Branch-and-Price

Jacques Desrosiers · Marco Lübbecke ·
Guy Desaulniers · Jean Bertrand Gauthier

Branch-and-Price

Springer

Jacques Desrosiers
GERAD and Département de sciences de la décision
HEC Montréal
Montréal, Canada

Guy Desaulniers
GERAD and Département de mathématiques et de génie industriel
Polytechnique Montréal
Montréal, Canada

Marco Lübbecke
GERAD and School of Business and Economics
RWTH Aachen University
Aachen, Germany

Jean Bertrand Gauthier
Mainz, Germany

ISBN 978-3-031-96916-4 ISBN 978-3-031-96917-1 (eBook)
https://doi.org/10.1007/978-3-031-96917-1

This Springer imprint is published by the registered company Springer Nature Switzerland AG
The registered company address is: Gewerbestrasse 11, 6330 Cham, Switzerland

About the authors

Jacques Desrosiers, born in 1950, is the second of a family of six children ($XY,\mathbf{XY},XX,XY,XY,XY$). His mother Gisèle, one of the few women of her time to hold a grade 11 diploma, was very good at mental arithmetic, while his father Raymond, excellent in a variety of sports such as golf ($+4$), baseball, bowling, and badminton, was a long-time traveling salesman.

Jacques obtained from Université de Montréal a bachelor's degree in mathematics in 1973, a master's degree in statistics in 1974, and a PhD (specialized in transportation) in 1979 under the supervision of Jacques A. Ferland and Jean-Marc Rousseau. Since 1978, he is a professor in the Department of Management Sciences at HEC Montréal.

Teenage Jacques was a member of the Laval Olympique athletics club: at the age of 15, he jumped 6.09 metres (20 feet less dust), which launched a five-year career in track and field. At the same age, and with the knowledge available to him at the time, he proved the Pythagorean theorem on a geometry homework assignment. This was one of the key events that pushed him towards mathematics. Jacques likes photography, a hobby he's been pursuing since 1972, the year he started to teach at high school, before teaching at college and university. He also loves red wines, especially those from Bordeaux. But he's not averse to Burgundy wines, as well as those from the Douro and Tuscany regions.

In 1993, he became partner in AD OPT Technologies which commercializes the *Altitude* software system for the management of air transport operations. In 1999, the company was listed on the Toronto Stock Exchange until 2004 when acquired by Kronos, a world leader in workforce management solutions. In 2019, the AD OPT division was sold to the aviation and transportation IT services provider IBS Software. The *Altitude* product line is powered by `GENCOL`, a state-of-the-art column generation solver developed at GERAD, the result of over 40 years of collaborative research among Montréal's universities in the field of applied mathematics and operations research. It's worth noting that the Natural Sciences and Engineering Research Council of Canada (NSERC) has been a major supporter of Jacques and his team. It has enabled a young university team to develop mathematical tools of much higher quality than those offered by the competition. The research, publications, and transfers carried out by Jacques during his career have been rewarded several times, notably in 1997 by the *Prix d'Excellence en Partenariat Innovateur* with his friend François Soumis, awarded jointly by NSERC and the Conference Board of Canada. This Synergy Award for Innovation recognizes outstanding research and development partnerships between academia and industry.

Marco Lübbecke was born on a Sunday in 1971. He spent his first years a dog-walking distance away from the former inner-German border without whose collapse he would have never met his wife Elisabeth.

Little Marco was thirsty for knowledge. His talent for explaining well showed early. However, his path to becoming a teaching awards-winning professor was winding. Marco is a first generation student, actually the first from his family who went to a higher school. When he was not admitted to the university subject he originally wanted to study, he enrolled in mathematics (which he liked in school). He graduated from the Technical University in Braunschweig, and stayed to earn a doctoral degree in mathematics in 2001. Marco applied for postdoctoral jobs in the US and Canada, but only received rejections. But why look far when everything good is so near? The combinatorial optimization and graph algorithms group welcomed Marco to Berlin. He completed his habilitation thesis in mathematics at the Technical University in 2007. A month before his contract (and so it seemed: his academic career) ended, the Technical University in Darmstadt appointed him as a visiting professor in discrete optimization. At the age of 39, Marco got his first and only permanent position when he was named full professor and head of the operations research group at RWTH Aachen University in 2010, only after five other individuals had declined that same offer. Also because of persistence, support, privilege, coincidence, and luck, Marco can live his passion (and pass it on). He is a very thankful person.

Marco is a father of six. For the 11km commute to the office he uses his bike. He likes photography (active) and music (passive), and he is not so bad at fixing things at home. Despite a chronic illness that may cost him control over his limbs one day, he started to play drums during the writing of this book. In 1983, Marco got his first computer, a Commodore C64. He was fascinated by this machine and he still thinks that limitations drive creativity. Apropos C. Marco's initiation in solvers was with `CPLEX` 3.0, and he was part of the first team that ever used `SCIP`'s column generation capabilities (with version 0.81). Since 2008 he has been leading the efforts of creating a decomposition based solver that automatically applies a reformulation and branch-price-and-cut to any mixed-integer program: the `GCG` project which is based on `SCIP`. Marco's research shows all facets of computational optimization: theory, structure, models, algorithms, implementations, experiments, and applications. He understands his work as truly interdisciplinary: he feels home in operations research, algorithmic discrete mathematics, and theoretical computer science, and more recently he has been making forays into engineering. Marco loves about mathematics that we exactly know the moment when we grasp something (and feel it when we do not understand it yet). He loves the pure beauty, the perfect organization, the waterproofness. And he loves the tears of joy in the eyes of practitioners when mathematical optimization helps making their lives easier.

Marco has no middle name. The copyright form of the first paper he published requested a middle initial, and he tried all 26 possibilities, "E." looked best to him.

Guy Desaulniers, born in 1964, grew up with his two brothers on the South shore of Montréal where he practiced various sports and developed his mathematical skills. Very young, he played hockey or baseball almost every day with his oldest brother and their friends. At 16, he started to play badminton intensively (up to 15 hours per week), won several Québec's university championships as a member of the Université de Montréal team, and became one of the best double players in Québec in 1990. A shoulder injury put an end to his badminton career. A few years later, he turned to running and his intense training allowed him to run a 5K in 20:08 and 10K in 42:38. Again, he injured himself, to an Achille's tendon this time, and stopped running. He got married to Lucie in 2000, became a proud father of two daugthers Alexane and Charlotte, and began a more sedentary life in Blainville, on the North shore of Montréal.

Guy always loved mathematics, a game in his view. At 3 years old, he started to learn mental arithmetic with his father Edouard; this helped him to be a first of the class student throughout his studies. During his high school fourth year, his math teacher, Madame Boisclair, asked him live to replace her to complete a lecture. He successfully took up the challenge and enjoyed the teaching experience. Later, during his undergraduate studies in mathematics, he often answered the questions of his classmates an hour before the exams. He loved sharing his passion with others, found it very rewarding, and decided to become a university professor.

Guy holds a BSc and a MSc in mathematics from Université de Montréal (1987, 1989), and a PhD in applied mathematics from Polytechnique Montréal (1993). His PhD thesis, supervised by François Soumis, focuses on trajectory optimization for autonomous vehicles. Surprisingly, it does not involve column generation. After his PhD, he occupied various professor and researcher positions during a difficult economical period in Québec and started to generate columns with François and Jacques. In 2000, he was hired as an associate professor in the Department of Applied Mathematics and Industrial Engineering at Polytechnique Montréal where he still teaches, conducts research, and supervises graduate students as a full professor since 2007. From 2015 to 2019, he was the Director of the GERAD, a world-renowned research center focusing on decision-making and data sciences that gathers more than 70 professors and 500 graduate students and post-docs. So far, Guy has supervised more than 55 MSc students, 35 PhD students, and 15 post-docs, and co-authored more than 125 papers in refereed journals and 10 book chapters.

His main research interests are on branch-and-price, mathematical decomposition methods, integer programming, and dynamic programming, applied to various problems arising in all modes of transportation, logistics, and personnel scheduling. He has been involved in many industrial research projects, especially with the optimization software companies AD OPT (IBS Software), Kronos (UKG) and Giro. Since 2019, Guy also acts as a scientific advisor at IvadoLabs, a company providing optimization and machine learning solutions to large Canadian organizations.

Jean Bertrand Gauthier, born in 1983, earned his bachelor's degree in actuarial science from Université du Québec à Montréal in 2010. He pursued higher education by specializing in the field of operations research at HEC Montréal where he obtained his PhD in 2016 under the supervision of Jacques Desrosiers. He then moved abroad to join the *Logistikmanagement* chair of Professor Stefan Irnich at Johannes Gutenberg University Mainz for a postdoctoral position. Moving to Germany proved to be a fulfilling experience, both professionally and personally, so he decided to settle down on that side of the Atlantic Ocean. Who knows if the plan to become a university professor would have come to fruition had he accepted the postdoctoral grant from IVADO in Montreal instead. Jean Bertrand was awarded a few gratifying prizes during his studies notably best master's thesis as well as best doctoral dissertation in their respective years of submission. He developed code for exact and heuristic methods to tackle various kinds of routing and network flow problems. His expertise includes linear programming, degeneracy, network models, and column generation. He has also been sharpening his LaTeX including PGF/TikZ skills since his graduate studies and considers himself a power user.

Jean Bertrand is an only child who happily spent hours on end playing with Lego. According to his mother's storytelling, he found the preparation time to go play outside at kindergarten far too long so he started helping his classmates tie their shoelaces. He might not have realized it then but he just might have tapped into the numerous positive rippling effects of lending a helping hand. Primary school was a breeze and so was high school despite some challenging periods during his teenage years. Books were an important part of his life already at a young age and this is still true to this day. Great examples of books he enjoys reading are those in which the story also serves to dig into the human psyche. This reminds him that psychology was an alternative program he contemplated. Even though he dismissed the admission offer, he is still genuinely interested in the wisdom of brilliant minds like Harper Lee, Jerome Salinger, Aldous Huxley and Nigel Barley.

Jean Bertrand was initiated to computer programming through Turing, an extinct descendant of Pascal. He found the possibility to command tasks from the computer exhilarating so much so that, in his spare time, he coded widgets that reproduce basic functionalities found in text editors. While he always found computers fascinating and *logical* quite literally, he did not think more of this experience until some fifteen years later when he learned C++ by himself. He laughs at the distant memory of arguing with the compiler because it refused to accept an int pointer for a method signed by a double pointer. *Just deference it to a float...* the irony is not lost on him that the logic for that was just waiting to be understood.

To my daughters, who mean so much to me: Amièle (1.1), Laurence (1.2), Léa (2.1), and Marie (2.2). To Alexandra (3.141592653589), for our endless and animated discussions. And to Karen for her unfailing support during this long journey.

– Jacques

Für père OR.

– Marco

To my wife Lucie, my daughters Alexane and Charlotte, and my mother Denise for their constant love. To François and Jacques, my mentors and good friends, who showed me the CG way and always supported me.

– Guy

À Louise sans qui j'accorderais un sens tout autre à l'importance des mots. À Bertrand qui a su partager sa logique des nombres avec moi. Für Hannah, die allein durch ihre Anwesenheit mein Leben optimiert.

– Jean Bertrand

Collectively, we dedicate this book to all column generation researchers around the world, friends and colleagues, students and programmers, without whom this book would certainly not exist.

Foreword

> The quality of a Dantzig-Wolfe reformulation can be appreciated, not only in terms of the optimal objective value of its linear relaxation, but also its feasible region. By considering complex structural constraints and integrality requirements in the pricing problem, we can imagine the objects it generates as *molecules* that are handled by the master problem. Of course, we can also identify those molecules in any integer solution of the original formulation, but with respect to its linear relaxation, it is like we are dealing with *atoms* that combine in any way, including ways that do not even respect chemical bonds.
>
> *Atoms vs. Molecules*
> François Soumis

The Dantzig-Wolfe reformulation and resolution by column generation allow complex constraints to be handled in subproblems, and often, to be separated into several subproblems, by, e.g., person, vehicle, scenario, etc. It is then possible to efficiently solve each of these subproblems, for example, by dynamic programming, integer programming, convex or non-convex non-linear programming. The exact resolution of the subproblems, compared to solving the linear relaxation of the complex constraints in the original problem, greatly reduces both the integrality gap and the size of the branch-and-bound tree to explore. This underpins the success of column generation to handle the many applications that follow.

The column generation algorithm was first used to tackle basic optimization problems such as cutting stock, multi-commodity flows, set partitioning/covering, bin packing, generalized assignment. These form the core of today's applications, with many more practical side constraints.

Regarding the applications, column generation now makes it possible to solve very large problems with good precision. We are successfully tackling weekly airline crew pairing problems of more than ten thousand flights and personalized monthly crew scheduling problems involving thousands of flight attendants requiring as many subproblems. The same applies to aircraft routing problems, covering thousands of flights while taking into account the maintenance requirements of each aircraft. We also solve public transit bus and driver scheduling problems for the world's largest cities, as well as similar problems for major rail and maritime networks, school transport, and on-demand transportation for disabled persons. Column generation is also utilized for solving freight transportation problems by truck, as well as scheduling of personnel working shifts in a wide range of domains: retail stores, healthcare systems, manufacturers, public safety, leisure and hospitality, and many others. It is also applied to problems concerning traffic assignment in road networks and in communication networks, where the subproblems generate the routes of users described by tens of thousands of origin-destination pairs.

The progress has been sufficiently significant for innovation to now focus on the simultaneous resolution of several transportation planning steps: crew pairings and monthly crew schedules, vehicle itineraries and crew schedules in air, rail and public transit applications. Progress has also made it possible to fairly rapidly deal with re-optimization problems that require the simultaneous treatment, over shorter time horizons, of what has been planned in several steps: for example, service timetables, vehicle itineraries, employee schedules, and user routes.

There is also a vast body of scientific literature on column-generation-based algorithms for solving vehicle routing problems with various side constraints: time windows, capacity limits or tour duration, heterogeneous fleet, pickup and delivery, time-dependent travel costs, electric vehicles with recharging, stochastic factors, etc. Many research projects focus on modeling resource constraints to get a good approximation of integer subproblems that can be solved in a reasonable time. The approximation is often refined during the solution process to improve the lower bound. In problems where networks may contain cycles, integrality gaps can be large, and a great deal of research develops specialized cuts to reduce them.

Column generation is now used by a large community of researchers. Looking at those who have cited the seminal paper *Time constrained routing and scheduling* which is cited more than 1400 times on Google Scholar Citation Index, there are 165 column-generation-based papers cited more than 100 times, some of them several hundred times. These papers have been written by around a hundred different teams.

Since 2006, Guy, Marco, and Jacques have organized and taught a five-day *School on Column Generation* multiple times (Montréal 2006, Darmstadt 2010, Paris 2014 and 2018) with up to 95 students each. They have used a lot of slides that evolved from one edition to the next. They were also involved in several editions of the *International Workshop on Column Generation* (Aussois 2008, Bromont 2012, Búzios 2016, Montréal 2023). Their experience has led to this book that meets the needs of the scientific community and, in particular, of new researchers wishing to enter the field. Hundreds of articles have been published on different aspects of the theory and applications. The notation used depends on the authors/applications and certain concepts sometimes have different names. It takes a lot of effort and time to go through all this literature. This book, which standardizes and simplifies notation while clarifying concepts, will enable students to acquire the basics much more quickly. What's more, this standardized language will make it easier to write papers that are clear and accessible to a broad community.

Montréal, October 2023 *François Soumis*

Preface

When one teaches, two learn.

Robert A. Heinlein

It must have been around the turn of the century that we were talking about writing a book on column generation. The first reference we could find, including a typo, is in an email from Jacques to Marco:

```
From: "Jacques Desrosiers" <jacques@crt.umontreal.ca>
To: <m.luebbecke@tu-bs.de>
Subject: Re: Happy New Year!
Date: Wed, 17 Jan 2001 08:42:46 -0500
[...]

ps : I am still interseted in ... the book.
```

A lot can happen in 15 years, and a lot *did* happen in 15 years, but no book writing. The need for a standard text on column generation was more pressing than ever, and it was clear to us that we were the ones to write it. A serious step forward was the setting up of a subversion repository. The first commit message reads

```
r4347 | luebbecke | 2015-01-14 21:21:50 +0100 | 2 lines
this is the beginning of [DDL1x], replace x by 5,6, or 7
```

As you can see, we were very optimistic about the duration of such a project. On January 26, 2015, Jacques had a very detailed outline of a book ready, entitled "branch-and-price," and it contained 16 chapters. The journey had finally started.

Integer (linear) programs are a standard way of formalizing a vast array of optimization problems in industry, services, management, science, and technology. By the logic of the underlying business problem, such models are often composed of independent building blocks that are kept together by, e.g., spatial, temporal, or financial constraints. Over the years, branch-and-price became a, likely *the* standard approach to solving such *structured* integer programs. The charm of the method lies in its ability to leverage algorithms for the building blocks by way of decomposition. Hundreds and hundreds of papers have been written on successful applications in logistics, transportation, production, energy, health care, education, politics, sports, etc. Besides collecting and unifying the literature, we wanted to share all our experience with the subject. One of the most rewarding side effects of writing what one knows is that it mercilessly reveals what one does *not* know. "Facts" that we took for granted (but which were wrong), details that were missing, papers that we have overlooked, formalisms that have never been spelled out. Yes, we learned a lot!

We assume the reader to be reasonably familiar with basic linear programming theory, in particular the concept of duality and a geometric understanding of the feasible region of a linear program. We recommend the fantastic book by Chvátal (1983) for both. Because we like puns, readers should also be familiar with non-basic theory. We expect them to have modeling experience with network, linear and integer linear programs. Essentials are presented in Chapter 1 (Linear and Integer Linear Programming).

This book is about solving integer linear programs by branch-and-price, i.e., column generation applied in every node of a search tree. Column Generation is an algorithm for solving large scale linear programs: as such, we have the dedicated Chapter 2 in which we see the similarities and differences with the primal simplex algorithm. The Dantzig-Wolfe decomposition is, in fact, a reformulation method: Chapter 3 presents the classical way for linear programs; this is based on the convexification approach of a sub-domain. Chapter 4 goes further, adapting it to integer linear programs, and also presenting the reformulation based on the discretization approach. Chapter 5 (Vehicle Routing and Crew Scheduling Problems) follows and gives access to important applications: we make use of what we have seen previously. Chapter 6 explores the Dual Point of View: it notably presents another decomposition method, the Lagrangian relaxation approach. We see its relationships with the Dantzig-Wolfe reformulation. A better understanding of duality leads to stabilization approaches. Chapter 7 (Branch-Price-and-Cut) presents how to handle various branching and cutting decisions to get integer solutions, indeed to solving the original model. The final chapter, Conclusion, is where we stop after several years of understanding, writing, classifying, etc. We tell the story of Montréal's `GENCOL` group since the early 80s and list some possible writing subjects for interested people.

It is our intention that there is something for everyone, for the beginner, for the practitioner, as well as for the seasoned researcher. Every chapter may have some surprising news or viewpoint also for the experts. In each one, we set out with a brief motivation and then cover the main theory in a few sections, sometimes interwoven with some illustrations. A section *Good to Know* follows with more advanced material that can be read after having worked through some examples. Then comes *More to Know* with even more advanced topics that may require that the previous sections are well digested.

The book is full of examples, every chapter contains a (long!) separate section. In contrast to the preceding theory, the examples demonstrate how column generation and branch-and-price is *lived*. A lot of ideas are transported there, and the reader should not skip them. *Reference notes* is the place to see historical perspectives and complementary lectures. Note, however, that we do not even try to cover the entire literature. There is a set of exercises in every chapter and we give all the answers but two (7.21 and 8.2). Sometimes, exercises provide new theory.

Our book contains (alas, only a few) hints about an implementation; these are marked in the margin with a keyboard logo. Some general remarks can be found at the very end. What our book does not contain is a single line, or even only an inspiration generated by some large language model or similar. All what we have written stems from natural intelligence. And we are proud of that.

We are four authors, with different views; compromises were necessary and some topics are missing. We had physically exhausting arguments about ways of presentations, assumptions, ... and they were among the most productive and enlightening events in our research careers. While our writing is in English, it should not be too hard to find evidence of the French Canadian culture of Jacques, Guy, and Jean Bertrand as well as the German one of Marco. Readers who know us very well can even differentiate our writing styles. Over the years, many colleagues asked us what source we would recommend for learning about column generation and branch-and-price. And now, here it is; you hold it in your hands. Enjoy reading and working with it as much as we enjoyed writing it.

We would like to acknowledge the help received from Katrin Heßler (Johannes Gutenberg-Universität Mainz), Elina Rönnberg (Linköping University), Frédéric Quesnel (Université du Québec à Montréal), and José Manuel Valério de Carvalho (Universidade do Minho) through their attentive reading of many chapters of the book. We are grateful to Jialin Yan from Springer for supporting us so constructively during the final production process.

Special thanks go to François Soumis (Polytechnique Montréal) for writing the Foreword and to Lina Vandal, a talented Montréal artist, for painting the cover illustration *Mathemagician*.

Anyhow, whatever, anyway... we could not possibly forget Karen, Elisabeth, Lucie, and Hannah: for all their words of encouragement.

This book contains many photos of friends and colleagues. All these were taken by ourselves. It would not have been possible to publish open access without the financial support that we received from the RWTH Aachen University, HEC Montréal, Polytechnique Montréal, and the GERAD Operations Research Center.

Knowlton, Aachen, Blainville, and Mainz

Jacques
Marco
Guy
Jean Bertrand

Contents

About the authors v

Foreword xi

Preface xiii

1 Linear and Integer Linear Programming 1
Introduction . . . 2
1.1 Polyhedral Theory . . . 3
1.2 Linear Programming . . . 6
Linear program . . . 6
Dual point of view . . . 7
Optimality conditions . . . 10
Sensitivity analysis . . . 11
1.3 Primal Simplex Algorithm . . . 14
Mechanics . . . 14
Sufficient optimality conditions . . . 15
Pricing problem . . . 16
Extreme ray direction . . . 17
Adjacent extreme point . . . 17
Initialization and pseudo-code . . . 18
Treatment of bounds on the variables . . . 21
1.4 Integer Linear Programming . . . 23
Strength of formulations . . . 24
Extended formulations . . . 25
Integrality property . . . 26
Total unimodularity . . . 27
Network flow problems . . . 27
1.5 Branch-and-Bound . . . 28
Optimality gap . . . 30
Variable fixing by reduced cost . . . 30
Illustration 1: Branch-and-bound experiment . . . 31
Branch-and-cut . . . 33
1.6 Good to Know . . . 33
Linear programs for the pricing . . . 34
Irrational data . . . 35
1.7 Reference notes . . . 37
Exercises . . . 39

2 Column Generation 43
Introduction . . . 45

2.1 Algorithm . . . 47
Master problem . . . 47
Pricing problem . . . 48
Iteration . . . 49
Initialization . . . 51
2.2 Good to Know . . . 53
Objects, representations, and encodings . . . 53
Several subproblems . . . 56
Pivot rules and column management . . . 57
Heuristic pricing . . . 59
2.3 More to Know . . . 60
Lower bounds . . . 60
Convergence . . . 63
Limited numeric precision . . . 65
Static and auxiliary variables . . . 66
Relaxed pricing, relaxed master . . . 68
Paving the way to a practical implementation . . . 69
2.4 Examples . . . 71
One-dimensional cutting stock problem . . . 71
Integer master problem . . . 71
Integer pricing problem . . . 72
Empty vs. zero cutting patterns . . . 73
Cutting stock problem with rolls of different widths . . . 74
Edge coloring problem . . . 76
Set covering master problem . . . 77
Pricing the matchings . . . 77
Sports scheduling . . . 78
Single depot vehicle scheduling problem . . . 80
Set partitioning master problem . . . 80
Shortest path pricing problem . . . 81
Zero schedule . . . 83
Vehicle routing and crew scheduling problems . . . 84
Aircraft routing with schedule synchronization . . . 86
Routes and schedules . . . 87
Integer master problem . . . 88
Integer pricing problems . . . 89
Practical observations . . . 90
2.5 Reference Notes . . . 91
Exercises . . . 95

3 Dantzig-Wolfe Decomposition for Linear Programming 103
Introduction . . . 105
3.1 Dantzig-Wolfe Reformulation . . . 105
Minkowski-Weyl theorem . . . 106
Master problem . . . 108

Polytope and polyhedral cone 110
3.2 Column Generation Approach 111
Pricing problem 112
Post-processing a solution of the master problem 114
Primal solution 114
Dual solution 114
Basic solution 115
On grouping the constraints 115
Block-diagonal structure 116
Lower and upper bounds 119
3.3 Good to Know 120
A set of extreme points and extreme rays 120
Extreme cases for the constraints in the reformulated set 120
All 120
None 121
Static variables 122
3.4 More to Know 123
Restricted compact formulation 124
Block-diagonal structure 127
Positive edge rule and Improved Primal Simplex algorithm 128
Benders decomposition of a linear program 130
Relaxed master, subproblem, and initialization 132
Decomposition theorem of a network flow solution 133
3.5 Examples 136
2D illustration 137
Compact formulation 137
Grouping of constraints 137
Extended formulation 138
Optimal primal-dual solutions for the compact formulation . 139
Column generation approach 140
Time constrained shortest path problem (*TCSPP*) 142
Compact formulation 142
Grouping of constraints 143
Extended formulation 144
Optimal primal-dual solutions for the compact formulation . 144
Column generation approach 145
Single depot vehicle scheduling problem: two compact formulations 148
Formulation with *origin-to-destination* arc 149
Formulation with *destination-to-origin* arc 151
Solving a sequence of restricted compact formulations 155
3.6 Reference Notes 158
Exercises 161

4 Dantzig-Wolfe Decomposition for Integer Linear Programming 173
Introduction 176

4.1 Reformulation by Convexification 177
Minkowski-Weyl theorem, again 177
Integer master problem 178
Polytope and polyhedral cone 181
Integer pricing problem 181
4.2 Reformulation by Discretization 183
Hilbert-Giles-Pulleyblank theorem 183
Integer master problem 185
Polytope and polyhedral cone 187
Two-dimensional illustrations 188
Illustration 1: Set $\mathcal{T}$ for a polytope 189
Illustration 2: Set $\mathcal{T}$ for a polyhedral cone 189
Illustration 3: Set $\mathcal{T}$ for a polyhedron 190
Illustration 4: Set $\mathcal{T}$ for another polyhedron 191
Integer pricing problem 191
Binary domain 193
Illustration 5: *TCSPP*: integrality 194
Some observations 196
Post-processing a solution of the master problem 197
4.3 Integrality Property: For or Against Virtue? 197
4.4 Block-diagonal Structure 200
Practical relevance 200
On grouping the constraints 201
Integer master problems (convexification and discretization) 202
Identical subproblems 204
Aggregation 205
Disaggregation 206
Some more observations 208
Lower and upper bounds 210
4.5 Good to Know 211
Non-linear encoding functions 212
Illustration 6: *TCSPP*: non-linear costs 215
Not all blocks are used 217
Bounded domains 218
Identical subproblems 221
Unbounded domains 222
Shared variables across all blocks 222
4.6 More to Know 224
Reverse engineering a compact formulation 224
Convexification 224
Discretization 225
Extended compact and subproblem formulations 230
Automatic grouping of the constraints for reformulation 232
Graph-based methods 233
Methods based on constraint classes 234

Evaluating a decomposition . . . 236
4.7 Examples . . . 238
Integrality property in the knapsack problem . . . 239
Binary knapsack problem . . . 239
Knapsack problem . . . 241
Integrality property in the cutting stock problem . . . 244
Weak compact formulation . . . 244
Network-based compact formulation . . . 246
Comparison of the four linear relaxations . . . 249
Reverse Dantzig-Wolfe . . . 250
TCSPP: nine reformulations . . . 251
TCSPP: time bounding . . . 254
Generalized assignment problem . . . 259
Binary tasks-to-machine (many-to-one) patterns . . . 261
Binary task-to-machines (one-to-many) patterns . . . 262
Multi-commodity maximum flow problem . . . 264
Scene selection problem . . . 267
Symmetry breaking constraints . . . 268
Dantzig-Wolfe reformulation . . . 268
Quality of the lower bounds . . . 269
Design of balanced student teams . . . 270
Quadratic transportation problem . . . 270
Set partitioning reformulation . . . 272
Can you decode a secret vote? . . . 275
Binary vote patterns . . . 277
Edge coloring problem: two compact formulations . . . 279
First compact, with an identical block-diagonal structure . . 279
Second compact, without a block-index . . . 281
Compact formulations are not created equal . . . 282
4.8 Reference Notes . . . 282
Exercises . . . 285

5 Vehicle Routing and Crew Scheduling Problems 295
Introduction . . . 297
5.1 Vehicle Routing Problem with Time Windows . . . 297
Problem statement . . . 298
Network structure . . . 298
Illustration 1: A network with four customers . . . 299
Mathematical formulations . . . 300
Compact arc-flow formulation . . . 300
Extended set-partitioning formulation . . . 301
Column generation . . . 304
Elementary shortest paths with time windows and capacity . 304
Labeling algorithm for the *ESPPTWC* . . . 306
Illustration 2: Improved dominance . . . 309

Illustration 3: Improved dominance (cont.) 310
Subproblem relaxations 312
SPPTWC relaxation 313
Illustration 4: Generated labels for the *SPPTWC* 313
ng-SPPTWC relaxation 314
Illustration 5: Generated labels for the *ng-SPPTWC* 317
Illustration 6: Comparative results 317
5.2 Elementary Shortest Path Problem with Resource Constraints 319
Mathematical formulation 320
Labeling algorithm 321
Illustration 7: Common resource extension functions 322
5.3 Examples 324
Simultaneous pickup and delivery problem 324
Network structure 324
Extended formulation 325
Pricing problem 325
Pickup and delivery problem with time windows 326
Network structure 327
Illustration 8: A network for the *PDPTW* 328
Extended formulation 328
Pricing problem 329
Easier-to-solve pricing problem 331
Crew pairing problem with base constraints 332
Network structure 334
Extended formulation 336
Pricing problems 337
5.4 Good to Know 339
Compact linear formulation for the *VRPTW* 339
Extended set partitioning formulation 342
5.5 Reference Notes 343
Exercises 345

6 Dual Point of View **351**
Introduction 353
6.1 Row Generation 355
Illustration 1: *TCSPP* 356
Alternative master problem 359
Illustration 2: *TCSPP* (cont.) 360
6.2 Lagrangian Relaxation 361
Lagrangian subproblem 361
Lagrangian function and Lagrangian dual problem 362
Block-diagonal structure 363
Illustration 3: *TCSPP* (cont.) 363
Dantzig-Wolfe reformulation vs. Lagrangian relaxation 365
Properties of the Lagrangian function 370

Illustration 4: *TCSPP* (cont.) 372
Optimality conditions 375
Illustration 5: Optimality check 375
Illustration 6: *TCSPP* (cont.) 376
Subgradient algorithm 377
Illustration 7: *TCSPP* (cont.) 381
Primal solutions (fractional and integer) 383
6.3 Good to Know 385
Dual-optimal inequalities 385
Illustration 8: Cutting stock problem 386
Dual-optimal boxes 388
How helpful can the dual information be? 390
Illustration 9: Cutting stock problem (cont.) 390
Illustration 10: *TCSPP* (cont.) 391
Illustration 11: Multi-depot vehicle scheduling problem . . . 393
6.4 More to Know 398
Stabilized column generation 398
Illustration 12: Multi-depot vehicle scheduling (cont.) . . . 402
Properties 403
Alternative/complementary stabilization methods 405
Learning more from the dual point of view 407
Initialization 407
Dual estimates 407
Anticipation 408
6.5 Examples 409
Generalized assignment problem 409
Relaxation of the semi-assignment constraints 409
Relaxation of the capacity restrictions on the machines . . . 410
Capacitated *p*-median problem 411
Lagrangian relaxation 413
Dantzig-Wolfe reformulation 413
Various Lagrangian subproblems 414
Vehicle routing problem with time windows 414
Cutting stock problem 415
Edge coloring problem 416
Symmetric traveling salesperson problem 418
Formulations 419
Held and Karp lower bounds 421
Lagrangian lower bounds 423
Dantzig-Wolfe reformulation 426
Balancing printed circuit board assembly line systems 427
Dantzig-Wolfe reformulation 427
Lagrangian relaxation 429
Some computational results 430
A few comments 431

Hybrid algorithm for a 2-dimensional problem 432
6.6 Reference Notes 434
Exercises 437

7 Branch-Price-and-Cut **443**
Introduction 445
7.1 Integrality Test 447
7.2 Cutting Planes 449
Cutting planes on the original variables 449
Integration into the master constraints 450
Integration into the subproblem constraints 452
Illustration 1: Clique cuts for the vertex coloring problem . . 452
Illustration 2: Capacity cuts for the *VRPTW* 454
Illustration 3: z-cuts within the *ISP* 457
Discussion 459
Cutting planes on the master variables 459
Illustration 4: Chvátal-Gomory cuts 461
Illustration 5: Clique cuts for the set partitioning problem . . 462
Discussion 464
7.3 Branching 464
Branching on the original variables 465
Illustration 6: x-branching for the *VRPTW* 466
Identical subproblems and disaggregation 469
Branching on the master variables 470
Branching on partitioning $\mathcal{X}$ with a single hyperplane 471
Illustration 7: λ-branching for the *VRPTW* 475
Branching on lower and upper bounds on the x-variables . . 476
Ryan-Foster rule for set partitioning master problems 480
Branching on inter-tasks 481
Some practical rules on the sum of binary λ-variables . . . 483
Illustration 8: λ-branching for the *VRPTW* (cont.) 485
Ranking the candidates 486
7.4 Convexification vs. discretization revisited 487
Illustration 9: What you eat is what you are 488
Final remarks 491
7.5 Good to Know 493
Arc-flow variable fixing by reduced cost 493
Acceleration techniques 494
Large number of column generation iterations 495
Most of the time spent solving the *RMP* 496
Most of the time spent solving the *ISP* 497
Large number of branch-and-bound nodes 499
7.6 More to Know 501
Primal heuristics 501
Chvátal-Gomory cuts of higher ranks 506

Beginner's guide to an implementation 507
7.7 Examples 509
Time constrained shortest path problem 510
Semi-assignment branching 514
Ryan-Foster branching for vertex coloring 516
Ryan-Foster branching for bin packing 518
Edge coloring and odd-circuit cuts 519
k–path cuts for the *VRPTW* 520
VRPTW and non-robust rounded capacity cuts 522
VRPTW and Chvátal-Gomory rank-1 cuts 525
Preferential bidding system 527
Branch-first, Cut-second strategy 531
7.8 Reference notes 536
Exercises 538

8 Conclusion **545**
Outroduction 546
8.1 GENCOL 546
First decade (1980s) 547
Second decade (1990s) 548
Third decade (2000s) 549
Fourth decade (2010s) 550
Fifth decade (2020s) 551
8.2 Branch-and-Price: *More to Know* 552
Exercises 553

Solutions **555**

References **641**

1

Linear and Integer Linear Programming

The final test of a theory
is its capacity to solve the problems which originated it.

Linear Programming and Extensions
George Bernard Dantzig

Abstract This first chapter recalls the main notions of linear programming, that is, the primal and dual formulations as well as necessary and sufficient optimality conditions. We also describe the primal simplex algorithm and discuss some aspects of integer linear programs.

Contents

Introduction 2
1.1 Polyhedral Theory 3
1.2 Linear Programming 6
Linear program 6
Dual point of view 7
Optimality conditions 10
Sensitivity analysis 11
1.3 Primal Simplex Algorithm 14
Mechanics 14
Sufficient optimality conditions 15
Pricing problem 16
Extreme ray direction 17
Adjacent extreme point 17
Initialization and pseudo-code 18
Treatment of bounds on the variables 21

J. Desrosiers et al., *Branch-and-Price*,
https://doi.org/10.1007/978-3-031-96917-1_1

1.4 Integer Linear Programming 23
Strength of formulations 24
Extended formulations 25
Integrality property 26
Total unimodularity 27
Network flow problems 27
1.5 Branch-and-Bound 28
Optimality gap 30
Variable fixing by reduced cost 30
Illustration 1: Branch-and-bound experiment 31
Branch-and-cut 33
1.6 Good to Know 33
Linear programs for the pricing 34
Irrational data 35
1.7 Reference notes 37
Exercises 39

Acronyms

OR	operations research	2
LP	linear program	6
LD	dual formulation of the *LP*	7
SP	subproblem, or simply pricing in the primal simplex algorithm	16
ILP	integer linear program	23
MILP	mixed-integer linear program	24
CMCFP	capacitated minimum cost flow problem	27
LB	best known lower bound on $z^\star_{ILP}$	29
UB	best known upper bound on $z^\star_{ILP}$	29

Introduction

Operations Research (OR) is concerned with analytical methods to help decision-makers construct and/or adjust their operational, tactical, and strategic plans. In this respect, optimization, in particular linear programming, is commonplace in plenty of industrial applications around the globe. Since we expect the reader to be more than acquainted with linear programming, its father, George Bernard Dantzig (1914–2005), should hardly need an introduction. This chapter therefore serves to formalize terminology under a common notation. Nevertheless, we like to think that there is something for even our most seasoned readers.

Let us start straight away by getting the obvious out of the way.

Definition 1.1. With an *optimization program*, we aim to find values for *decision variables* subject to restrictions given by *constraints* and *bounds*, i.e., the *feasible region*, such that the value of the *objective function* is optimized. A *solution* is a set of decision values, one for each variable, which may either be *feasible* or *infeasible* depending on whether or not these values are in the feasible region.

Moreover, a feasible solution is *optimal* if it yields the *optimal objective value* in which case we also call it an *optimizer* and the *optimum*. Observe that, with the articles a/an/the, we acknowledge the possible existence of multiple optimal solutions but a unique optimal objective value.

Notation. Throughout the text, vectors and matrices, respectively, use lower and upper case letters. Both are set in **bold** face. In particular, the matrix $\mathbf{A} = [\mathbf{a}_j]_{j\in\{1,\ldots,n\}}$ contains n column vectors, where such a vector $\mathbf{a}_j = [a_{ij}]_{i\in\{1,\ldots,m\}}$ contains m rows.

This indexing extends to subsets of rows and/or columns as follows. We use $\mathbf{A}_{i*}$ to refer to the i-th row of matrix $\mathbf{A}$. For an ordered subset $I \subseteq \{1,\ldots,m\}$ of row indices and an ordered subset $J \subseteq \{1,\ldots,n\}$ of column indices, we denote by $\mathbf{A}_{IJ}$ the sub-matrix of $\mathbf{A}$ containing the rows and columns respectively indexed by I and J. If the row subset is omitted by dropping the first index, it allows for standard linear programming notation like $\mathbf{A}_B\mathbf{x}_B$, the subset of basic columns of $\mathbf{A}$ indexed by B multiplied by the corresponding vector of basic variables $\mathbf{x}_B$. The index subset $N = \{1,\ldots,n\}\setminus B$ for non-basic columns is used analogously.

We denote by $\mathbf{I}_r$ the $r\times r$ identity matrix, by $\mathbf{0}$ (resp. $\mathbf{1}$) a vector/matrix of all zero (resp. one) entries of contextually appropriate dimension, and by $\mathbf{e}_i$ a standard unit-vector with a single non-zero entry of 1 in row i. The transpose operation is denoted $\mathbf{A}^\intercal$. For number sets like $\mathbb{Z}$, $\mathbb{Q}$, and $\mathbb{R}$, the notation $\mathbb{Z}_+$, $\mathbb{Q}_+$, and $\mathbb{R}_+$ refers to their non-negative counterparts, that is, these sets include zero.

1.1 Polyhedral Theory

We put several definitions regarding polyhedral theory here up-front. While they may seem unmotivated or out of place, postponing their presentations as needed would hinder the flow of ideas surrounding more advanced material.

Definition 1.2. A *hyperplane* $\mathcal{H} \subset \mathbb{R}^n$ is defined by a vector $\mathbf{f} \in \mathbb{R}^n$ and a scalar $f \in \mathbb{R}$ as $\mathcal{H} = \{\mathbf{x} \in \mathbb{R}^n \mid \mathbf{f}^\intercal\mathbf{x} = f\}$.

Definition 1.3. An *open half-space* is the set of all points below (resp. above) a hyperplane $\mathcal{H}$, i.e., $\mathcal{H}^- = \{\mathbf{x} \in \mathbb{R}^n \mid \mathbf{f}^\intercal\mathbf{x} < f\}$ (resp. $\mathcal{H}^+ = \{\mathbf{x} \in \mathbb{R}^n \mid \mathbf{f}^\intercal\mathbf{x} > f\}$). A half-space is *closed* if the said hyperplane also belongs to it, i.e., the strict inequalities are replaced by non-strict ones in the above definitions.

Definition 1.4. A *polyhedron* $\mathcal{P} \subseteq \mathbb{R}^n$ is a set of points satisfying a finite, say m, number of linear inequalities, i.e., $\mathcal{P} = \{\mathbf{x} \in \mathbb{R}^n \mid \mathbf{Q}\mathbf{x} \geq \mathbf{q}\}$, where $(\mathbf{Q},\mathbf{q}) \in \mathbb{R}^{m\times(n+1)}$. Moreover, a polyhedron $\mathcal{P}$ is said to be *rational* if there exists $(\mathbf{Q}',\mathbf{q}') \in \mathbb{Q}^{m\times(n+1)}$ with rational coefficients such that $\mathcal{P} = \{\mathbf{x} \in \mathbb{R}^n \mid \mathbf{Q}'\mathbf{x} \geq \mathbf{q}'\}$.

Throughout the book, we only consider rational polyhedra, see Irrational data in Section 1.6 (Good to Know), even when we write, e.g., $\mathcal{P} \subseteq \mathbb{R}^n$.

Definition 1.5. Given k points $\mathbf{a}_i \in \mathbb{R}^n$, $i \in \{1,\dots,k\}$, a *combination* $\sum_{i=1}^k \alpha_i \mathbf{a}_i$ is ...

- *linear* if $\alpha_i \in \mathbb{R}, \ \forall i \in \{1,\dots,k\}$;
- *affine* if $\alpha_i \in \mathbb{R}, \ \forall i \in \{1,\dots,k\}$, and $\sum_{i=1}^k \alpha_i = 1$;
- *conic* if $\alpha_i \geq 0, \ \forall i \in \{1,\dots,k\}$;
- *convex* if $\alpha_i \geq 0, \ \forall i \in \{1,\dots,k\}$, and $\sum_{i=1}^k \alpha_i = 1$.

Some people also know comic combinations and we find that funny.

Definition 1.6. A set of k points $\mathbf{a}_i \in \mathbb{R}^n$, $i \in \{1,\dots,k\}$, is ...

- *linearly independent* if $\sum_{i=1}^k \alpha_i \mathbf{a}_i = 0 \Leftrightarrow \alpha_i = 0, \forall i \in \{1,\dots,k\}$;
- *affinely independent* if $\{\mathbf{a}_i - \mathbf{a}_j\}_{i \in \{1,\dots,k\} \setminus \{j\}}$ is linearly independent for any j.

Definition 1.7. A point $\mathbf{p} \in \mathcal{P}$ is an *extreme point of* $\mathcal{P}$ if it cannot be expressed as a non-trivial convex combination of two points in $\mathcal{P}$, i.e., there do not exist $\mathbf{x}_1, \mathbf{x}_2 \in \mathcal{P}$, $\mathbf{x}_1 \neq \mathbf{x}_2$, and a scalar $\lambda \in (0,1)$ such that $\mathbf{p} = \lambda \mathbf{x}_1 + (1-\lambda)\mathbf{x}_2$.

Definition 1.8. Let $\mathcal{P}^{\mathbf{0}} = \{\mathbf{r} \in \mathbb{R}^n_+ \mid \mathbf{Q}\mathbf{r} \geq \mathbf{0}\}$. If $\mathcal{P}^{\mathbf{0}} \neq \emptyset$, a point $\mathbf{r} \in \mathcal{P}^{\mathbf{0}} \setminus \{\mathbf{0}\}$ is a *ray of* $\mathcal{P}$. Equivalently, a vector $\mathbf{r} \in \mathbb{R}^n_+ \setminus \{\mathbf{0}\}$ is a ray of $\mathcal{P}$ if for any vector $\mathbf{x} \in \mathcal{P}$, $\mathbf{x} + \theta \mathbf{r} \in \mathcal{P}$, $\forall \theta \geq 0$.

Note that when $\mathbf{r} \in \mathbb{R}^n_+$ is a ray of $\mathcal{P}$, then $\alpha \mathbf{r}$ is also a ray of $\mathcal{P}$ for any scalar $\alpha > 0$. That is, we can scale a ray to a representation we prefer.

Definition 1.9. A ray $\mathbf{r}$ of $\mathcal{P}$ is an *extreme ray of* $\mathcal{P}$ if it cannot be written as a non-trivial conic combination of two rays of $\mathcal{P}$, i.e., there do not exist $\mathbf{r}_1, \mathbf{r}_2 \in \mathcal{P}^{\mathbf{0}}$, with $\mathbf{r}_1 \neq \theta \mathbf{r}_2$ for any $\theta > 0$, and scalars $\lambda_1, \lambda_2 \geq 0$ such that $\mathbf{r} = \lambda_1 \mathbf{r}_1 + \lambda_2 \mathbf{r}_2$.

The theorem by Minkowski and Weyl, commonly used in the Dantzig-Wolfe reformulation of linear and integer linear programs, states that there are two equivalent representations of a polyhedron, the one seen above as the intersection of finitely many closed *half-spaces*, the other one using its *vertices* and *rays*.

Theorem 1.1. (Nemhauser and Wolsey, 1988, Theorem 4.8, p. 96) *Let* $\mathcal{P} = \{\mathbf{x} \in \mathbb{R}^n \mid \mathbf{Q}\mathbf{x} \geq \mathbf{q}\}$ *be a polyhedron with at least one extreme point. An equivalent description of* $\mathcal{P}$ *using its extreme points* $\{\mathbf{x}_p\}_{p \in P}$ *and extreme rays* $\{\mathbf{x}_r\}_{r \in R}$ *is*

$$\mathcal{P} = \left\{ \mathbf{x} \in \mathbb{R}^n \;\middle|\; \begin{array}{ll} \sum_{p \in P} \mathbf{x}_p \lambda_p + \sum_{r \in R} \mathbf{x}_r \lambda_r = \mathbf{x} & \\ \sum_{p \in P} \lambda_p = 1 & \\ \lambda_p \geq 0 & \forall p \in P \\ \lambda_r \geq 0 & \forall r \in R \end{array} \right\}. \tag{1.1}$$

Proof. We refer to Nemhauser and Wolsey (1988) for step by step proofs and rather concentrate on three important ideas.

1. Both sets P and R are finite.
2. By assumption, the set P is non-empty but R could be. Nemhauser and Wolsey subsume this under a feasible region in the first quadrant with a matrix $\mathbf{Q}$ of rank n, that is, $\mathbf{x} \geq \mathbf{0}$ is included in $\mathbf{Qx} \geq \mathbf{q}$.
3. Both descriptions are equivalent because they describe the same set of solutions:

   $\Rightarrow$ Any $\mathbf{x} \in \mathcal{P}$ can be written as a convex combination of the extreme points $\{\mathbf{x}_p\}_{p\in P}$ plus a conic combination of the extreme rays $\{\mathbf{x}_r\}_{r\in R}$, that is, there exist scalars $\{\lambda_p\}_{p\in P}$ and $\{\lambda_r\}_{r\in R}$ such that (1.1) holds for all $\mathbf{x} \in \mathcal{P}$.

   $\Leftarrow$ Any $\boldsymbol{\lambda}$ that satisfies (1.1) corresponds to an $\mathbf{x} \in \mathbb{R}^n$ by definition but also necessarily to an $\mathbf{x} \in \mathcal{P}$. □

Definition 1.10. A bounded polyhedron is called a *polytope*.

Definition 1.11. A *polyhedral cone* is a polyhedron $\mathcal{P}$ that is also a cone (that is, $\mathbf{x} \in \mathcal{P} \Rightarrow \alpha\mathbf{x} \in \mathcal{P},\ \forall \alpha \geq 0$). It has the form $\mathcal{P} = \{\mathbf{x} \in \mathbb{R}^n \mid \mathbf{Qx} \geq \mathbf{0}\}$, where the n-dimensional $\mathbf{0}$-vector is the unique extreme point.

Figure 1.1 shows in (a) a *polytope* with seven extreme points whereas three polyhedra (that are not polytopes are displayed on the right. The *polyhedron* in (b) exhibits three extreme points and two extreme rays whereas that in (c) comprises only one extreme point. That in (d) is a *polyhedral cone*.

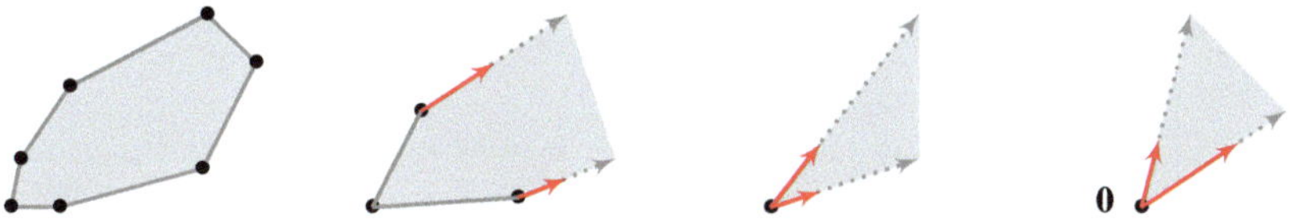

(a) Polytope (b) Polyhedron (c) Polyhedron (d) Polyhedral cone

Fig. 1.1: Extreme points and extreme rays.

Definition 1.12. The *convex hull* of a set of points $\mathcal{Y}$, denoted $\operatorname{conv}(\mathcal{Y})$, is the set of all convex combinations of $\mathcal{Y}$. It is the smallest convex set which contains $\mathcal{Y}$. Any polyhedron $\mathcal{P}$ is convex, thus we have $\mathcal{P} = \operatorname{conv}(\mathcal{P})$.

Figure 1.2 shows another polyhedron where there are three extreme points, namely, $(1,1)$, $(1,2)$, and $(2,1)$, but the set of extreme rays contains only a *single* element, say $(1,1)$. Moreover, the representation $(1,1)$ of this extreme ray is the same as that of the first extreme point. When necessary, we use a binary parameter to distinguish them: 1 for a vector representing a point and 0 for a vector representing a direction like a ray (see Schrijver, 1986, Corollary 7.1b). This is notably the case for the Dantzig-Wolfe decomposition approaches in the forthcoming chapters.

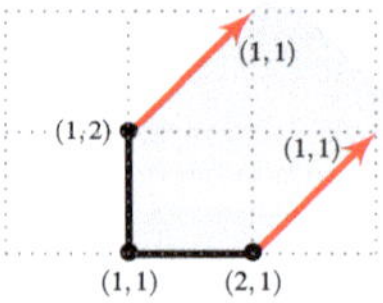

Fig. 1.2: Polyhedron with three extreme points represented as $\begin{bmatrix}(1,1)\\1\end{bmatrix}$, $\begin{bmatrix}(1,2)\\1\end{bmatrix}$, and $\begin{bmatrix}(2,1)\\1\end{bmatrix}$, and a single extreme ray represented as $\begin{bmatrix}(1,1)\\0\end{bmatrix}$.

1.2 Linear Programming

As calculus developed from the seventeenth century's need to solve problems of mechanics,
linear programming developed from the twentieth century's need to solve problems of management.

Vašek Chvátal (1983, p. 8)

This section recalls some fundamental results of linear programming, notably, the primal and dual formulations of a linear program, the duality principle, and a set of necessary and sufficient optimality conditions of a primal-dual solution pair.

Linear program

A *linear program* is an optimization program in which the objective function and the constraints are linear. We consider an $m \times n$ linear program, denoted *LP*, and expressed with inequality constraints as

$$\begin{aligned} z^{\star}_{LP} = \min \quad & \mathbf{c}^{\mathsf{T}}\mathbf{x} \\ \text{s.t.} \quad & \mathbf{A}\mathbf{x} \geq \mathbf{b} \\ & \mathbf{x} \geq \mathbf{0}, \end{aligned} \tag{1.2}$$

where $\mathbf{x}, \mathbf{c} \in \mathbb{R}^n$, $\mathbf{b} \in \mathbb{R}^m$, $\mathbf{A} \in \mathbb{R}^{m \times n}$. Recall that any equality constraint can be represented by two inequalities. Each constraint $\mathbf{A}_{i*}\mathbf{x} = \sum_{j=1}^{n} a_{ij}x_j \geq b_i$, $i \in \{1,\dots,m\}$, induces a closed half-space. The same is true for each lower bound $x_j \geq 0$, $j \in \{1,\dots,n\}$. The intersection of all these closed half-spaces forms the feasible region, denoted $\mathcal{A} = \{\mathbf{x} \in \mathbb{R}^n_+ \mid \mathbf{A}\mathbf{x} \geq \mathbf{b}\}$, which is a polyhedron by definition.

- An *extreme point* $\mathbf{x}_p$, $p \in P$, of $\mathcal{A}$ is *characterized by* (and can thus be identified with) a basic solution given by a partition of the variable indices $\{1,\dots,n\}$ into a subset B of size m and a subset N of size $n-m$. These sets induce vectors $\mathbf{x}_B$ of *basic* variables and $\mathbf{x}_N$ of *non-basic* variables. Then

$$\mathbf{x}_p = \begin{bmatrix} \mathbf{x}_B \\ \mathbf{x}_N \end{bmatrix} = \begin{bmatrix} \mathbf{x}_B \\ \mathbf{0} \end{bmatrix} = \begin{bmatrix} \mathbf{A}_B^{-1}\mathbf{b} \\ \mathbf{0} \end{bmatrix}, \qquad p \in P. \tag{1.3}$$

- An *extreme ray* $\mathbf{x}_r$, $r \in R$, is an *edge direction that extends to infinity*. It means that from a feasible extreme point, say $\mathbf{x}_p$, we have

$$\mathbf{x}_p + \theta \mathbf{x}_r \geq \mathbf{0}, \quad \mathbf{A}(\mathbf{x}_p + \theta \mathbf{x}_r) = \mathbf{b}, \quad \forall \theta > 0. \tag{1.4}$$

From this simple geometry, it follows that there are only three possible outcomes in optimizing the *LP*. It is *infeasible* if the feasible region is empty, i.e., $\mathcal{A} = \emptyset$. Otherwise, it is feasible in which case either there exists an optimizer which yields the *optimal* objective value or it is *unbounded* because the objective value can be improved indefinitely.

Proposition 1.1. *If the LP* (1.2) *has an optimal objective value $z^{\star}_{LP}$, then there exists at least one optimal solution $\mathbf{x}^{\star}_{LP}$ that is an extreme point $\mathbf{x}_p$, $p \in P$.*

Proof. Consider the level curve of the objective function $\mathbf{c}^{\mathsf{T}}\mathbf{x} = z^{\star}_{LP}$. The polyhedron $(\mathcal{A} \cap \{\mathbf{x} \in \mathbb{R}^n \mid \mathbf{c}^{\mathsf{T}}\mathbf{x} = z^{\star}_{LP}\})$ contains at least one extreme point $\mathbf{x}_p$. We show by contradiction that it must be an extreme point of $\mathcal{A}$ as well. Assume that $\mathbf{x}_p$ is not an extreme point of $\mathcal{A}$ and let $\mathbf{x}_p = \lambda \mathbf{x}_1 + (1-\lambda)\mathbf{x}_2$ be a convex combination for $\mathbf{x}_1, \mathbf{x}_2 \in \mathcal{A}$, $\mathbf{x}_1 \neq \mathbf{x}_2$. Since $\mathbf{c}^{\mathsf{T}}\mathbf{x}_1 \geq z^{\star}_{LP}$ and $\mathbf{c}^{\mathsf{T}}\mathbf{x}_2 \geq z^{\star}_{LP}$, they must be equal to $z^{\star}_{LP}$ as $\mathbf{c}^{\mathsf{T}}\mathbf{x}_p = z^{\star}_{LP}$ must hold. This would mean that $\mathbf{x}_1$ and $\mathbf{x}_2$ also belong to $(\mathcal{A} \cap \{\mathbf{x} \in \mathbb{R}^n \mid \mathbf{c}^{\mathsf{T}}\mathbf{x} = z^{\star}_{LP}\})$ which contradicts that $\mathbf{x}_p$ is an extreme point of $(\mathcal{A} \cap \{\mathbf{x} \in \mathbb{R}^n \mid \mathbf{c}^{\mathsf{T}}\mathbf{x} = z^{\star}_{LP}\})$. □

Proposition 1.2. *If the LP* (1.2) *has an unbounded objective value, then there exists at least one extreme ray $\mathbf{x}_r$, $r \in R$, and some vertex $\mathbf{x}_p$, $p \in P$, such that the points $\mathbf{x}_p + \theta \mathbf{x}_r \in \mathcal{A}$ improve the objective value indefinitely as $\theta \to \infty$.*

An immediate consequence is that we only need to search the extreme points and extreme rays to respectively find an optimal solution or detect unboundedness. Since this number is finite, we have a first brute-force approach.

Dual point of view

Duality theory states that we can view any linear program from a different perspective. In this respect, the formulation of the *LP* (1.2) is called *primal* and it can be associated with another one called *dual* which is denoted *LD*.The latter optimizes over $\boldsymbol{\pi}$ variables. The combination of these two programs is coined the *primal-dual* pair. Let us present them and explore the strength of their relationship in the remainder of this section. The primal is repeated on the left whereas the associated dual appears on the right:

$$\begin{array}{rll|rll} z^{\star}_{LP} = \min & \mathbf{c}^{\mathsf{T}}\mathbf{x} & & z^{\star}_{LD} = \max & \mathbf{b}^{\mathsf{T}}\boldsymbol{\pi} & \\ \text{s.t.} & \mathbf{A}\mathbf{x} \geq \mathbf{b} & [\boldsymbol{\pi}] & \text{s.t.} & \mathbf{A}^{\mathsf{T}}\boldsymbol{\pi} \leq \mathbf{c} & [\mathbf{x}] \\ & \mathbf{x} \geq \mathbf{0} & & & \boldsymbol{\pi} \geq \mathbf{0}. & \end{array} \tag{1.5}$$

Each variable in the primal is associated with a constraint in the dual. Each constraint in the primal is associated with a variable in the dual. In our notation we explicitly point to this relation by putting the respective variables in brackets after the constraints. Note how decision variables of one program appear as dual variables in the other. The sign of variables and type of constraints are deduced from one another in accordance with the optimization direction that is also inverted. The general recipe is given in Table 1.1 from which we establish a fundamental concept of duality stated in the following proposition.

primal minimization		$\longleftrightarrow$	dual maximization	
constraint	$\geq b_i$		variable	$\pi_i \geq 0$
	$\leq b_i$			$\pi_i \leq 0$
	$= b_i$			$\pi_i \in \mathbb{R}$
variable	$x_j \geq 0$		constraint	$\leq c_j$
	$x_j \leq 0$			$\geq c_j$
	$x_j \in \mathbb{R}$			$= c_j$

Table 1.1: Recipe to obtain primal-dual formulations.

Proposition 1.3. *The dual of the dual is the primal.*

Proof. We just have to apply the same transformation on the dual, first replacing $\max\ \mathbf{b}^\mathsf{T}\boldsymbol{\pi}$ by $-\min\ -\mathbf{b}^\mathsf{T}\boldsymbol{\pi}$:

$$\begin{aligned} z^\star_{LD} = -\min\quad & -\mathbf{b}^\mathsf{T}\boldsymbol{\pi} \\ \text{s.t.}\quad & -\mathbf{A}^\mathsf{T}\boldsymbol{\pi} \geq -\mathbf{c} \quad [\mathbf{x}] \\ & \boldsymbol{\pi} \geq \mathbf{0} \end{aligned}$$

$$\begin{aligned} z^\star_{LP} = -\max\quad & -\mathbf{c}^\mathsf{T}\mathbf{x} = \min\ \mathbf{c}^\mathsf{T}\mathbf{x} \\ \text{s.t.}\quad & -\mathbf{A}\mathbf{x} \leq -\mathbf{b} \qquad [\boldsymbol{\pi}] \\ & \mathbf{x} \geq \mathbf{0}. \end{aligned}$$

In particular, we interchange right-hand side coefficients with those of the objective function, multiply the system of constraints by -1, transpose the matrix $\mathbf{A}^\mathsf{T}$, and right-multiply everything by the dual variables $\mathbf{x}$ of the *LD*. □

Note 1.1 (Left-multiply, right-multiply, which is it?) As long as one follows matrix multiplication rules, either one is fine. We however find it convenient to present the dual with a right-multiplication by $\boldsymbol{\pi}$ (e.g., $\mathbf{b}^\mathsf{T}\boldsymbol{\pi}$ rather than $\boldsymbol{\pi}^\mathsf{T}\mathbf{b}$), because the same set of rules allows us to find the primal back. Moreover, the *LD* in (1.5) looks like an optimization program with decision variables on the right and constraints listed as rows. For virtually every other purpose, we use the left-multiplication because of notation elegance.

A linear program is either feasible or its domain is empty. The following proposition mathematically substantiates this statement. Indeed, it shows that we either find a feasible solution $\mathbf{x}$ or provide a certificate of infeasibility as $\boldsymbol{\pi}$.

Proposition 1.4 (Farkas' lemma). *Given* $\mathbf{A} \in \mathbb{R}^{m \times n}$ *and* $\mathbf{b} \in \mathbb{R}^m$, *then exactly one of the following assertions holds:*

1: There exists $\mathbf{x} \in \mathbb{R}^n$ *such that* $\mathbf{A}\mathbf{x} \geq \mathbf{b}$, $\mathbf{x} \geq \mathbf{0}$;
2: There exists $\boldsymbol{\pi} \in \mathbb{R}^m$ *such that* $\mathbf{A}^\mathsf{T}\boldsymbol{\pi} \leq \mathbf{0}$, $\boldsymbol{\pi} \geq \mathbf{0}$, $\mathbf{b}^\mathsf{T}\boldsymbol{\pi} > 0$.

Proof. The proof is left as an exercise, done by construction using the *Phase I* of the primal simplex algorithm (Exercise 1.3). □

Let us figure out what exactly this dual formulation *LD* gives us. Any feasible solution $\mathbf{x} \geq \mathbf{0}$ of cost $\bar{z}$ provides an upper bound on the optimal objective value, i.e., $z^\star_{LP} \leq \bar{z}$. It is also possible to derive a lower bound $\underline{z} \leq z^\star_{LP}$ by combining constraints, say $\boldsymbol{\pi}^\mathsf{T}\mathbf{A}\mathbf{x} \geq \boldsymbol{\pi}^\mathsf{T}\mathbf{b}$. If $\boldsymbol{\pi} \geq \mathbf{0}$, the inequality relations are preserved and this new constraint is obviously redundant. However, if every coefficient in the vector $\boldsymbol{\pi}^\mathsf{T}\mathbf{A}$ is less-than-or-equal to the corresponding one in $\mathbf{c}^\mathsf{T}$, that is, $\boldsymbol{\pi}^\mathsf{T}\mathbf{A} \leq \mathbf{c}^\mathsf{T}$, the right-hand side $\boldsymbol{\pi}^\mathsf{T}\mathbf{b}$ is one such lower bound: $\boldsymbol{\pi}^\mathsf{T}\mathbf{b} \leq \boldsymbol{\pi}^\mathsf{T}\mathbf{A}\mathbf{x} \leq \mathbf{c}^\mathsf{T}\mathbf{x}$.

The interpretation of the dual is immediate then: We want to find the largest such lower bound by finding an optimal combination of constraints over $\boldsymbol{\pi}$. This is known as *weak duality* and is captured in the following proposition. We subsequently show that the largest lower bound is in fact equal to the optimum, if it exists. This is known as *strong duality*.

Proposition 1.5 (Weak duality). *Given primal-dual linear programming formulations* (1.5), *if* $\mathbf{x}$ *is primal feasible and* $\boldsymbol{\pi}$ *is dual feasible, then*

$$\mathbf{b}^\mathsf{T}\boldsymbol{\pi} \leq z^\star_{LD} \leq z^\star_{LP} \leq \mathbf{c}^\mathsf{T}\mathbf{x}. \tag{1.6}$$

Proof. We have $\mathbf{b} \leq \mathbf{A}\mathbf{x}$, $\mathbf{x} \geq \mathbf{0}$, and $\mathbf{A}^\mathsf{T}\boldsymbol{\pi} \leq \mathbf{c}$, $\boldsymbol{\pi} \geq \mathbf{0}$. We left-multiply $\mathbf{b} \leq \mathbf{A}\mathbf{x}$ by $\boldsymbol{\pi}$ and right-multiply $\mathbf{A}^\mathsf{T}\boldsymbol{\pi} \leq \mathbf{c}$ by $\mathbf{x}$. This yields $\boldsymbol{\pi}^\mathsf{T}\mathbf{b} \leq \boldsymbol{\pi}^\mathsf{T}\mathbf{A}\mathbf{x} = (\mathbf{A}^\mathsf{T}\boldsymbol{\pi})^\mathsf{T}\mathbf{x} \leq \mathbf{c}^\mathsf{T}\mathbf{x}$, where the equality holds because of transpose properties. Since optimal values $z^\star_{LD} = \boldsymbol{\pi}^{\star^\mathsf{T}}\mathbf{b}$ and $z^\star_{LP} = \mathbf{c}^\mathsf{T}\mathbf{x}^\star_{LP}$ are as good as feasible ones, $\boldsymbol{\pi}^\mathsf{T}\mathbf{b} \leq \boldsymbol{\pi}^{\star^\mathsf{T}}\mathbf{b} \leq \mathbf{c}^\mathsf{T}\mathbf{x}^\star_{LP} \leq \mathbf{c}^\mathsf{T}\mathbf{x}$. Finally, the dot product $\boldsymbol{\pi}^\mathsf{T}\mathbf{b}$ is commutative. □

Corollary 1.1. *If* $\mathbf{c}^\mathsf{T}\mathbf{x} = \boldsymbol{\pi}^\mathsf{T}\mathbf{b}$ *for primal and dual feasible solutions* $\mathbf{x}$ *and* $\boldsymbol{\pi}$, *then these are optimal solutions for the LP and the LD, respectively.*

Corollary 1.2. *If the primal is unbounded, then the dual is infeasible (by weak duality,* $z^\star_{LD} \leq z^\star_{LP} = -\infty$, $\{\boldsymbol{\pi} \in \mathbb{R}^m \mid \mathbf{A}^\mathsf{T}\boldsymbol{\pi} \leq \mathbf{c}\} = \emptyset$, *and the LD has no feasible solution).*

Proposition 1.6 (Strong duality). *When either the LP or the LD has a finite optimal solution, then the other also has a finite optimal solution and their respective optima coincide:*

$$\exists \mathbf{x}^\star_{LP} \Leftrightarrow \exists \boldsymbol{\pi}^\star \quad and \quad \mathbf{b}^\mathsf{T}\boldsymbol{\pi}^\star = \mathbf{c}^\mathsf{T}\mathbf{x}^\star_{LP}. \tag{1.7}$$

Proof. This proof is left as an exercise, done by construction using an optimal basic solution given by the primal simplex algorithm (Exercise 1.8). □

Corollary 1.3. *There are four possible outcomes for the primal-dual formulations:*

1. primal feasible, dual feasible
2. primal feasible, dual infeasible
3. primal infeasible, dual feasible
4. primal infeasible, dual infeasible

Proof. Proposition 1.6 covers the first case and shows that both programs reach the same optimum. Corollary 1.2 covers the second and the third by mirroring the result. The fourth is established by recognizing that, with respect to Farkas' lemma (Proposition 1.4), primal infeasibility (i.e., $\mathbf{A}^\intercal\boldsymbol{\pi} \geq \mathbf{0}, \boldsymbol{\pi} \geq \mathbf{0}, \mathbf{b}^\intercal\boldsymbol{\pi} > 0$) and dual infeasibility (i.e., $\mathbf{Ax} \geq \mathbf{0}, \mathbf{x} \geq \mathbf{0}, \mathbf{c}^\intercal\mathbf{x} < 0$) can both be true simultaneously. □

Table 1.2 summarizes the primal-dual relationship we have established.

	Dual formulation		
Primal formulation	Feasible	Unbounded	Infeasible
Feasible: $z^\star_{LP} > -\infty$	$z^\star_{LD} = z^\star_{LP}$		
Unbounded: $z^\star_{LP} = -\infty$			$\{\boldsymbol{\pi} \in \mathbb{R}^m \mid \mathbf{A}^\intercal\boldsymbol{\pi} \leq \mathbf{c}\} = \emptyset$
Infeasible: $\{\mathbf{x} \in \mathbb{R}^n_+ \mid \mathbf{Ax} = \mathbf{b}\} = \emptyset$		$z^\star_{LD} = \infty$	$\{\boldsymbol{\pi} \in \mathbb{R}^m \mid \mathbf{A}^\intercal\boldsymbol{\pi} \leq \mathbf{c}\} = \emptyset$

Table 1.2: Relationship between the primal-dual formulations.

Finally, it can likewise be read from the dual perspective. Exercise 1.4 asks the reader to formulate a simultaneously infeasible primal-dual pair.

Optimality conditions

Necessary and sufficient optimality conditions are given in the next proposition.

Proposition 1.7 (Complementary slackness). *Given primal-dual linear programming formulations* (1.5)*, the primal-dual solution pair* $(\mathbf{x}, \boldsymbol{\pi})$ *is optimal if and only if* $\mathbf{x}$ *is primal feasible,* $\boldsymbol{\pi}$ *is dual feasible, and*

$$\begin{aligned} &\text{for all } j \in \{1,\dots,n\}: & c_j - \boldsymbol{\pi}^\intercal\mathbf{a}_j > 0 &\Rightarrow x_j = 0, \\ & & x_j > 0 &\Rightarrow c_j - \boldsymbol{\pi}^\intercal\mathbf{a}_j = 0; \\ &\text{for all } i \in \{1,\dots,m\}: & b_i - \mathbf{A}^\intercal_{i*}\mathbf{x} < 0 &\Rightarrow \pi_i = 0, \\ & & \pi_i > 0 &\Rightarrow b_i - \mathbf{A}^\intercal_{i*}\mathbf{x} = 0. \end{aligned} \tag{1.8}$$

Proof. We derive these conditions from Propositions 1.5–1.6 which give us $\boldsymbol{\pi}^\intercal\mathbf{b} = \boldsymbol{\pi}^\intercal\mathbf{Ax} = \mathbf{c}^\intercal\mathbf{x}$ for any optimal primal-dual pair $(\mathbf{x}, \boldsymbol{\pi})$. We can rewrite this as

$$(\mathbf{c}^\intercal - \boldsymbol{\pi}^\intercal\mathbf{A})\mathbf{x} = 0 \text{ and } \boldsymbol{\pi}^\intercal(\mathbf{Ax} - \mathbf{b}) = 0, \tag{1.9}$$

where $\mathbf{x}, \boldsymbol{\pi}, (\mathbf{c}^\intercal - \boldsymbol{\pi}^\intercal\mathbf{A})$, and $(\mathbf{Ax} - \mathbf{b})$ are non-negative vectors. Consequently, for each variable index $j \in \{1,\dots,n\}$, the product $(c_j - \boldsymbol{\pi}^\intercal\mathbf{a}_j)x_j$ must be zero such that at least one of these factors must be zero, i.e., either $x_j = 0$ or $c_j - \boldsymbol{\pi}^\intercal\mathbf{a}_j = 0$ or both. A similar result can be made for each constraint index $i \in \{1,\ldots,m\}$ using the components of $\boldsymbol{\pi}^\mathsf{T}(\mathbf{A}\mathbf{x}-\mathbf{b}) = 0$. □

It is obvious then that the interpretation of complementary slackness (1.8) is viscerally present in the structure of any optimal solution. While we are interested in an optimal solution $\mathbf{x}^\star_{LP}$, the previous propositions make it clear that a certificate of optimality for the *LP* necessarily comes as a primal-dual solution pair $(\mathbf{x}^\star_{LP}, \boldsymbol{\pi}^\star)$. Dual variables are so rooted in linear programming algorithms that it is convenient to list them even when only the primal formulation is presented. Accordingly, when we write about the dual variables in the following, we sometimes refer to their *values*, not the variables themselves. The reader will easily distinguish the respective meaning depending on the context so we do not introduce an extra notation.

Where do we go from here? For starters, we can already make one important observation: If $c_j - \boldsymbol{\pi}^\mathsf{T}\mathbf{a}_j > 0$ then variable x_j is useless. In fact, we could remove it from the model and achieve optimality just the same. How many such "useless" variables are there? Well, we can also observe that we are free to "test optimality" using any $\mathbf{x} \geq \mathbf{0}$ together with any $\boldsymbol{\pi} \geq \mathbf{0}$. With this in mind, let us first give names to the two expressions seen in (1.8).

Definition 1.13. Given *any* dual vector $\boldsymbol{\pi} \geq \mathbf{0}$, we call $\bar{c}_j = c_j - \boldsymbol{\pi}^\mathsf{T}\mathbf{a}_j$ the *reduced cost* of variable x_j, $j \in \{1,\ldots,n\}$, and let $\bar{\mathbf{c}} = [\bar{c}_j]_{j\in\{1,\ldots,n\}}$.

Definition 1.14. Given *any* primal vector $\mathbf{x} \geq \mathbf{0}$, let $s_i = b_i - \mathbf{A}^\mathsf{T}_{i*}\mathbf{x}$ be called the *slack* of constraint $i \in \{1,\ldots,m\}$. Depending on whether or not the slack is zero, we say that the constraint is *binding* (aka *tight*) or *non-binding*.

Given these, the complementary slackness conditions (1.8) take a simpler form:

$$
\begin{aligned}
\text{for all } j \in \{1,\ldots,n\}: \quad & \bar{c}_j > 0 \;\Rightarrow\; x_j = 0, \\
& x_j > 0 \;\Rightarrow\; \bar{c}_j = 0; \\
\text{for all } i \in \{1,\ldots,m\}: \quad & s_i < 0 \;\Rightarrow\; \pi_i = 0, \\
& \pi_i > 0 \;\Rightarrow\; s_i = 0,
\end{aligned}
\tag{1.10}
$$

which also gives us for an optimal primal-dual pair $(\mathbf{x}^\star_{LP}, \boldsymbol{\pi}^\star)$, $\sum_{j=1}^{n} \bar{c}_j x^\star_j = 0$ and $\sum_{i=}^{m} s_i \pi^\star_i = 0$, that is,

$$
\bar{\mathbf{c}}^\mathsf{T}\mathbf{x}^\star_{LP} = 0 \text{ and } \boldsymbol{\pi}^{\star\mathsf{T}}\mathbf{s} = 0. \tag{1.11}
$$

Sensitivity analysis

In economics, linear programs are used to compute profit maximizing production plans subject to scarce resources. The constraints are typically related to these resources while variables indicate how much of which product is produced on which resource. In this context, optimal dual values have a natural interpretation as *marginal profits* or *shadow prices* or *opportunity costs* (these mean all the same): For each resource the corresponding optimal dual value states by how much the profit could be theoretically increased if we had one more unit of this resource. On the other hand, if a constraint is not *binding*, we do not use all available units of the corresponding resource such that there is no value in having one more unit of it. We intuitively find complementary slackness in this interpretation.

One can derive this interpretation from a linear program with an optimal solution $(\mathbf{x}^\star_{LP}, \boldsymbol{\pi}^\star)$ of cost $z^\star_{LP}$ by keeping all parameters unchanged except one. Specifically, we consider a variation of a coefficient from either the right-hand side or the objective function. With respect to a right-hand side modification $\Delta b_i \in \mathbb{R}$ of b_i, $i \in \{1,\ldots,m\}$, the primal-dual formulations become

$$\begin{array}{ll|ll}
z^\star_{LP}(\Delta b_i) & & z^\star_{LD}(\Delta b_i) & \\
\quad = \min & \mathbf{c}^\mathsf{T}\mathbf{x} & \quad = \max & \mathbf{b}^\mathsf{T}\boldsymbol{\pi} + \Delta b_i \pi_i \\
\quad \text{s.t.} & \mathbf{A}\mathbf{x} \geq \mathbf{b} + \Delta b_i \mathbf{e}_i \quad [\boldsymbol{\pi}] & \quad \text{s.t.} & \mathbf{A}^\mathsf{T}\boldsymbol{\pi} \leq \mathbf{c} \quad [\mathbf{x}] \\
 & \mathbf{x} \geq \mathbf{0} & & \boldsymbol{\pi} \geq \mathbf{0}.
\end{array} \tag{1.12}$$

Looking at the dual, we see that $\boldsymbol{\pi}^\star$ remains feasible since only the objective function is modified. However, we do not know how $\mathbf{x}^\star_{LP}$ is changed: the primal program may have an optimal solution but the objective function can also be unbounded. By *weak duality*, the optimum is at least $\mathbf{b}^\mathsf{T}\boldsymbol{\pi}^\star + \Delta b_i \pi^\star_i = z^\star_{LP} + \Delta b_i \pi^\star_i$. It should be obvious that this interpretation is *local* to the current solution. With respect to the primal simplex algorithm we see next, most solvers can report a *sensitivity range* for which vector $\boldsymbol{\pi}^\star$ remains unchanged (and optimal) for the modified dual program, see Figure 1.3a.

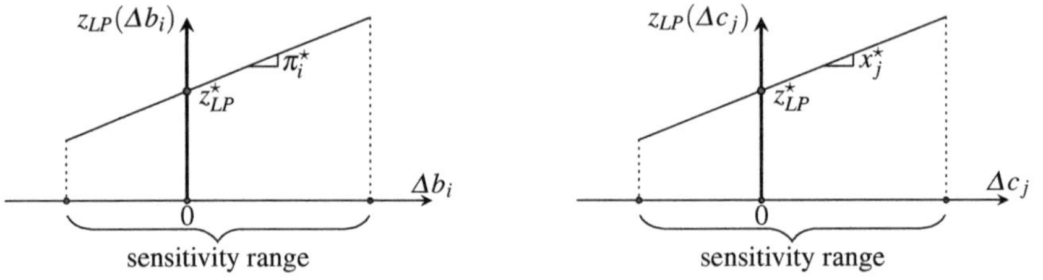

(a) Sensitivity analysis for coefficient b_i (b) Sensitivity analysis for coefficient c_j

Fig. 1.3: Variation of a coefficient from the right side or objective function.

We can do a similar study with a modification $\Delta c_j \in \mathbb{R}$ for the objective coefficient associated with column index c_j, $j \in \{1,\ldots,n\}$ (see Figure 1.3b):

$$
\begin{array}{ll|ll}
z^\star_{LP}(\Delta c_j) & & z^\star_{LD}(\Delta c_j) = & \\
\quad = \min \ \mathbf{c}^\mathsf{T}\mathbf{x} + \Delta c_j\, x_j & & \quad \max \quad \mathbf{b}^\mathsf{T}\boldsymbol{\pi} & \\
\quad \text{s.t.} \ \mathbf{A}\mathbf{x} \geq \mathbf{b} & [\boldsymbol{\pi}] & \quad \text{s.t.} \quad \mathbf{A}^\mathsf{T}\boldsymbol{\pi} \leq \mathbf{c} + \Delta c_j\, \mathbf{e}_j & [\mathbf{x}] \\
\quad \mathbf{x} \geq \mathbf{0} & & \quad \boldsymbol{\pi} \geq \mathbf{0}. &
\end{array}
\tag{1.13}
$$

Regarding the sensitivity analysis for the lower bounds, say $\mathbf{x} \geq \boldsymbol{\ell} = \mathbf{0}$, we combine the reduced cost expression $\bar{\mathbf{c}}^\mathsf{T} = \mathbf{c}^\mathsf{T} - \boldsymbol{\pi}^\mathsf{T}\mathbf{A}$ and the equality $\boldsymbol{\pi}^{\star\mathsf{T}}\mathbf{s} = 0$ (i.e., $\boldsymbol{\pi}^{\star\mathsf{T}}\mathbf{A}\mathbf{x}^\star_{LP} = \boldsymbol{\pi}^{\star\mathsf{T}}\mathbf{b}$) to rewrite $z^\star_{LP}$ in the following way:

$$
z^\star_{LP} = \mathbf{c}^\mathsf{T}\mathbf{x}^\star_{LP} = (\bar{\mathbf{c}}^\mathsf{T} + \boldsymbol{\pi}^{\star\mathsf{T}}\mathbf{A})\mathbf{x}^\star_{LP} = \boldsymbol{\pi}^{\star\mathsf{T}}\mathbf{b} + \bar{\mathbf{c}}^\mathsf{T}\mathbf{x}^\star_{LP} \tag{1.14}
$$

where $\bar{\mathbf{c}}^\mathsf{T}\mathbf{x}^\star_{LP} = 0$. However, it also writes as $\bar{\mathbf{c}}^\mathsf{T}\mathbf{x}^\star_{LP} = \sum_{j:x^\star_j = \ell_i} \bar{c}_j \ell_j + \sum_{j:x^\star_j > \ell_i} \bar{c}_j \ell_j$, where the second term is zero by the complementary slackness ($x_j > 0 \Rightarrow \bar{c}_j = 0$). Hence

$$
z^\star_{LP} = \boldsymbol{\pi}^{\star\mathsf{T}}\mathbf{b} + \sum_{j:x^\star_j = \ell_j (=0)} \bar{c}_j \ell_j. \tag{1.15}
$$

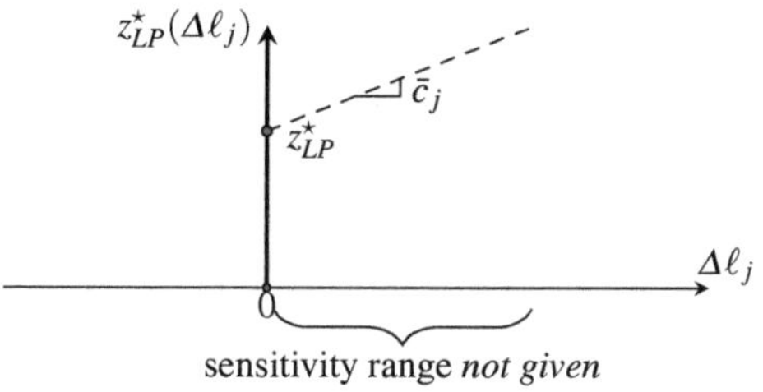

Fig. 1.4: Sensitivity analysis for a lower bound coefficient.

For an increase $\Delta\ell_j \in \mathbb{R}_+$ of the lower bound for variable $x_j = 0$, the objective function varies as

$$
z^\star_{LP}(\Delta\ell_j) = \boldsymbol{\pi}^{\star\mathsf{T}}\mathbf{b} + \bar{c}_j \Delta\ell_j = z^\star_{LP} + \bar{c}_j \Delta\ell_j. \tag{1.16}
$$

As above, we again face a line equation illustrated in Figure 1.4, but this time without having the sensitivity range computed by the commercial solvers. This can easily be done, see Exercise 1.6 (Sensitivity range for an increase of the lower bound).

Observe that the slopes $x^\star_j$, $\pi^\star_i$, and $\bar{c}_j$ are all non-negative for the three cases of the sensitivity analysis for the *LP* (1.2). This is not necessarily the same for other formulations: we always have $x^\star_j \geq 0$, but the last two varies according to the type of constraints ($=, \geq, \leq$) or the optimization criterion ($\min, \max$).

1.3 Primal Simplex Algorithm

In the primal simplex algorithm presented next, we move from one extreme point to the next *while always maintaining primal feasibility.* For each extreme point, we can derive corresponding dual values. We iterate until these values become dual feasible at which point we have achieved provable optimality.

For the primal simplex algorithm, we consider equality constraints only, that is, surplus and slack variables can be added in presence of inequalities. We also assume that the matrix $\mathbf{A}$ is of full row rank. This is called the *standard form* of the *LP*. The *new* vector of dual variables $\boldsymbol{\pi} \in \mathbb{R}^m$ associated with the equality constraints appears within brackets (see Exercise 1.7 Lost in translation?), and the dual formulation *LD* appears on the right of (1.17).

$$\begin{aligned} z_{LP}^\star = \min \quad & \mathbf{c}^\mathsf{T}\mathbf{x} \\ \text{s.t.} \quad & \mathbf{A}\mathbf{x} = \mathbf{b} \quad [\boldsymbol{\pi}] \\ & \mathbf{x} \geq \mathbf{0} \end{aligned} \qquad \Bigg| \qquad \begin{aligned} z_{LD}^\star = \max \quad & \mathbf{b}^\mathsf{T}\boldsymbol{\pi} \\ \text{s.t.} \quad & \mathbf{A}^\mathsf{T}\boldsymbol{\pi} \leq \mathbf{c} \quad [\mathbf{x}]. \end{aligned} \tag{1.17}$$

For a linear program expressed in standard form, the optimality conditions (1.8) simplify to

$$\textit{for all } j \in \{1,\ldots,n\}: \quad \begin{aligned} \bar{c}_j > 0 &\Rightarrow x_j = 0, \\ x_j > 0 &\Rightarrow \bar{c}_j = 0. \end{aligned} \tag{1.18}$$

Mechanics

The primal simplex algorithm relies on an important property of linear programs. If the *LP* (1.17) has a finite optimal objective value $z_{LP}^\star$, then there exists at least one optimal solution $\mathbf{x}_{LP}^\star$ given by an extreme point $\mathbf{x}_p, p \in P$. Otherwise, $z_{LP}^\star$ is unbounded and we have identified an improving direction, represented by an extreme ray $\mathbf{x}_r$, $r \in R$, that extends to infinity. From a geometric point of view, if $z_{LP}^\star > -\infty$, then the primal simplex algorithm moves from an extreme point to an adjacent one, indeed, from one basis to an adjacent one by changing a single column.

Let us rewrite the *LP* in terms of basic and non-basic variables indexed by B and N, respectively:

$$\begin{aligned} z_{LP}^\star = \min \quad & \begin{bmatrix} \mathbf{c}_B \\ \mathbf{c}_N \end{bmatrix}^\mathsf{T} \begin{bmatrix} \mathbf{x}_B \\ \mathbf{x}_N \end{bmatrix} = \mathbf{c}_B^\mathsf{T}\mathbf{x}_B + \mathbf{c}_N^\mathsf{T}\mathbf{x}_N \\ \text{s.t.} \quad & \begin{bmatrix} \mathbf{A}_B \ \mathbf{A}_N \end{bmatrix} \begin{bmatrix} \mathbf{x}_B \\ \mathbf{x}_N \end{bmatrix} = \mathbf{A}_B\mathbf{x}_B + \mathbf{A}_N\mathbf{x}_N = \mathbf{b} \\ & \mathbf{x}_B \geq \mathbf{0}, \quad \mathbf{x}_N \geq \mathbf{0}, \end{aligned} \tag{1.19}$$

where the matrix $\mathbf{A}_B$ is non-singular. Temporarily fixing $\mathbf{x}_N = \mathbf{0}$, the linear system $\mathbf{A}_B\mathbf{x}_B = \mathbf{b}$ is solved with $\mathbf{x}_B = \mathbf{A}_B^{-1}\mathbf{b}$ of cost $\mathbf{c}_B^\mathsf{T}\mathbf{A}_B^{-1}\mathbf{b}$.

To verify if this solution is optimal or not, the mathematical expression of $\mathbf{x}_B$, given by $\mathbf{x}_B = \mathbf{A}_B^{-1}\mathbf{b} - \mathbf{A}_B^{-1}\mathbf{A}_N\mathbf{x}_N$, is substituted in the objective function. This results in

$$\mathbf{c}_B^\top\mathbf{x}_B + \mathbf{c}_N^\top\mathbf{x}_N = \mathbf{c}_B^\top\mathbf{A}_B^{-1}\mathbf{b} + (\mathbf{c}_N^\top - \mathbf{c}_B^\top\mathbf{A}_B^{-1}\mathbf{A}_N)\mathbf{x}_N. \tag{1.20}$$

Observe that $\mathbf{c}_B^\top\mathbf{A}_B^{-1}$ appears twice in the above expression. It is rather computed only once as $\boldsymbol{\pi}^\top \in \mathbb{R}^m$ (called the vector of *simplex multipliers*), a particular choice for the dual values such that the coefficient of $\mathbf{x}_N$ in the objective function becomes the reduced cost vector $(\mathbf{c}_N^\top - \boldsymbol{\pi}^\top\mathbf{A}_N) = \bar{\mathbf{c}}_N^\top$. Left-multiplying the system of constraints by $\mathbf{A}_B^{-1}$, let $\bar{\mathbf{b}} = \mathbf{A}_B^{-1}\mathbf{b}$ and $\bar{\mathbf{A}}_N = \mathbf{A}_B^{-1}\mathbf{A}_N$, then the *LP* (1.19) becomes:

$$\begin{aligned} z_{LP}^\star = \boldsymbol{\pi}^\top\mathbf{b} + \min \quad & && \bar{\mathbf{c}}_N^\top\mathbf{x}_N \\ \text{s.t.} \quad & \mathbf{x}_B && + \bar{\mathbf{A}}_N\mathbf{x}_N = \bar{\mathbf{b}} \\ & \mathbf{x}_B \geq \mathbf{0}, && \quad \mathbf{x}_N \geq \mathbf{0}. \end{aligned} \tag{1.21}$$

This is also known as the *dictionary* representation.

Definition 1.15. Let the matrix $\mathbf{A}_B$ be non-singular. A *basic solution* $\mathbf{x}$ is given by

$$\mathbf{x} = \begin{bmatrix} \mathbf{x}_B \\ \mathbf{x}_N \end{bmatrix} = \begin{bmatrix} \mathbf{A}_B^{-1}\mathbf{b} \\ \mathbf{0} \end{bmatrix} = \begin{bmatrix} \bar{\mathbf{b}} \\ \mathbf{0} \end{bmatrix}. \tag{1.22}$$

The associated vector of *simplex multipliers* is $\boldsymbol{\pi}^\top = \mathbf{c}_B^\top\mathbf{A}_B^{-1}$.

Definition 1.16. A basic solution is *degenerate* if at least one of the basic variables takes value zero.

Sufficient optimality conditions

Given a basic solution $\mathbf{x} = \begin{bmatrix} \mathbf{x}_B \\ \mathbf{x}_N \end{bmatrix}$, where $\mathbf{x}_B = \bar{\mathbf{b}}$ and $\mathbf{x}_N = \mathbf{0}$, *sufficient conditions* for $\mathbf{x}$ to be optimal in (1.21) are $\bar{\mathbf{c}}_N \geq \mathbf{0}$, that is, $\bar{c}_j \geq 0$, $\forall j \in N$. By design of the primal simplex algorithm, observe that $\bar{\mathbf{c}}_B^\top = \mathbf{c}_B^\top - \boldsymbol{\pi}^\top\mathbf{A}_B = \mathbf{c}_B^\top - \mathbf{c}_B^\top\mathbf{A}_B^{-1}\mathbf{A}_B = \mathbf{0}^\top$. Hence, the stopping rule $\bar{\mathbf{c}}_N \geq \mathbf{0}$ for a finite objective value is equivalent to

$$\bar{c}_j \geq 0, \ \forall j \in \{1, \ldots, n\}, \ \text{ where } \ \bar{c}_j = 0, \ \forall j \in B. \tag{1.23}$$

These conditions are not all requested by the complementary slackness conditions: they are sufficient but not necessary. Indeed, only the *positive variables* are requested to have reduced costs of zero value in (1.18) whereas the primal simplex algorithm also imposes such a condition on *all* the basic variables, positive and degenerate.

Pricing problem

Given a basic solution $\mathbf{x}$ and corresponding dual multipliers $\boldsymbol{\pi}$, we verify whether the optimality conditions (1.23) are fulfilled. Otherwise, we identify a non-basic variable x_ℓ, $\ell \in N$, with a negative reduced cost ($\bar{c}_\ell < 0$). Since this value indicates the marginal impact of increasing the value of x_ℓ by one unit, the *Dantzig's rule* is to select a most promising one x_ℓ, i.e., one with the least reduced cost:

$$\ell \in \arg\min_{j \in N} \bar{c}_j. \tag{1.24}$$

Note 1.2 (Generic pricing.) The minimum reduced cost $\bar{c}_\ell$ is usually found by enumeration. Obviously dependent on the vector $\boldsymbol{\pi}$, it can also be formulated as an optimization program *SP* (the subproblem or pricing problem, see Linear programs for the pricing), whose value is given by

$$\bar{c}(\boldsymbol{\pi}) = \min_{j \in N} c_j - \boldsymbol{\pi}^\mathsf{T} \mathbf{a}_j. \tag{1.25}$$

The advanced reader can already observe that (1.25) is similar to the notation

- $\bar{c}(\boldsymbol{\pi})$ used in the subproblem of the column generation algorithm (Chapter 2),
- $\bar{c}(\boldsymbol{\pi}_\mathbf{b}, \pi_0)$ used in the subproblem of the Dantzig-Wolfe reformulation (Chapters 3 and 4),
- the value $LR(\boldsymbol{\pi}_\mathbf{b})$ utilized in the subproblem of the Lagrangian relaxation (Chapter 6).

In a very general sense, their purpose is always the same: *pricing the variables.*

Let the index-set N be further partitioned into $\{\ell, N'\}$. Formulation (1.21) becomes

$$\begin{aligned} z^\star_{LP} = \boldsymbol{\pi}^\mathsf{T}\mathbf{b} \, + \, \min \quad & \begin{bmatrix} \mathbf{0} \\ \bar{c}_\ell \\ \bar{\mathbf{c}}_{N'} \end{bmatrix}^\mathsf{T} \begin{bmatrix} \mathbf{x}_B \\ x_\ell \\ \mathbf{x}_{N'} \end{bmatrix} \\ \text{s.t.} \quad & \begin{bmatrix} \mathbf{I}_m & \bar{\mathbf{a}}_\ell & \bar{\mathbf{A}}_{N'} \end{bmatrix} \begin{bmatrix} \mathbf{x}_B \\ x_\ell \\ \mathbf{x}_{N'} \end{bmatrix} = \bar{\mathbf{b}} \\ & \mathbf{x}_B \geq \mathbf{0}, \; x_\ell \geq 0, \; \mathbf{x}_{N'} \geq \mathbf{0}. \end{aligned} \tag{1.26}$$

Since all non-basic variables are at zero, we can determine how much we can increase x_ℓ using a simple system of m equalities between the basic variables and x_ℓ:

$$\mathbf{x}_B + \bar{\mathbf{a}}_\ell x_\ell = \bar{\mathbf{b}}, \qquad \mathbf{x}_B \geq \mathbf{0}, \; x_\ell \geq 0, \tag{1.27}$$

hence the new $\mathbf{x}$-vector is given by

$$\begin{bmatrix} \mathbf{x}_B \\ x_\ell \\ \mathbf{x}_{N'} \end{bmatrix} = \begin{bmatrix} \bar{\mathbf{b}} \\ 0 \\ \mathbf{0} \end{bmatrix} + \begin{bmatrix} -\bar{\mathbf{a}}_\ell \\ 1 \\ \mathbf{0} \end{bmatrix} x_\ell. \tag{1.28}$$

System (1.27)–(1.28) capitalizes on the basis structure to modify the current solution. Using variable x_ℓ either yields infinite improvement or it trades places with exactly one currently basic variable. It is no surprise that we have nice adjectives for these two variables then: *entering* and *leaving*. This exchange of variable is called a *pivot*. Note that any variable with a negative reduced cost could be used in (1.27).

Extreme ray direction

If $\bar{\mathbf{a}}_\ell \leq \mathbf{0}$, the basic variables remain non-negative for every $x_\ell > 0$, that is,

$$x_\ell > 0 \quad \Rightarrow \quad \mathbf{x}_B = \bar{\mathbf{b}} - \bar{\mathbf{a}}_\ell x_\ell \geq \mathbf{0}, \tag{1.29}$$

and $z^\star_{LP} = -\infty$ as we have identified an extreme ray $\mathbf{x}_r = \begin{bmatrix} -\bar{\mathbf{a}}_\ell \\ 1 \\ \mathbf{0} \end{bmatrix}$, originating from the current basic solution. Note that because all data/coefficients are rational, $\mathbf{x}_r$ can be integer-scaled using an appropriate integer value for x_ℓ.

Adjacent extreme point

If $\bar{\mathbf{a}}_\ell \not\leq \mathbf{0}$, the maximum value of x_ℓ is restricted by its impact on $\mathbf{x}_B$. For all scalars $\bar{a}_{i\ell} > 0$, $i \in \{1,\ldots,m\}$, the corresponding components must remain non-negative: $\bar{b}_i - \bar{a}_{i\ell} x_\ell \geq 0$, i.e., $x_\ell \leq {}^{\bar{b}_i}/_{\bar{a}_{i\ell}}$. Hence x_ℓ is computed by the *minimum ratio*:

$$x_\ell = \min_{i \in \{1,\ldots,m\} | \bar{a}_{i\ell} > 0} \frac{\bar{b}_i}{\bar{a}_{i\ell}}. \tag{1.30}$$

- If $x_\ell > 0$, the solution moves on an edge direction from the current extreme point until it reaches an adjacent extreme point:

$$\mathbf{x}_p = \begin{bmatrix} \bar{\mathbf{b}} \\ 0 \\ \mathbf{0} \end{bmatrix} + \begin{bmatrix} -\bar{\mathbf{a}}_\ell \\ 1 \\ \mathbf{0} \end{bmatrix} x_\ell, \qquad x_\ell = \min_{i \in \{1,\ldots,m\} | \bar{a}_{i\ell} > 0} \frac{\bar{b}_i}{\bar{a}_{i\ell}}. \tag{1.31}$$

  At least one of the former basic variables decreases to zero and one such variable is chosen to leave the basis, say x_s, whereas x_ℓ enters it:

$$s \in \arg\min_{i \in \{1,\ldots,m\} | \bar{a}_{i\ell} > 0} \bar{b}_i / \bar{a}_{i\ell}. \tag{1.32}$$

- If $x_\ell = 0$, the extreme point remains the same—this is called a *degenerate pivot*—but the subset of basic variables is different: x_ℓ enters the basis at value zero and the variable x_s that leaves it is any one for which $\bar{a}_{s\ell} > 0$ and $\bar{b}_s = 0$, $s \in B$. This means we change a degenerate basis for another degenerate one.

Figure 1.5 illustrates the primal simplex algorithm. At every iteration, the pricing problem *SP* uses the simplex multipliers $\boldsymbol{\pi}$, computed in accordance with the current basis of the *LP*, to identify a variable x_ℓ of negative reduced cost $c(\boldsymbol{\pi})$, if any. Because the primal simplex maintains primal feasibility, the output of the algorithm is only one of these two cases:

- a *certificate of unboundedness* ($z^\star_{LP} = -\infty$) as an extreme ray $\mathbf{x}_r$, $r \in R$;
- a *certificate of optimality* as an extreme point $\mathbf{x}^\star_{LP} = \mathbf{x}^\star_p, p \in P$, together with the optimal value $z^\star_{LP} = \mathbf{c}^\intercal \mathbf{x}^\star_{LP}$ and an optimal dual vector $\boldsymbol{\pi}^\star$ such that $\mathbf{c}^\intercal \mathbf{x}^\star_{LP} = \mathbf{b}^\intercal \boldsymbol{\pi}^\star$.

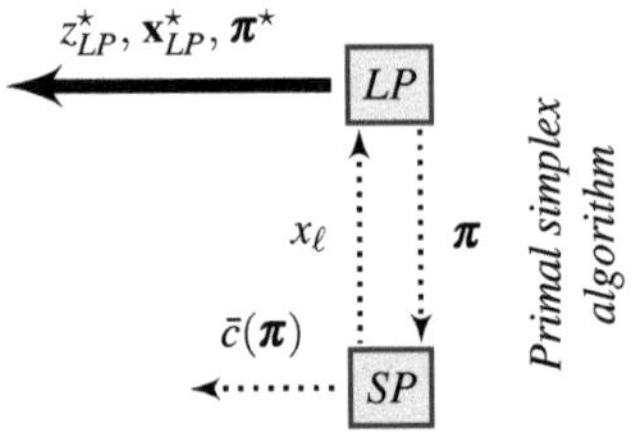

Fig. 1.5: Illustration of the primal simplex algorithm for solving the *LP* (1.17).

At every iteration of the algorithm, the *LP* sends the vector $\boldsymbol{\pi}$ of simplex multipliers to the pricing problem *SP* which returns the minimum (negative) reduced cost $\bar{c}(\boldsymbol{\pi})$ together with an entering variable x_ℓ. These actions appear as dotted lines in Figure 1.5. An optimal solution to the *LP*, here assumed finite, comes out as the primal-dual pair $(\mathbf{x}^\star_{LP}, \boldsymbol{\pi}^\star)$ together with its cost $z^\star_{LP}$. Although simple, this figure is later adapted to illustrate the column generation algorithm, the Dantzig-Wolfe reformulation of linear and integer linear programs, the later being finally compared to the Lagrangian relaxation of an integer linear program.

Initialization and pseudo-code

The reader might have realized that we assumed that a basis $\mathbf{A}_B$ is given at all times. The initialization of the primal simplex algorithm produces an initial basic solution to the *LP*. This is done using the so-called *Phase I* method.

An auxiliary problem is solved with m artificial variables, that is, $\mathbf{y} = [y_i]_{i=1,\ldots,m}$ that absorb the initial infeasibility of the *LP* by taking a trivial value $\mathbf{y} = \mathbf{b}$. Their presence is then discouraged by minimizing their sum. The *LP* is feasible if and only if $z^\star_{Phase\,I} = 0$. If so, we obtain an initial basic solution $\mathbf{x}$ that is also feasible for the *LP*; otherwise, $\mathcal{A} = \{\mathbf{x} \in \mathbb{R}^n \mid \mathbf{A}\mathbf{x} = \mathbf{b}\} = \emptyset$. The remainder of the solution process is sometimes called *Phase II*.

$$
\begin{aligned}
z^{\star}_{Phase\,I} = \min \quad & \mathbf{1}^{\mathsf{T}}\mathbf{y} \\
\text{s.t.} \quad & \mathbf{A}\mathbf{x} + \mathbf{y} = \mathbf{b} \\
& \mathbf{x},\ \mathbf{y} \geq \mathbf{0}.
\end{aligned} \tag{1.33}
$$

In practice, the two phases are merged using bloated objective coefficients on the artificial y-variables:

$$
\begin{aligned}
z^{\star}_{Phase\,I} = \min \quad & \mathbf{c}^{\mathsf{T}}\mathbf{x} + M\mathbf{1}^{\mathsf{T}}\mathbf{y} \\
\text{s.t.} \quad & \mathbf{A}\mathbf{x} + \mathbf{y} = \mathbf{b} \\
& \mathbf{x} \geq \mathbf{0}, \quad \mathbf{y} \geq \mathbf{0},
\end{aligned} \tag{1.34}
$$

where the notation of the cost coefficient M echoes the *big-M* concept. We would like this value to be as small as possible, but it should alas be big enough so as to guarantee that $\mathbf{y}$ is expelled from at least one optimal solution. The feasibility assumption now becomes more tricky to handle since it depends on a wise choice of M. That is, an optimal solution with $y_i > 0$ in the basis means one of two things: the value of M is too small or the *LP* is actually infeasible.

Note 1.3 (Full row-rank matrix.) The assumption that coefficient matrix is of full row rank is fulfilled whenever it contains at least one subset of m independent columns. This is the case in *Phase I* with matrix $\begin{bmatrix}\mathbf{A}\ \mathbf{I}_m\end{bmatrix}$.

Note 1.4 (Lower and upper bounds on the optimum.) Let $\begin{bmatrix}\mathbf{x}_B \\ \mathbf{x}_N\end{bmatrix}$ be a basic solution for the *LP* (1.17) with objective value $z_{LP} = \mathbf{c}_B^{\mathsf{T}}\mathbf{x}_B$ and corresponding simplex multipliers $\boldsymbol{\pi}^{\mathsf{T}} = \mathbf{c}_B^{\mathsf{T}}\mathbf{A}_B^{-1}$. If we assume that $z^{\star}_{LP}$ is finite, then there exists a positive integer number κ such that $\sum_{j=1}^{n} x_j \leq \kappa$ in every optimal solution. In that case, in addition to the current upper bound z_{LP} on $z^{\star}_{LP}$, we also have knowledge of a lower bound at every iteration, that is,

$$
z_{LP} + \kappa\bar{c}(\boldsymbol{\pi}) \leq z^{\star}_{LP} \leq z_{LP}. \tag{1.35}
$$

The lower bound can be generalized, see Exercise 1.9, to arbitrary dual values $\boldsymbol{\pi} \geq \mathbf{0}$ as

$$
\boldsymbol{\pi}^{\mathsf{T}}\mathbf{b} + \kappa\bar{c}(\boldsymbol{\pi}) \leq z^{\star}_{LP} \leq z_{LP}. \tag{1.36}
$$

Although not very exciting in practice for linear programs, the above expressions, or similar ones, appear within the following chapters on column generation, Dantzig-Wolfe decomposition, and Lagrangian relaxation.

We give the pseudo-code for the primal simplex algorithm with Dantzig's rule in Algorithm 1.1. It includes the expected input/output and the main loop which can be summarized in two parts: 1) the dual vector is defined from the current basis and passed as an argument to the pricing problem, and 2) the stopping rule based on the sufficient optimality conditions (1.23).

Algorithm 1.1: Primal simplex algorithm with Dantzig's rule.

input : *LP* in standard form (1.17), basic and non-basic index-sets B and N
output : Certificate of optimization

loop
  $\boldsymbol{\pi}^\top \leftarrow \mathbf{c}_B^\top \mathbf{A}_B^{-1}$ // $\mathcal{O}(m^3)$
  $\bar{\mathbf{b}} \leftarrow \mathbf{A}_B^{-1}\mathbf{b}$ // $\mathcal{O}(m^2)$
  $\bar{c}_j \leftarrow c_j - \boldsymbol{\pi}^\top \mathbf{a}_j, \forall j \in N$ // $\mathcal{O}(mn)$
  $\bar{c}(\boldsymbol{\pi}) \leftarrow \min_{j\in N} \bar{c}_j,\ \ell \leftarrow \arg\min_{j\in N} \bar{c}_j$ // $\mathcal{O}(n)$
  if $\bar{c}(\boldsymbol{\pi}) \geq 0$
    $\mathbf{x} \leftarrow [\mathbf{x}_B = \bar{\mathbf{b}}, \mathbf{x}_N = \mathbf{0}]$
    $z \leftarrow \mathbf{c}_B^\top \mathbf{x}_B$
    break *by optimality*
  $\bar{\mathbf{a}}_\ell \leftarrow \mathbf{A}_B^{-1}\mathbf{a}_\ell$ // $\mathcal{O}(m^2)$
  if $\bar{\mathbf{a}}_\ell \leq \mathbf{0}$
    $\mathbf{x} \leftarrow [\mathbf{x}_B = -\bar{a}_\ell, x_\ell = 1, \mathbf{x}_{N\setminus\{\ell\}} = \mathbf{0}],\ z \leftarrow -\infty,\ \{\mathbf{x} \in \mathbb{R}^n_+ \mid \mathbf{A}\mathbf{x} = \mathbf{b}\} = \emptyset$
    break *by unboundedness*
  $x_\ell \leftarrow \min_{i\in\{1,\dots,m\}\mid \bar{a}_{i\ell}>0} \bar{b}_i/\bar{a}_{i\ell},\ s \leftarrow \arg\min_{i\in\{1,\dots,m\}\mid \bar{a}_{i\ell}>0} \bar{b}_i/\bar{a}_{i\ell}$ // $\mathcal{O}(m)$
  $B \leftarrow B\setminus\{s\}\cup\{\ell\},\ N \leftarrow N\setminus\{\ell\}\cup\{s\}$ // $\mathcal{O}(1)$
return z, $\mathbf{x}$, and $\boldsymbol{\pi}$

Note 1.5 (Matrix operations everywhere.) We see in Algorithm 1.1 that from the dictionary data given in (1.21), only very little information is actually needed in an iteration: $\boldsymbol{\pi}^\top$, $\bar{\mathbf{b}}$, and $\bar{\mathbf{a}}_\ell$. All three vectors can be computed directly from the original data $\mathbf{A}$, $\mathbf{b}$, and $\mathbf{c}$. Even though on paper we write these computations using the basis inverse $\mathbf{A}_B^{-1}$, the inversion is never actually performed. Instead, the three vectors are computed by solving linear equation systems, all involving the same matrix $\mathbf{A}_B$ (or its transpose) which is LU factorized. The initial factorization costs $\mathcal{O}(m^3)$, but later using the factorization to solve the linear equation systems only costs $\mathcal{O}(m^2)$. Since the basis matrix changes only in one column per iteration, also the factorization needs only a small update. Efficient and sparsity maintaining proposals have been made, e.g., by Bartels and Golub (1969), Forrest and Tomlin (1972), and Reid (1982). This usually avoids the factorization when the basis changes, but over the iterations numerical errors may accumulate and finally, a re-factorization of the basis matrix must occur. Overall, computing only the necessary data via the solution of linear equation systems is subsumed under the name *revised* simplex method.

Note 1.6 (Efficiency vs. theory.) If all bases are non-degenerate, the objective function improves at every iteration and the primal simplex algorithm terminates after a finite number of steps. However, it can take $2^n - 1$ iterations in the worst case using Dantzig's pivot rule (Klee and Minty, 1972); see also Avis et al. (2008). In case of degeneracy, it can even cycle indefinitely (Hoffman, 1953). In practice, cycling is not an issue and the algorithm is quite efficient and only takes approximately $3m$ iterations:

> The most celebrated (and the most frequently quoted) source on this issue is Dantzig's 1963 book (p. 160):

Empirical experience with thousands of practical problems indicates that the number of iterations is usually close to the number of basic variables in the final set which were not present in the initial set. For an m-equation problem with m different variables in the final basic set, the number of iterations may run anywhere from m as a minimum, to 2m and rarely to 3m. The number is usually less than 3m/2 when there are less than 50 equations and 200 variables (to judge from informal empirical observations). Some believe that for a randomly chosen problem with fixed m, the number of iterations grows in proportion to n.

Dantzig's words are quite cautious, but they are also rather general and (perhaps deliberately) loosely defined. From the context it seems that he refers to the *Phase II* problem only, but it is not clear how the initial feasible vertex is obtained. Also, the specific pivoting rules used are not mentioned.

Shamir (1987)

Note 1.7 (One variant to rule them all.) Implementations of the algorithm exist in as many variants as the number of programmers who coded them. It suffices to have a look at the overwhelming list of options available in any modern solver to get a sense of the practical details necessary to create a successful solver. To name a few: pre-solve reductions, model structure exploits, partial pricing strategies, basis factorization updates, and numerical stability options. Yet, we follow the widespread usage of *the* instead of *a* primal simplex algorithm for every algorithm presented in this book.

Treatment of bounds on the variables

The *LP* (1.2) can be generalized to include lower and upper bounds on $\mathbf{x}$:

$$\begin{aligned} z^{\star}_{LP} = \min \quad & \mathbf{c}^{\mathsf{T}}\mathbf{x} \\ \text{s.t.} \quad & \mathbf{Ax} \geq \mathbf{b} \quad [\boldsymbol{\pi}] \\ & \boldsymbol{\ell} \leq \mathbf{x} \leq \mathbf{u}, \end{aligned} \tag{1.37}$$

and the complementary slackness optimality conditions take on a similar form.

Proposition 1.8 (Complementary slackness). *Given the linear programming formulation* (1.37)*, the primal-dual pair* $(\mathbf{x}, \boldsymbol{\pi})$ *is optimal if and only if* $\mathbf{x}$ *is primal feasible,* $\boldsymbol{\pi}$ *is dual feasible, and*

$$\begin{aligned} \textit{for all } j \in \{1,\dots,n\}: \quad & \bar{c}_j > 0 \;\Rightarrow\; x_j = \ell_j, \\ & \bar{c}_j < 0 \;\Rightarrow\; x_j = u_j, \\ & \ell_j < x_j < u_j \;\Rightarrow\; \bar{c}_j = 0; \\ \textit{for all } i \in \{1,\dots,m\}: \quad & s_i < 0 \;\Rightarrow\; \pi_i = 0, \\ & \pi_i > 0 \;\Rightarrow\; s_i = 0. \end{aligned} \tag{1.38}$$

Regarding the adaptation of the simplex algorithm given a basic solution $\begin{bmatrix} \mathbf{x}_B \\ \mathbf{x}_N \end{bmatrix}$, let $L = \{j \in N \mid x_j = \ell_j\}$ and $U = \{j \in N \mid x_j = u_j\}$. With this partition of N, the

LP (1.21) in standard form writes as

$$\begin{aligned} z^\star_{LP} = \boldsymbol{\pi}^\mathsf{T}\mathbf{b} + \min \quad & & \bar{\mathbf{c}}_L^\mathsf{T}\mathbf{x}_L &+ \bar{\mathbf{c}}_U^\mathsf{T}\mathbf{x}_U \\ \text{s.t.} \quad & \mathbf{x}_B + & \bar{\mathbf{A}}_L\mathbf{x}_L &+ \bar{\mathbf{A}}_U\mathbf{x}_U = \bar{\mathbf{b}} \\ & \mathbf{x}_B \geq \mathbf{0}, & \mathbf{x}_L \geq \mathbf{0}, & \quad \mathbf{x}_U \geq \mathbf{0}. \end{aligned} \tag{1.39}$$

The non-basic variables are either at their lower or upper bound and the pricing problem selects an entering variable x_ℓ according to the sign of $\bar{c}_\ell$: negative for $\ell \in L$, positive for $\ell \in U$. The stopping rule of the simplex algorithm for a finite objective $z^\star_{LP}$ becomes

$$\bar{c}_j \geq 0,\ \forall j \in L, \quad \bar{c}_j \leq 0,\ \forall j \in U, \quad \text{where } \bar{c}_j = 0,\ \forall j \in B. \tag{1.40}$$

Adapting (1.14)–(1.16), the sensitivity analysis for either $\Delta\ell_j \in \mathbb{R}$ or $\Delta u_j \in \mathbb{R}$ is easily done: $z^\star_{LP} = \mathbf{b}^\mathsf{T}\boldsymbol{\pi}^\star + \bar{\mathbf{c}}^\mathsf{T}\mathbf{x}^\star_{LP}$, which becomes, using the complementary conditions (1.38):

$$z^\star_{LP} = \mathbf{b}^\mathsf{T}\boldsymbol{\pi}^\star + \sum_{j:x^\star_j=\ell_j} \bar{c}_j\ell_j + \sum_{j:x^\star_j=u_j} \bar{c}_j u_j, \tag{1.41}$$

$$\text{hence} \qquad z^\star_{LP}(\Delta\ell_j) = z^\star_{LP} + \bar{c}_j\,\Delta\ell_j \quad \text{or} \quad z^\star_{LP}(\Delta u_j) = z^\star_{LP} + \bar{c}_j\,\Delta u_j. \tag{1.42}$$

Finally, a basic solution is *degenerate* if at least one of the basic variables is at its lower or upper bound.

Note 1.8 (Dantzig, Lemke, Beale, Ford, Fulkerson, Khachiyan, Karmarkar) These names refer to some of the *OR Fathers* of linear programming algorithms. Definitely, we already know the contribution of Dantzig (1963, 1990) who dates back his findings on the primal simplex to the summer of 1947. He is followed by both Lemke (1954) and Beale (1954) who designed a dual version that can be seen as the application of the primal method to the dual problem. Then come Dantzig et al. (1956) for the primal-dual method, a generalization of similar algorithms for network flow problems.

Khachiyan (1979) showed, finally, that linear programming problems can be solved in polynomial time with the ellipsoid method. Unfortunately, in practice, this algorithm is far from being competitive with the simplex methods. But it took only a few years to find a practical one with polynomial time complexity, an interior point algorithm designed by Karmarkar (1984).

Then, *why do we restrict our presentation to the primal simplex algorithm?* Simply because the column generation algorithm described in the next chapter and used to solve huge linear programs with gazillions of variables is the primal simplex algorithm, where the entering variables are not selected from an explicitly given set of variables, but pricing is done by solving an auxiliary optimization problem that searches the set of variables only implicitly.

Note 1.9 (A mathematical soap opera.) At this point, there are no timelines in which Dantzig's contribution to linear programming can be ignored. The same could be true for the astonishing work of Kantorovich (1939) but it appears to have been eclipsed by the Western literature.

Polyak (2002) ventures some explanations, the most prominent ones being political environment and language barrier. Despite Charnes and Cooper (1962) or Isbell and Marlow (1961) pointing out small technicalities, Schrijver (1986) confirms that Kantorovich's "Problem C" is indeed the general form of linear programming. In this groundbreaking work, Kantorovich solves the program by *resolving multipliers* and then also explains what we nowadays call sensitivity analysis by tiptoeing around capitalism. We can feel the tensions of the era in Koopmans (1962)'s assessment:

> Neither do I understand the preoccupation of Charnes and Cooper with temporal priority. Is the glory of American developers of linear programming in any way diminished if it now turns out that, unknown to them, important aspects of linear programming models and theory had been anticipated in another language and another economic environment, with which communication has been somewhat difficult? If an element of national pride is involved, can we not justifiably point to a conspicuous discrepancy in the time span between development and application of linear programming ideas and techniques in the two environments?

1.4 Integer Linear Programming

An *integer linear program ILP* literally shares the structure of a linear program but additionally requests that the variables $\mathbf{x}$ take on integer values:

$$\begin{aligned} z^\star_{ILP} = \min \quad & \mathbf{c}^\intercal \mathbf{x} \\ \text{s.t.} \quad & \mathbf{A}\mathbf{x} \geq \mathbf{b} \qquad [\boldsymbol{\pi}] \\ & \mathbf{x} \in \mathbb{Z}^n_+ . \end{aligned} \tag{1.43}$$

The integrality conditions make such optimization programs typically much harder to solve than linear programs, also complexity-wise: integer linear programming is $\mathcal{NP}$-hard. Like we did for linear programming, we postpone the description of an algorithm to the next section (Branch-and-Bound). Let us focus here on a few theoretical concepts we can establish.

When we drop the integrality requirements we obtain an optimization program which we know how to solve. In fact, the dual variables $\boldsymbol{\pi}$ are only meaningful in this so-called *linear relaxation* which we denote by *LP*. Even though an optimal solution $\mathbf{x}^\star_{LP}$ for the *LP* is likely fractional, it gives us a lower bound on the integer optimum, i.e., $z^\star_{LP} \leq z^\star_{ILP}$.

Note 1.10 (Mixed-integer linear programming.) If only some variables are required to be integer, we sometimes specifically call this a *mixed-integer linear program*

(*MILP*). We can write the integrality conditions on this subset of integer variables indexed by $I \subseteq \{1,\ldots n\}$ as $\mathbf{x} \in \mathbb{Z}_+^{|I|} \times \mathbb{R}_+^{n-|I|}$.

Given any minimization program (linear or otherwise), let $z_p^\star$ and $z_d^\star$ be primal and dual optima, respectively. While the construction of a dual program for the *ILP* is beyond the scope of this chapter (see Nemhauser and Wolsey, 1988), it is done in such a way that the notion of weak duality (e.g., Proposition 1.5) holds, i.e, $z_d \leq z_p$. However, in integer programming, strong duality (e.g., Proposition 1.6) does not hold in general, and the respective optima usually do not coincide.

Definition 1.17. The difference between the primal and dual optima $z_p^\star - z_d^\star \geq 0$ is called the *duality gap*.

Strength of formulations

Compared to modeling with linear programs, there is a *much* larger freedom in formulating an integer programming model for a given optimization problem. An intuition for this lies in the geometry that we see often in this book, e.g., in Figure 1.6: Together with the integrality requirement, a given set of integer points can be described by infinitely many different sets of linear constraints.

Definition 1.18. Given a set of points $\mathcal{Y} \subseteq \mathbb{Z}^n$, a polyhedron $\mathcal{P} \subseteq \mathbb{R}^n$ is called a *formulation* for $\mathcal{Y}$ if and only if $\mathcal{Y} = \mathcal{P} \cap \mathbb{Z}^n$.

The freedom of choosing a formulation makes modeling with *ILP*s an art, and it immediately brings up the question *how* we should model. First of all, the following formulation is always a wonderful option:

Definition 1.19. Given a polyhedron $\mathcal{P} = \{\mathbf{x} \in \mathbb{R}^n \mid \mathbf{A}\mathbf{x} \geq \mathbf{b}\}$, the *integer hull* of $\mathcal{P}$ is the convex hull of the integer points $\mathcal{P} \cap \mathbb{Z}^n$ it contains, i.e., $\mathrm{conv}(\mathcal{P} \cap \mathbb{Z}^n)$.

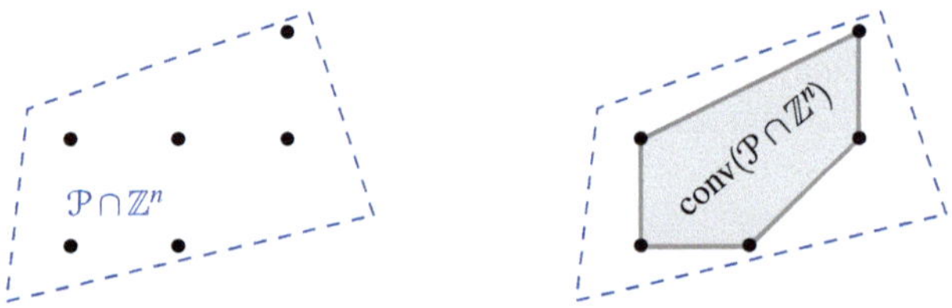

Fig. 1.6: Integer points of $\mathcal{P}$ vs. integer hull of $\mathcal{P}$.

The integer hull is itself a polyhedron, and each of its extreme points is integer by definition, see Figure 1.6. That is, also $\mathrm{conv}(\mathcal{Y})$ is a formulation for $\mathcal{Y} \subseteq \mathbb{Z}^n$. The interest in a linear description of the integer hull comes from the fact that optimizing a linear objective function over it ensures, when finite, the existence of an optimal integer solution as an extreme point. Alas, except for special cases, we usually do not know, let alone have access to such a perfect formulation for our optimization problems.

Definition 1.20. The difference between the integer optimum and that of the linear relaxation is called the *integrality gap*, i.e., $z^\star_{ILP} - z^\star_{LP}$.

For most of the *ILP*s in this book (and for most *ILP*s which are interesting in practice), the integrality gap is strictly positive. Given an *ILP*, the *relative integrality gap* $(z^\star_{ILP}-z^\star_{LP})/|z^\star_{ILP}|$ indicates how well the integer hull is approximated by the linear relaxation, in the direction of the objective function (undefined for $z^\star_{ILP} = 0$). Intuitively, the smaller the gap, the better. The standard notion of quality for *ILP* formulations reflects this.

Definition 1.21. Given a set of points $\mathcal{Y} \subseteq \mathbb{Z}^n$, and two formulations $\mathcal{P}_1, \mathcal{P}_2 \subseteq \mathbb{R}^n$ for $\mathcal{Y}$. We say that $\mathcal{P}_2$ is *stronger* than $\mathcal{P}_1$ if $\mathcal{P}_2 \subset \mathcal{P}_1$.

In our discussion of the branch-and-bound algorithm below, we see that a stronger model can result in less work to be done: The smaller the gap, intuitively, the more often one can prune by bound. A classical way of obtaining a stronger formulation is by adding inequalities that are redundant for the *ILP*.

Definition 1.22. Given a set of points $\mathcal{Y} \subseteq \mathbb{Z}^n$ and a formulation $\mathcal{P}_1 \subseteq \mathbb{R}^n$ for $\mathcal{Y}$. We call an inequality $\mathbf{q}^\intercal \mathbf{x} \geq q_0$ a *cutting plane* if $\mathcal{P}_2 = \mathcal{P}_1 \cap \{\mathbf{x} \in \mathbb{R}^n \mid \mathbf{q}^\intercal \mathbf{x} \geq q_0\}$ is a stronger formulation for $\mathcal{Y}$ than $\mathcal{P}_1$.

Also here, one usually restricts consideration in the direction of a particular objective function. There are very general proposals for cutting planes that work, in principle, for every *ILP*. For specialized algorithms like those in this book, also specialized cutting planes may work better. We discuss this in Chapter 7 (Branch-Price-and-Cut).

Extended formulations

Reformulation is an alternative, or complement, for obtaining stronger formulations. An idea is to start from a model, called *original* in this context, and go to a higher-dimensional space, i.e., use more variables. These may allow for different, "more expressive" ways of formulating constraints. Then project down to the space of the original variables.

Definition 1.23. Given a polyhedron $\mathcal{O} = \{\mathbf{x} \in \mathbb{R}^n \mid \mathbf{A}\mathbf{x} \geq \mathbf{b}\}$. A polyhedron $\mathcal{P} = \{(\mathbf{x}, \boldsymbol{\lambda}) \in \mathbb{R}^n \times \mathbb{R}^\ell \mid \mathbf{B}\mathbf{x} + \mathbf{D}\boldsymbol{\lambda} \geq \mathbf{d}\}$ is called an *extended formulation* of $\mathcal{O}$ if $\mathcal{O} = \text{proj}_{\mathbf{x}}(\mathcal{P})$, where $\text{proj}_{\mathbf{x}}(\mathcal{P})$ denotes the projection of $\mathcal{P}$ on the $\mathbf{x}$ variables, i.e., $\text{proj}_{\mathbf{x}}(\mathcal{P}) = \{\mathbf{x} \in \mathbb{R}^n \mid \exists \boldsymbol{\lambda} \in \mathbb{R}^\ell : (\mathbf{x}, \boldsymbol{\lambda}) \in \mathcal{P}\}$.

This definition extends to sets of integer points.

Definition 1.24. Given a set $\mathcal{Y} \subseteq \mathbb{Z}^n$ of integer points. Polyhedron $\mathcal{P}$ is an extended formulation of $\mathcal{Y}$ if $\mathcal{Y} = \text{proj}_{\mathbf{x}}(\mathcal{P}) \cap \mathbb{Z}^n$.

In order to understand the potential of extended formulations, consider the set $\mathcal{Y} \subseteq \mathbb{Z}^n$ of integer solutions of some *ILP*. The trivial extended formulation of $\mathcal{Y}$ is

$$\mathcal{P} = \{\boldsymbol{\lambda} \in [0,1]^{\mathcal{Y}} \mid \sum_{\mathbf{x} \in \mathcal{Y}} \mathbf{x}\lambda_{\mathbf{x}} = 1\}. \tag{1.44}$$

This is the linear relaxation of a binary program that selects exactly one integer point from $\mathcal{Y}$. Obviously, this (impractical) equivalent reformulation works for every *ILP*. The linear relaxation (1.44), however, is perfect as it describes $\text{conv}(\mathcal{Y})$.

A more useful extended formulation is given in Example 4.1 (Integrality property in the knapsack problem) for the binary knapsack problem; the projection is given in (4.105). We also devote the entire Chapter 4 (Dantzig-Wolfe Decomposition for Integer Linear Programming) to a clever reformulation technique that works for every *ILP*.

Integrality property

The remainder of this subsection deals with perfect formulations.

Definition 1.25. (Geoffrion, 1974, p. 89) The *ILP formulation* (1.43) possesses the *integrality property* if and only if its optimal objective value is equal to the optimal objective value of its linear relaxation, regardless of the coefficients in the objective function, i.e., $z^\star_{ILP} = z^\star_{LP}$, $\forall \mathbf{c} \in \mathbb{R}^n$.

Of course, one may obtain an integrality gap of zero for some objective function even though the property does not hold. We emphasize that the integrality property is an attribute of the model, not of the problem. A positive integrality gap does not rule out a different formulation (for the same problem) that possesses the integrality property. Note that the integrality property makes no guarantee about the integrality of the solution $\mathbf{x}^\star_{LP}$. Nevertheless, it is common practice to drop the integrality conditions on an *ILP* which possesses the integrality property. This implicitly introduces a bias in our choice of an algorithm in the sense that one expects the selected algorithm to output an optimal solution that naturally complies with the integer requirements of the *ILP*, see Exercise 1.11 (Integrality property and algorithm selection bias).

Proposition 1.9. *(Guignard, 2003, Definition 5.1) An integer linear program* (1.43) *whose solutions are in* $\mathcal{A} \cap \mathbb{Z}^n_+$*, where* $\mathcal{A} = \{\mathbf{x} \in \mathbb{R}^n_+ \mid \mathbf{A}\mathbf{x} \geq \mathbf{b}\}$*, possesses the integrality property if and only if* $\text{conv}(\mathcal{A} \cap \mathbb{Z}^n_+) = \mathcal{A}$.

Proof. $\Rightarrow$ Assume $\text{conv}(\mathcal{A} \cap \mathbb{Z}^n_+) \neq \mathcal{A}$. This implies that $\text{conv}(\mathcal{A} \cap \mathbb{Z}^n_+) \subset \mathcal{A}$ such that there exists at least one extreme point in $\mathcal{A}$ that is not in $\text{conv}(\mathcal{A} \cap \mathbb{Z}^n_+)$. It then suffices to assign objective coefficients that uniquely lead to this extreme point, i.e., $\exists \mathbf{c} \in \mathbb{R}^n \mid z^\star_{ILP} \neq z^\star_{LP}$.

$\Leftarrow$ Assume $\text{conv}(\mathcal{A} \cap \mathbb{Z}^n_+) = \mathcal{A}$. The equality implies that every extreme point of $\mathcal{A}$ is an integer vector. No matter the objective coefficients, at least one of these extreme points has the optimal objective value and $z^\star_{ILP} = z^\star_{LP}$, $\forall \mathbf{c} \in \mathbb{R}^n$. □

Definition 1.26. A polyhedron is called *integral* when every face contains an integer point. For all our purposes, this means that all extreme points have integer coordinates.

Total unimodularity

A matrix $\mathbf{A}$ is *totally unimodular* if each of its subdeterminants is 0, $+1$, or -1. Consequently, each entry in a totally unimodular matrix is also 0, $+1$, or -1. This fundamental definition ties into integer linear programming in the following way.

Theorem 1.2. *(Schrijver, 1986, Theorem 19.1) If matrix $\mathbf{A} \in \mathbb{Q}^{m\times n}$ is totally unimodular and vector $\mathbf{b} \in \mathbb{Q}^m$ is integer, then the polyhedron $\mathcal{P} = \{\mathbf{x} \in \mathbb{R}^n \mid \mathbf{Ax} \geq \mathbf{b}\}$ is integral.*

Proof. Every feasible basis $\mathbf{A}_B \subset \mathbf{A}$ of $\mathcal{P}$ is invertible and $\mathbf{A}_B^{-1}$ has integer $0, +1, -1$ coefficients. Every feasible extreme point solution $\mathbf{x} = \mathbf{A}_B^{-1}\mathbf{b}$ is thus also integer. □

With respect to integer programming, under the conditions in Theorem 1.2, we get an integer extreme point for free from solving the linear relaxation (if the optimum is finite, i.e., $z^\star_{LP} > -\infty$). Corollary 1.4 extends this result to the dual formulation.

Corollary 1.4. *(Schrijver, 1986, Corollary 19.1a) Given a pair of primal-dual formulations* $\min\{\mathbf{c}^\intercal\mathbf{x} \mid \mathbf{Ax} \geq \mathbf{b},\ \mathbf{x} \geq \mathbf{0}\} = \max\{\mathbf{b}^\intercal\boldsymbol{\pi} \mid \mathbf{A}^\intercal\boldsymbol{\pi} \leq \mathbf{c},\ \boldsymbol{\pi} \geq \mathbf{0}\}$ *with finite optimum $z^\star_{LP}$. If matrix $\mathbf{A}$ is totally unimodular and both vectors $\mathbf{b}$ and $\mathbf{c}$ are integer, then $z^\star_{LP}$ is integer and there exists an integer optimal solution pair $\mathbf{x}^\star_{LP}$ and $\boldsymbol{\pi}^\star$.*

Proof. Given $z^\star_{LP} > -\infty$, we know from Theorem 1.2 that there exists an integer optimal solution $\mathbf{x}^\star_{LP}$ to the primal formulation, which means that the optimum $z^\star_{LP} = \mathbf{c}^\intercal\mathbf{x}^\star_{LP}$ is also integer. By strong duality, this optimum holds for the dual formulation for which there exists an integer optimal solution $\boldsymbol{\pi}^\star$ since the transpose of a totally unimodular matrix is also totally unimodular. □

Network flow problems

A prime example of total unimodularity occurs in network flow problems. Consider the general form known as the *capacitated minimum cost flow problem* (*CMCFP*).It is defined on a directed graph $G = (N, A)$, where N is the set of nodes and A is the set of arcs. We associate with each node $i \in N$ an integer number b_i: if $b_i > 0$, node i is a *supply* node; if $b_i < 0$, node i is a *demand* node with a demand of $-b_i$; othetwise, $b_i = 0$ and node i is a *transshipment* node. We assume that $\sum_{i\in N} b_i = 0$, i.e., the demands and supplies are *balanced*. If $b_i = 0$, $\forall i \in N$, we also speak of a *circulation problem*. The cost vector is $\mathbf{c} = [c_{ij}]_{(i,j)\in A}$ and $\mathbf{x} = [x_{ij}]_{(i,j)\in A}$ represents the vector of flow variables bounded by non-negative vectors $\boldsymbol{\ell} = [l_{ij}]_{(i,j)\in A}$ and $\mathbf{u} = [u_{ij}]_{(i,j)\in A}$. An *arc-flow formulation* for the *CMCFP* is an integer linear program given by

$$
\begin{aligned}
z^{\star}_{ILP} = \min \quad & \sum_{(i,j)\in A} c_{ij}x_{ij} \\
\text{s.t.} \quad & \sum_{j:(i,j)\in A} x_{ij} - \sum_{j:(j,i)\in A} x_{ji} = b_i \quad [\pi_i] \quad \forall i \in N \\
& \ell_{ij} \le x_{ij} \le u_{ij} \qquad \forall (i,j) \in A \\
& x_{ij} \in \mathbb{Z}_+ \qquad \forall (i,j) \in A,
\end{aligned}
\tag{1.45}
$$

where the dual vector $\boldsymbol{\pi}$ again only makes sense for the linear relaxation. The equality constraints are known as *flow conservation* constraints. For each node $i \in N$, the equation reads as *outflow* minus *inflow* equals b_i. Matrix $\mathbf{A}$ is therefore of peculiar construction: it is a *node-arc incidence matrix* in which each column is associated with an arc $(i,j) \in A$ and contains exactly two non-zero entries: a $+1$ in row i and a -1 in row j.

It is well known that the incidence matrix of a directed graph is totally unimodular (and this property does not change if an identity matrix like for the bounds is appended). Thus, by Theorem 1.2 and Proposition 1.9, and assuming that the right-hand side $\mathbf{b} = [b_i]_{i\in N}$ as well as the bounds $\boldsymbol{\ell}$ and $\mathbf{u}$ are integer vectors, the arc-flow formulation (1.45) possesses the integrality property. It is therefore common practice to omit the integrality requirements on network flow models as all algorithms are designed to produce primal integer solutions. By Corollary 1.4, deriving a dual optimal integer solution from the latter is moreover possible if $\mathbf{c}$ is also an integer vector (see also Ahuja et al., 1993, Theorem 11.4, p. 413).

The particular structure of the incidence matrix simplifies e.g., reduced cost computations. As an example, this shows in the proof of the following little result.

Proposition 1.10. *Given a network flow problem on $G = (N,A)$ and an optimal dual solution $\boldsymbol{\pi}$, the translation $\pi_i + \alpha$, $\forall i \in N$, is also optimal for any scalar $\alpha \in \mathbb{R}$.*

Proof. We are given optimality conditions in Proposition 1.8 based on the reduced costs computed as $\bar{c}_{ij} = c_{ij} - \pi_i + \pi_j$, $\forall (i,j) \in A$. These remain identical for the proposed translation as $\bar{c}_{ij} = c_{ij} - (\pi_i + \alpha) + (\pi_j + \alpha)$, $\forall (i,j) \in A$. □

1.5 Branch-and-Bound

The standard solution approach to optimize the *ILP* (1.43) is the linear programming based *branch-and-bound* method. While linear programming algorithms have fathers, see Notes 1.8 and 1.9, this integer programming algorithm has mothers, namely Land and Doig (1960). The *linear relaxation* is solved first. If its optimal solution is not integer, we create at least two sub-domains, by adding constraints that (a) eliminate the fractional solution, but (b) keep all integer solutions to the *ILP* in the union of the sub-domains, preferably each of them in a single sub-domain (branching). This process is iterated and thus produces a search tree whose *root node* (node **0**) corresponds to the initial linear relaxation, with an optimal objective value $z^\star_{LP}$.

As we progress in the search tree, each new node is an *offspring* or *child* which contains some restricted linear relaxation that depends on the branching constraints introduced thus far. Depending on whether or not such a node ℓ has been solved (or "explored") already, let $z^{\star\ell}_{LP}$ denote the *node value* or the inherited one of its *parent*. Moreover, observe that $z^{\star\ell}_{LP}$ is monotonically increasing across the descendant line of a node. Let $\mathcal{L}$ denote the *live* (a.k.a. *active*) nodes of the tree, that is, those nodes that have yet to be explored. The most optimistic objective value for the *ILP*, say *LB*, is then simply a matter of keeping track of the linear relaxation with the minimal node value in $\mathcal{L}$, i.e.,

$$LB = \min_{\ell \in \mathcal{L}} z^{\star\ell}_{LP}. \tag{1.46}$$

The powerful *bound* aspect of the method allows us to discard entire sub-trees from the search. Further exploration of a node is not necessary in three cases which we call *pruning* by bound, by integrality, or by infeasibility. The node is removed from $\mathcal{L}$ then. Let *UB* be the objective value of the *incumbent solution*, the best integer solution encountered in the tree at any given time. Whenever a node yields an integer solution, it is pruned by integrality and the incumbent is updated as needed. A node is pruned by infeasibility if branching decisions induce an infeasible linear relaxation. Finally, for the nodes that remain, the following inequality holds

$$z^\star_{LP} \leq LB \leq z^\star_{ILP} \leq UB, \tag{1.47}$$

and states that the optimal objective value of the root node $z^\star_{LP}$ is not greater than the objective value of an optimal integer solution $z^\star_{ILP}$ which is itself sandwiched between lower and uppers bounds, as the objective values of the most optimistic linear relaxation *LB* and incumbent solution *UB*, respectively.

Whenever $z^{\star\ell}_{LP} \geq UB$, the associated node ℓ is pruned by bound because none of its offsprings can improve upon the incumbent solution. Consequently, all nodes in $\mathcal{L}$ can be reviewed for pruning whenever a new *UB* is identified. The branch-and-bound method terminates with a certificate of optimality for the *ILP* when $\mathcal{L} = \emptyset$.

The search tree may grow exponentially with branching. An indicator of the tree size is its *depth* (actually, its *height*). For any given node, we know its depth in the tree (also called the *level*), that is, the number of nodes in a descendant line starting from the root node at level 0. As branching is performed and child nodes are created, one question is in which order nodes should be explored. Three popular node selection rules are *depth-first*, *breadth-first*, and *best-first*.

- *Depth-first* acts in a last-in, first-out manner where the newest nodes are explored first; this diving in the tree is memory-friendly.
- *Breadth-first* acts in a first-in, first-out manner thus consistently progressing level by level.

- *Best-first* jumps around in the tree by selecting the next node according to the most promising lower bound.

After branching, the current basic solution is no longer feasible (by design), but usually still optimal. This is a perfect use case for the *dual* simplex method, see Lemke (1954) and Beale (1954) for their seminal contributions, and more recently, Koberstein and Suhl (2007).

Optimality gap

The *optimality gap* is computed as the difference between the objective value of the incumbent solution and the best we can hope for:

$$\textit{optimality gap} = UB - LB. \tag{1.48}$$

The *relative optimality gap* $(UB-LB)/|UB|$ (for $UB \neq 0$) is frequently used in practice to prematurely terminate an algorithm. For a fixed threshold $0 \leq \gamma < 1$, one may stop as soon as $UB - LB \leq \gamma\,|UB|$. This trades the optimality certificate in favor of a potential computation time reduction. If $\mathbf{c}^\intercal\mathbf{x} > 0$ for all feasible $\mathbf{x}$, an interpretation is that the resulting objective value is no worse than $1/(1-\gamma)\, z^\star_{ILP}$.

Note 1.11 (Role inversion of bounds if maximizing.) For a maximization program, the roles of *LB* and *UB* are inverted such that the relative optimality gap is computed as $|LB - UB|/|LB|$. This is why one speaks of primal bounds (coming from a feasible solution) and dual bounds (coming from a relaxation).

Variable fixing by reduced cost

The elimination of an integer variable (or *fixing it to zero*) can sometimes be done in a preprocessing step but also commonly using an optimal primal-dual solution pair $(\mathbf{x}^\star_{LP}, \boldsymbol{\pi}^\star)$ for the linear relaxation of the *ILP* (1.43) (see Nemhauser and Wolsey, 1988, p. 389). We assume known an upper bound *UB* on $z^\star_{ILP}$ and consider the lower bound $LB = z^\star_{LP} = \mathbf{c}^\intercal\mathbf{x}^\star_{LP} = \boldsymbol{\pi}^{\star\intercal}\mathbf{b}$.

Recall the sensitivity analysis for a variable x_j at its lower bound $\ell_j = 0$ in (1.16): $z^\star_{LP}(\Delta\ell_j) = \boldsymbol{\pi}^{\star\intercal}\mathbf{b} + \bar{c}_j\Delta\ell_j$, where $\bar{c}_j > 0$. Therefore, if the reduced cost $\bar{c}_j > UB - LB$ for some $j \in \{1,\ldots,n\}$, then $x^\star_j = 0$ in every optimal solution $\mathbf{x}^\star_{ILP}$ because $x^\star_j \geq 1$ would yield a contradiction on the upper bound, i.e., $\mathbf{c}^\intercal\mathbf{x}^\star_{ILP} \geq LB + \bar{c}_j > UB$. In this case, x_j can be fixed to 0 after solving the linear relaxation.

Observe that we only need a feasible dual solution to get *LB* and non-negative reduced costs. Consequently, the dual solution does not have to be optimal to apply this variable fixing technique. A version adapted to branch-and-price algorithms, where the column generation pricing problem consists in computing feasible paths in a network, is later presented in Chapter 7 (p. 493).

Illustration 1.1 **Branch-and-bound experiment**

Figure 1.7 depicts a branch-and-bound tree, where 25 linear relaxations have been solved, the branching decisions are not presented, and the node numbers are irrelevant. We reach a maximum level of 8 with node **9** being at level 3. We first take the opportunity to underscore that an integer objective value is not synonymous with an integer solution, see for instance $z_{LP}^{\star\mathbf{0}} = 7064$ and node **3** with an objective value of 7091. On the other hand, this particular application has integer objective coefficients. This implies that the optimal objective value is also integer, which explains why the node **29** is pruned by bound at $\lceil 7110.5 \rceil = 7111 \geq 7111$. Finally, node **30** is primal infeasible and the objective takes value $-\infty$.

This illustration is presented in the form of an exercise, answering one by one a series of questions.

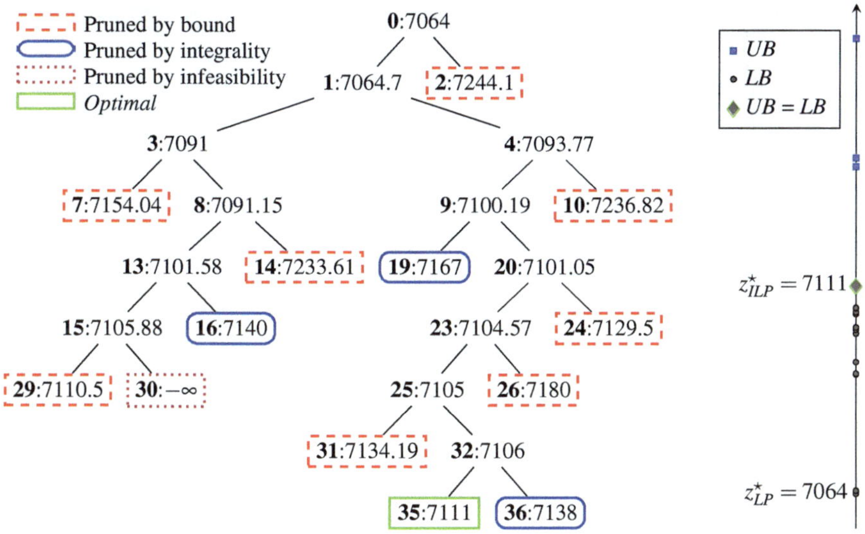

(a) Tree (b) Bounds

Fig. 1.7: A linear programming based branch-and-bound tree as well as lower and upper bounds evolution at a glance.

(a) Using the *best-first* rule, give the exploration order of the nodes.

► Recall that before being solved, every node inherits the objective value of its parent. The actual order is

0, 1, 2, 3, 4, 7, 8, 13, 14, 9, 10, 19, 20, 23,
24, 15, 16, 25, 26, 31, 32, 29, 30, 35, 36.

However, the root node produced two child nodes both initialized at $\tilde{z}_{LP} = z^{\star}_{LP}$. Whether one solves first the node **1** or **2** is then a matter of left or right internal workings. The same is true for every other node in the tree. Therefore, any order using a permutation of siblings such as **0**, **2**, **1**, **3**, **4**, ... is legitimate.

(b) Can the same question be answered for *depth-* and/or *breadth-first* rules? List pros and cons of *depth*, *breadth*, and *best-first* node selection rules.

▶ No. The *depth-first* rule would dive into node **2** or follow-up on nodes **1-3-7** depending once again on idiosyncratic coding details. The *breadth-first* rule would solve **1**, **2** and then **3**, **4** and then solve the hidden child nodes of **2** in order to complete the second level of the tree. Since we have no information after nodes **2** and **7**, we cannot conclude for either rules as the pruning would be impossible to predict.
By following successive branching decisions down a descendant line, the *depth-first* rule rapidly reaches leaf nodes whereby either an integer solution is found or the latest branching caused infeasibility, i.e., the effort is put on the upper bound. The *best-first* rule works on the other front by lifting the smallest lower bound. The *breadth-first* rule manages the tree size but does not actively seek progress on either bound.

(c) Venture an explanation regarding the absence of nodes **5** and **6**.

▶ After the node **2** is solved, the child nodes **5** and **6** are created right away from the fractional solution. Since they inherit $\tilde{z}_{LP} = 7244.1$, they are pruned at the same time as node **2**.

(d) Assuming an exact reproduction of the tree behavior, describe an implementation that would strictly produce the 25 nodes.

▶ The implementation would need to create child nodes only after the next node is selected. This implies quite a bit of bookkeeping since one would need to save the solution of the linear relaxation until the branching decision is applied. Furthermore, explored nodes have an ambivalent status depending on whether child nodes are already created or not.

(e) Give the largest tolerance for which the branch-and-bound method would have stopped at the integer solution 7140.

▶ Given the order listed in **(a)**, $LB = 7104.57$ when the integer solution 7140 is found which results in a relative optimality gap computed as

$$(7140 - 7104.57)/7140 \approx 0.0050.$$

Since we used the *best-first* node selection, the *largest* tolerance can be computed with respect to the objective value of the parent node, i.e.,

$$(7140 - 7101.58)/7140 \approx 0.0054.$$

For good measure, we should mention that this small tree is in no way representative of the expected benefits of a premature stopping rule.

(f) Compute the relative integrality gap.

▶ $(7111 - 7064)/7111 \approx 0.0066 < 1\,\%$.

Branch-and-cut

In the branch-and-bound algorithm, the number of nodes to explore in the search tree and, thus, the total computation time usually increase with the integrality gap. To achieve a better dual bound and reduce this gap, cutting planes can be added to the formulation.

In theory, only a limited number of them should be added to partially describe the convex hull of the feasible solution set $\mathcal{Y}$ around an optimal solution. However, given that the useful ones are not identified as such, they cannot be added a priori. On the other hand, we know families of inequalities, usually of exponential or even infinite cardinality, to which they belong. Given the solution $\mathbf{x}^{\star}_{LP}$ of a linear relaxation, we can then search in those inequality families, one or several cutting planes that cut off $\mathbf{x}^{\star}_{LP}$ from the domain of the current linear relaxation domain without removing any solution in $\mathcal{Y}$. This search is performed using a *separation algorithm.* Note that, interestingly, we can even discard feasible but non-optimal integer solutions, e.g., see the successive z-cuts restricting the objective value in Example 7.9 (Preferential bidding system).

The pure cutting plane algorithm (e.g. Gomory, 1963) solves a sequence of linear relaxations, each obtained from the previous one by adding cutting planes, until finding an integer optimal solution. Because it converges very slowly in general, it is rather merged with the branch-and-bound algorithm to yield the *branch-and-cut* algorithm (Padberg and Rinaldi, 1991).

After solving the linear relaxation at a node of the search tree that cannot be pruned, an exact or heuristic separation algorithm is called to search for cuts. If cuts are found in a limited computational effort, they are added to the relaxation which is then reoptimized. Otherwise, branching is performed to create child nodes. Given that the separation algorithm might be time-consuming, a compromise must be achieved between searching for cuts and exploring less nodes in the search tree.

Note 1.12 (Branch-and-cut in practice.) A lot of research and engineering went into implementations of the branch-and-cut algorithm. One could not solve as large *ILP*s as we do today without a choreography of preprocessing techniques, cutting plane strategies, primal heuristics, node selection rules, branching strategies, etc. We remark on some of these components in Chapter 7 (Branch-Price-and-Cut).

1.6 Good to Know

Here we examine some more advanced theoretical elements that may be overlooked on a first reading of the chapter.

Linear programs for the pricing

Consider the *LP* (1.17) in standard form and assume we are given a vector $\boldsymbol{\pi} \in \mathbb{R}^m$. The minimum reduced cost $\bar{c}(\boldsymbol{\pi}) = \min_{j\in\{1,\ldots,n\}} \bar{c}_j$ is usually found by enumeration. We can however formulate different linear programs for computing it.

Let $\mu = \min_{j\in\{1,\ldots,n\}} \bar{c}_j$ denote the least reduced cost. By linearizing the minimum function, the following formulation uses μ as a *variable unrestricted in sign.* It is maximized with respect to $\boldsymbol{\pi}$, that is, pushed as much as possible to the right until the least reduced cost value is encountered:

$$\begin{aligned} \bar{c}(\boldsymbol{\pi}) = \max \quad & \mu \\ \text{s.t.} \quad & \mu \leq \bar{c}_j \quad [y_j] \quad \forall j \in \{1,\ldots,n\}. \end{aligned} \tag{1.49}$$

In (1.49), y_j acts as a dual variable for the constraint $j \in \{1,\ldots,n\}$ and the dual program reads as

$$\begin{aligned} \bar{c}(\boldsymbol{\pi}) = \min \quad & \sum_{j\in\{1,\ldots,n\}} \bar{c}_j y_j \\ \text{s.t.} \quad & \sum_{j\in\{1,\ldots,n\}} y_j = 1 \quad [\mu] \\ & y_j \geq 0 \qquad \forall j \in \{1,\ldots,n\}. \end{aligned} \tag{1.50}$$

In (1.50), we are looking for a convex combination of variables, where an extreme point solution selects a single variable, one with the least reduced cost.

Alternatives to (1.49) use a vector $\mathbf{w}$ of positive weights $w_j > 0$, $\forall j \in \{1,\ldots,n\}$, e.g., $w_j = \|\mathbf{a}_j\|$. The dual formulation becomes

$$\begin{aligned} \bar{c}(\mathbf{w},\boldsymbol{\pi}) = \max \quad & \mu \\ \text{s.t.} \quad & \mu \leq \frac{\bar{c}_j}{w_j} \quad [y_j] \quad \forall j \in \{1,\ldots,n\}. \end{aligned} \tag{1.51}$$

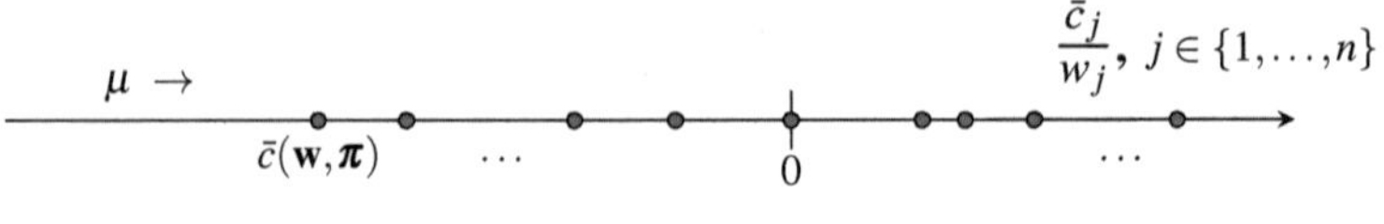

We propose two versions for the primal pricing problem.

The left one is derived from the above whereas the right one is obtained after rewriting $\mu \leq \frac{\bar{c}_j}{w_j}$ as $\mu w_j \leq \bar{c}_j$, $\forall j \in \{1,\ldots,n\}$:

$$\begin{array}{lll} \min & \sum_{j\in\{1,\ldots,n\}} \left(\frac{\bar{c}_j}{w_j}\right) y_j & \\ \text{s.t.} & \sum_{j\in\{1,\ldots,n\}} y_j = 1 & [\mu] \\ & y_j \geq 0, \ \forall j \in \{1,\ldots,n\} & \end{array} \quad \Bigg| \quad \begin{array}{lll} \min & \sum_{j\in\{1,\ldots,n\}} \bar{c}_j y_j & \\ \text{s.t.} & \sum_{j\in\{1,\ldots,n\}} w_j y_j = 1 & [\mu] \\ & y_j \geq 0, \ \forall j \in \{1,\ldots,n\}. & \end{array} \tag{1.52}$$

By construction in the primal simplex algorithm, the reduced cost of all the basic variables are set to 0 (i.e., $\bar{c}_j = 0, \forall j \in B$) by defining $\boldsymbol{\pi}^\mathsf{T} = \mathbf{c}_B^\mathsf{T}\mathbf{A}_B^{-1}$, but other variants exist.

For example, as suggested by the optimality conditions in Proposition 1.7 (*Complementary slackness*), we can impose $\bar{c}_j = 0$ only for the positive variables. Optimizing the reduced cost of the other variables results in the *improved primal simplex* algorithm (IPS) (El Hallaoui et al., 2011; Raymond et al., 2010b; Metrane et al., 2010). The pricing problem in its dual version writes as

$$\begin{aligned} \bar{c}(\boldsymbol{\pi}) = \max_{\boldsymbol{\pi},\mu} \quad & \mu \\ \text{s.t.} \quad & 0 = c_j - \boldsymbol{\pi}^\mathsf{T}\mathbf{a}_j \quad [y_j] \quad \forall j \in \{1,\ldots,n\} : x_j > 0 \\ & \mu \leq c_j - \boldsymbol{\pi}^\mathsf{T}\mathbf{a}_j \quad [y_j] \quad \forall j \in \{1,\ldots,n\} : x_j = 0. \end{aligned} \tag{1.53}$$

In this case, the entering variable is a convex combination of several variables, basic and non-basic, and only non-degenerate pivots occur.

Similarly, one can optimize the reduced cost of all the variables (see Exercise 1.10: this leads the *minimum mean cycle-canceling algorithm* (MMCC), well known for network flow problems (Goldberg and Tarjan, 1989; Radzik and Goldberg, 1994) and extended to linear programs in Gauthier et al. (2018); Gauthier and Desrosiers (2022). This algorithm also ensures non-degenerate pivots.

Irrational data

Let us take a look at the irrational polyhedron of Figure 1.8,

$$\mathcal{P} = \{x_1, x_2 \in \mathbb{R}_+ \mid \sqrt{3}x_1 - x_2 \geq 0,\ x_1 \geq 1\}. \tag{1.54}$$

This helps to understand the importance of rational values for $\mathbf{Q} \in \mathbb{R}^{m\times n}$ and $\mathbf{q} \in \mathbb{R}^m$ in the Definition 1.4 of $\mathcal{P} = \{\mathbf{x} \in \mathbb{R}^n \mid \mathbf{Qx} \geq \mathbf{q}\}$.

Because there is no integer solution that satisfies $x_1 \geq 1, x_2 \geq 0, \sqrt{3}x_1 - x_2 = 0$ ($\sqrt{3}x_1$ being irrational while x_2 is integer), we can be as close as we want to the line

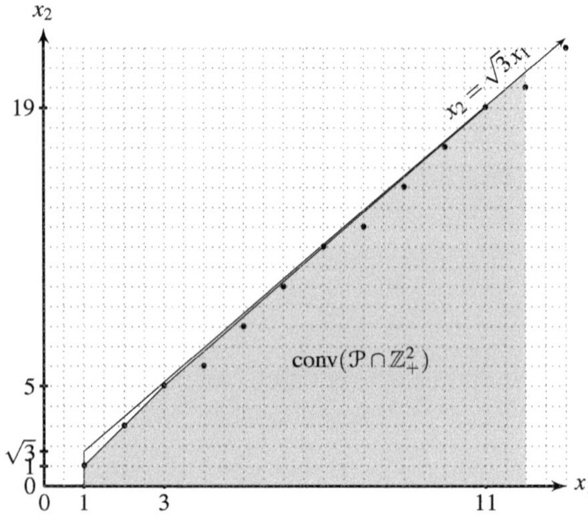

Fig. 1.8: Subset of points of the integer hull of $\mathcal{P}$ close to the line $x_2 = \sqrt{3}x_1$.

$x_2 = \sqrt{3}x_1$ by choosing a large positive integer x_1 and setting $x_2 = \lfloor \sqrt{3}x_1 \rfloor$, but we never hit this line. It also means that the integer hull of $\mathcal{P}$, that is, $\text{conv}(\mathcal{P} \cap \mathbb{Z}^2_+)$, is defined by an infinite number of inequalities and possesses an infinite number of extreme points. Consequently, $\text{conv}(\mathcal{P} \cap \mathbb{Z}^2_+)$ *is not a polyhedron.*

x_1	$\lfloor \sqrt{3}x_1 \rfloor = x_2$	$\sqrt{3}x_1$	slope
1	1	1.7320508	
3	5	5.1961524	2
11	19	19.0525589	1.75
41	71	71.0140831	1.733333333
153	265	265.0037736	1.732142857
362	627	627.0023923	1.732057416
1 351	2 340	2 340.0006410	1.732052578
2 131	3 691	3 691.0002709	1.732051282
5 042	8 733	8 733.0001718	1.732050842
18 817	32 592	32 592.0000460	1.732050817
29 681	51 409	51 409.0000195	1.732050810

Table 1.3: A subset of the extreme points of $\text{conv}(\mathcal{P} \cap \mathbb{Z}^2_+)$.

Apart from the extreme point $(1,0)$, Table 1.3 lists the first 10 extreme points of the upper part of $\text{conv}(\mathcal{P} \cap \mathbb{Z}^2_+)$, from $(1,1)$ to $(18\,817, 32\,592)$. Observe the decrease in the vertical distance $\sqrt{3}x_1 - x_2$ to the line equation from 0.7320508 to 0.0000195, and the convergence of the slope to $\sqrt{3} \approx 1.732050808$. Note that the last pair of coordinates $(29\,681, 51\,409)$ is potentially an extreme point, but only if the next selected pair (x_1,x_2) closer to the line produces a smaller slope than the current one (otherwise that point is on the edge joining two extreme points).

The impact on the following *ILP* is also interesting:

$$\begin{aligned} z^{\star}_{ILP} = \min \quad & \sqrt{3}x_1 - x_2 \\ \text{s.t.} \quad & \sqrt{3}x_1 - x_2 \geq 0 \\ & x_1 \geq 1,\ x_2 \geq 0,\ \textit{integer}. \end{aligned} \tag{1.55}$$

The *ILP* (1.55) is clearly feasible and its objective value $\sqrt{3}x_1 - x_2$ is bounded from below by 0. However, *it has no optimal integer solution.* This is indeed a sufficient reason to only use rational data in the definition of polyhedra and *ILP*s.

1.7 Reference notes

Introduction At the end of each chapter, we provide some pointers to the literature. The domain of this book grew so large over the past decades that we did not even try to be comprehensive in this endeavor. Instead, we primarily list articles and books that have some personal link to us authors or to the writing process of this book. That said, don't be sad if your favorite reference is not listed here, or your own paper. If you wish to have it included in a future edition of this book, just drop us a line.

Section 1.1 The reader can find more on the polyhedral theory in Nemhauser and Wolsey (1988, Part I., FOUNDATIONS 1.4) as well as the proof of the Minkowski and Weyl theorem (Theorem 4.8, p. 96). See also Ziegler (1995) for 11 chapters on *Lectures on Polytopes*.

Sections 1.2–1.3 Linear programming was first introduced by Leonid Kantorovich in 1939 (see Note 1.9, *A mathematical soap opera.*) in order to achieve a best possible organization and planning of production in the Soviet Union. It remained a secret (to Western eyes) until 1947, when Dantzig published the simplex algorithm, see the books Dantzig (1963, 1966) and the co-authored ones Dantzig and Thapa (1997, 2003). Linear programming was also rediscovered and applied to shipping problems by Koopmans (1942, 1947), an American economist who shared in 1975 with Kantorovich the Nobel Prize for Economics "for their contributions to the theory of optimum allocation of resources." See MacTutor for a biography.

Our preferred teaching books include Chvátal (1983) as well as Bertsimas and Tsitsiklis (1997), Griva et al. (2008), and Vanderbei (2020). Note that the sensitivity analysis results must be interpreted with caution when the optimal primal solution is degenerate, see for example Koltai and Tatay (2011) and references therein.

Sections 1.4–1.5. These two sections are on *Integer Linear Programming* and solution methods. We again suggest Nemhauser and Wolsey (1988) but also Schrijver

(1986), Bertsimas and Weismantel (2005), and Conforti et al. (2014). For a comprehensive book on the theory, algorithms, and applications of *Network Flows*, see Ahuja et al. (1993).

For various historical aspects on the above subjects, see Dantzig (1963, Chapter 2), Dantzig (1982, 1990), Lenstra et al. (1991), Nemhauser (1994), Bixby (2002, 2012), Jünger et al. (2010), Assad and Gass (2011), and Grötschel (2012). A wonderful resource is the *Subject to* podcast, *"a series of informal conversations with relevant figures in the fields of Operations Research, Combinatorial Optimization and Logistics, and they are hosted by Anand Subramanian (2023), an Associate Professor at Universidade Federal da Paraíba, Brazil."*

We could not make any practical use of this book, if there wasn't excellent solvers for linear and integer programs. Koch et al. (2022) report on the enormous progress in this area from 2001 to 2020. Let us quote the highlights of this paper:

> The article contains a comprehensive performance comparison of the virtual best *LP/MILP* solver from 2001 with the best solvers from 2020.
>
> From 2001 to 2020, hard- and software combined, *LP* solving got about 180× faster for instances that could be solved using the 2001 solvers.
>
> Solving *MILP* got about 1000× faster in the same period.
>
> Yet, the above numbers considerably underestimate the progress made, as many instances deemed unsolvable 20 years ago can now be solved within seconds.
>
> The article presents a deep dive into how *LP* and *MILP* solvers can be measured, how to evaluate without bias, and how to interpret the results.

These speedups from algorithmics and software even multiply with that from hardware (Bixby and Rothberg, 2007). But from the "trick question" by Bixby (2017) that we paraphrase as

> *To solve a mixed-integer program, would you rather*
> 1. use today's solution technology on a machine from 1991, or
> 2. use the 1991 solution technology on a machine from today?

(the "correct" answer is the first) remember that brain power beats machine power in this domain. When you proceed in this book, *solvers* alone will not suffice. One needs *frameworks* in which one can also create and modify the algorithms, tailor the implementation to the problem at hand. While there are several very good ones, our heart is with the SCIP Optimization Suite (Achterberg, 2009) for Marco and GENCOL for Jacques, Guy, and Jean Bertrand.

Good to Know A convexity constraint naturally appears in the pricing problem of the *minimum mean-cycle* algorithm (Goldberg and Tarjan, 1989) and the *improved primal simplex* algorithm (El Hallaoui et al., 2011). This appears to be also true for the *primal simplex* algorithm, see the pricing problem (1.50) and Gauthier et al. (2018).

Finally, other aspects of irrational data can be found in Schrijver (1986).

Exercises

1.1 George Dantzig

What was the dissertation subject of George Dantzig?

1.2 DDL1x

In the Preface, what is the meaning of `[DDL1x], replace x by 5, 6, or 7`?

1.3 Farkas' lemma

Give a proof for **Proposition 1.4**: Given $\mathbf{A} \in \mathbb{R}^{m \times n}$ and $\mathbf{b} \in \mathbb{R}^m$, then exactly one of the following assertions holds:

1: There exists $\mathbf{x} \in \mathbb{R}^n$ such that $\mathbf{Ax} \geq \mathbf{b}$, $\mathbf{x} \geq \mathbf{0}$;
2: There exists $\boldsymbol{\pi} \in \mathbb{R}^m$ such that $\mathbf{A}^\mathsf{T}\boldsymbol{\pi} \leq \mathbf{0}$, $\boldsymbol{\pi} \geq \mathbf{0}$, $\mathbf{b}^\mathsf{T}\boldsymbol{\pi} > 0$.

Hint: Firstly, rewrite $\mathbf{Ax} \geq \mathbf{b}$, $\mathbf{x} \geq \mathbf{0}$ as $\mathbf{Ax} - \mathbf{s} = \mathbf{b}$, $\mathbf{x}, \mathbf{s} \geq \mathbf{0}$. Secondly, apply the *Phase I* of the primal simplex algorithm.

1.4 Infeasible primal-dual pair of linear programs

We are interested in the relationship between the primal and dual formulations as expressed in Table 1.2. In particular, we want to conceive an example where both formulations are simultaneously infeasible.

(a) Show that there exists no such example with a single variable in the primal.
(b) Propose a primal-dual pair of infeasible linear programs.

1.5 Direction in the primal simplex algorithm

Verify that the direction $\begin{bmatrix} -\bar{\mathbf{a}}_\ell \\ 1 \\ \mathbf{0} \end{bmatrix}$ given in (1.28), either for reaching an adjacent extreme point or identifying an extreme ray, satisfies $\mathbf{A}\begin{bmatrix} -\bar{\mathbf{a}}_\ell \\ 1 \\ \mathbf{0} \end{bmatrix} = \mathbf{0}$.

1.6 Sensitivity range for an increase of the lower bound

For an increase $\Delta\ell_j \in \mathbb{R}_+$ of the lower bound for the non-basic variable $x^\star_\ell = 0$, show how to compute the sensitivity range in (1.16).

1.7 Lost in translation?

How do we ensure signed multipliers $\boldsymbol{\pi} \geq \mathbf{0}$ for the *original dual vector* when the *LP* (1.2) with inequality constraints is rewritten in standard form (1.17)?

1.8 Primal simplex and strong duality

Let the *LP* (1.2) with inequality constraints be rewritten in standard form (1.17). Show that an optimal basic solution given by the primal simplex algorithm satisfies Proposition 1.6 (*Strong duality*).

1.9 Lower bound on the optimum
Given a finite objective value $z^\star_{LP}$, arbitrary dual values $\boldsymbol{\pi} \geq \mathbf{0}$, and minimum reduced cost $\bar{c}(\boldsymbol{\pi}) \leq 0$, show that the optimum $z^\star_{LP}$ in (1.2) is bounded from below as in (1.36):

$$\boldsymbol{\pi}^\mathsf{T}\mathbf{b} + \kappa\bar{c}(\boldsymbol{\pi}) \leq z^\star_{LP}.$$

1.10 Optimizing π for maximizing the smallest reduced cost
Consider the following problem where $\boldsymbol{\pi} \in \mathbb{R}^m$ is unknown but *optimized* for maximizing the smallest reduced cost:

$$\max_{\boldsymbol{\pi}} \min_{j\in\{1,\ldots,n\}} \bar{c}_j = \max_{\boldsymbol{\pi}} \min_{j\in\{1,\ldots,n\}} c_j - \boldsymbol{\pi}^\mathsf{T}\mathbf{a}_j. \tag{1.56}$$

(a) Formulate this problem as a linear program.

(b) Formulate the dual of the linear program in (a).

1.11 Integrality property and algorithm selection bias
The integrality property is totally algorithm independent since it is only about having an integrality gap of zero (Definition 1.25). When solving the linear relaxation of an *ILP* having the integrality property using the primal simplex algorithm, justify whether or not we always terminate with an optimal integer solution.

1.12 Unbalanced capacitated minimum cost flow problem
Let the *CMCFP* (1.45) be defined on $G = (N,A)$, where the set of supply nodes is $S = \{i \in N \mid b_i > 0\}$, that of the demand nodes is $D = \{i \in N \mid b_i < 0\}$, and all lower bounds are set to zero, i.e., $\ell_{ij} = 0, \forall(i,j) \in A$. Note that even if the network is balanced or $\sum_{i\in S} b_i > \sum_{i\in D} -b_i$, the total supply from S can be insufficient to satisfy the total demand at D due to the set of upper bound restrictions $\{u_{ij}\}_{(i,j)\in A}$.

For any integer supply/demand values, show that by adding a single extra node, here denoted 0, and a set of arcs with appropriate cost and upper bound values, we obtain a *balanced* and *feasible* network that can provide a meaningful solution anyway.

1.13 Totally unimodular matrices with consecutive ones
To determine if a matrix $\mathbf{A} \in \mathbb{Z}^{m\times n}$ is totally unimodular, there exist sufficient conditions such as the following set:

(i) $a_{ij} \in \{-1,0,1\}, \forall i \in \{1,\ldots,m\}, j \in \{1,\ldots,n\}$;

(ii) $\sum_{i=1}^{m} |a_{ij}| \leq 2, \forall j \in \{1,\ldots,n\}$;

(iii) there exists a partition M_1 and M_2 of the row indices (M_2 can be empty) such that $\sum_{i\in M_1} a_{ij} = \sum_{i\in M_2} a_{ij}$ for any column $j \in \{1,\ldots,n\}$ with $\sum_{i=1}^{m} |a_{ij}| = 2$.

Consider an $m \times n$ binary matrix $\mathbf{A}$ for which the 1's are consecutive in each column, i.e., if $a_{ij} = a_{kj} = 1$ for $k > i+1$, then $a_{\ell j} = 1$ for all $\ell = i+1,\ldots,k-1$.

(a) Use the sufficient conditions (i)–(iii) to show that the determinant of $\mathbf{A}$ is 0, $+1$, or -1 when $m = n$. *Hint*: Consider the determinant of the matrix obtained by subtracting from each row of $\mathbf{A}$ except its first, its previous row.
(b) Using this result, prove that $\mathbf{A}$ is totally unimodular for any dimension $m \times n$.

1.14 Variable fixing at upper bound
Consider the *ILP* (1.43) with an additional upper bound $u_j \in \mathbb{Z}_+$ on the variable x_j. Let *UB* be an upper bound on $z^\star_{ILP}$ and $(\mathbf{x}^\star_{LP}, \boldsymbol{\pi}^\star)$ be an optimal primal-dual solution pair of the linear relaxation, with $x^\star_j = u_j$. Determine a sufficient condition on the reduced cost $\bar{c}_j$ of x_j that allows to fix x_j to its upper bound u_j.

1.15 Formulations for the minimum-weight perfect matching problem
Let $G = (N, E)$ be an undirected graph, where N is its node set and E its edge set. We assume that there are an even number $2n$ of nodes in N and that a weight $w_e \in \mathbb{R}$ is associated with each edge $e \in E$. A *perfect matching* in G is a subset M of n edges in E such that every node $i \in N$ is incident to exactly one edge in M. The *minimum-weight perfect matching problem* consists in finding a perfect matching M in G such that the sum of its edge weights is minimal.

The minimum-weight perfect matching problem can be formulated as the following *ILP*. Let x_e, $e \in E$, be a binary variable that takes value 1 if edge e is in the selected matching M and 0 otherwise. Furthermore, we denote by $\delta(i)$ the subset of edges incident to node $i \in N$. The *ILP* is

$$\begin{aligned} \min \quad & \sum_{e \in E} w_e x_e \\ \text{s.t.} \quad & \sum_{e \in \delta(i)} x_e = 1 && \forall i \in N \\ & x_e \in \{0,1\} && \forall e \in E. \end{aligned} \tag{1.57}$$

The graph G in Figure 1.9 is used as an example for some of the following questions.

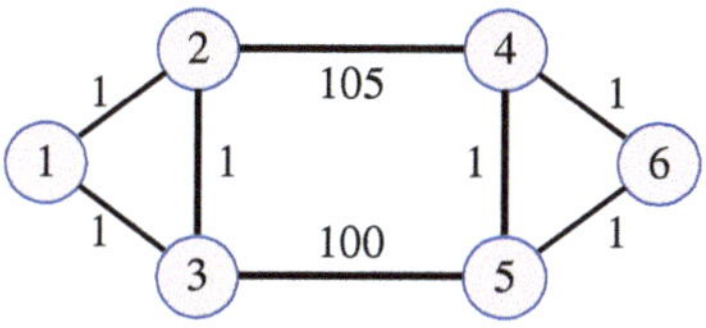

Fig. 1.9: Graph $G = (N, E)$ for exercise 1.15.

(a) For the example in Figure 1.9, find the optimal solutions of model (1.57) and its linear relaxation, and compute the relative integrality gap.
(b) For any subset of nodes S, let $E(S)$ be the subset of edges in E that have their two endpoints in S. The blossom inequalities (Edmonds, 1965)

$$\sum_{e \in E(S)} x_e \leq \frac{1}{2}(|S|-1), \qquad \forall S \subset N \text{ such that } |S| \text{ is odd} \tag{1.58}$$

specify that, for any subset of nodes S of odd cardinality, there are at most $\frac{1}{2}(|S|-1)$ edges that have their two endpoints in S. Show that these inequalities are valid for the feasible domain of (1.57), i.e., they are not violated by any feasible solution of (1.57).

(c) For the example in Figure 1.9, find a blossom inequality (1.58) that is violated by its optimal linear relaxation solution.

(d) Denote by $\mathcal{P}_1$ and $\mathcal{P}_2$ the feasible domains of the linear relaxations of (1.57) and (1.57)–(1.58), respectively. Show that formulation $\mathcal{P}_2$ is stronger than formulation $\mathcal{P}_1$.

(e) Because the number of blossom inequalities (1.58) is exponential in the number of nodes in N, these inequalities cannot be all added a priori and are rather generated as needed in a branch-and-cut algorithm. Given an optimal linear relaxation solution $\mathbf{x}^\star_{LP} = \{x^\star_e\}_{e \in E}$, we can use a separation algorithm that solves an *ILP* to determine if $\mathbf{x}^\star_{LP}$ violates a blossom inequality. Formulate this separation *ILP*.
Hint: A feasible solution to this *ILP* should identify a subset of nodes S and its objective function should aim at maximizing the left-hand side of (1.58) minus its right-hand side.

(f) Polynomial-time separation algorithms for the blossom inequalities also exist (see, e.g., Grötschel and Holland, 1985). They are based on the following equivalent version of these inequalities:

$$\sum_{e \in \delta(S)} x_e \geq 1, \qquad \forall S \subset N \text{ such that } |S| \text{ is odd} \tag{1.59}$$

where $\delta(S)$ is the subset of edges with an end node in S and the other in $N \setminus S$. Show that the inequalities (1.59) are equivalent to the inequalities (1.58).

2

Column Generation

> Le temps que l'on prend pour dire *Je t'aime*
> C'est le seul qui reste au bout de nos jours.
>
> *Gens du pays*
> Gaston Rochon / Gilles Vigneault

Abstract This chapter describes the classical column generation algorithm used to solve a linear program with a huge number of variables. Besides its main properties, we offer several practical hints to be included in any worthwhile implementation and present a few classical examples of its application.

Contents

Introduction . . . 45
2.1 Algorithm . . . 47
Master problem . . . 47
Pricing problem . . . 48
Iteration . . . 49
Initialization . . . 51
2.2 Good to Know . . . 53
Objects, representations, and encodings . . . 53
Several subproblems . . . 56
Pivot rules and column management . . . 57
Heuristic pricing . . . 59
2.3 More to Know . . . 60
Lower bounds . . . 60
Convergence . . . 63
Limited numeric precision . . . 65
Static and auxiliary variables . . . 66
Relaxed pricing, relaxed master . . . 68

J. Desrosiers et al., *Branch-and-Price*,
https://doi.org/10.1007/978-3-031-96917-1_2

Paving the way to a practical implementation 69
2.4 Examples 71
One-dimensional cutting stock problem 71
Integer master problem 71
Integer pricing problem 72
Empty vs. zero cutting patterns 73
Cutting stock problem with rolls of different widths 74
Edge coloring problem 76
Set covering master problem 77
Pricing the matchings 77
Sports scheduling 78
Single depot vehicle scheduling problem 80
Set partitioning master problem 80
Shortest path pricing problem 81
Zero schedule 83
Vehicle routing and crew scheduling problems 84
Aircraft routing with schedule synchronization 86
Routes and schedules 87
Integer master problem 88
Integer pricing problems 89
Practical observations 90
2.5 Reference Notes 91
Exercises 95

Acronyms

MP	master problem	47
IMP	integer master problem	47
RMP	restricted *MP*	48
SP	subproblem (or pricing problem)	48
ISP	integer subproblem	49
SP^k	subproblem for block $k \in K$	56
lb	lower bound on $z^\star_{MP}$	63
LB	best known lower bound on $z^\star_{MP}$	63
VRPTW	vehicle routing problem with time windows	63
$MP_\mathcal{Y}$	*MP* allowing for inadmissible columns	68
CSP	one-dimensional cutting stock problem	71
ISP^k	integer subproblem for block $k \in K$	74
ECP	edge coloring problem	76
SDVSP	single depot vehicle scheduling problem	80
SPP	set partitioning problem	80
MDVSP	multiple depot vehicle scheduling problem	85
ARPSS	aircraft routing problem with schedule synchronization	86

Introduction

In mathematical optimization, we often take an algorithmic approach to overcome modeling hurdles we do not know how to deal with directly. Examples are the integrality of variables or an exponential number of constraints. A typical procedure then is to first ignore these hurdles (relax integrality, drop constraints), and iteratively work towards respecting what was initially ignored (by branching, separating cuts) until it can be proven that we have found what we were originally looking for. Column generation is no different in this philosophy. Our hurdle here is that there are many—too many—variables in a linear program to cope with. Thus, we first drop them from the model and dynamically re-insert only a small subset of them until we can prove that the smaller model we constructed is sufficient to contain an optimal solution to the original model. For those of us who were not in love with optimization before, this is a good point to start.

In a linear program with gazillions of variables, the rationale behind needing only very few of the variables to prove optimality comes from the fact that in an extreme point solution almost all of the variables are at value zero (or at their bounds, to be precise). Not in the model or at value zero makes no difference then. The dynamic addition of variables to the linear program comes with adding column-coefficients to its constraint matrix, which gives the method its name. In practice, we see that the fraction of variables actually generated is astoundingly tiny.

Column generation is often mentioned in the context of the Dantzig-Wolfe decomposition (see Chapter 3) and sometimes even confused with it. However, the former lives completely independently of the latter and this is why we present the column generation algorithm first. We bring ourselves in the mood with a quote from George Nemhauser.

> *Column generation* refers to linear programming (*LP*) algorithms designed to solve problems in which there are a huge number of variables compared to the number of constraints and the simplex algorithm step of determining whether the current basic solution is optimal or finding a variable to enter the basis is done by solving an optimization problem rather than by enumeration.
>
> To the best of my knowledge, the idea of using column generation to solve linear programs was first proposed by Ford and Fulkerson (1958). However, I couldn't find the term *column generation* in that paper or the subsequent two seminal papers by Dantzig and Wolfe (1960) and Gilmore and Gomory (1961, 1963). The first use of the term that I could find was in Appelgren (1969), a paper with the title "A column generation algorithm for a ship scheduling problem".
>
> Ford and Fulkerson (1958) gave a formulation for a *multicommodity maximum flow* problem in which the variables represented path flows for each commodity. The commodities represent distinct origin-destination pairs and integrality of the flows is not required. This formulation needs a number of variables exponential in the size of the underlying network since the number of paths in a graph is exponential in the size of the network. What motivated them to propose this formulation? A more natural and smaller formulation in terms of the number of constraints plus the numbers of variables is easily obtained by using arc variables rather than path variables. Ford and Fulkerson observed that even with an exponential number of variables in the path formulation, the minimum reduced cost for each commodity could be calculated by solving a shortest path problem, which was already known to be

> an easy problem. Moreover the number of constraints in the path formulation is the number of arcs, while in the arc formulation it is roughly the (number of nodes)×(number of commodities) + number of arcs. Therefore the size of the basis in the path formulation is independent of the number of commodities and is significantly smaller when the number of commodities is large. This advantage in size they claimed might make it possible to solve instances with a large number of commodities with the simplex method. Modestly, they stated that they really had no idea whether the method would be practical since they had only solved a few small instances by hand. – Nemhauser (2012)

With respect to the origins of the method, Dantzig (1963) credits two 1958 works in his book. Independently of Ford and Fulkerson (1958), Jewell (1958) suggests (in an unpublished report) a very similar multi-commodity flow algorithm, which can also be seen as a precursor to column generation. We have also found that Manne (1958) proposes a large-scale linear programming model that explicitly considers all possible production schedules for the capacitated lot-sizing problem, and that Dzielinski and Gomory (1965) use column generation to handle that formulation.

The work of Kantorovich and Zalgaller (1951) has recently resurfaced from the Saint Petersburg (formerly Leningrad) State University. It contains an entire section dedicated to industrial cutting problems where the authors rely on dynamic pattern generation through what we today call dynamic programming. We share the enthusiasm of Uchoa and Sadykov (2024) who go as far as calling it *work zero* (YouTube for a video presentation hosted by Anand Subramanian).

We make one point clear right from the beginning. Even though the method was termed *generalized linear programming* in the early days (Magnanti et al., 1976), it never became competitive for solving linear programs, except for special cases. In addition, tailored implementations were designed to exploit matrix structures, but they did not perform better than the simplex method. By contrast, column generation is a real winner in the context of integer programs which makes it a *must-have* in the computational mixed-integer programming "bag of tricks."

Fig. 2.1: George Nemhauser (San Pedro de Atacama, Chile, 2013-06-12).

2.1 Algorithm

In essence, the column generation algorithm is the primal simplex algorithm with a small but fundamental difference in the pricing step: Rather than *explicitly* computing the reduced costs of a usually prohibitively large number of variables, one solves an auxiliary optimization program that *implicitly* searches for a variable of negative reduced cost, or proves that none exists.

Master problem

We would like to solve a linear program, which in our context is called the *master problem MP*. It contains a finite but typically huge number of variables indexed by the set $\mathcal{X}$:

$$\begin{aligned} z^\star_{MP} = \min \quad & \sum_{\mathbf{x}\in\mathcal{X}} c_{\mathbf{x}}\lambda_{\mathbf{x}} \\ \text{s.t.} \quad & \sum_{\mathbf{x}\in\mathcal{X}} \mathbf{a}_{\mathbf{x}}\lambda_{\mathbf{x}} \geq \mathbf{b} \quad [\boldsymbol{\pi}] \\ & \lambda_{\mathbf{x}} \geq 0 \qquad \forall \mathbf{x}\in\mathcal{X}, \end{aligned} \tag{2.1}$$

where $c_{\mathbf{x}} \in \mathbb{R}$, $\mathbf{a}_{\mathbf{x}}, \mathbf{b} \in \mathbb{R}^m$, $\forall \mathbf{x}\in\mathcal{X}$, and $m \ll |\mathcal{X}| < \infty$. We assume that the *MP* is feasible and $z^\star_{MP}$ is finite. The non-negative dual vector $\boldsymbol{\pi} = [\pi_i]_{i=1,\dots,m}$ associated with the inequality constraints appears as usual within brackets on the right.

Note 2.1 (Mind the bounds.) We underscore the non-negativity restriction on all λ-variables. Imposing upper bounds on any variable, say $\lambda_{\mathbf{x}} \leq u_{\mathbf{x}}$, can introduce adverse effects in column generation. We advice to steer clear of upper bounded variables as long as these effects are not well understood.

Note 2.2 (How about integrality?) In most of the examples in practice and in this book, a master problem results from the linear relaxation of an integer linear program which we actually wish to solve. The latter may come in different forms but we always call it the *integer master problem*, or *IMP* for short. As always, from solving the linear relaxation *MP* we obtain a lower bound $z^\star_{MP} \leq z^\star_{IMP}$.

Note 2.3 (Our choice of the index set $\mathcal{X}$.) Advanced readers with some knowledge about the column generation literature may wonder why we write $\lambda_{\mathbf{x}}$, $\mathbf{x}\in\mathcal{X}$, rather than, say, λ_j, $j \in J$. The reason is in anticipation that each master variable corresponds to a solution of a pricing problem. For example, $\mathcal{X} = \left\{\begin{bmatrix}\mathbf{x}_p\\1\end{bmatrix}\right\}_{p\in P} \cup \left\{\begin{bmatrix}\mathbf{x}_r\\0\end{bmatrix}\right\}_{r\in R}$ in a Dantzig-Wolfe reformulation (Chapter 3), the first set referring to the extreme points of a polyhedron whereas to its extreme rays for the second set. With the alternative notation, one requires a mapping to maintain this correspondence. To see what we mean, try to express that an index $j \in J$ is related to a solution $\mathbf{x}\in\mathcal{X}$, and vice versa.

As mentioned, we have no hope to directly solve the *MP* due to its sheer size. In practice, we see models with a number of variables no computer on earth could even store, let alone apply algorithms to optimize them. Therefore, we start working with a relatively small subset $\mathcal{X}' \subseteq \mathcal{X}$. The resulting model is a restriction of the *MP* and consequently called the *restricted master problem RMP*:

$$\begin{aligned} z_{RMP} = \min \quad & \sum_{\mathbf{x} \in \mathcal{X}'} c_{\mathbf{x}} \lambda_{\mathbf{x}} \\ \text{s.t.} \quad & \sum_{\mathbf{x} \in \mathcal{X}'} a_{i\mathbf{x}} \lambda_{\mathbf{x}} \geq b_i \quad [\pi_i] \quad \forall i \in \{1, \dots, m\} \\ & \lambda_{\mathbf{x}} \geq 0 \qquad\qquad\quad \forall \mathbf{x} \in \mathcal{X}' \subset \mathcal{X}, \end{aligned} \tag{2.2}$$

where the column-coefficients $\mathbf{a}_{\mathbf{x}} = [a_{i\mathbf{x}}]_{i=1,\dots,m}$ are here written in scalar form. We assume for the moment that $\mathcal{X}'$ is chosen such that the *RMP* (2.2) is feasible. We can therefore solve to optimality the *RMP* with an algorithm of our choice and obtain its optimum z_{RMP} and a primal-dual pair $(\boldsymbol{\lambda}, \boldsymbol{\pi})$ of optimal solutions. Note that $\boldsymbol{\lambda} = [\lambda_{\mathbf{x}}]_{\mathbf{x} \in \mathcal{X}'}$ can be trivially extended to $\boldsymbol{\lambda} = [\lambda_{\mathbf{x}}]_{\mathbf{x} \in \mathcal{X}}$, where $\lambda_{\mathbf{x}} = 0$, $\forall \mathbf{x} \in \mathcal{X} \setminus \mathcal{X}'$, and we obtain a feasible solution to the *MP* (2.1) as well. How would we know whether this solution is optimal for the master problem? The primal simplex algorithm would identify a variable $\lambda_{\mathbf{x}}$, $\mathbf{x} \in \mathcal{X} \setminus \mathcal{X}'$, of negative reduced cost or verify that none exists. Since an explicit search of $\mathcal{X}$ (or $\mathcal{X} \setminus \mathcal{X}'$ for that matter) is out of the question, this is where the *pricing problem* comes to the rescue. It is also known as the *subproblem*, *oracle*, or *column generator*. We denote it by *SP*.

Note 2.4 (Lightweight notation.) Again, we do not notationally distinguish between variables and their values. In addition, as we later iteratively solve the *RMP* for different subsets $\mathcal{X}'$, one would need to correctly write $z^{\star}_{RMP}(\mathcal{X}')$ to denote the optimum obtained. Similarly, this applies to primal and dual solutions. We choose to omit any superscript * or $'$ or other reference to optimality or iteration for the *RMP* or the *SP*. However, in the text we sometimes refer to "the current" iteration or solution.

Pricing problem

In order to prove that the *RMP* solution is optimal for the *MP*, the pricing problem needs to verify that the reduced costs $\bar{c}_{\mathbf{x}} = c_{\mathbf{x}} - \boldsymbol{\pi}^{\mathsf{T}} \mathbf{a}_{\mathbf{x}}$ of all $\lambda_{\mathbf{x}}$, $\mathbf{x} \in \mathcal{X}$, are non-negative. Recall that $\boldsymbol{\pi}$ is available from solving to optimality the *RMP*. In place of an explicit search, which is prohibitive, we rely on an optimization program with the reduced cost formula in the objective function to implicitly compute $\bar{c}(\boldsymbol{\pi})$ amongst all $\bar{c}_{\mathbf{x}}$, $\forall \mathbf{x} \in \mathcal{X}$:

$$\begin{aligned} \bar{c}(\boldsymbol{\pi}) = \min_{\mathbf{x} \in \mathcal{X}} \quad & c_{\mathbf{x}} - \sum_{i=1}^{m} \pi_i a_{i\mathbf{x}} \\ \text{s.t.} \quad & c_{\mathbf{x}} = c(\mathbf{x}) \\ & a_{i\mathbf{x}} = a_i(\mathbf{x}) \quad \forall i \in \{1, \dots, m\}. \end{aligned} \tag{2.3}$$

For the reduced cost computation, we assume that functions $c(\mathbf{x})$ and $\mathbf{a}(\mathbf{x}) = [a_i(\mathbf{x})]_{i=1,\dots,m}$ give us, from any $\mathbf{x} \in \mathcal{X}$, representative cost and column-coefficients for every variable $\lambda_{\mathbf{x}}$. We will see many specializations of these functions and how we interpret them later on.

If $\bar{c}_{\mathbf{x}} \geq 0, \forall \mathbf{x} \in \mathcal{X}$, then $\bar{c}(\boldsymbol{\pi}) \geq 0$ which proves optimality of the *MP*. Otherwise, from a minimizer $\mathbf{x} \in \mathcal{X} \setminus \mathcal{X}'$ we obtain the variable $\lambda_{\mathbf{x}}$ (with a negative reduced cost) to be added to the *RMP*, formally by adding $\mathbf{x}$ to $\mathcal{X}'$. Figure 2.2 shows the exchange of information between the (restricted) master problem and the subproblem; note the similarity to the primal simplex algorithm in Figure 1.5. We have stated the *SP* (2.3) very generally, but in Chapter 3, it is a linear program and thereafter almost always an integer linear program. The latter is actually the most relevant case for the entire book and we see this already in the examples in this chapter. We therefore often stress this case of having an integer subproblem by abbreviating it *ISP*.

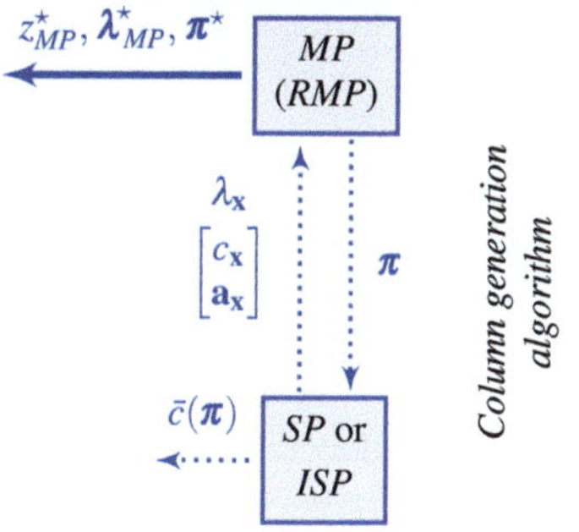

Fig. 2.2: Information flow of the column generation algorithm solving the *MP* (2.1).

Note 2.5 (Pricing known columns! Why not?) Even if the pricing problem aims at determining if there exists a variable $\lambda_{\mathbf{x}}, \mathbf{x} \in \mathcal{X} \setminus \mathcal{X}'$, with a negative reduced cost, it is defined over the whole set of variables indexed by $\mathcal{X}$. There are two reasons for this. First, for any optimal $\boldsymbol{\pi}$ to the current *RMP*, we already have $\bar{c}_{\mathbf{x}} \geq 0, \forall \mathbf{x} \in \mathcal{X}'$, which means that the columns in $\mathcal{X}'$ cannot be generated again. Second, excluding these variables from the subproblem's feasible domain is often a difficult task.

Iteration

What completes the column generation algorithm now lies right in front of us. If the *SP* solution proves optimality of the *MP* via $\bar{c}(\boldsymbol{\pi}) \geq 0$, we stop. In this case, the information of the last iteration gives us $z^\star_{MP} = z_{RMP}$, a primal solution $\boldsymbol{\lambda}^\star_{MP}$ where $\lambda^\star_{\mathbf{x}} = \lambda_{\mathbf{x}}, \forall \mathbf{x} \in \mathcal{X}'$, and $\lambda^\star_{\mathbf{x}} = 0, \forall \mathbf{x} \in \mathcal{X} \setminus \mathcal{X}'$, as well as a dual solution $\boldsymbol{\pi}^\star = \boldsymbol{\pi}$. Otherwise $\bar{c}(\boldsymbol{\pi}) < 0$, we then re-optimize the *RMP* with the added column $\mathbf{a}_{\mathbf{x}}$ of cost $c_{\mathbf{x}}$ to obtain a new $\boldsymbol{\lambda}$ and in particular a new $\boldsymbol{\pi}$ that is passed to the *SP*. Algorithm 2.1 summarizes this iterative process with a loop whose content we refer to as one column generation iteration.

Algorithm 2.1: The column generation algorithm.

---

input : *RMP* with feasible subset $\mathcal{X}' \subset \mathcal{X}$, *SP* (or *ISP*)
output : Optimal primal-dual solutions $\boldsymbol{\lambda}^\star_{MP}, \boldsymbol{\pi}^\star$ and optimum $z^\star_{MP}$ for the *MP*

loop
  Solve the *RMP* to obtain an optimal primal-dual solution $\boldsymbol{\lambda}_{RMP}, \boldsymbol{\pi}$ of cost z_{RMP}
  Solve the *SP* to obtain the minimum reduced cost $\bar{c}(\boldsymbol{\pi})$ with corresponding $\mathbf{x} \in \mathcal{X}$
  if $\bar{c}(\boldsymbol{\pi}) \geq 0$
    break by optimality of the *MP*
  Generate the variable $\lambda_{\mathbf{x}}$ with encoding $\begin{bmatrix} c_{\mathbf{x}} \\ \mathbf{a}_{\mathbf{x}} \end{bmatrix}$ via $\mathcal{X}' \leftarrow \mathcal{X}' \cup \{\mathbf{x}\}$
return $\boldsymbol{\lambda}_{RMP}$, $\boldsymbol{\pi}$, and z_{RMP}

Note 2.6 (Non-positivity of the minimum reduced cost.) Since $\bar{c}(\boldsymbol{\pi}) < 0$ as long as the *MP* is not solved to optimality, we only have to convince ourselves that the *SP* finds $\bar{c}(\boldsymbol{\pi}) = 0$ at the last column generation iteration. Considering that the *MP* is feasible by assumption, there must exist at least one variable $\lambda^\star_{\mathbf{x}} > 0$, $\mathbf{x} \in \mathcal{X}$, in any non-trivial solution. By the complementary slackness conditions (1.8), any positive λ-variable necessarily has a zero reduced cost such that $\bar{c}_{\mathbf{x}} = 0$ for this particular column. For an intuitive argument, if we use a simplex type method to optimize the *RMP*, any basic λ-variable has a zero reduced cost.

Note 2.7 (They talk to each other, Baby!) The dual values are the means of communication between the *RMP* and *SP*, see Figure 2.2. As the objective function of the subproblem reflects the reduced cost of a given $\lambda_{\mathbf{x}}$, $\mathbf{x} \in \mathcal{X}$, i.e., $\bar{c}_{\mathbf{x}} = c_{\mathbf{x}} - \sum_{i=1}^{m} \pi_i a_{i\mathbf{x}}$, we can put the interpretation we have learned in Section 1.2 on Sensitivity analysis to good use. Specifically, we expect the dual values, depending on their signs and magnitudes, to give a smaller or larger incentive or penalty to use a certain $\mathbf{x}$. That is, all else being equal, whether or not and by how much it is attractive to have a non-zero value $a_{i\mathbf{x}}$ depends on π_i. For example, in a vehicle routing problem where $a_{i\mathbf{x}}$ represents the number of times customer i is visited in route $\mathbf{x}$, minimizing the reduced cost with a large value for π_i encourages multiple visits at i.

Note 2.8 (Availability of the algorithm opens up modeling opportunities.) Our presentation reveals the two ingredients we need to make the column generation algorithm work: A restricted master problem (which is a linear program) and a corresponding subproblem. The two must be compatible in the sense that they refer to the same $\mathcal{X}$, of course. Some authors call the master problem a *column generation formulation* to reflect that they apply the column generation algorithm using the corresponding *RMP* and *SP*. We do not use this terminology in this book, but let us stress that having the *algorithm* available enables us to formulate a problem using different kinds of *models*. In this book's examples, we often see combinatorial concepts like subsets, configurations, sequences, etc., on which the variables of a model could be defined naturally. Without the column generation algorithm, it would often be impossible to solve these models.

Note 2.9 (The apple does not fall far from the tree.) Since the column generation algorithm is a variant of the primal simplex algorithm, it also inherits the flaws of the latter. We speak in particular of *cycling* and *degeneracy*.

- The phenomenon of cycling is a rather theoretical construct, but of course it could occur when solving the *RMP*, and it could even additionally occur when solving the *SP*. Interestingly, as long as we generate *new* columns from the *SP*, that is, $\mathbf{x} \in \mathcal{X} \setminus \mathcal{X}'$, we cannot cycle in the column generation iterations. Nevertheless, assuming the "standard" precautions of the simplex algorithm of Chapter 1, the column generation algorithm is correct, i.e., it finitely terminates with an optimal solution to the *MP*. Notable implementation decisions which can introduce cycling side-effects at the column generation level are to solve the *RMP* only heuristically and/or delete "non-used" columns.
- In contrast, degeneracy *is* a practical concern in the column generation algorithm as it is in the primal simplex algorithm. This is particularly true since the *MP*s that we solve are often linear relaxations of combinatorial optimization problems which can have highly degenerate solutions. Modern solvers handle degeneracy while solving a linear program. Yet, in the column generation algorithm, we control the iterations ourselves, and it may be us who have to deal with degeneracy. If the issue occurs, in the sense that significant stalling in the column generation process happens, soft instead of hard constraints as in Example 2.6 (Aircraft routing with schedule synchronization), classical perturbation of the right-hand side vector $\mathbf{b}$ or, more generally, dual variable stabilization techniques from Chapter 6 (Dual Point of View) can be effective. This anticipation of degeneracy potential is also a common background theme to smart decisions we should do regarding the columns we bring back from the *SP*: *several subproblems, parallelization, complementary pricing, heuristic pricing, etc.* These items are scattered throughout the rest of this chapter as we find opportunity to precise them. For now, we can already say that they highlight yet another inherited trait: the pricing problem is (often) the bottleneck.

Initialization

Algorithm 2.1 reflects the assumption that we are given an $\mathcal{X}' \subseteq \mathcal{X}$ such that the *RMP* is feasible. For particular applications, producing such a set can be very easy as in Example 2.1. In practice, a feasible solution may be available, e.g., from previously solving an *MP* with similar data. Or we have a specifically designed primal heuristic for our application. Lagrangian relaxation in Chapter 6 gives us good reasons to call the *SP* several times with various dual vectors to obtain initial variables for the *RMP*. Since the subproblem is always available, this is rather cheap to implement.

Even if heuristics like the above can often provide us with a feasible solution to the *RMP*, we have no guarantee of this. In fact, looking at the bigger picture where column generation is used within each node of a branch-and-bound tree (see Chapter 7), branching decisions sooner or later lead to infeasibility at some nodes. It is therefore useful to have a mechanism that can cope with this systematically. The classic way is to use artificial variables ($\mathbf{y} = [y_j]_{j=1,\ldots,m}$) as in a *Phase I* of the primal simplex algorithm, see Section 1.3 (p. 18) and Section 2.3 (p. 66). In Algorithm 2.2, the *RMP* is therefore always feasible but we verify in the exit condition whether there remains any positive artificial variable and return the appropriate certificate of optimization.

Algorithm 2.2: The column generation algorithm using artificial variables.

---

input : *RMP*, *SP* (or *ISP*), big-*M*
output : Certificate of optimization
initialization : $\mathcal{X}' \leftarrow \emptyset$, m artificial variables $\mathbf{y}$

loop
  $z_{RMP}, \boldsymbol{\lambda}_{RMP}, \mathbf{y}, \boldsymbol{\pi} \leftarrow RMP$
  $\bar{c}(\boldsymbol{\pi}), \mathbf{x}, c_{\mathbf{x}}, \mathbf{a}_{\mathbf{x}} \leftarrow SP$
  if $\bar{c}(\boldsymbol{\pi}) \geq 0$
    if $\exists j \in \{1,\ldots,m\} \mid y_j > 0$
      break *by infeasibility of the MP*
    else
      break *by optimality of the MP*
  $\mathcal{X}' \leftarrow \mathcal{X}' \cup \{\mathbf{x}\}$
return $\boldsymbol{\lambda}_{RMP}, \mathbf{y}, \boldsymbol{\pi}$, and z_{RMP}

Such an initialization is used in Example 3.2 where we can witness that the penalty cost may need some tuning. This is not the case for the following easy to implement procedure which is based on Farkas' Lemma, see Proposition 1.4. It states that either the domain of the *RMP* in the form of $\mathbf{Ax} \geq \mathbf{b}$, $\mathbf{x} \geq \mathbf{0}$, is non-empty or there is a vector $\boldsymbol{\pi} \geq \mathbf{0}$ with $\boldsymbol{\pi}^\mathsf{T}\mathbf{A} \leq \mathbf{0}$ and $\boldsymbol{\pi}^\mathsf{T}\mathbf{b} > 0$. Such a vector $\boldsymbol{\pi}$, a ray in the dual, proves infeasibility of the *RMP* as $\boldsymbol{\pi}^\mathsf{T}\mathbf{Ax} \geq \boldsymbol{\pi}^\mathsf{T}\mathbf{b}$ cannot be fulfilled. The dual ray $\boldsymbol{\pi}$ is typically provided by the linear programming solver when the *RMP* is infeasible. The idea is to add a column $\mathbf{a}_{\mathbf{x}}$ to $\mathbf{A}$ with $\boldsymbol{\pi}^\mathsf{T}\mathbf{a}_{\mathbf{x}} > 0$, thus *destroying* this proof of infeasibility with respect to $\boldsymbol{\pi}^\mathsf{T}\mathbf{A} \leq \mathbf{0}$. We can identify an $\mathbf{x} \in \mathcal{X}$ from which we obtain an $\mathbf{a}_{\mathbf{x}}$ with this property by solving

$$
\begin{aligned}
F(\boldsymbol{\pi}) = \max_{\mathbf{x}\in\mathcal{X}} \sum_{i=1}^{m} \pi_i a_{i\mathbf{x}} = -\min_{\mathbf{x}\in\mathcal{X}} \quad & -\sum_{i=1}^{m} \pi_i a_{i\mathbf{x}} \\
\text{s.t.} \quad & a_{i\mathbf{x}} = a_i(\mathbf{x}) \quad \forall i \in \{1,\ldots,m\}.
\end{aligned}
\tag{2.4}
$$

We see that we can mostly re-use the *SP* (2.3) except for temporarily setting all cost coefficients $c_{\mathbf{x}} = 0$. If $F(\boldsymbol{\pi}) > 0$, we have found a variable $\lambda_{\mathbf{x}}$ that reduces infeasibility so we add it to the *RMP* with its correct cost coefficient $c_{\mathbf{x}} = c(\mathbf{x})$. Otherwise, the *MP* is proven infeasible. Even though the mathematics is much older, the method we just described became known as *Farkas pricing* through the `SCIP` branch-price-and-cut framework (Achterberg, 2009). Exercise 2.20 asks the reader to propose a pseudo-code to initialize the column generation algorithm with Farkas pricing.

The work on multi-objective programs of Lin et al. (2017) compares these initialization methods. The authors observe that the big-M approach generally produces initial solutions with smaller objective values in less iterations although it does require an appropriate guesswork on M.

Let us finally mention that one can relax the *RMP* when this helps to easier initialize it. A typical example is to relax a set partitioning model to a set covering model. In Example 2.3 and Exercise 2.14, solving the relaxed *RMP* still leads to a feasible solution for the *MP*. And even if it does not, we at least obtain a lower bound on $z^\star_{MP}$.

As our knowledge in theory and understanding of the applications and contexts grow, we get a better feeling for advantages and disadvantages of using these methods in practice. We point in particular to the influence of artificial variables on the dual, see Section 2.3.

2.2 Good to Know

In addition to the core material, we feel that it is "good to know" some more background on what is important in actual implementations. In particular, the usefulness of heuristics cannot be stressed enough.

Objects, representations, and encodings

In the literature, one reads about negative reduced cost *columns* (even though the cost is defined on *variables*), or that the pricing problem returns a *pattern* or other *things*. Our brains easily "translate" between these concepts, and therefore they are often *identified*. We briefly discuss this here, because typically, the relations are not one-to-one. The discussion is not all that important, but it reminds us of how much we only implicitly convey in our language. If you need a proof of this, try to write a book on branch-and-price with co-authors!

The basics of mathematical model building apply, of course, also to our situation. In practical applications we would like to find crew schedules, cutting patterns, vehicle routes, etc. These are the *objects* the planner is interested in; we note that these objects belong to the end-user, they define the rules. It is part of the OR expert's job to create a column generator (*SP* as a mathematical model) that reflects the planning situation. What does this mean? An *SP* model is correct, if every object is *represented* by a solution in the model, and every solution of the model has an interpretation as an object. There can be many representations of the same object (e.g., in Example 4.1 where the same packing is represented by many paths). Also, a representation needs not uniquely identify an object (think of parallel machine scheduling, where an assignment of jobs to machines represents many schedules as the ordering of jobs on each machine may be unimportant).

We all know that coming up with a mathematical model can be a very creative task. In our context, the representations of objects need to be given exactly by the solutions $\mathbf{x} \in \mathcal{X}$ to the *SP*. In our forthcoming examples, these $\mathbf{x}$ can be subsets of nodes or paths in a network, solutions to linear and integer programs, etc. All of these can be presented in the form of vectors, so we henceforth assume that elements $\mathbf{x} \in \mathcal{X}$ are in fact vectors in an appropriate vector space. There can be infinitely many objects, but theory later tells us that we can restrict ourselves to finite $\mathcal{X}$ in all cases of interest in this book.

In general, the formulation of the *MP* does not (and needs not) contain all the detailed characteristics of an object that is available from its representation $\mathbf{x} \in \mathcal{X}$. Thus, typically, the cost $c_{\mathbf{x}}$ and in particular the column $\mathbf{a}_{\mathbf{x}}$ *encode* only the information that is relevant to formulate the objective function and the constraints of the *MP* (e.g., the flights operated in a weekly aircraft route but not the many events regarding flights, aircraft maintenance, crew schedules). Observe, again, that we index the λ-variables in the *MP* by the elements in $\mathcal{X}$, as these are—literally—the central piece of information.

Figure 2.3 sketches an example of an *aircraft routing problem* (Examples 2.5 and 2.6). The *object* we are interested in is an aircraft route together with the departure times of the flights. Its *representation* in the *SP* is given by an $\mathbf{x} \in \mathcal{X}$ which describes a directed path originating and ending at a (virtual) depot. This vector contains the values of the binary flow variables from which we derive the precise order in which the flights are operated, as well as the departure times. The corresponding *encoding* in the *RMP* (formulated as a set partitioning model) is given by the column-coefficients $\mathbf{a}_{\mathbf{x}}$ and its cost $c_{\mathbf{x}}$. The column $\mathbf{a}_{\mathbf{x}}$ typically indicates the set of operated flights as a binary vector, potentially (as in Example 2.6) also the departure times. A typical encoding for $c_{\mathbf{x}}$ includes the capital, fixed, and operational costs of an aircraft and revenues.

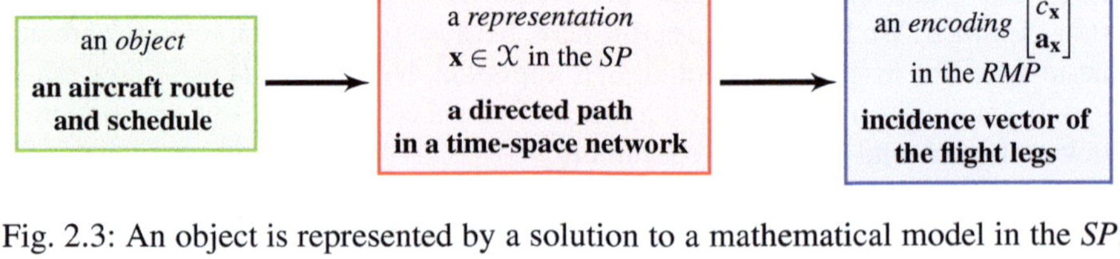

Fig. 2.3: An object is represented by a solution to a mathematical model in the *SP*; its encoding in the *MP* usually carries less detailed information.

Note 2.10 (About the encoding functions.) The cost $c_{\mathbf{x}}$ and column-coefficients $\mathbf{a}_{\mathbf{x}}$ are naturally defined as functions of $\mathbf{x} \in \mathcal{X}$, that is, $c_{\mathbf{x}} = c(\mathbf{x})$ and $a_{i\mathbf{x}} = a_i(\mathbf{x})$, $\forall i \in \{1,\ldots,m\}$. The general framework of the Dantzig-Wolfe decomposition for (integer) linear programming (see Chapters 3 and 4) requires these to preserve *vector addition* and *scalar multiplication.* In many real-life applications, the cost $c(\mathbf{x})$ is indeed non-linear and therefore does not fulfill this requirement. We can think of very complicated cost functions, e.g., for the airline crew schedules, which are negotiated by lawyers, not OR practitioners. Non-linear functions also appear for the computation of some column-coefficients when Chvátal-Gomory cuts are added to restrict an integer master problem formulation as these make use of *ceiling* or *floor* functions (see Chapter 7). We describe in Section 4.5 some conditions for which a reformulation using non-linear encoding functions holds. This may increase the difficulty in solving such non-linear pricing problems but the encoding always ends up in the *RMP* with scalars for the cost $c_{\mathbf{x}}$ and components of $\mathbf{a}_{\mathbf{x}}$.

Note 2.11 (Give me my objects back!) Once the *MP* is solved, we need to translate a $\boldsymbol{\lambda}^{\star}$-solution back to a plan the practitioner understands, i.e., in terms of objects. Remember that we do not directly work with objects in the *MP*, not even with the object's representation. We work with a potentially simplified encoding. We therefore need to first retrieve a representation $\mathbf{x} \in \mathcal{X}$ from an encoding $c_{\mathbf{x}}$ and $\mathbf{a}_{\mathbf{x}}$. This *decoding* can be trivial, e.g., if there is a bijection between representations and encodings like in the edge coloring problem (Example 2.3). In general, however, the decoding can be an expensive task as it amounts to finding a solution $\mathbf{x}$ to the *SP* given the input c and $\mathbf{a}$, i.e.,

$$\exists \mathbf{x} \in \mathcal{X} \quad \text{such that} \quad c(\mathbf{x}) = c, \quad a_i(\mathbf{x}) = a_i,\ \forall i \in \{1,\ldots,m\}. \tag{2.5}$$

Since the encoding functions are not necessarily injective, the returned representation needs not be unique which means it could even be different from the one that was generated in the first place. In an implementation, it is worthwhile to consider storing the representation $\mathbf{x} \in \mathcal{X}$ together with the encoding $(c_{\mathbf{x}}, \mathbf{a}_{\mathbf{x}})$. It may be easier and faster when retrieving the corresponding objects to show to the end-user.

Note 2.12 (Empty and zero objects.) In almost every application, we have the option to do nothing. For instance, in machine scheduling we could leave a machine just idle, i.e., assign it the empty schedule. Likewise, we might not use a vehicle (empty tour) or do not select any vertex from a graph (empty set). Generally, there exists an *empty object* which may have a cost associated with it, say, a fixed cost for providing the resource even if it is not used. Likewise, it may or may not be represented by a zero-vector. It is however useful to have a dummy that fulfills a mathematical interpretation of doing nothing, the *zero object* with a zero-cost and a zero-vector representation. We postpone its formal creation to Chapter 4 (Not all blocks are used, p. 217). We can however already see it in action in Examples 2.1 through 2.4: One-dimensional cutting stock problem, Cutting stock problem with rolls of different widths, Edge coloring problem, and Single depot vehicle scheduling problem.

Several subproblems

In many applications, we have several (different) column generators which are responsible for generating several (different) kinds of variables. Such a situation occurs in many classical optimization problems, one of which is the cutting stock problem with large rolls of different widths (Example 2.2). Another example appears in the airline industry if a heterogeneous aircraft fleet is used to service the flight legs: A column encodes a possible route for a specific aircraft type, and we have a specific column generator for each type. Similarly, we have several column generators in vehicle routing problems if vehicles have different capacities that may even influence customers they are allowed to visit.

For a finite set K of column generators, we reflect this in our notation by indexing $k \in K$ in the finite sets $\mathcal{X}^k$ as well as encoding functions $c^k(\mathbf{x}^k)$ and $\mathbf{a}^k(\mathbf{x}^k)$. In generic form, the *MP* is given by

$$
\begin{aligned}
z^\star_{MP} = \min \quad & \sum_{\mathbf{x}^1 \in \mathcal{X}^1} c_{\mathbf{x}^1} \lambda_{\mathbf{x}^1} + \ldots + \sum_{\mathbf{x}^{|K|} \in \mathcal{X}^{|K|}} c_{\mathbf{x}^{|K|}} \lambda_{\mathbf{x}^{|K|}} \\
\text{s.t.} \quad & \sum_{\mathbf{x}^1 \in \mathcal{X}^1} \mathbf{a}_{\mathbf{x}^1} \lambda_{\mathbf{x}^1} + \ldots + \sum_{\mathbf{x}^{|K|} \in \mathcal{X}^{|K|}} \mathbf{a}_{\mathbf{x}^{|K|}} \lambda_{\mathbf{x}^{|K|}} \geq \mathbf{b} \quad [\boldsymbol{\pi}] \\
& \lambda_{\mathbf{x}^1} \geq 0 \qquad \forall \mathbf{x}^1 \in \mathcal{X}^1 \\
& \qquad \ddots \qquad \vdots \\
& \lambda_{\mathbf{x}^{|K|}} \geq 0 \qquad \forall \mathbf{x}^{|K|} \in \mathcal{X}^{|K|}.
\end{aligned}
\tag{2.6}
$$

When solving the *MP* (2.6) with the column generation algorithm, the *RMP* again contains only subsets $\mathcal{X}'^k \subseteq \mathcal{X}^k$ for all $k \in K$. A bit more compactly, it reads as

$$
\begin{aligned}
z_{RMP} = \min \quad & \sum_{k \in K} \sum_{\mathbf{x}^k \in \mathcal{X}'^k} c_{\mathbf{x}^k} \lambda_{\mathbf{x}^k} \\
\text{s.t.} \quad & \sum_{k \in K} \sum_{\mathbf{x}^k \in \mathcal{X}'^k} a_{i\mathbf{x}^k} \lambda_{\mathbf{x}^k} \geq b_i \quad [\pi_i] \quad \forall i \in \{1, \ldots, m\} \\
& \lambda_{\mathbf{x}^k} \geq 0 \qquad \forall k \in K,\ \mathbf{x}^k \in \mathcal{X}'^k.
\end{aligned}
\tag{2.7}
$$

This implies that we must deal with $|K|$ generators identifying the respective columns to be added to the *RMP*. These different SP^k, $k \in K$, are

$$
\begin{aligned}
\bar{c}^k(\boldsymbol{\pi}) \quad = \quad \min_{\mathbf{x}^k \in \mathcal{X}^k} \quad & c_{\mathbf{x}^k} - \sum_{i=1}^{m} \pi_i a_{i\mathbf{x}^k} \\
\text{s.t.} \quad & c_{\mathbf{x}^k} = c^k(\mathbf{x}^k) \\
& a_{i\mathbf{x}^k} = a_i^k(\mathbf{x}^k) \quad \forall i \in \{1, \ldots, m\}.
\end{aligned}
\tag{2.8}
$$

Note, however, that all subproblems use the same vector $\boldsymbol{\pi} \geq \mathbf{0}$ of dual values. The column generation algorithm terminates when $\bar{c}^k(\boldsymbol{\pi}) \geq 0$ for *all* $k \in K$. Otherwise, at least one negative reduced cost variable is identified during the pricing step. In this case, a minimizer $\mathbf{x}^k$ from (2.8) is added to $\mathcal{X}'^k$, and variable $\lambda_{\mathbf{x}^k}$ with its cost $c_{\mathbf{x}^k}$ and column-coefficients $\mathbf{a}_{\mathbf{x}^k}$, is incorporated in the *RMP*.

Note 2.13 (Solving all the subproblems may be useless.) From our understanding of the communication between the *RMP* and subproblems, see Note 2.7, we may already anticipate what could go wrong with several column generators: the duals all communicate the *same* information to the different subproblems and these in turn may all react to incentives and penalties in the same way. For example, we may have different subproblems to produce routes for different types of vehicles. Presenting them all the same dual vector may lead to routes that all try to visit the same customers. It is obvious that we may have wasted a lot of time in solving all subproblems when in fact, only very little diversity in columns is produced. This is consistent with the findings of Goffin and Vial (2000) who relate the performance of the analytic center cutting plane method to the variance-covariance matrix of the selected columns: the performance increases with the selection of non-correlated columns.

Note 2.14 (Parallelization may not be as effective as you think.) When implementing a column generation algorithm with several subproblems, it is an obvious and advertised idea to solve these in parallel. As mentioned in Note 2.13, solving all pricing problems may be a waste of time and by solving all of them in parallel we only reach this conclusion faster. Instead, we might consider parallelizing the solution of each pricing problem and combine this with techniques we learn throughout the book.

Pivot rules and column management

In the primal simplex method, Dantzig's original suggestion is to fully consider all variables and pick one with most negative reduced cost. This classical rule implies that we solve the pricing problem to optimality. With regard to a practically efficient implementation, we cannot overemphasize that we do not always need to do so. In principle, *any* variable of negative reduced cost promises progress. This is all the more true in column generation where the pricing problem is typically much more of a computational burden than a simple enumeration, i.e., (2.3) vs. (1.25). Therefore, this step has received considerable attention in the literature, often the particular problem structure is exploited.

While a lot of alternative pivot rules were proposed for the primal simplex method, not many are transferred to the column generation algorithm but this can certainly be worth a thought. One suggestion is the *lambda pricing* rule (Bixby et al., 1992)

$$l(\boldsymbol{\pi}) = \min_{\mathbf{x}\in\mathcal{X}} \left\{ \frac{c_{\mathbf{x}}}{\boldsymbol{\pi}^\intercal \mathbf{a}_{\mathbf{x}}} \;\middle|\; \boldsymbol{\pi}^\intercal \mathbf{a}_{\mathbf{x}} > 0 \right\} \quad \text{where } c_{\mathbf{x}} \geq 0, \forall \mathbf{x} \in \mathcal{X}, \text{ is assumed.} \tag{2.9}$$

Clearly, the reduced cost $c_{\mathbf{x}} - \boldsymbol{\pi}^\intercal \mathbf{a}_{\mathbf{x}}$ is non-negative for all $\mathbf{x} \in \mathcal{X}$ if and only if $l(\boldsymbol{\pi}) \geq 1$. Yes, this is just a reformulation of the *SP* (2.3). However, in many applications where the column-coefficients $\mathbf{a}_{\mathbf{x}}$ are binary, a small ratio does not only prefer smaller cost coefficients but also more non-zero entries in $\mathbf{a_x}$. This is reported to be useful for set partitioning master problems. How would we solve the *SP* in (2.9)? We would not. Instead, this rule is used to select a variable in case there are several available from the *SP* (see next paragraph). Such a selection could be based on other pivot rules with good properties, like the *positive edge* rule (Towhidi et al., 2014) that almost surely avoids degenerate pivots (p. 128).

Besides these general implementation variants inherited from the primal simplex method, there are more opportunities in the column generation context. Very importantly, the algorithms for solving the *SP* or *ISP* are usually capable of producing more than one variable of negative reduced cost. This applies to combinatorial algorithms like dynamic programs and also to solving the subproblem by branch-and-bound. With this possibility, new questions arise: how many variables to generate per iteration, and which variables to actually use? It is a common mistake to base this selection solely on reduced cost. Indeed, in the light of Note 2.13, selecting those with smallest reduced cost is likely to give columns with many similar coefficients. A common heuristic to counteract this effect is called *complementary pricing*, see Example 2.4. It explicitly aims at generating a set of very diverse columns. When many variables are generated but not all are used, the (most promising of the) excessive ones can be kept in a *column pool* for later iterations, especially if these are computationally expensive to generate (Barnhart et al., 1998). We already know that we do not need to generate the most negative columns; the column pool even allows us to keep some variables with positive reduced cost.

The management of columns also pertains to the master problem. The growing number of variables can indeed slow down the re-optimization of the *RMP* such that one can think about removing inactive columns. For example, a simple *age* criterion can help identify such columns. Does the reader see the potential for column generation cycling as raised in Note 2.9? For large master problems, with thousands of rows, there may come a time when even the linear solver struggles to solve the *RMP*. In such a situation, we recommend to keep the size of the set $\mathcal{X}'$ under control, say no more than $3m$ variables.

Note 2.15 (Restricted master problem solvers.) As a final note concerning the restricted master problem, remember that its role is to provide dual solutions which guide the generation of columns. There are several options to re-optimize the *RMP*, primal and dual simplex methods, barrier methods (a.k.a. interior point methods), and more. One cannot generally say which one will perform best, but *each solver* provides different dual values, and the effect on solving the subproblems should be tested. It then also becomes obvious that we can even resort to approximations to solve the *RMP*, especially if it is computationally costly. Observe that there are two intertwined topics here: the speed with which we can solve the *RMP* and the usefulness of the dual values we provide to the pricing problem. The streams of research concerned with the latter topic are abundant and orbit around the fact that the dual values transferred from the *RMP* to the *SP* are critical to achieve stellar performance in column generation. Chapter 6 on the dual point of view provides significant insights to support this assertion and ultimately brings to light hybrid algorithms for column generation. In all cases, avoid imposing an upper bound on any variable, say $\lambda_{\mathbf{x}} \leq u_{\mathbf{x}}$, as this can introduce adverse effects in column generation. For example, the upper bounds of 1 on the binary variables of a set partitioning model should remain implicit in its linear relaxation. An intuitive explanation is that, in the simplex method with bounded variables (see complementary slackness conditions 1.8), a non-basic variable at its upper bound can have a negative reduced cost at optimality and may therefore be generated again, indefinitely.

Heuristic pricing

The pricing problem is a playground for experimentation and it offers large opportunities for speeding up the overall process. Actually, the entire viability of a column generation approach may depend on this. Indeed, since the pricing problem is usually solved very often, it had better be *fast*. As usual in OR, the meaning of fast cannot be taken at face value but rather in comparison to other components of the algorithm. A guideline can be to strive for balancing the computation time spent in the master and subproblems. That is, if the *SP* solves fast, in the sense that comparatively meaningful progress is made timely in the *RMP*, there is no need for tricks. Ultimately, it is almost certain that so-called heuristic pricing, in which we can make choices about which subset of variables we select for actually computing the reduced costs, and how we pick from that subset, should be included. This way, ideally, one needs to solve the pricing problem to optimality *only once*, in the last column generation iteration, to prove optimality of the *MP*.

The subproblem often exhibits very particular structure; for instance a *knapsack problem* or a *shortest path problem with resource constraints*, in many variants. We dedicate an important part of Chapter 5 (Vehicle Routing and Crew Scheduling Problems) to the latter; and entire research projects are devoted to designing, often heuristic, algorithms to effectively solve specific pricing problems. Such heuristics can be very elaborate and exploit a lot of problem knowledge and theory.

Generally speaking, one can think of a *cascade* of heuristics per iteration, starting with the computationally cheapest (this can be checking the variables in the column pool), and successively call more computationally expensive ones when previous heuristics in the cascade fail to deliver (enough) negative reduced cost variables.

When talking about algorithms in operations research, we do not have to decide between exact and heuristic approaches. An example for a symbiosis is to heuristically solve an integer pricing problem by prematurely terminating branch-and-bound, e.g., by imposing a time limit or a tolerance on solution quality. Another example is *partial pricing* in which one can use an exact algorithm to solve a restriction of it, i.e., to optimize over a well-defined subset of $\mathcal{X}$. Such a restriction could come from temporarily neglecting some pricing problems as suggested in Note 2.13 or from "user experience" regarding prevalent features we expect to see in optimal columns. In the former case, if we have many subproblems, one can e.g., cyclically walk through them, maybe in batches; or prioritize those which were successful in previous calls. Or quite the contrary: ask subproblems that have *not* delivered much. Partial pricing reduces similarity between generated columns since $\boldsymbol{\pi}$ multipliers are updated more often. In the latter case, for instance, in an airline crew pairing problem, the schedule lengths may range, say from one to five days, but most schedules last at most three days in practice. One could first use a restricted subproblem that generates schedules spanning at most three days rather than five. We refer to any and all these ideas as *heuristic pricing* in contrast to *exact pricing* for when we solve the *SP* to optimality. They are part of the toolbox of acceleration strategies in column generation (Desaulniers et al., 2002), see also Acceleration techniques.

Note 2.16 (Bury the hatchet.) Being adepts of an exact method in column generation does not reduce or take away from the importance of heuristics. Let us contribute once more in breaking the stigma sometimes associated with their usage by quoting what has been dismissed as lesser mathematics all the way back in 1977.

> Heuristic solution methods for integer programming have maintained a noticeably separate existence from algorithms. Algorithms have long constituted the more respectable side of the family, assuring an optimal solution in a finite number of steps. Methods that merely claim to be clever, and do not boast an entourage of supporting theorems and proofs, are accorded a lower status. Algorithms are conceived in analytic purity in the high citadels of academic research, heuristics are midwifed by expediency in the dark corners of the practitioner's lair.
>
> Recently, however, there has been a growing recognition that the algorithms are not always successful, and that their heuristic cousins deserve a chance to prove their mettle. Partly this comes from an emerging awareness that algorithms and heuristics are not as different as once supposed—algorithms, after all, are merely fastidious heuristics in which epsilons and deltas abide by the dictates of mathematical etiquette. It may even be said that algorithms exhibit a somewhat compulsive aspect, being denied the freedom that would allow an occasional inconsistency or an exception to ultimate convergence. (Unfortunately, ultimate convergence sometimes acquires a religious significance; it seems not to happen in this world.)
>
> The heuristic approach, robust and boisterous, may have special advantages in terrain too rugged or varied for algorithms. In fact, those who are fond of blurring distinctions suggest that an algorithm worth its salt is one with "heuristic power." – Glover (1977)

2.3 More to Know

This section contains more theoretical underpinnings for a practical implementation. Almost all relate to the dual in one way or another.

Lower bounds

As the *RMP* is a restriction of the *MP*, z_{RMP} iteratively approaches $z^\star_{MP}$ from above (and they finally meet). That is, from each feasible solution to the *RMP*, we trivially obtain an upper bound on the optimum $z^\star_{MP}$. Additionally having a lower bound enables us to evaluate solution quality. We establish a variety of such bounds more or less specialized depending on the structure of the *MP*.

The assumption regarding finiteness of $z^\star_{MP}$ implies that the variables coming from every subproblem attain finite values in any optimal solution $\boldsymbol{\lambda}^\star$ of the *MP* (2.6). Specifically,

$$\exists \kappa^k \quad \text{such that} \quad \forall \boldsymbol{\lambda}^\star: \quad \sum_{\mathbf{x}^k \in \mathcal{X}^k} \lambda^\star_{\mathbf{x}^k} \leq \kappa^k, \quad \forall k \in K. \tag{2.10}$$

We have such κ^k readily available in several applications, e.g., the number of vehicles at some depot k or rolls of type k to cut.

Proposition 2.1. *Given arbitrary dual values* $\boldsymbol{\pi} \geq \mathbf{0}$ *and minimum reduced costs* $\bar{c}^k(\boldsymbol{\pi})$, $\forall k \in K$, *then the optimum* $z^\star_{MP}$ *is bounded from below as*

$$\boldsymbol{\pi}^\intercal \mathbf{b} + \sum_{k \in K} \kappa^k \bar{c}^k(\boldsymbol{\pi}) \leq z^\star_{MP}, \qquad \forall \boldsymbol{\pi} \geq \mathbf{0}. \tag{2.11}$$

Proof. Let $\boldsymbol{\lambda}^\star$ denote an optimal solution to (2.6) and let the reduced cost be defined as usual as $\bar{c}_{\mathbf{x}^k} = c_{\mathbf{x}^k} - \boldsymbol{\pi}^\intercal \mathbf{a}_{\mathbf{x}^k}$, $\forall k \in K$, $\mathbf{x}^k \in \mathcal{X}^k$. We have

$$z^\star_{MP} = \sum_{k \in K} \sum_{\mathbf{x}^k \in \mathcal{X}^k} c_{\mathbf{x}^k} \lambda^\star_{\mathbf{x}^k} = \sum_{k \in K} \sum_{\mathbf{x}^k \in \mathcal{X}^k} (\bar{c}_{\mathbf{x}^k} + \boldsymbol{\pi}^\intercal \mathbf{a}_{\mathbf{x}^k}) \lambda^\star_{\mathbf{x}^k}. \tag{2.12}$$

We bound the two resulting double summations separately. First,

$$\sum_{k \in K} \sum_{\mathbf{x}^k \in \mathcal{X}^k} \bar{c}_{\mathbf{x}^k} \lambda^\star_{\mathbf{x}^k} \geq \sum_{k \in K} \bar{c}^k(\boldsymbol{\pi}) \sum_{\mathbf{x}^k \in \mathcal{X}^k} \lambda^\star_{\mathbf{x}^k} \geq \sum_{k \in K} \bar{c}^k(\boldsymbol{\pi}) \kappa^k, \tag{2.13}$$

where the left inequality holds because $\bar{c}_{\mathbf{x}^k} \geq \bar{c}^k(\boldsymbol{\pi})$, $\forall k \in K, \mathbf{x}^k \in \mathcal{X}^k$, by definition, and the right inequality uses the non-positivity of the minimum reduced cost and the bound (2.10). Second,

$$\sum_{k \in K} \sum_{\mathbf{x}^k \in \mathcal{X}^k} \boldsymbol{\pi}^\intercal \mathbf{a}_{\mathbf{x}^k} \lambda^\star_{\mathbf{x}^k} = \boldsymbol{\pi}^\intercal \sum_{k \in K} \sum_{\mathbf{x}^k \in \mathcal{X}^k} \mathbf{a}_{\mathbf{x}^k} \lambda^\star_{\mathbf{x}^k} \geq \boldsymbol{\pi}^\intercal \mathbf{b}, \tag{2.14}$$

where the inequality is tantamount to the feasibility of $\boldsymbol{\lambda}^\star$ in (2.6). □

Of course, the lower bound (2.11) holds in particular for an optimal dual solution to the *RMP*, so that by strong duality we have $\boldsymbol{\pi}^\intercal \mathbf{b} = z_{RMP}$ and therefore the following by-product.

Corollary 2.1. *Given optimal dual values* $\boldsymbol{\pi}$ *with objective value* $z_{RMP} = \boldsymbol{\pi}^\intercal \mathbf{b}$ *and minimum reduced costs* $\bar{c}^k(\boldsymbol{\pi})$, $\forall k \in K$, *then the optimum* $z^\star_{MP}$ *is bounded from below and above as*

$$z_{RMP} + \sum_{k \in K} \kappa^k \bar{c}^k(\boldsymbol{\pi}) \leq z^\star_{MP} \leq z_{RMP}. \tag{2.15}$$

There is an intuitive interpretation. The minimum reduced cost is the largest possible (potentially not realizable) per unit improvement in the objective function. We know from (2.10) that we cannot use more units than κ^k from the sum of all the variables coming from the pricing problem SP^k, $k \in K$. Note that at optimality of the MP, we have $\bar{c}^k(\boldsymbol{\pi}) = 0$, $\forall k \in K$, thus the lower bound finally reaches the upper bound and the interval collapses. The lower bound is, however, not monotonic over the iterations, see Figure 2.4. Finally, if we have only one subproblem, (2.10) and (2.15) reduce to

$$\exists \kappa \quad \text{such that} \quad \forall \boldsymbol{\lambda}^\star : \quad \sum_{\mathbf{x} \in \mathcal{X}} \lambda^\star_{\mathbf{x}} \leq \kappa \tag{2.16a}$$

$$z_{RMP} + \kappa \bar{c}(\boldsymbol{\pi}) \leq z^\star_{MP} \leq z_{RMP}. \tag{2.16b}$$

Note 2.17 (Lower bounds with heuristic pricing?) Our lower bounds above use the minimum reduced cost $\bar{c}^k(\boldsymbol{\pi})$ for each $k \in K$. Since $\bar{c}^k(\boldsymbol{\pi}) \leq 0$, all these expressions also hold for any under-approximation $\underline{c}^k(\boldsymbol{\pi}) \leq \bar{c}^k(\boldsymbol{\pi})$. In particular, the lower bound in (2.11) becomes

$$\boldsymbol{\pi}^\intercal \mathbf{b} + \sum_{k \in K} \kappa^k \underline{c}^k(\boldsymbol{\pi}) \leq z^\star_{MP}, \qquad \forall \boldsymbol{\pi} \geq \mathbf{0}. \tag{2.17}$$

Such a value $\underline{c}^k(\boldsymbol{\pi})$ can for example come from solving the ISP^k with a truncated branch-and-bound algorithm. In contrast, a lower bound is unavailable if we only have a primal heuristic solution $\mathbf{x}^k$ (including exact solutions of a restricted ISP^k) because the inequality $\bar{c}_{\mathbf{x}^k} \geq \bar{c}^k(\boldsymbol{\pi})$ is likely strict. Obviously, we also do not have a lower bound under partial pricing since we have no reduced cost information for at least one $k \in K$.

We conclude this topic with two more specialized lower bounds.

Corollary 2.2. *Let* $|K| = 1$ *and* $c_{\mathbf{x}} = 1$, $\forall \mathbf{x} \in \mathcal{X}$. *Given optimal dual values* $\boldsymbol{\pi}$ *with objective value* $z_{RMP} = \boldsymbol{\pi}^\intercal \mathbf{b}$ *and minimum reduced cost* $\bar{c}(\boldsymbol{\pi})$, *then* $z^\star_{MP}$ *is bounded from below as*

$$\frac{\boldsymbol{\pi}^\intercal \mathbf{b}}{1 - \bar{c}(\boldsymbol{\pi})} \leq z^\star_{MP}, \qquad \forall \boldsymbol{\pi} \geq \mathbf{0}. \tag{2.18}$$

Proof. Under the assumptions, it holds that $\sum_{\mathbf{x} \in \mathcal{X}} \lambda^\star_{\mathbf{x}} = z^\star_{MP}$ and the result follows from fixing $\kappa = z^\star_{MP}$ in (2.16b). Alternatively, a direct proof (Exercise 2.5) uses the fact that the vector $\frac{\boldsymbol{\pi}}{1-\bar{c}(\boldsymbol{\pi})}$ is dual feasible, hence $\frac{\boldsymbol{\pi}^\intercal \mathbf{b}}{1-\bar{c}(\boldsymbol{\pi})}$ is a lower bound on $z^\star_{MP}$ by Proposition 1.5 (weak duality). □

Corollary 2.3. (Farley, 1990) *Let* $|K| = 1$ *and* $c_{\mathbf{x}} \geq 0$, $\forall \mathbf{x} \in \mathcal{X}$. *Given optimal dual values* $\boldsymbol{\pi}$ *with objective value* $z_{RMP} = \boldsymbol{\pi}^\intercal \mathbf{b}$ *and minimum lambda price* $l(\boldsymbol{\pi})$, *then* $z^\star_{MP}$ *is bounded from below as*

$$l(\boldsymbol{\pi}) \cdot \boldsymbol{\pi}^\intercal \mathbf{b} \leq z^\star_{MP}, \qquad \forall \boldsymbol{\pi} \geq \mathbf{0}. \tag{2.19}$$

Proof. We show that the definition of $l(\boldsymbol{\pi})$ (2.9) implies that $l(\boldsymbol{\pi}) \cdot \boldsymbol{\pi}$ is dual feasible, i.e., $l(\boldsymbol{\pi}) \cdot \boldsymbol{\pi} \geq \mathbf{0}$ and $l(\boldsymbol{\pi}) \cdot \boldsymbol{\pi}^\intercal \mathbf{a}_\mathbf{x} \leq c_\mathbf{x}$, $\forall \mathbf{x} \in \mathcal{X}$, so (2.19) holds by weak duality. We have $l(\boldsymbol{\pi}) \geq 0$ such that $l(\boldsymbol{\pi}) \cdot \boldsymbol{\pi} \geq \mathbf{0}$ is obviously true. Furthermore, if $\boldsymbol{\pi}^\intercal \mathbf{a}_\mathbf{x} > 0$ for a given index $\mathbf{x}$, this is equivalent to $l(\boldsymbol{\pi}) \leq \frac{c_\mathbf{x}}{\boldsymbol{\pi}^\intercal \mathbf{a}_\mathbf{x}}$; otherwise, $l(\boldsymbol{\pi}) \cdot \boldsymbol{\pi}^\intercal \mathbf{a}_\mathbf{x} \leq 0 \leq c_\mathbf{x}$: hence, $l(\boldsymbol{\pi}) \cdot \boldsymbol{\pi}^\intercal \mathbf{a}_\mathbf{x} \leq c_\mathbf{x}$, $\forall \mathbf{x} \in \mathcal{X}$. □

Out of curiosity, compare with the publication date of the lambda pricing rule. Exercise 2.5 asks the reader to show that the lower bound (2.19) reduces to that of Corollary 2.2 if $c_\mathbf{x} = 1$, $\forall \mathbf{x} \in \mathcal{X}$.

Convergence

Having a lower bound on $z^\star_{MP}$ available at every iteration $t \geq 1$ of the column generation algorithm, say lb_t, we can compute LB, the best known lower bound observed so far, i.e.,

$$LB = \max_{t \geq 1}\{lb_1, \ldots, lb_t\}. \tag{2.20}$$

The column generation algorithm terminates exactly when $LB = z_{RMP}$. Figure 2.4 illustrates the convergence of the column generation algorithm on an instance of the classical *vehicle routing problem with time windows* (*VRPTW*) formulated as a set partitioning problem, see Chapter 5. We plot the development of the *RMP*'s objective value and of the lower bound from expression (2.16b) over the iterations. The precise problem definition is unimportant here; let us rather make some observations about this figure which are rather typical in column generation, and let us try to offer some first interpretations.

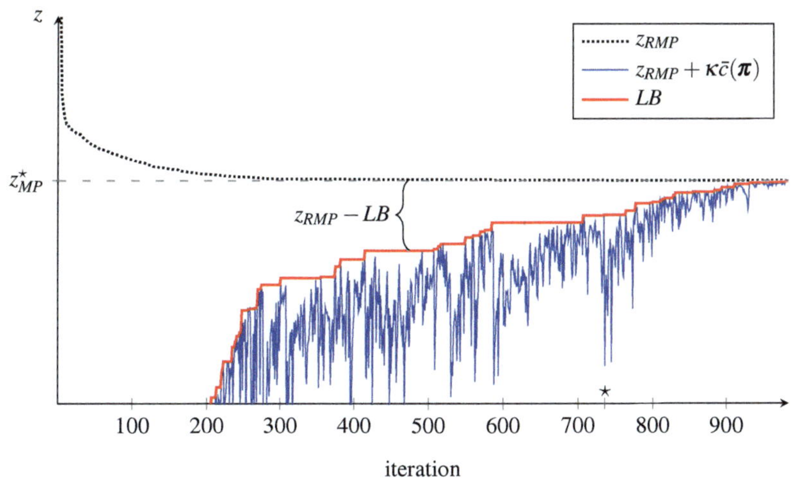

Fig. 2.4: Lower and upper bounds convergence and tailing-off effect.

1. The two bounds behave totally differently. The objective values z_{RMP} are non-increasing and give us a smooth upper bound plot. Concerning the lower bound, it does not even appear until about iteration 200, because we have restricted the z-axis plot range. As is the case here, it is not uncommon for the lower bound to be rather far away from the upper bound until the very last iterations of the column generation algorithm. The sequence of minimum reduced costs approaches zero only very late, and the worst case bound κ can be a poor estimate, and thus be large. In any case, we observe that the lower bound does not develop monotonically.
2. Even though the upper bound is monotone, we observe the initially satisfactory progress that gets smaller and smaller as we get closer to the optimum. In fact, there is comparably little progress for the better part of the iterations. This effect is known as *tailing-off*. For an empirical explanation attempt, we consult the raw data from which the plots are produced. We first observe that tailing-off is a matter of scale. Would we zoom on the beginning of the tail, we would observe a decrease in the objective value, yet progressively tiny, see Figure 6.17b. Often enough, the step-length is actually zero, i.e., we have degenerate pivots, a frequent case for the set partitioning formulation of the *VRPTW*. This manifests in *plateaus*, longer sequences of iterations during which the objective value does not improve at all. We highlight that an optimal solution is reached at iteration 736 (marked by a $\star$), although optimality is only proven at iteration 952. For a probabilistic explanation of the frequent occurrence of degenerate or small step-length pivots in the simplex method or the column generation algorithm, see Exercise 2.19 where we show (via an estimation using the *Hypergeometric* distribution) that the probability of such events is increasing with the number of basic variables modified by the entering variable.
3. Vanderbeck (2005, Chapter 12, p. 349) coins memorable names for other effects we see besides the plateaus and long tail, and they all relate to the dual solutions produced from the *RMP*.

    - In the early iterations with good progress on the objective value, the *RMP* is rather empty and often initialized with a bunch of artificial variables. Despite the apparent improvement, the generated variables typically do not reflect good solutions yet. This is related to the fact that in this phase also the dual variables are rather far away from their respective optimal values. This initial, only seemingly productive phase, is called *heading-in*.
    - From there on, the dual values alternate between interesting values (called the *bang-bang* phase) which is echoed in the lower bounds that go up and down erratically, called the *yo-yo* effect. While we can clearly see the yo-yo effect, how can we intuitively grasp bang-bang? Indeed, even a single negative reduced cost column entering a basis $\mathbf{A}_B$ while solving the *RMP* can largely modify the simplex multipliers $\boldsymbol{\pi}^\mathsf{T} = \mathbf{c}_B^\mathsf{T}\mathbf{A}_B^{-1}$. This new column has for consequence that we tend to go overboard in the other direction changing the dual values accordingly. For those familiar with the Lagrangian relaxation method, the same bang-bang phase is observed for the successive

Langrangian multipliers. An important question is therefore: *Can we better control the dual values within the column generation algorithm?* Several efficient options are presented in Chapter 6, Dual Point of View.

Note 2.18 (Early termination and integer optimum.) A better theoretical understanding of the effects observed here comes with our discussion about the dual point of view in Chapter 6. However, this is a perfect place to introduce *early termination* as a means to exploit the availability of lower bounds in the light of tailing-off. The linear programs we solve by column generation are usually relaxations of integer programs. These are solved for the lower bounds they provide. If we do not witness enough progress on the objective value, we may stop generating columns, use *LB* (2.20) as a valid lower bound, and start cutting and branching, see Chapter 7. A possible stopping criterion is to have reached a given relative optimality gap γ, i.e., to stop column generation when

$$\frac{z_{RMP} - LB}{|LB|} \leq \gamma \tag{2.21}$$

for $LB \neq 0$. If we know that $z^{\star}_{MP} \in \mathbb{Z}$, we can tighten the gap using $\lceil LB \rceil$ in the previous expression which is true because we terminate exactly when

$$\lceil LB \rceil \geq z_{RMP}. \tag{2.22}$$

Limited numeric precision

Numbers processed in a computer are represented with a limited amount of memory, say, 64 bits in the standard *double-precision floating-point format*. It is clear that irrational numbers cannot be stored in finite computer memory. However, even many rational numbers are not stored accurately. This holds for fractions that are periodical in binary or where the number of available significant digits is simply too small. The above referenced *double* format can hold up to fifteen significant decimal digits only (in total, before and after the decimal point). Computations under this *fixed-precision arithmetic* may propagate the small inaccuracies, and in fact the error accumulation can be dramatic.

It is an immediate consequence that in computer implementations, comparisons of numbers are usually not done *exactly*. Instead, for a given small *tolerance* $\varepsilon > 0$, two numbers a and b are considered equal when $|a-b| \leq \varepsilon$. In particular, all properties of a solution obtained from a mathematical optimization solver, like optimality, feasibility, and integrality only hold *within the tolerances*. As an example of particular importance to us, "negative reduced cost $\bar{c}_{\mathbf{x}} < 0$" translates to $\bar{c}_{\mathbf{x}} < -\varepsilon$. In a sense, a variable must have a *considerably* negative reduced cost to be considered attractive. Different solvers may use different tolerances, typical values range around $\varepsilon = 10^{-6}$, much larger than the actual machine-precision.

In the column generation algorithm, we alternatively solve the *MP* and *SP*. Thus, we need to ensure a compatibility of tolerances throughout all portions of the implementation. Effectively, the optimality tolerance $\varepsilon \geq 0$ used when solving the *RMP* should be the same tolerance used to stop pricing in Line 4 of Algorithm 2.1, i.e., when $\bar{c}(\boldsymbol{\pi}) \geq -\varepsilon$. Failing to do so may result in the generation of "useless" columns with reduced costs between $-\varepsilon$ and 0. This is a prime source for infinite loops.

Also the integrality tolerance can be a pitfall. The tolerance used in the solver of the *ISP* must be integrated into the method generating the column-coefficients. For instance, consider the *ISP* is subject to $c_{\mathbf{x}} = 1$ and $\mathbf{a}_{\mathbf{x}} = \mathbf{x}$, receives as input $\boldsymbol{\pi} = [1.001, -0.35]$, and outputs an optimal integer solution $\mathbf{x}^\star = [.999, 0.001]$ because we accept an integrality tolerance of 10^{-3}. We establish column-coefficients $\mathbf{a}_{\mathbf{x}}$ for the *MP* with respect to the given integrality tolerance of the *SP* as $\mathbf{a}_{\mathbf{x}} = [1, 0]$. It would even be wise to recompute the reduced cost:

$$\bar{c}(\boldsymbol{\pi}) = 1 - \begin{bmatrix}1.001, -0.35\end{bmatrix} \begin{bmatrix}0.999\\0.001\end{bmatrix} = 1 - 0.999649 > 0 \neq 1 - 1.001 < 0.$$

The total error induced by the integrality tolerance depends on the non-zero entries in $\mathbf{x}$ and the size of the corresponding values in $\boldsymbol{\pi}$. Moreover, it is obviously uncorrelated to the optimality tolerance ε.

Feasibility tolerances, i.e., the small amount by which constraints may be violated, can be of importance when degeneracy is counteracted by explicitly perturbing the right-hand sides by tiny (how tiny?) numbers. Specialized/combinatorial algorithms to solve the *ISP* may be a remedy, or libraries for *arbitrary-precision arithmetic*. The latter support true rational numbers and therefore guarantee exact numerical results. This comes with a trade-off of more expensive computations and pays off presumably only in mission critical applications.

Static and auxiliary variables

There are natural modeling situations, in which the master problem at hand, *MP* or *IMP*, contains variables that are not generated in the column generation process. We call these *static variables*. In the following model we denote them by $\mathbf{v}$:

$$\begin{aligned} z^\star_{MP} = \min \quad & \sum_{\mathbf{x} \in \mathcal{X}} c_{\mathbf{x}} \lambda_{\mathbf{x}} + \mathbf{c}_{\mathbf{v}}^\intercal \mathbf{v} \\ \text{s.t.} \quad & \sum_{\mathbf{x} \in \mathcal{X}} \mathbf{a}_{\mathbf{x}} \lambda_{\mathbf{x}} + \mathbf{A}_{\mathbf{v}} \mathbf{v} \geq \mathbf{b} && [\boldsymbol{\pi}] \\ & \mathbf{B}_{\mathbf{v}} \mathbf{v} \geq \mathbf{b}_{\mathbf{v}} && [\boldsymbol{\beta}] \\ & \mathbf{0} \leq \mathbf{v} \leq \mathbf{u}_{\mathbf{v}}, \\ & \lambda_{\mathbf{x}} \geq 0, && \forall \mathbf{x} \in \mathcal{X}, \end{aligned} \tag{2.23}$$

where $\mathbf{c_v}$, $\mathbf{A_v}$, $\mathbf{B_v}$, $\mathbf{b_v}$, and $\mathbf{u_v}$ are vectors and matrices of appropriate dimensions for their parameters. Since the static variables are part of the *MP*, but *cannot* be generated by the *SP* or *ISP*, they need to be *already present* in the *RMP* initially. Moreover, as the λ-variables play no part in constraints $\mathbf{B_v v} \geq \mathbf{b_v}$, the dual values $\boldsymbol{\beta}$ are irrelevant for the *SP*. Example 2.6 uses static variables t_{m_i} in the *IMP* (2.49).

In formulation (2.23), every feasible solution $(\boldsymbol{\lambda}, \mathbf{v})$ can be interpreted as a solution to the real problem at hand. The situation is different when we *purposefully* introduce static variables in the *MP* or *IMP* that *must not* be part of a feasible or optimal solution. Such *auxiliary variables* help us accomplishing several important technical tasks in the course of this book, so they deserve a separate presentation. Let us denote an auxiliary variable by y_j, $j \in J$. The *MP* takes the form

$$
\begin{aligned}
z^\star_{MP} = \min \quad & \sum_{\mathbf{x} \in \mathcal{X}} c_\mathbf{x} \lambda_\mathbf{x} + \boldsymbol{\delta}^\intercal \mathbf{y} \\
\text{s.t.} \quad & \sum_{\mathbf{x} \in \mathcal{X}} \mathbf{a}_\mathbf{x} \lambda_\mathbf{x} + \mathbf{S}\mathbf{y} \geq \mathbf{b} \qquad [\boldsymbol{\pi}] \\
& \mathbf{0} \leq \mathbf{y} \leq \mathbf{u}_\mathbf{y}, \\
& \lambda_\mathbf{x} \geq 0, \qquad \forall \mathbf{x} \in \mathcal{X},
\end{aligned}
\tag{2.24}
$$

where the auxiliary variables are accounted for in $\mathbf{y} = [y_j]_{j \in J}$, $\boldsymbol{\delta} = [\delta_j]_{j \in J}$, $\mathbf{S} = [\mathbf{s}_j]_{j \in J}$, and $\mathbf{u_y} = [u_j]_{j \in J}$. These parameters are *not* given from the problem description, but defined by the modeler; they can even dynamically change in the solution process. Of course, we can combine auxiliary variables with other static ones.

Auxiliary variables rightfully feel like they "do not belong" in the model: If any of them remains with a positive value in an optimal solution, this solution is not feasible for the problem at hand. So, of what use are they?

A well-known example is given by the *artificial variables* used in a *Phase I* initialization with parameters $\delta_j = M$, $u_j = \infty$, and $\mathbf{s}_j = \mathbf{e}_j$, $\forall j \in \{1, \dots, m\}$. We stress again that we are free to play with the big-*M* objective coefficients of artificial variables, and that they are not defined from problem data.

Let us interpret the property, that an optimal solution to the *MP* with at least one positive auxiliary variable implies that the solution in λ-variables alone is infeasible for the original problem. Stated differently, auxiliary variables induce a relaxation of the *MP*, such that optimality is achieved if

$$
\bar{c}(\boldsymbol{\pi}) \geq 0 \text{ and } y_j = 0, \ \forall j \in J. \tag{2.25}
$$

Furthermore, by duality, auxiliary variables imply a restriction on $\boldsymbol{\pi}$ in the dual formulation, that is,

$$
\mathbf{S}^\intercal \boldsymbol{\pi} \leq \boldsymbol{\delta} \quad [\mathbf{y}]. \tag{2.26}
$$

For the artificial variables, this becomes

$$
\pi_j \leq M \quad [y_j] \quad \forall j \in \{1, \dots, m\}. \tag{2.27}
$$

In other words, the M-penalties impose upper bounds on the dual variables. This restriction in the dual space is a reason why auxiliary variables may make the *RMP* easier to solve. We discuss this in more detail in Chapter 6.

A less obvious application of auxiliary variables arises in the use of a relaxed pricing problem (see below), where variables appear in the *MP*, thus relax it, and therefore constrain its dual in the way indicated above.

Relaxed pricing, relaxed master

When the pricing problem over $\mathcal{X}$ is difficult to solve exactly, we may rather solve a relaxation over a larger index set, say $\mathcal{Y} \supset \mathcal{X}$, when possible. A typical example is in vehicle routing problems, where each customer needs to be visited exactly once. This amounts to finding elementary paths in an appropriate network, one for each vehicle. Customarily, the *IMP* is formulated as a set partitioning problem, where *binary* columns encode whether in a route a customer is visited or not. However, the elementary shortest path problem to solve in the *ISP*, moreover constrained by time windows, vehicle capacity, etc., is strongly $\mathcal{NP}$-hard. Thus researchers rather solve more or less relaxed pricing problems that allow for the presence of cycles, that is, some customers are visited more than once in a path. For the non-elementary shortest path versions, pseudo-polynomial time algorithms are available. The corresponding columns then encode how often a customer is visited, thus column-coefficients become *non-negative integers*.

More generally, we replace the *MP* (2.1) by a relaxed formulation $MP_{\mathcal{Y}}$, temporarily allowing for *inadmissible* columns:

$$\begin{aligned} z^\star_{MP_{\mathcal{Y}}} = \min \quad & \sum_{\mathbf{x}\in\mathcal{X}} c_{\mathbf{x}}\lambda_{\mathbf{x}} + \sum_{\mathbf{x}\in\mathcal{Y}\setminus\mathcal{X}} c_{\mathbf{x}}\lambda_{\mathbf{x}} \\ \text{s.t.} \quad & \sum_{\mathbf{x}\in\mathcal{X}} \mathbf{a}_{\mathbf{x}}\lambda_{\mathbf{x}} + \sum_{\mathbf{x}\in\mathcal{Y}\setminus\mathcal{X}} \mathbf{a}_{\mathbf{x}}\lambda_{\mathbf{x}} \geq \mathbf{b} \quad [\boldsymbol{\pi}] \\ & \lambda_{\mathbf{x}} \geq 0, \forall \mathbf{x}\in\mathcal{X} \\ & \lambda_{\mathbf{x}} \geq 0, \forall \mathbf{x}\in\mathcal{Y}\setminus\mathcal{X}. \end{aligned} \tag{2.28}$$

This relaxation can be combined with heuristic strategies on both the master and pricing problems to support solving the enlarged $MP_{\mathcal{Y}}$ faster. The price to pay for the acceleration in solving the relaxed subproblem is a possible deterioration of the lower bound on $z^\star_{IMP}$, that is, $z^\star_{MP_{\mathcal{Y}}} \leq z^\star_{MP} \leq z^\star_{IMP}$. Since our goal is to solve the original *IMP*, we ultimately need to get rid of the extra variables $\lambda_{\mathbf{x}}$, $\mathbf{x} \in \mathcal{Y}\setminus\mathcal{X}$, e.g., by post-processing, or during the exploration of a branch-and-bound tree.

Paving the way to a practical implementation

Algorithm 2.1, the textbook column generation algorithm, is to solve to optimality the pricing problem in every iteration, add one variable to the *RMP* at a time, and re-solve the latter to optimality. In practice, however, when implemented this way, it is almost certain that this does not give satisfactory performance. The heavy-work and creativity actually begin *after* we have a functional textbook implementation.

Figure 2.5 compares a textbook implementation to two more practical ones ("simple" and "advanced"), which in particular make heavy use of heuristic pricing. The "advanced" one even aims at generating columns that are particularly helpful for an integer solution. We plot the development of bounds similar to Figure 2.4. Our comments complement Table 2.1 which lists additional information such as the number of columns generated, the percentage of time spent solving the *ISP*, the optimal objective value of the linear relaxation, and the relative integrality gap.

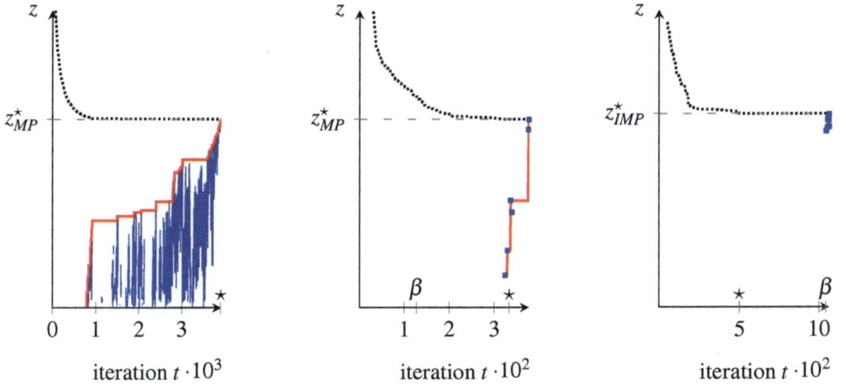

(a) 3 898 iterations (2 111.7 s) (b) 377 iterations (46.9 s) (c) 1 063 iterations (105.7 s)

Fig. 2.5: Column generation process at the root node for three implementations.

implementation	# iterations	# columns	time (s)	% *ISP*	$z^\star_{MP}$	% IG
textbook	3 898	3 897	2 111.7	98.8	5 705.18	2.61
simple	377	41 334	46.9	57.9	5 705.18	2.61
advanced	1 063	53 119	105.7	55.5	5 858	0

Table 2.1: Computational results at the root node for three implementations.

1. Let us first compare Figures 2.5a and 2.5b, textbook vs. using a few restricted pricing problems and better column management ("simple"). The z-axes are of the same scale and are again restricted in plot range. In the latter however, we

notice that the lower bounds are not always available because we use heuristic pricing. In fact, consulting the result log signals that *LB* is not available at all until iteration 127 (marked with a β), the first time we use exact pricing. From then on, we sporadically obtain new lower bounds and update the value of *LB* accordingly. The iterations on the t-axes on the other hand differ by one order of magnitude. What is more, perhaps countering intuition, we generate about 10 times more columns with heuristic pricing (41 334 vs. 3 897). When we additionally register that this nevertheless makes the whole process about 45 times faster (2 111.7 / 46.9), there hardly remains any convincing to do that judicious use of heuristics within the column generation algorithm yields tremendous results despite momentarily losing track of solution quality until exact pricing is turned on. Between the computation price per column or per iteration, it is unclear which one is better to control. Suffice it to say that aiming for a higher throughput on either one is a good target.

2. We see that the difference $z_{RMP} - LB$ still reaches 0 very late. We also underscore that not solving the *RMP* to optimality would not give an objective value z_{RMP}. This would cause holes but we would of course nevertheless be able to compute a non-increasing upper bound.
3. In Figure 2.5c, the "advanced" implementation, we manage to use exact pricing only nine times (out of 1 063 iterations). This all happens close to optimality of the *MP*, where not much can be discerned. The horizontal line marking the optimal objective value is slightly off compared to the first two figures. This is because all implementations use the idea of *relaxed pricing, relaxed master* but the "advanced" one is stronger, i.e., *less* inadmissible columns. We reach a better solution in the *MP* which is in fact by chance integer optimal for the *IMP*, that is, $\boldsymbol{\lambda}^\star_{MP}$ is binary, hence $z^\star_{MP} = z^\star_{IMP}$. We leave the reader to speculate on the computational cost of using exact pricing at every iteration. Since the solution process would be completely altered, we can say that even if we expect less iterations, the stronger pricing problem already warns us of a more expensive computation time. A rhetorical question that soon enough becomes laughable for larger instances. Aside from speeding up the process, we see that heuristics can be sufficient to reach very good solutions. As a case in point, we already attain the optimum about 500 iterations prior to the first use of exact pricing and take away that it is the proof of optimality that is a significant burden. Furthermore, our heuristically generated columns are constructed by keeping track of the current *RMP* solution in such a way that they are more likely to favor integrality when they are optimized in the *RMP*. There could exist a fractional solution with the same optimum but since we reach the latter with heuristics that foster integrality and only perform degenerate pivots from then on, we recall the words of Louis Pasteur (1822–1895): "*le hasard ne favorise que les esprits préparés.*"

2.4 Examples

In this section we present examples that exhibit several aspects of the column generation versatility. Our presentation of the *MP* (or the *integer master problem IMP*) usually precedes that of the *ISP*, the *integer subproblem*. This *ISP* is formulated to find mathematical representations of objects tied to the *IMP* and also comprises encoding functions. In these various examples, our objects are cutting patterns, matchings, and vehicle schedules/routes.

In Example 2.1, we start with the classical One-dimensional cutting stock problem proposed in the '60s where we cut paper rolls of identical size into smaller items. Example 2.2 generalizes it to different roll sizes, the Cutting stock problem with rolls of different widths. Example 2.3 describes the Edge coloring problem for finding the minimum number of colors such that no incident edges of a graph have the same color. Example 2.4 is based on a network flow problem: the Single depot vehicle scheduling problem. A set of trips to service is given, each with a known starting time and duration, and vehicle routes are generated by solving a shortest path problem on an acyclic network. It can be extended to the problem with several depots, and more generally, to a number of Vehicle routing and crew scheduling problems with various operating rules (Example 2.5). Amongst others, the vehicle routing problem with time windows has largely driven the research on theoretical aspects of column generation for integer programs. Finally, Example 2.6 presents an Aircraft routing with schedule synchronization that occurs in the long-term planning process of airlines.

Example 2.1 ***One-dimensional cutting stock problem***

✎ This is the first practical application of column generation, already done in the early '50s. There is a single pricing problem but we propose two formulations one of which uses the zero object.

The *one-dimensional cutting stock problem* (*CSP*) is a classical (maybe *the* classical) example in column generation. We are given large paper rolls ("raws") of width W as well as m small items of width $w_i < W$ and demand b_i, $i \in \{1,\ldots,m\}$. All problem data is positive. Each demand is of course integer and we assume each width is integer as well. The goal is to minimize the number of rolls to be cut into items such that the demand for each small item is satisfied.

Integer master problem

A standard formulation as an *integer master problem IMP* due to Kantorovich and Zalgaller (1951); Gilmore and Gomory (1961, 1963) is

$$
\begin{aligned}
z^{\star}_{IMP} = \min \quad & \sum_{\mathbf{x}\in\mathcal{X}} \lambda_{\mathbf{x}} \\
\text{s.t.} \quad & \sum_{\mathbf{x}\in\mathcal{X}} a_{i\mathbf{x}}\lambda_{\mathbf{x}} \geq b_i \qquad [\pi_i] \quad \forall i \in \{1,\dots,m\} \\
& \lambda_{\mathbf{x}} \in \mathbb{Z}_+ \qquad\qquad \forall \mathbf{x} \in \mathcal{X},
\end{aligned}
\tag{2.29}
$$

where $\mathcal{X}$ represents the set of *cutting patterns* (all possible ways of cutting a roll into items). Figure 2.6 illustrates a subset of these for three items of width 3, 4, and 5 cut from a roll of width 11. The non-negative integer variable $\lambda_{\mathbf{x}} \in \mathbb{Z}_+$ determines how many times cutting pattern $\mathbf{x}$ is used so we literally count the rolls we use in the objective function, i.e., $c_{\mathbf{x}} = 1$, $\forall \mathbf{x} \in \mathcal{X}$. Regarding $\mathbf{a}_{\mathbf{x}} = [a_{i\mathbf{x}}]_{i=1,\dots,m}$, the integer coefficient $a_{i\mathbf{x}} \in \mathbb{Z}_+$ specifies how often item i is obtained in cutting pattern $\mathbf{x}$. The linear relaxation *MP* of (2.29) is then solved by the column generation algorithm, using the non-negative vector $\boldsymbol{\pi} = [\pi_i]_{i=1,\dots,m}$ of dual variables. Finding an initial subset $\mathcal{X}' \subseteq \mathcal{X}$ of cutting patterns to form a feasible *RMP* is easy in this case. For instance, $\mathcal{X}'$ could contain all the singleton cutting patterns $\mathbf{e}_i$, $i \in \{1,\dots,m\}$, where only item i is cut exactly once.

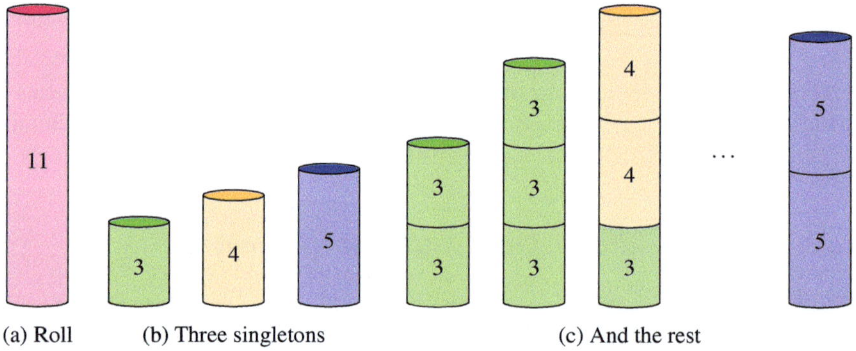

Fig. 2.6: Cutting patterns for three items of width 3, 4, and 5 in a roll of width 11.

Integer pricing problem

Solving the *RMP* provides us $\boldsymbol{\pi}$. Finding an improving cutting pattern $\mathbf{x}$ (if any) can then be done by solving the *ISP* as an *integer knapsack problem* with the first formulation we learn about in Martello and Toth (1990):

$$\bar{c}(\boldsymbol{\pi}) = \min_{\mathbf{x}\in\mathcal{X}} c_{\mathbf{x}} - \sum_{i=1}^{m} \pi_i a_{i\mathbf{x}} \quad \equiv 1 - \max \sum_{i=1}^{m} \pi_i x_i$$
$$\text{s.t.} \quad \sum_{i=1}^{m} w_i x_i \leq W \tag{2.30}$$
$$x_i \in \mathbb{Z}_+ \quad \forall i \in \{1,\ldots,m\}.$$

One can see that our encoding functions for the *ISP* reflect the understanding of what a column is for the *IMP*:

$$c_{\mathbf{x}} = 1 \quad \text{and} \quad a_{i\mathbf{x}} = x_i, \; \forall i \in \{1,\ldots,m\}. \tag{2.31}$$

This simplicity between the objects and encoding functions is because the objects $\mathcal{X} = \{\mathbf{x} \in \mathbb{Z}_+^m \mid \sum_{i=1}^m w_i x_i \leq W\}$ are literally defined in the same way as parameter $\mathbf{a}_{\mathbf{x}} = \mathbf{x}$. We underscore that this cannot be taken for granted, even for the same problem. We formulate the knapsack problem in two different ways in Example 4.1 and discuss how that impacts the cutting stock problem in Example 4.2.

Empty vs. zero cutting patterns

The *empty cutting pattern* $x_i = 0$, $\forall i \in \{1,\ldots,m\}$, is feasible in (2.30) but it incurs a cost of 1 in the objective function as (2.31) indicates. It corresponds to a roll that is not cut, and as such, it should not be counted in the objective function. This is actually mathematically ensured by the column generation algorithm because its reduced cost is 1 for any $\boldsymbol{\pi}$ so it cannot ever be *generated*. Nevertheless, we can compute the "real" contribution of not using a roll in the *IMP* using the zero object. The *zero cutting pattern* with zero-cost and zero-column-coefficients can easily be introduced with the use of a supplementary binary variable x_0 in the *ISP* formulated as

$$\bar{c}(\boldsymbol{\pi}) = \min \quad x_0 - \sum_{i=1}^{m} \pi_i x_i$$
$$\text{s.t.} \quad \sum_{i=1}^{m} w_i x_i \leq W x_0 \tag{2.32}$$
$$x_0 \in \{0,1\}$$
$$x_i \in \mathbb{Z}_+ \qquad \forall i \in \{1,\ldots,m\}.$$

As shown in Figure 2.7a, for any solution to the *ISP* (2.32), the cost encoding function now depends on x_0 whereas the column-coefficients are as before. Figures 2.7b and 2.7c correspond to solutions of formulation (2.30) where $x_0 = 1$ is implicit. The introduced zero cutting pattern appears in Figure 2.7d where $x_0 = 0$ forces $\mathbf{x} = \mathbf{0}$. The empty cutting pattern is dominated by the zero cutting pattern because its reduced cost is always $0 < 1$ for any $\boldsymbol{\pi}$, but finding either of them in the *ISP* results in the same conclusion: we have reached optimality. The same is also true if we turn to a situation where several pricing problems exist. We know it is hard to grasp its purpose at this stage but we can at least observe that the same problem can be solved with two different pricing problems. For the advanced reader, the zero object is not a question of efficiency but rather a mathematical construct we later use in a Dantzig-Wolfe reformulation to work with a *polyhedral cone*.

$$\begin{bmatrix} c_{\mathbf{x}} \\ [a_{i\mathbf{x}}]_{i=1,\dots,m} \end{bmatrix} = \begin{bmatrix} x_0 \\ [x_i]_{i=1,\dots,m} \end{bmatrix}$$
(a) Encoding functions

$$\begin{bmatrix} 1 \\ \mathbf{x} \end{bmatrix}$$
(b) $\mathbf{x} \neq \mathbf{0}$

$$\begin{bmatrix} 1 \\ \mathbf{0} \end{bmatrix}$$
(c) Empty

$$\begin{bmatrix} 0 \\ \mathbf{0} \end{bmatrix}$$
(d) Zero

Fig. 2.7: Cutting patterns in a roll.

Example 2.2 **Cutting stock problem with rolls of different widths**

We consider two modifications to the one-dimensional cutting stock problem. We now have rolls of different widths and measure the total trim loss rather than the number of rolls. We use an index k to differentiate the roll widths in their respective integer subproblems.

The *one-dimensional cutting stock problem using different roll widths* still aims to fulfill m demands b_i, $i \in \{1,\dots,m\}$, for items of width w_i but this time we may use various roll widths W^k, $k \in K$. We consider a different objective function that minimizes the total trim loss because it is a better indicator for the usage of raws.

For $k \in K$, let the integer variable $\lambda_{\mathbf{x}^k} \in \mathbb{Z}_+$ represent the number of times cutting pattern $\mathbf{x}^k = \begin{bmatrix} x_0^k \\ [x_i^k]_{i=1,\dots,m} \end{bmatrix}$ is used, where the binary variable x_0^k indicates if a roll of width W^k is cut or not. The *IMP* becomes

$$\begin{aligned} z^\star_{IMP} = \min \quad & \sum_{k\in K}\sum_{\mathbf{x}^k\in\mathcal{X}^k} c_{\mathbf{x}^k}\lambda_{\mathbf{x}^k} \\ \text{s.t.} \quad & \sum_{k\in K}\sum_{\mathbf{x}^k\in\mathcal{X}^k} a_{i\mathbf{x}^k}\lambda_{\mathbf{x}^k} = b_i \qquad [\pi_i] \quad \forall i \in \{1,\dots,m\} \\ & \lambda_{\mathbf{x}^k} \in \mathbb{Z}_+ \qquad\qquad \forall k \in K, \mathbf{x}^k \in \mathcal{X}^k, \end{aligned} \tag{2.33}$$

where $\mathcal{X}^k$ denotes the set of (representations of) cutting patterns for the roll width W^k (hence $\mathcal{X} = \cup_{k\in K}\mathcal{X}^k$). Observe the usage of equality constraints in (2.33): over-satisfying the demand would reduce the trim loss. Adapting (2.32) with index k and a restriction on item production, a formulation for the *ISP*^{k} becomes

$$\bar{c}^k(\boldsymbol{\pi}) = \min \quad c_{\mathbf{x}^k} - \sum_{i=1}^{m} \pi_i a_{i\mathbf{x}^k} \tag{2.34a}$$

$$\text{s.t.} \quad \sum_{i=1}^{m} w_i x_i^k \leq W^k x_0^k \tag{2.34b}$$

$$x_i^k \leq b_i \qquad \forall i \in \{1,\dots,m\} \tag{2.34c}$$

$$x_0^k \in \{0,1\} \tag{2.34d}$$

$$x_i^k \in \mathbb{Z}_+ \qquad \forall i \in \{1,\dots,m\} \tag{2.34e}$$

$$c_{\mathbf{x}^k} = W^k x_0^k - \sum_{i=1}^{m} w_i x_i^k \tag{2.34f}$$

$$a_{i\mathbf{x}^k} = x_i^k \qquad \forall i \in \{1,\dots,m\}. \tag{2.34g}$$

For any solution to (2.34), the cost $c_{\mathbf{x}^k}$ measures the trim loss of cutting pattern $\mathbf{x}^k$ whereas the column-coefficient $a_{i\mathbf{x}^k}$ indicates how many times item i is cut in $\mathbf{x}^k$. Observe that there is a $\mathbf{0}$-vector of dimension $m+1$ in every set $\mathcal{X}^k$, $k \in K$. With respect to the empty cutting pattern, a solution $\mathbf{x}^k = 0$ and $x_0^k = 1$ gives a "misleading" trim loss of W^k but nevertheless correctly terminates column generation since this value also corresponds to a positive reduced cost for any $\boldsymbol{\pi}$.

Note 2.19 (More than we bargain for.) Whenever it is interesting to take item i, the nature of optimization dictates that we take this item as much as possible. This is the notion of *maximal* cutting pattern which is illustrated in Figure 2.8 along with the trim loss given by a separated cylinder on the top.

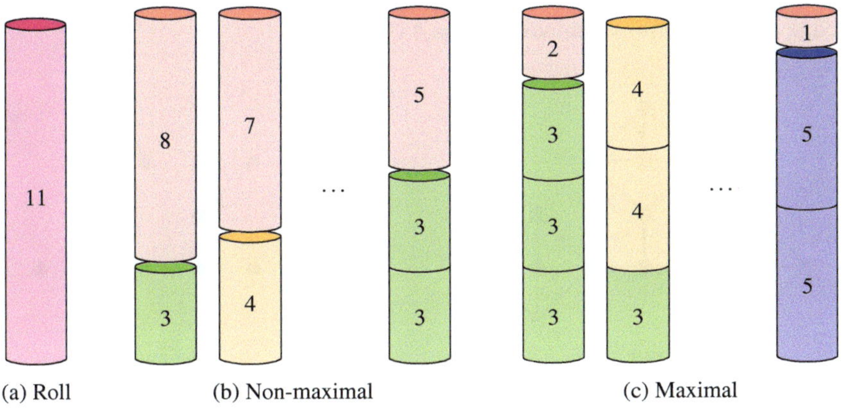

Fig. 2.8: Cutting patterns with trim loss.

Consequently, non-maximal patterns do not appear naturally when we optimize the *ISP* leading to potential over-packing with respect to the demand. Indeed, depending on problem data, it is possible that an item can be cut more times than it is needed, i.e., $\exists i \mid \lceil W^k/w_i \rceil > b_i$. In such a case, the constraints $x_i \leq b_i$, $\forall i \in \{1,\dots,m\}$, in formulation (2.34) intercept such cutting patterns that are obviously not suitable for an integer solution. We can therefore also point to *relaxed pricing, relaxed master* and predict that the absence of these constraints can weaken the objective value of the linear relaxation but does not compromise the integer optimum. However, since these are only bounds, they do not really make the *ISP* more challenging to solve. Exercise 2.16 exposes all of this on a simple instance where we minimize the number of rolls used.

Example 2.3 **Edge coloring problem**

✎ We discuss the notions of objects, representations, and encodings, and highlight that the modeler ties theory to practice.

The *edge coloring problem* (*ECP*) is defined on an undirected graph $G = (N, E)$, where N denotes the set of nodes and E the set of edges. We want to color the edges in such a way that no incident edges have the same color. The minimum number of colors required to color all the edges is called the *chromatic index* of the graph. A set of pairwise non-incident edges is called a *matching*. We observe that an edge coloring is a set of matchings, *each in a separate color*, that partitions the edges.

Figure 2.9a shows a planar graph with 16 vertices and 24 edges. We then show a possible matching (Figure 2.9b) as well as a *maximal matching* (Figure 2.9c) in which it is not possible to include an additional edge. Finally, we are able to draw an edge coloring with four colors by trial-and-error in Figure 2.9d. We have found the chromatic index of G since there are 7 edges in any maximal matching, hence an edge coloring requires at least $\lceil \frac{24}{7} \rceil = 4$ colors.

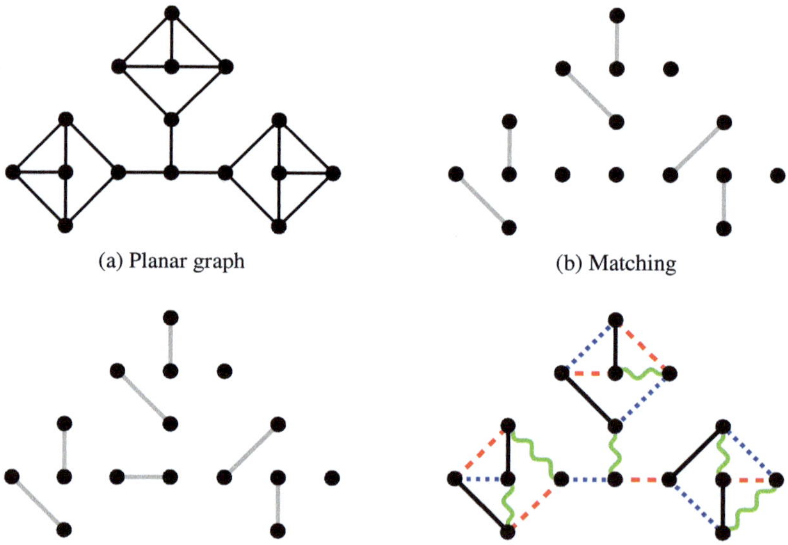

(a) Planar graph (b) Matching (c) Maximal matching (d) Four colors

Fig. 2.9: Coloring edges of a planar graph G with 16 vertices and 24 edges.

Set covering master problem

Nemhauser and Park (1991) and Mehrotra and Trick (1996) tackle the *ECP* by directly modeling the partitioning of the edge set into matchings. They observe that, in the model, the colors of the matchings are irrelevant. In fact, leaving the colors in the model induces a symmetry: any permutation of the colors in an edge coloring gives again an edge coloring with the same number of colors. Their set covering formulation therefore works only with variables for matchings which do not have a color. The *IMP* reads as

$$\begin{aligned} z^\star_{IMP} = \min \quad & \sum_{\mathbf{x}\in\mathcal{X}} \lambda_{\mathbf{x}} \\ \text{s.t.} \quad & \sum_{\mathbf{x}\in\mathcal{X}} \mathbf{a}_{\mathbf{x}}\lambda_{\mathbf{x}} \geq \mathbf{1} && [\boldsymbol{\pi}] \\ & \lambda_{\mathbf{x}} \text{ binary} && \forall \mathbf{x}\in\mathcal{X}, \end{aligned} \tag{2.35}$$

where $\mathcal{X}$ denotes the set of incidence vectors $\mathbf{x} = [x_e]_{e\in E}$ of matchings which specify, in binary, which edges of E are contained in a matching. The variable $\lambda_{\mathbf{x}}$ indicates if matching $\mathbf{x}$ is used or not, so we count the number of colors we use in the objective function, i.e., $c_{\mathbf{x}} = 1, \forall \mathbf{x} \in \mathcal{X}$. The covering constraints ensure that each edge in the set E appears in at least one matching and we here have that $\mathbf{a}_{\mathbf{x}} = \mathbf{x}, \forall \mathbf{x} \in \mathcal{X}$.

Note 2.20 (Covering constraints.) The reader may wonder why we do not write equality constraints, even though we want to *partition* the set of edges into matchings (or schedule the same number of games for each team). The nonchalant reason is because we can. Indeed, it is easy to turn a solution to (2.35) into a solution to the *ECP* of the same cost by dropping edges from the selected matchings until each edge is covered exactly once, hence the model is ultimately correct. In fact, the horizontal dotted edge (blue) in the middle of Figure 2.9d could also be solid (gray). The benefits and concerns regarding such manipulations are postponed to more advanced portions of the book, see Exercises 2.14–2.17 for an appetizer and Chapter 6 for dual incentives.

Pricing the matchings

Let $\delta(S) \subseteq E$ for $S \subseteq N$ denote the edges with one endpoint in S and the other one in $N \setminus S$. Such a set is called a *cut*. Let $\mathbf{x} = [x_e]_{e\in E}$. Then the set of matchings can be defined via

$$\mathcal{X} = \left\{ \mathbf{x} \in \{0,1\}^{|E|} \;\middle|\; \sum_{e\in\delta(\{i\})} x_e \leq 1, \ \forall i \in N \right\}, \tag{2.36}$$

where x_e takes value 1 if edge e is selected in the matching, and 0 otherwise. The set of constraints ensures that, for each node $i \in N$, at most one incident edge is selected. Using $\boldsymbol{\pi} = [\pi_e]_{e\in E}$, the non-negative dual values for the covering constraints in the linear relaxation of (2.35), together with the added binary variable x_0 to introduce the zero matching when $x_0 = 0$, the *ISP* reads as

$$\begin{aligned} \bar{c}(\boldsymbol{\pi}) = \min \quad & x_0 - \sum_{e \in E} \pi_e x_e \\ \text{s.t.} \quad & \sum_{e \in \delta(\{i\})} x_e \leq x_0 && \forall i \in N \\ & x_0 \in \{0,1\} \\ & x_e \in \{0,1\} && \forall e \in E \\ & c_{\mathbf{x}} = x_0 \\ & a_{e\mathbf{x}} = x_e && \forall e \in E. \end{aligned} \tag{2.37}$$

Note 2.21 (Artistic indices.) Observe that the objects, matchings with edges in E, are represented by binary vectors $\mathbf{x} = [x_e]_{e \in E}$ of cardinality $|E|$, which are identical to their encodings $\mathbf{a}_{\mathbf{x}} = [a_{e\mathbf{x}}]_{e \in E}$. The close relationship between these three is illustrated when indexing the λ-variables with objects:

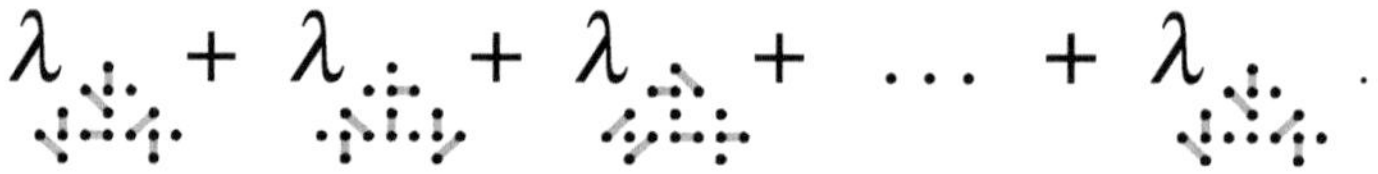

This makes it even more apparent why we may also understand an expression like "negative reduced cost matchings."

Note 2.22 (An object is an object, basta!) What we call an object only depends on the problem (and solution!) understanding of the end-user: does she want an edge coloring, or did she realize that a set of matchings suffices which she can color on her own? An experienced modeler recognizes such degree of freedom and exploits it when building a model. For example, it is a perfectly realistic situation that an end-user with a cutting stock problem wishes to obtain a detailed plan, how many orders to cut from exactly which roll. That is, in the end-user's perspective, the objects specify the precise roll. The modeler realizes the symmetry, builds a model without roll-indices, and thus adopts a perspective in which objects *do not* specify a roll. It is fair to say that the modeler introduced another layer of abstraction into our discussion about objects, representations, and encodings. It goes without saying that the modeler should not forget to return objects to the end-user in their perspective and language; when roll-indices are desired, these need to be provided, regardless of whether the employed mathematical model specifies them or not. As a practical remark this reminds us of the communication skills demanded from the OR expert when translating between the real and the model world.

Sports scheduling

An interesting application of the *ECP* is sports scheduling as described in Januario et al. (2016) and Ribeiro et al. (2023). For example, a *single round-robin tournament* is a competition in which each team plays against every other team exactly once. It is also said to be *compact* if every team plays once per round such that the number of rounds matches the chromatic index. This situation can be modeled on a complete graph, where the nodes correspond to the teams, the edges to the games, and the colors to the rounds in which the games are played.

Suppose we have been mandated to determine a competition schedule where games must occur every Saturday. Let us use the unusual matchup depicted in Figure 2.9. The optimal solution we found earlier is reproduced in Figure 2.10a but it does not mean anything to the sports league. They expect a schedule where every game appears on a calendar. In this case, we can associate any two vertices (teams) connected by a blue/dotted edge (game) with the first Saturday, see Figure 2.10b. The second Saturday with red/dashed, third with green/wavy, and fourth with black/solid.

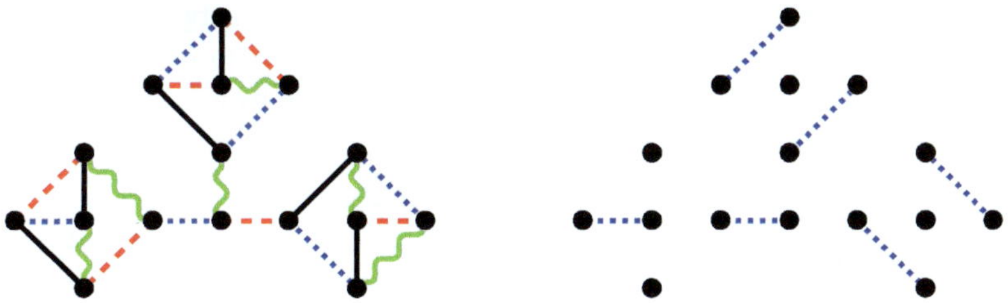

(a) Reproduction (b) First Saturday

Fig. 2.10: Four colors into a schedule.

What if the team represented by the right-most node cannot play on the second weekend? The right-most node has three adjacent vertices so we just need to reorder the colors to match the forbidden Saturday, in this case exchange red/dashed with black/solid.

What if the team represented by the left-most node also cannot play on the second weekend? This particular solution is conflicting with the additional constraint because the edge have different colors but the same date restriction. Figure 2.11a gives an alternative four color solution that fulfills both restrictions where we simply change three edge colors around the left-most node. Figure 2.11b confirms that neither teams associated with left-most and right-most nodes are playing on the second Saturday.

It may not always be easy nor even possible to match the chromatic index with additional restrictions. Manually fiddling with a solution trying to figure this out can rapidly become an error-prone burden. Can we state a constraint for the *IMP* that can handle both restrictions? We can model this as $\sum_{\mathbf{x}\in\mathcal{X}} r_{\mathbf{x}}\lambda_{\mathbf{x}} = 3$, where $r_{\mathbf{x}}$ is a parameter that takes value 1 if matching $\mathbf{x}$ includes the left-most or right-most nodes, 0 otherwise.

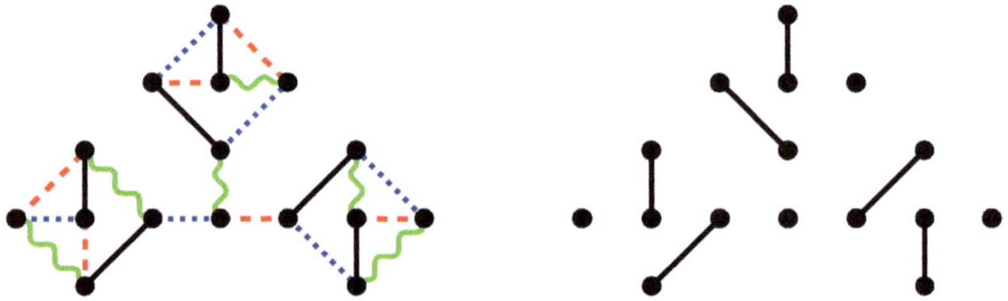

(a) Another four colors (b) Second Saturday

Fig. 2.11: Four colors into a schedule, take two.

Example 2.4 **Single depot vehicle scheduling problem**

> This example again proposes two formulations for the subproblem, the second bringing in the zero object. The *SP* turns out to be a shortest path problem on an acyclic network, easily solvable by a dynamic programming algorithm.

The *single depot vehicle scheduling problem* (*SDVSP*) involves a set N of n trips, each one defined by a pair of start time and location, and a pair of end time and location, that must be operated by a set of vehicles. The objects in this problem are *vehicle schedules*, each consisting in a sequence of trips and deadheads (empty moves for repositioning) which must start and end at the given depot. We assume a homogeneous fleet of v vehicles. The cost structure typically includes travel costs and a fixed cost per vehicle used. The problem consists in determining a set of schedules such that each trip $i \in N$ belongs to exactly one schedule, the vehicles used are available, and the total operating cost is minimized.

The *SDVSP* occurs in urban transit bus scheduling where vehicles are *buses*, trips are *bus trips*, and vehicle schedules are *bus schedules*. The goal is to determine a least-cost set of bus schedules. Obviously, it can be adapted to other contexts such as airline management, where vehicles are *aircraft*, trips are *flights*, and vehicle schedules are *aircraft schedules*.

Set partitioning master problem

Although the *SDVSP* can be formulated and easily solved as a network flow problem (see Exercise 2.8), for pedagogical reasons in this chapter on column generation, we use an alternative formulation for it, a set partitioning type problem (*SPP*) :

$$z^\star_{IMP} = \min \quad \sum_{\mathbf{x}\in\mathcal{X}} c_{\mathbf{x}}\lambda_{\mathbf{x}} \tag{2.38a}$$

$$\text{s.t.} \quad \sum_{\mathbf{x}\in\mathcal{X}} a_{i\mathbf{x}}\lambda_{\mathbf{x}} = 1 \qquad [\pi_i] \qquad \forall i \in N \tag{2.38b}$$

$$\sum_{\mathbf{x}\in\mathcal{X}} a_{o\mathbf{x}}\lambda_{\mathbf{x}} \leq v \qquad [\pi_o] \tag{2.38c}$$

$$\lambda_{\mathbf{x}} \in \{0,1\} \qquad \forall \mathbf{x} \in \mathcal{X}. \tag{2.38d}$$

where $\mathcal{X}$ denotes the set of (representations of) schedules in (2.38), and binary variable $\lambda_{\mathbf{x}}$ takes value 1 if and only if schedule $\mathbf{x} \in \mathcal{X}$ is selected. Such a schedule is encoded by its cost $c_{\mathbf{x}}$ and column $\mathbf{a}_{\mathbf{x}} = [a_{i\mathbf{x}}]_{i\in N}$, where the binary column-coefficient $a_{i\mathbf{x}}$ takes value 1 if bus trip i is operated in schedule $\mathbf{x}$, and 0 otherwise. Additionally, we have a constant parameter $a_{o\mathbf{x}} = 1$, $\forall \mathbf{x} \in \mathcal{X}$, where index o represents the depot from which the vehicles are originating. Given the huge number of schedules in $\mathcal{X}$, the linear relaxation *MP* of (2.38) is solved by the column generation algorithm, where the dual values $\boldsymbol{\pi} = [\pi_i]_{i\in N}$ and π_o are used.

Shortest path pricing problem

The *SP* can be modeled as a shortest path problem in an acyclic directed time-space network, where there is a node for each timetabled trip and the depot is represented by two nodes, an origin o and a destination d. We denote by N the set of trip nodes. The set of arcs is constructed as follows:

- We introduce an arc between trips i and j if and only if it is feasible to travel from the end of trip i to the start of trip j while satisfying the start/end time constraints. Such arcs are called *inter-trip arcs*, denoted by I.
- We need arcs from the origin o to each trip and from each trip to the destination d (back to the depot).
- Finally, we model the option of an empty schedule that operates no trips at all with arc (o,d). This arc may incur a cost c_{od} for an available vehicle that remains parked at the depot. In that case, this cost would be assigned to the slack variable in (2.38c).

This gives us the network $G = (V,A)$, with the node set $V = N \cup \{o,d\}$ and the arc set $A = I \cup (\{o\} \times N) \cup (N \times \{d\}) \cup \{(o,d)\}$, see Figure 2.12. A schedule is a binary solution $\mathbf{x} \in \mathcal{X}$ corresponding to an od-path in this network, and vice versa. Observe that we nonetheless do not need any integrality requirements to obtain such solutions.

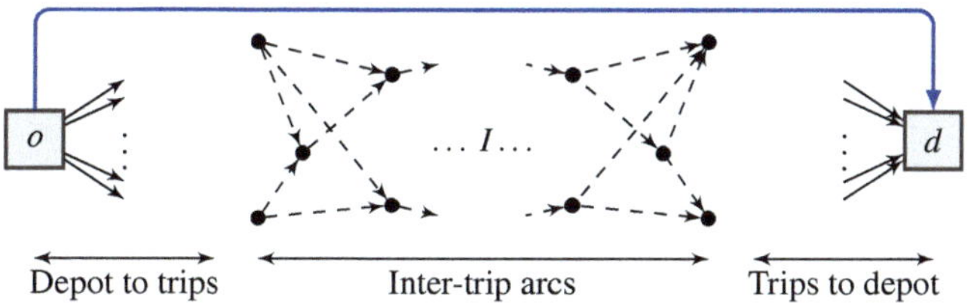

Fig. 2.12: Network $G = (V,A)$ with the option to model an empty schedule.

Let the variable x_{ij} represent whether arc $(i,j) \in A$ is used on a path or not. The path constraints that define $\mathcal{X}$ are

$$\begin{aligned}
\sum_{j:(o,j)\in A} x_{oj} &= 1 \\
\sum_{j:(i,j)\in A} x_{ij} - \sum_{j:(j,i)\in A} x_{ji} &= 0 \quad \forall i \in N \\
\sum_{i:(i,d)\in A} x_{id} &= 1 \\
x_{ij} &\geq 0 \quad \forall (i,j) \in A.
\end{aligned} \tag{2.39}$$

A vector $\mathbf{x} \in \mathcal{X}$ in (2.39) is encoded in the *IMP* (2.38) as

$$c_{\mathbf{x}} = \sum_{(i,j)\in A} c_{ij}x_{ij}, \;\; a_{i\mathbf{x}} = \sum_{j:(i,j)\in A} x_{ij}, \; \forall i \in N, \;\; \text{and} \;\; a_{o\mathbf{x}} = \sum_{j:(o,j)\in A} x_{oj} = 1. \tag{2.40}$$

Note 2.23 (A sum to one.) Since there is always one redundant constraint in a network flow problem, we can arbitrarily drop any one we please. This means that we can also modify said constraint with any transformation that respects flow conservation. Consequently, we can replace the $a_{o\mathbf{x}}$ expression by $a_{d\mathbf{x}} = \sum_{i:(i,d)\in A} x_{id} = 1$. Another possible replacement is to identify any subset of arc-flow variables that necessarily sum to one in any path-flow solution. For example, the subset of arcs on which a vehicle travels at some specified time value is acceptable because exactly one of these must be taken.

Combining (2.39) and (2.40) together with the dual values $\boldsymbol{\pi} = [\pi_i]_{i\in N}$ and π_o leads us to a formulation for the *SP* given by

$$\begin{aligned}
\bar{c}(\boldsymbol{\pi}, \pi_o) &= \min_{\mathbf{x}\in\mathcal{X}} \; c_{\mathbf{x}} - \sum_{i\in N} \pi_i a_{i\mathbf{x}} - \pi_o a_{o\mathbf{x}} \\
&= \min_{\mathbf{x}\in\mathcal{X}} \; \sum_{(i,j)\in A} c_{ij}x_{ij} - \sum_{i\in N} \pi_i \Big(\sum_{j:(i,j)\in A} x_{ij} \Big) - \pi_o \Big(\sum_{j:(o,j)\in A} x_{oj} \Big) \\
&= \min_{\mathbf{x}\in\mathcal{X}} \; \sum_{(i,j)\in A} \tilde{c}_{ij}x_{ij},
\end{aligned} \tag{2.41}$$

where grouping the objective coefficients of each x_{ij}-variable in (2.41) leads to an *adjusted cost* $\tilde{c}_{ij}$ for arc $(i,j) \in A$ in the *SP* (Exercise 2.7).

Note 2.24 (Reducing costs needs not lead to reduced costs.) One sometimes finds in the literature a misleading notion that $\tilde{c}_{ij}$ is the reduced cost $\bar{c}_{ij}$ of arc-flow variable x_{ij}. It is *not* and the reason is simple: variable x_{ij} actually belongs to the *SP*, so its reduced cost certainly cannot be computed with the dual variables of another program, namely the *MP*. The reader can also verify that formulating the *SDVSP* as a network flow problem in x_{ij}-variables does not lead to arc reduced costs computed as in (2.41), see Exercise 2.8.

Zero schedule

We again introduce the zero schedule by using a supplementary variable, in this case by adding to A the reverse arc (d,o) at zero-cost to form the network $G_{do} = (N, A_{do})$, where $A_{do} = A \cup \{(d,o)\}$, depicted in Figure 2.13.

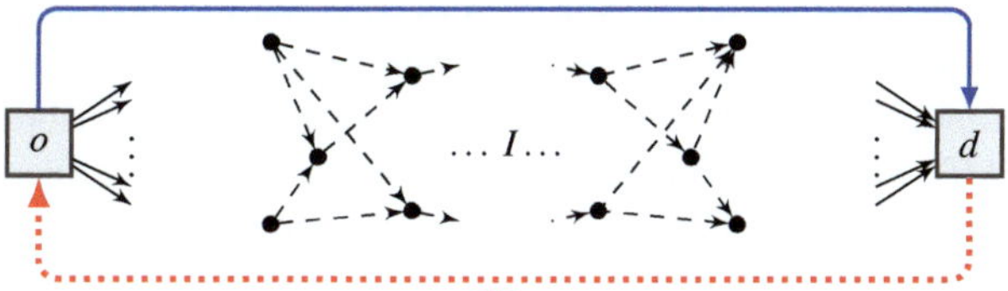

Fig. 2.13: Network G_{do} with arc (d,o) to model the zero schedule.

A formulation for the *SP* is given by

$$\bar{c}(\boldsymbol{\pi}, \pi_o) = \min \sum_{(i,j)\in A} \tilde{c}_{ij} x_{ij} \tag{2.42a}$$

$$\sum_{j:(o,j)\in A} x_{oj} = x_{do} \tag{2.42b}$$

$$\sum_{j:(i,j)\in A} x_{ij} - \sum_{j:(j,i)\in A} x_{ji} = 0 \qquad \forall i \in N \tag{2.42c}$$

$$\sum_{j:(j,d)\in A} x_{jd} = x_{do} \tag{2.42d}$$

$$x_{ij} \geq 0 \qquad \forall (i,j) \in A_{do} \tag{2.42e}$$

$$x_{do} \leq 1. \tag{2.42f}$$

A (binary) solution to (2.42) either gives a schedule represented by a unit-cycle composed of an *od*-path completed by the reverse arc with $x_{do} = 1$ or the zero schedule with $x_{do} = 0$ and $x_{ij} = 0$, $\forall (i,j) \in A$. To account for the zero schedule in the *IMP* (2.38), the encoding is slightly modified as

$$c_{\mathbf{x}} = \sum_{(i,j)\in A} c_{ij} x_{ij}, \; a_{i\mathbf{x}} = \sum_{j:(i,j)\in A} x_{ij}, \; \forall i \in N, \; \text{and} \; a_{o\mathbf{x}} = \sum_{j:(o,j)\in A} x_{oj} = x_{do}, \tag{2.43}$$

where only the column-coefficient $a_{o\mathbf{x}}$ is modified to no longer be the constant term 1.

Note 2.25 (Complementary/orthogonal columns.) The solution of a shortest path problem by dynamic programming provides a tree of shortest paths from the origin node o to all other nodes, see Ahuja et al. (1993, §4.3). Saving this solution tree is done via pointers to predecessors. Some implementations may even discard the destination node d to post-process it after resolution. A posteriori, those ending at a trip node can be completed to the depot. In Figure 2.14, there is only one shortest path of cost 3 which is o24d (in dashed pattern). Because nodes 2 and 3 are not connected to the destination node d (not end trips), the only additional path we can deduce is o25d (dotted completion) of cost $3 + 2 = 5$. It is common practice to retrieve several of these *od*-paths with negative reduced cost. For example, we can select a shortest one, remove the corresponding trip-nodes from N, and greedily select another path that covers different trips in a complementary way. This can also be done in the case of different pricing problems. In this fashion, we in fact produce *orthogonal columns* which likely make different improving contributions to the set partitioning *RMP*. Such columns can, as an appreciated side effect, even make a contribution to *integer* feasible solutions to the set partitioning model (2.38).

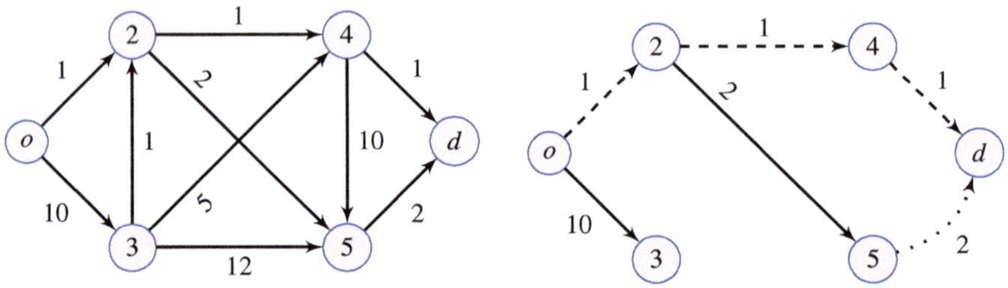

(a) Small network (b) Tree of shortest paths

Fig. 2.14: Solving the shortest path problem by dynamic programming.

Example 2.5 **Vehicle routing and crew scheduling problems**

✎ The purpose of this example is to shortly introduce a large class generalizing the above routing problem. Indeed, several routing and scheduling applications in the airline and rail industries are designed on time-space acyclic networks. Many are however on networks containing cycles, in practice much more difficult to solve.

Similarly to (2.38), a large variety of *vehicle routing and crew scheduling problems* can be formulated as set partitioning models where the right-hand side imposes that each *task* is performed exactly once (e.g., aircraft routing), or as set covering models where tasks are covered at least once (e.g., pilot schedules, where a pilot may travel as a passenger for repositioning), or as generalized partitioning/covering models, where the right-hand side is a positive integer vector asking for covering each task a certain number of times (e.g., flight attendant schedules and locomotive itineraries). Binary columns encode vehicle routes or crew schedules, where a unit-entry represents a task that is performed, see Chapter 5 for an extensive presentation. Figures 2.15 and 2.16 recollect our experience that the world of vehicle routing and crew scheduling is not without a certain fascination effect to children of all ages.

(a) 2-year-old Jacques (b) 7-month-old Jean Bertrand

Fig. 2.15: Initiation to vehicle routing.

Because vehicles and crews are intimately related, these models generalize the *SDVSP* in various ways. For example, we can account for heterogeneous fleets and/or several depots in vehicle routing, discriminate at a workforce level in crew scheduling where a preferential bidding system is in place, and also include a large number of operating rules. These relate mostly to maintenance schedules in vehicle fleet planning and to worker collective agreement regulations in crew scheduling. In recent years, computers and solution methods have become powerful enough to think about solving both types of problems simultaneously.

In such models, we can have several pricing problems and the columns are generated by solving constrained shortest path problems on appropriate networks, for example, on time-space networks where nodes represent locations at specific times while arcs represent activities such as trains, flight legs, deadheads, lunch and break periods, etc. The master problem manages the vehicle routes and/or crew schedules while a column generator handles constraints specific to a single route or schedule, for example, the time at which a customer is serviced or an aircraft maintenance is done, see Desrosiers et al. (1995) and Desaulniers et al. (1998a).

The *multiple depot vehicle scheduling problem* (*MDVSP*) is one such immediate generalization of the *SDVSP*, see Exercise 2.9. In this problem, we impose that a vehicle returns to its departing depot such that the *MDVSP* can no longer be formulated as a network flow problem. In fact, Bertossi et al. (1987) show that it is $\mathcal{NP}$-hard if it involves more than one depot. This gives rise to one pricing problem per depot, again formulated as a shortest path problem. The *IMP* remains a set partitioning model, with side constraints for the number of vehicles available at each depot. Due to its simplicity as there are no operating rules to integrate into the implementation, the *MDVSP* is often used to easily test acceleration strategies to overcome degeneracy in problems solved by the column generation algorithm (Oukil et al., 2007).

Fig. 2.16: Léa's perception of vehicle routing (Desrosiers et al., 1995) and, later on, airline crew scheduling.

Example 2.6 **Aircraft routing with schedule synchronization**

> ✎ In this example, we encode three types of information for an aircraft route: the cost, the operated flights, and the time at which they are scheduled. These time values are synchronized within an interval from one day to the other.

This *aircraft routing problem with schedule synchronization* (*ARPSS*) occurs in the long-term planning process of airlines (Ioachim et al., 1999). It involves operating, at minimum cost, a set of flights which present schedule flexibility on the departure time. For each flight, the data includes an identifier, the origin and destination airports, a duration, a minimum ground time at the destination and a time window on the departure. Waiting is permitted before and within the time windows at no cost. A flight is also characterized by the day of the week when it is operated. For flights with the same identifier that are flown on different weekdays, the departure has to be scheduled at the same time every day, thus imposing schedule synchronization.

Let N denote the set of flights to be operated during a week and M the set of group identifiers for flights that must respect the same departure time. Let $m_i \in M$ denote the group identifier of flight $i \in N$. Hence, all pairs of flights $i, j \in N$ with $m_i = m_j$ must have the same departure time. We assume that all time values are integer and that the departure time windows $[a_i, b_i]$ and $[a_j, b_j]$ of two such synchronized flights i and j satisfy $a_i = (a_j \bmod 1\,440)$ and $b_i = (b_j \bmod 1\,440)$, where 1 440 minutes correspond to 24 hours.

Let K denote the set of available aircraft which are assumed to be of different types, i.e., with different operating costs, seating capacities, etc. For each aircraft $k \in K$, define a network $G^k_{do} = (V^k, A^k)$, where V^k is the set of nodes and A^k the set of arcs (see Figure 2.13 for a similar representation). Let o^k and d^k denote the source and sink nodes in G^k_{do}, while N^k are the nodes representing flights compatible with aircraft k (e.g., with an appropriate capacity). To ease the presentation and avoid vector dimension inconsistencies, flights in $N \setminus N^k$ are also incorporated (as dummy nodes) in V^k. Hence, $V^k = N \cup \{o^k, d^k\}$. For $i \in N$, let d_i and s_i denote the duration and minimum ground time, respectively, and assume $a_{o^k} = b_{o^k} = 0$.

The arc set A^k contains three types of arcs: the return arc (d^k, o^k), the depot arcs, and the flight connection arcs. The depot arcs are (o^k, d^k) to account for an unused aircraft, and the pair $(o^k, i), (i, d^k)$, for all $i \in N^k$, to link each compatible flight with the source and sink nodes. A flight connection arc (i, j) for $i, j \in N^k$, corresponds to a feasible connection between a flight arriving at an airport and one departing from the same airport; hence $a_i + d_i + s_i \leq b_j$. For the sink nodes, let

$$a_{d^k} = 0 \text{ and } b_{d^k} = \max_{i \in N^k} b_i + d_i + s_i, \qquad \forall k \in K. \tag{2.44}$$

Finally, a cost c^k_{ij}, $(i, j) \in A^k$, depends on the objective of the optimization program. The capital and fixed costs for aircraft k are assigned to arc (d^k, o^k) while the operational cost of a flight $i \in N^k$ performed by aircraft k (which may include loss revenues for not assigning enough seating capacity) is assigned to all arcs with a tail at node i.

Routes and schedules

A route for aircraft k comprises the selected flights, their departure times, and the total cost. The k-th pricing problem uses the variables

- x^k_{ij}, $(i, j) \in A^k$: 1 if arc (i, j) is selected, 0 otherwise;
- x^k_i, $i \in N$: 1 if flight i is operated by aircraft k, 0 otherwise;
- t^k_i, $i \in N$: departure time of flight i if operated by aircraft k, 0 otherwise.

Feasible routes, where $x^k_{d^k o^k} = 1$, satisfy the following path and time constraints:

$$\sum_{j:(o^k, j) \in A^k} x^k_{o^k j} = \sum_{j:(j, d^k) \in A^k} x^k_{j d^k} = x^k_{d^k o^k} = 1 \tag{2.45a}$$

$$\sum_{j:(i,j) \in A^k} x^k_{ij} - \sum_{j:(j,i) \in A^k} x^k_{ji} = 0 \qquad \forall i \in N \tag{2.45b}$$

$$\sum_{j:(i,j) \in A^k} x^k_{ij} = x^k_i \in \{0, 1\} \qquad \forall i \in N \tag{2.45c}$$

$$x^k_{ij}(t^k_i + d_i + s_i - t^k_j) \leq 0 \qquad \forall (i, j) \in A^k \setminus (d^k, o^k) \tag{2.45d}$$

$$0 \leq t^k_i \leq b_i x^k_{d^k o^k} \qquad \forall i \in \{o^k, d^k\} \tag{2.45e}$$

$$a_i x^k_i \leq t^k_i \leq b_i x^k_i \qquad \forall i \in N \tag{2.45f}$$

$$x^k_{ij} \in \{0, 1\} \qquad \forall (i, j) \in A^k. \tag{2.45g}$$

Flow conservation for the path constraints appear in (2.45a) for the depot nodes o^k and d^k, and in (2.45b) for the flight nodes. Constraints (2.45c) indicate if flight i is operated or not by aircraft k. Time windows at the depot and flight nodes are given in (2.45e) and (2.45f), respectively. Non-linear constraints (2.45d) model the compatibility requirements between the flow and time variables: if flight j immediately follows flight i, both operated by aircraft k, then $x_{ij}^k = 1$ and departure time $t_j^k \geq t_i^k + d_i + s_i$ (the sum of arrival and minimum ground times of flight i). A solution to model (2.45) is in terms of

$$\mathbf{x}^k = [x_{ij}^k]_{(ij)\in A^k}, \ \mathbf{t}^k = [t_i^k]_{i\in N}, \text{ and } [x_i^k]_{i\in N}. \tag{2.46}$$

Note that the zero schedule for which the cost is zero reveals itself if $x_{d^k o^k}^k = 0$, where not only the flow variables are zero but also the time variables.

Although the objective function of the pricing problem is discussed later, we can already write the components of the column encoding for aircraft k as

$$c^k(\mathbf{x}^k,\mathbf{t}^k) = \sum_{(i,j)\in A^k} c_{ij}^k x_{ij}^k \qquad \text{(route cost)} \tag{2.47a}$$

$$a_i^k(\mathbf{x}^k,\mathbf{t}^k) = x_i^k = \sum_{j:(i,j)\in A^k} x_{ij}^k \qquad \forall i \in N \ \text{(selected flights)} \tag{2.47b}$$

$$b_i^k(\mathbf{x}^k,\mathbf{t}^k) = t_i^k \qquad \forall i \in N \ \text{(flight departure times)} \tag{2.47c}$$

$$a_o^k(\mathbf{x}^k,\mathbf{t}^k) = x_{d^k o^k}^k. \qquad \text{(aircraft availability)} \tag{2.47d}$$

Integer master problem

The *IMP* can be written as a set partitioning type formulation with additional same-departure-time constraints. Let $(\mathbf{x}_p^k, \mathbf{t}_p^k)$, $p \in P^k$, be a specific path-solution to the k-th pricing problem, where P^k is the set of feasible routes, and let λ_p^k denote the associated variable.

Recall that M is the set of group identifiers for flights that must respect the same departure time and $m_i \in M$ denotes the group identifier of flight $i \in N$, that is, all pairs of flights $i, j \in N$ with $m_i = m_j$ must have the same departure time. Figure 2.17 shows how we handle the synchronization constraints in the *IMP* with the help of translated time variables t_{m_i}, for $m_i \in M$ and $i \in N$.

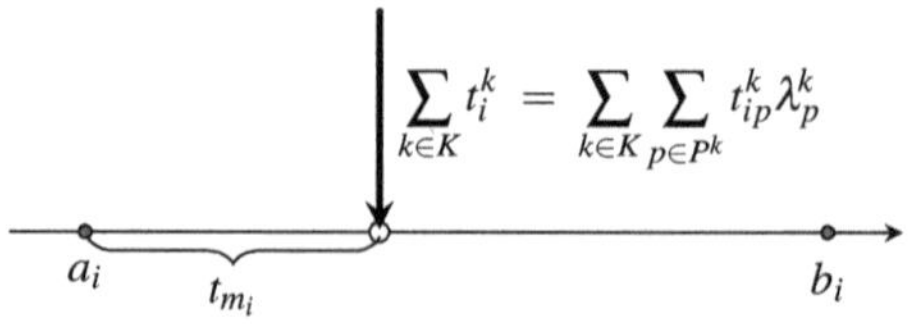

Fig. 2.17: Translated time variable t_{m_i}, $m_i \in M$ and $i \in N$.

For a flight $i \in N$, the variable $t_{m_i} \geq 0$ represents the slack between the beginning of the flight time window a_i and the optimized departure time, that is,

$$t_{m_i} = \sum_{k \in K} t_i^k - a_i, \quad \text{where } t_i^k = \sum_{p \in P^k} t_{ip}^k \lambda_p^k \ (= \sum_{p \in P^k} b_{ip}^k \lambda_p^k) \tag{2.48}$$

is a combination of the various path-solutions. Then the *IMP* reads as

$$
\begin{aligned}
z_{IMP}^\star = \min \; & \sum_{k \in K} \sum_{p \in P^k} c_p^k \lambda_p^k && && (2.49a)\\
\text{s.t.} \; & \sum_{k \in K} \sum_{p \in P^k} a_{ip}^k \lambda_p^k && = 1 \quad [\pi_i] \quad \forall i \in N && (2.49b)\\
& \sum_{k \in K} \sum_{p \in P^k} b_{ip}^k \lambda_p^k - t_{m_i} && = a_i \quad [\beta_i] \quad \forall i \in N && (2.49c)\\
& \sum_{p \in P^k} a_{op}^k \lambda_p^k && = 1 \quad [\pi_o^k] \quad \forall k \in K && (2.49d)\\
& \lambda_p^k && \geq 0 \quad \forall k \in K, p \in P^k && (2.49e)\\
& \sum_{p \in P^k} x_{ijp}^k \lambda_p^k && = x_{ij}^k \in \{0,1\} \quad \forall k \in K, (i,j) \in A^k. && (2.49f)
\end{aligned}
$$

The objective function (2.49a), the partitioning constraints (2.49b), and (2.49d) describe a minimum cost aircraft assignment problem. The synchronization constraints between different aircraft performing flights subject to same-departure-time constraints are given in (2.49c). We here assume that each flight in N must be synchronized with at least another one. Trivial bounds can be imposed on t_{m_i}, that is, $0 \leq t_{m_i} \leq b_i - a_i$ for any given flight $i \in N$ but since they are redundant with the constraints (2.45f) in the pricing problems, they do not appear in the above formulation.

Integer pricing problems

The linear relaxation *MP* is solved by column generation. Using the dual values $\boldsymbol{\pi} = [\pi_i]_{i \in N}$ for the partitioning constraints, $\boldsymbol{\beta} = [\beta_i]_{i \in N}$ for the synchronization constraints, and π_o^k, $k \in K$, for the aircraft availability constraints, the ISP^k reads

$$
\begin{aligned}
\bar{c}^k(\boldsymbol{\pi}, \boldsymbol{\beta}, \pi_o^k) &= \min_{(2.45),\,(2.47)} c^k(\mathbf{x}^k, \mathbf{t}^k) - \sum_{i \in N} \pi_i a_i^k(\mathbf{x}^k, \mathbf{t}^k) - \sum_{i \in N} \beta_i b_i^k(\mathbf{x}^k, \mathbf{t}^k) - \pi_o^k a_o^k(\mathbf{x}^k, \mathbf{t}^k)\\
&= \min_{(2.45)} \sum_{(i,j) \in A^k} c_{ij}^k x_{ij}^k - \sum_{i \in N} \pi_i x_i^k - \sum_{i \in N} \beta_i t_i^k - \pi_o^k x_{d^k o^k}^k,
\end{aligned} \tag{2.50}
$$

where we substitute the encoding functions (2.47) in the second line. A specialized dynamic programming algorithm is used to solve the ISP^k as a shortest path problem with time windows and linear node costs, where the time variables are pushed forward or backward depending on their dis-synchronized values (Ioachim et al., 1998).

Practical observations

- The time variables t_m, $m \in M$, are not generated in (2.49) but present in the *RMP* at the start. These are *static* variables, see Section 2.3, p. 66.
- To reduce numerical instability with the simplex software used to re-optimize the *RMP*, we redefine the $b_i^k(\mathbf{x}^k, \mathbf{t}^k)$ function (2.47c) to

$$b_i^k(\mathbf{x}^k, \mathbf{t}^k) = t_i^k - a_i x_i^k,\ \forall i \in N \quad \text{(relative departure times of flights).} \tag{2.51}$$

This new function transforms the departure time of a flight into the slack obtained with respect to the beginning of the time window. The column-coefficients

$$b_{ip}^k = t_{ip}^k - a_i x_{ip}^k, \quad i \in N, \tag{2.52}$$

are reduced to values less than or equal to 1 440 minutes (a day) instead of values of up to 10 080 minutes (a week). With this new interpretation of b_{ip}^k, the constraints (2.49c) are replaced by

$$\sum_{k \in K} \sum_{p \in P^k} b_{ip}^k \lambda_p^k - t_{m_i} = 0 \quad [\beta_i] \ \forall i \in N. \tag{2.53}$$

Ioachim et al. (1999) report an average decrease of 30 % on the computation time of the *MP* with this modified formulation.

- To further reduce the computation time, they introduce a tolerance on the right-hand side of the synchronization constraints (2.49c) or (2.53). Because the time data in their 12 test problems is given in multiples of 5 minutes, they use a tolerance of ± 2 minutes implemented with additional upper bounded static variables, see Exercise 2.18. Combined with a few other strategies (multiple generated columns and specialized branching rules), the average total computation time is reduced by 97.6 % when compared to their first implementation.
- The solver used for the ISP^k (2.50) can only generate schedules for which at least one of the time variables is either at its lower or upper bound. This is due to the positive or negative penalties on the time variables in (2.50). It may happen that the same path is generated more than once but with different time values. Convex combinations are then used to achieve synchronization. With respect to the *IMP*, the fractional values of the λ-variables are no concern because integrality is requested on the binary x-variables computed in (2.49f). This is an announcement of the Dantzig-Wolfe decomposition principle applied to integer linear programming in Chapter 4.

2.5 Reference Notes

Introduction Over the last 65 years, there have been many important contributors to column generation and branch-and-price. Ford and Fulkerson (1958) and Jewell (1958, 1966) have started the stream of research in the Western world, independently suggesting a similar multi-commodity flow algorithm considered as a precursor to column generation. This solution method is used by Paul Gilmore and Ralph Gomory for solving cutting stock problems (Gilmore and Gomory, 1961, 1963). At that time, only the linear relaxation of integer linear programs is solved to optimality. There were no known tools for obtaining optimal integer solutions yet, a fact mentioned by Vašek Chvátal in his book (Chvátal, 1983).

Kantorovich (1939) is often cited with respect to the cutting stock problem but was, until recently, mistakenly associated with a weak formulation. In this seminal paper, the author considers a pattern formulation but assumes that the number of such patterns is no concern. Much less known, the subsequent work of Kantorovich and Zalgaller (1951) tackles this issue. While it only exists in Russian and is still not readily available, it contains undeniable evidence of a column generation scheme devised to solve cutting problems of industrial nature by identifying patterns on-the-fly using "dynamic programming." Uchoa and Sadykov (2024) finally give due credit to this and correct the historical perspective.

Column generation is also used in the context of the Dantzig-Wolfe decomposition (Dantzig and Wolfe, 1960, 1961) when the extended reformulation in terms of the extreme points and extreme rays comprises too many variables, for example, see Dzielinski and Gomory (1965) to generate production schedules for the capacitated lot-sizing problem and Appelgren (1969, 1971) for a ship scheduling problem obtained from a Swedish shipowning company: *"Problems with about 40 ships and 50 cargoes are solved in about 2.5 minutes on an IBM 7090."*

In these early days, the book *"Optimization Theory for Large Systems"* by Lasdon (1970) is certainly one of the most important works on column generation, and more generally, on decomposition methods. Because of its implementation difficulties, slow convergence for obtaining optimal linear solutions, and lack of efficient methods for handling integrality of the decision variables, column generation is put aside after about ten years, researchers turning their heads towards *Lagrangian relaxation and subgradient algorithms* for the next decade, and still in use today, see for example Held and Karp (1970, 1971), Geoffrion (1974), Held et al. (1974), Shapiro (1979a), Fisher (1981), and Guignard (2003). We come back on some of these aspects in Chapter 6, Dual Point of View.

David Ryan from New Zealand is one of the major contributors in the 1980s. Together with Brian Foster, he notably proposes a branching rule for set partitioning formulations known as *Ryan-Foster branching* that can be applied together with the column generation algorithm (Ryan and Foster, 1981). His research includes urban and airline crew scheduling applications, e.g., Falkner and Ryan (1988), Ryan (1992), Butchers et al. (2001), and Weide et al. (2010), as well as solution strategies for highly degenerate set partitioning models, e.g., Ryan and Falkner (1988) and Ryan and Osborne (1988).

At the same time, on the other side of the earth, the *Montréal French Connection* takes another look at the column generation algorithm: François Soumis and Jacques Desrosiers, later joined by Pierre Hansen in the '90s, mainly in the field of vehicle routing and crew scheduling applications while developing various constrained shortest path algorithms as column generators, the first one being *"Plus court chemin avec contraintes d'horaires"* (Desrosiers et al., 1983) published in French, see Chapter 5. During the last forty years, more than two hundreds master and doctoral students together with skilled programmers have worked on various aspects of column generation at the GERAD and CIRRELT research centers, several of them becoming well-respected university researchers as well. Initiated in 1981 by François and Jacques, the `GENCOL` solver developed at GERAD is commercialized by GIRO for bus driver scheduling and AD OPT for optimizing air transportation activities.

In the 1990s, Cynthia Barnhart, Ellis Johnson, George Nemhauser, and Martin Savelsbergh propose the name *Branch-and-Price* to highlight the integration of column generation within a branch-and-bound tree (Barnhart et al., 1998). The Georgia Tech group is well known for its work on various airline problems, see for example Barnhart et al. (1995), Klabjan et al. (2001), Klabjan et al. (2002), Cohn and Barnhart (2003), and especially the survey paper Barnhart et al. (2003) with its large number of references on the usage of the column generation algorithm.

In an early paper, George Nemhauser also derives cutting planes defined on the master problem variables (Nemhauser and Park, 1991), which are later generalized, notably by Vanderbeck (2000, 2011) and Jepsen et al. (2008).

It is impossible to name all the past and current contributors to column generation and branch-and-price, although several are mentioned here and there in the book as we cover various subjects. However, let us mention some surveys that appeared over the years:

- Solomon and Desrosiers (1988), Desrosiers et al. (1995), Desaulniers et al. (1998a), and Feillet (2010) on constrained vehicle routing and crew scheduling problems;
- Vanderbeck (2005), Vanderbeck and Savelsbergh (2006), and Vanderbeck and Wolsey (2010) on reformulation and decomposition of integer programs.

Section 2.1 This section more or less follows the lines of Lübbecke and Desrosiers (2005). The index set $\mathcal{X}$ for the $\lambda_{\mathbf{x}}$-variables of the *MP* is a notation that easily transfers to the Dantzig-Wolfe decomposition principle for linear and integer linear programs (Chapters 3–4), where the reformulations are performed either by the convexification or discretization of the domain of the pricing problem. It is as well used with the Lagrangian relaxation (Chapter 6). Another introductory text is *"A primer in column generation"* (Desrosiers and Lübbecke, 2005), the first chapter of the book *"Column generation"* (Desaulniers et al., 2005). Observe in (2.3) that the functions $c(\mathbf{x})$ and $\mathbf{a}(\mathbf{x})$ for respectively the cost and column-coefficients for every variable $\lambda_{\mathbf{x}}$ are *always* computed in the pricing problem.

Initializing column generation is often done using artificial variables. Tobias Achterberg coins an alternative method as *Farkas pricing* in his PhD dissertation (Achterberg, 2007), under the supervision of Martin Grötschel at the Technical University of Berlin. He is the creator and first developer of `SCIP` (scipopt.org), actually recognized as one of the fastest non-commercial solvers for mixed-integer linear and non-linear programs. It is also a framework for branch-cut-and-price, regularly used at the *School on Column Generation*, e.g., Paris (2018, 2014) and Darmstadt (2010). The latest documentation update is Bestuzheva et al. (2021).

At the Université de Bordeaux, Sadykov and Vanderbeck (2021) develop `BaPCod` (swmath.org/software/9871), a generic branch-and-price code. It solves mixed-integer linear programs by the application of a Dantzig-Wolfe reformulation on an original or compact model, a technique that we see in the two forthcoming chapters. It is actually available under the name `Coluna` (atoptima.com), an open-source platform.

Good to Know Notes 2.13–2.14 explain that solving several subproblems in parallel might not be very efficient because each one is cannibalizing the others with the same dual information. However, there are several attempts, for example, see Basso and Ceselli (2022).

More to Know *Static and auxiliary variables* are widely used for soft constraints, a nice application in column generation being the Aircraft routing with schedule synchronization (Ioachim et al., 1999). They also come from a priori known or imposed dual constraints such as dual-optimal inequalities and within stabilized column generation for highly degenerate problems (Valério de Carvalho, 2005; Ben Amor and Valério de Carvalho, 2005; Ben Amor et al., 2006b; Ben Amor and Desrosiers, 2006; Oukil et al., 2007; Ben Amor et al., 2009; Gschwind and Irnich, 2016). These are also used in a crossover method to recover a primal basic optimal solution given an optimal primal-dual pair obtained via an interior point method (Ben Amor et al., 2006a): in essence, simply add very small dual boxes around optimal dual values, discard primal variables that must not be positive in any optimal solution, and re-solve the primal problem.

The shortest path problem with time windows, for which pseudo-polynomial time algorithms are available (Desrosiers et al., 1983; Desrochers, 1986; Desrochers and Soumis, 1988c,b), is an example of a *relaxed pricing* problem compared to the strongly $\mathcal{NP}$-hard elementary version (Feillet et al., 2004) as it allows for multiple visits at nodes, hence leading to a *relaxed master* problem (Desrosiers et al., 1984) where the undesired columns are eliminated during the branch-and-bound. The same applies for various elementary resource constrained shortest path problems, see Chapter 5.

The *VRPTW* computer code used for the results of Figures 2.4–2.5 and several subsequent ones has been designed by Stefan Irnich at Johannes Gutenberg University Mainz. The implementation has matured with major contributions from members of his team: Claudia Bode, Jean Bertrand Gauthier, Timo Gschwind, Katrin Heßler, Timo Hintsch, Ann-Kathrin Rothenbächer, David Sayah, Konrad Steiner, and Christian Tilk.

Examples

2.1 One-dimensional cutting stock problem and 2.2 Cutting stock problem with rolls of different widths. Starting with Gilmore and Gomory (1961, 1963), this problem has been extensively studied in conjunction with the column generation algorithm, see for example, Farley (1990); Vanderbeck (1999, 2000); Belov and Scheithauer (2002, 2006); Valério de Carvalho (2002); Ben Amor and Valério de Carvalho (2005); Vanderbeck (2011); Gondzio et al. (2013); Delorme et al. (2016). For pseudo-polynomial formulations for the cutting stock problem, see Delorme and Iori (2020).

2.4 Single depot vehicle scheduling problem. This is a pedagogical alternative to the cutting stock problem as an application of the column generation algorithm. It can be formulated as a network flow problem in an acyclic graph (Desrosiers et al., 1995, §2.1, p. 38) or, as proposed here, a set partitioning model whose *path-flow* variables are generated via the simplest label-setting algorithm, see Ahuja et al. (1993, §4.4 Shortest path problems in acyclic networks).

2.6 Aircraft routing with schedule synchronization. The ± 2 minutes tolerance on the synchronization constraints (Ioachim et al., 1999) for the flight departure times comes from Nicolas Bélanger, a programmer of the GENCOL team at the GERAD research center. It results in a drastic decrease of the computation times. Then, there was a fundamental question Jacques asked: *Can we do the same in the dual space?* This was the starting point of the stabilized column generation, with a research paper a month later (du Merle et al., 1999).

Exercises

2.19 Degenerate or small step-length pivots. The evaluation of the probabilities using the *Hypergeometric* distribution is new. These may explain the computational results obtained with the simplex algorithm applied to network flow problems, where the ratio of degenerate pivots to total pivots varies between 70 % and 90 %, see Ahuja et al. (1993, Figure 18.7).

Fig. 2.18: David Ryan (right) and his friend Michael Saunders from Stanford University (Tongariro National Park, New Zealand, 2004-09-12).

Exercises

2.1 Ralph Gomory

Who is Ralph Gomory, one of the coauthors of the first large-scale application of column generation (Gilmore and Gomory, 1961)? List a few of his main contributions to the field of operations research.

2.2 Finiteness of $\mathcal{X}$ and $z^\star_{MP}$

We here consider two aspects of finiteness.
(a) Justify that $\mathcal{X}$ in (2.1) must comprise a *finite* number of indices.
(b) Mathematically express in terms of the λ-variables that $z^\star_{MP}$ is bounded.

2.3 Cutting stock problem with rolls of different width

Consider the cutting stock problem with rolls of different width in Example 2.2.
(a) Show that if $W^{k_1} < W^{k_2}$, we can add $\sum_{i=1}^m w_i x_i^{k_2} \geq W^{k_1} + 1$ to the ISP^{k_2}.
(b) How to impose n^k available rolls of width W^k, for all $k \in K$? Do not forget the impact on the ISP^k.

2.4 Cutting stock problem with rolls of different width: a single subproblem

Reconsider the cutting stock problem with rolls of different width in Example 2.2. We have formulated the *MP* using one subproblem per roll width. In this exercise, we consider to formulate the *MP* using a single subproblem which corresponds to solving simultaneously all knapsack problems. The *IMP* reads as

$$
\begin{aligned}
z^\star_{IMP} = \min \quad & \sum_{\mathbf{x}\in\mathcal{X}} c_{\mathbf{x}} \lambda_{\mathbf{x}} \\
\text{s.t.} \quad & \sum_{\mathbf{x}\in\mathcal{X}} a_{i\mathbf{x}} \lambda_{\mathbf{x}} = b_i \qquad [\pi_i] \quad \forall i \in \{1,\dots,m\} \\
& \lambda_{\mathbf{x}} \in \mathbb{Z}_+ \qquad\qquad \forall \mathbf{x} \in \mathcal{X},
\end{aligned}
\tag{2.54}
$$

where $\mathcal{X} = \{\mathbf{x} = [\mathbf{x}^k]_{k\in K} \mid \sum_{i=1}^m w_i x_i^k \leq W^k, \forall k \in K\}$; $c_{\mathbf{x}} = \sum_{k\in K} W^k - \sum_{i=1}^m w_i x_i^k$; and $\mathbf{a}_{i\mathbf{x}} = \sum_{k\in K} x_i^k$, $\forall i \in \{1,\dots,m\}$. Note that $x_i^k \in Z_+$, $\forall i \in \{1,\dots,m\}, k \in K$, and $\mathbf{x} \in \mathbb{Z}_+^{m|K|}$.
(a) Formulate the *ISP*.
(b) Compare the *IMP* formulations with respect to solution symmetry, column interpretation, difficulty of solving the *MP* and *ISP*, fractional λ-variable.

2.5 Farley's lower bound

We place ourselves under the assumptions of Corollary 2.2, that is, $|K| = 1$ and $c_{\mathbf{x}} = 1, \forall \mathbf{x} \in \mathcal{X}$.
(a) Show that given any $\boldsymbol{\pi} \geq \mathbf{0}$, the vector $\bar{\boldsymbol{\pi}} = \frac{\boldsymbol{\pi}}{1-\bar{c}(\boldsymbol{\pi})}$ is dual feasible.
(b) Show that $\frac{\boldsymbol{\pi}^\intercal \mathbf{b}}{1-\bar{c}(\boldsymbol{\pi})} \leq z^\star_{MP}$ is a special case of Farley's bound $l(\boldsymbol{\pi}) \cdot \boldsymbol{\pi}^\intercal \mathbf{b} \leq z^\star_{MP}$.
(c) Show that given optimal dual values $\boldsymbol{\pi} \geq \mathbf{0}$ to the current *RMP* and $\bar{c}(\boldsymbol{\pi}) < 0$, the lower bound $\frac{\boldsymbol{\pi}^\intercal \mathbf{b}}{1-\bar{c}(\boldsymbol{\pi})} \leq z^\star_{MP}$ is strictly better than $z_{RMP} + \kappa\bar{c}(\boldsymbol{\pi}) \leq z^\star_{MP}$.

2.6 Sufficient optimality condition
Show that the termination condition $\bar{c}(\boldsymbol{\pi}) \geq 0$ is a sufficient but not necessary optimality condition. (*Hint:* A counter example is sufficient.)

2.7 Single depot vehicle scheduling problem: adjusted arc costs
For the *SDVSP* (Example 2.4), grouping the coefficients of each x_{ij}-variable in the objective function of the *SP* (2.41) leads to an adjusted arc cost $\tilde{c}_{ij}$, $(i,j) \in A$, computed as

$$\sum_{(i,j)\in A} \tilde{c}_{ij}x_{ij} = \sum_{(i,j)\in A} c_{ij}x_{ij} - \sum_{i\in N} \pi_i\Big(\sum_{j:(i,j)\in A} x_{ij}\Big) - \pi_o\Big(\sum_{j:(o,j)\in A} x_{oj}\Big)$$

Give the mathematical expressions for the adjusted arc cost in

- $\tilde{c}_{ij}$, $\forall i \in N, (i,j) \in A$;
- $\tilde{c}_{oj}$, $\forall (o,j) \in A$.

2.8 Single depot vehicle scheduling problem: arc-flow formulation
Let us consider the following network flow formulation for the *SDVSP* (Example 2.4), with non-negative integer variables x_{od} and x_{do}, binary variables x_{ij} for arcs $(i,j) \in I \cup (\{o\} \times N) \cup (N \times \{d\})$, and variables $x_i = 1$, $\forall i \in N$. Let $A = I \cup (\{o\} \times N) \cup (N \times \{d\}) \cup \{(o,d)\}$ and $A_{do} = A \cup \{(d,o)\}$.

$$\begin{aligned}
z^\star_{LP} = \min \quad & \sum_{(i,j)\in A} c_{ij}x_{ij} & & \\
\text{s.t.} \quad & \sum_{j:(i,j)\in A} x_{ij} = x_i & [\alpha_i] \quad & \forall i \in N \\
& -\sum_{j:(j,i)\in A} x_{ji} = -x_i & [\beta_i] \quad & \forall i \in N \\
& \sum_{j:(o,j)\in A} x_{oj} = x_{do} & [\beta_o] \quad & \\
& -\sum_{i:(i,d)\in A} x_{id} = -x_{do} & [\beta_d] \quad & \\
& x_i = 1 & & \forall i \in N \\
& 0 \leq x_{do} \leq v & & \\
& x_{ij} \geq 0 & & \forall (i,j) \in A.
\end{aligned} \tag{2.55}$$

The above formulation comprises $2n$ flow conservation equations for the trips and two more for the origin and destination depots. There is also a dual variable identified within bracket for every flow conservation equation.

(a) Draw the network, position the $2n+2$ dual variables on the appropriate nodes. *Hint*: There are $2n+2$ nodes: β_o obviously goes on node o, β_d on d, α_i on ...
(b) Give the mathematical expressions of the reduced cost for all the arcs: $\bar{c}_{od}$; $\bar{c}_{do}$; $\bar{c}_i$, $\forall i \in N$; $\bar{c}_{ij}$, $\forall (i,j) \in A$.

2.9 Multiple depot vehicle scheduling problem

The *MDVSP* often occurs in urban transit bus scheduling, as follows. Consider a bus timetable that specifies a set N of n bus trips to operate, each one defined by a pair of start time and location, and a pair of end time and location. Consider also a homogeneous fleet of vehicles that are housed in a set K of depots, where there are v^k vehicles available in depot $k \in K$. The problem consists of determining a set of bus schedules such that each bus trip $i \in N$ belongs to exactly one schedule, the vehicle availability is met, and the total operating cost is minimized. A schedule must start and end at the same depot (*depot restriction* constraints) and can be seen as a sequence of trips and deadheads. The cost structure includes travel costs and a fixed cost per bus used.

(a) Formulate the *MDVSP* using the following notation: the binary variable $\lambda_{\mathbf{x}^k}$ takes value 1 if and only if the schedule indexed by $\mathbf{x}^k \in \mathcal{X}^k$ is selected. Such a schedule has a cost $c_{\mathbf{x}^k}$ and it is represented by $\mathbf{a}_{\mathbf{x}^k} = [a_{i\mathbf{x}^k}]_{i \in N}$, where the binary coefficient $a_{i\mathbf{x}^k}$ takes value 1 if trip i is operated in schedule $\mathbf{x}^k$, and 0 otherwise.

(b) Associate a commodity to each depot by adapting the network $G_{do} = (N, A_{do})$ as seen on Figure 2.13 with an index k, that is, let the network $G^k_{do} = (N^k, A^k_{do})$ for depot k, where $N^k = N \cup \{o^k, d^k\}$, $A^k = I \cup (\{o^k\} \times N) \cup (N \times \{d^k\}) \cup \{(o^k, d^k)\}$, $A^k_{do} = A^k \cup \{(d^k, o^k)\}$, and arc (d^k, o^k) be utilized to count the number of buses used and also to model the zero-schedule. Formulate the *MDVSP* as a multi-commodity network flow model, with non-negative integer arc-variables based on Figure 2.19.

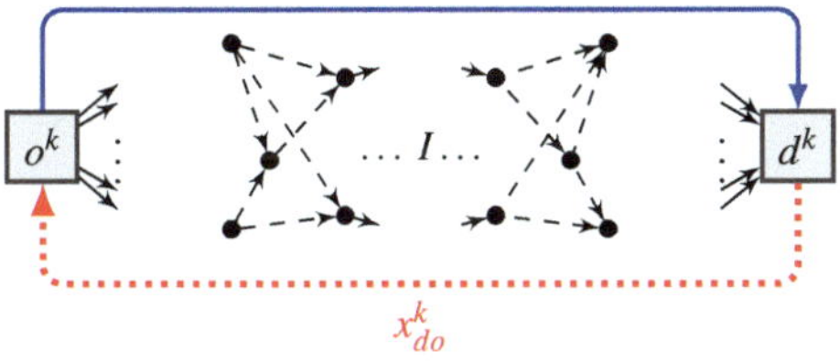

Fig. 2.19: Network $G^k_{do} = (N^k, A^k_{do})$ with arc (d^k, o^k) to model the zero schedule.

(c) Show that if we drop the *depot restriction* constraints, we can model this relaxed problem as a Single depot vehicle scheduling problem. Draw an appropriate *SDVSP* network for a 2-depot *MDVSP*. Using such a network, show how to find the minimum number of buses $v^\star$ for the *MDVSP*.

2.10 Maximum flow problem: arc-flow and path-flow formulations

Figure 2.20 depicts an instance of the maximum flow problem (Ford and Fulkerson, 1956) between the origin node o and destination node d on the network $G_{do} = (V, A_{do})$, where $V = N \cup \{o, d\}$ is the set of nodes, $A_{do} = A \cup \{(d, o)\}$ is the set of arcs, and each arc $(i, j) \in A$ has an upper bound u_{ij}.

(a) Give an arc-flow formulation.

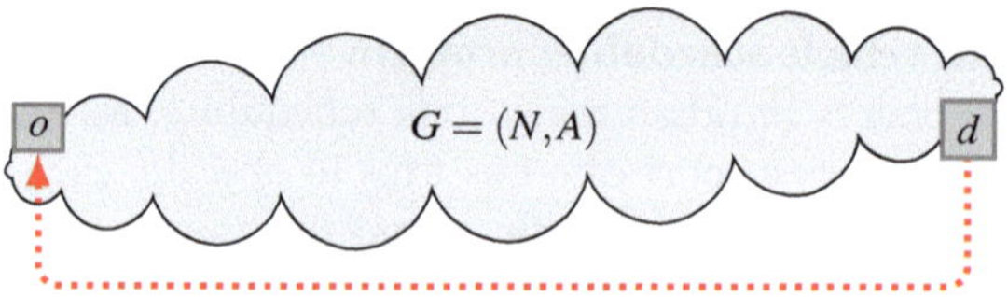

Fig. 2.20: Maximum flow on network $G_{do} = (V, A_{do})$.

(b) Propose a path-flow formulation solvable by column generation, that is, provide the *MP* and *SP* formulations, both being *maximization* programs. In the latter, provide the encoding functions $c_{\mathbf{x}}$ and $\mathbf{a}_{\mathbf{x}}$, where $\mathbf{x} = [x_{ij}]_{(i,j)\in A}$. In the former, use $\lambda_{\mathbf{x}}$ for the master variables and $\boldsymbol{\pi} = [\pi_{ij}]_{(i,j)\in A}$ for the dual variables.
(c) What are the possible values for the dual variables of the *RMP*?
(d) Show that there is only one possible value for $\bar{c}(\boldsymbol{\pi}) > 0$, computed here as the *maximum* reduced cost?
(e) Given $\boldsymbol{\pi}^\star$ at the last column generation iteration, give an x-solution for the *SP*.

2.11 Maximizing the smallest reduced cost within column generation
One desires to solve the following *MP* (in standard form) by the column generation algorithm but using a different pivoting rule:

$$\begin{aligned} z^\star_{MP} = \min \quad & \sum_{\mathbf{x}\in\mathcal{X}} c_{\mathbf{x}}\lambda_{\mathbf{x}} \\ \text{s.t.} \quad & \sum_{\mathbf{x}\in\mathcal{X}} \mathbf{a}_{\mathbf{x}}\lambda_{\mathbf{x}} = \mathbf{b} \quad [\boldsymbol{\pi}] \\ & \lambda_{\mathbf{x}} \geq 0 \qquad \forall \mathbf{x}\in\mathcal{X}. \end{aligned} \tag{2.56}$$

(a) Give a linear programming model that *optimizes* the dual vector $\boldsymbol{\pi} \in \mathbb{R}^m$ such that the smallest reduced cost is as large as possible. In other words, formulate the following *SP* as a linear program:

$$\bar{c}^\star = \max_{\boldsymbol{\pi}\in\mathbb{R}^m} \min_{\mathbf{x}\in\mathcal{X}} \bar{c}_{\mathbf{x}} = \max_{\boldsymbol{\pi}\in\mathbb{R}^m} \min_{\mathbf{x}\in\mathcal{X}} c_{\mathbf{x}} - \boldsymbol{\pi}^\mathsf{T}\mathbf{a}_{\mathbf{x}}. \tag{2.57}$$

(b) Write the dual formulation of the proposed linear program.
(c) Give an interpretation in the context of a network flow problem.

2.12 Tailing-off effect
Column generation is known to suffer from the tailing-off effect, that is, there is smaller and smaller incremental progress per iteration the closer we get to an optimal solution. Describe ways to lessen this effect in the context of solving an integer master problem.

2.13 Transforming a set partitioning problem into a set covering problem
Let the following set partitioning problem (*SPP*)

$$
\begin{aligned}
z^{\star}_{SPP} = \min \quad & \sum_{j=1}^{n} c_j x_j \\
\text{s.t.} \quad & \sum_{j=1}^{n} a_{ij} x_j = 1 \qquad \forall i \in \{1,\ldots,m\} \\
& x_j \in \{0,1\} \quad \forall j \in \{1,\ldots,n\},
\end{aligned}
\tag{2.58}
$$

with positive cost values, be feasible and define $n_j = \sum_{i=1}^{m} a_{ij}$ as the number of elements in the column indexed by j. Garfinkel and Nemhauser (1972, Theorem 1, p. 300) suggest selecting big-$M > \sum_{j=1}^{n} c_j$ and prove that the set covering problem (*SCP*) defined with costs

$$
C_j = c_j + M n_j, \qquad \forall j \in \{1,\ldots,n\}, \tag{2.59}
$$

has the same set of optimal solutions as the *SPP*. The proof goes as follows. Let $[x^{\star}_j]_{j=1,\ldots,n}$ be an optimal solution to the *SPP*, hence feasible for the *SCP*. Then

$$
\sum_{j=1}^{n} C_j x^{\star}_j = \sum_{j=1}^{n} c_j x^{\star}_j + M \sum_{j=1}^{n} n_j x^{\star}_j = \sum_{j=1}^{n} c_j x^{\star}_j + Mm \qquad < M(m+1). \tag{2.60}
$$

That is, any optimal *SPP* solution incurs a fixed cost Mm in the *SCP* formulation. On the other hand, any feasible solution $[x^{\bullet}_j]_{j=1,\ldots,n}$ to the *SCP* that over-covers the right-side cannot be optimal because it incurs a larger cost:

$$
\sum_{j=1}^{n} C_j x^{\bullet}_j = \sum_{j=1}^{n} c_j x^{\bullet}_j + M \sum_{j=1}^{n} n_j x^{\bullet}_j \geq \sum_{j=1}^{n} c_j x^{\bullet}_j + M(m+1) \qquad > M(m+1). \tag{2.61}
$$

► Assume that (2.58) is rather an integer master problem in column generation. Since $\sum_{\mathbf{x} \in \mathcal{X}} c_{\mathbf{x}}$ is then unknown, the above suggestion for big-M cannot be used. Derive a suitable one.

2.14 Set covering vs. set partitioning
In the formulation (2.35) of the edge coloring problem, we use a set covering problem (*SCP*) rather than a set partitioning one (*SPP*):

$$
\begin{aligned}
z^{\star}_{SCP} = \min \; & \sum_{\mathbf{x} \in \mathcal{X}} c_{\mathbf{x}} \lambda_{\mathbf{x}} \\
\text{s.t.} \; & \sum_{\mathbf{x} \in \mathcal{X}} \mathbf{a}_{\mathbf{x}} \lambda_{\mathbf{x}} \geq \mathbf{1} \; [\boldsymbol{\pi} \in \mathbb{R}^m_+] \\
& \lambda_{\mathbf{x}} \in \{0,1\}, \forall \mathbf{x} \in \mathcal{X}
\end{aligned}
\qquad \Bigg| \qquad
\begin{aligned}
z^{\star}_{SPP} = \min \; & \sum_{\mathbf{x} \in \mathcal{X}} c_{\mathbf{x}} \lambda_{\mathbf{x}} \\
\text{s.t.} \; & \sum_{\mathbf{x} \in \mathcal{X}} \mathbf{a}_{\mathbf{x}} \lambda_{\mathbf{x}} = \mathbf{1} \; [\boldsymbol{\pi} \in \mathbb{R}^m] \\
& \lambda_{\mathbf{x}} \in \{0,1\}, \forall \mathbf{x} \in \mathcal{X}.
\end{aligned}
\tag{2.62}
$$

Indeed, greater-than-or-equal-to-one inequalities are often used rather than equality constraints. When and why?

(a) In the *SCP* and *SPP* models, let $R_{\mathbf{x}}$, for $\mathbf{x} \in \mathcal{X}$, be the subset of rows covered by the binary column $\mathbf{a}_{\mathbf{x}}$ and let $c_{\mathbf{x}} = c(R_{\mathbf{x}})$ be the smallest cost for covering these constraints. Show that the *SPP* can be equivalently replaced by the *SCP* if for all subsets of rows $R'_{\mathbf{x}} \subseteq R_{\mathbf{x}}$, we have $c(R'_{\mathbf{x}}) \leq c(R_{\mathbf{x}})$.
(b) What are the primal and dual advantages of such a replacement?

2.15 Cutting stock problem: integer optimum bound
With respect to the One-dimensional cutting stock problem formulated with greater-than-or-equal constraints as in (2.29), the optimum $z^\star_{IMP}$ takes an integer value for the number of rolls cut. Given an optimal solution $\boldsymbol{\lambda}^\star_{MP}$ of cost $z^\star_{MP}$ to the linear relaxation, derive a valid upper bound on $z^\star_{IMP}$.

2.16 Cutting stock problem with equality constraints
Assume in the *CSP* of Example 2.1 that we want to satisfy the item demands exactly, that is, the *IMP* is formulated with equality constraints as

$$\begin{aligned} z^\star_{IMP} = \min \quad & \sum_{\mathbf{x} \in \mathcal{X}} \lambda_{\mathbf{x}} \\ \text{s.t.} \quad & \sum_{\mathbf{x} \in \mathcal{X}} a_{i\mathbf{x}} \lambda_{\mathbf{x}} = b_i \quad [\pi_i \in \mathbb{R}] \quad \forall i \in \{1, \ldots, m\} \\ & \lambda_{\mathbf{x}} \in \mathbb{Z}_+ \quad \forall \mathbf{x} \in \mathcal{X}, \end{aligned} \tag{2.63}$$

where $\mathcal{X} = \left\{ \mathbf{x} \in \mathbb{Z}^m_+ \;\middle|\; \sum_{i=1}^m w_i x_i \leq W \right\}$ is the set of solutions to the *ISP* as given by the integer knapsack problem in (2.30). Given a roll of width 11 and a unique item of width 2 with a demand of 4, answer the following questions:

(a) List all solutions of the *ISP*, i.e., all patterns $\mathbf{x} \in \mathcal{X}$.
(b) Express the *IMP* explicitly.
(c) Provide all optimal solutions to the *IMP*.
(d) List all the extreme points of $\text{conv}(\mathcal{X})$.
(e) Can we find an optimal integer solution to the *IMP* with these extreme points?
(f) Solve the *MP* by column generation starting with an artificial variable of cost big-M.
(g) What happens if we rather use

$$\mathcal{X} = \left\{ \mathbf{x} \in \mathbb{Z}^m_+ \;\middle|\; \sum_{i=1}^m w_i x_i \leq W,\, x_i \leq b_i, \forall i \in \{1, \ldots, m\} \right\}$$

as the set of solutions to the *ISP* which thus excludes patterns with excess production?

2.17 Edge coloring problem with equality constraints
Assume in the *ECP* of Example 2.3 that the *IMP* is formulated with equality constraints, that is, each edge is assigned exactly one color:

$$\begin{aligned} z_{IMP}^{\star} = \min \; & \sum_{\mathbf{x}\in\mathcal{X}} \lambda_{\mathbf{x}} \\ \text{s.t.} \; & \sum_{\mathbf{x}\in\mathcal{X}} x_e \lambda_{\mathbf{x}} = 1 \qquad [\pi_e \in \mathbb{R}] \quad \forall e \in E \\ & \lambda_{\mathbf{x}} \in \{0,1\} \qquad\qquad \forall \mathbf{x} \in \mathcal{X}, \end{aligned} \tag{2.64}$$

where $\mathcal{X} = \{\mathbf{x} \in \{0,1\}^{|E|} \mid \sum_{e\in\delta(\{i\})} x_e \leq 1, \; \forall i \in N\}$ is the set of solutions to the *ISP* as given by (2.37). Use the graph with five edges in Figure 2.21 to answer the following questions:

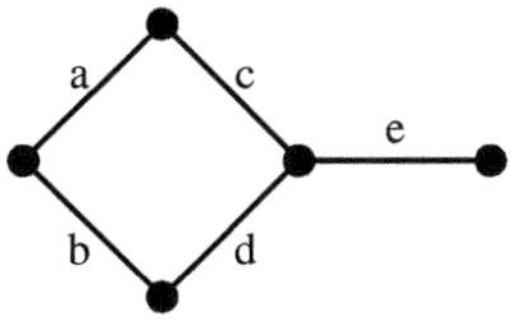

Fig. 2.21: Edge coloring on a five-edge graph.

(a) List all solutions of the *ISP*, i.e., all matchings $\mathbf{x} \in \mathcal{X}$.
(b) Express the *IMP* explicitly.
(c) Provide all optimal solutions to the *IMP*.
(d) List all the extreme points of $\text{conv}(\mathcal{X})$.
(e) Can we find an optimal integer solution to the *IMP* with these extreme points?
(f) Solve the *MP* by column generation starting with five artificial variables of cost big-M.

2.18 Tolerance on the schedule synchronization constraints
In Example 2.6, formulate the *IMP* to allow for a tolerance of ± 2 minutes on the synchronized time values.

2.19 Degenerate or small step-length pivots
The *Hypergeometric* distribution $H(n,K,N)$ is a discrete probability law describing the following model: We simultaneously draw n balls from an urn containing K winning balls amongst N, where $n < N$ and $K < N$. The random variable $X \sim H(n,K,N)$ counts the number of winning balls drawn. For $\max(0, n+K-N) \leq k \leq \min(n,K)$, the probability of picking exactly k winning balls is

$$\Pr(X = k) = \frac{\binom{K}{k}\binom{N-K}{n-k}}{\binom{N}{n}}, \tag{2.65}$$

with an expected value $E[X] = n \times \frac{K}{N}$.

Assume that the *RMP* comprises $N = 200$ basic variables, where $\mathbf{B}$ denotes the basis, out of which 40 are *degenerate*. If the entering variable $\lambda_{\mathbf{x}}$ with column $\mathbf{a}_{\mathbf{x}}$ potentially modifies $n = 20$ basic variables, i.e., $\bar{\mathbf{a}}_{\mathbf{x}} = \mathbf{B}^{-1}\mathbf{a}_{\mathbf{x}}$ contains 20 non-zero elements, let us calculate the probability of having a degenerate pivot using the distribution $H(20,20,200)$, where $K = 40 \times 50\,\%$, that is, we assume that a potentially modified degenerate variable increases half of the time.

Let the random variable $X \sim H(20,20,200)$ denote the number of degenerate variables that are asked to decrease with the selection of $\lambda_{\mathbf{x}}$ with $E[X] = 2$. A degenerate pivot occurs if $X \geq 1$ which has a probability of

$$\Pr(X \geq 1) = 1 - \Pr(X = 0) = 89.1\,\% \tag{2.66}$$

(a) Compute the probability of a degenerate pivot for $n = 5, 10, \ldots, 50$ non-zero elements in $\bar{\mathbf{a}}_{\mathbf{x}}$ using the distribution $X \sim H(n, 20, 200)$.

(b) Let $X \sim H(n, 20, 500)$, where $K = 20$ basic variables lead to a degenerate pivot or a relatively small step-length while $N = 500$ is the dimension of the basis. For $n = 5, 10, \ldots, 50$ non-zero elements in $\bar{\mathbf{a}}_{\mathbf{x}}$, compute $\Pr(X \geq 1)$.

2.20 Pseudo-code practice
Taking inspiration from the pseudo-code in Algorithm 2.2, write one that uses Farkas pricing as an initialization method.

3

Dantzig-Wolfe Decomposition for Linear Programming

> Tell me and I forget,
> teach me and I may remember,
> involve me and I learn.
>
> *Xunzi*
> Xun Kuang

Abstract This chapter describes the Dantzig-Wolfe decomposition principle applied to linear programming. This decomposition principle is no more and no less than a mathematical reformulation which uses the Minkowski-Weyl theorem to express some constraints under an alternative geometric interpretation. As the resulting reformulation typically contains a huge number of variables, we carry the essence of the column generation algorithm by deriving the master and pricing problems. We then oppose the given linear program to its reformulation and finally explore different examples.

Contents

Introduction 105
3.1 Dantzig-Wolfe Reformulation 105
Minkowski-Weyl theorem 106
Master problem 108
Polytope and polyhedral cone 110
3.2 Column Generation Approach 111
Pricing problem 112
Post-processing a solution of the master problem 114
Primal solution 114
Dual solution 114

J. Desrosiers et al., *Branch-and-Price*,
https://doi.org/10.1007/978-3-031-96917-1_3

Basic solution 115
On grouping the constraints 115
Block-diagonal structure 116
Lower and upper bounds 119
3.3 Good to Know 120
A set of extreme points and extreme rays 120
Extreme cases for the constraints in the reformulated set 120
All 120
None 121
Static variables 122
3.4 More to Know 123
Restricted compact formulation 124
Block-diagonal structure 127
Positive edge rule and Improved Primal Simplex algorithm 128
Benders decomposition of a linear program 130
Relaxed master, subproblem, and initialization 132
Decomposition theorem of a network flow solution 133
3.5 Examples 136
2D illustration 137
Compact formulation 137
Grouping of constraints 137
Extended formulation 138
Optimal primal-dual solutions for the compact formulation . 139
Column generation approach 140
Time constrained shortest path problem (*TCSPP*) 142
Compact formulation 142
Grouping of constraints 143
Extended formulation 144
Optimal primal-dual solutions for the compact formulation . 144
Column generation approach 145
Single depot vehicle scheduling problem: two compact formulations 148
Formulation with *origin-to-destination* arc 149
Formulation with *destination-to-origin* arc 151
Solving a sequence of restricted compact formulations 155
3.6 Reference Notes 158
Exercises 161

Acronyms

LP	linear program (original or compact formulation)	105
MP	master problem (extended formulation)	109
RMP	restricted *MP*	112
SP	subproblem	113
SP^k	subproblem for block $k \in K$	118

RLP	restricted *LP* (restricted compact formulation)	124
CMCFP	capacitated minimum cost flow problem	133
TCSPP	time constrained shortest path problem	142
lb	lower bound on $z^\star_{MP}$	146
LB	best known lower bound on $z^\star_{MP}$	148
SDVSP	single depot vehicle scheduling problem	148

Introduction

Given a linear program, a Dantzig-Wolfe decomposition acts on it by grouping the constraints in two subsets. The Minkowski-Weyl theorem is then used to reformulate one of these subsets and perform a substitution in the other and the objective function. Using this mathematical transformation, the linear program gives way to an equivalent formulation which is typically expressed with a huge number of variables. If solving this new formulation looks like a task fitted for the column generation algorithm, that is because it is one. The reformulation process even identifies the constraints of the pricing problem.

This chapter serves two purposes. On the one hand, we derive the master and pricing problems via a direct mathematical reformulation rather than by construction as we have seen in the examples of Chapter 2. On the other hand, since the linear program is already given, this is the beginning of an answer to why we should reformulate at all. Indeed, we plant the idea that there might be a structure to exploit. Chapter 4 picks up from there and hammers this point home.

3.1 Dantzig-Wolfe Reformulation

We would like to solve the following linear program *LP*

$$\begin{aligned} z^\star_{LP} = \min \quad & \mathbf{c}^\intercal \mathbf{x} \\ \text{s.t.} \quad & \mathbf{A}\mathbf{x} \geq \mathbf{b} \qquad [\boldsymbol{\sigma}_\mathbf{b}] \\ & \mathbf{D}\mathbf{x} \geq \mathbf{d} \qquad [\boldsymbol{\sigma}_\mathbf{d}] \\ & \mathbf{x} \in \mathbb{R}^n_+, \end{aligned} \tag{3.1}$$

with appropriate dimensions for all vectors and matrices. We assume that the *LP* is feasible and $z^\star_{LP}$ is finite. The dual variables are denoted by $\boldsymbol{\sigma}_\mathbf{b}, \boldsymbol{\sigma}_\mathbf{d} \geq \mathbf{0}$ to differentiate them from those of the forthcoming reformulation $\boldsymbol{\pi}_\mathbf{b}, \boldsymbol{\pi}_\mathbf{d} \geq \mathbf{0}$. In the context of this chapter we call this the *original* formulation with its *original* x-variables. The first step is to group the constraints, say $\mathbf{A}\mathbf{x} \geq \mathbf{b}$ and $\mathbf{D}\mathbf{x} \geq \mathbf{d}$, as listed in (3.1). Figure 3.1 depicts the domain of the *LP* (3.1) with respect to the two non-empty sets $\mathcal{A}$ and $\mathcal{D}$ that reflect this grouping as

$$\mathcal{A} = \{\mathbf{x} \in \mathbb{R}^n_+ \mid \mathbf{Ax} \geq \mathbf{b}\} \neq \emptyset \tag{3.2a}$$

$$\mathcal{D} = \{\mathbf{x} \in \mathbb{R}^n_+ \mid \mathbf{Dx} \geq \mathbf{d}\} \neq \emptyset. \tag{3.2b}$$

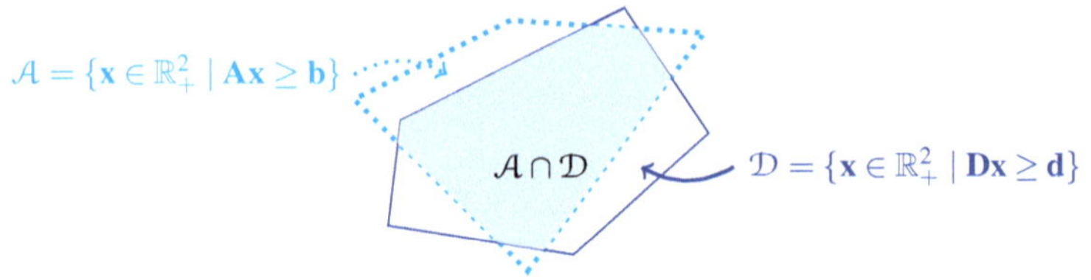

Fig. 3.1: Domain $\mathcal{A} \cap \mathcal{D} = \{\mathbf{x} \in \mathbb{R}^2_+ \mid \mathbf{Ax} \geq \mathbf{b}, \mathbf{Dx} \geq \mathbf{d}\}$ of the *LP*.

The next step is to express $\mathcal{D}$ in a different way. We do this by using a geometric result derived by Minkowski and Weyl. We then perform a substitution accordingly to obtain an equivalent optimization program. After solving the reformulation, we finally recover a solution for the original formulation by back substitution.

Minkowski-Weyl theorem

The theorem by Minkowski and Weyl states that there are two equivalent representations of a polyhedron. In linear programming, we are used to seeing a polyhedron defined by linear constraints, i.e., as the intersection of finitely many closed *half-spaces*. This is called the outer description or H-representation. Theorem 3.1 presents the alternative which Figure 3.2 informally illustrates on a 2-dimensional polytope: it can equivalently be represented using its *vertices*. This is called the inner description or V-representation. Although presented in Chapter 1, we repeat it here for convenience.

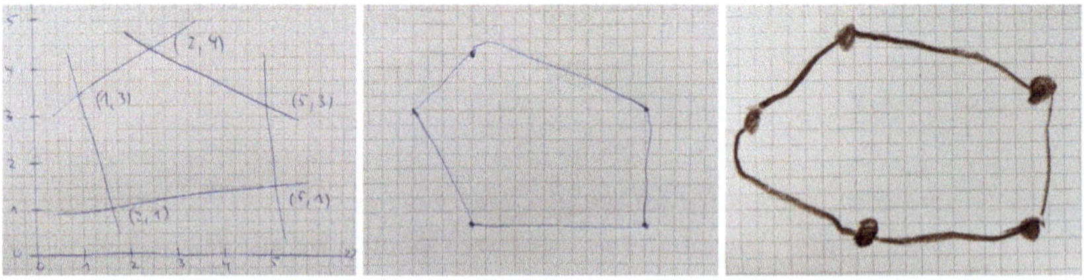

(a) 48-year-old Marco (b) 8-year-old Emil (c) 3-year-old Thea

Fig. 3.2: Drawing a polytope is child's play: Marco explains how to draw a polytope starting with $x_1 + 3x_2 \leq 14$ but it looks too complicated to Emil and Thea who settle on an easier idea.

Theorem 3.1. (Nemhauser and Wolsey, 1988, Theorem 4.8, p. 96) *Let* $\mathcal{P} = \{\mathbf{x} \in \mathbb{R}^n \mid \mathbf{Qx} \geq \mathbf{q}\}$ *be a polyhedron with at least one extreme point. An equivalent description of* $\mathcal{P}$ *using its extreme points* $\{\mathbf{x}_p\}_{p \in P}$ *and extreme rays* $\{\mathbf{x}_r\}_{r \in R}$ *is*

$$\mathcal{P} = \left\{ \mathbf{x} \in \mathbb{R}^n \;\middle|\; \begin{array}{llr} \sum_{p \in P} \mathbf{x}_p \lambda_p & + \sum_{r \in R} \mathbf{x}_r \lambda_r = \mathbf{x} & \\ \sum_{p \in P} \lambda_p = 1 & & \\ \lambda_p \geq 0 & & \forall p \in P \\ & \lambda_r \geq 0 & \forall r \in R \end{array} \right\}. \tag{3.3}$$

Proof. We refer to Nemhauser and Wolsey (1988) for step by step proofs and rather concentrate on three important ideas.

1. Both sets P and R are finite.
2. By assumption, the set P is non-empty but R could be. Nemhauser and Wolsey subsume this under a feasible region in the first quadrant with a matrix $\mathbf{Q}$ of rank n, that is, $\mathbf{x} \geq \mathbf{0}$ is included in $\mathbf{Qx} \geq \mathbf{q}$.
3. Both descriptions are equivalent because they describe the same set of solutions:

   $\Rightarrow$ Any $\mathbf{x} \in \mathcal{P}$ can be written as a convex combination of the extreme points $\{\mathbf{x}_p\}_{p \in P}$ plus a conic combination of the extreme rays $\{\mathbf{x}_r\}_{r \in R}$, that is, there exist scalars $\{\lambda_p\}_{p \in P}$ and $\{\lambda_r\}_{r \in R}$ such that (1.1) holds for all $\mathbf{x} \in \mathcal{P}$.

   $\Leftarrow$ Any $\boldsymbol{\lambda}$ that satisfies (1.1) corresponds to an $\mathbf{x} \in \mathbb{R}^n$ by definition but also necessarily to an $\mathbf{x} \in \mathcal{P}$. □

Observe that here and in what follows we adopt a simplified notation: for $p \in P$ and $r \in R$, we abbreviate $\lambda_{\mathbf{x}_p}$ and $\lambda_{\mathbf{x}_r}$ by λ_p and λ_r, respectively. The theorem is illustrated in Figure 3.3. In (a), we have an (unbounded) polyhedron and $\mathbf{x}$ is a convex combination of the three extreme points plus a conic combination of the upper ray. In (b), we have a polytope and $\mathbf{x}$ is a convex combination of three selected extreme points. In (c), we have a polyhedral cone and $\mathbf{x}$ is a convex combination of the single extreme point $\mathbf{0}$ plus a conic combination of the two extreme rays. Finally, in (d), we have in quadrants I, II, and IV, a polyhedron defined by a matrix that does not have full row rank so no $\mathbf{x}$ appears for substitution.

Note 3.1 (Multiple representations of $\mathbf{x}$*.)* From (3.3), any vector $\mathbf{x} \in \mathcal{P}$ uniquely reconstructs from a λ-combination but *the converse is not true*. This can be seen on Figure 3.3b where the two-dimensional vector $\mathbf{x}$ can be written as a convex combination of the extreme points $\{D, F, G\}$ as they form a triangle including $\mathbf{x}$, or as a convex combination of $\{C, D, F\}$ for the same reason. Moreover, an infinite number of convex combinations can be derived from the four extreme points $\{C, D, F, G\}$: we only have to take any convex combination of the first two ways! This is also true if extreme rays are involved.

Note 3.2 (Polytope and polyhedral cone.) In Theorem 3.1, there are two special cases for polyhedron $\mathcal{P}$ which are of interest to us: a polytope and a polyhedral

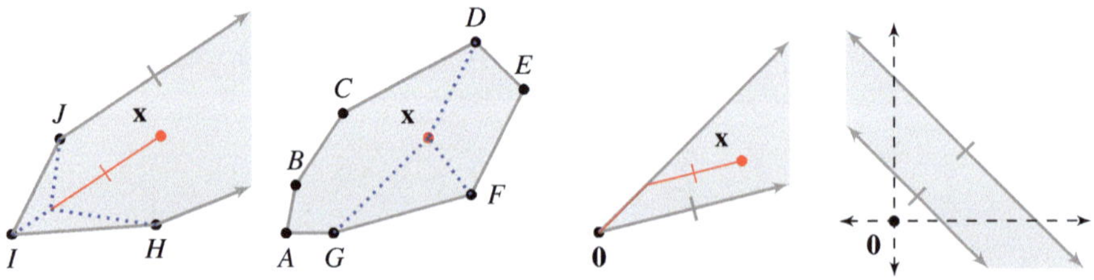

(a) Polyhedron (b) Polytope (c) Polyhedral cone (d) Not full row rank

Fig. 3.3: Illustration of the Minkowski-Weyl theorem.

cone. If $\mathcal{P}$ is a polytope (bounded polyhedron), i.e., $R = \emptyset$, it is straightforward to adapt our presentation by simply omitting that component from the substitution as in set (3.4)_left. If $\mathcal{P}$ is a polyhedral cone, i.e., $\mathbf{q} = \mathbf{0}$ and $\mathbf{x} \in \mathbb{R}^n_+$, the only extreme point is $\mathbf{0}$ such that the substitution can also be simplified as in set (3.4)_right.

$$
\begin{array}{r|l}
\begin{aligned}
\sum_{p \in P} \mathbf{x}_p \lambda_p &= \mathbf{x} \\
\sum_{p \in P} \lambda_p &= 1 \\
\lambda_p &\geq 0, \ \forall p \in P.
\end{aligned}
&
\begin{aligned}
\sum_{r \in R} \mathbf{x}_r \lambda_r &= \mathbf{x} \\
\\
\lambda_r &\geq 0, \ \forall r \in R.
\end{aligned}
\\
\text{Substitution for a polytope} & \text{Substitution for a polyhedral cone}
\end{array}
\tag{3.4}
$$

Note 3.3 (Nice rays!) Even though we speak of *the* extreme rays of a polyhedron $\mathcal{P}$, they are unique only up to scaling. In Note 3.9, for example, we normalize the rays for the sake of the argument. While in this chapter, the scaling may be a matter of taste or aesthetics, in the next chapter it becomes important, and in Chapter 7 it may even become a matter of performance. For this reason, we prefer to scale all extreme rays in this book to integer rays whose components are relatively prime. That is, we scale rays to their shortest possible integer representation. Note that scaling a ray to have integer components is possible because we deal only with rational data. We prefer American English over the British, but esthetics looks un-aesthetical.

Master problem

We remind the reader that, *a priori,* an extreme point and an extreme ray could have the same vector representation, see Figure 1.2. By using Theorem 3.1 on polyhedron $\mathcal{D}$ (3.2b), we consider its finite sets of extreme points $\{\mathbf{x}_p\}_{p \in P}$ and extreme rays $\{\mathbf{x}_r\}_{r \in R}$ which we gather into the set

$$
\mathcal{X} = \left\{ \begin{bmatrix} \mathbf{x}_p \\ 1 \end{bmatrix} \right\}_{p \in P} \cup \left\{ \begin{bmatrix} \mathbf{x}_r \\ 0 \end{bmatrix} \right\}_{r \in R} \tag{3.5}
$$

After substitution of $\mathbf{x}$ as per (3.3) into the objective function as well as the constraints of $\mathcal{A}$, i.e.,

$$\mathbf{c}^\mathsf{T}\mathbf{x} = \mathbf{c}^\mathsf{T}\Big(\sum_{p\in P}\mathbf{x}_p\lambda_p + \sum_{r\in R}\mathbf{x}_r\lambda_r\Big) \text{ and } \mathbf{A}\mathbf{x} = \mathbf{A}\Big(\sum_{p\in P}\mathbf{x}_p\lambda_p + \sum_{r\in R}\mathbf{x}_r\lambda_r\Big) \geq \mathbf{b}, \qquad (3.6)$$

we obtain an equivalent linear program expressed with λ-variables associated with the extreme points and extreme rays:

$$\begin{aligned}
z^\star_{LP} = \min \quad & \sum_{p\in P}(\mathbf{c}^\mathsf{T}\mathbf{x}_p)\lambda_p + \sum_{r\in R}(\mathbf{c}^\mathsf{T}\mathbf{x}_r)\lambda_r && && (3.7\text{a})\\
\text{s.t.} \quad & \sum_{p\in P}(\mathbf{A}\mathbf{x}_p)\lambda_p + \sum_{r\in R}(\mathbf{A}\mathbf{x}_r)\lambda_r \geq \mathbf{b} && [\boldsymbol{\pi}_\mathbf{b}] && (3.7\text{b})\\
& \sum_{p\in P}\lambda_p = 1 && [\pi_0] && (3.7\text{c})\\
& \lambda_p \geq 0 && \forall p \in P && (3.7\text{d})\\
& \lambda_r \geq 0 && \forall r \in R && (3.7\text{e})\\
& \sum_{p\in P}\mathbf{x}_p\lambda_p + \sum_{r\in R}\mathbf{x}_r\lambda_r = \mathbf{x} \in \mathbb{R}^n_+. && && (3.7\text{f})
\end{aligned}$$

The change of variables as per (3.3) obviously appears as a system of constraints, that is, (3.7c)–(3.7f). In particular, we refer to $\sum_{p\in P}\lambda_p = 1$ as the *convexity constraint*. Moreover, the last equation containing the original variables $\mathbf{x}$ reminds us of the relation to the original *LP*. Recall that the non-negativity constraints $\mathbf{x} \in \mathbb{R}^n_+$ in $\mathcal{A}$ are redundant and could be omitted altogether as they already appear in $\mathcal{D}$. However, with respect to the extreme points and extreme rays of the substitution, it should be clear from Figure 3.1 that *not each of them needs to be feasible for the MP*. Finally, note the different dual variables $\boldsymbol{\pi}_\mathbf{b}$ associated with the substituted constraints as well as π_0 for the additional convexity constraint.

The reformulation (3.7) lives in a higher dimensional space than the original one (3.1). For this reason, the latter is also referred to as the *compact* formulation, even though the number of its variables and constraints need not be polynomial in the size of the underlying problem instance. In contrast, the former is also called the *extended* formulation whereby the cardinalities of sets P and R are indeed typically exponential in the dimensions of the compact formulation. As the column generation algorithm appears to be quite suited for solving (3.7), let us call this extended formulation the *Dantzig-Wolfe master problem MP* whose optimal objective value $z^\star_{MP}$ is equal to $z^\star_{LP}$ by construction. As we transition to the column generation approach, let $c_\mathbf{x} = \mathbf{c}^\mathsf{T}\mathbf{x}$ and $\mathbf{a}_\mathbf{x} = \mathbf{A}\mathbf{x}$ specialize our generic cost and column-coefficient encoding functions $c(\mathbf{x})$ and $\mathbf{a}(\mathbf{x})$. We obtain the shorthand notation

$$\begin{bmatrix} c_p = c_{\mathbf{x}_p} = \mathbf{c}^\mathsf{T}\mathbf{x}_p \\ \mathbf{a}_p = \mathbf{a}_{\mathbf{x}_p} = \mathbf{A}\mathbf{x}_p \\ 1 \end{bmatrix}, \ \forall p \in P, \quad \text{and} \quad \begin{bmatrix} c_r = c_{\mathbf{x}_r} = \mathbf{c}^\mathsf{T}\mathbf{x}_r \\ \mathbf{a}_r = \mathbf{a}_{\mathbf{x}_r} = \mathbf{A}\mathbf{x}_r \\ 0 \end{bmatrix}, \ \forall r \in R, \qquad (3.8)$$

which allows us to rewrite the *MP* more concisely as

$$
\begin{aligned}
z^{\star}_{MP} = \min \quad & \sum_{p\in P} c_p\lambda_p + \sum_{r\in R} c_r\lambda_r && && (3.9a)\\
\text{s.t.} \quad & \sum_{p\in P} \mathbf{a}_p\lambda_p + \sum_{r\in R} \mathbf{a}_r\lambda_r \geq \mathbf{b} && [\boldsymbol{\pi}_\mathbf{b}] && (3.9b)\\
& \sum_{p\in P} \lambda_p = 1 && [\pi_0] && (3.9c)\\
& \lambda_p \geq 0 && \forall p \in P && (3.9d)\\
& \lambda_r \geq 0 && \forall r \in R && (3.9e)\\
& \sum_{p\in P} \mathbf{x}_p\lambda_p + \sum_{r\in R} \mathbf{x}_r\lambda_r = \mathbf{x}. && && (3.9f)
\end{aligned}
$$

Observe that the relation between the original x-variables and the λ-variables of the reformulation is otherwise not involved in any part of the optimization. It is customary to discard this system altogether while solving the *MP* so as to not incur the additional upkeep. The corresponding optimal solution $\mathbf{x}^{\star}_{MP}$ is then computed *a posteriori*, only once after the *MP* is solved.

Note 3.4 (Tracking the zero extreme point.) It often arises that $\mathbf{0}$ is an extreme point of $\mathcal{D}$ that we refer to explicitly as an index in the set P, i.e., $\lambda_\mathbf{0}$, $c_\mathbf{0} = 0$, and $\mathbf{a}_\mathbf{0} = \mathbf{0}$ while keeping (of course) the coefficient of 1 in the convexity constraint.

Note 3.5 (Encoding functions are scalar products.) Consider the *LP* formulated as

$$
\begin{aligned}
z^{\star}_{LP} = \min \quad & 5x_1 + 7x_2 + 8x_3 && (3.10a)\\
\text{s.t.} \quad & 2x_1 + 4x_2 + 5x_3 \geq 20 && (3.10b)\\
& 6x_1 + 5x_2 + 2x_3 \geq 10 && (3.10c)\\
& (x_1, x_2, x_3) \in \mathcal{X} \subset \mathbb{R}^3_+. && (3.10d)
\end{aligned}
$$

Suppose that one of the extreme points in $\mathcal{X}$ is given by $\mathbf{x}_p = (1,0,2)$ for some $p \in P$. The encoding functions for the cost and column-coefficients in the *MP* are computed as scalar products by the substitution in the objective function and the two constraints. Computing those values out, we can observe that this particular extreme point is infeasible for the *MP* when taken by itself, i.e., $12 < 20$.

$$
\begin{bmatrix} \mathbf{c}^\intercal\mathbf{x}_p \\ \mathbf{A}\mathbf{x}_p \end{bmatrix} \lambda_p = \begin{bmatrix} 5x_{1p} + 7x_{2p} + 8x_{3p} \\ 2x_{1p} + 4x_{2p} + 5x_{3p} \\ 6x_{1p} + 5x_{2p} + 2x_{3p} \end{bmatrix} \lambda_p = \begin{bmatrix} 21 \\ 12 \\ 10 \end{bmatrix} \lambda_p.
$$

Polytope and polyhedral cone

By Note 3.2, if we can recognize that $\mathcal{D}$ is a polytope or a polyhedral cone, then we can specialize the *MP* (3.9) as

$$\begin{array}{ll|ll}
\min & \sum_{p\in P} c_p\lambda_p & \min & \sum_{r\in R} c_r\lambda_r \\
\text{s.t.} & \sum_{p\in P} \mathbf{a}_p\lambda_p \geq \mathbf{b} \quad [\boldsymbol{\pi}_\mathbf{b}] & \text{s.t.} & \sum_{r\in R} \mathbf{a}_r\lambda_r \geq \mathbf{b} \quad [\boldsymbol{\pi}_\mathbf{b}] \\
& \sum_{p\in P} \lambda_p = 1 \quad [\pi_0] & & \\
& \lambda_p \geq 0, \ \forall p \in P & & \lambda_r \geq 0, \ \forall r \in R \\
& \sum_{p\in P} \mathbf{x}_p\lambda_p = \mathbf{x} & & \sum_{r\in R} \mathbf{x}_r\lambda_r = \mathbf{x}.
\end{array} \tag{3.11}$$

- On the left, we assume that $\mathcal{D}$ is bounded meaning that $R = \emptyset$.
- On the right, we assume that $\mathcal{D} = \{\mathbf{x} \in \mathbb{R}^n_+ \mid \mathbf{Dx} \geq \mathbf{0}\}$ is a polyhedral cone. The only extreme point $\mathbf{0}$ is dropped from the reformulation because the convexity constraint in (3.9) reduces to $\lambda_\mathbf{0} = 1$, thus allowing for a simplified *MP* without this constraint. Indeed, extreme point $\mathbf{0}$ can be dropped altogether since it has no contribution anywhere else ($c_\mathbf{0} = 0$ and $\mathbf{a}_\mathbf{0} = \mathbf{0}$).

  If the only extreme point differs from $\mathbf{0}$, the reformulation needs to be adjusted by the respective contributions but it can be dropped just the same. This obviously also means that we drop the dual variable π_0. Accordingly, we can amend the default input for the forthcoming *SP* (3.14) as $\bar{c}(\boldsymbol{\pi}_\mathbf{b}) = \min_{\mathbf{x}\in\mathcal{D}} (\mathbf{c}^\intercal - \boldsymbol{\pi}_\mathbf{b}^\intercal \mathbf{A})\mathbf{x}$. We remark that if $\mathbf{0}$ also belongs to $\mathcal{A}$, we then have a trivial feasible solution $\mathbf{0}$ and $z^\star_{LP} \leq 0$.

We reproduce Figure 3.3 using the associated column types in the set $\mathcal{X}$. Figure 3.4a is the general case with both extreme points and extreme rays. Figure 3.4b contains only extreme points. Finally, in Figure 3.4c, we have a polyhedral cone with the $\mathbf{0}$-vector as the single extreme point.

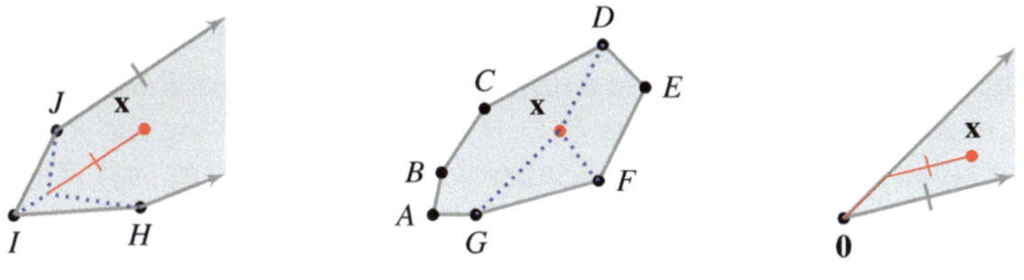

(a) $\mathcal{X} = \left\{\begin{bmatrix}\mathbf{x}_p\\1\end{bmatrix}\right\}_{p\in P} \cup \left\{\begin{bmatrix}\mathbf{x}_r\\0\end{bmatrix}\right\}_{r\in R}$ (b) $\mathcal{X} = \left\{\begin{bmatrix}\mathbf{x}_p\\1\end{bmatrix}\right\}_{p\in P}$ (c) $\mathcal{X} = \left\{\begin{bmatrix}\mathbf{0}\\1\end{bmatrix}\right\} \cup \left\{\begin{bmatrix}\mathbf{x}_r\\0\end{bmatrix}\right\}_{r\in R}$

Fig. 3.4: Column types in $\mathcal{X}$ for a polyhedron, a polytope, and a polyhedral cone.

3.2 Column Generation Approach

Figure 3.5 gives us the opportunity to refresh our memory about the column generation method for solving the *MP*. We indeed simply go over the concepts seen in Chapter 2 while adapting the notation with respect to the extended formulation. Most notably, we formalize the *SP* and derive lower bound formulas, where we trade the generic dual vector $\boldsymbol{\pi}$ for $\begin{bmatrix}\boldsymbol{\pi}_{\mathbf{b}} \\ \pi_0\end{bmatrix}$. In this process, we start to explore how the choice of the decomposed set $\mathcal{D}$ can have an impact on the resulting reformulation which leads us to two particularly important concepts: post-processing a solution of the master problem and a so-called block-diagonal structure.

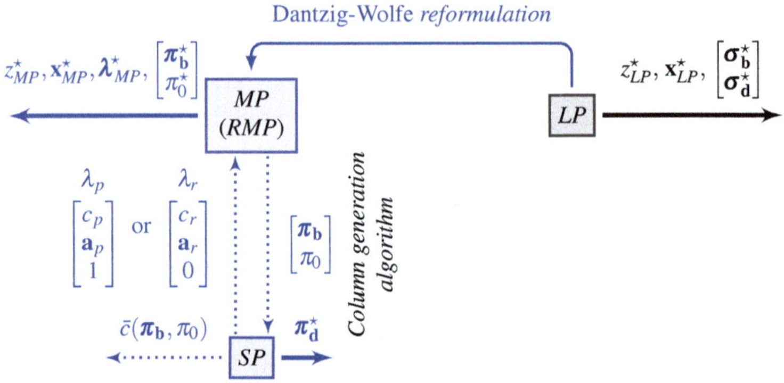

Fig. 3.5: Information flow of the column generation algorithm solving the *MP* (3.9), a Dantzig-Wolfe reformulation of the *LP* (3.1).

Pricing problem

From the *MP* (3.9), we can derive a restricted master problem *RMP* expressed with respect to relatively small subsets $P' \subset P$ and $R' \subset R$. Solving the *RMP* provides a primal solution $\boldsymbol{\lambda}$ of cost z_{RMP} together with dual values $(\boldsymbol{\pi}_{\mathbf{b}}, \pi_0)$. The latter are used in the subproblem to determine whether there remain any negative reduced cost variables. Formally, *we therefore have to find*

$$\min\{\min_{p\in P} \bar{c}_p, \min_{r\in R} \bar{c}_r\}, \tag{3.12}$$

where $\bar{c}_p$ and $\bar{c}_r$ respectively denote the reduced cost of variables λ_p and λ_r, i.e.,

$$\begin{aligned} \bar{c}_p &= c_p - \boldsymbol{\pi}_{\mathbf{b}}^\intercal \mathbf{a}_p - \pi_0 & \forall p \in P \\ \bar{c}_r &= c_r - \boldsymbol{\pi}_{\mathbf{b}}^\intercal \mathbf{a}_r & \forall r \in R. \end{aligned} \tag{3.13}$$

Yet, what we have available is the set $\mathcal{D} = \{\mathbf{x} \in \mathbb{R}^n_+ \mid \mathbf{D}\mathbf{x} \geq \mathbf{d}\}$. The combination of the Minkowski-Weyl theorem and simplex-type algorithms is the beauty of the reformulation. It relies on the existence of an alternative description for the set $\mathcal{D}$

(Minkowski-Weyl) and our capacity to identify an extreme point or extreme ray of the latter if needed (simplex-type solution). Accordingly, we define the *SP* as

$$\bar{c}(\boldsymbol{\pi}_{\mathbf{b}}, \pi_0) = -\pi_0 + \min_{\mathbf{x} \in \mathcal{D}} (\mathbf{c}^\mathsf{T} - \boldsymbol{\pi}_{\mathbf{b}}^\mathsf{T}\mathbf{A})\mathbf{x}, \tag{3.14}$$

that is,
$$\begin{aligned} \bar{c}(\boldsymbol{\pi}_{\mathbf{b}}, \pi_0) = \quad -\pi_0 + \min \quad & (\mathbf{c}^\mathsf{T} - \boldsymbol{\pi}_{\mathbf{b}}^\mathsf{T}\mathbf{A})\mathbf{x} \\ \text{s.t.} \quad & \mathbf{D}\mathbf{x} \geq \mathbf{d} \qquad [\boldsymbol{\pi}_{\mathbf{d}}] \\ & \mathbf{x} \in \mathbb{R}_+^n. \end{aligned} \tag{3.15}$$

We then ensure to solve it with an algorithm that returns an optimal solution in set $\mathcal{X}$, e.g., a simplex-type algorithm or an interior point algorithm with a crossover procedure. In this fashion, the output of the *SP almost matches* what we want in (3.12):

$$\bar{c}(\boldsymbol{\pi}_{\mathbf{b}}, \pi_0) = \begin{cases} -\infty & \text{if } c_r - \boldsymbol{\pi}_{\mathbf{b}}^\mathsf{T}\mathbf{a}_r < 0 \quad \text{for some } r \in R \\ c_p - \boldsymbol{\pi}_{\mathbf{b}}^\mathsf{T}\mathbf{a}_p - \pi_0 & \text{otherwise} \qquad\qquad \text{for some } p \in P. \end{cases} \tag{3.16}$$

If the *SP* selects an extreme point, then $\bar{c}(\boldsymbol{\pi}_{\mathbf{b}}, \pi_0)$ matches (3.12). Otherwise, any extreme ray with a negative reduced cost leads to $-\infty$ which means that one such extreme ray is picked arbitrarily and, moreover, that its corresponding reduced cost $\bar{c}_r$ is finite but does not necessarily translate to $\min_{r \in R}\{c_r - \boldsymbol{\pi}_{\mathbf{b}}^\mathsf{T}\mathbf{a}_r\}$.

Let us move to handling the generated variables as well as the stopping criterion, both of which revolve around checking the value of $\bar{c}(\boldsymbol{\pi}_{\mathbf{b}}, \pi_0)$:

- if $\bar{c}(\boldsymbol{\pi}_{\mathbf{b}}, \pi_0) < 0$ and *finite*, the *SP* identifies an extreme point $\mathbf{x}_p$, $p \in P \setminus P'$, and column $\begin{bmatrix} \mathbf{a}_p \\ 1 \end{bmatrix}$ of cost c_p is added to the *RMP*.
- if $\bar{c}(\boldsymbol{\pi}_{\mathbf{b}}, \pi_0) = -\infty$, the *SP* identifies an extreme ray $\mathbf{x}_r$, $r \in R \setminus R'$, and a column $\begin{bmatrix} \mathbf{a}_r \\ 0 \end{bmatrix}$ of cost c_r is added to the *RMP*.
- if $\bar{c}(\boldsymbol{\pi}_{\mathbf{b}}, \pi_0) \geq 0$, no negative reduced cost column exists and the column generation algorithm terminates with optimal solutions $\boldsymbol{\lambda}_{MP}^\star$ and $\mathbf{x}_{MP}^\star$ for the *MP*.

The *LP*, *MP*, *RMP*, and *SP* are all linear programs. By construction, we always have $z_{MP}^\star = z_{LP}^\star$ but the solutions $\mathbf{x}_{MP}^\star$ and $\mathbf{x}_{LP}^\star$ may very well differ from one another. The *MP* also provides $\boldsymbol{\lambda}_{MP}^\star$ whereas optimal dual values are only partially given by $\boldsymbol{\pi}_{\mathbf{b}}^\star$, see Figure 3.5. The missing part $\boldsymbol{\pi}_{\mathbf{d}}^\star$ is computed in (3.15) using the forthcoming Proposition 3.1 that also shows that the dual solutions are equivalent, $(\boldsymbol{\pi}_{\mathbf{b}}^\star, \boldsymbol{\pi}_{\mathbf{d}}^\star) \equiv (\boldsymbol{\sigma}_{\mathbf{b}}^\star, \boldsymbol{\sigma}_{\mathbf{d}}^\star)$. Example 3.2 illustrates, however, that the dual variables' individual values can be different.

Note 3.6 (Adjusted costs, again.) The interaction between the *MP* (3.9) and *SP* (3.15) means that we compute the reduced cost of any λ-variable from an element of $\mathcal{X}$ in the *SP*, i.e., $\bar{c}_p$ or $\bar{c}_r$ (3.13). With respect to variable x_j, $j \in \{1, \ldots, n\}$, of the *LP* (3.1), we echo Note 2.24 and adopt the convention of Dantzig and Thapa (2003, p. 287) to call an objective coefficient of the *SP*

$$\tilde{c}_j = c_j - \boldsymbol{\pi}_{\mathbf{b}}^{\mathsf{T}} \mathbf{A}_j \tag{3.17}$$

an *adjusted cost* in contrast to a reduced cost computed as

$$\bar{c}_j = c_j - \boldsymbol{\sigma}_{\mathbf{b}}^{\mathsf{T}} \mathbf{A}_j - \boldsymbol{\sigma}_{\mathbf{d}}^{\mathsf{T}} \mathbf{D}_j. \tag{3.18}$$

Post-processing a solution of the master problem

We remind ourselves that in spite of performing a Dantzig-Wolfe reformulation, our goal is to solve the original formulation, that is, find an optimal solution in terms of the original primal and dual variables $\mathbf{x}$ and $\begin{bmatrix} \boldsymbol{\sigma}_{\mathbf{b}} \\ \boldsymbol{\sigma}_{\mathbf{d}} \end{bmatrix}$, respectively. To this end, we assume that the *MP* is solved to optimality and break down our examination of solutions for the *LP* into primal, dual, and basic. In other words, since the extended and compact formulations are equivalent, an optimal solution to the *MP* that we post-process into one for the *LP* is also optimal.

Primal solution

An optimal solution $\mathbf{x}_{LP}^{\star}$ is obtained directly from the substitution in the Minkowski-Weyl Theorem 3.1:

$$\sum_{p \in P} \mathbf{x}_p \lambda_p^{\star} + \sum_{r \in R} \mathbf{x}_r \lambda_r^{\star} = \mathbf{x}_{MP}^{\star} \equiv \mathbf{x}_{LP}^{\star}. \tag{3.19}$$

Dual solution

From an optimal solution to the *MP*, we also recover optimal dual values $\boldsymbol{\pi}_{\mathbf{b}}^{\star}$ but still missing is $\boldsymbol{\pi}_{\mathbf{d}}^{\star}$. The following proposition fills this deficiency and finds an optimal dual solution for the *LP*.

Proposition 3.1. (Walker, 1969) *Let* $\begin{bmatrix} \boldsymbol{\pi}_{\mathbf{b}}^{\star} \\ \pi_0^{\star} \end{bmatrix}$ *be optimal dual values for the MP* (3.9) *and* $\boldsymbol{\pi}_{\mathbf{d}}^{\star}$ *be optimal dual values with optimum* $\bar{c}(\boldsymbol{\pi}_{\mathbf{b}}^{\star}, \pi_0^{\star})$ *in the SP* (3.15). *Then* $\begin{bmatrix} \boldsymbol{\pi}_{\mathbf{b}}^{\star} \\ \boldsymbol{\pi}_{\mathbf{d}}^{\star} \end{bmatrix}$ *is an optimal dual solution for the LP* (3.1).

Proof. For any dual optimal solution $\boldsymbol{\sigma}^{\star} = \begin{bmatrix} \boldsymbol{\sigma}_{\mathbf{b}}^{\star} \\ \boldsymbol{\sigma}_{\mathbf{d}}^{\star} \end{bmatrix}$, we have $\boldsymbol{\sigma}_{\mathbf{b}}^{\star\mathsf{T}} \mathbf{b} + \boldsymbol{\sigma}_{\mathbf{d}}^{\star\mathsf{T}} \mathbf{d} = z_{LP}^{\star} = z_{MP}^{\star} = \boldsymbol{\pi}_{\mathbf{b}}^{\star\mathsf{T}} \mathbf{b} + \pi_0^{\star}$ by strong duality in the *LP* and *MP*. We prove the dual feasibility of $\begin{bmatrix} \boldsymbol{\pi}_{\mathbf{b}}^{\star} \\ \boldsymbol{\pi}_{\mathbf{d}}^{\star} \end{bmatrix}$ for the *LP* and show that $\pi_0^{\star} = \boldsymbol{\pi}_{\mathbf{d}}^{\star\mathsf{T}} \mathbf{d}$.

With respect to the *SP*, whenever $\bar{c}(\boldsymbol{\pi}_{\mathbf{b}}, \pi_0)$ is finite for some $\mathbf{x}_p$, $p \in P$, we have equal primal-dual objective values $-\pi_0 + (\mathbf{c}^{\mathsf{T}} - \boldsymbol{\pi}_{\mathbf{b}}^{\mathsf{T}} \mathbf{A}) \mathbf{x}_p = -\pi_0 + \boldsymbol{\pi}_{\mathbf{d}}^{\mathsf{T}} \mathbf{d}$ as well as non-negative reduced costs for all its variables, i.e., $\bar{\mathbf{c}}^\mathsf{T} = (\mathbf{c}^\mathsf{T} - \boldsymbol{\pi}_\mathbf{b}^\mathsf{T}\mathbf{A}) - \boldsymbol{\pi}_\mathbf{d}^\mathsf{T}\mathbf{D} \geq \mathbf{0}^\mathsf{T}$. Exchanging $\boldsymbol{\pi}$ for $\boldsymbol{\sigma}$, the latter inequalities also corresponds to non-negative reduced costs in the *LP*. In particular, when the *MP* is solved to optimality, this means that $\begin{bmatrix}\boldsymbol{\pi}_\mathbf{b}^\star \\ \boldsymbol{\pi}_\mathbf{d}^\star\end{bmatrix}$ is dual feasible for the *LP*. Finally, we know from Note 2.6 that this occurs when $\bar{c}(\boldsymbol{\pi}_\mathbf{b}^\star, \pi_0^\star) = 0$, that is, $-\pi_0^\star + \boldsymbol{\pi}_\mathbf{d}^{\star\mathsf{T}}\mathbf{d} = 0$. □

Basic solution

There is no reason to assume that the recovered optimal solution $\mathbf{x}_{MP}^\star$ (3.19) is a basic solution for the original *LP*. We show here how to derive m independent columns to form a basis out of $\mathbf{x}_{MP}^\star$. Firstly, partition the x-variables of the primal solution into two subsets, that is, $\mathbf{x}_{MP}^\star = \{\mathbf{x}_L^\star, \mathbf{x}_F^\star\}$, where $\mathbf{x}_L^\star$ is the vector of variables at their *lower* bounds of zero and $\mathbf{x}_F^\star$ is the vector of positive ones (these are *free* to move upward or downward relatively to their actual values). Secondly, solve a linear program in which the variables $\mathbf{x}_L = \mathbf{0}$ are fixed and $\mathbf{x}_F \geq \mathbf{0}$ is optimized:

$$\begin{aligned} z_{LP}^\star = \min \quad & \mathbf{c}_F^\mathsf{T}\mathbf{x}_F + \mathbf{c}_L^\mathsf{T}\mathbf{x}_L \\ \text{s.t.} \quad & \mathbf{A}_F\mathbf{x}_F + \mathbf{A}_L\mathbf{x}_L \geq \mathbf{b} \\ & \mathbf{D}_F\mathbf{x}_F + \mathbf{D}_L\mathbf{x}_L \geq \mathbf{d} \\ & \mathbf{x}_F \geq \mathbf{0}, \quad \mathbf{x}_L = \mathbf{0}. \end{aligned} \tag{3.20}$$

Solving (3.20) by a simplex-type algorithm identifies a combination of the free variables satisfying the inequalities, some of them possibly taking a zero value. The basis may also comprise degenerate variables from index set L.

On grouping the constraints

Regardless of how the constraints of the compact formulation are grouped into $\mathcal{A}$ *and* $\mathcal{D}$, the *LP* (3.1) and its reformulation *MP* (3.9) are equivalent by construction, that is,

$$z_{MP}^\star = z_{LP}^\star. \tag{3.21}$$

For example, inverting the role of $\mathcal{A}$ and $\mathcal{D}$ means we have to manage a different set of extreme points and extreme rays, i.e., those of polyhedron $\mathcal{A}$, but we still obtain the same optimal objective value. We have established that we *can* reformulate the *LP* using the Minkowski-Weyl theorem but as the resulting *MP* gives an equivalent solution, let us now try to understand why we *should* under three segments.

- First of all, calling our attention to the several subproblems section of Chapter 2, we might be able to identify a so-called *block-diagonal* structure. This allows for a decentralized optimization strategy and is in fact the historical motivation of the decomposition (Dantzig, 1963, pp. 448–449). If such a structure exists, any grouping that combines blocks into larger ones sacrifices the potential for decentralization. We come back to the practical relevance of a block-diagonal structure in Section 4.4.

- Second, we have established that rather than enumerating extreme points and extreme rays, we can use the set $\mathcal{D}$ as is in the *SP* while being careful of its output. If we have a specialized algorithm that particularly excels at solving an optimization program with structure $\mathcal{D}$, the hope is that the huge number of variables in the *MP* can be mitigated by the efficiency with which we can solve the *SP*. One such notable structure is present in network flow problems for which integer solutions are easily attainable when solving in $\mathbf{x} \in \mathbb{R}^n_+$.
- Finally, grouping the constraints of the compact formulation in (3.2) must be jointly exhaustive but not necessarily mutually exclusive. This means that we obviously account for all constraints but allow for the possibility that some of them could be duplicated in both sets. Despite not usually being considered as "constraints," this is already the case for the non-negativity requirements. Furthermore, we are also allowed to integrate constraints that are redundant or implied from the problem. In this fashion, we may be able to create a notable structure in the *SP* (or *ISP*) while nonetheless preserving some duplicate constraints in the *MP*. While this seems like a lot of work to achieve an equivalent solution, Chapter 4 on integer linear programming picks up on these so-called *structured* constraints and aims to break free from the relation (3.21) by showing that not all groupings are created equal when we consider integrality requirements.

Block-diagonal structure

In many applications, matrix $\mathbf{D}$ has a block-diagonal structure, that is,

$$\mathbf{D} = \begin{bmatrix} \mathbf{D}^1 & & & \\ & \mathbf{D}^2 & & \\ & & \ddots & \\ & & & \mathbf{D}^{|K|} \end{bmatrix} \quad \text{and} \quad \mathbf{d} = \begin{bmatrix} \mathbf{d}^1 \\ \mathbf{d}^2 \\ \vdots \\ \mathbf{d}^{|K|} \end{bmatrix}, \tag{3.22}$$

where $\mathbf{D}^k$ and $\mathbf{d}^k$ are of compatible dimensions, for all $k \in K$. In that case, the *LP* writes as

$$
\begin{array}{lllllllll}
z^{\star}_{LP} = \min & \mathbf{c}^{1\mathsf{T}}\mathbf{x}^1 & + & \mathbf{c}^{2\mathsf{T}}\mathbf{x}^2 & + \ldots + & \mathbf{c}^{|K|\mathsf{T}}\mathbf{x}^{|K|} & & & \\
\text{s.t.} & \mathbf{A}^1\mathbf{x}^1 & + & \mathbf{A}^2\mathbf{x}^2 & + \ldots + & \mathbf{A}^{|K|}\mathbf{x}^{|K|} & \geq \mathbf{b} & [\boldsymbol{\sigma}_{\mathbf{b}}] & \\
& \mathbf{D}^1\mathbf{x}^1 & & & & & \geq \mathbf{d}^1 & [\boldsymbol{\sigma}_{\mathbf{d}^1}] & \\
& \mathbf{x}^1 \in \mathbb{R}^{n^1}_+ & & & & & & & \\
& & & \mathbf{D}^2\mathbf{x}^2 & & & \geq \mathbf{d}^2 & [\boldsymbol{\sigma}_{\mathbf{d}^2}] & (3.23)\\
& & & \mathbf{x}^2 \in \mathbb{R}^{n^2}_+ & & & & & \\
& & & & \ddots & & \vdots & \vdots & \\
& & & & & \mathbf{D}^{|K|}\mathbf{x}^{|K|} & \geq \mathbf{d}^{|K|} & [\boldsymbol{\sigma}_{\mathbf{d}^{|K|}}] & \\
& & & & & \mathbf{x}^{|K|} \in \mathbb{R}^{n^{|K|}}_+ & & &
\end{array}
$$

which can be concisely rewritten as

$$
\begin{aligned}
z^{\star}_{LP} = \min \quad & \sum_{k\in K} \mathbf{c}^{k\mathsf{T}}\mathbf{x}^k \\
\text{s.t.} \quad & \sum_{k\in K} \mathbf{A}^k\mathbf{x}^k \geq \mathbf{b} && [\boldsymbol{\sigma}_{\mathbf{b}}] \\
& \mathbf{D}^k\mathbf{x}^k \geq \mathbf{d}^k && [\boldsymbol{\sigma}_{\mathbf{d}^k}] \quad \forall k \in K \\
& \mathbf{x}^k \in \mathbb{R}^{n^k}_+ && \forall k \in K.
\end{aligned}
\tag{3.24}
$$

To take advantage of these blocks, let us group the constraints as

$$
\mathcal{A} = \left\{ \left\{ \mathbf{x}^k \in \mathbb{R}^{n^k}_+ \right\}_{k\in K} \;\middle|\; \sum_{k\in K} \mathbf{A}^k\mathbf{x}^k \geq \mathbf{b} \right\} \tag{3.25a}
$$

$$
\mathcal{D}^k = \left\{ \mathbf{x}^k \in \mathbb{R}^{n^k}_+ \mid \mathbf{D}^k\mathbf{x}^k \geq \mathbf{d}^k \right\}, \; \forall k \in K. \tag{3.25b}
$$

The constraints in $\mathcal{A}$ are said to be *global* or *linking* because, without them, we would have $|K|$ independent smaller problems. The set $\mathcal{D}^k$ is defined by *local* or *block* constraints which only involve the variables $\mathbf{x}^k$.

Let the set

$$
\mathcal{X} = \cup_{k\in K}\, \mathcal{X}^k, \;\; \text{where } \mathcal{X}^k = \left\{ \begin{bmatrix} \mathbf{x}^k_p \\ 1 \end{bmatrix} \right\}_{p\in P^k} \cup \left\{ \begin{bmatrix} \mathbf{x}^k_r \\ 0 \end{bmatrix} \right\}_{r\in R^k} \tag{3.26}
$$

be given by the union of the finite sets of extreme points and extreme rays, respectively indexed by P^k and R^k. With respect to each subsystem indexed by k, we adjust the notation in the change of variables (3.3) as

$$
\sum_{p\in P^k} \mathbf{x}^k_p \lambda^k_p + \sum_{r\in R^k} \mathbf{x}^k_r \lambda^k_r = \mathbf{x}^k, \; \sum_{p\in P^k} \lambda^k_p = 1, \; \lambda^k_p, \lambda^k_r \geq 0, \; \forall p \in P^k, r \in R^k, \tag{3.27}
$$

and define the cost and column parameters

$$\begin{aligned} c_p^k &= \mathbf{c}^{k\mathsf{T}}\mathbf{x}_p^k, & \mathbf{a}_p^k &= \mathbf{A}^k\mathbf{x}_p^k, & \forall p \in P^k, \\ c_r^k &= \mathbf{c}^{k\mathsf{T}}\mathbf{x}_r^k, & \mathbf{a}_r^k &= \mathbf{A}^k\mathbf{x}_r^k, & \forall r \in R^k. \end{aligned} \tag{3.28}$$

Similarly to (3.9), the *MP* then reads more concisely as

$$z_{MP}^\star = \min \sum_{k\in K}\sum_{p\in P^k} c_p^k\lambda_p^k + \sum_{k\in K}\sum_{r\in R^k} c_r^k\lambda_r^k \tag{3.29a}$$

$$\text{s.t.} \sum_{k\in K}\sum_{p\in P^k} \mathbf{a}_p^k\lambda_p^k + \sum_{k\in K}\sum_{r\in R^k} \mathbf{a}_r^k\lambda_r^k \geq \mathbf{b} \quad [\boldsymbol{\pi}_\mathbf{b}] \tag{3.29b}$$

$$\sum_{p\in P^k} \lambda_p^k = 1 \quad [\pi_0^k] \quad \forall k \in K \tag{3.29c}$$

$$\lambda_p^k \geq 0 \quad \forall k \in K, p \in P^k \tag{3.29d}$$

$$\lambda_r^k \geq 0 \quad \forall k \in K, r \in R^k \tag{3.29e}$$

$$\sum_{p\in P^k} \mathbf{x}_p^k\lambda_p^k + \sum_{r\in R^k} \mathbf{x}_r^k\lambda_r^k = \mathbf{x}^k \in \mathbb{R}_+^{n^k} \quad \forall k \in K. \tag{3.29f}$$

A block-diagonal structure yields several pricing problems, namely, one for each set $\mathcal{D}^k$, $k \in K$. Each SP^k uses the dual variable π_0^k associated with the k-th convexity constraint in (3.29):

$$\bar{c}^k(\boldsymbol{\pi}_\mathbf{b}, \pi_0^k) = -\pi_0^k + \min_{\mathbf{x}^k\in\mathcal{D}^k} (\mathbf{c}^{k\mathsf{T}} - \boldsymbol{\pi}_\mathbf{b}^\mathsf{T}\mathbf{A}^k)\mathbf{x}^k, \tag{3.30}$$

that is,

$$\begin{aligned} \bar{c}^k(\boldsymbol{\pi}_\mathbf{b}, \pi_0^k) = \quad -\pi_0^k + \min \quad & (\mathbf{c}^{k\mathsf{T}} - \boldsymbol{\pi}_\mathbf{b}^\mathsf{T}\mathbf{A}^k)\mathbf{x}^k \\ \text{s.t.} \quad & \mathbf{D}^k\mathbf{x}^k \geq \mathbf{d}^k \quad [\boldsymbol{\pi}_{\mathbf{d}^k}] \\ & \mathbf{x}^k \in \mathbb{R}_+^{n^k}. \end{aligned} \tag{3.31}$$

The column generation algorithm terminates when $\bar{c}^k(\boldsymbol{\pi}_\mathbf{b}, \pi_0^k) \geq 0$, for all $k \in K$. Otherwise, extreme points and extreme rays identified in (3.30) give rise to new columns that are added to the *RMP* in the appropriate block k. Indeed, observe the presence of index k in every column $\mathbf{a}_\mathbf{x}^k$. By construction in (3.28), these have all the same dimension, $(m \times n^k) \cdot (n^k \times 1) = m \times 1$, in the m linking constraints of (3.29b). With respect to the $|K|$ convexity constraints (3.29c), we have a 1/0 (point/ray) in the k-th row and 0 elsewhere.

Note 3.7 (Partial pricing for block-diagonal structure.) When we have many pricing problems, we may decide to call only a (small) fraction of them in each iteration, that is, follow a *partial pricing* strategy. Then unavoidably, we have to say *how* we select the subproblems to call. One can decide for those which were most successful in previous iterations. Or those that have not been called for many iterations.

When the pricing problems arise from a Dantzig-Wolfe reformulation, we have another option. The dual variable π_0^k corresponding to the (convexity constraint of) subproblem SP^k indicates, by how much the *RMP* objective function value could be (theoretically) improved, if we would increase the convexity constraint's right hand side by one unit. In other words, π_0^k is a predictor (admittedly, an optimistic one) for the usefulness of one more column that comes from SP^k. In partial pricing, we can select the most useful subproblems according to this prediction.

Note 3.8 (Identical subproblems.) In applications, it often happens that the data across all blocks is the same, i.e., $\mathbf{c}^k = \mathbf{c}$, $\mathbf{A}^k = \mathbf{A}$, $\mathbf{D}^k = \mathbf{D}$, and $\mathbf{d}^k = \mathbf{d}$, $\forall k \in K$. For example, this is the case for the single depot vehicle routing problem with homogeneous fleet. By definition of the SP^k (3.31), it is immediate that in this case we deal with identical pricing problems. What is not so immediate is how we can take advantage of this. We postpone the answer to Section 4.4 (p. 204).

Lower and upper bounds

The following proposition provides lower and upper bounds on $z^\star_{MP}$. By construction, they also apply to $z^\star_{LP}$.

Proposition 3.2. *Given optimal dual values* $(\boldsymbol{\pi}_\mathbf{b}, [\pi_0^k]_{k\in K})$ *with objective value* z_{RMP} *and minimum reduced costs* $\bar{c}^k(\boldsymbol{\pi}_\mathbf{b}, \pi_0^k)$, $\forall k \in K$, *then the optimum* $z^\star_{MP}$ *is bounded from below and above as*

$$z_{RMP} + \sum_{k\in K} \bar{c}^k(\boldsymbol{\pi}_\mathbf{b}, \pi_0^k) \le z^\star_{MP} = z^\star_{LP} \le z_{RMP}. \tag{3.32}$$

Proof. Based on the proof of Corollary 2.1, we mention two additional elements. If there exists some $k \in K$ such that $\bar{c}^k(\boldsymbol{\pi}_\mathbf{b}, \pi_0^k) = -\infty$, the left expression is obviously a lower bound. Otherwise, $\sum_{p\in P} \lambda_p^k = 1$ implies $\kappa^k = 1$, $\forall k \in K$. □

We adapt Proposition 2.1 for arbitrary dual values $\boldsymbol{\pi}_\mathbf{b} \ge \mathbf{0}$ and $\pi_0^k \in \mathbb{R}$, $\forall k \in K$, as

$$\boldsymbol{\pi}_\mathbf{b}^\intercal \mathbf{b} + \sum_{k\in K} \pi_0^k + \sum_{k\in K} \bar{c}^k(\boldsymbol{\pi}_\mathbf{b}, \pi_0^k) \le z^\star_{MP} = z^\star_{LP}, \qquad \forall \boldsymbol{\pi}_\mathbf{b} \ge \mathbf{0}, \pi_0^k \in \mathbb{R}, \forall k \in K. \tag{3.33}$$

Presented here in the context of the Dantzig-Wolfe decomposition, this lower bound on $z^\star_{LP}$ historically only appears in the '70s as the *Lagrangian dual bound*, see Chapter 6.

With respect to Note 2.17 on the availability of lower bounds with heuristic pricing, we can still only compute a lower bound in (3.33) if we have under-approximations $\underline{c}^k(\boldsymbol{\pi}_\mathbf{b}, \pi_0^k) \le \bar{c}^k(\boldsymbol{\pi}_\mathbf{b}, \pi_0^k)$, $\forall k \in K$.

3.3 Good to Know

We first derive the set $\mathcal{X}$ for a particular domain $\mathcal{D}$ and next discuss two extreme cases for the grouping of the constraints. We then examine the use of *static* variables in the *MP*, as opposed to the generated ones.

A set of extreme points and extreme rays

Figure 3.6 shows an interesting polyhedron for the domain $\mathcal{D}$ of the *SP*. Although it depicts five 2-dimensional vectors, the set of extreme points comprises $(1,1)$, $(0,2)$, and $(4,3)$ whereas there is only the single vector $(1,1)$ in the set of extreme rays. As a technical detail, observe that the latter vector is common to both sets. Therefore, $\mathcal{X} = \left\{ \begin{bmatrix} (1,1) \\ 1 \end{bmatrix}, \begin{bmatrix} (0,2) \\ 1 \end{bmatrix}, \begin{bmatrix} (4,3) \\ 1 \end{bmatrix}, \begin{bmatrix} (1,1) \\ 0 \end{bmatrix} \right\}$ and $|\mathcal{X}| = 4$.

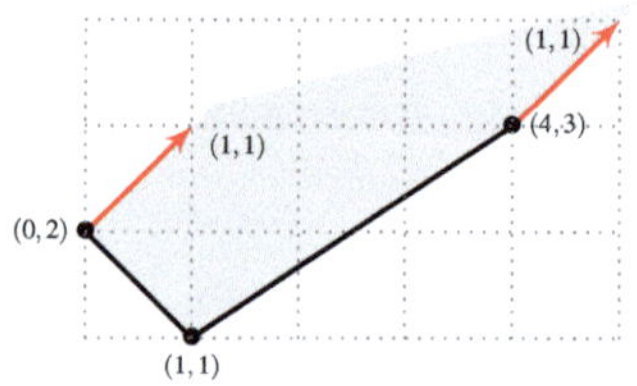

Fig. 3.6: A polyhedron with three extreme points and a single extreme ray.

Extreme cases for the constraints in the reformulated set

There are two extreme cases for the grouping of constraints in which $\mathcal{D}$ either contains *all* or *none* of the constraints of the *LP*. The first solves the *LP* in the *SP* whereas the *MP* is the *LP* in the second case.

All

The domain of the *SP* is defined by all the constraints of the *LP*, that is,

$$\mathcal{A} = \{\mathbf{x} \in \mathbb{R}^n_+\} \tag{3.34a}$$

$$\mathcal{D} = \{\mathbf{x} \in \mathbb{R}^n_+ \mid \mathbf{A}\mathbf{x} \geq \mathbf{b},\ \mathbf{D}\mathbf{x} \geq \mathbf{d}\}. \tag{3.34b}$$

Then, the *MP* (3.9) has only the convexity constraint whereas the objective function of the *SP* (3.15) is the same as that of the *LP*, except for $-\pi_0$, the constant term. Therefore, we directly find both the primal and dual optimal solutions $\mathbf{x}^\star$ and $(\boldsymbol{\sigma}_\mathbf{b}^\star, \boldsymbol{\sigma}_\mathbf{d}^\star)$, respectively, from solving the *SP*. Depending on the stopping criterion used, the column generation algorithm takes one or two iterations (Exercise 3.10).

None

Inverting the roles of $\mathcal{A}$ and $\mathcal{D}$ seen in the case *all* gives

$$\mathcal{A} = \{\mathbf{x} \in \mathbb{R}_+^n \mid \mathbf{Ax} \geq \mathbf{b},\ \mathbf{Dx} \geq \mathbf{d}\} \tag{3.35a}$$
$$\mathcal{D} = \{\mathbf{x} \in \mathbb{R}_+^n\}. \tag{3.35b}$$

The domain of the *SP* is a polyhedral cone formed by the extreme point $\mathbf{0}$ and n extreme rays represented by *orthogonal unit-vectors* $\mathbf{x}_r = \mathbf{e}_r$, $\forall r \in \{1,\ldots,n\}$, with a 1 in position r and 0 otherwise. Therefore, any non-negative solution $\mathbf{x} \in \mathcal{D}$ can be expressed as

$$\mathbf{x} = \begin{bmatrix} x_1 \\ \vdots \\ x_n \end{bmatrix} = \mathbf{0} + \sum_{r\in\{1,\ldots,n\}} \mathbf{e}_r \lambda_r, \qquad \lambda_r \geq 0,\ \forall r \in \{1,\ldots,n\}, \tag{3.36}$$

or equivalently by $x_r = \lambda_r$, $\forall r \in \{1,\ldots,n\}$. Consequently, there is no need for a change of variables from $\mathbf{x}$ to $\boldsymbol{\lambda}$. Let us now face an *RMP* which comprises a subset $R' \subset \{1,\ldots,n\}$ of the original x-variables. Assume that we know a dual solution $(\boldsymbol{\pi}_\mathbf{b}, \boldsymbol{\pi}_\mathbf{d})$ to that *RMP* which we use in the *SP* to find a variable of negative reduced cost, if any:

$$\bar{c}(\boldsymbol{\pi}_\mathbf{b}, \boldsymbol{\pi}_\mathbf{d}) = \min_{\mathbf{x}\in\mathbb{R}_+^n} \quad \mathbf{c}^\mathsf{T}\mathbf{x} - \boldsymbol{\pi}_\mathbf{b}^\mathsf{T}\mathbf{Ax} - \boldsymbol{\pi}_\mathbf{d}^\mathsf{T}\mathbf{Dx}. \tag{3.37}$$

As long as there remains an extreme ray with a negative reduced cost, the *SP* returns an unbounded optimum $\bar{c}(\boldsymbol{\pi}_\mathbf{b}, \boldsymbol{\pi}_\mathbf{d}) = -\infty$. At every iteration, we therefore add to the *RMP* an arbitrary x_r amongst all those with a negative reduced cost and proceed until the column generation algorithm reaches $\bar{c}(\boldsymbol{\pi}_\mathbf{b}, \boldsymbol{\pi}_\mathbf{d}) \geq 0$.

Note 3.9 (Primal simplex algorithm reloaded.) To overcome the arbitrary choice of the negative reduced cost entering variables when $\bar{c}(\boldsymbol{\pi}_\mathbf{b}, \boldsymbol{\pi}_\mathbf{d}) = -\infty$ in (3.37), we add the *normalizing* constraint $\mathbf{1}^\mathsf{T}\mathbf{x} = 1$ to the *SP*:

$$\begin{aligned} \bar{c}(\boldsymbol{\pi}_\mathbf{b}, \boldsymbol{\pi}_\mathbf{d}) = \min \quad & (\mathbf{c}^\mathsf{T} - \boldsymbol{\pi}_\mathbf{b}^\mathsf{T}\mathbf{A} - \boldsymbol{\pi}_\mathbf{d}^\mathsf{T}\mathbf{D})\mathbf{x} \\ \text{s.t.} \quad & \mathbf{1}^\mathsf{T}\mathbf{x} = 1 \qquad [\mu] \\ & \mathbf{x} \in \mathbb{R}_+^n, \end{aligned} \tag{3.38}$$

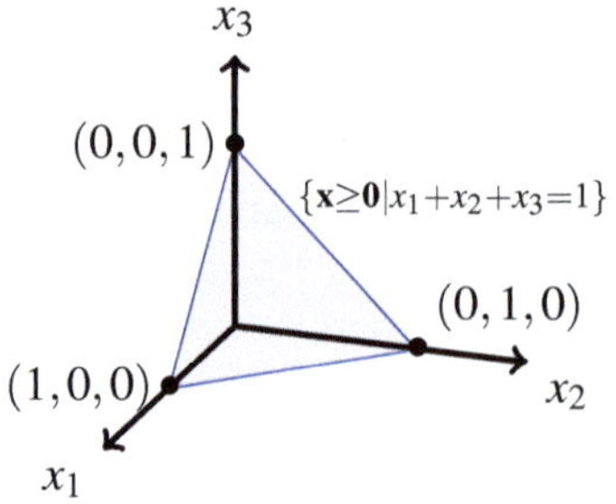

where $\mu \in \mathbb{R}$ is the dual variable associated with this added constraint. In this bounded *SP*, an extreme ray $\mathbf{x}_r = \mathbf{e}_r$, $r \in \{1,\ldots,n\}$, is replaced by an equivalent extreme point

$$\mathbf{x}_p = \mathbf{e}_p, \quad p \in \{1,\ldots,n\}$$

in a one-to-one correspondence. The objective function is $(\mathbf{c}^\mathsf{T} - \boldsymbol{\pi}_\mathbf{b}^\mathsf{T}\mathbf{A} - \boldsymbol{\pi}_\mathbf{d}^\mathsf{T}\mathbf{D})\mathbf{x} = \bar{\mathbf{c}}^\mathsf{T}\mathbf{x}$, hence

$$\begin{aligned} \bar{c}(\boldsymbol{\pi}_\mathbf{b},\boldsymbol{\pi}_\mathbf{d}) = \min \quad & \sum_{j=1}^{n} \bar{c}_j x_j \\ \text{s.t.} \quad & \sum_{j=1}^{n} x_j = 1 \quad [\mu] \\ & x_j \geq 0 \qquad \forall j \in \{1,\ldots,n\}. \end{aligned} \tag{3.39}$$

An extreme point solution to (3.39) selects exactly one variable at value 1. This strategy effectively reproduces the primal simplex algorithm using Dantzig's rule (see Section 1.6, p. 34) with the detour of a Dantzig-Wolfe decomposition.

Static variables

For some applications, we can identify a grouping of constraints such that some variables are *not* present in the reformulated set. We also qualify such variables as *static* (see Section 2.3) because keeping them as is in the *MP* is more efficient than performing an additional reformulation and generating columns for them. Consider that the set of original variables is partitioned into x- and v-variables and the *LP* writes as

$$\begin{aligned} z^\star_{LP} = \min \quad & \mathbf{c}^\mathsf{T}\mathbf{x} + \mathbf{c}_\mathbf{v}^\mathsf{T}\mathbf{v} \\ \text{s.t.} \quad & \mathbf{A}\mathbf{x} + \mathbf{A}_\mathbf{v}\mathbf{v} \geq \mathbf{b} \quad [\boldsymbol{\sigma}_\mathbf{b}] \\ & \mathbf{B}_\mathbf{v}\mathbf{v} \geq \mathbf{b}_\mathbf{v} \quad [\boldsymbol{\sigma}_\mathbf{B}] \\ & \mathbf{0} \leq \mathbf{v} \leq \mathbf{u}_\mathbf{v} \\ & \mathbf{D}\mathbf{x} \geq \mathbf{d} \quad [\boldsymbol{\sigma}_\mathbf{d}] \\ & \mathbf{x} \geq \mathbf{0}, \end{aligned} \tag{3.40}$$

with $\mathcal{D} = \{\mathbf{x} \in \mathbb{R}^n_+ \mid \mathbf{D}\mathbf{x} \geq \mathbf{d}\}$, $\mathcal{D}_\mathbf{v} = \{\mathbf{v} \in \mathbb{R}^{n_\mathbf{v}}_+ \mid \mathbf{B}_\mathbf{v}\mathbf{v} \geq \mathbf{b}_\mathbf{v}, \mathbf{v} \leq \mathbf{u}_\mathbf{v}\}$, and appropriate dimensions for vectors and matrices. We could define two subproblems with domains $\mathbf{x} \in \mathcal{D}$ and $\mathbf{v} \in \mathcal{D}_\mathbf{v}$. This is not necessarily a good idea. For example, if $\mathbf{B}_\mathbf{v}\mathbf{v} \geq \mathbf{b}_\mathbf{v}$ is absent, we would replace $2n_\mathbf{v}$ (easy) bound constraints $\mathbf{0} \leq \mathbf{v} \leq \mathbf{u}_\mathbf{v}$ by $2^{n_\mathbf{v}}$ extreme points.

We rather perform a Dantzig-Wolfe reformulation with the grouping

$$\mathcal{A} = \{\mathbf{x} \in \mathbb{R}^n_+, \mathbf{v} \in \mathbb{R}^{n_\mathbf{v}}_+ \mid \mathbf{A}\mathbf{x} + \mathbf{A}_\mathbf{v}\mathbf{v} \geq \mathbf{b}, \mathbf{B}_\mathbf{v}\mathbf{v} \geq \mathbf{b}_\mathbf{v}, \mathbf{v} \leq \mathbf{u}_\mathbf{v}\} \tag{3.41a}$$

$$\mathcal{D} = \{\mathbf{x} \in \mathbb{R}^n_+ \mid \mathbf{D}\mathbf{x} \geq \mathbf{d}\}. \tag{3.41b}$$

The *MP* is obtained by substituting $\mathbf{x}$ in terms of the elements of $\mathcal{X}$, as in (3.5):

$$\begin{aligned}
z^\star_{MP} = \min \quad & \sum_{p\in P} \mathbf{c}^\mathsf{T}\mathbf{x}_p\lambda_p + \sum_{r\in R} \mathbf{c}^\mathsf{T}\mathbf{x}_r\lambda_r + \mathbf{c}_\mathbf{v}^\mathsf{T}\mathbf{v} & & \\
\text{s.t.} \quad & \sum_{p\in P} \mathbf{A}\mathbf{x}_p\lambda_p + \sum_{r\in R} \mathbf{A}\mathbf{x}_r\lambda_r + \mathbf{A}_\mathbf{v}\mathbf{v} \geq \mathbf{b} & [\boldsymbol{\pi}_\mathbf{b}] & \\
& \mathbf{B}_\mathbf{v}\mathbf{v} \geq \mathbf{b}_\mathbf{v} & [\boldsymbol{\sigma}_\mathbf{B}] & \\
& \mathbf{0} \leq \mathbf{v} \leq \mathbf{u}_\mathbf{v} & & \\
& \sum_{p\in P} \lambda_p = 1 & [\pi_0] & \\
& \lambda_p \geq 0 & & \forall p \in P \\
& \lambda_r \geq 0 & & \forall r \in R \\
& \sum_{p\in P} \mathbf{x}_p\lambda_p + \sum_{r\in R} \mathbf{x}_r\lambda_r = \mathbf{x} \in \mathbb{R}^n_+. & &
\end{aligned} \tag{3.42}$$

The v-variables are unaffected by the substitution. Hence the λ-variables are dynamically generated as usual as opposed to these *static* v-variables in the *MP*.

Static variables notably arise if we want to exploit a particular structure only shared by the original x-variables. For example, inventory routing problems can be modeled using *dynamic routing variables* and *static inventory variables* (Desaulniers et al., 2016b). Another classic example arises in network design problems with static design variables (Frangioni and Gorgone, 2014).

3.4 More to Know

In this section on advanced topics, we derive a *restricted compact formulation* whose purpose is to directly solve the *LP* by activating its rows and variables in accordance with solutions of the *SP*. The exploited structure incidentally appears in the *positive edge rule* for the primal simplex algorithm and allows to relate the Dantzig-Wolfe decomposition method to the *improved primal simplex* algorithm as well as the *minimum mean cycle-canceling* algorithm. We also present the *Benders decomposition for linear programming*: it turns out to be equivalent to a Dantzig-Wolfe reformulation of the dual problem. We finally prove a fundamental theorem in network flow theory with the help of a Dantzig-Wolfe reformulation, that is, the *decomposition of an arc-flow solution* into a combination of directed path-flows and directed cycle-flows.

Restricted compact formulation

Assume the standard form with equality constraints (1.17) for the *LP*, with the possible addition of slack, surplus, and costly artificial variables for initialization. It can be a viable strategy to deal with the very large size of a linear program by dynamically expanding its formulation as needed. It turns out that the Dantzig-Wolfe decomposition can help in this context. Let us develop a so-called *column activation* approach. The idea is to use a *restricted compact formulation*, denoted *RLP*,

$$
\begin{aligned}
z_{RLP} = \min \quad & \sum_{j \in J} c_j x_j && && (3.43a)\\
\text{s.t.} \quad & \sum_{j \in J} \mathbf{a}_j x_j = \mathbf{b} && [\boldsymbol{\sigma}_\mathbf{b}] && (3.43b)\\
& \sum_{j \in J} \mathbf{d}_j x_j = \mathbf{d} && [\boldsymbol{\sigma}_\mathbf{d}] && (3.43c)\\
& x_j \geq 0 && \forall j \in J, && (3.43d)
\end{aligned}
$$

where J is hopefully a small subset of $\{1,\ldots,n\}$ and the pair $(\boldsymbol{\sigma}_\mathbf{b}, \boldsymbol{\sigma}_\mathbf{d})$ denotes the current dual values. We assume given the grouping of the constraints as

$$
\begin{aligned}
\mathcal{A} &= \{\mathbf{x} \in \mathbb{R}^n_+ \mid \mathbf{A}\mathbf{x} = \mathbf{b}\} && (3.44a)\\
\mathcal{D} &= \{\mathbf{x} \in \mathbb{R}^n_+ \mid \mathbf{D}\mathbf{x} = \mathbf{d}\}. && (3.44b)
\end{aligned}
$$

In a Dantzig-Wolfe reformulation, we use a solution $\mathbf{x} = [x_j]_{j=1,\ldots,n}$ from the *SP* to generate a $\lambda_\mathbf{x}$-variable for the *MP*. In this approach, there is no reformulation, we rather use that $\mathbf{x}$-solution to activate all positive x_j-variables in the *RLP*.

Note 3.10 (Is this approach tractable?) The number of variables in the *SP* remains very large but should be much easier to solve than the *LP* since it only contains the structured constraints (3.43c). Consequently, such a solution approach is only interesting if the number of activated x-variables is rather small compared to that in the compact formulation, and additionally, if $z^\star_{LP}$ is a close approximation of $z^\star_{ILP}$ in the case of an integer linear compact formulation.

Such a technique has been used to solve large-scale linear multi-commodity flow problems (Löbel, 1998) and the cutting stock problem (Valério de Carvalho, 1999, 2002). In these applications, the *LP* is written in terms of arc-flow variables and the *SP*s are shortest path problems, a well-known structure. When a subproblem generates a negative reduced cost path, the utilized arc-flow variables are activated in the *RLP* as well as the newly active flow conservation equations. This process allows the implicit combination of arcs into paths, without explicitly having to generate these paths. Example 3.4 presents an application of this procedure.

More formally, to find an optimal solution for the *LP*, Algorithm 3.1 outlines that we can iteratively solve the *RLP* and *SP* in which we input $\boldsymbol{\pi}_\mathbf{b} = \boldsymbol{\sigma}_\mathbf{b}$ and $\pi_0 = \boldsymbol{\sigma}_\mathbf{d}^\intercal \mathbf{d}$, (see also Figure 3.7). The equality system in $\mathcal{D}$ (3.44b) gives us the opportunity to recall the *SP* with amended linear relations and parameters:

$$\begin{aligned} \bar{c}(\boldsymbol{\sigma}_\mathbf{b}, \boldsymbol{\sigma}_\mathbf{d}^\top \mathbf{d}) = -\boldsymbol{\sigma}_\mathbf{d}^\top \mathbf{d} + \min \quad & (\mathbf{c}^\top - \boldsymbol{\sigma}_\mathbf{b}^\top \mathbf{A})\mathbf{x} \\ \text{s.t.} \quad & \mathbf{D}\mathbf{x} = \mathbf{d} \qquad [\boldsymbol{\pi}_\mathbf{d}] \\ & \mathbf{x} \in \mathbb{R}^n_+. \end{aligned} \tag{3.45}$$

Algorithm 3.1: The column activation algorithm.

input : *RLP* with feasible subset $J \subset \{1,\ldots,n\}$, *SP*

output : Optimal primal-dual solutions $\mathbf{x}^\star_{LP}$, $\begin{bmatrix} \boldsymbol{\sigma}^\star_\mathbf{b} \\ \boldsymbol{\sigma}^\star_\mathbf{d} \end{bmatrix}$ and optimum $z^\star_{LP}$ for the *LP*

loop
- $\mathbf{x}_{RLP}$, $\boldsymbol{\sigma}_\mathbf{b}$, $\boldsymbol{\sigma}_\mathbf{d}$, $z_{RLP} \leftarrow RLP$
- $\bar{c}(\boldsymbol{\sigma}_\mathbf{b}, \boldsymbol{\sigma}_\mathbf{d}^\top \mathbf{d})$, $\mathbf{x}$, $\boldsymbol{\pi}_\mathbf{d} \leftarrow SP$
- **if** $\bar{c}(\boldsymbol{\sigma}_\mathbf{b}, \boldsymbol{\sigma}_\mathbf{d}^\top \mathbf{d}) \geq 0$
  - **break** by optimality of the *LP*
- $J \leftarrow J \cup \{j \in \{1,\ldots,n\} \mid x_j > 0\}$

return $\mathbf{x}_{RLP}$, $\begin{bmatrix} \boldsymbol{\sigma}_\mathbf{b} \\ \boldsymbol{\pi}_\mathbf{d} \end{bmatrix}$, and z_{RLP}

We show that the set J changes at every iteration until optimality is proven by Proposition 3.3. Indeed, by (3.33), we can compute the *Lagrangian dual bound* from arbitrary dual values $\boldsymbol{\pi}_\mathbf{b}$ and π_0, say $\boldsymbol{\sigma}_\mathbf{b}$ and $\boldsymbol{\sigma}_\mathbf{d}^\top \mathbf{d}$ respectively, i.e.,

$$\boldsymbol{\sigma}_\mathbf{b}^\top \mathbf{b} + \boldsymbol{\sigma}_\mathbf{d}^\top \mathbf{d} + \bar{c}(\boldsymbol{\sigma}_\mathbf{b}, \boldsymbol{\sigma}_\mathbf{d}^\top \mathbf{d}) \leq z^\star_{LP}. \tag{3.46}$$

For optimal dual values to the *RLP*, we always have $\bar{c}(\boldsymbol{\sigma}_\mathbf{b}, \boldsymbol{\sigma}_\mathbf{d}^\top \mathbf{d}) \leq 0$. This is because the *RLP* is a restriction of the original formulation which gives us an upper bound

$$z^\star_{LP} \leq z_{RLP} = \sum_{j \in J} c_j x_j = \boldsymbol{\sigma}_\mathbf{b}^\top \mathbf{b} + \boldsymbol{\sigma}_\mathbf{d}^\top \mathbf{d} \tag{3.47}$$

by strong duality. We also have non-negative reduced costs on the activated variables, i.e., $\bar{c}_j = c_j - \boldsymbol{\sigma}_\mathbf{b}^\top \mathbf{a}_j - \boldsymbol{\sigma}_\mathbf{d}^\top \mathbf{d}_j \geq 0$, $\forall j \in J$. Furthermore, since $\mathbf{D}\mathbf{x} = \mathbf{d}$, we can rewrite the objective function of the *SP* as $\bar{c}(\boldsymbol{\sigma}_\mathbf{b}, \boldsymbol{\sigma}_\mathbf{d}^\top \mathbf{d}) = (\mathbf{c}^\top - \boldsymbol{\sigma}_\mathbf{b}^\top \mathbf{A} - \boldsymbol{\sigma}_\mathbf{d}^\top \mathbf{D})\mathbf{x}$. Therefore, when its value is negative for a solution $\mathbf{x} \in \mathcal{D}$, there exists at least one non-activated variable x_j, $j \notin J$, that is positive.

Proposition 3.3. *Let* $\mathbf{x}_{RLP}$ *and* $\begin{bmatrix} \boldsymbol{\sigma}_\mathbf{b} \\ \boldsymbol{\sigma}_\mathbf{d} \end{bmatrix}$ *be optimal primal-dual solutions to the current RLP* (3.43) *with objective value* z_{RLP}. *If* $\boldsymbol{\pi}_\mathbf{d}$ *is an optimal dual solution to the SP* (3.45) *with* $\bar{c}(\boldsymbol{\sigma}_\mathbf{b}, \boldsymbol{\sigma}_\mathbf{d}^\top \mathbf{d}) = 0$, *then* $\mathbf{x}_{RLP} \in \mathbb{R}^{|J|}_+$ *(embedded in* $\mathbb{R}^n_+$*) and* $\begin{bmatrix} \boldsymbol{\sigma}_\mathbf{b} \\ \boldsymbol{\pi}_\mathbf{d} \end{bmatrix}$ *form a pair of optimal primal-dual solutions for the LP* (3.1).

Proof. If $\bar{c}(\boldsymbol{\sigma}_\mathbf{b}, \boldsymbol{\sigma}_\mathbf{d}^\top \mathbf{d}) = 0$, we have that $z^\star_{LP} = z_{RLP} = \sum_{j \in J} c_j x_j = \boldsymbol{\sigma}_\mathbf{b}^\top \mathbf{b} + \boldsymbol{\sigma}_\mathbf{d}^\top \mathbf{d}$ by (3.46)–(3.47), where the solution $\mathbf{x}_{RLP}$ extended with $x_j = 0$, $\forall j \in \{1,\ldots,n\} \setminus J$,

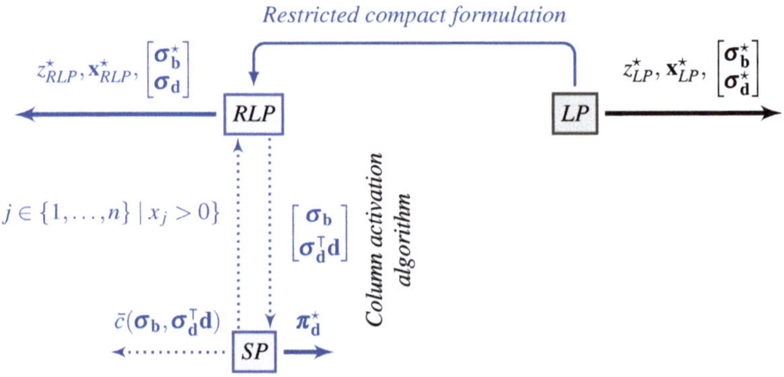

Fig. 3.7: Information flow of the column activation algorithm solving the *RLP* (3.43), a restriction of the *LP* (1.17) in standard form.

is primal feasible for the *LP*. In the *SP*, this condition means that we have equal primal-dual objective values

$$-\boldsymbol{\sigma}^\mathsf{T}_{\mathbf{d}}\mathbf{d} + (\mathbf{c}^\mathsf{T} - \boldsymbol{\sigma}^\mathsf{T}_{\mathbf{b}}\mathbf{A})\mathbf{x} = -\boldsymbol{\sigma}^\mathsf{T}_{\mathbf{d}}\mathbf{d} + \boldsymbol{\pi}^\mathsf{T}_{\mathbf{d}}\mathbf{d} = 0 \tag{3.48}$$

and non-negative reduced costs

$$(\mathbf{c}^\mathsf{T} - \boldsymbol{\sigma}^\mathsf{T}_{\mathbf{b}}\mathbf{A}) - \boldsymbol{\pi}^\mathsf{T}_{\mathbf{d}}\mathbf{D} \geq \mathbf{0}^\mathsf{T}. \tag{3.49}$$

Since these inequalities also correspond to non-negative reduced costs in the *LP*, the pair $(\boldsymbol{\sigma}_{\mathbf{b}}, \boldsymbol{\pi}_{\mathbf{d}})$ is dual feasible which implies that $z^\star_{LP} = \boldsymbol{\sigma}^\mathsf{T}_{\mathbf{b}}\mathbf{b} + \boldsymbol{\pi}^\mathsf{T}_{\mathbf{d}}\mathbf{d}$. There is no contradiction with the leading value for $z^\star_{LP}$ since $\boldsymbol{\sigma}^\mathsf{T}_{\mathbf{d}}\mathbf{d} = \boldsymbol{\pi}^\mathsf{T}_{\mathbf{d}}\mathbf{d}$ in (3.48). □

Note 3.11 (Surprising dual.) We underscore that, in the *column activation algorithm*, only the dual multipliers $\boldsymbol{\pi}_{\mathbf{d}}$ combined with $\boldsymbol{\sigma}_{\mathbf{b}}$ are guaranteed to be optimal for the *LP* even though $\boldsymbol{\sigma}_{\mathbf{d}}$ allows us to compute the optimal objective value $z^\star_{LP}$ correctly. Indeed, whenever we find an optimal degenerate solution for the *RLP* (3.43c), we remark that the corresponding dual values $\boldsymbol{\sigma}_{\mathbf{d}}$ may not be optimal for the *LP*. Consider the *LP* given as

$$\begin{array}{llll|lll} \min\limits_{\mathbf{x}\geq\mathbf{0}} & 0x_1 & +0x_2 & +0x_3 & -5x_4 & & \\ \text{s.t.} & x_1 & & & & =1 & [\sigma_1] \\ & & x_2 & & & =2 & [\sigma_2] \\ & & & x_3 & +x_4 & =0 & [\sigma_3], \end{array} \tag{3.50}$$

where $\mathcal{D} = \{\mathbf{x} \in \mathbb{R}^4_+ \mid x_3 + x_4 = 0\}$. Given $J = \{1,2,3\}$, an optimal solution to the *RLP* is the primal degenerate basic $\mathbf{x}_{RLP} = (1,2,0)$ and dual $\boldsymbol{\sigma}^\mathsf{T} = \begin{bmatrix}0 & 0 & 0\end{bmatrix}$ with $\bar{c}_4 = -5 < 0$. The *SP* given by $\bar{c}(\boldsymbol{\sigma}_{\mathbf{b}}, \boldsymbol{\sigma}^\mathsf{T}_{\mathbf{d}}\mathbf{d})$ reads as

$$\bar{c}(\begin{bmatrix}\sigma_1\\ \sigma_2\end{bmatrix}, \sigma_3 \cdot 0) = -0 + \min_{\mathbf{x}\geq\mathbf{0}} 0x_1 + 0x_2 + 0x_3 - 5x_4 \quad \text{s.t.} \;\; x_3 + x_4 = 0 \;\; [\pi_3],$$

where an optimal primal-dual solution $\mathbf{x}_{SP} = (0,0,0,0)$, $\pi_3 = -5$ has cost 0. Hence $\mathbf{x}_{RLP} = (1,2,0)$ embedded in $\mathbb{R}^4_+$ is optimal for the *LP* with appropriate optimal dual vector $\begin{bmatrix}\sigma_1 \ \sigma_2 \ \pi_3\end{bmatrix} = \begin{bmatrix}0 \ 0 \ -5\end{bmatrix}$ giving $\bar{c}_1 = 0$, $\bar{c}_2 = 0$, $\bar{c}_3 = 5$, and $\bar{c}_4 = 0$.

Note 3.12 (Homogeneous system.) Assume that $\mathbf{d} = \mathbf{0}$ in the set $\mathcal{D}$ (3.44b). Since $\boldsymbol{\pi}_{\mathbf{d}}^{\mathsf{T}}\mathbf{d} = 0$, $\forall \boldsymbol{\pi}_{\mathbf{d}}$, we can work out a simplification like in the polyhedral cone case where we drop the only extreme point $\mathbf{0}$ and the convexity constraint, see right side of (3.11). That is, we search for negative cost extreme rays in the *SP*

$$\begin{aligned} \bar{c}(\boldsymbol{\sigma}_{\mathbf{b}}) = \min \quad & (\mathbf{c}^{\mathsf{T}} - \boldsymbol{\sigma}_{\mathbf{b}}^{\mathsf{T}}\mathbf{A})\mathbf{x} \\ \text{s.t.} \quad & \mathbf{D}\mathbf{x} = \mathbf{0} \qquad [\boldsymbol{\pi}_{\mathbf{d}}] \\ & \mathbf{x} \in \mathbb{R}_+. \end{aligned} \tag{3.51}$$

Furthermore, the *RLP* can be solved using all the constraints (3.43b) derived from $\mathcal{A}$ but only those from $\mathcal{D}$ that are activated, that is, only those with non-zero coefficients on the left-hand side of the constraints (3.43c).

Note 3.13 (Row and column management.) The initialization of the *RLP* for a suitable set J can be done with a Phase I. With this in mind, we can generalize the idea of row activation in $\mathcal{D}$ seen in Note 3.12. In this case, we have to process the dual variables associated with the rows in $\mathcal{D}$ to obtain a complete dual vector $\boldsymbol{\sigma}_{\mathbf{d}}$. For each omitted row, we test whether it is satisfied or not and respectively assign 0 or big-M dual values. The zero-case does not modify the value of the dual objective function while the other one requires an artificial variable of big-M cost per unit to penalize primal infeasibility. In the same vein, once we have activated a lot of variables, it can be reasonable to deactivate some that are no longer used.

Block-diagonal structure

We easily adapt the *column activation* approach to a block-diagonal structure (3.24) with the *RLP* as

$$\begin{aligned} z_{RLP} = \min \quad & \sum_{k \in K} \sum_{j \in J^k} c_j^k x_j^k \\ \text{s.t.} \quad & \sum_{k \in K} \sum_{j \in J^k} \mathbf{a}_j^k x_j^k = \mathbf{b} \qquad [\boldsymbol{\sigma}_{\mathbf{b}}] \\ & \sum_{j \in J^k} \mathbf{d}_j^k x_j^k = \mathbf{d}^k \qquad [\boldsymbol{\sigma}_{\mathbf{d}^k}] \quad \forall k \in K \\ & x_j^k \geq 0 \qquad \forall k \in K,\ j \in J^k \subset \{1,\ldots,n^k\}, \end{aligned} \tag{3.52}$$

and the SP^k, $k \in K$, as

$$\begin{aligned}\bar{c}^k(\boldsymbol{\sigma}_{\mathbf{b}}, \boldsymbol{\sigma}_{\mathbf{d}^k}^{\mathsf{T}}\mathbf{d}^k) = -\boldsymbol{\sigma}_{\mathbf{d}^k}^{\mathsf{T}}\mathbf{d}^k + \min \quad & (\mathbf{c}^{k\mathsf{T}} - \boldsymbol{\sigma}_{\mathbf{b}}^{\mathsf{T}}\mathbf{A}^k)\mathbf{x}^k \\ \text{s.t.} \quad & \mathbf{D}^k\mathbf{x}^k = \mathbf{d}^k \qquad [\boldsymbol{\pi}_{\mathbf{d}^k}] \\ & \mathbf{x}^k \in \mathbb{R}_+.\end{aligned} \tag{3.53}$$

The arguments used in the proof of Proposition 3.3 still apply. A lower bound expression with arbitrary dual values writes as

$$\boldsymbol{\sigma}_{\mathbf{b}}^{\mathsf{T}}\mathbf{b} + \sum_{k\in K} \boldsymbol{\sigma}_{\mathbf{d}^k}^{\mathsf{T}}\mathbf{d}^k + \sum_{k\in K} \bar{c}^k(\boldsymbol{\sigma}_{\mathbf{b}}, \boldsymbol{\sigma}_{\mathbf{d}^k}^{\mathsf{T}}\mathbf{d}^k) \le z_{LP}^\star,$$

and optimality is achieved when $\bar{c}^k(\boldsymbol{\sigma}_{\mathbf{b}}, \boldsymbol{\sigma}_{\mathbf{d}^k}^{\mathsf{T}}\mathbf{d}^k) = 0$, $\forall k \in K$. At this point, the solution $\{\mathbf{x}_{RLP}^k \in \mathbb{R}_+^{|J^k|}\}_{k\in K}$, in which each vector is embedded in $\mathbb{R}_+^{n^k}$, is an optimal primal solution to the *LP* (3.24) with objective value z_{RLP} whereas an optimal dual one is given by $\boldsymbol{\sigma}_{\mathbf{b}}^\star$ and $\boldsymbol{\pi}_{\mathbf{d}^k}^\star$, $\forall k \in K$.

Positive edge rule and Improved Primal Simplex algorithm

The structure with right-hand side vector $\mathbf{d} = \mathbf{0}$ in (3.43c) implicitly occurs in the primal simplex algorithm each time the solution is degenerate. Assume again the standard form (1.17) for the *LP*, this time writing it in terms of the basic and non-basic variables (1.19), and recall the formulation (1.21) after the left-multiplication of the system of constraints by the basis inverse $\mathbf{A}_B^{-1}$:

$$\begin{aligned} z_{LP}^\star = \mathbf{c}_B^{\mathsf{T}}\bar{\mathbf{b}} + \min \quad & \qquad \bar{\mathbf{c}}_N^{\mathsf{T}}\mathbf{x}_N \\ \text{s.t.} \quad & \mathbf{x}_B + \bar{\mathbf{A}}_N\mathbf{x}_N = \bar{\mathbf{b}} \qquad [\boldsymbol{\sigma}] \\ & \mathbf{x}_B \ge \mathbf{0}, \qquad \mathbf{x}_N \ge \mathbf{0}, \end{aligned} \tag{3.54}$$

where $\bar{\mathbf{b}} = \mathbf{A}_B^{-1}\mathbf{b}$, $\bar{\mathbf{c}}_N^{\mathsf{T}} = (\mathbf{c}_N^{\mathsf{T}} - \mathbf{c}_B^{\mathsf{T}}\mathbf{A}_B^{-1}\mathbf{A}_N)$, and $\bar{\mathbf{A}}_N = \mathbf{A}_B^{-1}\mathbf{A}_N$. We can rewrite this *dictionary* representation by separating $\mathbf{x}_\mathbf{B} = \bar{\mathbf{b}} \ge \mathbf{0}$ into positive and zero parts. We do this by reorganizing the rows such that $\mathbf{x}_\mathbf{B} = \begin{bmatrix}\mathbf{x}_1\\ \mathbf{x}_2\end{bmatrix} = \begin{bmatrix}\bar{\mathbf{b}}_1 > \mathbf{0}\\ \bar{\mathbf{b}}_2 = \mathbf{0}\end{bmatrix}$. We then rewrite (3.54) by letting every matrix/vector of the system reflect this:

$$\begin{aligned} z_{LP}^\star = \mathbf{c}_1^{\mathsf{T}}\bar{\mathbf{b}}_1 + \min \quad & \mathbf{0}^{\mathsf{T}}\mathbf{x}_1 + \mathbf{0}^{\mathsf{T}}\mathbf{x}_2 + \bar{\mathbf{c}}_N^{\mathsf{T}}\mathbf{x}_N \\ \text{s.t.} \quad & \mathbf{x}_1 \qquad\qquad + \bar{\mathbf{A}}_{1N}\mathbf{x}_N = \bar{\mathbf{b}}_1 \qquad [\boldsymbol{\sigma}_1] \\ & \qquad\quad \mathbf{x}_2 + \bar{\mathbf{A}}_{2N}\mathbf{x}_N = \mathbf{0} \qquad [\boldsymbol{\sigma}_2] \\ & \mathbf{x}_1 \ge \mathbf{0}, \quad \mathbf{x}_2 \ge \mathbf{0}, \quad \mathbf{x}_N \ge \mathbf{0}. \end{aligned} \tag{3.55}$$

For a non-optimal basic solution $\mathbf{x}_B$, we assume that none of the non-basic variables $\mathbf{x}_N$ are yet activated. Then the current dual values are $\boldsymbol{\sigma}_1 = \mathbf{0}$ and $\boldsymbol{\sigma}_2 = \mathbf{0}$, and the *SP* (3.51) becomes

$$
\begin{aligned}
\bar{c}(\boldsymbol{\sigma}_1) = \min \quad & && \bar{\mathbf{c}}_N^{\mathsf{T}}\mathbf{x}_N \\
\text{s.t.} \quad & \mathbf{x}_2 &+& \bar{\mathbf{A}}_{2N}\mathbf{x}_N = \mathbf{0} \qquad [\boldsymbol{\pi}_2] \\
& \mathbf{x}_2 \geq \mathbf{0}, && \mathbf{x}_N \geq \mathbf{0}.
\end{aligned} \tag{3.56}
$$

With respect to the *SP* (3.56), let us examine three pricing strategies.

1. Dantzig's rule in the *primal simplex* algorithm selects a non-basic variable x_ℓ, $\ell \in N$, with minimum negative reduced cost $\bar{c}_\ell < 0$, if any. This is independent of the constraints $\mathbf{x}_2 = -\bar{\mathbf{a}}_{2\ell}\mathbf{x}_\ell \geq \mathbf{0}$, where $\bar{\mathbf{a}}_{2\ell} = \mathbf{A}_{2B}^{-1}\mathbf{a}_\ell$. If any coefficient of $\bar{\mathbf{a}}_{2\ell}$ is positive, we have a degenerate pivot because $\mathbf{x}_2$ cannot decrease.
2. The *positive edge* rule (Raymond et al., 2010a; Towhidi et al., 2014) also selects a single negative reduced cost non-basic variable x_ℓ, but with $\bar{\mathbf{a}}_{2\ell} = \mathbf{0}$, hence ensuring a non-degenerate pivot. Such a variable is called *compatible with the row-partition of* (3.55). The selection of x_ℓ is done without explicitly computing $\bar{\mathbf{a}}_{2j}$, $\forall j \in N$. Let $\mathbf{v}$ be a random vector for which all components are non-zero. If $\bar{\mathbf{a}}_{2j} = \mathbf{0}$, then $\mathbf{v}^{\mathsf{T}}\bar{\mathbf{a}}_{2j} = \mathbf{0}$. Otherwise $\bar{\mathbf{a}}_{2j} \neq 0$ and $\mathbf{v}^{\mathsf{T}}\bar{\mathbf{a}}_{2j} = 0 \Leftrightarrow \mathbf{v} \perp \bar{\mathbf{a}}_{2j}$, that is, if and only if the two vectors are orthogonal, which has probability zero for a continuous random vector. Define $\mathbf{w}^{\mathsf{T}} = \mathbf{v}^{\mathsf{T}}\mathbf{A}_{2B}^{-1}$. Then for any variable x_j, we have $\mathbf{v}^{\mathsf{T}}\bar{\mathbf{a}}_{2j} = \mathbf{v}^{\mathsf{T}}\mathbf{A}_{2B}^{-1}\mathbf{a}_j = \mathbf{w}^{\mathsf{T}}\mathbf{a}_j$, and one can use $\mathbf{w}^{\mathsf{T}}\mathbf{a}_j = 0$ for a compatibility-test using the original vector $\mathbf{a}_j$. (This is similar to replacing $\mathbf{c}_B^{\mathsf{T}}(\mathbf{A}_B^{-1}\mathbf{a}_j)$ by $(\mathbf{c}_B^{\mathsf{T}}\mathbf{A}_B^{-1})\mathbf{a}_j$ in the reduced cost formula of x_j, where $\mathbf{c}_B^{\mathsf{T}}\mathbf{A}_B^{-1} = \boldsymbol{\pi}^{\mathsf{T}}$ is computed once and next used directly on $\mathbf{a}_j$, $\forall j \in N$.)
3. The third strategy involves the incompatible variables (those with $\bar{\mathbf{a}}_{2\ell} \neq \mathbf{0}$), amongst which are $\mathbf{x}_2$. The *Improved Primal Simplex* algorithm (IPS) (El Hallaoui et al., 2011; Raymond et al., 2010b; Metrane et al., 2010) uses the *SP* (3.56) to select a negative reduced cost extreme ray $\mathbf{x}_r = \begin{bmatrix}\mathbf{x}_{2r} \\ \mathbf{x}_{Nr}\end{bmatrix}$ that satisfies $\mathbf{x}_2 + \bar{\mathbf{A}}_{2N}\mathbf{x}_N = \mathbf{0}$, hence leading to a non-degenerate pivot. To avoid unboundedness, we can scale the rays in (3.56) with a normalizing constraint, e.g., the sum of the variables equals 1 (Dantzig and Thapa, 2003, §10.2).

Note 3.14 (Minimum mean cycle-canceling algorithm (MMCC).) We can define the *SP* (3.56) not only in terms of the zero-variables $\mathbf{x}_2$ and $\mathbf{x}_N$ but also $\mathbf{x}_1 = \bar{\mathbf{b}}_1 > \mathbf{0}$. It suffices to make the change of variables

$$
\mathbf{x}_1 = \bar{\mathbf{b}}_1 + (\mathbf{y}_1^+ - \mathbf{y}_1^-), \qquad \mathbf{y}_1^+ \geq \mathbf{0},\ \mathbf{0} \leq \mathbf{y}_1^- \leq \bar{\mathbf{b}}_1, \tag{3.57}
$$

in the *LP* (3.55), or more cleverly in the original formulation that becomes

$$
\begin{aligned}
z_{LP}^{\star} = \mathbf{c}_1^{\mathsf{T}}\bar{\mathbf{b}}_1 + \min \quad & \mathbf{c}_1^{\mathsf{T}}\mathbf{y}_1^+ - \mathbf{c}_1^{\mathsf{T}}\mathbf{y}_1^- + \mathbf{c}_2^{\mathsf{T}}\mathbf{x}_2 + \mathbf{c}_N^{\mathsf{T}}\mathbf{x}_N && \text{(3.58a)} \\
\text{s.t.} \quad & \mathbf{y}_1^- \leq \bar{\mathbf{b}}_1 && \text{(3.58b)} \\
& \mathbf{A}_1\mathbf{y}_1^+ - \mathbf{A}_1\mathbf{y}_1^- + \mathbf{A}_2\mathbf{x}_2 + \mathbf{A}_N\mathbf{x}_N = \mathbf{0} && \text{(3.58c)} \\
& \mathbf{y}_1^+ \geq \mathbf{0}, \quad \mathbf{y}_1^- \geq \mathbf{0}, \quad \mathbf{x}_2 \geq \mathbf{0}, \quad \mathbf{x}_N \geq \mathbf{0}. && \text{(3.58d)}
\end{aligned}
$$

In this fashion, the *SP* is solved on (3.58c)–(3.58d) with a normalizing constraint whereas the step size of the non-degenerate pivot is controlled by the upper bound constraints (3.58b). It is noteworthy that this algorithm does not revolve around basic solutions. It has originally been developed for network flow problems for which the solutions of the *SP* are cycles of minimum average cost (Goldberg and Tarjan, 1989). It is known to be strongly polynomial since its inception but tighter complexity results have been derived in Radzik and Goldberg (1994) and more recently in Gauthier et al. (2015). The algorithm can be ported to arbitrary linear programs in which case the complexity is shown to be pseudo-polynomial (Gauthier and Desrosiers, 2022). Finally, Gauthier et al. (2018) present a generic framework in which the IPS and MMCC algorithms are extreme cases ensuring non-degenerate pivots. The various cases depend on the number of positive variables for which the change of variables is performed: none in IPS, all in MMCC.

Benders decomposition of a linear program

"Benders decomposition is Dantzig-Wolfe decomposition applied to the dual. Under this approach, the number of variables is reduced at the expense of usually adding many new inequalities. Analogous to generating columns of the D-W master only when needed, the inequalities of the Benders master are generated only when needed." – *Dantzig and Thapa (2003, §10.3)*

Benders decomposition (Benders, 1962) is often used for reformulating a *mixed-integer linear program MILP*, for example one with two linked sets of variables such as

$$\begin{aligned} z^{\star}_{MILP} = \min \quad & \mathbf{p}^{\intercal}\mathbf{y} + \mathbf{q}^{\intercal}\mathbf{x} \\ \text{s.t.} \quad & \mathbf{P}\mathbf{y} + \mathbf{Q}\mathbf{x} \geq \mathbf{b} \\ & \mathbf{y} \in \mathcal{Y} \\ & \mathbf{x} \geq \mathbf{0}, \end{aligned} \tag{3.59}$$

where $\mathcal{Y} \subseteq \mathbb{Z}^m_+$ is a set of integer points. Indeed, we deal with complicating variables instead of complicating constraints. For a given feasible $\mathbf{y} \in \mathcal{Y}$, the resulting *linear* program is much easier to solve:

$$\begin{aligned} \min_{\mathbf{x} \geq \mathbf{0} \,|\, \mathbf{y}} \quad & \mathbf{q}^{\intercal}\mathbf{x} \\ \text{s.t.} \quad & \mathbf{Q}\mathbf{x} \geq \mathbf{b} - \mathbf{P}\mathbf{y} \quad [\boldsymbol{\sigma}]. \end{aligned} \tag{3.60}$$

We can leverage this "easy" program idea by rewriting formulation (3.59) as a bi-level optimization program, where the lower-level receives $\mathbf{y}$-values as input:

$$z^{\star}_{MILP} = \min_{\mathbf{y} \in \mathcal{Y}} \left\{ \mathbf{p}^{\intercal}\mathbf{y} + \min_{\mathbf{x} \geq \mathbf{0} | \mathbf{y}} \{ \mathbf{q}^{\intercal}\mathbf{x} \mid \mathbf{Q}\mathbf{x} \geq \mathbf{b} - \mathbf{P}\mathbf{y} \} \right\} \tag{3.61a}$$

$$= \min_{\mathbf{y}\in\mathcal{Y}} \left\{ \mathbf{p}^\mathsf{T}\mathbf{y} + \max_{\boldsymbol{\sigma}\geq\mathbf{0}|\mathbf{y}} \{(\mathbf{b}-\mathbf{P}\mathbf{y})^\mathsf{T}\boldsymbol{\sigma} \mid \mathbf{Q}^\mathsf{T}\boldsymbol{\sigma} \leq \mathbf{q}\} \right\}. \tag{3.61b}$$

Benders decomposition revolves around the dual passage to (3.61b). We do this to obtain a fixed polyhedron $\{\boldsymbol{\sigma} \in \mathbb{R}^m_+ \mid \mathbf{Q}^\mathsf{T}\boldsymbol{\sigma} \leq \mathbf{q}\}$, independent of a specific $\mathbf{y}$, and thus be able to use the Minkowski-Weyl theorem. This leads us to an exact solution method which iteratively introduces refining cuts in $\mathbf{y}$. That is, we can derive a so-called *Benders master problem* from (3.61b) and use a cutting plane algorithm on its relaxation. Let us give more detailed explanations in the forthcoming comparison with the Dantzig-Wolfe decomposition.

For the record, we could even have integrality restrictions on $\mathbf{x}$ in (3.61a) but would only be able to dualize the linear relaxation. This in turn implies that we can adapt Benders decomposition with a branching mechanism to solve a more general starting formulation. Benders decomposition is now widely used for linear, non-linear, integer, stochastic, multi-stage, bi-level, and other optimization programs, see Rahmaniani et al. (2017) for a survey. The L-shaped method (Van Slyke and Wets, 1969), taking its name from what the program being solved looks like, is to the Benders decomposition what the column generation algorithm is to the Dantzig-Wolfe decomposition.

To show the equivalence with the Dantzig-Wolfe decomposition in the context of *maximizing* a linear program, we use the simplified primal-dual pair (assumed feasible)

$$\begin{array}{rll|rll} z^\star_{LP} = \displaystyle\max_{\boldsymbol{\pi}_\mathbf{b}\geq\mathbf{0},\boldsymbol{\pi}_\mathbf{d}\geq\mathbf{0}} & \mathbf{b}^\mathsf{T}\boldsymbol{\pi}_\mathbf{b} + \mathbf{d}^\mathsf{T}\boldsymbol{\pi}_\mathbf{d} & & z^\star_{LD} = \displaystyle\min_{\mathbf{x}\geq\mathbf{0}} & \mathbf{c}^\mathsf{T}\mathbf{x} & \\ \text{s.t.} & \mathbf{A}^\mathsf{T}\boldsymbol{\pi}_\mathbf{b} + \mathbf{D}^\mathsf{T}\boldsymbol{\pi}_\mathbf{d} \leq \mathbf{c} & [\mathbf{x}] & \text{s.t.} & \mathbf{A}\mathbf{x} \geq \mathbf{b} & [\boldsymbol{\pi}_\mathbf{b}] \\ & & & & \mathbf{D}\mathbf{x} \geq \mathbf{d} & [\boldsymbol{\pi}_\mathbf{d}], \end{array} \tag{3.62}$$

for which we know that a Dantzig-Wolfe reformulation of the *LD* gives the *MP* (3.9). Using the same grouping of constraints in $\mathcal{A}$ and $\mathcal{D}$, the *LP* can be rewritten as a bi-level program in which the lower-level receives $\boldsymbol{\pi}_\mathbf{b}$-values as input:

$$z^\star_{LP} = \max_{\boldsymbol{\pi}_\mathbf{b}\geq\mathbf{0}} \left\{ \mathbf{b}^\mathsf{T}\boldsymbol{\pi}_\mathbf{b} + \max_{\boldsymbol{\pi}_\mathbf{d}\geq\mathbf{0}|\boldsymbol{\pi}_\mathbf{b}} \{\mathbf{d}^\mathsf{T}\boldsymbol{\pi}_\mathbf{d} \mid \mathbf{D}^\mathsf{T}\boldsymbol{\pi}_\mathbf{d} \leq \mathbf{c} - \mathbf{A}^\mathsf{T}\boldsymbol{\pi}_\mathbf{b}\} \right\} \tag{3.63a}$$

$$= \max_{\boldsymbol{\pi}_\mathbf{b}\geq\mathbf{0}} \left\{ \mathbf{b}^\mathsf{T}\boldsymbol{\pi}_\mathbf{b} + \min_{\mathbf{x}\geq\mathbf{0}|\boldsymbol{\pi}_\mathbf{b}} \{(\mathbf{c} - \mathbf{A}^\mathsf{T}\boldsymbol{\pi}_\mathbf{b})^\mathsf{T}\mathbf{x} \mid \mathbf{D}\mathbf{x} \geq \mathbf{d}\} \right\}. \tag{3.63b}$$

For any optimal solution $\boldsymbol{\pi}^\star_\mathbf{b}$, there cannot be an extreme ray $\mathbf{x}_r$, $r \in R$, of the minimization program for which $(\mathbf{c}^\mathsf{T} - \boldsymbol{\pi}^{\star\mathsf{T}}_\mathbf{b}\mathbf{A})\mathbf{x}_r = c_r - \mathbf{a}_r^\mathsf{T}\boldsymbol{\pi}^\star_\mathbf{b} < 0$, otherwise the *LP* in (3.62) is *infeasible*. Therefore, the *feasibility cuts* $\mathbf{a}_r^\mathsf{T}\boldsymbol{\pi}_\mathbf{b} \leq c_r$, $\forall r \in R$, are imposed on $\boldsymbol{\pi}_\mathbf{b}$ and there only remains to consider extreme points of the (inner) minimization program given $\boldsymbol{\pi}_\mathbf{b}$ as input:

$$z^\star_{LP} = \max_{\boldsymbol{\pi}_\mathbf{b} \geq \mathbf{0}} \left\{ \mathbf{b}^\intercal \boldsymbol{\pi}_\mathbf{b} + \min_{\{\mathbf{x}_p\}_{p \in P} | \boldsymbol{\pi}_\mathbf{b}} \{c_p - \mathbf{a}_p^\intercal \boldsymbol{\pi}_\mathbf{b}\} \right\} \tag{3.64}$$
$$\text{s.t.} \quad \mathbf{a}_r^\intercal \boldsymbol{\pi}_\mathbf{b} \leq c_r \quad \forall r \in R.$$

Because $\min_{\{\mathbf{x}_p\}_{p \in P} | \boldsymbol{\pi}_\mathbf{b}} \{c_p - \mathbf{a}_p^\intercal \boldsymbol{\pi}_\mathbf{b}\} = \max_{\pi_0 \in \mathbb{R}} \pi_0$ s.t. $\pi_0 \leq c_p - \mathbf{a}_p^\intercal \boldsymbol{\pi}_\mathbf{b}$, $\forall p \in P$, we obtain the so-called Benders master problem:

$$\begin{aligned} z^\star_{LP} = \max_{\boldsymbol{\pi}_\mathbf{b} \geq \mathbf{0}, \pi_0 \in \mathbb{R}} \quad & \mathbf{b}^\intercal \boldsymbol{\pi}_\mathbf{b} + \pi_0 \\ \text{s.t.} \quad & \mathbf{a}_p^\intercal \boldsymbol{\pi}_\mathbf{b} + \pi_0 \leq c_p \quad [\lambda_p] \quad \forall p \in P \\ & \mathbf{a}_r^\intercal \boldsymbol{\pi}_\mathbf{b} \leq c_r \quad [\lambda_r] \quad \forall r \in R, \end{aligned} \tag{3.65}$$

which indeed corresponds to the dual of the *MP* (3.9). Moreover, the constraints indexed by $p \in P$ are known as *optimality cuts*, a name we can better appreciate upon substituting $\mathbf{b}^\intercal \boldsymbol{\pi}_\mathbf{b} + \pi_0 = \mu \in \mathbb{R}$:

$$\begin{aligned} z^\star_{LP} = \max_{\boldsymbol{\pi}_\mathbf{b} \geq \mathbf{0}, \mu \in \mathbb{R}} \quad & \mu \\ \text{s.t.} \quad & \mu \leq c_p - (\mathbf{a}_p - \mathbf{b})^\intercal \boldsymbol{\pi}_\mathbf{b} \quad [\lambda_p] \quad \forall p \in P \\ & 0 \leq c_r - \mathbf{a}_r^\intercal \boldsymbol{\pi}_\mathbf{b} \quad [\lambda_r] \quad \forall r \in R. \end{aligned} \tag{3.66}$$

Relaxed master, subproblem, and initialization

The literature refers to both formulations (3.65) and (3.66) as the Benders master problem. From our knowledge of the Dantzig-Wolfe decomposition, it is immediate that we can instantiate a *relaxed Benders master problem* with subsets $P' \subset P$ and $R' \subset R$. Furthermore, initialization is straightforward since we can start with no constraints at all and some arbitrary $\boldsymbol{\pi}_\mathbf{b} \geq \mathbf{0}$. We then generate constraints by solving (simplex-type) the subproblem, that is, the inner optimization program in (3.63b):

$$\bar{c}(\boldsymbol{\pi}_\mathbf{b}) = \min \{(\mathbf{c} - \mathbf{A}^\intercal \boldsymbol{\pi}_\mathbf{b})^\intercal \mathbf{x} \mid \mathbf{D}\mathbf{x} \geq \mathbf{d}, \mathbf{x} \geq \mathbf{0}\} = \min_{\mathbf{x} \in \mathcal{D}} (\mathbf{c} - \mathbf{A}^\intercal \boldsymbol{\pi}_\mathbf{b})^\intercal \mathbf{x} \tag{3.67}$$

and converting the output $\mathbf{x} \in \{\mathbf{x}_p\}_{p \in P}$ or $\mathbf{x} \in \{\mathbf{x}_r\}_{r \in R}$ into an appropriate type of cut with parameters $\mathbf{a}_\mathbf{x}$ and $c_\mathbf{x}$. We terminate with an optimal solution to the Benders master problem when $\bar{c}(\boldsymbol{\pi}_\mathbf{b}) \geq 0$, the *LP* (3.62) being a maximization program.

Note 3.15 (What is what?) Observe that the *LP* rewritten as (3.63b) is now solved with $\boldsymbol{\pi}_\mathbf{b}$ and $\mathbf{x}$ but without $\boldsymbol{\pi}_\mathbf{d}$ variables. At the risk of creating confusion, it is only fair to put this back in the perspective of the more general model (3.59), i.e., $\boldsymbol{\pi}_\mathbf{b}$ plays the role of $\mathbf{y}$, $\boldsymbol{\pi}_\mathbf{d}$ that of $\mathbf{x}$, and $\mathbf{x}$ that of $\boldsymbol{\sigma}$.

Decomposition theorem of a network flow solution

Any network flow solution can be decomposed into a limited number of paths and cycles. Although the classical constructive proof for the existence of such a decomposition is very simple (Ahuja et al., 1993, Theorem 3.5), it does not generalize to solutions of an arbitrary linear program. Gauthier et al. (2014) show that a similar result indeed does hold with a proof based on a Dantzig-Wolfe reformulation. We present the edulcorated version of this proof within the scope of the capacitated minimum cost flow problem (*CMCFP*) as described in Chapter 1 (Network flow problem, p. 27). Let us repeat the canonical formulation (1.45) for convenience.

$$\begin{aligned} z^\star_{LP} = \min \quad & \sum_{(i,j)\in A} c_{ij}x_{ij} \\ \text{s.t.} \quad & \sum_{j:(i,j)\in A} x_{ij} - \sum_{j:(j,i)\in A} x_{ji} = b_i \quad \forall i \in N \\ & 0 \le \ell_{ij} \le x_{ij} \le u_{ij} \qquad \forall (i,j) \in A. \end{aligned} \tag{3.68}$$

Proposition 3.4 (Flow Decomposition Theorem). *Any arc-flow solution* $\mathbf{x}^0$ *to the CMCFP* (3.68) *can be represented as a combination of directed paths and directed cycles—though not necessarily uniquely—with the following properties:*

(a) Every directed path with positive flow connects a supply node to a demand node.
(b) At most $|A|+|N|$ *directed paths and directed cycles have positive flow amongst which at most* $|A|$ *directed cycles.*
(c) If $b_i = 0$, $\forall i \in N$, *only directed cycles have positive flow on circulation* $\mathbf{x}^0$.

Proof. We first transform the *CMCFP* into a circulation problem. We then perform a Dantzig-Wolfe reformulation on the latter with a pricing problem whose solutions can be interpreted as paths or cycles of the original problem. The remaining part uses linear programming arguments to show that any arc-flow solution $\mathbf{x}^0$ can be represented in the master problem with a limited number of variables.

1. We obtain a convenient representation of a circulation problem in two steps. First, we partition the set of nodes in three subsets:

   - the *supply* nodes $S = \{i \in N \mid b_i > 0\}$;
   - the *demand* nodes $D = \{i \in N \mid b_i < 0\}$;
   - the *transshipment* nodes $N \setminus \{S \cup D\} = \{i \in N \mid b_i = 0\}$.

   Second, Figure 3.8 shows how we augment this partitioned network with nodes o and d. Transshipment nodes are only visually omitted whereas we count five supply nodes $s_1,\dots,s_5$ in S and four demand nodes $d_1,\dots,d_4$ in D. Nodes o and d respectively connect supply and demand nodes with appropriately directed arcs. We create the loop with arc (d,o). We denote this augmented network by $G_{do} = (N_{do}, A_{do})$, where $N_{do} = N \cup \{o,d\}$ and $A_{do} = A \cup \{(d,o)\} \cup \{(o,j)_{j\in S}\} \cup \{(i,d)_{i\in D}\}$.

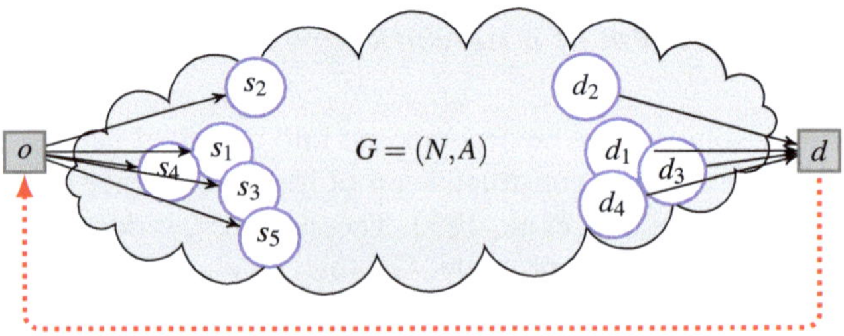

Fig. 3.8: Circulation problem on the network G_{do}.

The supply and demand requirements are respectively transferred on the zero-cost arc sets $\{(o, j) \mid j \in S\}$ and $\{(i,d) \mid i \in D\}$, i.e., $\ell_{oj} = u_{oj} = b_j$, $j \in S$, and $\ell_{id} = u_{id} = -b_i$, $i \in D$. Lower and upper bounds on x_{do} are given as $\ell_{do} = u_{do} = \sum_{j \in S} b_j = -\sum_{i \in D} b_i$, that is, both are equal to the total supply or demand. The circulation problem is given by

$$
\begin{aligned}
z^{\star}_{LP} = \min \quad & \sum_{(i,j) \in A_{do}} c_{ij} x_{ij} \\
\text{s.t.} \quad & \sum_{j:(i,j) \in A_{do}} x_{ij} - \sum_{j:(j,i) \in A_{do}} x_{ji} = 0 \quad && \forall i \in N_{do} \\
& 0 \leq \ell_{ij} \leq x_{ij} \leq u_{ij} && \forall (i,j) \in A_{do}.
\end{aligned} \tag{3.69}
$$

2. We then perform a Dantzig-Wolfe reformulation by grouping the constraints as

$$
\mathcal{A} = \{\{x_{ij} \in \mathbb{R}_+\}_{(i,j) \in A_{do}} \mid \ell_{ij} \leq x_{ij} \leq u_{ij}\}
$$

$$
\mathcal{D} = \left\{ \{x_{ij} \in \mathbb{R}_+\}_{(i,j) \in A_{do}} \;\middle|\; \sum_{j:(i,j) \in A_{do}} x_{ij} - \sum_{j:(j,i) \in A_{do}} x_{ji} = 0, \;\; \forall i \in N_{do} \right\}.
$$

The set $\mathcal{D}$ describes a flow circulation without upper bounds. It actually forms a polyhedral cone that can be described in terms of the extreme point $\mathbf{0}$ and a finite number of extreme rays, i.e., $\mathcal{X} = \left\{ \left[\begin{smallmatrix} \mathbf{0} \\ 1 \end{smallmatrix}\right] \right\} \cup \left\{ \left[\begin{smallmatrix} \mathbf{x}_r \\ 0 \end{smallmatrix}\right] \right\}_{r \in R}$ On the one hand, the extreme point $\mathbf{0}$ can be discarded from the reformulation as it has no contribution in the objective function of the *MP* nor its constraint set. On the other hand, every extreme ray is associated with a directed cycle in G_{do}. These directed cycles with positive flow can be represented using a unit-flow and are next interpreted in terms of directed paths and directed cycles in the original network G:

- a G_{do}-cycle with $x_{do} = 1$: a *directed path* in G from a supply to a demand node;
- a G_{do}-cycle with $x_{do} = 0$: a *directed cycle* in G.

Define $\mathcal{P}$ and $\mathcal{C}$ to be respectively the sets of paths and cycles in G, where these are understood to be directed even though we omit the precision.

Let $\mathbf{x} = [x_{ij}]_{(i,j)\in A}$ be the flow variables with respectively the lower and upper bounds $\boldsymbol{\ell} = [\ell_{ij}]_{(i,j)\in A}$ and $\mathbf{u} = [u_{ij}]_{(i,j)\in A}$. Define also $\mathbf{x}_S = [x_{oj}]_{j\in S}$ and $\mathbf{x}_D = [x_{id}]_{i\in D}$ together with $\mathbf{b}_S = [b_j]_{j\in S}$ and $\mathbf{b}_D = [b_i]_{i\in D}$.

Any non-zero solution $[\mathbf{x}, \mathbf{x}_S, \mathbf{x}_D, x_{do}]^\mathsf{T}$ to $\mathcal{D}$ can be written as *a conic combination of the extreme rays* $\mathbf{x}_r$, $r \in R$, on G_{do}, that is, equivalently on G in terms of the supply-demand paths $[\mathbf{x}_p, \mathbf{x}_{Sp}, \mathbf{x}_{Dp}, 1]^\mathsf{T}$, $p \in \mathcal{P}$, and cycles $[\mathbf{x}_c, \mathbf{0}, \mathbf{0}, 0]^\mathsf{T}$, $c \in \mathcal{C}$:

$$\begin{bmatrix} \mathbf{x} \\ \mathbf{x}_S \\ \mathbf{x}_D \\ x_{do} \end{bmatrix} = \sum_{p\in\mathcal{P}} \begin{bmatrix} \mathbf{x}_p \\ \mathbf{x}_{Sp} \\ \mathbf{x}_{Dp} \\ 1 \end{bmatrix} \lambda_p + \sum_{c\in\mathcal{C}} \begin{bmatrix} \mathbf{x}_c \\ \mathbf{0} \\ \mathbf{0} \\ 0 \end{bmatrix} \lambda_c, \quad \lambda_p, \lambda_c \geq 0, \forall p \in \mathcal{P}, c \in \mathcal{C}. \tag{3.71}$$

Let $c_p = \mathbf{c}^\mathsf{T}\mathbf{x}_p$, $p \in \mathcal{P}$, be the cost of a path and $c_c = \mathbf{c}^\mathsf{T}\mathbf{x}_c$, $c \in \mathcal{C}$, be the cost of a cycle. The *MP* is then given as

$$\begin{aligned}
z^\star_{MP} = \min \quad & \sum_{p\in\mathcal{P}} c_p\lambda_p + \sum_{c\in\mathcal{C}} c_c\lambda_c && && \text{(3.72a)}\\
\text{s.t.} \quad & \sum_{p\in\mathcal{P}} \mathbf{x}_p\lambda_p + \sum_{c\in\mathcal{C}} \mathbf{x}_c\lambda_c = \mathbf{x} && && \text{(3.72b)}\\
& \sum_{p\in\mathcal{P}} \mathbf{x}_{Sp}\lambda_p = \mathbf{x}_S \ (= \mathbf{b}_S) && && \text{(3.72c)}\\
& \sum_{p\in\mathcal{P}} \mathbf{x}_{Dp}\lambda_p = \mathbf{x}_D \ (= -\mathbf{b}_D) && && \text{(3.72d)}\\
& \sum_{p\in\mathcal{P}} \lambda_p = x_{do} && && \text{(3.72e)}\\
& \lambda_p \geq 0 && \forall p \in \mathcal{P} && \text{(3.72f)}\\
& \lambda_c \geq 0 && \forall c \in \mathcal{C} && \text{(3.72g)}\\
& \mathbf{x} \in [\boldsymbol{\ell}, \mathbf{u}]. && && \text{(3.72h)}
\end{aligned}$$

3. In the above reformulation, the constraint (3.72e) is redundant and as such can be removed. Indeed, $x_{do} = \mathbf{1}^\mathsf{T}\mathbf{b}_S = -\mathbf{1}^\mathsf{T}\mathbf{b}_D$ is implied by the flow conservation equations of the supply nodes (3.72c) or demand nodes (3.72d). Since any solution $\mathbf{x}^0$ to the *CMCFP* (3.68) obviously satisfies the bound constraints, we can describe this solution with a basis of size $|A| + |S| + |D|$ in respectively (3.72b), (3.72c), and (3.72d), where $|S| + |D| \leq |N|$ nodes. Therefore, counting only the positive λ-variables, there exists a representation for $\mathbf{x}^0$ that uses at most $|A| + |N|$ paths and cycles, amongst which at most $|A|$ cycles (λ_c-variables).

   In the case of a circulation problem for which $b_i = 0$, $\forall i \in N$, there are no paths involved, that is, no λ_p-variables because $S = D = \emptyset$, and $\mathbf{x}^0$ can be written as a combination of at most $|A|$ cycles. □

Fig. 3.9: François Soumis, a lead researcher on column generation and Dantzig-Wolfe decomposition at GERAD in Montréal (Tromsø, Norway, 2010-06-23).

3.5 Examples

Example 3.1 illustrates the mechanics of a reformulation on a two-dimensional program. Example 3.2 solves the linear relaxation of a Time constrained shortest path problem (*TCSPP*) for which the search for an optimal integer solution is postponed to Chapters 4 and 7. In Example 3.3, we present two compact formulations of the Single depot vehicle scheduling problem. The second facilitates the reformulation in terms of extreme rays. Finally, Example 3.4 uses the *SP* of a Dantzig-Wolfe reformulation to activate rows and variables of the compact formulation.

There is something mechanical in solving a Dantzig-Wolfe reformulation, see Figure 3.5. Starting with a given compact formulation, we next establish a grouping of constraints from which we derive an extended formulation. We then usually solve this reformulation by column generation and finally do a back substitution to obtain a solution for the compact formulation. We enumerate these elements concisely as

- Compact formulation (*LP*)
- Grouping of constraints ($\mathcal{A}$ and $\mathcal{D}$)
- Extended formulation (*MP*)
- Column generation approach (*RMP* and *SP*)
- Optimal primal-dual solutions for the compact formulation ($\mathbf{x}^{\star}_{LP}, \boldsymbol{\sigma}^{\star}_{\mathbf{b}}$ and $\boldsymbol{\sigma}^{\star}_{\mathbf{d}}$).

If the *MP* can be solved directly, we obviously also obtain a primal solution for the compact formulation. Otherwise, we use column generation with the *SP* which also gives us a dual solution for the compact formulation at optimality of the *MP*. We follow this pattern in Examples 3.1 and 3.2. From there on, we may skip and/or blur together some of these elements in our presentation.

Example 3.1 ***2D illustration***

✎ This is a visual illustration of a Dantzig-Wolfe reformulation which is solved directly and again by column generation.

We use a simple 2-dimensional linear program to hone our skills on performing a Dantzig-Wolfe reformulation. The simplicity of the problem allows us to solve it in three different ways: the *LP* and the *MP* directly as well as by column generation. We insist that solving the *MP* directly is *only* possible if we can conceivably enumerate all extreme points and extreme rays. The visual interpretation of this small problem obviously grants us this.

Compact formulation

Consider the following 2-dimensional linear program with dual variables appearing within brackets as usual:

$$\begin{aligned}
z^\star_{LP} = \min \quad & x_1 + x_2 \\
\text{s.t.} \quad & -x_1 + 4x_2 \leq 8 \quad [\sigma_1] \\
& 3x_1 + 4x_2 \geq 24 \quad [\sigma_2] \\
& x_1 \leq 10 \quad [\sigma_3] \\
& x_2 \geq 2 \quad [\sigma_4] \\
& x_2 \leq 6 \quad [\sigma_5] \\
& x_1, \; x_2 \geq 0.
\end{aligned} \tag{3.73}$$

The unique primal optimal solution to the *LP* is $\mathbf{x}^\star_{LP} = (4,3)$ with $z^\star_{LP} = 7$. This is easy to verify given the four extreme points $\{(4,3), (10,4.5), (10,2), (16/3,2)\}$. We also have that $\boldsymbol{\sigma}^{\star\mathsf{T}} = \left[-1/16,\ 5/16,\ 0,\ 0,\ 0\right]$ is the unique optimal dual solution.

Grouping of constraints

Figure 3.10 depicts the feasible domain where the sets $\mathcal{A}$ and $\mathcal{D}$ are given by

$$\mathcal{A} = \{x_1, x_2 \geq 0 \mid 3x_1 + 4x_2 \geq 24,\ x_1 \leq 10,\ x_2 \leq 6\} \tag{3.74a}$$
$$\mathcal{D} = \{x_1, x_2 \geq 0 \mid -x_1 + 4x_2 \leq 8,\ x_2 \geq 2\}. \tag{3.74b}$$

As a simple courtesy, we inform the reader that we ask to redo this example while inverting the roles of $\mathcal{A}$ and $\mathcal{D}$ in Exercise 3.6. Likewise in Exercise 3.7 where we rather want to maximize the objective function.

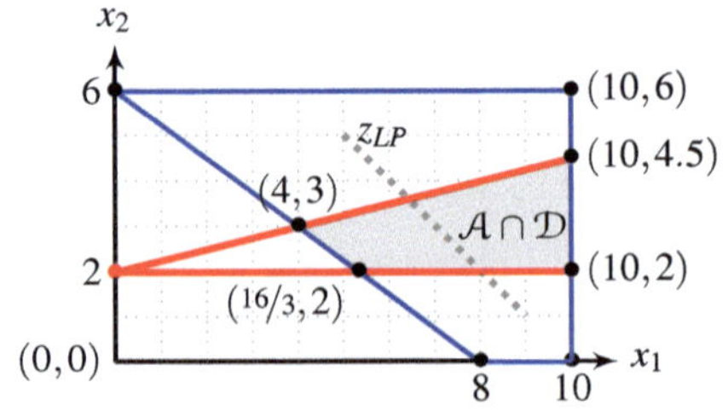

Fig. 3.10: Domain of the *LP* (3.73) as the shaded intersection of $\mathcal{A}$ and $\mathcal{D}$. A level curve of the objective function appears as a dotted line.

Extended formulation

The polyhedron $\mathcal{D}$ comprises a single extreme point $(0,2)$ and two extreme rays: the first has a positive slope $1/4$ while the second is horizontal, see Figure 3.11(right). The latter two are represented by the Euclidean vectors $(4,1)$ and $(1,0)$ which correspond to the illustrated vectors $(0,2) \to (4,3)$ and $(0,2) \to (1,2)$ in the unbounded domain $\mathcal{D}$.

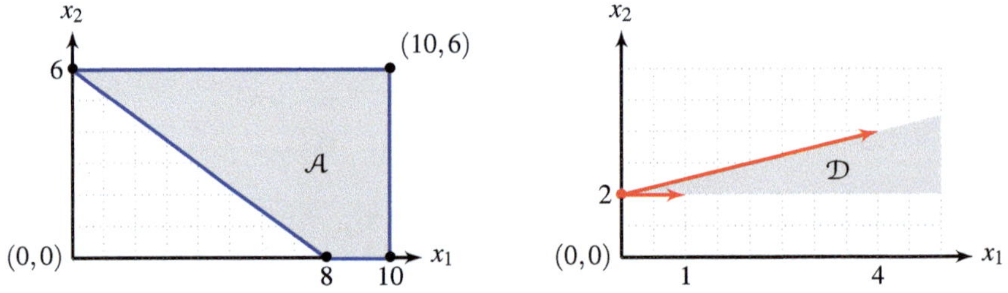

Fig. 3.11: Sets $\mathcal{A}$ and $\mathcal{D}$ for the *LP* (3.73).

Using the Minkowski-Weyl Theorem 3.1 on $\mathcal{D}$, let λ_1 be used for the extreme point $\mathbf{x}_1 = (0,2)$ whereas λ_2 and λ_3 are used for the extreme rays $\mathbf{x}_2 = (4,1)$ and $\mathbf{x}_3 = (1,0)$, respectively. After substitution in the constraints of $\mathcal{A}$ and the objective function, the *MP* comprises four constraints, where the row indices correspond to those of the dual variables $\boldsymbol{\pi}_\mathbf{b}^\mathsf{T} = [\pi_2, \pi_3, \pi_5]$, π_0 being reserved for the convexity constraint (here involving only variable λ_1):

$$
\begin{array}{llllllll}
\mathbf{x}: & & (0,2) & & (4,1) & & (1,0) & \\
z^\star_{MP} & = \min & 2\lambda_1 & + & 5\lambda_2 & + & \lambda_3 & \\
& \text{s.t.} & 8\lambda_1 & + & 16\lambda_2 & + & 3\lambda_3 \geq 24 & [\pi_2] \\
& & & & 4\lambda_2 & + & \lambda_3 \leq 10 & [\pi_3] \\
& & 2\lambda_1 & + & \lambda_2 & & \leq 6 & [\pi_5] \\
& & \lambda_1 & & & & = 1 & [\pi_0] \\
& & \lambda_1, & & \lambda_2, & & \lambda_3 \geq 0 & \\
& & (0,2)\lambda_1 & + & (4,1)\lambda_2 & + & (1,0)\lambda_3 = (x_1,x_2). &
\end{array}
\tag{3.75}
$$

Because $c_{\mathbf{x}} = x_1 + x_2$ in (3.73), the costs of the λ-variables are easily computed, e.g., $c_{(4,1)} = 4+1 = 5$ for λ_2. For the column coefficients in the constraints, the computation is done analogously. For example, for row index 2, $a_{2\mathbf{x}} = 3x_1 + 4x_2$, hence $a_{2(0,2)} = 8$, $a_{2(4,1)} = 16$, and $a_{2(1,0)} = 3$.

Optimal primal-dual solutions for the compact formulation

Solving the *MP* (3.75), we find

- a primal λ-solution: $\boldsymbol{\lambda}^\star_{MP} = (1,1,0)$ with objective value $z^\star_{MP} = 7 = z^\star_{LP}$;
- a primal x-solution: $\mathbf{x}^\star_{MP} = (0,2)\lambda^\star_1 + (4,1)\lambda^\star_2 + (1,0)\lambda^\star_3 = (4,3) = \mathbf{x}^\star_{LP}$;
- a partial dual solution: $\boldsymbol{\pi}^{\star\mathsf{T}}_{\mathbf{b}} = [\pi^\star_2, \pi^\star_3, \pi^\star_5] = [5/16, 0, 0]$ together with $\pi^\star_0 = -1/2$.

To complete the dual solution, we need to define and solve the pricing problem. This *SP*, a linear program with the dual vector $\boldsymbol{\pi}_{\mathbf{d}} = \begin{bmatrix}\pi_1 \\ \pi_4\end{bmatrix}$, writes as

$$
\begin{aligned}
\bar{c}(\boldsymbol{\pi}_{\mathbf{b}}, \pi_0) = -\pi_0 + \min_{\mathbf{x}\in\mathcal{D}} \; & c_{\mathbf{x}} - \begin{bmatrix}\pi_2 \\ \pi_3 \\ \pi_5\end{bmatrix}^{\mathsf{T}} \mathbf{a}_{\mathbf{x}} \\
\text{s.t. } \; & c_{\mathbf{x}} = x_1 + x_2 \\
& \mathbf{a}_{\mathbf{x}} = \begin{bmatrix} 3x_1 + 4x_2 \\ x_1 \\ x_2 \end{bmatrix},
\end{aligned}
$$

that is,

$$
\begin{array}{lll}
\bar{c}(\boldsymbol{\pi}_{\mathbf{b}}, \pi_0) = -\pi_0 + \min & (1 - 3\pi_2 - \pi_3)x_1 + (1 - 4\pi_2 - \pi_5)x_2 & \\
\text{s.t.} & -x_1 + 4x_2 \leq 8 & [\pi_1] \\
& x_2 \geq 2 & [\pi_4] \\
& x_1 \geq 0. &
\end{array}
\tag{3.76}
$$

Note that the objective coefficients $(1-3\pi_2-\pi_3)$ and $(1-4\pi_2-\pi_5)$ in the *SP* are *adjusted* costs of the variables x_1 and x_2. Their reduced costs can rather be derived from the *LP* as $\bar{c}_1=(1+\sigma_1-3\sigma_2-\sigma_3)$ and $\bar{c}_2=(1-4\sigma_1-4\sigma_2-\sigma_4-\sigma_5)$.

Solving the *SP* (3.76) for $\boldsymbol{\pi}^\star_{\mathbf{b}}$ and $\pi^\star_0$, that is,

$$\begin{aligned}\bar{c}(\boldsymbol{\pi}^\star_{\mathbf{b}},\pi^\star_0)=\frac{1}{2}\ +\ \min\ & \frac{1}{16}x_1-\frac{4}{16}x_2 \\ \text{s.t.}\ & -x_1+4x_2\leq 8 \quad [\pi_1] \\ & x_2\geq 2 \quad [\pi_4] \\ & x_1\geq 0,\end{aligned} \tag{3.77}$$

we identify the extreme point $(0,2)$, the obvious optimal value $\bar{c}(\boldsymbol{\pi}^\star_{\mathbf{b}},\pi^\star_0)=0$, and the requested dual values $\pi^\star_1=-1/16$ and $\pi^\star_4=0$. We have confirmed in this reformulation that $z^\star_{MP}=z^\star_{LP}$. Moreover, as both primal and dual optimal solutions of the *LP* are unique, we have $\mathbf{x}^\star_{MP}=\mathbf{x}^\star_{LP}$ and $\boldsymbol{\pi}^\star=\boldsymbol{\sigma}^\star$.

Column generation approach

We now solve the *MP* (3.75) by column generation.

Iteration 1. Let us first discard the last equation in terms of λ- and x-variables while introducing the surplus variable s_2 in the row indexed by 2 and slack variables s_3 and s_5 in rows indexed by 3 and 5, respectively. Because the slack variables can be basic, the initialization requires only two artificial variables, say y_2 and y_0 with big-M here equal to 100:

$$\begin{aligned}z_{RMP}=\min\ & 100y_2+100y_0 \\ \text{s.t.}\ & y_2-s_2=24 \quad [\pi_2] \\ & s_3=10 \quad [\pi_3] \\ & s_5=6 \quad [\pi_5] \\ & y_0=1 \quad [\pi_0] \\ & y_2,\ y_0,\ s_2,\ s_3,\ s_5\geq 0.\end{aligned}$$

The first basic solution takes the four values $(y_2,s_3,s_5,y_0)=(24,10,6,1)$ with dual vector $\boldsymbol{\pi}^\intercal_{\mathbf{b}}=[\pi_2,\pi_3,\pi_5,\pi_0]=[100,0,0,100]$ and $z_{RMP}=2500$. Solving the *SP* (3.76) with these dual values, that is,

$$\begin{aligned}\bar{c}(\boldsymbol{\pi}_{\mathbf{b}},\pi_0)=-100\ +\ \min\ & -299x_1-399x_2 \\ \text{s.t.}\ & -x_1+4x_2\leq 8 \quad [\pi_1] \\ & x_2\geq 2 \quad [\pi_4] \\ & x_1\geq 0,\end{aligned}$$

returns $\bar{c}(\boldsymbol{\pi}_\mathbf{b}, \pi_0) = -\infty$ with one of the two extreme rays $\mathbf{x}_2$ or $\mathbf{x}_3$, see Figure 3.11(right). Indeed, introducing the slack variable s_1, surplus s_4, and artificial y_4 into the above formulation, the *SP* writes as

$$\begin{array}{llllllll} \bar{c}(\boldsymbol{\pi}_\mathbf{b}, \pi_0) = -100 + \min & -299x_1 & -399x_2 & +100y_4 & & & & \\ \text{s.t.} & -x_1 & +4x_2 & & +s_1 & & =8 & [\pi_1] \\ & & x_2 & +y_4 & & -s_4 & =2 & [\pi_4] \\ & x_1, & & y_4, & s_1, & s_4 & \geq 0. & \end{array}$$

Firstly, there are various ways to reach the extreme point $\mathbf{x}_1 = \begin{bmatrix} x_1 \\ x_2 \end{bmatrix} = \begin{bmatrix} 0 \\ 2 \end{bmatrix}$ of cost -198, where the basis $\begin{bmatrix} -1 & 4 \\ 0 & 1 \end{bmatrix}$ and its inverse are identical. This provides the dual vector

$$\boldsymbol{\pi}_\mathbf{d}^\intercal = [\pi_1, \pi_4] = [-299, -399] \begin{bmatrix} -1 & 4 \\ 0 & 1 \end{bmatrix} = [299, -1695].$$

Secondly, both variables s_1 and s_4 then lead to negative cost extreme rays, i.e.,

$$\begin{bmatrix} x_1 \\ x_2 \end{bmatrix} = \begin{bmatrix} 0 \\ 2 \end{bmatrix} - \begin{bmatrix} -1 & 4 \\ 0 & 1 \end{bmatrix} \begin{bmatrix} 1 \\ 0 \end{bmatrix} s_1 = \begin{bmatrix} 0 \\ 2 \end{bmatrix} + \begin{bmatrix} 1 \\ 0 \end{bmatrix} s_1, \text{ with } \bar{c}_{s_1} = -299;$$

$$\begin{bmatrix} x_1 \\ x_2 \end{bmatrix} = \begin{bmatrix} 0 \\ 2 \end{bmatrix} - \begin{bmatrix} -1 & 4 \\ 0 & 1 \end{bmatrix} \begin{bmatrix} 0 \\ -1 \end{bmatrix} s_4 = \begin{bmatrix} 0 \\ 2 \end{bmatrix} + \begin{bmatrix} 4 \\ 1 \end{bmatrix} s_4, \text{ with } \bar{c}_{s_4} = -1695.$$

Let us select the extreme ray $\mathbf{x}_2$ identified by $(4,1)$ to add in the *RMP* as λ_2.

Iteration 2. The second *RMP* to solve, where we omit artificial variable y_2 simply to make room in the presentation, writes as

$$\begin{array}{lllllllll} & \mathbf{x}: & & & & & (4,1) & & \\ z_{RMP} & = \min & 100y_0 & & & & +\ 5\lambda_2 & & \\ & \text{s.t.} & & -\ s_2 & & & +\ 16\lambda_2 & = 24 & [\pi_2] \\ & & & & s_3 & & +\ 4\lambda_2 & = 10 & [\pi_3] \\ & & & & & s_5 & +\ \lambda_2 & = \ 6 & [\pi_5] \\ & & y_0 & & & & & = \ 1 & [\pi_0] \\ & & y_0, & s_2, & s_3, & s_5, & \lambda_2 & \geq 0. & \end{array}$$

The second basic solution takes the four values $(\lambda_2, s_3, s_5, y_0) = (1.5, 4, 4.5, 1)$ with dual vector $[\pi_2, \pi_3, \pi_5, \pi_0] = [5/16, 0, 0, 100]$ and $z_{RMP} = 107.5$.

Solving the *SP* (3.76) with these dual values, that is,

$$
\begin{aligned}
\bar{c}(\boldsymbol{\pi}_{\mathbf{b}}, \pi_0) = -100 \quad + \quad \min \quad & \frac{1}{16}x_1 \quad - \frac{5}{16}x_2 \\
\text{s.t.} \quad & -x_1 \quad + 4x_2 \leq 8 \quad [\pi_1] \\
& \qquad\qquad x_2 \geq 2 \quad [\pi_4] \\
& x_1 \geq 0,
\end{aligned}
$$

returns $\bar{c}(\boldsymbol{\pi}_{\mathbf{b}}, \pi_0) = -100.625$ with the extreme point $\mathbf{x}_1 = (0,2)$.

Iteration 3. Discarding the slack variable s_2, the third *RMP* is given by

$$
\begin{array}{llcccccccl}
\mathbf{x}: & & & & & (4,1) & & (0,2) & & \\
z_{RMP} = \min & 100y_0 & & & + & 5\lambda_2 & + & 2\lambda_1 & & \\
\text{s.t.} & & & & & 16\lambda_2 & + & 8\lambda_1 \geq 24 & & [\pi_2] \\
& & s_3 & & + & 4\lambda_2 & & = 10 & & [\pi_3] \\
& & & s_5 & + & \lambda_2 & + & 2\lambda_1 = \;\; 6 & & [\pi_5] \\
& y_0 & & & & & + & \lambda_1 = \;\; 1 & & [\pi_0] \\
& y_0, & s_3, & s_5, & & \lambda_2, & & \lambda_1 \geq 0. & &
\end{array}
$$

The third basic solution takes the four values $(\lambda_1, \lambda_2, s_3, s_5) = (1,1,2,3)$ with dual vector $[\pi_2, \pi_3, \pi_5, \pi_0] = [5/16, 0, 0, -1/2]$ and $z_{RMP} = 7$. The *SP* is exactly the one given in (3.77). It returns $\bar{c}(\boldsymbol{\pi}_{\mathbf{b}}, \pi_0) = 0$ and proves the optimality of this solution.

Example 3.2 ***Time constrained shortest path problem*** *(TCSPP)*

✎ A Dantzig-Wolfe reformulation and the solution process, step by step.

Consider the network $G = (N, A)$ depicted in Figure 3.12, where N is the set of six nodes and A is the set of ten arcs. Besides a cost c_{ij} attributed to each arc $(i, j) \in A$, there is also a traversal time t_{ij}. The problem is to find a shortest path from node 1 to node 6 such that its total traversal time does not exceed 14 time units. This example first appears in Ahuja et al. (1993, p. 599) within a chapter devoted to Lagrangian relaxation. It is later used in Desrosiers and Lübbecke (2005) where the Dantzig-Wolfe decomposition principle is applied to decompose an integer linear programming formulation of it.

Compact formulation

Let x_{ij} be a binary flow variable for $(i, j) \in A$. The integer linear programming formulation *ILP* provided in (3.78) for this *time constrained shortest path problem* (*TCSPP*) is as follows: one unit of flow must leave the source node 1 and enter the sink node 6, while flow conservation must hold at all other nodes. The restriction of at most 14 time units appears in (3.78e).

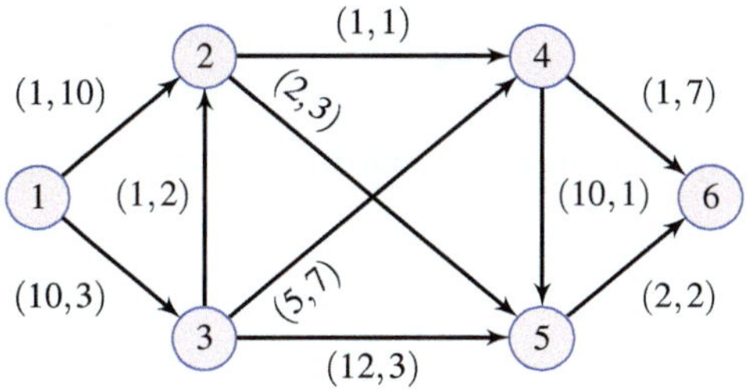

Fig. 3.12: Network $G=(N,A)$ with (c_{ij},t_{ij}) values, $\forall(i,j)\in A$.

$$
\begin{aligned}
z^\star_{ILP} = \min \quad & \sum_{(i,j)\in A} c_{ij}x_{ij} && && \text{(3.78a)}\\
\text{s.t.} \quad & \sum_{j:(1,j)\in A} x_{1j} = 1 && [\sigma_1] && \text{(3.78b)}\\
& \sum_{j:(i,j)\in A} x_{ij} - \sum_{j:(j,i)\in A} x_{ji} = 0 && [\sigma_i] \quad \forall i\in\{2,\dots,5\} && \text{(3.78c)}\\
& -\sum_{i:(i,6)\in A} x_{i6} = -1 && [\sigma_6] && \text{(3.78d)}\\
& \sum_{(i,j)\in A} t_{ij}x_{ij} \le 14 && [\sigma_7] && \text{(3.78e)}\\
& x_{ij}\in\{0,1\} && \forall(i,j)\in A. && \text{(3.78f)}
\end{aligned}
$$

By replacing the binary constraints by $x_{ij}\ge 0$, $\forall(i,j)\in A$, we obtain the linear relaxation with associated dual variables σ_i, $\forall i\in 1,\dots,7$. One can easily compute an optimal solution for the *LP* as $z^\star_{LP}=7$ with positive flow variables $x^\star_{12}=0.8$, $x^\star_{13}=x^\star_{32}=0.2$, $x^\star_{25}=x^\star_{56}=1$ and dual vector $\boldsymbol{\sigma}^{\star\mathsf{T}}=[0,-21,-16,-24,-29,-35,-2]$.

Grouping of constraints

Although we see in the next chapter why this is not (only) a good idea, let us select a grouping of the constraints based on the network flow structure of (3.78):

$$
\begin{aligned}
\mathcal{A} &= \{\mathbf{x}\in[0,1]^{|A|} \mid \text{(3.78e)}\} && \text{(3.79a)}\\
\mathcal{D} &= \{\mathbf{x}\in[0,1]^{|A|} \mid \text{(3.78b)–(3.78d)}\}. && \text{(3.79b)}
\end{aligned}
$$

The constraints defining $\mathcal{D}$ express a path structure from node 1 to node 6. By inspection, we find nine paths corresponding to the set $\{\mathbf{x}_p\}_{p\in P}$ of extreme points of $\mathcal{D}$ and no extreme rays. An extreme point $\mathbf{x}_p$ is a vector of dimension 10 (the number of arcs in A), that is, $\mathbf{x}_p=[x_{ijp}]_{(i,j)\in A}$, $p\in P$. Hence, we interpret from the Minkowski-Weyl Theorem 3.1 that any arc-flow solution $\mathbf{x}\in\mathcal{D}$ can be represented as a convex combination of these extreme points, that is,

$$\sum_{p\in P} \mathbf{x}_p\lambda_p = \mathbf{x}, \qquad \sum_{p\in P} \lambda_p = 1, \qquad \lambda_p \geq 0,\ \forall p \in P, \tag{3.80}$$

where the first equality written component-wise for the arc-flow variable vector $\mathbf{x}$ is $\sum_{p\in P} x_{ijp}\lambda_p = x_{ij}$, $\forall(i,j) \in A$.

Extended formulation

The substitution of (3.80) in the objective function of (3.78) and in the constraints defining $\mathcal{A}$ results in a *MP* with only two constraints: the traversal time constraint from $\mathcal{A}$ (with an associated dual variable π_7) and the convexity constraint (with the dual variable π_0). It is formulated as

$$\begin{aligned} z^\star_{MP} = \min \ & \sum_{p\in P} c_p\lambda_p \\ \text{s.t.}\ & \sum_{p\in P} t_p\lambda_p \leq 14 && [\pi_7] \\ & \sum_{p\in P} \lambda_p = 1 && [\pi_0] \\ & \lambda_p \geq 0 && \forall p \in P \\ & \sum_{p\in P} x_{ijp}\lambda_p = x_{ij} && \forall(i,j) \in A, \end{aligned} \tag{3.81}$$

where c_p is the cost of path p whereas t_p corresponds to its traversal time. These coefficients are respectively the contribution of the extreme point (path) $\mathbf{x}_p$ to the objective function and the traversal time constraint of $\mathcal{A}$:

$$c_p = \sum_{(i,j)\in A} c_{ij}x_{ijp} \quad \text{and} \quad t_p = \sum_{(i,j)\in A} t_{ij}x_{ijp}. \tag{3.82}$$

Let λ_{12456} be the variable associated with path $1 \to 2 \to 4 \to 5 \to 6$, and similarly for the other eight path variables. Since $x^\star_{ij}$ can be computed *a posteriori* when the $\lambda^\star_p$-variables are known, there is no need to keep the relation between the x_{ij}- and λ_p-variables while solving the *MP* and (3.81) becomes

$$\begin{aligned} \min\ & 3\lambda_{1246} + 14\lambda_{12456} + 5\lambda_{1256} + 13\lambda_{13246} + 24\lambda_{132456} + 15\lambda_{13256} + 16\lambda_{1346} + 27\lambda_{13456} + 24\lambda_{1356} \\ \text{s.t.}\ & 18\lambda_{1246} + 14\lambda_{12456} + 15\lambda_{1256} + 13\lambda_{13246} + 9\lambda_{132456} + 10\lambda_{13256} + 17\lambda_{1346} + 13\lambda_{13456} + 8\lambda_{1356} \leq 14 \\ & \lambda_{1246} + \lambda_{12456} + \lambda_{1256} + \lambda_{13246} + \lambda_{132456} + \lambda_{13256} + \lambda_{1346} + \lambda_{13456} + \lambda_{1356} = 1 \\ & \lambda_{1246},\ \lambda_{12456},\ \lambda_{1256},\ \lambda_{13246},\ \lambda_{132456},\ \lambda_{13256},\ \lambda_{1346},\ \lambda_{13456},\ \lambda_{1356} \geq 0. \end{aligned}$$

Optimal primal-dual solutions for the compact formulation

An optimal solution to this *MP* is $\lambda^\star_{13256} = 0.2$ and $\lambda^\star_{1256} = 0.8$ with $z^\star_{MP} = 7$. Projecting back to the original variables with $x^\star_{ij} = \sum_{p\in P} x_{ijp}\lambda^\star_p$, it corresponds to the arc-flows $x^\star_{12} = 0.8$, $x^\star_{13} = x^\star_{32} = 0.2$, and $x^\star_{25} = x^\star_{56} = 1$ previously found while solving the *LP*. Branching decisions are needed to recover integrality, the subject of Chapter 7. On Figure 3.13, each arc has a width proportional to its flow value and it is easy to validate that flow conservation is satisfied.

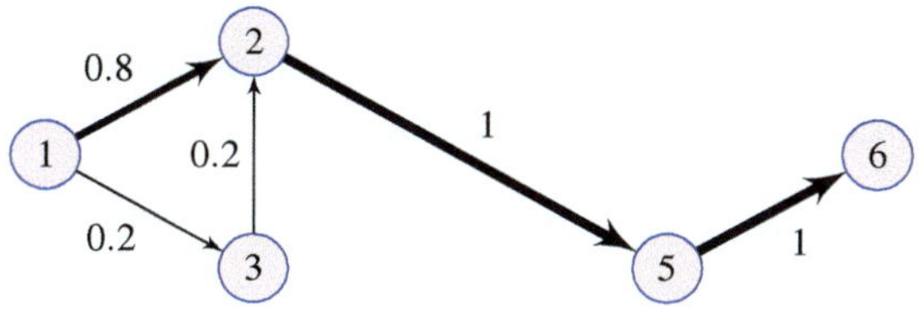

Fig. 3.13: Optimal arc-flow solution as a combination of paths 13256 and 1256.

Moreover, $\pi^\star_7 = -2 = \sigma^\star_7$ while $\pi^\star_0 = 35$. Our goal is now to find the missing dual values for $\pi^\star_i$, $i \in \{1,\ldots,6\}$, to obtain an equivalent primal-dual optimal solution to the *LP*. To do this, we simply have to solve the *SP* with input $(\pi^\star_7, \pi^\star_0) = (-2, 35)$:

$$
\begin{aligned}
\bar{c}(-2,35) = -35 + \min \quad & \sum_{(i,j)\in A} (c_{ij} + 2t_{ij}) x_{ij} \\
\text{s.t.} \quad & \sum_{j:(1,j)\in A} x_{1j} = 1 \quad [\pi_1] \\
& \sum_{j:(i,j)\in A} x_{ij} - \sum_{j:(j,i)\in A} x_{ji} = 0 \quad [\pi_i] \quad \forall i \in \{2,\ldots,5\} \\
& - \sum_{i:(i,6)\in A} x_{i6} = -1 \quad [\pi_6] \\
& x_{ij} \ge 0 \quad \forall (i,j) \in A,
\end{aligned}
$$

that is, $\boldsymbol{\pi}^{\star\mathsf{T}}_{\mathbf{d}} = [29, 8, 13, 5, 0, -6]$. This program is a network flow problem and, as such, naturally gives integer primal solutions but an infinite number of equivalent dual solutions. Indeed, by Proposition 1.10, we can translate an optimal dual solution by any scalar. In this case, $\boldsymbol{\pi}^\star_{\mathbf{d}}$ is a translation by $+29$ of

$$\boldsymbol{\sigma}^{\star\mathsf{T}}_{\mathbf{d}} = [0, -21, -16, -24, -29, -35]$$

found while directly solving the original *LP*. An alternative optimal dual vector to $\boldsymbol{\sigma}^\star = \begin{bmatrix} \boldsymbol{\sigma}^\star_{\mathbf{d}} \\ \sigma^\star_7 \end{bmatrix}$ is therefore given by $\boldsymbol{\pi}^\star = \begin{bmatrix} \boldsymbol{\pi}^\star_{\mathbf{d}} \\ \pi^\star_7 \end{bmatrix}$, where $\pi^\star_7 = \sigma^\star_7$.

Column generation approach

The *MP* (3.81) comprises nine λ-variables and two constraints with dual variables π_7 and π_0. During the column generation algorithm, an iteration consists of:

- Optimizing the *RMP* to determine the objective value z_{RMP} as well as the dual values π_7 and π_0;
- Finding a variable λ_p, $p \in P$, with a negative reduced cost $\bar{c}_p$ (if any), that is,

$$\bar{c}_p = c_p - \pi_7 t_p - \pi_0 = \sum_{(i,j)\in A} c_{ij} x_{ijp} - \pi_7 \sum_{(i,j)\in A} t_{ij} x_{ijp} - \pi_0 \ < 0. \qquad (3.83)$$

This amounts to solving the *SP*, a shortest path problem with an *adjusted cost* $(c_{ij} - \pi_7 t_{ij})$ for every arc $(i, j) \in A$, defined as

$$\bar{c}(\pi_7, \pi_0) = -\pi_0 + \min_{\mathbf{x}\in\mathcal{D}} \sum_{(i,j)\in A} (c_{ij} - \pi_7 t_{ij}) x_{ij}. \qquad (3.84)$$

Table 3.1 gives the five iterations for solving the *MP*, where we report for the *RMP*: a primal solution, the objective value z_{RMP}, as well as dual values π_0 and π_7, and regarding the *SP*: the minimum reduced cost $\bar{c}(\pi_7, \pi_0)$ given by path p of cost c_p and traversal time t_p. Finally, the last column shows the lower bound lb on $z^\star_{LP}$ computed as

$$lb = z_{RMP} + \bar{c}(\pi_7, \pi_0) = 14\,\pi_7 + \pi_0 + \bar{c}(\pi_7, \pi_0). \qquad (3.85)$$

Take a look at the sequence of generated paths. We first generate the least cost path but its duration is too long. From there on, the process adapts by generating a new path that leverages the cost or time component depending on the opportunity cost given by π_7 and π_0.

	RMP				*SP*				
t	primal solution	z_{RMP}	π_0	π_7	$\bar{c}(\pi_7, \pi_0)$	p	c_p	t_p	lb
1	$y_0 = 1, s_7 = 14$	100.0	100.00	0.00	−97.0	1246	3	18	3.00
2	$y_0 = 0.22, \lambda_{1246} = 0.78$	24.6	100.00	−5.39	−32.9	1356	24	8	−8.33
3	$\lambda_{1246} = 0.6, \lambda_{1356} = 0.4$	11.4	40.80	−2.10	−4.8	13256	15	10	6.60
4	$\lambda_{1246} = \lambda_{13256} = 0.5$	9.0	30.00	−1.50	−2.5	1256	5	15	6.50
5	$\lambda_{13256} = 0.2, \lambda_{1256} = 0.8$	7.0	35.00	−2.00	0.0	-	-	-	7.00
	$x^\star_{12} = 0.8$, $x^\star_{13} = x^\star_{32} = 0.2$, and $x^\star_{25} = x^\star_{56} = 1$								

Table 3.1: Iterations of the column generation algorithm while solving the *MP*.

Iteration 1. We initialize the column generation algorithm with a *Phase I*. To do so, we need only one artificial variable y_0 with a relatively large cost, say 100, for the convexity constraint while an added non-negative slack variable s_7 can be basic for the time constraint. We do not have any path variables yet and the *RMP* contains two constraints.

$$\begin{aligned} z_{RMP} = \min\ & 100 y_0 \\ \text{s.t.}\quad & s_7 = 14 \quad [\pi_7] \\ & y_0 = 1 \quad [\pi_0] \\ & y_0,\ s_7 \geq 0. \end{aligned}$$

This *RMP* is solved by inspection: $y_0 = 1$, $s_7 = 14$, $z_{RMP} = 100$, and the dual variables are $\pi_0 = 100$ and $\pi_7 = 0$. The *SP* returns path 1246 with a reduced cost $\bar{c}(\pi_7, \pi_0) = -97$, cost 3 and duration 18. The lower bound is $lb = 3$.

Iteration 2. The *RMP* contains three variables: y_0, s_7, and λ_{1246}.

$$\begin{array}{llllll} z_{RMP} = \min & 100y_0 + & 3\lambda_{1246} & & & \\ \text{s.t.} & & 18\lambda_{1246} & + s_7 & = 14 & [\pi_7] \\ & y_0 + & \lambda_{1246} & & = 1 & [\pi_0] \\ & y_0, & \lambda_{1246}, & s_7 & \geq 0. & \end{array}$$

An optimal solution is $z_{RMP} = 24.6$ with $y_0 = 0.22$ and $\lambda_{1246} = 0.78$, which is still infeasible for the *RMP* as $y_0 > 0$. The dual variables are $\pi_0 = 100$ and $\pi_7 = -5.39$. Solving the *SP* gives path 1356 with a reduced cost -32.9, cost 24, and duration 8. The lower bound *lb* is worse than the previous one as $-8.33 < 3$.

Iteration 3. The third *RMP* contains four variables: y_0, s_7, λ_{1246}, and λ_{1356}.

$$\begin{array}{lllllll} z_{RMP} = \min & 100y_0 + & 3\lambda_{1246} + & 24\lambda_{1356} & & & \\ \text{s.t.} & & 18\lambda_{1246} + & 8\lambda_{1356} & + s_7 & = 14 & [\pi_7] \\ & y_0 + & \lambda_{1246} + & \lambda_{1356} & & = 1 & [\pi_0] \\ & y_0, & \lambda_{1246}, & \lambda_{1356}, & s_7 & \geq 0. & \end{array}$$

Solving it provides $z_{RMP} = 11.4$ with $\lambda_{1246} = 0.6$ and $\lambda_{1356} = 0.4$. This is feasible for the *RMP*, thus ending the *Phase I* and the artificial variable is removed. The dual variables are $\pi_0 = 40.80$ and $\pi_7 = -2.10$. Solving the *SP* generates path 13256 with a reduced cost -4.8, cost 15, and duration 10. We obtain a better lower bound $lb = 6.6 > 3$.

Iteration 4. The *RMP* writes as

$$\begin{array}{lllllll} z_{RMP} = \min & 3\lambda_{1246} + & 24\lambda_{1356} + & 15\lambda_{13256} & & & \\ \text{s.t.} & 18\lambda_{1246} + & 8\lambda_{1356} + & 10\lambda_{13256} & + s_7 & = 14 & [\pi_7] \\ & \lambda_{1246} + & \lambda_{1356} + & \lambda_{13256} & & = 1 & [\pi_0] \\ & \lambda_{1246}, & \lambda_{1356}, & \lambda_{13256}, & s_7 & \geq 0. & \end{array}$$

We find $z_{RMP} = 9.0$ with $\lambda_{1246} = \lambda_{13256} = 0.5$ while $\pi_0 = 30.00$ and $\pi_7 = -1.50$. Solving the *SP* finds path 1256 with a reduced cost -2.5, cost 5, and duration 15. The lower bound $lb = 6.5 < 6.6$ again does not improve on the best one available.

Iteration 5. The *RMP* contains four path variables:

$$\begin{array}{llllllll} z_{RMP} = \min & 3\lambda_{1246} + & 24\lambda_{1356} + & 15\lambda_{13256} + & 5\lambda_{1256} & & & \\ \text{s.t.} & 18\lambda_{1246} + & 8\lambda_{1356} + & 10\lambda_{13256} + & 15\lambda_{1256} & + s_7 & = 14 & [\pi_7] \\ & \lambda_{1246} + & \lambda_{1356} + & \lambda_{13256} + & \lambda_{1256} & & = 1 & [\pi_0] \\ & \lambda_{1246}, & \lambda_{1356}, & \lambda_{13256}, & \lambda_{1256}, & s_7 & \geq 0. & \end{array}$$

We finally find $z_{RMP} = 7.0$ with $\lambda_{13256} = 0.2$, $\lambda_{1256} = 0.8$, $\pi_0 = 35.00$ and $\pi_7 = -2.00$. Solving the *SP* yields $\bar{c}(\pi_7, \pi_0) = 0$ and, therefore, the current solution to the *RMP* is optimal for the *MP*. Note that $lb = 7.0$ corresponds to the current upper bound z_{RMP}.

Figure 3.14 shows the evolution of the lower and upper bounds on $z^\star_{MP}$, where the upper bounds are only depicted when the solutions are also feasible for the *RMP*, i.e., starting at iteration 3. The integer optimum $z^\star_{ILP} = 13$ of the *ILP* (3.78) appears for comparison only. Finally, the best known lower bound *LB* is not depicted but would be updated at each iteration with the largest available lower bound.

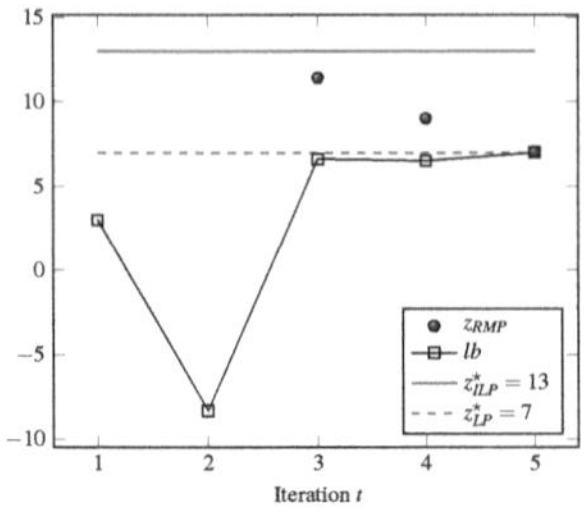

Fig. 3.14: Lower and upper bounds evolution in the column generation algorithm.

Note 3.16 (Master solutions: direct vs. column generation.) In total, four path variables are generated during the column generation process. We land back on the same primal solution as solving the *MP* directly. In general, this would be a purely coincidental observation not to mention it is practically irrelevant as we only solve the *MP* via column generation. It is however an opportunity to sharpen our intuition. Indeed, in this case, it is not surprising that we reach an identical master solution although the answer is not as trivial as one might conclude at first thought. Jumping to this conclusion based on the fact that the optimal solution of the compact formulation is unique is only half the answer. Indeed, Note 3.1 tells us that there could be infinitely many equivalent solutions in λ-variables for some $\mathbf{x}$. The other half of the answer therefore comes from the fact that the optimal solution for the *MP* uses only two λ-variables.

Example 3.3 ***Single depot vehicle scheduling problem: two compact formulations***

✎ Two linear programming compact formulations leading to the same *MP*, obviously with different *SP*s. Surprisingly, neither of them is the intuitive pricing problem!

Let us revisit the Single depot vehicle scheduling problem (Example 2.4) where the given *SP* (2.41) is a *shortest path problem defined from origin node o to destination node d*. This is intuitively interesting but such a pricing problem should be derived mathematically. We propose two compact formulations for the *SDVSP* on which we apply some Dantzig-Wolfe reformulations. They both provide us similar but different ways to derive the set partitioning model (2.38) introduced in Chapter 2.

Recall the acyclic network $G = (V,A)$ in Figure 2.12, with node set $V = N \cup \{o,d\}$ and arc set $A = I \cup (\{o\} \times N) \cup (N \times \{d\}) \cup \{(o,d)\}$, where I is the set of *inter-trip arcs* and arc (o,d) is used for the unused vehicles remaining at the depot. The first compact formulation uses the network G whereas the second adds the reverse arc (d,o), i.e., $G_{do} = (V, A_{do})$ with $A_{do} = A \cup \{(d,o)\}$. Both formulations are network flow models. Figures 3.15 and 3.16 respectively capture their essence in the corresponding presentation.

Formulation with *origin-to-destination* arc

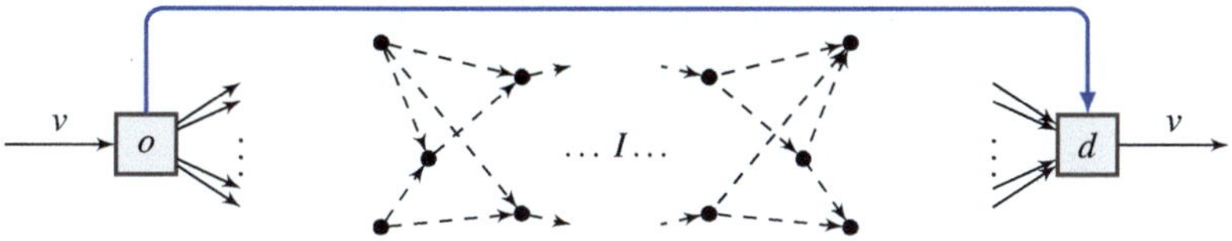

Fig. 3.15: Network G with v available vehicles at the depot.

Let the non-negative (integer) variable x_{ij} be the flow through arc $(i,j) \in A$. The first model uses the variable x_{od} as a slack variable for the unused vehicles remaining at the depot, with an eventual parking cost, see Figure 3.15. A linear programming network flow formulation is

$$z^\star_{LP} = \min \sum_{(i,j)\in A} c_{ij} x_{ij} \tag{3.86a}$$

$$\text{s.t.} \sum_{j:(i,j)\in A} x_{ij} = 1 \qquad [\sigma_i] \qquad \forall i \in N \tag{3.86b}$$

$$\sum_{j:(o,j)\in A} x_{oj} = v \qquad [\beta_o] \tag{3.86c}$$

$$\sum_{j:(i,j)\in A} x_{ij} - \sum_{j:(j,i)\in A} x_{ji} = 0 \qquad [\beta_i] \qquad \forall i \in N \tag{3.86d}$$

$$- \sum_{j:(j,d)\in A} x_{jd} = -v \qquad [\beta_d] \tag{3.86e}$$

$$x_{ij} \geq 0 \qquad \forall (i,j) \in A. \tag{3.86f}$$

The constraints (3.86b) ensure that every trip in N is operated once whereas path constraints (3.86c)–(3.86e) impose that v vehicles are available at the depot to flow through the network G. Except for arc (o,d), the flow on the arcs is *implicitly limited* to one unit due to the first set of constraints. An important observation is that the model (3.86) corresponds to a network comprising $2|N|+2$ nodes, i.e., two for every trip $i \in N$ and two for the depot at nodes o and d. Such a network representation, *different* from that of Figure 3.15, is derived in Exercise 2.8; see also forthcoming

Note 3.18. Our reformulation of the *LP* (3.86) groups the constraints as

$$\mathcal{A} = \{\mathbf{x} \in \mathbb{R}_+^{|A|} \mid (3.86\text{b})\} \tag{3.87a}$$

$$\mathcal{D} = \{\mathbf{x} \in \mathbb{R}_+^{|A|} \mid (3.86\text{c})\text{–}(3.86\text{e})\}. \tag{3.87b}$$

In $\mathcal{D}$, the flow is now implicitly bounded by v units on every arc of the acyclic network G. In fact, an extreme point of $\mathcal{D}$ is an od-path with exactly v units of flow such that $\mathcal{X} = \left\{\begin{bmatrix}\mathbf{x}_p\\1\end{bmatrix}\right\}_{p\in P}$.

Such a path either uses the single arc (o,d) with $x_{od} = v$ (no trips are operated) or it comprises at least two arcs and $x_{od} = 0$ (some trips are operated). The substitution of the convex combination of the extreme points in the constraints defining $\mathcal{A}$ and the objective function is done as usual. We however reserve λ and π for later use by writing the *MP* in terms of primal θ- and dual γ-variables as

$$\begin{aligned}
z^\star_{MP} = \min \quad & \sum_{p\in P} C_p\theta_p \\
\text{s.t.} \quad & \sum_{p\in P} v_{ip}\theta_p = 1 \quad [\gamma_i] \quad \forall i \in N \\
& \sum_{p\in P} \theta_p = 1 \quad [\gamma_0] \\
& \theta_p \geq 0 \quad \forall p \in P \\
& \sum_{p\in P} \mathbf{x}_p\theta_p = \mathbf{x},
\end{aligned} \tag{3.88}$$

where C_p is the cost of v units of flow on path $\mathbf{x}_p$, and $v_{ip} = v$ if $\mathbf{x}_p$ operates trip $i \in N$, 0 otherwise. Taking into account the dual values γ_i, $\forall i \in N$, and γ_0, retrieved from the solution of an *RMP*, the *SP* solves a shortest path problem sending v units from o to d, with the minimum reduced cost value here given by

$$\bar{C}(\boldsymbol{\pi}, \pi_0) = \min \quad C_p - \sum_{i\in N} \gamma_i v_{ip} - \gamma_0 \tag{3.89a}$$

$$\text{s.t.} \quad \sum_{j:(o,j)\in A} x_{oj} = v \tag{3.89b}$$

$$\sum_{j:(i,j)\in A} x_{ij} - \sum_{j:(j,i)\in A} x_{ji} = 0 \quad \forall i \in N \tag{3.89c}$$

$$- \sum_{j:(j,d)\in A} x_{jd} = -v \tag{3.89d}$$

$$x_{ij} \geq 0 \quad \forall (i,j) \in A \tag{3.89e}$$

$$C_p = \sum_{(i,j)\in A} c_{ij}x_{ij} \tag{3.89f}$$

$$v_{ip} = \sum_{j:(i,j)\in A} x_{ij} \quad \forall i \in N. \tag{3.89g}$$

To obtain the linear relaxation of a set partitioning problem from (3.88), we use the change of variables

$$\lambda_p = v\theta_p, \quad \forall p \in P, \tag{3.90}$$

which leads to the forthcoming *MP* with

- scaled cost coefficients $c_p = C_p/v, \forall p \in P$;
- binary column coefficients $a_{ip} = v_{ip}/v, \forall i \in N$;
- dual variables $\pi_i = \gamma_i, \forall i \in N$, and $\pi_0 = \gamma_0/v$;
- v available vehicles at the depot:

$$\begin{aligned} z^\star_{MP} = \min \quad & \sum_{p\in P} c_p\lambda_p \\ \text{s.t.} \quad & \sum_{p\in P} a_{ip}\lambda_p = 1 \quad [\pi_i] \quad \forall i \in N \\ & \sum_{p\in P} \lambda_p = v \quad [\pi_0] \\ & \lambda_p \geq 0 \quad \forall p \in P \\ & \sum_{p\in P} \left(\frac{\mathbf{x}_p}{v}\right) \lambda_p = \mathbf{x}. \end{aligned} \tag{3.91}$$

If the cost of an unused vehicle is zero ($c_{od} = 0$), the λ-variable associated with the empty schedule ($x_{od} = v$ and a zero-flow everywhere else) can be omitted and the depot constraint becomes a less-than-or-equal constraint ($\sum_{p\in P} \lambda_p \leq v$). Although mathematically accurate, the above derivation of the *MP* as the linear relaxation of a set partitioning-type model (when the last set of constraints is removed) is rather laborious. The next one is much simpler, obviously derived from a different compact formulation.

Formulation with *destination-to-origin* arc

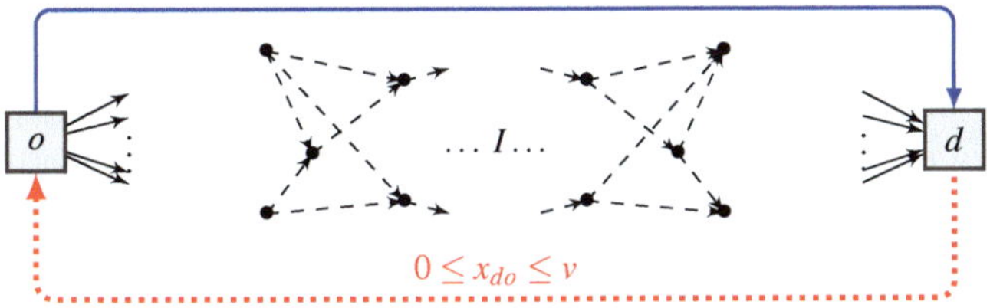

Fig. 3.16: Network G_{do} with v available vehicles at the depot.

Let x_{ij} be the flow through arc $(i, j) \in A_{do}$. The second flow model uses the zero-cost variable $x_{do} \leq v$ and its slack accounts for the unused vehicles (Figure 3.16). A linear programming network flow formulation is

$$
\begin{aligned}
z^\star_{LP} = \min \quad & \sum_{(i,j)\in A} c_{ij} x_{ij} && && (3.92a)\\
\text{s.t.} \quad & \sum_{j:(i,j)\in A} x_{ij} = 1 && [\sigma_i] \quad \forall i \in N && (3.92b)\\
& x_{do} \leq v && [\sigma_{do}] && (3.92c)\\
\sum_{j:(i,j)\in A_{do}} x_{ij} - & \sum_{j:(j,i)\in A_{do}} x_{ji} = 0 && [\beta_i] \quad \forall i \in N \cup \{o,d\} && (3.92d)\\
& x_{ij} \geq 0 && \forall (i,j) \in A_{do}. && (3.92e)
\end{aligned}
$$

Constraints (3.92b) ensure that every trip in N is operated exactly once. The second set (3.92c) makes at most v vehicles available at the depot whereas (3.92d) provides the $n+2$ flow conservation equations at every node $i \in N \cup \{o,d\}$. Except for arcs (o,d) and (d,o), the flow on the arcs is *implicitly limited* to one unit by the first set of constraints. We group the partitioning constraints and the upper bound on x_{do} in $\mathcal{A}$ whereas $\mathcal{D}$ contains the flow conservation equations:

$$
\begin{aligned}
\mathcal{A} &= \{\mathbf{x} \in \mathbb{R}_+^{|A_{do}|} \mid (3.92b)\text{–}(3.92c)\} && (3.93a)\\
\mathcal{D} &= \{\mathbf{x} \in \mathbb{R}_+^{|A_{do}|} \mid (3.92d)\}. && (3.93b)
\end{aligned}
$$

The set $\mathcal{D}$ is a polyhedral cone with the unique extreme point $\mathbf{0}$ and the set R of extreme rays. Such a ray is a cycle formed by a path from o to d and arc (d,o). We represent it with a unit-flow on the selected arcs. The reformulation writes as

$$
\begin{aligned}
z^\star_{MP} = \min \quad & 0\lambda_{\mathbf{0}} + \sum_{r\in R} c_r \lambda_r && && (3.94a)\\
\text{s.t.} \quad & 0\lambda_{\mathbf{0}} + \sum_{r\in R} a_{ir} \lambda_r = 1 && [\pi_i] \quad \forall i \in N && (3.94b)\\
& \sum_{r\in R} b_r \lambda_r \leq v && [\pi_{do}] && (3.94c)\\
& \lambda_{\mathbf{0}} = 1 && [\pi_0] && (3.94d)\\
& \lambda_r \geq 0 && \forall r \in R && (3.94e)\\
& \sum_{r\in R} \mathbf{x}_r \lambda_r = \mathbf{x}. && && (3.94f)
\end{aligned}
$$

Discarding the extreme point $\mathbf{0}$ from the reformulation, the *MP* no longer comprises the convexity constraint (3.94d) and gives the linear relaxation of the set partitioning-type model as

$$
\begin{aligned}
z^\star_{MP} = \min \quad & \sum_{r\in R} c_r \lambda_r && && (3.95a)\\
\text{s.t.} \quad & \sum_{r\in R} a_{ir} \lambda_r = 1 && [\pi_i] \quad \forall i \in N && (3.95b)\\
& \sum_{r\in R} b_r \lambda_r \leq v && [\pi_{do}] && (3.95c)
\end{aligned}
$$

$$\lambda_r \geq 0 \qquad \forall r \in R \tag{3.95d}$$

$$\sum_{r \in R} \mathbf{x}_r \lambda_r = \mathbf{x}. \tag{3.95e}$$

Using the dual variables π_i, $\forall i \in N$, and π_{do} retrieved from the *MP* (3.95), the *SP* turns out to be a network flow *circulation* problem for which an optimal negative reduced cost cycle, if any, is *represented* using a unit-flow on x_{do}, that is,

$$\bar{c}(\boldsymbol{\pi}, \pi_{do}) = \min \quad c_r - \sum_{i \in N} \pi_i a_{ir} - \pi_{do} b_r \tag{3.96a}$$

$$\text{s.t.} \quad \sum_{j:(i,j) \in A_{od}} x_{ij} - \sum_{j:(j,i) \in A_{do}} x_{ji} = 0 \qquad \forall i \in N \cup \{o, d\} \tag{3.96b}$$

$$x_{ij} \geq 0 \qquad \forall (i,j) \in A_{do} \tag{3.96c}$$

$$c_r = \sum_{(i,j) \in A} c_{ij} x_{ij} \tag{3.96d}$$

$$a_{ir} = \sum_{j:(i,j) \in A} x_{ij} \qquad \forall i \in N \tag{3.96e}$$

$$b_r = x_{do}. \tag{3.96f}$$

Solving (3.96), we set $x_{do} = 1$ for any negative reduced cost extreme ray $\mathbf{x}_r$, $r \in R$, while accordingly computing c_r and a_{ir}, $\forall i \in N$. Otherwise, we find the zero-cost extreme point $\mathbf{0}$ for which $\bar{c}_{\mathbf{0}} = 0$, and the algorithm terminates, see Exercise 3.11 (c).

Note 3.17 (Taking advantage of the new interpretation.) In practice, we do not directly solve the *SP* (3.96) but rather find a shortest path from o to d and complete such a path with $x_{do} = 1$. This provides a negative reduced cost cycle, if any. Much more interestingly, we find in this way the true minimum reduced cost $\bar{c}_r$, $r \in R$ (scaled with $b_r = 1$). We hence mathematically take advantage of the new interpretation, replacing every extreme point by a corresponding extreme ray, and thus facilitate the writing of the Dantzig-Wolfe reformulation (3.95).

Note 3.18 (Start and end trip nodes.) The *SP* (3.96) uses the network G_{do}. The arc-flow formulation (3.92) rather uses a network comprising $2|N| + 2$ nodes, one per equality constraint in (3.92b) and (3.92d). Indeed, every trip is represented by two nodes for the start and end of service; these are connected by an arc and the task is completed when the arc is traversed. For trip $i \in N$, this arc is associated with variable $x_i = 1$, see Figure 3.17 and Exercise 2.8. The cost of such an arc can be set to zero or to the fixed cost incurred to cover that trip. Since each trip has to be covered exactly once, the sum of the costs of all the trips is a constant.

Note 3.19 (Aircraft routing.) The network representation with two nodes for a trip is in fact natural in the airline industry, for example, a flight Montréal – Frankfurt is represented by the two city-nodes. Figure 3.18 illustrates such a simplified network for an airline company that operates 15 daily flights between cities A, B, and C. In (a), we have the instance data to be read as *from:to:departure time:flight duration*.

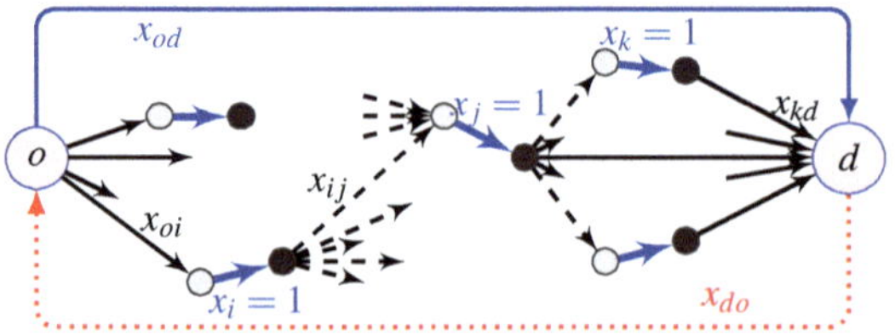

Fig. 3.17: Network for the arc-flow formulation (3.92) of the *SDVSP*.

In (b), we present these flights on a time-space network, where a node identifier indicates a city (A, B, or C), the activity number (1, 2, ...) within the airport, and the type (D for departure, A for arrival). The time is increasing from top to bottom, with a night arc in each city from the end of a day to the beginning of the next one. In general, for a set of n flights, the network is polynomial in size, that is, $2n$ nodes (two per flight) and $3n$ arcs (one per flight [solid] plus one for each ground task [dashed]). Moreover, any solution to this minimum cost flow problem can be decomposed into a set of directed cycles by Proposition 3.4. The goal is to determine the minimum number of (identical) aircraft needed by finding appropriate routing, see Exercises 3.14–3.15.

```
A:B:8h00:0h50
B:A:8h15:2h13
C:A:8h45:3h17
A:C:9h00:1h55
B:C:9h09:0h32
B:C:9h34:0h32
A:B:10h55:0h50
A:C:11h25:1h55
C:B:11h27:1h33
B:A:12h08:2h13
C:B:12h30:1h33
C:B:15h07:1h33
B:A:15h32:2h13
C:B:16h00:1h33
B:C:16h10:0h32
```

(a) Raw flight data

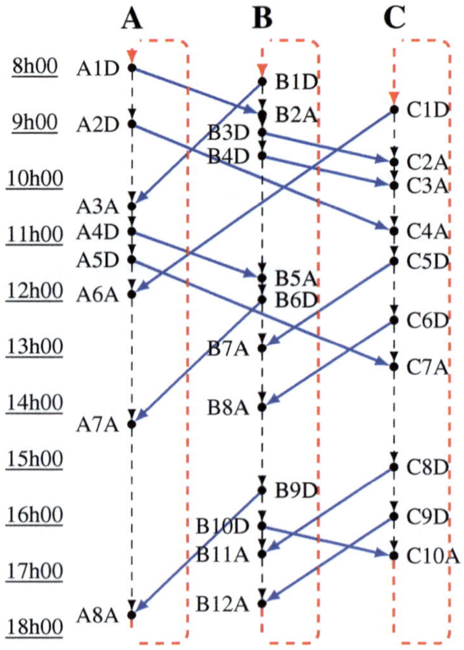

(b) Time-space network

Fig. 3.18: Instance with 15 daily flights between cities A, B, and C.

Example 3.4 **Solving a sequence of restricted compact formulations**

✎ An example of row and column activation using the *restricted compact formulation* theory presented on p. 124.

We consider again the Time constrained shortest path problem (*TCSPP*) with an alternative formulation to (3.78) on a network augmented by the no-cost and no-time arc $(d,o) = (6,1)$, that is, $G_{do} = (N, A_{do})$, where $A_{do} = A \cup \{(6,1)\}$, see Figure 3.19.

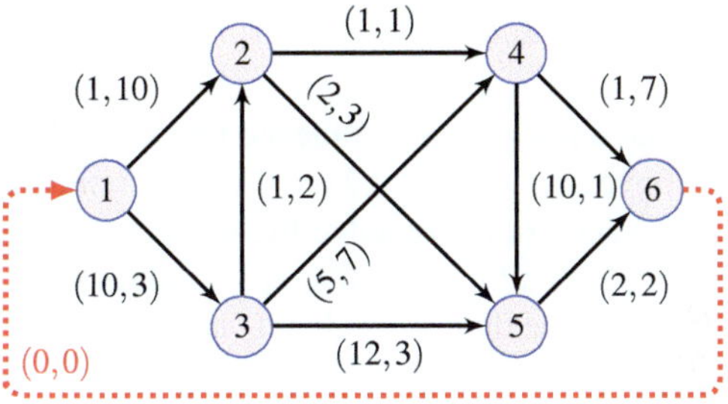

Fig. 3.19: Network G_{do} with parameters (c_{ij}, t_{ij}), $\forall (i,j) \in A_{do}$.

The linear relaxation of this alternative compact formulation is given by

$$z^\star_{LP} = \min \sum_{(i,j)\in A} c_{ij} x_{ij} \tag{3.97a}$$

$$\text{s.t.} \quad \sum_{j:(i,j)\in A_{do}} x_{ij} - \sum_{j:(j,i)\in A_{do}} x_{ji} = 0 \qquad [\sigma_i] \qquad \forall i \in \{1,\dots,6\} \tag{3.97b}$$

$$\sum_{(i,j)\in A} t_{ij} x_{ij} \leq 14 \qquad [\sigma_7] \tag{3.97c}$$

$$x_{61} = 1 \qquad [\sigma_{61}] \tag{3.97d}$$

$$x_{ij} \geq 0 \qquad \forall (i,j) \in A_{do}. \tag{3.97e}$$

The flow conservation equations in (3.97b) describe a circulation, the second imposes the traversal time restriction (3.97c) while constraint $x_{61} = 1$ (with the associated dual variable denoted σ_{61}) implicitly asks for a single unit from source node $o = 1$ to terminal node $d = 6$ by the flow conservation equations. We use the grouping of constraints

$$\mathcal{A} = \{\mathbf{x} \in \mathbb{R}_+^{|A_{do}|} \mid \text{(3.97c)–(3.97d)}\} \tag{3.98a}$$

$$\mathcal{D} = \{\mathbf{x} \in \mathbb{R}_+^{|A_{do}|} \mid \text{(3.97b)}\}. \tag{3.98b}$$

The *SP* derived from $\mathcal{D}$ becomes a network flow circulation problem with the extreme rays encoded using a unit-flow on the associated cycles, and the single extreme point $\mathbf{0}$, that is, $\mathcal{X} = \left\{\begin{bmatrix}\mathbf{0}\\1\end{bmatrix}\right\} \cup \left\{\begin{bmatrix}\mathbf{x}_r\\0\end{bmatrix}\right\}_{r\in R}$ The *MP* writes as

$$
\begin{aligned}
z^\star_{MP} = \min \quad & 0\lambda_\mathbf{0} + \sum_{r\in R} c_r \lambda_r \\
\text{s.t.} \quad & 0\lambda_\mathbf{0} + \sum_{r\in R} t_r \lambda_r \le 14 && [\pi_7] \\
& \sum_{r\in R} \lambda_r = 1 && [\pi_{61}] \\
& \lambda_\mathbf{0} = 1 && [\pi_0] \\
& \lambda_r \ge 0 && \forall r \in R \\
& \sum_{r\in R} x_{ijr}\lambda_r = x_{ij} && \forall (i,j) \in A_{do},
\end{aligned}
\tag{3.99}
$$

where c_r is the cost of the unit-flow cycle indexed by $r \in R$ whereas t_r is its traversal time. Obviously, the convexity constraint $\lambda_\mathbf{0} = 1$ can be removed. Rather than solving this *MP* by column generation, we solve the original *LP* (3.97) by sequentially activating in the restricted *LP* all the positive x_{ij}-variables identified by the pricing problem defined in terms of the dual variables σ_7 and σ_{61} as

$$
\bar{c}(\sigma_7, \sigma_{61}) = \min_{\mathbf{x}\in\mathcal{D}} \sum_{(i,j)\in A} (c_{ij} - \sigma_7 t_{ij}) x_{ij} - \sigma_{61} x_{61}. \tag{3.100}
$$

In this example, the *SP* is solved as a shortest path problem from 1 to 6 after which arc $(6,1)$ is added, that is, $x_{61} = 1$. We reach optimality in four iterations detailed below. Figure 3.20 highlights that at each iteration, the *RLP* corresponds to a subnetwork of G_{do} that depends on the nodes and arcs that have been activated.

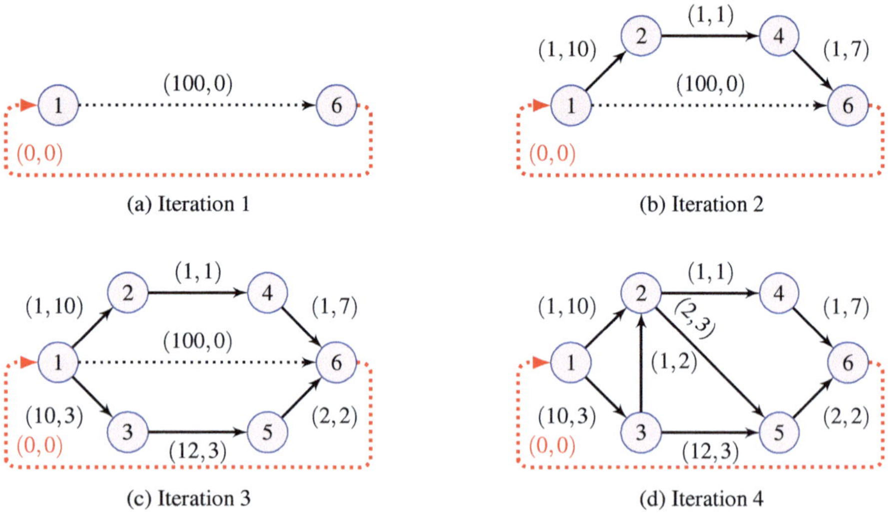

(a) Iteration 1
(b) Iteration 2
(c) Iteration 3
(d) Iteration 4

Fig. 3.20: Subnetworks of G_{do}.

Iteration 1. The algorithm starts with a *RLP* comprising the two constraints in $\mathcal{A}$, that is, $0 \le 14$ and the single variable $x_{61} = 1$. Moreover, because x_{61} appears in the flow conservation constraints for nodes 1 and 6, these are activated from $\mathcal{D}$, and we need an artificial variable, say y_{16} with cost 100 and zero traversal time, to satisfy them:

$$
\begin{array}{rrrrl}
z_{RLP} = \min & 100y_{16} & & & \\
\text{s.t.} & y_{16} & -\ x_{61} & = 0 & [\sigma_1] \\
 & -y_{16} & +\ x_{61} & = 0 & [\sigma_6] \\
 & & & 0 \le 14 & [\sigma_7] \\
 & & x_{61} & = 1 & [\sigma_{61}] \\
 & y_{16} \ge 0. & & &
\end{array}
$$

Solving the above *RLP* gives $z_{RLP} = 100$ with $y_{16} = x_{61} = 1$, $\sigma_{61} = 100$, and $\sigma_7 = 0$ for the inactive constraint. The *SP* finds $\bar{c}(\sigma_7, \sigma_{61}) = -97$ with cycle 12461 of cost 3 and duration 18. The variables x_{12}, x_{24}, and x_{46} are added to the *RLP* and the flow conservation equations at nodes 2 and 4 are activated.

Iteration 2. The *RLP* contains five variables and writes as

$$
\begin{array}{rrrrrrrl}
z_{RLP} = \min & 100y_{16} & +10x_{12} & +x_{24} & +x_{46} & & & \\
\text{s.t.} & y_{16} & +x_{12} & & & -x_{61} & = 0 & [\sigma_1] \\
 & & -x_{12} & +x_{24} & & & = 0 & [\sigma_2] \\
 & & & -x_{24} & +x_{46} & & = 0 & [\sigma_4] \\
 & -y_{16} & & & -x_{46} & +x_{61} & = 0 & [\sigma_6] \\
 & & 10x_{12} & +x_{24} & +7x_{46} & & \le 14 & [\sigma_7] \\
 & & & & & x_{61} & = 1 & [\sigma_{61}] \\
 & y_{16}, & x_{12}, & x_{24}, & x_{46} & & \ge 0. &
\end{array}
$$

Solving it gives $z_{RLP} = 24.6$ with $y_{16} = 0.22$, $x_{12} = x_{24} = x_{46} = 0.78$, and $x_{61} = 1$. This is still infeasible for the *LP*. The relevant dual values for the *SP* are $\sigma_7 = -5.39$ and $\sigma_{61} = 100$. The *SP* returns cycle 13561 of reduced cost -32.9, cost 24, and duration 8. The variables x_{13}, x_{35}, and x_{56} are added to the *RLP* while the constraints at nodes 3 and 5 are activated.

Iteration 3. The *RLP* now contains eight variables and corresponds to

$$
\begin{array}{rrrrrrrrrrl}
z_{RLP} = \min & 100y_{16} & +x_{12} & +10x_{13} & +x_{24} & +12x_{35} & +x_{46} & +2x_{56} & & & \\
\text{s.t.} & y_{16} & +x_{12} & +x_{13} & & & & & -x_{61} & = 0 & [\sigma_1] \\
 & & -x_{12} & & +x_{24} & & & & & = 0 & [\sigma_2] \\
 & & & -x_{13} & & +x_{35} & & & & = 0 & [\sigma_3] \\
 & & & & -x_{24} & & +x_{46} & & & = 0 & [\sigma_4] \\
 & & & & & -x_{35} & & +x_{56} & & = 0 & [\sigma_5] \\
 & -y_{16} & & & & & -x_{46} & -x_{56} & +x_{61} & = 0 & [\sigma_6] \\
 & & 10x_{12} & +3x_{13} & +x_{24} & +3x_{35} & +7x_{46} & +2x_{56} & & \le 14 & [\sigma_7] \\
 & & & & & & & & x_{61} & = 1 & [\sigma_{61}] \\
 & y_{16}, & x_{12}, & x_{13}, & x_{24}, & x_{35}, & x_{46}, & x_{56} & & \ge 0. &
\end{array}
$$

Solving it provides $z_{RLP} = 11.4$ with $x_{12} = x_{24} = x_{46} = 0.6$ and $x_{13} = x_{35} = x_{56} = 0.4$. This is feasible for the *LP*. Hence it ends the *Phase I* and we can discard y_{16} from the *RLP*. The dual variables to be forwarded to the *SP* are $\sigma_7 = -2.10$ and $\sigma_{61} = 40.80$. The *SP* finds cycle 132561 of reduced cost -4.8, cost 15, and duration 10. It identifies two additional variables to add to the *RLP*: x_{32} and x_{25}.

Iteration 4. The *RLP* contains nine flow variables whose formulation is given by

$$
\begin{array}{rlllllllllll}
z_{RLP} = \min & x_{12} & +10x_{13} & +x_{24} & +2x_{25} & +12x_{32} & +12x_{35} & +x_{46} & +2x_{56} & & & \\
\text{s.t.} & x_{12} & +x_{13} & & & & & & & -x_{61} & = 0 & [\sigma_1] \\
 & -x_{12} & & +x_{24} & +x_{25} & -x_{32} & & & & & = 0 & [\sigma_2] \\
 & & -x_{13} & & & +x_{32} & +x_{35} & & & & = 0 & [\sigma_3] \\
 & & & -x_{24} & & & & +x_{46} & & & = 0 & [\sigma_4] \\
 & & & & -x_{25} & & -x_{35} & & +x_{56} & & = 0 & [\sigma_5] \\
 & & & & & & & -x_{46} & -x_{56} & +x_{61} & = 0 & [\sigma_6] \\
 & 10x_{12} & +3x_{13} & +x_{24} & +3x_{25} & +2x_{32} & +3x_{35} & +7x_{46} & +2x_{56} & & \leq 14 & [\sigma_7] \\
 & & & & & & & & & x_{61} & = 1 & [\sigma_{61}] \\
 & x_{12}, & x_{13}, & x_{24}, & x_{25}, & x_{32}, & x_{35}, & x_{46}, & x_{56} & & \geq 0. &
\end{array}
$$

Solving it yields $z_{RLP} = 7$ with $x_{12} = 0.8$, $x_{13} = x_{32} = 0.2$ and $x_{25} = x_{56} = 1$. The relevant dual values for the *SP* are $\sigma_7 = -2$ and $\sigma_{61} = 35$. The *SP* finds $\bar{c}(-2, 35) = 0$ and the current solution to the *RLP* is optimal for the *LP*. It was found in four iterations, one less compared to the column generation algorithm depicted in Table 3.1. The original variables x_{34} and x_{45} have not been activated in the *RLP*.

3.6 Reference Notes

Section 3.1 Section 3.1 Several references (in German, English, and French) to the work of Minkowski and Weyl are available in Schrijver (1986) who attributes it as well to Farkas. See also Charnes and Cooper (1958). Note again that the Dantzig-Wolfe decomposition principle (Dantzig and Wolfe, 1960, 1961) is in fact a *reformulation* of the original formulation *LP*, not a solution method.

Section 3.2 We found the middle name of almost all cited authors, sometimes via a personal exchange. Figure 3.21 is a record of one such communication with Warren E. Walker, see Proposition 3.1 (Walker, 1969). We welcome any addition or correction to this unabbreviated bibliography. That being said, we will promptly honor any wish to suppress such a detail.

We recall that the lower bound (3.33) on $z^{\star}_{LP}$ using arbitrary dual values, here developed in the context of the Dantzig-Wolfe decomposition, is better known as the *Lagrangian dual bound* (see Section 6.2, Proposition 6.1 and, more generally, relation (6.18) for a linear program with a block-diagonal structure).

Von: Warren Walker - LR <w.e.walker@tudelft.nl>
Datum: Donnerstag, 24. Sept. 2020 um 18:37 Uhr
Betreff: RE: middle initial
An: Marco Lübbecke <marco.luebbecke@rwth-aachen.de>
Cc: Jacques Desrosiers <jacques.desrosiers@gerad.ca>, Warren Walker - LR <w.e.walker@tudelft.nl>

Dear Prof. Lübbecke:

Thank you for your kind e-mail message. I have no problem sharing my middle name. It is Elliott.

Would it be possible to send me the paper that cites my 1969 letter? I would love to see the citation (that was more than 50 years ago!).

Sincerely,

Warren Elliott Walker

Professor (Emeritus) of Policy Analysis
Faculty of Technology, Policy and Management
Faculty of Aerospace Engineering
Delft University of Technology
P.O. Box 5015
2600 GA Delft
The Netherlands

2020-09-24 18:12 UTC+01:00 Marco Lübbecke <marco.luebbecke@rwth-aachen.de>

> Dear Dr. Walker,
>
> Jacques Desrosiers (Cc) and myself with co-authors write a book on branch-and-price. We also cite your 1969 letter to the editor of Operations Research. In all our references, we list all the authors with all their first names, when we can.
>
> Is it too impolite to ask what the middle initial "E." means? I would appreciate it if you would reveal this to us, but I fully understand if you prefer not to share.
>
> Stay in good health, all the best, Marco
>
> --
> Prof. Dr. Marco Lübbecke
> RWTH Aachen University
> Chair of Operations Research
> Kackertstrasse 7
> D-52072 Aachen
> Germany

Fig. 3.21: Warren E? Walker.

Good to Know Static variables are widely used: as artificial variables or for a known initial feasible solution, for perturbation of the right-hand side vector and soft constraints, as a support for dual information in dual-optimal inequalities (Section 6.3) and stabilization techniques (Section 6.4), for shared variables across blocks (p. 222), etc.

More to Know Further readings on the solution of the *LP* by using a series of restricted compact problems can be found in Mamer and McBride (2000) and Sadykov and Vanderbeck (2013). More generally, pseudo-polynomial arc-flow formulations allow to develop tighter formulations for mixed-integer programs, see Valério de Carvalho (2002) and Delorme and Iori (2020). For such a *LP*, the subproblem solutions are re-expressed in the variables of the arc-flow formulation and added to the restricted compact along with the newly active constraints.

By separating rows of a linear program before pricing, some of the dual variables can be optimized rather than take fixed values. Some papers based on this concept are: the *positive edge* (PE) rule (Raymond et al., 2010a; Towhidi et al., 2014; Omer et al., 2015b) implemented in COIN-OR's Clp in 2015 (see Figure 3.22), the *improved primal simplex* (IPS) algorithm (Raymond et al., 2010b; Metrane et al., 2010; El Hallaoui et al., 2011; Omer et al., 2015a; Omer and Soumis, 2015), the *dynamic constraint aggregation* (DCA) methods (El Hallaoui et al., 2005, 2008, 2010; Yaakoubi et al., 2020), the *integral simplex using decomposition* algorithm (ISUD) (Zaghrouti et al., 2014; Rosat et al., 2017b,a,c), the integral column generation heuristic that combines ISUD and column generation to solve large-scale set partitioning problems (Tahir et al., 2019) and its distributed version (Foutlane et al., 2022), the *minimum mean cycle-canceling* algorithm for linear programs (Gauthier and Desrosiers, 2022), and, for a general framework, *vector space decomposition* by Gauthier et al. (2018). For the relationships between IPS, DCA, and PE, see Gauthier et al. (2016).

De: Jérémy Omer <jeremy.omer@xx.xx>
Objet: Rép : Dual-guided Pivot Rules for Linear Programming (using Clp)
Date: 25 juin 2015 à 13:25:30 UTC-4
À: John Forrest <john.forrest@fastercoin.com>
Cc: Francois Soumis <francois.soumis@gerad.ca>, Mehdi Towhidi <mehdi.towhidi@gerad.ca>, Jacques Desrosiers <jacques.desrosiers@gerad.ca>

Dear John,

Thank you very much for your feedback. I think that the way you chose to include positive edge in Clp is the right one and I am looking forward to using the version of Clp that includes positive edge. I personnally do not have any views on how to word the announcement and I trust that you will find a better wording than I would.

Sincerely,

Jeremy Omer

2015-06-25 4:30 GMT-04:00 John Forrest <john.forrest@fastercoin.com>:

> Jeremy,
>
> I have got round to putting Positive Edge coding into Clp (trunk). It does not seem to be worth making it default as Clp is quite cunning on whether to use dual or primal – based on how degenerate it looks – but it does improve time on many problems.
>
> What I have done is to use "psi" as trigger in stand-alone Clp – so if user sets – psi 0.5 they get Positive Edge. The user can use PE just for dual or primal, but that is probably only useful when I get round to using in Cbc.
>
> I will be putting code into svn later today and will announce tomorrow (probably). Have you any views on how to word the announcement?
> I have added reference in Clp papers.
>
> If user in stand-alone Clp types
> psi??
> then they get
> psi : Two-dimension pricing factor for Positive edge. The Positive Edge criterion has been added to select incoming variables to try and avoid degenerate moves. Variables not in promising set have their infeasibility weight multiplied by psi so 0.01 would mean that if there were any promising variables, then they would always be chosen, while 1.0 effectively switches algorithm off. There are two ways of switching on this feature. One way is to set psi positive and then the Positive Edge criterion will be used for Primal and Dual. The other way is to select pesteep in dualpivot choice (for example), then the absolute value of psi is used - default 0.5. Until this settles down it is only implemented in clp.
>
> Code donated by Jeremy Omer.
> See Towhidi, M., Desrosiers, J., Soumis, F., The positive edge criterion within COIN-OR's CLP.
> and
> Omer, J., Towhidi, M., Soumis, F., The positive edge pricing rule for the dual simplex.
>
> John Forrest

Fig. 3.22: *Positive edge* officially becomes part of COIN-OR's Clp in 2015.

If the Benders subproblems are integer programs, standard duality theory cannot be used to derive the cuts. Let us mention a few alternatives, although more can be found in the survey by Rahmaniani et al. (2017). If the subproblem variables are binary, Laporte and Louveaux (1993) propose lower-bounding functions instead of optimality cuts. Such constraints enforce a change to the current master problem solution, or the acceptance of its associated cost. Sherali and Fraticelli (2002) apply the reformulation-linearization technique or lift-and-project cuts to solve the discrete subproblem and show that these cuts can be expressed in terms of the master problem variables. Logic-based Benders decomposition (Hooker and Ottosson, 2003) relies on an "inference dual" subproblem to find the tightest bound on the cost of the current master problem solution, which is then used to derive a cut to be added to the Benders master problem. This approach allows to handle subproblems of any type (e.g., with integrality requirements or non-linear constraints, and even cast as a constraint programming model) but the cuts must be tailored to the problem at hand. Branch-and-bound procedures can also be devised, a natural choice being to impose branching decisions on the subproblem variables. As this may yield a very large search tree because such decisions have a local impact on the model, Hamdouni et al. (2007) rather impose decisions on the master problem variables, even if these already take integer values. This is well suited if a few costly binary variables can be positive in a master problem solution. This has a great impact on the solution cost which leads to efficient pruning strategies.

Examples

3.2 Time constrained shortest path problem (*TCSPP*). Jacques started using this example from Ahuja et al. (1993) in a *Column Generation School* held at the University of Buenos Aires (2004), upon the invitation of Irene Loiseau.

3.3 Single depot vehicle scheduling problem: two compact formulations. The Single depot vehicle scheduling problem is a good alternative to the cutting stock problem when teaching the column generation algorithm. It is a network flow problem for which decomposition is quite natural. By the way, amongst the three networks presented, only the last one in Figure 3.17 corresponds to the arc-flow formulation (3.92).

Fig. 3.23: Jacques, Gilbert Laporte, and François Soumis at the *Pierre Laurin research award* ceremony (HEC Montréal, Canada, November 1984).

Exercises

3.1 Philip Wolfe

Everybody knows George B. Dantzig for his numerous contributions to linear programming. Who is Philip Wolfe, coauthor in Dantzig and Wolfe (1960, 1961)? Where did he study? Was he a student of Dantzig? A colleague?

3.2 Column generation vs. Dantzig-Wolfe decomposition

How does column generation relate to the Dantzig-Wolfe decomposition?

3.3 Bounded variables in the compact formulation
All Dantzig-Wolfe reformulations we have derived naturally respect the warning present in Note 2.1. That is, none of the λ-variables have explicit upper bounds, only the implicit ones imposed by the convexity constraints or the structural ones derived from the set $\mathcal{A}$. This remains true even when the variables of the original formulation are upper bounded. Let us do the exercise on the *LP* given as

$$\begin{aligned} z^{\star}_{LP} = \min \quad & \mathbf{c}^{\mathsf{T}}\mathbf{x} \\ \text{s.t.} \quad & \mathbf{A}\mathbf{x} \geq \mathbf{b} \qquad [\boldsymbol{\sigma}_{\mathbf{b}}] \\ & \mathbf{D}\mathbf{x} \geq \mathbf{d} \qquad [\boldsymbol{\sigma}_{\mathbf{d}}] \\ & \boldsymbol{\ell} \leq \mathbf{x} \leq \mathbf{u} \\ & \mathbf{x} \in \mathbb{R}^n, \end{aligned} \tag{3.101}$$

where $\mathbf{D}$ is of full row rank. Derive a Dantzig-Wolfe reformulation with the grouping of constraints

$$\mathcal{A} = \{\mathbf{x} \in \mathbb{R}^n \mid \mathbf{A}\mathbf{x} \geq \mathbf{b}\} \neq \emptyset \tag{3.102a}$$
$$\mathcal{D} = \{\mathbf{x} \in \mathbb{R}^n \mid \mathbf{D}\mathbf{x} \geq \mathbf{d},\, \boldsymbol{\ell} \leq \mathbf{x} \leq \mathbf{u}\} \neq \emptyset. \tag{3.102b}$$

3.4 Duplicated cost and column variables
Functions $c_{\mathbf{x}}$ and $\mathbf{a}_{\mathbf{x}}$ often appear as variables in the formulation of the *SP* because they are indeed computed and optimized in the pricing problem. We here present a mathematical justification for this choice.

Let the *LP* be formulated as follows, where in addition to the variables $\mathbf{x} \geq \mathbf{0}$, we have also duplicated ones, $c_{\mathbf{x}} \in \mathbb{R}$ and $\mathbf{a}_{\mathbf{x}} \in \mathbb{R}^m$:

$$\begin{aligned} z^{\star}_{LP} = \min \quad & c_{\mathbf{x}} \\ \text{s.t.} \quad & \mathbf{a}_{\mathbf{x}} \geq \mathbf{b} \\ & \mathbf{D}\mathbf{x} \geq \mathbf{d} \\ & c_{\mathbf{x}} = \mathbf{c}^{\mathsf{T}}\mathbf{x} \\ & \mathbf{a}_{\mathbf{x}} = \mathbf{A}\mathbf{x} \\ & \mathbf{x} \in \mathbb{R}^n_+. \end{aligned} \tag{3.103}$$

Let the grouping of the constraints be

$$\mathcal{A} = \left\{ \begin{bmatrix} c_{\mathbf{x}} \\ \mathbf{a}_{\mathbf{x}} \\ \mathbf{x} \end{bmatrix} \in \mathbb{R}^{m+1} \times \mathbb{R}^n_+ \;\middle|\; \mathbf{a}_{\mathbf{x}} \geq \mathbf{b} \right\} \tag{3.104a}$$

$$\mathcal{D} = \left\{ \begin{bmatrix} c_{\mathbf{x}} \\ \mathbf{a}_{\mathbf{x}} \\ \mathbf{x} \end{bmatrix} \in \mathbb{R}^{m+1} \times \mathbb{R}^n_+ \;\middle|\; \mathbf{D}\mathbf{x} \geq \mathbf{d},\; c_{\mathbf{x}} = \mathbf{c}^{\mathsf{T}}\mathbf{x},\; \mathbf{a}_{\mathbf{x}} = \mathbf{A}\mathbf{x} \right\}. \tag{3.104b}$$

(a) Formulate the *MP*.
(b) Formulate the *SP*.

3.5 Dual formulations for the *MP*

Dropping the constraint relating the x- and λ-variables, write the dual formulation, denoted *DMP*, of

(a) the *MP* (3.9) using $|P|+|R|$ constraints (drop $\sum_{p\in P}\mathbf{x}_p\lambda_p+\sum_{r\in R}\mathbf{x}_r\lambda_r=\mathbf{x}$).

(b) the *MP* (3.29) issued from a Dantzig-Wolfe reformulation of the compact formulation (3.24) exhibiting a block-diagonal structure using $\sum_{k\in K}(|P^k|+|R^k|)$ constraints (drop $\sum_{p\in P^k}\mathbf{x}_p^k\lambda_p^k+\sum_{r\in R^k}\mathbf{x}_r^k\lambda_r^k=\mathbf{x}^k,\ \forall k\in K$).

3.6 2D illustration: role inversion

Redo Example 3.1 while inverting the role of $\mathcal{A}$ and $\mathcal{D}$, that is,

$$\mathcal{A}=\{x_1,x_2\geq 0 \mid -x_1+4x_2\leq 8,\ x_2\geq 2\} \quad (3.105a)$$
$$\mathcal{D}=\{x_1,x_2\geq 0 \mid 3x_1+4x_2\geq 24,\ x_1\leq 10,\ x_2\leq 6\}. \quad (3.105b)$$

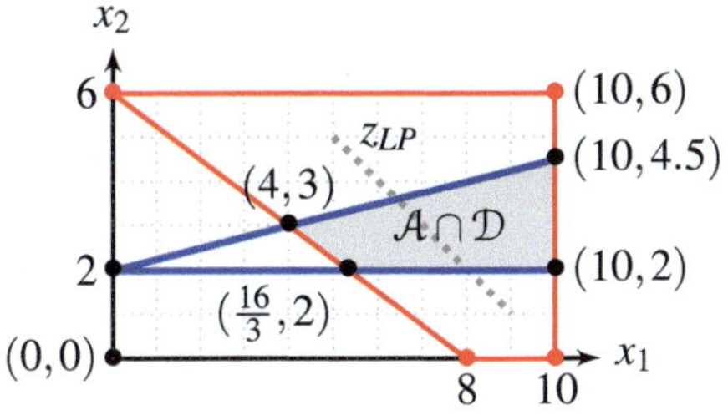

$$\begin{array}{llllll}
z_{LP}^\star = \min & x_1 & + \ x_2 & & & \\
\text{s.t.} & -x_1 & +4x_2 & \leq & 8 & [\sigma_1]\\
 & 3x_1 & +4x_2 & \geq & 24 & [\sigma_2]\\
 & x_1 & & \leq & 10 & [\sigma_3]\\
 & & x_2 & \geq & 2 & [\sigma_4]\\
 & & x_2 & \leq & 6 & [\sigma_5]\\
 & x_1 & & \geq & 0. &
\end{array}$$

(a) Describe the set $\mathcal{X}$ in terms of the extreme points and extreme rays of $\mathcal{D}$.
(b) Give the mathematical expressions of $c_\mathbf{x}$ and $\mathbf{a}_\mathbf{x}$.
(c) Perform a Dantzig-Wolfe reformulation of the *LP*; state the *MP* and *SP*.
(d) Solve the *MP* directly and by column generation.
(e) Verify that $\boldsymbol{\lambda}_{MP}^\star$ leads to the optimal primal solution $\mathbf{x}^\star=(4,3)$ and that the *MP* together with the *SP* provide $\boldsymbol{\pi}^\star=(-1/16,5/16,0,0,0)$ for the *LP*.

3.7 2D illustration: maximization

Redo Example 3.1 while *maximizing* the objective function:

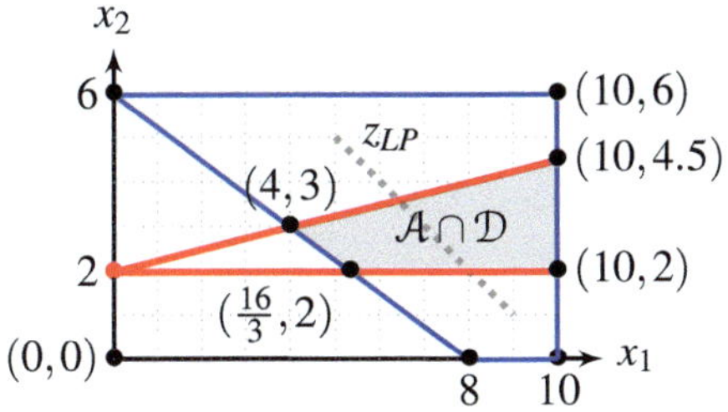

$$\begin{array}{llllll}
z_{LP}^\star = \max & x_1 & + \ x_2 & & & \\
\text{s.t.} & -x_1 & +4x_2 & \leq & 8 & [\sigma_1]\\
 & 3x_1 & +4x_2 & \geq & 24 & [\sigma_2]\\
 & x_1 & & \leq & 10 & [\sigma_3]\\
 & & x_2 & \geq & 2 & [\sigma_4]\\
 & & x_2 & \leq & 6 & [\sigma_5]\\
 & x_1 & & \geq & 0. &
\end{array}$$

(a) Solve the *LP* to identify the optimal primal and dual solutions $\mathbf{x}_{LP}^\star$ and $\boldsymbol{\sigma}^\star$ together with $z_{LP}^\star$.

(b) Show that another Dantzig-Wolfe reformulation provides the same results with

$$\mathcal{A} = \{x_1, x_2 \geq 0 \mid 3x_1 + 4x_2 \geq 24,\ x_1 \leq 10,\ x_2 \leq 6\} \tag{3.106a}$$
$$\mathcal{D} = \{x_1, x_2 \geq 0 \mid -x_1 + 4x_2 \leq 8,\ x_2 \geq 2\}. \tag{3.106b}$$

- Write and solve the *MP* to derive the optimal primal solutions $\boldsymbol{\lambda}^\star_{MP}$ and $\mathbf{x}^\star_{MP}$, and the partial dual vector $\boldsymbol{\pi}^\star_{\mathbf{b}}$.
- Write the *SP*. Verify next with the use of the *SP* the finding of $\boldsymbol{\pi}^\star_{\mathbf{d}}$.

3.8 Minimum reduced cost of zero at optimality of the *MP*
At optimality of the *MP*, show that $\bar{c}(\boldsymbol{\pi}_{\mathbf{b}}, \pi_0) = 0$. Recall that the *MP* needs not be solved by a simplex-type algorithm.

3.9 The Dantzig-Wolfe lower bound does not depend on π_0
Consider a single *SP* in (3.32). Show that the lower bound $z_{RMP} + \bar{c}(\boldsymbol{\pi}_{\mathbf{b}}, \pi_0)$ on $z^\star_{MP}$ is independent of π_0.

3.10 All constraints in the pricing problem
Given is $\mathcal{D} = \{\mathbf{x} \in \mathbb{R}^n_+ \mid \mathbf{Ax} \geq \mathbf{b}, \mathbf{Dx} \geq \mathbf{d}\}$ whereas $\mathcal{A} = \{\mathbf{x} \in \mathbb{R}^n_+\}$. Assume a Dantzig-Wolfe reformulation based on $\mathcal{D}$, as on p. 120.

(a) Formulate the *MP*.
(b) Formulate the *SP*.
(c) Describe the column generation process, starting with an artificial solution. How many times is the *RMP* solved? The *SP*?

3.11 Generating variable $\lambda_\mathbf{0}$
Given that $\mathbf{0}$ belongs to the set of extreme points of $\mathcal{D}^k$, $k \in K$, that is, $\mathbf{0} \in \{\mathbf{x}^k_p\}_{p \in P^k}$, show how variable $\lambda^k_\mathbf{0}$ can be generated for the *RMP*, although $c^k_\mathbf{0} = 0$ and $\mathbf{a}^k_\mathbf{0} = \mathbf{0}$.

(a) Consider a single pricing problem ($|K| = 1$).
(b) Consider $|K| > 1$ pricing problems.
(c) Consider that $\mathcal{D}^k$ is a polyhedral cone, for all $k \in K$.

3.12 Time constrained shortest path problem: duality
The *MP* (3.81) in Example 3.2 is given by

$$\begin{aligned}
\min\ & 3\lambda_{1246} + 14\lambda_{12456} + 5\lambda_{1256} + 13\lambda_{13246} + 24\lambda_{132456} + 15\lambda_{13256} + 16\lambda_{1346} + 27\lambda_{13456} + 24\lambda_{1356} \\
\text{s.t.}\ & 18\lambda_{1246} + 14\lambda_{12456} + 15\lambda_{1256} + 13\lambda_{13246} + 9\lambda_{132456} + 10\lambda_{13256} + 17\lambda_{1346} + 13\lambda_{13456} + 8\lambda_{1356} \leq 14 \\
& \lambda_{1246} + \lambda_{12456} + \lambda_{1256} + \lambda_{13246} + \lambda_{132456} + \lambda_{13256} + \lambda_{1346} + \lambda_{13456} + \lambda_{1356} = 1 \\
& \lambda_{1246},\ \lambda_{12456},\ \lambda_{1256},\ \lambda_{13246},\ \lambda_{132456},\ \lambda_{13256},\ \lambda_{1346},\ \lambda_{13456},\ \lambda_{1356} \geq 0.
\end{aligned}$$

The dual, written in terms of $[\pi_7, \pi_0]$, is therefore

$$
\begin{array}{rll}
\max & 14\pi_7 + \pi_0 & \\
\text{s.t.} & 18\pi_7 + \pi_0 \le 3 & [\lambda_{1246}] \\
& 14\pi_7 + \pi_0 \le 14 & [\lambda_{12456}] \\
& 15\pi_7 + \pi_0 \le 5 & [\lambda_{1256}] \\
& 13\pi_7 + \pi_0 \le 13 & [\lambda_{13246}] \\
& 9\pi_7 + \pi_0 \le 24 & [\lambda_{132456}] \\
& 10\pi_7 + \pi_0 \le 15 & [\lambda_{13256}] \\
& 17\pi_7 + \pi_0 \le 16 & [\lambda_{1346}] \\
& 13\pi_7 + \pi_0 \le 27 & [\lambda_{13456}] \\
& 8\pi_7 + \pi_0 \le 24 & [\lambda_{1356}] \\
& \pi_7 \le 0,\ \pi_0 \in \mathbb{R}. &
\end{array}
\tag{3.107}
$$

(a) Draw the feasible region of the dual, indicate the improving direction for the objective function, and identify the optimal solution $(\pi_7^\star, \pi_0^\star)$.

(b) Because of the convexity constraint ($\sum_{p\in P} \lambda_p = 1$), the duration constraint can also be written as $\sum_{p\in P} t_p\lambda_p \le 14(\sum_{p\in P} \lambda_p)$ and we have an alternative formulation for the *MP* and the corresponding dual written in terms of π_7 and μ:

$$
\begin{array}{rlll}
\min & \sum_{p\in P} c_p\lambda_p & & \\
\text{s.t.} & \sum_{p\in P} (t_p - 14)\lambda_p \le 0 & [\pi_7] & \\
& \sum_{p\in P} \lambda_p = 1 & [\mu] & \\
& \lambda_p \ge 0 & \forall p \in P &
\end{array}
\quad\Bigg|\quad
\begin{array}{rll}
\max & \mu & \\
\text{s.t.} & 4\pi_7 + \mu \le 3 & [\lambda_{1246}] \\
& \mu \le 14 & [\lambda_{12456}] \\
& \pi_7 + \mu \le 5 & [\lambda_{1256}] \\
& -\pi_7 + \mu \le 13 & [\lambda_{13246}] \\
& -5\pi_7 + \mu \le 24 & [\lambda_{132456}] \\
& -4\pi_7 + \mu \le 15 & [\lambda_{13256}] \\
& 3\pi_7 + \mu \le 16 & [\lambda_{1346}] \\
& -\pi_7 + \mu \le 27 & [\lambda_{13456}] \\
& -6\pi_7 + \mu \le 24 & [\lambda_{1356}] \\
& \pi_7 \le 0, \mu \in \mathbb{R}. &
\end{array}
\tag{3.108}
$$

Draw the dual region of the alternative *MP*, indicate the improving direction for the objective function, and identify the optimal solution $(\pi_7^\star, \mu^\star)$.

3.13 Time constrained shortest path problem: circulation pricing problem
We consider again Example 3.2 with the compact formulation (3.78) which we modify using a network augmented by arc $(d,o) = (6,1)$, that is, $G_{do} = (N, A_{do})$, where $A_{do} = A \cup \{(6,1)\}$, as illustrated in Figure 3.19. The linear relaxation of the new compact formulation (3.97) is

$$
z_{LP}^\star = \min \sum_{(i,j)\in A_{do}} c_{ij}x_{ij} \tag{3.109a}
$$

$$
\text{s.t.} \quad \sum_{j:(i,j)\in A_{do}} x_{ij} - \sum_{j:(j,i)\in A_{do}} x_{ji} = 0 \quad [\sigma_i] \quad \forall i \in \{1,\dots,6\} \tag{3.109b}
$$

$$\sum_{(i,j)\in A_{do}} t_{ij}x_{ij} \leq 14 \quad [\sigma_7] \tag{3.109c}$$
$$x_{61} = 1 \quad [\sigma_{61}] \tag{3.109d}$$
$$x_{ij} \geq 0 \quad \forall (i,j) \in A_{do}. \tag{3.109e}$$

The flow conservation equations in (3.109b) describe a circulation, the second imposes the traversal time restriction (3.109c) while constraint $x_{61} = 1$ (with the associated dual variable denoted σ_{61}) implicitly asks for a single unit from source node $o = 1$ to terminal node $d = 6$ by the flow conservation equations.

We apply a Dantzig-Wolfe reformulation to the above formulation using the grouping of constraints

$$\mathcal{A} = \left\{ \mathbf{x} \in \mathbb{R}_+^{|A_{do}|} \;\middle|\; (3.109c)\text{–}(3.109d) \right\} \tag{3.110a}$$
$$\mathcal{D} = \left\{ \mathbf{x} \in \mathbb{R}_+^{|A_{do}|} \;\middle|\; (3.109b) \right\}. \tag{3.110b}$$

(a) Describe the set $\{\mathbf{x}_p\}_{p\in P}$ of extreme points of $\mathcal{D}$ and $\{\mathbf{x}_r\}_{r\in R}$ of extreme rays.
(b) Write the formulations for the *MP* and *SP*.

3.14 Aircraft routing

Recall Note 3.19 on the routing of aircraft to operate 15 flight legs, each one being obviously covered exactly once. Consider the data and network provided in Figure 3.18. The network $G = (N,A)$ comprises 30 nodes in set N, two per flight leg. We also find 45 arcs in set A, out of which the 15 flight arcs (diagonal). The other 30 arcs are ground arcs (dashed) at the airports, where aircraft are stationed until their next flight assignment. There are three special ground arcs, one per city: these are the night arcs (dashed) going from the end of the day to the beginning of the next, allowing for a repeated schedule. Let the corresponding three sets of arcs be denoted $Flight$, $Ground$, and $Night \subset Ground$. Finally, let $A = Flight \cup Ground$.

This aircraft routing problem is similar to the Single depot vehicle scheduling problem of Examples 2.4 and 3.3, except that there is no depot although some aircraft are obviously available at the start of the day in each city. The mathematical formulation of this minimum cost flow problem is composed of the objective function for counting the number of aircraft, the flow conservation constraints with net flow equal to zero for all nodes, and the bounds for all flow variables:

$$\begin{aligned} z^\star_{LP} = \min \quad & \sum_{(i,j)\in A} c_{ij}x_{ij} \\ \text{s.t.} \quad & \sum_{j:(i,j)\in A} x_{ij} - \sum_{j:(j,i)\in A} x_{ji} = 0 \quad \forall i \in N \\ & \ell_{ij} \leq x_{ij} \leq u_{ij} \quad \forall (i,j) \in A. \end{aligned} \tag{3.111}$$

(a) For all $(i,j) \in A$, provide adequate numerical values for
- the c_{ij}-coefficients in the objective function,
- the lower and upper bounds ℓ_{ij} and u_{ij}, respectively.

(b) Solve the *LP* (3.111) and describe a solution in terms of cycles.

(c) An optimal solution requires one aircraft at the beginning of the day in city **C**. Give the full *cycle* for this aircraft. How many legs does it operate?

(d) How to handle cities in different time zones? Assume that the time zone associated with each city is available, say `A:+3, B:+2, C:+2` where the signed number follows the *coordinated universal time* format. Moreover, the departure time in Figure 3.18a should be understood as *local departure time*.

(e) Show that this aircraft routing problem can be transformed into a Single depot vehicle scheduling problem by sketching an appropriate network.

(f) For a set of $|Flight| = n$ flights, the proposed network is polynomial in size, that is, $2n$ nodes and $3n$ arcs. Show how to decrease the number of nodes and arcs.

3.15 Aircraft routing: reformulation and column generation
Let the aircraft routing problem of Note 3.19 be formulated as

$$\begin{aligned} z^\star_{LP} = \min \quad & \sum_{(i,j)\in Night} x_{ij} \\ \text{s.t.} \quad & x_{ij} = 1 \quad [\sigma_{ij}] \quad \forall (i,j) \in Flight \\ & \sum_{j:(i,j)\in A} x_{ij} - \sum_{j:(j,i)\in A} x_{ji} = 0 \quad [\sigma_i] \quad \forall i \in N \\ & x_{ij} \geq 0 \quad \forall (i,j) \in Ground, \end{aligned} \tag{3.112}$$

and apply a Dantzig-Wolfe reformulation with the grouping of constraints

$$\mathcal{A} = \left\{ \mathbf{x} \in \mathbb{R}^{|A|}_+ \,\middle|\, x_{ij} = 1,\ \forall (i,j) \in Flight \right\} \tag{3.113a}$$

$$\mathcal{D} = \left\{ \mathbf{x} \in \mathbb{R}^{|A|}_+ \,\middle|\, \sum_{j:(i,j)\in A} x_{ij} - \sum_{j:(j,i)\in A} x_{ji} = 0,\ \forall i \in N \right\}. \tag{3.113b}$$

Give the formulations for the *MP* and *SP*.

3.16 Linear relaxation of the cutting stock problem
Recall the One-dimensional cutting stock problem seen in Example 2.1. Given is a large set K of paper rolls of identical width W and m demands b_i, $i \in \{1,\dots,m\}$, for small items of width w_i. The goal is to satisfy the demand using a minimum number of rolls. For $k \in K$, let x^k_0 be a binary variable indicating if roll k is selected or not and x^k_i be the number of times item i is cut in roll k. Consider the following integer linear programming formulation, where the dual variables are only used in the *LP*:

$$z^\star_{ILP} = \min \quad \sum_{k\in K} x^k_0 \tag{3.114a}$$

$$\text{s.t.} \quad \sum_{k\in K} x^k_i \geq b_i \qquad [\sigma_i \geq 0] \qquad \forall i \in \{1,\dots,m\} \tag{3.114b}$$

$$\sum_{i=1}^{m} w_i x_i^k \leq W x_0^k \qquad [\sigma_0^k \leq 0] \quad \forall k \in K \tag{3.114c}$$

$$x_i^k \geq 0 \qquad \forall k \in K,\ i \in \{0,\dots,m\} \tag{3.114d}$$

$$x_0^k \in \{0,1\} \qquad \forall k \in K \tag{3.114e}$$

$$x_i^k \in \mathbb{Z}_+ \qquad \forall k \in K,\ i \in \{1,\dots,m\}. \tag{3.114f}$$

We are interested in solving and analyzing the linear relaxation of (3.114) by dropping the binary and integrality requirements on $\mathbf{x}^k = [x_i^k]_{i=0,\dots,m}$, $\forall k \in K$.

(a) Prove that $z_{LP}^{\star} = \sum_{i=1}^{m} w_i b_i / W$. *Hint*: Firstly show that $\sum_{i=1}^{m} w_i b_i / W$ is a lower bound; secondly, find appropriate values for all $\mathbf{x}^k$, $k \in K$, to reach that bound.

(b) Show that $\sigma_i^{\star} = w_i / W$, $\forall i \in \{1,\dots,m\}$ is part of an optimal dual solution.

(c) We know that performing a Dantzig-Wolfe reformulation using any grouping of the constraints of a linear program leads to $z_{MP}^{\star} = z_{LP}^{\star}$. Define the following grouping of the constraints for the *LP*:

$$\mathcal{A} = \left\{ \{\mathbf{x}^k \in \mathbb{R}_+^{m+1}\}_{k \in K} \;\middle|\; \sum_{k \in K} x_i^k = b_i,\ \forall i \in \{1,\dots,m\} \right\} \tag{3.115a}$$

$$\mathcal{D}^k = \left\{ \mathbf{x}^k \in \mathbb{R}_+^{m+1} \;\middle|\; \sum_{i=1}^{m} w_i x_i^k \leq W x_0^k,\ x_0^k \leq 1 \right\},\ \forall k \in K. \tag{3.115b}$$

Give *three ways* on how to improve on $z_{LP}^{\star}$, a lower bound on $z_{ILP}^{\star}$.

3.17 PS, IPS, and MMCC: a dual point of view

The *primal simplex* (**PS**) as well as the *improved primal simplex* (**IPS**) and the *minimum mean cycle-canceling* (**MMCC**) algorithms can be reproduced with various pricing strategies used in a Dantzig-Wolfe reformulation, see p. 128. We here examine a dual point of view. Assume that the *LP* is given in standard form as

$$\begin{aligned} z_{LP}^{\star} = \min \quad & \mathbf{c}^{\mathsf{T}}\mathbf{x} \\ \text{s.t.} \quad & \mathbf{A}\mathbf{x} = \mathbf{b} \qquad [\boldsymbol{\pi} \in \mathbb{R}^m] \\ & \mathbf{x} \in \mathbb{R}_+^n. \end{aligned} \tag{3.116}$$

Given any feasible solution $\mathbf{x}^0$, we partition the variables indexed by $J = \{1,\dots,n\}$ in two subsets:

- F for positive (or free) variables: $F = \{j \in J \mid x_j^0 > 0\}$;
- L for variables at zero (lower bound): $L = \{j \in J \mid x_j^0 = 0\}$.

A variable x_j indexed in F can increase or decrease in value relatively to x_j^0 whereas one indexed in L can only increase. We use the following change of variables to transpose this into *forward* variable y_j and *backward* variables y_{j+n} as seen in Figure 3.24

$$x_j = x_j^0 + y_j - y_{j+n}, \quad 0 \le y_j \le r_j^0,\ 0 \le y_{j+n} \le r_{j+n}^0 \qquad \forall j \in J. \tag{3.117}$$

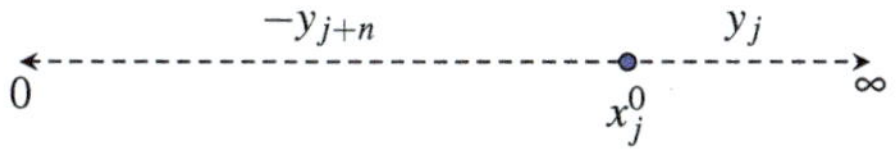

Fig. 3.24: Forward and backward variables for x_j as part of a feasible solution $\mathbf{x}^0$.

The parameters r_j^0 and r_{j+n}^0 are upper bounds that depend on those of variable x_j. By construction, they are zero for backward variables associated with variables indexed in L, i.e., $r_{j+n} = 0, \forall j \in L$. We see as much if we specialize (3.117) according to partition $J = F \cup L$:

$$x_j = \begin{cases} x_j^0 + y_j - y_{j+n}, & y_j, y_{j+n} \ge 0,\ y_{j+n} \le x_j^0, \quad \forall j \in F \\ x_j^0 + y_j, & y_j \ge 0 \qquad\qquad\qquad\quad\ \forall j \in L. \end{cases} \tag{3.118}$$

This notation is in keeping with (3.57) in which the superscripts $+/-$ indicate forward and backward possibilities. The contribution of y_j to the objective function is $c_j y_j$ whereas that of y_{j+n} is $-c_j y_{j+n}$.

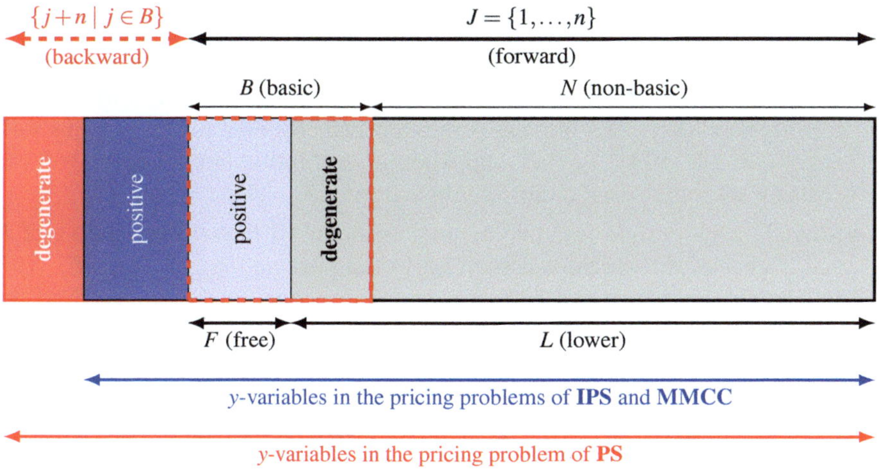

Fig. 3.25: Various subsets of the y-variables for a basic solution $\mathbf{x}^0$.

Figure 3.25 relates these sets to those we know when $\mathbf{x}^0$ is a basic solution, i.e., $F \subseteq B$ and $L \supseteq N$. Both **PS** and **IPS** work with a basic solution but **MMCC** has no such restriction. Given the feasible primal solution $\mathbf{x}^0$, the three algorithms use the same stopping rule: $\mu = 0$ in (3.119) but *try to verify* different optimality conditions with their respective sets of dual constraints. Each formulation comprises the same objective function $\bar{c}(\boldsymbol{\pi})$ maximizing μ over the possible values of $\boldsymbol{\pi}$ but different sets of dual constraints with appropriate y-variables within brackets. Given $\mathbf{x}^0$, variable μ optimizes the smallest reduced cost and optimality is reached if it equals zero.

$$
\begin{array}{llll}
& \bar{c}(\boldsymbol{\pi}) = \max_{\boldsymbol{\pi} \in \mathbb{R}^m} \mu & \text{subject to:} & \\
\textbf{PS} & 0 = c_j - \boldsymbol{\pi}^\mathsf{T} \mathbf{a}_j & [y_j \in \mathbb{R}] & \forall j \in B \\
& \mu \leq c_j - \boldsymbol{\pi}^\mathsf{T} \mathbf{a}_j & [y_j \geq 0] & \forall j \in N \\
\textbf{IPS} & 0 = c_j - \boldsymbol{\pi}^\mathsf{T} \mathbf{a}_j & [y_j \in \mathbb{R}] & \forall j \in F \\
& \mu \leq c_j - \boldsymbol{\pi}^\mathsf{T} \mathbf{a}_j & [y_j \geq 0] & \forall j \in L \\
\textbf{MMCC} & \mu \leq -(c_j - \boldsymbol{\pi}^\mathsf{T} \mathbf{a}_j) & [y_{j+n} \geq 0] & \forall j \in F \\
& \mu \leq \quad c_j - \boldsymbol{\pi}^\mathsf{T} \mathbf{a}_j & [y_j \geq 0] & \forall j \in J.
\end{array}
\tag{3.119}
$$

- On the one hand, **PS** imposes a zero reduced cost for all basic variables, that is, the equality constraints $0 = c_j - \boldsymbol{\pi}^\mathsf{T}\mathbf{a}_j$, $\forall j \in B$. The primal formulation in y-variables then replaces the associated $y_j \in \mathbb{R}$ by $y_j - y_{j+n}$ with $y_j, y_{j+n} \geq 0$, $\forall j \in B$. On the other hand, when $\mu = 0$, it also verifies the non-negativity of the reduced cost for non-basic variables (set N). These two sets of constraints provide sufficient optimality conditions as the backward variables for the degenerate variables are not part of any necessary and sufficient optimality conditions and indeed induce degenerate pivots.
- In conformity with the complementary slackness (necessary and sufficient) optimality conditions, **IPS** imposes a zero reduced cost only for the positive variables, $0 = c_j - \boldsymbol{\pi}^\mathsf{T}\mathbf{a}_j$, $\forall j \in F$, and when $\mu = 0$, fulfills non-negative reduced costs for the variables at their lower bound (set L).
- **MMCC** aims for non-negative reduced costs for all forward variables (set J) and all backward variables associated to the positive variables (set F). This follows the necessary and sufficient optimality conditions established on the residual problem (Gauthier et al., 2014).

By taking the dual formulation for each of the three linear programs in (3.119), show that this leads to the pricing strategies in y-variables of respectively **PS**, **IPS**, and **MMCC**.

Note that even though **IPS** and **MMCC** use the same y-variables, y_j, $y_{j+n} \geq 0$, $\forall j \in F$, and $y_j \geq 0$, $\forall j \in L$, or equivalently $y_{j+n} \geq 0$, $\forall j \in F$, and $y_j \geq 0$, $\forall j \in J$, their pricing problems may have different optima. Furthermore, because they are both centered on necessary and sufficient optimality conditions, the set of solutions in the pricing problem of **MMCC** must be a superset of that of **IPS**. Indeed, **MMCC** is less restrictive with the inequality constraints on μ, optimizing all reduced cost values rather than fixing some at zero.

4

Dantzig-Wolfe Decomposition for Integer Linear Programming

Ignoramus et ignorabimus

Über die Grenzen des Naturerkennens
Emil du Bois-Reymond

Wir müssen wissen.
Wir werden wissen.

Retirement address to the Society of German Scientists and Physicians
David Hilbert

Abstract This chapter extends the Dantzig-Wolfe decomposition principle to integer linear programs. Indeed, we capitalize on the fact that we can request partial integrality requirements of the compact formulation in the pricing problem. We investigate two related, but different approaches in deriving the integer master and integer pricing problems. The first is based on the convexification of the reformulated domain, as in the classical Dantzig-Wolfe decomposition for linear programming. The second is based on its discretization, taking into account also interior points of the reformulated domain. In the context of integer linear programming, both of these provide more than only mathematical reformulations. In particular, we show that solving the linear relaxation of the integer master problem may provide a better bound than that of the compact formulation.

Contents

Introduction 176
4.1 Reformulation by Convexification 177
Minkowski-Weyl theorem, again 177
Integer master problem 178

J. Desrosiers et al., *Branch-and-Price*,
https://doi.org/10.1007/978-3-031-96917-1_4

Polytope and polyhedral cone 181
Integer pricing problem 181
4.2 Reformulation by Discretization 183
Hilbert-Giles-Pulleyblank theorem 183
Integer master problem 185
Polytope and polyhedral cone 187
Two-dimensional illustrations 188
Illustration 1: Set $\mathcal{T}$ for a polytope 189
Illustration 2: Set $\mathcal{T}$ for a polyhedral cone 189
Illustration 3: Set $\mathcal{T}$ for a polyhedron 190
Illustration 4: Set $\mathcal{T}$ for another polyhedron 191
Integer pricing problem 191
Binary domain 193
Illustration 5: *TCSPP*: integrality 194
Some observations 196
Post-processing a solution of the master problem 197
4.3 Integrality Property: For or Against Virtue? 197
4.4 Block-diagonal Structure 200
Practical relevance 200
On grouping the constraints 201
Integer master problems (convexification and discretization) 202
Identical subproblems 204
Aggregation 205
Disaggregation 206
Some more observations 208
Lower and upper bounds 210
4.5 Good to Know 211
Non-linear encoding functions 212
Illustration 6: *TCSPP*: non-linear costs 215
Not all blocks are used 217
Bounded domains 218
Identical subproblems 221
Unbounded domains 222
Shared variables across all blocks 222
4.6 More to Know 224
Reverse engineering a compact formulation 224
Convexification 224
Discretization 225
Extended compact and subproblem formulations 230
Automatic grouping of the constraints for reformulation 232
Graph-based methods 233
Methods based on constraint classes 234
Evaluating a decomposition 236
4.7 Examples 238
Integrality property in the knapsack problem 239

Binary knapsack problem . . . 239
Knapsack problem . . . 241
Integrality property in the cutting stock problem . . . 244
Weak compact formulation . . . 244
Network-based compact formulation . . . 246
Comparison of the four linear relaxations . . . 249
Reverse Dantzig-Wolfe . . . 250
TCSPP: nine reformulations . . . 251
TCSPP: time bounding . . . 254
Generalized assignment problem . . . 259
Binary tasks-to-machine (many-to-one) patterns . . . 261
Binary task-to-machines (one-to-many) patterns . . . 262
Multi-commodity maximum flow problem . . . 264
Scene selection problem . . . 267
Symmetry breaking constraints . . . 268
Dantzig-Wolfe reformulation . . . 268
Quality of the lower bounds . . . 269
Design of balanced student teams . . . 270
Quadratic transportation problem . . . 270
Set partitioning reformulation . . . 272
Can you decode a secret vote? . . . 275
Binary vote patterns . . . 277
Edge coloring problem: two compact formulations . . . 279
First compact, with an identical block-diagonal structure . . . 279
Second compact, without a block-index . . . 281
Compact formulations are not created equal . . . 282
4.8 Reference Notes . . . 282
Exercises . . . 285

Acronyms

ILP	integer linear program (original or compact formulation)	177
LP	linear relaxation of the *ILP*	177
IMP	integer master problem (convexification)	178
MP	linear relaxation of the *IMP*	179
RMP	restricted *MP*	181
ISP	integer subproblem	181
$\ddot{IMP}$	integer master problem (discretization)	185
$\ddot{MP}$	linear relaxation of the $\ddot{IMP}$	186
ISP^k	integer subproblem for block $k \in K$	203
BKP	binary knapsack problem	239
KP	knapsack problem	241
CSP	one-dimensional cutting stock problem	244
TCSPP	time constrained shortest path problem	251

GAP	generalized assignment problem	259
SSP	scene selection problem	267
VRT	variable reduction technique	268
IQP	integer quadratic program	270

Introduction

Almost all the applications using the Dantzig-Wolfe decomposition principle found in the literature are in fact based on integer programming models. Amongst these, bin packing and cutting stock, edge coloring, generalized assignment, aircraft routing and scheduling, crew pairing, traffic assignment in satellite communication systems, etc. We bring attention to Chapter 2 in which we already have some integer pricing problems such as the knapsack problem in the cutting stock problem (p. 71).

This chapter differentiates itself from the traditional Dantzig-Wolfe decomposition by focusing on the integrality conditions present in the compact formulation. Specifically, we impose that the subproblem produces solutions that respect them. We describe two approaches to produce a reformulation which differ in how the subproblem's domain is represented: *convexification* and *discretization*. The former uses the Minkowski-Weyl theorem seen in Chapter 3 whereas the latter uses a theorem by Hilbert that has been proven in our context by Giles and Pulleyblank.

Either way, given an integer linear program, a Dantzig-Wolfe reformulation produces an equivalent *integer master problem* whose linear relaxation is typically solved by the column generation algorithm. This incidentally means that solving the reformulation is done by combining branch-and-bound with column generation. The famous *branch-and-price* is coined. This chapter focuses on the root node where we show how and why the linear relaxation of an extended formulation is likely stronger than that of its compact counterpart. We postpone the branch-and-bound and technical aspects induced by the usage of column generation to Chapter 7.

We bring the reader in the mood by quoting François Soumis' evolving analogy since 1981.

> *Atoms vs. Molecules*
> The quality of a Dantzig-Wolfe reformulation can be appreciated, not only in terms of the optimal objective value of its linear relaxation, but also its feasible region. By considering complex structural constraints and integrality requirements in the pricing problem, we can imagine the objects it generates as *molecules* that are handled by the master problem. Of course, we can also identify those molecules in any integer solution of the original formulation, but with respect to its linear relaxation, it is like we are dealing with *atoms* that combine in any way, including ways that do not even respect chemical bonds.
>
> The pricing problem is a subcontracted chemist who may be costly to call but I hope that it pays off in the master problem by providing some structure for the type of solutions we are looking for. This cost can also often be offset because we can split the contract by vehicle, person, or sub-system.
>
> Moreover, as the objective function does not influence the structure of a molecule, this interpretation of the reformulated domain applies even if there is no such function.

4.1 Reformulation by Convexification

We would like to solve the following integer linear program *ILP*, which we again call the *original* or *compact* formulation in our context:

$$\begin{aligned} z^\star_{ILP} = \min \quad & \mathbf{c}^\mathsf{T}\mathbf{x} \\ \text{s.t.} \quad & \mathbf{A}\mathbf{x} \geq \mathbf{b} & [\boldsymbol{\sigma}_\mathbf{b}] \\ & \mathbf{D}\mathbf{x} \geq \mathbf{d} & [\boldsymbol{\sigma}_\mathbf{d}] \\ & \mathbf{x} \in \mathbb{Z}^n_+, \end{aligned} \tag{4.1}$$

with appropriate dimensions for all vectors and matrices. Our presentation assumes that all variables are required to be integer but it can be generalized to mixed-integer linear programs. As before, we assume that (4.1) is feasible and $z^\star_{ILP}$ finite. The dual vectors $\boldsymbol{\sigma}_\mathbf{b}$ and $\boldsymbol{\sigma}_\mathbf{d}$ are used in the linear relaxation *LP*. Define the non-empty sets $\mathcal{A}$ and $\mathcal{D}$ that group the constraints in two subsets:

$$\mathcal{A} = \{\mathbf{x} \in \mathbb{Z}^n_+ \mid \mathbf{A}\mathbf{x} \geq \mathbf{b}\} \neq \emptyset \tag{4.2a}$$

$$\mathcal{D} = \{\mathbf{x} \in \mathbb{Z}^n_+ \mid \mathbf{D}\mathbf{x} \geq \mathbf{d}\} \neq \emptyset. \tag{4.2b}$$

There are *two* differences with the previous chapter: integrality requirements in $\mathcal{A}$ and $\mathcal{D}$. Their presence in the former is mandatory since we want to respect the original restrictions. They are optional in the latter although it is precisely how the *magic* happens.

Note 4.1 (Infeasible compact.) Formally, the non-empty nature of $\mathcal{A}$ and $\mathcal{D}$ in (4.2) is a consequence of the feasibility of the compact formulation, that is, $\mathcal{A} \cap \mathcal{D} \neq \emptyset$. If the compact formulation is indeed infeasible, then $\mathcal{A} \cap \mathcal{D} = \emptyset$ which either means at least one of the sets is empty or they have no common solution. We can detect this at the *SP* level if $\mathcal{D} = \emptyset$ and at the *MP* level otherwise, i.e., $\mathcal{A} \cap \mathcal{D} = \emptyset$. This logic holds in practice where we iteratively solve linear relaxations eventually subject to branching decisions, see Algorithm 2.2, Exercise 2.20, and of course Chapter 7. That is, we detect infeasibility in linear relaxations and then interpret what it means for integrality.

Minkowski-Weyl theorem, again

Let us remind ourselves of what happens in Chapter 3. The reformulation lies in a higher dimensional space where the variables convey the vertex-description of the polyhedron $\{\mathbf{x} \in \mathbb{R}^n \mid \mathbf{D}\mathbf{x} \geq \mathbf{d}\}$ given by the Minkowski-Weyl Theorem 3.1.

However, we cannot apply this theorem directly to $\mathcal{D}$ (4.2b), because the latter is not a polyhedron but an integer set. Instead, we consider its convexification, that is, we replace $\mathcal{D}$ by $\text{conv}(\mathcal{D})$ which is a polyhedron. The set of integer solutions is not changed, and in particular the vertices of $\text{conv}(\mathcal{D})$ are integer. This allows us to write the Minkowski-Weyl Theorem in a specialized form.

Theorem 4.1. *Let* $\mathcal{Q}=\mathcal{P}\cap\mathbb{Z}^n$ *be a non-empty integer set, where* $\mathcal{P}=\{\mathbf{x}\in\mathbb{R}^n \mid \mathbf{Qx}\geq\mathbf{q}\}$ *is a polyhedron with at least one extreme point. An equivalent description of* $\mathcal{Q}$ *using the extreme points* $\{\mathbf{x}_p\}_{p\in P}$ *and extreme rays* $\{\mathbf{x}_r\}_{r\in R}$ *of* $\text{conv}(\mathcal{Q})$ *is*

$$\mathcal{Q}=\left\{\mathbf{x}\in\mathbb{Z}^n \;\middle|\; \begin{array}{lll} \sum_{p\in P}\mathbf{x}_p\lambda_p + \sum_{r\in R}\mathbf{x}_r\lambda_r = \mathbf{x} & & \\ \sum_{p\in P}\lambda_p = 1 & & \\ \lambda_p \geq 0 & & \forall p\in P \\ & \lambda_r \geq 0 & \forall r\in R \end{array}\right\}. \tag{4.3}$$

Proof. The proof of Theorem 3.1 adapts without difficulty to the polyhedron $\text{conv}(\mathcal{Q})$ since the underlying polyhedron $\mathcal{P}$ fulfills a stronger assumption that it also contains at least one integer point. We restate both directions for this equivalence.

- $\Rightarrow$ Any integer point $\mathbf{x}\in\mathcal{Q}$ can be written as a convex combination of the finitely many extreme points $\{\mathbf{x}_p\}_{p\in P}$ of $\text{conv}(\mathcal{Q})$ plus a conic combination of the finitely many extreme rays $\{\mathbf{x}_r\}_{r\in R}$ of $\text{conv}(\mathcal{Q})$, that is, there exist scalars $\{\lambda_p\}_{p\in P}$ and $\{\lambda_r\}_{r\in R}$ such that (4.3) holds for all $\mathbf{x}\in\mathcal{Q}$.
- $\Leftarrow$ Any combination $\boldsymbol{\lambda}$ that satisfies (4.3) corresponds to an $\mathbf{x}\in\mathbb{Z}^n$ by definition but also necessarily to an $\mathbf{x}\in\mathcal{Q}$. □

Integer master problem

By using Theorem 4.1 on the integer set $\mathcal{D}$, we consider the finite sets of extreme points $\{\mathbf{x}_p\}_{p\in P}$ and extreme rays $\{\mathbf{x}_r\}_{r\in R}$ of $\text{conv}(\mathcal{D})$ which, again, we gather into the set

$$\mathcal{X}=\left\{\begin{bmatrix}\mathbf{x}_p\\1\end{bmatrix}\right\}_{p\in P}\cup\left\{\begin{bmatrix}\mathbf{x}_r\\0\end{bmatrix}\right\}_{r\in R} \tag{4.4}$$

We obtain a Dantzig-Wolfe reformulation of the *ILP* (4.1) by substituting (4.3) into $\mathbf{c}^\intercal\mathbf{x}$ and the constraints of $\mathcal{A}$. The resulting integer linear program typically contains a huge number of variables and we therefore brace ourselves for column generation by referring to the following formulation as the *integer master problem IMP*:

$$\begin{array}{rllll} z^\star_{IMP} = \min & \sum_{p\in P} c_p\lambda_p + \sum_{r\in R} c_r\lambda_r & & & \\ \text{s.t.} & \sum_{p\in P}\mathbf{a}_p\lambda_p + \sum_{r\in R}\mathbf{a}_r\lambda_r \geq \mathbf{b} & & [\boldsymbol{\pi}_\mathbf{b}] & \\ & \sum_{p\in P}\lambda_p = 1 & & [\pi_0] & \\ & \lambda_p \geq 0 & & & \forall p\in P \\ & & \lambda_r \geq 0 & & \forall r\in R \\ & \sum_{p\in P}\mathbf{x}_p\lambda_p + \sum_{r\in R}\mathbf{x}_r\lambda_r = \mathbf{x}\in\mathbb{Z}^n_+, & & & \end{array} \tag{4.5}$$

where the encoding functions are defined as

$$\begin{aligned} c_p &= c_{\mathbf{x}_p} = \mathbf{c}^\mathsf{T}\mathbf{x}_p, & \mathbf{a}_p &= \mathbf{a}_{\mathbf{x}_p} = \mathbf{A}\,\mathbf{x}_p, && \forall p \in P \\ c_r &= c_{\mathbf{x}_r} = \mathbf{c}^\mathsf{T}\mathbf{x}_r, & \mathbf{a}_r &= \mathbf{a}_{\mathbf{x}_r} = \mathbf{A}\,\mathbf{x}_r, && \forall r \in R. \end{aligned} \tag{4.6}$$

As always, the dual variables $\boldsymbol{\pi}_\mathbf{b} \geq \mathbf{0}$ and $\pi_0 \in \mathbb{R}$ are only used in the linear relaxation *MP*. By construction, (4.1) and (4.5) are equivalent integer linear programs, thus $z^\star_{IMP} = z^\star_{ILP}$ even though $\mathbf{x}^\star_{IMP}$ and $\mathbf{x}^\star_{ILP}$ may differ. The *IMP* also finds $\boldsymbol{\lambda}^\star_{IMP}$, a vector of variables *that is not required to be integer*. We trivially have $z^\star_{MP} \leq z^\star_{IMP} = z^\star_{ILP}$. Let us state in a formal proposition a more general result which positions $z^\star_{LP}$ among these optima.

Proposition 4.1. *Given the ILP* (4.1), *let the LP be its linear relaxation and the MP be the linear relaxation of its reformulation IMP* (4.5). *Then,*

$$z^\star_{LP} \leq z^\star_{MP} \leq z^\star_{IMP} \;\; (= z^\star_{ILP}). \tag{4.7}$$

Proof. In the minimization context, this follows from comparing the feasible domains of the respective programs:

$$\{\mathbf{x} \in \mathbb{R}^n_+ \mid \mathbf{Ax} \geq \mathbf{b},\, \mathbf{Dx} \geq \mathbf{d}\} \supseteq \{\mathbf{x} \in \mathbb{R}^n_+ \mid \mathbf{Ax} \geq \mathbf{b}\} \cap \operatorname{conv}(\mathcal{D}) \supseteq \mathcal{A} \cap \mathcal{D}, \tag{4.8}$$

where the last inclusion is always proper except in the absurd case in which all domains reduce to a single solution. □

Note 4.2 (Proper inclusions and equal optima.) We underline that the equality $z^\star_{MP} = z^\star_{IMP}$ can nonetheless occur despite this proper quality as it depends on where the objective function leads us. The same can be said for the equality $z^\star_{LP} = z^\star_{MP}$. Let us come back to this observation in Section 4.3.

Figure 4.1 illustrates the various sets involved in Proposition 4.1, where the dots on the domains correspond to 2-dimensional integer points. There are five feasible integer points for the *ILP*, the dots in $\mathcal{A} \cap \mathcal{D}$, and obviously at least one of them is an optimal solution $\mathbf{x}^\star_{ILP}$ (depending on the objective function $\mathbf{c}^\mathsf{T}\mathbf{x}$). The remaining integer points are then all infeasible for the *ILP*, but we can see that eight respect the constraints in $\mathcal{D}$. We cannot stress enough that these observations are in general inaccessible so trying to reduce the number of optimal candidates to four (the point in the middle certainly cannot be optimal for the *ILP*) is a misplaced good intention. Back to the Minkowski-Weyl Theorem 4.1, we can observe that an optimal solution for the *ILP* can indeed be expressed using only the extreme points of $\operatorname{conv}(\mathcal{D})$ in a convex combination. Furthermore, the different variable space of the *MP* compared to that of the compact formulation is not visible here because we only see the projection. Nonetheless, it is interesting to note that the extreme points of the linear relaxation of the *ILP* (seven of them) lose all meaning compared to those of the *MP* (six of them). Ultimately, we can see that the *MP*'s domain in Figure 4.1d is a superset of the *ILP*'s and a subset of the *LP*'s which are in this case also both proper. As

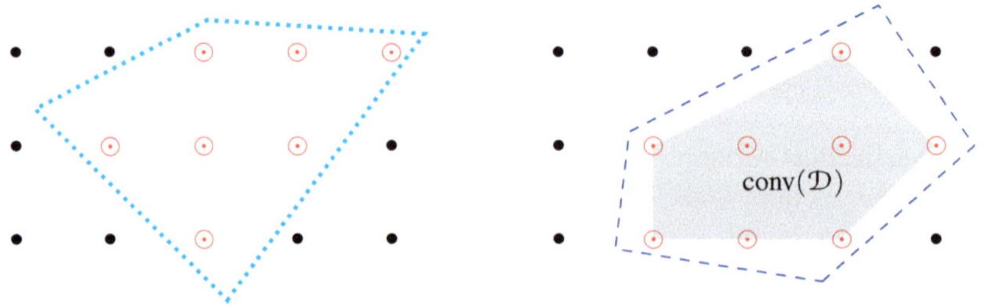

(a) Integer set $\mathcal{A} = \{\mathbf{x} \in \mathbb{Z}^2_+ \mid \mathbf{Ax} \geq \mathbf{b}\}$

(b) Integer set $\mathcal{D} = \{\mathbf{x} \in \mathbb{Z}^2_+ \mid \mathbf{Dx} \geq \mathbf{d}\}$ and its integer hull $\text{conv}(\mathcal{D})$

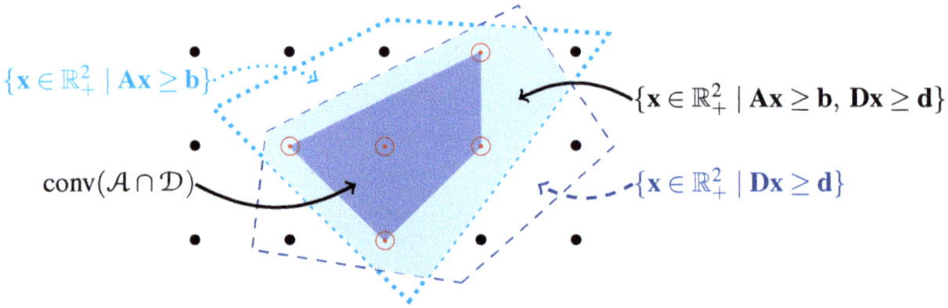

(c) Domain $\{\mathbf{x} \in \mathbb{R}^2_+ \mid \mathbf{Ax} \geq \mathbf{b}, \mathbf{Dx} \geq \mathbf{d}\}$ of the *LP* in light color compared to integer hull $\text{conv}(\mathcal{A} \cap \mathcal{D})$ of the *ILP* in dark color

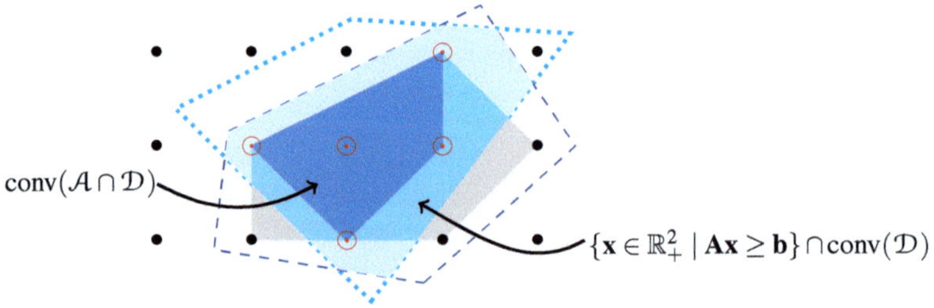

(d) Domain $\{\mathbf{x} \in \mathbb{R}^2_+ \mid \mathbf{Ax} \geq \mathbf{b}\} \cap \text{conv}(\mathcal{D})$ of the *MP* compared to that of the *LP* in light color and integer hull $\text{conv}(\mathcal{A} \cap \mathcal{D})$ of the *ILP* in dark color

Fig. 4.1: Reformulation based on the convexification of the integer set $\mathcal{D}$.

additional food for thought on column generation, the beauty of the reformulation by convexification is that we find ourselves working with an implicit description of the integer hull of $\mathcal{D}$. Similar observations are derived if we interchange the role of $\mathcal{A}$ and $\mathcal{D}$, see Exercise 4.2.

Polytope and polyhedral cone

For future referencing, let us consider two special cases for the domain $\mathcal{D}$ for which the *IMP* specializes to

$$\begin{array}{ll|ll}
\min & \sum_{p\in P} c_p\lambda_p & \min & \sum_{r\in R} c_r\lambda_r \\
\text{s.t.} & \sum_{p\in P} \mathbf{a}_p\lambda_p \geq \mathbf{b} \quad [\boldsymbol{\pi}_\mathbf{b}] & \text{s.t.} & \sum_{r\in R} \mathbf{a}_r\lambda_r \geq \mathbf{b} \quad [\boldsymbol{\pi}_\mathbf{b}] \\
 & \sum_{p\in P} \lambda_p = 1 \quad [\pi_0] & & \\
 & \lambda_p \geq 0, \ \forall p \in P & & \lambda_r \geq 0, \ \forall r \in R \\
 & \sum_{p\in P} \mathbf{x}_p\lambda_p = \mathbf{x} \in \mathbb{Z}^n_+ & & \sum_{r\in R} \mathbf{x}_r\lambda_r = \mathbf{x} \in \mathbb{Z}^n_+.
\end{array} \tag{4.9}$$

- On the left, $\text{conv}(\mathcal{D})$ is bounded and hence $R = \emptyset$ so the λ_r-variables are simply removed.
- On the right, $\text{conv}(\mathcal{D})$ is a polyhedral cone and, as seen in Section 3.1 (p. 110), the single extreme point $\mathbf{0}$ and the convexity constraint $\lambda_\mathbf{0} = 1$ are removed.

Integer pricing problem

Figure 4.2 summarizes the Dantzig-Wolfe decomposition based on the *convexification* of $\mathcal{D}$, i.e., $\text{conv}(\mathcal{D})$, we have performed on the *ILP*. On the right, optimizing the latter yields $z^\star_{ILP}$ and $\mathbf{x}^\star_{ILP}$ whereas optimizing the *LP* gives $z^\star_{LP}$, $\mathbf{x}^\star_{LP}$, $\boldsymbol{\sigma}^\star_\mathbf{b}$ and $\boldsymbol{\sigma}^\star_\mathbf{d}$. On the left, the *ILP* is reformulated into the equivalent *IMP* by substitution. The latter is solved by branch-price-and-cut (Chapter 7), where the root node hosts the *MP*, its linear relaxation. In a branch-and-bound/branch-and-cut solver for the *ILP* (right side), we already know how to derive branching and cutting decisions that revolve around the fractional x-values which must be integer. That same logic applies for the *IMP* (left side) where we indeed request integrality on the x-variables. We also point to Exercise 7.3 in which the reader must adapt a branch-and-bound algorithm to reach integer optimality with sub-optimal solutions for the linear relaxations.

Observe that we again differentiate the dual vectors $\boldsymbol{\pi}^\star_\mathbf{b}$ for the *MP* and $\boldsymbol{\sigma}^\star_\mathbf{b}$ for the *LP* despite them being associated with the same set of constraints. It is the same reason as in Chapter 3: they are two different linear programs although it is perhaps more evident this time around as Proposition 4.1 establishes that they are not even necessarily equivalent as they may reach different optima. If the *MP* is solved by column generation, it is done by coordinating the *RMP* (a linear program) and the *ISP* (an integer linear program), where the absence of a $\boldsymbol{\pi}_\mathbf{d}$ output can be noted.

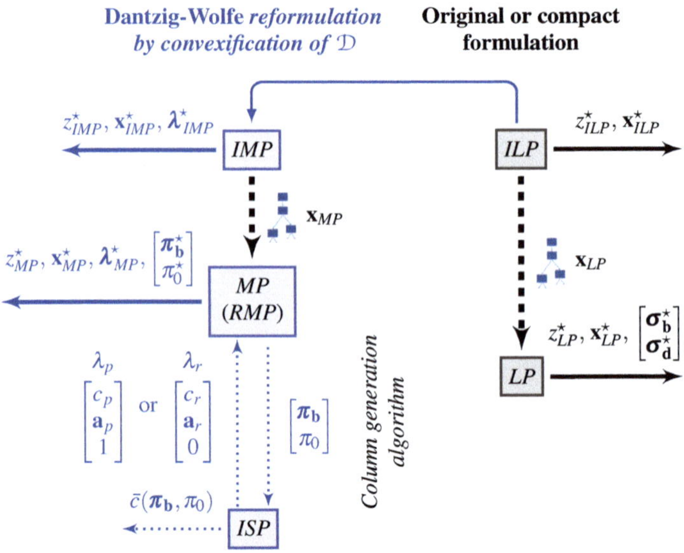

Fig. 4.2: Information flow of the column generation algorithm solving the *MP*, the linear relaxation of the *IMP* (4.5), a Dantzig-Wolfe reformulation by convexification of the *ILP* (4.1).

Let us formally define the last piece of this picture: the *ISP*. In principle, we could determine whether there remains any negative reduced cost variable by solving

$$\min\left\{\min_{p\in P}\{c_p - \boldsymbol{\pi}_{\mathbf{b}}^\intercal \mathbf{a}_p - \pi_0\},\ \min_{r\in R}\{c_r - \boldsymbol{\pi}_{\mathbf{b}}^\intercal \mathbf{a}_r\}\right\}. \tag{4.10}$$

Since we wish to avoid enumerating the content of $\mathcal{X} = \left\{\begin{bmatrix}\mathbf{x}_p\\ 1\end{bmatrix}\right\}_{p\in P} \cup \left\{\begin{bmatrix}\mathbf{x}_r\\ 0\end{bmatrix}\right\}_{r\in R}$, we rather define the integer linear subproblem *ISP* as

$$\bar{c}(\boldsymbol{\pi}_{\mathbf{b}}, \pi_0) = -\pi_0 + \min_{\mathbf{x}\in\mathcal{D}} (\mathbf{c}^\intercal - \boldsymbol{\pi}_{\mathbf{b}}^\intercal \mathbf{A})\mathbf{x} \tag{4.11}$$

while ensuring that we use an algorithm returning a solution $\mathbf{x} \in \mathcal{D}$ which is actually an extreme point (e.g., optimal) or extreme ray (unbounded) of conv($\mathcal{D}$). Accordingly, we expect the output of the *ISP* to be

$$\bar{c}(\boldsymbol{\pi}_{\mathbf{b}}, \pi_0) = \begin{cases} -\infty & \text{if } c_r - \mathbf{a}_r^\intercal \boldsymbol{\pi}_{\mathbf{b}} < 0 \quad \text{for some } r \in R \\ c_p - \boldsymbol{\pi}_{\mathbf{b}}^\intercal \mathbf{a}_p - \pi_0 & \text{otherwise} \qquad\qquad\ \ \text{for some } p \in P. \end{cases} \tag{4.12}$$

4.2 Reformulation by Discretization

Let us underline once more how convexification lets the λ-variables in the *IMP* (4.5) be fractional so that they can be combined to form integrality on $\mathbf{x}$. These variables correspond to extreme points and extreme rays of conv($\mathcal{D}$). For example, reconsider Figure 4.1b where any of the three non-extreme integer points in conv($\mathcal{D}$) could be optimal for the *IMP*. Once again, it is in general irrelevant that we can intersect with $\mathcal{A}$, as seen in Figure 4.1c, to reduce this number of points from three to two; otherwise we may as well just enumerate $\mathcal{A} \cap \mathcal{D}$ by brute-force.

Discretization acknowledges non-extreme integer points of conv($\mathcal{D}$) directly in the reformulation. We call these extra points *interior*, even when they may lie on the boundary, in which case they are interior to the respective face. This implies that we deal with a larger set of λ-variables, compared to convexification, but we can impose integrality on them. We do this by using a theorem of Hilbert (1890), proven in our context by Giles and Pulleyblank (1979), to obtain a representation of $\mathcal{D}$. That is, even though this is not a polyhedron, the theorem provides a representation for the points that matter. We also mention that reading the proof is worthwhile as we reuse its content afterwards. Remember our assumption from Note 3.3 that our rays are scaled to integers. This is why the theorem assumes rational data.

Hilbert-Giles-Pulleyblank theorem

Theorem 4.2. (Nemhauser and Wolsey, 1988, Theorem 6.1, p. 104) *Let* $\mathcal{Q} = \mathcal{P} \cap \mathbb{Z}^n$ *be a non-empty integer set, where* $\mathcal{P} = \{\mathbf{x} \in \mathbb{R}^n \mid \mathbf{Q}\mathbf{x} \geq \mathbf{q}\}$ *is a rational polyhedron with at least one extreme point. An equivalent description of* $\mathcal{Q}$ *using a finite subset* $\{\mathbf{x}_p\}_{p\in\ddot{P}}$ *of its integer points and the (integer-scaled) extreme rays* $\{\mathbf{x}_r\}_{r\in\ddot{R}}$ *of* conv($\mathcal{Q}$) *is*

$$\mathcal{Q} = \left\{ \mathbf{x} \in \mathbb{Z}^n \;\middle|\; \begin{array}{lll} \sum_{p\in\ddot{P}} \mathbf{x}_p\lambda_p + \sum_{r\in\ddot{R}} \mathbf{x}_r\lambda_r = \mathbf{x} & & \\ \sum_{p\in\ddot{P}} \lambda_p = 1 & & \\ \lambda_p \in \{0,1\} & & \forall p \in \ddot{P} \\ & \lambda_r \in \mathbb{Z}_+ & \forall r \in \ddot{R} \end{array} \right\}. \tag{4.13}$$

Proof. The proof shows that an appropriate finite subset $\mathcal{T} = \{\mathbf{x}_p\}_{p\in\ddot{P}} \subseteq \mathcal{Q}$ exists so that (4.13) is indeed equivalent. Since the proof is based on Minkowski-Weyl, we only prove one direction of the equivalence: Any integer point $\mathbf{x} \in \mathcal{Q}$ can be written as exactly one element of $\mathcal{T}$ plus a non-negative integer combination of the finitely many extreme rays (scaled to be integer) $\{\mathbf{x}_r\}_{r\in\ddot{R}}$ of conv($\mathcal{Q}$), that is, there exist integer scalars $\{\lambda_p\}_{p\in\ddot{P}}$ and $\{\lambda_r\}_{r\in\ddot{R}}$ such that (4.13) holds for any $\mathbf{x} \in \mathcal{Q}$.

Let $\{\mathbf{x}_p\}_{p\in P}$ and $\{\mathbf{x}_r\}_{r\in R}$ be the sets of extreme points and extreme rays of conv($\mathcal{Q}$), respectively. Had we not assumed our extreme rays to be integer-scaled anyway, we would have to consider this here at the latest. We use the set $\{\mathbf{x}_r\}_{r\in\ddot{R}}$ instead of $\{\mathbf{x}_r\}_{r\in R}$ for the (shortest) representatives with integer coordinates. By the Minkowski-Weyl Theorem 4.1, we can write $\mathcal{Q}$ as

$$\left\{\mathbf{x}\in\mathbb{Z}^n \,\middle|\, \sum_{p\in P}\mathbf{x}_p\alpha_p+\sum_{r\in\ddot{R}}\mathbf{x}_r\beta_r=\mathbf{x},\ \sum_{p\in P}\alpha_p=1,\alpha_p\geq 0,\forall p\in P,\beta_r\geq 0,\forall r\in\ddot{R}\right\}. \quad (4.14)$$

We separate the fractional and integral portions of β_r, $r\in\ddot{R}$, to rewrite (4.14) as

$$\underbrace{\begin{array}{l}\sum_{p\in P}\mathbf{x}_p\alpha_p+\sum_{r\in\ddot{R}}\mathbf{x}_r(\beta_r-\lfloor\beta_r\rfloor)\\[2ex] \sum_{p\in P}\alpha_p=1,\alpha_p\geq 0,\forall p\in P,\ \beta_r\geq 0,\forall r\in\ddot{R}\end{array}}_{\text{some integer vector } \mathbf{x}_q,\ q\in\ddot{P}} \quad + \quad \underbrace{\begin{array}{l}\sum_{r\in\ddot{R}}\mathbf{x}_r\lfloor\beta_r\rfloor \quad = \mathbf{x}\\[2ex] \beta_r\geq 0,\forall r\in\ddot{R}.\end{array}}_{\text{integer combination of } \mathbf{x}_r,\ r\in\ddot{R}} \quad (4.15)$$

Remember that we only care for integer solutions $\mathbf{x}\in\mathbb{Z}^n$. It is obvious that the system within the right underbrace yields integer points. With respect to the system within the left underbrace, it has a bounded domain because the extreme rays are limited by

$$0\leq\beta_r-\lfloor\beta_r\rfloor<1, \qquad \forall r\in\ddot{R}. \quad (4.16)$$

We can therefore define a finite subset $\mathcal{T}$ that contains all these points $\{\mathbf{x}_q\}_{q\in\ddot{P}}$, i.e.,

$$\mathcal{T}=\left\{\mathbf{x}_q\in\mathcal{Q} \,\middle|\, \begin{array}{llll}\sum_{p\in P}\mathbf{x}_p\alpha_p & + & \sum_{r\in\ddot{R}}\mathbf{x}_r\tau_r=\mathbf{x}_q & \\ \sum_{p\in P}\alpha_p=1 & & & \\ \alpha_p\geq 0 & & & \forall p\in P \\ & & 0\leq\tau_r<1 & \forall r\in\ddot{R}\end{array}\right\}. \quad (4.17)$$

It is irrelevant that (4.17) is not even linear programming material ($\tau_r<1$) as it only serves to show the existence of $\mathcal{T}=\{\mathbf{x}_p\}_{p\in\ddot{P}}$. By definition of extreme points, we necessarily have $\ddot{P}\supseteq P$, where the additional points come from non-trivial linear combinations. Ultimately, given an integer-scaled description of the extreme rays $\{\mathbf{x}_r\}_{r\in\ddot{R}}$, there indeed exists $\ddot{P}\supseteq P$ such that (4.13) holds by picking *exactly one* index $q\in\ddot{P}$ and letting $\lambda_r=\lfloor\beta_r\rfloor$, $\forall r\in\ddot{R}$. □

While the reformulation by discretization differentiates itself from convexification by the possibility that we may have $\{\mathbf{x}_p\}_{p\in P}\neq\{\mathbf{x}_p\}_{p\in\ddot{P}}$ for the relevant points, we have $\{\mathbf{x}_r\}_{r\in R}=\{\mathbf{x}_r\}_{r\in\ddot{R}}$ by our assumption of (shortest) integer scaling of the extreme rays, see Note 3.3. We could therefore refrain from the extra notion $\ddot{R}$, but we keep it in order to remind us later *where we come from.*

Note 4.3 (Applied mathematics.) From his experience building the Mars Science Laboratory at Jet Propulsion Laboratory, Adam Steltzner has this to say on the subject of engineering versus pure mathematics:

> [us engineers] are not paid to do things right, we're paid to do them just right enough.

Operations research certainly shares some of that philosophy. Indeed, because all our algorithms run on computer code, any model we use sooner or later must fit into that scope to be solved. What this means for us is that for all practical purposes, we could have written the entire book using $\mathbb{Q}$ instead of $\mathbb{R}$ even though discretization is the only substitution which in theory requires data in the rational set. Recall Note 2.16 or Section Limited numeric precision in Chapter 2 for more proof of this.

Integer master problem

Let us now move on to the formal Dantzig-Wolfe decomposition by discretization on the given *ILP* (4.1) as well as the usual grouping of the constraints

$$\mathcal{A} = \{\mathbf{x} \in \mathbb{Z}^n_+ \mid \mathbf{Ax} \geq \mathbf{b}\} \neq \emptyset \tag{4.18a}$$
$$\mathcal{D} = \{\mathbf{x} \in \mathbb{Z}^n_+ \mid \mathbf{Dx} \geq \mathbf{d}\} \neq \emptyset. \tag{4.18b}$$

By using the Hilbert-Giles-Pulleyblank Theorem 4.2 on $\mathbf{x} \in \mathcal{D}$, we consider an integer-scaled description of its extreme rays $\{\mathbf{x}_r\}_{r\in\ddot{R}}$ and a corresponding finite set of integer points $\{\mathbf{x}_p\}_{p\in\ddot{P}}$, $\ddot{P} \supseteq P$, which we gather into the set

$$\mathcal{X} = \left\{\begin{bmatrix}\mathbf{x}_p\\ 1\end{bmatrix}\right\}_{p\in\ddot{P}} \cup \left\{\begin{bmatrix}\mathbf{x}_r\\ 0\end{bmatrix}\right\}_{r\in\ddot{R}}. \tag{4.19}$$

A reformulation of the *ILP* (4.1) is obtained by substituting (4.13) for $\mathbf{x} \in \mathbb{Z}^n_+$ in its objective function and the constraints of $\mathcal{A}$. This yields an equivalent integer linear program, denoted *IMP̈*, which reads as

$$z^\star_{IM\ddot{P}} = \min \sum_{p\in\ddot{P}} c_p\lambda_p + \sum_{r\in\ddot{R}} c_r\lambda_r \tag{4.20a}$$
$$\text{s.t.} \sum_{p\in\ddot{P}} \mathbf{a}_p\lambda_p + \sum_{r\in\ddot{R}} \mathbf{a}_r\lambda_r \geq \mathbf{b} \qquad [\boldsymbol{\pi}_\mathbf{b}] \tag{4.20b}$$
$$\sum_{p\in\ddot{P}} \lambda_p = 1 \qquad [\pi_0] \tag{4.20c}$$
$$\lambda_p \in \{0,1\} \qquad \forall p \in \ddot{P} \tag{4.20d}$$
$$\lambda_r \in \mathbb{Z}_+ \qquad \forall r \in \ddot{R} \tag{4.20e}$$
$$\sum_{p\in\ddot{P}} \mathbf{x}_p\lambda_p + \sum_{r\in\ddot{R}} \mathbf{x}_r\lambda_r = \mathbf{x} \in \mathbb{Z}^n_+, \tag{4.20f}$$

where $c_j = \mathbf{c}^\top \mathbf{x}_j$ and $\mathbf{a}_j = \mathbf{A}\mathbf{x}_j$, $j \in \ddot{P} \cup \ddot{R}$. Despite the reformulation *IMP̈* (4.20) being different from the *IMP* (4.5) because it ensures the integrality of the $\boldsymbol{\lambda}_{IM\ddot{P}}$-vector in addition to that on the $\mathbf{x}$-vector, solving its linear relaxation *MP̈* is not different from solving the *MP* because they both have the same *RMP*.

Proposition 4.2. *The linear relaxations of the Dantzig-Wolfe reformulations by discretization* (4.20) *or convexification* (4.5) *of* $\mathcal{D}$ *give the same bound, i.e.,* $z^\star_{M\ddot{P}} = z^\star_{MP}$ *and* $\mathbf{x}^\star_{M\ddot{P}} = \mathbf{x}^\star_{MP}$.

Proof. The sets of extreme rays $\{\mathbf{x}_r\}$ of $\operatorname{conv}(\mathcal{D})$ indexed by $\ddot{R}$ and R are identical by assumption. Moreover, any point $\mathbf{x} \in \{\mathbf{x}_p\}_{p\in\ddot{P}} \subseteq \mathcal{D}$ can be written as a convex combination of the extreme points $\{\mathbf{x}_p\}_{p\in P}$ plus a conic combination of the extreme rays $\{\mathbf{x}_r\}_{r\in R}$ of $\operatorname{conv}(\mathcal{D})$. Hence $z^\star_{M\ddot{P}} = z^\star_{MP} = \min\{\mathbf{c}^\top\mathbf{x} \mid \mathbf{A}\mathbf{x} \geq \mathbf{b}, \mathbf{x} \in \operatorname{conv}(\mathcal{D})\}$. Finally, $\mathbf{x}^\star_{M\ddot{P}} = \mathbf{x}^\star_{MP} = \sum_{p\in P'} \mathbf{x}_p\lambda_p + \sum_{r\in R'} \mathbf{x}_r\lambda_r$ in the final *RMP*. □

Note 4.4 (We are all cut from the same cloth.) It feels strange that *different* integer master problems with therefore *different* relaxations (the master problems) should lead to *the same* restricted master problem. In order to understand this, remember that both use *the same ISP* and this is only able to produce extreme points and rays, regardless of whether $\ddot{P} \supset P$ or not (and again, $\{\mathbf{x}_r\}_{r\in R} = \{\mathbf{x}_r\}_{r\in\ddot{R}}$). With this in mind, the distinction between convexification and discretization rather becomes a conceptual discussion than a real practical difference. We revisit this thought in Chapter 7. Of course, this is not to say that three different persons, say from Montréal, Bordeaux, and Rio, use their machines and favorite solvers for the *ISP* and *RMP*, and reach different optimal solutions $\mathbf{x}^\star_{JD}$, $\mathbf{x}^\star_{FV}$, and $\mathbf{x}^\star_{EU}$, but this is a different story.

Proposition 4.3. *The integrality requirements on* $\mathbf{x}$ *are redundant in the IMP̈* (4.20) *(if the integrality conditions on the variables of* $\mathcal{D}$ *are the same as in* $\mathcal{A}$*), i.e.,*

$$
\begin{aligned}
z^\star_{IM\ddot{P}} = \min \quad & \sum_{p\in\ddot{P}} c_p\lambda_p + \sum_{r\in\ddot{R}} c_r\lambda_r \\
\text{s.t.} \quad & \sum_{p\in\ddot{P}} \mathbf{a}_p\lambda_p + \sum_{r\in\ddot{R}} \mathbf{a}_r\lambda_r \geq \mathbf{b} && [\boldsymbol{\pi}_\mathbf{b}] \\
& \sum_{p\in\ddot{P}} \lambda_p = 1 && [\pi_0] \\
& \lambda_p \in \{0,1\} && \forall p \in \ddot{P} \\
& \lambda_r \in \mathbb{Z}_+ && \forall r \in \ddot{R}.
\end{aligned}
\tag{4.21}
$$

Proof. We prove that an optimal integer solution $\boldsymbol{\lambda}^\star_{IM\ddot{P}}$ to (4.21) implies integrality on $\mathbf{x}$, i.e.,

$$
\begin{aligned}
\boldsymbol{\lambda}^{\star}_{IM\ddot{P}} \in \mathbb{Z}_{+}^{|\ddot{P}|+|\ddot{R}|} &\Rightarrow \sum_{p\in\ddot{P}} \mathbf{x}_p\lambda_p^{\star} \in \mathbb{Z}_{+}^{n} \qquad \text{and} \qquad \sum_{r\in\ddot{R}} \mathbf{x}_r\lambda_r^{\star} \in \mathbb{Z}_{+}^{n} \\
&\Rightarrow \sum_{p\in\ddot{P}} \mathbf{x}_p\lambda_p^{\star} + \sum_{r\in\ddot{R}} \mathbf{x}_r\lambda_r^{\star} = \mathbf{x} \in \mathbb{Z}_{+}^{n}.
\end{aligned} \tag{4.22}
$$

By assumption, $\mathbf{x}_p$, $p \in \ddot{P}$, is integer and obviously $\mathbf{x}_p\lambda_p^{\star}$ remains so for binary $\lambda_p^{\star}$. This is likewise true for integer-scaled $\mathbf{x}_r$, $r \in \ddot{R}$, and non-negative integer $\lambda_r^{\star}$. □

Note 4.5 (Interior integer points.) We sometimes refer only to the interior integer points specified by $\ddot{P}$. Let us denote their subset by

$$\ddot{I} = \ddot{P} \setminus P. \tag{4.23}$$

If $\mathcal{D}$ is bounded, then $\ddot{I}$ contains all interior integer points of conv($\mathcal{D}$). We thus expect the cardinality of $\ddot{I}$ to be much larger than that of P. If $\mathcal{D}$ is unbounded, then $\ddot{I}$ contains a finite subset of interior integer points whose cardinality increases with the integer-scale of the extreme rays (see the Two-dimensional illustrations on p. 188).

Note 4.6 (Discretization is not enumeration.) The Hilbert-Giles-Pulleyblank Theorem shows that $|\mathcal{T}| < \infty$ such that the reformulation $IM\ddot{P}$ contains a finite number of integer variables. Just like in the Minkowski-Weyl Theorem, it only matters that this alternative representation *exists*. In particular, discretization is not an explicit enumeration of all the integer points in the set $\mathcal{D}$, even in the bounded case, see Illustration 4.1. Indeed, we can use column generation to service the reformulation thus producing variables as needed. That being said, we should turn to enumeration if and when it is tractable in which case one would obviously try to find a minimal subset $\mathcal{T} \subseteq \mathcal{D}$.

Polytope and polyhedral cone

We can still consider the same two special cases for the domain $\mathcal{D}$ (4.18b) that yield very similar $IM\ddot{P}$ reformulations than in the convexification models (4.9):

$$
\begin{array}{ll|ll}
\min & \sum_{p\in\ddot{P}} c_p\lambda_p & \min & \sum_{r\in\ddot{R}_0} c_r\lambda_r \\
\text{s.t.} & \sum_{p\in\ddot{P}} \mathbf{a}_p\lambda_p \geq \mathbf{b} \quad [\boldsymbol{\pi}_\mathbf{b}] & \text{s.t.} & \sum_{r\in\ddot{R}_0} \mathbf{a}_r\lambda_r \geq \mathbf{b} \quad [\boldsymbol{\pi}_\mathbf{b}] \\
& \sum_{p\in\ddot{P}} \lambda_p = 1 \quad [\pi_0] & & \\
& \lambda_p \in \{0,1\},\ \forall p \in \ddot{P} & & \lambda_r \in \mathbb{Z}_{+},\ \forall r \in \ddot{R}_0 \\
& \sum_{p\in\ddot{P}} \mathbf{x}_p\lambda_p = \mathbf{x} \in \mathbb{Z}_{+}^{n} & & \sum_{r\in\ddot{R}_0} \mathbf{x}_r\lambda_r = \mathbf{x} \in \mathbb{Z}_{+}^{n}\,.
\end{array} \tag{4.24}
$$

- On the left, we assume that $\mathcal{D}$ is bounded, so is $\text{conv}(\mathcal{D})$, meaning that $\ddot{R} = \emptyset$ and $\mathcal{T} = \{\mathbf{x}_p\}_{p\in\ddot{P}} = \mathcal{D}$. The substitution of $\sum_{p\in\ddot{P}} \mathbf{x}_p \lambda_p = \mathbf{x}$ leads to a *binary master problem* that selects exactly one of the integer points of $\mathcal{D}$.
- On the right, we assume that $\text{conv}(\mathcal{D})$ is a polyhedral cone such that vector $\mathbf{0}$ is the unique extreme point as $\mathcal{D} = \{\mathbf{x} \in \mathbb{Z}^n_+ \mid \mathbf{D}\mathbf{x} \geq \mathbf{0}\}$. From $\mathcal{T} = \{\mathbf{x}_p\}_{p\in\ddot{P}}$ defined by (4.17), we have $\ddot{I} = \ddot{P} \setminus \{\mathbf{0}\}$ which gives us

$$\mathcal{X} = \left\{ \begin{bmatrix} \mathbf{0} \\ 1 \end{bmatrix} \right\} \cup \left\{ \begin{bmatrix} \mathbf{x}_p \\ 1 \end{bmatrix} \right\}_{p\in\ddot{I}} \cup \left\{ \begin{bmatrix} \mathbf{x}_r \\ 0 \end{bmatrix} \right\}_{r\in\ddot{R}}. \tag{4.25}$$

The definition of $\ddot{R}_\mathbf{0}$ comes from discarding the $\mathbf{0}$-vector from the reformulation and then treating every interior integer point in $\{\mathbf{x}_p\}_{p\in\ddot{I}}$ as a ray. Indeed, for any integer scalar $\lambda_p \in \mathbb{Z}_+$, $p \in \ddot{I}$, we have that $\mathbf{x}_p\lambda_p \in \mathbb{Z}^n_+$ and $\mathbf{D}(\mathbf{x}_p\lambda_p) \geq \mathbf{0}$, hence $\mathbf{x}_p\lambda_p \in \mathcal{D}$ as well as $\sum_{p\in\ddot{I}} \mathbf{x}_p\lambda_p \in \mathcal{D}$. Consequently, let

$$\ddot{R}_\mathbf{0} = \ddot{I} \cup \ddot{R}, \qquad \mathcal{X} = \left\{ \begin{bmatrix} \mathbf{x}_r \\ 0 \end{bmatrix} \right\}_{r\in\ddot{R}_\mathbf{0}}, \tag{4.26}$$

and express the substitution as

$$\sum_{r\in\ddot{R}_\mathbf{0}} \mathbf{x}_r\lambda_r = \mathbf{x}, \quad \lambda_r \in \mathbb{Z}_+, \ \forall r \in \ddot{R}_\mathbf{0}. \tag{4.27}$$

Note 4.7 (Hilbert basis.) For the friends of interesting mathematical notions, we remark that there is a smallest set of integer rays whose non-negative integer combinations yield all integer points contained in a polyhedral cone. It is unique and it is called the *Hilbert basis*.

Two-dimensional illustrations

Using 2-dimensional illustrations, let us examine the set $\mathcal{T} = \{\mathbf{x}_p\}_{p\in\ddot{P}}$ defined by (4.17) depending on whether the set $\mathcal{D}$ (4.18b) is bounded or not.

- If it is bounded, we realize in Illustration 4.1 that $\mathcal{T} = \mathcal{D}$ is the only complete representation.
- In the unbounded case, $\mathcal{T}$ contains all the extreme points of $\text{conv}(\mathcal{D})$, that is, $\{\mathbf{x}_p\}_{p\in P} \subseteq \mathcal{T}$, and typically *some other points*. Illustrations 4.2 (polyhedral cone), 4.3 (polyhedron with a single, non-zero, extreme point), and 4.4 (three extreme points and two extreme rays) help us underscore that we really only need $\mathcal{T}$ to exist (i.e., it does not have to be minimal) in the proof of Theorem 4.2.

Other illustrations can be found in Nemhauser and Wolsey (1988, p. 105) and Vanderbeck and Wolsey (2010, Example 2, p. 437).

Illustration 4.1 **Set $\mathcal{T}$ for a polytope**

If $\mathcal{D}$ is bounded, then $\ddot{R} = \emptyset$ and

$$\mathcal{T} = \left\{ \mathbf{x} \in \mathcal{D} \subset \mathbb{Z}^n_+ \;\middle|\; \sum_{p \in P} \mathbf{x}_p \alpha_p = \mathbf{x}, \sum_{p \in P} \alpha_p = 1, \alpha_p \geq 0, \forall p \in P \right\} = \mathcal{D}, \tag{4.28}$$

i.e., $\mathcal{T}$ is composed of not only the extreme points $\{\mathbf{x}_p\}_{p \in P}$ of $\text{conv}(\mathcal{D})$ but *all* its integer points, see Figure 4.3.

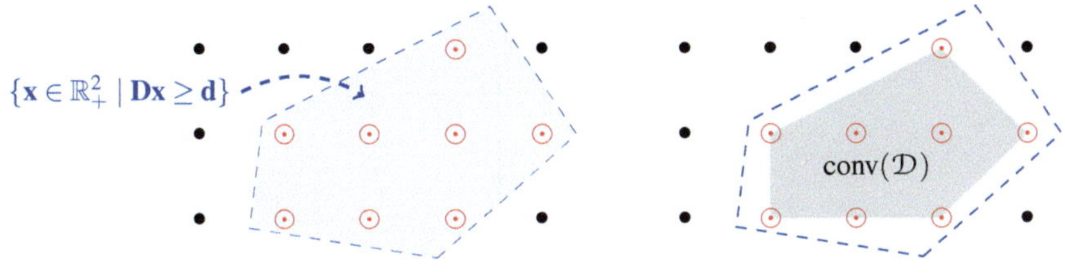

(a) Integer set $\mathcal{D} = \{\mathbf{x} \in \mathbb{Z}^2_+ \mid \mathbf{D}\mathbf{x} \geq \mathbf{d}\} \subset \mathbb{Z}^2_+$ (b) $\mathcal{T} = \text{conv}(\mathcal{D}) \cap \mathbb{Z}^2_+ = \mathcal{D}$

Fig. 4.3: Bounded set $\mathcal{D} \subset \mathbb{Z}^2_+$: $\mathcal{T}$ comprises all its integer points.

Illustration 4.2 **Set $\mathcal{T}$ for a polyhedral cone**

In Figure 4.4, $\text{conv}(\mathcal{D}) = \{x_1, x_2 \geq 0 \mid 5x_1 - 2x_2 \geq 0,\ -x_1 + 3x_2 \geq 0\}$, with integer-scaled extreme rays $(3,1)$ and $(2,5)$. The 13 integer points in set $\mathcal{T}$ come from the single extreme point $\mathbf{0}$ and the 12 interior ones. Using the definition (4.26), we have $|\ddot{R}_\mathbf{0}| = 14$ from $|\ddot{I}| = 12$ integer points and $|\ddot{R}| = 2$ extreme rays. Observe that 8 interior points identified by a $\star$, indeed rays in (4.26), are redundant given ray $(1,1)$. The Hilbert basis is thus $\{(1,1),(2,1),(1,2),(2,4),(3,1),(2,5)\}$.

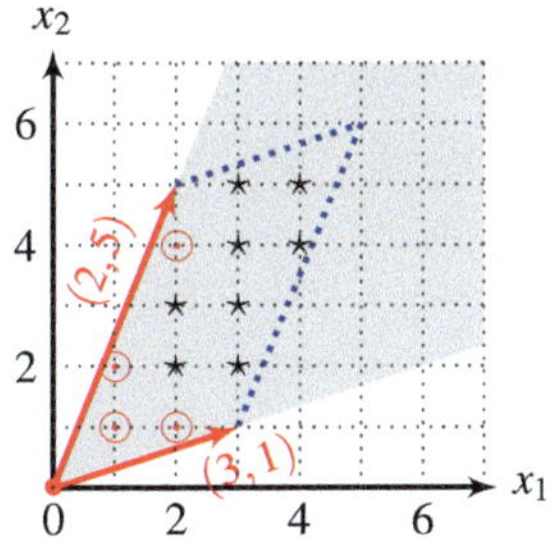

Fig. 4.4: Given the integer-scaled extreme rays $\{(3,1),(2,5)\}$, $\mathcal{T}$ comprises the extreme point $(0,0)$ and 12 interior integer points.

Illustration 4.3 **Set $\mathcal{T}$ for a polyhedron**

In Figure 4.5a, the unbounded set

$$\text{conv}(\mathcal{D}) = \{x_1, x_2 \geq 0 \mid -x_1 + 4x_2 \leq 8,\ x_2 \geq 2\}$$

(also used in Figure 3.11) is a polyhedron with extreme point $(0,2)$ and integer extreme rays $(1,0)$ and $(4,1)$. These rays are integer-scaled differently in Figures 4.5b, that is, multiplied by 2. This has an impact on the number of points in $\mathcal{T}$: on the left, $|\mathcal{T}| = 1$ whereas $|\mathcal{T}| = 4$ on the right.

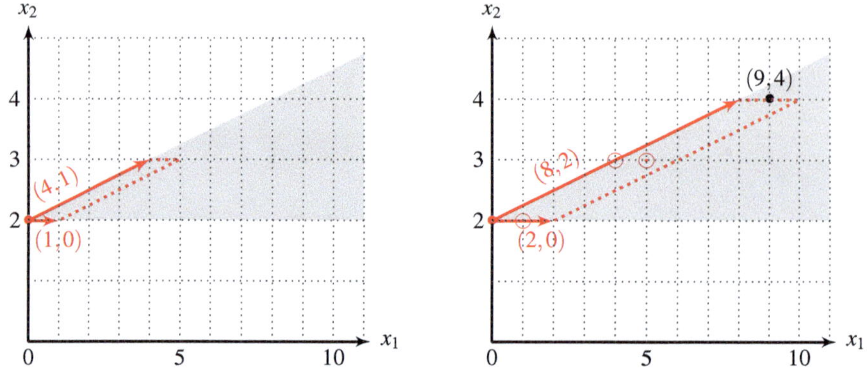

(a) Integer-scaled rays $\{(1,0),(4,1)\}$

(b) Integer-scaled rays $\{(2,0),(8,2)\}$

Fig. 4.5: Different extreme ray scalings give different discretization sets $\mathcal{T}$ for unbounded polyhedron conv($\mathcal{D}$) defined by the extreme point $(0,2)$ and two extreme rays: $\mathcal{T} = \{(0,2)\}$ in (a), $\mathcal{T} = \{(0,2),(1,2),(4,3),(5,3)\}$ in (b).

Using expression (4.15) on Figure 4.5b, we write $(9,4)$ as the sum of point $(1,2)$ indexed in $\ddot{P}$ plus the non-negative integer combination $0(2,0)+1(8,2)$ of the two integer rays indexed in $\ddot{R}$. To do so, we first decompose $(9,4)$ according to the Minkowski-Weyl Theorem 3.1, as a convex combination of the single extreme point $(0,2)$ with $\alpha_1 = 1$ plus a conic combination of the two extreme rays with $\beta_1 = \frac{1}{2}$, $\beta_2 = 1$. Next we separate the fractional and integral portions of $\beta_1 = \frac{1}{2} + 0$ and $\beta_2 = 0 + 1$ as in Theorem 4.2, finding $(1,2) \in \mathcal{T}$ and the non-negative integer combination of the rays $0(2,0)+1(8,2)$.

$$\begin{aligned}(9,4) &= (0,2) + \tfrac{1}{2}(2,0) + 1(8,2) \\ &= \underbrace{(0,2) + \tfrac{1}{2}(2,0) + 0(8,2)}_{(1,2) \in \mathcal{T}} + \underbrace{0(2,0) \quad + \quad 1(8,2)}_{\text{integer combination of } \mathbf{x}_r,\ r \in \ddot{R}} = (1,2) + 1(8,2).\end{aligned}$$

These observations about the potential number of interior points needed provide another reason for us to scale rays to the shortest possible integer vector.

Illustration 4.4 **Set $\mathcal{T}$ for another polyhedron**

Figure 4.6 presents another unbounded case, this time with three extreme points and two extreme rays for $\text{conv}(\mathcal{D}) \subset \mathbb{Z}_+^2$. The 43 integer points in set $\mathcal{T}$ come from the three extreme points and the 40 interior ones, including point $(3,7)$ on an edge. However, many of these interior points, indeed 29 identified by a $\star$, are useless in $\mathcal{T}$ as they can be obtained using the other 14 points and a non-negative integer combination of the two integer-scaled extreme rays.

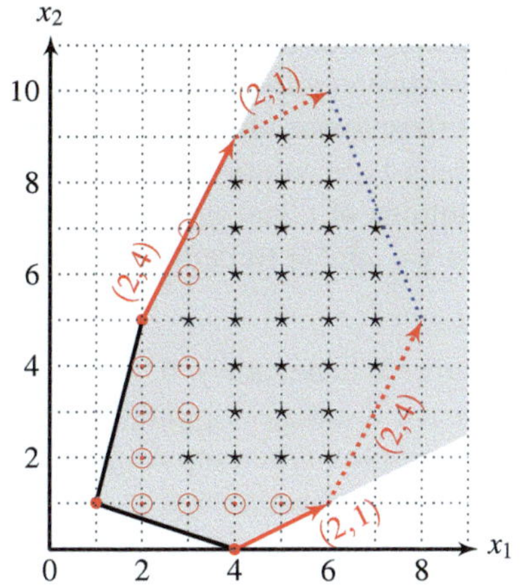

Fig. 4.6: Integer set $\mathcal{T} \subset \text{conv}(\mathcal{D}) \subset \mathbb{Z}_+^2$ derived from the three extreme points $(4,0)$, $(1,1)$, and $(2,5)$ and the two integer-scaled extreme rays $(2,1)$ and $(2,4)$.

Integer pricing problem

Proposition 4.2 is reflected in the way we handle the pricing problem in discretization because we can use the same pricing problem as in convexification, that is,

$$\bar{c}(\boldsymbol{\pi}_{\mathbf{b}}, \pi_0) = -\pi_0 + \min_{\mathbf{x} \in \mathcal{D}} (\mathbf{c}^\mathsf{T} - \boldsymbol{\pi}_{\mathbf{b}}^\mathsf{T} \mathbf{A})\mathbf{x}. \tag{4.29}$$

Recall that we impose that the *ISP* gives us a point $\mathbf{x}_p$, $p \in P$, with a finite objective value or a ray $\mathbf{x}_r$, $r \in R$, that leads to unboundedness. Let us try to match this to the set of variables we must formally consider in discretization:

$$\min \left\{ \min_{p \in \ddot{P}} \bar{c}_{\mathbf{x}_p}, \min_{r \in \ddot{R}} \bar{c}_{\mathbf{x}_r} \right\}. \tag{4.30}$$

With respect to the extreme rays, without our assumption that $\{\mathbf{x}_r\}_{r\in R} = \{\mathbf{x}_r\}_{r\in \ddot{R}}$, it could be possible that

$$\min_{r\in \ddot{R}} \bar{c}_{\mathbf{x}_r} \neq \min_{r\in R} \bar{c}_{\mathbf{x}_r} \tag{4.31}$$

because the scale influences the objective value. However, this difference disappears in (4.29) since the scale is irrelevant for unboundedness, i.e., $\bar{c}(\boldsymbol{\pi}_{\mathbf{b}}, \pi_0) = -\infty$ for all negative reduced cost extreme rays. Practically speaking, this is the place where we have to cast an extreme ray into an appropriate integer scale for the reformulation.

With respect to the extreme points, if $\bar{c}(\boldsymbol{\pi}_{\mathbf{b}}, \pi_0)$ is finite, then we have a point $\mathbf{x}_p$, $p \in P \subseteq \ddot{P}$. Since every interior point of $\mathcal{D}$ is dominated with respect to objective values by at least one extreme point of $\text{conv}(\mathcal{D})$, this strategy cannot produce an integer point indexed in $\ddot{I}$. Unfortunately, this also means that we cannot meaningfully conceive an algorithm that differentiates interior points. For instance, we could drop the necessity to ensure extreme-type solutions in the pricing problem but this just arbitrarily trades integer solutions of equal cost with one another. Note 4.8 underscores that the integer points we appear to be missing for the reformulation can be recovered in branching.

Note 4.8 (Structural deficiency.) Whenever $\ddot{I} \neq \emptyset$, the corresponding interior integer points possibly needed to reach λ-integer optimality are typically not actively, at least not by algorithms with simplex-type outputs, generated at the root node. Degraeve and Jans (2007) refer to this phenomenon as *structural deficiency*. While the authors restrict their analysis to binary master problems, it immediately applies to general *IMP*s. The thing is, binary master problems is perhaps the setting in which this disparity between the columns we need and those we get appears just barely out of reach. An example of such a situation occurs for the capacitated lot-sizing problem (Manne, 1958) solved by column generation (Dzielinski and Gomory, 1965). The pricing problem generates only the production plans that satisfy the Wagner-Whitin property (Wagner and Whitin, 1958), but it is known that an optimal integer solution does not necessarily satisfy this property. Jans (2010) mentions two other cases with structural deficiency: the capacitated facility location problem and the split delivery vehicle routing problem with time windows. With respect to more general integer master problems, we very intuitively recognize that non-extreme objects are needed, e.g., the cutting stock problem (see Exercise 2.16). We see in Chapter 7 how to recover these neglected integer points of $\mathcal{D}$ with branching and cutting decisions.

The conclusion is that there is no difference, at the root node, in the solution process of the linear relaxations $\mathit{M\ddot{P}}$ or *MP*, see Figures 4.7 and 4.2 where the *RMP* is the same and $\mathbf{x}^{\star}_{M\ddot{P}} = \mathbf{x}^{\star}_{MP} = \sum_{p\in P'} \mathbf{x}_p\lambda_p + \sum_{r\in R'} \mathbf{x}_r\lambda_r$. What shows up in this information flow is that both x- and λ-variables can be candidates for branching and cutting decisions. In comparison, we can only branch on the x-variables in the reformulation by convexification as illustrated for the *ILP*.

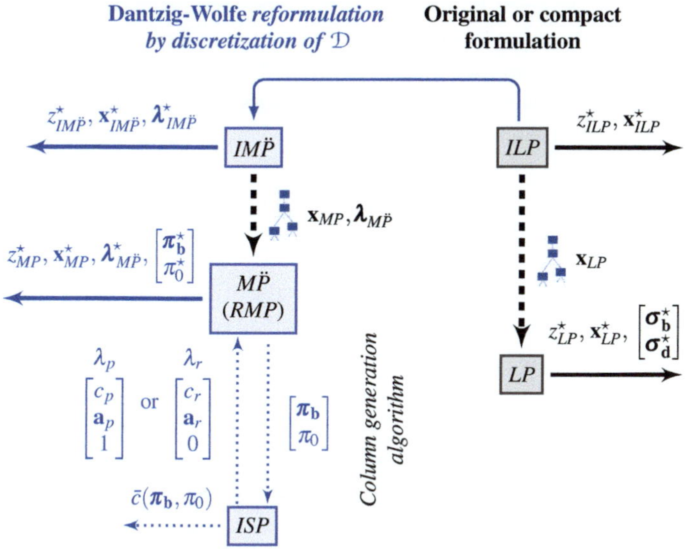

Fig. 4.7: Information flow of the column generation algorithm solving the $M\ddot{P}$, the linear relaxation of the $IM\ddot{P}$ (4.20), a Dantzig-Wolfe reformulation by discretization of the *ILP* (4.1).

Binary domain

In the important special case $\mathcal{D} \subseteq \{0,1\}^n$, there are no extreme rays ($\ddot{R} = \emptyset$), nor interior integer points ($\ddot{I} = \emptyset$ and $\ddot{P} = P$). The *IMP* issued from convexification allows to impose binary requirements on the λ-variables, as in the $IM\ddot{P}$. Several large-scale applications belong to this class, in particular, some reformulations that give rise to set partitioning and set covering models such as the *binary cutting stock problem*, the *multiple depot vehicle scheduling problem*, and the *edge coloring problem*.

Proposition 4.4. *If there is a binary condition on every variable of the pricing problem ($\mathcal{D} \subseteq \{0,1\}^n$), we can equivalently write the convexification model with integrality on the λ-variables only, i.e.,*

$$\begin{aligned}
z^\star_{IMP} = \min \quad & \sum_{p\in P} c_p\lambda_p \\
\text{s.t.} \quad & \sum_{p\in P} \mathbf{a}_p\lambda_p \geq \mathbf{b} && [\boldsymbol{\pi}_{\mathbf{b}}] \\
& \sum_{p\in P} \lambda_p = 1 && [\pi_0] \\
& \lambda_p \in \{0,1\} && \forall p \in P.
\end{aligned} \tag{4.32}$$

Proof. By assumption, we have that $\mathcal{D}$ is bounded so this model is an immediate adaptation of (4.9) in which we replace $\sum_{p\in P}\mathbf{x}_p\lambda_p = \mathbf{x} \in \{0,1\}^n$ by $\lambda_p \in \{0,1\}$,

$\forall p \in P$. This is correct since $P = \ddot{P}$, so it is the same as a discretization model which ensures integrality on $\mathbf{x}$ by Proposition 4.3. The equivalence comes from the fact that integrality on $\mathbf{x}$ also implies that on $\boldsymbol{\lambda}$. Indeed, binary vector $\mathbf{x}^{\star}_{IMP}$ is an extreme point of conv$(\mathcal{D})$. Because there are no interior integer points, there do not exist any integer convex combinations, hence $\mathbf{x}^{\star}_{IMP} \in \{0,1\}^n \Leftrightarrow \boldsymbol{\lambda}^{\star}_{IMP} \in \{0,1\}^{|P|}$. □

Illustration 4.5 TCSPP: **integrality**

We reconsider the time constrained shortest path problem as described in Example 3.2. Recall that we are looking for a least-cost path on network $G = (N,A)$, as depicted in Figure 4.8, no longer than 14 time units.

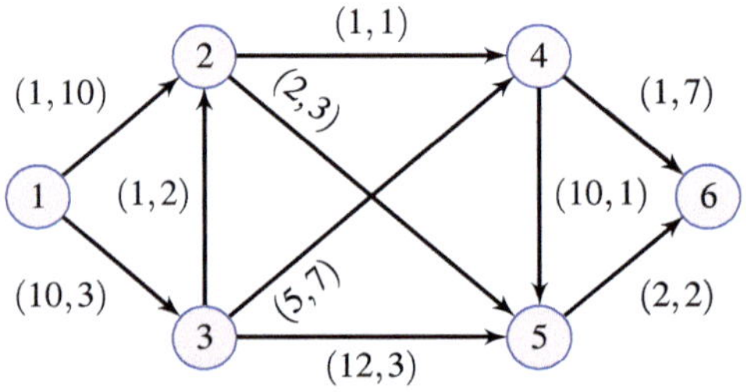

Fig. 4.8: Network $G = (N,A)$ with (c_{ij}, t_{ij}) values, $\forall (i,j) \in A$.

The compact formulation is

$$z^{\star}_{ILP} = \min \sum_{(i,j)\in A} c_{ij} x_{ij} \tag{4.33a}$$

$$\text{s.t.} \quad \sum_{j:(1,j)\in A} x_{1j} = 1 \quad [\sigma_1] \tag{4.33b}$$

$$\sum_{j:(i,j)\in A} x_{ij} - \sum_{j:(j,i)\in A} x_{ji} = 0 \quad [\sigma_i] \quad \forall i \in \{2,\ldots,5\} \tag{4.33c}$$

$$-\sum_{i:(i,6)\in A} x_{i6} = -1 \quad [\sigma_6] \tag{4.33d}$$

$$\sum_{(i,j)\in A} t_{ij} x_{ij} \leq 14 \quad [\sigma_7] \tag{4.33e}$$

$$x_{ij} \in \{0,1\} \quad \forall (i,j) \in A. \tag{4.33f}$$

Following the same grouping of constraints but taking into account the integrality restrictions from the compact formulation in $\mathcal{A}$ leads to

$$\mathcal{A} = \left\{ \mathbf{x} \in \{0,1\}^{|A|} \;\middle|\; \sum_{(i,j)\in A} t_{ij} x_{ij} \leq 14 \right\} \tag{4.34a}$$

$$\mathcal{D} = \left\{ \mathbf{x} \in \{0,1\}^{|A|} \;\middle|\; \sum_{j:(i,j)\in A} x_{ij} - \sum_{j:(j,i)\in A} x_{ji} = [+1\ 0\ 0\ 0\ 0\ -1]_i, \forall i \in N \right\}. \quad (4.34b)$$

Because $\mathcal{D}$ describes a network flow problem which possesses the integrality property, we can literally reuse model (3.81) while simply switching to an integer master problem

$$\begin{aligned} z^\star_{IMP} = \min \; & \sum_{p\in P} c_p\lambda_p \\ \text{s.t.} \; & \sum_{p\in P} t_p\lambda_p \le 14 && [\pi_7] \\ & \sum_{p\in P} \lambda_p = 1 && [\pi_0] \\ & \lambda_p \in \{0,1\} && \forall p \in P, \end{aligned} \quad (4.35)$$

where we use convexification notation and $\boldsymbol{\lambda}$-integrality because $\mathcal{D}$ is on a binary domain such that there is no difference with discretization by Proposition 4.4. Recall also that integrality on $\mathbf{x}$ is implied from $\boldsymbol{\lambda}$ in discretization so we also drop the relation.

In Example 4.5 on the Generalized assignment problem, we derive a Dantzig-Wolfe reformulation of a binary program which simplifies back into the compact formulation. It is a consequence of Proposition 4.5 extended to a block-diagonal structure. (See also Exercise 4.21 on Useless Dantzig-Wolfe reformulations for two additional cases.)

Proposition 4.5. *If* $\mathcal{D} = \{\mathbf{x} \in \{0,1\}^n \mid \sum_{j=1}^n x_j = 1\}$, *then the IMP is no more and no less than the ILP.*

Proof. By the Minkowski-Weyl Theorem 4.1, we reformulate $\text{conv}(\mathcal{D}) = \{\mathbf{x} \ge \mathbf{0} \mid \sum_{j=1}^n x_j = 1\}$ which is the convex hull of the n unit vectors in $\mathbb{R}^n_+$. The set of all the extreme points of $\text{conv}(\mathcal{D})$ is therefore $\{\mathbf{e}_1, \ldots, \mathbf{e}_n\}$.

We here observe that $|P| = n$ and express $\mathbf{x}$ as

$$\sum_{j=1}^n \mathbf{e}_j\lambda_j = \mathbf{x} \in \{0,1\}^n, \quad \sum_{j=1}^n \lambda_j = 1, \quad \lambda_j \ge 0,\ j = 1,\ldots,n. \quad (4.36)$$

Writing the first equation component-wise, we get $\lambda_1 = x_1, \ldots, \lambda_n = x_n$, that is, we simply renamed the x-variables. Finally, the introduced convexity constraint is exactly the original constraint in $\mathcal{D}$. $\square$

Some observations

Table 4.1 summarizes key elements of the reformulation by discretization. The first case is the general model whereas others come from the simplifications based on the nature of the set $\mathcal{D} = \{\mathbf{x} \in \mathbb{Z}^n_+ \mid \mathbf{Dx} \geq \mathbf{d}\}$. We describe the index sets of the λ-variables with respect to P, $\ddot{I}$, and $\ddot{R}$, give a reference to the reformulation model, and list their type. We then use it to make some observations and draw a few words of comparison with convexification.

Nature of $\mathcal{D}$	Index sets		*IMP̈*	λ_p	λ_r
$\mathcal{D} \subseteq \mathbb{Z}^n_+$	$p \in P \cup \ddot{I},\ r \in \ddot{R}$		(4.20)	$\in \{0,1\}$	$\in \mathbb{Z}_+$
$\mathcal{D}$ *bounded*	$p \in P \cup \ddot{I}$		(4.24)_left	$\in \{0,1\}$	
$\mathcal{D} \subseteq \{0,1\}^n$	$p \in P$		(4.32)	$\in \{0,1\}$	
$\text{conv}(\mathcal{D})$ *polyhedral cone*		$r \in \ddot{I} \cup \ddot{R}$	(4.24)_right		$\in \mathbb{Z}_+$

Table 4.1: Various results for a Dantzig-Wolfe reformulation of the *ILP* (4.1) based on the discretization of $\mathcal{D} = \{\mathbf{x} \in \mathbb{Z}^n_+ \mid \mathbf{Dx} \geq \mathbf{d}\}$.

1. In the general case ($\mathcal{D} \subseteq \mathbb{Z}^n_+$), the set $\ddot{P} \cup \ddot{R}$ is usually larger than $P \cup R$, i.e., $\ddot{R}$ is the same as R but $\ddot{P} = P \cup \ddot{I}$, where $\ddot{I} \neq \emptyset$ most of the time. This is particularly true in the bounded case, where P is only associated with the extreme points of the polytope $\text{conv}(\mathcal{D})$ whereas $\ddot{I}$ corresponds to *all its interior integer points.*
2. A Dantzig-Wolfe reformulation by convexification requires $\mathbf{x}$ integer. The discretization approach shows that there also *exists* a reformulation with $\boldsymbol{\lambda}$ integer. The interior integer points are those that allow this. Within a column generation context, the potentially much larger size of $\ddot{I}$ is not worrying since we consider them implicitly. In fact, we never even have meaningful access to them when we solve the *MP̈*. Indeed, recall that the *RMP* is the same for both the convexification and discretization of $\mathcal{D}$ (assuming $\{\mathbf{x}_r\}_{r\in R} = \{\mathbf{x}_r\}_{r\in \ddot{R}}$). The differences in the approaches therefore materialize in branching and cutting decisions.
3. In this respect, we have no choice but to derive decisions based on what the integrality requirements are. On $\mathbf{x}_{MP}$, on $\boldsymbol{\lambda}_{M\ddot{P}}$, on both? Chapter 7 concentrates on deriving appropriate decisions while exposing technical challenges. We underscore here that the question of whether we can impose integrality on $\boldsymbol{\lambda}$ is one with plenty of history. We answer it affirmatively and it implies the discretization point of view.
4. In writing the *MP̈*, an upper bound of 1 on the λ_p-variables appears naturally and it is thus tempting to impose that in the model. Observe that these bounds are redundant since they are implied by the convexity constraint. This gives us the opportunity to echo Note 2.15 and warn against using these upper bounds.

Post-processing a solution of the master problem

We have already seen in Chapter 3 that the projection back to $\mathbf{x}$ is dropped while solving the *RMP*s. This of course remains true for a Dantzig-Wolfe reformulation of an integer linear program, regardless of whether we apply a convexification or discretization of $\mathcal{D}$. The fact that we are ultimately looking for an integer solution, however, implies that this projection is done in a post-processing of a solution of the master problem. We do not only figure out the values for the x-variables but also test whether integrality on $\mathbf{x}$ is fulfilled.

If the solution of the final *RMP* is integer, i.e.,

$$\sum_{p\in P'} \mathbf{x}_p\lambda_p^\star + \sum_{r\in R'} \mathbf{x}_r\lambda_r^\star = \mathbf{x}_{MP}^\star \in \mathbb{Z}_+^n, \tag{4.37}$$

we are done as it obviously constitutes an optimal integer solution $\mathbf{x}_{IMP}^\star \equiv \mathbf{x}_{IM\ddot{P}}^\star \equiv \mathbf{x}_{ILP}^\star$ as well. Otherwise, $\mathbf{x}_{MP}^\star \notin \mathbb{Z}_+^n$ is fractional despite having integer vectors $\mathbf{x}_p, \mathbf{x}_r \in \mathbb{Z}_+^n$, for all $p \in P$ and $r \in R$. In convexification, a cutting or branching decision is required to progress in the branch-and-bound algorithm, see Chapter 7.

In discretization, we have seen in Proposition 4.3 that integrality on $\boldsymbol{\lambda}$ implies that on $\mathbf{x}$. This means that testing whether a solution is integer can be reduced to checking the values of $\boldsymbol{\lambda}$. It also means that it is *sufficient* to derive branching and cutting decisions on λ-variables. We stress that while we *can* neglect the relation to $\mathbf{x}$ entirely as in (4.21) (*not necessary*), it is likely more interesting to keep it in mind in post-processing because of opportunities for cutting and branching decisions on x-variables that would otherwise vanish.

4.3 Integrality Property: For or Against Virtue?

Evaluating the success of a reformulation essentially lies in comparing how much it can be solved faster than the compact formulation. We first observe that a Dantzig-Wolfe reformulation is a detour if the compact formulation has an integrality gap of zero. Indeed, we consider that the latter is already simple to solve since little branch-and-bound "if any" is needed as $z_{ILP}^\star = z_{LP}^\star$.

Proposition 4.6. *If the ILP* (4.1) *has an integrality gap of zero, then*

$$z_{LP}^\star = z_{MP}^\star = z_{IMP}^\star = z_{ILP}^\star. \tag{4.38}$$

Proof. Proposition 4.1 establishes $z_{LP}^\star \le z_{MP}^\star \le z_{IMP}^\star = z_{ILP}^\star$ for any grouping of the constraints. By assumption, we have that $z_{ILP}^\star = z_{LP}^\star$ which leads to $z_{MP}^\star = z_{IMP}^\star$. □

Note 4.9 (Integrality property of the ILP transfers to the IMP.) Furthermore, if the formulation of the *ILP* has the integrality property (Definition 1.25 and Proposition 1.9), the integrality gap of zero in Proposition 4.6 obviously transfers for any objective coefficients $\mathbf{c} \in \mathbb{R}^n$. That is to say that the formulation of the *IMP*, where the $\mathbf{x}_p$ and $\mathbf{x}_r$ are fixed parameters, also possesses the integrality property. For example, if we apply a Dantzig-Wolfe reformulation on the *capacitated minimum cost flow problem*, the *mixed-integer* linear programming reformulation also has the integrality property, despite λ-variables possibly taking fractional values. Indeed, no matter the way we group the constraints of the compact formulation, the x-domain of the arc-flow variables in the *ILP* remains the same in the *IMP*.

Ultimately, it is fair to assume that the *ILP* has a non-negligible integrality gap such that it is likely too difficult to solve with current algorithms/technology. The possibility to obtain a stronger linear relaxation than that of the compact formulation is quite interesting because reducing the integrality gap can significantly help our chances in completing the branch-and-bound search. In general, the quality of the lower bound $z^\star_{MP}$ with respect to $z^\star_{ILP}$ is influenced by the information available in $\mathcal{D}$. Figure 4.9 is a good place to start the explanation. Observe that some integer solutions of $\mathcal{D}$ are allowed in the *MP* despite the fact that they are infeasible for the *IMP*. In contrast, fractional solutions in the *LP* are shaved off in chunks from the *MP*. Moreover, we hand-pick an objective function whose level curve shows best- and worst-case scenarios of a reformulation. In the best-case (level curve moving to the left), we are very lucky and reach integer optimality at the point marked by a $\star$, directly in the *MP*. In the worst-case (level curve moving to the right), we are rather unlucky and reach a lower bound $z^\star_{MP}$ that is no better than $z^\star_{LP}$ as the level curve is parallel to the bottom-right edge of the *LP*'s domain. We underscore that in practice, we do expect a better linear relaxation from the *MP* regardless of the objective coefficients. In this case, there is only one other level curve parallel to the bottom-left edge of the *LP*'s domain with improvement going downwards for which $z^\star_{LP} = z^\star_{MP}$.

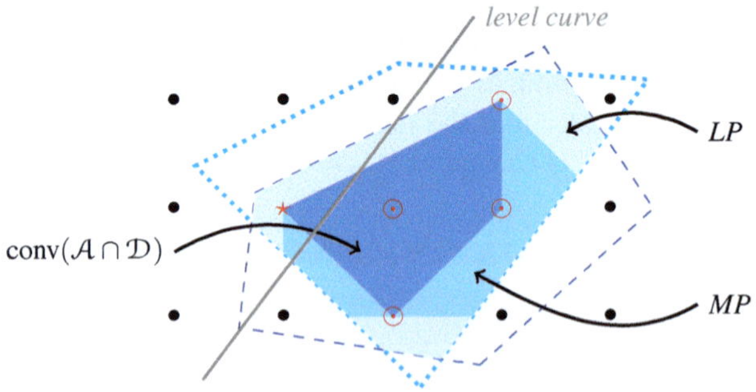

Fig. 4.9: Trade-off of a reformulation with respect to the domains of the *LP* and *MP*.

Since we control $\mathcal{D}$, the pricing problem can be as easy or as difficult as we wish. In theory, we can solve the *ILP* in the *ISP* but we rather look for a grouping of constraints $\mathcal{A}$ and $\mathcal{D}$ with a best trade-off between the quality of the lower bound and the ease with which the *ISP* can be solved. This is of course easier said than done since it is not at all clear how this trade-off looks like. As such, it is hard to turn a blind eye on the efficiency we can achieve in the bottleneck operation. In particular, we may be tempted to look for an *ISP* with a formulation that possesses the integrality property. Proposition 4.7 warns us that leveraging on this idea defeats the purpose of trying to capitalize on integrality. In other words, a necessary condition to obtain a better lower bound than $z^\star_{LP}$ is that the *formulation* of the *ISP* must *not* have the integrality property.

Proposition 4.7. *If the formulation of the ISP* (4.11) *possesses the integrality property, then the linear relaxation of the IMP is no better than that of the ILP, i.e.,*

$$z^\star_{LP} = z^\star_{MP} \leq z^\star_{ILP}. \tag{4.39}$$

Proof. By Proposition 1.9, the formulation of the *ISP* (4.11) possesses the integrality property if $\text{conv}(\mathcal{D}) = \{\mathbf{x} \in \mathbb{R}^n_+ \mid \mathbf{Dx} \geq \mathbf{d}\}$, where $\mathcal{D} = \{\mathbf{x} \in \mathbb{Z}^n_+ \mid \mathbf{Dx} \geq \mathbf{d}\}$. The feasible regions of the *MP* and *LP* are therefore the same, i.e.,

$$\{\mathbf{x} \in \mathbb{R}^n_+ \mid \mathbf{Ax} \geq \mathbf{b}\} \cap \text{conv}(\mathcal{D}) = \{\mathbf{x} \in \mathbb{R}^n_+ \mid \mathbf{Ax} \geq \mathbf{b}, \mathbf{Dx} \geq \mathbf{d}\},$$

such that their optimal objective values are equal. □

Note 4.10 (Property means property.) The word *property* is a little bit ambiguous because it is charged with the meaning of *attribute* or *quality*, the latter being familiarly understood as a positive trait. Historically, this is exactly what people were looking for! Make no mistake, the integrality property is first and foremost a mathematical characterization of a matrix. It turns out to be fruitful in integer linear programming when the formulation is solved heads on, but *it is not necessarily a desirable property for the formulation of the ISP.*

While these observations can be enlightening, it does not make selecting the sets $\mathcal{A}$ and $\mathcal{D}$ any less of a dilemma. On the one hand, we would like the reformulation to be meaningful in that a *reduction of the solution space* is achieved leading to a potentially smaller integrality gap $z^\star_{IMP} - z^\star_{MP} \leq z^\star_{ILP} - z^\star_{LP}$. On the other hand, since solving the *ISP* is a bottleneck operation, we would like it to expose some *structural properties* that allow it to be solved in *reasonable* time in spite of the integrality requirements. Indeed, the trade-off between the discarded fractional solutions and the difficulty of solving the integer pricing problem appears to be complex and sometimes impossible to predict. Example 4.3 analyzes the impact of different reformulations on the master and pricing problems and the resulting lower bounds.

We conclude by playing devil's advocate on the integrality property for the formulation of the *ISP*. There are two well-known papers for fundamental problems where the integrality property is exploited: the *traveling salesperson problem* (Held and Karp, 1970) and the *multi-commodity maximum flow problem* (Ford and Fulkerson, 1958). This allows us to underscore two things:

1. The linear relaxation of the *ILP* may already be pretty good so the pricing bottleneck may become the only critical issue, see Example 6.4 on the Symmetric traveling salesperson problem.
2. A block-diagonal structure (Sections 3.2 and 4.4) helps to reduce the bottleneck strain because we rather handle $|K|$ smaller independent problems, as in Example 4.6 on the Multi-commodity maximum flow problem.

4.4 Block-diagonal Structure

The block-diagonal structure of Section 3.2 is extended to the *ILP* given as

$$\begin{aligned} z^\star_{ILP} = \min \quad & \sum_{k \in K} \mathbf{c}^{k\intercal} \mathbf{x}^k \\ \text{s.t.} \quad & \sum_{k \in K} \mathbf{A}^k \mathbf{x}^k \geq \mathbf{b} && [\boldsymbol{\sigma}_\mathbf{b}] \\ & \mathbf{D}^k \mathbf{x}^k \geq \mathbf{d}^k && [\boldsymbol{\sigma}_{\mathbf{d}^k}] \quad \forall k \in K \\ & \mathbf{x}^k \in \mathbb{Z}_+^{n^k} && \forall k \in K, \end{aligned} \tag{4.40}$$

the main difference being that $\mathcal{D}^k = \{\mathbf{x}^k \in \mathbb{Z}_+^{n^k} \mid \mathbf{D}^k\mathbf{x}^k \geq \mathbf{d}^k\}$, $k \in K$, is defined for integer vectors $\mathbf{x}^k$, i.e., $\mathcal{D}^k$ is a discrete set.

Practical relevance

The lightweight notation of a formulation for the *ILP* with a single block serves the academic purpose particularly well. In the remainder of this section, we directly use models obtained by convexification or discretization with a notation that can become cumbersome. To convince the reader not to be deterred by the mathematics, we list some large-scale applications from the literature to underscore the practical relevance of a block-diagonal structure:

a) 3 to 6 time periods in the inventory routing problem (Desaulniers et al., 2016b);
b) 6 and 9 aircraft types for daily schedules of a heterogeneous fleet at European and American carriers, respectively (Desaulniers et al., 1997d);
c) 14 subproblems, one for each starting day of the week for round-trip crew schedules at the two Air France crew bases (Desaulniers et al., 1997a);
d) 2 to 20 blocks in the automated Dantzig-Wolfe reformulation computational study of mixed-integer linear programs (Bergner et al., 2015);

e) 18 to 108 pilots for the construction of personalized monthly schedules at Air Canada (Gamache et al., 1998);

f) 26 locomotive types and 171 critical locomotives in the problem of assigning locomotives to train-segments at CN North America (Ziarati et al., 1997) – see Example 7.10 Branch-first, Cut-second strategy;

g) 55 to 1131 crew members, each with its own pricing problem due to personal desiderata for the construction of monthly schedules at Air France (Gamache et al., 1999). As mentioned in Note 2.13, it is in practice too time consuming to solve all these ($\mathcal{NP}$-hard) subproblems. Particularly in this application, a partial pricing strategy is used, that is, only 15 of these are solved at every column generation iteration with the same set of dual values.

On grouping the constraints

We know that block-diagonal structures like (4.40) very naturally arise from the definition of an optimization problem, where otherwise independent entities (like vehicles, warehouses, containers) are linked by coordinating constraints. The latter are like a conductor who takes care that the individual players in an orchestra sound well together. An index set K represents the entities and thus induces the blocks. Variables that appear in the constraints that belong to one block, obviously cannot appear in constraints of other blocks. In other words, whenever two variables with different indices in K appear in the same constraint, this constraint must be a coordinating one. In the `Coluna` (atoptima.com) branch-and-price solver, the set K is called the *axis*. Figure 4.10a shows the form of the corresponding coefficient matrix of the model. It is called the *single-bordered*, or more precisely, the *single-row-bordered* block-diagonal form. As we have mentioned before, such a model structure is particularly well suited for a Dantzig-Wolfe reformulation.

$$\begin{bmatrix} \mathbf{A}^1 & \mathbf{A}^2 & \cdots & \mathbf{A}^{|K|} \\ \mathbf{D}^1 & & & \\ & \mathbf{D}^2 & & \\ & & \ddots & \\ & & & \mathbf{D}^{|K|} \end{bmatrix} \qquad \begin{bmatrix} \mathbf{S}^1 & \mathbf{D}^1 & & & \\ \mathbf{S}^2 & & \mathbf{D}^2 & & \\ \vdots & & & \ddots & \\ \mathbf{S}^{|K|} & & & & \mathbf{D}^{|K|} \end{bmatrix} \qquad \begin{bmatrix} \mathbf{S} & \mathbf{A}^1 & \mathbf{A}^2 & \cdots & \mathbf{A}^{|K|} \\ \mathbf{S}^1 & \mathbf{D}^1 & & & \\ \mathbf{S}^2 & & \mathbf{D}^2 & & \\ \vdots & & & \ddots & \\ \mathbf{S}^{|K|} & & & & \mathbf{D}^{|K|} \end{bmatrix}$$

(a) Single-row-bordered (b) Single-column-bordered (c) Double-bordered

Fig. 4.10: Forms of (coefficient) matrices interesting for decompositions

Related block-diagonal forms are with a *column-border* as in Figure 4.10b, or *double-bordered* like in Figure 4.10c. The latter is also known as *arrowhead form.* We see such forms in Shared variables across all blocks (p. 222), where it is shown that they are amenable to a Dantzig-Wolfe reformulation as well. Also, a Benders decomposition of a linear program may apply (p. 130).

When it comes to grouping the constraints for a decomposition, all we need is the set K, or in other words, we need to identify the blocks. What at first sounds like little freedom, can be non-obvious in some models. Think, for instance, of multi-product lot-sizing problems, in which we decide about production and inventory quantities over time. One can take the perspective of an individual product and follow it along the time periods, or take the view of a single time period that sees all products simultaneously. This gives (at least) two ways, how blocks, i.e., the index set K, could be defined. One can decompose along the product axis, or along the time axis.

The matrices in Figure 4.11 hint at this freedom of grouping constraints into blocks. Besides relying on our problem and model knowledge, we can use algorithms to find the blocks for us. The interested reader can learn about the basics of how this works in Automatic grouping of the constraints for reformulation (p. 232).

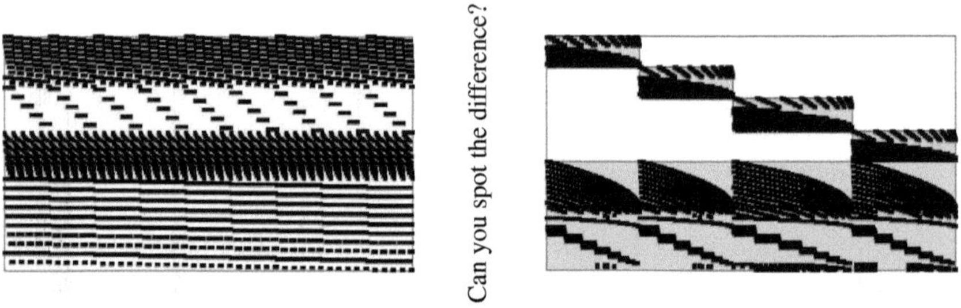

Fig. 4.11: T-shirt logo of the 2012 CG-Workshop (Bromont, Canada).

Integer master problems (convexification and discretization)

In the convexification approach, the substitution for block k is based on

$$\begin{aligned}
&\sum_{p\in P^k} \mathbf{x}_p^k \lambda_p^k + \sum_{r\in R^k} \mathbf{x}_r^k \lambda_r^k = \mathbf{x}^k \in \mathbb{Z}_+^{n^k} \\
&\sum_{p\in P^k} \lambda_p^k = 1 \\
&\lambda_p^k \geq 0, \quad \lambda_r^k \geq 0 \qquad \forall p \in P^k, r \in R^k
\end{aligned} \tag{4.41}$$

where P^k and R^k are the index sets of the extreme points and extreme rays of $\operatorname{conv}(\mathcal{D}^k)$. Similarly to (3.29), the reformulation then reads as

$$z_{IMP}^\star = \min \sum_{k\in K} \sum_{p\in P^k} c_p^k \lambda_p^k + \sum_{k\in K} \sum_{r\in R^k} c_r^k \lambda_r^k \tag{4.42a}$$

$$\text{s.t.} \sum_{k\in K} \sum_{p\in P^k} \mathbf{a}_p^k \lambda_p^k + \sum_{k\in K} \sum_{r\in R^k} \mathbf{a}_r^k \lambda_r^k \geq \mathbf{b} \quad [\boldsymbol{\pi}_\mathbf{b}] \tag{4.42b}$$

$$\sum_{p\in P^k} \lambda_p^k = 1 \qquad [\pi_0^k] \qquad \forall k\in K \tag{4.42c}$$

$$\lambda_p^k \geq 0 \qquad \forall k\in K, p\in P^k \tag{4.42d}$$

$$\lambda_r^k \geq 0 \qquad \forall k\in K, r\in R^k \tag{4.42e}$$

$$\sum_{p\in P^k} \mathbf{x}_p^k \lambda_p^k + \sum_{r\in R^k} \mathbf{x}_r^k \lambda_r^k = \mathbf{x}^k \in \mathbb{Z}_+^{n^k} \qquad \forall k\in K. \tag{4.42f}$$

Regarding the discretization approach, the substitution for block k is based on

$$\begin{aligned}
&\sum_{p\in \ddot{P}^k} \mathbf{x}_p^k \lambda_p^k + \sum_{r\in \ddot{R}^k} \mathbf{x}_r^k \lambda_r^k = \mathbf{x}^k \in \mathbb{Z}_+^{n^k} \\
&\sum_{p\in \ddot{P}^k} \lambda_p^k = 1 \\
&\lambda_p^k \in \{0,1\} \qquad \forall p \in \ddot{P}^k \\
&\lambda_r^k \in \mathbb{Z}_+ \qquad \forall r \in \ddot{R}^k,
\end{aligned} \tag{4.43}$$

where $\ddot{P}^k$ and $\ddot{R}^k$ are defined as in the Hilbert-Giles-Pulleyblank Theorem 4.2, and again, we assume $\{\mathbf{x}_r^k\}_{r\in\ddot{R}^k} = \{\mathbf{x}_r^k\}_{r\in R^k}$, $\forall k \in K$. The reformulation then reads as

$$z^\star_{IM\ddot{P}} = \min \sum_{k\in K} \sum_{p\in \ddot{P}^k} c_p^k \lambda_p^k + \sum_{k\in K} \sum_{r\in \ddot{R}^k} c_r^k \lambda_r^k \tag{4.44a}$$

$$\text{s.t.} \sum_{k\in K} \sum_{p\in \ddot{P}^k} \mathbf{a}_p^k \lambda_p^k + \sum_{k\in K} \sum_{r\in \ddot{R}^k} \mathbf{a}_r^k \lambda_r^k \geq \mathbf{b} \quad [\boldsymbol{\pi}_\mathbf{b}] \tag{4.44b}$$

$$\sum_{p\in \ddot{P}^k} \lambda_p^k = 1 \qquad [\pi_0^k] \qquad \forall k\in K \tag{4.44c}$$

$$\lambda_p^k \in \{0,1\} \qquad \forall k\in K, p\in \ddot{P}^k \tag{4.44d}$$

$$\lambda_r^k \in \mathbb{Z}_+ \qquad \forall k\in K, r\in \ddot{R}^k \tag{4.44e}$$

$$\sum_{p\in \ddot{P}^k} \mathbf{x}_p^k \lambda_p^k + \sum_{r\in \ddot{R}^k} \mathbf{x}_r^k \lambda_r^k = \mathbf{x}^k \in \mathbb{Z}_+^{n^k} \qquad \forall k\in K, \tag{4.44f}$$

where the relation (4.44f) can be dropped altogether by Proposition 4.3. In the post-processing, note that we can also work with $\sum_{p\in\ddot{P}^k} \mathbf{x}_p^k \lambda_p^k \in \mathbb{Z}_+^{n^k}$ and $\sum_{r\in\ddot{R}^k} \mathbf{x}_r^k \lambda_r^k \in \mathbb{Z}_+^{n^k}$, for all $k \in K$.

In either approach, we can derive without much fanfare that, for each block $k \in K$, we output extreme points or extreme rays from the ISP^k which writes no differently than (3.30) except for integrality restrictions in $\mathcal{D}^k$, that is,

$$\bar{c}^k(\boldsymbol{\pi}_\mathbf{b}, \pi_0^k) = -\pi_0^k + \min_{\mathbf{x}^k\in\mathcal{D}^k} (\mathbf{c}^{k\mathsf{T}} - \boldsymbol{\pi}_\mathbf{b}^\mathsf{T}\mathbf{A}^k)\mathbf{x}^k. \tag{4.45}$$

Identical subproblems

An important special case of the compact formulation with a block-diagonal structure (4.40), known as *identical subproblems*, occurs if the data across all blocks is identical, i.e.,

$$\mathbf{c}^k = \mathbf{c}, \ \ \mathbf{A}^k = \mathbf{A}, \ \ \mathbf{D}^k = \mathbf{D}, \ \ \mathbf{d}^k = \mathbf{d}, \qquad \forall k \in K. \tag{4.46}$$

Several applications are later presented, for example,

- The One-dimensional cutting stock problem where identical rolls are cut (p. 244);
- The Vehicle Routing Problem with Time Windows where identical vehicles are routed and scheduled (p. 297).

Given the parameters in (4.46), the *ILP* is formulated as

$$\begin{aligned} z^\star_{ILP} = \min \quad & \mathbf{c}^\intercal (\sum_{k \in K} \mathbf{x}^k) \\ \text{s.t.} \quad & \mathbf{A}(\sum_{k \in K} \mathbf{x}^k) \geq \mathbf{b} && [\boldsymbol{\sigma}_\mathbf{b}] \\ & \mathbf{D}\mathbf{x}^k \geq \mathbf{d} && [\boldsymbol{\sigma}_{\mathbf{d}^k}] \quad \forall k \in K \\ & \mathbf{x}^k \in \mathbb{Z}^n_+ && \forall k \in K, \end{aligned} \tag{4.47}$$

where the cost vector $\mathbf{c}^\intercal$ and matrix $\mathbf{A}$ of the linking constraints factorize outside the sum. With the obvious grouping

$$\mathcal{A} = \left\{ \{\mathbf{x}^k \in \mathbb{Z}^n_+\}_{k \in K} \;\middle|\; \mathbf{A}(\sum_{k \in K} \mathbf{x}^k) \geq \mathbf{b} \right\} \tag{4.48a}$$

$$\mathcal{D}^k = \{\mathbf{x}^k \in \mathbb{Z}^n_+ \mid \mathbf{D}\mathbf{x}^k \geq \mathbf{d}\}, \qquad \forall k \in K, \tag{4.48b}$$

and the resulting ISP^k (4.45) that are all the same in objective functions and domains except for the independent dual value π_0^k, this means that the respective variables of the $|K|$ subproblems live in isomorphic subspaces and in fact could formally live in the same subspace. For the identical sets in (4.48b), we therefore choose one representative

$$\mathbf{x}^k \in \mathcal{D} = \{\mathbf{x} \in \mathbb{Z}^n_+ \mid \mathbf{D}\mathbf{x} \geq \mathbf{d}\}, \qquad \forall k \in K, \tag{4.49}$$

which gives us a single pricing problem later defined in (4.58). In the convexification approach, we use extreme points and extreme rays of conv$(\mathcal{D})$, that is, $\mathcal{X} = \left\{ \begin{bmatrix} \mathbf{x}_p \\ 1 \end{bmatrix} \right\}_{p \in P} \cup \left\{ \begin{bmatrix} \mathbf{x}_r \\ 0 \end{bmatrix} \right\}_{r \in R}$, to express $\mathbf{x}^k \in \text{conv}(\mathcal{D})$ as

$$\sum_{p \in P} \mathbf{x}_p \lambda_p^k + \sum_{r \in R} \mathbf{x}_r \lambda_r^k = \mathbf{x}^k \in \mathbb{Z}^n_+, \quad \sum_{p \in P} \lambda_p^k = 1, \quad \lambda_p^k, \lambda_r^k \geq 0, \ \forall p \in P, r \in R. \tag{4.50}$$

The discretization approach rather uses

$$\mathcal{X} = \left\{ \begin{bmatrix} \mathbf{x}_p \\ 1 \end{bmatrix} \right\}_{p \in \ddot{P}} \cup \left\{ \begin{bmatrix} \mathbf{x}_r \\ 0 \end{bmatrix} \right\}_{r \in \ddot{R}} \tag{4.51}$$

with binary λ_p-variables and non-negative integer λ_r-variables. Let us continue with the former approach, pointing out modifications as needed for the latter. Independent of index $k \in K$, our shorthand encoding functions $c_{\mathbf{x}} = \mathbf{c}^\mathsf{T}\mathbf{x}$ and $\mathbf{a}_{\mathbf{x}} = \mathbf{A}\mathbf{x}$, $\forall \mathbf{x} \in \mathcal{X}$, are still distinguishable by subscript p or r. The *IMP* becomes

$$\begin{aligned}
z^\star_{IMP} = \min \; & \sum_{p \in P} c_p (\sum_{k \in K} \lambda_p^k) && + \sum_{r \in R} c_r (\sum_{k \in K} \lambda_r^k) \\
\text{s.t.} \; & \sum_{p \in P} \mathbf{a}_p (\sum_{k \in K} \lambda_p^k) && + \sum_{r \in R} \mathbf{a}_r (\sum_{k \in K} \lambda_r^k) \geq \mathbf{b} \quad [\boldsymbol{\pi}_\mathbf{b}] \\
& \sum_{p \in P} \lambda_p^k = 1 && \qquad [\pi_0^k] \quad \forall k \in K \\
& \lambda_p^k \geq 0, && \qquad \lambda_r^k \geq 0 \qquad \forall k \in K, p \in P, r \in R \\
& \sum_{p \in P} \mathbf{x}_p \lambda_p^k && + \sum_{r \in R} \mathbf{x}_r \lambda_r^k = \mathbf{x}^k \in \mathbb{Z}^n_+ \quad \forall k \in K.
\end{aligned} \tag{4.52}$$

Aggregation

Since the λ_p^k-variables for different k refer to the same extreme point $\mathbf{x}_p$, we collapse them into one variable (the same applies to the λ_r^k-variables, of course). Let

$$\lambda_p = \sum_{k \in K} \lambda_p^k, \; \forall p \in P, \quad \text{and} \quad \lambda_r = \sum_{k \in K} \lambda_r^k, \; \forall r \in R. \tag{4.53}$$

The aggregation of the λ^k-variables implies the *aggregated convexity constraint*

$$\sum_{p \in P} \lambda_p = \sum_{p \in P} \sum_{k \in K} \lambda_p^k = \sum_{k \in K} (\sum_{p \in P} \lambda_p^k) = \sum_{k \in K} (1) = |K|. \tag{4.54}$$

Adding this redundant constraint with dual variable $\pi_{\text{agg}} \in \mathbb{R}_+$ to the *IMP* (4.52) and substituting the aggregated λ_p, λ_r-variables, index k disappears from the objective function and the constraints in $\mathcal{A}$, but remains elsewhere:

$$z^\star_{IMP} = \min \sum_{p \in P} c_p \lambda_p + \sum_{r \in R} c_r \lambda_r \tag{4.55a}$$

$$\text{s.t.} \sum_{p \in P} \mathbf{a}_p \lambda_p + \sum_{r \in R} \mathbf{a}_r \lambda_r \geq \mathbf{b} \quad [\boldsymbol{\pi}_\mathbf{b}] \tag{4.55b}$$

$$\sum_{p \in P} \lambda_p = |K| \quad [\pi_{\text{agg}}] \tag{4.55c}$$

$$\lambda_p \geq 0, \quad \lambda_r \geq 0 \quad \forall p \in P, r \in R \tag{4.55d}$$

$$\lambda_p = \sum_{k\in K} \lambda_p^k, \quad \lambda_r = \sum_{k\in K} \lambda_r^k \qquad \forall p \in P, r \in R \tag{4.55e}$$

$$\sum_{p\in P} \lambda_p^k = 1 \qquad \forall k \in K \tag{4.55f}$$

$$\lambda_p^k \geq 0, \qquad \lambda_r^k \geq 0 \qquad \forall k \in K, p \in P, r \in R \tag{4.55g}$$

$$\sum_{p\in P} \mathbf{x}_p \lambda_p^k + \sum_{r\in R} \mathbf{x}_r \lambda_r^k = \mathbf{x}^k \in \mathbb{Z}_+^n \qquad \forall k \in K. \tag{4.55h}$$

Disaggregation

The next proposition shows that solving the linear relaxation obtained from imposing $\mathbf{x}^k \geq \mathbf{0}$ in (4.55h) rather than $\mathbf{x}^k \in \mathbb{Z}_+^n$ can be done on a simplified formulation.

Proposition 4.8. *The linear relaxation of the IMP* (4.55) *can be solved using the linear program defined by* (4.55a)–(4.55d) *only.*

Proof. Let the *MP* be defined accordingly as

$$z_{MP}^\star = \min \sum_{p\in P} c_p \lambda_p + \sum_{r\in R} c_r \lambda_r \tag{4.56a}$$

$$\text{s.t.} \sum_{p\in P} \mathbf{a}_p \lambda_p + \sum_{r\in R} \mathbf{a}_r \lambda_r \geq \mathbf{b} \qquad [\boldsymbol{\pi}_\mathbf{b}] \tag{4.56b}$$

$$\sum_{p\in P} \lambda_p = |K| \qquad [\pi_{\text{agg}}] \tag{4.56c}$$

$$\lambda_p \geq 0, \qquad \lambda_r \geq 0 \qquad \forall p \in P, r \in R. \tag{4.56d}$$

The constraints in (4.55e)–(4.55g) can be fulfilled, without modifying the objective value, by infinitely many disaggregated values for the λ_p^k, λ_r^k-variables. In particular,

$$\lambda_p^k = \tfrac{\lambda_p}{|K|}, \quad \lambda_r^k = \tfrac{\lambda_r}{|K|}, \qquad \forall k \in K, p \in P, r \in R. \tag{4.57}$$

Then, from $\sum_{p\in P} \mathbf{x}_p \lambda_p^k + \sum_{r\in R} \mathbf{x}_r \lambda_r^k = \mathbf{x}^k$, $\forall k \in K$, in (4.55h), we easily deduce that $\mathbf{x}^k \geq \mathbf{0}$, $\forall k \in K$. □

If the *MP* (4.56) is solved by column generation, the pricing problem that also mutes index k from the aggregated convexity constraint writes as

$$\bar{c}(\boldsymbol{\pi}_\mathbf{b}, \pi_{\text{agg}}) = -\pi_{\text{agg}} + \min_{\mathbf{x}\in\mathcal{D}} (\mathbf{c}^\intercal - \boldsymbol{\pi}_\mathbf{b}^\intercal \mathbf{A})\mathbf{x}. \tag{4.58}$$

With respect to the linear relaxation, this proof of existence for the disaggregated values means that our work is done. In reality, disaggregation already reminds us of the follow-up work: we have to reach integrality. It is therefore part of the post-processing on solutions in which we compute values for the x-variables and test whether integrality requirements hold. In this respect, the way we choose to disaggregate values has an obvious influence and it makes sense to consider λ^k-values that match our expectations regarding integrality. For instance, Proposition 4.9 states that there is no need to deal with $|K|$ fractions for the extreme rays.

Proposition 4.9. *In the disaggregation, we can consolidate the use of all extreme rays into a single block.*

Proof. Since $\lambda_r = \sum_{k\in K} \lambda_r^k$ and $\lambda_r^k \geq 0$, $\forall k \in K,\ r \in R$, a single index k can be selected, say

$$\lambda_r^k = \begin{cases} \lambda_r, & \text{for } k = 1 \\ 0, & \text{for } k \neq 1 \end{cases} \qquad \forall r \in R, \tag{4.59}$$

where we arbitrarily pick the first index. □

Discretization offers a tailored alternative disaggregation scheme for the extreme points as well. The formulation of the *IMP̈* with aggregated variables is

$$z^\star_{IM\ddot{P}} = \min \sum_{p\in\ddot{P}} c_p\lambda_p + \sum_{r\in\ddot{R}} c_r\lambda_r \tag{4.60a}$$

$$\text{s.t.} \sum_{p\in\ddot{P}} \mathbf{a}_p\lambda_p + \sum_{r\in\ddot{R}} \mathbf{a}_r\lambda_r \geq \mathbf{b} \tag{4.60b}$$

$$\sum_{p\in\ddot{P}} \lambda_p = |K| \tag{4.60c}$$

$$\lambda_p \in \mathbb{Z}_+, \qquad \lambda_r \in \mathbb{Z}_+ \qquad \forall p \in \ddot{P}, r \in \ddot{R} \tag{4.60d}$$

$$\lambda_p = \sum_{k\in K} \lambda_p^k, \qquad \lambda_r = \sum_{k\in K} \lambda_r^k \qquad \forall p \in \ddot{P}, r \in \ddot{R} \tag{4.60e}$$

$$\sum_{p\in\ddot{P}} \lambda_p^k = 1 \qquad \forall k \in K \tag{4.60f}$$

$$\lambda_p^k \in \{0,1\} \qquad \forall k \in K, p \in \ddot{P} \tag{4.60g}$$

$$\lambda_r^k \in \mathbb{Z}_+ \qquad \forall k \in K, r \in \ddot{R} \tag{4.60h}$$

$$\sum_{p\in\ddot{P}} \mathbf{x}_p\lambda_p^k + \sum_{r\in\ddot{R}} \mathbf{x}_r\lambda_r^k = \mathbf{x}^k \in \mathbb{Z}_+^n \qquad \forall k \in K. \tag{4.60i}$$

Proposition 4.10. *The integrality requirements on* $\mathbf{x}^k$ *and* λ_p^k, λ_r^k*-variables are redundant in the IMP̈* (4.60) *(if the integrality conditions on the variables of* $\mathcal{D} = \mathcal{D}^k$*,* $\forall k \in K$*, are the same as in* $\mathcal{A}$*), i.e.,*

$$z^\star_{IM\ddot{P}} = \min \sum_{p\in\ddot{P}} c_p\lambda_p + \sum_{r\in\ddot{R}} c_r\lambda_r \tag{4.61a}$$

$$\text{s.t.} \sum_{p\in\ddot{P}} \mathbf{a}_p\lambda_p + \sum_{r\in\ddot{R}} \mathbf{a}_r\lambda_r \geq \mathbf{b} \tag{4.61b}$$

$$\sum_{p\in\ddot{P}} \lambda_p = |K| \tag{4.61c}$$

$$\lambda_p \in \mathbb{Z}_+, \qquad \lambda_r \in \mathbb{Z}_+ \qquad \forall p \in \ddot{P}, r \in \ddot{R}. \tag{4.61d}$$

Proof. We prove that an optimal integer solution $\boldsymbol{\lambda}^{\star}_{IM\ddot{P}}$ to (4.61) implies integrality on $\mathbf{x}^k$ and λ^k_p, λ^k_r, $\forall k \in K$, $p \in \ddot{P}$, $r \in \ddot{R}$, i.e.,

$$\boldsymbol{\lambda}^{\star}_{IM\ddot{P}} \in \mathbb{Z}_+^{|\ddot{P}|+|\ddot{R}|} \Rightarrow \sum_{p\in\ddot{P}} \mathbf{x}_p\lambda^k_p + \sum_{r\in\ddot{R}} \mathbf{x}_r\lambda^k_r \in \mathbb{Z}^n_+, \; \forall k \in K.$$

By Proposition 4.3, for each $k \in K$, if λ^k_p, $p \in \ddot{P}$, and λ^k_r, $r \in \ddot{R}$, are integers, then $\sum_{p\in\ddot{P}} \mathbf{x}_p\lambda^k_p + \sum_{r\in\ddot{R}} \mathbf{x}_r\lambda^k_r$ is also integer. By Proposition 4.9 (which also holds in discretization), λ_r can be consolidated into a single block such that if $\lambda_r \in \mathbb{Z}_+$, then λ^k_r is integer for all $k \in K$.

With respect to λ_p, $\forall p \in \ddot{P}$, the system

$$\begin{aligned} \lambda_p = \sum_{k\in K} \lambda^k_p \quad & \forall p \in \ddot{P} \\ \sum_{p\in\ddot{P}} \lambda^k_p = 1 \quad & \forall k \in K \\ \lambda^k_p \geq 0 \quad & \forall k \in K, p \in \ddot{P}, \end{aligned} \tag{4.62}$$

is the set of constraints of a zero-cost balanced transportation problem with supply $\lambda_p \in \mathbb{R}_+$, $\forall p \in \ddot{P}$, and a unit demand $\forall k \in K$. The system can be solved by hand, greedily filling in the demands, one by one. If the λ_p-variables are already integer, we trivially obtain a binary solution for the λ^k_p-variables. □

Proposition 4.10 makes it clear that both aggregated *IMP* (4.55) and *IM$\ddot{P}$* (4.60) have the same linear relaxation. Furthermore, compared to the disaggregated solution (4.57), the above proof gives us an alternative disaggregation rule that matches the integrality requirements on λ-variables in discretization. These values are likely not all that useful for anticipating the integrality requirements on $\mathbf{x}^k$, $k \in K$, in convexification except if Proposition 4.4 regarding binary x-variables holds.

Note 4.11 (Lexicographic ordering of the extreme points.) Because the $\mathbf{x}_p$ can be generated in various sequences, and in particular since static disaggregation rules like (4.57) produce a highly symmetric solution, Vanderbeck (2011) suggests a dynamic numbering of the positive λ_p-variables in the solution to the final *RMP* based on the lexicographic order of the corresponding vectors. Exercise 4.9 asks to prove a mathematical expression for computing λ^k_p-values in the solution of the transportation problem (4.62). This alternative disaggregation rule to the simple average (4.57) can prove useful for branching in Chapter 7.

Some more observations

To complement this analysis on Identical subproblems, we recall in Table 4.2 some results on the reformulation (4.61) of the *ILP* (4.47) with identical blocks based on the discretization approach. Compared to Table 4.1, we still do not have λ_p^k- and λ_r^k-variables but we also omit the fourth entry on the polyhedral cone. The former omission comes from Proposition 4.10 whereas the latter comes from the forthcoming Proposition 4.11 which states that, in that case, we would in fact never write the compact formulation with a block-diagonal structure if it is well-thought. This is indeed what happens for the Single depot vehicle scheduling problem in (3.92) and the Network-based compact formulation for the *CSP* in (4.120).

Nature of $\mathcal{D}$	Index sets	λ_p	λ_r
$\mathcal{D} \subseteq \mathbb{Z}_+^n$	$p \in P \cup \ddot{I},\ r \in \ddot{R}$	$\in \mathbb{Z}_+$	$\in \mathbb{Z}_+$
$\mathcal{D}$ *bounded*	$p \in P \cup \ddot{I}$	$\in \mathbb{Z}_+$	
$\mathcal{D} \subseteq \{0,1\}^n$	$p \in P$	$\in \mathbb{Z}_+$	

Table 4.2: Various results for a Dantzig-Wolfe reformulation (4.61) of the *ILP* (4.47) with *identical* subproblems based on the discretization of $\mathcal{D}$ (4.49).

Proposition 4.11. *Given identical subproblems, if* conv$(\mathcal{D})$ *derived from* (4.49) *is a polyhedral cone, then the compact formulation can be written without index k.*

Proof. The following proof in discretization is valid in convexification, replacing as needed $\ddot{R}_\mathbf{0}$ by R, and $\lambda_r \in \mathbb{Z}_+$ by $\lambda_r \geq 0$.

Because conv$(\mathcal{D})$ is a polyhedral cone and we assume non-negative variables, we have that $\mathcal{D} = \{\mathbf{x} \in \mathbb{Z}_+^n \mid \mathbf{D}\mathbf{x} \geq \mathbf{0}\}$ which gives us $\mathcal{X} = \left\{\begin{bmatrix}\mathbf{0}\\1\end{bmatrix}\right\} \cup \left\{\begin{bmatrix}\mathbf{x}_r\\0\end{bmatrix}\right\}_{r\in\ddot{R}_\mathbf{0}}$. In that case, the $IM\ddot{P}$ (4.60) becomes

$$
\begin{aligned}
z^\star_{IM\ddot{P}} = \min\ & 0\lambda_\mathbf{0} + \sum_{r\in\ddot{R}_\mathbf{0}} c_r\lambda_r && && (4.63a)\\
\text{s.t.}\ & \mathbf{0}\lambda_\mathbf{0} + \sum_{r\in\ddot{R}_\mathbf{0}} \mathbf{a}_r\lambda_r \geq \mathbf{b} && && (4.63b)\\
& \lambda_\mathbf{0} = |K| && && (4.63c)\\
& && \lambda_r \in \mathbb{Z}_+ & \forall r \in \ddot{R}_\mathbf{0} && (4.63d)\\
& \lambda_\mathbf{0} = \sum_{k\in K} \lambda_\mathbf{0}^k, && \lambda_r = \sum_{k\in K} \lambda_r^k & \forall r \in \ddot{R}_\mathbf{0} && (4.63e)\\
& \lambda_\mathbf{0}^k = 1, && \lambda_r^k \in \mathbb{Z}_+ & \forall k \in K, r \in \ddot{R}_\mathbf{0} && (4.63f)\\
& \mathbf{0}\lambda_\mathbf{0}^k + \sum_{r\in\ddot{R}_\mathbf{0}} \mathbf{x}_r\lambda_r^k = \mathbf{x}^k \in \mathbb{Z}_+^n && & \forall k \in K. && (4.63g)
\end{aligned}
$$

In the above, all convexity constraints are useless ($\lambda_\mathbf{0}^k = 1, \forall k \in K$) as well as their redundant sum ($\lambda_\mathbf{0} = \sum_{k\in K} \lambda_\mathbf{0}^k$ and $\lambda_\mathbf{0} = |K|$) and are thus removed:

$$
\begin{aligned}
z^{\star}_{IM\ddot{P}} = \min \quad & \sum_{r\in\ddot{R}_0} c_r\lambda_r && && (4.64a)\\
\text{s.t.} \quad & \sum_{r\in\ddot{R}_0} \mathbf{a}_r\lambda_r \geq \mathbf{b} && && (4.64b)\\
& \lambda_r \in \mathbb{Z}_+ && \forall r\in\ddot{R}_\mathbf{0} && (4.64c)\\
& \lambda_r = \sum_{k\in K}\lambda_r^k && \forall r\in\ddot{R}_\mathbf{0} && (4.64d)\\
& \lambda_r^k \in \mathbb{Z}_+ && \forall k\in K, r\in\ddot{R}_\mathbf{0} && (4.64e)\\
& \sum_{r\in\ddot{R}_0}\mathbf{x}_r\lambda_r^k = \mathbf{x}^k \in \mathbb{Z}^n_+ && \forall k\in K. && (4.64f)
\end{aligned}
$$

Moreover, given a λ_r-solution, we have by Proposition 4.9 that $\forall r \in \ddot{R}_\mathbf{0}$, $\lambda_r^k = \lambda_r$ for $k = 1$ and $\lambda_r^k = 0$ for $k \neq 1$. The formulation of the *IMP̈* thus simplifies to

$$
\begin{aligned}
z^{\star}_{IM\ddot{P}} = \min \quad & \sum_{r\in\ddot{R}_0} c_r\lambda_r && && (4.65a)\\
\text{s.t.} \quad & \sum_{r\in\ddot{R}_0} \mathbf{a}_r\lambda_r \geq \mathbf{b} && && (4.65b)\\
& \lambda_r \in \mathbb{Z}_+ && \forall r\in\ddot{R}_\mathbf{0} && (4.65c)\\
& \sum_{r\in\ddot{R}_0}\mathbf{x}_r\lambda_r = \mathbf{x}^1 \in \mathbb{Z}^n_+. && && (4.65d)
\end{aligned}
$$

Because the index of $\mathbf{x}^1$ is the only remaining block reference, we can mute it entirely such that the reformulation does not need index k. This is indeed the same as (4.24)_right coming by construction from the compact formulation

$$z^{\star}_{ILP} = \min\ \mathbf{c}^{\intercal}\mathbf{x} \quad \text{s.t. } \mathbf{Ax} \geq \mathbf{b},\ \mathbf{Dx} \geq \mathbf{0},\ \mathbf{x} \in \mathbb{Z}^n_+, \tag{4.66}$$

on which a Dantzig-Wolfe reformulation is performed using $\mathcal{A} = \{\mathbf{x} \in \mathbb{Z}^n_+ \mid \mathbf{Ax} \geq \mathbf{b}\}$ and $\mathcal{D} = \{\mathbf{x} \in \mathbb{Z}^n_+ \mid \mathbf{Dx} \geq \mathbf{0}\}$. □

Lower and upper bounds

With respect to lower and upper bounds on $z^{\star}_{MP}$, our recognizable expressions resurface with a particular attention to optimal objective values of the compact formulation, or lack thereof.

Proposition 4.12. *Given optimal dual values* $(\boldsymbol{\pi}_\mathbf{b}, [\pi_0^k]_{k\in K})$ *with objective value* z_{RMP} *and minimum reduced costs* $\bar{c}^k(\boldsymbol{\pi}_\mathbf{b}, \pi_0^k)$, $\forall k \in K$, *then the optimum* $z^{\star}_{MP}$ *is bounded from below and above as*

$$z_{RMP} + \sum_{k\in K} \bar{c}^k(\boldsymbol{\pi}_\mathbf{b}, \pi_0^k) \leq z^{\star}_{MP} \leq z_{RMP}. \tag{4.67}$$

The proof of Proposition 3.2 continues to hold in the context of integer linear programming. Also still valid are the comments regarding relaxed pricing solutions, i.e, $\underline{c}^k(\boldsymbol{\pi}_\mathbf{b}, \pi_0^k) \leq \bar{c}^k(\boldsymbol{\pi}_\mathbf{b}, \pi_0^k)$, as well as arbitrary dual values $\boldsymbol{\pi}_\mathbf{b} \geq \mathbf{0}$ and $\pi_0^k \in \mathbb{R}$, $\forall k \in K$, i.e.,

$$\boldsymbol{\pi}_\mathbf{b}^\intercal \mathbf{b} + \sum_{k \in K} \pi_0^k + \sum_{k \in K} \bar{c}^k(\boldsymbol{\pi}_\mathbf{b}, \pi_0^k) \leq z_{MP}^\star, \qquad \forall \boldsymbol{\pi}_\mathbf{b} \geq \mathbf{0}, \pi_0^k \in \mathbb{R}, k \in K. \tag{4.68}$$

Observe however that $z_{LP}^\star$ and $z_{ILP}^\star$ are respectively incomparable to the lower and upper bounds given by (4.67) despite Proposition 4.1 providing $z_{LP}^\star \leq z_{MP}^\star \leq z_{ILP}^\star$. In fact, an *upper bound on the linear relaxation's optimal objective value* z_{RMP} is incomparable to any *integer upper bound*, let alone the integer optimum. While there is little reason to care for $z_{LP}^\star$, the optimal objective value $z_{ILP}^\star$ is of primary interest. We bridge the missing information in Chapter 7 when we embed the $z_{MP}^\star$ in a branch-and-bound algorithm.

Finally, we conclude this section with a generalization of Proposition 4.7 for the *ILP* with a block-diagonal structure.

Proposition 4.13. *If the formulation of each* ISP^k (4.45), $k \in K$, *possesses the integrality property, then the linear relaxation of the IMP is no better than that of the ILP, i.e.,*

$$z_{LP}^\star = z_{MP}^\star \leq z_{ILP}^\star. \tag{4.69}$$

Proof. By Proposition 1.9, the formulation of the ISP^k possesses the integrality property if $\operatorname{conv}(\mathcal{D}^k) = \{\mathbf{x}^k \in \mathbb{R}_+^{n^k} \mid \mathbf{D}^k\mathbf{x}^k \geq \mathbf{d}^k\}$, where $\mathcal{D}^k = \{\mathbf{x}^k \in \mathbb{Z}_+^{n^k} \mid \mathbf{D}^k\mathbf{x}^k \geq \mathbf{d}^k\}$. The feasible regions of the *MP* and *LP* are therefore the same, i.e.,

$$\left\{ \{\mathbf{x}^k \in \mathbb{R}^{n^k}\}_{k \in K} \;\middle|\; \sum_{k \in K} \mathbf{A}^k\mathbf{x}^k \geq \mathbf{b} \right\} \cap \left\{ \prod_{k \in K} \operatorname{conv}(\mathcal{D}^k) \right\}$$
$$= \left\{ \{\mathbf{x}^k \in \mathbb{R}^{n^k}\}_{k \in K} \;\middle|\; \sum_{k \in K} \mathbf{A}^k\mathbf{x}^k \geq \mathbf{b}, \mathbf{D}^k\mathbf{x}^k \geq \mathbf{d}^k, \forall k \in K \right\},$$

such that their optimal objective values are equal. □

4.5 Good to Know

In this section, we first examine the impact of non-linear functions in the compact formulation and show that the discretization approach can manage that for bounded domain in pricing problems. We next propose a way to write a compact formulation so as to allow unused blocks in (4.40). We also examine a block-diagonal structure with additional variables that are shared by all the blocks and derive an appropriate reformulation.

Non-linear encoding functions

Let us quote what we have said in Note 2.10 regarding our encoding functions:

> The cost $c_{\mathbf{x}}$ and column-coefficients $\mathbf{a}_{\mathbf{x}}$ are naturally defined as functions of $\mathbf{x} \in \mathcal{X}$, that is, $c_{\mathbf{x}} = c(\mathbf{x})$ and $a_{i\mathbf{x}} = a_i(\mathbf{x})$, $\forall i \in \{1,\ldots,m\}$. The general framework of the Dantzig-Wolfe decomposition for (integer) linear programming (see Chapters 3 and 4) requires these to preserve *vector addition* and *scalar multiplication*. In many real-life applications, the cost $c(\mathbf{x})$ is indeed non-linear and therefore does not fulfill this requirement. We can think of very complicated cost functions, e.g., for the airline crew schedules, which are negotiated by lawyers, not OR practitioners. Non-linear functions also appear for the computation of some column-coefficients when Chvátal-Gomory cuts are added to restrict an integer master problem formulation as these make use of *ceiling* or *floor* functions (see Chapter 7). We describe in Section 4.5 some conditions for which a reformulation using non-linear encoding functions holds. This may increase the difficulty in solving such non-linear pricing problems but the encoding always ends up in the *RMP* with scalars for the cost $c_{\mathbf{x}}$ and components of $\mathbf{a}_{\mathbf{x}}$.

The encoding functions preserve *vector addition* and *scalar multiplication* if we have $\forall \mathbf{x}_1, \mathbf{x}_2 \in \mathcal{X}$ and $\alpha_1, \alpha_2 \in \mathbb{R}$:

$$\begin{aligned} c(\alpha_1\mathbf{x}_1 + \alpha_2\mathbf{x}_2) &= c(\alpha_1\mathbf{x}_1) + c(\alpha_2\mathbf{x}_2) \\ a_i(\alpha_1\mathbf{x}_1 + \alpha_2\mathbf{x}_2) &= a_i(\alpha_1\mathbf{x}_1) + a_i(\alpha_2\mathbf{x}_2) \qquad \forall i \in \{1,\ldots,m\}. \end{aligned} \tag{4.70}$$

This is generally required because a Dantzig-Wolfe reformulation is rooted by the Minkowski-Weyl or Hilbert-Giles-Pulleyblank theorems to express $\mathcal{D}$ or its sibling $\operatorname{conv}(\mathcal{D})$ in a different but equivalent way: any solution $\mathbf{x} \in \mathcal{D}$ can be expressed as a combination of vectors scaled by λ-variables, and vice versa. Obviously, the linear functions $c(\mathbf{x}) = \mathbf{c}^\mathsf{T}\mathbf{x}$ and $\mathbf{a}(\mathbf{x}) = \mathbf{A}\mathbf{x}$ satisfy this property.

Without loss of generality, assume non-linearity is present in only $c(\mathbf{x})$ or $\mathbf{a}(\mathbf{x})$ and that we can solve the *ISP* to optimality regardless. In either case, we consider a combination of points from the *ISP* as obtained in the *RMP* and conclude that decisions we can make from a solution λ_{RMP} (optimal or otherwise) in the branch-and-bound algorithm are compromised. If $c(\mathbf{x})$ is non-linear, the cost evaluation of such a combination is likely erroneous. If $\mathbf{a}(\mathbf{x})$ is non-linear, the evaluation of the left-hand side contribution of such a combination is likely erroneous and maybe even infeasible for the constraints in $\mathcal{A}$.

One exception to this general rule occurs if $\operatorname{conv}(\mathcal{D})$ (4.2b) is a polytope, which means that $\mathcal{D}$ is a finite set of integer points since all variables must be integer. In this case, the discretization approach yields an integer linear reformulation in which we pick one and only one λ_p-variable, $p \in \ddot{P}$. A careful examination of the proof of Theorem 4.2 tells us that we do not actually need convexification then because $R = \ddot{R} = \emptyset$ such that $\mathcal{T}$ directly interprets as the entire set of integer solutions, i.e., $\mathcal{T} = \mathcal{D}$. Interestingly, this means that non-linear coefficients $c(\mathbf{x})$ or $\mathbf{a}(\mathbf{x})$ we compute for the objective function in the *ISP* are also evaluated correctly in any binary solution $\boldsymbol{\lambda}_{IM\ddot{P}}$. *Under these conditions, our subcontracted chemist can identify the molecules and even still price them properly for the master problem.* Let us state this formally in Proposition 4.14.

Proposition 4.14. *Let $c(\mathbf{x})$ and $\mathbf{a}(\mathbf{x})$, $\mathbf{x} \in \mathcal{D}$, be non-linear functions. If $\mathcal{D}$ is a finite set of integer $\mathbf{x}$-points, then the IM$\ddot{P}$ (4.24)_left is an integer linear program that correctly substitutes the compact formulation.*

Proof. We first prove that a binary solution of the extended formulation is also valid for the compact formulation and then that branch-and-bound terminates exactly. Finally that it also holds in column generation.

1. For an optimal solution $\mathbf{x}^\star_{IM\ddot{P}} = \sum_{p \in \ddot{P}} \mathbf{x}_p \lambda^\star_p$, where $\sum_{p \in \ddot{P}} \lambda^\star_p = 1$, $\lambda^\star_p \in \{0,1\}$, $\forall p \in \ddot{P}$, the following holds for precisely one variable, say $\lambda^\star_q = 1$:

$$c(\mathbf{x}^\star_{IM\ddot{P}}) = c(\sum_{p \in \ddot{P}} \mathbf{x}_p \lambda^\star_p) = c(\mathbf{x}_q \lambda^\star_q) = c(\mathbf{x}_q) = c_q = c_q \lambda^\star_q = \sum_{p \in \ddot{P}} c_p \lambda^\star_p, \qquad (4.71)$$

that is to say that the non-linear cost $c(\sum_{p \in \ddot{P}} \mathbf{x}_p \lambda^\star_p)$ is indeed correctly evaluated at a single integer point indexed in $\ddot{P}$, the same being obviously true for

$$\mathbf{a}(\mathbf{x}^\star_{IM\ddot{P}}) = \mathbf{a}(\sum_{p \in \ddot{P}} \mathbf{x}_p \lambda^\star_p) = \sum_{p \in \ddot{P}} \mathbf{a}_p \lambda^\star_p. \qquad (4.72)$$

2. Assuming that the *IM$\ddot{P}$* is solved by branch-and-bound, there only remains to recognize that the objective values of the linear relaxations we obtain progress monotonously throughout child nodes of the tree, i.e., $z^\star_{MP} \le z^\star_{IM\ddot{P}}$. We have this by convexity of the linear objective function $\sum_{p \in \ddot{P}} c_p \lambda_p$ irrespective of the fact that the coefficients c_p may come from a non-linear expression.
3. Solving the *IM$\ddot{P}$* by column generation implies that we must be able to solve *exactly* the *ISP* over $\mathcal{D}$ and a non-linear objective function. □

What does not transpire in this proof is that fractional solutions of the extended formulation can lose all meaning when projected back to the compact formulation. That is, if we project a fractional $\boldsymbol{\lambda}_{M\ddot{P}}$-solution into an $\mathbf{x}$-solution, it suffices to take another look at (4.70) to realize that what we have found cannot be interpreted correctly. Remarkably, this misinterpretation likely remains even if a non-binary combination of λ-variables yields $\mathbf{x}$ integer. This means that we have to derive branching and cutting decisions until we reach a binary $\boldsymbol{\lambda}_{M\ddot{P}}$-solution for which we know the projection back onto the compact formulation is correct. Furthermore, some notions we have gotten accustomed to may lose their footing. For instance, we cannot take for granted that the adjusted cost $\tilde{c}_j$, $j \in \{1,\ldots,n\}$, in the pricing problem makes sense, see Note 3.6. That is, the non-linearity in encoding functions $c(\mathbf{x})$ or $\mathbf{a}(\mathbf{x})$ may make it impossible to separate the objective coefficients per variable x_j, $j \in \{1,\ldots,n\}$, in the reduced cost expression of the pricing problem, i.e.,

$$\bar{c}(\boldsymbol{\pi}_\mathbf{b}, \pi_0) = -\pi_0 + \min_{\mathbf{x} \in \mathcal{D}} c(\mathbf{x}) - \boldsymbol{\pi}^\intercal_\mathbf{b} \mathbf{a}(\mathbf{x}). \qquad (4.73)$$

Proposition 4.14 cannot be generalized to a polyhedron (unbounded domain) conv($\mathcal{D}$). Even in this case, the monotonous under-approximation of objective values in the *IM$\ddot{P}$* is still guaranteed. The issue is rather that when we reach an integer

λ-solution, its projection back onto the compact formulation can still yield an unpredictable objective value. Indeed, for the *ILP* with a *linear* cost function, we have

$$\mathbf{c}^\mathsf{T}\mathbf{x}^\star_{IM\ddot{P}} = \mathbf{c}^\mathsf{T}\Big(\sum_{p\in\ddot{P}} \mathbf{x}_p\lambda^\star_p + \sum_{r\in\ddot{R}} \mathbf{x}_r\lambda^\star_r\Big) = \sum_{p\in\ddot{P}} (\mathbf{c}^\mathsf{T}\mathbf{x}_p)\lambda^\star_p + \sum_{r\in\ddot{R}} (\mathbf{c}^\mathsf{T}\mathbf{x}_r)\lambda^\star_r, \tag{4.74}$$

whereas the cost distribution does not in general hold for a *non-linear* one $c(\cdot)$ because

$$c(\mathbf{x}^\star_{IM\ddot{P}}) \neq \sum_{p\in\ddot{P}} c(\mathbf{x}_p)\lambda^\star_p + \sum_{r\in\ddot{R}} c(\mathbf{x}_r)\lambda^\star_r. \tag{4.75}$$

For example, consider Figure 4.5b: as we write $(9,4)$ as the sum of the point $(1,2)$ indexed in $\ddot{P}$ plus the non-negative integer combination $0(2,0)+1(8,2)$ of the two integer rays indexed in $\ddot{R}$, the two-dimensional quadratic cost function $c(x,y) = x^2+y^2$ results in

$$(81+16) = 97 \neq (1+4)+(64+4) = 73.$$

Moreover, there might not be a unique representation of $\mathbf{x}^\star_{IM\ddot{P}}$ as a combination of the λ-variables because this combination depends on the scaling adopted for the integer rays.

An unpractical possibility to solve such a model exactly would be to reach either infeasibility or integrality at every leaf without ever being able to prune based on linear relaxations. The idea may become relevant if we can bound the error on the real objective value, e.g., through Taylor series or similar tools, which would give us back pruning capabilities. Specifically, can we find ε such that the difference between the original objective value and the one computed in the $M\ddot{P}$ is bounded from above in

$$c\Big(\sum_{p\in P'} \mathbf{x}_p\lambda_p\Big) - \sum_{p\in P'} c_p\lambda_p \overset{?}{\leq} \varepsilon. \tag{4.76}$$

Note 4.12 (A perspective with non-linear constraints.) One could of course also see the encoding functions $c(\mathbf{x})$ and $\mathbf{a}(\mathbf{x})$ as constraints of the pricing problem in which case it is the domain that is non-linear. As long as we can solve this non-linearity exactly, the conclusion obviously remains the same regarding discretization. This implies that we may be able to handle non-linear constraints of the compact formulation by confining them to the pricing problem. Figure 4.12 sketches an arbitrary non-linear domain for the pricing problem, say $\mathbf{x} \in \mathcal{X}$, and the consequence this has on the extended formulation. In that case, we do not respect the Minkowski-Weyl conditions required to apply Theorems 3.1, 4.1, and 4.2, so we do not have an equivalent description of the substituted set, i.e., $\mathcal{A} \cap \text{conv}(\mathcal{X}) \neq \mathcal{A} \cap \mathcal{X}$, where $\mathcal{A} = \{\mathbf{x} \in \mathbb{Z}^n \mid \mathbf{A}\mathbf{x} \geq \mathbf{b}\}$. Convexification introduces "feasible" solutions (integer or fractional) in the master problem that are actually infeasible in $\mathcal{X}$ even if they are integer. A Dantzig-Wolfe reformulation by convexification therefore cannot treat these problematic solutions whereas one by discretization rejects them correctly upon branching as we are only allowed to pick one point in $\mathcal{X}$.

(a) Finite set of integer solutions $\mathcal{X}$ (b) Minkowski-Weyl conditions are not met

Fig. 4.12: A set $\mathcal{X} \subset \mathbb{Z}^2_+$ whose convexification creates undesirable points in the $M\ddot{P}$.

Note 4.13 (More non-linear cases.) The proof also holds for a maximization in which case monotonous progress is guaranteed by *concavity* of the linear objective function. It immediately generalizes to a block-diagonal structure where we pick one and only one λ^k_p-variable per block $k \in K$, $p \in \ddot{P}^k$. The same is true for identical subproblems where $\lambda_p = \sum_{k \in K} \lambda^k_p$ in the reformulation (4.60) takes integer values.

Note 4.14 (Non-linear mixed-integer conjecture.) The extent to which non-linear encoding functions hold in a Dantzig-Wolfe reformulation by discretization on a mixed-integer compact formulation is an open question. In other words, a polytope $\text{conv}(\mathcal{D})$ defined from a mixed-integer compact formulation is not necessarily synonymous with a finite set of integer points. A conjecture is that all variables influenced by non-linearity must take finite values. We point back to Example 2.6 in which we allow convex combinations of λ_p-variables under non-linear timing restrictions. Formally, only combinations with the same routing but a different schedule are considered integer feasible, that is, the t-variables are continuous but have no impact on routing cost or feasibility, e.g., 0.25(9h) + 0.75(14h) = 12h45. If the time variables had an influence on cost or traveling patterns, the story would be much different.

An integer program with a quadratic objective function is given in Example 4.8. It has a block-diagonal structure with identical data in some blocks such that we use aggregation to get two integer pricing problems. In the mean time, let us prepare the ground with Illustration 4.6.

Illustration 4.6 TCSPP: **non-linear costs**

To spice things up, let us pick up on Illustration 4.5 and introduce non-linear elements, in the objective function of the compact formulation, in two different ways:

(a) The cost on each arc $(i, j) \in A$ is given by $\cos(c^2_{ij})$ instead;
(b) The objective function is given by $\sin(\sum_{(i,j) \in A} c_{ij} x_{ij})$ instead.

The specific cosine and sine functions we use are not important as one would be hard-pressed to find any real-world meaning for them in a routing context. Let us rather concentrate on whether or not we can handle this non-linearity. Should the reader attempt to work out the solutions before seeing the answers, the idea is to write the compact formulation, confirm that we can perform a Dantzig-Wolfe reformulation using $\mathcal{D}$, and if so solve the extended formulation to see how the linear relaxation behaves.

(a) For each arc $(i,j) \in A$, we simply apply the non-linear function $\cos(c_{ij}^2)$ to the listed arc costs c_{ij} which gives $\cos(10^2) \approx 0.8623$ for arc $(1,3)$.

$(i,j) \in A$	(1,2)	(1,3)	(2,4)	(3,2)	(2,5)	(3,4)	(3,5)	(5,6)	(4,5)	(4,6)
c_{ij}	1	10	1	1	2	5	12	2	10	1
$\cos(c_{ij}^2)$	0.5403	0.8623	0.5403	0.5403	-0.6536	0.9912	0.8711	-0.6536	0.8623	0.5403

Each *coefficient* comes from a non-linear expression but the compact formulation remains entirely linear, i.e.,

$$z_{ILP}^\star = \min \sum_{(i,j)\in A} \cos(c_{ij}^2) x_{ij} \qquad \text{s.t. } \mathbf{x} \in \mathcal{A} \cap \mathcal{D}. \tag{4.77}$$

Computing the cost of a path is no different than we are use to, e.g., the cost of path 1246 is $c_{1246} = 0.5403 + 0.5403 + 0.5403 = 1.6209$ as listed in Table 4.3.

$p \in P$	1246	1256	12456	13246	**13256**	132456	1346	13456	1356	
c_p	3	5	14	13	**15**	24	16	27	24	original
c_p	1.6209	-0.7670	1.2893	2.4832	**0.0953**	2.1516	2.3938	2.0622	1.0798	non-linear
t_p	18	15	14	13	**10**	9	17	13	8	

Table 4.3: Path parameters c_p and t_p using a non-linear function on *arc* costs.

It is thus immediate that we can use the extended formulation we already know. The optimal integer solution is $\lambda_{13256}^\star = 1$ at cost $z_{IM\ddot{P}}^\star = 0.0953$. The linear relaxation is $\lambda_{1256} = 0.8$ and $\lambda_{13256} = 0.2$ at cost $z_{MP}^\star = -0.5945$. Furthermore, this value means the same in the compact formulation, i.e., $x_{12} = 0.8$, $x_{25} = x_{56} = 1$, $x_{13} = 0.2$, and $x_{32} = 0.2$ which can be evaluated indifferently as

$$\begin{aligned} &.8(.5403 - .6536 - .6536) + .2(.8623 + .5403 - .6536 - .6536) \\ = &.8(.5403) - .6536 - .6536 + .2(.8623) + .2(.5403) = -.5945. \end{aligned}$$

(b) The compact formulation now contains a non-linear objective function but remains subject to the usual constraints, i.e.,

$$z_{NLP}^\star = \min \sin\Big(\sum_{(i,j)\in A} c_{ij} x_{ij} \Big) \qquad \text{s.t. } \mathbf{x} \in \mathcal{A} \cap \mathcal{D}. \tag{4.78}$$

The domain conv($\mathcal{D}$) is a polytope that contains only binary solutions so we can perform a Dantzig-Wolfe reformulation by Proposition 4.14. This also results in the extended formulation we already know but it is worth to write out the pricing problem that takes over the non-linear objective function, i.e.,

$$\begin{aligned}\bar{c}(\pi_7, \pi_0) &= -\pi_0 + \min_{\mathbf{x}\in\mathcal{D}} \ \sin\Big(\sum_{(i,j)\in A} c_{ij}x_{ij}\Big) - \pi_7 \sum_{(i,j)\in A} t_{ij}x_{ij} \\ &= -\pi_0 + \min_{p\in P} \ c_p - \pi_7 t_p.\end{aligned}$$

We can observe that it no longer makes sense to speak of the adjusted cost $\tilde{c}_{ij}$ of an arc $(i, j) \in A$. The cost of each path is computed in Table 4.4 with the cost of path 1246 now calculated as $\sin(1+1+1) = 0.1411$.

$p \in P$	1246	1256	12456	13246	13256	**132456**	1346	13456	**1356**	
c_p	3	5	14	13	15	**24**	16	27	**24**	original
c_p	0.1411	-0.9589	0.9906	0.4202	0.6503	**-0.9056**	-0.2879	0.9564	**-0.9056**	non-linear
t_p	18	15	14	13	10	**9**	17	13	**8**	

Table 4.4: Path parameters c_p and t_p using a non-linear function on *path* costs.

An optimal integer solution is $\lambda^\star_{1356} = 1$ at cost $z^\star_{IM\ddot{P}} = -0.9056$. The linear relaxation is $\lambda_{1256} = 0.8571$ and $\lambda_{1356} = 0.1429$ at cost $z^\star_{MP} = -0.9513$. Converting this solution to the compact formulation yields $x_{12} = x_{25} = 0.8571$, $x_{56} = 1$, and $x_{13} = x_{35} = 0.1429$. We observe that the real objective value is not equal to what we establish with the extended formulation, i.e.,

$$\begin{aligned}&.8571(-.9589) + .1429(-.9056) = -.9513 \\ &\neq \sin[1(.8571) + 2(.8571) + 2(1) + 10(.1429) + 12(.1429)] = .9903.\end{aligned}$$

In this case, it is even largely greater than the integer optimum. What is important to remember is that, since we are minimizing, linear relaxations of the extended formulation always provide a lower bound on integer optimality, e.g., $-0.9513 \leq -0.9056$.

Not all blocks are used

We have seen in Chapter 2 that we can do column generation without any reference to a Dantzig-Wolfe reformulation. All we need is a master problem and a *suitable* pricing problem that is able to generate master variables of negative reduced cost, if there are any. With the material in the present chapter, we understand that a Dantzig-Wolfe reformulation is the theoretical link between the *IMP* and *ISP*, the link that explains what *suitable* means. When we discuss branching decisions and cutting planes in Chapter 7, this link proves to be useful again, and we may need a compact formulation from which we derive, by an appropriate reformulation, the master and pricing problems. Since we know (almost) all the constraints, *reversing* a Dantzig-Wolfe reformulation is not so difficult. Think about how you would do it! We return to this in Section 4.6 More to Know.

There is a small twist, however, in the very common context of block-diagonal models, i.e., the presence of several *ISP*s. In many applications, not all pricing problems necessarily contribute to the final master problem solution. When we formulate the *ISP*s, and the *ILP* as well, we must therefore actively construct this possibility. The intuition is right: we must introduce a binary variable that *enables/disables* a pricing problem. This relates to the zero-object we have briefly discussed in Note 2.12. Indeed, we have already done this by incorporating to the *ISP*

- x_0 in (2.32) for the One-dimensional cutting stock problem;
- x_0^k in (2.34) for the Cutting stock problem with rolls of different widths;
- x_0 in (2.37) for the Edge coloring problem.

Bounded domains

Assume that we need to write a compact formulation *ILP* given the following information:

- $|K|$ blocks with bounded domains

$$\mathcal{D}^k = \left\{ \mathbf{x}^k \in \mathbb{Z}_+^{n^k} \;\middle|\; \mathbf{D}^k \mathbf{x}^k \geq \mathbf{d}^k \right\} \qquad \forall k \in K \tag{4.79}$$

- a set of linking constraints

$$\mathcal{A} = \left\{ \left\{ \mathbf{x}^k \in \mathbb{Z}_+^{n^k} \right\}_{k \in K} \;\middle|\; \sum_{k \in K} \mathbf{A}^k \mathbf{x}^k \geq \mathbf{b} \right\} \tag{4.80}$$

- two cost components per block k: the linear portion $\mathbf{c}^{k\intercal} \mathbf{x}^k$ and a fixed cost $c_0^k \geq 0$ for using the block ($\mathbf{x}^k \neq \mathbf{0}$);

As expected, we use a binary variable in each block, say x_0^k taking value 1 if block k is used, otherwise 0. We do however need one more trick: since every block k has a bounded domain $\mathcal{D}^k$, we can trivially introduce adequate upper bounds $\mathbf{x}^k \leq \mathbf{u}^k$ that do not modify the feasible region. An appropriate compact formulation is then given by

$$z_{ILP}^{\star} = \min \quad \sum_{k \in K} c_0^k x_0^k + \mathbf{c}^{k\intercal} \mathbf{x}^k \tag{4.81a}$$

$$\text{s.t.} \quad \sum_{k \in K} \mathbf{A}^k \mathbf{x}^k \geq \mathbf{b} \tag{4.81b}$$

$$\mathbf{D}^k\mathbf{x}^k \geq \mathbf{d}^k x_0^k \qquad \forall k \in K \tag{4.81c}$$

$$\mathbf{x}^k \leq \mathbf{u}^k x_0^k \qquad \forall k \in K \tag{4.81d}$$

$$x_0^k \in \{0,1\} \qquad \forall k \in K \tag{4.81e}$$

$$\mathbf{x}^k \in \mathbb{Z}_+^{n^k} \qquad \forall k \in K, \tag{4.81f}$$

where $x_0^k \in \{0,1\}$ implies that (4.81c)–(4.81f) either results in $\mathbf{x}^k = \mathbf{0}$ or corresponds to $\mathcal{D}^k$ (4.79).

*Note 4.15 (0*$0\mathbf{0}$*, and nothing to do with James Bond.)* We note the following small technicality although it is not restrictive by any means. If $\mathbf{0} \notin \mathcal{D}^k$, any integer solution $\mathbf{x}^k \neq \mathbf{0}$ implies $x_0^k = 1$ and vice versa. If $\mathbf{0} \in \mathcal{D}^k$, the solution $x_0^k = 1$, $\mathbf{x}^k = \mathbf{0}$ co-exists with the solution $x_0^k = 0$, $\mathbf{x}^k = \mathbf{0}$ but the latter is always preferred since it yields a cost $0 \leq c_0^k$. We can therefore see that if $c_0^k = 0$ and $\mathbf{0} \in \mathcal{D}^k$, there is no need to construct an artificial zero-object although we can do so anyway.

Let us now take a look at the two Dantzig-Wolfe reformulations *IMP* and *IMP̈* using the grouping of constraints

$$\mathcal{A}_0 = \left\{ \left\{ \begin{bmatrix} x_0^k \\ \mathbf{x}^k \end{bmatrix} \in \{0,1\} \times \mathbb{Z}_+^{n^k} \right\}_{k \in K} \middle| \sum_{k \in K} \mathbf{A}^k\mathbf{x}^k \geq \mathbf{b} \right\} \tag{4.82a}$$

$$\mathcal{D}_0^k = \left\{ \begin{bmatrix} x_0^k \\ \mathbf{x}^k \end{bmatrix} \in \{0,1\} \times \mathbb{Z}_+^{n^k} \middle| \mathbf{D}^k\mathbf{x}^k \geq \mathbf{d}^k x_0^k,\ \mathbf{x}^k \leq \mathbf{u}^k x_0^k \right\}, \qquad \forall k \in K. \tag{4.82b}$$

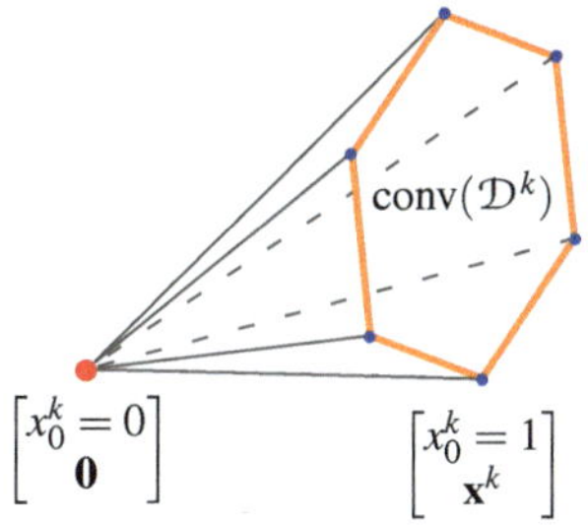

Fig. 4.13: Illustration of conv($\mathcal{D}_0^k$).

Block k is illustrated in Figure 4.13: the polytope conv($\mathcal{D}^k$) derived from (4.79) is replaced by conv($\mathcal{D}_0^k$) that includes one additional extreme point, $\begin{bmatrix} x_0^k \\ \mathbf{x}^k \end{bmatrix} = \begin{bmatrix} 0 \\ \mathbf{0} \end{bmatrix}$. Let us first examine the reformulation based on convexification. The set of extreme points of conv($\mathcal{D}_0^k$), here denoted $\mathcal{X}_0^k$, is

$$\mathcal{X}_0^k = \left\{ \begin{bmatrix} 0 \\ \mathbf{0} \end{bmatrix} \right\} \cup \left\{ \begin{bmatrix} 1 \\ \mathbf{x}_p^k \end{bmatrix} \right\}_{p \in P^k}, \tag{4.83}$$

where P^k is the index-set of the extreme points of $\operatorname{conv}(\mathcal{D}^k)$. For the λ^k-variables of the reformulation, let the lower-index be 0 for the zero-extreme point and $p \in P^k$ for the others. The *IMP* becomes

$$
\begin{aligned}
z^\star_{IMP} = \min \sum_{k\in K} \begin{bmatrix} c_0^k & \mathbf{c}^{k\mathsf{T}} \end{bmatrix} \begin{bmatrix} 0 \\ \mathbf{0} \end{bmatrix} \lambda_0^k &+ \sum_{k\in K}\sum_{p\in P^k} \begin{bmatrix} c_0^k & \mathbf{c}^{k\mathsf{T}} \end{bmatrix} \begin{bmatrix} 1 \\ \mathbf{x}_p^k \end{bmatrix} \lambda_p^k \\
\text{s.t.} \sum_{k\in K} \begin{bmatrix} 0 & \mathbf{A}^k \end{bmatrix} \begin{bmatrix} 0 \\ \mathbf{0} \end{bmatrix} \lambda_0^k &+ \sum_{k\in K}\sum_{p\in P^k} \begin{bmatrix} 0 & \mathbf{A}^k \end{bmatrix} \begin{bmatrix} 1 \\ \mathbf{x}_p^k \end{bmatrix} \lambda_p^k \geq \mathbf{b} \\
\lambda_0^k &+ \sum_{p\in P^k} \lambda_p^k = 1 && \forall k \in K \\
\lambda_0^k \geq 0, & \qquad \lambda_p^k \geq 0 && \forall k \in K, p \in P^k \\
\begin{bmatrix} 0 \\ \mathbf{0} \end{bmatrix} \lambda_0^k &+ \sum_{p\in P^k} \begin{bmatrix} 1 \\ \mathbf{x}_p^k \end{bmatrix} \lambda_p^k = \begin{bmatrix} x_0^k \\ \mathbf{x}^k \end{bmatrix} && \forall k \in K \\
& \begin{bmatrix} x_0^k \\ \mathbf{x}^k \end{bmatrix} \in \{0,1\} \times \mathbb{Z}_+^{n^k} \ \forall k \in K.
\end{aligned}
$$

Because $\sum_{p\in P^k} \lambda_p^k = x_0^k$, we obtain a simplified program by treating λ_0^k as a binary slack variable in the convexity constraint of block k, i.e., $1 - \lambda_0^k = \sum_{p\in P^k} \lambda_p^k \in \{0,1\}$. The *IMP*, where we compute the vector product for the coefficients of the λ_p^k-variables, is

$$
\begin{aligned}
z^\star_{IMP} = \min \quad & \sum_{k\in K}\sum_{p\in P^k} (c_0^k + c_p^k)\lambda_p^k && && (4.84a) \\
\text{s.t.} \quad & \sum_{k\in K}\sum_{p\in P^k} \mathbf{a}_p^k \lambda_p^k \geq \mathbf{b} && && (4.84b) \\
& \sum_{p\in P^k} \lambda_p^k \leq 1 && \forall k \in K && (4.84c) \\
& \lambda_p^k \geq 0 && \forall k \in K, p \in P^k && (4.84d) \\
& \sum_{p\in P^k} \lambda_p^k = x_0^k \in \{0,1\} && \forall k \in K && (4.84e) \\
& \sum_{p\in P^k} \mathbf{x}_p^k \lambda_p^k = \mathbf{x}^k \in \mathbb{Z}_+^{n^k} && \forall k \in K. && (4.84f)
\end{aligned}
$$

Observe that in an optimal integer solution, an activated block k has a binding convexity constraint in $\operatorname{conv}(\mathcal{D}^k)$, i.e., $\sum_{p\in P^k} \lambda_p^k = 1$ in (4.84e), otherwise 0 if it is deactivated. This also implies that the fixed cost c_0^k is adequately measured by the objective function even if $\mathbf{x}^k = \mathbf{0}$ in which case the model ignores $c_0^k \geq 0$ using the kth less-than-or-equal convexity constraint. Interestingly, solving the linear relaxation of (4.84) means that a block k can be used but still be non-binding in the convexity constraint, that is, $0 < \sum_{p\in P^k} \lambda_p^k < 1$.

In the discretization approach, the constraints (4.84e)–(4.84f) can be removed altogether because the binary condition on λ_p^k ensures those on x_0^k and $\mathbf{x}^k$ by Proposition 4.10. We do note that variable x_0^k is completely useless even in the post-processing because any binary decision can be analogously done on $\sum_{p \in \ddot{P}^k} \lambda_p^k$:

$$\begin{aligned} z^\star_{IM\ddot{P}} = \min \quad & \sum_{k \in K} \sum_{p \in \ddot{P}^k} (c_0^k + c_p^k)\lambda_p^k \\ \text{s.t.} \quad & \sum_{k \in K} \sum_{p \in \ddot{P}^k} \mathbf{a}_p^k \lambda_p^k \geq \mathbf{b} \\ & \sum_{p \in \ddot{P}^k} \lambda_p^k \leq 1 && \forall k \in K \\ & \lambda_p^k \in \{0,1\} && \forall k \in K, p \in \ddot{P}^k. \end{aligned} \tag{4.85}$$

Identical subproblems

Let us again focus on the discretization case and assume that we also have $|K|$ *identical bounded subproblems* from the *ILP* (4.81), i.e., $\mathbf{A}^k = \mathbf{A}$, $\mathbf{D}^k = \mathbf{D}$, $\mathbf{d}^k = \mathbf{d}$, $\mathbf{c}^k = \mathbf{c}$, and non-negative fixed cost $c_0^k = c_0$, $\forall k \in K$. Then we perform aggregation in the *IM$\ddot{P}$* (4.85), or equivalently adapt (4.61), as

$$z^\star_{IM\ddot{P}} = \min \quad \sum_{p \in \ddot{P}} (c_0 + c_p)\lambda_p \tag{4.86a}$$

$$\text{s.t.} \quad \sum_{p \in \ddot{P}} \mathbf{a}_p \lambda_p \geq \mathbf{b} \qquad [\boldsymbol{\pi}_\mathbf{b}] \tag{4.86b}$$

$$\sum_{p \in \ddot{P}} \lambda_p \leq |K| \qquad [\pi_{\text{agg}}] \tag{4.86c}$$

$$\lambda_p \in \mathbb{Z}_+ \qquad \forall p \in \ddot{P}. \tag{4.86d}$$

Note 4.16 (Got enough blocks?) Since $z^\star_{IM\ddot{P}}$ is finite, there exists κ such that $\mathbf{1}^\intercal \boldsymbol{\lambda}^\star_{IM\ddot{P}} \leq \kappa$, for any optimal solution $\boldsymbol{\lambda}^\star_{IM\ddot{P}}$. This matches the definition of κ in (2.16a). If $|K| \geq \kappa$, the aggregated convexity constraint (4.86c) is redundant such that we can drop it from the model and assume that $\pi_{\text{agg}} = 0$ until further notice. This is the case in many applications if for example the number of vehicles to schedule or rolls to cut is a priori known to be sufficiently large. However, it often comes back in the *RMP* with branching decisions of the form

$$\sum_{p \in \ddot{P}} \lambda_p \leq \lfloor \mathbf{1}^\intercal \boldsymbol{\lambda}^\star_{M\ddot{P}} \rfloor \quad \text{and} \quad \sum_{p \in \ddot{P}} \lambda_p \geq \lfloor \mathbf{1}^\intercal \boldsymbol{\lambda}^\star_{M\ddot{P}} \rfloor + 1. \tag{4.87}$$

In fact, dropping the aggregated convexity constraint does not serve the implementation since it eventually requires two constraints to be expressed.

What we can rather do is dynamically modify the bounds of a *static variable*

$$v \in [0, \kappa] \quad \text{in} \quad \sum_{p \in \ddot{P}} \lambda_p - v = 0. \tag{4.88}$$

Note 4.17 (A long way...) Observe that we have come a long way from the aggregated discretization model (4.60) even after omitting extreme rays. We have seen some such simplified models and even created them intuitively for instance in Example 2.2, Cutting stock problem with rolls of different widths. In retrospective, it is a good place to underscore that these intuitive *column generation models* often rely on discretization whether knowingly or not, that is, we must be careful as to what is the meaning of $\mathcal{X}$ in the extended formulation.

Unbounded domains

Because $z^\star_{IMP}$ is finite, we can always argue that $\mathcal{D}^k$, hence $\operatorname{conv}(D^k)$ and $\operatorname{conv}(D^k_0)$ as well, is bounded by an appropriate large hyperbox with big-M values, that is, $\mathbf{u}^k = \mathbf{1}M$, $\forall k \in K$. Therefore, the above development can be applied by imposing these bounds in the compact formulation and deriving a Dantzig-Wolfe reformulation where only extreme points are present.

The alternative scenario in which we would interpret a solution $\mathbf{x}^k \in \mathcal{D}^k$, for which $\mathbf{x}^k < \mathbf{u}^k$ is not fulfilled, as an extreme ray of $\operatorname{conv}(\mathcal{D}^k)$ is not easy to conceive because we need two points to define a ray. This can however be done if $\operatorname{conv}(\mathcal{D}^k)$ is a polyhedral cone where the *zero*-extreme point is implicitly given. In that case the set $\mathcal{X}^k_0$ becomes

$$\mathcal{X}^k_0 = \left\{ \begin{bmatrix} 0 \\ \mathbf{0} \end{bmatrix} \right\} \cup \left\{ \begin{bmatrix} 1 \\ \mathbf{0} \end{bmatrix} \right\} \cup \left\{ \begin{bmatrix} 1 \\ \mathbf{x}^k_r \end{bmatrix} \right\}_{r \in R^k}, \tag{4.89}$$

and the fixed cost c^k_0 is only meaningful for $\mathbf{x}^k_r \neq \mathbf{0}$.

Shared variables across all blocks

Consider the following *ILP* with a block-diagonal structure plus an extra component called the *shared* (or *linking*) variables $\mathbf{y} \in \mathbb{Z}^n_+$:

$$\begin{aligned} z^\star_{ILP} = \min \quad & \mathbf{s}^\intercal \mathbf{y} + \sum_{k \in K} \mathbf{c}^{k\intercal} \mathbf{x}^k \\ \text{s.t.} \quad & \mathbf{S}\mathbf{y} + \sum_{k \in K} \mathbf{A}^k \mathbf{x}^k \geq \mathbf{b} \\ & \mathbf{S}^k \mathbf{y} + \mathbf{D}^k \mathbf{x}^k \geq \mathbf{d}^k \quad \forall k \in K \\ & \mathbf{y} \in \mathbb{Z}^n_+, \quad \mathbf{x}^k \in \mathbb{Z}^{n^k}_+ \quad \forall k \in K. \end{aligned} \tag{4.90}$$

We recognize that we can treat the y-variables as static by adding the constraints $\mathbf{y}=\mathbf{y}^k$, $\forall k \in K$, to (4.90) in anticipation of a reformulation over $|K|$ pricing problems:

$$\begin{aligned}
z^{\star}_{ILP} = \min \quad & \mathbf{s}^{\mathsf{T}}\mathbf{y} + \sum_{k\in K} \mathbf{c}^{k\mathsf{T}}\mathbf{x}^k && && \text{(4.91a)}\\
\text{s.t.} \quad & \mathbf{S}\mathbf{y} + \sum_{k\in K} \mathbf{A}^k\mathbf{x}^k \geq \mathbf{b} && && \text{(4.91b)}\\
& \mathbf{y} - \mathbf{y}^k = 0 && \forall k \in K && \text{(4.91c)}\\
& \mathbf{y} \in \mathbb{Z}^n_+ && && \text{(4.91d)}\\
& \mathbf{S}^k\mathbf{y}^k + \mathbf{D}^k\mathbf{x}^k \geq \mathbf{d}^k && \forall k \in K && \text{(4.91e)}\\
& \mathbf{y}^k \in \mathbb{Z}^n_+, \quad \mathbf{x}^k \in \mathbb{Z}^{n^k}_+ && \forall k \in K. && \text{(4.91f)}
\end{aligned}$$

Let the grouping of the constraints be

$$\mathcal{A} = \left\{ \mathbf{y} \in \mathbb{Z}^n_+, \left\{ \begin{bmatrix} \mathbf{y}^k \\ \mathbf{x}^k \end{bmatrix} \in \mathbb{Z}^n_+ \times \mathbb{Z}^{n^k}_+ \right\}_{k\in K} \;\middle|\; \mathbf{S}\mathbf{y} + \sum_{k\in K} \mathbf{A}^k\mathbf{x}^k \geq \mathbf{b}, \mathbf{y} = \mathbf{y}^k, \forall k \in K \right\} \quad \text{(4.92a)}$$

$$\mathcal{D}^k = \left\{ \begin{bmatrix} \mathbf{y}^k \\ \mathbf{x}^k \end{bmatrix} \in \mathbb{Z}^n_+ \times \mathbb{Z}^{n^k}_+ \;\middle|\; \mathbf{S}^k\mathbf{y}^k + \mathbf{D}^k\mathbf{x}^k \geq \mathbf{d}^k \right\}, \qquad \forall k \in K. \quad \text{(4.92b)}$$

For the ease of presentation, assume that $\mathcal{D}^k$ is bounded, $\forall k \in K$. We then express $\begin{bmatrix} \mathbf{y}^k \\ \mathbf{x}^k \end{bmatrix}$ as a convex combination of the extreme points of $\text{conv}(\mathcal{D}^k)$:

$$\begin{bmatrix} \mathbf{y}^k \\ \mathbf{x}^k \end{bmatrix} = \sum_{p\in P^k} \begin{bmatrix} \mathbf{y}^k_p \\ \mathbf{x}^k_p \end{bmatrix} \lambda^k_p, \quad \sum_{p\in P^k} \lambda^k_p = 1, \qquad \lambda^k_p \geq 0, \ \forall p \in P^k. \quad \text{(4.93)}$$

The Dantzig-Wolfe reformulation becomes

$$\begin{aligned}
z^{\star}_{IMP} = \min \quad & \mathbf{s}^{\mathsf{T}}\mathbf{y} + \sum_{k\in K}\sum_{p\in P^k} c^k_p\lambda^k_p && && \text{(4.94a)}\\
\text{s.t.} \quad & \mathbf{S}\mathbf{y} + \sum_{k\in K}\sum_{p\in P^k} \mathbf{a}^k_p\lambda^k_p \geq \mathbf{b} \quad [\boldsymbol{\pi}_{\mathbf{b}}] && && \text{(4.94b)}\\
& \mathbf{y} - \sum_{p\in P^k} \mathbf{y}^k_p\lambda^k_p = 0 \quad [\boldsymbol{\pi}^k_{\mathbf{y}}] && \forall k \in K && \text{(4.94c)}\\
& \sum_{p\in P^k} \lambda^k_p = 1 \quad [\pi^k_0] && \forall k \in K && \text{(4.94d)}\\
& \mathbf{y} \in \mathbb{Z}^n_+ && && \text{(4.94e)}\\
& \lambda^k_p \geq 0 && \forall k \in K, p \in P^k && \text{(4.94f)}\\
& \sum_{p\in P^k} \mathbf{y}^k_p\lambda^k_p = \mathbf{y}^k \in \mathbb{Z}^{n^k}_+ && \forall k \in K && \text{(4.94g)}
\end{aligned}$$

$$\sum_{p \in P^k} \mathbf{x}_p^k \lambda_p^k = \mathbf{x}^k \in \mathbb{Z}_+^{n^k} \qquad \forall k \in K, \tag{4.94h}$$

with the ISP^k written as

$$\begin{aligned} \bar{c}(\boldsymbol{\pi}_\mathbf{b}, \boldsymbol{\pi}_\mathbf{y}^k, \pi_0^k) = -\pi_0^k + \min \quad & c_\mathbf{x}^k - \boldsymbol{\pi}_\mathbf{b}^\intercal \mathbf{a}_\mathbf{x}^k + \boldsymbol{\pi}_\mathbf{y}^{k\intercal} \mathbf{y}^k \\ \text{s.t.} \quad & \begin{bmatrix} \mathbf{y}^k \\ \mathbf{x}^k \end{bmatrix} \in \mathcal{D}^k \\ & c_\mathbf{x}^k = \mathbf{c}^{k\intercal} \mathbf{x}^k, \;\; \mathbf{a}_\mathbf{x}^k = \mathbf{A}^k \mathbf{x}^k. \end{aligned} \tag{4.95}$$

Handling the linking variables in this way is similar to what is known as *Lagrangian decomposition* (Guignard and Kim, 1987).

4.6 More to Know

In Chapter 7, we assume that we can branch on the x- or λ-variables as we wish. In this section we justify the assumption that we can always work with a compact formulation in x-variables even though we may be given an extended formulation only in λ-variables. We call this *reverse engineering a compact formulation* or *reverse Dantzig-Wolfe decomposition.* There may be different *ILP* formulations to start from, and we remark that a reformulation can capitalize on this. We conclude by showing how to algorithmically group the constraints for a reformulation.

Reverse engineering a compact formulation

We have drawn several parallels and pointed to sometimes subtle differences between intuitive models we constructed in Chapter 2 and those we derived via a Dantzig-Wolfe reformulation. Let us formally describe the process of deriving a compact formulation if and when it is apparently not given. We consider both convexification and discretization models.

Convexification

Given an extended formulation and the pricing problem, all we have to realize is that any information we need to rewrite an equivalent compact formulation is readily available. In particular, it is useful to remind ourselves that $c_p = \mathbf{c}^\intercal \mathbf{x}_p$, $\mathbf{a}_p = \mathbf{A}\mathbf{x}_p$, $\forall p \in P$, and likewise for extreme rays. The relationship given by (4.96) could have been presented much earlier and reminds us that both of these formulations always co-exist.

$$
\begin{aligned}
z^\star_{IMP} = \min \; & \sum_{p\in P} c_p\lambda_p + \sum_{r\in R} c_r\lambda_r \\
\text{s.t.} \; & \sum_{p\in P} \mathbf{a}_p\lambda_p + \sum_{r\in R} \mathbf{a}_r\lambda_r \geq \mathbf{b} \\
& \sum_{p\in P} \lambda_p = 1 \\
& \lambda_p \geq 0, \quad \lambda_r \geq 0 \qquad \forall p \in P, r \in R \\
& \sum_{p\in P} \mathbf{x}_p\lambda_p + \sum_{r\in R} \mathbf{x}_r\lambda_r = \mathbf{x} \in \mathbb{Z}^n_+ \\
\text{with } & \{\mathbf{x}_p\}_{p\in P} \text{ and } \{\mathbf{x}_r\}_{r\in R} \text{ derived from } \mathcal{D} = \{\mathbf{x} \in \mathbb{Z}^n_+ \mid \mathbf{D}\mathbf{x} \geq \mathbf{d}\}
\end{aligned}
\qquad
\begin{aligned}
z^\star_{ILP} = \min \; & \mathbf{c}^\mathsf{T}\mathbf{x} \\
\text{s.t.} \; & \mathbf{A}\mathbf{x} \geq \mathbf{b} \\
& \mathbf{D}\mathbf{x} \geq \mathbf{d} \\
& \mathbf{x} \in \mathbb{Z}^n_+
\end{aligned}
\tag{4.96}
$$

It is not really more difficult to reverse a convexification model where a block-diagonal structure is present, see the *ILP* (4.40) and *IMP* (4.42). Now that the reader understands what we mean by reverse Dantzig-Wolfe, let us tackle the discretization case.

Discretization

Let us assume for the rest of the presentation that the given $\ddot{IMP}$ comprises only one pricing problem defined over $\mathcal{D} = \{\mathbf{x} \in \mathbb{Z}^n_+ \mid \mathbf{D}\mathbf{x} \geq \mathbf{d}\}$. Moreover, $c_p = \mathbf{c}^\mathsf{T}\mathbf{x}_p$ and $\mathbf{a}_p = \mathbf{A}\mathbf{x}_p, \forall p \in \ddot{P}$, and likewise for rays when used. The case of discretization is at first sight slightly more subtle to reverse because it may happen that some constraints have been simplified away (whether knowingly or not). For instance, from the compact formulation (4.81) with a block-diagonal structure, we have derived the very concise and yet equivalent discretization model (4.86) which assumes $|K|$ identical subproblems, i.e., $\mathbf{A}^k = \mathbf{A}$, $\mathbf{D}^k = \mathbf{D}$, $\mathbf{d}^k = \mathbf{d}$, $\mathbf{c}^k = \mathbf{c}$, and non-negative fixed cost $c^k_0 = c_0$, $\forall k \in K$. Using Note 4.16 to also get rid of the aggregated convexity constraint, we put these side-by-side in Proposition 4.15.

Observe that the number of blocks $|K|$ that is assumed to be large enough is optimized by design in the extended formulation $\ddot{IMP}$ (via a static variable with dynamic bounds in $[0, \kappa]$) whereas expressing the compact formulation *ILP* forces us to explicitly decide how many blocks we want to deal with.

Proposition 4.15. (Villeneuve et al., 2005) *Consider an extended formulation by discretization* (4.97)*_left in λ-variables without a convexity constraint as well as a single pricing problem in x-variables. Then, the compact formulation* (4.97)*_right that uses x^k-variables in a block-diagonal structure with identical data across all blocks which need not all be used is equivalent.*

$$
\begin{aligned}
z^{\star}_{IM\ddot{P}} = \min \quad & \sum_{p\in\ddot{P}} (c_0 + c_p)\lambda_p \\
\text{s.t.} \quad & \sum_{p\in\ddot{P}} \mathbf{a}_p \lambda_p \geq \mathbf{b} \\
& \lambda_p \in \mathbb{Z}_+ \quad \forall p \in \ddot{P} \\
& \textit{with } \{\mathbf{x}_p\}_{p\in\ddot{P}} \textit{ derived from } \mathcal{D} = \{\mathbf{x}\in\mathbb{Z}^n_+ \mid \mathbf{D}\mathbf{x}\geq\mathbf{d}\}
\end{aligned}
\qquad
\begin{aligned}
z^{\star}_{ILP} = \min \quad & \sum_{k\in K} c_0 x^k_0 + \mathbf{c}^{\mathsf{T}}\mathbf{x}^k \\
\text{s.t.} \quad & \sum_{k\in K} \mathbf{A}\mathbf{x}^k \geq \mathbf{b} \\
& \mathbf{D}\mathbf{x}^k \geq \mathbf{d}x^k_0 & \forall k \in K \\
& \mathbf{x}^k \leq \mathbf{u}x^k_0 & \forall k \in K \\
& x^k_0 \in \{0,1\} & \forall k \in K \\
& \mathbf{x}^k \in \mathbb{Z}^n_+ & \forall k \in K
\end{aligned}
\tag{4.97}
$$

What is important to understand here is that whenever we have an extended formulation $IM\ddot{P}$ where (part of) the substitution system does not appear, it remains easy to recover a compact formulation. This is the result of Villeneuve et al. (2005) that proves the existence of an *ILP*. Their proof in fact is restricted to a bounded domain $\mathcal{D}$ to allow reversing non-linear functions $c(\mathbf{x})$ and $\mathbf{a}(\mathbf{x})$ as we have considered in Proposition 4.14 in which case we could for instance read $\sum_{k\in K} c_0 x^k_0 + c(\mathbf{x}^k)$ in the compact formulation. In particular, our point is that it is sufficient to implicitly recover the substitution system simply by accepting that we know a Dantzig-Wolfe reformulation on the compact formulation which indeed leads to the given extended formulation. Let us underline that this existence neither implies that the compact formulation we propose is unique, nor that it is even the most 'compact' one.

Propositions 4.16 and 4.17 are evidence of this since they provide, under their respective assumptions, alternative compact formulations that do not even have a block-diagonal structure. The first is based on the polyhedral cone whereas the second is based on the integrality property. These are in fact generalizations of what we have seen in Example 3.3 on the Single depot vehicle scheduling problem: two compact formulations. Using Proposition 4.15, we can derive a compact formulation with $|K|$ identical vehicles, yielding one pricing problem per vehicle $k \in K$. Yet there is no such index k in the *ILP* (3.92), a network flow circulation problem (Figure 3.16 is reproduced in Figure 4.14).

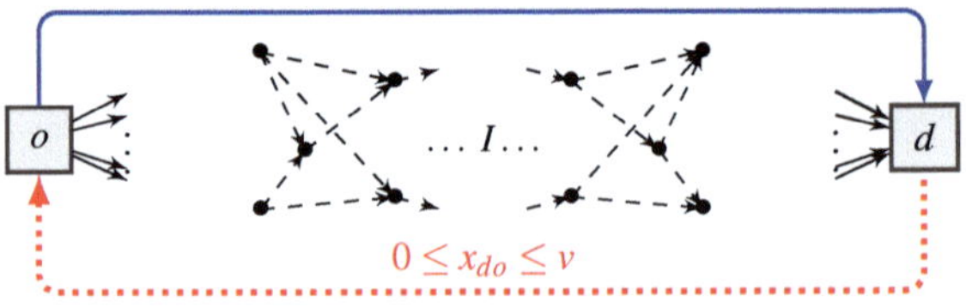

Fig. 4.14: Network G_{do} with v identical vehicles available at the depot.

With respect to the former proposition, the od-shortest path problem has been transformed into a circulation problem on the network G_{do} and the extreme rays scaled to one unit. For the latter proposition, the formulation we use to solve the circulation problem does have the integrality property and we also obtain a compact formulation without index k where x_0 plays the same role as x_{do}.

Proposition 4.16. *Consider an extended formulation by discretization* (4.98)_*left in λ-variables without a convexity constraint as well as a single pricing problem in x-variables with a domain $\mathcal{D} = \{\mathbf{x} \in \mathbb{Z}^n_+ \mid \mathbf{D}\mathbf{x} \geq \mathbf{0}\}$ whose convex hull* $\operatorname{conv}(\mathcal{D})$ *is a polyhedral cone. Then, the compact formulation* (4.98)_*right that also uses x-variables and does not have a block-diagonal structure is equivalent.*

$$
\begin{aligned}
z^\star_{IM\ddot{P}} = \min \; & \sum_{r \in \ddot{R}_\mathbf{0}} c_r \lambda_r \\
\text{s.t.} \; & \sum_{r \in \ddot{R}_\mathbf{0}} \mathbf{a}_r \lambda_r \geq \mathbf{b} \\
& \lambda_r \in \mathbb{Z}_+ \quad \forall r \in \ddot{R}_\mathbf{0} \\
& \text{with } \{\mathbf{x}_r\}_{r \in \ddot{R}_\mathbf{0}} \text{ derived from } \mathcal{D} = \{\mathbf{x} \in \mathbb{Z}^n_+ \mid \mathbf{D}\mathbf{x} \geq \mathbf{0}\}
\end{aligned}
\qquad
\begin{aligned}
z^\star_{ILP} = \min \; & \mathbf{c}^\mathsf{T}\mathbf{x} \\
\text{s.t.} \; & \mathbf{A}\mathbf{x} \geq \mathbf{b} \\
& \mathbf{D}\mathbf{x} \geq \mathbf{0} \\
& \mathbf{x} \in \mathbb{Z}^n_+
\end{aligned}
\tag{4.98}
$$

Proof. Recall the definition of $\ddot{R}_\mathbf{0} = \ddot{I} \cup \ddot{R}$ in (4.26) where the $\mathbf{0}$-vector is removed from the $IM\ddot{P}$. The result then follows from Proposition 4.9. □

Proposition 4.17. *Consider an extended formulation by discretization* (4.99)_*left in λ-variables without a convexity constraint as well as a single pricing problem in x-variables. Then, the compact formulation* (4.99)_*right that also uses x-variables and does not have a block-diagonal structure is equivalent if and only if the following conditions are satisfied:*

1. *The ISP formulation in $\mathbf{x} \in \mathcal{D} = \{\mathbf{x} \in \mathbb{Z}^n_+ \mid \mathbf{D}\mathbf{x} \geq \mathbf{d}\}$ has the integrality property;*
2. $\mathbf{0} \notin \mathcal{A} = \{\mathbf{x} \in \mathbb{Z}^n_+ \mid \mathbf{A}\mathbf{x} \geq \mathbf{b}\}$ *or* $z^\star_{IM\ddot{P}} \leq 0$.

$$
\begin{aligned}
z^\star_{IM\ddot{P}} = \min \; & \sum_{p \in \ddot{P}} (c_0 + c_p) \lambda_p \\
\text{s.t.} \; & \sum_{p \in \ddot{P}} \mathbf{a}_p \lambda_p \geq \mathbf{b} \\
& \lambda_p \in \mathbb{Z}_+ \quad \forall p \in \ddot{P} \\
& \text{with } \{\mathbf{x}_p\}_{p \in \ddot{P}} \text{ derived from } \mathcal{D} = \{\mathbf{x} \in \mathbb{Z}^n_+ \mid \mathbf{D}\mathbf{x} \geq \mathbf{d}\}, \\
& \text{the ISP formulation having the integrality property}
\end{aligned}
\qquad
\begin{aligned}
z^\star_{ILP} = \min \; & c_0 x_0 + \mathbf{c}^\mathsf{T}\mathbf{x} \\
\text{s.t.} \; & \mathbf{A}\mathbf{x} \geq \mathbf{b} \\
& \mathbf{D}\mathbf{x} \geq \mathbf{d} x_0 \\
& \mathbf{x} \leq \mathbf{u} x_0 \\
& x_0 \in \mathbb{Z}_+ \\
& \mathbf{x} \in \mathbb{Z}^n_+
\end{aligned}
\tag{4.99}
$$

Proof. Observe first that the given $IM\ddot{P}$ is formulated with variables λ_p, $\forall p \in \ddot{P}$. Hence the set $\mathcal{D}$ contains a finite number of integer points such that the values of $\mathbf{x}$ are implicitly upper bounded, say by $\mathbf{u}$. We start by deriving a Dantzig-Wolfe reformulation of the ILP using the grouping of constraints

$$
\mathcal{A}(x_0) = \left\{ \begin{bmatrix} x_0 \\ \mathbf{x} \end{bmatrix} \in \mathbb{Z}^{n+1}_+ \;\middle|\; \mathbf{A}\mathbf{x} \geq \mathbf{b} \right\} \tag{4.100a}
$$

$$\mathcal{D}(x_0) = \left\{ \begin{bmatrix} x_0 \\ \mathbf{x} \end{bmatrix} \in \mathbb{Z}_+^{n+1} \;\middle|\; \mathbf{Dx} \geq \mathbf{d}x_0,\ \mathbf{x} \leq \mathbf{u}x_0 \right\}. \tag{4.100b}$$

We then show that it can be simplified into the $IM\ddot{P}$ if and only if the two conditions are satisfied.

By Theorem 4.2, every point of $\mathcal{D}(x_0)$ can be expressed using a finite set $\left\{ \begin{bmatrix} x_{0r} \\ \mathbf{x}_r \end{bmatrix} \right\}_{r \in \ddot{R}_0}$ of integer-scaled rays of the polyhedral cone $\operatorname{conv}(\mathcal{D}(x_0))$, see (4.24)_right. A Dantzig-Wolfe reformulation based on the discretization of $\mathcal{D}(x_0)$ is therefore

$$\begin{aligned}
\min \quad & 0\lambda_{\left[\begin{smallmatrix}0\\\mathbf{0}\end{smallmatrix}\right]} && + && \sum_{r\in\ddot{R}_0} c_r\lambda_r \\
\text{s.t.} \quad & \mathbf{0}\lambda_{\left[\begin{smallmatrix}0\\\mathbf{0}\end{smallmatrix}\right]} && + && \sum_{r\in\ddot{R}_0} \mathbf{a}_r\lambda_r \geq \mathbf{b} \\
& \lambda_{\left[\begin{smallmatrix}0\\\mathbf{0}\end{smallmatrix}\right]} = 1, && && \lambda_r \in \mathbb{Z}_+ \qquad\qquad \forall r \in \ddot{R}_0 \\
& \begin{bmatrix} 0 \\ \mathbf{0} \end{bmatrix}\lambda_{\left[\begin{smallmatrix}0\\\mathbf{0}\end{smallmatrix}\right]} && + && \sum_{r\in\ddot{R}_0} \begin{bmatrix} x_{0r} \\ \mathbf{x}_r \end{bmatrix}\lambda_r = \begin{bmatrix} x_0 \\ \mathbf{x} \end{bmatrix} \in \mathbb{Z}_+^{n+1},
\end{aligned} \tag{4.101}$$

where the encoding functions are $c_r = c_0 x_0 + \mathbf{c}^\mathsf{T}\mathbf{x}_r$, $\mathbf{a}_r = \mathbf{A}\mathbf{x}_r$, $\forall r \in \ddot{R}_0$. Note that we explicitly keep the convexity constraint $\lambda_{\left[\begin{smallmatrix}0\\\mathbf{0}\end{smallmatrix}\right]} = 1$ for future reference to the unique zero-extreme point of the polyhedral cone but the constraints in x_0 and $\mathbf{x}$ can be dropped.

$\Leftarrow$ We start with Condition *1*. By assumption, we have $\mathcal{D} = \mathcal{D}(1)$ since the upper bounds $\mathbf{u}$ are implicit. Figure 4.15 depicts the polyhedral cone $\operatorname{conv}(\mathcal{D}(x_0))$ and makes the previous statement obvious.

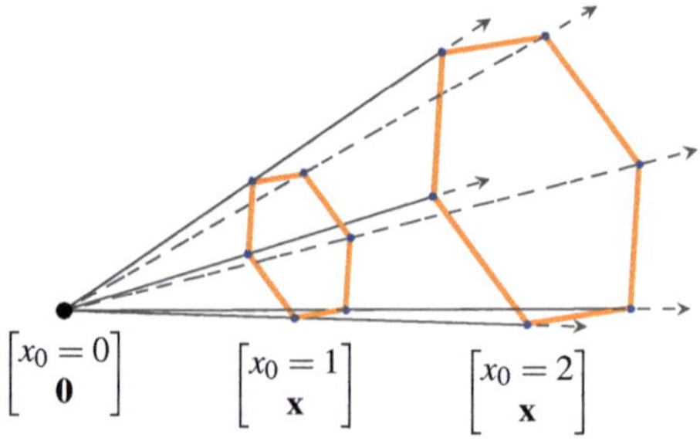

Fig. 4.15: Polyhedral cone $\operatorname{conv}(\mathcal{D}(x_0))$.

By the integrality property of the *ISP* formulation defined on $\mathbf{Dx} \geq \mathbf{d}$, the extreme points of the polytope

$$\{\mathbf{x} \in \mathbb{R}_+^n \mid \mathbf{Dx} \geq \mathbf{d},\ \mathbf{x} \leq \mathbf{u}\} = \operatorname{conv}(\mathcal{D}) \tag{4.102}$$

are *integer* just like those of conv($D(1)$). Thus, the set $\left\{\begin{bmatrix}1\\ \mathbf{x}_p\end{bmatrix}\right\}_{p\in\ddot{P}}$ of integer points of $\mathcal{D}(1)$ is in a one-to-one correspondence with the set $\{\mathbf{x}_p\}_{p\in\ddot{P}}$ of the λ_p-variables in the $IM\ddot{P}$.

We need to show that any integer-scaled ray in the reformulation (4.101) can be written using the points of $\mathcal{D}(1)$. In fact, apart from the single extreme point, all integer points of conv($\mathcal{D}(x_0)$) are rays, those indexed in $\ddot{R}_\mathbf{0}$ being a subset. Hence it is sufficient to show that all points of $\mathcal{D}(x_0)$, $x_0 \geq 2$, write in terms of $\mathcal{D}(1)$.

This is true for the extreme points of $\left\{\begin{bmatrix}x_0\\ \mathbf{x}\end{bmatrix} \in \mathbb{R}^{n+1}_+ \;\middle|\; \mathbf{D}\mathbf{x} \geq \mathbf{d}x_0,\ \mathbf{x} \leq \mathbf{u}x_0\right\}$ given by $\left\{x_0\begin{bmatrix}1\\ \mathbf{x}_p\end{bmatrix}\right\}_{p\in P}$, hence integer for $x_0 \in \mathbb{Z}_+$. Let $\ddot{P}_{adj} \subset \ddot{P}\times\ddot{P}$ denote the set of indices (p,p') of two adjacent integer points of $\mathcal{D}(1)$.

- The points of $\mathcal{D}(2)$ are either twice a point of $\mathcal{D}(1)$, i.e., $2\begin{bmatrix}1\\ \mathbf{x}_p\end{bmatrix}$, $\forall p \in \ddot{P}$, or twice a fractional point *halfway between two adjacent integer points* of $\mathcal{D}(1)$:

$$2\times\left(\frac{1}{2}\begin{bmatrix}1\\ \mathbf{x}_p\end{bmatrix}+\frac{1}{2}\begin{bmatrix}1\\ \mathbf{x}_{p'}\end{bmatrix}\right)=\begin{bmatrix}1\\ \mathbf{x}_p\end{bmatrix}+\begin{bmatrix}1\\ \mathbf{x}_{p'}\end{bmatrix}\in\mathcal{D}(2), \qquad \forall (p,p')\in\ddot{P}_{adj}.$$

- Similarly, $\begin{bmatrix}3\\ \mathbf{x}\end{bmatrix}$ writes as three times a point of $\mathcal{D}(1)$, or three times a fractional point *either at* $1/3$ *or at* $2/3$ *in between two adjacent integer points* of $\mathcal{D}(1)$:

$$3\times\left(\frac{2}{3}\begin{bmatrix}1\\ \mathbf{x}_p\end{bmatrix}+\frac{1}{3}\begin{bmatrix}1\\ \mathbf{x}_{p'}\end{bmatrix}\right)=2\begin{bmatrix}1\\ \mathbf{x}_p\end{bmatrix}+\begin{bmatrix}1\\ \mathbf{x}_{p'}\end{bmatrix}\in\mathcal{D}(3), \qquad \forall (p,p')\in\ddot{P}_{adj};$$

$$3\times\left(\frac{1}{3}\begin{bmatrix}1\\ \mathbf{x}_p\end{bmatrix}+\frac{2}{3}\begin{bmatrix}1\\ \mathbf{x}_{p'}\end{bmatrix}\right)=\begin{bmatrix}1\\ \mathbf{x}_p\end{bmatrix}+2\begin{bmatrix}1\\ \mathbf{x}_{p'}\end{bmatrix}\in\mathcal{D}(3), \qquad \forall (p,p')\in\ddot{P}_{adj}.$$

- More generally, the points of $\mathcal{D}(x_0)$, $x_0 \geq 2$, are of two types:

$$\begin{bmatrix}x_0\\ \mathbf{x}\end{bmatrix}=\begin{cases} x_0\begin{bmatrix}1\\ \mathbf{x}_p\end{bmatrix} & \forall p\in\ddot{P}\\[2ex] (x_0-\alpha)\begin{bmatrix}1\\ \mathbf{x}_p\end{bmatrix}+\alpha\begin{bmatrix}1\\ \mathbf{x}_{p'}\end{bmatrix} & \forall (p,p')\in\ddot{P}_{adj},\ \alpha\in\{1,\dots,x_0-1\}.\end{cases}$$

  As $\ddot{P}_{adj} \subset \ddot{P}\times\ddot{P}$, it is therefore sufficient to know the points indexed by $\ddot{P}$ to write every point of $\mathcal{D}(x_0)$, $x_0 \geq 2$.

Condition **2** comprises two expressions, each one independently showing that the introduced zero-vector can be removed from the reformulation (4.101).

- If $\mathbf{0} \notin \mathcal{A}$, then an optimal solution to (4.101) writes in terms of positive λ_r-variables and $\lambda_{\begin{bmatrix}0\\ \mathbf{0}\end{bmatrix}}$ is useless.
- The second condition $z^\star_{IM\ddot{P}} \leq 0$ in (4.99)_left is treated in two parts. If $z^\star_{IM\ddot{P}} < 0$, the zero-vector cannot be optimal in the reformulation (4.101). If $z^\star_{IM\ddot{P}} = 0$, there exists an optimal solution $\boldsymbol{\lambda}^\star_{IM\ddot{P}} \geq \mathbf{0}$, the same being in (4.101) in terms of the λ_r-variables, $r \in \ddot{R}_\mathbf{0} = \ddot{P}$, and the zero-vector can be removed.

Conditions *1* and *2* being sufficient, the zero-vector can be removed from the reformulation (4.101) and the index-set $\ddot{R}_0$ replaced by $\ddot{P}$: this is the *IM*$\ddot{P}$.

$\Rightarrow$ The proof is done by contradiction on either Condition *1* or Condition *2*.

1. Assume that the *ISP* does not possess the integrality property.
Then there exists at least one rational point $\mathbf{x}_\bullet \in \{\mathbf{x} \in \mathbb{R}^n_+ \mid \mathbf{D}\mathbf{x} \geq \mathbf{d}\}$ that is not in $\operatorname{conv}(\mathcal{D})$, hence cannot be written neither as a convex combination of the extreme points of $\{\mathbf{x}_p\}_{p\in P}$. nor using the larger set $\{\mathbf{x}_p\}_{p\in\ddot{P}}$. Multiplying $\mathbf{x}_\bullet$ by a sufficiently large positive integer number x_0, $\begin{bmatrix} x_{0r} \\ \mathbf{x}_r \end{bmatrix} = \begin{bmatrix} x_0 \\ x_0\mathbf{x}_\bullet \end{bmatrix}$ is an integer-scaled ray indexed by $r \in \ddot{R}_0$, but it cannot be written using the points of $\mathcal{D}(1)$. We can then modify the objective coefficients $\mathbf{c}$ so that $x_0\mathbf{x}_\bullet$ is integer optimal for the pricing problem defined on $\mathcal{D}(x_0)$.
To sum up, the absence of the integrality property invalidates the one-to-one correspondence between $\ddot{P}$ and $\ddot{R}_0$.

2. Assume that $\mathbf{0} \in \mathcal{A}$ and $z^\star_{IM\ddot{P}} > 0$.
By construction, the zero-vector $\begin{bmatrix} 0 \\ \mathbf{0} \end{bmatrix}$ belongs to $\mathcal{D}(x_0)$ and it is also feasible in $\mathcal{A}(x_0)$ as $\mathbf{0}$ is in $\mathcal{A}$. As such, the minimum objective value in the reformulation (4.101) is less than or equal to 0, obviously in contradiction with $z^\star_{IM\ddot{P}} > 0$.

In both cases of contradiction, the reformulation (4.101) of the *ILP* is not equivalent to the given *IM*$\ddot{P}$, hence these two conditions are also necessary. □

Note 4.18 (Lost in reduction.) In finding a compact formulation without index k from Propositions 4.16 or 4.17, it may be that we lose track of the essence of the problem. This is notably the case if the x-variables can be simplified from the *ILP* model thus giving no way to recover an actual solution for the real problem. This surprisingly happens in Example 4.10 (Edge coloring problem: two compact formulations) for which the second *ILP* is useless except for its objective value.

Extended compact and subproblem formulations

Imagination, creativity, thinking outside the box, knowledge may help in choosing a compact formulation *ILP* to start with. For example, any finite set $\mathcal{D} = \{\mathbf{x} \in \mathbb{R}^n_+ \mid \mathbf{D}\mathbf{x} \geq \mathbf{d}\} \cap \mathbb{Z}^n_+$ can be represented by an acyclic state-space network, say $G_\mathcal{D}$, in which each solution corresponds to a path from the source node to the destination node, and vice versa, see Nemhauser and Wolsey (1988, p. 312) and Vanderbeck (2000, p. 117). This might be interesting since an *ISP* formulation based on the network $G_\mathcal{D}$ possesses the integrality property. It should be noted that the number of nodes grows exponentially with the row size of $\mathbf{D}$ and linearly with the L_1-norm of the right-hand side $\mathbf{d}$.

While the alternative to solve over $\mathcal{D}$ by branch-and-bound is always there, we consider here the possible advantage of using the network $G_{\mathcal{D}}$ which has the integrality property. A full description of $G_{\mathcal{D}}$ means that every path it contains is feasible so the graph size may rapidly become a burden to treat. If enumeration is tractable, we can imagine that this may indeed be a decent alternative. Otherwise, dynamic programming (Bellman, 1957) may also be a worthy option. A major factor for this potential superiority is that dynamic programming typically does not need to explore the entire state-space. In fact, one often embeds some portions of that state-space onto the arcs of a much smaller network. The algorithm then filters out (partial) paths based on infeasibility or dominance conditions. This is illustrated in Example 4.1, where we present an alternative formulation for the knapsack problem which has the integrality property (Gilmore and Gomory, 1966).

Putting this in the perspective of a Dantzig-Wolfe reformulation and more specifically for the formulation of the integer pricing problem, we can compare the linear relaxations of different compact *ILP*s and their corresponding integer master reformulations. Furthermore, we now have different algorithms to solve the various formulations of the *ISP*s. This is illustrated in Example 4.2 (Integrality property in the cutting stock problem), where two compact formulations are compared. The first has a block-diagonal structure with identical subproblems, the second has no block-index. Given their integer master reformulation (the same $\ddot{IMP}$ for both *ILPs*) and their subproblem formulations and properties ($c_0 = 1$ in the first, $c_0 = 0$ and polyhedral cone in the second), the reader can easily find back the two original *ILPs*, rather different, in terms of the x-variables of their respective *ISPs*. Early applications of this *extended ISP formulation* can be found in Eppen and Martin (1987) to reformulate multi-item capacitated lot-sizing problems and Lavoie et al. (1988) for the crew pairing problem at Air France.

We can also use an *ILP* formulation based on the flow variables of $G_{\mathcal{D}}$ and, without any reformulation, apply the column activation algorithm we have devised in Section 3.4 (Restricted compact formulation) to solve the *LP* by iteratively filling a restricted model (*RLP*). That is, one generates paths with the *ISP* based on network $G_{\mathcal{D}}$ and activates the corresponding arc-variables as well as the newly active flow conservation constraints. As mentioned previously in Section 3.4, such a technique has been used to solve the cutting stock problem (Valério de Carvalho, 2002).

Dynamic programming may also be able to handle non-linearity and continuous variables. In such cases, we cannot speak of a one-to-one correspondence between paths and solutions. The *VRPTW* where we consider continuous time variables fits this bill in Chapter 5 and Section 5.4 describes the state-space network whose solutions corresponds to what we find by dynamic programming. By construction, we find earliest-visit even though any other visiting time t_i feasible with respect to time windows on a path is indeed feasible in the *ISP*. In general, we can summarize that a dynamic programming algorithm starts with a trivial strategy and recursively considers a more complex one until the problem is solved. One can describe this recursion on a state-space network in which every path from the start node to the destination node corresponds to a solution to the problem.

Automatic grouping of the constraints for reformulation

Throughout the preceding two chapters, we cared about describing the grouping of the *ILP* constraints into (the definition of) the sets $\mathcal{A}$ and $\mathcal{D}$. This grouping of constraints itself is sometimes also called a *decomposition* (with an overload of the notion). So far, we took the grouping for granted. However, it is not always given or even only known. The *ILP* model may not be formally stated on paper, but a specific instance could be presented only as a file, e.g., in LP or MPS format. Variables and constraints may be only generically numbered, without any explanation of their meaning, just the "flat" model. Actually, this is the standard situation in which a solver "sees" the *ILP*.

Figure 4.16a shows the coefficient matrix of instance ns1778858, which is a general MILP taken from the publicly available benchmark library MIPLIB (Koch et al., 2011). Each black dot represents a non-zero entry in the matrix, reminding us again of how sparse practical optimization models are. The model is listed as "unknown application," it is anonymously donated to the collection. In Figure 4.16b, we see "the same" matrix, only with rows and columns re-arranged. Here, very clearly, a structure becomes apparent, a small border and 16 blocks, shaded to support visibility. With this information, and of course, cost coefficients and right-hand sides re-arranged accordingly, we can apply a Dantzig-Wolfe reformulation to this model. Note that re-ordering the matrix does not change the model, nor its feasible region.

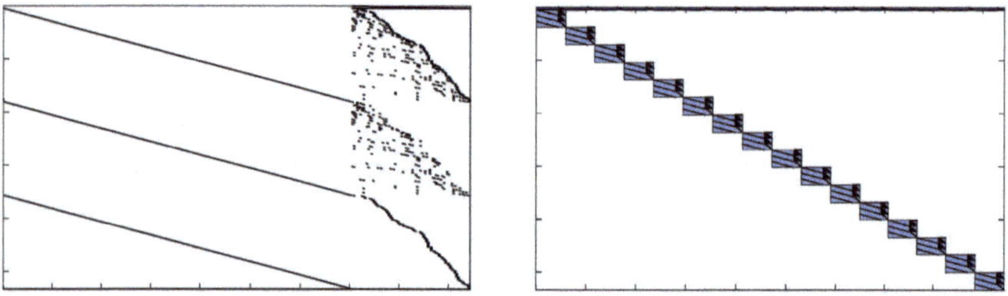

(a) Original matrix (b) Permuted matrix

Fig. 4.16: Permuting the constraint matrix of MIPLIB instance ns1778858.

Permuting the rows and columns of a matrix such that a "model structure becomes visible" is called *structure detection*. It is not as well-defined as it sounds. There are different forms into which one and the same matrix could be permuted, see Figures 4.10 and 4.17. Even when the form is given, the number of blocks is not "clear," as different constraints can go to the border or to the blocks, see Figure 4.19. Usually, we "don't know" what the modeler had in mind. Even though this knowledge can be helpful, the following shows that we can also do without it.

(a) Original matrix (b) Single-bordered form (c) Double-bordered form

Fig. 4.17: Coefficient matrix of `b2c1s1` permuted into two different forms.

Graph-based methods

It is natural to represent a matrix $\mathbf{A} = [a_{ij}]_{i\in\{1,\dots,m\},j\in\{1,\dots,n\}}$ as a bipartite graph. We introduce a set of nodes $I = \{r_1,\dots,r_m\}$, one for each row, and a set of nodes $J = \{c_1,\dots,c_n\}$, one for each column. There is an edge between two nodes $r_i \in I$ and $c_j \in J$ for every non-zero element $a_{ij} \neq 0$ of $\mathbf{A}$, see Figure 4.18. Let us restrict attention to single-bordered block-diagonal forms which are the most relevant for a Dantzig-Wolfe reformulation. It is important to see that for a given such form, if we would remove all nodes from I that correspond to the border, we would be left with a number of connected components, each of which represents one block.

$$
\begin{array}{c} \\ r_1 \\ r_2 \\ r_3 \\ r_4 \\ r_5 \\ r_6 \end{array}
\begin{array}{c}
\begin{array}{cccccc} c_1 & c_2 & c_3 & c_4 & c_5 & c_6 \end{array} \\
\begin{bmatrix}
1 & 1 & & 1 & & \\
1 & & 1 & 1 & & 1 \\
 & 1 & & & & \\
1 & & 1 & & & \\
 & & & 1 & & 1 \\
 & & & 1 & 1 & 1
\end{bmatrix}
\end{array}
$$

(a) A matrix ...

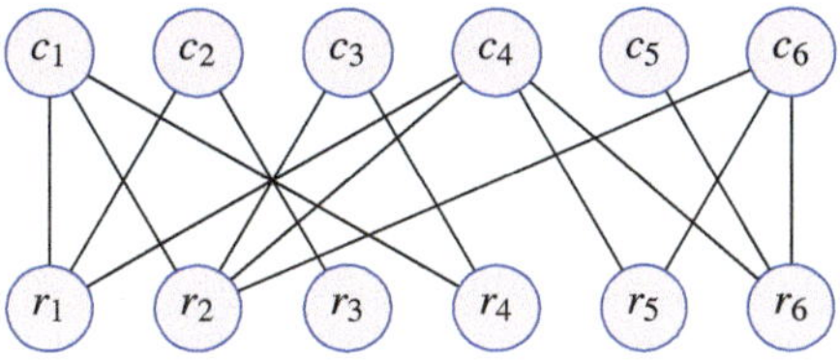

(b) ... and its representation

Fig. 4.18: Bipartite graph representation of a matrix.

This gives the following idea, based on graph partitioning: First decide about a number $|K|$ of blocks we want to see. Then delete as few nodes from I as possible (and the incident edges) such that the remaining graph partitions into $|K|$ connected components. This effectively minimizes the number of constraints in $\mathcal{A}$, the border. In this basic *vertex* $|K|$*-separator problem*, nothing is said about whether all nodes are equally important or how large the resulting connected components are. Variants that use node weights or impose balancing constraints exist, e.g., for preferring certain types of constraints over others (see below). Note that we could remove nodes corresponding to either rows or columns or both, depending on whether we are looking for linking constraints, linking variables, or both. An alternative representation of a matrix is via a *hypergraph*, in which nodes represent the non-zero elements, and all the entries in a row (or a column) are represented by a hyperedge. This leads to hypergraph partitioning problems, but we do not follow up on this here. All the above problems are $\mathcal{NP}$-hard. However, the problem of identifying a block-diagonal matrix structure is generally interesting, e.g., in numerical linear algebra for parallelizing the solution of linear equation systems. Also because of this, several good heuristic graph partitioning implementations are publicly available.

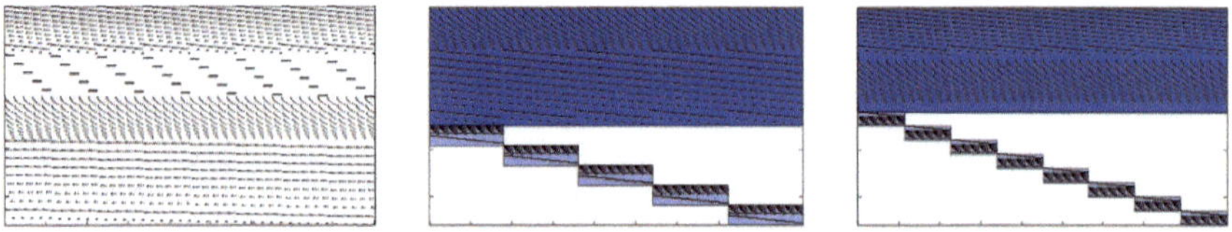

(a) Original matrix (b) Permuted to 5 blocks (c) Permuted to 8 blocks

Fig. 4.19: Permuting the matrix of `10teams` into different numbers of blocks.

A drawback of the above is that one has to specify the number $|K|$ of blocks upfront. When one assumes that blocks are all identical, one can come up with good guesses based on the content of constraints (Wang and Ralphs, 2013), but this is not satisfactory in general. Khaniyev et al. (2018) suggest a graph clustering approach that borrows ideas from community detection. Intuitively speaking, a graph has a *community structure* when there are subgraphs, the communities, in which the graph density is (much) higher than in between the communities. In the graph clustering language, there should be many *intra-cluster* edges and few *inter-cluster* edges. A classical measure to quantify the quality of a clustering is *modularity*. In order to remove edges that run between clusters altogether, one can eliminate nodes (from I, and the incident edges). The removed nodes correspond to rows in the border, the clusters correspond to blocks. Since the clustering is driven by maximizing, e.g., the modularity, the number of resulting blocks is part of the output.

Let us mention that using the graph representation, one can even identify identical blocks: whenever the two subgraphs representing two blocks are *isomorphic*, the blocks are identical (given that the respective cost coefficients and right-hand sides coincide as well). Also other forms can be detected with graph based methods, e.g., *staircase forms*, see Figure 4.20.

Methods based on constraint classes

When you compare how we *extracted* a structure from a coefficient matrix to the way the structure *gets into* the model, there is a mismatch. It is unlikely that a human modeler would think of incidences between variables and constraints when creating a model. Instead, they express the *semantic* elements of a problem, casted in linear constraints. Typically, these constraints come in groups of similar meaning, e.g., one group for assigning items, one for capacities, one for implications, etc. Modelers give names to constraints which belong together. Look at any model in this book and you see these groups. The constraints in one semantic group are

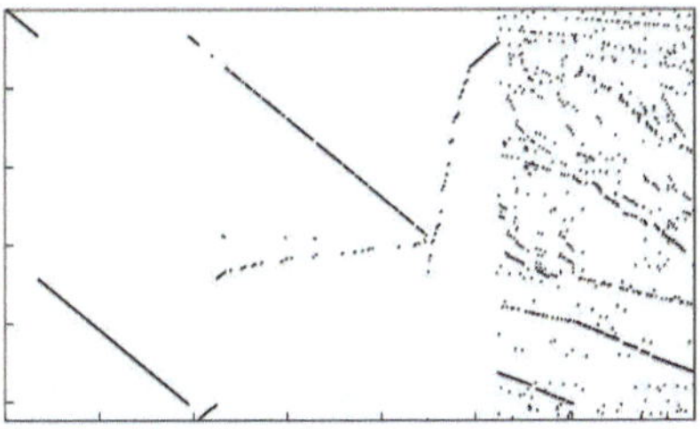

(a) Original matrix

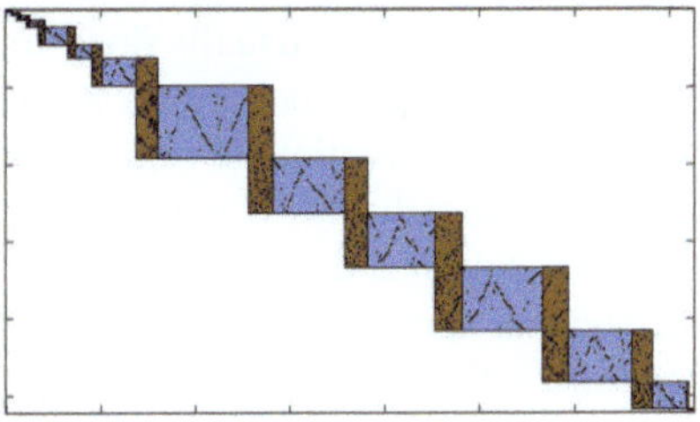

(b) Permuted matrix

Fig. 4.20: Coefficient matrix of `ic97_potential` permuted to a staircase form.

typically generated from the elements in some index set, but more importantly, all constraints in a group are of the same *type*, i.e., of knapsack type, set packing type, variable upper bound type, etc. We refer to the `MIPLIB` (Gleixner et al., 2021) for a description of the various types of linear constraints, of which there are not so many. Such a classification into types happens in standard MILP solvers usually already at the preprocessing stage. However, we can classify constraints not only according to their type, but also based on other attributes like the number (and type) of variables they contain, coefficients and right-hand sides, their names (if given), etc. Note that one could similarly classify the variables.

Assume that we have identified a number of classes $I_1,\ldots,I_\ell \subseteq I$ of constraints which together partition I, that is, every constraint is part of exactly one class. In the models in this book, then two cases occur for a class: either, *all* constraints in the class go to the master problem, that is, are included in $\mathcal{A}$. Or, the constraints (perfectly) *distribute* over the blocks in $\mathcal{D}^1,\ldots,\mathcal{D}^{|K|}$. You may ask yourself, well, how else? But again, we "see" this because we have the model on paper. If we wouldn't have this information, in particular no information about an index set K, we needed to extract it. Based on the previous observations, here is one way how.

Since we do not know which of the classes are entirely assigned to $\mathcal{A}$, we may try *every subset* of $\{I_1,\ldots,I_\ell\}$, of which there are 2^ℓ many. When ℓ is not too large, and in this book we often have a rather small one-digit number of ℓ constraint classes, this number is manageable. We tentatively assign a subset $I_\mathcal{A} \subseteq \{I_1,\ldots,I_\ell\}$ of entire constraint classes "to the border" $\mathcal{A}$ and check whether the constraints in the *remaining* classes $I_\mathcal{D} = \{I_1,\ldots,I_\ell\} \setminus I_\mathcal{A}$ can be distributed to more than one block. We already know how to accomplish the latter. We consider the bipartite graph representing the submatrix corresponding to the constraints (and variables) in $I_\mathcal{D}$ and check its number of connected components. This is an easy problem and can be done via a quick graph search. The components reveal the blocks and their number, potentially only one. The entire procedure gives us many *candidates* for a decomposition and we have to choose one. This is discussed below. For the record: The structure in Figure 4.16 was detected using the method just described, and we produced the t-shirt logo in Figure 4.11 with a hypergraph partitioning algorithm.

Let us remark that it usually does not make sense to test *every* subset of constraint classes. Maybe one is looking for particular decompositions, or problem knowledge is available, or theoretical considerations from below apply, and the number of candidates can be reduced. One can also classify variables. And a better grouping of constraints could be done by taking the different classes of variables into account as well. Such an approach is taken in a different context by Salvagnin (2016), applying a *partition refinement* algorithm. Its potential for structure detection for reformulations is not fully exploited to date.

Finally note that things would be much easier if we had access to the index sets from which variables and constraints are generated. When a model is formulated using, e.g., an algebraic modeling language, this information is available. Annotations in models could help, too. It is the ever recurring theme: the more information you have, the more you can exploit.

Evaluating a decomposition

Looking for different matrix forms, asking for different numbers of blocks, selecting different groups of constraints to go into the master problem, one can easily produce hundreds or thousands of candidate decompositions for a given *ILP*. Not all of them are suited for a successful Dantzig-Wolfe reformulation, but how would we know? What does it even mean, "suited," because in theory they all "work." We are in need of a quality measure for decompositions, a *score*.

The first that comes to mind is the dual bound. While for linear programs, literally every decomposition leads to an equivalent reformulation, c.f. $z^\star_{MP} = z^\star_{LP}$ in (3.21), the situation is different for integer programs. From the proof of Proposition 4.1, in particular (4.8), it is clear that the definition of $\mathcal{D}$ (or $\mathcal{D}^1, \ldots, \mathcal{D}^{|K|}$ for that matter) can make a difference in terms of the dual bound obtained from the reformulation. However, the *best* dual bound is not a reasonable goal because we can always obtain it by reformulating *all* constraints. We also need to take the computational resources into account that have to be invested to compute that bound.

Estimating the runtime it needs to solve a given *ILP* is a difficult undertaking, however, and this is no easier for an *IMP*. From experience and also from experiments, some *proxy measures* were suggested that are considered beneficial for the performance of the column generation algorithm. These include the following.

- *Properties of the border.*
  Intuitively, fewer constraints in the border (that is, in the definition of $\mathcal{A}$) may speed-up the re-optimization of the *RMP* and may give a better dual bound because more constraints go into the definition of the blocks. Bergner et al. (2015) give a penalty linear in the size of the border. Khaniyev et al. (2018) experiment with exponential decay: The penalty for adding ever more constraints to the border is diminishing with a larger border size.
  Observe that for some classical models, the decomposition from the literature uses relatively large borders, see e.g., Figure 4.21.

- *Properties of the blocks.*
  For computational reasons, one rather wants to have a larger number of smaller blocks instead of a few large ones. Whether or not having many blocks counteracts the quality of the dual bound is not clear. Intuitively, a large integrality gap in the blocks may hint at a good dual bound obtained from the reformulation, but computing the gap is expensive and cannot be directly used as an objective in, e.g., the graph-based methods. At the same time, a large integrality gap is detrimental to computation time for the blocks.
  Khaniyev et al. (2018) aim for as many blocks as possible, a balanced size is preferred, and ideally, blocks should be identical. Their main evaluation measure for the quality of the blocks is modularity.

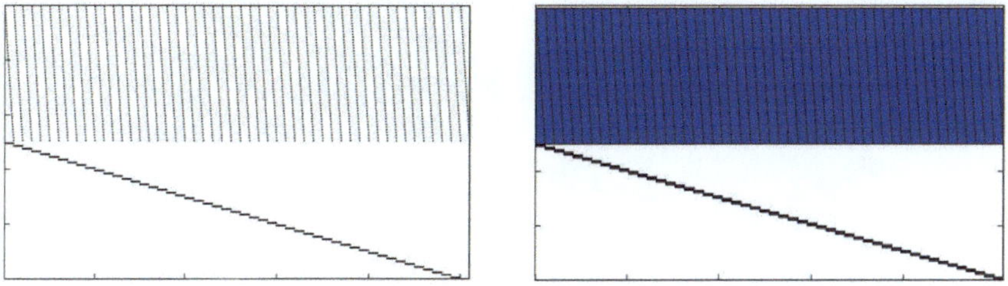

(a) Original matrix (b) "Permuted" matrix

Fig. 4.21: A textbook example (bin packing) in which a large border is "correct."

Combining these goals, small border and many blocks, gives *visually appealing* decompositions (Bergner et al., 2015), like in Figure 4.16. But don't be deluded by the beauty. It is certainly fascinating that we can discover (or rather "enforce") a previously unknown structure in general MILPs. However, only because we have nice pictures, it does not mean that there is a structure in the sense of this book: the subproblems may just also be general MILPs, only smaller; nothing "special" to exploit. We call these models "unstructured," and the graph-based methods may be a good way to deal with them. Yet, this book is almost exclusively about "structured" models, those which contain subproblems for which we have a good oiled machinery ready to go. In this case, the methods based on constraint (and variable) classes may work best. We may look specifically for model structure like knapsacks or networks that should go into the subproblem(s). On the other hand, e.g., semi-assignment constraints should go into the border (and if possible, nothing else), because this leads to set partitioning master problems for which we have, e.g., special branching rules at hand (see the Ryan-Foster rule, p. 480). In fact, Proposition 4.5 says that, in terms of the dual bound, it never pays off to reformulate semi-assignment constraints. We may want to put constraints in the blocks that destroy the integrality property in the pricing, see Section 4.3. Identical blocks are a winner because we can perform an aggregation (p. 205).

Some remarks at the close: We cannot speak of "the" decomposition of an *ILP*—there are many. Seemingly small changes to a decomposition may lead to very different behavior in the subsequent column generation process, both in terms of dual bound quality and running time (Bergner et al., 2015). Bastubbe et al. (2018) have computational results that suggest that randomly grouping constraints does *not* yield a good quality in terms of dual bound. And almost unavoidably these days, one can learn a good structure (Basso and Ceselli, 2023) and whether (or not) to use it in a Dantzig-Wolfe reformulation (Kruber et al., 2017).

Fig. 4.22: Fabio Furini and Alberto Ceselli (Montréal, Canada, 2023-05-17).

Most structure detection concepts described here (and others) are implemented in the generic branch-and-price solver `GCG` (Gamrath and Lübbecke, 2010) which is based on `SCIP` (scipopt.org) and available from the same web site.

We would like to emphasize that the thoughts in this subsection significantly expand the applicability of the methods presented in this book. Once the grouping of (variables and) constraints is decided, all reformulations, column generation, later cutting and branching, all "tricks," can be generically implemented. This does not imply that tailoring the approach to a particular application would not help, quite to the contrary. However, this reminds us that, *in principle*, branch-and-price is a general method to solve general MILPs. Whether it is successful or not hinges, in particular, on the question whether we find a *good* decomposition or not.

4.7 Examples

This section presents a selection of examples. In Example 4.1, we formulate the knapsack problem in two different ways, the first having the integrality property whereas the second not. Example 4.2 examines two different models of the cutting stock problem. The first features an *ISP* formulation that does not possess the integrality property whereas the second one does. Both are derived from alternative formulations of the knapsack problem. Example 4.3 presents nine reformulations for the time constrained shortest path problem in which an upper bound on the arrival time is imposed on every node in Example 4.4. This is followed by Example 4.5 where we study the generalized assignment problem and then the multi-commodity maximum flow problem in Example 4.6. In Examples 4.7 to 4.9, we exploit properties of the knapsack problem formulations for three applications. Finally, Example 4.10 revisits the edge coloring problem with two compact formulations.

Example 4.1 ***Integrality property in the knapsack problem***

✎ This example underscores that the integrality property is associated with the *formulation* or *model* used, not with the problem statement.

We analyze the knapsack problem starting with the binary version. To appreciate how the integrality property plays out, we use two integer linear programming formulations one of which being based on network flows.

Generically, we are given a set of m items and a knapsack of capacity W. Let u_i be the utility coefficient for item i and let w_i be its size. We also assume that the sizes and the knapsack capacity are integer numbers. We use a small instance whose data is listed in Table 4.5.

item	i	1	2	3	4
utility	u_i	45	18	20	8
size	w_i	5	3	2	1
capacity	W	7			

Table 4.5: Knapsack capacity and potential items to pack.

Binary knapsack problem

The *binary knapsack problem* (*BKP*) consists of selecting a subset of items to put in the knapsack, i.e., $\sum_{i=1}^{m} w_i > W$, so as to maximize the total utility of the selected items.

Binary linear programming formulation. For $i \in \{1,\ldots,m\}$, let the binary variable x_i take value 1 if item i is selected, 0 otherwise:

$$\begin{aligned} z^{\star}_{ILP} = \max \quad & \sum_{i=1}^{m} u_i x_i \\ \text{s.t.} \quad & \sum_{i=1}^{m} w_i x_i \leq W \\ & x_i \in \{0,1\} \quad \forall i \in \{1,\ldots,m\}. \end{aligned} \tag{4.103}$$

This integer linear program appears as the pricing problem in the One-dimensional cutting stock problem (Example 2.1). The *formulation* (4.103) *for the BKP does not possess the integrality property* because $z^\star_{LP} > z^\star_{ILP}$ for various instances, see Exercise 4.11.

Network arc-flow formulation. The *BKP* can also be formulated as a longest path problem on an acyclic network whose node set depends on the capacity W of the knapsack. In this construction, any arc-flow solution inherently fulfills the capacity restriction which therefore only leaves us to interpret what we take in the knapsack. Figure 4.23 presents such a network for the data in Table 4.5.

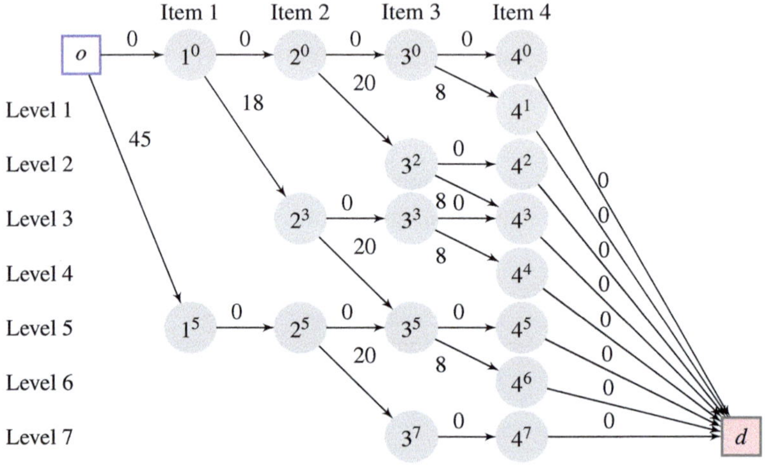

Fig. 4.23: Network for a binary knapsack problem (capacity of 7 and 4 items).

In general, such an acyclic network can be drawn with several columns of nodes. It has one column corresponding to each item as well as two extra columns for a source node o and a sink node d. The column of item i has $W+1$ level-nodes: $i^0, i^1, \ldots, i^W$, where node i^ℓ indicates that the selected items amongst $1,2,\ldots,i$ have consumed $\ell \le W$ units of the available capacity. For each item $i \in \{1,\ldots,m-1\}$, a node i^ℓ has at most two outgoing arcs:

- $(i^\ell, (i+1)^\ell)$ with utility $c_{i^\ell,(i+1)^\ell} = 0$ and item $(i+1)$ is not selected;
- $(i^\ell, (i+1)^{\ell+w_{i+1}})$ with utility $c_{i^\ell,(i+1)^{\ell+w_{i+1}}} = u_{i+1}$ and item $(i+1)$ is selected.

For the last item $i = m$, we connect all the level-nodes m^ℓ to the sink node d with arcs of zero-utility. The source node o has exactly two incident arcs, $(o, 1^0)$ of zero-utility and $(o, 1^{w_1})$ of utility u_1. The network is reduced by removing all the nodes (and their incident arcs) without predecessors. Let it be denoted by $G = (V, A)$ with nodes in $V = N \cup \{o, d\}$, where the set N of appropriate level-nodes are connected by arcs in A. The longest path model of the *BKP* reads as

$$
\begin{aligned}
z_{ILP}^{\star} = \max \quad & \sum_{(i,j)\in A} c_{ij}x_{ij} \\
\text{s.t.} \quad & \sum_{j:(i,j)\in A} x_{ij} - \sum_{j:(j,i)\in A} x_{ji} = \begin{cases} 1 & i=o \\ 0 & \forall i \in N \\ -1 & i=d \end{cases} \\
& x_{ij} \in \{0,1\} \ \forall (i,j) \in A.
\end{aligned}
\tag{4.104}
$$

Formulation (4.104) possesses the integrality property because $z_{LP}^{\star} = z_{ILP}^{\star}$ for any cost coefficients. The binary value x_i, used to determine if item i is selected or not, is computed afterwards and is expressed in terms of the arc-flow variables:

$$
x_i = \sum_{\ell \in \{1,\ldots,W\}:((i-1)^{\ell-w_i},i^{\ell})\in A} x_{(i-1)^{\ell-w_i},i^{\ell}}, \quad \forall i \in \{1,\ldots,m\}, \tag{4.105}
$$

that is, $x_i = 1$ if one of the diagonal arcs from the preceding item $(i-1)$ is used in the *od*-path.

Knapsack problem

The above can be generalized to the *knapsack problem* (*KP*) in which we can place multiple copies of an item in the knapsack. In the following formulation, the integer variable x_i represents the number of times item i is selected:

$$
\begin{aligned}
z_{ILP}^{\star} = \max \quad & \sum_{i=1}^{m} u_i x_i \\
\text{s.t.} \quad & \sum_{i=1}^{m} w_i x_i \le W \\
& x_i \in \mathbb{Z}_+ \quad \forall i \in \{1,\ldots,m\}.
\end{aligned}
\tag{4.106}
$$

The *formulation* (4.106) *for the KP does not possess the integrality property*, a special case being the binary version in (4.103). An alternative linear programming formulation is to use a longest path problem. There are several ways to construct more or less dense networks. Figure 4.24 depicts one such way using the data from Table 4.5. Observe that any *od*-path still provides a capacity feasible knapsack solution.

Given this relatively low-density acyclic network $G = (V,A)$, where $V = N \cup \{o,d\}$ with the appropriate sets of level-nodes N and arc set A. The longest path formulation of the knapsack problem reads as

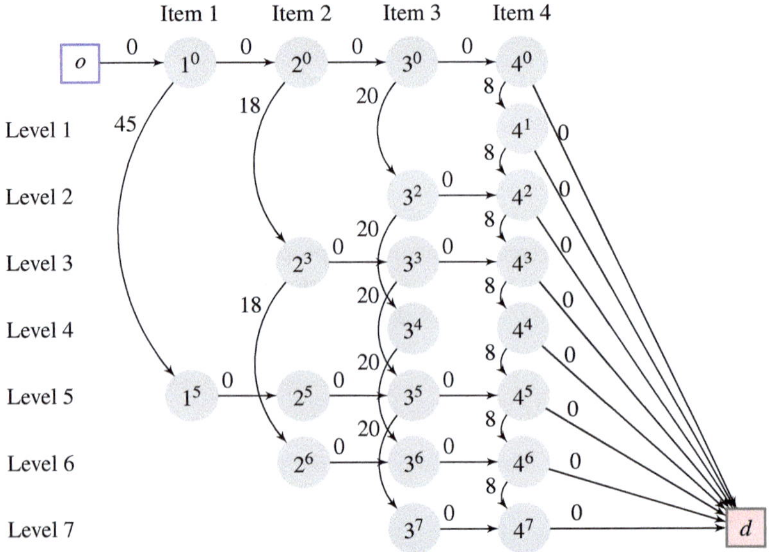

Fig. 4.24: Low-density network for a knapsack problem (capacity of 7 and 4 items).

$$
\begin{aligned}
z^{\star}_{ILP} = \max \quad & \sum_{(i,j)\in A} c_{ij}x_{ij} \\
\text{s.t.} \quad & \sum_{j:(i,j)\in A} x_{ij} - \sum_{j:(j,i)\in A} x_{ji} = \begin{cases} 1 & i = o \\ 0 & \forall i \in N \\ -1 & i = d \end{cases} \\
& x_{ij} \in \{0,1\} \quad \forall (i,j) \in A.
\end{aligned}
\tag{4.107}
$$

This formulation possesses the integrality property because $z^{\star}_{LP} = z^{\star}_{ILP}$, for any set of cost coefficients. As before, the number of times item i is selected in a pattern is computed *a posteriori*, in this case as

$$
x_i = \sum_{\ell:(i^{\ell}, i^{\ell+w_i})\in A} x_{i^{\ell}, i^{\ell+w_i}}, \qquad \forall i \in \{1,\ldots,m\}. \tag{4.108}
$$

There is more to say about these two formulations. Because the first formulation (4.106) does not possess the integrality property while the second (4.107) does, we may think that, in a Dantzig-Wolfe reformulation, the lower bound $z^{\star}_{MP}$ on $z^{\star}_{IMP}$ achieved by formulating the *ISP* with the first will be better than the second one. This is not true because both *ISP*s are *equivalent* integer formulations.

Note 4.19 (High-density network.) Contrary to the *BKP* and Figure 4.23, we can take advantage of the fact that we are allowed to pack the same item several times by discarding the packing ordering implied in Figure 4.24. We can thus define the *KP* by a longest path problem on a much denser multi-arc acyclic network. We see this in Figure 4.25 where we have only one dimension for the knapsack capacity. For each item, we use arcs starting at each capacity consumption which remains feasible. For example, item 2 takes $w_2 = 3$ units and grants utility $u_2 = 18$ until level 4 inclusively. Trivial dominance can then be applied to significantly reduce the number of arcs. This gives the type of networks used by Valério de Carvalho (1999).

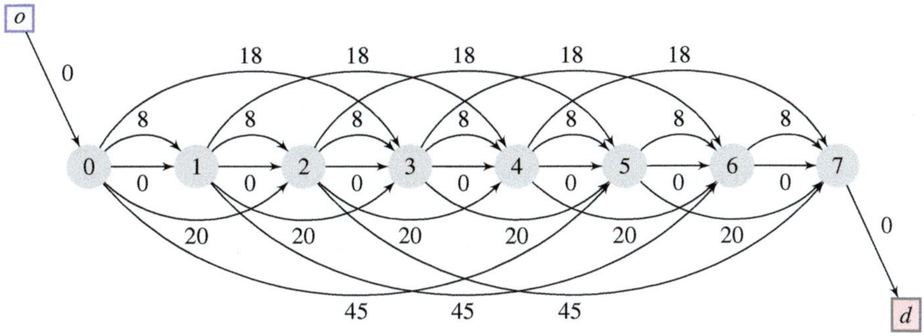

Fig. 4.25: High-density network for a knapsack problem (capacity of 7 and 4 items).

Fig. 4.26: José Manuel Valério de Carvalho (Montréal, Canada, 2023-05-18).

Example 4.2 ***Integrality property in the cutting stock problem***

✎ We here compare the reformulations of two compact formulations for the cutting stock problem (*CSP*). In the first, the *ISP* formulation does not possess the integrality property while it does in the second. They both nevertheless provide the same lower bound on $z^\star_{ILP}$.

We establish in Proposition 4.1 that we hope to exploit integrality in the pricing problem by reaching a better bound than $z^\star_{LP}$. We then establish in Proposition 4.7 that an *ISP* whose formulation has the integrality property only reaches $z^\star_{MP} = z^\star_{LP}$. With this in mind, we consider the *CSP* in which the *ISP* is a knapsack problem. With our new found understanding from Example 4.1 of how the integrality property plays out in the latter, we consider two different compact formulations for the *CSP*. We then perform a Dantzig-Wolfe reformulation on both of them and realize that they reach the same bound. Let us make sense of what appears contradictory. To this end, we recall the formulation seen in Example 2.1:

$$\begin{aligned} z^\star_{IMP} = \min \quad & \sum_{\mathbf{x}\in\mathcal{X}} \lambda_{\mathbf{x}} \\ \text{s.t.} \quad & \sum_{\mathbf{x}\in\mathcal{X}} a_{i\mathbf{x}}\lambda_{\mathbf{x}} \geq b_i \qquad \forall i \in \{1,\ldots,m\} \\ & \lambda_{\mathbf{x}} \in \mathbb{Z}_+ \quad \forall \mathbf{x} \in \mathcal{X}, \end{aligned} \tag{4.109}$$

where $\mathcal{X} = \{\mathbf{x} \in \mathbb{Z}^m_+ \mid \sum_{i=1}^m w_i x_i \leq W\}$ denotes the set of cutting patterns, $\lambda_{\mathbf{x}}$ determines how many times pattern $\mathbf{x}$ is used in a solution, and parameter $a_{i\mathbf{x}} \in \mathbb{Z}_+$ denotes how many times item i is obtained from cutting pattern $\mathbf{x}$.

Weak compact formulation

Let $\mathbf{b} = [b_i]_{i=1,\ldots,m}$ and consider now an alternative formulation for the *CSP* which uses indexed commodities in set K, one per roll. For $k \in K$, define x^k_0 as a binary variable taking value 1 if roll k is used and 0 otherwise, let $\mathbf{x}^k = [x^k_i]_{i=1,\ldots,m}$, where x^k_i denotes the number of times item i is cut in roll k. This formulation, denoted ILP_K, reads as

$$\begin{aligned} z^\star_{ILP_K} = \min \quad & \sum_{k\in K} x^k_0 \\ \text{s.t.} \quad & \sum_{k\in K} \mathbf{x}^k \geq \mathbf{b} \\ & \sum_{i\in\{1,\ldots,m\}} w_i x^k_i \leq W x^k_0 \qquad \forall k \in K \\ & x^k_0 \in \{0,1\} \quad \forall k \in K \\ & x^k_i \in \mathbb{Z}_+ \qquad \forall k \in K,\, i \in \{1,\ldots,m\}. \end{aligned} \tag{4.110}$$

We say that this formulation is weak because its linear relaxation gives very poor information on integer optimality. That is, the optimal number of rolls utilized is trivially equal to the *total item demand* divided by the *width of a roll*, see Exercise 3.16:

$$z^\star_{LP_K} = \frac{1}{W} \sum_{i\in\{1,\ldots,m\}} w_i b_i. \tag{4.111}$$

This lower bound on the number of rolls can be improved in a trivial way to $\lceil z^\star_{LP_K} \rceil$ but also by applying on the ILP_K a Dantzig-Wolfe reformulation with an *ISP* formulation that does not possess the integrality property.

We start by observing that the ILP_K has a block-diagonal structure isomorphic in K which we rewrite as

$$\begin{aligned} z^\star_{ILP_K} = \min \quad & \sum_{k\in K} x^k_0 \\ \text{s.t.} \quad & \sum_{k\in K} \mathbf{x}^k \geq \mathbf{b} \\ & \begin{bmatrix} x^k_0 \\ \mathbf{x}^k \end{bmatrix} \in \mathcal{D}_K \quad \forall k \in K, \end{aligned} \tag{4.112}$$

where

$$\mathcal{D}_K = \left\{ \begin{bmatrix} x_0 \\ \mathbf{x} \end{bmatrix} \in \{0,1\} \times \mathbb{Z}^m_+ \;\middle|\; \sum_{i=1}^{m} w_i x_i \leq W x_0 \right\}. \tag{4.113}$$

$\mathcal{D}_K$ is a polytope such that $\mathcal{X} = \{\mathbf{x}_p\}_{p\in P}$ is only defined by the set of extreme points. The set P denotes the index-set of cutting patterns $\begin{bmatrix} x_{0p} \\ \mathbf{x}_p \end{bmatrix}_{p\in P}$, with $x_{0p} \in \{0,1\}$ and $\mathbf{x}_p = [x_{ip}]_{i=1,\ldots,m}$, where x_{ip} is the number of times item i is cut in pattern p. For $k \in K$, express $\begin{bmatrix} x^k_0 \\ \mathbf{x}^k \end{bmatrix} \in \mathcal{D}_K$ as

$$\begin{bmatrix} \sum_{p\in P} x_{0p}\lambda^k_p \\ \sum_{p\in P} \mathbf{x}_p \lambda^k_p \end{bmatrix} = \begin{bmatrix} x^k_0 \\ \mathbf{x}^k \end{bmatrix}, \quad \sum_{p\in P} \lambda^k_p = 1, \quad \lambda^k_p \geq 0, \forall p \in P, \tag{4.114}$$

and substitute in the objective function and demand constraints of (4.112) to obtain the IMP_K:

$$z^\star_{IMP_K} = \min \quad \sum_{k\in K} \sum_{p\in P} x_{0p} \lambda^k_p \tag{4.115a}$$

$$\text{s.t.} \quad \sum_{k\in K} \sum_{p\in P} \mathbf{x}_p \lambda^k_p \geq \mathbf{b} \tag{4.115b}$$

$$\sum_{p\in P} \lambda^k_p = 1 \qquad \forall k \in K \tag{4.115c}$$

$$\lambda^k_p \geq 0 \qquad \forall k \in K,\ p \in P \tag{4.115d}$$

$$\sum_{p \in P} x_{0p} \lambda_p^k = x_0^k \in \{0,1\} \qquad \forall k \in K \tag{4.115e}$$

$$\sum_{p \in P} \mathbf{x}_p \lambda_p^k = \mathbf{x}^k \in \mathbb{Z}_+^m \qquad \forall k \in K. \tag{4.115f}$$

To eliminate the symmetry due to the identical subproblems, we use the aggregation procedure described on p. 204, where $\lambda_p = \sum_{k \in K} \lambda_p^k$:

$$\begin{aligned} z^\star_{MP_K} = \min \quad & \sum_{p \in P} x_{0p} \lambda_p \\ \text{s.t.} \quad & \sum_{p \in P} \mathbf{x}_p \lambda_p \geq \mathbf{b} \\ & \sum_{p \in P} \lambda_p = |K| \\ & \lambda_p \geq 0 \qquad \forall p \in P. \end{aligned} \tag{4.116}$$

The *ISP* based on $\mathcal{D}_K$ (4.113) is the one already seen in (2.32) where we constructed the zero-cutting pattern by design, see also Not all blocks are used p. 217. The aggregated convexity constraint is redundant because we can always use the zero-cutting pattern (index by $p = \mathbf{0}$) at cost 0 as a slack variable. Furthermore, if at any point we come to generate this pattern, we also know that optimality of the *MP* is reached because $c - \boldsymbol{\pi}^\intercal \mathbf{0} \geq 0$, $\forall c \geq 0$. This means that correctly representing its cost $x_{0p} = 0 \neq 1$ is irrelevant. These comments lead to this simplified formulation identical to the *MP* derived from (4.109)

$$\begin{aligned} z^\star_{MP_K} = \min \quad & \sum_{p \in P} \lambda_p \\ \text{s.t.} \quad & \sum_{p \in P} \mathbf{x}_p \lambda_p \geq \mathbf{b} \\ & \lambda_p \geq 0 \quad \forall p \in P. \end{aligned} \tag{4.117}$$

Finally, because $\mathcal{D}_K$ does not possess the integrality property, one expects a potentially better lower bound $z^\star_{MP_K}$ than the one evaluated by $z^\star_{LP_K}$, i.e.,

$$z^\star_{LP_K} \leq z^\star_{MP_K} \leq z^\star_{ILP_K}. \tag{4.118}$$

Network-based compact formulation

We next examine an integer linear programming compact formulation for the *CSP* with a network flow structure, denoted ILP_{NF}. Additionally, the cutting patterns are associated with extreme rays rather than extreme points. In this case, a Dantzig-Wolfe reformulation also leads to the linear relaxation of the *IMP* (4.109), thus obtained in a simpler way compared to the reformulation we have derived starting with the weak compact formulation (4.110).

Consider the network in Figure 4.24, where $V = N \cup \{o, d\}$ with the appropriate sets of level nodes N and the arc set A used for the longest path model of the knapsack problem (4.107). We add arc (d, o) to A to form a circulation network denoted $G_{do} = (V, A_{do})$, where $A_{do} = A \cup \{(d, o)\}$, see Figure 4.27.

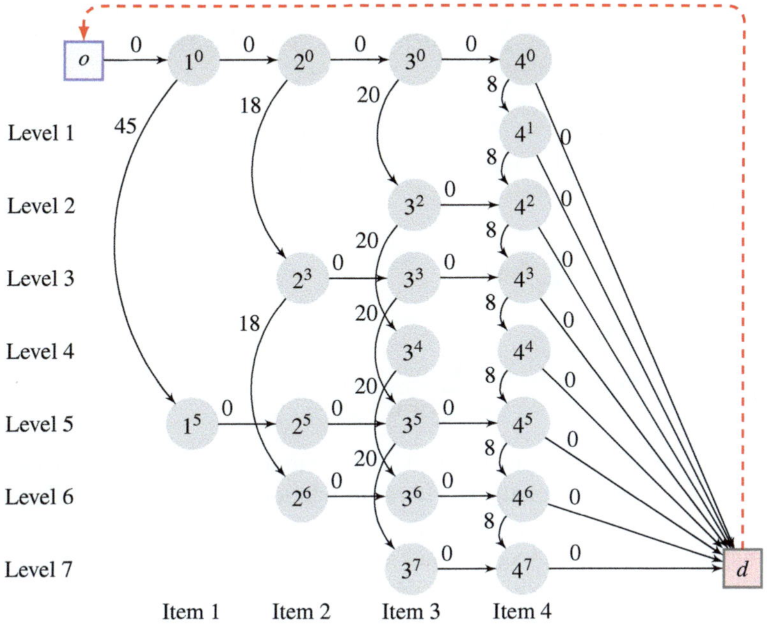

Fig. 4.27: Network G_{do} for the formulation ILP_{NF} (capacity of 7 and 4 items).

Let $\mathbf{x} = [x_{ij}]_{(i,j) \in A_{do}}$ be a vector of non-negative integer flow variables. The network-based formulation of the *CSP* is composed of the objective function counting the total number of patterns used on arc (d, o), the requested demands b_i, $i \in \{1, \dots, m\}$, and the flow conservation constraints on all nodes of V. From (4.108), we have

$$x_i = \sum_{\ell:(i^\ell, i^{\ell+w_i}) \in A_{do}} x_{i^\ell, i^{\ell+w_i}} \tag{4.119}$$

and the compact formulation ILP_{NF} of the integer linear program is

$$\begin{aligned} z^\star_{ILP_{NF}} = \min \quad & x_{do} \\ \text{s.t.} \quad & \sum_{\ell:(i^\ell, i^{\ell+w_i}) \in A_{do}} x_{i^\ell, i^{\ell+w_i}} \geq b_i \quad \forall i \in \{1, \dots, m\} \\ & \sum_{j:(i,j) \in A_{do}} x_{ij} - \sum_{j:(j,i) \in A_{do}} x_{ji} = 0 \quad \forall i \in V \\ & x_{ij} \in \mathbb{Z}_+ \quad \forall (i, j) \in A_{do}. \end{aligned} \tag{4.120}$$

To replicate the network-based knapsack subproblem, we use the grouping

$$\mathcal{A}_{NF} = \left\{ \mathbf{x} \in \mathbb{Z}_+^{|A_{do}|} \,\middle|\, \sum_{\ell:(i^\ell, i^{\ell+w_i}) \in A_{do}} x_{i^\ell, i^{\ell+w_i}} \geq b_i, \ \forall i \in \{1,\dots,m\} \right\} \tag{4.121a}$$

$$\mathcal{D}_{NF} = \left\{ \mathbf{x} \in \mathbb{Z}_+^{|A_{do}|} \,\middle|\, \sum_{j:(i,j)\in A_{do}} x_{ij} - \sum_{j:(j,i)\in A_{do}} x_{ji} = 0, \ \forall i \in V \right\}. \tag{4.121b}$$

Observe that $\mathbf{x} \in \mathbb{Z}_+^{|A_{do}|}$ in (4.121b) defines an *integer* cutting pattern. Given the dual values $\pi_i \geq 0$, $i \in \{1,\dots,m\}$, for the m constraints in $\mathcal{A}_{NF}$, the formulation of the *ISP* is given by

$$\begin{aligned} \min_{\mathbf{x}\in\mathcal{D}_{NF}} \quad & c(\mathbf{x}) - \sum_{i=1}^{m} \pi_i a_i(\mathbf{x}) \\ & c(\mathbf{x}) = x_{do} \\ & a_i(\mathbf{x}) = x_i = \sum_{\ell:(i^\ell, i^{\ell+w_i})\in A} x_{i^\ell, i^{\ell+w_i}} \quad \forall i \in \{1,\dots,m\}. \end{aligned} \tag{4.122}$$

Because the formulation of the *ISP* (4.122) is a circulation problem, it possesses the integrality property. A Dantzig-Wolfe reformulation IMP_{NF} of the ILP_{NF} (4.120) is constructed as follows. Any $\mathbf{x} \in \mathcal{D}_{NF}$ can be expressed as a convex combination of the unique extreme point $\mathbf{x_0} = \mathbf{0}$ and a conic combination of the extreme rays $\mathbf{x}_r$, $r \in R$. Therefore, the convexity constraint $\lambda_\mathbf{0} = 1$ is discarded from the reformulation of the IMP_{NF} and as $\lambda_\mathbf{0}$ does not contribute to the objective function nor the demand constraints, we have

$$\sum_{r\in R} \mathbf{x}_r \lambda_r = \mathbf{x}, \qquad \lambda_r \geq 0, \forall r \in R. \tag{4.123}$$

The representation of an extreme ray $\mathbf{x}_r$, indeed a cutting pattern, can be done with a single unit of flow in the circulation network G_{do}, yielding by the flow conservation constraints, $c_r = x_{do,r} = 1$ for all $r \in R$. The substitution results in

$$z^\star_{IMP_{NF}} = \min \quad \sum_{r\in R} \lambda_r \tag{4.124a}$$

$$\text{s.t.} \quad \sum_{r\in R} a_{ir}\lambda_r \geq b_i \qquad \forall i \in \{1,\dots,m\} \tag{4.124b}$$

$$\lambda_r \geq 0 \qquad \forall r \in R \tag{4.124c}$$

$$\sum_{r\in R} x_{ij,r}\lambda_r = x_{ij} \in \mathbb{Z}_+ \qquad \forall (i,j) \in A_{do}, \tag{4.124d}$$

where

$$a_{ir} = x_{ir} = \sum_{\ell:(i^\ell, i^{\ell+w_i})\in A} x_{i^\ell, i^{\ell+w_i}, r} \tag{4.125}$$

encodes the number of times item i is obtained in pattern r. Taking the linear relaxation while relaxing the last constraints involving both the x- and λ-variables that are no longer needed leads to the linear relaxation of the *IMP* (4.109), where the zero-cutting pattern is explicitly excluded, i.e., $R \equiv P \setminus \{\mathbf{0}\}$.

$$\begin{aligned} z^\star_{MP_{NF}} = \min \quad & \sum_{r\in R} \lambda_r \\ \text{s.t.} \quad & \sum_{r\in R} a_{ir}\lambda_r \geq b_i \quad \forall i \in \{1,\dots,m\} \\ & \lambda_r \geq 0 \quad \forall r \in R. \end{aligned} \tag{4.126}$$

Comparison of the four linear relaxations

The linear programs MP_K (4.117) and MP_{NF} (4.126) being exactly the same, we arrive at the inevitable conclusion that they reach the same lower bound on the optimal number of cutting patterns, i.e.,

$$z^\star_{LP_K} \leq z^\star_{MP_K} = z^\star_{MP_{NF}} = z^\star_{LP_{NF}}, \tag{4.127}$$

where the last equality comes from the integrality property in the formulation of the *ISP* (4.122) defined on $\mathcal{D}_{NF}$ while the inequality comes from the general result of Proposition 4.1 for a Dantzig-Wolfe reformulation. In fact, one typically expects $z^\star_{LP_K} \ll z^\star_{MP_K}$ as per the weakness of the information provided by the linear relaxation. For example, consider an instance with an integer demand b for a single item of width $w = \frac{W}{2} + \varepsilon$ such that no two items can be cut in a roll. The optimal number of rolls is $z^\star_{ILP_K} = b$ while we have from (4.111)

$$z^\star_{LP_K} = (\frac{W}{2} + \varepsilon)\frac{b}{W} = \frac{b}{2} + \frac{\varepsilon b}{W} \approx \frac{b}{2} = \frac{z^\star_{ILP_K}}{2} \tag{4.128}$$

for a small $\varepsilon > 0$. That is an integrality gap of 50 % in strong contrast with that of zero for the linear relaxation of the extended formulations (IMP_K and IMP_{NF}) which is even integer optimal. The cutting stock problem is one of the few for which the following definition is applicable.

Definition 4.1. (Baum, 1977; Baum and Trotter, 1982) Given $\mathbf{A} \in \mathbb{Z}^{m\times n}_+$ and $\mathbf{b} \in \mathbb{Z}^m_+$, an integer linear program of the form $\min\{\mathbf{1}^\mathsf{T}\mathbf{x} \mid \mathbf{A}\mathbf{x} \geq \mathbf{b}, \mathbf{x} \in \mathbb{Z}^n_+\}$ is said to have the *integer round-up property* if the integrality gap is smaller than one, i.e., $z^\star_{ILP} = \lceil z^\star_{LP} \rceil$.

Although the integer round-up property is often observed for the presented extended formulations of this problem, it is not always the case. However, Scheithauer and Terno (1995) conjecture that the *modified* integer round-up property, in which the integrality gap difference is smaller than two ($z^\star_{ILP} \leq \lceil z^\star_{LP} \rceil + 1$), always holds. Rietz et al. (2002) present various families of instances and construct an instance with the largest known gap of 7/6. Figure 4.28 summarizes these results.

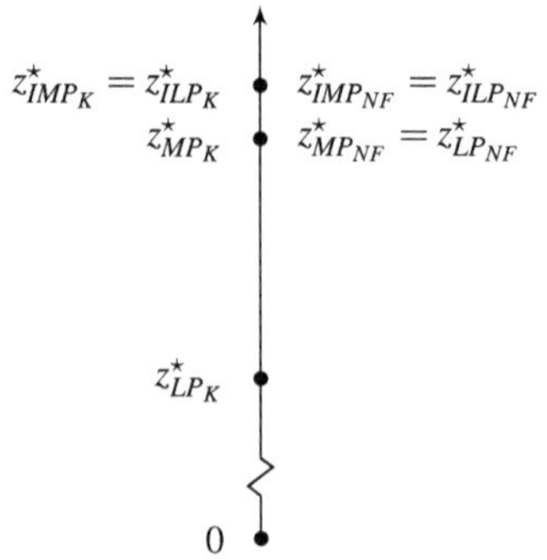

Fig. 4.28: Lower bounds on the minimum number of rolls.

Reverse Dantzig-Wolfe

Performing a Dantzig-Wolfe reformulation on the ILP_K (4.112) with $|K|$ identical blocks leads us back to the *IMP* (4.109). This should be no surprise considering we can also reverse the latter using Proposition 4.15. The alternative network-based compact formulation ILP_{NF} (4.120) which also yields the same *IMP* (4.109) allows us to recall Proposition 4.16.

Furthermore, if we are given the classical formulation *IMP* and the domain of the pricing problem is formulated as an od-path on G (rather than a circulation on G_{do}), that is, the grouping in (4.121) is defined on arc set A as

$$\mathcal{A} = \left\{ \mathbf{x} \in \mathbb{Z}_+^{|A|} \;\middle|\; \sum_{\ell:(i^\ell, i^{\ell+w_i}) \in A} x_{i^\ell, i^{\ell+w_i}} \geq b_i, \; \forall i \in \{1, \ldots, m\} \right\} \tag{4.129a}$$

$$\mathcal{D} = \left\{ \mathbf{x} \in \mathbb{Z}_+^{|A|} \;\middle|\; \sum_{j:(i,j) \in A} x_{ij} - \sum_{j:(j,i) \in A} x_{ji} = \left\{ \begin{array}{rl} 1 & \text{for } i = o \\ 0 & \forall i \in N \\ -1 & \text{for } i = d \end{array} \right. \right\}, \tag{4.129b}$$

then, Proposition 4.17 can be used because

- the *ISP* formulation possesses the integrality property and
- the zero-vector is not a solution to the covering constraints (i.e., $\mathbf{0} \notin \mathcal{A}$).

Finally, observe that variable x_0 is the flow variable x_{do}, that is,

$$\mathcal{D}(x_{do}) = \left\{ \begin{bmatrix} x_{do} \\ \mathbf{x} \end{bmatrix} \in \mathbb{Z}_+^{n+1} \;\middle|\; \sum_{j:(i,j) \in A} x_{ij} - \sum_{j:(j,i) \in A} x_{ji} = \left\{ \begin{array}{rl} x_{do} & \text{for } i = o \\ 0 & \forall i \in N \\ -x_{do} & \text{for } i = d \end{array} \right. \right\} \tag{4.130}$$

and any upper bounding constraint such as $\mathbf{x} \leq \mathbf{1} x_{do}$ is absent from (4.130) as $x_{do} = 0 \Rightarrow \mathbf{x} = \mathbf{0}$. Indeed, we once again derive back the network-based compact formulation ILP_{NF} (4.120).

Example 4.3 TCSPP: nine reformulations

✎ Given the original *ILP*, we examine different subproblems that give various lower bounds $\lceil z^\star_{MP} \rceil$ on $z^\star_{ILP}$.

Let us observe the impact of various groupings for $\mathcal{A}$ and $\mathcal{D}$ on the optimal objective value of the *MP*. To achieve this, we base ourselves on the *TCSPP* seen in Example 3.2. Recall the formulation of the *ILP* as

$$z^\star_{ILP} = \min \sum_{(i,j)\in A} c_{ij}x_{ij} \tag{4.131a}$$

$$\text{s.t.} \quad \sum_{j:(1,j)\in A} x_{1j} = 1 \qquad [\sigma_1] \tag{4.131b}$$

$$\sum_{j:(i,j)\in A} x_{ij} - \sum_{j:(j,i)\in A} x_{ji} = 0 \qquad [\sigma_i] \quad \forall i \in \{2,\ldots,5\} \tag{4.131c}$$

$$-\sum_{i:(i,6)\in A} x_{i6} = -1 \qquad [\sigma_6] \tag{4.131d}$$

$$\sum_{(i,j)\in A} t_{ij}x_{ij} \leq 14 \qquad [\sigma_7] \tag{4.131e}$$

$$x_{ij} \in \{0,1\} \qquad \forall (i,j) \in A. \tag{4.131f}$$

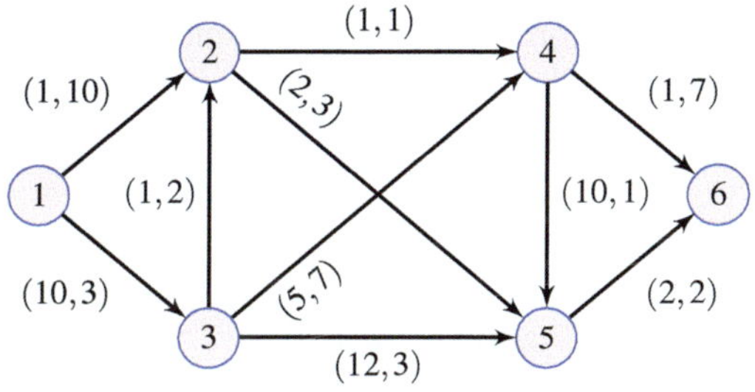

Fig. 4.29: Network $G = (N,A)$ with (c_{ij},t_{ij}) values, $\forall (i,j) \in A$.

We already know that our initial choice of $\mathcal{D}$ in Example 3.2, here denoted $\mathcal{D}_1$, expresses a path structure from node 1 to 6 on the network G of Figure 4.29:

$$\mathcal{A}_1 = \{\mathbf{x} \in \{0,1\}^{|A|} \mid (4.131e)\} \tag{4.132a}$$

$$\mathcal{D}_1 = \{\mathbf{x} \in \{0,1\}^{|A|} \mid (4.131b)\text{–}(4.131d)\}. \tag{4.132b}$$

The corresponding lower bound on $z^\star_{ILP} = 13$ only reaches $z^\star_{MP} = z^\star_{LP} = 7$ because the associated *ISP* formulation is the shortest path network flow formulation (3.84) which possesses the integrality property.

We propose the following eight alternatives for $\mathcal{D}$, where we use different subsets of constraints and sometimes combine them with constraints implied from the original *ILP* (Exercise 4.13 asks for the corresponding sets for $\mathcal{A}$):

$$\mathcal{D}_2 = \mathcal{D}_1 \cap \{3 \le \sum_{(i,j)\in A} x_{ij} \le 5\} \qquad \mathcal{D}_3 = \{\mathbf{x} \in \{0,1\}^{|A|} \mid \sum_{(i,j)\in A} t_{ij}x_{ij} \le 14\}$$

$$\mathcal{D}_4 = \mathcal{D}_3 \cap \{3 \le \sum_{(i,j)\in A} x_{ij} \le 5\} \qquad \mathcal{D}_5 = \mathcal{D}_3 \cap \{\sum_{j=2}^{3} x_{1j} = \sum_{i=4}^{5} x_{i6} = 1\}$$

$$\mathcal{D}_6 = \mathcal{D}_4 \cap \mathcal{D}_5 \qquad \mathcal{D}_7 = \mathcal{D}_6 \cap \{\sum_{i:(i,j)\in A} x_{ij} \le 1, \forall j \in \{2,\ldots,5\}\}$$

$$\mathcal{D}_8 = \mathcal{D}_6 \cap \{\sum_{j:(i,j)\in A} x_{ij} \le 1, \forall i \in \{2,\ldots,5\}\} \quad \mathcal{D}_9 = \{\mathbf{x} \in \{0,1\}^{|A|} \mid (4.131b)\text{–}(4.131e)\}.$$

- $\mathcal{D}_2$ is a restriction of $\mathcal{D}_1$ obtained by forcing the total number of selected arcs to be between three and five.
- $\mathcal{D}_3$ is defined by the duration constraint over binary variables.
- $\mathcal{D}_4$ is a restriction of $\mathcal{D}_3$ obtained by again limiting the number of selected arcs.
- $\mathcal{D}_5$ is another restriction of $\mathcal{D}_3$ this time obtained by imposing the selection of one arc from the source node and one to the sink node.
- $\mathcal{D}_6$ combines the two previous conditions with the intersection of $\mathcal{D}_4$ and $\mathcal{D}_5$.
- $\mathcal{D}_7$ and $\mathcal{D}_8$ are two different restrictions of $\mathcal{D}_6$: for $\mathcal{D}_7$, an additional constraint for every node 2 to 5 enforces a limit of one unit on its inflows; $\mathcal{D}_8$ is defined similarly using the outflows.
- $\mathcal{D}_9$ corresponds to the extreme case where *all constraints* appear in the *ISP*.

Table 4.6 presents the size of the *MP* in terms of the number of rows times the number of columns, the optimal objective value $z^\star_{MP}$, and its rounded up value for these nine variations of $\mathcal{D}$ (see also Figure 4.30).

$\mathcal{D}$	*MP* size	$z^\star_{MP}$	$\lceil z^\star_{MP} \rceil$
$\mathcal{D}_1$	2×9	7	7
$\mathcal{D}_2$	2×9	7	7
$\mathcal{D}_3$	7×278	9	9
$\mathcal{D}_4$	7×217	9	9
$\mathcal{D}_5$	5×55	9	9
$\mathcal{D}_6$	5×47	9	9
$\mathcal{D}_7$	5×32	13	13
$\mathcal{D}_8$	5×31	11.25	12
$\mathcal{D}_9$	1×6	13	13

Table 4.6: Lower bounds $z^\star_{MP}$ and $\lceil z^\star_{MP} \rceil$ on $z^\star_{ILP}$ for nine variations of $\mathcal{D}$.

For $\mathcal{D}_1$, we have already seen that the *MP* (3.81) comprises 2 rows and 9 path-variables (or 9 extreme points satisfying the path-flow constraints) with $z^\star_{MP} = 7$. The set $\mathcal{D}_2$ has identical measures as its sibling $\mathcal{D}_1$. This could have been predicted since the added constraints $3 \leq \sum_{(i,j)\in A} x_{ij} \leq 5$ are redundant with respect to the network structure, that is, all 9 paths are in this range.

For $\mathcal{D}_3$, the *ISP* uses a formulation for solving a binary knapsack problem that does not possess the integrality property (see p. 239): the lower bound increases from 7 to 9. There are 278 extreme points (again, 10-dimensional vectors): solving the 7×278 *MP* (one convexity constraint and six flow conservation equations) results in an optimal fractional solution with an objective value $z^\star_{MP} = 9$. For $\mathcal{D}_4$, the number of selected arcs indeed has an influence on the feasible set. Despite the reduction of the feasible space, the lower bound unfortunately remains at 9.

For the sets $\mathcal{D}_5$ and $\mathcal{D}_6$, an optimal fractional solution is found for both *MP*s, also with $z^\star_{MP} = 9$. The sets $\mathcal{D}_7$ and $\mathcal{D}_8$ respectively yield lower bounds of 13 and 11.25 which is integer optimal for the former and rounds up to one unit below the optimal objective value for the latter.

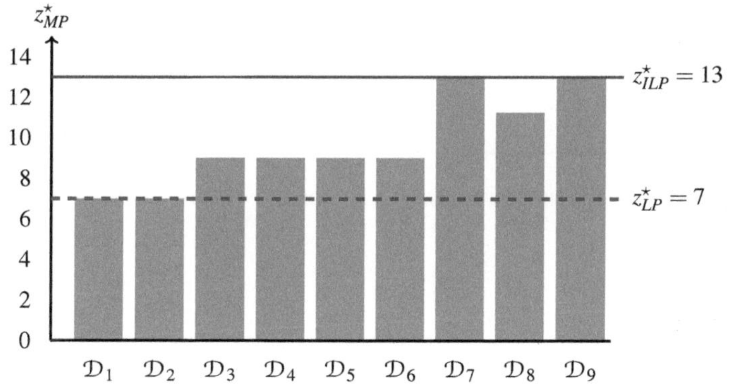

Fig. 4.30: Lower bound $z^\star_{MP}$ on $z^\star_{ILP}$ for various definitions of the set $\mathcal{D}$.

It is a good place to remind the reader of what we discussed *on grouping the constraints* (p. 115). The constraints $3 \leq \sum_{(i,j)\in A} x_{ij} \leq 5$ are redundant with respect to the original *ILP*. Adding them in the set $\mathcal{D}_2$ has no influence yet it does reduce the number of extreme points in $\mathcal{D}_4$. Unfortunately, this does not translate in an improved bound. In contrast, the inflow bound constraint in $\mathcal{D}_7$ allows us to reach the optimal integer bound using only five constraints in the *MP*, that is, (4.131c) plus the convexity constraint.

Finally, what if we restrict solutions in $\mathcal{D}$ to satisfy $4 \leq \sum_{(i,j)\in A} x_{ij} \leq\ 4$. Intuitively, this amounts to discarding short paths (arc-length 3) that are likely too expensive (as *expressways with tolls*) to be optimal and long paths (arc-length 5) that are likely unable to meet the time restriction. Whether such a guess compromises the *IMP* or not is of course hard to predict in general which reminds us of the heuristic pricing section seen in column generation.

As a case in point, observe that we are lucky in discarding path 1-2-4-6 of length 3, cost 3, and time 18 since it is being discarded for the "wrong" reason. In contrast, we are unlucky with path 1-2-5-6 of length 3, cost 5, and time 15 because it is needed to reach optimality for the *MP*. However, since the integer optimal path is correctly kept, we nonetheless terminate with a correct solution. To sum up, any rule of thumb should *efficiently* filter out objects and, if it is never turned off, must *empirically* observe no false negatives with respect to integrality. Table 4.7 classifies the nine paths according to this rule from the point of view of the *MP* and *IMP*. We have four indicators 'N', 'P', 'FN', and 'FP' which respectively express that a path is correctly discarded, correctly kept, mistakenly discarded, and mistakenly kept.

path	length	cost	time	*MP*	*IMP*
1246	3	3	18	N	N
1256	3	5	15	FN	N
12456	4	14	14	FP	FP
13246	4	13	13	FP	P
13256	4	15	10	P	FP
132456	5	24	9	N	N
1346	3	16	17	N	N
13456	4	27	13	FP	FP
1356	3	24	8	N	N

Table 4.7: False negatives and false positives from a rule of thumb.

*Example 4.4 TCSPP: **time bounding***

If you don't manage the important constraints in the *ISP*, here the time component, the reformulation is useless.

Consider again the *TCSPP* from Example 3.2. We already have an upper bound of 14 units of time on node 6 within an optimal path from the origin to the destination. We now also impose an upper bound on all other nodes. These bounds, denoted u_i, $i \in \{1,\dots,6\}$, are respectively 0, 9, 3, 5, 12, and 14 (see Figure 4.31).

Note 4.20 exposes simple ideas to filter out *obvious* portions of the input that cannot be part of an optimal integer solution. For pedagogical reasons, we later assume not to perform such a preprocessing step and examine some compact formulations and reformulations.

Note 4.20 (Preprocessing.) Looking at arc $(1,2)$, we see that the travel time ($t_{12} = 10$) is larger than the upper bound at the arrival node ($u_2 = 9$): this arc cannot be used in any feasible integer solution. Indeed, every *arc* (i, j) for which $t_{ij} > u_j$ can be discarded, so is the case for $(3,4)$. We can also eliminate every *node* for which

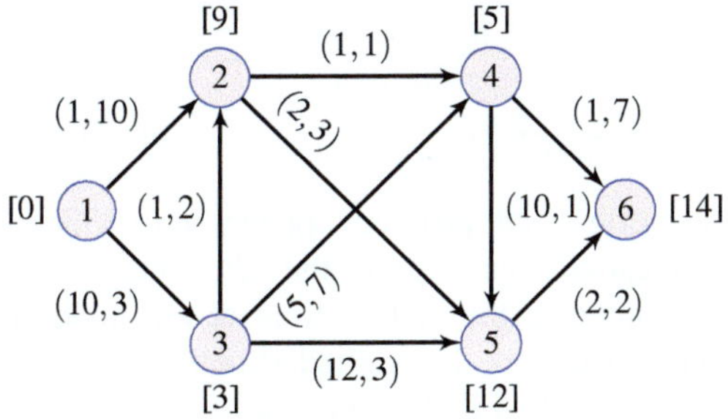

Fig. 4.31: Network with (c_{ij}, t_{ij}) values, $\forall (i, j) \in A$, and upper bounds $[u_i]$, $\forall i \in N$.

the earliest arrival time exceeds its upper bound: node 4 is discarded. This can be seen by computing the earliest arrival times by solving a shortest time path tree from node 1. After this preprocessing is performed, we get the simplified network seen in Figure 4.32 which contains only two paths, the optimal one being 13256 with a cost of 15.

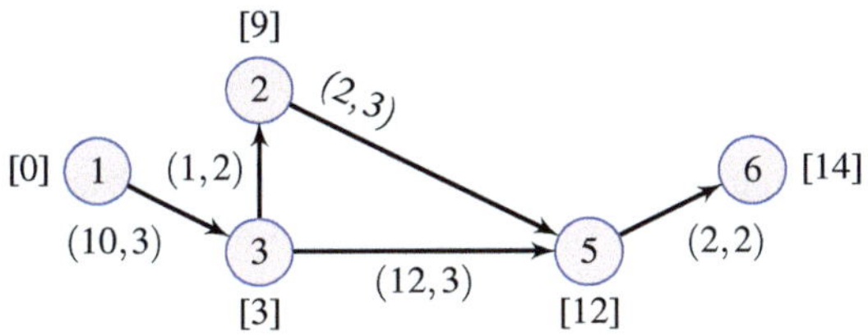

Fig. 4.32: Remaining network after preprocessing.

A non-linear program *NLP* has been proposed by Desrosiers et al. (1983). It involves the usual arc-flow variables $\mathbf{x} = [x_{ij}]_{(i,j)\in A}$ and the time variables $\mathbf{t} = [t_i]_{i\in N}$:

$$z^{\star}_{NLP} = \min \sum_{(i,j)\in A} c_{ij} x_{ij} \tag{4.133a}$$

$$\text{s.t.} \sum_{j:(1,j)\in A} x_{1j} = 1 \tag{4.133b}$$

$$\sum_{j:(i,j)\in A} x_{ij} - \sum_{j:(j,i)\in A} x_{ji} = 0 \qquad \forall i \in \{2,\ldots,5\} \tag{4.133c}$$

$$\sum_{i:(i,6)\in A} x_{i6} = 1 \tag{4.133d}$$

$$x_{ij} \in \{0,1\} \qquad \forall (i,j) \in A \tag{4.133e}$$

$$t_1 = 0 \tag{4.133f}$$

$$0 \leq t_i \leq u_i(\sum_{j:(j,i)\in A} x_{ji}) \qquad \forall i \in \{2,\ldots,6\} \tag{4.133g}$$

$$x_{ij}(t_i + t_{ij} - t_j) \leq 0 \qquad \forall (i,j) \in A. \tag{4.133h}$$

A formulation of the shortest path problem appears in (4.133a)–(4.133e), where the flow variables are requested binary due to the additional constraints in the model. The time variables computing the arrival time at the nodes follow in (4.133f)–(4.133g), where a node i that is not visited (binary in-flow $\sum_{j:(j,i)\in A} x_{ji} = 0$) is assigned a zero value ($t_i = 0$). Finally, we find in (4.133h) the precedence constraints on each arc, a non-linear representation of $x_{ij} = 1 \Rightarrow t_i + t_{ij} \leq t_j$ (otherwise $x_{ij} = 0$ and the non-linear expression is trivially satisfied). Dynamic programming algorithms are in practice used for solving such time window constrained shortest path problems using 2-dimensional labels, that is, time and cost. More about this is presented in Chapter 5 (Vehicle Routing and Crew Scheduling Problems).

To apply a Dantzig-Wolfe reformulation based on (4.133), constraints (4.133h) are linearized as

$$t_i + t_{ij} \leq t_j + M(1 - x_{ij}), \qquad \forall (i,j) \in A, \tag{4.134}$$

where M is a large constant. Let the *ILP* be given by (4.133a)–(4.133g) and (4.134). As expected from using a big-M model, even the tightest value typically leads to a poor linear relaxation which only gets even poorer with larger values. Table 4.8 presents some increasing relatively large values for M: we observe that $z^\star_{LP}$ decreases as more flow goes on path 1246 of attractive cost 3 but infeasible duration 18.

$z^\star_{LP}$	3.900	3.422	3.205	3.040	3.020
big-M	25	50	100	500	1000

Table 4.8: Optimal objective value $z^\star_{LP}$ as a function of big-M.

That is to say that with the grouping of constraints as

$$\mathcal{A} = \left\{ \begin{bmatrix} \mathbf{x} \\ \mathbf{t} \end{bmatrix} \in \{0,1\}^{|A|} \times \mathbb{R}_+^{|N|} \;\middle|\; \text{(4.133f)–(4.133g) and (4.134)} \right\} \tag{4.135a}$$

$$\mathcal{D} = \left\{ \begin{bmatrix} \mathbf{x} \\ \mathbf{t} \end{bmatrix} \in \{0,1\}^{|A|} \times \mathbb{R}_+^{|N|} \;\middle|\; \text{(4.133b)–(4.133d)} \right\}, \tag{4.135b}$$

the *ISP* formulation on set $\mathcal{D}$ is a shortest path problem *without time constraints*, as such it possesses the integrality property but the quality of $z^\star_{MP} = z^\star_{LP}$ in the reformulation decreases as M increases. Although we know that this grouping is certainly not the best one, let us take a look at the *IMP* and *ISP*.

For convenience, we first re-write constraints (4.133g) and (4.134) as

$$t_i - u_i(\sum_{j:(j,i)\in A} x_{ji}) \leq 0 \qquad \forall i \in \{2,\ldots,6\} \tag{4.136}$$

$$t_i - t_j + Mx_{ij} \leq M - t_{ij} \qquad \forall (i,j) \in A. \tag{4.137}$$

Secondly, because the time variables are not reformulated in $\mathcal{D}$, they appear as *static* ones in the *IMP* together with the nine path-variables λ_p, $p \in P$, where P is the index set of the extreme points of $\mathrm{conv}(\mathcal{D})$:

$$\begin{aligned}
z^\star_{IMP} = \min \quad & \sum_{p\in P} c_p\lambda_p \\
\text{s.t.} \quad & t_1 = 0 \\
& t_i + \sum_{p\in P} a_{ip}\lambda_p \leq 0 && [\alpha_i \leq 0] \quad \forall i \in \{2,\ldots,6\} \\
& t_i - t_j + \sum_{p\in P} b_{ijp}\lambda_p \leq M - t_{ij} && [\beta_{ij} \leq 0] \quad \forall (i,j) \in A \\
& \sum_{p\in P} \lambda_p = 1 && [\pi_0 \in \mathbb{R}] \\
& \lambda_p \geq 0 && \forall p \in P \\
& \sum_{p\in P} x_{ijp}\lambda_p = x_{ij} \in \{0,1\} && \forall (i,j) \in A.
\end{aligned} \tag{4.138}$$

The dual variables $\boldsymbol{\alpha} = [\alpha_i]_{i=2,\ldots,6}$, $\boldsymbol{\beta} = [\beta_{ij}]_{(i,j)\in A}$, and π_0 are involved in the *ISP* given as

$$\bar{c}(\boldsymbol{\alpha},\boldsymbol{\beta},\pi_0) = -\pi_0 + \min_{\mathbf{x}\in\mathcal{D}} \quad c_{\mathbf{x}} + \sum_{i=2}^{6} \alpha_i a_{i\mathbf{x}} + \sum_{(i,j)\in A} \beta_{ij} b_{ij\mathbf{x}} \tag{4.139a}$$

$$\text{s.t.} \quad c_{\mathbf{x}} = \sum_{(i,j)\in A} c_{ij}x_{ij} \tag{4.139b}$$

$$a_{i\mathbf{x}} = -u_i(\sum_{j:(j,i)\in A} x_{ji}) \qquad \forall i \in \{2,\ldots,6\} \tag{4.139c}$$

$$b_{ij\mathbf{x}} = Mx_{ij} \qquad \forall (i,j) \in A. \tag{4.139d}$$

Even if the travel times are integer, there is no guarantee that optimal time variables take integer values. These can be computed a posteriori for the optimal path 13256 using the precedence relations, e.g., the earliest arrival times, or a small cost can be imposed in the objective function as in min $0.01\sum_{i\in N} t_i + \sum_{(i,j)\in A} c_{ij}x_{ij}$.

In a second attempt, we rather use the compact formulation

$$z^\star_{ILP} = \min \quad \sum_{(i,j)\in A} c_{ij}x_{ij} \tag{4.140a}$$

$$\text{s.t.} \quad \sum_{j:(1,j)\in A} x_{1j} = 1 \tag{4.140b}$$

$$\sum_{j:(i,j)\in A} x_{ij} - \sum_{j:(j,i)\in A} x_{ji} = 0 \qquad \forall i \in \{2,\ldots,5\} \tag{4.140c}$$

$$\sum_{i:(i,6)\in A} x_{i6} = 1 \tag{4.140d}$$

$$x_{ij} \in \{0,1\} \qquad \forall (i,j) \in A \tag{4.140e}$$

$$t_1 = 0 \tag{4.140f}$$

$$0 \le t_i \le M(\sum_{j:(j,i)\in A} x_{ji}) \qquad \forall i \in \{2,\ldots,6\} \tag{4.140g}$$

$$t_i + t_{ij} \le t_j + M(1 - x_{ij}) \qquad \forall (i,j) \in A \tag{4.140h}$$

$$t_i \le u_i \qquad \forall i \in \{2,\ldots,6\}, \tag{4.140i}$$

where (4.140g) is later used in the *ISP* to impose a zero time value on nodes that are not visited while (4.140i) appears in the *IMP* as upper bounds. The impact of $M \ge 25$ on the objective value of the linear relaxation is similar to that of the first model, see Table 4.9:

$z^\star_{LP}$	3.800	3.400	3.200	3.040	3.020
big-M	25	50	100	500	1000

Table 4.9: Optimal objective value $z^\star_{LP}$ as a function of big-M (version 2).

With the grouping given as

$$\mathcal{A} = \left\{ \begin{bmatrix} \mathbf{x} \\ \mathbf{t} \end{bmatrix} \in \{0,1\}^{|A|} \times \mathbb{R}_+^{|N|} \;\middle|\; (4.140\text{i}) \right\} \tag{4.141a}$$

$$\mathcal{D} = \left\{ \begin{bmatrix} \mathbf{x} \\ \mathbf{t} \end{bmatrix} \in \{0,1\}^{|A|} \times \mathbb{R}_+^{|N|} \;\middle|\; (4.140\text{b})\text{–}(4.140\text{d}),\ (4.140\text{f})\text{–}(4.140\text{h}) \right\}, \tag{4.141b}$$

the *IMP* reads as

$$z^\star_{IMP} = \min \sum_{p\in P} c_p \lambda_p \tag{4.142a}$$

$$\text{s.t.} \sum_{p\in P} t_{ip}\lambda_p \le u_i \qquad [\pi_i \le 0] \quad \forall i \in \{2,\ldots,6\} \tag{4.142b}$$

$$\sum_{p\in P} \lambda_p = 1 \qquad [\pi_0 \in \mathbb{R}] \tag{4.142c}$$

$$\lambda_p \ge 0 \qquad \forall p \in P \tag{4.142d}$$

$$\sum_{p\in P} x_{ijp}\lambda_p = x_{ij} \in \{0,1\} \qquad \forall (i,j) \in A, \tag{4.142e}$$

where $c_p = \sum_{(i,j)\in A} c_{ij} x_{ijp}$ is the cost of path p and t_{ip} is the arrival time at node i along that path. The chosen values for big-M are here so large that all possible paths are valid in the *ISP*. Dropping integrality, the *MP* reads as

$$\begin{array}{llllllllllll}
\min & 3\lambda_{1246} & +14\lambda_{12456} & +5\lambda_{1256} & +13\lambda_{13246} & +24\lambda_{132456} & +15\lambda_{13256} & +16\lambda_{1346} & +27\lambda_{13456} & +24\lambda_{1356} & & \\
\text{s.t.} & 10\lambda_{1246} & +10\lambda_{12456} & +10\lambda_{1256} & +5\lambda_{13246} & +5\lambda_{132456} & +5\lambda_{13256} & & & & \leq & 9 \\
& & & & 3\lambda_{13246} & +3\lambda_{132456} & +3\lambda_{13256} & +3\lambda_{1346} & +3\lambda_{13456} & +3\lambda_{1356} & \leq & 3 \\
& 11\lambda_{1246} & +11\lambda_{12456} & & +6\lambda_{13246} & +6\lambda_{132456} & & +10\lambda_{1346} & +10\lambda_{13456} & & \leq & 5 \\
& & 12\lambda_{12456} & +13\lambda_{1256} & & +7\lambda_{132456} & +8\lambda_{13256} & & +11\lambda_{13456} & +6\lambda_{1356} & \leq & 12 \\
& 18\lambda_{1246} & +14\lambda_{12456} & +15\lambda_{1256} & +13\lambda_{13246} & +9\lambda_{132456} & +10\lambda_{13256} & +17\lambda_{1346} & +13\lambda_{13456} & +8\lambda_{1356} & \leq & 14 \\
& \lambda_{1246} & +\lambda_{12456} & +\lambda_{1256} & +\lambda_{13246} & +\lambda_{132456} & +\lambda_{13256} & +\lambda_{1346} & +\lambda_{13456} & +\lambda_{1356} & = & 1 \\
& \lambda_{1246}, & \lambda_{12456}, & \lambda_{1256}, & \lambda_{13246}, & \lambda_{132456}, & \lambda_{13256}, & \lambda_{1346}, & \lambda_{13456}, & \lambda_{1356} & \geq & 0.
\end{array}$$

Interestingly, the *ISP* defined in terms of $\boldsymbol{\pi} = [\pi_i]_{i=2,\ldots,6}$ and π_0 as

$$\begin{aligned}
\bar{c}(\boldsymbol{\pi}, \pi_0) = \min \quad & c_{\mathbf{x}} - \sum_{i=2}^{6} \pi_i t_{i\mathbf{x}} \\
\text{s.t.} \quad & \begin{bmatrix} \mathbf{x} \\ \mathbf{t} \end{bmatrix} \in \mathcal{D} \\
& c_{\mathbf{x}} = \sum_{(i,j)\in A} c_{ij} x_{ij} \\
& t_{i\mathbf{x}} = t_i \qquad \forall i \in \{2,\ldots,6\},
\end{aligned} \tag{4.143}$$

naturally computes the earliest arrival time at the visited nodes because of the positive penalties $-\pi_i \geq 0, \forall i \in \{2,\ldots,6\}$, on the time variables. The solution given in Example 3.2 is again optimal for the *MP*: $\lambda^{\star}_{1256} = 0.8$ and $\lambda^{\star}_{13256} = 0.2$ with the lower bound $z^{\star}_{MP} = 7$ on $z^{\star}_{ILP} = z^{\star}_{IMP} = 15$. This is an improvement on the objective values $z^{\star}_{LP}$ of Table 4.9 but, as the time upper bounds (4.140i) only appear in $\mathcal{A}$ (4.141a), once again, these are not at all used during the solution of the *ISP*, an important lack of information in a *time constrained* shortest path problem!

Example 4.5 ***Generalized assignment problem***

✎ Two block-diagonal structures appear in the *ILP*. The classical choice of blocks results in what we expect from a Dantzig-Wolfe reformulation where each subproblem builds specialized *binary* patterns that are assembled by the master problem. Surprisingly, using the other structure leads back to the compact formulation.

The *generalized assignment problem* (*GAP*) seeks a least-cost assignment of n tasks to a set K of machines such that each task is assigned exactly once, subject to capacity restrictions on the machines (Savelsbergh, 1997). A compact formulation is

$$z^{\star}_{ILP} = \min \quad \sum_{k\in K}\sum_{i=1}^{n} c^k_i x^k_i \tag{4.144a}$$

$$\text{s.t.} \quad \sum_{k\in K} x^k_i = 1 \qquad [\sigma_i] \quad \forall i \in \{1,\ldots,n\} \tag{4.144b}$$

$$\sum_{i=1}^{n} b_i^k x_i^k \leq b^k \qquad [\beta^k] \quad \forall k \in K \tag{4.144c}$$

$$x_i^k \in \{0, 1\} \qquad \forall k \in K,\ i \in \{1,\ldots,n\}, \tag{4.144d}$$

where c_i^k is the cost of assigning task i to machine k, b_i^k is the capacity used when task i is assigned to machine k, and b^k is the capacity of machine k. The binary variable x_i^k takes value 1 if and only if task i is assigned to machine k. This formulation has a block-diagonal structure over the machines which are all linked by the *semi-assignment constraints* (4.144b), see Figure 4.33.

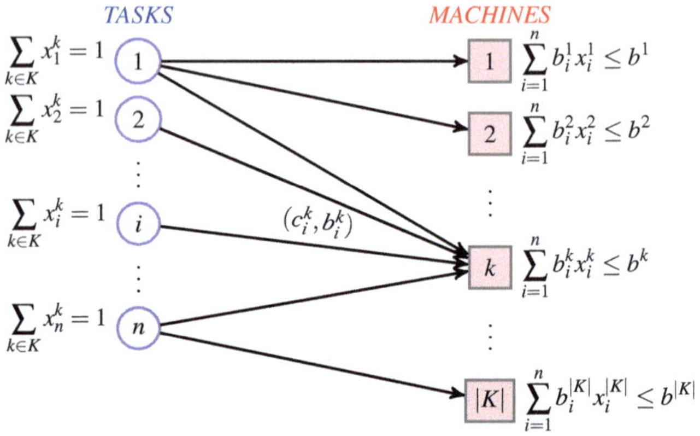

Fig. 4.33: Network representation of the *GAP*.

Fig. 4.34: Martin Savelsbergh (Bertinoro, Italy, 2005-06-23), the one who coined the tag-name *Branch-and-Price* with his colleagues (Barnhart et al., 1998).

Binary tasks-to-machine (many-to-one) patterns

Let $\mathbf{x} = [\mathbf{x}^k]_{k \in K}$, where

$$\mathbf{x}^k = [x_i^k]_{i=1,\ldots,n} = \begin{bmatrix} x_1^k \\ \vdots \\ x_n^k \end{bmatrix} \tag{4.145}$$

and group the constraints as

$$\mathcal{A} = \left\{ \left\{ \mathbf{x}^k \in \{0,1\}^n \right\}_{k \in K} \;\middle|\; \sum_{k \in K} x_i^k = 1,\ \forall i \in \{1,\ldots,n\} \right\} \tag{4.146a}$$

$$\mathcal{D}^k = \left\{ \mathbf{x}^k \in \{0,1\}^n \;\middle|\; \sum_{i=1}^{n} b_i^k x_i^k \le b^k \right\}, \quad \forall k \in K. \tag{4.146b}$$

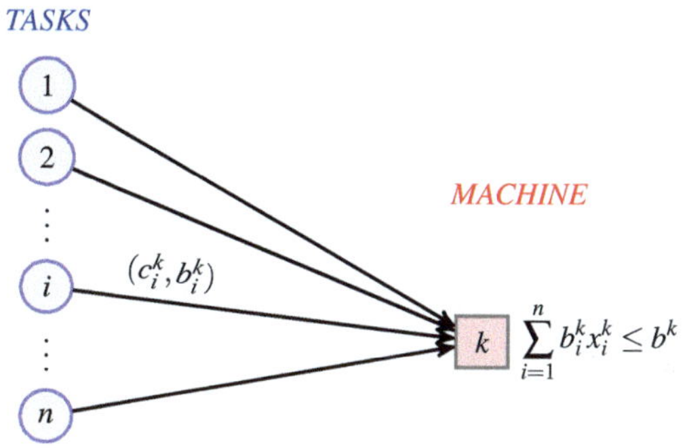

Fig. 4.35: Network representation of the *binary tasks-to-machine patterns* in $\mathcal{D}^k$.

For $k \in K$, the *binary tasks-to-machine patterns* $\mathbf{x}_p^k = [x_{ip}^k]_{i=1,\ldots,n}$, $p \in P^k$, illustrated in Figure 4.35, are used to reformulate (4.144) into the *IMP*

$$\begin{aligned}
z_{IMP}^{\star} = \min \quad & \sum_{k \in K} \sum_{p \in P^k} c_p^k \lambda_p^k \\
\text{s.t.} \quad & \sum_{k \in K} \sum_{p \in P^k} a_{ip}^k \lambda_p^k = 1 && [\pi_i] \quad \forall i \in \{1,\ldots,n\} \\
& \sum_{p \in P^k} \lambda_p^k = 1 && [\pi_0^k] \quad \forall k \in K \\
& \lambda_p^k \ge 0 && \forall k \in K,\ p \in P^k \\
& \sum_{p \in P^k} \mathbf{x}_p^k \lambda_p^k = \mathbf{x}^k \in \{0,1\}^n && \forall k \in K,
\end{aligned} \tag{4.147}$$

where $c_p^k = \sum_{i=1}^{n} c_i^k x_{ip}^k$ is the cost of pattern p on machine k whereas $a_{ip}^k = x_{ip}^k$ is equal to 1 if and only if task i is assigned to machine k in pattern p. Relaxing the binary requirements on $\mathbf{x}^k$ in (4.147), $\forall k \in K$, the *MP* is solved by the column generation algorithm, where an ISP^k turns out to be a binary knapsack problem:

$$\begin{aligned}
\bar{c}^k(\boldsymbol{\pi}, \pi_0^k) = \quad & -\pi_0^k + \min_{\mathbf{x} \in \mathcal{D}^k} c_{\mathbf{x}}^k - \sum_{i=1}^{n} \pi_i a_{i\mathbf{x}}^k \\
& \text{s.t. } c_{\mathbf{x}}^k = \sum_{i=1}^{n} c_i^k x_i^k \\
& a_{i\mathbf{x}}^k = x_i^k \qquad \forall i \in \{1,\ldots,n\},
\end{aligned}$$

$$\begin{aligned}
\text{that is, } \bar{c}^k(\boldsymbol{\pi}, \pi_0^k) = \quad & -\pi_0^k + \min \quad \sum_{i=1}^{n} (c_i^k - \pi_i) x_i^k \\
& \text{s.t.} \quad \sum_{i=1}^{n} b_i^k x_i^k \leq b^k \\
& x_i^k \in \{0,1\} \quad \forall i \in \{1,\ldots,n\}.
\end{aligned} \tag{4.148}$$

Because formulation (4.148) does not possess the integrality property, $z_{MP}^\star$ is greater-than-or-equal to $z_{LP}^\star$ in (4.144), i.e., $z_{LP}^\star \leq z_{MP}^\star \leq z_{ILP}^\star$.

Binary task-to-machines (one-to-many) patterns

In the following, we rather write the vector of variables as $\mathbf{x} = [\mathbf{x}_i]_{i=1,\ldots,n}$, where

$$\mathbf{x}_i = [x_i^k]_{k \in K} = \begin{bmatrix} x_i^1 & \cdots & x_i^{|K|} \end{bmatrix}. \tag{4.149}$$

Consider the Dantzig-Wolfe reformulation where we group the constraints as

$$\mathcal{A} = \left\{ \left\{ \mathbf{x}_i \in \{0,1\}^{|K|} \right\}_{i \in \{1,\ldots,n\}} \;\middle|\; \sum_{i=1}^{n} b_i^k x_i^k \leq b^k, \ \forall k \in K \right\} \tag{4.150a}$$

$$\mathcal{D}_i = \{ \mathbf{x}_i \in \{0,1\}^{|K|} \mid \sum_{k \in K} x_i^k = 1 \}, \qquad \forall i \in \{1,\ldots,n\}, \tag{4.150b}$$

that is, the domain $\mathcal{D}_i$ of the ISP_i contains exactly one of the semi-assignment constraints (4.144b), see Figure 4.36. Solutions to $\mathcal{D}_i$ are the extreme points of $\operatorname{conv}(\mathcal{D}_i) = \{\mathbf{x}_i \geq \mathbf{0} \mid \sum_{k \in K} x_i^k = 1\}$, the convex hull of the $|P| = |K|$ unit row-vectors in $\mathbb{R}_+^{|K|}$. These are $\{\mathbf{e}_1^\top, \ldots, \mathbf{e}_{|K|}^\top\}$ and the ISP_i has the integrality property (select machine k for task i ($x_i^k = 1$) with the smallest adjusted cost). Therefore, $z_{MP}^\star = z_{LP}^\star$ and the literature never considers this decomposition scheme compared to the previous one that provides in general a much better lower bound.

However, Proposition 4.5 extended to a block-diagonal structure tells us much more: *the IMP is the original ILP*. For the sake of completeness of this example, let us derive the reformulation. The index set P being the same as K, the Minkowski-Weyl substitution on $\operatorname{conv}(\mathcal{D}_i)$ writes as

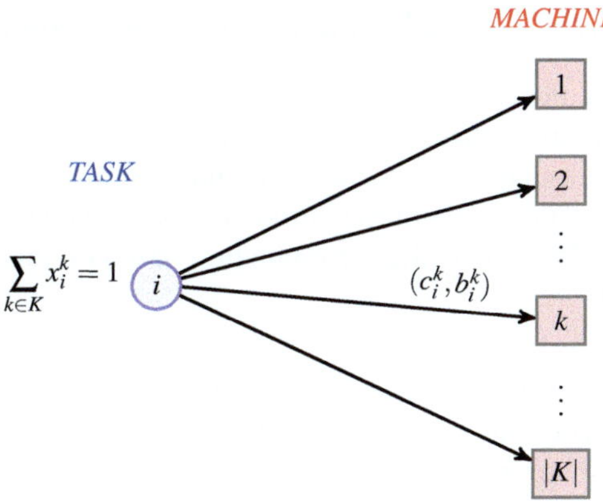

Fig. 4.36: Network representation of the *binary task-to-machines patterns* in $\mathcal{D}_i$.

$$\sum_{k\in K} \mathbf{e}_k^\top \lambda_{ik} = \mathbf{x}_i = \begin{bmatrix} x_i^1 & \cdots & x_i^{|K|} \end{bmatrix} \qquad \forall i \in \{1,\ldots,n\} \tag{4.151a}$$

$$\sum_{k\in K} \lambda_{ik} = 1 \qquad \forall i \in \{1,\ldots,n\} \tag{4.151b}$$

$$\lambda_{ik} \geq 0 \qquad \forall i \in \{1,\ldots,n\},\ k \in K. \tag{4.151c}$$

Writing the first equations component-wise, we get $\lambda_{ik} = x_i^k$, $\forall i \in \{1,\ldots,n\}$, $k \in K$. The reformulation becomes

$$z^\star_{IMP} = \min \quad \sum_{k\in K}\sum_{i=1}^{n} c_i^k \lambda_{ik} \tag{4.152a}$$

$$\text{s.t.} \quad \sum_{i=1}^{n} b_i^k \lambda_{ik} \leq b^k \qquad \forall k \in K \tag{4.152b}$$

$$\sum_{k\in K} \lambda_{ik} = 1 \qquad \forall i \in \{1,\ldots,n\} \tag{4.152c}$$

$$\lambda_{ik} = x_i^k \in \{0,\, 1\} \qquad \forall i \in \{1,\ldots,n\},\ k \in K, \tag{4.152d}$$

where the convexity constraints (4.151b) become the semi-assignment constraints. *Et voilà*, this is the *ILP* model from which we started! Obviously, we again have $z^\star_{MP} = z^\star_{LP}$.

Since we used only properties of the semi-assignment constraints, we conclude that it never pays to reformulate these. This applies, e.g., to the textbook models of bin packing, facility location, p-median, parallel machine scheduling, vertex coloring, vehicle routing, etc.

Example 4.6 ***Multi-commodity maximum flow problem***

✎ In this problem, the capacity constraints on the arcs link the flow of the various commodities whereas the flow conservation constraints display an obvious block-diagonal structure. The number of constraints of the latter type is several orders of magnitude larger than that of the former which makes a Dantzig-Wolfe reformulation based on this observation an ideal choice.

The *multi-commodity maximum flow problem* (*MCMFP*) is defined on a directed network $G = (N, A)$. The commodities are each ruled by their own flow conservation constraints but they share the same network. Each commodity $k \in K$ is defined by its own origin and its own destination in N, that is, $K = \{(o,d) \in N \times N, o \neq d\}$. For example, in an internet network, a message between a given pair of nodes defines a specific commodity. The goal is to send the maximum flow from all the origins to all destinations while satisfying the shared capacity u_{ij} on each arc $(i, j) \in A$.

Let x_{ij}^{od} denote the integer flow of commodity (o,d) on arc $(i, j) \in A$ and let the unbounded integer variable $x_0^{od} \geq 0$ measure the total flow from o to d. An integer linear programming formulation is

$$z_{ILP}^{\star} = \max \sum_{(o,d)\in K} x_0^{od} \tag{4.153a}$$

$$\text{s.t.} \sum_{(o,d)\in K} x_{ij}^{od} \leq u_{ij} \qquad \forall (i,j) \in A \tag{4.153b}$$

$$\sum_{j:(i,j)\in A} x_{ij}^{od} - \sum_{j:(j,i)\in A} x_{ji}^{od} = \begin{cases} x_0^{od} & \text{for } i = o \\ 0 & \forall i \in N, i \neq o,d \\ -x_0^{od} & \text{for } i = d \end{cases} \quad \forall (o,d) \in K \tag{4.153c}$$

$$x_0^{od}, x_{ij}^{od} \in \mathbb{Z}_+ \qquad \forall (o,d) \in K, (i,j) \in A. \tag{4.153d}$$

The *MCMFP* is known to be $\mathcal{NP}$-hard for $|K| \geq 2$ (Even et al., 1976), where the reduction is shown for the feasibility version, thus independent of the objective function. The proposed grouping keeps the capacity constraints in the *IMP* and moves the $|K|$-block structure into the pricing. Let $\mathbf{x}^{od} = [x_{ij}^{od}]_{(i,j)\in A}$, $\forall (o,d) \in K$.

$$\mathcal{A} = \left\{ \left\{ \begin{bmatrix} x_0^{od} \\ \mathbf{x}^{od} \end{bmatrix} \in \mathbb{Z}_+^{|A|+1} \right\}_{(o,d)\in K} \middle| \; (4.153\text{b}) \right\} \tag{4.154a}$$

$$\mathcal{D}^{od} = \left\{ \begin{bmatrix} x_0^{od} \\ \mathbf{x}^{od} \end{bmatrix} \in \mathbb{Z}_+^{|A|+1} \middle| \; (4.153\text{c})^{od} \right\}, \qquad \forall (o,d) \in K. \tag{4.154b}$$

This example is presented in the form of an exercise, answering one by one a series of questions. Assume that G has $|N| = 1\,000$ nodes and $|A| = 10\,000$ arcs and we want to send flow *from every node to every other*, i.e., $|K| = |N|(|N| - 1) = 999\,000$.

(a) *The ILP formulation* (4.153) *comprises how many constraints and variables?*

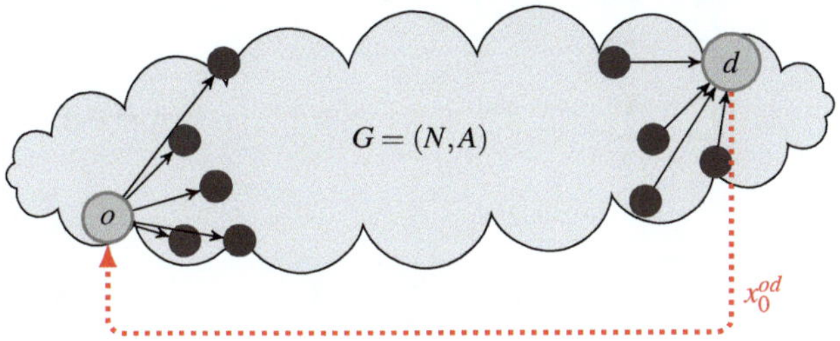

Fig. 4.37: Network structure of $\mathcal{D}^{od}$ for commodity $(o,d) \in K$.

► Capacity constraints in (4.153b) comprise one constraint per arc in A (10 000). Constraints in (4.153c) comprise one constraint per node in N (1 000) times the number of commodities $|K|$ (999 000), that is, 999 000 000. The total *number of constraints* is therefore 999 010 000.

The number of variables is given by the number of arcs in A (10 000) times the number of commodities $|K|$ (999 000), plus the 999 000 variables x_0^{od}, that is, a total of 9 990 999 000 *variables*.

(b) *Give the set* $\mathcal{X}^{od}$ *of extreme points and extreme rays of* $\text{conv}(\mathcal{D}^{od})$.

► On the one hand, $\mathcal{D}^{od}$ describes a network circulation structure. On the second hand, $\text{conv}(\mathcal{D}^{od})$ is a polyhedral cone and, here omitting the binary identifier,

$$\mathcal{X}^{od} = \left\{ \begin{bmatrix} 0 \\ \mathbf{0} \end{bmatrix} \right\} \cup \left\{ \begin{bmatrix} x_{0,r}^{od} \\ \mathbf{x}_r^{od} \end{bmatrix} \right\}_{r \in R^{od}}, \tag{4.155}$$

where $x_{0,r}^{od}$ can be scaled to 1 for any extreme ray indexed by $r \in R^{od}$. By flow conservation, this is also the case for the positive components of $\mathbf{x}_r^{od}$, that is, $x_{ij,r}^{od} = 1$ if arc (i, j) belongs to the od-path, 0 otherwise.

(c) *Formulate the IMP and give the row-size and column-size of the MP.*

► Discarding the zero extreme points and the convexity constraints for all commodities, we use the variables $\lambda_r^{od} \geq 0$, $\forall (o,d) \in K$, $r \in R^{od}$, for the Dantzig-Wolfe reformulation:

$$x_{ij}^{od} = \sum_{r \in R^{od}} x_{ij,r}^{od} \lambda_r^{od} \quad \text{and} \quad x_0^{od} = \sum_{r \in R^{od}} x_{0,r}^{od} \lambda_r^{od} = \sum_{r \in R^{od}} \lambda_r^{od}, \tag{4.156}$$

also given as $\begin{bmatrix} 1 \\ \mathbf{x}^{od} \end{bmatrix} = \sum_{r \in R^{od}} \begin{bmatrix} 1 \\ \mathbf{x}_r^{od} \end{bmatrix} \lambda_r^{od}$ in vector form.

The *IMP* writes as

$$
\begin{aligned}
z^\star_{IMP} = \max & \sum_{(o,d)\in K}\sum_{r\in R^{od}} \lambda_r^{od} \\
\text{s.t.} & \sum_{(o,d)\in K}\sum_{r\in R^{od}} x_{ij,r}^{od}\lambda_r^{od} \le u_{ij} \quad [\pi_{ij}] && \forall (i,j)\in A \\
& \lambda_r^{od} \ge 0 && \forall (o,d)\in K, r\in R^{od} \\
& \sum_{r\in R^{od}} \begin{bmatrix} 1 \\ \mathbf{x}_r^{od}\end{bmatrix}\lambda_r^{od} = \begin{bmatrix} x_0^{od} \\ \mathbf{x}^{od}\end{bmatrix} \in \mathbb{Z}_+^{|A|+1} && \forall (o,d)\in K.
\end{aligned} \tag{4.157}
$$

Discarding the *integrality constraints* on the 9 990 999 000 x-variables, the *MP* comprises *only* $|A|$ capacity constraints, that is, 10 000 constraints. This is less than 0.001 % of the number of constraints of the original *ILP* (4.153).

The precise number of λ-variables is unknown but huge, that is, the number of extreme rays of $\mathcal{D}^{od}$ times the number of commodities $|K|$. Therefore, the *MP* is solved by column generation.

(d) *Formulate the pricing problems. How would you solve them? Give the stopping criterion for the column generation algorithm.*

► Let $\boldsymbol{\pi} = [\pi_{ij}]_{(i,j)\in A}$. Because $x_{0,r}^{od} = 1$ for all rays indexed by $r \in R^{od}$, every pricing problem is an od-shortest path problem:

$$\bar{c}^{od}(\boldsymbol{\pi}) = \max \, (1 - \sum_{(i,j)\in A} \pi_{ij}x_{ij}^{od}) = 1 - \min \sum_{(i,j)\in A} \pi_{ij}x_{ij}^{od} \tag{4.158a}$$

$$\text{s.t.} \sum_{j:(i,j)\in A} x_{ij}^{od} - \sum_{j:(j,i)\in A} x_{ji}^{od} = \begin{cases} 1 & \text{for } i = o \\ 0 & \forall i \in N, i \neq o,d \\ -1 & \text{for } i = d \end{cases} \tag{4.158b}$$

$$x_{ij}^{od} \ge 0 \qquad \forall (i,j) \in A. \tag{4.158c}$$

The column generation algorithm stops when $\bar{c}^{od}(\boldsymbol{\pi}) \le 0$, $\forall (o,d) \in K$. While we can solve (4.158) with any linear programming solver, there is an abundance of literature on the shortest path problem. In our case, we have a directed graph on which there can be no negative cycles since all objective coefficients are non-negative, i.e., $\boldsymbol{\pi} \ge \mathbf{0}$.

The *all-pairs shortest path problem* can be solved by the Floyd-Warshall algorithm (Floyd, 1962) that runs in $O(|N|^3)$ time, see Ahuja et al. (1993, §5.6). This effectively means that we obtain $\bar{c}^{od}(\boldsymbol{\pi})$, $\forall (o,d) \in K$, but this might not be a good strategy. Indeed, we have seen in Note 2.13 that pricing problems generate columns that compete with each other such that this may not be the most efficient use of computing resources.

An alternative is to solve the *single-source shortest path problem* using Dijkstra's algorithm (Dijkstra, 1959) that runs in $O(|N|^2)$ time, see Ahuja et al. (1993, §4.5). The resulting tree of shortest paths contains solutions between a node and every other. This means that we are *de facto* solving $|N| - 1$ pricing problems in one go, i.e., finding $\bar{c}^{oj}(\boldsymbol{\pi})$, $\forall j \neq o \in N$. We can eventually analyze these paths to keep a diversified subset for the *RMP*. Despite Note 2.14 warning us about parallelization ineffectiveness, this alternative seems like a promising compromise because there is barely any overhead compared to solving for only (o,d). We have also yet to mention Dial's implementation (Ahuja et al., 1993, §4.6) which can improve the bottleneck of Dijkstra's algorithm in practice.

A more recent complexity result for the all-pairs variant, we deal with is $O(|A||N| + |N|^2 \log\log|N|)$ from Pettie (2004) whereas for the single-source variant, that of $O(|A| + |N|\log|N|)$ by Fredman and Tarjan (1984, 1987) still stands. It is unlikely that the outcome of this discussion changes with theoretical complexity improvements on the shortest path problem unless one can derive a factor much smaller than $|N|$ between both variants. These results do however suggest that we should pay closer attention to the number of arcs in the graph. One should obviously carry out some experiments especially for instances with a smaller number of commodities, i.e., $|K| \ll |N|(|N|-1)$.

(e) *How does $z^\star_{MP}$ compare to $z^\star_{LP}$?*

▶ Because the shortest path formulation in (4.158) possesses the integrality property, we have $z^\star_{LP} = z^\star_{MP}$.

Example 4.7 **Scene selection problem**

✎ Exploiting the structure helps a lot! Tightening the *ILP* with additional constraints to avoid symmetry difficulties is much less efficient than reformulating the original problem with an *ISP* whose formulation does not possess the integrality property.

This *scene selection problem* (*SSP*) is taken from Van Hentenryck (2002) for which various symmetry breaking formulations are evaluated in Jans and Desrosiers (2010). Given a cast of actors forming the set A and various scenes to shoot forming the set N, a movie producer must decide when shootings occur. We know that

- Each scene requires the presence of a specific subset of actors: the parameter $a_{ij} = 1$ indicates if actor $j \in A$ is needed for scene $i \in N$, 0 otherwise;
- There can be at most W scenes shot each day during up to m days of shootings;
- Actor j is paid c_j for each day of presence.

Let the binary variables $x_i^k = 1$ if scene i is shot on day $k \in K = \{1,\ldots,m\}$ and $y_j^k = 1$ if actor j appears on day k. The objective is to minimize the total cost:

$$z^\star_{ILP} = \min \quad \sum_{k\in K}\sum_{j\in A} c_j y_j^k \tag{4.159a}$$

$$\text{s.t.} \quad \sum_{k\in K} x_i^k = 1 \qquad \forall i \in N \tag{4.159b}$$

$$\sum_{i\in N} x_i^k \le W \qquad \forall k \in K \tag{4.159c}$$

$$a_{ij} x_i^k \le y_j^k \qquad \forall k \in K,\ i \in N,\ j \in A \tag{4.159d}$$

$$x_i^k \in \{0,1\} \qquad \forall k \in K,\ i \in N \tag{4.159e}$$

$$y_j^k \in \{0,1\} \qquad \forall k \in K,\ j \in A. \tag{4.159f}$$

In model (4.159),

- each scene i is shot on one day: *assignment* constraints (4.159b);
- at most W scenes are shot on each day k: *capacity* constraints (4.159c);
- if scene i is shot on day k, then $x_i^k = 1$ and the required actors are present: *actor* constraints (4.159d).

Technically, it is sufficient to impose the binary restrictions on the x-variables. The following analysis and results are based on $|A| = 11$, $|N| = 19$, $W = 5$, and $m = 5$.

Symmetry breaking constraints

In Exercise 4.14, the reader is asked to show that the objective value $z_{LP}^\star$ of the linear relaxation of (4.159) is equal to $\sum_{j\in A} c_j$, a weak lower bound on the optimal objective value $z_{ILP}^\star$. This formulation contains a lot of symmetry. For any solution, we can obtain an equivalent one by permuting the days. We can decrease the symmetry using a *variable reduction technique* (*VRT*):

- Take one scene at random, say scene one, and impose that it is shot on day 1.
- For scene two, if it is not shot on day one, then it is on day two, hence impose $x_2^3 = x_2^4 = x_2^5 = 0$ and $x_2^1 + x_2^2 = 1$. We continue this for scenes 3 and 4.
- The remainder of the scenes can be shot on any day, hence for scene 5 and the following ones, we cannot reduce the number of variables.

For *VRT*, let us consider three different orders of the scenes. The first keeps the *original* numbering. Next, for each scene, we evaluated the total cost for the actors needed in it. In the *LowHigh* order, the scenes are numbered from the lowest total cost to the highest, whereas the numbering is reversed in the *HighLow* order.

Dantzig-Wolfe reformulation

For $k \in K$, let $\mathbf{x}^k = [x_i^k]_{i=1,\dots,|N|}$ and $\mathbf{y}^k = [y_j^k]_{j=1,\dots,|A|}$. We group the constraints as

$$\mathcal{A} = \left\{ \left\{ \begin{bmatrix} \mathbf{x}^k \\ \mathbf{y}^k \end{bmatrix} \in \{0,1\}^{|N|+|A|} \right\}_{k\in K} \;\middle|\; (4.159\text{b}) \right\} \tag{4.160a}$$

$$\mathcal{D}^k = \left\{ \begin{bmatrix} \mathbf{x}^k \\ \mathbf{y}^k \end{bmatrix} \in \{0,1\}^{|N|+|A|} \;\middle|\; (4.159\text{c})\text{–}(4.159\text{d}) \right\}, \quad \forall k \in K, \tag{4.160b}$$

where the domain of the ISP^k defines *binary scene selection patterns* which can be enumerated exhaustively for the chosen numbers of scenes, actors, and days.

Quality of the lower bounds

In Table 4.10 and Figure 4.38, we report some results for the linear relaxation of the various compact formulations, that is, for the *ILP* alone and for the *ILP* + *VRT* strategies, as well as for the *IMP* based on (4.160). We report the computation time in seconds to calculate the linear programming solution with the default setting for symmetry breaking in `CPLEX 11.2` and the integrality gap (*IP* gap (%)) compared to the optimal objective value $z^{\star}_{ILP} = 334\,144$.

	time (s)	Linear relaxation	*IP* gap (%)
ILP	4.00	137 739.00	58.8
ILP + *VRT* (*Original*)	5.79	177 075.57	47.0
ILP + *VRT* (*LowHigh*)	8.53	157 388.55	52.9
ILP + *VRT* (*HighLow*)	2.75	219 013.83	34.5
IMP	1.34	330 405.41	1.1

Table 4.10: Results for the *SSP*, where $z^{\star}_{ILP} = 334\,144$.

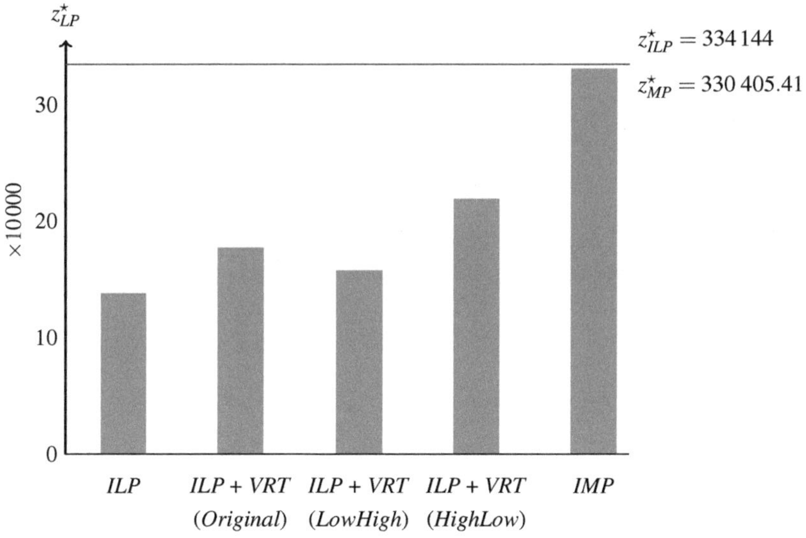

Fig. 4.38: Quality of various linear programming relaxations for the *SSP*.

The results bring to light three important considerations.

1. This is an example of a formulation where the various linear relaxation values of *ILP* + *VRT* improve on that of *ILP*. In fact, the integrality gap improves from

52.9 % for *VRT* (*LowHigh*) to 47.0 % for *VRT* (*Original*) and 34.5 % for *VRT* (*HighLow*) compared to 58.8 % for the *ILP* alone.

2. The linear relaxation values of *ILP* + *VRT* depend on the order of the input data. This can partially be explained as follows. In *VRT*, we fix the first variable: $x_1^1 = 1$. This forces the actors needed for that first scene to be present and we are *fully charged for the cost of that scene*. For the second scene, there are two possibilities: either it is in the first day or in the second, so we induce less fractional values for the decision variables. We see indeed that the order *HighLow* performs substantially better in terms of $z^\star_{LP}$ value compared to *LowHigh*.

3. For the *IMP* (or $\mathit{IM}\dot{P}$ based on discretization), the order of the input data has no impact on the linear relaxation $z^\star_{MP}$, and the integrality gap falls to only 1.1 %. More importantly, there are no more symmetry issues due to the permutation of the days while solving the *MP*.

Example 4.8 **Design of balanced student teams**

We start with an integer quadratic program with symmetry issues as well as fractional solutions almost systematically when the integrality requirements are relaxed. The reformulation turns out to be a set partitioning model, much easier to manage.

Students of a same class must be partitioned into teams in such a way that each team provides a good representation of the class composition. In general, when facing this kind of problem, we can fall back upon descriptive statistical measures where the goal is to reproduce statistics observed for the population in each sample. From the school administration, we are given, for each student, a list of attributes such as age, gender, highest grade, field of studies, country of origin, etc. We may even be given relative importance for these attributes. This gives us a way to score each partition with a *weighted sum of squared deviations*. This particular measure leads to an integer quadratic program (*IQP*).

Quadratic transportation problem

Assume the presence of 26 students within a group of a *Branch-and-Price Course*, namely Amièle, Bob, Charlotte, Dominic, ..., Xavier, Yolaine, and Zoe. Each team must comprise 4 or 5 students. Each student being assigned to a single team, we face the *supply-demand* constraints of the classical transportation problem (see Figure 4.39).

Let n denote the number of students, K the set of teams, and n_k the number of students in team $k \in K$. Let x_{ik} be a binary variable that takes value 1 if student $i \in \{1,\dots,n\}$ is assigned to team k, 0 otherwise.

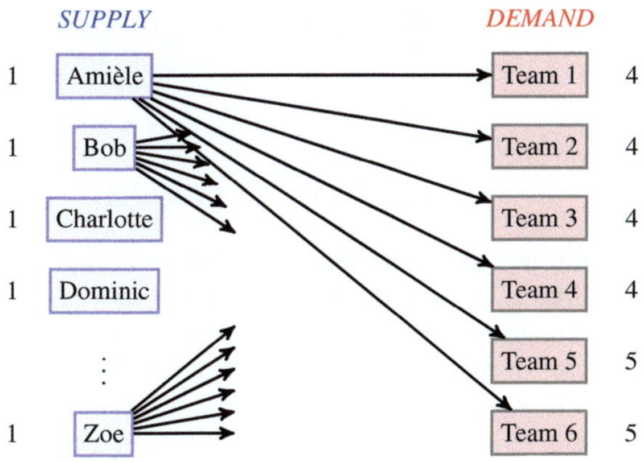

Fig. 4.39: *Supply-demand* for the design of balanced student teams.

Following Figure 4.39, the set of transportation constraints is

$$\begin{aligned} \sum_{k\in K} x_{ik} &= 1 && \forall i \in \{1,\dots,n\} \\ \sum_{i=1}^{n} x_{ik} &= n_k && \forall k \in K \\ x_{ik} &\in \{0,1\} && \forall k \in K, i \in \{1,\dots,n\}. \end{aligned} \tag{4.161}$$

This is an easy structure but the difficulties arise from scoring the possible assignments *while optimizing the objective function*. For every attribute $q \in Q$, we have

- the personal statistic of student i denoted by s_{iq}, for all $i \in \{1,\dots,n\}$;
- the target value t_q, a parameter computed a priori by some well-defined rule such as the average or percentage taken over the class, that is, $\sum_{i=1}^{n} s_{iq}/n$;
- the weight factor w_q.

The contribution of each statistic q to the objective function of students present in team k is calculated by averaging their scores as

$$t_{kq} = \sum_{i=1}^{n} \frac{s_{iq}x_{ik}}{n_k} \tag{4.162}$$

Using vector notation, we have the target-vector $\mathbf{t} = [t_q]_{q\in Q}$ and the score-vector $\mathbf{t}_k = [t_{kq}]_{q\in Q}$ of team k. Let A be the set of arcs in the transportation problem structure (4.161). Incorporating the objective function in (4.163), the goal is to partition the students into teams such that the corresponding weighted sum of squared deviations between the scores and targets reaches its minimum:

$$z^{\star}_{IQP} = \min \sum_{k\in K}\sum_{q\in Q} w_q\Big(\sum_{i=1}^{n} \frac{s_{iq}x_{ik}}{n_k} - t_q\Big)^2 \tag{4.163a}$$

$$\text{s.t.} \quad \sum_{k \in K} x_{ik} = 1 \qquad \forall i \in \{1,\ldots,n\} \tag{4.163b}$$

$$\sum_{i=1}^{n} x_{ik} = n_k \qquad \forall k \in K \tag{4.163c}$$

$$x_{ik} \in \{0,1\} \qquad \forall (i,k) \in A. \tag{4.163d}$$

Solving this quadratic transportation problem while relaxing the binary requirements on the x-variables is, in most cases, useless because we proportionally adjust the "presence" x_{ik} of team members in computing the score-vector $\mathbf{t}_k$ (4.162). For example, if the number of men in the class is 17 out of 26 (65.4 %), this amounts to a target of 2.6155 in a team of 4, or 3.2692 in a team of 5. These target values are impossible to reach in any binary solution but easy in a fractional one. Hence, a solution to the relaxation of (4.163) is in general so fractional that even finding a single integer solution by branch-and-bound in a reasonable amount of time is hard. Moreover, the model suffers from symmetry induced by the identifiers of the teams with the same size.

Set partitioning reformulation

The proposed Dantzig-Wolfe reformulation by discretization of this compact formulation allows us to overcome these difficulties by capitalizing on the following observation: If we *know* all team members T_k for some team k, it is easy to compute the contribution to the objective function of any attribute q. We see this clearly in any integer solution where we have $T_k = \{i \mid x_{ik} = 1\}$ such that

$$t_{kq} = \sum_{i=1}^{n} \frac{s_{iq} x_{ik}}{n_k} = \frac{\sum_{i \in T_k} s_{iq}}{|T_k|}. \tag{4.164}$$

Figure 4.40 displays four out of the six teams in a 2-dimensional attribute space, where Amièle, Bob, Dominic, and Zoe belong to the first team. Assignments for teams 3 and 5 have been omitted to lighten the content. The score-vector $\mathbf{t}_1 \in \mathbb{R}^2$ for team 1 is here computed as the average of statistic-vectors $\mathbf{s}_1, \mathbf{s}_2, \mathbf{s}_4$, and $\mathbf{s}_{26}$ whereas the target-vector for the class is computed as $\mathbf{t} = \sum_{i=1}^{26} \mathbf{s}_i / 26$.

We use the grouping of the constraints

$$\mathcal{A} = \left\{ \mathbf{x} \in \{0,1\}^{|A|} \;\middle|\; \sum_{k \in K} x_{ik} = 1,\ \forall i \in \{1,\ldots,n\} \right\} \tag{4.165a}$$

$$\mathcal{D}^k = \left\{ \mathbf{x} \in \{0,1\}^{|A|} \;\middle|\; \sum_{i=1}^{n} x_{ik} = n_k \right\} \qquad \forall k \in K, \tag{4.165b}$$

and the convexification and discretization approaches are equivalent by Proposition 4.4. As $\mathcal{D}^k$ is bounded for all k, Proposition 4.14 applies and allows us to use the quadratic cost function. We then rely on the enumeration of all solutions.

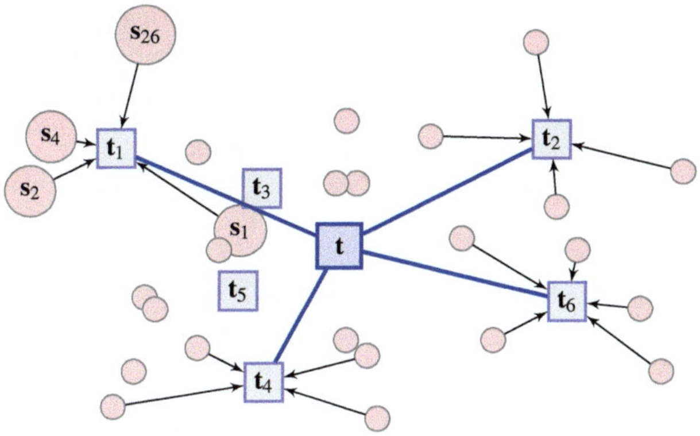

Fig. 4.40: Partition of 17 students (out of 26) vs. target-vector **t**.

Figure 4.41 illustrates the *binary students-to-team patterns* in $\mathcal{D}^k$. Note that it is identical to the *tasks-to-machine* decomposition already used for the Generalized assignment problem, the alternative *teams-to-student patterns* giving back the compact formulation by Proposition 4.5 extended to a block-diagonal structure.

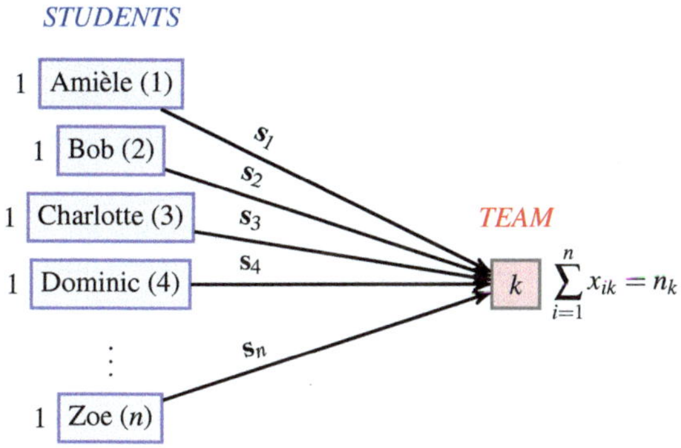

Fig. 4.41: Illustration of the *binary students-to-team patterns* in $\mathcal{D}^k$, where the statistic-vectors $\mathbf{s}_i$, $i \in \{1,\ldots,n\}$, of the selected students are used to compute the score-vector $\mathbf{t}_k$ for team k and then the non-linear values $(t_{kq} - t_q)^2$, $q \in Q$.

Since all subproblems with the same team size n_k are isomorphic, we use aggregation. Let $\ell \in L$ denote the distinct team sizes. The total number of teams is

$$|P| = \sum_{\ell \in L} |P^\ell|, \quad \text{where } |P^\ell| = \binom{n}{\ell}. \tag{4.166}$$

Then, one needs to select a number of teams amongst $|P|$ that partition the n students such that the sum of the weighted distances between their team-vector scores and the target-vector score is minimal. For a team $p \in P^\ell$ of size ℓ, we denote by

- $\lambda_p^\ell \in \{0,1\}$: 1 if the team is chosen, 0 otherwise;
- $a_{ip}^\ell \in \{0,1\}$: 1 if student i belongs to the team, 0 otherwise ($\sum_{i=1}^n a_{ip}^\ell = \ell$);
- $c_p^\ell = \sum_{q\in Q} w_q (t_{pq}^\ell - t_q)^2$, where $t_{pq}^\ell = \sum_{i=1}^n s_{iq}\, a_{ip}^\ell/\ell$ is the score for attribute q.

The quadratic program (4.163) is reformulated as a set partitioning model:

$$
\begin{aligned}
z^\star_{IMP} = \min \quad & \sum_{\ell\in L}\sum_{p\in P^\ell} c_p^\ell \lambda_p^\ell && && (4.167a)\\
\text{s.t.} \quad & \sum_{\ell\in L}\sum_{p\in P^\ell} a_{ip}^\ell \lambda_p^\ell = 1 && \forall i \in \{1,\ldots,n\} && (4.167b)\\
& \sum_{p\in P^\ell} \lambda_p^\ell \le m^\ell && \forall \ell \in L && (4.167c)\\
& \lambda_p^\ell \in \{0,1\} && \forall \ell \in L,\ p\in P^\ell, && (4.167d)
\end{aligned}
$$

where $\sum_{p\in P^\ell} \lambda_p^\ell$ is the number of selected teams of size ℓ, and the parameter m^ℓ is either given or derived from the data. As a result of the aggregation process on identical subproblems (see p. 204), model (4.167) does not suffer from the duplication of teams of the same type.

In our instance involving 26 students, we have teams of sizes 4 and 5, hence $L = \{4,5\}$ such that $|P|^4 = 14\,950$, $|P|^5 = 65\,780$, for a total of $|P| = 80\,730$, a relatively small number for modern set partitioning solvers. We also derive $m^4 = \lfloor 26/4 \rfloor = 6$ and $m^5 = \lfloor 26/5 \rfloor = 5$. In this particular instance, we can use equality constraints with $m^4 = 4$ and $m^5 = 2$ in (4.167c) if we realize that the only integer partition is four teams of 4 students and two of 5. Assume given the enumerated teams of 4 and 5 students, here indexed by $p \in P^4$ and $p' \in P^5$, respectively. The set partitioning model writes as

$$
\begin{aligned}
z^\star_{IMP} = \min \quad & \sum_{p\in P^4} c_p\lambda_p + \sum_{p'\in P^5} c_{p'}\lambda_{p'} && (4.168a)\\
\text{s.t.} \quad & \sum_{p\in P^4} \mathbf{a}_p\lambda_p + \sum_{p'\in P^5} \mathbf{a}_{p'}\lambda_{p'} = \mathbf{1} && (4.168b)\\
& \sum_{p\in P^4} \lambda_p = 4, \quad \sum_{p'\in P^5} \lambda_{p'} = 2 && (4.168c)\\
& \lambda_p,\ \lambda_{p'} \in \{0,1\} \quad \forall p\in P^4,\ p'\in P^5, && (4.168d)
\end{aligned}
$$

where, for all $p \in P^4$ and $p' \in P^5$, $\mathbf{a}_p$ and $\mathbf{a}_{p'}$ are binary vectors of dimension 26, the number of non-zero entries in $\mathbf{a}_p$ is 4, that of $\mathbf{a}_{p'}$ is 5. Every student is assigned exactly once in (4.168b); four teams of 4 students are selected and two of size 5 in (4.168c). An optimal solution to the *IMP* (4.168) is composed of 6 teams.

The number of admissible teams can be reduced by discarding any that does not satisfy certain *practical* rules. For example, as the number of men in our class is 17 out of 26 (65.4 %), this amounts to a target of 2.6155 in a team of 4, so we can decide to only accept teams with 2 or 3 men and this reduces the number of admissible teams of size 4 to 11 016. We do the same for a team of 5, accepting only teams with 3 or 4 men: the number decreases to 45 900. Full enumeration is tractable and the *IMP* (4.168) can be solved directly with any modern solver: it involves 56 916 team-variables and $26+2$ constraints. Note that using methods to heuristically reduce the number of variables treated leads to potentially suboptimal solutions. In the realm of practical rules, this comment becomes a gray area. Imagine that an integer solution with only 1 man in a team exists and that it even has a better objective value than what we have found with respect to our practical rule regarding restricted gender distribution. What do we conclude about optimality then? If such a team composition would be rejected anyway to avoid giving rise to what is ultimately an "unwarranted" sentiment of injustice, then said practical rule is really a constraint of the problem.

Finally, in a 2-semester academic program, assume the teams are broken after the first semester as in Exercise 4.15. How would you deal with such a situation?

Example 4.9 **Can you decode a secret vote?**

✎ In this problem, we have no objective function but a Dantzig-Wolfe reformulation can still be of service. Recall the analogy *atoms vs. molecules* on p. 176.

Twenty shareholders are voting to elect the president of a strategic working group within six external candidates ($k \in K = \{a, \ldots, f\}$). Each shareholder (or voter) owns a share percentage p_i, $i \in N = \{1, \ldots, 20\}$. The rules stipulate that each shareholder i casts one vote, with weight p_i, towards a particular candidate k. Moreover, a round system provision is established if no candidate receives at least 50 % of votes. Table 4.11 details the ownership distribution of the shareholders and the outcome of the vote. Amongst other things, no candidate has received more than 50 %, candidate f received the highest score 35.9 % while candidate a received only 4.5 %.

Can you decode the results, that is, finding fully or partially who voted for who?

Stated otherwise and putting the moral high ground aside, is there an efficient way to put persuasion talent to service for the second round?

We are looking for an integer programming formulation, *where there is no objective function*. One such set of constraints that comes to mind is based on the flow formulation of the generalized assignment problem (4.144), see Figure 4.42 and system (4.169), where x_i^k is a binary variable taking value 1 if shareholder i votes for candidate k, 0 otherwise, whereas b^k is the total percentage of votes obtained by candidate k:

Shareholder i	1	2	3	4	5	6	7	8	9	10
% Shares (p_i)	14.3	13.2	12.4	8.4	7.8	6.2	5.7	5.5	4.5	4.2
Shareholder i	11	12	13	14	15	16	17	18	19	20
% Shares (p_i)	3.6	3.1	2.7	2.4	1.5	1.4	1.3	1.1	0.4	0.3
Candidate k	a	b	c	d	e	f				
Total % Votes (b^k)	4.5	11.1	13.8	17.3	17.4	35.9				

Table 4.11: Percentage shares of every shareholder and first round election results.

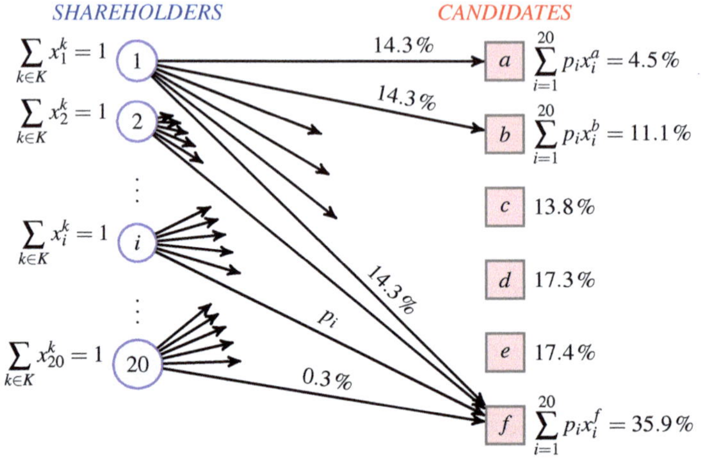

Fig. 4.42: Illustration of the secret vote formulation.

$$
\begin{aligned}
\sum_{k\in K} \quad x_i^k &= 1 && \forall i \in N \\
\sum_{i\in N} p_i x_i^k &= b^k && \forall k \in K \\
x_i^k &\in \{0, 1\} && \forall k \in K,\ i \in N.
\end{aligned}
\tag{4.169}
$$

Although valid, this set of constraints is indeed rather weak in the sense that, while solving the linear relaxation, there is nothing to account for the voting rule imposed by the binary requirements within the knapsack constraints $\sum_{i=1}^{20} p_i x_i^k = b^k$, $\forall k \in K$. Let us take a look at Figure 4.43, the solution obtained from the linear relaxation of (4.169).

Obviously, voter 3 with 12.4 % shares, for which this solution splits his vote between candidate a (0.11290) and c (0.88710), cannot vote for candidate a in any integer solution because he or she received only 4.5 % of the votes. This means that some x-variables can *a priori* be set to zero, indeed the first 8 shareholders for a, the first 3 for b, and the first for c. However, much more can be done.

		Candidates						
Voters	Shares (%)	*a*	*b*	*c*	*d*	*e*	*f*	Row sum
1	14.3						1	1
2	13.2					1		1
3	12.4	0.11290		0.88710				1
4	8.4				1			1
5	7.8			0.17940	0.43590		0.38462	1
6	6.2						1	1
7	5.7		1					1
8	5.5				1			1
9	4.5		1					1
10	4.2						1	1
11	3.6						1	1
12	3.1		0.29032				0.70968	1
13	2.7					1		1
14	2.4					0.62500	0.37500	1
15	1.5						1	1
16	1.4			1				1
17	1.3	1						1
18	1.1	1						1
19	0.4	1						1
20	0.3	1						1
	Results (%)	4.5	11.1	13.8	17.3	17.4	35.9	100

Fig. 4.43: Fractional solution to the secret vote formulation.

Binary vote patterns

A Dantzig-Wolfe reformulation of (4.169) can help. For $k \in K$, let $\mathbf{x}^k = [x_i^k]_{i=1,\ldots,20}$ and group the constraints as

$$\mathcal{A} = \left\{ \left\{ \mathbf{x}^k \in \{0,1\}^{20} \right\}_{k \in K} \middle| \sum_{k \in K} x_i^k = 1,\ \forall i \in \{1,\ldots,20\} \right\} \tag{4.170a}$$

$$\mathcal{D}^k = \left\{ \mathbf{x}^k \in \{0,1\}^{20} \middle| \sum_{i=1}^{20} p_i x_i^k = b^k \right\}, \qquad \forall k \in K. \tag{4.170b}$$

For $k \in K$, the *vote patterns* in $\mathcal{D}^k$, i.e., the binary vectors

$$\mathbf{x}_p^k = [x_{ip}^k]_{i=1,\ldots,20},\ p \in \ddot{P}^k,$$

are used to reformulate (4.169) by substituting for all $i \in \{1,\ldots,20\}$,

$$\begin{aligned} \sum_{p \in \ddot{P}^k} x_{ip}^k \lambda_p^k &= x_i^k \\ \sum_{p \in \ddot{P}^k} \lambda_p^k &= 1 \\ \lambda_p^k &\in \{0,1\} \quad \forall p \in \ddot{P}^k. \end{aligned} \tag{4.171}$$

For example, for candidate a who received 4.5 % of the percentage of votes, there are only $|\mathcal{D}^a| = 10$ possible binary patterns, and the associated 20-dimensional points appear in Table 4.12, with zero-entries for the first 8 voters (with more than 4.5 % of the shares).

Voters	Shares (%)										
9	4.5	1									
10	4.2		1								
11	3.6										
12	3.1			1	1						
13	2.7					1	1	1			
14	2.4								1		
15	1.5					1				1	
16	1.4			1			1		1	1	1
17	1.3									1	1
18	1.1				1			1			1
19	0.4						1	1	1		1
20	0.3		1		1	1		1	1	1	1

Table 4.12: The ten binary vote patterns in $\mathcal{D}^a$.

The reformulation of (4.169) only contains the constraints of a relatively small-sized set partitioning problem with $20+6$ rows and $\sum_{k\in K} |\ddot{P}^k|$ columns:

$$\begin{aligned} \sum_{k\in K} \sum_{p\in \ddot{P}^k} x^k_{ip} \lambda^k_p &= 1 && \forall i \in \{1,\ldots,20\} \\ \sum_{p\in \ddot{P}^k} \lambda^k_p &= 1 && \forall k \in K \\ \lambda^k_p &\in \{0,1\} && \forall k \in K,\ p \in \ddot{P}^k. \end{aligned} \tag{4.172}$$

In the above model, every shareholder i is voting exactly once and one pattern indexed in $\ddot{P}^k$ is selected for each candidate k. In a full enumeration model, we immediately see that a 0-1 branching decision on a fractional λ^k_p is natural. In a column generation context, this is a more complex matter.

An integer solution to such a set of constraints (4.172) provides one possible way to vote, but this might not be the actual one. Hence an enumeration process is required, preferably with a statistical analysis performed on all feasible solutions.

In the above example, it is not possible to decode the vote, even partially, with the actual information. In Exercise 4.16, we examine additional information that might help decoding such a vote. For example, the six candidates are not external but rather amongst the twenty shareholders, or as in the technical paper by Jaumard and Soumis (1986) at the origin of this application, we are also given the number of shareholders who voted for each candidate.

Example 4.10 **Edge coloring problem: two compact formulations**

✎ One extended set covering formulation, two compacts.

Recall the Edge coloring problem, Example 2.3, which is defined on an undirected graph $G=(N,E)$, where N denotes the set of nodes and E the set of edges. Let $\delta(\{i\}) \subseteq E$ denote the subset of edges incident to $i \in N$, and $E(S) \subseteq E$ for $S \subseteq N$ the edges with both endpoints in S. Moreover, we say that $S \subseteq N$ is an *odd set* if $|S| \geq 3$ and is odd.

First compact. We are given the $\ddot{IMP}$ (2.35) (a set covering formulation) and the binary *ISP* (2.37) that finds matchings of minimum reduced cost. We design, with the help of Proposition 4.15, a compact formulation *ILP* with several blocks, not all of them being used. We also propose a grouping of the constraints such that the Dantzig-Wolfe reformulation leads to this $\ddot{IMP}$.

Second compact. We next provide an alternative "compact" model, this time without a block-index. We use a linear description of the matching polytope and Proposition 4.17 that benefits from the integrality property of the pricing problem.

First compact, with an identical block-diagonal structure

Let κ denote an upper bound on the *chromatic index*, that is, the minimum number of colors required to color all the edges such that no incident edges have the same color. Vizing's theorem states that the edges of any graph with maximum degree Δ can be colored with at most $\Delta+1$ colors (Vizing, 1964). For $k \in \{1,\ldots,\Delta+1\}$, let $x_0^k = 1$ if color k is selected, 0 otherwise. Let the binary variable x_e^k take value 1 if edge e is colored by color k, 0 otherwise.

Starting with the structure of the *ISP* (2.37) reproduced below, where $\boldsymbol{\pi} \geq \mathbf{0}$ is the dual vector of the linear relaxation of the set covering formulation (2.35),

$$\bar{c}(\boldsymbol{\pi}) = \min \; x_0 - \sum_{e \in E} \pi_e x_e \tag{4.173a}$$

$$\text{s.t.} \quad \sum_{e \in \delta(\{i\})} x_e \leq x_0 \qquad \forall i \in N \tag{4.173b}$$

$$x_0,\, x_e \in \{0,1\} \qquad \forall e \in E \tag{4.173c}$$

$$c_{\mathbf{x}} = x_0,\; a_{e\mathbf{x}} = x_e \qquad \forall e \in E. \tag{4.173d}$$

a compact formulation with $\kappa = \Delta+1$ blocks (Proposition 4.15) is given by

$$
\begin{aligned}
z^\star_{ILP} = \min \quad & \sum_{k=1}^{\kappa} x_0^k \\
\text{s.t.} \quad & \sum_{k=1}^{\kappa} x_e^k \geq 1 && \forall e \in E \\
& \sum_{e \in \delta(\{i\})} x_e^k \leq x_0^k && \forall k \in \{1,\ldots,\kappa\},\ i \in N \\
& x_0^k \in \{0,1\} && \forall k \in \{1,\ldots,\kappa\} \\
& x_e^k \in \{0,1\} && \forall k \in \{1,\ldots,\kappa\},\ e \in E.
\end{aligned}
\tag{4.174}
$$

To derive the $IM\ddot{P}$ (2.35) from the above compact formulation, let $\mathbf{x}^k = [x_e^k]_{e \in E}$, $\forall k \in \{1,\ldots,\kappa\}$, and group the constraints as

$$
\mathcal{A} = \left\{ \left\{ \begin{bmatrix} x_0^k \\ \mathbf{x}^k \end{bmatrix} \in \{0,1\}^{|E|+1} \right\}_{k \in K} \middle| \sum_{k=1}^{\kappa} x_e^k \geq 1,\ \forall e \in E \right\} \tag{4.175a}
$$

$$
\mathcal{D}^k = \left\{ \begin{bmatrix} x_0^k \\ \mathbf{x}^k \end{bmatrix} \in \{0,1\}^{|E|+1} \middle| \sum_{e \in \delta(\{i\})} x_e^k \leq x_0^k,\ \forall i \in N \right\}, \quad \forall k \in \{1,\ldots,\kappa\}. \tag{4.175b}
$$

Here we face κ identical binary domains $\mathcal{D}^k$, one per possible color, where, for $\mathbf{x} = [x_e]_{e \in E}$,

$$
\begin{bmatrix} x_0^k \\ \mathbf{x}^k \end{bmatrix} \in \mathcal{D} = \left\{ \begin{bmatrix} x_0 \\ \mathbf{x} \end{bmatrix} \in \{0,1\}^{|E|+1} \middle| \sum_{e \in \delta(\{i\})} x_e \leq x_0,\ \forall i \in N \right\}, \ \forall k \in \{1,\ldots,\kappa\}. \tag{4.176}
$$

Let $\ddot{P} (= P)$ be the index set of extreme points of $\text{conv}(\mathcal{D})$. The zero-vector $\begin{bmatrix} 0 \\ \mathbf{0} \end{bmatrix}$ (indexed by $p = 0$) can be removed from $\mathcal{X} = \left\{ \begin{bmatrix} x_{0p} \\ \mathbf{x}_p \end{bmatrix} \right\}_{p \in \ddot{P}}$, where $\mathbf{x}_p = [x_{ep}]_{e \in E}$. This results in less-than-or-equal-to-1 inequalities for the κ convexity constraints. Aggregation of the λ_p^k-variables into $\lambda_p = \sum_{k=1}^{\kappa} \lambda_p^k$ leads to $\sum_{p \in \ddot{P} \setminus \{0\}} \lambda_p \leq \kappa$, a redundant constraint according to the definition of κ. The (aggregated) $IM\ddot{P}$ reads as

$$
\begin{aligned}
z^\star_{IM\ddot{P}} = \min \quad & \sum_{p \in \ddot{P} \setminus \{0\}} \lambda_p \\
\text{s.t.} \quad & \sum_{p \in \ddot{P} \setminus \{0\}} x_{ep} \lambda_p \geq 1 && \forall e \in E \\
& \lambda_p \text{ binary} && \forall p \in \ddot{P} \setminus \{0\}.
\end{aligned}
$$

Note that the λ_p-variables should be in $\mathbb{Z}_+$ by Proposition 4.10 but they are restricted to binary values as it is useless to over cover the edges while minimizing the number of matchings.

Second compact, without a block-index

Edmonds (1965) observes that for any matching, the induced subgraph on any odd cardinality vertex subset S has at most $(|S|-1)/2$ edges. In fact, this observation is already all that is needed to fully describe the matching polytope, that is, the convex hull of incidence vectors of matchings

$$\operatorname{conv}\left(\left\{\mathbf{x}\in\{0,1\}^{|E|}\;\middle|\;\sum_{e\in\delta(\{i\})} x_e \le 1\right\}\right), \tag{4.177}$$

see Nemhauser and Wolsey (1988, Part III.2 Matching) for a comprehensive study. A *linear* description of the matching polytope is therefore

$$\sum_{e\in\delta(\{i\})} x_e \le 1 \qquad \forall i\in N \tag{4.178a}$$

$$\sum_{e\in E(S)} x_e \le \frac{1}{2}(|S|-1) \qquad \text{for all odd sets } S\subseteq N, \tag{4.178b}$$

$$0\le x_e\le 1 \qquad \forall e\in E, \tag{4.178c}$$

where the constraints (4.178b) are called the *odd set* or *blossom* constraints and the bounds $0\le x_e\le 1$ replace the binary condition $x_e\in\{0,1\}$ because of the integrality property. Let (4.178) describe the *ISP*'s domain $\mathcal{D}$ such that its formulation fulfills Condition ***1*** of Proposition 4.17. Deriving

$$\mathcal{A}=\{\mathbf{x}\in\{0,1\}^{|E|}\mid x_e\ge 1,\ \forall e\in E\} \tag{4.179}$$

from the *IM$\ddot{P}$* formulation, obviously $\mathbf{0}\notin\mathcal{A}$ such that Condition ***2*** is also satisfied. A natural upper bound on any edge variable is obviously $x_e\le 1$. Therefore, using $c_0=1$, $\mathbf{c}=\mathbf{0}$, $\mathbf{A}=\mathbf{I}$ (i.e., $\mathbf{a_x}=\mathbf{x}$), and $\mathbf{u}=\mathbf{1}$, the following "compact" formulation (4.180) in integer x-variables with an exponential number of constraints is derived from Proposition 4.17:

$$z^\star_{ILP}=\min\ x_0 \tag{4.180a}$$

$$\text{s.t.}\ x_e\ge 1 \qquad \forall e\in E \tag{4.180b}$$

$$\sum_{e\in\delta(\{i\})} x_e\le x_0 \qquad \forall i\in N \tag{4.180c}$$

$$\sum_{e\in E(S)} x_e\le\frac{1}{2}(|S|-1)x_0 \qquad \forall S\subseteq N: |S|\ge 3,\ \text{odd} \tag{4.180d}$$

$$x_e\le x_0 \qquad \forall e\in E \tag{4.180e}$$

$$x_0, x_e\in\mathbb{Z}_+ \qquad \forall e\in E. \tag{4.180f}$$

It gives back the set covering formulation (2.35) in a Dantzig-Wolfe reformulation using the grouping of constraints

$$\mathcal{A}(x_0) = \left\{ \begin{bmatrix} x_0 \\ \mathbf{x} \end{bmatrix} \in \mathbb{Z}_+^{n+1} \;\middle|\; (4.180b) \right\} \tag{4.181a}$$

$$\mathcal{D}(x_0) = \left\{ \begin{bmatrix} x_0 \\ \mathbf{x} \end{bmatrix} \in \mathbb{Z}_+^{n+1} \;\middle|\; (4.180c)\text{–}(4.180e) \right\}. \tag{4.181b}$$

Compact formulations are not created equal

Recall Note 4.18 where we express concerns about a compact formulation without index k derived from Proposition 4.17: we *may* lose track of the essence of the problem. This actually is the case with the above compact (4.180). Although its reformulation with a scaling using $x_0 = 1$ for the extreme rays of $\mathcal{D}(x_0)$ results in the set covering formulation (2.35), where the matchings are well identified, *none are available in this ILP*.

If we pay attention to the x_e-variables in (4.180b), we realize they all necessarily take value 1. Moreover, Vizing's theorem tells us that $x_0^\star$ is either Δ or $\Delta + 1$, which means that (4.180c) is tantamount to $\Delta \le x_0 \le \Delta + 1$. Finally, the left-hand side of the odd set constraints in (4.180d) simply counts the number of edges in $E(S)$. We can rewrite another model, simplified, with these observations as

$$z_{ILP}^\star = \min \; x_0 \tag{4.182a}$$

$$x_0 \ge \frac{2|E(S)|}{|S|-1} \qquad \forall S \subseteq N : |S| \ge 3, \text{ odd} \tag{4.182b}$$

$$x_0 \in \{\Delta, \Delta + 1\}. \tag{4.182c}$$

This (third) compact formulation (4.182) has a single variable. An optimal solution does establish the chromatic index but does not give an actual edge coloring. Intuitively, if any constraint in (4.182b) is violated by $x^0 = \Delta$, we trivially establish that an optimal solution is $x_0^\star = \Delta + 1$. We can then even derive an edge coloring in polynomial time (Vizing, 1964). The entire difficulty therefore lies in deciding whether $x_0 = \Delta$ is feasible which requires the verification of *all* constraints in (4.182b). If the optimum is indeed $x_0^\star = \Delta$, this knowledge does not make it any easier to assign colors to edges. In finding the second compact formulation without block-index, we just threw the baby out with the bath water because we lost track of the essence of the real problem.

4.8 Reference Notes

This chapter is the bridge between two algorithms: Column Generation (Chapter 2) and Branch-Price-and-Cut (Chapter 7) in which we see how to deal with solutions that come from linear relaxations until integrality is fulfilled.

Introduction The analogy *atomes vs. molécules* developed by François Soumis originally started as *grains de sable vs. briques*. The former better reflecting that the objects we generate with the pricing problem are 'complex/structured' and of various 'shapes.'

Section 4.1 The convexification of the reformulated domain $\mathcal{D}$ naturally extends the Dantzig-Wolfe decomposition principle to integer linear programs, see for example Barnhart et al. (1998), Desrosiers and Lübbecke (2005), and Lübbecke and Desrosiers (2005).

Section 4.2 The discretization approach ensures that branching decisions can also be done on the λ-variables, in addition to the x-variables already present in the compact formulation. Discretization is generalized to mixed-integer linear programs in that one applies convexification to the continuous part (Vanderbeck and Savelsbergh, 2006). Johnson (1989) has an early paper on the subject. See also the PhD dissertation by Vanderbeck (1994).

Section 4.3 The integrality property of an *ILP formulation* is first defined in Geoffrion (1974). For years, the quality of the lower bound induced by it is essentially investigated in conjunction with the Lagrangian relaxation method (see Fisher (1981) and Guignard (2003) amongst others), a decomposition approach that can be seen as the dual point of view of the Dantzig-Wolfe decomposition. We come back on this aspect in Chapter 6.

Section 4.4 Most of the industrial applications involve block-diagonal structures, that is, several (and different) pricing problems to build up columns of various types. Examples reported on p. 200 range from two to more than a thousand such blocks. Other applications are listed in Lübbecke and Desrosiers (2005), notably in vehicle routing and crew scheduling, see Desrosiers et al. (1995) and Desaulniers et al. (1998a). Proposition 4.11 is new, exploiting the fact that $\text{conv}(\mathcal{D})$ is a polyhedral cone. In that case, a set of identical subproblems allows for a compact formulation written without block indices.

Good to Know The material in Not all blocks are used is new and largely used in the so-called reverse Dantzig-Wolfe decomposition (see More to Know). Clausen et al. (2022) inspired the formulations on the Shared variables across all blocks.

More to Know There are three main results in this section that are devoted to the reverse Dantzig-Wolfe decomposition. Proposition 4.15 is due to Daniel Villeneuve (1999), later published in Villeneuve et al. (2005). Together with Éric Gélinas and Norbert Lingaya, Daniel is one of the architects of the `GENCOL` solver used by the Montréal companies AD OPT for solving routing and scheduling airline problems and GIRO for optimizing bus driver schedules. This proposition regarding the existence of a compact formulation with a set of identical blocks was derived the same day as the block-diagonal structure used in the *Unified framework for deterministic time constrained vehicle routing and crew scheduling problems* (Desaulniers et al., 1998a). The two other results, Propositions 4.16 and 4.17, are new and show that compact formulations without block-index are also possible, depending on the properties of the pricing problem. These are two late answers to François Vanderbeck asking/waiting/hoping for such results.

Finding structures in matrices is an old topic. The graph-based methods go back to Ferris and Horn (1998) and to Aykanat et al. (2004) (using hypergraphs). Bergner et al. (2015) pioneered the graph partitioning ideas in the context of (automatic) Dantzig-Wolfe reformulation of general MILPs. Khaniyev et al. (2018) describe graph clustering for structure detection and suggest several quality scores, in particular modularity. The methods based on constraint and variable classes are documented in code (only): they are part of the general decomposition solver GCG (Gamrath and Lübbecke, 2010). GCG is also used for the machine learning based methods for detecting and exploiting decompositions (Basso and Ceselli, 2023; Kruber et al., 2017).

Examples

4.1 Integrality property in the knapsack problem. The book *Knapsack Problems: Algorithms and Computer Implementations* (Martello and Toth, 1990) is a must.

4.2 Integrality property in the cutting stock problem. As already mentioned in the previous chapter, flow-based formulations allow to develop tight formulations for mixed-integer programs, see Valério de Carvalho (2002), Delorme et al. (2016), Delorme and Iori (2020) and de Lima et al. (2022). The *reverse Dantzig-Wolfe* paragraphs shows that there are alternatives to the compact formulation with a block-diagonal structure. Here we used Proposition 4.15 on the one hand and Proposition 4.16 exploiting the property of a polyhedral cone on the other hand.

4.3 *TCSPP*: nine reformulations. The more the reformulated domain $\mathcal{D}$ contains constraints information, the better is the lower bound $z^\star_{MP}$ on $z^\star_{ILP}$. The trade-off analysis of this problem originally designed in Ahuja et al. (1993, p. 599) explores the impact of a few constraints up to all, the last being irrelevant as it solves the *ILP* within the *ISP*.

4.5 Generalized assignment problem. The reformulation of the semi-assignment constraints giving back the compact formulation is an observation due to Krunal Kishor Patel (fall 2021), a PhD student at Polytechnique Montréal. We wrote Proposition 4.5 accordingly.

4.6 Multi-commodity maximum flow problem. Ford and Fulkerson (1958), "*A suggested computation for maximal multicommodity network flows*," is a seminal paper on the Dantzig-Wolfe decomposition approach, without the name. Wang (2018a,b) are complementary papers on applications, mathematical formulations, and solution methods for various multi-commodity network flow problems.

(*Reinventing the wheel, several times.*) The Floyd-Warshall algorithm is an example of dynamic programming, and was published in its currently recognized form by Robert Floyd in 1962. However, it is essentially the same as algorithms previously published by Bernard Roy in 1959 and also by Stephen Warshall in 1962 for finding the transitive closure of a graph, and is closely related to Kleene's algorithm (published in 1956) for converting a deterministic finite automaton into a regular expression. The modern formulation of the

algorithm as three nested for-loops was first described by Peter Ingerman, also in 1962.
– Wikipedia

(*Robert W Floyd.*) [P]eople often assume that my books are in error or incomplete when I refer to Bob as "Robert W Floyd," since the indexes to my books give a full middle name for almost everybody else. The truth is that he was indeed born with another middle name ["Willoughby"], but he had it legally changed to "W"—just as President Truman's middle name was simply "S". Bob liked to point out that "W." is a valid abbreviation for "W".
– Knuth (2003)

4.8 Design of balanced student teams. The material of this section is inspired by Desrosiers et al. (2005). The difficulties arising from a quadratic objective function are bypassed by the reformulation approach resulting in a set partitioning model.

4.9 Can you decode a secret vote? Work on this problem began in 1985, following a confidential (and deciphered) request. The main results of the mathematical analysis are available in Jaumard and Soumis (1986).

4.10 Edge coloring problem: two compact formulations. To our knowledge, the second compact formulation based on the polyhedral description of the matching polytope by Edmonds (1965) is new. This is an application of Proposition 4.17 requiring the integrality property of the *ISP* formulation. In theory, the convex hull of any integer linear problem can be described by a set of linear constraints, possibly an exponential number, hence there should be an original formulation without a block-index. However, it does not ensure the presence of relevant information to build a viable solution.

Exercises
4.9 Identical subproblems: lexicographic ordering of the extreme points. Formula (4.184) for the computation of the disaggregated λ_p^k-values is due to Vanderbeck and Wolsey (2010, eq. (13.51)).

Exercises

4.1 Hermann Minkowski

When and where did Hermann Minkowski live? List a few of his scientific contributions.

4.2 Alternative decomposition of the 2D illustration

In Figure 4.1, the Dantzig-Wolfe reformulation of the illustrated 2-dimensional *ILP* is based on the convexification of $\mathcal{D}$. The domain of the *MP* is therefore given by $\{\mathbf{x} \in \mathbb{R}^2_+ \mid \mathbf{Ax} \geq \mathbf{b}\} \cap \text{conv}(\mathcal{D})$.
Assume that the reformulation is rather based on the convexification of $\mathcal{A}$. Compare the domain $\{\mathbf{x} \in \mathbb{R}^2_+ \mid \mathbf{Dx} \geq \mathbf{d}\} \cap \text{conv}(\mathcal{A})$ of the reformulated *MP* to the integer hull $\text{conv}(\mathcal{A} \cap \mathcal{D})$ of the *ILP*.

4.3 2D convexification practice
Given below are the respective integer sets $\mathcal{A}$ and $\mathcal{D}$, where the outer border describes the linear relaxation. Complete the exercise by plotting

(a) the complete integer linear program and its linear relaxation.
(b) the convex hull of $\mathcal{D}$ and the feasible region of the *MP* for a Dantzig-Wolfe reformulation based on conv($\mathcal{D}$).
(c) the convex hull of $\mathcal{A}$ and the feasible region of the *MP* for a Dantzig-Wolfe reformulation based on conv($\mathcal{A}$).

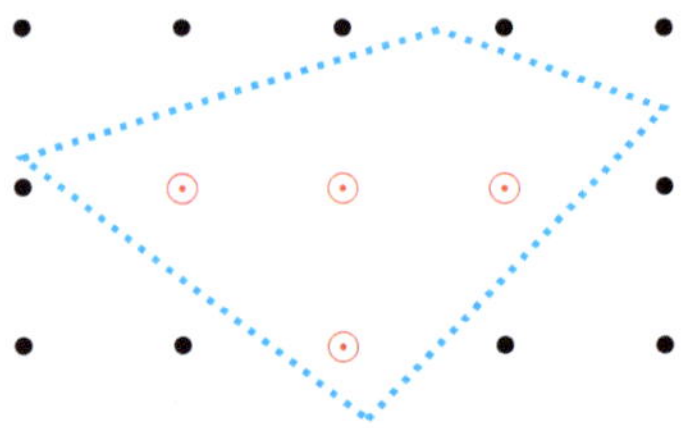
(a) Integer set $\mathcal{A} = \{\mathbf{x} \in \mathbb{Z}^2_+ \mid \mathbf{Ax} \geq \mathbf{b}\}$

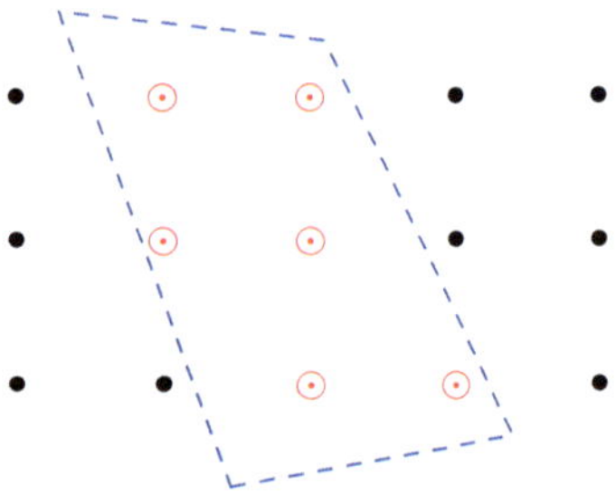
(b) Integer set $\mathcal{D} = \{\mathbf{x} \in \mathbb{Z}^2_+ \mid \mathbf{Dx} \geq \mathbf{d}\}$

Fig. 4.44: Integer sets for Exercise 4.3.

4.4 Reformulation by discretization
(a) In Figure 4.45a, we use the integer scaled rays $(1,0)$ and $(4,1)$. Write points $(17,5)$ and $(9,4)$ according to the Giles-Pulleyblank expression (4.15).
(b) In Figure 4.45b we use the integer scaled rays $(2,0)$ and $(8,2)$. Write points $(6,3)$ and $(7,3)$ according to the Giles-Pulleyblank expression (4.15).

4.5 Trick question
Both reformulations of the compact formulation by Theorems 4.1 (Minkowski-Weyl) and 4.2 (Hilbert-Giles-Pulleyblank) make use of the convexification of $\mathcal{D}$, the domain of the *ISP*. In practice, do we really perform such a convexification?

4.6 Alternative expression for the Dantzig-Wolfe lower bound
Show that the Dantzig-Wolfe lower bound $z_{RMP} + \sum_{k\in K} \bar{c}^k(\boldsymbol{\pi}_\mathbf{b}, \pi_0^k) \leq z^\star_{MP}$ in Proposition 4.12 is independent of all dual values π_0^k, $k \in K$.

4.7 Optimality test for the compact formulation
Given dual values $\boldsymbol{\pi}_\mathbf{b} \geq \mathbf{0}$ and $\pi_0 \in \mathbb{R}$ with respect to the linear relaxation of the *IMP* (4.5), let an optimal solution $\mathbf{x}_p$, $p \in P$, to the *ISP* (4.11) be feasible for the single block compact formulation *ILP* (4.1). Show that $\mathbf{x}_p$ is optimal for the *ILP* if $\boldsymbol{\pi}_\mathbf{b}^\intercal(\mathbf{b} - \mathbf{a}_p) = 0$.

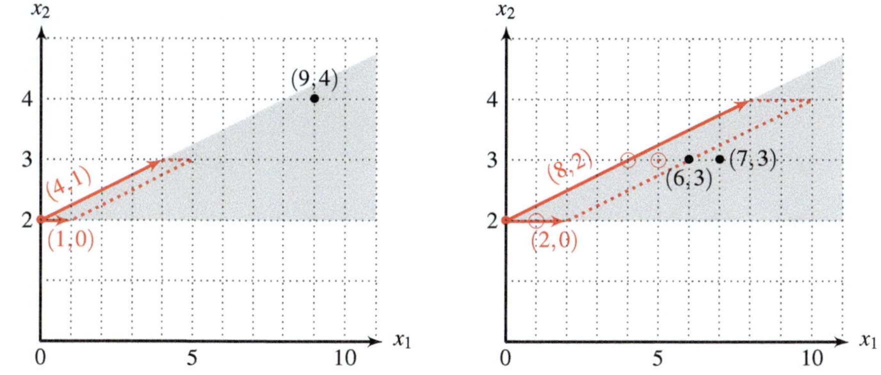

(a) Integer-scaled rays $\{(1,0),(4,1)\}$ (b) Integer-scaled rays $\{(2,0),(8,2)\}$

Fig. 4.45: Different extreme ray scalings give different discretization sets.

4.8 Identical subproblems: solving an aggregated compact formulation
Let the block constraints $\mathbf{D}\mathbf{x}^k \geq \mathbf{d}$, $\forall k \in K$, in the compact formulation (4.47) be summed to $\mathbf{D}(\sum_{k\in K}\mathbf{x}^k) \geq |K|\mathbf{d}$ and let $\mathbf{y} = \sum_{k\in K}\mathbf{x}^k$. Show that solving the *MP* (4.56) in Proposition 4.8 is equivalent to solving the *MP* of a Dantzig-Wolfe reformulation of the following *ILP*

$$\begin{aligned} z^\star_{ILP} = \min \quad & \mathbf{c}^\intercal \mathbf{y} \\ \text{s.t.} \quad & \mathbf{A}\mathbf{y} \geq \mathbf{b} \\ & \mathbf{D}\mathbf{y} \geq |K|\mathbf{d} \\ & \mathbf{y} \in \mathbb{Z}^n_+, \end{aligned} \tag{4.183}$$

using the grouping $\mathcal{A}_\mathbf{y} = \{\mathbf{y} \in \mathbb{Z}^n_+ \mid \mathbf{A}\mathbf{y} \geq \mathbf{b}\}$ and $\mathcal{D}_\mathbf{y} = \{\mathbf{y} \in \mathbb{Z}^n_+ \mid \mathbf{D}\mathbf{y} \geq |K|\mathbf{d}\}$.

4.9 Identical subproblems: lexicographic ordering of the extreme points.
Assume that the numbering of the positive aggregated λ_p-variables in the solution to the final *RMP* is given by the lexicographic order of the $\mathbf{x}_p$-vectors, say from 1 to m. Show that a solution to the transportation problem (4.62) of dimension $m \times |K|$ is given by recursively computing for $p = 1,\ldots,m$:

$$\lambda^k_p = \min\left\{1,\, \lambda_p - \sum_{j=1}^{k-1} \lambda^j_p,\, (k - \sum_{i=1}^{p-1} \lambda_i)^+\right\}, \qquad \text{for } k = 1,\ldots,|K|. \tag{4.184}$$

4.10 Not all blocks are used
From Section 4.5, let the *ILP* (4.81) be reproduced below

$$z^\star_{ILP} = \min \quad \sum_{k\in K} c^k_0 x^k_0 + \mathbf{c}^{k\intercal}\mathbf{x}^k \tag{4.185a}$$

$$\text{s.t.} \quad \sum_{k\in K} \mathbf{A}^k\mathbf{x}^k \geq \mathbf{b} \tag{4.185b}$$

$$\mathbf{D}^k\mathbf{x}^k \geq \mathbf{d}^k x_0^k \qquad \forall k \in K \qquad (4.185c)$$
$$\mathbf{x}^k \leq \mathbf{u}^k x_0^k \qquad \forall k \in K \qquad (4.185d)$$
$$x_0^k \in \{0,1\} \qquad \forall k \in K \qquad (4.185e)$$
$$\mathbf{x}^k \in \mathbb{Z}_+^{n^k} \qquad \forall k \in K, \qquad (4.185f)$$

and rather consider the following grouping of the constraints:

$$\mathcal{A}_\mathbf{0} = \left\{ \left\{ \begin{bmatrix} x_0^k \\ \mathbf{x}^k \end{bmatrix} \in \{0,1\} \times \mathbb{Z}_+^{n^k} \right\}_{k \in K} \,\middle|\, \sum_{k \in K} \mathbf{A}^k \mathbf{x}^k \geq \mathbf{b} \right\} \qquad (4.186a)$$

$$\mathcal{D}_\mathbf{0}^k = \left\{ \begin{bmatrix} x_0^k \\ \mathbf{x}^k \end{bmatrix} \in \mathbb{R}_+ \times \mathbb{Z}_+^{n^k} \,\middle|\, \mathbf{D}^k\mathbf{x}^k \geq \mathbf{d}^k x_0^k,\ \mathbf{x}^k \leq \mathbf{u}^k x_0^k \right\}, \qquad \forall k \in K. \qquad (4.186b)$$

Compared to the grouping in (4.82), observe that $x_0^k \in \{0,1\}$ in $\mathcal{D}_\mathbf{0}^k$ is replaced by $x_0^k \geq 0$. Indeed, we have replaced the polytope $\text{conv}(\mathcal{D}^k)$ by a polyhedral cone with the newly created zero-vector as the single extreme point, see Figure 4.46. Show that the resulting Dantzig-Wolfe reformulation leads to an *IMP* formulation identical to that of (4.84).

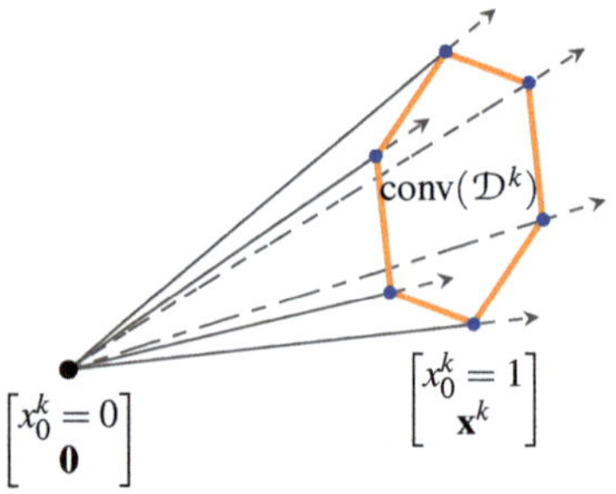

Fig. 4.46: Illustration of $\text{conv}(\mathcal{D}_0^k)$.

4.11 Binary knapsack problem

Show that the formulation (4.103) of the binary knapsack problem does not possess the integrality property using the data in Table 4.13.

i	1	2	3	4
u_i	20	36	18	4
w_i	2	4	3	1
$W = 7$				

Table 4.13: Knapsack capacity and potential items to pack.

4.12 Cutting stock problem: λ-integrality

The reformulation *IMP* (4.109) requests λ-integrality whereas the reformulations IMP_K (4.115) and IMP_{NF} (4.124) do not. How can we validate the λ-integrality of the Gilmore-Gomory formulation using a Dantzig-Wolfe reformulation?

4.13 Time constrained shortest path problem: nine reformulations

(a) For the grouping of the constraints of the *TCSPP* in Example 4.3, write the sets $\mathcal{A}_1$ to $\mathcal{A}_9$, the counterparts of $\mathcal{D}_1$ to $\mathcal{D}_9$.

(b) For the reformulation based on the extreme case set $\mathcal{D}_9$ for which all constraints appear in the *ISP*, write the *IMP*.

4.14 Reformulation of the scene selection problem

(a) In the *ILP* (4.159) of Example 4.7, show that $z^\star_{LP} = \sum_{j\in A} c_j$.

(b) Describe the set of extreme points arising from the proposed grouping (4.160). Give an interpretation/description of a possible *day*-pattern. Recall that this Dantzig-Wolfe reformulation does not suffer from any symmetry due to the permutation of the days (p. 270).

(c) Formulate the $\ddot{IMP}$ based on discretization.

(d) Formulate the *ISP*.

(e) How would you enumerate all the patterns of Example 4.7?

(f) How does $z^\star_{MP}$ compare to $z^\star_{LP}$?

4.15 Design of balanced student teams

Consider Example 4.8 in which an optimal solution to the *IMP* (4.168) is composed of 6 teams. Let the corresponding column-vectors be denoted $\mathbf{b}_1, \ldots, \mathbf{b}_6$.

(a) Assume that in a 2-semester academic program, the teams are broken at the end of the first semester such that no members of a team are together during the second semester. Given the six column-vectors, propose a model to build the new teams.

(b) For our instance, can you provide a model that computes simultaneously the teams for the two semesters?

4.16 Secret ballot

In the following, we examine situations where additional knowledge might help decode the vote in Example 4.9. For each, enumerate the set of patterns in $\mathcal{D}^a$. Assume that we know a more precise share distribution with two decimal digits given in Table 4.14.

Shareholder i	1	2	3	4	5	6	7	8	9	10
Shares (p_i %)	14.33	13.24	12.42	8.43	7.77	6.21	5.66	5.51	4.51	4.24
Shareholder i	11	12	13	14	15	16	17	18	19	20
Shares (p_i %)	3.61	3.07	2.66	2.37	1.49	1.41	1.32	1.13	0.36	0.26

Table 4.14: Percentage shares (two decimal digits).

(a) Table 4.15 gives the first round election results with either two decimal digits or rounded to one decimal place (using a half-up rounding rule, i.e., a decimal number $0.05 \leq x < 0.15$ is rounded to $x = 0.1$). Modify the model (4.169) to account for these situations.

Candidate k	a	b	c	d	e	f	
Votes (b_2^k %)	4.48	11.14	13.76	17.29	17.44	35.89	*(Two decimal digits)*
Votes (b^k %)	4.5	11.1	13.8	17.3	17.4	35.9	*(Rounded)*

Table 4.15: First round results.

(b) Every candidate is a shareholder whom voted for him or herself. From candidate *a* to *f*, the respective shareholders are 16, 11, 15, 3, 8, and 13.

(c) We know the number of shareholders who voted for each candidate. From candidate *a* to *f*, the respective numbers are 2, 3, 4, 3, 4, and 4.

4.17 Cutting stock with rolls of different widths: compact formulation
For the Cutting stock problem with rolls of different widths, Example 2.2:
(a) Give a compact formulation *ILP*.
(b) Propose a grouping of the constraints such that the Dantzig-Wolfe reformulation leads to the *IMP* (2.33), where the λ-variables take non-negative integer values.

4.18 Aircraft routing with schedule synchronization: compact formulation
Given the domain $\mathcal{D}^k$ in (2.45) for the ISP^k, the master problem (2.49) as well as the parameters and encoding functions defined in (2.46)–(2.48), propose a compact formulation for the Aircraft routing with schedule synchronization, Example 2.6.

4.19 Single depot vehicle scheduling problem: compact formulations
Given are:
(1) the Single depot vehicle scheduling problem of Example 2.4 formulated as the following *IMP*, a set partitioning model,

$$z^\star_{IMP} = \min \quad \sum_{\mathbf{x}\in\mathcal{X}} c_\mathbf{x}\lambda_\mathbf{x} \tag{4.187a}$$

$$\text{s.t.} \quad \sum_{\mathbf{x}\in\mathcal{X}} a_{i\mathbf{x}}\lambda_\mathbf{x} = 1 \qquad [\pi_i] \qquad \forall i \in N \tag{4.187b}$$

$$\sum_{\mathbf{x}\in\mathcal{X}} b_\mathbf{x}\lambda_\mathbf{x} \leq v \qquad [\pi_v] \tag{4.187c}$$

$$\lambda_\mathbf{x} \in \{0,1\} \qquad \forall \mathbf{x} \in \mathcal{X}, \tag{4.187d}$$

where v identical vehicles are available at the depot and $\mathcal{X}$ is the index set of the extreme points of the pricing problem (4.188);
(2) the *ISP* formulated as a unit-flow shortest path problem from o to d given the dual values $\boldsymbol{\pi} = [\pi_i]_{i\in N}$ and π_v,

$$\bar{c}(\boldsymbol{\pi}, \pi_v) = \min \quad c_{\mathbf{x}} - \sum_{i \in N} \pi_i a_{i\mathbf{x}} - \pi_v b_{\mathbf{x}} \tag{4.188a}$$

$$\text{s.t.} \quad \sum_{j:(o,j) \in A} x_{oj} = 1 \tag{4.188b}$$

$$\sum_{j:(i,j) \in A} x_{ij} - \sum_{j:(j,i) \in A} x_{ji} = 0 \qquad \forall i \in N \tag{4.188c}$$

$$- \sum_{i:(i,d) \in A} x_{id} = -1 \tag{4.188d}$$

$$x_{ij} \geq 0 \qquad \forall (i,j) \in A \tag{4.188e}$$

$$c_{\mathbf{x}} = \sum_{(i,j) \in A} c_{ij} x_{ij} \tag{4.188f}$$

$$a_{i\mathbf{x}} = \sum_{j:(i,j) \in A} x_{ij} \qquad \forall i \in N \tag{4.188g}$$

$$b_{\mathbf{x}} = \sum_{j:(o,j) \in A} x_{oj} \tag{4.188h}$$

on the network $G = (N, A)$ depicted in Figure 4.47, where arc (o, d) is used if some vehicles remain at the depot, with a possible parking cost.

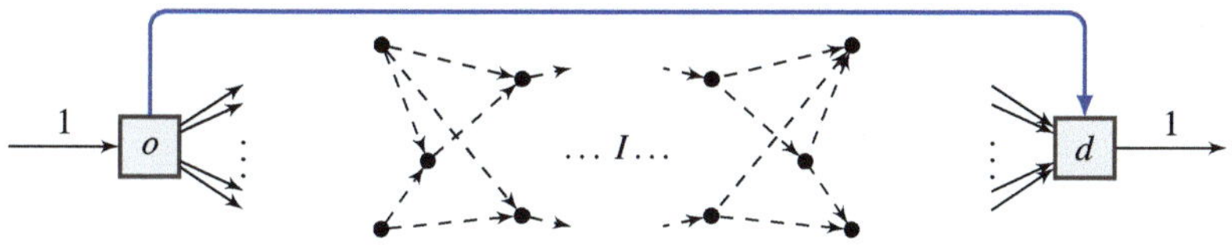

Fig. 4.47: Network $G = (N, A)$ for the *od*-shortest path pricing problem.

(a) Using Proposition 4.15, propose a compact formulation with a block-diagonal structure with identical data across all blocks which need not all be used.

(b) Using Proposition 4.17, give a compact formulation without a block-index. It is known that this *ILP*, a network flow problem, possesses the integrality property. Does the *IMP* reformulation, a set partitioning model, also has the integrality property?

4.20 Multiple depot vehicle scheduling problem

A multi-commodity network flow model for the *MDVSP* (Ribeiro and Soumis, 1994), where commodity k is associated with depot k, is

$$z^{\star}_{ILP} = \min \sum_{k \in K} \sum_{(i,j) \in A^k_{do}} c^k_{ij} x^k_{ij} \tag{4.189a}$$

$$\text{s.t.} \quad \sum_{k \in K} \sum_{j:(i,j) \in A^k} x^k_{ij} = 1 \qquad \forall i \in N \tag{4.189b}$$

$$x^k_{do} \leq v^k \qquad \forall k \in K \tag{4.189c}$$

$$\sum_{j:(i,j)\in A^k_{do}} x^k_{ij} - \sum_{j:(j,i)\in A^k_{do}} x^k_{ji} = 0 \qquad \forall k \in K, i \in N^k \tag{4.189d}$$

$$x^k_{ij} \in \mathbb{Z}_+ \qquad \forall k \in K, (i,j) \in A^k_{do}, \tag{4.189e}$$

where the network $G^k_{do} = (N^k, A^k_{do})$ for depot k is represented as in Figure 4.48, with

- $N^k = N \cup \{o^k, d^k\}$,
- $A^k = I \cup (\{o^k\} \times N) \cup (N \times \{d^k\}) \cup \{(o^k, d^k)\}$,
- $A^k_{do} = A^k \cup \{(d^k, o^k)\}$, and
- arc (d^k, o^k) is utilized to count the number of buses used.

Note that $x^k_{ij} \in \{0,1\}$, $\forall k \in K, (i,j) \in A^k \setminus \{(d^k, o^k)\}$, by constraints (4.189b).

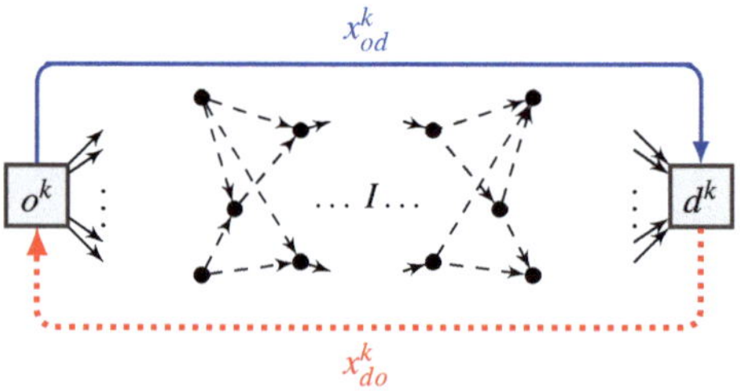

Fig. 4.48: Network $G^k_{do} = (N^k, A^k_{do})$.

(a) What is the row-size of the *ILP*?
(b) Propose a grouping of the constraints in the form $\mathcal{A}$ and $\mathcal{D}$.
(c) Describe the set $\mathcal{X}$ of extreme points and extreme rays of conv($\mathcal{D}$).
(d) Formulate the *IMP*. Give the row-size of the *MP*.
(e) Formulate the *ISP*.
(f) How would you solve the *ISP*?
(g) How does $z^\star_{MP}$ compare to $z^\star_{LP}$?
(h) Assume that there is a vehicle cost c_v, large compared to the traveling costs. How would you initialize the column generation algorithm for solving the *MP*? *Suggestion:* Aggregate all depots into one and solve a Single depot vehicle scheduling problem (*SDVSP*). Formulate this network flow problem. How is an optimal solution used in the *IMP*? – Strategy used in Oukil et al. (2007).

4.21 Useless Dantzig-Wolfe reformulations

Proposition 4.5 shows that if we reformulate $\mathcal{D} = \{\mathbf{x} \in \{0,1\}^n \mid \sum_{j=1}^n x_j = 1\}$, then the *IMP* is no more, no less than the *ILP*. Are there other situations where a Dantzig-Wolfe reformulation gives back the original formulation?

5

Vehicle Routing and Crew Scheduling Problems

Not everything that counts can be counted,
and not everything that can be counted counts.

Informal Sociology: A Casual Introduction to Sociological Thinking
William Bruce Cameron

Abstract In this chapter we illustrate the application of column generation to a variety of vehicle routing and crew scheduling problems, starting with the classical vehicle routing problem with time windows. In particular, we describe dynamic programming algorithms for solving shortest path problems with resource constraints.

Contents

Introduction . . . 297
5.1 Vehicle Routing Problem with Time Windows . . . 297
Problem statement . . . 298
Network structure . . . 298
Illustration 1: A network with four customers . . . 299
Mathematical formulations . . . 300
Compact arc-flow formulation . . . 300
Extended set-partitioning formulation . . . 301
Column generation . . . 304
Elementary shortest paths with time windows and capacity . 304
Labeling algorithm for the *ESPPTWC* . . . 306
Illustration 2: Improved dominance . . . 309
Illustration 3: Improved dominance (cont.) . . . 310
Subproblem relaxations . . . 312
SPPTWC relaxation . . . 313

J. Desrosiers et al., *Branch-and-Price*,
https://doi.org/10.1007/978-3-031-96917-1_5

Illustration 4: Generated labels for the *SPPTWC* 313
ng-SPPTWC relaxation 314
Illustration 5: Generated labels for the *ng-SPPTWC* 317
Illustration 6: Comparative results 317
5.2 Elementary Shortest Path Problem with Resource Constraints 319
Mathematical formulation 320
Labeling algorithm 321
Illustration 7: Common resource extension functions 322
5.3 Examples 324
Simultaneous pickup and delivery problem 324
Network structure 324
Extended formulation 325
Pricing problem 325
Pickup and delivery problem with time windows 326
Network structure 327
Illustration 8: A network for the *PDPTW* 328
Extended formulation 328
Pricing problem 329
Easier-to-solve pricing problem 331
Crew pairing problem with base constraints 332
Network structure 334
Extended formulation 336
Pricing problems 337
5.4 Good to Know 339
Compact linear formulation for the *VRPTW* 339
Extended set partitioning formulation 342
5.5 Reference Notes 343
Exercises 345

Acronyms

SPPRC	shortest path problem with resource constraints	297
ESPPRC	elementary shortest path problem with resource constraints	297
VRPTW	vehicle routing problem with time windows	297
ILP	original or compact integer formulation	300
ISP^k	integer subproblem for block $k \in K$	301
IMP	extended integer formulation	302
MP	linear relaxation of the *IMP*	304
ISP	integer subproblem	304
ESPPTWC	elementary shortest path problem with time windows and capacity	304
REFs	resource extension functions	306
SPPTWC	shortest path problem with time windows and capacity	312
ng-SPPTWC	shortest *ng*-path problem with time windows and capacity	312
SPDP	simultaneous pickup and delivery problem	324

PDPTW	pickup and delivery problem with time windows	324
CPPBC	air crew pairing problem with base constraints	324
SPPTW	shortest path problem with time windows	343
VRPB	vehicle routing problem with backhauls	347

Introduction

Vehicle routing problems are amongst the most studied problems in operations research and their most common variants (the traveling salesperson problem, the capacitated vehicle routing problem, and the vehicle routing problem with time windows) have served for decades as guinea pigs for testing several new algorithmic ideas (such as branch-and-bound and cutting planes) that are now widely used in numerous mathematical programming algorithms. In particular, a school bus routing problem was the first problem modeled as an integer program and solved by an exact column-generation-based algorithm (see Desrosiers et al., 1984), i.e., a branch-and-price algorithm. Soon after, the development of such algorithms following this framework spread to crew scheduling problems, namely, public transit bus driver scheduling and aircrew scheduling. Nowadays, state-of-the-art branch-and-price algorithms have been designed for a wide variety of vehicle routing and crew scheduling problems. These problems can often be formulated using set partitioning, set covering, or generalized set covering models where the set partitioning or covering constraints impose that the tasks (flights, customers, trains, etc.) be covered adequately by a set of columns encoding, e.g., vehicle routes, bus itineraries, crew schedules, or ship routes (see Desrosiers et al., 1995; Desaulniers et al., 1998a). These problems share a common feature: they possess an underlying network structure such that, in a column generation context, the pricing problem is often a *shortest path problem with resource constraints* (*SPPRC*).

The main goal of this chapter is to illustrate the application of column generation to vehicle and crew scheduling problems, by presenting a variety of examples, formulating them, and discussing how column generation can be applied to solve their linear relaxations. In particular, we describe dynamic programming algorithms for solving the *SPPRC* and the elementary *SPPRC* (*ESPPRC*). The discussion about how integer solutions can be computed is postponed to Chapter 7, where branch-price-and-cut is presented.

5.1 Vehicle Routing Problem with Time Windows

The *vehicle routing problem with time windows* (*VRPTW*) is one of the first problems for which a branch-and-price algorithm was developed at the beginning of the 1990s. Since then, numerous papers describing improved algorithms have been published, see Desaulniers et al. (2014); Costa et al. (2019).

Problem statement

The *VRPTW* can be stated as follows. Consider a sufficient number of identical vehicles, each with a capacity Q, that are housed in a single depot. These vehicles are used to deliver merchandise to a set of customers C. With each customer $i \in C$, we associate a demand $q_i \leq Q$ that must be satisfied by a vehicle in a single visit, and a start of service time window $[a_i, b_i]$. Time windows are said to be hard in the following sense. On the one hand, a vehicle can arrive at a customer i earlier than a_i because it can wait to start service. On the other hand, it cannot arrive later than b_i. With the depot, we also associate a time window $[\bar{a}, \bar{b}]$ representing the planning horizon ($\bar{a} < \bar{b}$). Let $c_{ij} \geq 0$ be the travel cost between locations i and j, and $t_{ij} \geq 0$ the corresponding travel time, where i and j are either the depot or a customer. Without loss of generality, we assume that if i is a customer, then t_{ij} includes the service time at i. The objective of the *VRPTW* consists in determining a set of feasible vehicle routes such that each customer is visited exactly once and the total travel cost is minimized. A route is feasible if it starts and ends at the depot, visits each customer at most once (i.e., the route is elementary), the sum of the demands of the visited customers does not exceed Q, and their time windows are respected.

To reflect practice, we assume that

- all parameters Q, q_i, a_i, b_i, and t_{ij} are integers;
- the total travel time (including some service time) of any cycle through a subset of at least two locations is positive;
- the travel times t_{ij} satisfy the triangle inequality.

Definition 5.1. We say that the travel times satisfy the *triangle inequality* if

$$t_{ik} + t_{kj} \geq t_{ij}, \qquad \forall (i,k,j) \in \Lambda,$$

where Λ is the set of all triplets of distinct locations (customers or depot). That is, it is at least as fast to travel directly from i to j than have an in-between stop at k. For travel times proportional to Euclidean distances, this is always true.

Network structure

We model the *VRPTW* using the following network $G = (N, A)$. The node set $N = C \cup \{o, d\}$ contains a node for each customer in C and two nodes o and d representing the depot at the beginning and the end of the routes, respectively. With nodes o and d, we associate the fictitious demands $q_o = q_d = 0$ and the time windows $[a_o, b_o] = [a_d, b_d] = [\bar{a}, \bar{b}]$. Furthermore, we set $c_{od} = t_{od} = 0$ on arc (o, d). The arc set A contains all arcs $(i, j) \in (N \setminus \{d\}) \times (N \setminus \{o\})$ such that $q_i + q_j \leq Q$ and $a_i + t_{ij} \leq b_j$, i.e., those that can possibly be part of a feasible route.

Illustration 5.1 **A network with four customers**

Figure 5.1 provides an example of a network G. In this example, $C = \{1,2,3,4\}$, the time windows $[a_i, b_i]$ are given beside the nodes, and the travel costs and times (c_{ij}, t_{ij}) are specified beside the arcs. We assume that $Q = 5$, $q_1 = q_3 = 1$, and $q_2 = q_4 = 2$. Note that arc $(4,2) \notin A$ because t_{42} is such that $a_4 + t_{42} > b_2$. Other arcs are discarded for similar reasons. A feasible route corresponds to a path from o to d. However, not all o-d paths in G are feasible. For instance, path $o234d$ is not feasible even if all arcs $(o,2), (2,3), (3,4), (4,d) \in A$. Indeed, following this path, the earliest start of service time at customer 3 is $8 + 5 = 13$, which makes it impossible to reach customer 4 before $b_4 = 18$ (i.e., $13 + 6 \not\leq 18$). Observe that G is not acyclic. Indeed, both arcs $(2,3)$ and $(3,2)$ exist and form a cycle. Moreover, there exist non-elementary paths (for example, path $o1343d$) that respect the time windows and the capacity constraint. Obviously, this path cannot be part of a feasible *VRPTW* solution given that each customer must be visited exactly once.

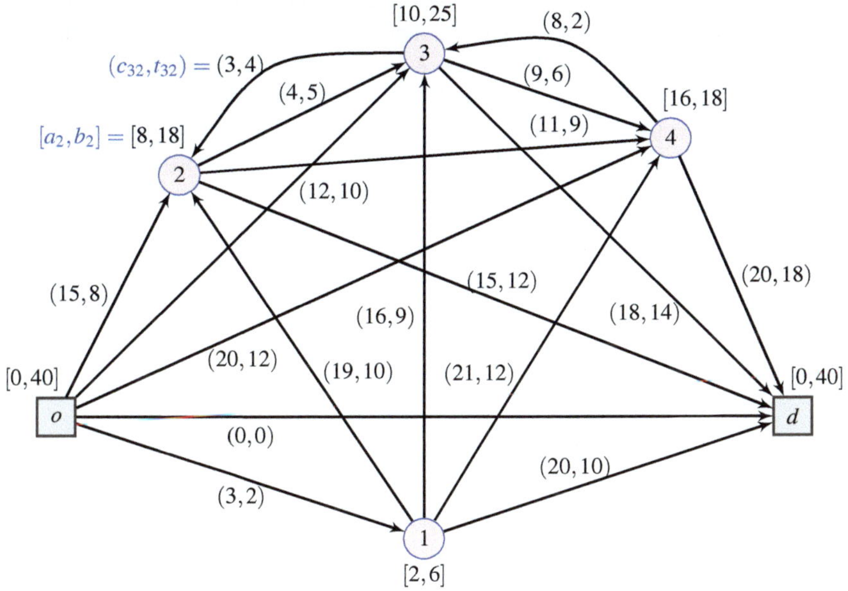

Fig. 5.1: A network $G = (N,A)$ for the *VRPTW*.

The optimal solution to this example is formed of the two feasible routes $o32d$ and $o14d$ with respective costs 30 and 44, for a total cost of 74. The total demand of route $o32d$ is 3 and that of route $o14d$ is 3, both respecting the vehicle capacity $Q = 5$.

Furthermore, using the earliest service start times, a feasible schedule for the first route indicates to start service at times 10 and 14 at customers 3 and 2, respectively, after leaving the depot o at time 0 and before returning to the depot d at time 26. Similarly, the earliest times 0, 2, 16 and 34 at nodes o, 1, 4 and d, respectively, define a schedule that respects the time windows along the second route. Observe that, with the latter schedule, the vehicle needs to wait two units of time between customers 1 and 4.

Mathematical formulations

Various formulations can be proposed for the *VRPTW*. In the following, we present a compact fromulation and an extended one.

Compact arc-flow formulation

Let K be a set of vehicles that is sufficiently large to be unconstraining. With each arc $(i,j) \in A$ and each vehicle $k \in K$, we associate the binary variable x_{ij}^k that takes value 1 if arc (i,j) is traversed by vehicle k and 0 otherwise. Furthermore, with each node $i \in N$ and each vehicle $k \in K$, we define a continuous variable t_i^k that indicates the start of service time at node i if this node is visited by vehicle k or takes value 0 otherwise. This start of service time corresponds to the departure time from the depot for $i = o$ and to the arrival time at the depot for $i = d$.

The *VRPTW* can be formulated as the following mixed-*ILP*:

$$z_{ILP}^{\star} = \min \sum_{k \in K} \sum_{(i,j) \in A} c_{ij} x_{ij}^k \tag{5.1a}$$

$$\text{s.t.} \quad \sum_{k \in K} \sum_{j:(i,j) \in A} x_{ij}^k = 1 \qquad \forall i \in C \tag{5.1b}$$

$$\sum_{j:(i,j) \in A} x_{ij}^k - \sum_{j:(j,i) \in A} x_{ji}^k = \begin{cases} 1 & \text{for } i = o \\ 0 & \forall i \in C \\ -1 & \text{for } i = d \end{cases} \qquad \forall k \in K \tag{5.1c}$$

$$\sum_{i \in C} \sum_{j:(i,j) \in A} q_i x_{ij}^k \leq Q \qquad \forall k \in K \tag{5.1d}$$

$$a_i \leq t_i^k \leq b_i \qquad \forall k \in K, i \in \{o,d\} \tag{5.1e}$$

$$a_i \Big(\sum_{j:(i,j) \in A} x_{ij}^k \Big) \leq t_i^k \leq b_i \Big(\sum_{j:(i,j) \in A} x_{ij}^k \Big) \qquad \forall k \in K, i \in C \tag{5.1f}$$

$$x_{ij}^k (t_i^k + t_{ij} - t_j^k) \leq 0 \qquad \forall k \in K, (i,j) \in A \tag{5.1g}$$

$$x_{ij}^k \in \{0,1\} \qquad \forall k \in K, (i,j) \in A. \tag{5.1h}$$

Constraints (5.1b) ensure that each customer is visited exactly once. Constraint set (5.1c) defines the structure of a path from o to d for each vehicle $k \in K$. We impose vehicle capacity for each vehicle through constraints (5.1d), and express time windows at the depots and the customers through (5.1e) and (5.1f), respectively. Finally, constraints (5.1g) link the arc-flow variables with the time variables. Such a constraint states that if $x_{ij}^k = 1$, then $t_j^k \geq t_i^k + t_{ij}$; otherwise $x_{ij}^k = 0$ and it is trivially satisfied. These constraints prevent the formation of cycles because the time along a cycle is assumed to increase.

As presented, constraints (5.1g) are non-linear but they can be linearized as follows: $t_i^k + t_{ij} \leq t_j^k + M(1 - x_{ij}^k), \forall k \in K, (i,j) \in A$, where M is a large constant that can be set to $b_j - a_i$. Moreover, we can discard all the constraints for which $b_i + t_{ij} \leq a_j$, $(i,j) \in A$, as they are always satisfied.

Extended set-partitioning formulation

Model (5.1) has a block-diagonal structure (see Section 4.4), where the set $\mathcal{D}^k$ is defined by the constraints (5.1c)–(5.1h) for each vehicle $k \in K$. To derive an extended formulation, we apply a Dantzig-Wolfe reformulation using the convexification of the ISP^k domain. Let

$$\mathcal{A} = \left\{ \left\{ \mathbf{x}^k \in \{0,1\}^{|A|} \right\}_{k \in K} \;\middle|\; \sum_{k \in K} \sum_{j:(i,j) \in A} x_{ij}^k = 1,\ \forall i \in C \right\} \tag{5.2a}$$

$$\mathcal{D}^k = \left\{ \begin{bmatrix} \mathbf{x}^k \\ \mathbf{t}^k \end{bmatrix} \in \{0,1\}^{|A|} \times \mathbb{R}^{|C|+2} \;\middle|\; \mathbf{D}^k \begin{bmatrix} \mathbf{x}^k \\ \mathbf{t}^k \end{bmatrix} \geq \mathbf{d}^k \right\}, \qquad \forall k \in K, \tag{5.2b}$$

where the set of constraints $\mathbf{D}^k \begin{bmatrix} \mathbf{x}^k \\ \mathbf{t}^k \end{bmatrix} \geq \mathbf{d}^k$ denotes the constraints (5.1c)–(5.1g) associated with vehicle k. Observing that all domains $\mathcal{D}^k$ are identical, we can aggregate them into a single one as discussed in Section 4.4 and omit index k. The resulting domain $\mathcal{D}$ is defined by the following constraints:

$$\sum_{j:(i,j) \in A} x_{ij} - \sum_{j:(j,i) \in A} x_{ji} = \begin{cases} 1 & \text{for } i = o \\ 0 & \forall i \in C \\ -1 & \text{for } i = d \end{cases} \tag{5.3a}$$

$$\sum_{i \in C} \sum_{j:(i,j) \in A} q_i x_{ij} \leq Q \tag{5.3b}$$

$$a_i \leq t_i \leq b_i \qquad \forall i \in \{o,d\} \tag{5.3c}$$

$$a_i\Big(\sum_{j:(i,j) \in A} x_{ij}\Big) \leq t_i \leq b_i\Big(\sum_{j:(i,j) \in A} x_{ij}\Big) \qquad \forall i \in C \tag{5.3d}$$

$$x_{ij}(t_i + t_{ij} - t_j) \leq 0 \qquad \forall (i,j) \in A \tag{5.3e}$$

$$x_{ij} \in \{0,1\} \qquad \forall (i,j) \in A. \tag{5.3f}$$

This domain is bounded and, thus, its convex hull can be described using only extreme points. As in the previous chapter, we denote by P, the index set of these extreme points. The extreme point indexed by $p \in P$ is denoted $(\mathbf{x}_p, \mathbf{t}_p)$. It specifies a route and a schedule, i.e., a start of service time at every visited node along this route (the service start time at every other node is set to zero).

Note 5.1 (The nature of the extreme points.) An extreme point $(\mathbf{x}_p, \mathbf{t}_p)$, $p \in P$, is such that the subvector $\mathbf{x}_p$ defines a single path $(i_0 = o, i_1, \ldots, i_m, i_{m+1} = d)$ in network G visiting m customer nodes. For all $i \in C \setminus \{i_1, \ldots, i_m\}$, $t_{ip} = 0$. The other time components t_{ip}, $i \in \{i_1, \ldots, i_m\}$, represent an extreme point of the domain defined by

$$a_i \leq t_i \leq b_i \qquad \forall i \in \{i_0, i_1, \ldots, i_m, i_{m+1}\} \tag{5.4a}$$

$$t_{i_j} + t_{i_j, i_{j+1}} \leq t_{i_{j+1}} \qquad \forall j \in \{0, 1, \ldots, m\}. \tag{5.4b}$$

Therefore, each route can induce multiple extreme points that differ by their schedule. One such schedule is given by the earliest start of service time at each node according to the chosen route. Another is obtained by considering the latest service start times. Several others derived from a mix of earliest and latest start times may also yield an extreme point.

For an extreme point $(\mathbf{x}_p, \mathbf{t}_p)$, $p \in P$, let

$$c_p = c(\mathbf{x}_p, \mathbf{t}_p) = \sum_{(i,j) \in A} c_{ij} x_{ij} \tag{5.5}$$

be its cost and

$$a_{ip} = a_i(\mathbf{x}_p, \mathbf{t}_p) = \sum_{j:(i,j) \in A} x_{ij} \tag{5.6}$$

be a binary parameter defined for each customer $i \in C$ that indicates if the route associated with $(\mathbf{x}_p, \mathbf{t}_p)$ visits or not customer i.

Convexifying $\mathcal{D}$ to apply a Dantzig-Wolfe reformulation to model (5.1) yields the following *IMP*, see formulation (4.55) adapted to the bounded case:

$$z^\star_{IMP} = \min \quad \sum_{p \in P} c_p \lambda_p \tag{5.7a}$$

$$\text{s.t.} \quad \sum_{p \in P} a_{ip} \lambda_p = 1 \qquad \forall i \in C \tag{5.7b}$$

$$\sum_{p \in P} \lambda_p = |K| \tag{5.7c}$$

$$\lambda_p \geq 0 \qquad \forall p \in P \tag{5.7d}$$

$$\lambda_p = \sum_{k \in K} \lambda_p^k \qquad \forall p \in P \tag{5.7e}$$

$$\sum_{k \in K} \lambda_p^k = 1 \qquad \forall k \in K \tag{5.7f}$$

$$\lambda_p^k \geq 0 \qquad \forall k \in K, p \in P \tag{5.7g}$$

$$\sum_{p \in P} \mathbf{x}_p \lambda_p^k = \mathbf{x}^k \in \{0,1\} \qquad \forall k \in K, \tag{5.7h}$$

where the variables λ_p^k, $p \in P$, $k \in K$, are the weights associated with the extreme points $(\mathbf{x}_p, \mathbf{t}_p)$ in the convex combination for vehicle k and the variables λ_p, $p \in P$, are the sums of these weights per vehicle. Constraints (5.7b) impose a single visit to each customer, whereas constraints (5.7c) ensure that each vehicle is assigned to a route-schedule, possibly the empty one represented by arc (o,d).

Note 5.2 (The time components are not encoded.) The encoding $(c_p, \mathbf{a}_p)$ of every extreme point $(\mathbf{x}_p, \mathbf{t}_p)$ does not depend on the time components. Therefore, instead of considering the complete set of extreme points $(\mathbf{x}_p, \mathbf{t}_p)$, $p \in P$, it is sufficient to consider a subset of it, indexed by $\hat{P} \subseteq P$, that contains a single extreme point $(\mathbf{x}_{\hat{p}}, \mathbf{t}_{\hat{p}})$ representing all extreme points $(\mathbf{x}_p, \mathbf{t}_p)$, $p \in P$, such that $\mathbf{x}_p = \mathbf{x}_{\hat{p}}$. For instance, $\mathbf{t}_{\hat{p}}$ may be the schedule formed of all the earliest start of service times for the route represented by $\mathbf{x}_{\hat{p}}$. Furthermore, given that all $\mathbf{x}$ variables are binary, constraints (5.7d)–(5.7h) imply that all variables λ_p, $p \in \hat{P}$, are also binary by Proposition 4.4. Finally, since $|K|$ is assumed unrestrictive and there is a zero-cost empty route-schedule, constraint (5.7c) can be omitted. However, they are usually kept and used in the branching tree, see Note 4.16.

These observations yield the following equivalent *IMP*:

$$z^\star_{IMP} = \min \quad \sum_{p \in \hat{P}} c_p \lambda_p \tag{5.8a}$$

$$\text{s.t.} \quad \sum_{p \in \hat{P}} a_{ip} \lambda_p = 1 \qquad [\pi_i] \qquad \forall i \in C \tag{5.8b}$$

$$\sum_{p \in P} \lambda_p = |K| \qquad [\pi_o] \tag{5.8c}$$

$$\lambda_p \in \{0,1\} \qquad \forall p \in \hat{P}, \tag{5.8d}$$

where the dual variables $\pi_i \in \mathbb{R}$, $\forall i \in C$, and π_o appear in the linear relaxation. For ease of notation, the dual variable associated with the aggregated convexity constraint (5.8c) is here denoted π_o as we count the number of vehicles used as those exiting the origin depot.

Note 5.3 (How about encoding the time components?) This simplification of the *IMP* to binary λ-variables is possible because the time components $\mathbf{t}_p$ of the original extreme points $(\mathbf{x}_p, \mathbf{t}_p)$, $p \in P$, do not play a direct role in the *IMP*. They are only useful to ensure that every route associated with an extreme point indexed in $\hat{P}$ respects the time windows. This simplification is not possible when the column encodings involve time components, for example, in vehicle routing applications with vehicle synchronization constraints (Example 2.6) or with time-dependent delivery costs. For the sake of conciseness, we consider in the following that each extreme point $(\mathbf{x}_p, \mathbf{t}_p)$, $p \in \hat{P}$, is only associated with a route (and not a schedule). We also call $\hat{P}$ the set of feasible routes and designate a route by the index $p \in \hat{P}$ of its corresponding extreme point, i.e., we sometimes say route p instead of the route represented by the extreme point $(\mathbf{x}_p, \mathbf{t}_p)$ or associated with the λ_p-variable.

The linear relaxation of model (5.8) is known to be much tighter, in general, than the linear relaxation of model (5.1), see Proposition 4.1. Because of this advantage, the state-of-the-art is to solve this extended formulation of the *VRPTW* by branch-price-and-cut, see Chapter 7. Below, we discuss how to solve the *MP* issued from (5.8) by column generation.

Column generation

In practice, the *MP* contains a very large number of variables, namely, one per feasible route or, equivalently, one per extreme point of the convex hull of the domain $\mathcal{D}$ defined by constraints (5.3). Column generation must, thus, be applied to solve it. In this case, the *ISP* is an elementary shortest path problem with time windows and capacity (*ESPPTWC*). This *ISP* is typically solved using a dynamic programming algorithm, more specifically, a labeling algorithm. Given that solving elementary shortest path problems can be highly time-consuming, relaxations of the *ISP* are often used to ease the route generation process. Below, we define the *ISP*, describe a labeling algorithm for solving it, and discuss two *ISP* relaxations.

Elementary shortest path problem with time windows and capacity

Given a vector $\boldsymbol{\pi} = [\pi_i]_{i \in C}$ of dual values associated with constraint set (5.8b) and $\pi_o = 0$ because the aggregated convexity constraint is obviously not binding (see Note 4.16), the *ISP* consists of finding a feasible route, i.e., its associated extreme point $(\mathbf{x}_p, \mathbf{t}_p)$, $p \in \hat{P}$, with a negative reduced cost. The reduced cost of a route $p \in \hat{P}$ is given by

$$\bar{c}_p = -\pi_o + c_p - \sum_{i \in C} \pi_i a_{ip}, \tag{5.9}$$

or, equivalently,

$$\bar{c}_p = -\sum_{j:(o,j) \in A} \pi_o x_{ojp} + \sum_{(i,j) \in A} c_{ij} x_{ijp} - \sum_{i \in C} \sum_{j:(i,j) \in A} \pi_i x_{ijp} = \sum_{(i,j) \in A} (c_{ij} - \pi_i) x_{ijp}. \tag{5.10}$$

We denote by $\tilde{c}_{ij} = c_{ij} - \pi_i$ the *adjusted cost* of arc $(i, j) \in A$ and write the objective function of the *ISP* as

$$\min \sum_{(i,j) \in A} \tilde{c}_{ij} x_{ij}. \tag{5.11}$$

Note that, for each vehicle $k \in K$, the capacity constraint (5.1d) can be written similarly to the time window constraints (5.1e)–(5.1g). Indeed, dropping index k as well as introducing a variable t_i^{load} for each node $i \in N$ that indicates the load accumulated up to node i, vehicle capacity can be modeled using the following constraints:

$$0 \leq t_i^{load} \leq Q \qquad \forall i \in \{o,d\} \tag{5.12a}$$

$$0 \leq t_i^{load} \leq Q(\sum_{j:(i,j)\in A} x_{ij}) \qquad \forall i \in C \tag{5.12b}$$

$$x_{ij}(t_i^{load} + q_j - t_j^{load}) \leq 0 \qquad \forall (i,j) \in A. \tag{5.12c}$$

Consequently, the *ISP* can be formulated as

$$\bar{c}(\boldsymbol{\pi}) = \min \sum_{(i,j)\in A} \tilde{c}_{ij} x_{ij} \tag{5.13a}$$

$$\text{s.t.} \quad \sum_{j:(i,j)\in A} x_{ij} - \sum_{j:(j,i)\in A} x_{ji} = \begin{cases} 1 & \text{for } i = o \\ 0 & \forall i \in C \\ -1 & \text{for } i = d \end{cases} \tag{5.13b}$$

$$0 \leq t_i^{load} \leq Q \qquad \forall i \in \{o,d\} \tag{5.13c}$$

$$0 \leq t_i^{load} \leq Q(\sum_{j:(i,j)\in A} x_{ij}) \qquad \forall i \in C \tag{5.13d}$$

$$x_{ij}(t_i^{load} + q_j - t_j^{load}) \leq 0 \qquad \forall (i,j) \in A \tag{5.13e}$$

$$a_i \leq t_i^{time} \leq b_i \qquad \forall i \in \{o,d\} \tag{5.13f}$$

$$a_i(\sum_{j:(i,j)\in A} x_{ij}) \leq t_i^{time} \leq b_i(\sum_{j:(i,j)\in A} x_{ij}) \qquad \forall i \in C \tag{5.13g}$$

$$x_{ij}(t_i^{time} + t_{ij} - t_j^{time}) \leq 0 \qquad \forall (i,j) \in A \tag{5.13h}$$

$$x_{ij} \in \{0,1\} \qquad \forall (i,j) \in A, \tag{5.13i}$$

where we now denote by t_i^{time} the time variable at node $i \in N$. In this formulation, the objective function (5.13a) aims at minimizing the reduced cost of the selected path-variable in the *MP*. Constraints (5.13b), (5.13f)–(5.13i) are equivalent to (5.1c), (5.1e)–(5.1h) for a given vehicle k, whereas constraints (5.13c)–(5.13e) enforce vehicle capacity. Note that, compared to (5.3), the domain of this *ISP* formulation contains additional load variables. Therefore, its extreme points involve additional components which express a loading schedule. As for the time components, these load components are not involved in the column encoding of the *VRPTW*.

Note 5.4 (Cycle elimination.) Observe that model (5.13) does not allow cycles in a feasible path because all time variables t_i^{time}, $i \in N$, must take a unique value and all cycles in G are assumed to have a positive duration. We purposely model the capacity constraint similarly to the time window constraints to highlight their common structure. They are generally called *resource constraints* (see Section 5.2) and the *ISP* is, thus, an *ESPPRC*, more precisely, an *ESPPTWC* in this case.

The *ISP* is typically not solved using an algorithm based on model (5.13). It is rather solved using a labeling algorithm applied on the underlying network G as discussed next.

Labeling algorithm for the *ESPPTWC*

As previously stated, the *ESPPTWC* consists of finding a least-cost *o-d* path in network G amongst those respecting the time windows, the vehicle capacity, and the elementary requirements. The cost of a path is computed as the sum of the adjusted costs $\tilde{c}_{ij}$ of its arcs (i,j) and is, thus, equal to the reduced cost of the corresponding route variable with respect to the current dual solution $\boldsymbol{\pi}$. Given that the adjusted costs $\tilde{c}_{ij}$, $(i,j) \in A$, can be negative, the cycles in G can have a negative cost and be attractive. Consequently, the labeling algorithm for solving the *ESPPTWC* must include an explicit mechanism to forbid them.

A labeling algorithm is a dynamic programming algorithm where a partial path from o to a node $j \in N$ and its attributes are represented by a multi-dimensional vector, called a *label*, associated with node j. Starting from an initial label at source node o, the algorithm creates iteratively partial paths by extending through the network the partial paths previously created using *resource extension functions* (REFs). After every extension, the feasibility of the newly created partial path is checked against so-called *resource windows* and, when deemed infeasible, its corresponding label is discarded from the extension process. To avoid enumerating all feasible *o-d* paths, a *dominance rule* can be applied to keep only Pareto-optimal labels, ensuring that all prefixes of at least one optimal *o-d* path are preserved. Before providing the pseudo-code of a labeling algorithm, we discuss in detail these concepts for the *ESPPTWC*.

Each partial path $p = (i_0 = o, i_1, \ldots, i_m = j)$ from o up to a node $j \in N$ is represented by a label

$$E_p = \left(T_p^{rCost}, T_p^{time}, T_p^{load}, [T_p^{cust_i}]_{i \in C}\right) \tag{5.14}$$

at node j. Its components are defined as follows:

$T_p^{rCost} = \sum_{\ell=1}^{m} \tilde{c}_{i_{\ell-1},i_\ell}$ is the reduced cost;

$T_p^{time} = T_{pm}^{time}$ is the earliest feasible time computed recursively as

$$\begin{aligned} T_{p0}^{time} &= a_{i_0} \\ T_{p\ell}^{time} &= \max\{a_{i_\ell}, T_{p,\ell-1}^{time} + t_{i_{\ell-1},i_\ell}\}, \qquad \ell = 1,\ldots,m; \end{aligned}$$

$T_p^{load} = \sum_{\ell=1}^{m} q_{i_\ell}$ is the accumulated load;

$T_p^{cust_i}$, $i \in C$, is an indicator equal to 1 if customer i is visited, i.e., if there exists $\ell \in \{1,\ldots,m\}$ such that $i_\ell = i$, and 0 otherwise.

All components beside the cost component are needed to check the feasibility of path p. They are associated with so-called *resources* (time, load, number of visits to each customer) which are consumed along the arcs and must respect constraints expressed as resource windows at each visited node. The resource window at node $j \in N$ is given by the corresponding time window $[a_j, b_j]$ for the time resource, $[0, Q]$ for the load resource, and $[0, 1]$ for each customer resource. The above partial path p represented by label $E_p = \left(T_p^{rCost}, T_p^{time}, T_p^{load}, [T_p^{cust_i}]_{i \in C}\right)$ is feasible if and only if

$$T_p^{time} \in [a_j, b_j] \tag{5.15a}$$

$$T_p^{load} \in [0, Q] \tag{5.15b}$$

$$T_p^{cust_i} \in [0, 1], \quad \forall i \in C. \tag{5.15c}$$

For convenience, the reduced cost is also considered as a resource without windows except at the depot node d, where a zero upper bound can be imposed. We denote by $\mathcal{R}$ the set of all resources, i.e.,

$$\mathcal{R} = \{rCost, time, load, [cust_i]_{i \in C}\}. \tag{5.16}$$

The initial label E_0 at node o represents the single-node partial path $p_0 = (o)$ and is set to

$$E_0 = \left(0, a_o, 0, [0]_{i \in C}\right). \tag{5.17}$$

To create new partial path $p' = (i_0 = o, i_1, \ldots, i_m = j, h)$ from path $p = (i_0, i_1, \ldots, i_m)$, label E_p is extended along arc $(j, h) \in A$. This extension yields a new label

$$E_{p'} = \left(T_{p'}^{rCost}, T_{p'}^{time}, T_{p'}^{load}, [T_{p'}^{cust_i}]_{i \in C}\right),$$

where

$$T_{p'}^{rCost} = f_{jh}^{rCost}(T_p^{rCost}) = T_p^{rCost} + \tilde{c}_{jh} \tag{5.18a}$$

$$T_{p'}^{time} = f_{jh}^{time}(T_p^{time}) = \max\{a_h, T_p^{time} + t_{jh}\} \tag{5.18b}$$

$$T_{p'}^{load} = f_{jh}^{load}(T_p^{load}) = T_p^{load} + q_h \tag{5.18c}$$

$$T_{p'}^{cust_i} = f_{jh}^{cust_i}(T_p^{cust_i}) = \begin{cases} T_p^{cust_i} + 1 & \text{if } h = i \\ T_p^{cust_i} & \text{otherwise} \end{cases} \quad \forall i \in C, \tag{5.18d}$$

and the functions $f_{jh}^r(\cdot)$, $r \in R$, $(j, h) \in A$, are the REFs. Label $E_{p'}$ is discarded if deemed infeasible according to (5.15).

When applied iteratively on network G, the extension process presented above can generate all feasible o-d paths. In practice, this might not be possible because there may exist a very large number of feasible paths. To avoid enumerating all of them, a dominance rule is applied. Such a rule, typically, compares pairs of labels associated with the same node and is based on a dominance definition. Before presenting it, let us define what is a feasible extension of a path.

Definition 5.2. Let $p = (i_0 = o, i_1, \ldots, i_m)$ be a feasible partial path starting at node o. A *feasible extension* of p is a path $\chi = (j_1, j_2, \ldots, j_\ell)$ such that $i_m = j_1$ and the path $p' = (i_0 = o, i_1, \ldots, i_m = j_1, j_2, \ldots, j_\ell)$ obtained by concatenating p and χ is feasible.

The $\oplus$ symbol is used to denote the concatenation of two paths, i.e., $p' = p \oplus \chi$ in Definition 5.2. When searching for a feasible shortest o-d path, a partial path p starting in node o can be discarded from the search if it can be proven that every path $p \oplus \chi$ resulting from any of its feasible extensions χ is not a shortest path amongst all feasible paths reaching the end node of $p \oplus \chi$, or at least not a unique one. In general, the following more restrictive dominance definition is adopted.

Definition 5.3. Let p and p' be two feasible partial paths from o to $j \in N$. Label E_p (resp., path p) is said to *dominate* label $E_{p'}$ (resp., path p') if, for every feasible (single- or multiple-arc) extension χ' of p', $p \oplus \chi'$ is feasible and $\bar{c}_{p \oplus \chi'} \leq \bar{c}_{p' \oplus \chi'}$.

Clearly, if E_p dominates $E_{p'}$ but not the opposite, label $E_{p'}$ can be safely discarded because no feasible extension of p' can yield a better path than the best path resulting from an extension of p. On the other hand, if both labels E_p and $E_{p'}$ dominate each other, then one of them must be kept while the other can be discarded.

Because Definition 5.3 cannot be used in practice (finding all feasible extensions of p' is often intractable), we rather rely on *sufficient* conditions to identify dominated labels. Given that all REFs $f^r_{jh}(\cdot)$ defined in (5.18) are non-decreasing functions, the following conditions are sufficient:

$$T_p^{rCost} \leq T_{p'}^{rCost} \tag{5.19a}$$

$$T_p^{time} \leq T_{p'}^{time} \tag{5.19b}$$

$$T_p^{load} \leq T_{p'}^{load} \tag{5.19c}$$

$$T_p^{cust_i} \leq T_{p'}^{cust_i}, \qquad \forall i \in C. \tag{5.19d}$$

The dominance rule is then stated as:

If conditions (5.19) *hold, then label E_p dominates label $E_{p'}$.*

When not all conditions (5.19) are satisfied, we say that label E_p does not dominate label $E_{p'}$ because we cannot prove the opposite with these conditions.

The number of feasible paths enumerated by the labeling process depends on the capability of the dominance rule to identify dominated labels. As proposed by Feillet et al. (2004), this rule can be strengthened by introducing the notion of unreachability and revising the definition of the customer resources. This notion is motivated by the fact that, in practice, the time windows of the customers are, typically, spread throughout the planning horizon. Therefore, when the time component T_p^{time} of a label E_p representing a path p is relatively large, it becomes infeasible to extend this path to the customer nodes with relatively early time windows, which are then deemed unreachable. In this case, the customer components $T_p^{cust_i}$ associated with these unreachable customers i are set to 1 to ease dominance.

Illustration 5.2 **Improved dominance**

Consider the network of Figure 5.1 and the two paths $p = (o,1,3)$ and $p' = (o,2,3)$ which both end at node 3. Let us assume that $\pi_2 = 20$ and $\pi_3 = 32$. Therefore, $\tilde{c}_{o1} = 3$, $\tilde{c}_{o2} = 15$, $\tilde{c}_{13} = -6$, and $\tilde{c}_{23} = -16$. The labels representing p and p' are

$$E_p = (-3,11,2,[1,0,1,0]) \text{ and } E_{p'} = (-1,13,3,[0,1,1,0]).$$

According to the dominance rule (5.19), E_p does not dominate $E_{p'}$ because the condition $T_p^{cust_1} \leq T_{p'}^{cust_1}$ is violated. In fact, this is the only violated condition, which is part of the dominance rule to ensure that customer 1 can be visited in a feasible extension of path p if this is possible for path p'. However, we can observe that no feasible extension of p' visits customer 1. Given that all travel times are non-negative, this can be deduced from the inequality $T_{p'}^{time} = 13 > b_1 = 6$. In this case, we say that customer 1 is unreachable from path p' and, therefore, E_p dominates $E_{p'}$.

For the *VRPTW*, we define the notion of unreachability as follows.

Definition 5.4. Let p be a feasible partial path from o to j that is represented by label $E_p = \left(T_p^{rCost}, T_p^{time}, T_p^{load}, [T_p^{cust_i}]_{i \in C}\right)$. Under the assumption previously made that the travel times satisfy the triangle inequality, customer $i \in C$ is said to be *unreachable from p* if and only if at least one of the four following conditions does not hold:
$T_p^{cust_i} = 1;\quad (j,i) \notin A;\quad T_p^{time} + t_{ji} > b_i;\quad T_p^{load} + q_i > Q.$

Note that the assumption on the travel times is necessary for the second and third conditions to be valid. Indeed, if the triangle inequality on the travel times does not hold then it might be possible to use a multiple-arc subpath to reach node i from node j even if $(j,i) \notin A$ or to reach node i from node j in less than t_{ji} time units.

To strengthen the dominance rule (5.19), we replace the set of customer resources by a set of unreachable customer resources, i.e., the resource set becomes

$$\mathcal{R} = \{rCost, time, load, [uCust_i]_{i \in C}\}, \tag{5.20}$$

where $uCust_i$ denotes the unreachability resource for customer i. A partial path p is then represented by a label

$$E_p = \left(T_p^{rCost}, T_p^{time}, T_p^{load}, [T_p^{uCust_i}]_{i \in C}\right), \tag{5.21}$$

where the component $T_p^{uCust_i}$, $i \in C$, takes value 1 if customer i is unreachable from p, and 0 otherwise. These new resources are treated as follows in the labeling algorithm. First, their resource windows are all equal to $[0,1]$ at every node $j \in N$. The label feasibility conditions (5.15c) are thus replaced by

$$T_p^{uCust_i} \in [0,1], \quad \forall i \in C. \tag{5.22}$$

Second, in the initial label E_0, all components $T_0^{uCust_i}$, $i \in C$, are set to 0. Third, when extending a label E_p along an arc $(j,h) \in A$ to create a new label $E_{p'}$, the computation of the customer resource components (5.18d) is replaced by

$$\begin{aligned} T_{p'}^{uCust_i} &= f_{jh}^{uCust_i}(T_p^{time}, T_p^{load}, T_p^{uCust_i}) \\ &= \begin{cases} T_p^{uCust_i} + 1 & \text{if } h = i \\ \max\{T_p^{uCust_i}, U_{jh}^i(T_p^{time}, T_p^{load})\} & \text{otherwise} \end{cases} \quad \forall i \in C, \end{aligned} \tag{5.23}$$

where the function $U_{jh}^i(\cdot)$ returns 1 if customer i is unreachable from p' because of the time windows or the capacity constraint, and 0 otherwise, i.e.,

$$U_{jh}^i(T_p^{time}, T_p^{load}) = \begin{cases} 1 \text{ if } f_{jh}^{time}(T_p^{time}) + t_{hi} > b_i \text{ or } f_{jh}^{load}(T_p^{load}) + q_i > Q \\ 0 \text{ otherwise.} \end{cases} \tag{5.24}$$

Finally, the conditions (5.19d) in the dominance rule are replaced by

$$T_p^{uCust_i} \le T_{p'}^{uCust_i}, \qquad \forall i \in C. \tag{5.25}$$

Illustration 5.3 **Improved dominance (cont.)**

Returning to the example presented above with paths $p = (o,1,3)$ and $p' = (o,2,3)$, label $E_p = (-3, 11, 2, [1,0,1,0])$ remains unchanged whereas the label $E_{p'}$ becomes $E_{p'} = (-1, 13, 3, [1,1,1,0])$ because customer 1 cannot be reached from node 3. The updated dominance rule indicates that label E_p dominates label $E_{p'}$, which can then be discarded.

In Algorithm 5.1, we provide the pseudo-code of a labeling algorithm for the *ESPPTWC* that uses the ingredients discussed above. We assume that $t_{ij} > 0$ for all arcs $(i,j) \in A$ such that $i \in C$ (which is the case if all service times are positive). Under this assumption and despite the fact that the adjusted costs can be negative, Algorithm 5.1 is a label-setting algorithm because it extends the labels in increasing order of their time component, ensuring that only permanent labels are extended. Note that this order guarantees a minimum number of label extensions.

Let us describe this pseudo-code. At each node $i \in N$, the labels are stored in two disjoint sets: $\mathcal{P}_i$ contains the labels that have been processed (i.e., extended) along the arcs leaving node i and $\mathcal{U}_i$ those that are unprocessed. The initialization creates empty sets putting only the initial label E_0 in the set of unprocessed labels at node o. The main while loop (Steps 1 to 10) is executed until there are no more labels to extend. At each iteration, it selects an unprocessed label with a minimum time component in Step 2. This label is denoted E_p and is associated with node j. This label is then extended in Step 4 along each arc leaving node j before being transferred to the set of processed labels $\mathcal{P}_j$ in Step 10. For every extension of E_p along an arc (j,h), the new label $E_{p'}$ is first checked for feasibility in Step 5. If feasible, it is compared to the labels in $\mathcal{U}_h \cup \mathcal{P}_h$ to determine if it is dominated (Step 6). If so, $E_{p'}$ is discarded and the algorithm moves on to the next extension. Otherwise, we add $E_{p'}$ to the set $\mathcal{U}_h$ and filter out dominated labels D, if any (Steps 8 and 9). Finally, once the main loop is completed, a shortest feasible o-d path is returned in Step 12 if one exists, i.e., if $\mathcal{P}_d \neq \emptyset$, as tested in Step 11. In the next subsection, we present an application of this algorithm to a relaxation of the *ESPPTWC*.

Algorithm 5.1: A labeling algorithm for the *ESPPTWC*.

input : $G = (N, A)$
output : Least reduced cost path $p \in \hat{P}$
initialization : $\mathcal{U}_o \leftarrow \{E_O\}$, $\mathcal{U}_i \leftarrow \emptyset, \forall i \in N \setminus \{o\}$, $\mathcal{P}_i \leftarrow \emptyset, \forall i \in N$

1 **while** $\bigcup_{i \in N} \mathcal{U}_i \neq \emptyset$
2     $j, E_p \leftarrow$ Pick unprocessed label `// min. time` $\underset{i \in N, E_q \in \mathcal{U}_i}{\arg\min} \{T_q^{time}\}$
3   **for** $(j,h) \in A$
4         $E_{p'} \leftarrow$ Extend label E_p to node h `// e.g.` (5.18a)`--`(5.18c) `and` (5.23)
5   **if** $E_{p'}$ *is feasible* `// e.g.` (5.15a)`,` (5.15b) `and` (5.22)
6             $D \leftarrow$ Check dominance between labels $\{E_{p'}\}$ and $\mathcal{U}_h \cup \mathcal{P}_h$ `// optional`
7   **if** $D = \emptyset$
8                 $D \leftarrow$ Check dominance between labels $\mathcal{U}_h$ and $\{E_{p'}\}$ `// optional`
9                 $\mathcal{U}_h \leftarrow (\mathcal{U}_h \setminus D) \cup \{E_{p'}\}$
10     $\mathcal{P}_j \leftarrow \mathcal{P}_j \cup \{E_p\}$, $\mathcal{U}_j \leftarrow \mathcal{U}_j \setminus \{E_p\}$
11 **if** $\mathcal{P}_d \neq \emptyset$
12   **return** path p such that $E_p \in \underset{E_q \in \mathcal{P}_d}{\arg\min}\, T_q^{rCost}$

Other variants of Algorithm 5.1 can be devised by changing, for example, the order in which the labels are extended (Lines 2 or 3), when dominance is executed, or between which pairs of labels it is applied (Lines 6/8). For hard-to-solve *ESPPTWC* instances, many feasible labels are generated, yielding a large number of comparisons to perform in the dominance procedure which, then, becomes a highly critical bottleneck operation. Indeed, consider Algorithm 5.2 in which we perform a pairwise dominance check with labels from each set. Observe that one of the sets always contains only one label $E_{p'}$ and that an obvious overlap in the work effort appears with respect to $\mathcal{U}_h$. Furthermore, the computational burden increases exponentially with the number of generated labels. A more intricate dominance check procedure along with well-thought-out data structures (see, e.g., Sadykov et al., 2021) can therefore impact significantly the efficiency of the labeling algorithm. Such technical details are, however, out of the scope of this book. We can nonetheless observe an obvious trade-off from such a tactical decision: during the time that dominance is postponed we have to manage more labels, in particular generate extensions that would otherwise be dominated.

Note 5.5 (Unreachability and feasibility check.) When unreachability resources are used to impose path elementarity, their values can be exploited to avoid performing extensions that would yield infeasible labels. Indeed, in Step 3, all arcs (j,h) such that $h \in C$ and $T_p^{uCust_h} = 1$ can be skipped. In this case, there is no need to perform the feasibility check in Step 5 if $h \in C$ as it is equivalent to testing if $T_p^{uCust_h} = 1$.

Algorithm 5.2: Dominance check.

input : $\mathcal{L}_d, \mathcal{L}_c$
output : Set of labels in $\mathcal{L}_d$ that are dominated by at least one label in $\mathcal{L}_c$
initialization : $D \leftarrow \emptyset$

for $l_d \in \mathcal{L}_d$
  for $l_c \in \mathcal{L}_c$
    if *l_d is dominated by l_c* // e.g. (5.19a)--(5.19c), (5.25)
      $D \leftarrow D \cup \{l_d\}$
      break
return D

Note 5.6 (More acceleration tricks.) As surveyed in Costa et al. (2019), several other acceleration strategies have been developed to solve the *ESPPTWC* (and its variants) and generate columns more rapidly. Among others, we highlight bounded bidirectional labeling (see, e.g., Righini and Salani, 2006; Tilk et al., 2017, and Exercise 5.12), decremental state-space relaxation (see, e.g., Boland et al., 2006; Righini and Salani, 2008; Martinelli et al., 2014; Contardo et al., 2015), completion bounds (see, e.g., Lübbecke, 2005; Baldacci et al., 2008; Martinelli et al., 2014), and heuristic pricing (see, e.g., Fukasawa et al., 2006; Desaulniers et al., 2008; Contardo et al., 2015). For details, the interested readers are invited to consult the above references.

Subproblem relaxations

When the time windows are wide and the vehicle capacity loose, the *ESPPTWC* can be highly time-consuming and using it as an *ISP* may become impractical despite the strong bounds it might yield. To overcome this difficulty, the *ESPPTWC* is often replaced by an easier-to-solve relaxation. The most popular *ISP* relaxations (see Desaulniers et al., 2014) in the context of the *VRPTW* allow the presence of cycles in the paths, i.e., a customer can be visited more than once in a feasible path. These relaxations differ by the rules defining the allowed cycles. In this section, we present two such relaxations, namely, the shortest path problem with time windows and a capacity constraint (*SPPTWC*) and the shortest *ng*-path problem with time windows and a capacity constraint (*ng-SPPTWC*).

When the *ISP* can generate paths with cycles, the *IMP* obtained from a Dantzig-Wolfe reformulation differs from formulation (5.8) in two ways.

- First, the set of feasible routes $\hat{P}$ is augmented to include paths with the allowed cycles. However, these additional paths must satisfy the time windows and vehicle capacity. In this respect, multiple visits to the same customer induce multiple start of service times at this customer (one for each visit) as well as multiple deliveries of the customer demand. Note that it is not possible to cycle infinitely because of the time windows and vehicle capacity.

- Second, parameter a_{ip}, $i \in C$, $p \in \hat{P}$, is redefined to count the number of visits (possibly more than one) to customer i in route p. Observe that routes with cycles cannot be part of a feasible solution to (5.8). Indeed, when a route $p \in \hat{P}$ visits a customer $i \in C$ more than once, $a_{ip} \geq 2$ and the constraint (5.8b) associated with i ensures that $\lambda_p = 0$ in any integer solution. Nevertheless, when generated, such a variable can take a positive value (not larger than $1/a_{ip}$) in an *MP* solution, possibly weakening the lower bound $z^{\star}_{MP}$.

Note 5.7 (Reduced cost of a path variable.) Even when a_{ip} can be larger than one for a customer $i \in C$, the reduced cost of variable λ_p still writes as

$$\bar{c}_p = c_p - \sum_{i \in C} \pi_i a_{ip}. \tag{5.26}$$

Consequently, when solving the *ISP*, the dual value π_i must be subtracted each time that customer i is visited to correctly compute the reduced cost of a path-variable with cycles.

SPPTWC relaxation

The first *ISP* relaxation that has been proposed for the *VRPTW* is called the *SPPTWC*. It simply allows all cycles. For instance, in the example network of Figure 5.1, the path $(o,1,3,2,3,d)$, which contains the cycle $(3,2,3)$, becomes feasible. Its earliest service start times are 0, 2, 11, 15, 20, and 34, indicating that service starts twice at customer 3, namely, at times 11 and 20. The total load of this path is 5, including the demand of customer 3 twice. Using the *SPPTWC* as the *ISP* instead of the *ESPPTWC* drastically reduces the complexity of the *ISP*. Indeed, there exist pseudo-polynomial labeling algorithms for solving the *SPPTWC* such as the one discussed below.

To solve the *SPPTWC*, the labeling Algorithm 5.1 is modified as follows:

- no unreachable customer resources are considered in the label definition;
- REF (5.23) is omitted from the label extension function in Step 4;
- condition (5.22) is removed from the feasibility check in Step 5;
- condition (5.25) is removed from the dominance rule in Steps 6 and 8.

Illustration 5.4 **Generated labels for the *SPPTWC***

The application of this modified algorithm to the *SPPTWC* example of Figure 5.1 yields the labels given in Table 5.1 assuming that $\pi_1 = 22$, $\pi_2 = 20$, $\pi_3 = 32$, and $\pi_4 = 25$. In this table, each line corresponds to the extension of a label $E_p \in \mathcal{U}_j$, the one listed in the first column (xLbl), along each arc $(j,h) \in A$ leaving the node $j \in N$ associated with this label. The first line reports the labels created by extending the initial label $E_0 \in \mathcal{U}_o$. Each new label $E_{p'}$ resulting from an extension along an arc

(j,h) is given in the column nLbl under Node h. For example, label $E_1 = (3,2,1)$ in column nLbl under Node 1 is obtained by extending E_0 along arc $(o,1)$. The next column F/I indicates if $E_{p'}$ is feasible (F) or infeasible (I). Finally, if this label turns out to be dominated later, the last column dLbl under the same node specifies a label that dominates it. Otherwise, a dash (-) appears in this column. The other lines provide the same information for the extensions of each non-dominated label, except those associated with the sink node d (no arcs leave node d). These lines follow the order in which the labels are extended, i.e., the labels with the smallest time component first.

Table 5.1 indicates that a total of 51 labels are generated, excluding label E_0. Fifteen of them are associated with infeasible partial paths either because of the time windows, the capacity constraint or both. Not many labels are dominated except at node d. This is not unusual for such a small example when the travel times satisfy the triangle inequality. Nevertheless, dominance helps to avoid the enumeration of a few feasible *o*-*d* paths. At node d, 7 of the 17 created labels are shown to be dominated. It shows that if network G is enlarged with additional nodes and arcs and d is not the sink node, the dominance rule would be effective at eliminating feasible paths passing through this node. Note that applying dominance at the sink node is not helpful as the labels at this node do not have to be extended. On the other hand, keeping several negative reduced cost labels may be of interest to generate more than one path per column generation iteration.

An optimal path is associated with a least-cost feasible label at node d, namely, label E_{51} (in bold) with reduced cost -62 in this example. To retrieve the corresponding path, we must find the predecessor labels and their associated nodes, until reaching source node o. Thus, every label must also store a pointer to its predecessor label. For label E_{51}, the sequence of predecessor labels (also in bold) is E_{25} (at node 3), E_{16} (at node 2), E_7 (at node 3 again), E_1 (at node 1), and E_0 (at node o). Reversing the order of these nodes yields path $(o,1,3,2,3,d)$ with the earliest service start times 0, 2, 11, 15, 20, and 34 (according to the label time components) and a total load of 5.

ng-SPPTWC relaxation

Introduced by Baldacci et al. (2012), the current state-of-the-art *ISP* relaxation in the context of the *VRPTW* is the *ng-SPPTWC* relaxation which allows the presence of certain cycles in the generated routes.

Let $NG_i \subseteq C$, $i \in C$, be a subset of customers called the *neighborhood* of node i. Such a subset contains i and typically its ρ closest customers, where ρ is a predefined parameter. A feasible *o*-*d* *ng*-path respects the time windows and the capacity constraint, and may contain a cycle $(j_1, j_2, \ldots, j_m = j_1)$ if and only if there exists a node j_h, $h \in \{2,\ldots,m-1\}$, such that $j_1 \notin NG_{j_h}$. Indeed, such a cycle is allowed because j_1 is not in the neighborhood of j_h, which means that travel from j_1 to j_h, and back to j_1 induces a relatively long detour. Therefore, any path containing this

	Node 1			Node 2			Node 3			Node 4			Node d		
xLbl	nLbl	F/I	dLbl	nLbl	F/I	dLbl	nLbl	F/I	dLbl	nLbl	F/I	dLbl	nLbl	F/I	dLbl
$\mathbf{E_0 = (0,0,0)}$	$\mathbf{E_1 = (3,2,1)}$	F	-	$E_2 = (15,8,2)$	F	-	$E_3 = (12,10,1)$	F	-	$E_4 = (20,16,2)$	F	-	$E_5 = (0,0,0)$	F	-
$E_1 = (3,2,1)$				$E_6 = (0,12,3)$	F	-	$\mathbf{E_7 = (-3,11,2)}$	F	-	$E_8 = (2,16,3)$	F	E_{14}	$E_9 = (1,12,1)$	F	E_5
$E_2 = (15,8,2)$							$E_{10} = (-1,13,3)$	F	E_7	$E_{11} = (6,17,4)$	F	E_{17}	$E_{12} = (10,20,2)$	F	E_5
$E_3 = (12,10,1)$				$E_{13} = (-17,14,3)$	F	-				$E_{14} = (-11,16,3)$	F	-	$E_{15} = (-2,24,1)$	F	-
$E_7 = (-3,11,2)$				$\mathbf{E_{16} = (-32,15,4)}$	F	-				$E_{17} = (-26,17,4)$	F	-	$E_{18} = (-17,25,2)$	F	-
$E_6 = (0,12,3)$							$E_{19} = (-16,17,4)$	F	-	$E_{20} = (-9,21,5)$	I	-	$E_{21} = (-5,24,3)$	F	-
$E_{13} = (-17,14,3)$							$E_{22} = (-33,19,4)$	F	-	$E_{23} = (-26,23,5)$	I	-	$E_{24} = (-22,26,3)$	F	-
$E_{16} = (-32,15,4)$							$\mathbf{E_{25} = (-48,20,5)}$	F	-	$E_{26} = (-41,24,6)$	I	-	$E_{27} = (-37,27,4)$	F	-
$E_4 = (20,16,2)$							$E_{28} = (3,18,3)$	F	-				$E_{29} = (15,34,2)$	F	E_5
$E_{14} = (-11,16,3)$							$E_{30} = (-28,18,4)$	F	-				$E_{31} = (-16,34,3)$	F	E_{18}
$E_{17} = (-26,17,4)$							$E_{32} = (-43,19,5)$	F	-				$E_{33} = (-31,35,4)$	F	E_{27}
$E_{19} = (-16,17,4)$				$E_{34} = (-45,21,6)$	I	-				$E_{35} = (-39,23,6)$	I	-	$E_{36} = (-30,31,4)$	F	E_{27}
$E_{28} = (3,18,3)$				$E_{37} = (-26,22,5)$	I	-				$E_{38} = (-20,24,5)$	I	-	$E_{39} = (-11,32,3)$	F	E_{18}
$E_{30} = (-28,18,4)$				$E_{40} = (-57,22,6)$	I	-				$E_{41} = (-51,24,6)$	I	-	$E_{42} = (-42,32,4)$	F	-
$E_{22} = (-33,19,4)$				$E_{43} = (-62,23,6)$	I	-				$E_{44} = (-56,25,6)$	I	-	$E_{45} = (-47,33,4)$	F	-
$E_{32} = (-43,19,5)$				$E_{46} = (-72,23,7)$	I	-				$E_{47} = (-66,25,7)$	I	-	$E_{48} = (-57,33,5)$	F	-
$E_{25} = (-48,20,5)$				$E_{49} = (-77,24,7)$	I	-				$E_{50} = (-71,26,7)$	I	-	$\mathbf{E_{51} = (-62,34,5)}$	F	-

Table 5.1: Labels generated for the *SPPTWC*.

cycle is probably costly and has less chance to be selected in an optimal solution of the *MP*.

Note 5.8 (A compromise.) The *ng-SPPTWC* is identical to the *SPPTWC* if $NG_i = \{i\}$ for all $i \in C$ and to the *ESPPTWC* if $NG_i = C$ for all $i \in C$. Consequently, when $2 \leq \rho \leq |C| - 1$, the *ng-SPPTWC* offers a compromise between the *SPPTWC* and the *ESPPTWC*.

To solve the *ng-SPPTWC*, the labeling Algorithm 5.1 is modified as follows.

- First, all unreachable customer resources are replaced by new customer resources that enforce the *ng*-restrictions. In a label, we denote by T^{ngCust_i}, $i \in C$, the corresponding resource components which is equal to 0 if the label can be extended directly to node i and to 1 if it cannot because this extension would result in a forbidden cycle. The resource windows associated with all these resources are set to $[0,1]$ at each node in N and the label feasibility conditions (5.22) in Step 5 are replaced by

$$T_p^{ngCust_i} \in [0,1], \quad \forall i \in C. \tag{5.27}$$

- Furthermore, when extending a label E_p along an arc $(j,h) \in A$ in Step 4, the computation of the customer resource components (5.18d) for the new label $E_{p'}$ is replaced by

$$T_{p'}^{ngCust_i} = f_{jh}^{ngCust_i}(T_p^{ngCust_i}) = \begin{cases} 0 & \text{if } i \notin NG_h \\ T_p^{ngCust_i} + 1 & \text{if } h = i \\ T_p^{ngCust_i} & \text{otherwise} \end{cases} \quad \forall i \in C, \tag{5.28}$$

  where $NG_d = \emptyset$.
- Finally, the conditions (5.25) are replaced by

$$T_p^{ngCust_i} \leq T_{p'}^{ngCust_i}, \qquad \forall i \in C \cap NG_h, \tag{5.29}$$

  in the dominance rule applied in Steps 6 and 8.

Observe that, according to (5.28), $T_{p'}^{ngCust_i} = 0$ if $i \notin NG_h$ for every label $E_{p'}$ associated with a node $h \in C$. Consequently, the comparisons for the customers $i \notin NG_h$ can be omitted from the conditions (5.29). This highlights the advantage in terms of computational efficiency that can result from using the *ng-SPPTWC* as an *ISP* instead of the *ESPPTWC*. Indeed, if $\rho \ll |C|$, then there are much higher chances to find dominated labels for the *ng-SPPTWC* because the dominance rule involves $\rho + 3$ comparisons instead of $|C| + 3$.

Note 5.9 (Memory leaks.) Instead of using binary customer resources to enforce the *ng*-requirements, we can also use a subset of visited customers that cannot be the next one to be visited. For a path p and its associated label E_p (with customer components $T_p^{ngCust_i}$, $i \in C$), this subset, which is called the *memory* of path p and denoted M_p, is defined as $M_p = \{i \in C \mid T_p^{ngCust_i} = 1\}$. In a memory-based labeling algorithm, the customer components are not needed because the memory of a path can be directly built as follows in the label extension process. Let E_p be a label associated with a node $j \in N$ and having a memory M_p. When extending E_p along an arc $(j,h) \in A$ to create a new path p', memory $M_{p'}$ is obtained by adding h to M_p and forgetting all customers that are not in NG_h, i.e.,

$$M_{p'} = (M_p \cup \{h\}) \cap NG_h. \tag{5.30}$$

This set, thus, contains the customers where path p' cannot be directly extended without violating the *ng*-restrictions. With this memory representation, the dominance condition (5.29) is replaced by

$$M_p \subseteq M_{p'}, \tag{5.31}$$

clearly specifying that all direct extensions of path p that are forbidden due to the *ng*-restrictions must also be forbidden for path p'.

Illustration 5.5 **Generated labels for the *ng-SPPTWC***

Consider again the network of Figure 5.1 to define a *ng-SPPTWC* with $\pi_1 = 22$, $\pi_2 = 20$, $\pi_3 = 32$, $\pi_4 = 25$ (i.e., the same dual values as above), $NG_1 = \{1,3\}$, $NG_2 = \{2,3\}$, $NG_3 = \{3,4\}$, and $NG_4 = \{3,4\}$ (thus, $\rho = 1$). Given these neighborhoods, the cycles $(3,2,3)$ and $(3,4,3)$ are infeasible, whereas cycle $(2,3,2)$ is feasible because $2 \notin NG_3$. Therefore, the *SPPTWC* optimal solution computed above, namely path $(o,1,3,2,3,d)$, is infeasible for this *ng-SPPTWC*. Indeed, the memories of its first subpaths are $M_o = \emptyset$, $M_{o1} = \{1\}$, $M_{o13} = \{3\}$, $M_{o132} = \{2,3\}$, indicating that subpath $(o,1,3,2)$ cannot be extended to node 3 without creating a forbidden cycle. The *ng-SPPTWC* optimal path is $(o,1,3,2,d)$ with a reduced cost of -37 (see Exercise 5.14).

Illustration 5.6 **Comparative results**

To conclude this section on the *ISP* relaxations, we present some computational results to assess the quality of the lower bounds that can be obtained using different relaxations and the time required to compute them. We have selected a small subset of six representative *VRPTW* instances from the well-known benchmark data sets of Solomon (1987).

All these instances contain 100 customers which are randomly located around the depot. The instances in class R1 have relatively narrow time windows, whereas those in class R2 have larger time windows, offering many options to cycle. For each of these six instances, we solved by column generation the *MP* obtained using five different *ISP*s, namely, the *SPPTWC*, the *ng-SPPTWC* with $\rho = 5, 10, 15$, and the *ESPPTWC*. In fact, the *SPPTWC* and *ng-SPPTWC* are enhanced with 2-cycle elimination constraints which forbid cycles of the type (j,h,j) and can yield improved lower bounds. The column generation algorithm applies several state-of-the-art acceleration techniques such as tabu search heuristic pricing and bidirectional labeling (see Desaulniers et al., 2014). All tests were run on an `Intel Core i7-4770` processor clocked at 3.40GHz running the Linux x86-64 operating system. The *RMP*s were solved using `IBM CPLEX` version 12.6.

		SPPTWC		*ng*-SPPTWC ($\rho = 5$)			*ng*-SPPTWC ($\rho = 10$)			*ng*-SPPTWC ($\rho = 15$)			ESPPTWC	
Instance	$z^\star_{IMP}$	T(s)	$z^\star_{MP}$	T(s)	$z^\star_{MP}$	GP(%)	T(s)	$z^\star_{MP}$	GP(%)	T(s)	$z^\star_{MP}$	GP(%)	T(s)	$z^\star_{MP}$	GP(%)
R104	971.5	7	949.1	8	955.2	27.2	9	956.1	31.3	11	956.2	31.7	11	956.9	34.8
R108	932.1	8	907.2	11	911.9	16.1	12	912.8	22.5	12	912.9	22.9	15	913.5	25.3
R110	1068.0	8	1048.5	8	1055.2	34.4	7	1055.6	36.4	8	1055.6	36.4	9	1055.6	36.4
R202	1029.6	66	1009.8	86	1016.9	35.9	68	1021.7	60.1	69	1022.2	62.6	139	1022.2	62.6
R203	870.8	87	846.5	193	857.4	44.9	264	861.8	63.0	156	864.4	73.7	3650	866.9	84.0
R205	949.8	63	916.6	89	935.2	56.0	85	938.1	64.8	76	938.9	67.2	232	938.9	67.2

Table 5.2: Comparative results for different *ISP* relaxations.

Table 5.2 reports the results of these experiments. For each instance, it provides the optimal value of the instance $z^\star_{IMP}$ and, for each algorithm (identified by the *ISP* used), the time T in seconds required to solve the *MP*, and the lower bound achieved $z^\star_{MP}$. For the last four algorithms, we also indicate the integrality gap closed GP in percentage obtained by using a stronger *ISP*. For an *ISP A*, this gap is computed as

$$GP(A) = \frac{z^\star_{MP}(A) - z^\star_{MP}(SPPTWC)}{z^\star_{IMP} - z^\star_{MP}(SPPTWC)}. \tag{5.32}$$

To further highlight the impact of allowing cycles on the lower bound achieved, Figure 5.2 specifies, for each *ISP A* other than *ESPPTWC* and each instance, the ratio of the lower bound achieved using *A* to that obtained using *ESPPTWC*, i.e., $z^\star_{MP}(A)/z^\star_{MP}(ESPPTWC)$.

These results clearly show that the lower bounds provided by the *SPPTWC* can be quite weak compared to those derived with the *ESPPTWC*. Using the latter *ISP* can, however, yield much larger computation times for the instances with large time windows (R2 instances). For the R1 instances where cycles are often not possible, using this *ISP* does not have a major impact on the computation times. For the R2 instances, the *ng-SPPTWC* with $\rho = 15$ offers a good compromise between lower bound quality and computation time. Observe, however, that, for the R203 instance, a much better lower bound can be achieved with the *ESPPTWC* but at the expense of a very large computation time.

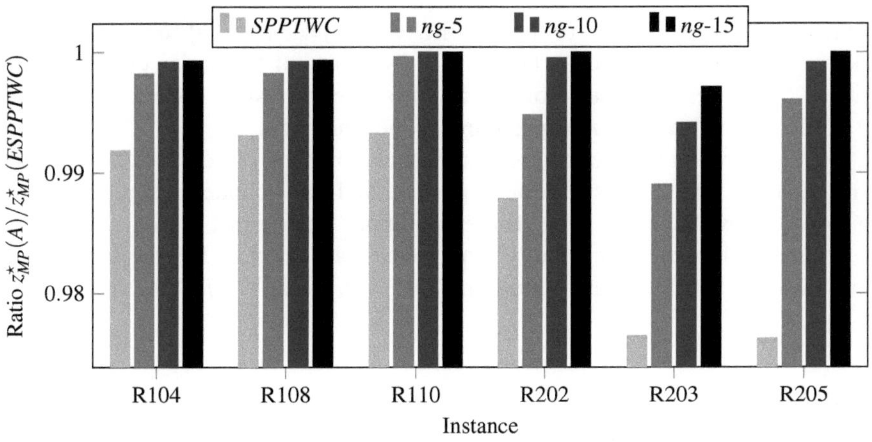

Fig. 5.2: Quality of the lower bounds achieved compared to the *ESPPTWC*.

5.2 Elementary Shortest Path Problem with Resource Constraints

Besides the *VRPTW*, many vehicle routing and crew scheduling problems are solved by branch-price-and-cut algorithms nowadays (for a survey on vehicle routing applications, see Costa et al., 2019). For most of them, the underlying *ISP* is modeled as a *SPPRC* or a *ESPPRC*. In this section, we start by describing these shortest path problems and labeling algorithms that can be used to solve them.

The *ESPPTWC* is a special case of the *ESPPRC*, where the set of resources is $\mathcal{R} = \{rCost, time, load, [cust_i]_{i \in C}\}$, their consumption is limited by resource windows (5.15) at every node of the network, and their REFs are defined by (5.18). These REFs are relatively simple: each one depends on a single resource value. More complex REFs that depend on more than one resource value can be devised such as those defined in (5.23) which allow to extend the unreachable customer resources $uCust_i$, $i \in C$, using both *time* and *load* resources, and as discussed later, those used to extend the reduced cost resource in certain personnel scheduling applications where the cost is a complex function defined by a collective agreement. Such general resources and corresponding REFs have led to the following definition of the *ESPPRC*.

Let $G = (N, A)$ be a network with node set N and arc set A. Set N contains a source node o and a sink node d. Let $\mathcal{R}$ be a set of resources, including a reduced cost resource denoted *rCost*. For each resource $r \in \mathcal{R}$ and each node $i \in N$, a resource window $[a_i^r, b_i^r]$ (possibly unrestrictive) specifies the values that can be taken by resource r at node i. In particular, $[a_o^{rCost}, b_o^{rCost}] = [0, 0]$ imposes that the cost of a path containing only the source node o is equal to 0. Furthermore, for each arc $(i, j) \in A$ and resource $r \in \mathcal{R}$, a REF $f_{ij}^r(\mathbf{t}_i)$ provides a lower bound on the value that resource r can take when extended along arc (i, j) in function of the values of all the resources at node i, i.e., of the vector $\mathbf{t}_i = \left[t_i^{r'}\right]_{r' \in \mathcal{R}}$, where $t_i^{r'}$ denotes the value taken by resource r' at node i. The *ESPPRC* consists of finding an elementary o-d path in G that respects the resource windows at each visited node and minimizes t_d^{rCost}, i.e., the value of its *rCost* resource at the sink node.

As it is common in practice, we assume that along any cycle W in G, there exists a resource r^+ whose consumption is positive and has bounded resource windows $[a_i^{r^+}, b_i^{r^+}]$ at every node $i \in W$. These conditions ensure that a path cannot cycle indefinitely when the elementarity requirements are relaxed. For the *VRPTW*, the time resource satisfies these conditions, that is, $t_{ij} > 0$, $\forall (i,j) \in A$, and $0 \le a_i \le b_i < \infty$, $\forall i \in N$.

Mathematical formulation

The *ESPPRC* can be formulated as a non-linear mixed-integer program that involves two types of variables. For each arc $(i,j) \in A$, there is a binary arc-flow variable that is equal to 1 if (i,j) belongs to the solution, and 0 otherwise. For each resource $r \in \mathcal{R}$ and each node $i \in N$, a continuous variable t_i^r indicates the value of resource r at node i if the computed path visits node i (otherwise t_i^r is meaningless and simply set to 0).

With this notation, the *ESPPRC* can be expressed as

$$\min \quad t_d^{rCost} \tag{5.33a}$$

$$\text{s.t.} \quad \sum_{j:(i,j) \in A} x_{ij} - \sum_{j:(j,i) \in A} x_{ji} = \begin{cases} 1 & \text{for } i = o \\ 0 & \forall i \in N \setminus \{o,d\} \\ -1 & \text{for } i = d \end{cases} \tag{5.33b}$$

$$a_i^r \le t_i^r \le b_i^r \qquad \forall i \in \{o,d\}, r \in \mathcal{R} \tag{5.33c}$$

$$a_i^r \Big(\sum_{j:(i,j) \in A} x_{ij} \Big) \le t_i^r \le b_i^r \Big(\sum_{j:(i,j) \in A} x_{ij} \Big) \qquad \forall i \in N \setminus \{o,d\}, r \in \mathcal{R} \tag{5.33d}$$

$$x_{ij} \left(f_{ij}^r(\mathbf{t}_i) - t_j^r \right) \le 0 \qquad \forall (i,j) \in A, r \in \mathcal{R} \tag{5.33e}$$

$$x_{ij} \in \{0,1\} \qquad \forall (i,j) \in A. \tag{5.33f}$$

The objective function (5.33a) minimizes the total path cost, or more precisely, the path reduced cost assuming that an appropriate adjusted cost $\tilde{c}_{ij}$ is given for all $(i,j) \in A$. Flow conservation constraints (5.33b) ensure a path structure from o to d. The resource windows are expressed through constraints (5.33c)–(5.33d), whereas non-linear inequalities (5.33e) guarantee that the values taken by the resources at every visited node respect the lower bounds provided by the REFs. Binary requirements (5.33f) complete this formulation. The assumption on the r^+ resource mentioned above ensures that any node $i \in N$ can be visited at most once because $t_i^{r^+}$ can take a single value in a feasible solution.

The *SPPRC* is obtained by omitting the elementarity requirements from the definition of the *ESPPRC*, yielding a relaxation that allows feasible paths with cycles. However, when cycles are possible, model (5.33) is not valid because each variable t_i^{r+}, $i \in N$, can take a single value. Instead, in this case, it is possible to devise an arc-flow model based on a time-discretized network, where each node in $N \setminus \{o,d\}$ is represented by a set of nodes, namely, one for each feasible service start time at this node (see Section Good to Know). On the other hand, when cycles are not possible, for example, if network G is acyclic like in the air crew pairing problem discussed later in this chapter, the *SPPRC* is equivalent to the *ESPPRC*. In this case, model (5.33) is valid for the *SPPRC*.

Note 5.10 (Falling short of a resource window lower bound.) Constraints (5.33e) indicate that each REF $f_{ij}^r(\mathbf{t}_i)$ provides only a lower bound on the resource value t_j^r whenever arc (i,j) is used. Therefore, it does not mean that the solution is infeasible if $f_{ij}^r(\mathbf{t}_i) < a_j^r$ as t_j^r can still take a value in $[a_j^r, b_j^r]$ and satisfy the corresponding constraint (5.33e). To avoid falling short of the resource window lower bound, the REF $f_{ij}^r(\mathbf{t}_i)$ can always be replaced by $\max\{a_j^r, f_{ij}^r(\mathbf{t}_i)\}$ as done in (5.18b).

Labeling algorithm

Under the assumption that all REFs f_{ij}^r, $r \in R$, $(i,j) \in A$, are non-decreasing with respect to each of their components, the *ESPPRC* can be solved by a labeling algorithm obtained by generalizing Algorithm 5.1. In fact, two modifications are needed.

- First, the resource set $\mathcal{R}$ can be arbitrary, yielding the following label definition. A label E_p representing a partial path $p = (i_0 = o, i_1, \ldots, i_m = j)$ from source node o to a node $j \in N$ contains $|\mathcal{R}|$ components, one for each resource in $\mathcal{R}$. It writes as $E_p = (T_p^r)_{r\in\mathcal{R}}$, where T_p^r specifies the consumption of resource r along path p which is computed recursively using REFs. Typically, $\mathcal{R}$ contains a reduced cost resource, one unreachable node resource for each node that cannot be traversed twice in the same path, and possibly additional resources.
- Second, in Step 2, the next label to extend is selected according to a given resource component. If the consumption of this resource is always non-decreasing along any path, then the algorithm remains a label-setting algorithm. Otherwise, it is rather a label-correcting algorithm as extended labels may be dominated subsequently.

In this algorithm, we set the initial label E_0 at node o to $E_0 = (a_o^r)_{r\in\mathcal{R}}$. A label $E_p = (T_p^r)_{r\in\mathcal{R}}$ residing at node $j \in N$ that is extended along an arc $(j,h) \in A$ yields a label $E_{p'} = (T_{p'}^r)_{r\in\mathcal{R}}$, where

$$T_{p'}^r = f_{jh}^r(E_p), \qquad \forall r \in \mathcal{R}. \tag{5.34}$$

It is assumed that, for al $r \in \mathcal{R}$, the function f^r_{jh} guarantees the satisfaction of the window lower bound a^r_h for resource r like the REF f^{time}_{jh} in (5.18b). Label $E_{p'}$ is considered feasible if $T^r_{p'} \le b^r_h, \forall r \in \mathcal{R}$. Otherwise, it is discarded. Finally, given two labels $E_p = \left(T^r_p\right)_{r\in\mathcal{R}}$ and $E_{p'} = \left(T^r_{p'}\right)_{r\in\mathcal{R}}$ associated with two partial paths ending at the same node, sufficient conditions for label E_p to dominate label $E_{p'}$ according to Definition 5.3 are

$$T^r_p \le T^r_{p'}, \qquad \forall r \in \mathcal{R}. \tag{5.35}$$

For solving the *SPPRC*, the same labeling algorithm can be applied. However, no resources are needed to ensure path elementarity.

Illustration 5.7 **Common resource extension functions**

The above dominance rule in (5.35) is valid only under the assumption that the REFs are non-decreasing with respect to each of their components. Such REFs are very common in practical applications. As illustrated in Figure 5.3, the following four functions f_1 to f_4 of a variable t are non-decreasing, where $a \ge 0$ and b are real numbers:

$$f_1(t) = t + a \qquad\qquad f_2(t) = t - a$$
$$f_3(t) = \max\{b, t+a\} \qquad\qquad f_4(t) = b.$$

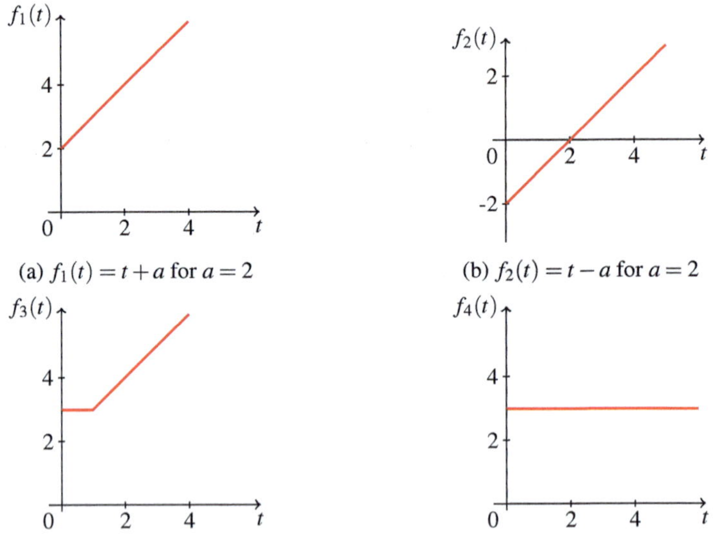

(a) $f_1(t) = t + a$ for $a = 2$ (b) $f_2(t) = t - a$ for $a = 2$

(c) $f_3(t) = \max\{b, t+a\}$ for $a = 2$ and $b = 3$ (d) $f_4(t) = b$ for $b = 3$

Fig. 5.3: Examples of non-decreasing functions f_1, f_2, f_3, and f_4.

- Function f_1 can be used to cumulate a quantity such as the load onboard a vehicle, see REF (5.18c).
- Function f_2, although non-decreasing with a slope equal to 1, can model the decrease of a quantity like the residual capacity of the battery of an electric vehicle when the vehicle is moving or the residual loading capacity of a vehicle when picking up some freight at a customer.
- As for function f_3, it can be applied to enforce satisfying the lower bound of a resource window whenever the resource consumption is not sufficient to reach it, as proposed for the start of service time in the *VRPTW*, see REF (5.18b).
- Finally, function f_4 can be used as a reset function where a quantity is set to a predetermined value, for example, the residual capacity of a battery may be set to its full capacity after a full recharge. Other applications occur in crew scheduling, for example in the airline industry, where the number of flights performed, the working time, and the flying time must be reset to zero at the beginning of each day.

Note 5.11 (Composition of non-decreasing functions.) The function resulting from a composition of non-decreasing functions is always non-decreasing, see Exercise 5.16. For example, function $f_3(t) = \max\{b, t+a\}$ is a composition of the non-decreasing functions $g(t) = b$, $h(t) = t + a$ and $\max\{x, y\}$ and is, therefore, non-decreasing. Observe also that the function $\min\{x, y\}$ is also non-decreasing. This property allows defining non-decreasing REFs to accommodate almost every real-life situation.

Fig. 5.4: Marius Solomon (Banff, Canada, 2004-05-16).

5.3 Examples

To illustrate different ways of modeling with resource constraints, we present three additional examples of vehicle routing and crew scheduling problems, namely,

- the simultaneous pickup and delivery problem (*SPDP*),
- the pickup and delivery problem with time windows (*PDPTW*),
- the air crew pairing problem with base constraints (*CPPBC*).

Each of these problems can be modeled using a compact arc-flow formulation which can be reformulated by applying a Dantzig-Wolfe decomposition. By the sake of conciseness, we only present the reformulation (the *IMP*) for each problem and we focus on the *ISP* by describing its underlying network structure and the resources considered.

Example 5.1 **Simultaneous pickup and delivery problem**

✎ We present an application where the value of a resource depends on the value taken by another one.

In the *SPDP*, least-cost routes for an unlimited fleet of identical capacitated vehicles housed in a single depot must be determined to service a set C of customers exactly once each. As opposed to the *VRPTW*, the set C is divided in two disjoint subsets:

- C^P contains the so-called *pickup* customers, where pickups of merchandise destined to the depot must be performed;
- C^D contains the *delivery* customers, where deliveries of merchandise from the depot must be made.

The quantity delivered or picked up at customer $i \in C$ is denoted q_i. Note that, in practice, a customer can request both a delivery and a pickup, but it is considered here as two separate customers. A travel cost c_{ij} is incurred when traveling from location i to location j.

Let P be the set of feasible routes. A feasible route $p \in P$ starts and ends at the depot, is elementary, and can visit pickup and delivery customers in any order as long as the vehicle capacity Q is respected. Let C_p^P and C_p^D be the (possibly empty) subsets of pickup and delivery customers visited in route p, respectively. The vehicle assigned to this route leaves the depot with a load of $\sum_{i \in C_p^D} q_i \leq Q$ and returns to it with a load of $\sum_{i \in C_p^P} q_i \leq Q$. When visiting a customer $i \in C_p^D$, its load decreases; when visiting a customer $i \in C_p^P$, its load increases and must not exceed Q.

Network structure

The feasible routes in the *SPDP* can be modeled as paths in a network $G = (N, A)$ similar to the one presented for the *VRPTW* (see Figure 5.1). Node set $N = C \cup \{o, d\}$

contains a source node o, a destination node d, and one node for each customer in C. Arc set A contains all arcs $(i,j) \in (N \setminus \{d\}) \times (N \setminus \{o\})$ except those that cannot be traversed in any feasible route due to vehicle capacity, i.e., the arcs (i,j) that satisfy one of the following conditions:

1. $i,j \in C^P$ and $q_i + q_j > Q$;
2. $i,j \in C^D$ and $q_i + q_j > Q$;
3. $i \in C^P$, $j \in C^D$ and $q_i + q_j > Q$.

The last case is infeasible because any quantity delivered after a pickup is onboard the vehicle when the pickup occurs.

Extended formulation

Like the *VRPTW*, the *SPDP* can be modeled using a compact arc-flow formulation with a block-diagonal structure. Applying a Dantzig-Wolfe reformulation to this formulation and aggregating the identical *ISP*s yield the following extended set partitioning formulation of the *SPDP*:

$$z^{\star}_{IMP} = \min \quad \sum_{p \in \hat{P}} c_p \lambda_p \tag{5.36a}$$

$$\text{s.t.} \quad \sum_{p \in \hat{P}} a_{ip} \lambda_p = 1 \qquad [\pi_i] \qquad \forall i \in C \tag{5.36b}$$

$$\lambda_p \in \{0,1\} \qquad \forall p \in \hat{P}, \tag{5.36c}$$

where $\hat{P}$ is the set of feasible routes (each with arbitrary feasible time and load schedules), c_p is the cost of route $p \in \hat{P}$, a_{ip}, $i \in C$, $p \in \hat{P}$, is a binary parameter indicating whether or not customer i is visited in route p, and λ_p, $p \in \hat{P}$, is a binary variable equal to 1 if route p is selected and 0 otherwise. This model is indeed identical to model (5.8). However, set $\hat{P}$ does not contain the same routes in both models because the route feasibility rules in the *VRPTW* and in the *SPDP* differ.

Pricing problem

There is a single *ISP* which is an *ESPPRC* defined on network G. Setting $\pi_o = 0$ for notational convenience, the reduced cost $\bar{c}_p$ of a route $p \in P$ is given by

$$\bar{c}_p = c_p - \sum_{i \in C} a_{ip} \pi_i = \sum_{(i,j) \in A} (c_{ij} - \pi_i) x_{ijp}, \tag{5.37}$$

where $x_{ijp} = 1$ if arc $(i,j) \in A$ is traversed by path p and 0 otherwise. Consequently, the adjusted cost of an arc $(i,j) \in A$ is equal to $\tilde{c}_{ij} = c_{ij} - \pi_i$.

To ensure route feasibility and compute the route reduced cost, we use resources. The resource set is defined as

$$\mathcal{R} = \{rCost, loadP, maxL, [cust_i]_{i \in C}\}. \tag{5.38}$$

Because the resources *rCost* and $cust_i$, $i \in C$, are the same as for the *VRPTW*, we focus on the two resources *loadP* and *maxL* that are used to impose the vehicle capacity constraint. The meaning of these resources is:

- *loadP*: the load collected at the pickup customers;
- *maxL*: the maximum load on board the vehicle since the start of the route.

The resource windows for these two resources are $[0, Q]$ on all nodes in N. Their REFs along any arc $(j,h) \in A$ are as follows:

$$f_{jh}^{loadP}(\mathbf{t}_j) = \begin{cases} t_j^{loadP} + q_h & \text{if } h \in C^P \\ t_j^{loadP} & \text{otherwise} \end{cases} \tag{5.39}$$

$$f_{jh}^{maxL}(\mathbf{t}_j) = \begin{cases} \max\{t_j^{maxL}, t_j^{loadP} + q_h\} & \text{if } h \in C^P \\ t_j^{maxL} + q_h & \text{if } h \in C^D \\ t_j^{maxL} & \text{otherwise,} \end{cases} \tag{5.40}$$

where $\mathbf{t}_j = \left[t_j^r\right]_{r \in R}$ is the vector of the resource values at node j. REF (5.39) is straightforward. However, REF (5.40) is more complex and ensues from the following reasoning.

- On the one hand, if $h \in C^P$, then the maximum load on board up to node h was either achieved before node h (t_j^{maxL}) or it is achieved at node h in which case it corresponds to the total load picked up so far ($t_j^{loadP} + q_h$).
- On the other hand, if $h \in C^D$, then this maximum load is the maximum load at node j, increased by the quantity delivered because this quantity comes from the depot and was, therefore, on board when the maximum load was achieved.

Unlike several relatively simple REFs, the REF (5.40) for resource *maxL* does not only depend on the resource variable t_j^{maxL} but also on t_j^{loadP}. Nevertheless, the REFs for all resources in $\mathcal{R}$, including (5.39)–(5.40), are non-decreasing with respect to each of their components. Therefore, the ESPPRC labeling algorithm described above can be used to solve this *ISP*.

Example 5.2 ***Pickup and delivery problem with time windows***

✎ In addition to the standard *time* and *capacity* constraints, this routing problem imposes *pairing* and *precedence* constraints for route feasibility, i.e., a pair of pickup and delivery nodes must be serviced by the same vehicle and a delivery can only be performed after its corresponding pickup.

In the *PDPTW*, an unlimited fleet of vehicles must be routed to service a set $U = \{1,2,\ldots,n\}$ of n transportation requests, where each request $u \in U$ is defined by a pickup location u^+, a pickup time window $[a_{u^+}, b_{u^+}]$, a delivery location u^-, a delivery time window $[a_{u^-}, b_{u^-}]$, and a quantity q_u to transport from u^+ to u^-. All vehicles are identical: they have the same capacity Q and are associated with a single depot. Departures and arrivals at the depot are restricted by a time window $[\bar{a}, \bar{b}]$. The travel cost and time between two locations i and j (pickup points, delivery points, or depot) are denoted c_{ij} and t_{ij}, respectively, where any service time at i is assumed to be included in t_{ij}.

The goal of the *PDPTW* is to build a set of feasible vehicle routes such that each request is serviced in exactly one route and the total travel cost is minimized. A route is said to be feasible if it starts and ends at the depot, services each request at most once, meets the time windows at the visited locations, respects the vehicle capacity at all times, and, for each serviced request, performs its pickup before its delivery. Note that several requests can be onboard a vehicle at the same time as long as vehicle capacity is respected, i.e., a route can visit a second pickup location u_2^+ between a first pickup location u_1^+ and its corresponding delivery location u_1^-.

The pickup and delivery structure in the *PDPTW* differs from that of the *SPDP*. In the latter, the delivery locations receive merchandise from the depot and the merchandise collected at the pickup locations is destined to the depot. In the *PDPTW*, the pickup and delivery locations are paired and the vehicles leave and return to the depot empty.

As for the *VRPTW*, assume that

- all parameters Q, q_i, a_i, b_i, and t_{ij} are integer;
- the total travel time of any cycle is positive;
- the matrix of the travel times t_{ij} satisfy the triangle inequality. Under this last assumption, a *PDPTW* instance can be feasible only if, for each request $u \in U$, $a_{u^+} + t_{u^+u^-} \leq b_{u^-}$.

Network structure

For the *PDPTW*, the feasible routes correspond to paths in the following network $G = (N, A)$. The node set $N = U^+ \cup U^- \cup \{o, d\}$ contains a pair of source and sink nodes o and d to represent the depot at the start and the end of a route, respectively. For both nodes, define $[\bar{a}, \bar{b}]$ as their time window. Furthermore, node sets U^+ and U^- contain the request pickup and delivery locations, respectively.

The arc set A contains all arcs $(i, j) \in (N \setminus \{d\}) \times (N \setminus \{o\})$ except those for which we can show that they cannot be traversed in any feasible route due to the pickup and delivery precedence constraints (conditions 1–3 below), the vehicle capacity (conditions 4 and 5), or the time windows (conditions 6–10), i.e., the arcs (i, j) that satisfy one of the following conditions:

1. $i = u^-$ and $j = u^+$ for a given $u \in U$;
2. $i = o$ and $j = u^-$ for a given $u \in U$;
3. $i = u^+$ and $j = d$ for a given $u \in U$;

4. $i = u^+$, $j = v^+$ for $u, v \in U$, and $q_u + q_v > Q$;
5. $i = u^-$, $j = v^-$ for $u, v \in U$, and $q_u + q_v > Q$;
6. $a_i + t_{ij} > b_j$;
7. $i = u^+$, $j \in \{v^+, v^-\}$ for $u, v \in U$ with $u \neq v$, and $\max\{a_i + t_{ij}, a_j\} + t_{ju^-} > b_{u^-}$;
8. $i \in \{u^+, u^-\}$, $j = v^+$ for $u, v \in U$ with $u \neq v$, and $a_i + t_{ij} + t_{jv^-} > b_{v^-}$;
9. $i \in \{u^+, u^-\}$, $j = v^-$ for $u, v \in U$ with $u \neq v$, and $\max\{a_{v^+} + t_{v^+i}, a_i\} + t_{ij} > b_j$;
10. $i = u^-$, $j \in \{v^+, v^-\}$ for $u, v \in U$ with $u \neq v$, and $a_{u^+} + t_{u^+i} + t_{ij} > b_j$.

Case 7 is infeasible because there is not enough time to perform delivery u^- when the corresponding pickup u^+ is immediately followed by the pickup or delivery j. Similarly, case 8 is infeasible because there is not enough time to reach the delivery location v^- when the corresponding pickup location v^+ is visited immediately after performing pickup or delivery j. Note that this case can only occur if $a_i + t_{ij} > a_j$ (otherwise, $a_j + t_{jv^-} > b_{v^-}$ and the problem would be infeasible). Cases 9 and 10 are deduced similarly, but taken into account that the pickup location u^+ (case 9) or v^+ (case 10) must be visited prior to u^- or v^-, respectively.

Illustration 5.8 **A network for the *PDPTW***

An example of a network G is presented in Figure 5.5, where time windows are provided beside the nodes and travel times beside the arcs. The travel times between two request nodes are assumed to be symmetric, i.e., $t_{ij} = t_{ji}$ for request nodes $i, j \in N$. Note that arcs may be bidirectional (with arrows at both ends) or monodirectional. The arcs in A are represented by solid lines. The dotted arcs do not belong to A because they meet at least one of the five time-related conditions 6–10 stated above. For instance, the arcs $(3^+, 2^+)$, $(2^+, 3^+)$ and $(1^+, 2^-)$ satisfy condition 6 ($70 + 35 > 60$), condition 7 ($\max\{40 + 35, 70\} + 20 > 90$), and condition 9 ($\max\{40 + 25, 30\} + 35 > 90$), respectively. Observe that, in a feasible route, the requests can be intertwined in various ways: for example, the routes $o1^+1^-3^+3^-d$, $o1^+3^+3^-1^-d$, and $o1^+3^+1^-3^-d$ are all feasible if $q_1 + q_3 \leq Q$.

Extended formulation

The *PDPTW* can also be formulated using an extended set partitioning model. Let

- $\hat{P}$ be the set of feasible routes, each associated with a feasible time schedule and a feasible loading schedule;
- c_p the cost of route $p \in \hat{P}$;
- a_{up}, $u \in U$, $p \in \hat{P}$, a binary parameter equal to 1 if route p performs request u and 0 otherwise;
- λ_p, $p \in \hat{P}$, a binary variable equal to 1 if route p is selected and 0 otherwise.

The set partitioning *IMP* is given by

$$z^\star_{IMP} = \min \quad \sum_{p \in \hat{P}} c_p \lambda_p \tag{5.41a}$$

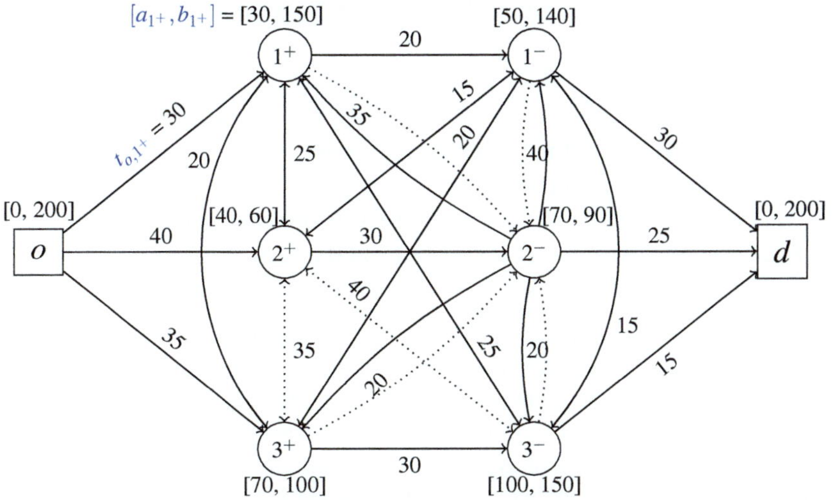

Fig. 5.5: A network $G = (N, A)$ for the *PDPTW*.

$$\text{s.t.} \quad \sum_{p \in \hat{P}} a_{up} \lambda_p = 1 \qquad [\pi_u] \qquad \forall u \in U \tag{5.41b}$$

$$\lambda_p \in \{0, 1\} \qquad \forall p \in \hat{P}. \tag{5.41c}$$

This model differs from the set partitioning model (5.8) of the *VRPTW* by the set of feasible routes $\hat{P}$ but also by the fact that the set partitioning constraints are defined for each request in U and not for each node in $U^+ \cup U^-$.

Pricing problem

Because there is a single depot and the vehicles are all identical, there is a single *ISP* which is an *ESPPRC* defined on network G. To define the adjusted arc costs, consider the reduced cost $\bar{c}_p$ of a route $p \in \hat{P}$:

$$\bar{c}_p = c_p - \sum_{u \in U} a_{up} \pi_u = \sum_{(i,j) \in A} (c_{ij} - 0.5\pi_{u(i)}) x_{ijp}, \tag{5.42}$$

where x_{ijp} is equal to 1 if arc $(i, j) \in A$ is included in path p and 0 otherwise, $u(i)$ is the request associated with node i and $\pi_{u(i)} = 0$ if $i = o$. The adjusted cost of arc $(i, j) \in A$ is, thus, $\tilde{c}_{ij} = c_{ij} - 0.5\pi_{u(i)}$. Remark that the dual variable π_u, $u \in U$, may have been subtracted, for example, only on the arcs with tail node u^+ or only on the arcs with head node u^-.

To enforce route feasibility and compute the route reduced cost, we use the following resource set:

$$\mathcal{R} = \{rCost, time, load, [req_u]_{u\in U}, [on_u]_{u\in U}, [notOn_u]_{u\in U}\}. \tag{5.43}$$

These resources are respectively

$rCost$: Reduced cost;
$time$: Earliest service start time at the current node;
$load$: Load onboard the vehicle;
req_u: Number of times (0 or 1) that request $u \in U$ has been performed;
on_u: Equal to 1 if request $u \in U$ is onboard and 0 otherwise;
$notOn_u$: Equal to 1 if request $u \in U$ is not onboard and 0 otherwise.

The last two resources are used to ensure that 1) any picked up item is delivered afterwards, 2) a delivery cannot be performed unless the corresponding pickup has, and 3) a delivery cannot be realized more than once. Because the *rCost* and *time* resources are identical to those in the *VRPTW*, we concentrate on the other resources. The resource windows per node type for these resources are given in Table 5.3. The columns on_v and $notOn_v$ indicate the windows when $v \neq u$.

Node type	$load$	req_u	on_u	on_v	$notOn_u$	$notOn_v$
source o	$[0,Q]$	$[0,1]$	$[0,0]$	$[0,0]$	$[1,1]$	$[1,1]$
sink d	$[0,Q]$	$[0,1]$	$[0,0]$	$[0,0]$	$[1,1]$	$[1,1]$
pickup u^+	$[0,Q]$	$[0,1]$	$[1,1]$	$[0,1]$	$[0,0]$	$[0,1]$
delivery u^-	$[0,Q]$	$[0,1]$	$[0,0]$	$[0,1]$	$[1,1]$	$[0,1]$

Table 5.3: Resource windows for the *PDPTW*.

Given that $\mathbf{t}_i = \left[t_i^r\right]_{r\in\mathcal{R}}$ denotes the vector of the resource values at node i, the REFs for these resources are given by

$$f_{ij}^{load}(\mathbf{t}_i) = \begin{cases} t_i^{load} + q_{u(j)} & \text{if } j \in U^+ \\ t_i^{load} - q_{u(j)} & \text{if } j \in U^- \\ t_i^{load} & \text{otherwise} \end{cases} \tag{5.44}$$

$$f_{ij}^{req_u}(\mathbf{t}_i) = \begin{cases} t_i^{req_u} + 1 & \text{if } j = u^+ \\ t_i^{req_u} & \text{otherwise} \end{cases} \qquad \forall u \in U \tag{5.45}$$

$$f_{ij}^{on_u}(\mathbf{t}_i) = \begin{cases} t_i^{on_u} + 1 & \text{if } j = u^+ \\ t_i^{on_u} - 1 & \text{if } j = u^- \\ t_i^{on_u} & \text{otherwise} \end{cases} \qquad \forall u \in U \tag{5.46}$$

$$f_{ij}^{notOn_u}(\mathbf{t}_i) = \begin{cases} t_i^{notOn_u} + 1 & \text{if } j = u^- \\ t_i^{notOn_u} - 1 & \text{if } j = u^+ \\ t_i^{notOn_u} & \text{otherwise} \end{cases} \qquad \forall u \in U. \tag{5.47}$$

Because all REFs are non-decreasing, the *ISP* can be solved using the *ESPPRC* labeling algorithm described above (p. 321). It can be proven that the conditions on the resources on_u and $notOn_u$, $u \in U$, in the dominance rule (5.35) are equivalent to the relation

$$O_p = O_{p'}, \tag{5.48}$$

where O_ℓ denotes the set of onboard requests at the end node of a path ℓ.

Easier-to-solve pricing problem

When solving the *ISP*, dominance is rather limited as defined above because condition (5.48) is difficult to satisfy. This condition is necessary to ensure that every feasible extension of path p' is also feasible for path p as imposed by the dominance Definition 5.3. Indeed, if there exists a request $u \in U$ such that $u \in O_p \setminus O_{p'}$, then any feasible extension of p' visiting u^- must visit u^+ beforehand which makes it infeasible for p. Furthermore, if there exists $u \in U$ such that $u \in O_{p'} \setminus O_p$, then any feasible extension of p' must visit u^- without visiting u^+, which again yields an infeasible extension of p.

It is possible to increase label elimination by relaxing Definition 5.3 (i.e., by considering a less restrictive dominance condition) as follows.

Definition 5.5. Let p and p' be two feasible partial paths between node o and a node $j \in N$ which are represented by labels E_p and $E_{p'}$, respectively. Label E_p is said to *dominate* label $E_{p'}$ if, for every feasible extension χ' of p', there exists a feasible extension χ of p such that $\bar{c}_{p\oplus\chi} \leq \bar{c}_{p'\oplus\chi'}$.

Compared to Definition 5.3, Definition 5.5 allows to use an extension χ of p that differs from χ' to find a path $p \oplus \chi$ that dominates $p' \oplus \chi'$. Nevertheless, like Definition 5.3, this new definition cannot be directly applied in practice as it is often impossible to enumcrate all feasible extensions of p'. However, for the *PDPTW*, Røpke and Cordeau (2009) define sufficient conditions for identifying dominated labels according to Definition 5.5. To do so, they slightly modify the *ISP* cost structure as follows such that it satisfies the triangle inequality on the adjusted costs at the delivery nodes. For each request $u \in U$, let

$$\delta_u = \max_{(i,j)\in A_{u^-}} \{0, \tilde{c}_{ij} - (\tilde{c}_{iu^-} + \tilde{c}_{u^-j})\} \tag{5.49}$$

be the maximum violation of the triangle inequality at node u^-, where

$$A_{u^-} = \{(i,j) \in A \mid (i,u^-),(u^-,j) \in A\}. \tag{5.50}$$

Given that, in any feasible route, pickup node u^+ must be visited if delivery node u^- is, it is possible to increase by δ_u the costs of the arcs leaving u^- and decrease those of the arcs leaving u^+ without changing the total reduced cost of a route. This operation yields the following adjusted costs:

$$\tilde{c}_{ij} = \begin{cases} \tilde{c}_{ij} + \delta_{u(i)} & \text{if } i \in U^- \\ \tilde{c}_{ij} - \delta_{u(i)} & \text{if } i \in U^+ \\ \tilde{c}_{ij} & \text{otherwise.} \end{cases} \tag{5.51}$$

After this transformation, one can easily verify that $\tilde{c}_{iu^-} + \tilde{c}_{u^-j} \geq \tilde{c}_{ij}$ for each request $u \in U$ and arc $(i, j) \in A_{u^-}$. Note that this transformation must be recomputed at every column generation iteration because the adjusted arc costs depend on the dual values of the *MP* constraints.

Now, let us consider two paths p and p', both ending at a node $j \in N$, and a feasible extension χ' of p' that contains a single delivery node $u^- \in U^-$ (preceded by node i and succeeded by node j) such that $u \in O_{p'} \setminus O_p$. Then, the extension χ obtained from χ' by replacing the arcs (i, u^-) and (u^-, j) by the single arc (i, j) might be considered as a feasible extension of p (if all resource constraints are verified) whose cost does not exceed that of χ'. When χ' contains several delivery nodes $u^- \in U^-$ such that $u \in O_{p'} \setminus O_p$, then this arc replacement process can be performed several times to obtain a feasible extension χ.

Consequently, with the dominance Definition 5.5, the condition (5.48) of the dominance rule (5.35) can be replaced by

$$O_p \subseteq O_{p'}. \tag{5.52}$$

This condition is equivalent to omitting the condition on the resource *notOn* from the dominance rule (5.35). Note that this resource still needs to be considered in the labels as its resource windows are required to filter out infeasible paths.

Finally, remark that, to apply the technique proposed in this section, all arcs that can be part of a feasible path must be in set A. This condition must be met at all times when searching for an integer solution to the compact formulation.

Example 5.3 **Crew pairing problem with base constraints**

✎ This problem introduces multiple resources to properly compute the reduced cost of a pairing.

The *CPPBC* is an optimization problem solved while planning the operations of a major airline. Given a weekly flight schedule to be repeated over several consecutive weeks, it consists of finding least-cost crew pairings such that each scheduled flight is covered by an anonymous crew and the work is relatively well distributed amongst different crew bases to which the crew members are assigned. The computed pairings are subsequently assigned to individual crew members in a monthly crew rostering step.

The definition of the *CPPBC* varies from one airline to another as it depends on various rules imposed by the airline, the aviation authorities, and the employees' collective agreement. We present one relatively simple variant of this problem. The flight schedule is composed of a set of flights F to be operated during a generic week by aircraft of the same type. Each flight $f \in F$ is defined by an origin and a

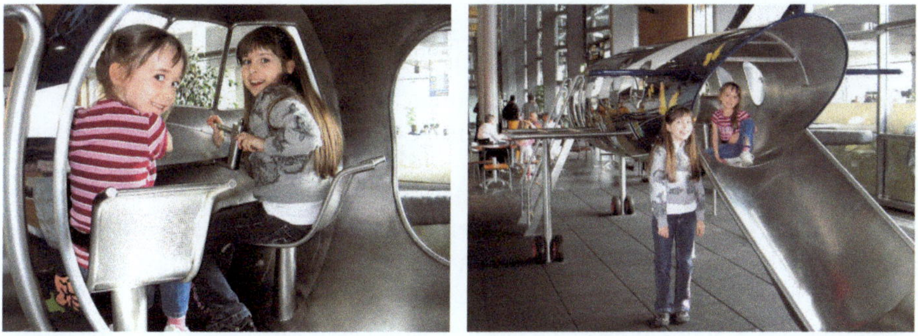

Fig. 5.6: Alexane and Charlotte training at OSL to become pilots or flight attendants.

destination airport, and a schedule indicating the departure day and time as well as the arrival day and time (from which the flight duration l_f can be deduced). Each flight must be covered by one crew. The crew members are typically assigned to different crew bases (airports). The set of these bases is denoted by B.

To cover the flights in F, crew pairings must be determined. A pairing is a sequence of one or several duties that starts and ends at the same crew base. A duty is a sequence of flights separated by connections and corresponds to a work day for a crew. A duty can start and end at different airports. When a pairing contains several duties, they are separated by rest periods. A rest period cannot occur at the base to which a crew is assigned. A pairing is feasible if it satisfies the following constraints:

- It spans a maximum of $\bar{n}^D$ days (e.g., $\bar{n}^D = 4$) with $\bar{n}^D \leq 7$;
- The duration (span) of a *duty* falls in the interval $[\underline{t}^{DTY}, \bar{t}^{DTY}]$;
- The duration of a *connection* between two consecutive flights in a duty falls in the interval $[\underline{t}^{CNX}, \bar{t}^{CNX}]$;
- The duration of a *rest* period between two consecutive duties falls in the interval $[\underline{t}^{RST}, \bar{t}^{RST}]$.

Denote by $\hat{P}$ the set of feasible pairings. Note that crews can also be repositioned from one airport to another by deadheading (i.e., travel as passengers) on certain flights. We omit this possibility in the considered *CPPBC* variant.

A solution to the *CPPBC* is composed of a set of feasible pairings that covers each flight in F exactly once. Because the flight schedule is repeated weekly for a certain period of time, this solution must be reproducible one week after the other. Hence, a pairing can start before the end of the week (e.g., on Saturday if the week is from Monday to Sunday) and end after (e.g., on Tuesday).

The cost of a solution is given by the sum of the costs of the selected pairings. Let F_p be the subset of flights covered in a pairing $p \in \hat{P}$. Denote by $l_p = \sum_{f \in F_p} l_f$ its total flying time and by s_p its total span, i.e., the difference between the arrival time of the last flight in pairing p and the departure time of its first flight. The cost

c_p of pairing p is equal to:

$$c_p = \max\{l_p, \alpha s_p\}, \tag{5.53}$$

where $0 < \alpha < 1$ is a parameter value (e.g., $\alpha = 1/4$) dictated by the collective agreement to ensure a fair pay if the pairing contains too long inactivity periods (connections and rests).

Finally, given that the number of available crew members varies at each base $b \in B$ (typically, it depends on the number of flights incident to airport b), the distribution of the total flying time amongst the bases should be aligned as much as possible with the crew availability at each base. To model this, let L_b be a *soft maximum* on the total flying time in the pairings assigned to base $b \in B$. This maximum can be violated at the following costs: any excess of at most μ_b units is penalized at a rate of ω_b^1 per unit, whereas any additional excess is penalized at a larger rate of $\omega_b^2 > \omega_b^1$, yielding a piecewise-linear penalty function.

Network structure

Let W be the set of days in the week. The feasible pairings that start on day $w \in W$ from base $b \in B$ can be represented as paths in the following acyclic time-space network $G^{wb} = (N^{wb}, A^{wb})$, where N^{wb} and A^{wb} denote its node and arc sets. In a time-space network, each node (except possibly the source and sink nodes) is associated with a time and a location, and each arc models, in general, a movement in time, in space, or in both dimensions. For simplicity, assume that there are no overnight flights in F. Thus, each flight operates on a single day.

Let W_w be the subset of days that can be spanned by a pairing starting on day w. For example, if $w = Saturday$ and $\bar{n}^D = 3$, then $W_w = \{Saturday, Sunday, Monday\}$. The set N^{wb} contains a *source* node and a *sink* node that represent the start and the end of a pairing, respectively. Furthermore, for each flight $f \in F$ operated on a day in W_w, there are up to three nodes in set N^{wb}:

- a *departure* node,
- an *arrival* node,
- a *to-flight* node.

The first two nodes always exist and their meaning is obvious. Their associated locations are the flight departure and arrival airports, respectively. The time associated with the *departure* node is the flight departure time, but that associated with the *arrival* node is its arrival time plus the minimal connection time $\underline{t}^{CNX}$. The *to-flight* node is an auxiliary node that offers the opportunity to end a rest period by starting a duty with the corresponding flight or to extend this period. It is associated with the flight departure airport and time, and exists only if this airport is not the base b. Figure 5.7 depicts the nodes and arcs associated with a single non-base airport.

The arc set $A^{wb} = \{A^{FLY}, A^{RST}, A^{SOD}, A^{CNX}, A^{WT}, A^{SOP}, A^{EOP}\}$ contains seven disjoint subsets of arc types given respectively as

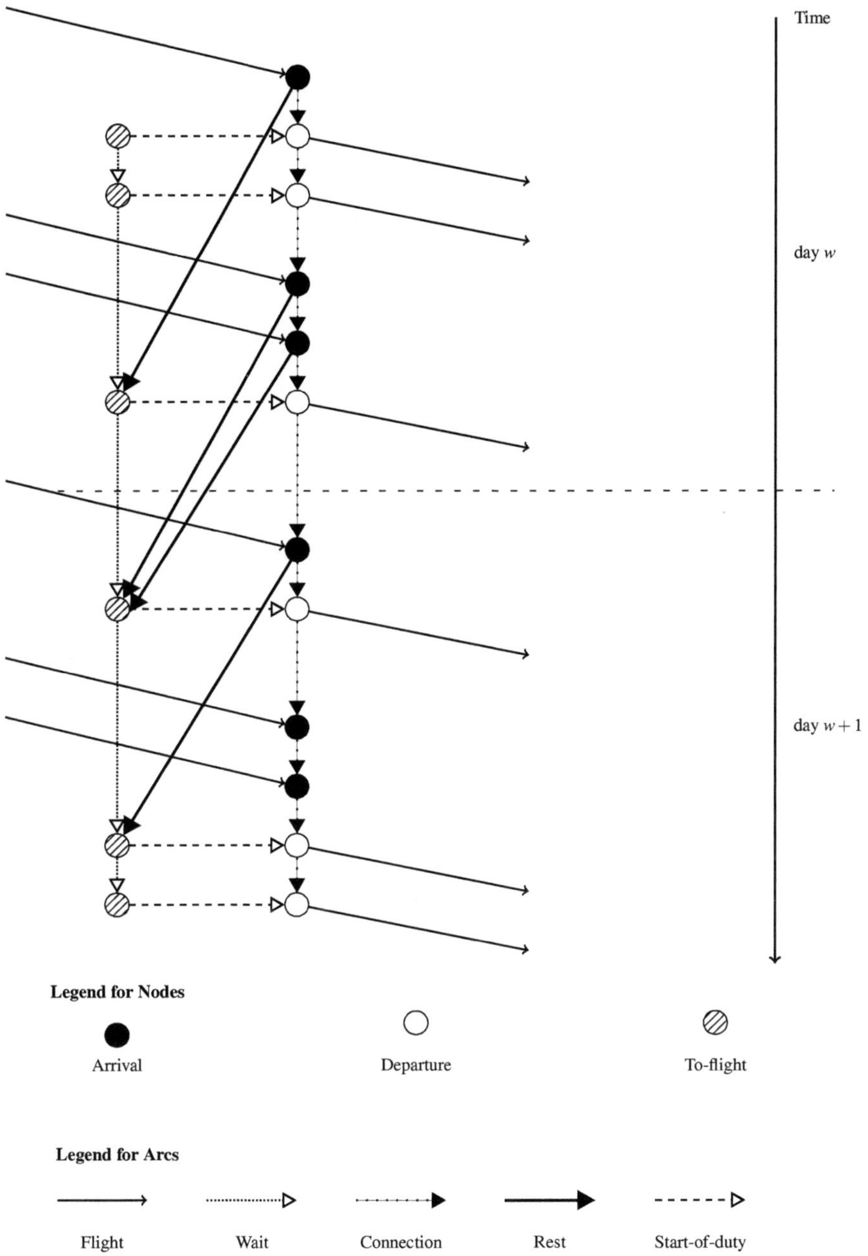

Fig. 5.7: Subnetwork of G^{wb} describing the nodes and arcs of a non-base airport.

- Each flight operated on a day in W_w is represented by a *flight* arc that links its departure and arrival nodes.
- For each flight that does not end at base b, there is a *rest* arc connecting its arrival node to the earliest to-flight node associated with the flight arrival airport such that the minimum and maximum rest period durations $\underline{t}^{RST}$ and $\bar{t}^{RST}$ are satisfied. This arc may not exist if there is no such to-flight node.
- For each flight that does not start at base b, there is a *sod (start-of-duty)* arc between its to-flight and departure nodes.
- The departure and arrival nodes associated with each airport are linked together in chronological order by a chain of *connection* arcs. These arcs allow the extension of a connection over the minimum connection duration.
- The to-flight nodes associated with each airport other than base b are linked together in chronological order by a chain of *wait* arcs. These arcs allow the extension of a rest period over the minimum rest duration.
- There is a *sop (start-of-pairing)* arc between the source node and each departure node at base b on day w.
- There is an *eop (end-of-pairing)* arc between each arrival node at base b (on any day in W_w) and the sink node.

Three parameters are associated with each arc $(i, j) \in A^{wb}$:

- E_{ij}: elapsed time (difference between the times associated with nodes j and i);
- F_{ij}: flying time;
- $f(i, j)$: associated flight.

The elapsed time of *sop* and *eop* arcs is equal to zero. The flying time of every arc except the *flight* arcs is equal to zero. The associated flight is relevant only for the *flight* arcs.

Extended formulation

In this section, an extended formulation (an *IMP*) is directly proposed for the *CPPBC*. For each flight $f \in F$ and each pairing $p \in \hat{P}$, let a_{fp} be a binary parameter indicating whether pairing p covers or not flight f. For each day $w \in W$ and base $b \in B$, we denote by $\hat{P}^{wb} \subset \hat{P}$ the subset of pairings starting on day w from base b. For each pairing $p \in \hat{P}^{wb}$, we define a binary variable λ_p^{wb} that is equal to 1 if p is selected in the solution and 0 otherwise. Furthermore, for each base $b \in B$, we define two non-negative integer static variables y_b^1 and y_b^2 that indicate the amount of flying time assigned to base b in excess of L_b and $L_b + \mu_b$, respectively.

Given this notation, the proposed *IMP* is the following set partitioning model with side constraints:

$$z_{IMP}^{\star} = \min \quad \sum_{w \in W} \sum_{b \in B} \sum_{p \in \hat{P}^{wb}} c_p \lambda_p^{wb} + \sum_{b \in B} \left(\omega_b^1 y_b^1 + \omega_b^2 y_b^2 \right) \tag{5.54a}$$

$$\text{s.t.} \quad \sum_{w\in W}\sum_{b\in B}\sum_{p\in \hat{P}^{wb}} a_{fp}\lambda_p^{wb} = 1 \qquad [\pi_f] \quad \forall f\in F \tag{5.54b}$$

$$\sum_{w\in W}\sum_{p\in \hat{P}^{wb}} l_p\lambda_p^{wb} - y_b^1 - y_b^2 \le L_b \qquad [\pi_b] \quad \forall b\in B \tag{5.54c}$$

$$\lambda_p^{wb} \in \{0,1\} \qquad \forall w\in W, b\in B, p\in \hat{P}^{wb} \tag{5.54d}$$

$$0 \le y_b^1 \le \mu_b \qquad \forall b\in B \tag{5.54e}$$

$$y_b^2 \ge 0 \qquad \forall b\in B. \tag{5.54f}$$

The objective function (5.54a) minimizes the sum of the pairing costs and the penalties incurred for not respecting the maximum flying time at each base. The set partitioning constraints (5.54b) ensure that each flight is included in exactly one pairing. The soft base constraints (5.54c) allow the computation of the excess flying time assigned to each base and its breakdown in the y_b^1 and y_b^2 variables. The variable domains are expressed by (5.54d)–(5.54f).

Pricing problems

There is one integer subproblem ISP^{wb} for each day $w \in W$ and each base $b \in B$ that allows the generation of pairings in $\hat{P}^{wb}$, i.e., starting on day w from base b. It is a *SPPRC* defined on the time-space network G^{wb} described above. It aims at finding feasible pairings with a negative reduced cost, where the reduced cost $\bar{c}_p$ of a pairing $p \in \hat{P}^{wb}$ is given by

$$\begin{aligned}\bar{c}_p &= c_p - \sum_{f\in F} a_{fp}\pi_f - l_p\pi_b \\ &= \max\{l_p, \alpha s_p\} - \sum_{f\in F_p} \pi_f - \sum_{f\in F_p} l_f\pi_b \\ &= \max\{\sum_{f\in F_p} (l_f(1-\pi_b) - \pi_f), \alpha s_p - \sum_{f\in F_p} (\pi_f + l_f\pi_b)\}.\end{aligned} \tag{5.55}$$

To ensure feasibility of all source-to-sink paths in network G^{wb}, we impose resource constraints. Unconstrained resources are also used to correctly compute the pairing reduced cost. Seven resources are defined in

$$\mathcal{R} = \{rCost,\ rCostF,\ rCostS,\ dtyMin,\ dtyMax,\ cnxMin,\ rstMax\}, \tag{5.56}$$

where

rCost: Reduced cost;
rCostF: Total flying time minus cumulated duals; first component in (5.55);
rCostS: Span minus cumulated duals; second component in (5.55);
dtyMin: Time remaining to reach the minimum duty duration $\underline{t}^{DTY}$;
dtyMax: Current duty duration;

$cnxMax$: Current connection duration;

$rstMax$: Current rest duration.

The two resources $rCostF$ and $rCostS$ compute the reduced cost of the partial pairing according to each component in (5.55) as if the other component was irrelevant. Its true reduced cost is given by the resource $rCost$ whose value is obtained by taking the maximum value of these two resources.

The resource windows for the $rCost$, $rCostF$ and $rCostS$ resources are set to $[0,0]$ at the source node; they are unconstraining (neither lower nor upper bound) at all other nodes. The resource windows associated with the other resources are given in Table 5.4. Observe that the windows for resource $dtyMin$ at the to-flight and sink nodes are $[0,0]$ because these nodes can only be reached after completing a duty and, in this case, there should be no time remaining to reach the minimum duty duration. Furthermore, for resource $dtyMax$, the window upper bound at all nodes except the source node accounts for the minimum connection time $\underline{t}^{CNX}$ that is automatically added to the last *flight* arc of a duty.

Node type	$dtyMin$	$dtyMax$	$cnxMax$	$rstMax$
source	$[0,0]$	$[0,0]$	$[0,0]$	$[0,0]$
sink	$[0,0]$	$[0,\bar{t}^{DTY}+\underline{t}^{CNX}]$	$[0,\bar{t}^{CNX}]$	$[0,\bar{t}^{RST}]$
departure	$[0,\underline{t}^{DTY}]$	$[0,\bar{t}^{DTY}+\underline{t}^{CNX}]$	$[0,\bar{t}^{CNX}]$	$[0,\bar{t}^{RST}]$
arrival	$[0,\underline{t}^{DTY}]$	$[0,\bar{t}^{DTY}+\underline{t}^{CNX}]$	$[0,\bar{t}^{CNX}]$	$[0,\bar{t}^{RST}]$
to-flight	$[0,0]$	$[0,\bar{t}^{DTY}+\underline{t}^{CNX}]$	$[0,\bar{t}^{CNX}]$	$[0,\bar{t}^{RST}]$

Table 5.4: Resource windows for an *ISP* of the *CPPBC*.

The REFs of the resources in $\mathcal{R}$ for each arc $(i,j) \in A^{wb}$ are defined as

$$f_{ij}^{rCost}(\mathbf{t}_i) = \max\{f_{ij}^{costF}(\mathbf{t}_i), f_{ij}^{costS}(\mathbf{t}_i)\} \tag{5.57a}$$

$$f_{ij}^{costF}(\mathbf{t}_i) = \begin{cases} t_i^{costF} + F_{ij}(1-\pi_b) - \pi_{f(i,j)} & \text{if } (i,j) \in A^{FLY} \\ t_i^{costF} & \text{otherwise} \end{cases} \tag{5.57b}$$

$$f_{ij}^{costS}(\mathbf{t}_i) = \begin{cases} t_i^{costS} + \alpha E_{ij} - \pi_{f(i,j)} - F_{ij}\pi_b & \text{if } (i,j) \in A^{FLY} \\ t_i^{costS} + \alpha E_{ij} & \text{otherwise} \end{cases} \tag{5.57c}$$

$$f_{ij}^{dtyMin}(\mathbf{t}_i) = \begin{cases} \underline{t}^{DTY} & \text{if } (i,j) \in A^{SOP} \cup A^{SOD} \\ \max\{t_i^{dtyMin} - E_{ij}, 0\} & \text{if } (i,j) \in A^{FLY} \cup A^{CNX} \\ t_i^{dtyMin} & \text{otherwise} \end{cases} \tag{5.57d}$$

$$f_{ij}^{dtyMax}(\mathbf{t}_i) = \begin{cases} t_i^{dtyMax} + E_{ij} & \text{if } (i,j) \in A^{FLY} \cup A^{CNX} \\ 0 & \text{otherwise} \end{cases} \tag{5.57e}$$

$$f_{ij}^{cnxMax}(\mathbf{t}_i) = \begin{cases} \underline{t}^{CNX} & \text{if } (i,j) \in A^{FLY} \\ t_i^{cnxMax} + E_{ij} & \text{if } (i,j) \in A^{CNX} \\ 0 & \text{otherwise} \end{cases} \tag{5.57f}$$

$$f_{ij}^{rstMax}(\mathbf{t}_i) = \begin{cases} \underline{t}^{CNX} + E_{ij} & \text{if } (i,j) \in A^{RST} \\ t_i^{rstMax} + E_{ij} & \text{if } (i,j) \in A^{WT} \\ 0 & \text{otherwise,} \end{cases} \tag{5.57g}$$

where $\mathbf{t}_i = \left[t_i^r\right]_{r \in R}$ is the vector of the resource values at node i. Let us comment on these resource extension functions. REFs (5.57b)–(5.57c) ensure that both components in (5.55) are well computed, whereas REF (5.57a) computes the maximum. Observe that the latter REF does not depend on t_i^{rCost} but rather on t_i^{costF} and t_i^{costS}. Next, REFs (5.57e)–(5.57g) are straightforward: each allows the computation of the consumption of the corresponding resource along the arcs and resets the resource value to 0, $\underline{t}^{CNX}$ or $\underline{t}^{RST}$ when a duty, a connection or a rest starts, respectively. Finally, REFs (5.57d) are used to compute the time remaining to reach the minimum duty time $\underline{t}^{DTY}$. Traversing a *rest* or an *eop* arc (indicating the end of a duty) when this time is positive is not feasible because of the resource windows $[0,0]$ for $dtyMin$ on the destination node of such an arc (either a to-flight node or the sink node).

Notice that all REFs (5.57) are non-decreasing functions with respect to each of their components. Therefore, the ISP^{wb} for each base $b \in B$ and day $w \in W$ can be formulated using the *SPPRC* model (5.33) (recall that the *ESPPRC* is equivalent to a *SPPRC* when the underlying network is acyclic) and, thus, the labeling algorithm described above for the *SPPRC* can be used to solve it.

5.4 Good to Know

In this final section, we show how it is possible to derive a compact *linear* arc-flow formulation for the *VRPTW* by embedding the time information in the structure of the network.

Compact linear formulation for the VRPTW

To do so, we replace the network $G = (N,A)$ described before (p. 298) by a *time-expanded network* $G^\Delta = (N^\Delta, A^\Delta)$ where each node $i \in N \setminus \{o\}$ is represented by $b_i - a_i + 1$ copy nodes in N^Δ, one for each integer time unit in time window $[a_i, b_i]$. For node o, a single copy associated with time a_o is required. This origin node is denoted $\bar{o}$. For each node $\ell \in N^\Delta$, we denote by ν_ℓ and τ_ℓ its corresponding node in N and its associated time, respectively, and we write $\ell = \langle \nu_\ell, \tau_\ell \rangle$.

Let $N_i^\Delta = \{\ell \in N^\Delta \mid v_\ell = i\}$ be the set of nodes in N^Δ representing node $i \in N$. This allows us to internalize the time information on the arc connection, that is, $t_{v_h,v_\ell} = \tau_\ell - \tau_h$, $(h,\ell) \in N_i^\Delta \times N_j^\Delta$. In the arc set A^Δ, every arc $(i,j) \in A$ is replaced by the set of arcs

$$A_{ij}^\Delta = \{(h,\ell) \in N_i^\Delta \times N_j^\Delta \mid t_{v_h,v_\ell} = t_{ij} \vee (t_{v_h,v_\ell} > t_{ij} \wedge \tau_\ell = a_j)\}, \tag{5.58}$$

which means we only keep arc (h,ℓ) if $t_{v_h,v_\ell} = t_{ij}$ or $t_{v_h,v_\ell} > t_{ij} \wedge \tau_\ell = a_j$, that is, each arc echoes the original travel time plus a waiting time but only if the latter is necessary and minimal. All arcs in A_{ij}^Δ have the same cost c_{ij}. By construction, every path from the source node $\bar{o}$ to any sink node $\ell \in N_d^\Delta$ satisfies the time windows at the visited nodes.

Figure 5.8 illustrates such a subset derived from arc $(1,4)$ in the example network G of Figure 5.1. Recall that $[a_1,b_1] = [2,6]$, $[a_4,b_4] = [16,18]$ and $t_{14} = 12$. Therefore, if service starts at customer 1 at time 2 (node $\langle 1,2\rangle$), then the vehicle has to wait two time units before the time window lower bound at customer 4, yielding the arc $(\bar{h},\bar{\ell}) \in A_{14}^\Delta$ with $\bar{h} = \langle 1,2\rangle$ and $\bar{\ell} = \langle 4,16\rangle$. Similarly, one time unit of waiting must occur if service starts at time 3, yielding the arc $(\bar{h},\bar{\ell}) \in A_{14}^\Delta$ with $\bar{h} = \langle 1,3\rangle$ and $\bar{\ell} = \langle 4,16\rangle$. For the other arcs $(h,\ell) \in A_{14}^\Delta$, no waiting occurs, i.e., $\tau_h + t_{14} = \tau_\ell$. Note that some nodes and their adjacent arcs may be removed from the sets N^Δ and A^Δ because they cannot be reached from the source node $\bar{o}$. For instance, node $\langle 1,3\rangle$ and arc $(\langle 1,3\rangle,\langle 4,16\rangle)$ can be discarded in the example. We do note that in general, the difficulty of identifying such nodes and arcs increases with the size of the network and the width of the time windows.

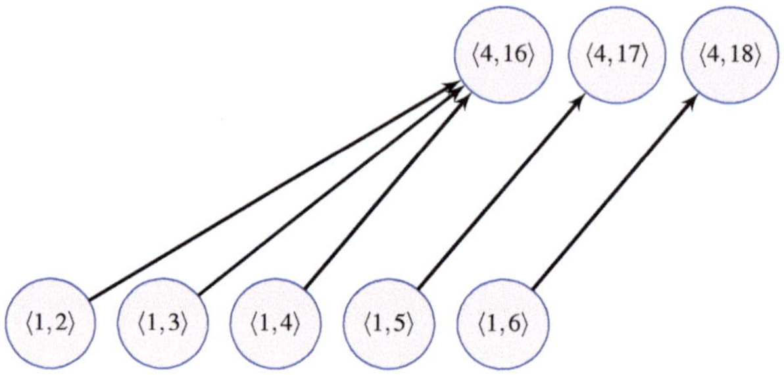

Fig. 5.8: Node subsets N_1^Δ and N_4^Δ, and arc subset A_{14}^Δ.

Observe also that G^Δ is an *acyclic network*. However, a path in G^Δ may visit more than one node representing the same customer if the travel time and the time windows allow it. For instance, in the example of Figure 5.1, the path o343d with associated start of service times 0, 10, 16, 18, and 32 is represented in the corresponding network G^Δ by the path going through the nodes $\bar{o}$, $\langle 3,10\rangle$, $\langle 4,16\rangle$, $\langle 3,18\rangle$ and $\langle d,32\rangle$. Such a path is non-elementary as it visits more than once the same customer and is thus infeasible for the *VRPTW*.

Using the binary arc-flow variables $x^k_{h\ell}$, $(h,\ell) \in A^\Delta, k \in K$, the *VRPTW* is formulated in a compact form, without any time variables nor time-related constraints, as

$$
\begin{aligned}
z^\star_{ILP} = \min \quad & \sum_{k\in K} \sum_{(h,\ell)\in A^\Delta} c_{v_h v_\ell} x^k_{h\ell} && && (5.59a)\\
\text{s.t.} \quad & \sum_{k\in K} \sum_{h\in N^\Delta_i} \sum_{\ell:(h,\ell)\in A^\Delta} x^k_{h\ell} = 1 && \forall i \in C && (5.59b)\\
& \sum_{\ell:(\bar{o},\ell)\in A^\Delta} x^k_{\bar{o}\ell} = 1 && \forall k \in K && (5.59c)\\
& \sum_{\ell:(h,\ell)\in A^\Delta} x^k_{h\ell} - \sum_{\ell:(\ell,h)\in A^\Delta} x^k_{\ell h} = 0 && \forall k \in K, i \in C, h \in N^\Delta_i && (5.59d)\\
& \sum_{h\in N^\Delta_d} \sum_{\ell:(\ell,h)\in A^\Delta} x^k_{\ell h} = 1 && \forall k \in K && (5.59e)\\
& \sum_{(h,\ell)\in A^\Delta} q_{v_h} x^k_{h\ell} \le Q && \forall k \in K && (5.59f)\\
& x^k_{h\ell} \in \{0,1\} && \forall k \in K, (h,\ell) \in A^\Delta. && (5.59g)
\end{aligned}
$$

Constraints (5.59b) impose that each customer $i \in C$ be visited exactly once and that its start of service time be within its time window. Consequently, they ensure that all selected paths are elementary. For each vehicle $k \in K$, constraints (5.59c)–(5.59e) define a path structure from the source node $\bar{o}$ to any destination node in N^Δ_d. They also ensure that any path respects the time windows of the visited customers. Finally, the capacity constraints for each vehicle k are expressed through inequalities (5.59f).

Note that, like for the Binary knapsack problem (p. 239), the capacity constraints can also be modeled directly in the network structure by using three-dimensional nodes $\ell = \langle v_\ell, \tau_\ell, \zeta_\ell \rangle$, where $\zeta_\ell \in \{0,1,\dots,Q\}$ represents the load accumulated up to node v_ℓ. In this augmented network, nodes h and ℓ are linked by an arc if and only if $(\langle v_h, \tau_h\rangle, \langle v_\ell, \tau_\ell\rangle)$ belongs to the arc set A^Δ defined above and $\zeta_h + q_{v_\ell} = \zeta_\ell$. This alternative modeling is useful to better understand the labeling algorithms presented earlier. However, because it is not required to derive such a linear arc-flow model, we continue our discussion with model (5.59) which is based on the time-expanded network G^Δ.

Observe that constraints (5.59c)–(5.59g) are separable by vehicle $k \in K$. Therefore, model (5.59) has a block-diagonal structure. Unlike model (5.1), these constraints are linear (except for the integrality requirements) and, thus, permit to derive an extended formulation via a Dantzig-Wolfe reformulation.

Extended set partitioning formulation

To apply a Dantzig-Wolfe reformulation to model (5.59), we exploit the block-diagonal structure in index k. We define $\mathcal{D}^k = \{\mathbf{x}^k \in \{0,1\}^{|A^\Delta|} \mid \mathbf{D}^k\mathbf{x}^k \geq \mathbf{d}^k\}$, where $\mathbf{D}^k\mathbf{x}^k \geq \mathbf{d}^k$ represents the constraints (5.59c)–(5.59f) associated with vehicle k, and the following *elementarity constraints*

$$\sum_{(h,\ell)\in A^\Delta:v_h=i} x^k_{hl} \leq 1 \qquad \forall i \in C. \tag{5.60}$$

These additional constraints are not necessary in model (5.59) because they are redundant with constraints (5.59b) but they are needed here to guarantee that the feasible solutions in $\mathcal{D}^k$ correspond to elementary routes. Observing that all domains $\mathcal{D}^k$ are identical, we can aggregate them into a single one and omit index k. The resulting domain $\mathcal{D}$ is defined by the following constraints:

$$\sum_{\ell:(\bar{o},\ell)\in A^\Delta} x_{\bar{o}\ell} = 1 \tag{5.61a}$$

$$\sum_{\ell:(h,\ell)\in A^\Delta} x_{h\ell} - \sum_{\ell:(\ell,h)\in A^\Delta} x_{\ell h} = 0 \qquad \forall i \in C, h \in N^\Delta_i \tag{5.61b}$$

$$\sum_{h\in N^\Delta_d} \sum_{\ell:(\ell,h)\in A^\Delta} x_{\ell h} = 1 \tag{5.61c}$$

$$\sum_{(h,\ell)\in A^\Delta} q_{v_h} x_{h\ell} \leq Q \tag{5.61d}$$

$$\sum_{(h,\ell)\in A^\Delta:v_h=i} x_{h\ell} \leq 1 \qquad \forall i \in C \tag{5.61e}$$

$$x_{h\ell} \in \{0,1\} \qquad \forall (h,\ell) \in A^\Delta. \tag{5.61f}$$

This domain is bounded and, thus, its convex hull can be described using only extreme points. We can show that every solution in $\mathcal{D}$ is an extreme point of $conv(\mathcal{D})$ and, thus, there is a bijection between the set of extreme points of $conv(\mathcal{D})$ and the set of feasible routes. Consequently, reformulating the compact formulation (5.59) by convexification or by discretization yields the same *IMP*. As in the previous chapter, we denote by P the index set of these extreme points.

For a feasible route (an extreme point) $\mathbf{x}_p$, $p \in P$, let $c_p = c_{\mathbf{x}_p} = \sum_{(h,\ell)\in A^\Delta} c_{v_h v_\ell} x_{h\ell}$ be its cost and $a_{ip} = a_{i\mathbf{x}_p} = \sum_{(h,\ell)\in A^\Delta:v_h=i} x_{h\ell}$, $i \in C$, be a binary parameter indicating if it visits or not customer i. Using the Dantzig-Wolfe decomposition principle, model (5.59) can be reformulated as the following set partitioning model:

$$z^\star_{IMP} = \min \quad \sum_{p\in P} c_p \lambda_p \tag{5.62a}$$

$$\text{s.t.} \quad \sum_{p\in P} a_{ip}\lambda_p = 1 \qquad [\pi_i] \qquad \forall i \in C \tag{5.62b}$$

$$\lambda_p \in \{0,1\} \qquad \forall p \in P, \tag{5.62c}$$

where, for each route $p \in P$, the binary variable λ_p indicates whether or not it is part of the solution, and the dual variables $\pi_i \in \mathbb{R}$, $\forall i \in C$, appear in the linear relaxation. Note that this reformulation should include constraint $\sum_{p \in P} \lambda_p \leq |K|$ but it is not necessary given that the number of vehicles $|K|$ is assumed to be sufficiently large.

5.5 Reference Notes

Section 5.1 The first column generation algorithm for the *multiple traveling salesperson problem with time windows* was devised by Desrosiers et al. (1984) for a school bus application. In that paper, the columns are generated by solving a shortest path problem with time windows on nodes (*SPPTW*), using a *label-correcting* algorithm (Desrosiers et al., 1983). It led the way to the resource constrained shortest path problem (Desrochers, 1988) as we know it today. Many algorithms have been developed to tackle this particular pricing problem at the core of numerous vehicle routing problem variants as surveyed by Costa et al. (2019).

As one of the simplest variants, the *VRPTW* turned out to play a central role in testing new ideas (around twenty papers proposing different column generation algorithms for this problem have been published so far). The first is due to Desrochers et al. (1992a), who exploited the possibility to use a subproblem relaxation, namely, the *SPPTWC* instead of the *ESPPTWC*. By introducing the notion of customer unreachability, Feillet et al. (2004) have succeeded to design the first efficient algorithm that relies on the *ESPPTWC* to yield strong lower bounds.

Nevertheless, this *ISP* remains too difficult to solve when the time windows and vehicle capacity are not enough constraining to sufficiently limit the possibility to cycle. At the beginning of the 2010s, the *ng-SPPTWC* was introduced by Baldacci et al. (2011, 2012) as an *ISP* that offers a good compromise between lower bound quality and computational effort to solve it.

Fig. 5.9: Stefan Røpke announcing the solution of Solomon's instance R208.

Research on the *VRPTW* has benefited from a common playground, namely, the well-known benchmark instances introduced by Solomon (1987). This dataset includes 56 varied 100-customer instances that differ by how the customers are geographically distributed in a squared region (randomly, in clusters, or a mix of both) and how wide are the time windows. Desrochers et al. (1992a) were able to solve only 7 of these 56 instances. It took 25 years before all 56 instances were solved to optimality by Pecin et al. (2017a). Note that, at the 2012 Column Generation workshop (Bromont, Quebec), Stefan Røpke reported solving instance R208, famous for being the only remaining open one at the time, but he never published this result. Since then, researchers have started to tackle the instances of Gehring and Homberger which have a structure similar to those of Solomon but with 200 to 1000 customers (see Gehring and Homberger, 1999; Pecin et al., 2017a; Sadykov et al., 2021).

Section 5.2 Desrochers and Soumis (1988c) developed the first *label-setting* algorithm for the *SPPTW*. This study was immediately followed by another one on the *SPPRC* by Desrochers (1988). This seminal work was, however, rejected for publication (see Figure 5.10). For a survey on the *SPPRCs*, the reader is referred to Irnich and Desaulniers (2005). There are thereafter plenty of noteworthy contributions to the canon of knowledge on the *ESPPRC* such as the first bi-directional algorithm of Righini and Salani (2006).

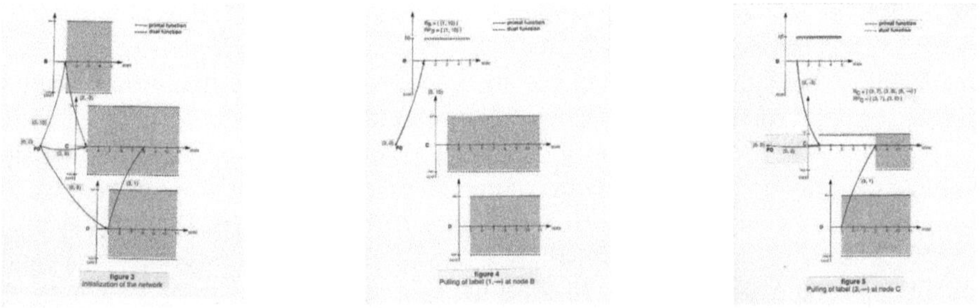

Fig. 5.10: A few steps into the *SPPRC* (Desrochers, 1988).

Examples

5.1 Simultaneous pickup and delivery problem. This example is inspired by Desaulniers et al. (1998a, §3).

5.2 Pickup and delivery problem with time windows. Dumas et al. (1991) proposed the first algorithm for the *PDPTW*. This paper contains the main results of Yvan's master thesis (1983–1985) under the supervision of Jacques Desrosiers: *Confection de routes de véhicules pour le transport de plusieurs origines à plusieurs destinations*, Université de Montréal. The improved dominance rule was, much later, devised by Røpke and Cordeau (2009).

5.3 Crew pairing problem with base constraints. Crew scheduling in air transportation has also been a fertile ground for the development of cutting-edge branch-and-price algorithms. Lavoie et al. (1988) were the first to devise one for the crew pairing problem. Several more were developed including those of Desaulniers et al. (1997a) and Vance et al. (1997). Recently, Quesnel et al. (2017) presented a study focusing on the *crew pairing problem with base constraints*. Finally, Desaulniers et al. (2020c) solve the largest published industrial monthly crew pairing instance with 46 588 flights. Embedded in a rolling-horizon procedure with relatively large time periods, their algorithm combines column generation with the *dynamic constraint aggregation* algorithm for set partitioning problems (El Hallaoui et al., 2005, 2008) that efficiently manages degeneracy of the *RMP*.

Early surveys on Dantzig-Wolfe decomposition applied to vehicle routing and crew scheduling problems include Solomon and Desrosiers (1988), Desrosiers et al. (1995), and Desaulniers et al. (1998a).

Exercises

5.1 Martin Desrochers

Who is Martin Desrochers? List a few of his scientific contributions.

5.2 Zero-objects for the *VRPTW*

Consider the compact arc-flow formulation (5.1) for the Vehicle Routing Problem with Time Windows and the actual grouping of the constraints (5.2), where the $|K|$ available vehicles are all used.

(a) Propose a compact formulation in which we require the possibility of not using all of them.
(b) Write an appropriate grouping of the constraints.
(c) Give the *MP* resulting from a Dantzig-Wolfe reformulation.

5.3 Two-index arc-flow formulation for the *VRPTW*

Propose a two-index arc-flow model for the *VRPTW*. See the description of the network structure on p. 298.

5.4 Tightness of the linear relaxations for the *VRPTW*

The linear relaxation of the *IMP* (5.8) can be tighter than that of the compact formulation *ILP* (5.1) (considering the linearized version of constraints (5.1g)). To illustrate this, compute an optimal solution to the linear relaxation of both models (5.8) and (5.1) for the following small example and compare their values. Set $C = \{1,2,3\}$ contains three customers with $q_i = 10$ for all $i \in C$. Assume that the time windows are not constraining at all, $Q = 15$ and $|K| = 3$. Furthermore, let $c_{ij} = 1$ for all arcs $(i,j) \in A$ except arc (o,d) for which $c_{od} = 0$.

5.5 Dominance rule for the *ESPPTWC*

Under the assumption that all REFs $f^r_{jh}(\cdot)$ defined in (5.18) are non-decreasing with respect to each of their variables, show that (5.19) are sufficient dominance conditions.

5.6 Dominance of labels by a dominated label
Prove that, if the newly created label $E_{p'}$ associated with a node $h \neq d$ is dominated in Step 6 of Algorithm 5.1, then it cannot dominate any label $E_{p''} \in \mathcal{U}_h \cup \mathcal{P}_h$, unless $E_{p'} = E_{p''}$.

5.7 Dominance of processed labels by a non-dominated label
Under the assumption that $t_{uv} > 0$ for all arcs $(u, v) \in A$ such that $v \in C$, prove that, in Step 8 of Algorithm 5.1, the newly created label $E_{p'}$ associated with a node $h \neq d$ cannot dominate any processed label $E_{p''} \in \mathcal{P}_h$.

5.8 Omitting the *load* resource in the *SPPTWC*
In some *VRPTW* instances, the time windows are much more constraining than the vehicle capacity. Therefore, to speed up the solution of the *ISP*, one possibility is to omit the *load* resource and add infeasible path cuts when load-infeasible routes are part of a *MP* solution. If this resource was not considered in the *SPPTWC* of Illustration 5.4, which labels listed in Table 5.1 would be additionally dominated or not generated?

5.9 Revisiting the time-constrained shortest path problem
(a) Formulate the time-constrained shortest path problem of Example 3.2 as a *SPPRC* by defining the resource set, the resource windows, and the REFs.
(b) Solve this example using a labeling algorithm.

5.10 Two-cycle elimination
How would you modify the labeling algorithm proposed for solving the *SPPTWC* if 2-cycles were not allowed? A 2-cycle is a sequence of three nodes (i, j, i) starting and ending at the same node i.

5.11 Backward labeling for the *ESPPTWC*
Algorithm 5.1 for the *ESPPTWC* is qualified as a forward labeling algorithm because it extends partial paths from the source node o towards the destination node d in network G. A backward labeling algorithm works in the opposite direction: it extends partial paths from node d towards node o using backward REFs. Define the main components (label definition, resource windows, initial label, REFs, dominance rule) of a backward labeling algorithm.

5.12 Bidirectional labeling for the *ESPPTWC*
When feasible *od*-paths can contain a relatively large number of nodes, it may be advantageous to use a bidirectional labeling algorithm. In such an algorithm, labeling is performed in both forward and backward directions. Labels are extended forwardly from node o until reaching the halfway point of a monotonic resource such as the time. Labels are also extended backwardly from node d until reaching the same halfway point for the corresponding backward resource. When all labels have been extended, forward and backward labels associated with the same node are merged to form complete *od*-paths. Given a forward label

$$E_p = \left(T_p^{rCost}, T_p^{time}, T_p^{load}, [T_p^{cust_i}]_{i \in C}\right)$$

and a backward label (as defined in the solution to Exercise 5.11)

$$E_p^{bw} = (T_p^{bw\_rCost}, T_p^{bw\_time}, T_p^{bw\_rCap}, [T_p^{bw\_cust_i}]_{i \in C})$$

associated with the same node $j \in C$, specify the conditions that must be met to ensure that the *od*-path $p \oplus p'$ is feasible.

5.13 Unreachable customers in the *ng-SPPTWC*
How can we integrate the usage of unreachable customers in the labeling algorithm proposed for solving the *ng-SPPTWC*?

5.14 Applying the *ng-SPPTWC* **labeling algorithm**
Apply the *ng-SPPTWC* labeling algorithm to solve the *ng-SPPTWC* instance defined in Illustration 5.5. Consider the unreachability definition and the dominance rule proposed in the solution to Exercise 5.13.

5.15 Non-decreasing REFs assumption for the *ESPPRC*
The proposed generalized labeling Algorithm 5.1 for the *ESPPRC* relies on the assumption that the REFs f_{ij}^r, $r \in R$, $(i, j) \in A$, are non-decreasing with respect to each of their variables. Why would this algorithm be invalid if some of these REFs were not non-decreasing?

5.16 Composition and addition of non-decreasing functions
Let $f(\mathbf{x}), g(\mathbf{x}) : \mathbb{R}^n \mapsto \mathbb{R}^n$ be non-decreasing functions with respect to each of their variables.

(a) Show that the composition $f \circ g(\mathbf{x}) = f(g(\mathbf{x}))$ is a non-decreasing function.
(b) Show that the sum $(f+g)(\mathbf{x}) = f(\mathbf{x}) + g(\mathbf{x})$ is a non-decreasing function.

5.17 Customers with both a pickup and a delivery
In the *SPDP*, assume that a customer i requests both a pickup of q_i^P and a delivery of q_i^D. Would it be valid to use a single node for this customer which is a pickup node with demand $q_i^P - q_i^D$ if $q_i^P \geq q_i^D$ and a delivery node with demand $q_i^D - q_i^P$ otherwise?

5.18 Unreachable customers for the *SPDP*
To speed up the labeling algorithm for solving the pricing problem of the *SPDP*, we can replace the customer resources by unreachable customer resources. What would be the conditions to declare a customer unreachable?

5.19 *SPDP* **is a special case of the** *PDP*
Model the *SPDP* as a pickup and delivery problem (*PDP*), i.e., a *PDPTW* without time windows.

5.20 The vehicle routing problem with backhauls
In the vehicle routing problem with backhauls (*VRPB*), the set of customers C is partitioned in two disjoint subsets: subset C^D contains customers requesting a delivery and subset C^P, those requesting a pickup. All delivered items originate from a single depot and all picked up items must be transported to this depot. A route is

feasible if all visited delivery customers are serviced before all visited pickup customers, and the capacity of the vehicle is satisfied all along the route. Each customer must be visited exactly once. The *VRPB* can be formulated as the *IMP* (5.8) (with the appropriate set of routes $\hat{P}$) and solved by a branch-and-price algorithm. How would you model the *ISP*? Describe the underlying network, the adjusted arc costs, and the resources used. Use the following notation: customer demand q_i, vehicle capacity Q, arc cost c_{ij}, and dual variable π_i.

5.21 Resource validity for the ***PDPTW***
The *ISP* defined for the *PDPTW* involves the subset of resources

$$\{load, [req_u]_{u \in U}, [on]_{u \in U}, [notOn]_{u \in U}\}$$

that are constrained by the resource windows described in Table 5.3 and extended using the REFs (5.44)–(5.47). Show that the resources req_u, on_u, and $notOn_u$ for $u \in U$ ensure that each request u is picked up at most once, delivered at most once, and, when picked up, it is delivered afterwards.

5.22 Unreachable requests for the ***PDPTW***
To speed up the labeling algorithm for solving the pricing problem of the *PDPTW*, we can replace the request resources by unreachable request resources. What would be the conditions to declare a request unreachable?

5.23 Dominance only for same onboard requests in the ***PDPTW***
Show the equivalence between (5.48) and the conditions on the resources on_u and $notOn_u$, $u \in U$, in the dominance rule (5.35).

5.24 Pairing feasibility constraints
In the proposed *CPPBC* model, which pairing feasibility constraints are directly taken into account by the structure of the networks G^{wb}, $w \in W, b \in B$?

5.25 Minimum and maximum resource constraints for the ***CPPBC***
In the proposed *ISP* for the *CPPBC*, two distinct resources *dtyMin* and *dtyMax* are used to impose that the duration of a duty falls in the interval $[\underline{t}^{DTY}, \bar{t}^{DTY}]$.

(a) Why is it not sufficient to only use resource *dtyMax* (that tracks the current duty duration) to handle both minimum and maximum duty duration constraints?
(b) In the labeling algorithm for solving an *ISP*, consider two labels E_p and $E_{p'}$ associated with the same node such that $T_p^{dtyMax} < \underline{t}^{DTY}$ and $T_{p'}^{dtyMax} < \underline{t}^{DTY}$. Show that, according to the dominance conditions (5.35), E_p cannot dominate $E_{p'}$ unless $T_p^{dtyMax} = T_{p'}^{dtyMax} < \underline{t}^{DTY}$ or $\underline{t}^{DTY} \leq T_p^{dtyMax} \leq T_{p'}^{dtyMax}$.

5.26 New pairing constraints for the ***CPPBC***
How would you model the following constraints in the *ISP* for the *CPPBC*?

(a) The total flying time in each duty should not exceed a maximum time $\bar{t}^{FLY}$.
(b) The duration of a rest before the last duty of a pairing can be less than $\underline{t}^{RST}$ but not less than $\underline{\underline{t}}^{RST}$, where $\underline{\underline{t}}^{RST} < \underline{t}^{RST}$. Hint: A new type of arcs can be useful.

6

Dual Point of View

There is a crack in everything
That's how the light gets in.

Anthem
Leonard Cohen

Abstract In this chapter, we dive into the dual space: row generation rather than column generation, the Lagrangian relaxation method and its strong relationship with the Dantzig-Wolfe decomposition, and ways we borrow from the dual point of view in accelerating the column generation algorithm.

Contents

Introduction 353
6.1 Row Generation 355
Illustration 1: *TCSPP* 356
Alternative master problem 359
Illustration 2: *TCSPP* (cont.) 360
6.2 Lagrangian Relaxation 361
Lagrangian subproblem 361
Lagrangian function and Lagrangian dual problem 362
Block-diagonal structure 363
Illustration 3: *TCSPP* (cont.) 363
Dantzig-Wolfe reformulation vs. Lagrangian relaxation 365
Properties of the Lagrangian function 370
Illustration 4: *TCSPP* (cont.) 372
Optimality conditions 375
Illustration 5: Optimality check 375
Illustration 6: *TCSPP* (cont.) 376
Subgradient algorithm 377

J. Desrosiers et al., *Branch-and-Price*,
https://doi.org/10.1007/978-3-031-96917-1_6

Illustration 7: *TCSPP* (cont.) 381
Primal solutions (fractional and integer) 383
6.3 Good to Know 385
Dual-optimal inequalities 385
Illustration 8: Cutting stock problem 386
Dual-optimal boxes 388
How helpful can the dual information be? 390
Illustration 9: Cutting stock problem (cont.) 390
Illustration 10: *TCSPP* (cont.) 391
Illustration 11: Multi-depot vehicle scheduling problem . . . 393
6.4 More to Know 398
Stabilized column generation 398
Illustration 12: Multi-depot vehicle scheduling (cont.) . . . 402
Properties 403
Alternative/complementary stabilization methods 405
Learning more from the dual point of view 407
Initialization 407
Dual estimates 407
Anticipation 408
6.5 Examples 409
Generalized assignment problem 409
Relaxation of the semi-assignment constraints 409
Relaxation of the capacity restrictions on the machines . . . 410
Capacitated *p*-median problem 411
Lagrangian relaxation 413
Dantzig-Wolfe reformulation 413
Various Lagrangian subproblems 414
Vehicle routing problem with time windows 414
Cutting stock problem 415
Edge coloring problem 416
Symmetric traveling salesperson problem 418
Formulations 419
Held and Karp lower bounds 421
Lagrangian lower bounds 423
Dantzig-Wolfe reformulation 426
Balancing printed circuit board assembly line systems 427
Dantzig-Wolfe reformulation 427
Lagrangian relaxation 429
Some computational results 430
A few comments 431
Hybrid algorithm for a 2-dimensional problem 432
6.6 Reference Notes 434
Exercises 437

Acronyms

ILP	integer linear program (original or compact formulation)	355
IMP	integer master problem (extended integer formulation)	355
MP	linear relaxation of the *IMP*	355
DMP	dual formulation of the *MP*	355
RMP	restricted *MP*	355
ISP	integer subproblem	356
TCSPP	time constrained shortest path problem	356
AMP	alternative *MP*	360
LDP	Lagrangian dual problem of the *ILP*	363
ISP^k	integer subproblem for block $k \in K$	366
LP	linear relaxation of the *ILP*	369
LR	Lagrangian function	371
LB	best known lower bound on $z^\star_{ILP}$	376
UB	best known upper bound on $z^\star_{ILP}$	376
DOI	dual-optimal inequality	386
DDOI	deep dual-optimal inequality	386
CSP	cutting stock problem	386
MP_δ	dual-boxed *MP*	388
DMP_δ	dual-boxed *DMP*	388
MDVSP	multiple depot vehicle scheduling problem	393
SDVSP	single depot vehicle scheduling problem	394
MP_{stab}	stabilized *MP*	399
GAP	generalized assignment problem	409
CpMP	capacitated *p*-median problem	412
VRPTW	vehicle routing problem with time windows	414
SPPTWC	shortest path problem with time windows and capacity	415
ECP	edge coloring problem	416
TSP	traveling salesperson problem	418
ATSP	asymmetric traveling salesperson problem	418
STSP	symmetric traveling salesperson problem	418
SEC	subtour elimination constraints	419
BC, AD	Before Computers, After Dantzig	421
MSTP	minimum spanning tree problem	422
PCB	printed circuit board	427
MILP	mixed-integer linear program	439

Introduction

Several topics that we have seen in this book relate in some form to duality, for instance, optimality criteria in linear programming, or the communication between the master and subproblem(s) in the column generation algorithm. We interpret the material of previous chapters in the light of duality and see that this opens new doors. The Lagrangian relaxation method for computing lower bounds for an integer linear program is important in this context. It is said to be dual to the Dantzig-Wolfe decomposition, and we clarify this statement. Most importantly, embracing this dual perspective enables us to improve the column generation algorithm in practice. Most ideas have a geometric interpretation and are thus very visual, so we work with many illustrations.

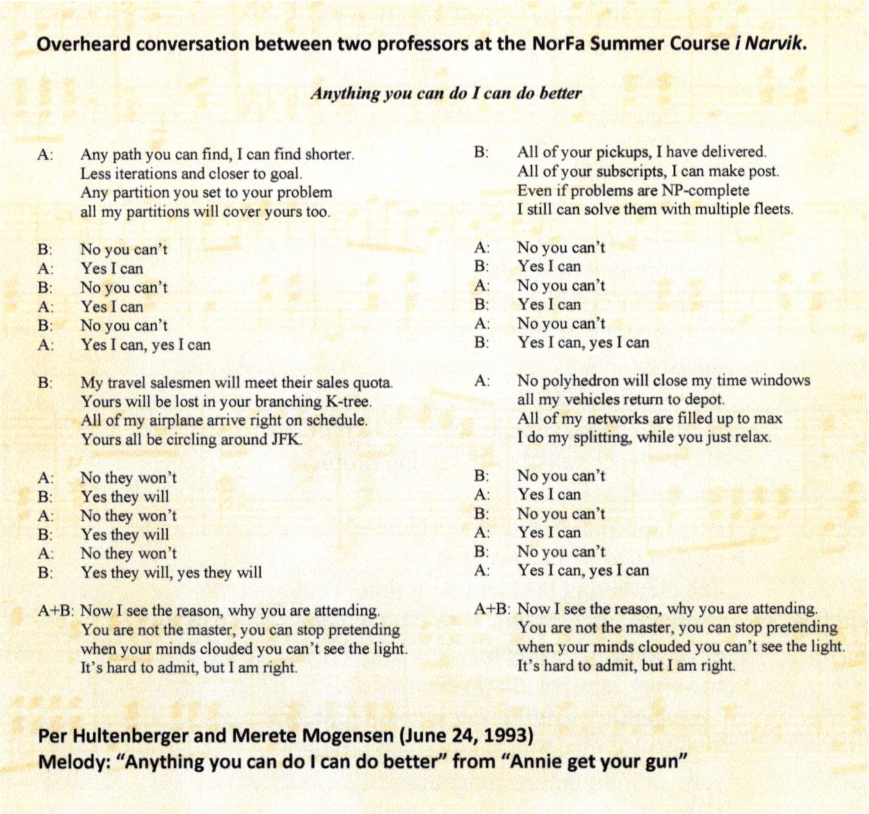

Overheard conversation between two professors at the NorFa Summer Course *i Narvik*.

Anything you can do I can do better

A: Any path you can find, I can find shorter.
Less iterations and closer to goal.
Any partition you set to your problem
all my partitions will cover yours too.

B: No you can't
A: Yes I can
B: No you can't
A: Yes I can
B: No you can't
A: Yes I can, yes I can

B: My travel salesmen will meet their sales quota.
Yours will be lost in your branching K-tree.
All of my airplane arrive right on schedule.
Yours all be circling around JFK.

A: No they won't
B: Yes they will
A: No they won't
B: Yes they will
A: No they won't
B: Yes they will, yes they will

A+B: Now I see the reason, why you are attending.
You are not the master, you can stop pretending
when your minds clouded you can't see the light.
It's hard to admit, but I am right.

B: All of your pickups, I have delivered.
All of your subscripts, I can make post.
Even if problems are NP-complete
I still can solve them with multiple fleets.

A: No you can't
B: Yes I can
A: No you can't
B: Yes I can
A: No you can't
B: Yes I can, yes I can

A: No polyhedron will close my time windows
all my vehicles return to depot.
All of my networks are filled up to max
I do my splitting, while you just relax.

B: No you can't
A: Yes I can
B: No you can't
A: Yes I can
B: No you can't
A: Yes I can, yes I can

A+B: Now I see the reason, why you are attending.
You are not the master, you can stop pretending
when your minds clouded you can't see the light.
It's hard to admit, but I am right.

Per Hultenberger and Merete Mogensen (June 24, 1993)
Melody: "Anything you can do I can do better" from "Annie get your gun"

Fig. 6.1: Overheard conversation between two professors.

Figure 6.1 reports in an artistic way on a conversation between professors Oli Madsen and Jacques Desrosiers at the NorFa Summer Course *i Narvik* in 1993. The first favors the dual methods, the second the primal ones. *If all you have is a hammer, everything looks like a nail*, and these two researchers have different hammers. Just in the *Montréal spirit*, the spirit of collaboration, we write this chapter hoping that together they have an even stronger hammer.

6.1 Row Generation

We are (still!) interested in solving the *ILP* given by the compact formulation

$$
\begin{aligned}
z^\star_{ILP} = \min \quad & \mathbf{c}^\intercal \mathbf{x} \\
\text{s.t.} \quad & \mathbf{Ax} \geq \mathbf{b} \qquad [\boldsymbol{\sigma}_\mathbf{b}] \\
& \mathbf{Dx} \geq \mathbf{d} \qquad [\boldsymbol{\sigma}_\mathbf{d}] \\
& \mathbf{x} \in \mathbb{Z}^n_+ .
\end{aligned} \tag{6.1}
$$

We assume that (6.1) is feasible and $z^\star_{ILP}$ is finite. Again, the dual vectors $\boldsymbol{\sigma}_\mathbf{b}$ and $\boldsymbol{\sigma}_\mathbf{d}$ are only meaningful in the linear relaxation. We recall that in conducting a Dantzig-Wolfe reformulation, we divide the constraints in two parts, reflected by the non-empty sets

$$\mathcal{A} = \{\mathbf{x} \in \mathbb{Z}^n_+ \mid \mathbf{Ax} \geq \mathbf{b}\} \tag{6.2a}$$

$$\mathcal{D} = \{\mathbf{x} \in \mathbb{Z}^n_+ \mid \mathbf{Dx} \geq \mathbf{d}\}. \tag{6.2b}$$

Using the convexification of $\mathcal{D}$, we derive the *IMP* (4.5), whose linear relaxation gives us the *MP*:

$$
\begin{aligned}
z^\star_{MP} = \min \quad & \sum_{p \in P} c_p \lambda_p + \sum_{r \in R} c_r \lambda_r \\
\text{s.t.} \quad & \sum_{p \in P} \mathbf{a}_p \lambda_p + \sum_{r \in R} \mathbf{a}_r \lambda_r \geq \mathbf{b} \qquad [\boldsymbol{\pi}_\mathbf{b}] \\
& \sum_{p \in P} \lambda_p = 1 \qquad [\pi_0] \\
& \lambda_p \geq 0 \qquad \forall p \in P \\
& \lambda_r \geq 0 \qquad \forall r \in R.
\end{aligned} \tag{6.3}
$$

For later reference, we also state its dual and refer to it as the *DMP*:

$$
\begin{aligned}
z^\star_{DMP} = \max \quad & \mathbf{b}^\intercal \boldsymbol{\pi}_\mathbf{b} + \pi_0 \\
\text{s.t.} \quad & \mathbf{a}_p^\intercal \boldsymbol{\pi}_\mathbf{b} + \pi_0 \leq c_p \qquad [\lambda_p] \quad \forall p \in P \\
& \mathbf{a}_r^\intercal \boldsymbol{\pi}_\mathbf{b} \leq c_r \qquad [\lambda_r] \quad \forall r \in R \\
& \boldsymbol{\pi}_\mathbf{b} \geq \mathbf{0}, \ \pi_0 \in \mathbb{R}.
\end{aligned} \tag{6.4}
$$

We ask a question which is central to this chapter: While solving the *MP* by column generation, a primal algorithm, what happens in the dual space? The *RMP* is a restriction of the *MP*, with variables missing. The dual of the *RMP* therefore misses constraints, and is thus a relaxation of the *DMP*. As the column generation algorithm iteratively refines an inner approximation of the *MP*'s feasible region, it iteratively refines an outer approximation of the *DMP*'s feasible region at the same time. The column generation algorithm in the primal is a *cutting plane algorithm* in the dual (Cheney and Goldstein, 1959; Kelley, 1960).

Note 6.1 (Separation problem.) The subproblem finds a master variable of negative reduced cost, if there is any. That is, $c_p - \mathbf{a}_p^\mathsf{T}\boldsymbol{\pi}_\mathbf{b} - \pi_0 < 0$ must hold for a λ_p-variable, similarly for the λ_r-variables, $c_r - \mathbf{a}_r^\mathsf{T}\boldsymbol{\pi}_\mathbf{b} < 0$. This is equivalent to $\mathbf{a}_p^\mathsf{T}\boldsymbol{\pi}_\mathbf{b} + \pi_0 > c_p$ or $\mathbf{a}_r^\mathsf{T}\boldsymbol{\pi}_\mathbf{b} > c_r$, that is, a constraint in the dual which is violated by $(\boldsymbol{\pi}_\mathbf{b}, \pi_0)$. In our interpretation, the *ISP* thus acts as a *separation problem* in the dual: to identify a constraint that is violated by the current dual solution, or to prove that none exists.

Illustration 6.1 TCSPP

We illustrate the dual cutting plane algorithm using our long standing Time constrained shortest path problem (*TCSPP*). The problem is to find a shortest path from node 1 to 6 such that its total traversal time does not exceed 14 time units, see Figure 6.2.

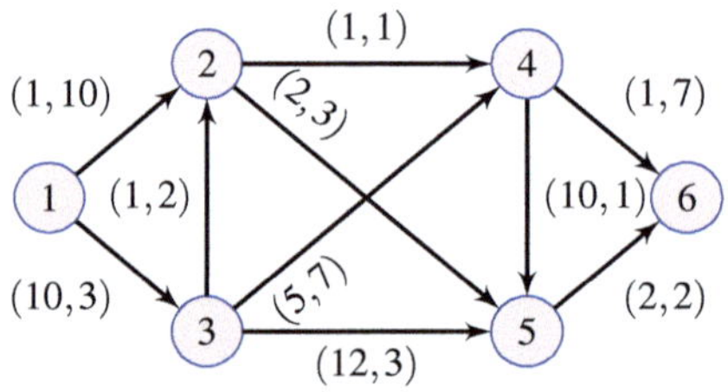

Fig. 6.2: Network $G = (N,A)$ with (c_{ij}, t_{ij}) values, $\forall (i,j) \in A$.

Table 6.1 shows again the five column generation iterations to solve the *MP*. The subproblem's domain is bounded, so only extreme points can be generated.

	RMP				*ISP*				
t	*RMP* solution	z_{RMP}	π_0	π_7	$\bar{c}(\pi_7, \pi_0)$	p	c_p	t_p	lb_t
1	$y_0 = 1$	100.0	100.00	0.00	−97.0	1246	3	18	3.0
2	$y_0 = 0.22, \lambda_{1246} = 0.78$	24.6	100.00	−5.39	−32.9	1356	24	8	−8.3
3	$\lambda_{1246} = 0.6, \lambda_{1356} = 0.4$	11.4	40.80	−2.10	−4.8	13256	15	10	6.6
4	$\lambda_{1246} = \lambda_{13256} = 0.5$	9.0	30.00	−1.50	−2.5	1256	5	15	6.5
5	$\lambda_{13256} = 0.2, \lambda_{1256} = 0.8$	7.0	35.00	−2.00	0.0	-	-	-	7.0

Table 6.1: Column generation iterations: solving the *MP*.

We mirror these iterations by following the generated sequence and *observe* what simultaneously happens in the dual space. As per Note 6.1, at every column generation iteration in the primal, the variable generated by the *ISP* corresponds to a violated constraint in the dual of the *RMP*, i.e., the *relaxed DMP*. Therefore, an optimal path p with cost c_p and column coefficient t_p translates into a cut expressed by $t_p\pi_7 + \pi_0 \leq c_p$. We can see in Figure 6.3 how accumulating these cuts iteratively outer approximates the *DMP*'s feasible region. The dots indicate the extreme points of the current dual feasible region (unbounded on the left and bottom). We mark the respective optimal dual solution in each iteration with a white dot. Let us stress again that these are a by-product of solving the *RMP*, and explicitly solving the relaxed *DMP* is not necessary.

(a) We initially present both, the *RMP* and relaxed *DMP*:

$$\begin{array}{llrcll|llrcll}
z_{RMP} = \min & 100y_0 & & & & & \max & 14\pi_7 + & \pi_0 & & & \\
\text{s.t.} & & s_7 & = & 14 & [\pi_7] & \text{s.t.} & & \pi_0 & \leq & 100 & [y_0] \\
& y_0 & & = & 1 & [\pi_0] & & \pi_7 & & \leq & 0 & [s_7] \\
& y_0, & s_7 & \geq & 0 & & & \pi_7 \in \mathbb{R}, & \pi_0 \in \mathbb{R}. & & &
\end{array}$$

Observe that $\pi_7 \leq 0$ whereas $\pi_0 \leq 100$ is due to the big-M cost of the artificial variable y_0 in the convexity constraint of the primal formulation. The optimal dual solution $(\pi_7, \pi_0) = (0, 100)$ gives $z_{RMP} = 100$. From path $p = 1246$, we add constraint $18\pi_7 + \pi_0 \leq 3$ to the relaxed *DMP*.

(b) In the second iteration, the relaxed *DMP* thus reads as

$$\begin{array}{lrcll}
z_{RMP} = \max\ 14\pi_7 + & \pi_0 & & & \\
\text{s.t.} & \pi_0 & \leq & 100 & [y_0] \\
18\pi_7 + & \pi_0 & \leq & 3 & [\lambda_{1246}] \\
\pi_7 \leq 0, & \pi_0 \in \mathbb{R}. & & &
\end{array}$$

The optimal objective value decreases to $z_{RMP} = {}^{221}/_{9}$ at $(-{}^{97}/_{18}, 100)$. The primal solution using $y_0 = 0.22$ is still infeasible for the original problem. This is reflected by π_0 being at its upper bound value 100. From path $p = 1356$, we add the dual constraint $8\pi_7 + \pi_0 \leq 24$.

(c) This leads to the next dual problem:

$$\begin{array}{lrcll}
z_{RMP} = \max\ 14\pi_7 + & \pi_0 & & & \\
\text{s.t.} & \pi_0 & \leq & 100 & [y_0] \\
18\pi_7 + & \pi_0 & \leq & 3 & [\lambda_{1246}] \\
8\pi_7 + & \pi_0 & \leq & 24 & [\lambda_{1356}] \\
\pi_7 \leq 0, & \pi_0 \in \mathbb{R}. & & &
\end{array}$$

The optimal objective value falls again, to $z_{RMP} = 11.4$ at $(-2.1, 40.8)$. From path $p = 13256$, we add constraint $10\pi_7 + \pi_0 \leq 15$ to the dual.

(d) The fourth dual problem, more and more constrained, is

$$\begin{array}{lrcll}
z_{RMP} = \max\ 14\pi_7 + & \pi_0 & & & \\
\text{s.t.} & \pi_0 & \leq & 100 & [y_0] \\
18\pi_7 + & \pi_0 & \leq & 3 & [\lambda_{1246}] \\
8\pi_7 + & \pi_0 & \leq & 24 & [\lambda_{1356}] \\
10\pi_7 + & \pi_0 & \leq & 15 & [\lambda_{13256}] \\
\pi_7 \leq 0, & \pi_0 \in \mathbb{R}. & & &
\end{array}$$

The maximum objective value again decreases to $z_{RMP} = 9$ at $(-1.5, 30)$. From path $p = 1256$, we add constraint $15\pi_7 + \pi_0 \leq 5$.

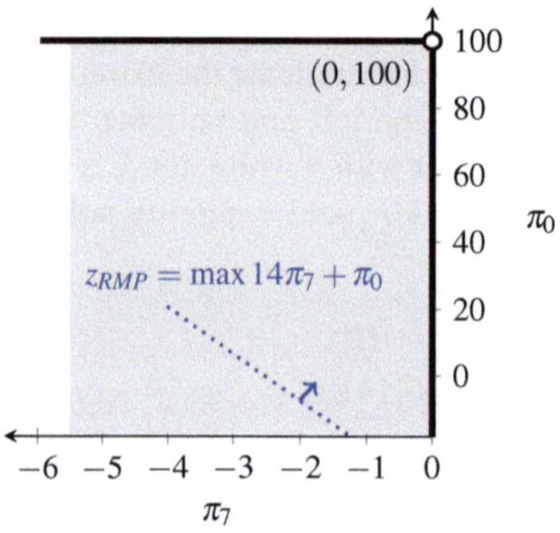

(a) **Iter. 1.** $z_{RMP} = 100$ at $(0, 100)$

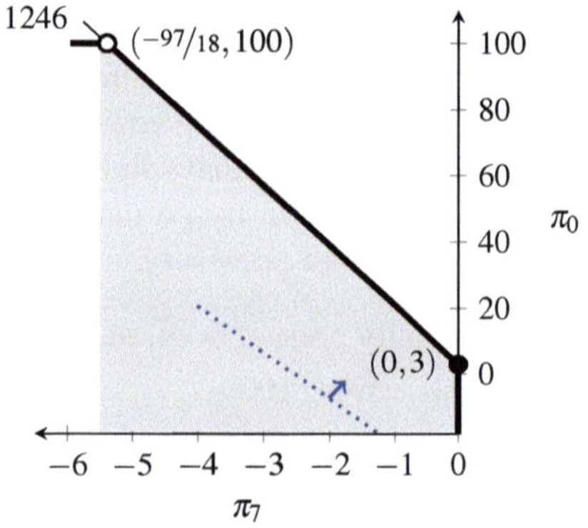

(b) **Iter. 2.** $z_{RMP} = 221/9$ at $(-97/18, 100)$

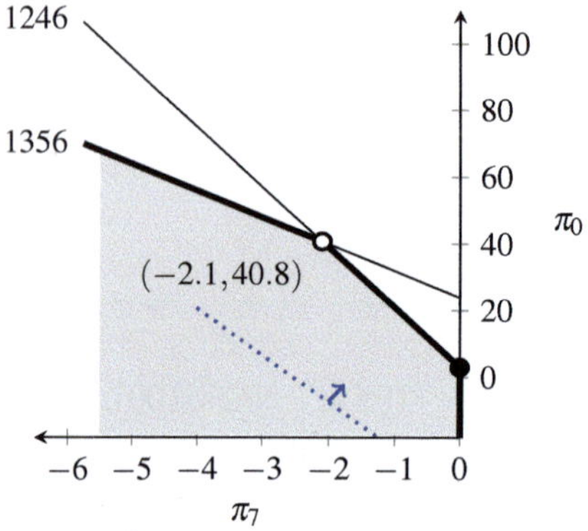

(c) **Iter. 3.** $z_{RMP} = 11.4$ at $(-2.1, 40.8)$

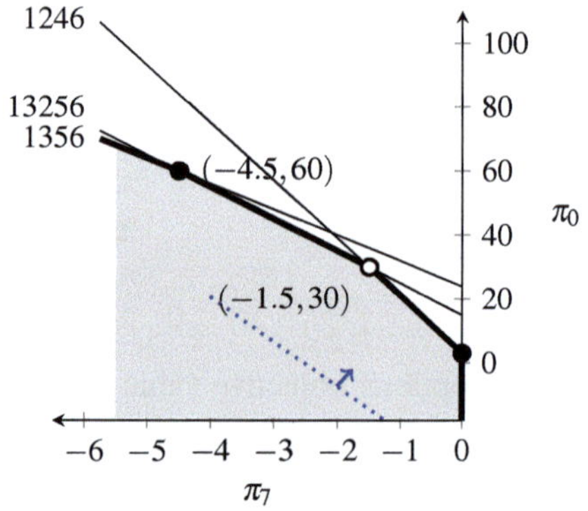

(d) **Iter. 4.** $z_{RMP} = 9$ at $(-1.5, 30)$

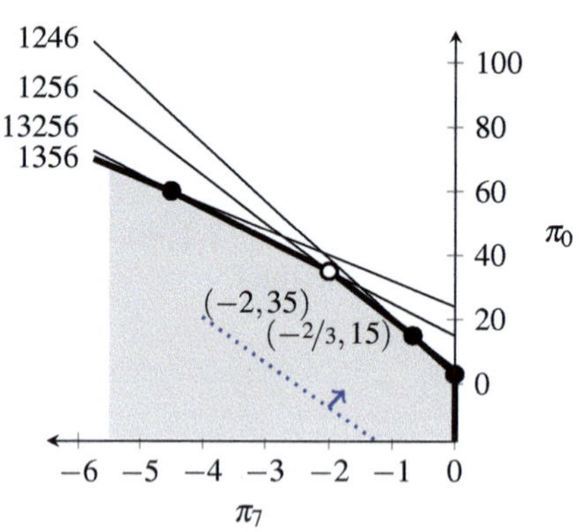

(e) **Iter. 5.** $z^{\star}_{MP} = 7$ at $(-2, 35)$

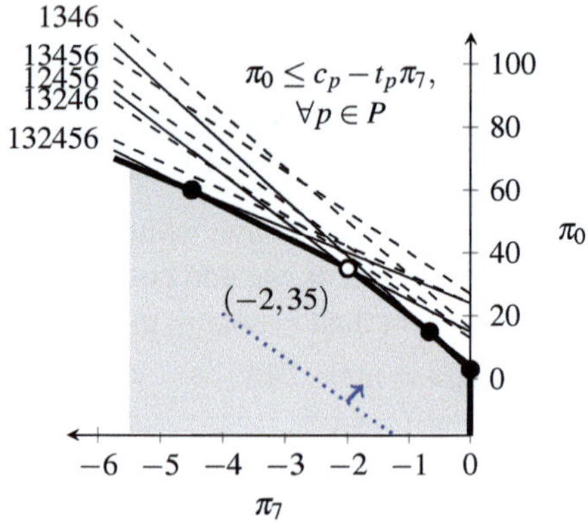

(f) All of the nine constraints

Fig. 6.3: Outer approximating the *DMP*'s feasible region.

(e) The fifth and final dual problem in this illustration is

$$
\begin{aligned}
z_{RMP} = \max\ & 14\pi_7 + \pi_0 \\
\text{s.t.}\ & \pi_0 \leq 100 \quad [y_0] \\
& 18\pi_7 + \pi_0 \leq 3 \quad [\lambda_{1246}] \\
& 8\pi_7 + \pi_0 \leq 24 \quad [\lambda_{1356}] \\
& 10\pi_7 + \pi_0 \leq 15 \quad [\lambda_{13256}] \\
& 15\pi_7 + \pi_0 \leq 5 \quad [\lambda_{1256}] \\
& \pi_7 \leq 0,\ \pi_0 \in \mathbb{R},
\end{aligned}
$$

where $z_{RMP} = 7$ at $(-2, 35)$. There is no more column of negative reduced cost, thus no violated constraint in the dual, and $z^{\star}_{DMP} = z^{\star}_{MP} = 7$.

(f) The last sub-figure (f) shows all the nine dual constraints. The five ones that were not generated are labeled and appear as dashed lines. Observe that all of them are, as expected by the fact that the *MP* is solved to optimality, redundant for the feasible region of the *DMP*.

Note 6.2 (Hold on copycat.) We could have produced this illustration by directly solving the *DMP* using the *ISP* as a separation algorithm to find violated dual constraints. In fact, we would find exactly the same sequence of iterations and dual solutions as when solving the *MP* by the column generation algorithm. This is purely coincidental since we have such a small example. In general, equivalent optimal solutions in the master and/or subproblems, limited numeric precision, and runtime behavior can all influence unpredictably independent solving of the *MP* and *DMP*. While there is no trivial answer as to which one would be faster to solve, the presence of degeneracy may give incentive to solve the *DMP*. The primal is indeed often very degenerate, in particular for the set partitioning/covering master problems we see in practice. Degeneracy in the dual appears when there are many primal solutions of the same cost, in particular many optimal solutions. This can for instance happen when primal variables have identical cost, which is usually not the case in practice. On the flip side, as the number of constraints grows in the relaxed *DMP*, it becomes increasingly difficult to solve, proportionally more so than with the growing number of generated columns in the *RMP*.

Alternative master problem

We next present a slightly different *MP*. A reason for this mathematical development is that the forthcoming Lagrangian function has strong ties to the dual of this alternative master problem. In the *DMP* (6.4), let us use the change of variables

$$\mu = \mathbf{b}^{\intercal}\boldsymbol{\pi}_{\mathbf{b}} + \pi_0 \tag{6.5}$$

in both the objective function and the constraints indexed by $p \in P$. Hence, the value of the objective function is now expressed as the single dual variable μ:

$$
\begin{aligned}
\max \quad & \mu \\
\text{s.t.} \quad & (\mathbf{a}_p - \mathbf{b})^\intercal \boldsymbol{\pi}_\mathbf{b} + \mu \le c_p \quad [\lambda_p] \quad \forall p \in P \\
& \mathbf{a}_r^\intercal \boldsymbol{\pi}_\mathbf{b} \le c_r \quad [\lambda_r] \quad \forall r \in R \\
& \boldsymbol{\pi}_\mathbf{b} \ge \mathbf{0}, \ \mu \in \mathbb{R}.
\end{aligned}
\tag{6.6}
$$

Dualizing, we obtain an alternative primal formulation for the *MP*, denoted by *AMP*:

$$z^\star_{AMP} = \min \quad \sum_{p \in P} c_p \lambda_p + \sum_{r \in R} c_r \lambda_r \tag{6.7a}$$

$$\text{s.t.} \quad \sum_{p \in P} (\mathbf{a}_p - \mathbf{b}) \lambda_p + \sum_{r \in R} \mathbf{a}_r \lambda_r \ge \mathbf{0} \quad [\boldsymbol{\pi}_\mathbf{b}] \tag{6.7b}$$

$$\sum_{p \in P} \lambda_p = 1 \quad [\mu] \tag{6.7c}$$

$$\lambda_p \ge 0, \quad \lambda_r \ge 0 \quad \forall p \in P, r \in R. \tag{6.7d}$$

A careful examination of the two primal formulations in (6.3) and (6.7) shows that we can derive the latter from the former by using the convexity constraint in order to move the right-hand side vector $\mathbf{b}$ to the left of the constraints as $-\mathbf{b}(\sum_{p \in P} \lambda_p)$.

The introduced change of variables (6.5) results in an affine transformation between the two dual domains. Thus, there is a one-to-one correspondence between their extreme points. Since we accordingly transform the objective function, this correspondence also holds for their respective objective values and, obviously, $z^\star_{AMP} = z^\star_{MP}$.

Note 6.3 (Consequences of a finite optimal objective value.) There are a few technical implications of a finite $z^\star_{ILP}$: in the *ILP*, the sum of the x-variables is bounded, or equivalently, $\mathbf{x} \le \mathbf{u}$ for a sufficiently large upper bound vector $\mathbf{u}$. Moreover, in the *MP* and *AMP* (with finite $z^\star_{MP}$ and $z^\star_{AMP}$, respectively), $\sum_{r \in R} \lambda^\star_r$ has to be finite in any optimal solution $\boldsymbol{\lambda}^\star$ (note that $\sum_{p \in P} \lambda^\star_p$ is already equal to 1). Therefore, from a dual point of view, for any optimal dual solution $\boldsymbol{\pi}^\star_\mathbf{b}$, there cannot be an extreme ray $r \in R$, for which the corresponding $\lambda^\star_r$ has negative reduced cost:

$$\forall \boldsymbol{\pi}^\star_\mathbf{b}, \qquad c_r - \mathbf{a}_r^\intercal \boldsymbol{\pi}^\star_\mathbf{b} \ge 0, \ \forall r \in R. \tag{6.8}$$

Illustration 6.2 TCSPP **(cont.)**

Figure 6.4 presents side-by-side the dual domains for the two Dantzig-Wolfe master problems, together with their respective objective functions in dashed lines.

- For the *MP* on the left, we draw the constraints $\pi_0 \le c_p - \mathbf{a}_p^\intercal \boldsymbol{\pi}_\mathbf{b}, \forall p \in P$, see (6.4) with objective function $z^\star_{MP} = \max\ \mathbf{b}^\intercal \boldsymbol{\pi}_\mathbf{b} + \pi_0$.
- For the *AMP* on the right, we use $\mu \le c_p + (\mathbf{b} - \mathbf{a}_p)^\intercal \boldsymbol{\pi}_\mathbf{b}, \forall p \in P$, from (6.6), where the objective function is $z^\star_{AMP} = \max\ \mu$.
- In both cases, we obtain $\pi^\star_7 = -2$ and $z^\star_{MP} = z^\star_{AMP} = 7$.

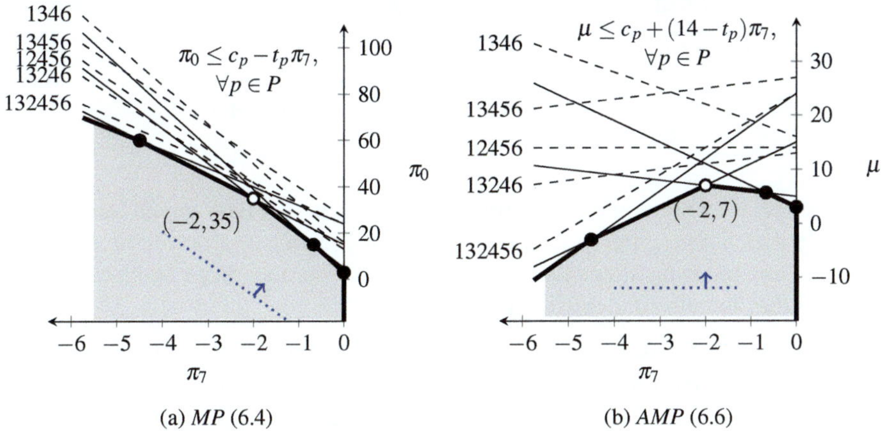

Fig. 6.4: Dual domains of the two equivalent Dantzig-Wolfe master problems.

6.2 Lagrangian Relaxation

A Lagrangian relaxation of an integer linear program is a common approach to calculate a *lower bound* on $z^\star_{ILP}$, usually better and never worse than the one derived from the linear relaxation $z^\star_{LP}$. As in a Dantzig-Wolfe reformulation, a Lagrangian relaxation relies on observing two different roles in the constraints of the *ILP*. On the one hand, *structured* constraints can be exploited, say, because they describe a problem formulation for which we have a tailored algorithm available. On the other hand, *complicating* constraints which may prevent the direct application of such an algorithm.

Lagrangian subproblem

In order to perform a Lagrangian relaxation of the *ILP* (6.1), we first appropriately divide its constraints into the *complicating* ones, say in

$$\mathcal{A} = \{\mathbf{x} \in \mathbb{Z}^n_+ \mid \mathbf{Ax} \ge \mathbf{b}\}, \tag{6.9}$$

and the *structured* ones, say in

$$\mathcal{D} = \{\mathbf{x} \in \mathbb{Z}^n_+ \mid \mathbf{Dx} \ge \mathbf{d}\}. \tag{6.10}$$

We could completely relax $\mathbf{Ax} \ge \mathbf{b}$, but instead we penalize their violation $(\mathbf{b} - \mathbf{Ax})$ in the objective function via so-called *Lagrangian multipliers* $\boldsymbol{\pi}_\mathbf{b} \ge \mathbf{0}$. This yields the *Lagrangian subproblem*, an integer linear program we denote by *ISP* and present in concise and expanded form:

$$\begin{aligned} LR(\boldsymbol{\pi}_{\mathbf{b}}) &= \min_{\mathbf{x}\in\mathcal{D}} \; \mathbf{c}^\top\mathbf{x} + \boldsymbol{\pi}_{\mathbf{b}}^\top(\mathbf{b}-\mathbf{A}\mathbf{x}) \\ &= \boldsymbol{\pi}_{\mathbf{b}}^\top\mathbf{b} + \min \; (\mathbf{c}^\top - \boldsymbol{\pi}_{\mathbf{b}}^\top\mathbf{A})\mathbf{x} \\ & \qquad \text{s.t.} \qquad \mathbf{D}\mathbf{x} \geq \mathbf{d} \\ & \qquad\qquad\qquad \mathbf{x} \in \mathbb{Z}_+^n. \end{aligned} \tag{6.11}$$

Since this is a common misconception, let us repeat that Lagrangian relaxation is precisely that, a *relaxation* of the *ILP*, and *not* an equivalent reformulation. The notion of relaxation is justified because the *ISP* (6.11) not only drops some constraints from the *ILP*, but also underestimates $z_{ILP}^\star$.

Proposition 6.1 (Weak Lagrangian duality). *Solving the ISP* (6.11) *for any* $\boldsymbol{\pi}_{\mathbf{b}} \geq \mathbf{0}$ *yields a lower bound* $LR(\boldsymbol{\pi}_{\mathbf{b}})$ *on* $z_{ILP}^\star$, *called a* Lagrangian bound, *that is,*

$$LR(\boldsymbol{\pi}_{\mathbf{b}}) \leq z_{ILP}^\star, \quad \forall \boldsymbol{\pi}_{\mathbf{b}} \geq \mathbf{0}. \tag{6.12}$$

Proof. Obviously, if $\boldsymbol{\pi}_{\mathbf{b}} = \mathbf{0}$, $LR(\mathbf{0}) \leq z_{ILP}^\star$ because the set $\mathbf{A}\mathbf{x} \geq \mathbf{b}$ is discarded from the *ILP* (6.1). We show that this remains true for all $\boldsymbol{\pi}_{\mathbf{b}} \geq \mathbf{0}$ by definition of the *ISP*:

$$\begin{aligned} LR(\boldsymbol{\pi}_{\mathbf{b}}) &\stackrel{\text{def}}{=} \min \{\mathbf{c}^\top\mathbf{x} + \boldsymbol{\pi}_{\mathbf{b}}^\top(\mathbf{b}-\mathbf{A}\mathbf{x}) \mid \mathbf{x}\in\mathcal{D}\} \\ &\leq \min \{\mathbf{c}^\top\mathbf{x} + \boldsymbol{\pi}_{\mathbf{b}}^\top(\mathbf{b}-\mathbf{A}\mathbf{x}) \mid \mathbf{A}\mathbf{x}\geq\mathbf{b},\, \mathbf{x}\in\mathcal{D}\} \leq z_{ILP}^\star. \end{aligned} \tag{6.13}$$

The first inequality comes from adding $\mathbf{A}\mathbf{x} \geq \mathbf{b}$ to the set of constraints while the second is from the term $\boldsymbol{\pi}_{\mathbf{b}}^\top(\mathbf{b}-\mathbf{A}\mathbf{x}) \leq \mathbf{0}$ added to the objective function in (6.1). □

In our presentation, we use non-negative Lagrangian multipliers. Regarding their sign, however, standard duality concepts apply: For equality constraints, the multipliers are unrestricted in sign; and for less-than-or-equal constraints, they are non-positive. After all, the term $\boldsymbol{\pi}_{\mathbf{b}}^\top(\mathbf{b}-\mathbf{A}\mathbf{x})$ should be a penalty term.

Note 6.4 (Are we French, or what?) Several major publications on Lagrangian relaxation stick closer to the namesake and use *Lagrangean relaxation* instead. Google, however, reveals that the majority of scientific literature is not francophone, and we do not start a revolution here. (See Exercise 6.1 on the nationality of Joseph-Louis Lagrange.)

Lagrangian function and Lagrangian dual problem

The Lagrangian subproblem's optimal objective value $LR(\boldsymbol{\pi}_{\mathbf{b}})$, and thus the quality of the lower bound on $z_{ILP}^\star$, clearly depends on the choice of $\boldsymbol{\pi}_{\mathbf{b}}$. We call

$$LR : \mathbb{R}_+^m \to \mathbb{R}, \quad \boldsymbol{\pi}_{\mathbf{b}} \mapsto LR(\boldsymbol{\pi}_{\mathbf{b}}), \tag{6.14}$$

the *Lagrangian function*. Because we are interested in the best possible Lagrangian bound, we need to maximize the Lagrangian function over $\boldsymbol{\pi}_{\mathbf{b}} \geq \mathbf{0}$. This *master* optimization program in $\boldsymbol{\pi}_\mathbf{b}$ is called the *Lagrangian dual problem* of the *ILP* with respect to the set of constraints $\mathbf{Ax} \geq \mathbf{b}$. We denote it by *LDP*:

$$z^\star_{LDP} = \max_{\boldsymbol{\pi}_\mathbf{b} \geq \mathbf{0}} LR(\boldsymbol{\pi}_\mathbf{b}) \quad = \max_{\boldsymbol{\pi}_\mathbf{b} \geq \mathbf{0}} \left\{ \boldsymbol{\pi}_\mathbf{b}^\mathsf{T}\mathbf{b} + \min_{\mathbf{x}\in\mathcal{D}}(\mathbf{c}^\mathsf{T} - \boldsymbol{\pi}_\mathbf{b}^\mathsf{T}\mathbf{A})\mathbf{x} \right\}. \tag{6.15}$$

Block-diagonal structure

As several times before, the case of a single subproblem can be generalized for an integer linear program with a block-diagonal structure. Let the *ILP* be given as

$$\begin{aligned} z^\star_{ILP} = \min \quad & \sum_{k\in K} \mathbf{c}^{k\mathsf{T}}\mathbf{x}^k \\ \text{s.t.} \quad & \sum_{k\in K} \mathbf{A}^k\mathbf{x}^k \geq \mathbf{b} \quad [\boldsymbol{\sigma}_\mathbf{b}] \\ & \mathbf{x}^k \in \mathcal{D}^k = \{\mathbf{x}^k \in \mathbb{Z}_+^{n^k} \mid \mathbf{D}^k\mathbf{x}^k \geq \mathbf{d}^k\} \quad \forall k \in K, \end{aligned} \tag{6.16}$$

where the dual vector $\boldsymbol{\sigma}_\mathbf{b} \geq \mathbf{0}$ is only meaningful in the linear relaxation. We relax the inequalities $\sum_{k\in K} \mathbf{A}^k\mathbf{x}^k \geq \mathbf{b}$ and penalize their violation in the objective function with the vector of *Lagrangian multipliers* $\boldsymbol{\pi}_\mathbf{b} \geq \mathbf{0}$; so the *ISP* becomes

$$\begin{aligned} LR(\boldsymbol{\pi}_\mathbf{b}) = \min \quad & \sum_{k\in K} \mathbf{c}^{k\mathsf{T}}\mathbf{x}^k + \boldsymbol{\pi}_\mathbf{b}^\mathsf{T}(\mathbf{b} - \sum_{k\in K} \mathbf{A}^k\mathbf{x}^k) \\ \text{s.t.} \quad & \mathbf{x}^k \in \mathcal{D}^k, \quad \forall k \in K. \end{aligned} \tag{6.17}$$

Setting aside $\boldsymbol{\pi}_\mathbf{b}^\mathsf{T}\mathbf{b}$, we see that $LR(\boldsymbol{\pi}_\mathbf{b})$ is separable in index k. The *Lagrangian bound* from Proposition 6.1 can therefore be computed as

$$LR(\boldsymbol{\pi}_\mathbf{b}) = \boldsymbol{\pi}_\mathbf{b}^\mathsf{T}\mathbf{b} + \sum_{k\in K} \min_{\mathbf{x}^k\in\mathcal{D}^k} (\mathbf{c}^{k\mathsf{T}} - \boldsymbol{\pi}_\mathbf{b}^\mathsf{T}\mathbf{A}^k)\mathbf{x}^k \leq z^\star_{ILP}, \qquad \forall \boldsymbol{\pi}_\mathbf{b} \geq \mathbf{0}. \tag{6.18}$$

We ultimately arrive at the *LDP* given by

$$z^\star_{LDP} = \max_{\boldsymbol{\pi}_\mathbf{b} \geq \mathbf{0}} LR(\boldsymbol{\pi}_\mathbf{b}) = \max_{\boldsymbol{\pi}_\mathbf{b} \geq \mathbf{0}} \left\{ \boldsymbol{\pi}_\mathbf{b}^\mathsf{T}\mathbf{b} + \sum_{k\in K} \min_{\mathbf{x}^k\in\mathcal{D}^k} (\mathbf{c}^{k\mathsf{T}} - \boldsymbol{\pi}_\mathbf{b}^\mathsf{T}\mathbf{A}^k)\mathbf{x}^k \right\}. \tag{6.19}$$

Illustration 6.3 TCSPP **(cont.)**

Let us apply the Lagrangian relaxation method on the *time constrained shortest path problem* and take a look at the Lagrangian bounds.

$$z^\star_{ILP} = \min \qquad \sum_{(i,j)\in A} c_{ij}x_{ij} \tag{6.20a}$$

$$\text{s.t.} \quad \sum_{j:(1,j)\in A} x_{1j} = 1 \quad [\sigma_1] \tag{6.20b}$$

$$\sum_{j:(i,j)\in A} x_{ij} - \sum_{j:(j,i)\in A} x_{ji} = 0 \quad [\sigma_i] \quad \forall i \in \{2,\ldots,5\} \tag{6.20c}$$

$$-\sum_{i:(i,6)\in A} x_{i6} = -1 \quad [\sigma_6] \tag{6.20d}$$

$$\sum_{(i,j)\in A} t_{ij}x_{ij} \leq 14 \quad [\sigma_7] \tag{6.20e}$$

$$x_{ij} \in \{0,1\} \quad \forall (i,j) \in A. \tag{6.20f}$$

Recall that an optimal integer solution to (6.20) is given by path 13246 of duration 13 and cost $z^\star_{ILP} = 13$, whereas the linear relaxation gives a lower bound of $z^\star_{LP} = 7$.

We replicate the grouping (3.79) seen in our Dantzig-Wolfe reformulation by keeping the path structure from node 1 to 6 in

$$\mathcal{D} = \{\mathbf{x} \in \{0,1\}^{|A|} \mid (6.20\text{b})\text{–}(6.20\text{d})\}. \tag{6.21}$$

We then relax the duration constraint $\sum_{(i,j)\in A} t_{ij}x_{ij} \leq 14$ and penalize its violation in the objective function with the *Lagrangian multiplier* $\pi_7 \leq 0$:

$$LR(\pi_7) = \min_{\mathbf{x}\in\mathcal{D}} \sum_{(i,j)\in A} c_{ij}x_{ij} + \pi_7(14 - \sum_{(i,j)\in A} t_{ij}x_{ij}). \tag{6.22}$$

By Proposition 6.1, formulation (6.22) is a relaxation of (6.20). This is obvious when $\pi_7 = 0$, since this amounts to removing the duration constraint from (6.20). Computing $LR(0) = 3$, we obtain the shortest path 1246 of duration $18 > 14$, indeed infeasible for (6.20).

When $\pi_7 \ll 0$, say $\pi_7 = -100$, the penalized time component $\pi_7(14 - \sum_{(i,j)\in A} t_{ij}x_{ij})$ in (6.22) dominates the cost component $\sum_{(i,j)\in A} c_{ij}x_{ij}$ and solving the *ISP* with $\pi_7 = -100$ yields path 1356 with the large cost 24 compared to $z^\star_{ILP} = 13$. In that case, the lower bound becomes the poor $LR(-100) = 24 + (-100)(14-8) = -576 \ll 13$.

In order to gain some intuition on the largest Lagrangian bound, let us plot $LR(\pi_7)$ for values of π_7 in a region where we expect the maximum, say $-5 \leq \pi_7 \leq 0$, for values $\{-5, -4.5, -4, \ldots, -0.5, 0\}$. We observe in Figure 6.5 what appears to be a concave curve, with the maximum obtained at coordinates $(-2, 7)$ for our selection of π_7-values.

Note 6.5 (Duals, duals, duals.) We have seen variables (or multipliers) that correspond to the same constraints several times, dual variables for the original *LP*, dual variables for the *MP*, Lagrangian multipliers in the Lagrangian subproblem. The latter are *also* called dual variables in the literature, and penalizing violated constraints in the objective function is sometimes referred to as *dualizing* them. Even though the variables all have similar roles, and e.g., their sign is derived via arguments common to all of them, they usually do *not* have identical values in the different

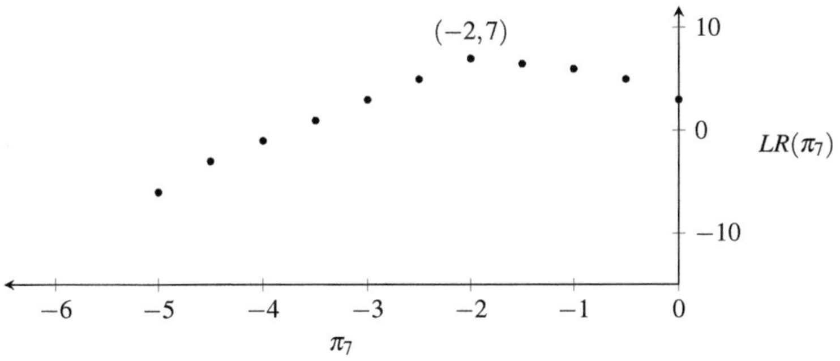

Fig. 6.5: Several values of $LR(\pi_7)$.

contexts, and should thus not be confused. We reflect this in our notation so the reason we use $\boldsymbol{\pi}_{\mathbf{b}}$ interchangeably is because of our leading claim that Lagrangian relaxation and Dantzig-Wolfe decomposition are dual from one another.

Dantzig-Wolfe reformulation vs. Lagrangian relaxation

Our development of the Lagrangian relaxation method is in large analogy to our presentation of the Dantzig-Wolfe decomposition. We make identical assumptions on the original *ILP*, and use an identical grouping of its constraints. We purposefully denote the Lagrangian multipliers $\boldsymbol{\pi}_{\mathbf{b}}$ like the dual variables of the Dantzig-Wolfe master problem, see (6.3) and (6.7). Figure 6.6 gives the big picture of what is coming, assuming a single bounded subproblem. We are already familiar with the left part of it, and we now establish the right part, justifying our choices above. We show the following:

- the formulations of the Dantzig-Wolfe and Lagrangian subproblems, respectively given by (4.11) and (6.11), are equivalent;
- the *LDP* (6.15) and *MP* (6.3) (or the alternative formulation *AMP* (6.7)) give an identical lower bound on $z^\star_{ILP}$, i.e., $z^\star_{LDP} = z^\star_{MP}$ $(= z^\star_{AMP})$.

To begin with, the Lagrangian bound (6.18) looks familiar since we derived it, in a slightly different way, as (3.33) for linear programs and (4.68) for integer linear programs. For all $\boldsymbol{\pi}_{\mathbf{b}} \geq \mathbf{0}$ and $\pi_0^k \in \mathbb{R}$, $\forall k \in K$, recall the ISP^k in a Dantzig-Wolfe reformulation

$$\bar{c}^k(\boldsymbol{\pi}_{\mathbf{b}}, \pi_0^k) = -\pi_0^k + \min_{\mathbf{x}^k \in \mathcal{D}^k} (\mathbf{c}^{k\intercal} - \boldsymbol{\pi}_{\mathbf{b}}^\intercal \mathbf{A}^k)\mathbf{x}^k, \quad k \in K. \tag{6.23}$$

Contrasting this with the Lagrangian subproblem (6.17), we see

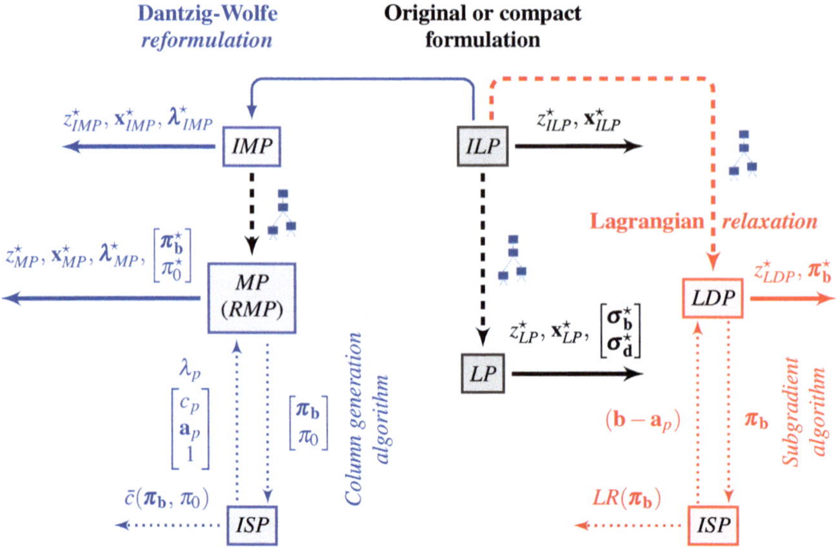

Fig. 6.6: Information flow of a Dantzig-Wolfe reformulation vs. a Lagrangian relaxation of the *ILP* with classical solution methods. Domain $\mathcal{D}$ is assumed bounded.

$$
\begin{aligned}
LR(\boldsymbol{\pi}_{\mathbf{b}}) &= \boldsymbol{\pi}_{\mathbf{b}}^{\mathsf{T}}\mathbf{b} + \sum_{k\in K} \min_{\mathbf{x}^k\in\mathcal{D}^k} (\mathbf{c}^{k\mathsf{T}} - \boldsymbol{\pi}_{\mathbf{b}}^{\mathsf{T}}\mathbf{A}^k)\mathbf{x}^k \\
&= \boldsymbol{\pi}_{\mathbf{b}}^{\mathsf{T}}\mathbf{b} + \sum_{k\in K} \left(\bar{c}^k(\boldsymbol{\pi}_{\mathbf{b}}, \pi_0^k) + \pi_0^k\right)
\end{aligned}
\tag{6.24}
$$

In other words, $\min_{\mathbf{x}^k\in\mathcal{D}^k}(\mathbf{c}^{k\mathsf{T}} - \boldsymbol{\pi}_{\mathbf{b}}^{\mathsf{T}}\mathbf{A}^k)\mathbf{x}^k = \bar{c}^k(\boldsymbol{\pi}_{\mathbf{b}}, \pi_0^k) + \pi_0^k$ is computed in the Lagrangian subproblem ISP^k whereas it is $\bar{c}^k(\boldsymbol{\pi}_{\mathbf{b}}, \pi_0^k)$ in the Dantzig-Wolfe one. These optimization programs are obviously equivalent: they have an identical domain $\mathcal{D}^k$ and only differ in the objective function by the constant term π_0^k.

Concerning lower bounds, let us consider a single subproblem (see Exercises 6.4 and 6.5 for several pricing problems). We rewrite the Lagrangian subproblem (6.11) using a convexification of $\mathcal{D}$, that is, we replace $\mathcal{D}$ by conv$(\mathcal{D})$. This does not change its set of integer solutions. As introduced for a Dantzig-Wolfe reformulation of the *ILP*, let P and R be the index sets of extreme points and extreme rays of conv$(\mathcal{D})$, respectively, and $\mathcal{X} = \left\{\begin{bmatrix}\mathbf{x}_p\\ 1\end{bmatrix}\right\}_{p\in P} \cup \left\{\begin{bmatrix}\mathbf{x}_r\\ 0\end{bmatrix}\right\}_{r\in R}$. We also reuse the concise notation

$$
\begin{aligned}
c_p &= \mathbf{c}^{\mathsf{T}}\mathbf{x}_p, \quad & \mathbf{a}_p &= \mathbf{A}\mathbf{x}_p \qquad & \forall p\in P \\
c_r &= \mathbf{c}^{\mathsf{T}}\mathbf{x}_r, \quad & \mathbf{a}_r &= \mathbf{A}\mathbf{x}_r \qquad & \forall r\in R.
\end{aligned}
\tag{6.25}
$$

For a given $\boldsymbol{\pi}_{\mathbf{b}}$, solving the *ISP* (6.11) with an algorithm that returns an optimal solution $\mathbf{x}$ that is either an extreme point or an extreme ray of conv$(\mathcal{D})$, the obtained

Lagrangian bound is

$$LR(\boldsymbol{\pi}_\mathbf{b}) = \begin{cases} -\infty & \text{if } c_r - \mathbf{a}_r^\intercal \boldsymbol{\pi}_\mathbf{b} < 0 \quad \text{for some } r \in R \\ c_p + (\mathbf{b} - \mathbf{a}_p)^\intercal \boldsymbol{\pi}_\mathbf{b} & \text{otherwise} \qquad\qquad\quad \text{for some } p \in P. \end{cases} \tag{6.26}$$

Proposition 6.2. *Given that $z^\star_{ILP}$ is finite, the largest Lagrangian bound is equal to the optimal objective value of the MP, that is, $z^\star_{LDP} = z^\star_{MP}$.*

Proof. The *LDP* is a linear max-min problem. We prove the equality $z^\star_{LDP} = z^\star_{MP}$ by rewriting the *LDP* as a linear program and showing that the latter is the dual of the *AMP*, that is, the alternative *MP*.

For a given $\boldsymbol{\pi}_\mathbf{b} \geq \mathbf{0}$, a lower bound $LR(\boldsymbol{\pi}_\mathbf{b})$ on $z^\star_{ILP}$ is either infinite $(-\infty)$ or finite by (6.26). Looking for a better bound than the useless $-\infty$, we further restrain $\boldsymbol{\pi}_\mathbf{b}$ with $c_r - \mathbf{a}_r^\intercal \boldsymbol{\pi}_\mathbf{b} \geq 0, \forall r \in R$, which we know must hold for any optimal $\boldsymbol{\pi}^\star_\mathbf{b}$, see Note 6.3. The *LDP* is thus restated as

$$\begin{aligned} z^\star_{LDP} = \max_{\boldsymbol{\pi}_\mathbf{b} \geq \mathbf{0}} \quad & \min_{p \in P} \; c_p + (\mathbf{b} - \mathbf{a}_p)^\intercal \boldsymbol{\pi}_\mathbf{b} \\ \text{s.t.} \quad & \qquad\qquad c_r - \mathbf{a}_r^\intercal \boldsymbol{\pi}_\mathbf{b} \geq 0 \quad \forall r \in R. \end{aligned} \tag{6.27}$$

We reformulate the inner optimization $\min_{p \in P} \; c_p + (\mathbf{b} - \mathbf{a}_p)^\intercal \boldsymbol{\pi}_\mathbf{b}$ as

$$\begin{aligned} \max \quad & \mu \\ \text{s.t.} \quad & \mu \leq c_p + (\mathbf{b} - \mathbf{a}_p)^\intercal \boldsymbol{\pi}_\mathbf{b} \quad \forall p \in P. \end{aligned} \tag{6.28}$$

Combining (6.27) and (6.28), we arrive at a reformulation of the *LDP* as a linear program in $\mu \in \mathbb{R}$ and $\boldsymbol{\pi}_\mathbf{b} \geq \mathbf{0}$, where the corresponding dual λ-variables associated with the constraints appear in brackets:

$$\begin{aligned} z^\star_{LDP} = \max \quad & & & \mu & & \\ \text{s.t.} \quad & (\mathbf{a}_p - \mathbf{b})^\intercal \boldsymbol{\pi}_\mathbf{b} & + \; & \mu \leq c_p & [\lambda_p] \quad & \forall p \in P \\ & \mathbf{a}_r^\intercal \boldsymbol{\pi}_\mathbf{b} & & \leq c_r & [\lambda_r] \quad & \forall r \in R \\ & \boldsymbol{\pi}_\mathbf{b} \geq \mathbf{0}, & & \mu \in \mathbb{R}. & & \end{aligned} \tag{6.29}$$

We here recognize (6.6), the dual formulation of the alternative Dantzig-Wolfe master problem (6.7), and hence, $z^\star_{LDP} = z^\star_{AMP} = z^\star_{MP}$. □

Note 6.6 (Another proof.) Instead of using (6.28), we formulate the inner optimization $\min_{p \in P} \; c_p + (\mathbf{b} - \mathbf{a}_p)^\intercal \boldsymbol{\pi}_\mathbf{b}$ as a different linear program. This time we keep the constant term $\mathbf{b}^\intercal \boldsymbol{\pi}_\mathbf{b}$ as an offset outside the optimization program:

$$\begin{aligned} \mathbf{b}^\intercal \boldsymbol{\pi}_\mathbf{b} + \min_{p \in P} \; c_p - \mathbf{a}_p^\intercal \boldsymbol{\pi}_\mathbf{b} = \mathbf{b}^\intercal \boldsymbol{\pi}_\mathbf{b} + \max \quad & \pi_0 \\ \text{s.t.} \quad & \pi_0 \leq c_p - \mathbf{a}_p^\intercal \boldsymbol{\pi}_\mathbf{b} \quad \forall p \in P. \end{aligned} \tag{6.30}$$

Combining (6.27) and (6.30), the *LDP* is rewritten as

$$
\begin{aligned}
z^\star_{LDP} = \max \quad & \mathbf{b}^\intercal \boldsymbol{\pi}_\mathbf{b} && + \pi_0 \\
\text{s.t.} \quad & \mathbf{a}_p^\intercal \boldsymbol{\pi}_\mathbf{b} && + \pi_0 \leq c_p \quad [\lambda_p] \quad \forall p \in P \\
& \mathbf{a}_r^\intercal \boldsymbol{\pi}_\mathbf{b} && \leq c_r \quad [\lambda_r] \quad \forall r \in R \\
& \boldsymbol{\pi}_\mathbf{b} \geq \mathbf{0}, \ \pi_0 \in \mathbb{R}.
\end{aligned}
\tag{6.31}
$$

We recognize (6.4), the dual formulation of the *MP* (6.3), hence $z^\star_{LDP} = z^\star_{MP}$. □

The *MP* and *LDP* compute equivalent vectors $\boldsymbol{\pi}^\star_\mathbf{b}$, a consequence of which we would like to state explicitly:

Proposition 6.3. *An optimal dual solution $(\boldsymbol{\pi}^\star_\mathbf{b}, [\pi_0^{\star k}]_{k\in K})$ to the MP in a Dantzig-Wolfe reformulation solves the LDP, i.e., $LR(\boldsymbol{\pi}^\star_\mathbf{b}) = z^\star_{LDP}$. Conversely, optimal Lagrangian multipliers $\boldsymbol{\pi}^\star_\mathbf{b}$ to the LDP are dual optimal for the MP although we are missing $\boldsymbol{\pi}_0^{\star k}$, $\forall k \in K$.*

Proof. Both directions are proven from our knowledge that $z^\star_{LDP} = z^\star_{MP}$ in Proposition 6.2 and their relationship in (6.24).
$\Rightarrow$ By assumption, $\bar{c}^k(\boldsymbol{\pi}^\star_\mathbf{b}, \pi_0^{\star k}) = 0$, $\forall k \in K$, and $z^\star_{MP} = \boldsymbol{\pi}^{\star\intercal}_\mathbf{b}\mathbf{b} + \sum_{k\in K} \pi_0^{\star k}$. We then have $LR(\boldsymbol{\pi}^\star_\mathbf{b}) = z^\star_{LDP} = z^\star_{MP}$.
$\Leftarrow$ By assumption, $LR(\boldsymbol{\pi}^\star_\mathbf{b}) = z^\star_{LDP}$. We see from the independent terms in $z^\star_{LDP} = z^\star_{MP} = \boldsymbol{\pi}^{\star\intercal}_\mathbf{b}\mathbf{b} + \sum_{k\in K} \pi_0^{\star k}$ that there exist some dual values π_0^k, $\forall k \in K$, such that $\boldsymbol{\pi}^\star_\mathbf{b}$ is indeed optimal. □

Corollary 6.1. *If we have $|K|$ identical subproblems or if $|K| = 1$, we can deduce optimal dual values $\pi_0^{\star k}$, $\forall k \in K$, from $\boldsymbol{\pi}^\star_\mathbf{b}$.*

Proof. For $|K| = 1$, we have $LR(\boldsymbol{\pi}^\star_\mathbf{b}) = z^\star_{LDP} = z^\star_{MP} = \mathbf{b}^\intercal\boldsymbol{\pi}^\star_\mathbf{b} + \pi_0^\star$ from which we deduce $\pi_0^\star = LR(\boldsymbol{\pi}^\star_\mathbf{b}) - \mathbf{b}^\intercal\boldsymbol{\pi}^\star_\mathbf{b}$. For identical subproblems, we rather deduce aggregated $\pi_{\text{agg}} = LR(\boldsymbol{\pi}^\star_\mathbf{b}) - \mathbf{b}^\intercal\boldsymbol{\pi}^\star_\mathbf{b}$ or, as we do for the primal λ-variables in Proposition 4.8, disaggregated values $\pi_0^{k\star} = \frac{LR(\boldsymbol{\pi}^\star_\mathbf{b}) - \mathbf{b}^\intercal\boldsymbol{\pi}^\star_\mathbf{b}}{|K|}$, $\forall k \in K$. □

At face value, solving a Lagrangian relaxation provides incomplete dual information and no primal information at all when compared to a Dantzig-Wolfe reformulation. More specifically, what we find is an optimizer $\boldsymbol{\pi}^\star_\mathbf{b}$ but we do not get values for π_0^k, $k \in K$, nor a solution $\mathbf{x}$. Indeed, the solutions $\mathbf{x}^k$, $k \in K$, that we find for $LR(\boldsymbol{\pi}^\star_\mathbf{b})$ are very likely meaningless on their own for the compact formulation as they do not need to satisfy the complicating constraints in $\mathcal{A}$.

Some questions that come to mind in light of this observation are:

- How useful is it to only have optimal Lagrangian multipliers $\boldsymbol{\pi}^\star_\mathbf{b}$?
- How important are the dual values π_0^k, $k \in K$?
- Is a Lagrangian relaxation easier to solve than a Dantzig-Wolfe reformulation?
- Can we easily find a primal solution $\mathbf{x}$ (even if only fractional) for the compact formulation in a Lagrangian relaxation?
- Can we leverage optimal Lagrangian multipliers $\boldsymbol{\pi}^\star_\mathbf{b}$ in a Dantzig-Wolfe reformulation?

The reader can find answers to these questions in the rest of the chapter. In particular, an intuitive primal construction is presented in Section Primal solutions (fractional and integer) whereas Note 6.18 uses column generation to reproduce the columns we need. Let us continue with our comparison for now.

In a Dantzig-Wolfe reformulation, Propositions 4.1 and 4.7 express the relations between the lower bound $z^\star_{MP}$ and optimal objective values of the *ILP* and its linear relaxation *LP*, respectively given by $z^\star_{ILP}$ and $z^\star_{LP}$. Obviously, we have the same relations regarding $z^\star_{LDP}$.

Corollary 6.2. *Let the LP be the linear relaxation of the ILP* (6.1). *Then*

(a) The lower bound $z^\star_{LDP}$ may improve on $z^\star_{LP}$, that is, $z^\star_{LP} \leq z^\star_{LDP} \leq z^\star_{ILP}$.
(b) If the Lagrangian subproblem formulation (6.11) *has the integrality property, then $z^\star_{LP} = z^\star_{LDP}$.*

Note again that *(b)* only states a sufficient condition. It may happen that the *ISP* formulation does not have the integrality property and the bound obtained from a Lagrangian relaxation (or Dantzig-Wolfe reformulation) is still not better than $z^\star_{LP}$. This nicely completes our knowledge about bounds as

$$z^\star_{LP} \leq z^\star_{LDP} = z^\star_{MP} \leq z^\star_{IMP} = z^\star_{ILP}, \tag{6.32}$$

see also Figure 6.6. Note that $LR(\boldsymbol{\pi}_\mathbf{b}) \geq z^\star_{LP}$ does *not* need to hold in general, see e.g., Illustration 6.3 where $LR(0) = 3 < z^\star_{LP} = 7$, or $LR(-100) = -576$.

Note 6.7 (Together forever.) By Proposition 6.2, we cannot help it: Dantzig-Wolfe reformulating an integer linear program, and solving to optimality the resulting master problem by the column generation algorithm, we always also compute the best Lagrangian bound. Even better, by Proposition 6.1 we implicitly compute Lagrangian bounds in every iteration, which is very useful for early termination, see Note 2.18.

Note 6.8 (How to decompose?) We have no general theory yet concerning how to split the set of constraints of an *ILP* to obtain a good decomposition in the Dantzig-Wolfe reformulation. We do not even know what "good" in this context generally means. Yet, this is a natural question, and we discussed some practical considerations in Chapter 4 in Automatic grouping of the constraints for reformulation. With the relation to Lagrangian relaxation in place, we have another source of information and experience at hand: The motivation for identifying *complicating*, or master constraints, is exactly the same in both worlds.

Note 6.9 (I can see clearly now.) Some students have trouble deriving/memorizing the adjusted costs formula $\tilde{\mathbf{c}}$ (cf. Note 3.6) of the subproblem's x-variables in the Dantzig-Wolfe *ISP*. As a playful meaning on the song words of Johnny Nash (1972), the objective function of the Lagrangian *ISP* (6.11) comes to help here as

$$\mathbf{c}^\intercal\mathbf{x} + \boldsymbol{\pi}_\mathbf{b}^\intercal(\mathbf{b} - \mathbf{A}\mathbf{x}) = \boldsymbol{\pi}_\mathbf{b}^\intercal\mathbf{b} + (\mathbf{c}^\intercal - \boldsymbol{\pi}_\mathbf{b}^\intercal\mathbf{A})\mathbf{x} = \boldsymbol{\pi}_\mathbf{b}^\intercal\mathbf{b} + \tilde{\mathbf{c}}\mathbf{x} \tag{6.33}$$

produces the expression $\tilde{\mathbf{c}} = \mathbf{c}^\mathsf{T} - \boldsymbol{\pi}_\mathbf{b}^\mathsf{T}\mathbf{A}$ very naturally. This is particularly true when adding further constraints to the original *ILP* what we do in Chapter 7: There is one dual variable associated with each new constraint in $\mathcal{A}$.

Properties of the Lagrangian function

Let us recollect a few notions from calculus that allow us to describe the essential properties of the Lagrangian function in the forthcoming propositions. Let $f : I \to \mathbb{R}$ denote a continuous real-valued function defined on a convex open set $I \subseteq \mathbb{R}^n$.

Definition 6.1. A function f is *concave* if, for all vectors $\mathbf{x}_1, \mathbf{x}_2 \in I$ and any scalar $\alpha \in [0, 1]$,

$$f(\alpha\mathbf{x}_1 + (1-\alpha)\mathbf{x}_2) \geq \alpha f(\mathbf{x}_1) + (1-\alpha) f(\mathbf{x}_2). \tag{6.34}$$

Definition 6.2. The *partial derivatives* of f, denoted by $\partial f / \partial x_i$, are derivatives with respect to one variable x_i, $i \in \{1, \ldots, n\}$, where all others are kept constant. The *gradient* of f, denoted by ∇f, evaluated at $\mathbf{x}$, is the column vector

$$\nabla f(\mathbf{x}) = \left(\frac{\partial f(\mathbf{x})}{\partial x_1}, \frac{\partial f(\mathbf{x})}{\partial x_2}, \ldots, \frac{\partial f(\mathbf{x})}{\partial x_n} \right)^\mathsf{T}. \tag{6.35}$$

Proposition 6.4. *Given a concave function f, differentiable at $\mathbf{x}_1 \in I$, the* first-order approximation $f(\mathbf{x}_1) + (\mathbf{x} - \mathbf{x}_1)^\mathsf{T} \nabla f(\mathbf{x}_1)$ *(the first term of the Taylor series at $\mathbf{x}_1$) is an overestimation of $f(\mathbf{x})$, that is,*

$$f(\mathbf{x}) \leq f(\mathbf{x}_1) + (\mathbf{x} - \mathbf{x}_1)^\mathsf{T} \nabla f(\mathbf{x}_1), \qquad \forall \mathbf{x} \in I. \tag{6.36}$$

Definition 6.3. Given a concave function f, a *subgradient* at $\mathbf{x}_1 \in I$ is a vector $\mathbf{s} \in \mathbb{R}^n$ that replaces the gradient while also satisfying this overestimation, that is,

$$f(\mathbf{x}) \leq f(\mathbf{x}_1) \; + \; (\mathbf{x} - \mathbf{x}_1)^\mathsf{T} \mathbf{s}, \qquad \forall \mathbf{x} \in I. \tag{6.37}$$

Definition 6.4. The *subdifferential* $\partial f(\mathbf{x}_1)$ of f at $\mathbf{x}_1 \in I$ is the set of all subgradients at $\mathbf{x}_1$. Given a concave function f, if the subdifferential $\partial f(\mathbf{x}_1)$ contains a single element, f is differentiable in $\mathbf{x}_1$ and that element is the gradient $\nabla f(\mathbf{x}_1)$. Otherwise, f is *subdifferentiable* in $\mathbf{x}_1$.

In Figure 6.7, we illustrate these notions on one-dimensional functions f and g. The function f on the left is differentiable everywhere and the gradient is just the derivative f'. Any value $f(x)$ is overestimated by the first-order approximation at x_1. The function g on the right is not differentiable at x_1, but subdifferentiable. A subgradient s at x_1 is the *slope of a tangent* to g at x_1. The subdifferential is given by the interval $[s_a, s_b]$, where s_a and s_b are the one-sided limits

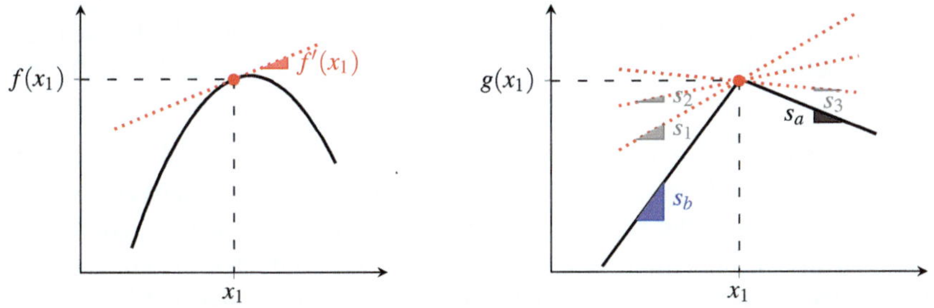

(a) Gradient of f: tangent sloped by $f'(x_1)$ (b) Subgradients of g: tangents sloped by $s \in [s_a, s_b]$

Fig. 6.7: Differentiable and subdifferentiable functions with their tangents at x_1.

$$s_a = \lim_{x \to x_1^+} \frac{g(x) - g(x_1)}{x - x_1}, \qquad s_b = \lim_{x \to x_1^-} \frac{g(x) - g(x_1)}{x - x_1}. \tag{6.38}$$

Proposition 6.5. *On the domain over which it is finite, the Lagrangian function LR is piecewise linear, continuous, concave, and subdifferentiable.*

Proof. $LR(\boldsymbol{\pi}_\mathbf{b})$ is finite if $\boldsymbol{\pi}_\mathbf{b} \in \{\boldsymbol{\pi}_\mathbf{b} \geq \mathbf{0} \mid c_r - \boldsymbol{\pi}_\mathbf{b}^\intercal \mathbf{a}_r \geq 0, \forall r \in R\}$. Hence, there exists a finite set of extreme points $\{\mathbf{x}_p\}_{p \in P}$ such that $LR(\boldsymbol{\pi}_\mathbf{b}) = \min_{p \in P} c_p + \boldsymbol{\pi}_\mathbf{b}^\intercal(\mathbf{b} - \mathbf{a}_p)$. Thus, LR is the pointwise minimum of a finite family of affine functions in $\boldsymbol{\pi}_\mathbf{b}$, and as such it is piecewise linear, continuous, and concave. From the latter two, sudifferentiability follows, cf. Figure 6.8. □

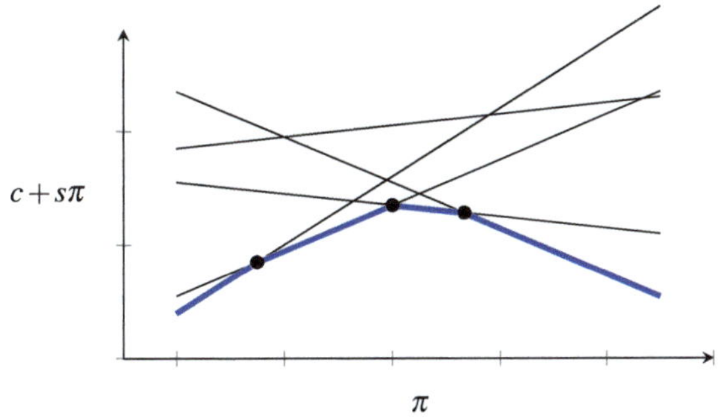

Fig. 6.8: Illustrating the proof of Proposition 6.5: the graph of the function resulting from a pointwise minimum of a finite family of affine functions. It also shows the coordinates where this function is not differentiable.

Definition 6.5. The *hypograph* of a function $f : S \to \mathbb{R}$, $S \subseteq \mathbb{R}^n$, is the set of points lying on or below its graph: $\text{hyp}(f) = \{(\mathbf{x}, y) \in S \times \mathbb{R} \mid y \leq f(\mathbf{x})\} \subseteq \mathbb{R}^{n+1}$. Similarly, the set of points on or above the function's graph is its *epigraph*, denoted $\text{epi}(f)$. The intersection $\text{hyp}(f) \cap \text{epi}(f) = \{(\mathbf{x}, y) \in S \times \mathbb{R} \mid y = f(\mathbf{x})\}$ is the graph of f.

Note 6.10 (Polyhedral function.) We have presented the traditional way of deriving the properties of the Lagrangian function. If you look back, alternatively, we can see that we can maximize the Lagrangian function by solving a linear program, the dual of the *AMP*, see Proposition 6.2 and Figure 6.4b. The domain of this linear program is a polyhedron, actually, the *hypograph* of the Lagrangian function *LR*, so the graph of *LR* is exactly the boundary of that hypograph. As such it is continuous, piecewise linear, concave, and subdifferentiable. (Optimizing the *ILP* in maximization form would result in "the graph of *LR* is the boundary of the corresponding *epigraph*.")

Given the remarkable structure of the Lagrangian function, we also note that solving the *ISP* to optimality trivially identifies a subgradient.

Proposition 6.6. *Let* $\mathbf{x}_1$ *be an optimal solution to the ISP* (6.11) *for a given* $\boldsymbol{\pi}_{\mathbf{b},1} > \mathbf{0}$, *i.e.,* $LR(\boldsymbol{\pi}_{\mathbf{b},1}) = c_1 + \boldsymbol{\pi}_{\mathbf{b},1}^{\mathsf{T}}(\mathbf{b} - \mathbf{a}_1)$. *Then,* $(\mathbf{b} - \mathbf{a}_1)$ *is a subgradient of LR at* $\boldsymbol{\pi}_{\mathbf{b},1}$.

Proof. We show that $(\mathbf{b} - \mathbf{a}_1)$ satisfies Definition 6.3. For any vector of multipliers $\boldsymbol{\pi}_{\mathbf{b}} \geq \mathbf{0}$, we have:

$$\begin{aligned} LR(\boldsymbol{\pi}_{\mathbf{b}}) &= c_p + \boldsymbol{\pi}_{\mathbf{b}}^{\mathsf{T}}(\mathbf{b} - \mathbf{a}_p) \quad \text{for some optimal extreme point } \mathbf{x}_p, p \in P \\ &\leq c_1 + \boldsymbol{\pi}_{\mathbf{b}}^{\mathsf{T}}(\mathbf{b} - \mathbf{a}_1) \\ &= c_1 + \boldsymbol{\pi}_{\mathbf{b},1}^{\mathsf{T}}(\mathbf{b} - \mathbf{a}_1) - \boldsymbol{\pi}_{\mathbf{b},1}^{\mathsf{T}}(\mathbf{b} - \mathbf{a}_1) + \boldsymbol{\pi}_{\mathbf{b}}^{\mathsf{T}}(\mathbf{b} - \mathbf{a}_1) \\ &= LR(\boldsymbol{\pi}_{\mathbf{b},1}) \; + \; (\boldsymbol{\pi}_{\mathbf{b}} - \boldsymbol{\pi}_{\mathbf{b},1})^{\mathsf{T}}(\mathbf{b} - \mathbf{a}_1). \end{aligned}$$

Given $\boldsymbol{\pi}_{\mathbf{b}}$, the proof starts with an optimal extreme point $\mathbf{x}_p$ of $\text{conv}(\mathcal{D})$ which leads to an inequality when compared with $\mathbf{x}_1$. Adding $\boldsymbol{\pi}_{\mathbf{b},1}^{\mathsf{T}}(\mathbf{b} - \mathbf{a}_1) - \boldsymbol{\pi}_{\mathbf{b},1}^{\mathsf{T}}(\mathbf{b} - \mathbf{a}_1) = 0$ to the right-hand side maintains the equality that is written in terms of $LR(\boldsymbol{\pi}_{\mathbf{b},1})$ in the last equation. □

Illustration 6.4 TCSPP **(cont.)**

Given any $\pi_7 \leq 0$, formulation (6.22) models a network flow problem for which the value $LR(\pi_7)$ is attained at one of the nine extreme points $\mathbf{x}_p = [x_{ijp}]_{(i,j) \in A}$, $p \in P$, which correspond to the nine possible paths from node 1 to 6. Therefore, we can reformulate (6.22) as

$$LR(\pi_7) = \min_{p \in P} \sum_{(i,j) \in A} c_{ij} x_{ijp} + \pi_7 (14 - \sum_{(i,j) \in A} t_{ij} x_{ijp}). \tag{6.39}$$

Recall that $c_p = \sum_{(i,j) \in A} c_{ij} x_{ijp}$ is the cost of the incidence vector $\mathbf{x}_p$ of path p, whereas its duration is given by $t_p = \sum_{(i,j) \in A} t_{ij} x_{ijp}$. With this notation, we write the Lagrangian function (6.39) as

$$LR(\pi_7) = \min_{p \in P} \quad c_p + (14 - t_p)\pi_7, \tag{6.40}$$

where we denote the affine function (or just "line") with intercept c_p and slope $(14 - t_p)$ by

$$\mu_p(\pi_7) = c_p + (14 - t_p)\pi_7. \tag{6.41}$$

Table 6.2 exhibits the nine line equations. The reader can easily verify the values obtained previously for $LR(0) = 3$ and $LR(-100) = -576$. Indeed, for $\pi_7 = 0$, the Lagrangian function (6.40) reaches its minimum with path 1246 whereas path 1356 is used for $\pi_7 = -100$.

p	1246	1256	12456	13246	13256	132456	1346	13456	1356
c_p	3	5	14	13	15	24	16	27	24
t_p	18	15	14	13	10	9	17	13	8
$\mu_p(\pi_7)$	$3-4\pi_7$	$5-\pi_7$	14	$13+\pi_7$	$15+4\pi_7$	$24+5\pi_7$	$16-3\pi_7$	$27+\pi_7$	$24+6\pi_7$
$\mu_p(0)$	**3**	5	14	13	15	24	16	27	24
$\mu_p(-100)$	403	105	14	−87	−385	−476	316	−73	**−576**

Table 6.2: Line equations $\mu_p(\pi_7) = c_p + (14 - t_p)\pi_7, \forall p \in P$.

As shown in Figure 6.9, finding an optimal extreme point $\mathbf{x}_p$ for a given $\pi_7 \leq 0$ is easy *if we can draw* these lines. Amongst them moreover materializes the Lagrangian function LR as the pointwise minimum of this family of functions that are affine in π_7. As expected from Proposition 6.5, it is piecewise linear, continuous, concave, and subdifferentiable. In particular, we see that it is composed of the four line segments determined by the paths 1246, 1256, 13256, and 1356:

$$LR(\pi_7) = \begin{cases} 3 - 4\pi_7 & -2/3 \leq \pi_7 \leq 0 \\ 5 - \pi_7 & -2 \leq \pi_7 \leq -2/3 \\ 15 + 4\pi_7 & -4.5 \leq \pi_7 \leq -2 \\ 24 + 6\pi_7 & \pi_7 \leq -4.5. \end{cases} \tag{6.42}$$

Some examples of what we can compute: Not only is path 1246 the optimizer for $\pi_7 = 0$ but on the whole interval $[-2/3, 0]$; for $\pi_7 = -1$ it is path 1256, for $\pi_7 = -4$ we find path 13256. We have a boundary point at $\pi_7 = 0$ and three breakpoints $\pi_7 \in \{-4.5, -2, -2/3\}$. For instance, the breakpoint $\pi_7 = -4.5$ is obtained from intersecting lines $\mu_{13256}(\pi_7) = 15 + 4\pi_7$ and $\mu_{1356}(\pi_7) = 24 + 6\pi_7$. Evidently, for $\pi_7 = -4.5$ both paths 13256 and 1356 are optimal. The subdifferentials $[s_a, s_b]$ for the three breakpoints are

breakpoint	−4.5	−2	−2/3
subdifferential	[4, 6]	[−1, 4]	[−4, −1]

.

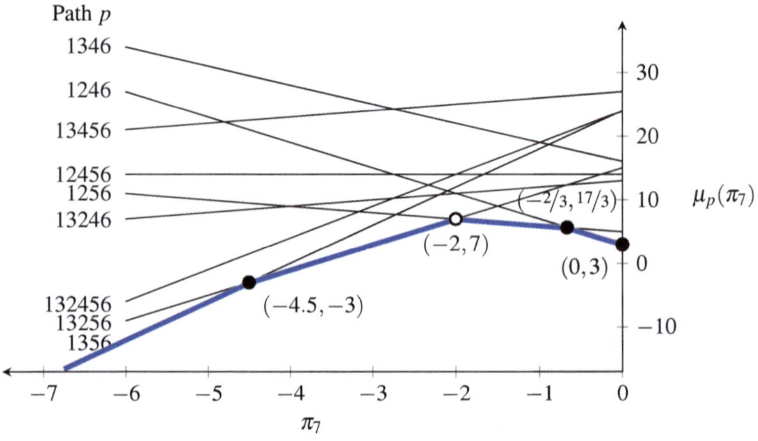

Fig. 6.9: The Lagrangian function LR, $\pi_7 \leq 0$.

We can even read off the optimal Lagrangian multiplier $\pi_7^\star = -2$ which yields $z_{LDP}^\star = LR(-2) = 7 = z_{MP}^\star$, in line with Proposition 6.2. Its proof in fact already alerted us of what we would observe here: Figures 6.4b and 6.9 both reveal exactly the same Lagrangian functions. We could have even constructed it in Illustration 6.1 had we solved the (dual of the) *AMP* instead of the (dual of the) *MP*. That the paths generated by column generation define *the entire* Lagrangian function is of course neither desired nor does it generally happen; it is just a coincidence due to our small *TCSPP* example.

Note 6.11 (Heuristic vs. relaxed pricing.) In practice, we may choose not to solve the Lagrangian subproblem to optimality, but only heuristically, or a relaxation instead. In the first case, we may not attain (we stay above) the pointwise minimum of the affine functions and only obtain an outer approximation of the Lagrangian function. In the second case, we may go below the pointwise minimum and cut away parts of the Lagrangian function. Therefore, while both variants may help solving the subproblem faster, using the former we do not obtain a valid lower bound on $z_{ILP}^\star$, but for the latter we do. All is well, we arrive at the same conclusion for column generation respectively in Note 2.17 and Section Relaxed pricing, relaxed master.

Note 6.12 (Row generation, again.) Kelley (1960) suggests to maximize a concave function using an outer approximation by linear inequalities. In this context, the suggestion of Cheney and Goldstein (1959) is fundamentally no different. Maximizing the Lagrangian function is an application case for this algorithm which is why in the literature *Kelley-Cheney-Goldstein's cutting plane method* is sometimes related to column generation. Note the years these papers were published: around the same time as column generation was designed by Dantzig and Wolfe (1960). It is unsurprising then that the algorithm we see in Section Subgradient algorithm faces the same hurdles:

> Although one can concoct more complicated criteria for terminating the process, such complications are probably not worth the effort. A more difficult problem is to obtain accuracy near the termination of the process [at which point] the successive cutting planes generally tend to become more and more parallel to one another. As this occurs, the determination of successive $\mathbf{t}_k$ becomes more and more difficult due to the limited precision with which the computations must be performed. Thus, there is a point where continued computation would cause the process to degenerate completely, oscillate or converge on the wrong solution.
>
> – Kelley (1960)

Optimality conditions

We first state a necessary and sufficient optimality condition for the *LDP*. We then observe that it is possible that we solve the *ILP* by chance.

Proposition 6.7. *A vector of Lagrangian multipliers* $\boldsymbol{\pi}_\mathbf{b} \geq \mathbf{0}$ *is optimal for the LDP* (6.15) *if and only if the subdifferential at* $\boldsymbol{\pi}_\mathbf{b}$ *contains a zero-subgradient.*

Proof. Trivial by the properties of the Lagrangian function established in Proposition 6.5. □

Proposition 6.8. *Given an arbitrary vector of Lagrangian multipliers* $\boldsymbol{\pi}_\mathbf{b} \geq \mathbf{0}$ *and an optimizer* $\hat{\mathbf{x}}$ *for* $LR(\boldsymbol{\pi}_\mathbf{b})$ (6.11), *if* $\hat{\mathbf{x}}$ *is feasible for the ILP* (6.1) *and* $\boldsymbol{\pi}_\mathbf{b}^\top(\mathbf{b} - \mathbf{A}\hat{\mathbf{x}}) = 0$, *then* $\hat{\mathbf{x}}$ *is an optimal solution to the ILP.*

Proof. The integer solution $\hat{\mathbf{x}}$ provides an upper bound, $z^\star_{ILP} \leq \mathbf{c}^\top\hat{\mathbf{x}}$. By Proposition 6.1, we also have a lower bound $LR(\boldsymbol{\pi}_\mathbf{b}) = \mathbf{c}^\top\hat{\mathbf{x}} + \boldsymbol{\pi}_\mathbf{b}^\top(\mathbf{b} - \mathbf{A}\hat{\mathbf{x}}) \leq z^\star_{LDP} \leq z^\star_{ILP}$, where $\boldsymbol{\pi}_\mathbf{b}^\top(\mathbf{b} - \mathbf{A}\hat{\mathbf{x}}) = 0$ holds. Hence, $z^\star_{LDP} = z^\star_{ILP} = \mathbf{c}^\top\hat{\mathbf{x}}$ and $\mathbf{x}^\star_{ILP} = \hat{\mathbf{x}}$. □

There is a very important difference between the relaxation of equality and inequality constraints in Proposition 6.8. In the former case, a successful feasibility test for the *ILP*, i.e., $\mathbf{A}\hat{\mathbf{x}} = \mathbf{b}$ implies that the second test condition $\boldsymbol{\pi}_\mathbf{b}^\top(\mathbf{b} - \mathbf{A}\hat{\mathbf{x}}) = 0$ holds as well. In the latter case, we simply cannot shortcut the second condition. We also remind the reader that the above result was already derived in the context of Dantzig-Wolfe, see Exercise 4.7.

Illustration 6.5 **Optimality check**

Consider the following integer linear program:

$$\begin{aligned} z^\star_{ILP} = \min \quad & -2x_1 - 3x_2 \\ \text{s.t.} \quad & x_1 + 4x_2 \leq 5 \\ & x_1, \quad x_2 \in \{0,1\}, \end{aligned} \tag{6.43}$$

whose optimal solution is given by $x_1^\star = x_2^\star = 1$ with $z^\star_{ILP} = -5$. The Lagrangian relaxation of the redundant constraint $x_1 + 4x_2 \leq 5$ with the multiplier $\pi \leq 0$ gives the Lagrangian subproblem

$$LR(\pi) = \min_{x_1,x_2 \in \{0,1\}} -2x_1 - 3x_2 + \pi(5 - x_1 - 4x_2). \qquad (6.44)$$

For $\pi = -1$, we get $LR(-1) = -6$ at $(1, 0)$. This is a lower bound on $z^{\star}_{ILP}$, point $(1,0)$ satisfies the relaxed constraint but is not an optimal solution for the *ILP*. Indeed, the slack variable for $x_1 + 4x_2 \leq 5$ takes value 4, hence $\pi(5 - x_1 - 4x_2) = \pi(4) = -4 \neq 0$.

Note 6.13 (Once in a blue moon.) Proposition 6.7 fortunately agrees with intuitive first order condition and concavity. Although we can have a zero-subgradient in the *ISP* (if the Lagrangian function is flat on top), it is however very rarely observed in practice. The primal perspective adds to this intuition. Indeed, we know from the Dantzig-Wolfe decomposition that an optimal solution to the *MP* is typically a non-trivial convex combination of extreme points, or for set partitioning formulations, complementary columns from one or several subproblems. This implies that optimality of the *LDP* is typically attained at a breakpoint for which one would have to compute the subdifferential.

If Proposition 6.8 establishes optimality of the *ILP* for given $\boldsymbol{\pi}_{\mathbf{b}}$ and optimizer $\hat{\mathbf{x}}$ for $LR(\boldsymbol{\pi}_{\mathbf{b}})$, then Proposition 6.7 also re-establishes optimality of the *LDP* for said $\boldsymbol{\pi}_{\mathbf{b}}$ but not vice versa. Finding that the subdifferential at $\boldsymbol{\pi}_{\mathbf{b}}$ contains a zero-subgradient gives the lower bound $z^{\star}_{LDP} \leq z^{\star}_{ILP}$, most of the time with a strict inequality. Although the corresponding optimizer $\hat{\mathbf{x}}$ is integer in the *ISP*, we could also check its feasibility for the *ILP* and if $\boldsymbol{\pi}_{\mathbf{b}}^{\mathsf{T}}(\mathbf{b} - \mathbf{A}\hat{\mathbf{x}}) = 0$ holds. Typically, we do not bother with this test as we assume an optimal solution to the *ILP* is found with the exploration of a branch-and-price tree, see the forthcoming Chapter 7.

Illustration 6.6 TCSPP **(cont.)**

Figure 6.10 plots the Lagrangian function of the *TCSPP*. We display the best upper bound obtained from one of the four paths defining $LR(\pi_7)$, i.e., the feasible path 13256 of duration 10 gives $UB = c_{13256} = 15$. We also display the optimal objective values of the *ILP* and *LDP*, $z^{\star}_{ILP} = 13$ and $z^{\star}_{LDP} = 7$. Observe that none of the non-dominated extreme points yield a zero-subgradient. However, all subgradients contained in a subdifferential are obtained via convex combinations. In particular, for the breakpoint $\pi_7 = -2$, we have

$$\alpha(14 - t_{13256}) + (1 - \alpha)(14 - t_{1256}) = 5\alpha - 1 \in [-1, 4], \ \forall \alpha \in [0, 1],$$

where $t_{13256} = 10$ and $t_{1256} = 15$ (see Table 6.2). Solving this expression to zero yields the fractional combination of paths 13256 and 1256 with $\alpha^{\star} = 0.2$, already seen in Figure 3.13. The integrality gap is $z^{\star}_{ILP} - z^{\star}_{LDP} = 6$ and it is obvious we should not hold our breath to close the relative optimality gap $(UB - LB)/UB = 53.3\,\%$ using only the *LDP*. The branch-and-bound search tree is the forthcoming step in Chapter 7.

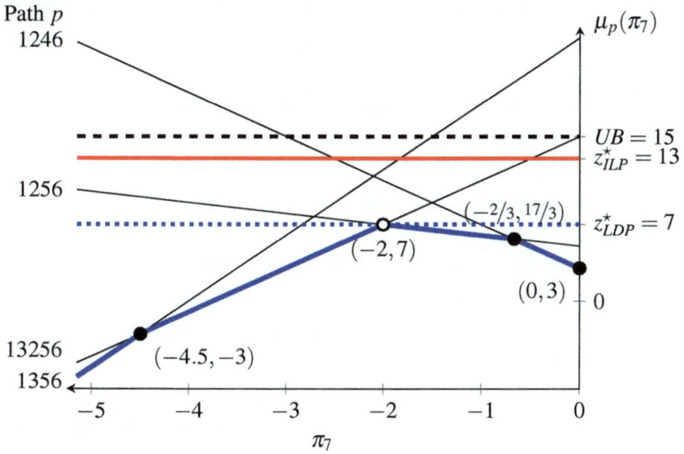

Fig. 6.10: Optimality observations on the Lagrangian function *LR*, $\pi_7 \leq 0$.

Subgradient algorithm

A Dantzig-Wolfe reformulation and a Lagrangian relaxation share common ground, but they are not identical. Traditionally, two completely different algorithms are used for their solution. In particular, even though both approaches arrive at the same lower bound on $z^\star_{ILP}$, the sequences of values for the dual variables and Lagrangian multipliers, respectively, may be different. In order to find out whether this is actually the case, as well as whether and how we can make use of this, we move on to get acquainted with the *subgradient algorithm*, the traditional method to solve the *LDP*. It is one of the easiest (also to implement) and thus, one of the most popular methods to obtain (near-)optimal Lagrangian multipliers.

The basic idea is almost a gradient ascent method for maximizing a differentiable function f. There, in order to move from one solution $\mathbf{x}_1$ to a new one $\mathbf{x}_1 + \theta\mathbf{y}$ using the direction $\mathbf{y}$ and step size $\theta > 0$, we use the *first-order Taylor approximation*

$$f(\mathbf{x}_1 + \theta\mathbf{y}) \approx f(\mathbf{x}_1) + \theta\nabla f(\mathbf{x}_1)^\intercal\mathbf{y}. \tag{6.45}$$

Taking $\mathbf{y} = \nabla f(\mathbf{x}_1)$ of steepest ascent, we obtain

$$f(\mathbf{x}_1 + \theta\mathbf{y}) \approx f(\mathbf{x}_1) + \theta\|\nabla f(\mathbf{x}_1)\|^2, \tag{6.46}$$

where $\|\mathbf{y}\|^2 = \mathbf{y}^\intercal\mathbf{y} = \sum_{i=1}^n y_i^2$ denotes the square of the Euclidean norm. For a small step size $\theta > 0$ in the direction given by the gradient, $f(\mathbf{x}_1 + \theta\mathbf{y})$ *increases* by approximately the positive amount $\theta\|\nabla f(\mathbf{x}_1)\|^2$. A canonical choice to determine an appropriate step size goes by the name of *line search* whereby we optimize the step size in the selected direction, i.e., $\arg\max_\theta f(\mathbf{x}_1 + \theta\mathbf{y})$. Mathematically, we see that both the direction ($\mathbf{y}$) and step size (θ) contribute to the convergence of the process. Of course, if f is concave, we know we have a global optimum whenever we find a local one. Figure 6.11 sketches a concave quadratic function ($-x^2$) for which one could, from any initial point x^0, compute exactly both the gradient ($-2x_0$) and step size ($\max_\theta -(x_0 + \theta(-2x_0))^2$) to land on $x^\star = 0$ in one iteration.

In practice, it is often unreasonable to solve for θ exactly. Two notable complicating factors are that f has no closed-form expression or that optimization is subject to constraints. Instead, we use a parameterized value θ that decreases gradually towards zero at which point we terminate the process with the best objective value identified. What is a good initial value and how fast should we decrease it? For the sake of convergence, a lot of ink has been spilled on this subject, e.g., Armijo (1966); Wolfe (1969); Bertsekas (1995). In a nutshell, the step size should be neither too small, nor too large. In the first case, the algorithm essentially remains at the same point and therefore makes no progress. In the second case, it might overshoot over a local optimum and oscillate back and forth from there on. The idea is therefore to find a compromise for the step size that allows us to "carefully approach a (local) optimum." The precise definition of this compromise is open to experimental trials because the nature of the function can largely exacerbate convergence issues in the sense that even with the same step sizes, some regions behave much more nicely than others.

On our example, for any fixed step size (e.g., $\theta = 0.25$), following the gradient towards the maximum attained at $x^\star = 0$ from the right at $x_0 = 1$ impacts the function value much more significantly than from the left at $x_0 = -0.5$, that is,

$$\begin{aligned} \nabla f(1) = -2(1) = -2 \quad &\text{and} \quad f(1 + 0.25(-2)) - f(1) = 0.75; \\ \nabla f(-0.5) = -2(-0.5) = 1 \quad &\text{and} \quad f(-0.5 + 0.25(1)) - f(-0.5) = 0.1875. \end{aligned}$$

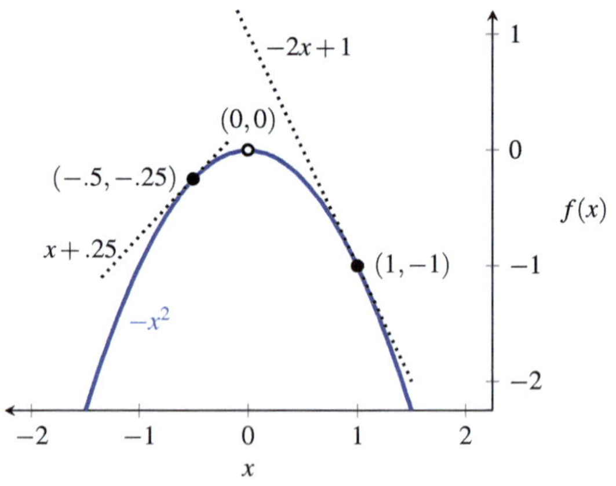

Fig. 6.11: Gradient ascent on function $f(x) = -x^2$ with first derivative $f'(x) = -2x$.

We now apply this to the Lagrangian function *LR*. A classical assumption is that the domain $\mathcal{D}$ of the Lagrangian subproblem is bounded in order to ensure $LR(\boldsymbol{\pi}_{\mathbf{b}}) > -\infty$. In theory, this can always be achieved by imposing (sufficiently large) upper bounds $\mathbf{u}$ on $\mathbf{x}$ in the *ISP* since we assume $z^{\star}_{ILP}$ to be finite, see Note 6.3. Recall by Proposition 6.6 that any extreme point $\mathbf{x}_p$ identified by the *ISP* gives us a subgradient and therefore the direction $(\mathbf{b} - \mathbf{a}_p)$. Following the latter with some step size θ, the next Lagrangian value is overestimated by

$$LR(\boldsymbol{\pi}_{\mathbf{b}} + \theta(\mathbf{b} - \mathbf{a}_p)) \leq LR(\boldsymbol{\pi}_{\mathbf{b}}) + \theta \|\mathbf{b} - \mathbf{a}_p\|^2. \tag{6.47}$$

The subgradient algorithm starts with a given $\boldsymbol{\pi}_{\mathbf{b},1} \geq \mathbf{0}$ and iteratively determines $\boldsymbol{\pi}_{\mathbf{b},t+1}$ for $t \geq 1$. Let $\mathbf{x}_t$ denote an optimal solution to $LR(\boldsymbol{\pi}_{\mathbf{b},t})$ and θ_t denote the step size. Then $\boldsymbol{\pi}_{\mathbf{b},t+1} = [\boldsymbol{\pi}_{\mathbf{b},t} + \theta_t(\mathbf{b} - \mathbf{a}_t)]^+$, that is,

$$\forall i \in \{1, \ldots, m\}, \; \pi_{i,t+1} = \max\{0, \pi_{i,t} + \theta_t(b_i - a_{i,t})\}. \tag{6.48}$$

There only remains to determine an appropriate step size. This discussion follows the same theme as in non-linear optimization for which we simply recall the aforementioned line search. Let us frame the discussion of the step size with respect to the subgradient algorithm. In the first case (too small), it essentially finds the same extreme point and does not converge. In the second case (too large), it might overshoot an optimal dual vector, and then oscillate. For so-called *diminishing step size rules* that satisfy $\theta_t \to 0$ and $\sum_{j=1}^{t} \theta_j \to \infty$ as t tends to infinity, Shapiro (1979a,b) proves convergence to an optimal solution. We list the Polyak step (Polyak, 1967) as a more practical alternative to the basic first one while bringing attention to Monique Guignard's assessment:

> Practical convergence of the subgradient method is unpredictable. For some problems, convergence is quick and fairly reliable, while other problems tend to produce erratic behavior of the multiplier sequence, or of the Lagrangean value, or both. In a "good" case, one will usually observe a saw-tooth pattern in the Lagrangean value for the first iterations, followed by a roughly monotonic improvement and asymptotic convergence to a value that is hopefully the optimal Lagrangean bound. In "bad" cases, the saw-tooth pattern continues, or, worse, the Lagrangean value keeps deteriorating. – Guignard (2003)

1. We can choose $\theta_t = 1/t, t \geq 1$, that satisfies the above conditions.
2. A more involved alternative is the *Polyak step*. Assume that $LR(\boldsymbol{\pi}_{\mathbf{b}})$ is given by the hyperplane induced by $LR(\boldsymbol{\pi}_{\mathbf{b},1})$, i.e.,

$$LR(\boldsymbol{\pi}_{\mathbf{b}}) = c_1 + \boldsymbol{\pi}_{\mathbf{b}}^{\intercal}(\mathbf{b} - \mathbf{a}_1). \tag{6.49}$$

Given the next iterate $\boldsymbol{\pi}_{\mathbf{b},2} = \boldsymbol{\pi}_{\mathbf{b},1} + \theta_1(\mathbf{b} - \mathbf{a}_1)$ and the overestimation of the Lagrangian function with

$$LR(\boldsymbol{\pi}_{\mathbf{b},1} + \theta_1(\mathbf{b} - \mathbf{a}_1)) \leq LR(\boldsymbol{\pi}_{\mathbf{b},1}) + (\theta_1(\mathbf{b} - \mathbf{a}_1))^{\intercal}(\mathbf{b} - \mathbf{a}_1) \tag{6.50}$$

from Proposition 6.6, an upper limit on θ that makes sense comes from considering that we should in principle never exceed $z^{\star}_{LDP}$:

Fig. 6.12: Monique Guignard (San Pedro de Atacama, Chile, 2013-06-12).

$$LR(\boldsymbol{\pi}_{\mathbf{b},1}) + (\theta_1(\mathbf{b} - \mathbf{a}_1))^\intercal(\mathbf{b} - \mathbf{a}_1) \leq z^\star_{LDP} \Leftrightarrow \theta \leq \frac{z^\star_{LDP} - LR(\boldsymbol{\pi}_{\mathbf{b},1})}{\|\mathbf{b} - \mathbf{a}_1\|^2}. \quad (6.51)$$

We see that the step size tends to increase with the distance to optimality in the numerator but is mitigated by the steepness of the subgradient in the denominator. Moreover, there is no issue with a division by zero because such a subgradient implies optimality has been reached, see Proposition 6.7. In practice, of course, we do not know $z^\star_{LDP}$ but rather use an upper bound UB on $z^\star_{ILP}$ which is updated each time an improving integer solution $\mathbf{x}^\bullet$ is found, that is, $UB = \mathbf{c}^\intercal\mathbf{x}^\bullet$. These values are such that $z^\star_{LDP} \leq z^\star_{ILP} \leq UB$. Hopefully, the quality of the lower bound induced by set $\mathcal{D}$ is such that UB is not too far from $z^\star_{LDP}$, otherwise UB may largely overestimate $z^\star_{LDP} = z^\star_{MP}$ and thus induce large step sizes when we would perhaps already be close to the linear relaxation optimality. The final expression for the step size is

$$\theta_t = \varepsilon_t \frac{UB - LR(\boldsymbol{\pi}_{\mathbf{b},t})}{\|\mathbf{b} - \mathbf{a}_t\|^2}, \qquad t \geq 1, \quad (6.52)$$

where ε_t is a parameter that hopefully handles convergence issues by influencing the size of θ_t. For instance, an initial value is received as input and it is divided by 2 each time the best known Lagrangian bound does not improve sufficiently in a certain number of consecutive iterations. In Illustration 6.7, we derive this initial value analytically. It is however usually the fruit of trial and error in which $\varepsilon = 2$ is a common default value.

Obviously, if $LB = UB$, we reach optimality of the *LDP* which in turn even means that $\mathbf{x}^\bullet$ is integer optimal for the *ILP*. We can also reach optimality of the *LDP* by finding an optimal solution $\mathbf{x}_t$ to the *ISP* with a subgradient of zero (Proposition 6.7), i.e., $\mathbf{b} - \mathbf{a}_t = \mathbf{0}$. In that case, $z^\star_{LDP} \leq z^\star_{ILP}$. Since these are unlikely to happen and we otherwise have no alternative optimality criterion to verify, we instead resort to terminate the algorithm with respect to parameters such as a limit T on the number of iterations and/or a tolerance on θ, see Algorithm 6.1. This makes the subgradient algorithm a *heuristic* which typically does not find the maximum of the Lagrangian function but rather a *good lower approximation* of $z^\star_{LDP}$. Take notice that there is no primal solution output. Indeed, the solution $\mathbf{x}_t$ associated with the best (known) lower bound is in all likelihood meaningless on its own for the compact formulation, see Figure 6.10 and Note 6.13.

Algorithm 6.1: The subgradient algorithm with the Polyak step.

input : *LDP* (6.15), *ISP* (6.11); maximum number of iterations T, ε, $\hat{\boldsymbol{\pi}}_\mathbf{b}$, *UB*
output : Approximation of $z^\star_{LDP}$ with associated $\boldsymbol{\pi}_\mathbf{b}$
initialization : $t \leftarrow 1$, $LB \leftarrow -\infty$, $\boldsymbol{\pi}_{\mathbf{b},t} \leftarrow \hat{\boldsymbol{\pi}}_\mathbf{b}$

loop
  $\mathbf{x}_t, LR(\boldsymbol{\pi}_{\mathbf{b},t}) \leftarrow ISP$
  if $\mathbf{x}_t$ *feasible for the ILP and* $\mathbf{c}^\top \mathbf{x}_t < UB$
    $UB \leftarrow \mathbf{c}^\top \mathbf{x}_t$
  if $LR(\boldsymbol{\pi}_{\mathbf{b},t}) > LB$
    $LB \leftarrow LR(\boldsymbol{\pi}_{\mathbf{b},t})$, $\boldsymbol{\pi}_\mathbf{b} \leftarrow \boldsymbol{\pi}_{\mathbf{b},t}$
  if $LB = UB$
    break by optimality of the *ILP*
  if $\mathbf{b} - \mathbf{a}_t = \mathbf{0}$
    break by optimality of the *LDP*
  if $t = T$
    break by stopping rule
  if *LB constant over 3 consecutive iterations*
    $\varepsilon \leftarrow \varepsilon/2$
  $\theta \leftarrow \varepsilon \frac{UB - LR(\boldsymbol{\pi}_{\mathbf{b},t})}{\|\mathbf{b} - \mathbf{a}_t\|^2}$
  $\boldsymbol{\pi}_{\mathbf{b},t+1} \leftarrow [\boldsymbol{\pi}_{\mathbf{b},t} + \theta(\mathbf{b} - \mathbf{a}_t)]^+$
  $t \leftarrow t + 1$
return *LB* and $\boldsymbol{\pi}_\mathbf{b}$

Illustration 6.7 TCSPP **(cont.)**

Let us run the subgradient algorithm with the Polyak step of $\theta_t = \varepsilon_t \frac{UB - LR(\pi_7)}{\|14 - t_p\|^2}$. We analyze what happens in the first couple of iterations and derive an appropriate initial value for ε in the process. Initializing at $UB = \max_{p \in P} c_p = 27$ and $\pi_7 = 0$,

- The first iteration yields $LR(0) = 3$ for $p = 1246$, see Figure 6.9, with $c_p = 3$ and $t_p = 18$ infeasible for the *ILP*. We obtain

$$\theta = \varepsilon \frac{27-3}{(14-18)^2} \text{ and } \pi_7 = 0 + \theta \cdot (14-18) = -6\varepsilon.$$

- For any $\varepsilon > 0.75$, we allow $\pi_7 < -4.5$ (see Figure 6.9) for which the optimal path identified in the *ISP* is $p = 1356$ with $c_p = 24$ and $t_p = 8$ at $LR(\pi_7) = 24 + 6\pi_7$. As that path is feasible for the *ILP*, we update $UB = 24$ and compute the step size and next iterate. We obtain

$$\theta = \varepsilon \frac{24-(24-36\varepsilon)}{(14-8)^2} = \varepsilon^2 \text{ and } \pi_7 = -6\varepsilon + 6\varepsilon^2.$$

- Since $\pi_7 \leq 0$, we avoid cycling with $-6\varepsilon + 6\varepsilon^2 < 0$ which solves to $\varepsilon < 1$. A decent initial parameter value for this instance is therefore $\varepsilon \in (0.75, 1)$ with larger values theoretically providing smoother convergence.

Table 6.3 reports the results with $\varepsilon = 0.9$. The plots in Figure 6.13 show the evolution of the Lagrangian multiplier π_7 as well as the upper and lower bounds over the 80 iterations. We can associate what we see in (a) with bang-bang and (b) with the yo-yo effect. Putting numerical precision aside, we see that the relations in (6.32) are fulfilled by $z^\star_{LP} = 7 \leq z^\star_{LDP} = z^\star_{MP} = 7 \leq z^\star_{ILP} = 13 \leq UB = 15$, which reminds us that we have *not* solved the original problem.

t	π_7	$14-t_p$	$LR(\pi_7)$	LB	UB	ε	θ
1	0.000000	−4	3.000000	3.000000	27	0.900000	1.350000
2	−5.400000	6	−8.400000	3.000000	24	0.900000	0.810000
3	−0.540000	−4	5.160000	5.160000	24	0.900000	1.059750
4	−4.779000	6	−4.674000	5.160000	24	0.900000	0.716850
5	−0.477900	−4	4.911600	5.160000	24	0.900000	1.073723
6	−4.772790	6	−4.636740	5.160000	24	0.900000	0.715919
7	−0.477279	−4	4.909116	5.160000	24	0.450000	0.536931
8	−2.625003	4	4.499986	5.160000	15	0.450000	0.295313
9	−1.443752	−1	6.443752	6.443752	15	0.450000	3.850312
10	−5.294064	6	−7.764381	6.443752	15	0.450000	0.284555
11	−3.586735	4	0.653060	6.443752	15	0.450000	0.403508
12	−1.972704	−1	6.972704	6.972704	15	0.450000	3.612283
13	−5.584987	6	−9.509924	6.972704	15	0.450000	0.306374
14	−3.746743	4	0.013028	6.972704	15	0.450000	0.421509
15	−2.060709	4	6.757165	6.972704	15	0.450000	0.231830
16	−1.133390	−1	6.133390	6.972704	15	0.225000	1.994987
17	−3.128377	4	2.486492	6.972704	15	0.225000	0.175971
18	−2.424492	4	5.302031	6.972704	15	0.225000	0.136378
19	−1.878981	−1	6.878981	6.972704	15	0.225000	1.827229
20	−3.706211	4	0.175157	6.972704	15	0.112500	0.104237
			...				
70	−2.000651	4	6.997395	6.999992	15	0.000110	0.000055
			...				
80	−2.000102	4	6.999593	6.999992	15	0.000014	0.000007

Table 6.3: Subgradient iterations initialized at $UB = 27$, $\pi_7 = 0$, and $\varepsilon = 0.9$ which is halved every 3 consecutive iterations without improvement; stopping rule $\theta < 10^{-5}$.

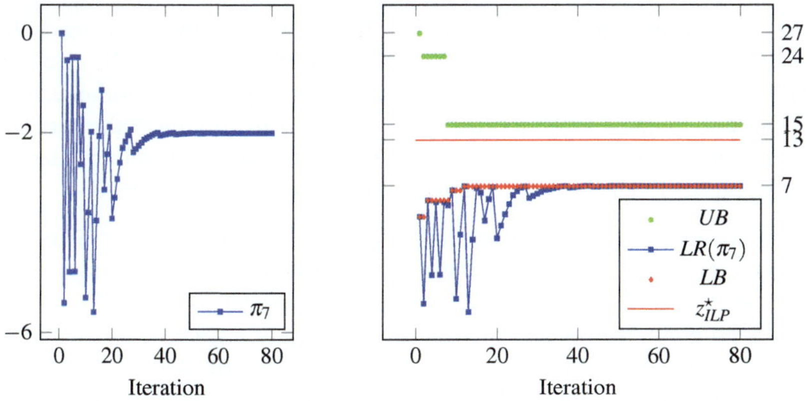

(a) Lagrangian multiplier π_7

(b) Upper and lower bounds

Fig. 6.13: Selection of plots for the evolution in Table 6.3.

Note 6.14 (Pedagogical example.) We may not have mentioned it enough but our *TCSPP* is pedagogical from the start. In practice, extreme points most likely cannot be enumerated nor can we describe the piecewise linear Lagrangian function explicitly. The specific initial value for $\varepsilon = 0.9$ is not as important as understanding that this parameter should have a pertinent scale for the problem at hand. The fact that we are unable to reach precise optimality despite being able to describe the optimized function analytically shows an obvious limitation of the subgradient algorithm.

Primal solutions (fractional and integer)

Apart from convergence issues, one of the main objections against the subgradient algorithm is that it exploits only local information to update $\boldsymbol{\pi}_{\mathbf{b}}$, that is, only the *one* last extreme point found by solving the Lagrangian subproblem. This is canonical to the expression of the Lagrangian dual problem (6.15)

$$z^\star_{LDP} = \max_{\boldsymbol{\pi}_{\mathbf{b}} \geq \mathbf{0}} LR(\boldsymbol{\pi}_{\mathbf{b}}) \quad = \max_{\boldsymbol{\pi}_{\mathbf{b}} \geq \mathbf{0}} \left\{ \boldsymbol{\pi}_{\mathbf{b}}^\mathsf{T} \mathbf{b} + \min_{\mathbf{x} \in \mathcal{D}} (\mathbf{c}^\mathsf{T} - \boldsymbol{\pi}_{\mathbf{b}}^\mathsf{T} \mathbf{A}) \mathbf{x} \right\},$$

whose optimizers are a vector $\boldsymbol{\pi}^\star_{\mathbf{b}}$ and a corresponding integer extreme point $\mathbf{x}_p$ of $\operatorname{conv}(\mathcal{D})$, for some $p \in P$, the latter being almost certainly infeasible for the compact formulation. Since we ultimately rather want to find an integer solution $\mathbf{x}^\star_{ILP}$ for the primal, a legitimate question is what can we really do with only a set of optimal dual multipliers $\boldsymbol{\pi}^\star_{\mathbf{b}}$ and this extreme point?

The truth is not much for obtaining an optimal integer solution. Interestingly however, for some applications, it is possible to compute, in a reasonable time, a heuristic but feasible integer solution for the *ILP* by transforming an integer solution of the *ISP* or using information gathered while solving it. This is the case for the *VRPTW* where, at every column generation iteration, a greedy algorithm can easily find from the computed *ISP* shortest path tree a set of disjoint paths partitioning the customers to service. See also Examples 6.4 Symmetric traveling salesperson problem and 6.5 Balancing printed circuit board assembly line systems.

Alternatively, Figure 6.14 shows that collecting the computed subgradients $(\mathbf{b} - \mathbf{a}_p)$ and their costs c_p can give us access to a primal solution, namely, $\mathbf{x}^\star_{MP}$, by solving the corresponding linear program (a restricted master problem). To efficiently solve this program, the optimal multiplier vector $\boldsymbol{\pi}^\star_{\mathbf{b}}$ or a good approximation of it $\hat{\boldsymbol{\pi}}_{\mathbf{b}}$ can be exploited in a dual stabilization method, as described next in various ways in Good to Know and More to Know. Since solving the Lagrangian dual problem is equivalent to solving the dual of the *MP* or *AMP* of a Dantzig-Wolfe reformulation, the solution $\mathbf{x}^\star_{MP}$ might not be integer. Therefore, to obtain an integer optimizer $\mathbf{x}^\star_{ILP}$, some additional work is required, that is, branching and cutting on the x- or λ-variables must be performed, as discussed in Chapter 7 (Branch-Price-and-Cut).

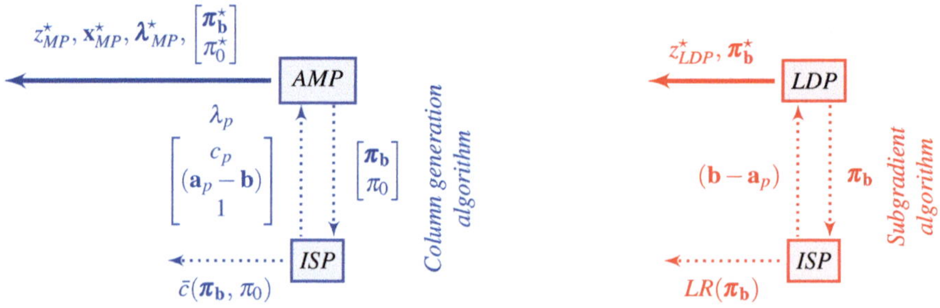

Fig. 6.14: *AMP* vs. *LDP*, where domain $\mathcal{D}$ is assumed bounded.

This makes it clear that, by contrast, the column generation algorithm uses a more elaborate update strategy for the dual values. It makes use of *all* extreme points gathered during the solution process by solving a linear program at every iteration. A reason for preferring a Lagrangian relaxation over a Dantzig-Wolfe reformulation was the sheer size of the linear programs that need to be solved, which was out of the question until some thirty years ago. With modern linear programming solvers this drawback is gone, and the lower bound obtained is the same. That is, in principle, the entire master problem class/implementation is not needed in a Lagrangian relaxation. However, in anticipation of wanting to find an (integer) solution for the compact formulation, one could say that it is unreasonable to simply discard all the pricing problem solutions we have found along the way. The only question that remains is whether or not we routinely solve a restricted master problem to update our multipliers. At the end of the day, we are left with a different sequence of multipliers and perhaps, most importantly, our understanding of the Lagrangian function.

6.3 Good to Know

We have established strong ties between the Lagrangian dual problem (6.15) and the Dantzig-Wolfe master problem (6.3). Not only do both provide the same lower bound on $z^\star_{ILP}$, also algorithmically, we solve both by iteratively updating the values of dual variables/Lagrangian multipliers. However, our initial interpretation of the column generation algorithm as being a cutting plane algorithm in the dual and our ensuing observations only rely on having a master and (integer) subproblem given. In the following, we thus abstract from the Dantzig-Wolfe decomposition and investigate *what we can learn from duality* for the column generation algorithm in general. Actually, this could have been an alternative title for this section.

Free from any mention of the sets $\mathcal{A}$ and $\mathcal{D}$, we recall the master problem from Chapter 2 and its dual for handy reference. In our further presentation, we assume equality constraints in the primal which implies that the dual variables are unrestricted in sign:

$$
\begin{array}{rll|l}
z^\star_{MP} = \min & \sum_{\mathbf{x}\in\mathcal{X}} c_{\mathbf{x}}\lambda_{\mathbf{x}} & & \max \ \mathbf{b}^\intercal\boldsymbol{\pi} \\
\text{s.t.} & \sum_{\mathbf{x}\in\mathcal{X}} \mathbf{a}_{\mathbf{x}}\lambda_{\mathbf{x}} = \mathbf{b} & [\boldsymbol{\pi}] & \text{s.t.} \ \mathbf{a}_{\mathbf{x}}^\intercal\boldsymbol{\pi} \le c_{\mathbf{x}} \quad [\lambda_{\mathbf{x}}] \quad \forall \mathbf{x}\in\mathcal{X} \\
 & \lambda_{\mathbf{x}} \ge 0 & \forall \mathbf{x}\in\mathcal{X} & \quad \boldsymbol{\pi}\in\mathbb{R}^m.
\end{array}
\tag{6.53}
$$

Solving the *MP* by the column generation algorithm involves repeatedly solving the *ISP*

$$
\begin{aligned}
\bar{c}(\boldsymbol{\pi}) = \quad & \min_{\mathbf{x}\in\mathcal{X}} \quad c_{\mathbf{x}} - \boldsymbol{\pi}^\intercal\mathbf{a}_{\mathbf{x}} \\
& \text{s.t.} \quad c_{\mathbf{x}} = c(\mathbf{x}) \\
& \qquad \mathbf{a}_{\mathbf{x}} = \mathbf{a}(\mathbf{x}).
\end{aligned}
\tag{6.54}
$$

The dual variables are the secret stars in the column generation algorithm. They determine which columns are generated and when we stop. In a Lagrangian relaxation, we explicitly work on these dual variables. In the remainder of this chapter, we thus concentrate on this dual perspective, in particular by more carefully choosing which dual variable values we work with in the column generation algorithm.

Dual-optimal inequalities

With our interpretation of column generation as iteratively constraining the dual, it immediately comes to mind that *further* constraining the dual can help in practice. A prominent and simple example is the set partitioning problem that can rather be solved as a set covering problem, often without compromising optimality: Replacing equality with inequality constraints in the primal, we restrain the dual vector $\boldsymbol{\pi}$, which is unrestricted in sign, to non-negative values $\boldsymbol{\pi} \geq \mathbf{0}$. Intuitively, as in any cutting plane algorithm, we wish to faster arrive at a description of the *relevant portion* of the Lagrangian function. This desire motivates the following general concept.

A *dual-optimal inequality* (*DOI*) cuts off part of the dual space by imposing a restriction on the dual variables that is satisfied by *all* optimal dual solutions. Such a constraint, indexed by j in a set J and expressed by

$$\mathbf{s}_j^\top \boldsymbol{\pi} \leq \delta_j, \qquad [y_j], \tag{6.55}$$

appears in the *RMP* as a non-negative auxiliary variable y_j:

$$\begin{aligned} z_{RMP} = \min \quad & \sum_{\mathbf{x} \in \mathcal{X}'} c_{\mathbf{x}} \lambda_{\mathbf{x}} + \sum_{j \in J} \delta_j y_j \\ \text{s.t.} \quad & \sum_{\mathbf{x} \in \mathcal{X}'} \mathbf{a}_{\mathbf{x}} \lambda_{\mathbf{x}} + \sum_{j \in J} \mathbf{s}_j y_j = \mathbf{b} \quad [\boldsymbol{\pi}] \\ & y_j \geq 0 \qquad \forall j \in J \\ & \lambda_{\mathbf{x}} \geq 0, \qquad \forall \mathbf{x} \in \mathcal{X}'. \end{aligned} \tag{6.56}$$

A *deep dual-optimal inequality* (*DDOI*) is even more restrictive in the sense that it only has to be satisfied by at least one optimal dual solution. Obviously, adding several *DOI*s in the formulation can be done without reservation, but one has to be careful not to cut away optimality when mixing *DDOI*s: these need to share at least one common dual optimal solution. Otherwise, the dual formulation is too constrained and, thus, yields a weaker primal lower bound.

Note 6.15 (Primal recovery procedure.) We may find a primal solution with positive y-variables that are absent from the original *MP*, and thus actually infeasible. To resolve this situation, one needs to implement a primal recovery procedure (Gschwind and Irnich, 2016) to swap these out for λ-variables. In a first step, a heuristic is applied in the hope of finding an alternative same-cost solution with $\mathbf{y} = \mathbf{0}$. In the case of *DOI*s or compatible *DDOI*s, this first step is always sufficient. For incompatible *DDOI*s or by extension arbitrary dual inequalities (a kind of over-stabilization), a second step is required to restore primal feasibility.

Illustration 6.8 **Cutting stock problem**

We present, without proof, three sets of *DOI*s for the linear relaxation of the cutting stock master problem (*CSP*, Example 2.1). Recall that the dual value $\pi_i^\star$ represents the marginal impact on $z_{MP}^\star$, the objective value in the linear relaxation of (2.29), of an augmentation of the demand b_i for item $i \in \{1, \ldots, m\}$ by one unit.

- *There always is* a non-negative optimal dual solution, that is, $\boldsymbol{\pi} \geq \mathbf{0}$. Therefore, the demand constraints can be written with greater-than-or-equal inequalities.

- Instead of cutting item $i \in \{1,\ldots,m\}$ from a roll, we can cut a set $S \subset \{1,\ldots,m\}$ of smaller items, as long as their total widths do not exceed the width of i. This does not increase the number of rolls (the objective value), which can be expressed as a *DOI*. Any optimal dual solution to the *MP* satisfies

$$w_i \geq \sum_{\ell \in S} w_\ell \quad \Rightarrow \quad \pi_i \geq \sum_{\ell \in S} \pi_\ell. \tag{6.57}$$

- As there are exponentially many constraints of the type (6.57), one may restrict to small cardinalities of S. In particular, for $|S| = 1$, we obtain the *ranking constraints*

$$w_1 \geq w_2 \geq \ldots \geq w_m \quad \Rightarrow \quad \pi_1 \geq \pi_2 \geq \ldots \geq \pi_m. \tag{6.58}$$

  Using $m-1$ successive ranking inequalities,

$$\pi_{i+1} - \pi_i \leq 0, \qquad [y_i] \qquad \forall i \in \{1,2,\ldots,m-1\}, \tag{6.59}$$

  the *MP* derived from the formulation (2.29) becomes:

$$\begin{aligned}
z^\star_{MP} = \min & \sum_{\mathbf{x}\in\mathcal{X}} \lambda_\mathbf{x} \\
\text{s.t.} & \sum_{\mathbf{x}\in\mathcal{X}} a_{1\mathbf{x}}\lambda_\mathbf{x} - y_1 \geq b_1 \quad [\pi_1] \\
& \sum_{\mathbf{x}\in\mathcal{X}} a_{2\mathbf{x}}\lambda_\mathbf{x} + y_1 - y_2 \geq b_2 \quad [\pi_2] \\
& \sum_{\mathbf{x}\in\mathcal{X}} a_{3\mathbf{x}}\lambda_\mathbf{x} + y_2 - y_3 \geq b_3 \quad [\pi_3] \\
& \qquad \vdots \qquad\qquad\qquad \vdots \\
& \sum_{\mathbf{x}\in\mathcal{X}} a_{m\mathbf{x}}\lambda_\mathbf{x} - y_{m-1} \geq b_m \quad [\pi_m] \\
& \lambda_\mathbf{x} \geq 0 \qquad \forall \mathbf{x} \in \mathcal{X} \\
& y_1, \quad y_2, \quad \ldots \quad y_{m-1} \geq 0.
\end{aligned} \tag{6.60}$$

Note 6.16 (Exchange vectors.) We notice in the primal *MP* (6.60) a remarkable structure in the column coefficients of the added zero-cost y-variables. Setting one of these variables to value one can be interpreted as replacing an item i by the smaller or equal width item $i+1$ in any cutting pattern containing i. As an effect in the column generation algorithm, for any generated cutting pattern that uses item i, we *implicitly* generate *another* feasible pattern without extra work. Such y-variables, respectively, their coefficient columns, are called *exchange vectors.* They are also explicitly used in a straightforward way to get rid of the basic y-variables so as to reconstruct an optimal primal solution for the *MP*, see Valério de Carvalho (2005); Ben Amor et al. (2006b).

Dual-optimal boxes

While dual-optimal inequalities impose *structural* constraints on the dual, we now aim at explicitly prescribing *values* to the dual variables. Ultimately, we would like to constrain their values to the vicinity of an optimal solution. To this end, a *dual box* $[\delta_{1i}, \delta_{2i}]$ imposes an interval on a dual variable π_i, $i \in \{1,\dots,m\}$, or more concisely in vector form for all π-variables,

$$\boldsymbol{\delta}_1 \leq \boldsymbol{\pi} \leq \boldsymbol{\delta}_2. \tag{6.61}$$

Let us denote the dual formulation of the master problem (6.53) in which these constraints are added by $DMP_{\boldsymbol{\delta}}$:

$$\begin{aligned} z^\star_{DMP_{\boldsymbol{\delta}}} = \max \quad & \mathbf{b}^\intercal \boldsymbol{\pi} \\ \text{s.t.} \quad & \mathbf{a}_\mathbf{x}^\intercal \boldsymbol{\pi} \leq \ c_\mathbf{x} \quad [\lambda_\mathbf{x}] \quad \forall \mathbf{x} \in \mathcal{X} \\ & -\boldsymbol{\pi} \leq -\boldsymbol{\delta}_1 \quad [\mathbf{y}_1] \\ & \boldsymbol{\pi} \leq \ \boldsymbol{\delta}_2 \quad [\mathbf{y}_2]. \end{aligned} \tag{6.62}$$

Dualizing, we obtain $2m$ auxiliary variables as surplus ($\mathbf{y}_1$) and slack ($\mathbf{y}_2$) variables in the primal formulation

$$\begin{aligned} z^\star_{MP_{\boldsymbol{\delta}}} = \min \quad & \sum_{\mathbf{x} \in \mathcal{X}} c_\mathbf{x} \lambda_\mathbf{x} - \boldsymbol{\delta}_1^\intercal \mathbf{y}_1 + \boldsymbol{\delta}_2^\intercal \mathbf{y}_2 \\ \text{s.t.} \quad & \sum_{\mathbf{x} \in \mathcal{X}} \mathbf{a}_\mathbf{x} \lambda_\mathbf{x} - \mathbf{y}_1 + \mathbf{y}_2 = \mathbf{b} \quad [\boldsymbol{\pi}] \\ & \mathbf{y}_1 \geq \mathbf{0}, \quad \mathbf{y}_2 \geq \mathbf{0} \\ & \lambda_\mathbf{x} \geq 0, \qquad \forall \mathbf{x} \in \mathcal{X}, \end{aligned} \tag{6.63}$$

where we likewise denote the *dual-boxed master problem* as $MP_{\boldsymbol{\delta}}$. Given the above, four linear programming programs have been identified:

- the *MP* and *DMP* with respective primal and dual solutions denoted $\boldsymbol{\lambda}^\star$ and $\boldsymbol{\pi}^\star$;
- the $MP_{\boldsymbol{\delta}}$ and $DMP_{\boldsymbol{\delta}}$ with respective solutions denoted $(\boldsymbol{\lambda}^\star_{\boldsymbol{\delta}}, \mathbf{y}_1^\star, \mathbf{y}_2^\star)$ and $\boldsymbol{\pi}^\star_{\boldsymbol{\delta}}$.

The primal vector $\boldsymbol{\lambda}^\star_{\boldsymbol{\delta}}$ is an optimal solution for the *MP* if $\mathbf{y}_1^\star = \mathbf{y}_2^\star = \mathbf{0}$, a condition easily verifiable by inspection. Our goal is to establish *a priori* conditions on $[\boldsymbol{\delta}_1, \boldsymbol{\delta}_2]$ for which solving the $MP_{\boldsymbol{\delta}}$ also solves the *MP*. There are three cases based on the possible values of $\boldsymbol{\pi}^\star$:

1. Obviously, if $[\boldsymbol{\delta}_1, \boldsymbol{\delta}_2]$ does not include any optimal vector $\boldsymbol{\pi}^\star$, then $\boldsymbol{\pi}^\star_{\boldsymbol{\delta}}$ is not feasible for the *DMP* as the dual boxes are too restrictive in the $DMP_{\boldsymbol{\delta}}$, nor does the $MP_{\boldsymbol{\delta}}$ correspond to the *MP* as it is too relaxed.
2. If $[\boldsymbol{\delta}_1, \boldsymbol{\delta}_2]$ contains an optimal solution $\boldsymbol{\pi}^\star$, in which case we call $[\boldsymbol{\delta}_1, \boldsymbol{\delta}_2]$ a *dual-optimal box*, then $\boldsymbol{\pi}^\star_{\boldsymbol{\delta}}$ solves the *DMP*, but unfortunately there is no guarantee that $\boldsymbol{\lambda}^\star_{\boldsymbol{\delta}}$ is an optimal solution for the *MP*, see Proposition 6.9 and Note 6.17.

3. If there exists a dual vector $\boldsymbol{\pi}^\star$ in the interior of $[\boldsymbol{\delta}_1, \boldsymbol{\delta}_2]$, that is, $\boldsymbol{\delta}_1 < \boldsymbol{\pi}^\star < \boldsymbol{\delta}_2$, then $\boldsymbol{\pi}^\star_{\boldsymbol{\delta}}$ still solves the *DMP* and $\boldsymbol{\lambda}^\star_{\boldsymbol{\delta}}$ additionally solves the *MP*, see Proposition 6.10.

Proposition 6.9. *If there exists an optimal solution $\boldsymbol{\pi}^\star$ for the DMP such that $\boldsymbol{\pi}^\star \in [\boldsymbol{\delta}_1, \boldsymbol{\delta}_2]$, then $\boldsymbol{\pi}^\star_{\boldsymbol{\delta}}$ is optimal for the DMP and $z^\star_{MP_{\boldsymbol{\delta}}} = z^\star_{DMP_{\boldsymbol{\delta}}} = z^\star_{DMP} = z^\star_{MP}$.*

Proof. By construction, $\boldsymbol{\pi}^\star_{\boldsymbol{\delta}}$ is feasible for the *DMP*, hence $\mathbf{b}^\top \boldsymbol{\pi}^\star_{\boldsymbol{\delta}} \leq \mathbf{b}^\top \boldsymbol{\pi}^\star$. The converse is also true as $\boldsymbol{\pi}^\star$ is feasible for the $DMP_{\boldsymbol{\delta}}$, hence $\mathbf{b}^\top \boldsymbol{\pi}^\star \leq \mathbf{b}^\top \boldsymbol{\pi}^\star_{\boldsymbol{\delta}}$. Their optimal objective values are therefore equal to one another and, moreover, equal to those of their respective primal formulation by strong duality. □

Note 6.17 (Boundary issues.) Writing out mathematically the primal objective functions gives some indication on the boundary issues:

$$\sum_{\mathbf{x} \in \mathcal{X}} c_{\mathbf{x}} \lambda^\star_{\boldsymbol{\delta}\mathbf{x}} - \boldsymbol{\delta}_1^\top \mathbf{y}_1^\star + \boldsymbol{\delta}_2^\top \mathbf{y}_2^\star = \sum_{\mathbf{x} \in \mathcal{X}} c_{\mathbf{x}} \lambda^\star_{\mathbf{x}}. \tag{6.64}$$

Comparing the left term to the right one, we can see that in the $MP_{\boldsymbol{\delta}}$, the objective value can be achieved using y-variables that are otherwise not existing in the *MP*, hence $(\mathbf{y}_1^\star, \mathbf{y}_2^\star)$ may differ from $(\mathbf{0}, \mathbf{0})$. For example, this is the case if the slack variables $\mathbf{y}_2^\star = \mathbf{b}$ for a positive right-hand side vector $\mathbf{b}$, together with $\boldsymbol{\lambda}^\star_{\boldsymbol{\delta}} = \mathbf{0}$, $\mathbf{y}_1^\star = \mathbf{0}$, and $\boldsymbol{\pi}^\star_{\boldsymbol{\delta}} = \boldsymbol{\delta}_2$.

It is quite unfortunate that imposing dual-optimal boxes in the $MP_{\boldsymbol{\delta}}$ is only guaranteed to produce an optimal solution $\boldsymbol{\pi}^\star$ for the *DMP*. The following proposition is stronger in its implication since it provides a sufficient condition on $[\boldsymbol{\delta}_1, \boldsymbol{\delta}_2]$ to ensure that solving the $MP_{\boldsymbol{\delta}}$ also solves the *MP*.

Proposition 6.10. *If there exists an optimal dual solution $\boldsymbol{\pi}^\star$ for the DMP such that $\boldsymbol{\delta}_1 < \boldsymbol{\pi}^\star < \boldsymbol{\delta}_2$, then $\boldsymbol{\lambda}^\star_{\boldsymbol{\delta}}$ is optimal for the MP as well.*

Proof. It suffices to show that $\mathbf{y}_1^\star = \mathbf{y}_2^\star = \mathbf{0}$. Firstly, note that $\boldsymbol{\pi}^\star$ is also optimal for the $DMP_{\boldsymbol{\delta}}$ and that any optimal solution for the $DMP_{\boldsymbol{\delta}}$ is optimal for the *DMP*. Secondly, by Proposition 1.7, every primal optimal solution for the $MP_{\boldsymbol{\delta}}$ satisfies the complementary slackness conditions with every dual optimal solution for the $DMP_{\boldsymbol{\delta}}$; therefore, consider the pair of primal and dual optimal solutions $(\boldsymbol{\lambda}^\star_{\boldsymbol{\delta}}, \mathbf{y}_1^\star, \mathbf{y}_2^\star)$ and $\boldsymbol{\pi}^\star$. Because $\boldsymbol{\delta}_1 < \boldsymbol{\pi}^\star < \boldsymbol{\delta}_2$, we must have $\mathbf{y}_1^\star = \mathbf{y}_2^\star = \mathbf{0}$ by complementary slackness in (6.62), and consequently, $\boldsymbol{\lambda}^\star_{\boldsymbol{\delta}}$ is optimal for the *MP*. □

Note 6.18 (Primal complement to Proposition 6.3.) Optimal Lagrangian multipliers $\boldsymbol{\pi}^\star_{\mathbf{b}}$ to the *LDP* (6.15) are also dual optimal for the *MP* by Proposition 6.3. Proposition 6.10 furnishes us with a bulletproof way to obtain a primal solution $\boldsymbol{\lambda}^\star$ and even complete the missing dual information $\pi_0^{\star k}$, $k \in K$. That is, we solve the $MP_{\boldsymbol{\delta}}$ with relatively small dual-optimal boxes around $\boldsymbol{\pi}^\star_{\mathbf{b}}$. Observe that we can easily initialize the $RMP_{\boldsymbol{\delta}}$ with $\mathcal{X}' = \emptyset$ as the y-variables ensure feasibility which then gives dual values $\boldsymbol{\pi}$ at either end of the dual box.

Exercises 6.9 and 6.11 are in the spirit of Note 6.18. The first considers whether it is interesting to use the pricing problem $\bar{c}^k(\boldsymbol{\pi}_{\mathbf{b}}^\star, \pi_0^k)$ to generate the pertinent columns in an empty *RMP* instead. The second is a crossover method to recover a basic solution to a linear program from an optimal dual solution, say coming from an interior-point algorithm.

How helpful can the dual information be?

Dual boxes theoretically enable us to constrain the *DMP* around any, in particular optimal, solution. Intuitively, such information should help the column generation process, especially since we are solving a relaxed primal formulation. However, the proof of the pudding is in the eating. In the three upcoming illustrations, we set dual boxes around some optimal $\boldsymbol{\pi}^\star$ in a symmetric and uniform manner, that is,

$$\boldsymbol{\pi}^\star - \boldsymbol{\delta} \leq \boldsymbol{\pi} \leq \boldsymbol{\pi}^\star + \boldsymbol{\delta}, \quad \boldsymbol{\delta} > \mathbf{0}. \tag{6.65}$$

However, everything that follows immediately generalizes to individual boxes on dual variables. Whether we should possess such perfect dual information prior to solving anything is an obvious and pressing question. We postpone an answer to the next section when we are convinced that it is worth the effort.

Illustration 6.9 **Cutting stock problem (cont.)**

The *CSP* is an example in which we have perfect dual information *if there is no loss*. In that case, $z_{IMP}^\star = z_{MP}^\star = \frac{\sum_{i=1}^m w_i b_i}{W}$, and for the linear relaxation of (2.29),

$$\pi_i^\star = \frac{w_i}{W}, \qquad \forall i \in \{1,\ldots,m\}. \tag{6.66}$$

This happens in the *triplet instances* for which each roll is cut into exactly three items. Table 6.4 reports the computational results of Ben Amor et al. (2006b) on ten randomly generated instances with $m = 501$. We compare four formulations: the standard one with a unit-demand for each of the 501 items; the classical one where items of identical width are aggregated, leading to fewer constraints (on average 194.3); formulation (6.60) with the added ranking constraints; and that with dual-optimal boxes of radius $\delta = 10^{-2}$.

The number of constraints in the *MP* is given by m'. The computation time in seconds is followed by the percent reduction *over the previous formulation.* We then list the proportion taken by the *RMP* and *ISP*, e.g., 522.4 (90 % of 579.4) seconds are spent solving the *RMP* in Unit-demand. The last column indicates the number of column generation iterations. All numbers are averages over the test set.

As expected, the classical aggregation of the demand for identical items is by far preferable to the formulation with unit-demands. The added ranking inequalities have only a small impact on the resolution of the *MP* while the dual-optimal

Formulation	m'	time (s)	Reduction (%)	RMP/ISP (%)	# iterations
Unit-demand	501	579.6	-	90 / 10	465.2
Aggregation of identical items	194.3	20.7	96.4	12 / 88	124.2
Ranking inequalities	194.3	19.8	5.8	11 / 89	113.3
Dual-optimal boxes	194.3	7.9	59.5	6 / 94	12.2

Table 6.4: Average results for 10 randomly generated 501-triplet instances.

boxes reduce the number of column generation iterations by an order of magnitude compared to the classical formulation (at the expense of a larger per-iteration computation time).

Illustration 6.10 TCSPP **(cont.)**

Reconsider our Dantzig-Wolfe reformulation of the *time constrained shortest path problem*, see Illustration 6.1 on p. 356. Before we solve the *MP* by the column generation algorithm, we impose the dual-optimal box $-2.1 \leq \pi_7 \leq -1.9$. That is, initially, the *RMP* contains the artificial variable y_0 in the convexity constraint, surplus variable y_1 with cost coefficient 2.1, and slack variable y_2 with cost coefficient -1.9, both in the duration constraint:

$$\begin{aligned} z_{RMP_\delta} = \min\ & 100y_0 + 2.1y_1 - 1.9y_2 \\ \text{s.t. } & -y_1 + y_2 \leq 14 \quad [\pi_7] \\ & y_0 = 1 \quad [\pi_0] \\ & y_0 \geq 0,\quad y_1 \geq 0,\quad y_2 \geq 0, \end{aligned} \tag{6.67}$$

1. In the first iteration, the artificial variable $y_0 = 1$ and the slack variable $y_2 = 14$ constitute a solution with $z_{RMP_\delta} = 73.4$. The corresponding dual solution is $\pi_7 = -1.9$ and $\pi_0 = 100$, at the boundary of the dual box. Path 1256 is generated (at cost 5 with duration 15) and the lower bound already reaches $lb = 6.9$.
2. In the second iteration,

$$\begin{aligned} z_{RMP_\delta} = \min\ & 100y_0 + 5\lambda_{1256} + 2.1y_1 - 1.9y_2 \\ \text{s.t. } & 15\lambda_{1256} - y_1 + y_2 \leq 14 \quad [\pi_7] \\ & y_0 = 1 \quad [\pi_0] \\ & y_0 \geq 0,\ \lambda_{1256} \geq 0,\quad y_1 \geq 0,\quad y_2 \geq 0, \end{aligned} \tag{6.68}$$

and $\lambda_{1256} = 1$ and surplus variable $y_1 = 1$ define an optimal primal solution (yet infeasible for the original problem). It provides an upper bound of $z_{RMP_\delta} = 7.1$ on $z^\star_{MP} = 7$ and the dual optimal solution is $\pi_7 = -2.1$ and $\pi_0 = 36.5$, again at the boundary of the box. The subproblem generates path 13256 (at cost 15 with duration 10) and the lower bound decreases to $lb = 6.6$.

3. Solving the third RMP_δ,

$$
\begin{aligned}
z_{RMP_\delta} = \min\ & 100y_0 + 5\lambda_{1256} + 15\lambda_{13256} + 2.1y_1 - 1.9y_2 \\
\text{s.t.}\ & 15\lambda_{1256} + 10\lambda_{13256} - y_1 + y_2 \le 14\ [\pi_7] \\
& y_0 = 1\ [\pi_0] \\
& y_0 \ge 0,\ \lambda_{1256} \ge 0,\ \lambda_{13256} \ge 0,\ y_1 \ge 0,\ y_2 \ge 0,
\end{aligned}
\tag{6.69}
$$

gives $\lambda_{1325} = 0.2$, $\lambda_{1256} = 0.8$, and $z_{RMP_\delta} = 7$. This is an optimal solution for the MP_δ because the reduced cost provided by the solution of the subproblem is zero. Since $y_1^\star = y_2^\star = 0$, we also have in hand an optimal solution for the *MP*, an expected conclusion as per Proposition 6.10 and the fact that a dual optimal value $\pi_7^\star = -2$ for the *MP* indeed lies in the interior of the dual box. The column generation algorithm terminates in three iterations rather than five, see Table 6.5.

	RMP				*ISP*				
t	RMP_δ solution	z_{RMP_δ}	π_0	π_7	$\bar{c}(\pi_7, \pi_0)$	p	c_p	t_p	lb
1	$y_0 = 1, y_2 = 14$	73.4	100.0	−1.9	−66.5	1256	5	15	6.9
2	$\lambda_{1256} = 1, y_1 = 1$	7.1	36.5	−2.1	−0.5	13256	15	10	6.6
3	$\lambda_{13256} = 0.2, \lambda_{1256} = 0.8$	7.0	35.0	−2.0	0.0				7.0

Table 6.5: Dual-optimal interval $-2.1 \le \pi_7 \le -1.9$.

The dual point of view explains the faster convergence of this dual box method. By enforcing $\pi_7 \in [-2.1, -1.9]$ around a dual optimal value $\pi_7^\star = -2$, we focus on the *relevant portion* of the Lagrangian function *LR* in Figure 6.15. Fed with dual values in the vicinity of the optimal dual solution, the *ISP* (when solved to optimality) can no longer generate certain dual cutting planes (here corresponding to paths 1246 and 1356) that describe the Lagrangian function in an *irrelevant* region.

Out of curiosity, let us also impose a dual box on $\pi_0^\star = 35$. If we add such dual information rather than using the artificial variable y_0 in the convexity constraint, for example $\pi_0 \in [34.99, 35.01]$, the first RMP_δ reads as

$$
\begin{aligned}
z_{RMP_\delta} = \min\ & +2.1y_1 - 1.9y_2 - 34.99y_3 + 35.01y_4 \\
\text{s.t.}\ & -y_1 + y_2 \le 14\ [\pi_7] \\
& -y_3 + y_4 = 1\ [\pi_0] \\
& y_1 \ge 0,\ y_2 \ge 0,\ y_3 \ge 0,\ y_4 \ge 0.
\end{aligned}
\tag{6.70}
$$

This does not change the number of column generation iterations as there are again the same two path-variables to generate, but the first solution gives $y_2 = 14$, $y_4 = 1$, and $z_{RMP_\delta} = 8.41$ much closer to the optimal objective value $z_{MP}^\star = 7$ than 73.4.

Here is an interesting observation that may have gone unnoticed regarding initializing the RMP_δ with good dual information: *the smaller the dual boxes around* $\boldsymbol{\pi}^\star$*, the closer the initial objective value* z_{RMP_δ} *is to* $\mathbf{b}^\intercal\boldsymbol{\pi}^\star$.

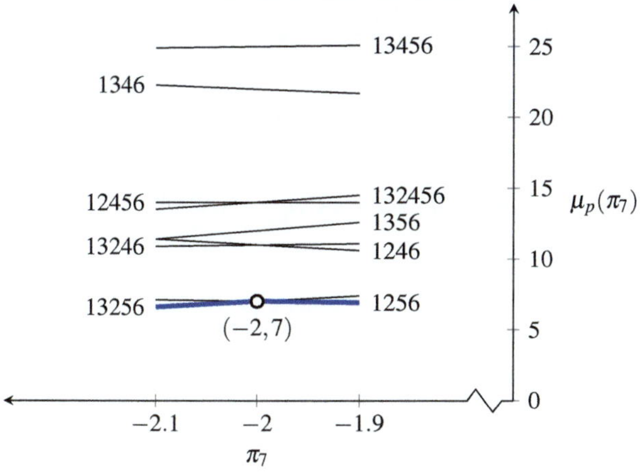

Fig. 6.15: The Lagrangian function *LR*, $-2.1 \leq \pi_7 \leq -1.9$.

Proposition 6.11. *Trivial feasibility of the* RMP_δ *for* (6.63) *can be achieved by setting the slack variable* $y_{i2} = b_i$ *if* $b_i \geq 0$ *or the surplus variable* $y_{i1} = -b_i$ *if* $b_i < 0$, $\forall i \in \{1, \ldots, m\}$. *Assuming symmetric dual-optimal boxes, the objective value of such a solution is* $z_{RMP_\delta} = \mathbf{b}^\mathsf{T}\boldsymbol{\pi}^\star + \sum_{i=1}^m \delta_i |b_i|$.

Proof. We have $z_{RMP_\delta} = \sum_{i=1}^m (-\delta_i y_{i1} + \delta_i y_{i2})$.
For each $i \in \{1, \ldots, m\}$, the contribution in the objective depends on the sign of b_i.
If $b_i = 0$, both products evaluate to zero.
If $b_i > 0$, we have $(\pi_i^\star + \delta_i) b_i = \pi_i^\star b_i + \delta_i b_i$.
If $b_i < 0$, we have $-(\pi_i^\star - \delta_i)(-b_i) = \pi_i^\star b_i + \delta_i(-b_i)$.
Summing the m terms, $z_{RMP_\delta} = \mathbf{b}^\mathsf{T}\boldsymbol{\pi}^\star + \sum_{i=1}^m \delta_i |b_i|$. □

Illustration 6.11 **Multi-depot vehicle scheduling problem**

✎ It's easier to find an optimal solution when you know where it hides. (*JD*)

In this illustration, we repeat the experiment of imposing a dual box around optimal dual values on a much larger problem. The *multi-depot vehicle scheduling problem* (*MDVSP*) is a natural extension of the single depot version (Example 2.4), where each vehicle is allowed to depart from any depot but must return to it at the end of the route. For this application, we use a multi-commodity network flow model as the compact formulation, where each depot has its own set of flow conservation constraints thus duplicating the number of inter-customer arcs for each depot whereas the customer visits are fulfilled by assignment constraints, see Exercises 2.9 and 4.20.

Assume that the network $G_{do}^k = (N^k, A_{do}^k)$ for depot k is represented in Figure 6.16 similarly to that of Figure 2.13, where $N^k = N \cup \{o^k, d^k\}$, $A_{do}^k = A^k \cup \{(d^k, o^k)\}$, and arc (d^k, o^k) is utilized to count the number of buses used.

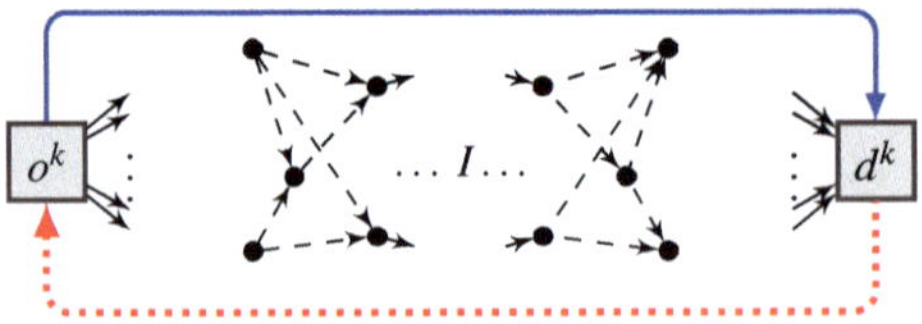

Fig. 6.16: Network $G^k_{do} = (N^k, A^k_{do})$.

A multi-commodity formulation (Ribeiro and Soumis, 1994), the commodity k being associated with depot k, is

$$z^\star_{ILP} = \min \quad \sum_{k\in K}\sum_{(i,j)\in A^k_{do}} c^k_{ij}x^k_{ij} \tag{6.71a}$$

$$\text{s.t.} \quad \sum_{k\in K}\sum_{j:(i,j)\in A^k} x^k_{ij} = 1 \quad [\sigma_i] \qquad \forall i \in N \tag{6.71b}$$

$$\sum_{j:(i,j)\in A^k_{do}} x^k_{ij} - \sum_{j:(j,i)\in A^k_{do}} x^k_{ji} = 0 \quad [\sigma^k_i] \qquad \forall k \in K, i \in N^k \tag{6.71c}$$

$$0 \le x^k_{d^k o^k} \le v^k \quad [\sigma^k_0] \qquad \forall k \in K \tag{6.71d}$$

$$x^k_{ij} \in \{0,1\} \qquad \forall k \in K, (i,j) \in A^k. \tag{6.71e}$$

The *MDVSP* has been shown to be $\mathcal{NP}$-hard for $|K| \ge 2$ (Bertossi et al., 1987). For $|K| = 1$, the *MDVSP* simplifies to the *SDVSP*, formulated as a network flow problem and solvable in polynomial time. This is also the case for $|K| \ge 2$ when the objective function corresponds to the minimization of the number of vehicles used to perform the n trips, that is, $\min \sum_{k\in K} x^k_{d^k o^k}$. Integrality of $x^k_{d^k o^k}$, $\forall k \in K$, follows from the binary requirements on x^k_{ij}, $\forall (i,j) \in A^k$.

From the compact formulation (6.71), we see a block-diagonal structure giving subproblems separable per depot (structural constraints) and the partitioning constraints (complicating ones). Let $\mathbf{x}^k = [x^k_{ij}]_{(i,j)\in A^k}$, $\forall k \in K$. Grouping the constraints as

$$\mathcal{A} = \left\{ \left\{ \begin{bmatrix} x^k_{d^k o^k} \\ \mathbf{x}^k \end{bmatrix} \in \mathbb{Z}_+ \times \{0,1\}^{|A^k|} \right\}_{k\in K} \middle| \text{ (6.71b) and (6.71d)} \right\} \tag{6.72a}$$

$$\mathcal{D}^k = \left\{ \begin{bmatrix} x^k_{d^k o^k} \\ \mathbf{x}^k \end{bmatrix} \in \mathbb{Z}_+ \times \{0,1\}^{|A^k|} \middle| \text{ (6.71c)} \right\}, \quad \forall k \in K, \tag{6.72b}$$

the zero-vector is discarded from the kth set of indices, the extreme rays are integer-scaled with $x^k_{d^k o^k} = 1$, and we have $\mathcal{X}^k = \left\{ \begin{bmatrix} 1 \\ \mathbf{x}^k_r \end{bmatrix} \right\}_{r\in R^k}$, $\forall k \in K$. The *IMP* formulation is therefore given as

$$z^\star_{IMP} = \min \sum_{k\in K} \sum_{\mathbf{x}^k \in \mathcal{X}^k} c_{\mathbf{x}^k} \lambda_{\mathbf{x}^k} \tag{6.73a}$$

$$\text{s.t.} \sum_{k\in K} \sum_{\mathbf{x}^k \in \mathcal{X}^k} a_{i\mathbf{x}^k} \lambda_{\mathbf{x}^k} = 1 \qquad [\pi_i] \qquad \forall i \in N \tag{6.73b}$$

$$\sum_{\mathbf{x}^k \in \mathcal{X}^k} \lambda_{\mathbf{x}^k} \leq v^k \qquad [\pi_0^k] \qquad \forall k \in K \tag{6.73c}$$

$$\lambda_{\mathbf{x}^k} \in \{0,1\} \qquad \forall k \in K, \mathbf{x}^k \in \mathcal{X}^k. \tag{6.73d}$$

The binary variable $\lambda_{\mathbf{x}^k}$ takes value 1 if and only if schedule $\mathbf{x}^k \in \mathcal{X}^k$ is selected. Such a schedule has a cost $c_{\mathbf{x}^k}$ and is represented by vector $\mathbf{a}_{\mathbf{x}^k} = [a_{i\mathbf{x}^k}]_{i\in N}$, where binary coefficient $a_{i\mathbf{x}^k}$ takes value 1 if trip i is operated in schedule $\mathbf{x}^k$, 0 otherwise. The coefficient of one for the $\lambda_{\mathbf{x}^k}$-variables in (6.73c) comes from $x^k_{d^k o^k} = 1$. For all intents and purposes, these are aggregated convexity constraints per depot $k \in K$ which we indeed find if we derive a Dantzig-Wolfe reformulation using extreme points instead.

The first row of Table 6.6 reports computational results on an instance comprising 4 depots and 800 bus trips to service (Ben Amor et al., 2009). With a dual box of radius ∞, this reference row corresponds to the original *MP* solved by column generation. The optimal objective value of 1 915 589.5 is obtained after 4 178 seconds of computing time. The algorithm called the pricing step 509 times from which 37 579 columns are generated. The last column indicates that almost one million simplex iterations (pivots) are performed in the successive *RMP*s. Finally, the initial solution is obtained via artificial variables with a million-cost each.

Dual boxes of radius δ	$z^\star_{MP_\delta}$	Initial objective value	time (s)	# iterations	# columns	# pivots
∞	**1 915 589.5**	800 000 000.0	4 178	509	37 579	926 161
100	1 915 589.5	2 035 590.5	836	119	9 368	279 155
10	1 915 589.5	1 927 590.5	118	35	2 789	40 599
1	1 915 589.5	1 916 790.5	52	20	1 430	8 744
0.1	1 915 589.5	1 915 710.5	48	19	1 333	8 630
0.01	**1 915 589.1**	1 915 602.5	37	17	1 145	6 288

Table 6.6: Dual-optimal boxes on an instance with 800 trips and 4 depots.

Figure 6.17 shows the convergence to optimality of the objective value. On the left, we underline that the order of magnitude difference in the objective value is correct. We therefore have a steady rapid decline in the first 50 iterations followed by more moderate improvements in the following 150, and finally a tail *seems to appear* for some 300 iterations. On the right, a closer inspection of this tail rather shows sustained decrease in the objective values albeit at an obvious much slower rate. To put these plots into perspective, the optimal objective value is 3 % of what is attained at iteration 109 whereas it is 61 % at iteration 209. By iteration 309, we stand at 96 % and the objective value continues to slowly decrease towards the optimal value.

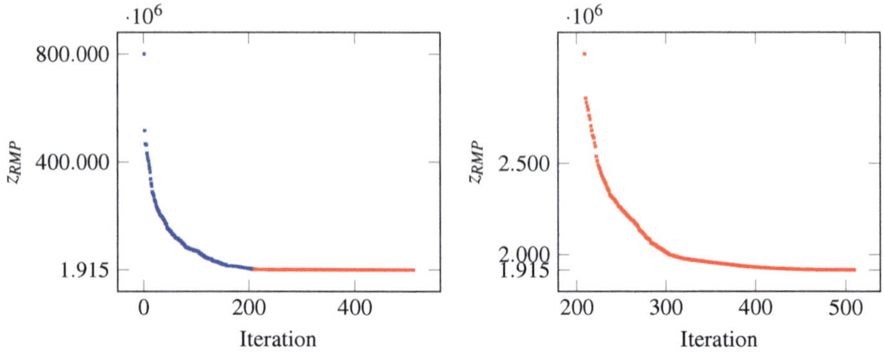

(a) Complete process (b) Zoom on tail

Fig. 6.17: The root node of an instance with 800 trips and 4 depots solved by the column generation algorithm; 509 iterations.

The second line of Table 6.6 reports the results obtained with a radius of 100 units around the optimal set of dual values obtained from the above reference run and every line that follows divides the allocated radius by 10. The impacts on the solution process are numerous. For one, we clearly identify initial objective values that are much closer to optimality, the small difference with Proposition 6.11 coming from the fact that there is no dual box associated with the four depot constraints. The computation time also decreases consistently as a result of cumulative economy in the various components of the column generation algorithm. Each *RMP* is re-optimized faster which can be explained by having identified fewer yet more relevant columns from the subproblems. This transfers to the number of column generation iterations since we improve towards optimality faster.

Figure 6.18 illustrates the convergence of the column generation algorithm on these various dual-boxed *RMP*s. The dual boxes of radius $\delta = 10$ are used as a yardstick to compare the scale of the convergence rate.

Finally, we have to underscore the last line with $\delta = 0.01$ which takes 17 iterations and is more than 100 times faster than the reference run. We find a solution of cost 1 915 589.1, indeed 0.4 units *less* than the optimal objective value of the *MP*. This means that the dual boxes are too restrictive, hence some positive y-variables appear in the solution of the MP_{δ}. Interestingly, this serves as additional warning regarding Note 6.17. Not only can we theoretically not use a closed interval but in practice, issues with numerical precision forbid too tiny intervals. In Illustration 6.9, we are able to reach optimality with 2 significant digits. It however proved to be too much to ask for this time around because of the much larger absolute objective values here. Fortunately, the silver lining is twofold. First, we already achieve very

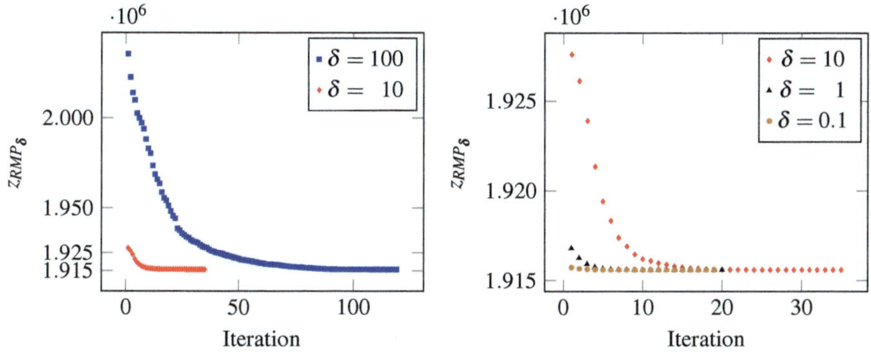

Fig. 6.18: The root node of an instance with 800 trips and 4 depots solved by column generation and dual-optimal boxes (yardstick radius $\delta = 10$; 35 iterations).

acceptable performance at $\delta = 1$. Second, a method that allows us to dynamically refine the radius and center of the dual boxes is in the making.

Fig. 6.19: Celso Ribeiro and Jacques (Angra dos Reis, Brazil, 2009-10-26).

Note 6.19 (What helps the integer program needs not help the linear program.) It may happen that initializing the *RMP* with the columns representing an integer solution can be worse than starting from scratch. We now know that every column can be interpreted as putting a bound on a linear combination on the dual variables; and this can be simply irrelevant/confusing and may lead the column generation algorithm astray. Moreover, a primal (optimal) integer solution usually forms only a small part of an initial basis $\mathbf{A}_B$ that needs to be completed with the columns of some artificial variables having a large cost. Since the simplex multipliers are computed as $\boldsymbol{\pi}^\mathsf{T} = \mathbf{c}_B^\mathsf{T}\mathbf{A}_B^{-1}$, the initial dual vector contains rather irrelevant penalty information which does not help. The column generation algorithm still spends its time searching for the columns producing the adequate dual values such that the solution process stops on the non-negative reduced cost optimality criterion.

6.4 More to Know

We have raised the fact that several algorithms are similar in spirit in that they follow a sequence of multipliers to achieve (near-)optimality. In particular, consider Figure 6.14 where vectors $(\mathbf{b} - \mathbf{a}_p)$, $p \in P$, are collected with respect to various multipliers. The co-existence of these methods does not end here; we hope that our *exposé* makes it natural to think of hybridization.

In fact, we have already established one such hybrid. Recall that in general the Lagrangian relaxation neither solves the *ILP*, nor does it even provide a primal solution. The dual point of view however gives us a simple mechanism to establish a primal solution using dual-optimal boxes in the Dantzig-Wolfe master problem.

In this section, we present more advanced hybridization ideas starting with stabilization of the dual values and move on to miscellaneous topics that may benefit from hybrid thinking.

Stabilized column generation

We have made at least two observations that motivate to *explicitly control* the update of the dual values in the column generation algorithm. One is the subgradient algorithm *overshooting* an optimal dual solution, and the resulting oscillation of dual variable values over the iterations, see Figure 6.13. It is quite understandable that we may see a similarly unstable behavior of the dual variable values in the column generation algorithm. This is particularly true when we work with (very) degenerate *RMP*s, modifying the basis with columns useless in the primal solution but sometimes largely modifying the dual values. Our other observation is the computational effectiveness of putting a dual box around a *good* dual solution, or in other words, to try to keep or steer away from *unimportant dual regions*.

Propositions 6.9 and 6.10 already hint at how we could proceed if we encounter instability of the dual values:

- Put a box around some dual solution and solve the boxed master problem.
- If the optimal dual solution is in the interior of the box, stop.
- Otherwise, relocate the box, and repeat.

This is indeed *stabilization* in a nutshell although we have yet to establish how to relocate the box. Marsten (1975) and Marsten et al. (1975) lay some ground rules for this relocation but their pioneering *boxstep method* lacks some of the self-adjusting flexibility we present in the sequel.

We define *stabilized column generation* as solving the *MP* (6.53) with a series of stabilized problems.

- Each one is defined as a linear programming approximation of the *MP*, indeed a *relaxation*, and it is solved by column generation. Such a formulation, denoted MP_{stab}, revolves around a *stability center* $\hat{\boldsymbol{\pi}}$ which acts as a *guess* (eventually very educated) for an optimal dual solution $\boldsymbol{\pi}^\star$.
- Around $\hat{\boldsymbol{\pi}}$, we define in a way similar to that of the $DMP_{\boldsymbol{\delta}}$ some relatively small dual boxes $[\boldsymbol{\delta}_1, \boldsymbol{\delta}_2]$ such that $\boldsymbol{\delta}_1 < \hat{\boldsymbol{\pi}} < \boldsymbol{\delta}_2$. These boxes define the *trust region*, hereafter denoted Ψ.
- By Proposition 6.10, we know that if there exists some $\boldsymbol{\pi}^\star \in \operatorname{int}(\Psi)$, then the $MP_{\boldsymbol{\delta}}$ solves the *MP*. We design the MP_{stab} with a penalty function in such a way that it honors this at zero penalty whenever there exists some $\boldsymbol{\pi}^\star \in \operatorname{int}(\Psi)$ but otherwise allows to *move outside the box* at a certain penalty cost.
- The optimal dual solution of the current MP_{stab} becomes the stability center of the next.

An example of a *penalty function* for variable π_i is illustrated in Figure 6.20, with a 3-piece linear concave function. Globally, this is formulated in the dual space as

$$\boldsymbol{\delta}_1 - \mathbf{w}_1 \leq \boldsymbol{\pi} \leq \boldsymbol{\delta}_2 + \mathbf{w}_2, \qquad \mathbf{w}_1, \mathbf{w}_2 \geq \mathbf{0}, \tag{6.74}$$

with the penalty terms $-\boldsymbol{\varepsilon}_1^\mathsf{T}\mathbf{w}_1 - \boldsymbol{\varepsilon}_2^\mathsf{T}\mathbf{w}_2$ in the objective function, where parameters $\boldsymbol{\varepsilon}_1, \boldsymbol{\varepsilon}_2 > \mathbf{0}$ by design. Observe that the trust region is not penalized:

$$-\,(\boldsymbol{\varepsilon}_1)^\mathsf{T}\mathbf{w}_1 \;-\; (\boldsymbol{\varepsilon}_2)^\mathsf{T}\mathbf{w}_2 \quad \begin{cases} < 0 & \Leftrightarrow \;\boldsymbol{\pi} \notin \Psi \\ = 0 & \Leftrightarrow \;\boldsymbol{\pi} \in \Psi \end{cases}. \tag{6.75}$$

Stabilization therefore comes from the fact that we are requesting substantial evidence that taking a value outside the trust region is beneficial.
The dual of the MP_{stab}, denoted DMP_{stab}, reads as

$$
\begin{alignat}{6}
z_{DMP_{stab}} = \max\;\; & \mathbf{b}^\mathsf{T}\boldsymbol{\pi} && -\boldsymbol{\varepsilon}_1^\mathsf{T}\mathbf{w}_1 && -\boldsymbol{\varepsilon}_2^\mathsf{T}\mathbf{w}_2 && && && \tag{6.76a}\\
\text{s.t.}\;\; & \mathbf{a}_\mathbf{x}^\mathsf{T}\boldsymbol{\pi} && && && \leq c_\mathbf{x} && [\lambda_\mathbf{x}] && \forall \mathbf{x} \in \mathcal{X} \tag{6.76b}\\
& -\boldsymbol{\pi} && -\mathbf{w}_1 && && \leq -\boldsymbol{\delta}_1 && [\mathbf{y}_1] && \tag{6.76c}\\
& \boldsymbol{\pi} && && -\mathbf{w}_2 && \leq \boldsymbol{\delta}_2 && [\mathbf{y}_2] && \tag{6.76d}\\
& && \mathbf{w}_1 \geq \mathbf{0}, && \mathbf{w}_2 \geq \mathbf{0}, && && && \tag{6.76e}
\end{alignat}
$$

and the primal formulation MP_{stab} is

$$
\begin{alignat}{5}
z_{MP_{stab}} = \min\;\; & \sum_{\mathbf{x}\in\mathcal{X}} c_\mathbf{x}\lambda_\mathbf{x} && -\boldsymbol{\delta}_1^\mathsf{T}\mathbf{y}_1 && +\boldsymbol{\delta}_2^\mathsf{T}\mathbf{y}_2 && && \tag{6.77a}\\
\text{s.t.}\;\; & \sum_{\mathbf{x}\in\mathcal{X}} \mathbf{a}_\mathbf{x}\lambda_\mathbf{x} && -\mathbf{y}_1 && +\mathbf{y}_2 = \mathbf{b} && [\boldsymbol{\pi}] && \tag{6.77b}\\
& \lambda_\mathbf{x} \geq 0 && && && && \forall \mathbf{x} \in \mathcal{X} \tag{6.77c}
\end{alignat}
$$

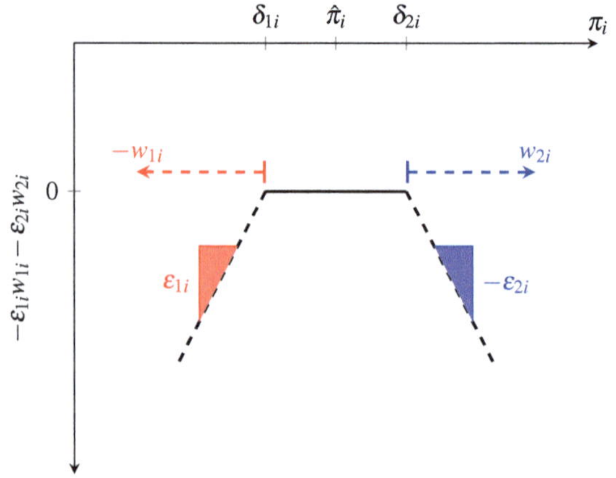

Fig. 6.20: Dual ε_i-penalties on values outside the dual box $\delta_{i1} \leq \pi_i \leq \delta_{i2}$.

$$\mathbf{y}_1 \leq \boldsymbol{\varepsilon}_1 \qquad [-\mathbf{w}_1 \leq \mathbf{0}] \tag{6.77d}$$

$$\mathbf{y}_2 \leq \boldsymbol{\varepsilon}_2 \quad [-\mathbf{w}_2 \leq \mathbf{0}] \tag{6.77e}$$

$$\mathbf{y}_1 \geq \mathbf{0}, \quad \mathbf{y}_2 \geq \mathbf{0}. \tag{6.77f}$$

Note 6.20 (Classes of primal variables.) Observe how auxiliary slack/surplus variables are structurally no different than the ε-bounded ones used for perturbation of the right-hand side vector **b** or the artificial variables used in a Phase I. In an implementation which uses the latter to ensure feasibility of the *RMP*, incorporating dual boxes should not be a hassle granted variable classes are well organized. This is a good time to make sure one uses *object-oriented programming* for their implementation.

Let us confirm that if the interior of the trust region contains an optimal solution for the *DMP*, then the stabilized problem solves the *MP*.

Corollary 6.3. *If there exists an optimal dual solution $\boldsymbol{\pi}^\star$ for the DMP such that $\boldsymbol{\delta}_1 < \boldsymbol{\pi}^\star < \boldsymbol{\delta}_2$, then, with respect to solving the MP_{stab}, $\boldsymbol{\lambda}^\star_{stab}$ and $\boldsymbol{\pi}^\star_{stab}$ are respectively primal and dual optimal solutions for the MP as well, for any $\boldsymbol{\varepsilon}_1, \boldsymbol{\varepsilon}_2 > \mathbf{0}$.*

Proof. By assumption, $\text{int}(\Psi)$ corresponds to open dual-optimal boxes so we know from Proposition 6.10 that the $MP_{\boldsymbol{\delta}}$ solves the *MP*, i.e., it finds optimal primal-dual solutions $\boldsymbol{\lambda}^\star_{\boldsymbol{\delta}}$ and $\boldsymbol{\pi}^\star_{\boldsymbol{\delta}} \in \text{int}(\Psi)$ for the *MP*. Since these are primal-dual feasible for the MP_{stab}, we know they must also be optimal because this already gives us $z^\star_{MP} = \mathbf{b}^\intercal \boldsymbol{\pi}$ for $\boldsymbol{\lambda}^\star_{stab} \equiv \boldsymbol{\lambda}^\star_{\boldsymbol{\delta}}$ and $\boldsymbol{\pi}^\star_{stab} \equiv \boldsymbol{\pi}^\star_{\boldsymbol{\delta}}$ along with the non-positive penalty term of zero as expected from (6.75), i.e., $\mathbf{w}^\star_1 = \mathbf{w}^\star_2 = \mathbf{0}$ in $-\boldsymbol{\varepsilon}_1\mathbf{w}_1 - \boldsymbol{\varepsilon}_2\mathbf{w}_2 = 0, \forall \boldsymbol{\varepsilon}_1, \boldsymbol{\varepsilon}_2 > \mathbf{0}$. □

Note 6.21 (Smooth penalty function.) Proposition 6.10 requires an open interval for the dual boxes to ensure we terminate with primal and dual solutions for the *MP*.

Since stabilization is based on dual optimal boxes, it means that we mathematically require the penalty function to be 'smooth' at $\boldsymbol{\pi}_i \in (\delta_{i1}, \delta_{i2})$, $\forall i \in \{1, \dots, m\}$, i.e., the trust region cannot reduce to a single point for any dual value.

As described above, the objective function in the DMP_{stab} penalizes the components of $\boldsymbol{\pi}$ when they lie outside the trust region $[\boldsymbol{\delta}_1, \boldsymbol{\delta}_2]$. In the primal formulation MP_{stab}, these $\boldsymbol{\delta}$-bounds appear in the objective function as the coefficients of the surplus ($\mathbf{y}_1$) and slack ($\mathbf{y}_2$) variables while the dual penalties $\boldsymbol{\varepsilon}_1$ and $\boldsymbol{\varepsilon}_2$ act as perturbation parameters on vector $\mathbf{b}$. Perturbation is known to be beneficial for coping with degeneracy in practice, see Dantzig (1963, §10-2, p. 231). We ultimately want the sequence of stabilized master problems to be solved faster than the original *MP*. In Illustrations 6.9–6.11, we have seen convincingly quicker solution processes when initialized with relatively small dual-optimal boxes. Otherwise, when the trust region does not contain an optimal dual solution, we solve an MP_{stab}, a relaxation of the *MP*, with a primary focus on the given trust region, at the same time being guided as to *how to move away* from the current dual boxes.

Let $s \geq 0$ index the sequence of stabilized master problems MP^s_{stab} solved. For each of them, we have as input

- the stability center $\hat{\boldsymbol{\pi}}^s$,
- the penalties $\boldsymbol{\varepsilon}^s_1$ and $\boldsymbol{\varepsilon}^s_2$,
- the trust region $\Psi^s = [\boldsymbol{\delta}^s_1, \boldsymbol{\delta}^s_2]$, where $\boldsymbol{\delta}^s_1 < \hat{\boldsymbol{\pi}}^s < \boldsymbol{\delta}^s_2$;

and as output

- a primal-dual pair $(\boldsymbol{\lambda}^s, \mathbf{y}^s_1, \mathbf{y}^s_2)$ and $(\boldsymbol{\pi}^s, \mathbf{w}^s_1, \mathbf{w}^s_2)$,
- with objective value z^s.

Formulation MP^s_{stab} is equivalent to the *MP* if and only if $\mathbf{y}^s_1 = \mathbf{y}^s_2 = \mathbf{0}$ so the (partly hidden) stopping criterion of the stabilized column generation algorithm is

$$\bar{c}(\boldsymbol{\pi}^s) = 0 \text{ and } \mathbf{y}^s_1 = \mathbf{y}^s_2 = \mathbf{0}. \tag{6.78}$$

Algorithm 6.2 outlines the mechanics of the stabilization process which terminates upon optimality of the *MP*. At every so-called *stabilized iteration*, we solve the MP^s_{stab} by column generation and thus reach $\bar{c}(\boldsymbol{\pi}^s) = 0$. We then check whether $\mathbf{y}^s_1 = \mathbf{y}^s_2 = \mathbf{0}$ holds true. Otherwise, we iterate with updated stabilization inputs.

In Algorithm 6.2, the trust region and penalty parameters are updated dynamically. We select the trust region to contain the current dual solution (in the beginning, an estimate), and solve the MP^s_{stab}. Component wise, if $\boldsymbol{\pi}^s$ lies in the trust region, reduce its width and augment the ε-penalties. Otherwise, relocate the stability center, recenter and enlarge the trust region, and decrease the penalties. Alternatively, the update (also called a *serious step*) can be performed each time a dual solution $\boldsymbol{\pi}$ improves upon the previously best known lower bound $\mathbf{b}^\mathsf{T}\boldsymbol{\pi} + \bar{c}(\boldsymbol{\pi}) > LB$, or when a number of iterations confirms that some dual boxes need to be changed. Finally note that, as $\mathbf{y}_1 = \mathbf{y}_2 = \mathbf{0}$ at the last stabilized iteration, the constraints (6.76c)–(6.76e) with the final parameters in the DMP_{stab} are in fact deep dual-optimal inequalities.

Algorithm 6.2: The stabilized column generation algorithm.

input : MP_{stab} (6.77), ISP (6.54), $\hat{\boldsymbol{\pi}}$, $\boldsymbol{\varepsilon}_1$, $\boldsymbol{\varepsilon}_2$, $\boldsymbol{\delta}_1$, $\boldsymbol{\delta}_2$
output : Optimal primal-dual solutions $\boldsymbol{\lambda}^\star_{MP}$ and $\boldsymbol{\pi}^\star$ for the MP, and optimum $z^\star_{MP}$
initialization : $s \leftarrow 0$

loop
    z^s, $(\boldsymbol{\lambda}^s, \mathbf{y}_1^s, \mathbf{y}_2^s)$, $(\boldsymbol{\pi}^s, \mathbf{w}_1^s, \mathbf{w}_2^s) \leftarrow MP^s_{stab}$
    if $\mathbf{y}_1^s = \mathbf{y}_2^s = \mathbf{0}$
        break by optimality of the MP
    else
        $\boldsymbol{\varepsilon}_1$, $\boldsymbol{\varepsilon}_2$, $\boldsymbol{\delta}_1$, $\boldsymbol{\delta}_2 \leftarrow$ Update penalties and radii according to $\boldsymbol{\pi}^s$
        $\hat{\boldsymbol{\pi}} \leftarrow \boldsymbol{\pi}^s$
        $s \leftarrow s+1$

return primal-dual pair $\boldsymbol{\lambda}^s$ and $\boldsymbol{\pi}^s$ with objective value z^s

Illustration 6.12 **Multi-depot vehicle scheduling problem (cont.)**

Let us illustrate this stabilization process on the *multi-depot vehicle scheduling problem*. As before, the instance comprises 4 depots and 800 trips to service and the dual estimates are computed by solving a network flow problem using an aggregation of the four depots into a single one (see Exercise 4.20).

As reported in Table 6.7, a standard implementation of the column generation algorithm requires 509 iterations, generates 37 579 columns, and runs in 4 178 seconds, where $3149/4178 = 75\,\%$ of the computation time is devoted to solving the *RMP*. The stabilized column generation is about 10 times faster, taking only 112 iterations, generating 4 749 columns in 439 seconds. Moreover, the total computation time splits half and half between solving the *RMP* and the pricing problems, respectively in 216 and 223 seconds.

	time (s)			Column generation	
	Total	*RMP*	*SP*	# iterations	# columns
Standard	4 178	3 149 75 %	1 029 25 %	509	37 579
Stabilized	439	216 49 %	223 51 %	112	4 749
Reduction	89 %	93 %	78 %	78 %	87 %

Table 6.7: Standard and stabilized results of the column generation algorithm.

In Figure 6.21(a), we see that every stabilized linear program, each one being a relaxation of the *MP*, takes only a few column generation iterations: 29, 18, 8, 8, 10, 4, 5, 4, 4, 5, etc. Visually, we discern six updates of the parameters when $z_{RMP_{stab}}$ clearly jumps, but there are actually eleven such updates when we consult the data.

In Figure 6.21(b), we find that the first approximate problem solved after an update is in general more difficult than the following ones.

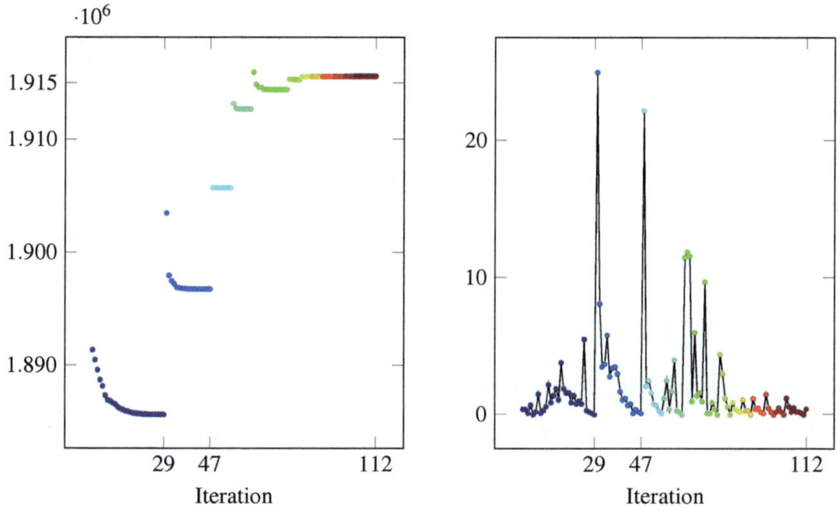

(a) Values of $z_{RMP_{stab}}$ (b) Computation times (sec) for the RMP_{stab}

Fig. 6.21: Stabilization process.

Properties

Propositions 6.12–6.14 prove that stabilized column generation ultimately solves the *MP*. We can support the purpose of each proposition with what we see (and do not see) in Figure 6.21a:

6.12 Every MP^s_{stab} is a relaxation of the *MP* and, as such, it provides a lower bound z^s on $z^\star_{MP}$. This is the last objective value at the end of each column generation process. However, neglecting the penalty terms from the dual objective value may yield a better lower bound as $\mathbf{b}^\intercal \boldsymbol{\pi}^s$;

6.13 Every (except maybe last) MP^s_{stab} strictly improves the lower bound $\mathbf{b}^\intercal \boldsymbol{\pi}^s$ of its predecessor. *It is easy to confuse this with the strict increase we see in the sequence* z^s;

6.14 The sequence of MP^s_{stab} reaches $z^\star_{MP}$ in a finite number of stabilized iterations.

Proposition 6.12. *Solving the* MP^s_{stab}, *lower bounds on* $z^\star_{MP}$ *are*

$$z^s \leq \mathbf{b}^\intercal \boldsymbol{\pi}^s \leq z^\star_{MP}. \tag{6.79}$$

Proof. The vector $\boldsymbol{\pi}^s$ is feasible for the *DMP* and hence it supplies the lower bound $\mathbf{b}^\intercal\boldsymbol{\pi}^s$ on $z^\star_{MP}$ by Proposition 1.5 (weak duality), disregarding the non-positive penalty computed in $z^s = \mathbf{b}^\intercal\boldsymbol{\pi}^s - (\boldsymbol{\varepsilon}_1^s)^\intercal\mathbf{w}_1^s - (\boldsymbol{\varepsilon}_2^s)^\intercal\mathbf{w}_2^s$. □

The next proposition states that if the MP^s_{stab} does not find an optimal dual solution to the *MP*, then $\mathbf{b}^\intercal\boldsymbol{\pi}^{s+1}$ strictly improves the lower bound on $z^\star_{MP}$.

Proposition 6.13. *For $s \geq 0$, if $\boldsymbol{\pi}^s$ is not optimal for the DMP, then*

$$\mathbf{b}^\intercal\boldsymbol{\pi}^{s+1} > \mathbf{b}^\intercal\boldsymbol{\pi}^s. \tag{6.80}$$

Proof. With respect to the MP^{s+1}_{stab}, the dual vector $\boldsymbol{\pi}^s = \hat{\boldsymbol{\pi}}^{s+1} \in \Psi^{s+1}$ always gives a penalty (6.75) evaluation of zero. Hence, we have by optimality of $\boldsymbol{\pi}^{s+1}$ that

$$\mathbf{b}^\intercal\boldsymbol{\pi}^{s+1} - (\boldsymbol{\varepsilon}_1^{s+1})^\intercal\mathbf{w}_1^{s+1} - (\boldsymbol{\varepsilon}_2^{s+1})^\intercal\mathbf{w}_2^{s+1} \geq \mathbf{b}^\intercal\boldsymbol{\pi}^s, \tag{6.81}$$

where the penalty is either negative or zero.

- If it is negative, we have $\boldsymbol{\pi}^{s+1} \notin \Psi^{s+1}$ and $\mathbf{b}^\intercal\boldsymbol{\pi}^{s+1} > \mathbf{b}^\intercal\boldsymbol{\pi}^s$.
- If it is zero, we have $\boldsymbol{\pi}^{s+1} \in \Psi^{s+1}$ and $\mathbf{b}^\intercal\boldsymbol{\pi}^{s+1} \geq \mathbf{b}^\intercal\boldsymbol{\pi}^s$. We show that the strict inequality holds by considering 1) $\boldsymbol{\pi}^{s+1} \in \mathrm{int}(\Psi^{s+1})$ and 2) $\boldsymbol{\pi}^{s+1} \notin \mathrm{int}(\Psi^{s+1})$:

1) By complementary slackness, $\mathbf{y}_1^s = \mathbf{y}_2^s = \mathbf{0}$ and $\boldsymbol{\pi}^{s+1}$ is optimal for the *DMP*. As such, $\mathbf{b}^\intercal\boldsymbol{\pi}^{s+1} > \mathbf{b}^\intercal\boldsymbol{\pi}^s$ because $\boldsymbol{\pi}^s$ is not optimal for the *DMP* by assumption.
2) Assume that $\mathbf{b}^\intercal\boldsymbol{\pi}^{s+1} = \mathbf{b}^\intercal\boldsymbol{\pi}^s$, then $\boldsymbol{\pi}^s$ is also optimal for the MP^{s+1}_{stab}. By Corollary 6.3 $\boldsymbol{\pi}^s = \hat{\boldsymbol{\pi}}^{s+1} \in \mathrm{int}(\Psi^{s+1})$ is also optimal for the *DMP*, a contradiction to the suboptimality assumption of $\boldsymbol{\pi}^s$. Since at least one component of $\boldsymbol{\pi}^{s+1}$ is at the boundary of Ψ^{s+1}, we have $\boldsymbol{\pi}^{s+1} \neq \boldsymbol{\pi}^s = \hat{\boldsymbol{\pi}}^{s+1}$ and $\mathbf{b}^\intercal\boldsymbol{\pi}^{s+1} > \mathbf{b}^\intercal\boldsymbol{\pi}^s$ as requested. □

Note 6.22 (Only one more to go.) This last particular case reminds us of Proposition 6.9 and Note 6.17 in which solving the MP_δ with dual-optimal boxes yields an optimal solution for the *DMP* but not for the *MP*. By design, if $\boldsymbol{\pi}^s$ is optimal for the *DMP* but $\boldsymbol{\lambda}^s$ is not optimal for the *MP*, then only one more stabilized iteration is required.

Proposition 6.14. *The strictly increasing sequence $\{\mathbf{b}^\intercal\boldsymbol{\pi}^s\}_{s\geq 0}$ reaches $z^\star_{MP}$ in a finite number of stabilized iterations.*

Proof. Consider the strictly increasing sequence $\{\mathbf{b}^\intercal\boldsymbol{\pi}^s\}_{s\geq 0}$ bounded from above by $z^\star_{MP}$ and converging to $\mathbf{b}^\intercal\boldsymbol{\pi}^\star_{stab} \leq z^\star_{MP}$. By contradiction, we show that the equality holds. Let $\mathcal{L}_{stab} = \{\boldsymbol{\pi} \mid \mathbf{b}^\intercal\boldsymbol{\pi} = \mathbf{b}^\intercal\boldsymbol{\pi}^\star_{stab}\}$ be the level set of value $\mathbf{b}^\intercal\boldsymbol{\pi}^\star_{stab}$. For s large enough, $\boldsymbol{\pi}^\star_{stab} \in \mathcal{L}_{stab} \cap \mathrm{int}(\Psi^s)$, and by optimality of $\boldsymbol{\pi}^s$, we have $\mathbf{b}^\intercal\boldsymbol{\pi}^\star_{stab} \leq \mathbf{b}^\intercal\boldsymbol{\pi}^s$.

Assuming $\mathbf{b}^\intercal\boldsymbol{\pi}^\star_{stab} < z^\star_{MP}$ means that $\boldsymbol{\pi}^\star_{stab}$ is not optimal for the *DMP*. If $\boldsymbol{\pi}^s$ is optimal for the *DMP*, then $\mathbf{b}^\intercal\boldsymbol{\pi}^\star_{stab} < \mathbf{b}^\intercal\boldsymbol{\pi}^s$, a contradiction on the limit of the sequence. Otherwise, $\boldsymbol{\pi}^s$ is not optimal for the *DMP* and $\mathbf{b}^\intercal\boldsymbol{\pi}^\star_{stab} \leq \mathbf{b}^\intercal\boldsymbol{\pi}^s < \mathbf{b}^\intercal\boldsymbol{\pi}^{s+1}$ by Proposition 6.13, contradicting again the limit of the sequence. □

Alternative/complementary stabilization methods

Alternative/complementary stabilization methods exist, let us mention a few.

- *Bundle methods*
  There are more elaborate proposals for penalty functions, e.g., piecewise linear functions with five pieces. These penalize even more, the further one moves away from the stability center. Readers coming from non-linear optimization recognize this as an approximation to a quadratic penalty function that is regularly used there. For instance, the *bundle methods* (Lemaréchal, 1978; Lemaréchal et al., 1995; Kiwiel, 1999; Frangioni, 2002) are therefore interesting alternatives to solve the *MP* as they have built-in stabilizing mechanism. While piecewise linear functions only require linear programming solvers, quadratic ones obviously must use quadratic solvers. The interested reader can consult the various mathematical programming solvers of Gurobi Optimization (gurobi.com). For an analysis of alternative stabilizing functions and performance evaluation on large-scale multi-depot vehicle scheduling problems and simultaneous vehicle and crew scheduling problems, see Ben Amor et al. (2009). Comparisons are also presented for five applications in Briant et al. (2008): cutting stock (which includes bin packing), vertex coloring, capacitated vehicle routing, multi-item lot sizing, and traveling salesperson.
- *Over-stabilization*
  Gschwind and Irnich (2016) use dual inequalities that are not necessarily (deep) *DOI*s (see Note 6.15). Their point is that almost any dual guess helps. After a recovery procedure on the primal variables, if any y-variable remains positive, the corresponding dual-inequality is removed from the RMP_{stab} and another stabilized iteration occurs. Otherwise, the current solution $\boldsymbol{\lambda}^\star_{stab}$ is optimal for the *MP*.
- *Dual variable smoothing*
  If we want to avoid large jumps in dual variable values from one iteration to the next, we may dampen the oscillation effect by shortening the step length. More precisely, at iteration t of the column generation algorithm, the dual vector $\tilde{\boldsymbol{\pi}}^t$ used in the *ISP* instead of $\boldsymbol{\pi}^t$ retrieved from the solution of the RMP^t is computed using previous or even all the previous dual vectors. In particular, Wentges (1997) proposes the smoothing rule

  $$\tilde{\boldsymbol{\pi}}^t = \alpha\hat{\boldsymbol{\pi}} + (1-\alpha)\boldsymbol{\pi}^t = \hat{\boldsymbol{\pi}} + (1-\alpha)(\boldsymbol{\pi}^t - \hat{\boldsymbol{\pi}}), \qquad t \geq 1, \tag{6.82}$$

  where $\alpha \in [0,1)$ and $\hat{\boldsymbol{\pi}}$ is the current stability center: this amounts to taking a step size $1-\alpha$ away from $\hat{\boldsymbol{\pi}}$ in the direction of $\boldsymbol{\pi}^t$. Neame (1999) rather uses

  $$\tilde{\boldsymbol{\pi}}^t = \alpha\tilde{\boldsymbol{\pi}}^{t-1} + (1-\alpha)\boldsymbol{\pi}^t, \qquad t \geq 1, \tag{6.83}$$

  that weighs in the previous dual values, see Pessoa et al. (2013). Solving the *ISP* may produce columns that are already present in the RMP^t, hence introducing a kind of noise in the column generation algorithm. In practice, this is easily manageable. However, optimizing $\bar{c}(\hat{\boldsymbol{\pi}}^t)$ might not yield a negative reduced cost column, even if one exists for $\boldsymbol{\pi}^t$. This is called a *mis-price*. In that case, we can confirm convergence to optimality by verifying that $\bar{c}(\boldsymbol{\pi}^t)$ is also equal to zero. Pessoa et al. (2018) suggest a parameter-free dual variable stabilization scheme that combines subgradient ideas with dual variable smoothing.

- *Interior-point algorithms*
  A way to get dual values that are not associated with primal degenerate basic solutions is to use the *analytic center cutting plane method* (ACCPM) initiated by Goffin and Vial (1990)–not to be confused with *Anaesthesia Critical Care & Pain Medicine* in a Google search. Goffin et al. (1992), Goffin et al. (1993), and Goffin and Vial (2002) make additional theoretical developments. At every iteration of the column generation algorithm, the method computes the analytic center (or an approximation of it) of the polyhedron defined by a lower bound constraint on the value of the objective function and the set of dual constraints (see the solution of Exercise 3.5 for dual formulations of the *MP* with a single subproblem and a block-diagonal structure). This polyhedron is called the *localization set*. Optimality cuts are derived from extreme point solutions of the *ISP*; feasibility cuts correspond to extreme rays (see Note 6.3). Dual boxes can also be added to restrain the dual domain. Note that a finite lower bound is obtained in (6.18) only when all pricing problems produce extreme point solutions for a given dual vector. It also means that they all have to be solved. For more, see Gondzio et al. (1996), the brief tutorial of Péton and Vial (2001), and the website www.maths.ed.ac.uk/~gondzio/software/accpm.

  Another way to achieve the centralization of the dual values is to generate several extreme points of the dual polyhedron and compute a convex combination of these (Rousseau et al., 2007). Alternatively, we can directly solve the *RMP* with an interior-point algorithm: by relaxing the optimality tolerance, it can be solved rather quickly, see Gondzio et al. (2013); Munari and Gondzio (2013). Let us quote the beginning of Gondzio et al. (2016)'s abstract:

  > The primal-dual column generation method (PDCGM) is a general-purpose column generation technique that relies on the primal-dual interior point method to solve the restricted master problems. The use of this interior point method variant allows to obtain suboptimal and well-centered dual solutions which naturally stabilizes the column generation process.

- *Dynamic separation of aggregated rows*
  To conclude, let us mention a recent stream of research. Costa et al. (2022) propose a new stabilization framework which relies on the dynamic generation of aggregated rows from the *MP*. Huge computation time reductions are observed with respect to a standard column generation algorithm on three applications, namely, the vehicle routing problem with time windows, the bin packing problem with conflicts, and the multi-person pose estimation problem.

Learning more from the dual point of view

When writing this book we were tempted several times to forward reference to this chapter. This was the case for statements that acquire a different quality or can be better motivated when viewed through the dual lens. We encourage the reader to go back through the chapters with the new knowledge in mind. We close this chapter with three more topics that allude to where the dual point of view can lead us.

Initialization

Even in the initialization phase, we have good reasons to use the Lagrangian relaxation to create a subset of initial columns. Looking back at Figure 6.14, we can use it to collect $(\mathbf{b} - \mathbf{a}_p)$ vectors, $\forall p \in P'$, and reuse these to optimize the *RMP*.

- In the initial phase, the subgradient algorithm for solving the *LDP* (6.15) does not suffer from degeneracy difficulties encountered in the standard column generation. Moreover, solving the Lagrangian subproblem *ISP* may be *on average* faster than solving it in a Dantzig-Wolfe reformulation. This is the case for some *VRPTW* instances, where initializing $\boldsymbol{\pi}_\mathbf{b}$ to $\mathbf{0}$ and slowly increasing its values in the subgradient algorithm keeps the adjusted costs at positive values and prevents the presence of cycles within the generated paths as opposed to the big-*M* initial values in the column generation algorithm which favors these cycles, hence largely increasing the number of labels in dynamic programming algorithms (Kohl, 1995). (This was indeed the first incentive to start the research at the GERAD center on controlling the values of the dual variables.)
- In the final phase, construct a relevant dual box-constrained or stabilized master problem, and proceed to find both primal and dual solutions to the *MP* (6.3).

Dual estimates

In very special cases like Illustration 6.9 on the *cutting stock problem*, we may have perfect dual information available, but this is unlikely in general. Next best, stabilized column generation makes successively improving guesses of *trustworthy* dual values. Regardless of which column generation variant we use, starting with good dual information helps generating good columns early, see Illustrations 6.10 and 6.11. It is therefore natural to ask for *dual heuristics* that produce good dual estimates.

The literature here is clearly insufficient. One option is to solve the linear relaxation of the *ILP*, or that of an approximation as seen for the *MDVSP* in Illustration 6.12. However, problem specific methods are probably more effective. Here is a suggestion for vehicle routing problems, given a feasible integer solution. An estimate of π_i for customer i in a set partitioning *MP* formulation is calculated by removing that customer from the set of routes and then calculating the savings of such a removal. A complementary estimate considers visiting this customer once more, this time computing how much it costs to insert it in an existing route, if possible, and otherwise the price of a new vehicle. These two estimates provide a dual box.

Finally, we should be aware of dual estimates that come as a side-effect. For example, when we have artificial or slack and surplus variables in the *MP*, their cost coefficients induce bounds on dual variables and should thus be chosen wisely, see (6.61)–(6.63). If applicable in the respective context, the penalty of an artificial variable should approximate the dual value in the corresponding primal constraint (as in the *MDVSP* example before). We can turn this knowledge into an advantage by replacing the usually large big-M penalties by better estimates. In any case, a smaller penalty for artificial variables usually tighter constrains the dual space.

Anticipation

The dual point of view brings to light alternative algorithms to compute dual solutions to the *MP*. But this co-existence of algorithms is not the end; actually, as we may see different dual values from different methods, a combination of approaches seems natural. Such a hybrid idea may be useful in a Dantzig-Wolfe reformulation when the *RMP* is difficult or time consuming to solve relatively to the *ISP*. One can experiment with obtaining a new dual solution from a few iterations of the subgradient algorithm instead of re-optimizing a degenerate *RMP*. When we have several pricing problems, we may modify the dual values slightly in between calling two different pricing problems, so as to increase the chance that more diverse columns are generated.

Consider for example a vehicle routing problem with 6 customers and assume that path p found by the *ISP* visits customers 1, 4, and 5. In a subgradient step with direction

$$\mathbf{b} - \mathbf{a}_p = \begin{bmatrix}1\\1\\1\\1\\1\\1\end{bmatrix} - \begin{bmatrix}1\\0\\0\\1\\1\\0\end{bmatrix} = \begin{bmatrix}0\\1\\1\\0\\0\\1\end{bmatrix},$$

we modify the incentives (dual values) to visit customers in the next iteration by (6.48). The new dual values are unchanged for customers 1, 4, and 5 whereas they are increased for customers 2, 3, and 6. In a Dantzig-Wolfe reformulation, it means that we can anticipate without solving the *MP* which customers would be favored in the next pricing iteration. This can be simulated in the *ISP* by simply removing from the network customers 1, 4, and 5 and reusing the same dual values.

More on such a *complementary column generation* strategy for set partitioning formulations can be found in Ghoniem and Sherali (2009).

6.5 Examples

In Examples 6.1 (Generalized assignment problem), 6.2 (Capacitated p-median problem), and 6.3 (Various Lagrangian subproblems), we convince ourselves that decomposition choices have the same theoretical foundations in a Lagrangian relaxation and in a Dantzig-Wolfe reformulation.

In Examples 6.4 (Symmetric traveling salesperson problem) and 6.5 (Balancing printed circuit board assembly line systems) we illustrate that, in some applications, integer solutions can be recovered rather easily from the linear relaxation such that we may be quite near integer optimality without the need for a master problem.

Finally, Example 6.6 (Hybrid algorithm for a 2-dimensional problem) presents a hybrid algorithm that combines Lagrangian relaxation information with a Dantzig-Wolfe master problem.

Example 6.1 ***Generalized assignment problem***

✎ A classical example where we compare the impact of two groupings of the constraints of the compact formulation.

Recall that the *generalized assignment problem* (*GAP*) seeks the assignment at minimum cost of n tasks to a set K of machines such that each task is assigned exactly once, subject to capacity restrictions on the machines. The compact integer linear programming formulation seen in Example 4.5 is

$$\begin{aligned} z^\star_{ILP} = \min \quad & \sum_{k\in K}\sum_{i=1}^{n} c_i^k x_i^k \\ \text{s.t.} \quad & \sum_{k\in K} x_i^k = 1 && \forall i \in \{1,\ldots,n\} \\ & \sum_{i=1}^{n} b_i^k x_i^k \le b^k && \forall k \in K \\ & x_i^k \in \{0,1\} && \forall k \in K,\ i \in \{1,\ldots,n\}, \end{aligned} \tag{6.84}$$

where c_i^k is the cost of assigning task i to machine $k \in K$, b_i^k is the capacity used when task i is assigned to machine k, and b^k is the capacity of machine k. The binary variable x_i^k takes value 1 if and only if task i is assigned to machine k. Let us examine a Lagrangian relaxation of either the semi-assignment constraints or the capacity restrictions on the machines. Feel free to compare these with both Dantzig-Wolfe reformulations we have previously derived.

Relaxation of the semi-assignment constraints

Consider now the option where the n semi-assignment constraints are relaxed in the objective function with the vector of Lagrangian multipliers $\boldsymbol{\pi}$, where $\pi_i \in \mathbb{R}$,

$\forall i \in \{1,\ldots,n\}$. The new objective function writes as

$$\sum_{k\in K}\sum_{i=1}^{n} c_i^k x_i^k + \sum_{i=1}^{n} \pi_i (1 - \sum_{k\in K} x_i^k) \;=\; \sum_{i=1}^{n} \pi_i + \sum_{k\in K}\sum_{i=1}^{n} (c_i^k - \pi_i) x_i^k.$$

The Lagrangian subproblem becomes

$$\begin{aligned} LR(\boldsymbol{\pi}) = \sum_{i=1}^{n} \pi_i + \sum_{k\in K} \min\ & \sum_{i=1}^{n} (c_i^k - \pi_i) x_i^k \\ \text{s.t.}\ & \sum_{i=1}^{n} b_i^k x_i^k \le b^k \qquad \forall k \in K \\ & x_i^k \in \{0,1\} \quad \forall k \in K,\ i \in \{1,\ldots,n\}. \end{aligned} \tag{6.85}$$

It has a block-diagonal structure separable per machine $k \in K$, where the kth optimization problem turns out to be a binary knapsack problem. Figure 6.22 depicts a network representation of the information we need for such a pricing problem, namely adjusted costs and capacity consumption with respect to machine k. This formulation does not possess the integrality property and the best lower bound can be better than that of the linear relaxation of (6.84), that is, $z^\star_{LP} \le z^\star_{LDP} \le z^\star_{ILP}$.

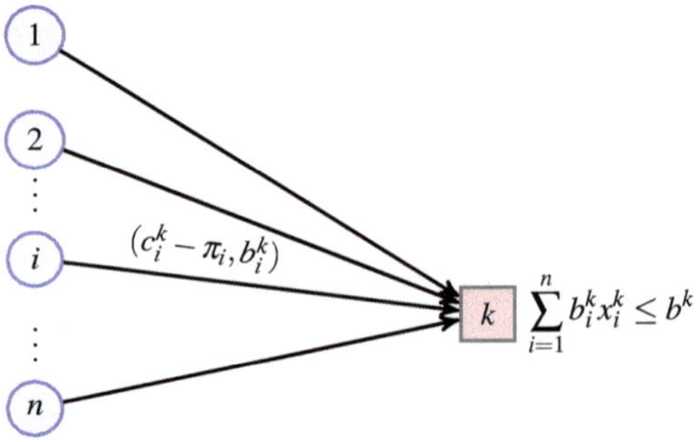

Fig. 6.22: Lagrangian relaxation of the semi-assignment constraints.

Relaxation of the capacity restrictions on the machines

Let $\boldsymbol{\omega}$ be a $|K|$-dimensional vector of Lagrangian multipliers associated with the relaxed capacity constraints on the machines, where $\omega^k \le 0$, $\forall k \in K$. The new objective function writes as

$$\sum_{k\in K}\sum_{i=1}^{n} c_i^k x_i^k + \sum_{k\in K} \omega^k (b^k - \sum_{i=1}^{n} b_i^k x_i^k) \;=\; \sum_{k\in K} \omega^k b^k + \sum_{i=1}^{n}\sum_{k\in K} (c_i^k - \omega^k b_i^k) x_i^k. \tag{6.86}$$

Reversing the summation order, the Lagrangian subproblem derived from (6.84) becomes

$$LR(\boldsymbol{\omega}) = \sum_{k\in K} \omega^k b^k + \sum_{i=1}^{n} \min \sum_{k\in K} (c_i^k - \omega^k b_i^k) x_i^k$$
$$\text{s.t.} \sum_{k\in K} x_i^k = 1 \qquad \forall i \in \{1,\ldots,n\} \tag{6.87}$$
$$x_i^k \in \{0,1\} \qquad \forall i \in \{1,\ldots,n\},\ k \in K.$$

The Lagrangian subproblem (6.87) is easily solvable by hand as it is separable per item i. Once again, we depict the required information for the pricing problem in Figure 6.23 using a network representation. For each item, select machine $k \in K$ with the least assignment cost, without considering the capacity of the machines, i.e., $x_i^k = 1$ for the selected machine k:

$$LR(\boldsymbol{\omega}) = \sum_{k\in K} \omega^k b^k + \sum_{i=1}^{n} \min_{k\in K}(c_i^k - \omega^k b_i^k). \tag{6.88}$$

The formulation of the *ISP* (6.87) possesses the integrality property such that the best lower bound is equal to that of the linear relaxation of (6.84), that is, $z_{LP}^\star = z_{LDP}^\star \leq z_{ILP}^\star$.

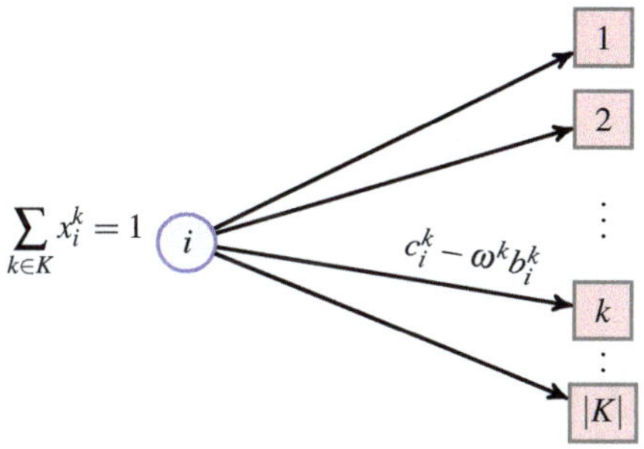

Fig. 6.23: Lagrangian relaxation of the capacity restrictions on the machines.

Example 6.2 **Capacitated *p*-median problem**

✎ Another classic problem for which we present both the Lagrangian and Dantzig-Wolfe approaches.

The *p-median problem* consists in partitioning a set of nodes $N = \{1,\ldots,n\}$ into p clusters and choosing, amongst a predefined subset $K \subseteq N$, a *median node* in each one while minimizing the sum of the distances between each node and the median of its cluster (Figure 6.24). If the medians provide a service and each node representing a customer requires that service, we introduce capacity constraints, that is, the total demand of a cluster must not exceed the capacity of its median node. This yields the *capacitated p-median problem* (*CpMP*), an important topic within distribution logistics and supply chain applications. As a generalization of the p-median problem, the *CpMP* is $\mathcal{NP}$-hard (Kariv and Hakimi, 1979).

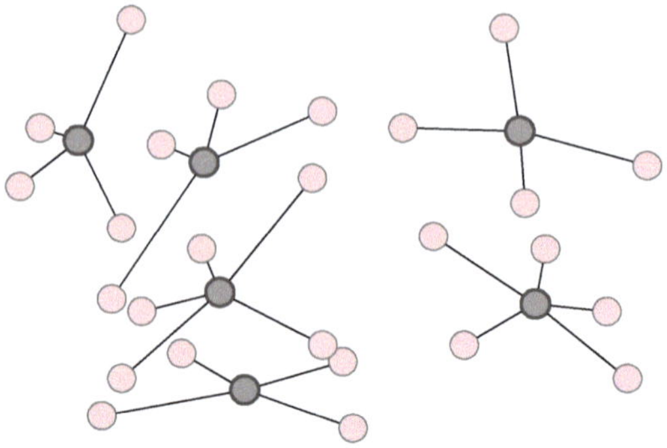

Fig. 6.24: Partition of 32 nodes into 6 clusters.

The presentation of the Lagrangian relaxation and Dantzig-Wolfe reformulation is inspired by Ceselli and Righini (2005). The notation is

N : set of customers to be served; d_i, integer demand of customer $i \in N$;
$K \subseteq N$: set of potential medians; D^k, integer capacity of median $k \in K$;
y^k, binary variable taking value 1 if median k is selected, 0 otherwise;
$c_i^k \geq 0$: cost for customer i when served from median k, with $c_k^k = 0, \forall k$;
x_i^k, binary variable taking value 1 if i is served from k, 0 otherwise.

$$
\begin{aligned}
z_{ILP}^{\star} = \min \quad & \sum_{k \in K} \sum_{i \in N} c_i^k x_i^k && && \text{(6.89a)} \\
\text{s.t.} \quad & \sum_{k \in K} x_i^k = 1 && \forall i \in N && \text{(6.89b)} \\
& \sum_{k \in K} y^k = p && && \text{(6.89c)} \\
& \sum_{i \in N} d_i x_i^k - D^k y^k \leq 0 && \forall k \in K && \text{(6.89d)} \\
& x_i^k \in \{0,1\} && \forall k \in K, i \in N && \text{(6.89e)} \\
& y^k \in \{0,1\} && \forall k \in K. && \text{(6.89f)}
\end{aligned}
$$

The objective function (6.89a) minimizes the sum of the assignment costs. Partitioning of the customers into p clusters appears in (6.89b)–(6.89c). The set of constraints (6.89d) imposes that the capacity of every selected median is not exceeded and also forbids the assignment of customers to unselected medians.

Lagrangian relaxation

To perform a Lagrangian relaxation of the *ILP*, we divide its constraints into the *linking* ones ($\mathcal{A}$) and the *structured* ones ($\mathcal{D}^k$, $\forall k \in K$). Let $\mathbf{x}^k = [x_i^k]_{i\in N}$ be the assignment vector for median k. The grouping we use is therefore

$$\mathcal{A} = \left\{ \left\{ \begin{bmatrix} \mathbf{x}^k \\ y^k \end{bmatrix} \in \{0,1\}^{n+1} \right\}_{k\in K} \;\middle|\; \sum_{k\in K} x_i^k = 1, \forall i \in N;\; \sum_{k\in K} y^k = p \right\} \tag{6.90a}$$

$$\mathcal{D}^k = \left\{ \begin{bmatrix} \mathbf{x}^k \\ y^k \end{bmatrix} \in \{0,1\}^{n+1} \;\middle|\; \sum_{i\in N} d_i x_i^k \le D^k y^k \right\}, \qquad \forall k \in K. \tag{6.90b}$$

Relaxing the constraints in $\mathcal{A}$ with multipliers $\boldsymbol{\pi} = [\pi_i \in \mathbb{R}]_{i\in N}$ for (6.89b) and $\gamma \in \mathbb{R}$ for (6.89c), the Lagrangian subproblem is separable per index k:

$$\begin{aligned} LR(\boldsymbol{\pi},\gamma) = \min \quad & \sum_{k\in K}\sum_{i\in N} c_i^k x_i^k + \sum_{i\in N} \pi_i \Big(1 - \sum_{k\in K} x_i^k\Big) + \gamma\Big(p - \sum_{k\in K} y^k\Big) \\ \text{s.t.} \quad & \begin{bmatrix} \mathbf{x}^k \\ y^k \end{bmatrix} \in \mathcal{D}^k, \qquad \forall k \in K \\ = \quad & \sum_{i\in N} \pi_i + \gamma p + \sum_{k\in K} \min_{\left[\begin{smallmatrix} \mathbf{x}^k \\ y^k \end{smallmatrix}\right] \in \mathcal{D}^k} \sum_{i\in N} (c_i^k - \pi_i) x_i^k - \gamma y^k. \end{aligned} \tag{6.91}$$

The ISP^k over $\mathcal{D}^k$ is a binary knapsack problem for $y^k = 1$, otherwise $\mathbf{x}^k = \mathbf{0}$. As such, it does not possess the integrality property and $z_{LP}^\star \le z_{LDP}^\star \le z_{ILP}^\star$, with strict inequalities in practice.

Dantzig-Wolfe reformulation

The same lower bound can be obtained via a Dantzig-Wolfe reformulation, that is, $z_{MP}^\star = z_{LDP}^\star$. Let $\mathcal{X}^k = \left\{ \begin{bmatrix} \mathbf{x}_p^k \\ y_p^k \end{bmatrix} \right\}_{p\in P^k}$ denote the set of extreme points of $\operatorname{conv}(\mathcal{D}^k)$, a polytope. Adapting formulation (4.42) with $c_p^k = \sum_{i\in N} c_i^k x_{ip}^k$, $\forall k \in K$, $p \in P^k$, the *IMP* writes as

$$z_{IMP}^\star = \min \sum_{k\in K}\sum_{p\in P^k} c_p^k \lambda_p^k \tag{6.92a}$$

$$\text{s.t.} \quad \sum_{k\in K}\sum_{p\in P^k} \mathbf{x}_p^k \lambda_p^k = \mathbf{1} \qquad [\boldsymbol{\pi}] \tag{6.92b}$$

$$\sum_{k\in K}\sum_{p\in P^k} y_p^k \lambda_p^k = p \qquad [\gamma] \tag{6.92c}$$

$$\sum_{p\in P^k} \lambda_p^k = 1 \qquad [\pi_0^k] \quad \forall k \in K \tag{6.92d}$$

$$\lambda_p^k \geq 0 \qquad \forall k \in K, p \in P^k \tag{6.92e}$$

$$\sum_{p\in P^k} \begin{bmatrix}\mathbf{x}_p^k \\ y_p^k\end{bmatrix} \lambda_p^k = \begin{bmatrix}\mathbf{x}^k \\ y^k\end{bmatrix} \in \{0,1\}^{n+1} \qquad \forall k \in K. \tag{6.92f}$$

For all $k \in K$, the following pricing problem $\bar{c}^k(\boldsymbol{\pi}, \gamma, \pi_0^k)$ is obviously equivalent to $\min_{\left[\begin{smallmatrix}\mathbf{x}^k \\ y^k\end{smallmatrix}\right]\in\mathcal{D}^k} \sum_{i\in N}(c_i^k - \pi_i)x_i^k - \gamma y^k$ in (6.91), that is, adapted from (4.45),

$$\bar{c}^k(\boldsymbol{\pi}, \gamma, \pi_0^k) = -\pi_0^k + \min_{\left[\begin{smallmatrix}\mathbf{x}^k \\ y^k\end{smallmatrix}\right]\in\mathcal{D}^k} \sum_{i\in N}(c_i^k - \pi_i)x_i^k - \gamma y^k. \tag{6.93}$$

Finally, let us mention that all the equations of the *IMP* (6.92) can be transformed into inequalities: the partitioning constraints become covering ones ($\geq \mathbf{1}$) given the assumed non-negative costs; because of the existence of the extreme point $\left[\begin{smallmatrix}\mathbf{0} \\ 1\end{smallmatrix}\right]$ for every k, the associated λ-variables can be seen as slack variables and hence discarded, resulting in $\leq p$ in (6.92c) and ≤ 1 in (6.92d). This implies restricted dual values in both Lagrangian and Dantzig-Wolfe approaches: $\boldsymbol{\pi} \geq \mathbf{0}$, $\gamma \leq 0$, and $\pi_0^k \leq 0$, $\forall k \in K$.

Example 6.3 ***Various Lagrangian subproblems***

✎ On the easiness of formulating the Lagrangian subproblem.

Vehicle routing problem with time windows

Consider the compact formulation (5.1) of the *VRPTW*, for which is given a set of $|K|$ identical vehicles. As in Kohl and Madsen (1997), let us relax with multipliers $\boldsymbol{\pi}_\mathbf{b} \in \mathbb{R}^{|C|}$ the constraints (5.1b) ensuring that each customer is visited exactly once, that is, $\sum_{k\in K}\sum_{j:(i,j)\in A} x_{ij}^k = 1$, $\forall i \in C$. The Lagrangian subproblem then reads as

$$\begin{aligned} LR(\boldsymbol{\pi}_\mathbf{b}) = \min \quad & \sum_{k\in K}\sum_{(i,j)\in A} c_{ij}x_{ij}^k + \sum_{i\in C}\pi_i(1 - \sum_{k\in K}\sum_{j:(i,j)\in A} x_{ij}^k) \\ \text{s.t.} \quad & \mathbf{x}^k \in \mathcal{D}^k, \qquad \forall k \in K, \end{aligned} \tag{6.94}$$

where $\mathcal{D}^k$ denotes the set of path, capacity, and time constraints (5.1c)–(5.1g) associated with vehicle k. Because all domains are identical, we can use a single one, say $\mathcal{D}$ (5.3), and rearrange the terms:

$$LR(\boldsymbol{\pi}_\mathbf{b}) = \sum_{i\in C}\pi_i + |K|\left(\min_{\mathbf{x}\in\mathcal{D}} \sum_{(i,j)\in A} c_{ij}x_{ij} - \sum_{i\in C}\sum_{j:(i,j)\in A} \pi_i x_{ij}\right). \tag{6.95}$$

Because $\sum_{j:(o,j)\in A} x_{oj} = 1$ is ensured by design of the *od*-path in $\mathcal{D}$, set C can be extended to $C \cup \{o\}$ with dual value $\pi_o = 0$. The Lagrangian subproblem becomes

$$LR(\boldsymbol{\pi}_{\mathbf{b}}) = \sum_{i\in C} \pi_i + |K| \left(\min_{\mathbf{x}\in\mathcal{D}} \sum_{(i,j)\in A} (c_{ij} - \pi_i) x_{ij} \right). \tag{6.96}$$

In their implementation, Kohl and Madsen (1997) solve the shortest path problem with time windows and a capacity constraint (*SPPTWC*), a relaxation of the elementary version. They find optimal Lagrangian multipliers using a bundle method (see various references on p. 405).

Cutting stock problem

We first consider the formulation (4.112) for the *CSP* which uses indexed commodities in the set K, one per roll:

$$\begin{aligned} z^{\star}_{ILP_K} = \min \quad & \sum_{k\in K} x_0^k \\ \text{s.t.} \quad & \sum_{k\in K} x_i^k \geq b_i \qquad \forall i \in \{1,\dots,m\} \\ & \begin{bmatrix} x_0^k \\ \mathbf{x}^k \end{bmatrix} \in \mathcal{D}_K \quad \forall k \in K, \end{aligned} \tag{6.97}$$

where

$$\mathcal{D}_K = \left\{ \begin{bmatrix} x_0 \\ \mathbf{x} \end{bmatrix} \in \{0,1\} \times \mathbb{Z}_+^m \;\middle|\; \sum_{i=1}^m w_i x_i \leq W x_0 \right\}. \tag{6.98}$$

Let us relax with multipliers $\boldsymbol{\pi}_{\mathbf{b}} \in \mathbb{R}_+^m$ the demand constraints $\sum_{k\in K} \mathbf{x}^k \geq \mathbf{b}$. Similarly to what we have seen above for the *VRPTW* in presence of $|K|$ identical blocks, the Lagrangian subproblem reads as

$$\begin{aligned} LR(\boldsymbol{\pi}_{\mathbf{b}}) = \min \quad & \sum_{k\in K} x_0^k + \sum_{i=1}^m \pi_i (b_i - \sum_{k\in K} x_i^k) \\ \text{s.t.} \quad & \begin{bmatrix} x_0^k \\ \mathbf{x}^k \end{bmatrix} \in \mathcal{D}_K, \qquad \forall k \in K, \\ = \quad & \sum_{i=1}^m \pi_i b_i + |K| \left(\min_{\mathbf{x}\in\mathcal{D}_K} x_0 - \sum_{i=1}^m \pi_i x_i \right). \end{aligned} \tag{6.99}$$

For $x_0 = 0$, all x-variables take value 0. Otherwise, $LR(\boldsymbol{\pi}_{\mathbf{b}})$ is an integer knapsack problem that writes as

$$LR(\boldsymbol{\pi}_{\mathbf{b}})|_{x_0=1} = \sum_{i=1}^m \pi_i b_i + |K| \left(1 - \max_{\sum_{i=1}^m w_i x_i \leq W} \sum_{i=1}^m \pi_i x_i \right). \tag{6.100}$$

This Lagrangian subproblem does not possess the integrality property and the best lower bound is greater-than-or-equal to the optimal value of the linear relaxation of (6.97), that is, $\max_{\boldsymbol{\pi}_\mathbf{b}} LR(\boldsymbol{\pi}_\mathbf{b}) = z^\star_{LDP_K} \geq z^\star_{LP_K}$.

We next consider the network-based formulation (4.120):

$$\begin{aligned}
z^\star_{ILP_{NF}} = \min \quad & x_{do} && && \text{(6.101a)}\\
\text{s.t.} \quad & \sum_{\ell:(i^\ell, i^{\ell+w_i}) \in A_{do}} x_{i^\ell, i^{\ell+w_i}} \geq b_i && \forall i \in \{1,\dots,m\} && \text{(6.101b)}\\
& \sum_{j:(i,j)\in A_{do}} x_{ij} - \sum_{j:(j,i)\in A_{do}} x_{ji} = 0 && \forall i \in V && \text{(6.101c)}\\
& x_{ij} \in \mathbb{Z}_+ && \forall (i,j) \in A_{do}. && \text{(6.101d)}
\end{aligned}$$

Relaxing the demand constraints (6.101b) with non-negative multipliers $\boldsymbol{\pi}_\mathbf{b}$, the Lagrangian subproblem defined on the single block $\mathcal{D}_{NF} = \{\mathbf{x} \in \mathbb{Z}_+^{|A_{do}|} \mid \text{(6.101c)}\}$ becomes

$$\begin{aligned}
LR(\boldsymbol{\pi}_\mathbf{b}) = & \min_{\mathbf{x}\in\mathcal{D}_{NF}} x_{do} + \sum_{i=1}^{m} \pi_i \Big(b_i - \sum_{\ell:(i^\ell, i^{\ell+w_i}) \in A_{do}} x_{i^\ell, i^{\ell+w_i}}\Big)\\
= & \sum_{i=1}^{m} \pi_i b_i + \min_{\mathbf{x}\in\mathcal{D}_{NF}} x_{do} - \sum_{i=1}^{m} \sum_{\ell:(i^\ell, i^{\ell+w_i}) \in A_{do}} \pi_i x_{i^\ell, i^{\ell+w_i}}.
\end{aligned} \tag{6.102}$$

This Lagrangian subproblem is defined on a polyhedral cone whose extreme rays can a priori be scaled to one unit ($x_{do} = 1$): it then becomes an easy shortest path problem on an acyclic network (illustrated in Figure 4.27). The best lower bound is equal to the optimal objective value of the linear relaxation of (6.101), that is, $z^\star_{LDP_{NF}} = z^\star_{LP_{NF}}$.

As seen in Example 4.2 Integrality property in the cutting stock problem (see Figure 4.28), both lower bounds are equal ($z^\star_{LDP_K} = z^\star_{LDP_{NF}}$) but obtained in a different way.

Edge coloring problem

We consider two compact formulations for which we derive the Lagrangian subproblems. From Examples 2.3 and 4.10, we recall the description and notation used. The *ECP* is defined on an undirected graph $G = (N, E)$, where N denotes the set of nodes and E the set of edges. We want to color the edges in such a way that no incident edges have the same color. Let $\delta(\{i\}) \subseteq E$ be the subset of edges incident to $i \in N$, and $E(S) \subseteq E$ for $S \subseteq N$ the edges with both endpoints in S. We say that $S \subseteq N$ is an *odd set* if $|S| \geq 3$ and odd.

First compact. Vizing's theorem states that the edges of any graph with maximum degree Δ can be colored with at most $\Delta + 1$ colors (Vizing, 1964). The compact formulation (4.174) with $\kappa = \Delta + 1$ blocks is given by

$$
\begin{aligned}
z^\star_{ILP} = \min \quad & \sum_{k=1}^{\kappa} x_0^k \\
\text{s.t.} \quad & \sum_{k=1}^{\kappa} x_e^k \geq 1 && \forall e \in E \\
& \sum_{e \in \delta(\{i\})} x_e^k \leq x_0^k && \forall k \in \{1,\dots,\kappa\},\ i \in N \\
& x_0^k \in \{0,1\} && \forall k \in \{1,\dots,\kappa\} \\
& x_e^k \in \{0,1\} && \forall k \in \{1,\dots,\kappa\},\ e \in E.
\end{aligned} \tag{6.103}
$$

Let $\mathbf{x}^k = [x_e^k]_{e \in E}$, $\forall k \in \{1,\dots,\kappa\}$. Relaxing in the objective function the $|E|$ edge-covering constraints with non-negative multipliers $\boldsymbol{\pi} = [\pi_e]_{e \in E}$, we obtain

$$
\begin{aligned}
LR(\boldsymbol{\pi}) = \min \quad & \sum_{k=1}^{\kappa} x_0^k + \sum_{e \in E} \pi_e \Big(1 - \sum_{k=1}^{\kappa} x_e^k\Big) \\
\text{s.t.} \quad & \begin{bmatrix} x_0^k \\ \mathbf{x}^k \end{bmatrix} \in \mathcal{D}, \ \forall k \in \{1,\dots,\kappa\},
\end{aligned} \tag{6.104}
$$

where $\mathcal{D} = \left\{ \begin{bmatrix} x_0 \\ \mathbf{x} \end{bmatrix} \in \{0,1\}^{|E|+1} \ \middle|\ \sum_{e \in \delta(\{i\})} x_e \leq x_0, \forall i \in N \right\}$ as in (4.176). Rearranging the terms and noting the κ identical blocks, we get the Lagrangian subproblem

$$
LR(\boldsymbol{\pi}) = \sum_{e \in E} \pi_e + \kappa \left(\min_{\left[\begin{smallmatrix} x_0 \\ \mathbf{x} \end{smallmatrix}\right] \in \mathcal{D}} x_0 - \sum_{e \in E} \pi_e x_e \right). \tag{6.105}
$$

The best lower bound $z^\star_{LDP}$ is either equal to Δ or greater than Δ. In the latter case, we can easily find a set of $\Delta + 1$ colors; in the former, one way to determine the number of colors (and matchings) is to solve the set covering formulation (2.35).

Second compact. We here use the linear description (4.178) of the matching polytope (Edmonds, 1965) and recall the *ILP* (4.180):

$$
\begin{aligned}
z^\star_{ILP} = \min \quad & x_0 && && (6.106a) \\
\text{s.t.} \quad & x_e \geq 1 && \forall e \in E && (6.106b) \\
& \sum_{e \in \delta(\{i\})} x_e \leq x_0 && \forall i \in N && (6.106c) \\
& \sum_{e \in E(S)} x_e \leq \frac{1}{2}(|S| - 1) x_0 && \forall S \subseteq N : |S| \geq 3, \text{ odd} && (6.106d) \\
& x_e \leq x_0 && \forall e \in E && (6.106e) \\
& x_0, x_e \in \mathbb{Z}_+ && \forall e \in E, && (6.106f)
\end{aligned}
$$

where x_0 is a non-negative integer variable. Relaxing again the $|E|$ edge-covering constraints with non-negative multipliers $\boldsymbol{\pi} = [\pi_e]_{e \in E}$, we obtain

$$LR(\boldsymbol{\pi}) = \min_{\left[\begin{smallmatrix} x_0 \\ \mathbf{x} \end{smallmatrix}\right] \in \mathcal{D}} x_0 + \sum_{e \in E} \pi_e (1 - x_e) = \sum_{e \in E} \pi_e + \min_{\left[\begin{smallmatrix} x_0 \\ \mathbf{x} \end{smallmatrix}\right] \in \mathcal{D}} x_0 - \sum_{e \in E} \pi_e x_e, \tag{6.107}$$

where

$$\mathcal{D} = \left\{ \begin{bmatrix} x_0 \\ \mathbf{x} \end{bmatrix} \in \mathbb{Z}_+^{|E|+1} \;\middle|\; \text{(6.106c)–(6.106e)} \right\}. \tag{6.108}$$

In that case, $\operatorname{conv}(\mathcal{D})$ is a polyhedral cone with the integrality property. For any $\boldsymbol{\pi} \geq \mathbf{0}$, $(x_0 = 0, \mathbf{x} = \mathbf{0})$ is feasible for the minimization problem over $\mathcal{D}$ with objective value zero; the minimum is therefore less-than-or-equal to zero. If negative, it comes from an optimal extreme ray with objective value $-\infty$ which also transfers to $LR(\boldsymbol{\pi})$. Otherwise we reach the optimal lower bound $z^\star_{LDP} = \sum_{e \in E} \pi^\star_e$, the one given by the linear relaxation of the set covering model (4.180). This might be quite difficult to obtain with the subgradient algorithm that is stopped after a given number of iterations.

Can we find finite lower bounds as in (6.105)? The answer is positive but requires two steps. In the first, we apply a Dantzig-Wolfe reformulation using the grouping $\mathcal{D}$ (6.108) and $\mathcal{A} = \left\{ \begin{bmatrix} x_0 \\ \mathbf{x} \end{bmatrix} \in \mathbb{Z}_+^{|E|+1} \mid x_e \geq 1, \forall e \in E \right\}$. The *IMP* is the set covering model (2.35), where a column comes from an extreme ray solution of $\operatorname{conv}(\mathcal{D})$ scaled to one unit, i.e., with $x_0 = 1$. At that point, no finite lower bound is available. In the second step, we turn ourselves to the column generation algorithm and the way to derive a lower bound on $z^\star_{MP}$. Proposition 2.1 used with a single block together with the condition $\sum_{r \in R} \lambda_r \leq \kappa$ on the number of colors results in

$$\sum_{e \in E} \pi_e + \kappa \left(\min_{\left[\begin{smallmatrix} x_0 = 1 \\ \mathbf{x} \end{smallmatrix}\right] \in \mathcal{D}} 1 - \sum_{e \in E} \pi_e x_e \right) \leq z^\star_{MP}, \tag{6.109}$$

with x_0 a priori fixed to 1 in $\mathcal{D}$ (6.108). For any non-negative $\boldsymbol{\pi}$, this is the lower bound (6.105) derived from the first compact formulation.

Example 6.4 Symmetric traveling salesperson problem

✎ Lagrangian relaxation and Dantzig-Wolfe reformulation in hiding.

The *traveling salesperson problem* (*TSP*) is a cornerstone combinatorial problem in which one must visit in a tour a set $N = \{1, \ldots, n\}$ of nodes (or cities) exactly once at minimal cost. In the *asymmetric* variant (*ATSP*), we are given a directed network $G = (N, A)$, where $A = \{(i, j) \mid i \neq j \in N\}$ denotes the set of *arcs*. In the *symmetric* variant (*STSP*), we are given an undirected network $G = (N, E)$, where $E = \{(i, j) \mid i < j \in N\}$ denotes the set of *edges*. The cost c_{ij} is defined either on A or E as needed.

Formulations

We consider first the Dantzig-Fulkerson-Johnson formulation (Dantzig et al., 1954) for the asymmetric case and refer the reader to Öncan et al. (2009) for an analysis and comparison of 24 different models. Let x_{ij} be a binary variable taking value 1 if arc $(i, j) \in A$ is taken in the tour and 0 otherwise. Assuming $n \geq 3$, we have

$$
\begin{aligned}
z^\star_{ATSP} = \min \quad & \sum_{(i,j)\in A} c_{ij}x_{ij} && && (6.110a)\\
\text{s.t.} \quad & \sum_{j:(k,j)\in A} x_{kj} = 1 && \forall k \in N && (6.110b)\\
& \sum_{i:(i,k)\in A} x_{ik} = 1 && \forall k \in N && (6.110c)\\
& x_{ij} \in \{0,1\} && \forall (i,j) \in A && (6.110d)\\
& \sum_{(i,j)\in A: i,j\in S} x_{ij} \leq |S| - 1 && \forall S \subset N : |S| \geq 2. && (6.110e)
\end{aligned}
$$

The assignment constraints (6.110b)–(6.110c) ensure that each node $k \in N$ is visited exactly once while (6.110e) are the $O(2^n)$ subtour elimination constraints (*SEC*).

In the *symmetric* variant, the cost between two nodes is the same in each direction. Let the *degree* of a node be the number of edges incident to it and x_{ij} a binary variable taking value 1 if edge $(i, j) \in E$ is taken, 0 otherwise. Assuming $n \geq 4$, the formulation includes the two-degree constraints (6.111b) in an undirected tour as well as the *SEC* (6.111d):

$$
\begin{aligned}
z^\star_{STSP} = \min \quad & \sum_{(i,j)\in E} c_{ij}x_{ij} && && (6.111a)\\
\text{s.t.} \quad & \sum_{i:(i,k)\in E} x_{ik} + \sum_{j:(k,j)\in E} x_{kj} = 2 && \forall k \in N && (6.111b)\\
& x_{ij} \in \{0,1\} && \forall (i,j) \in E && (6.111c)\\
& \sum_{(i,j)\in E: i,j\in S} x_{ij} \leq |S| - 1 && \forall S \subset N : |S| \geq 3. && (6.111d)
\end{aligned}
$$

Aside from dividing the number of variables by two, this formulation also drastically reduces solution symmetry because it naturally imposes the $n(n-1)/2$ *SEC* of cardinality 2, i.e., any undirected tour in (6.111) has two equivalent directed representations in (6.110). Figure 6.25 illustrates this statement with a significantly better lower bound (3800 vs. 3418) for the edge formulation while relaxing the *SEC* in both (6.110) and (6.111). Observe also the number of subtours of cardinality 2 in the arc-flow formulation. An edge one is therefore to be preferred when dealing with a symmetric *TSP*.

Note 6.23 (Redundant SEC.) Wolsey (1998) shows that the *SEC* on $\bar{S} = N \setminus S$ are redundant such that we can reduce (6.111d) by half as

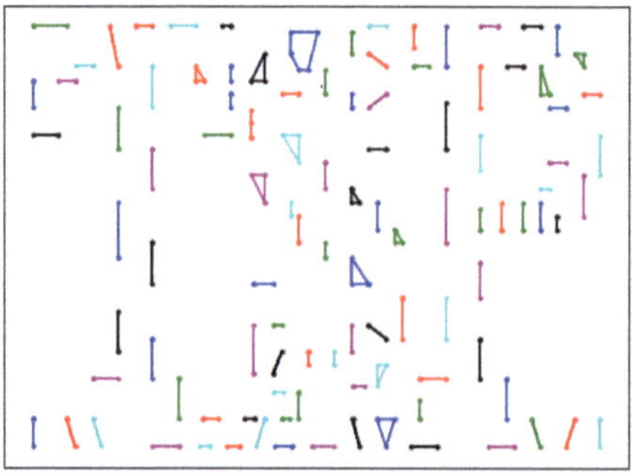

(a) Arc-flow: cost 3418, 105 subtours

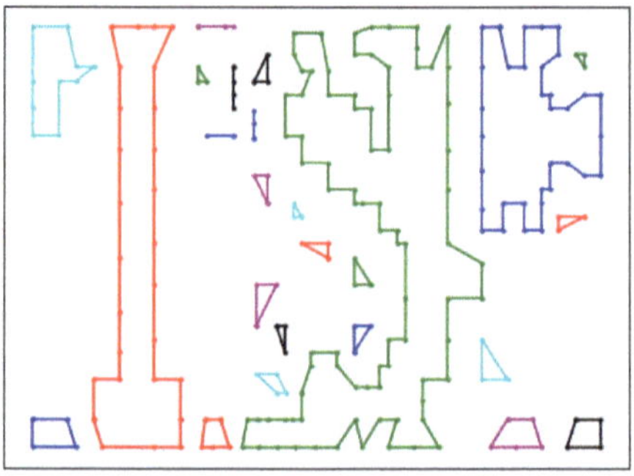

(b) Edge-flow: cost 3800, 25 subtours

Fig. 6.25: Lower bounds using arc-flow and edge-flow formulations without *SEC* for the symmetric instance `tsp225`.

$$\sum_{(i,j)\in E:i,j\in S} x_{ij} \le |S|-1 \qquad \forall S \subset N : 3 \le |S| \le \lfloor n/2 \rfloor.$$

In any solution (fractional or integer), the sum of the two-degree constraints over the elements of S, split by end point sets, writes as

$$\begin{aligned} 2|S| &= \sum_{k\in S}\left(\sum_{i:(i,k)\in E} x_{ik} + \sum_{j:(k,j)\in E} x_{kj}\right) \\ &= \sum_{k\in S}\left(\sum_{i\in S:(i,k)\in E} x_{ik} + \sum_{i\in \bar{S}:(i,k)\in E} x_{ik} + \sum_{j\in S:(k,j)\in E} x_{kj} + \sum_{j\in \bar{S}:(k,j)\in E} x_{kj}\right). \end{aligned} \tag{6.112}$$

Similarly, we have

$$2\sum_{(i,j)\in E:i,j\in S} x_{ij} = \sum_{k\in S}\left(\sum_{i\in S:(i,k)\in E} x_{ik} + \sum_{j\in S:(k,j)\in E} x_{kj}\right).$$

Taking the difference and interchanging the roles of S and $\bar{S}$ in the split sum yields

$$\begin{aligned} 2|S| - 2\sum_{(i,j)\in E:i,j\in S} x_{ij} &= \sum_{k\in S}\left(\sum_{i\in \bar{S}:(i,k)\in E} x_{ik} + \sum_{j\in \bar{S}:(k,j)\in E} x_{kj}\right) \\ &= \sum_{k\in \bar{S}}\left(\sum_{i\in S:(i,k)\in E} x_{ik} + \sum_{j\in S:(k,j)\in E} x_{kj}\right) \\ &= 2|\bar{S}| - 2\sum_{(i,j)\in E:i,j\in \bar{S}} x_{ij}. \end{aligned} \tag{6.113}$$

Since $\sum_{(i,j)\in E:i,j\in S} x_{ij} \leq |S|-1 \Leftrightarrow |S|-\sum_{(i,j)\in E:i,j\in S} x_{ij} \geq 1$, we indeed have

$$|S| - \sum_{(i,j)\in E:i,j\in S} x_{ij} \geq 1 \;\Leftrightarrow\; |\bar{S}| - \sum_{(i,j)\in E:i,j\in \bar{S}} x_{ij} \geq 1.$$

While halving the number of constraints sounds impressive, remember that 2^{n-1} does not change the order of magnitude. The floor function intervenes if n is odd in which case $|S| = (n-1)/2$ implies $|\bar{S}| = (n+1)/2$. Note that the reduction also applies to the *ATSP*.

With exponentially many *SEC*, formulation (6.111) is never solved heads on even for medium-sized instances. Interestingly, we observe in practice that we only need "few" of them to reach a strong lower bound on integer optimality but we typically close the gap using a variety of alternative cut types, see Section 6.6 (Reference Notes).

According to Applegate et al. (2003), branch-and-cut is the only exact method tractable for instances larger than 100 vertices. At the time of writing, their implementation called `Concorde` is deserving its name since it is the fastest *TSP* solver in existence (math.uwaterloo.ca/tsp/concorde). It has been used to obtain the optimal solutions to the full set of 110 TSPLIB instances, the largest having 85 900 cities.

Note 6.24 (BC and AD.) A historical record of the pioneering work of Dantzig et al. (1954) can still be found online (math.uwaterloo.ca/tsp/methods/dfj). Interestingly, we read there that the authors use a computer to solve a linear relaxation and then manually identify violated cuts for integer solutions. The repercussions of this groundbreaking idea extend nowadays far beyond the confines of the *TSP*. Laporte (2006) reminds us of this era with the idioms *Before Computers* and *After Dantzig* in A Short History of the Traveling Salesman Problem.

Held and Karp lower bounds

Picking up on the symmetric *TSP* solved by branch-and-cut using the formulation with *SEC* (6.111), we take a look at Held and Karp (1970, 1971) which derive a sharp lower bound on the optimum. The method exploits basic graph theory properties, so let us recall some definitions.

Given an undirected graph $G = (N, E)$, a *tree* T is a subgraph of G in which any two nodes are connected by exactly one path. T is connected and acyclic. It becomes disconnected if any edge is removed whereas a (simple) cycle is formed if any edge is added. If T has n nodes, then it contains $n-1$ edges. A *spanning tree* is a tree covering all nodes of N. Given weighted edges, a *minimum spanning tree* is a spanning tree with minimal cost.

A *1-tree* is a slight variant of a spanning tree on G which allows a single cycle at a given node of N, say η. Let $N_\eta = N \setminus \{\eta\}$ and $E_\eta \subset E$ denote the set of edges where neither endpoints are η. A 1-tree $\mathbf{T}_\eta$ is obtained by building a spanning tree (formed of $n-2$ edges) in the subgraph $G_\eta = (N_\eta, E_\eta)$ and then adjoining any two edges incident to η. It therefore contains n edges of E. Given weighted edges, a *minimum 1-tree* is a 1-tree with minimal cost. Since any *TSP* tour is a 1-tree in which each node has degree 2, the following proposition gives us access to an infinite amount of lower bounds.

Proposition 6.15. *(Held and Karp, 1970, §1) The cost of a minimum 1-tree is a lower bound on the optimal objective value $z^\star_{STSP}$ of the TSP. This bound is tight if such a minimum 1-tree is a tour, i.e., we have an optimal TSP tour.*

Proof. For any $\boldsymbol{\pi} = [\pi_k \in \mathbb{R}]_{k\in N}$, the transformation on edge cost $C_{ij} = c_{ij} - \pi_i - \pi_j$, $\forall (i,j) \in E$, is invariant for the *TSP* because all tours have their cost modified by $-2\sum_{k\in N} \pi_k$. It does however potentially modify the minimum 1-tree we can extract from G. With respect to the transformed costs, the value of an optimal tour is then greater-than-or-equal to the cost of a minimum 1-tree, that is,

$$z^\star_{STSP} - 2\sum_{k\in N} \pi_k \geq \min_{\mathbf{T}_\eta} \sum_{(i,j)\in\mathbf{T}_\eta} C_{ij} = \min_{\mathbf{T}_\eta} \sum_{(i,j)\in\mathbf{T}_\eta} c_{ij} - \sum_{k\in N} \pi_k d_{k\mathbf{T}_\eta}, \tag{6.114}$$

where $\mathbf{T}_\eta$ indexes the minimum 1-trees with respect to an arbitrary node say $\eta \in N$, $\sum_{(i,j)\in\mathbf{T}_\eta} C_{ij}$ is the transformed cost of $\mathbf{T}_\eta$, and $d_{k\mathbf{T}_\eta}$ is the degree of node k in said 1-tree. Reorganizing the terms, we have

$$\min_{\mathbf{T}_\eta} \sum_{(i,j)\in\mathbf{T}_\eta} c_{ij} + \sum_{k\in N} \pi_k (2 - d_{k\mathbf{T}_\eta}) \leq z^\star_{STSP}. \tag{6.115}$$

If $\mathbf{T}_\eta$ is a tour, then $d_{k\mathbf{T}_\eta} = 2$, $\forall k \in N$, such that the bound is tight. □

Let the *Held-and-Karp function* be defined as

$$HK : \mathbb{R}^{|N|} \mapsto \mathbb{R}, \quad \boldsymbol{\pi} \mapsto \min_{\mathbf{T}_\eta} \sum_{(i,j)\in\mathbf{T}_\eta} c_{ij} + \sum_{k\in N} \pi_k (2 - d_{k\mathbf{T}_\eta}). \tag{6.116}$$

Compared to (6.14)–(6.15), the similarities with the Lagrangian function are striking. We want to maximize $HK(\boldsymbol{\pi})$ whose every point is a lower bound on the optimal objective value $z^\star_{STSP}$, that is,

$$z^\star_{HK} = \max_{\boldsymbol{\pi}\in\mathbb{R}^n} HK(\boldsymbol{\pi}) = \max_{\boldsymbol{\pi}\in\mathbb{R}^n} \left\{ \min_{\mathbf{T}_\eta} \sum_{(i,j)\in\mathbf{T}_\eta} c_{ij} + \sum_{k\in N} \pi_k (2 - d_{k\mathbf{T}_\eta}) \right\}. \tag{6.117}$$

We show next that (6.117) can indeed be derived from a Lagrangian relaxation applied to an *ILP* with an appropriate grouping of constraints. Moreover, the combinatorial subproblem we solve in $\boldsymbol{\pi}$ can be solved *efficiently*. There are indeed a lot of algorithmic options nowadays to solve the minimum spanning tree problem (*MSTP*). One of the very accessible ones is the Jarník-Prim-Dijkstra algorithm developed concurrently by Jarník (1930); Prim (1957); Dijkstra (1959). Kruskal (1956)'s algorithm is another popular choice. Interestingly, Bader and Cong (2006) show that the most ancient suggestion by Borůvka (1926) lends itself particularly well to parallelization. Their algorithm therefore easily solves the minimum spanning forest problem that consists of several disjoint spanning trees.

Lagrangian lower bounds

Assuming symmetric cost, the following is an integer linear programming formulation for the *MSTP*. Let it take place on subgraph $G_\eta = (N_\eta, E_\eta)$ for some arbitrary node $\eta \in N$ such that we have $|N_\eta| = n-1$:

$$\begin{aligned} \min \quad & \sum_{(i,j)\in E_\eta} c_{ij}x_{ij} \\ \text{s.t.} \quad & \sum_{(i,j)\in E_\eta} x_{ij} = n-2 \\ & \sum_{(i,j)\in E_\eta : i,j\in S} x_{ij} \le |S|-1 \quad \forall S \subset N_\eta : |S| \ge 3 \\ & x_{ij} \in \{0,1\} \quad \forall (i,j) \in E_\eta, \end{aligned} \tag{6.118}$$

which we can show possesses the integrality property (Edmonds, 1971), see also Ahuja et al. (1993, §13.8 Minimum spanning trees and linear programming). We can then derive a compact formulation for the symmetric *TSP* by encapsulating the *MSTP* into tour solutions:

$$z^\star_{ILP} = \min \sum_{(i,j)\in E_\eta} c_{ij}x_{ij} + \sum_{i:(i,\eta)\in E} c_{i\eta}x_{i\eta} + \sum_{j:(\eta,j)\in E} c_{\eta j}x_{\eta j} \tag{6.119a}$$

$$\text{s.t.} \sum_{i:(i,\eta)\in E} x_{i\eta} + \sum_{j:(\eta,j)\in E} x_{\eta j} = 2 \tag{6.119b}$$

$$\sum_{(i,j)\in E_\eta} x_{ij} = n-2 \tag{6.119c}$$

$$\sum_{(i,j)\in E_\eta : i,j\in S} x_{ij} \le |S|-1 \qquad \forall S \subset N_\eta : |S| \ge 3 \tag{6.119d}$$

$$x_{ij} \in \{0,1\} \qquad \forall (i,j) \in E \tag{6.119e}$$

$$\sum_{i:(i,k)\in E} x_{ik} + \sum_{j:(k,j)\in E} x_{kj} = 2 \qquad \forall k \in N. \tag{6.119f}$$

In the above formulation, we have

- constraints (6.119c)–(6.119d) forcing the identification of a spanning tree in subgraph $G_\eta = (N_\eta, E_\eta)$;
- constraint (6.119b) creating a 1-tree at node η out of the latter;

- constraints (6.119f) imposing the 1-tree to have degree 2 on all nodes of N (conveniently including redundant node η);
- the objective function (6.119a) ensuring that both the spanning tree and 1-tree are minimal, hence a minimum cost traveling salesperson tour.

Note 6.25 (No missing SEC.) Formulations (6.111) and (6.119) are slightly different but they are indeed equivalent with respect to integrality. That is, constraint (6.119c) is redundant and the *SEC* (6.119d) over strict subsets of N_η are sufficient. Whether or not they also reach the same objective values for their respective linear relaxations is unclear. Finally, we cannot use the halving observation in Note 6.23 because formulation (6.118) on the *MSTP* does not have the degree constraints. In fact, it also does not work in (6.119) where the degree constraints and the *SEC* are not on the same arc set.

By relaxing the degree constraints (6.119f) into the objective function with Lagrangian multipliers $\boldsymbol{\pi} = [\pi_k]_{k\in N}$, where $\pi_\eta = 0$, we get a pricing problem whose domain

$$\mathcal{D} = \{\mathbf{x} \in \{0,1\}^{|E|} \mid (6.119\text{b})\text{–}(6.119\text{d})\} \tag{6.120}$$

is the set of 1-trees. We take notice that the subproblem domain features a block-diagonal structure in N_η and η, one for identifying a spanning tree on subgrah G_η, the other for adding two edges incident to η. The Lagrangian subproblem reads as

$$\begin{aligned} LR(\boldsymbol{\pi}) &= \min_{\mathbf{x}\in\mathcal{D}} \sum_{(i,j)\in E_\eta} c_{ij}x_{ij} + \sum_{i:(i,\eta)\in E} c_{i\eta}x_{i\eta} + \sum_{j:(\eta,j)\in E} c_{\eta j}x_{\eta j} + \sum_{k\in N} \pi_k\Big(2 - \sum_{i:(i,k)\in E} x_{ik} - \sum_{j:(k,j)\in E} x_{kj}\Big) \\ &= 2\sum_{k\in N} \pi_k + \min_{\mathbf{x}\in\mathcal{D}} \sum_{(i,j)\in E} (c_{ij} - \pi_i - \pi_j)x_{ij}. \end{aligned} \tag{6.121}$$

We see here that the adjusted costs in $(i, j) \in E$ are the same as those in the transformation used in the proof of Proposition 6.15 such that Held and Karp's function is a Lagrangian relaxation, i.e., $LR(\boldsymbol{\pi}) = HK(\boldsymbol{\pi})$.

We are here reminded that the *ISP* (6.121) does not need to be solved by linear programming as long as the outputs are extreme points of $\text{conv}(\mathcal{D})$. We also take note that the *STSP* belongs to the $\mathcal{NP}$-complete class whereas the *MSTP* is solvable in polynomial time. Because the Lagrangian subproblem in (6.121) possesses the integrality property, the best lower bound $z^\star_{LDP}$ is equal to $z^\star_{LP}$, the value of the linear relaxation of the *ILP* (6.119).

We obtain an optimality certificate if, at any point, we find $\boldsymbol{\pi}$ for which the identified minimum 1-tree has degree 2. Otherwise, we terminate when the multiplier sequence cannot be modified meaningfully, i.e., at any iteration $t \geq 1$, we iterate by $\boldsymbol{\pi}_{t+1} = \boldsymbol{\pi}_t + \theta_t(\mathbf{b} - \mathbf{a}_t)$, θ_t being a step size applied to the degree vector

$$(\mathbf{b} - \mathbf{a}_t) = [(2 - d_{k,t})]_{k\in N}, \tag{6.122}$$

where $d_{k,t} = \sum_{k:(i,k)\in E} x_{ik,t} + \sum_{k:(k,j)\in E} x_{kj,t}$ is the degree at node k in the current 1-tree solution.

Figure 6.26 plots the solution process of this procedure which terminates with a lower bound of $LB = 3876.25$ eventually as good as $z^{\star}_{LDP} = z^{\star}_{LP}$. Furthermore, the closer $(\mathbf{b} - \mathbf{a}_t)$ is to $\mathbf{0}$, the easier it is to patch up the corresponding 1-tree into a tour by reconnecting the nodes according to, e.g., a depth-first sequence of the edges. We can even further post-process that tour with heuristic neighborhood operators like 2-opt, see Kindervater and Savelsbergh (1997). The objective value of such a tour then gives us *UB* in (6.52). The ordinate also shows the optimum as well as the *SEC* relaxations. Figure 6.27 compares the solution we get as upper bound to an optimal one. The integrality gap is less than 1 % whereas the optimality gap is around 5.7 %. The former gap supports the claim of a 'sharp lower bound', that is, *LB* is actually not that far off from $z^{\star}_{ILP}$ which suggests that we expect $(\mathbf{b} - \mathbf{a}_t)$ to flirt with $\mathbf{0}$ as t increases. A reasonable explanation is that the pricing problem always satisfies almost *all SEC* constraints.

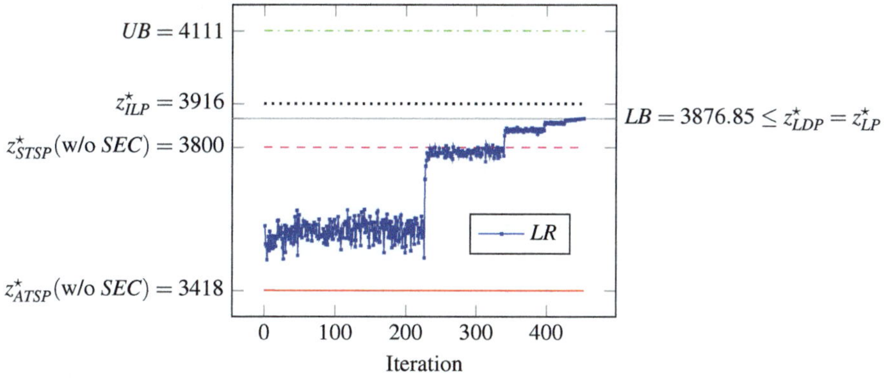

Fig. 6.26: Held and Karp lower bounds for the symmetric instance `tsp225`.

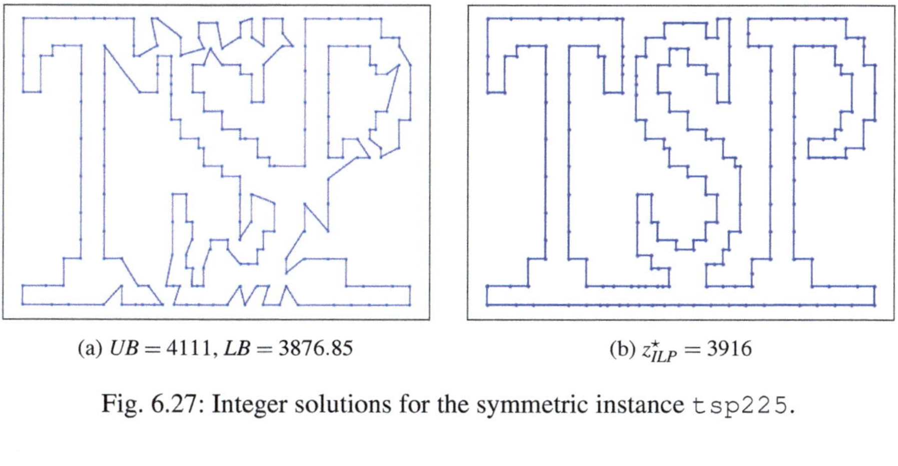

(a) $UB = 4111$, $LB = 3876.85$

(b) $z^{\star}_{ILP} = 3916$

Fig. 6.27: Integer solutions for the symmetric instance `tsp225`.

Dantzig-Wolfe reformulation

Held and Karp (1970, §3) also suggest to solve the linear relaxation of a Dantzig-Wolfe reformulation by *"A column-generation technique"*. They group the constraints in (6.119) as

$$\mathcal{A} = \{\mathbf{x} \in \{0,1\}^{|E|} \mid (6.119\text{f})\} \tag{6.123a}$$

$$\mathcal{D} = \{\mathbf{x} \in \{0,1\}^{|E|} \mid (6.119\text{b})\text{–}(6.119\text{d})\} \tag{6.123b}$$

and we know that $z^{\star}_{MP} = z^{\star}_{LP}$ by the integrality property of the *ISP* defined over the 1-tree domain $\mathcal{D}$. Indeed, the authors describe in their *linear program* (2), p. 1141, the *alternative MP* derived from

$$\begin{aligned} z^{\star}_{IMP} = \min \quad & \sum_{p\in P} c_p\lambda_p \\ \text{s.t.} \quad & \sum_{p\in P} (\mathbf{a}_p - \mathbf{b})\lambda_p = \mathbf{0} && [\boldsymbol{\pi}_{\mathbf{b}}] \\ & \sum_{p\in P} \lambda_p = 1 && [\mu] \\ & \lambda_p \in \{0,1\} && \forall p \in P, \end{aligned} \tag{6.124}$$

where $(\mathbf{a}_p - \mathbf{b}) = [(d_{kp} - 2)]_{k\in N}$, and $d_{kp} = \sum_{k:(i,k)\in E} x_{ikp} + \sum_{k:(k,j)\in E} x_{kjp}$ is the degree at node k in the extreme point $\mathbf{x}_p \in \text{conv}(\mathcal{D})$, $p \in P$ (see Figure 6.14). Because many nodes have degree 2 in such a 1-tree solution, this results in low density for the coefficient matrix.

> This column-generation procedure was programmed for the IBM/360 using standard means of avoiding error propagation: i.e., double-precision floating-point arithmetic, periodic basis reinversion, and checks for zero. The program was able to solve most problems with $n = 12$ and some problems with $13 \leq n \leq 20$. On larger problems, convergence was always too slow to permit optimal solutions to be reached. This slow convergence is consistent with the behavior of other column-generation techniques, which in practice yield good approximate solutions, rather than strictly optimum ones [Gilmore and Gomory (1963)].
>
> – Held and Karp (1970, p. 1146)

As in the `Concorde` implementation, various cuts in the x-variables can be added as needed to the linear relaxation of (6.124), always keeping the *ISP* a minimum 1-tree problem. A huge and predictable drawback is degeneracy of the basic solutions during the column generation process as the final integer solution is a single column (an optimal 1-tree). There are many questions not yet answered as never tested. Amongst these are:

- Can we benefit from a hybrid algorithm, for example, subgradient followed by a linear programming algorithm?
- Is one of the stabilization techniques a viable approach?
- Would constraints aggregation help?
- Should the dual master rather be solved?

Example 6.5 **Balancing printed circuit board assembly line systems**

✎ We here face a single block *ISP* formulation for which a feasible integer solution for the *ILP* is easily derived. We examine both the Dantzig-Wolfe and Lagrangian solution approaches.

Computerized machines are used to produce printed circuit boards (*PCB*s). We present an integer linear model inspired by Lapierre et al. (2000) for such a production on an assembly line made up of consecutive non-identical machines. This *ILP* optimizes the total cycle time while considering different assembly time values if the components are located at different slot positions on the machines. The following notation is used:

N : set of *PCB* types; p_i, proportion of *PCB*s of type $i \in N$;
J_i, set of components needed by type i; $J = \cup_{i \in N} J_i$;

K : set of machines; S^k, set of slots on machine $k \in K$; $S = \cup_{k \in K} S^k$;

t_{js} : time required to insert component j when located at slot s;

x_{js} : binary variable taking value 1 if component j is assigned to slot s, 0 otherwise;

t_i : assembly time for a *PCB* of type i, where

$$t_i = \max_{k \in K} \sum_{j \in J_i} \sum_{s \in S^k} t_{js} x_{js}, \qquad \forall i \in N. \qquad (6.125)$$

We assume that the total number of components $|J|$ is equal to $|S|$, the total number of slots on the machines; this may require dummy variables. The *ILP* minimizing the total cycle time to produce all *PCB* types writes as

$$z^\star_{ILP} = \min \quad \sum_{i \in N} p_i t_i \qquad (6.126a)$$

$$\text{s.t.} \quad t_i - \sum_{j \in J_i} \sum_{s \in S^k} t_{js} x_{js} \geq 0 \qquad \forall i \in N, k \in K. \qquad (6.126b)$$

$$\sum_{s \in S} x_{js} = 1 \qquad \forall j \in J \qquad (6.126c)$$

$$\sum_{j \in J} x_{js} = 1 \qquad \forall s \in S \qquad (6.126d)$$

$$x_{js} \in \{0,1\} \qquad \forall j \in J, s \in S. \qquad (6.126e)$$

Every constraint (6.126b) computes the assembly time for a *PCB* of type i on machine k. These constraints are the linear version of (6.125) in a minimization context. Constraints (6.126c) ensure that every component j is located on a slot s while (6.126d) prevents such a slot from having more than one component.

Dantzig-Wolfe reformulation

Consider the grouping of the constraints

$$\mathcal{A} = \{\mathbf{x} \in \{0,1\}^{|J|\times|S|} \mid (6.126b)\} \tag{6.127a}$$

$$\mathcal{D} = \{\mathbf{x} \in \{0,1\}^{|J|\times|S|} \mid (6.126c)\text{–}(6.126d)\} \tag{6.127b}$$

and observe that the time variables are not involved in these two sets, both being defined on $\mathbf{x} \in \{0,1\}^{|J|\times|S|}$. In fact, t_i, $\forall i \in N$, appear as static variables in the reformulation. Let us express $\mathbf{x}$ as a convex combination of the extreme points of $\operatorname{conv}(\mathcal{D})$.

The *IMP* becomes

$$z^\star_{IMP} = \min \sum_{i\in N} p_i t_i \tag{6.128a}$$

$$\text{s.t. } t_i - \sum_{j\in J_i}\sum_{s\in S^k} t_{js}\Big(\sum_{p\in P} x_{js,p}\lambda_p\Big) \geq 0 \quad [\pi_i^k] \quad \forall i \in N, k \in K \tag{6.128b}$$

$$\sum_{p\in P} \lambda_p = 1 \quad [\pi_0] \tag{6.128c}$$

$$\lambda_p \geq 0 \quad \forall p \in P \tag{6.128d}$$

$$\sum_{p\in P} x_{js,p}\lambda_p = x_{js} \in \{0,1\} \quad \forall j \in J, s \in S, \tag{6.128e}$$

where the dual vector $\boldsymbol{\pi} = [\pi_i^k]_{i\in N, k\in K}$ takes non-negative values whereas $\pi_0 \in \mathbb{R}$. The pricing problem writes as

$$\bar{c}(\boldsymbol{\pi}, \pi_0) = -\pi_0 + \min_{\mathbf{x}\in\mathcal{D}} \sum_{i\in N}\sum_{k\in K} \pi_i^k\Big(\sum_{j\in J_i}\sum_{s\in S^k} t_{js}x_{js}\Big). \tag{6.129}$$

The set $\mathcal{D}$ contains the constraints of an assignment problem, the *ISP* (6.129) possesses the integrality property, and $z^\star_{MP} = z^\star_{LP}$. For any $\boldsymbol{\pi} \geq \mathbf{0}$ obtained from the solution of the *RMP*, we compute a lower bound, indeed the Lagrangian dual bound given by

$$LR(\boldsymbol{\pi}) = \min_{\mathbf{x}\in\mathcal{D}} \sum_{i\in N}\sum_{k\in K} \pi_i^k\Big(\sum_{j\in J_i}\sum_{s\in S^k} t_{js}x_{js}\Big) \leq z^\star_{ILP}. \tag{6.130}$$

An interesting observation about $\boldsymbol{\pi}$ comes from the time variables. Because t_i is obviously positive, its reduced cost is zero in every solution for the *RMP*, that is,

$$\sum_{k\in K} \pi_i^k = p_i, \qquad \forall i \in N. \tag{6.131}$$

Even more interesting, for any solution $\mathbf{x}_p$, $p \in P$, to (6.129), the time variables are computed using (6.125) so as to satisfy a posteriori the time constraints (6.126b). Incidentally, this provides a feasible solution for the *ILP*, hence an upper bound *UB* at every iteration of the column generation algorithm.

Lagrangian relaxation

Lapierre et al. (2000) rather propose the Lagrangian relaxation approach, combined with the subgradient algorithm. Given $\boldsymbol{\pi} \geq \mathbf{0}$ for the relaxation of the time constraints (6.126b), the *ISP* reads as

$$\begin{aligned} LR(\boldsymbol{\pi}) &= \min_{\mathbf{x} \in \mathcal{D}} \sum_{i \in N} p_i t_i + \sum_{i \in N} \sum_{k \in K} \pi_i^k \Big(\sum_{j \in J_i} \sum_{s \in S^k} t_{js} x_{js} - t_i\Big) \\ &= \min_{\mathbf{x} \in \mathcal{D}} \sum_{i \in N} \Big(p_i - \sum_{k \in K} \pi_i^k\Big) t_i + \sum_{i \in N} \sum_{k \in K} \pi_i^k \Big(\sum_{j \in J_i} \sum_{s \in S^k} t_{js} x_{js}\Big). \end{aligned} \tag{6.132}$$

The variables t_i, $\forall i \in N$, are not restricted in any way by the assignment constraints in $\mathcal{D}$ and they are optimized independently of each other. If $(p_i - \sum_{k \in K} \pi_i^k)$ is negative, $t_i = \infty$; if positive, $t_i = -\infty$. Consequently, we must have $p_i - \sum_{k \in K} \pi_i^k = 0$, $\forall i \in N$, in any optimal solution, as already found in (6.131) from solving the *RMP*. *Under this condition*, all the time variables disappear from the Lagrangian subproblem:

$$\begin{aligned} LR(\boldsymbol{\pi}) = \min_{\mathbf{x} \in \mathcal{D}} \quad & \sum_{i \in N} \sum_{k \in K} \pi_i^k \Big(\sum_{j \in J_i} \sum_{s \in S^k} t_{js} x_{js}\Big) \\ \text{s.t.} \quad & \sum_{k \in K} \pi_i^k = p_i \qquad \forall i \in N. \end{aligned} \tag{6.133}$$

Hence, the lower bounds $LR(\boldsymbol{\pi})$ computed from (6.129) and (6.133) are the same. To complete the theoretical part, let us verify that the *LDP* is the dual of the *MP*, the linear relaxation of the *IMP* (6.128).

$$\begin{aligned} z_{LDP}^{\star} &= \max_{\boldsymbol{\pi} \geq \mathbf{0}} \left\{ \min_{p \in P} \sum_{i \in N} \sum_{k \in K} \pi_i^k \Big(\sum_{j \in J_i} \sum_{s \in S^k} t_{js} x_{js,p}\Big) \,\middle|\, \sum_{k \in K} \pi_i^k = p_i,\ \forall i \in N \right\} \\ &= \max_{\boldsymbol{\pi} \geq \mathbf{0}} \pi_0 \\ &\quad \text{s.t. } \pi_0 \leq \sum_{i \in N} \sum_{k \in K} \pi_i^k \Big(\sum_{j \in J_i} \sum_{s \in S^k} t_{js} x_{js,p}\Big) \qquad [\lambda_p \geq 0] \quad \forall p \in P \\ &\qquad\quad \sum_{k \in K} \pi_i^k = p_i \qquad [t_i \in \mathbb{R}] \quad \forall i \in N \\ &= \max_{\boldsymbol{\pi} \geq \mathbf{0}} \pi_0 \\ &\quad \text{s.t. } \pi_0 - \sum_{i \in N} \sum_{k \in K} \pi_i^k \Big(\sum_{j \in J_i} \sum_{s \in S^k} t_{js} x_{js,p}\Big) \leq 0 \qquad [\lambda_p \geq 0] \quad \forall p \in P \\ &\qquad\quad \sum_{k \in K} \pi_i^k = p_i \qquad [t_i \in \mathbb{R}] \quad \forall i \in N. \end{aligned} \tag{6.134}$$

Dualizing formulation (6.134), with $\lambda_p \geq 0, \forall p \in P$, and $t_i \in \mathbb{R}, \forall i \in N$, we recognize the objective function $\min \sum_{i \in N} p_i t_i$, the convexity constraint $\sum_{p \in P} \lambda_p = 1$ derived from π_0, and the optimality conditions on the reduced costs of the π_i^k-variables, i.e.,

$$\sum_{p \in P}(\sum_{j \in J_i}\sum_{s \in S^k} t_{js}x_{js,p})\lambda_p - t_i) \leq 0, \qquad \forall i \in N, k \in K.$$

Some computational results

The experimentation is conducted with four types of *PCB*s ($|N| = 4$) involving more than a hundred components and up to five machines ($1 \leq |K| \leq 5$). The manufacturer provided detailed information on each component, e.g., its size and which machines can handle it (Lapierre et al., 2000).

Vector $\boldsymbol{\pi}_1$ is initialized to $\pi^k_{i,1} = p_i/|K|$, $\forall i \in N$, $k \in K$, which gives equal multipliers to each machine, and the subgradient direction is computed as

$$-\mathbf{a}_t = \left[\sum_{j \in J_i}\sum_{s \in S^k} t_{js}x_{js,t}\right]_{i \in N, k \in K}$$

at iteration t. The parameter ε_t in the step-size expression (6.52), that is,

$$\theta_t = \varepsilon_t \frac{UB - LR(\boldsymbol{\pi}_t)}{||\mathbf{a}_t||^2}, \qquad t \geq 1,$$

starts at 2 and is divided by 2 when the lower bound is not improved during five iterations. The algorithm is stopped when the lower and upper bounds have not improved for 15 iterations. In Exercise 6.14, the reader is asked to compute the π-values from one iteration to the next while satisfying the normalizing constraints (6.131) and to show that they always remain positive in this implementation.

The solution approach is first tested on a 3-machine assembly line and the results appear in Figure 6.28. A cycle time solution of 58.46 hours is obtained at iteration 7 but the stopping criterion is only reached at iteration 30. The gap between the best solution and the lower bound (57.26 hours) is only 2.10 %. The computation time is 8.63 minutes on an IBM PC 286, a computer that can now only be found in a museum. Current computers would only require a few seconds.

The authors also examined various configurations and allocation of the *PCB* components to the machines. We present in Table 6.8 the results for an assembly line with 3 to 5 machines, where obviously the cycle time decreases with the added machines. They also report that *"in the worst case, the machine is busy 97.18 % of the time and, on average, the machines are busy more than 99 % of the time."*

Number of machines	cycle time (hours)	# iterations	Gap (%)	CPU time (min)
3	58.46	30	2.10	8.63
4	44.74	24	2.01	6.70
5	36.04	16	2.49	4.62

Table 6.8: Best cycle time solutions for 3 to 5 machines.

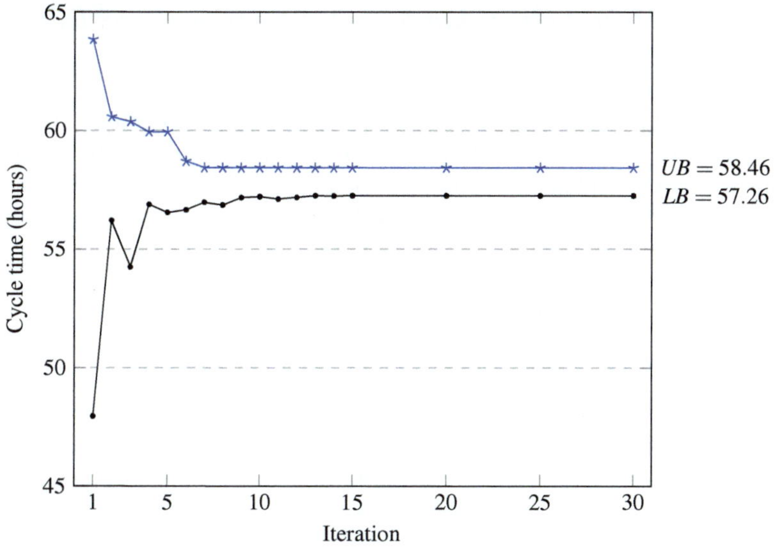

Fig. 6.28: Lower and upper bounds for a 3-machine assembly line.

A few comments

1. At optimality of the *RMP*, it is very likely that, for some combination of *PCB*-type and machine indexes, the time constraint (6.128b) is not binding, that is,

$$\exists i \in N, k \in K : \sum_{j \in J_i} \sum_{s \in S^k} t_{js}\Big(\sum_{p \in P} x_{js,p}\lambda_p\Big) < t_i, \tag{6.135}$$

   with the associated dual variable $\pi_i^k = 0$. Such a zero-value for a Lagrangian multiplier cannot occur in the implemented subgradient algorithm because they always remain positive, see the solution of Exercise 6.14.

2. At some point in the iterative process, the $\boldsymbol{\pi}$-vector should be re-initialized according to the x, t-variable values giving the best upper bound.
   For $i \in N$, let $K_i^=$ and $K_i^>$ denote the sets of machines with binding and non-binding time constraints (6.128b), respectively. Then $\pi_i^k = 0$, $\forall k \in K_i^>$, and use an equal distribution for the machines in binding ones, i.e., $\pi_i^k = \frac{p_i}{|K_i^=|}$, $\forall k \in K_i^=$.

3. Subgradient iterations are here similar to solving the *RMP* with a single column, determining the time values and an upper bound, but not retrieving the dual values. It would be simpler to solve the *RMP*; its row-dimension is less-than-or-equal to $4 \times 5 + 1$ and the column-dimension is the number of iterations.

Example 6.6 **Hybrid algorithm for a 2-dimensional problem**

✎ A sequence of subgradient iterations followed by column generation.

This example finds a primal solution in the x-variables of the compact formulation in addition to the dual values. It starts with a Lagrangian relaxation and ends with a Dantzig-Wolfe reformulation. Figure 6.29 depicts the domain of the *ILP* formulated in (6.136) with the signed dual variables within brackets for the *LP*:

$$\begin{aligned}
z^\star_{ILP} = \min \quad & x_1 + x_2 \\
\text{s.t.} \quad & -x_1 + 4x_2 \leq 8 & [\sigma_1 \leq 0] \\
& 3x_1 + 4x_2 \geq 24 & [\sigma_2 \geq 0] \\
& x_1 \leq 10 & [\sigma_3 \leq 0] \\
& x_2 \geq 2 & [\sigma_4 \geq 0] \\
& x_2 \leq 6 & [\sigma_5 \leq 0] \\
& x_1, \quad x_2 \in \mathbb{Z}_+.
\end{aligned} \tag{6.136}$$

Apart from the integrality requirements, this is the system (3.73) considered in Example 3.1. The optimal solution to the *ILP* is $(x_1^\star, x_2^\star) = (4,3)$ with $z^\star_{ILP} = z^\star_{LP} = 7$.

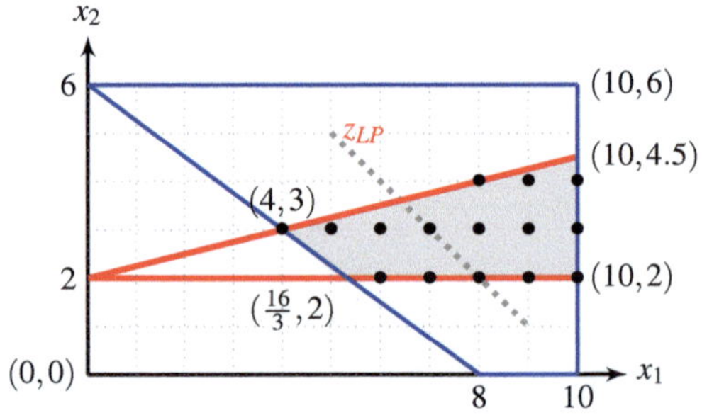

Fig. 6.29: The domain of (6.136). The objective function appears as a dotted line.

Let us relax the constraints $-x_1 + 4x_2 \leq 8$ and $x_2 \geq 2$ in the objective function:

$$x_1 + x_2 + \pi_1(8 + x_1 - 4x_2) + \pi_4(2 - x_2),$$

where $\pi_1 \leq 0$ and $\pi_4 \geq 0$ are Lagrangian multipliers. The Lagrangian subproblem reads as

$$\begin{aligned}
LR(\pi_1, \pi_4) = 8\pi_1 + 2\pi_4 + \min \quad & (1 + \pi_1)x_1 + (1 - 4\pi_1 - \pi_4)x_2 \\
\text{s.t.} \quad & (x_1, x_2) \in \mathcal{D},
\end{aligned} \tag{6.137}$$

where $\mathcal{D} = \{x_1, x_2 \in \mathbb{Z}_+ \mid 3x_1 + 4x_2 \geq 24,\ x_1 \leq 10,\ x_2 \leq 6\}$.

Observe on the right side that the *ISP* formulation (6.137) possesses the integrality property, the four extreme points of conv($\mathcal{D}$) having integer coordinates. These are the only possible optimal solutions to $LR(\pi_1, \pi_4)$, all being infeasible for the *ILP*. In that case, it is impossible to derive an upper bound on $z^\star_{ILP}$ by solving the Lagrangian subproblem.

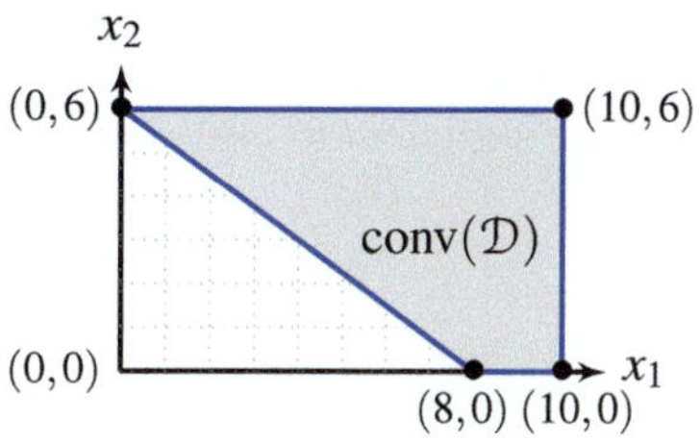

Let us start the solution process with the zero-valued Lagrangian multipliers such that the first lower bound on $z^\star_{ILP}$ is

$$LR(0,0) = \min_{\mathbf{x}\in\mathcal{D}} \; x_1 + x_2 = 6$$

obtained at $(0,6)$. We next compute the direction given by the subgradient

$$(8 + x_1 - 4x_2, 2 - x_2)\Big|_{(0,6)} = (-16, -4).$$

Using the arbitrary step size $\theta = 1/16$, we get $(\pi_1, \pi_4) = (-1, 0)$ by modifying $(0,0)$ by $1/16(-16,-4)$ while respecting the signed multipliers ($\pi_1 \leq 0$ and $\pi_4 \geq 0$). The Lagrangian subproblem (6.137) becomes

$$LR(-1,0) = -8 + \min_{\mathbf{x}\in\mathcal{D}} \; 5x_2,$$

and by inspection of the extreme points, the lower bound $LR(-1,0) = -8$ is obtained using either $(8,0)$ or $(10,0)$. With the first, the subgradient is $(16,2)$ and asks for a movement to the right. Let us use $\theta = 1/32$ on both multipliers which yields the next vector $(\pi_1, \pi_4) = (-1,0) + 1/32(16,2) = (-1/2, 1/16)$. The Lagrangian subproblem becomes

$$LR(-1/2, 1/16) = -31/8 + \min_{\mathbf{x}\in\mathcal{D}} \; 1/2 \cdot x_1 + 47/16 \cdot x_2$$

which solves to $-31/8 + 4 = 1/8$ at the extreme point $(8,0)$. The subgradient $(16,2)$ is the same as before and we again move both multipliers to the right using $\theta = 1/32$: $(\pi_1, \pi_4) = (-1/2, 1/16) + 1/32\,(16,2) = (0, 1/8)$, such that

$$LR(0, 1/8) = 1/4 + \min_{\mathbf{x}\in\mathcal{D}} \; x_1 + 7/8 \cdot x_2 = 1/4 + 42/8 = 5.5$$

is optimized at point $(0,6)$. The subgradient is $(-16,-4)$ and we reduce the step size to $\theta = 1/64$: $(\pi_1, \pi_4) = (0, 1/8) + 1/64\,(-16,-4) = (-1/4, 1/16)$, such that

$$LR(-1/4, 1/16) = -15/8 + \min_{\mathbf{x}\in\mathcal{D}} \; 3/4 \cdot x_1 + 31/16 \cdot x_2 = -15/8 + 6 = 4.125$$

is optimized at $(8,0)$. This is where we stop with this strategy, for which the sequence of lower bounds is $6, -8, 0.125, 5.5, 4.125 \leq z^\star_{LP} = 7$. We complete the solution process with the Dantzig-Wolfe decomposition. The domain of the *MP* is composed of the two relaxed constraints $-x_1+4x_2 \leq 8$ and $x_2 \geq 2$ and the convexity one. We already found the extreme points $(0,6)$ and $(8,0)$ and, in this example, these are sufficient to initialize the set of columns of the *RMP*:

$$
\begin{array}{rrrcrll}
\textit{Extreme points:} & & (0,6) & & (8,0) & & \\
z^\star_{RMP} = & \min & 6\lambda_1 & + & 8\lambda_2 & & \\
& \text{s.t.} & 24\lambda_1 & - & 8\lambda_2 & \leq 8 & [\pi_1] \\
& & 6\lambda_1 & & & \geq 2 & [\pi_4] \\
& & \lambda_1 & + & \lambda_2 & = 1 & [\pi_0] \\
& & \lambda_1, & & \lambda_2 & \geq 0, &
\end{array}
\tag{6.138}
$$

where the coefficients in the objective function and the constraints are computed by

$$
c_{\mathbf{x}} = x_1 + x_2 \qquad \text{and} \qquad \mathbf{a}_{\mathbf{x}} = \begin{bmatrix} -x_1+4x_2 \\ x_2 \\ 1 \end{bmatrix}. \tag{6.139}
$$

The *ISP* writes as

$$
\begin{aligned}
\bar{c}(\pi_1,\pi_4,\pi_0) &= \min_{\mathbf{x}\in\mathcal{D}} \; c_{\mathbf{x}} - [\pi_1,\pi_4,\pi_0]\mathbf{a}_{\mathbf{x}} \\
&= -\pi_0 + \min_{\mathbf{x}\in\mathcal{D}} \; (1+\pi_1)x_1 + (1-4\pi_1-\pi_4)x_2.
\end{aligned}
\tag{6.140}
$$

Solving the *RMP* (6.138) results in $z_{RMP} = 7$ obtained from $(\lambda_1,\lambda_2) = (0.5,0.5)$ and $(\pi_1,\pi_4,\pi_0) = (-1/16,\, 0,\, 7.5)$. For this dual vector, the *ISP* becomes

$$
\bar{c}(-1/16,\, 0,\, 7.5) = -7.5 + \min_{\mathbf{x}\in\mathcal{D}} 15/16\, x_1 + 5/4\, x_2 = 0
$$

at both extreme points $(0,6)$ and $(8,0)$. The column generation algorithm is terminated with the *MP* solved at optimality. It provides the dual values $(\pi_1,\pi_4) = (-1/16,\, 0)$ and the primal solution $(x_1,x_2) = 0.5(0,6)+0.5(8,0) = (4,3)$ that happens to be integer, hence also optimal for the *ILP*.

6.6 Reference Notes

Introduction Per Olov Lindberg invited Oli Madsen and Jacques Desrosiers to teach at the first column generation school, 1993 (at the NorFa Summer Course *i Narvik*). As Oli, P.O. was (and still is) an adept of the dual side (Google Scholar).

Section 6.1 The close relationship between the dual of the Alternative Dantzig-Wolfe Master Problem (6.6) and the Lagrangian function (6.29) is revealed in Figures 6.4b and 6.9. Such a relation is exploited by Held and Karp (1970), see Example 6.4 (Symmetric traveling salesperson problem).

Section 6.2 Lagrangian relaxation is a classical method for the optimization of integer linear programs, e.g., Geoffrion (1974), Shapiro (1979a), Nemhauser and Wolsey (1988), and Guignard (2003). Fisher (1981) surveys various solution approaches that have used one way or another the Lagrangian relaxation/subgradient approach. It includes a historical narrative and points at some practical aspects to deal with in solving the original problem. Magnanti et al. (1976) establish the equivalence between the Lagrangian dual problem (*LDP*) and Dantzig-Wolfe master problem (*MP*) both having been derived from a compact formulation.

On a similar note, Barahona and Anbil (2000) present the *volume algorithm*, a more elaborate subgradient method that is able to produce a primal solution in addition to a dual one; small infeasibilities may remain, though. In that respect, it almost matches the capabilities of the Dantzig-Wolfe master problem.

Good to Know Dual-optimal inequalities (*DOI*s) for column generation have been introduced by Valério de Carvalho (2005) while early use of dual boxes in the *Boxstep method* is due to Marsten (1975) and Marsten et al. (1975). *DOI*s are application-specific constraints which reduce the dual solution space that column generation searches over. Because they do not remove any optimal dual solutions, we are assured to terminate with a 'correct' optimal dual solution but we may eventually have to perform a primal recovery procedure to flush out any positive column associated with the *DOI*s (Valério de Carvalho, 2005; Ben Amor et al., 2006b). Additional results appear in Gschwind and Irnich (2016); Yarkony et al. (2020); Haghani et al. (2022).

More to Know The stabilized column generation algorithm follows the lines of du Merle et al. (1999) and Ben Amor et al. (2006b). Extensive computational experiments can be found in Oukil et al. (2007), Ben Amor et al. (2009), and Pessoa et al. (2013, 2018). Relations (6.82) and (6.83) are basically *statistical smoothing*, as used for example in time series forecasting, market analysis, image processing, etc., see Simonoff (1996).

For a linear program with degenerate solutions, either we provide good dual information to the stabilized column generation algorithm, or we can modify in the primal the representation of these solutions, from a basic one to a unique one by only considering the columns associated to the positive variables. Algorithms issued from that stream of research are the *dynamic constraint aggregation* for set partitioning problems (El Hallaoui et al., 2005, 2008, 2010) and the *improved primal simplex* (Metrane et al., 2010; El Hallaoui et al., 2011; Omer et al., 2015a), or more generally, Gauthier et al. (2016, 2018).

Exploiting the knowledge of integer solutions is a current area of research. The *integral simplex using decomposition* is a method that efficiently solves set partitioning problems: it iteratively moves through a sequence of integer solutions, decreasing the cost at each iteration, see the early work of Rönnberg and Larsson (2009, 2014), and more recently, Rosat et al. (2017b,a,c); Zaghrouti et al. (2018), and, in the context of column generation, Tahir et al. (2019).

Examples
6.3 Vehicle routing problem with time windows. Additional usage of Lagrangian

duality for this problem is available in Kallehauge et al. (2006). The authors use a stabilized cutting-plane algorithm for solving the Lagrangian problem. This is embedded within a branch-and-bound search, where strong valid inequalities are used at the master problem level. The result is a *Lagrangian Branch-Price-and-Cut* algorithm for the *VRPTW*.

6.4 Symmetric traveling salesperson problem. The *SEC* are facet defining for a complete graph (Lawler et al., 1985). In fact, non-negativity, upper bounds, subtours, 2-matchings, combs, clique trees, envelopes, path/star, path trees, and crowns are all facet-inducing, see the *survey of facial results for the traveling salesman polytope* by Ruland and Rodin (1998). Further readings include Nemhauser and Wolsey (1988, Part II.2.3 Valid Inequalities for the Symmetric Traveling Salesman Polytope and Part II.6.3 The Symmetric Traveling Salesman Problem).

Finally, we must once again point out that Held and Karp (1970) present within the same paper the Lagrangian relaxation/subgradient approach applied to the *STSP*, but also, much less known, the Dantzig-Wolfe reformulation/column generation counterpart.

Exercises

6.11 A crossover method for finding a basic $\mathbf{x}^{\star}_{LP}$. This is taken from Ben Amor et al. (2006a), where fixing variables to zero allows to accelerate the recovery of an optimal linear programming basis when we are only given an optimal dual solution. This incidentally occurs when solving linear programs by interior point algorithms.

Fixing arc-variables in routing and scheduling problems solved using the column generation algorithm can be found in Irnich et al. (2010).

6.15 Warehouse location problem is inspired by Ahuja et al. (1993, Exercise 16.9 Facility location), indeed derived from Erlenkotter (1978).

6.16 Dual inequalities for a two-echelon vehicle routing problem is derived from Mhamedi et al. (2022) who introduced these so-called *transfer inequalities*. This exercise highlights how they are implemented and how they can be interpreted.

Fig. 6.30: Maria Grazia Speranza and Oli Madsen (Izmir, Turkey, 2009-05-27).

Exercises

6.1 Joseph-Louis Lagrange

Is Lagrange French or Italian? Where was he born?

6.2 Nature of the *LDP*

Is the *LDP* (6.15) an integer or a linear program?

6.3 About the *AMP*

Let the *ILP* (6.1) be formulated with a bounded domain $\mathcal{D}$ in (6.2b). Assume obtained a dual vector $\boldsymbol{\pi}_{\mathbf{b}}$, optimal or not, while solving by column generation the restricted version of the alternative master problem

$$
\begin{aligned}
z^\star_{AMP} = \min \quad & \sum_{p\in P} c_p \lambda_p && && \text{(6.141a)}\\
\text{s.t.} \quad & \sum_{p\in P} (\mathbf{a}_p - \mathbf{b})\lambda_p \geq \mathbf{0} && [\boldsymbol{\pi}_{\mathbf{b}}] && \text{(6.141b)}\\
& \sum_{p\in P} \lambda_p = 1 && [\mu] && \text{(6.141c)}\\
& \lambda_p \geq 0 && \forall p \in P, && \text{(6.141d)}
\end{aligned}
$$

whose dual formulation is

$$
\begin{aligned}
z^\star_{AMP} = \max \quad & \mu \\
\text{s.t.} \quad & (\mathbf{a}_p - \mathbf{b})^\intercal \boldsymbol{\pi}_{\mathbf{b}} + \mu \leq c_p \quad [\lambda_p] \quad \forall p \in P \qquad \text{(6.142)}\\
& \boldsymbol{\pi}_{\mathbf{b}} \geq \mathbf{0},\ \mu \in \mathbb{R}.
\end{aligned}
$$

Although $\boldsymbol{\pi}_{\mathbf{b}}$ is associated with a right-side given by a $\mathbf{0}$-vector, show that it plays the same role as in the *MP* in the computation of
(a) the reduced cost of a λ_p-variable;
(b) the lower bound result: $\boldsymbol{\pi}_{\mathbf{b}}^\intercal \mathbf{b} + \min_{\mathbf{x}\in\mathcal{D}} \mathbf{c}^\intercal \mathbf{x} - \boldsymbol{\pi}_{\mathbf{b}}^\intercal \mathbf{A}\mathbf{x} \leq z^\star_{MP}$.

6.4 *AMP* formulations for a block-diagonal structure

For the *ILP* with a block-diagonal structure (6.16), provide the primal and dual formulations of the *AMP*.

6.5 Dantzig-Wolfe vs. Lagrange for a block-diagonal structure

For the *ILP* with a block-diagonal structure (6.16), prove Proposition 6.2: *Given that* $z^\star_{ILP}$ *is finite in* (6.16)*, the largest Lagrangian bound is equal to the optimal objective value of the MP, that is,* $z^\star_{LDP} = z^\star_{MP}$. Provide a proof relying on the (a) *MP*, (b) *AMP*.

6.6 *VRPTW*: Dantzig-Wolfe vs. Lagrange

It has been observed that at the beginning of the solution process of the *VRPTW* using column generation for the *MP* or subgradient for the *LDP*, *the latter is faster on average per iteration* for the solution of their respective subproblems, i.e., $\bar{c}(\boldsymbol{\pi}_{\mathbf{b}}, \pi_0)$ and $LR(\boldsymbol{\pi}_{\mathbf{b}})$. Both subproblems are solving a constrained shortest path problem with customer time windows and vehicle capacity. Explain such a behavior.

6.7 Yes or No? And why?
For each question, provide a justification for the answer. As needed, you may refer to Figure 6.6, where equivalent problems appear at the same level.

(a) Can the primal simplex algorithm solve a bounded integer linear program?
(b) Can the column generation algorithm solve an integer linear program?
(c) Can a Dantzig-Wolfe reformulation solve an integer linear program? a linear program?
(d) Can a Lagrangian relaxation solve an integer linear program? a linear program?

6.8 Ready to answer?
(a) How useful is it to only have optimal Lagrangian multipliers $\boldsymbol{\pi}_{\mathbf{b}}^{\star}$?
(b) How important are the dual values π_0^k, $k \in K$?
(c) Is a Lagrangian relaxation easier to solve than a Dantzig-Wolfe reformulation?
(d) Can we easily find a primal solution for the compact formulation in a Lagrangian relaxation?
(e) Can we leverage optimal Lagrangian multipliers $\boldsymbol{\pi}_{\mathbf{b}}^{\star}$ in a Dantzig-Wolfe reformulation?

6.9 Using optimal Lagrangian multipliers with dual boxes
Given optimal Lagrangian multipliers $\boldsymbol{\pi}_{\mathbf{b}}^{\star}$ for the *LDP*, can we easily derive a primal solution for the *MP* using the *RMP* initialized with $\mathcal{X}' = \emptyset$ and the pricing problem $\bar{c}^k(\boldsymbol{\pi}_{\mathbf{b}}^{\star}, \pi_0^k)$, where π_0^k may take any appropriate value?

6.10 Stabilization parameters: initialization and update
The MP_{stab} is constructed in such a way that we respect the properties of the $MP_{\boldsymbol{\delta}}$ whenever $\hat{\boldsymbol{\pi}} = \boldsymbol{\pi}^{\star}$. The formulation we propose in (6.77) uses the dual box (6.74) and the penalty function (6.75). How does it work out when we initialize the RMP_{stab} with yet no generated columns ($\mathcal{X}' = \emptyset$) and dual values $\hat{\boldsymbol{\pi}} = \boldsymbol{\pi}^{\star}$, i.e.,

$$\begin{aligned}
z_{RMP^0_{stab}} = \min \quad & \sum_{\mathbf{x}\in\mathcal{X}'} c_{\mathbf{x}}\lambda_{\mathbf{x}} - \boldsymbol{\delta}_1^{0\mathsf{T}}\mathbf{y}_1 + \boldsymbol{\delta}_2^{0\mathsf{T}}\mathbf{y}_2 \\
\text{s.t.} \quad & \sum_{\mathbf{x}\in\mathcal{X}'} \mathbf{a}_{\mathbf{x}}\lambda_{\mathbf{x}} - \mathbf{y}_1 + \mathbf{y}_2 = \mathbf{b} && [\boldsymbol{\pi}] \\
& \lambda_{\mathbf{x}} \geq 0 && \forall \mathbf{x} \in \mathcal{X}' \\
& \mathbf{y}_1 \leq \boldsymbol{\varepsilon}_1^0 && [-\mathbf{w}_1 \leq \mathbf{0}] \\
& \mathbf{y}_2 \leq \boldsymbol{\varepsilon}_2^0 && [-\mathbf{w}_2 \leq \mathbf{0}] \\
& \mathbf{y}_1 \geq \mathbf{0}, \quad \mathbf{y}_2 \geq \mathbf{0}.
\end{aligned}$$

(a) Discuss with respect to the parameter values $\boldsymbol{\delta}$ and $\boldsymbol{\varepsilon}$.
(b) By design, we ensure convergence by updating $\boldsymbol{\delta}$ in a way that it always reflects an optimal dual solution of the MP_{stab}. Suggest a self-adjustable rule to update $\boldsymbol{\delta}$ and $\boldsymbol{\varepsilon}$ by also considering that $\hat{\boldsymbol{\pi}}$ may not be that good of an approximation of an optimal dual solution $\boldsymbol{\pi}^{\star}$.

6.11 A crossover method for finding a basic $\mathbf{x}^\star_{LP}$
Let the vector $\boldsymbol{\delta} > \mathbf{0} \in \mathbb{R}^m$ restrict the vector $\boldsymbol{\pi}$ to lie within the dual-optimal box $[\boldsymbol{\pi}^\star - \boldsymbol{\delta}, \boldsymbol{\pi}^\star + \boldsymbol{\delta}]$ for the *LP*

$$\begin{aligned} z^\star_{LP} = \min \quad & \mathbf{c}^\intercal \mathbf{x} \\ \text{s.t.} \quad & \mathbf{A}\mathbf{x} = \mathbf{b} \quad [\boldsymbol{\pi}] \\ & \mathbf{x} \geq \mathbf{0} \end{aligned} \tag{6.143}$$

and let the *relaxed primal problem* and *restricted dual problem* be given as

$$\begin{array}{llll} \min & \mathbf{c}^\intercal \mathbf{x} - (\boldsymbol{\pi}^\star - \boldsymbol{\delta})^\intercal \mathbf{y}_1 + (\boldsymbol{\pi}^\star + \boldsymbol{\delta})^\intercal \mathbf{y}_2 & & \\ \text{s.t.} & \mathbf{A}\mathbf{x} - \quad \mathbf{y}_1 + \quad \mathbf{y}_2 = \mathbf{b} & [\boldsymbol{\pi}] & \\ & \mathbf{x} \geq \mathbf{0}, \quad \mathbf{y}_1 \geq \mathbf{0}, \quad \mathbf{y}_2 \geq \mathbf{0} & & \end{array}$$

$$\begin{array}{lrll} \max & \mathbf{b}^\intercal \boldsymbol{\pi} & & \\ \text{s.t.} & \mathbf{A}^\intercal \boldsymbol{\pi} \leq & \mathbf{c} & [\mathbf{x}] \\ & -\boldsymbol{\pi} \leq & -(\boldsymbol{\pi}^\star - \boldsymbol{\delta}) & [\mathbf{y}_1] \\ & \boldsymbol{\pi} \leq & (\boldsymbol{\pi}^\star + \boldsymbol{\delta}) & [\mathbf{y}_2]. \end{array}$$

Show that x_j can be *fixed to zero* if its reduced cost computed with respect to $\boldsymbol{\pi}^\star$ is large enough, that is, if $c_j - \sum_{i=1}^m \pi^\star_i a_{ij} > \sum_{i=1}^m \delta_i |a_{ij}|$.

6.12 Symmetric *TSP*: a compact formulation without block-index k
Given are:

- the *ISP* constraints (6.119b)–(6.119e) for the 1-tree problem;
- the two-degree constraints (6.119f) derived from the *IMP* (6.124).

Justify that, in the proposed *ILP* (6.119), there is neither the set of variables x^k_0 nor the block-index k as in Proposition 4.15.

6.13 Asymmetric and symmetric *TSP*: flow conservation constraints
Show that the flow conservation constraints are satisfied in the *ATSP* formulation (6.110) but not in the *STSP* ones, neither (6.111) nor (6.119).

6.14 Printed circuit boards: Lagrangian multipliers
In the column generation algorithm for solving the *RMP*, $\boldsymbol{\pi} \geq \mathbf{0}$ satisfies the normalizing constraints (6.131) as the result of the complementary slackness conditions. This is not guaranteed in the subgradient algorithm from one iteration to the next.

(a) Show how to compute the vector $\boldsymbol{\pi}_{t+1}$ derived from the usual expression $\boldsymbol{\pi}_{t+1} = \boldsymbol{\pi}_t + \theta_t(-\mathbf{a}_t)$, where $-\mathbf{a}_t = \left[\sum_{j \in J_i} \sum_{s \in S_k} t_{js} x_{js,t}\right]_{i \in N, k \in K}$ is the subgradient direction.
(b) For all $t \geq 1$, the subgradient direction $-\mathbf{a}_t$ only contains non-negative values and $\theta_t > 0$. Explain how some π-values may decrease.
(c) Show that the π-multipliers always remain positive in the implementation of the subgradient algorithm of Lapierre et al. (2000).

6.15 Warehouse location problem
Consider a mixed-integer linear program formulation (*MILP*) for determining the location of a number of warehouses, a variation of the Capacitated p-median problem, where a fixed cost is incurred for opening a warehouse. The notation used is:

N : set of customers to be served; d_i, demand of customer $i \in N$;
K : set of warehouses; D^k and F^k, capacity and fixed cost for warehouse $k \in K$; y^k, binary variable taking value 1 if warehouse k is selected, 0 otherwise;
$c_i^k \geq 0$: cost for customer i when served from warehouse k; x_i^k, fraction of the demand of customer i served from warehouse k.

$$z^\star_{MILP} = \min \sum_{k\in K}\sum_{i\in N} c_i^k x_i^k + \sum_{k\in K} F^k y^k \tag{6.144a}$$

$$\text{s.t.} \quad \sum_{k\in K} x_i^k = 1 \quad \forall i \in N \tag{6.144b}$$

$$\sum_{i\in N} d_i x_i^k - D^k y^k \leq 0 \quad \forall k \in K \tag{6.144c}$$

$$0 \leq x_i^k \leq 1 \quad \forall k \in K, i \in N \tag{6.144d}$$

$$y^k \in \{0,1\} \quad \forall k \in K. \tag{6.144e}$$

The first set of constraints (6.144b) states that the full demand of customer i has to be served by the warehouses. If $y^k = 1$ in (6.144c), then warehouse k is selected and its capacity D^k is available for serving the demand d_i of the assigned customers; otherwise $y^k = 0$ such that no customers can be served by this warehouse. Figure 6.31 illustrates the formulation, indeed very similar to the *GAP* structure.

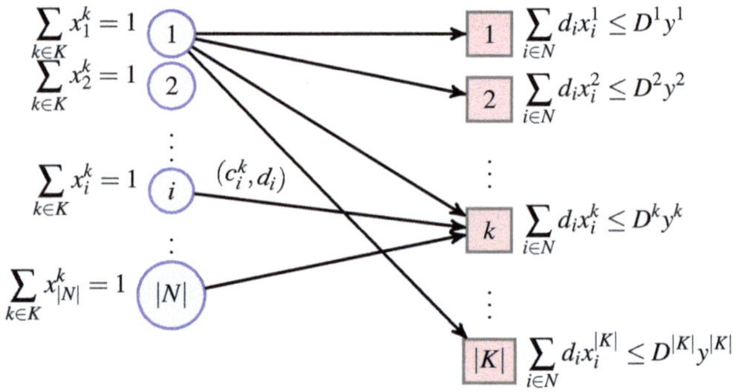

Fig. 6.31: Illustration of the warehouse location problem.

(a) Formulate the Lagrangian subproblem obtained by relaxing the first set of constraints (6.144b) with Lagrangian multipliers $\boldsymbol{\pi} = [\pi_i \in \mathbb{R}]_{i\in N}$. Does it possess the integrality property?

(b) Formulate the Lagrangian subproblem obtained by relaxing the second set of constraints (6.144c) with Lagrangian multipliers $\boldsymbol{\omega} = [\omega^k \leq 0]_{k\in K}$. Does it possess the integrality property?

(c) Assume now that $x_i^k \in \{0,1\}, \forall i \in N, k \in K$. These binary variables indicate that every customer i must be served from a single warehouse in the resulting *ILP*. What is the impact on $LR(\boldsymbol{\pi})$ and $LR(\boldsymbol{\omega})$?

6.16 Dual inequalities for a two-echelon vehicle routing problem

Consider a two-echelon vehicle routing problem with time windows that consists in determining first-echelon routes (from depots to satellites) for large vehicles and second-echelon routes (from satellites to customers) operated by smaller vehicles. All the freight to be delivered by a second-echelon route must be supplied by a single first-echelon route in time, i.e., before it starts from the satellite. The objective is to minimize the total routing cost of both echelons.

We model this problem using an extended formulation that relies on the following notation:

N : set of customers to be served;

K : set of feasible first-echelon routes of capacity Q^1; $C^k > 0$, cost of route $k \in K$; y^k, binary variable taking value 1 if route k is selected, 0 otherwise;

R : set of feasible second-echelon routes; $c_r > 0$, cost of route $r \in R$; d_r, total demand delivered by route $r \in R$; a_{ir}, binary parameter equal to 1 if route $r \in R$ visits customer $i \in N$;

R^k : subset of second-echelon routes that can be supplied by first-echelon route $k \in K$; λ_r^k, binary variable taking value 1 if route $r \in R^k$ is selected and supplied by route $k \in K$, 0 otherwise.

The integer master problem is

$$z^\star_{IMP} = \min \sum_{k\in K} C^k y^k + \sum_{k\in K}\sum_{r\in R^k} c_r \lambda_r^k \tag{6.145a}$$

$$\text{s.t.} \quad \sum_{k\in K}\sum_{r\in R^k} a_{ir}\lambda_r^k = 1 \qquad [\psi_i \in \mathbb{R}] \quad \forall i \in N \tag{6.145b}$$

$$\sum_{r\in R^k} d_r \lambda_r^k \le Q^1 y^k \qquad [\pi^k \le 0] \quad \forall k \in K \tag{6.145c}$$

$$y^k \in \{0,1\} \qquad \forall k \in K \tag{6.145d}$$

$$\lambda_r^k \in \{0,1\} \qquad \forall k \in K, r \in R^k. \tag{6.145e}$$

Constraints (6.145b) ensure that each customer is visited by exactly one second-echelon route, whereas constraints (6.145c) link the first- and second-echelon route variables. Indeed, the demand delivered by a second-echelon route can only be supplied by a first-echelon route if this route is selected and the total demand delivered by the second-echelon routes assigned to a first-echelon route cannot exceed capacity Q^1.

In practice, the set K is not very large so we assume that all first-echelon routes are enumerated a priori. This is not the case for the second-level routes in the sets R^k, $\forall k \in K$, that need to be generated as needed by column generation.

In the following questions, we focus on the linear relaxation of (6.145), i.e., on the *MP* whose dual variables are listed in brackets.

(a) Let $k_1, k_2 \in K$, $k_1 \neq k_2$, be two distinct first-echelon routes such that $R^{k_1} \subseteq R^{k_2}$. Mhamedi et al. (2022) prove that

$$\pi^{k_2} \leq \pi^{k_1} \tag{6.146}$$

is a deep dual-optimal inequality (*DDOI*), i.e., it preserves at least one optimal dual solution. Provide an informal explanation for this by looking at constraints (6.145c).

(b) Denoting by $u^{k_1k_2} \geq 0$ a static primal variable associated with dual inequality (6.146), how would you modify the *MP* to impose this inequality?

(c) Give a practical interpretation of $u^{k_1k_2}$. How can this variable help to speed up column generation?

(d) Let $F = \{(k_1, k_2) \in K^2 \mid k_1 \neq k_2, R^{k_1} \subseteq R^{k_2}\}$. Then, the inequalities

$$\pi^{k_2} \leq \pi^{k_1}, \qquad \forall (k_1, k_2) \in F, \tag{6.147}$$

form a set of *DDOI*s (Mhamedi et al., 2022). Considering their associated non-negative primal variables $u^{k_1k_2}$, $\forall (k_1, k_2) \in F$, write the modified constraints (6.145c) that take these dual inequalities into account.

(e) Assume that we solve model (6.145) augmented with the set of *DDOI*s (6.147), in their primal version (d), and get an integer solution $(\tilde{\boldsymbol{\lambda}}, \tilde{\mathbf{y}}, \tilde{\mathbf{u}})$. What can you say about the existence of an equivalent *feasible* integer solution obtained by transferring load between first-echelon routes according to the values of the u-variables
   - if none of them take a positive value?
   - if a single one takes a positive value?
   - if two or more take a positive value?

7

Branch-Price-and-Cut

> In theory there is no difference between theory and practice, while in practice there is.
>
> *The Yale Literary Magazine*
> Benjamin Brewster (1882)

Abstract Our aim from the very beginning is to solve an integer linear program. Yet, all that we have algorithmically accomplished so far is to solve the linear relaxation of a Dantzig-Wolfe reformulation. When the solution is fractional, we need to cut it, or branch. From a bird's eye view, all that remains to do is to use the column generation algorithm to solve the relaxation, and possibly strengthen it by additional valid inequalities, in every node of the branch-and-bound tree. Zooming in, we see that some subtleties have to be respected or can even be exploited.

Contents

Introduction . . . 445
7.1 Integrality Test . . . 447
7.2 Cutting Planes . . . 449
Cutting planes on the original variables . . . 449
Integration into the master constraints . . . 450
Integration into the subproblem constraints . . . 452
Illustration 1: Clique cuts for the vertex coloring problem . . 452
Illustration 2: Capacity cuts for the *VRPTW* . . . 454
Illustration 3: z-cuts within the *ISP* . . . 457
Discussion . . . 459
Cutting planes on the master variables . . . 459
Illustration 4: Chvátal-Gomory cuts . . . 461
Illustration 5: Clique cuts for the set partitioning problem . . 462
Discussion . . . 464

J. Desrosiers et al., *Branch-and-Price*,
https://doi.org/10.1007/978-3-031-96917-1_7

7.3 Branching 464
Branching on the original variables 465
Illustration 6: x-branching for the *VRPTW* 466
Identical subproblems and disaggregation 469
Branching on the master variables 470
Branching on partitioning $\mathcal{X}$ with a single hyperplane 471
Illustration 7: λ-branching for the *VRPTW* 475
Branching on lower and upper bounds on the x-variables . . 476
Ryan-Foster rule for set partitioning master problems 480
Branching on inter-tasks 481
Some practical rules on the sum of binary λ-variables . . . 483
Illustration 8: λ-branching for the *VRPTW* (cont.) 485
Ranking the candidates 486
7.4 Convexification vs. discretization revisited 487
Illustration 9: What you eat is what you are 488
Final remarks 491
7.5 Good to Know 493
Arc-flow variable fixing by reduced cost 493
Acceleration techniques 494
Large number of column generation iterations 495
Most of the time spent solving the *RMP* 496
Most of the time spent solving the *ISP* 497
Large number of branch-and-bound nodes 499
7.6 More to Know 501
Primal heuristics 501
Chvátal-Gomory cuts of higher ranks 506
Beginner's guide to an implementation 507
7.7 Examples 509
Time constrained shortest path problem 510
Semi-assignment branching 514
Ryan-Foster branching for vertex coloring 516
Ryan-Foster branching for bin packing 518
Edge coloring and odd-circuit cuts 519
k–path cuts for the *VRPTW* 520
VRPTW and non-robust rounded capacity cuts 522
VRPTW and Chvátal-Gomory rank-1 cuts 525
Preferential bidding system 527
Branch-first, Cut-second strategy 531
7.8 Reference notes 536
Exercises 538

Acronyms

ILP integer linear program 446

IMP	integer master problem (convexification)	446
MP	linear relaxation of the *IMP*	446
$\mathit{IM\ddot{P}}$	integer master problem (discretization)	446
$\mathit{M\ddot{P}}$	linear relaxation of the $\mathit{IM\ddot{P}}$	447
RMP	restricted master problem (convexification & discretization)	447
LP	linear relaxation of the *ILP*	449
ISP	integer subproblem	450
VRPTW	vehicle routing problem with time windows	451
ISP^k	integer subproblem for block $k \in K$	454
MDVSP	multiple depot vehicle scheduling problem	484
ESPPRC	elementary shortest path problem with resource constraints	456
LB	best known lower bound on $z^\star_{ILP}$	457
UB	best known upper bound on $z^\star_{ILP}$	457
lb	lower bound on $z^\star_{ILP}$	458
CPP	crew pairing problem	481
ESPPTWC	elementary shortest path problem with time windows and capacity	485
VRP	vehicle routing problem	494
AMP	alternative *MP*	496
DCA	dynamic constraint aggregation	496
MILP	mixed-integer linear program	498
LNS	large neighborhood search	504
SPPRC	shortest path problem with resource constraints	481
API	application programming interface	508
TCSPP	time constrained shortest path problem	510
ECP	edge coloring problem	519
TSPTW	traveling salesperson problem with time windows	521
REFs	resource extension functions	524

Introduction

We are almost at the end of our book on *branch-and-price*. What is left is this chapter which, incidentally, has almost the same title. Some hundred pages ago, we set out to solve integer programs, and what we have accomplished so far is solving a relaxation by column generation. In other words, we have solved the root node of a branch-and-bound tree. When we do this in every node of the tree, we obtain branch-and-price.

Note 7.1 (What's in a name?) The name of *branch-and-price* is well-accepted in the literature (it somehow "won" against *IP column generation*). There is less unanimity when cutting planes come into play. One finds branch-and-cut-and-price, branch-cut-and-price, and branch-price-and-cut. We can only guess that the former two are motivated by integrating the established *branch-and-cut* with pricing. In our view, doing column generation or not changes so fundamentally, e.g., an implementation that we think of rather complementing *branch-and-price* by cutting. This is why, the reader has figured it from this chapter's title, we stick to branch-price-and-cut.

One could continue the discussion whether this enumeration should be comma separated, like *branch, price, and cut*, and whether or not one likes the Oxford comma. But this is a maths book and not one about language.

We wish to (finally!) solve the compact formulation *ILP* which we repeat as

$$
\begin{aligned}
z^\star_{ILP} = \min \quad & \mathbf{c}^\intercal \mathbf{x} & \\
\text{s.t.} \quad & \mathbf{A}\mathbf{x} \geq \mathbf{b} & [\boldsymbol{\sigma}_\mathbf{b}] \\
& \mathbf{D}\mathbf{x} \geq \mathbf{d} & [\boldsymbol{\sigma}_\mathbf{d}] \\
& \mathbf{x} \in \mathbb{Z}^n_+ .
\end{aligned}
\tag{7.1}
$$

We assume this *ILP* to be feasible with finite optimum $z^\star_{ILP}$, as before. In Chapter 4 (Dantzig-Wolfe Decomposition for Integer Linear Programming), we learned about two reformulations of model (7.1) using the grouping of constraints

$$\mathcal{A} = \{\mathbf{x} \in \mathbb{Z}^n_+ \mid \mathbf{A}\mathbf{x} \geq \mathbf{b}\} \neq \emptyset \tag{7.2a}$$

$$\mathcal{D} = \{\mathbf{x} \in \mathbb{Z}^n_+ \mid \mathbf{D}\mathbf{x} \geq \mathbf{d}\} \neq \emptyset . \tag{7.2b}$$

The first reformulation is by *convexification* of $\mathcal{D}$, and we arrive at the following equivalent integer linear program *IMP* (7.3):

$$
\begin{aligned}
z^\star_{IMP} = \min \quad & \sum_{p \in P} c_p \lambda_p + \sum_{r \in R} c_r \lambda_r & & \\
\text{s.t.} \quad & \sum_{p \in P} \mathbf{a}_p \lambda_p + \sum_{r \in R} \mathbf{a}_r \lambda_r \geq \mathbf{b} & [\boldsymbol{\pi}_\mathbf{b}] & \\
& \sum_{p \in P} \lambda_p = 1 & [\pi_0] & \\
& \lambda_p \geq 0 & & \forall p \in P \\
& \lambda_r \geq 0 & & \forall r \in R \\
& \sum_{p \in P} \mathbf{x}_p \lambda_p + \sum_{r \in R} \mathbf{x}_r \lambda_r = \mathbf{x} \in \mathbb{Z}^n_+ , & &
\end{aligned}
\tag{7.3}
$$

where P and R denote the finite index sets for the extreme points and extreme rays of $\operatorname{conv}(\mathcal{D})$, respectively. The integrality requirements are imposed on the x-variables of the *ILP*, hence we can define cutting planes and branching decisions using the values of $\mathbf{x}^\star_{MP}$, see Figure 4.2. The above formulation reproduces the T-shirt advertisement for the 2018 School on Column Generation held in Paris, where the primal variables are represented by a *Tour Eiffel*. The dual ones for the linear relaxation, represented by an *Arc de Triomphe*, were subsequently incorporated.

The second reformulation, given by *discretization* of $\mathcal{D}$, gives us another equivalent integer linear program, denoted $IM\ddot{P}$, which reads as

$$
\begin{aligned}
z^\star_{IM\ddot{P}} = \min \quad & \sum_{p \in \ddot{P}} c_p \lambda_p \quad + \quad \sum_{r \in \ddot{R}} c_r \lambda_r \\
\text{s.t.} \quad & \sum_{p \in \ddot{P}} \mathbf{a}_p \lambda_p \quad + \quad \sum_{r \in \ddot{R}} \mathbf{a}_r \lambda_r \geq \mathbf{b} && [\boldsymbol{\pi}_\mathbf{b}] \\
& \sum_{p \in \ddot{P}} \lambda_p \qquad\qquad\qquad = 1 && [\pi_0] \\
& \qquad \lambda_p \in \{0,1\} && \forall p \in \ddot{P} \\
& \qquad\qquad\qquad \lambda_r \in \mathbb{Z}_+ && \forall r \in \ddot{R} \\
& \sum_{p \in \ddot{P}} \mathbf{x}_p \lambda_p \quad + \quad \sum_{r \in \ddot{R}} \mathbf{x}_r \lambda_r = \mathbf{x} \in \mathbb{Z}^n_+,
\end{aligned}
\tag{7.4}
$$

where $\ddot{P} \supseteq P$ and $\ddot{R}$ denote the finite index sets of integer points and integer-scaled extreme rays of $\operatorname{conv}(\mathcal{D})$, respectively. Even though the *IM$\ddot{P}$* (7.4) is different from the *IMP* (7.3), solving its linear relaxation *M$\ddot{P}$* is not different from solving the *MP* by Proposition 4.2. However, with more variables required to be integer, we have additional options to impose cutting and branching decisions, also on the variable values $\boldsymbol{\lambda}^\star_{M\ddot{P}}$, see Figure 4.7.

A word on the icon:
the first node is the root,
followed by a cut, and next,
a 2-side branching decision.

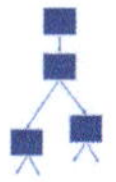

7.1 Integrality Test

It may be an obvious statement, but it has to be said: When we have optimally solved the master problem in the root node, a relaxation, and hold an integer solution in our hands, we are done. But what does it even mean, to have an integer solution?

Consider some node of the branch-and-price tree. The easiest is that the solution of the final *RMP* is integer. We denote this solution by $\boldsymbol{\lambda}^\star_{RMP} \in \mathbb{R}^{|P'|+|R'|}_+$ as it covers both cases, convexification and discretization. That is, with $\boldsymbol{\lambda}^\star_{RMP} \in \mathbb{Z}^{|P'|+|R'|}_+$ we imply $\boldsymbol{\lambda}^\star_{MP} \in \mathbb{Z}^{|P|+|R|}_+$ or $\boldsymbol{\lambda}^\star_{M\ddot{P}} \in \mathbb{Z}^{|\ddot{P}|+|\ddot{R}|}_+$, respectively, in which case also the solution in $\mathbf{x}^\star_{MP}$ is integer (see Propositions 4.2 and 4.3). The converse is not generally true: the vector

$$
\mathbf{x}^\star_{MP} = \sum_{p \in P'} \mathbf{x}_p \lambda^\star_p + \sum_{r \in R'} \mathbf{x}_r \lambda^\star_r \tag{7.5}
$$

can still be integer even when $\boldsymbol{\lambda}^\star_{RMP}$ is fractional. Therefore, both options should be checked, integrality of master and original variables, and maybe in this order:

1. If $\boldsymbol{\lambda}^\star_{RMP}$ is integer, so is $\mathbf{x}^\star_{MP}$.
2. Otherwise, test if $\mathbf{x}^\star_{MP}$ is integer.

There is a nasty case which is practically very relevant: when the master problem uses aggregated λ-variables in presence of $|K|$ identical subproblems, integrality of $\mathbf{x}^\star_{MP}$ in (7.5) needs not imply integrality of the disaggregated variables

$$\mathbf{x}^{k\star}_{MP} = \sum_{p \in P^{k\prime}} \mathbf{x}_p \lambda^{k\star}_p + \sum_{r \in R^{k\prime}} \mathbf{x}_r \lambda^{k\star}_r, \qquad \forall k \in K, \tag{7.6}$$

we are really interested in. If this happens, we have to perform a disaggregation from $\boldsymbol{\lambda}^\star_{RMP}$ to $\{\boldsymbol{\lambda}^{k\star}_{RMP}\}_{k\in K}$ first, e.g., by solving a zero-cost balanced transportation problem as in (4.62), and then project back to the original variable values $\{\mathbf{x}^{k\star}_{MP}\}_{k\in K}$ (see Disaggregation in Chapter 4). Only then do we perform the integrality test:

1. Perform a disaggregation of $\boldsymbol{\lambda}^\star_{RMP}$ into $\{\boldsymbol{\lambda}^{k\star}_{RMP}\}_{k\in K}$ and derive $\{\mathbf{x}^{k\star}_{MP}\}_{k\in K}$.

2a. If $\boldsymbol{\lambda}^\star_{RMP}$ is integer, the solution $\{\boldsymbol{\lambda}^{k\star}_{RMP}\}_{k\in K}$ to (4.62) is integer; so is $\{\mathbf{x}^{k\star}_{MP}\}_{k\in K}$.

2b. Otherwise, test if the solution $\{\mathbf{x}^{k\star}_{MP}\}_{k\in K}$ is integer.

In case the integrality tests fail, you need to read this chapter on cutting and branching decisions on original and master variables. And also when the integrality tests do not fail, it is just very nice to read this chapter.

Note 7.2 (Integer only at second glance.) Integrality of a solution can also be a matter of implementation or luck. Consider the polyhedron in Figure 7.1. It is described using two extreme points in set $\{(1,1),(2,1)\}$ and two extreme rays, again in the same set $\{(1,1),(2,1)\}$. In the convex combination of the extreme points with respective weights 0.4 and 0.6 plus 1.6 times the first extreme ray and 0.4 times the second, we reach the coordinates $(4,3) = 0.4(1,1)+0.6(2,1)+1.6(1,1)+0.4(2,1)$, which is a fractional λ-representation. However, it need not be. We could as well use the second extreme point and the first extreme ray with respective weights 1 and 2, that is, $1(2,1)+2(1,1) = (4,3)$, which is an integer solution in the λ-variables.

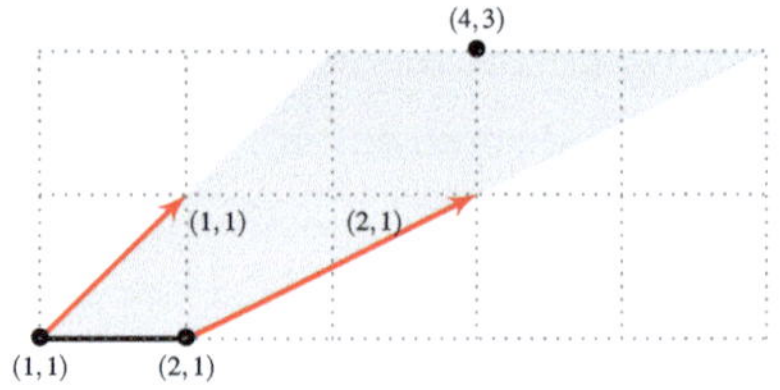

Fig. 7.1: A polyhedron with two extreme points and two extreme rays.

Referring back to Illustration 4.3, we also see that the integrality of a master solution may depend on the scaling of the rays. Avoiding unnecessary fractional solutions is our true motivation for scaling rays to their shortest possible integer representative.

7.2 Cutting Planes

Both Dantzig-Wolfe reformulations provide us with effectively the same master relaxation. It is potentially stronger than the standard linear relaxation of *ILP* (7.1). *Cutting planes*, or simply *cuts*, are a way to strengthen a relaxation even further.

Definition 7.1. Given an integer program $\min\{\mathbf{c}^\mathsf{T}\mathbf{x} \mid \mathbf{Q}\mathbf{x} \geq \mathbf{q}, \mathbf{x} \in \mathbb{Z}^n_+\}$ and its linear relaxation $\min\{\mathbf{c}^\mathsf{T}\mathbf{x} \mid \mathbf{Q}\mathbf{x} \geq \mathbf{q}, \mathbf{x} \in \mathbb{R}^n_+\}$,

1. an inequality $\mathbf{a}^\mathsf{T}\mathbf{x} \geq a_0$ is *valid* for $\mathcal{Q} := \{\mathbf{x} \in \mathbb{Z}^n_+ \mid \mathbf{Q}\mathbf{x} \geq \mathbf{q}\}$ if

$$\mathbf{a}^\mathsf{T}\bar{\mathbf{x}} \geq a_0, \qquad \forall \bar{\mathbf{x}} \in \mathcal{Q},$$

   that is, every integer point is feasible; and
2. a valid inequality $\mathbf{a}^\mathsf{T}\mathbf{x} \geq a_0$ for $\mathcal{Q}$ is a *cutting plane* if

$$\mathbf{a}^\mathsf{T}\mathbf{x}^* < a_0, \qquad \text{for some } \mathbf{x}^* \in \{\mathbf{x} \in \mathbb{R}^n_+ \mid \mathbf{Q}\mathbf{x} \geq \mathbf{q}\},$$

   that is, at least one fractional point becomes infeasible.

Most likely, when you formulate an *ILP* like (7.1), the literature knows several cutting planes for (a variant of) this model, and they are naturally formulated in the original x-variables. We explore this case first.

Cutting planes on the original variables

We start from the compact formulation (7.1) and assume that we know a set of cutting planes for it, say $\mathbf{F}\mathbf{x} \geq \mathbf{f}$. With these additional inequalities, we associate a dual vector $\boldsymbol{\sigma}_\mathbf{f} \geq \mathbf{0}$ in the *LP*. The strengthened *ILP* formulation reads as

$$\begin{aligned}
z^\star_{ILP} = \min \quad & \mathbf{c}^\mathsf{T}\mathbf{x} && && \text{(7.7a)}\\
\text{s.t.} \quad & \mathbf{A}\mathbf{x} \geq \mathbf{b} && [\boldsymbol{\sigma}_\mathbf{b}] && \text{(7.7b)}\\
& \mathbf{D}\mathbf{x} \geq \mathbf{d} && [\boldsymbol{\sigma}_\mathbf{d}] && \text{(7.7c)}\\
& \mathbf{F}\mathbf{x} \geq \mathbf{f} && [\boldsymbol{\sigma}_\mathbf{f}] && \text{(7.7d)}\\
& \mathbf{x} \in \mathbb{Z}^n_+. && && \text{(7.7e)}
\end{aligned}$$

We consider two cases for the Dantzig-Wolfe reformulation: the integration of $\mathbf{F}\mathbf{x} \geq \mathbf{f}$ either in $\mathcal{A}$ or in $\mathcal{D}$, that is, in the master or the subproblem constraints.

Integration into the master constraints

In the first case, the grouping of the constraints becomes

$$\mathcal{A} = \{\mathbf{x} \in \mathbb{Z}^n_+ \mid \mathbf{Ax} \geq \mathbf{b},\ \mathbf{Fx} \geq \mathbf{f}\} \tag{7.8a}$$
$$\mathcal{D} = \{\mathbf{x} \in \mathbb{Z}^n_+ \mid \mathbf{Dx} \geq \mathbf{d}\}. \tag{7.8b}$$

The cuts $\mathbf{Fx} \geq \mathbf{f}$ are reformulated in the same way as those of $\mathbf{Ax} \geq \mathbf{b}$, that is, as

$$\sum_{p\in P} \mathbf{Fx}_p\lambda_p + \sum_{r\in R} \mathbf{Fx}_r\lambda_r \geq \mathbf{f} \tag{7.9}$$

with dual vector $\boldsymbol{\pi}_\mathbf{f}$. Using the transformations $\mathbf{f}_p = \mathbf{Fx}_p$ for all $p \in P$ and $\mathbf{f}_r = \mathbf{Fx}_r$ for all $r \in R$, the *MP* reads as

$$\begin{aligned}
z^\star_{MP} = \min \quad & \sum_{p\in P} c_p\lambda_p + \sum_{r\in R} c_r\lambda_r && && (7.10a)\\
\text{s.t.} \quad & \sum_{p\in P} \mathbf{a}_p\lambda_p + \sum_{r\in R} \mathbf{a}_r\lambda_r \geq \mathbf{b} && [\boldsymbol{\pi}_\mathbf{b}] && (7.10b)\\
& \sum_{p\in P} \mathbf{f}_p\lambda_p + \sum_{r\in R} \mathbf{f}_r\lambda_r \geq \mathbf{f} && [\boldsymbol{\pi}_\mathbf{f}] && (7.10c)\\
& \sum_{p\in P} \lambda_p = 1 && [\pi_0] && (7.10d)\\
& \lambda_p \geq 0 && \forall p \in P && (7.10e)\\
& \lambda_r \geq 0 && \forall r \in R && (7.10f)\\
& \sum_{p\in P} \mathbf{x}_p\lambda_p + \sum_{r\in R} \mathbf{x}_r\lambda_r = \mathbf{x} \in \mathbb{R}^n_+. && && (7.10g)
\end{aligned}$$

This can be adapted for special cases such as bounded $\mathcal{D}$ (in which case $R = \emptyset$) and conv($\mathcal{D}$) a polyhedral cone (zero-extreme point removed and $r \in R$). Given dual values $\boldsymbol{\pi}_\mathbf{b}, \boldsymbol{\pi}_\mathbf{f} \geq \mathbf{0}$ and $\pi_0 \in \mathbb{R}$, the reduced cost of a λ-variable is computed with a "slightly longer" column vector now. The adapted *ISP* reads as

$$\begin{aligned}
\bar{c}(\boldsymbol{\pi}_\mathbf{b}, \boldsymbol{\pi}_\mathbf{f}, \pi_0) = -\pi_0 + & \min_{\mathbf{x}\in\mathcal{D}} \ c_\mathbf{x} - \boldsymbol{\pi}_\mathbf{b}^\intercal \mathbf{a}_\mathbf{x} - \boldsymbol{\pi}_\mathbf{f}^\intercal \mathbf{f}_\mathbf{x} \\
& \ \text{s.t.} \ \ c_\mathbf{x} = \mathbf{c}^\intercal\mathbf{x},\ \mathbf{a}_\mathbf{x} = \mathbf{Ax},\ \mathbf{f}_\mathbf{x} = \mathbf{Fx},
\end{aligned} \tag{7.11}$$

$$= -\pi_0 + \min_{\mathbf{x}\in\mathcal{D}} \ (\mathbf{c}^\intercal - \boldsymbol{\pi}_\mathbf{b}^\intercal\mathbf{A} - \boldsymbol{\pi}_\mathbf{f}^\intercal\mathbf{F})\mathbf{x}. \tag{7.12}$$

Comparing this *ISP* (7.12) to the standard subproblem formulation (4.11), we observe that the subproblem's domain $\mathcal{D}$ formally does not change. Having cutting (and later branching) decisions on the original variables, integrated into the reformulation in this way, is sometimes called *robust* or *compatible*. This is attractive because a (possibly combinatorial, or otherwise specialized) pricing algorithm that works for the *ISP* (4.11) can still be applied to solve the *ISP* (7.12). While this is formally true, read the following note where the devil is in the details.

Note 7.3 (A misinterpretation.) It is an advantage that formally "only the objective function of the pricing problem changes" when cutting (or branching) decisions are formulated on original variables, and reformulated as master constraints. However, in practice, this may introduce new variables in the objective function of the pricing problem. Consider a cutting plane that is (also) defined on original variables that have no cost associated with them in the compact formulation. In the reduced cost objective function of the *ISP*, these variables may have a non-zero adjusted cost coefficient, and thus variables may suddenly "appear." When the *ISP* is not solved as an MILP anyway, this *does* potentially change its character.

The *VRPTW* serves as an example, where the subproblem can be solved using a dynamic program that works on the arcs of the underlying network, that is, on the x_{ij}-variables. Formulating a cutting plane in terms of the time variables, say $\sum_{i \in C} f_i t_i \geq T$ with dual value denoted $\pi_T \geq 0$ in (7.10c), introduces the time variables in the objective function of the *ISP* (5.13)

$$\min_{\begin{bmatrix}\mathbf{x}\\ \mathbf{t}\end{bmatrix} \in \mathcal{D}} \sum_{(i,j) \in A} (c_{ij} - \pi_i) x_{ij} - \pi_T \sum_{i \in C} f_i t_i \tag{7.13}$$

and, depending on the sign of the f-coefficients, the *ISP* may require an alternative dynamic programming algorithm for solving the constrained shortest path problem with linear penalties on nodes (Ioachim et al., 1998).

Note 7.4 (Changes to the column generation algorithm.) One does not add cutting planes to an integer program all at once. Instead, *separation algorithms* are used to either find one (or several) inequalities that are violated by the current fractional solution, or to prove that no such inequality exists. Often in the literature, such an algorithm is described together with the valid inequalities $\mathbf{Fx} \geq \mathbf{f}$, and it is naturally called on a fractional solution $\mathbf{x}^\star_{MP}$. Algorithm 7.1 describes how this very harmonically combines with the column generation algorithm.

Algorithm 7.1: Column generation algorithm with separation of cutting planes on the original variables

---

initialize the *RMP* as usual
loop
  solve the *MP* to optimality via column generation to obtain $\boldsymbol{\lambda}^\star_{MP}$ and $\mathbf{x}^\star_{MP}$;
  call separation algorithms on $\mathbf{x}^\star_{MP}$;
  if *this produces a cut* $\mathbf{f}^\intercal \mathbf{x} \geq f_0$
    add $\sum_{p \in P'} \mathbf{f}_p \lambda_p + \sum_{r \in R'} \mathbf{f}_r \lambda_r \geq f_0$ to the *RMP* with dual variable π_{f_0};
    incorporate π_{f_0} in the objective function of the *ISP* (7.12);
  else
    break;

Integration into the subproblem constraints

In the second case, the grouping of the constraints is

$$\mathcal{A} = \{\mathbf{x} \in \mathbb{Z}_+^n \mid \mathbf{Ax} \geq \mathbf{b}\} \tag{7.14a}$$

$$\mathcal{D} = \{\mathbf{x} \in \mathbb{Z}_+^n \mid \mathbf{Dx} \geq \mathbf{d},\ \mathbf{Fx} \geq \mathbf{f}\}. \tag{7.14b}$$

Formally, $\mathcal{D}$ is modified (it may remain unchanged), and also conv($\mathcal{D}$) may change, with different extreme points and extreme rays. We still index them by $p \in P$ and $r \in R$. With this, the *ISP* writes as before, i.e.,

$$\bar{c}(\boldsymbol{\pi}_\mathbf{b}, \pi_0) = -\pi_0 + \min_{\mathbf{x} \in \mathcal{D}} \mathbf{c}^\mathsf{T}\mathbf{x} - \boldsymbol{\pi}_\mathbf{b}^\mathsf{T}\mathbf{Ax}, \tag{7.15}$$

where $\boldsymbol{\pi}_\mathbf{b}$ and π_0 come from solving the linear relaxation of the *IMP* (7.3). However, this notation hides that the pricing algorithm may change.

On the positive side, we have no need to use any reformulation of cutting planes, and no extra dual variables to respect in the pricing problem. Also, if we solve the *ISP* as an integer program anyway, this may even help speeding up its solution. Note, however, that when the current fractional master solution should be cut off, all master variables λ_j, $j \in P' \cup R'$, that are incompatible with the cuts, i.e., $\mathbf{Fx}_j \ngeq \mathbf{f}$, must be eliminated from the *RMP*.

Illustration 7.1 **Clique cuts for the vertex coloring problem**

Similar to the *edge coloring problem* seen in Example 2.3, consider the *vertex coloring problem* in which we are given an undirected graph $G = (N, E)$, where N denotes the set of nodes (or vertices) and E the set of edges. We wish to assign a color (from a set K, *in Anlehnung an den berühmten Kodak Kodacolor Fotofilm*, or as in *The Kolors*, an Italian rock band) to every node such that no two neighbors receive the same color. The total number of colors should be minimized (the minimum is the so-called *chromatic number* of the graph). A textbook model for this problem uses assignment variables $x_i^k \in \{0,1\}$ to represent whether node $i \in N$ is colored in color $k \in K$ or not. For the counting of colors, we use the binary variable x_0^k for color k.

$$z_{ILP}^\star = \min \quad \sum_{k \in K} x_0^k \tag{7.16a}$$

$$\text{s.t.} \quad \sum_{k \in K} x_i^k = 1 \qquad \forall i \in N \tag{7.16b}$$

$$x_i^k + x_j^k \leq x_0^k \qquad \forall k \in K,\ (i,j) \in E \tag{7.16c}$$

$$x_0^k, x_i^k \in \{0,1\} \qquad \forall k \in K,\ i \in N. \tag{7.16d}$$

The conflict inequalities (7.16c) have a block-diagonal form, one block for every color. This suggests to Dantzig-Wolfe reformulate these constraints. For $k \in K$, let $\mathbf{x}^k = [x_i^k]_{i=0,1,\ldots,|N|}$. Define the non-empty sets $\mathcal{A}$ and $\mathcal{D}^k$ that group the constraints as

$$\mathcal{A} = \left\{ \left\{ \mathbf{x}^k \in \{0,1\}^{|N|+1} \right\}_{k\in K} \;\middle|\; \sum_{k\in K} x_i^k = 1,\ \forall i \in N \right\} \tag{7.17a}$$

$$\mathcal{D}^k = \left\{ \mathbf{x}^k \in \{0,1\}^{|N|+1} \;\middle|\; x_i^k + x_j^k \le x_0^k,\ \forall (i,j) \in E \right\},\quad \forall k \in K. \tag{7.17b}$$

For every color $k \in K$ that is used ($x_0^k = 1$), every extreme point of $\text{conv}(\mathcal{D}^k)$ is an incidence vector of a subset of nodes, all of which are pairwise not neighbors, that is, an independent set in color k. We index these extreme points by $p \in P^k$ (and eliminate the zero solution with $x_0^k = 0$) to obtain the *MP* written as

$$\begin{aligned}
z_{MP}^\star = \min \quad & \sum_{k\in K} \sum_{p\in P^k} \lambda_p^k && && \text{(7.18a)}\\
\text{s.t.} \quad & \sum_{k\in K} \sum_{p\in P^k} a_{ip}^k \lambda_p^k = 1 && [\pi_i] && \forall i \in N \quad \text{(7.18b)}\\
& \sum_{p\in P^k} \lambda_p^k \le 1 && [\pi_0^k] && \forall k \in K \quad \text{(7.18c)}\\
& \lambda_p^k \ge 0 && && \forall k \in K, p \in P^k. \quad \text{(7.18d)}
\end{aligned}$$

Here $a_{ip}^k = x_{ip}^k$ is a binary parameter representing whether node $i \in N$ is contained in the independent set represented by the extreme point $\mathbf{x}_p^k = [x_{ip}^k]_{i=0,1,\dots,|N|}$ or not. The model selects at most one independent set per color such that every node appears in exactly one of them.

We observe that the linear relaxation of the compact formulation (7.16) is very weak: we can, e.g., color every node half in color $k = 1$ and half in $k = 2$, i.e., $x_i^1 = x_i^2 = 0.5$ for all $i \in N$. This implies $x_0^1 = x_0^2 = 1$ (and all other x_0^k-variables at zero, for $k \notin \{1,2\}$), and an optimal objective function value of $z_{LP}^\star = 2$. The information we obtain from this is: we need at least two colors to color all the nodes, which is completely useless for any graph that contains at least one edge.

A natural strengthening of the *ILP* model (7.16) uses *cliques* in G. These are subsets $Q \subseteq N$ of nodes, all of which are pairwise neighbors. In an integer solution, every color can be used at most once for the nodes in Q. A fractional solution, however, can color half of the nodes of Q in this color. A *clique inequality* forbids this defect:

$$\sum_{i\in Q} x_i^k \le 1 \text{ for every clique } Q \subseteq N, \text{ and every color } k \in K. \tag{7.19}$$

As before, if we express the original x-variables as convex combination of incidence vectors of independent sets, we obtain in the master problem the reformulated cutting planes

$$\sum_{i\in Q} \sum_{p\in P^k} a_{ip}^k \lambda_p^k \le 1 \text{ for every clique } Q \subseteq N, \text{ and every color } k \in K. \tag{7.20}$$

We read in this clique inequality for Q that we cannot (fractionally) select more than one independent set p that shares a vertex with Q. Since every independent set can share at most one vertex with Q anyway, this inequality is always fulfilled by every λ-solution. All our efforts were correct but in vain! Note that for a given color k, we also could have added the cutting plane (7.19) to $\mathcal{D}^k$. This may help speeding up the solution of the ISP^k. However, this does not strengthen the reformulation either because these inequalities are, of course, fulfilled in any binary solution to the pricing problem.

Note 7.5 (Too strong for you.) This illustration points to a general weakness of adding classical cutting planes in the original variables. A Dantzig-Wolfe reformulation *IMP* can be so strong already that adding further inequalities (in the original variables) does not strengthen the *MP* relaxation any further. Experiments by Witt (2019) support this intuition. At the same time, this is a reminder of the potentially substantial gain in strength when we apply a Dantzig-Wolfe reformulation: *all* cutting planes known for the independent set problem—and these are *many!*—are automatically satisfied via the reformulation.

Illustration 7.2 **Capacity cuts for the** ***VRPTW***

Consider the following two-index arc-flow formulation of the *VRPTW* (illustrated in Figure 7.2), similar to that of the solution of Exercise 5.3. Let C be the set of customers to be served and define $N = C \cup \{o,d\}$ as the set of nodes and $A_{do} = A \cup \{(d,o)\}$ as the set of arcs in the network $G_{do} = (N, A_{do})$. Let the set of resources be $\mathcal{R} = \{time, load\}$. For $r = time$, set $a_i^{time} = a_i$, $b_i^{time} = b_i$, $\forall i \in N$, and $t_{ij}^{time} = t_{ij}$, $\forall (i,j) \in A$. For $r = load$, set $a_i^{load} = 0$, $b_i^{load} = Q$, $\forall i \in N$, and $t_{ij}^{load} = q_j$, $\forall (i,j) \in A$. We also assume that all cost, time, and load parameters are integer and κ is the number of available identical vehicles. The mixed-*ILP* writes as

$$z^\star_{ILP} = \min \sum_{(i,j)\in A_{do}} c_{ij}x_{ij} \tag{7.21a}$$

$$\text{s.t.} \quad \sum_{j:(i,j)\in A} x_{ij} = 1 \qquad \forall i \in C \tag{7.21b}$$

$$x_{do} \le \kappa \tag{7.21c}$$

$$\sum_{j:(i,j)\in A} x_{ij} - \sum_{j:(j,i)\in A} x_{ji} = \begin{cases} x_{do} & \text{for } i = o \\ 0 & \forall i \in C \\ -x_{do} & \text{for } i = d \end{cases} \tag{7.21d}$$

$$a_i^r\Big(\sum_{j:(i,j)\in A} x_{ij}\Big) \le t_i^r \le b_i^r\Big(\sum_{j:(i,j)\in A} x_{ij}\Big) \qquad \forall r \in \mathcal{R}, i \in N \tag{7.21e}$$

$$x_{ij}(t_i^r + t_{ij}^r - t_j^r) \le 0 \qquad \forall r \in \mathcal{R}, (i,j) \in A \tag{7.21f}$$

$$x_{ij} \in \{0,1\} \qquad \forall (i,j) \in A \tag{7.21g}$$

$$x_{do} \ge 0, \text{ integer}. \tag{7.21h}$$

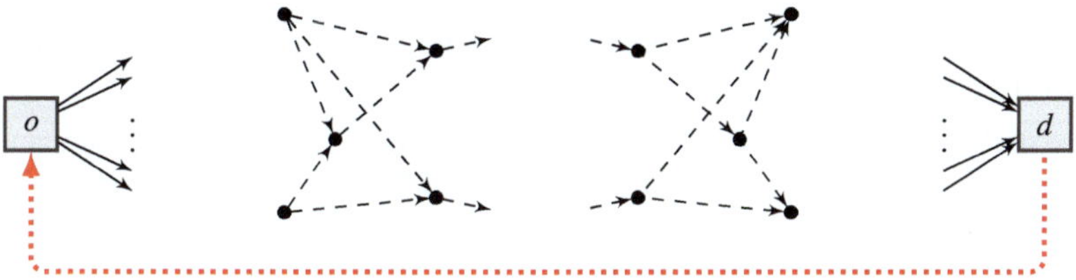

Fig. 7.2: Network structure of the *VRPTW*, where $x_{do} \leq \kappa$.

Using the grouping of the constraints

$$\mathcal{A} = \left\{ \begin{bmatrix} \mathbf{x} \\ x_{do} \end{bmatrix} \in \{0,1\}^{|A|} \times \mathbb{Z}_+ \;\middle|\; (7.21b) – (7.21c) \right\} \tag{7.22a}$$

$$\mathcal{D} = \left\{ \begin{bmatrix} \mathbf{x} \\ x_{do} \\ \mathbf{t} \end{bmatrix} \in \{0,1\}^{|A|} \times \mathbb{Z}_+ \times \mathbb{R}^{|\mathcal{R}||N|} \;\middle|\; (7.21d) – (7.21f) \right\}, \tag{7.22b}$$

we obtain the linear relaxation of the classical set partitioning model as the *MP*:

$$z^\star_{MP} = \min \quad \sum_{r \in R} c_r \lambda_r \tag{7.23a}$$

$$\text{s.t.} \quad \sum_{r \in R} a_{ir} \lambda_r = 1 \qquad [\pi_i] \qquad \forall i \in C \tag{7.23b}$$

$$\sum_{r \in R} \lambda_r \leq \kappa \qquad [\pi_\kappa] \tag{7.23c}$$

$$\lambda_r \geq 0 \qquad \forall r \in R, \tag{7.23d}$$

with dual variables $\pi_i \in \mathbb{R}$, $\forall i \in C$, and $\pi_\kappa \leq 0$. Note that the generated columns correspond to extreme rays of $\text{conv}(\mathcal{D})$.

If $z^\star_{MP}$ is fractional and we know that $z^\star_{ILP} \in \mathbb{Z}$, then we can add $\mathbf{c}^\mathsf{T}\mathbf{x} \geq z'$ (where $z' = \lceil z^\star_{MP} \rceil$) as a cut to the *ILP*. In the *MP* (7.23), this translates to

$$\sum_{r \in R} c_r \lambda_r \geq z' \qquad [\pi_z] \tag{7.24}$$

where, after reoptimizing the *RMP*, the associated dual π_z takes value 1 whereas $\pi_i = 0$, $\forall i \in C$, and $\pi_\kappa = 0$. This is not of great dual value! However, when such a cut $\mathbf{c}^\mathsf{T}\mathbf{x} \geq \lceil z^\star_{MP} \rceil$ can be applied to dynamically restrict the domain of the *ISP*, it turns out to be extremely powerful, see Illustration 7.3 (z-cuts within the *ISP*) and Example 7.9 (Preferential bidding system). Nevertheless, the mathematical expression of this cut in the *ILP*, $\sum_{(i,j) \in A_{do}} c_{ij} x_{ij} \geq c$, is the same as any cut in the x-variables written as $\sum_{(i,j) \in A_{do}} f_{ij} x_{ij} \geq f$. If we can correctly implement the former, we can also for the latter.

If the number of vehicles x_{do} is optimized on arc (d,o), say using a large cost c_{do}, then $x_{do} \geq \lceil x^\star_{do} \rceil$ is a valid cut for an optimal fractional value $x^\star_{do}$. In the *IMP*, this translates as $\sum_{r\in R}\lambda_r \geq \lceil x^\star_{do}\rceil$. Otherwise, branching decisions on x_{do} can be imposed in the search tree.

One of the most commonly used family of cuts for vehicle routing problems is the *rounded capacity cuts* that were initially proposed for the capacitated vehicle routing problem (Laporte et al., 1985). They are defined as follows. Let $S \subseteq C$ be a subset of customers such that $|S| \geq 2$ and denote by $\delta^-(S) = \{(i,j) \in A \mid i \notin S, j \in S\}$ the subset of arcs entering S. A lower bound $LB(S)$ on the number of vehicles required to service the customers in S is given by $LB(S) = \lceil \sum_{i\in S} q_i/Q\rceil$. Consequently, in a feasible solution to the *VRPTW*, the total vehicle flow entering subset S must be at least equal to $LB(S)$. This yields the following rounded capacity cuts:

$$\sum_{(i,j)\in\delta^-(S)} x_{ij} \geq LB(S) \qquad \forall S \subseteq C \text{ such that } |S| \geq 2, \tag{7.25}$$

that can be rewritten in terms of the λ-variables as follows:

$$\sum_{r\in R} n_{Sr}\lambda_r \geq LB(S) \qquad [\pi_S] \qquad \forall S \subseteq C \text{ such that } |S| \geq 2, \tag{7.26}$$

where n_{Sr} indicates the number of times that route $r \in R$ enters subset S and π_S is the associated dual variable. In the *ISP*, the adjusted cost $\tilde{c}_{ij}$ of arc $(i,j) \in A_{do}$ becomes

$$\tilde{c}_{ij} = \begin{cases} c_{ij} - \pi_K & \text{for } (i,j) = (d,o) \\ c_{ij} - \sum_{S\in\mathcal{S}_{ij}} \pi_S & \forall (i,j) \in A \text{ such that } i = o \\ c_{ij} - \pi_i - \sum_{S\in\mathcal{S}_{ij}} \pi_S & \forall (i,j) \in A \text{ such that } i \neq o, \end{cases} \tag{7.27}$$

where $\mathcal{S}_{ij}$ is the set of subsets S such that $(i,j) \in \delta^-(S)$ (i.e., (i,j) enters set S). Obviously, only the subsets S for which a capacity cut has been generated need to be considered.

To illustrate the potential impact of the rounded capacity cuts, we use the *VRPTW* example presented in Illustration 5.1 with $C = \{1,2,3,4\}$, $q_1 = q_3 = 1$, $q_2 = q_4 = 2$, and $Q = 5$. Applying column generation to solve the *MP* (with an *ESPPRC* pricing problem), we obtain the optimal solution

$$\lambda^\star_{132} = \lambda^\star_{134} = \lambda^\star_{24} = 0.5 \text{ and } \lambda^\star_r = 0 \text{ for all other routes } r \in R,$$

with a cost $z^\star_{MP} = 65.5$. In terms of the x-variables, this solution writes as:

$$x^\star_{do} = 1.5, x^\star_{o1} = x^\star_{13} = x^\star_{4d} = 1, x^\star_{o2} = x^\star_{24} = x^\star_{32} = x^\star_{34} = x^\star_{2d} = 0.5,$$
$$x^\star_{ij} = 0 \text{ for all other arcs } (i,j) \in A_{do}.$$

For the set $S = C$, we compute that $\sum_{(i,j)\in\delta^-(S)} x^\star_{ij} = \sum_{r\in R} n_{Sr}\lambda^\star_r = 1.5$ and $LB(S) = \lceil \frac{6}{5} \rceil = 2$. Therefore, the corresponding rounded capacity inequality (7.26) is violated ($1.5 \not\geq 2$). Adding this cut to the *MP* and re-optimizing, it produces an improved dual bound $z^\star_{MP} = 73.5$ attained at solution:

$$\lambda_1^\star = \lambda_{132}^\star = \lambda_{24}^\star = \lambda_{34}^\star = 0.5 \text{ and } \lambda_r^\star = 0 \text{ for all other routes } r \in R.$$

Given that the optimal value $z_{IMP}^\star = 74$, we observe that this single cut closes 94.1% $(= (73.5 - 65.5)/(74 - 65.5))$ of the integrality gap. Furthermore, adding the cut $\mathbf{c}^\mathsf{T}\mathbf{x} \geq \lceil z_{MP}^\star \rceil$ on the objective function fully closes the gap.

This result is impressive but adding rounded capacity cuts is not always so effective. Indeed, it depends on the tightness of the capacity constraint with respect to other route feasibility constraints. For instance, very narrow time windows in the *VRPTW* may heavily restrict route feasibility and increase the number of vehicles required. In this case, because the average number of customers per route is small, the capacity constraint is not very tight and the rounded capacity cuts are typically not violated by an optimal *MP* solution. At the opposite, wide time windows suggest that vehicle capacity might play a major role and, thus, rounded capacity cuts may help to tighten the linear relaxation. Given that there is an exponential number of rounded capacity inequalities, they should be generated dynamically using one or several separation algorithms (see Lysgaard et al., 2004). Nevertheless, it seems a good practice to always add a priori to the *MP* the inequality (7.26) for $S = C$.

To conclude this illustration, let us highlight two weaknesses of the rounded capacity cuts. First, the right-hand side $LB(S)$ is only a lower bound on the number of vehicles required to serve the customers in S, even if we only consider the vehicle capacity constraint. In fact, computing this number exactly amounts to solving an $\mathcal{NP}$-hard bin packing problem, which is too time-consuming to be solved in practice. On the other hand, $LB(S)$ can easily be computed and taken into account in the cut separation algorithms. Second, the rounded capacity cuts are robust cutting planes because they are defined on the x-variables of the two-index compact formulation of the *VRPTW* and do not change the nature of *ISP*. However, the notion of paths is not explicit in this formulation and the left-hand side of (7.25) does not count the number of vehicles used to serve the customers in S but rather the number of times that a vehicle enters into S, counting a vehicle multiple times if it enters set S more than once. This translates into coefficients n_{Sr} in (7.26) that may take a value greater than 1 even if the corresponding variable λ_r is clearly associated with a single vehicle. In Example 7.7, we present a non-robust version of the rounded capacity cuts where these coefficients are replaced by binary coefficients indicating whether route r enters subset S at least once or not.

Illustration 7.3 **z-cuts within the *ISP***

The First Cut is the Deepest.

Song by Cat Stevens (1967)

Assume we impose within the pricing problem the additional bound constraints

$$LB \leq \mathbf{c}^\mathsf{T}\mathbf{x} \leq UB. \tag{7.28}$$

We call these *z-cuts* because they impose restrictions on the value of the objective function. The aim of these cuts is to remove not only the current fractional point, but also non-optimal *integer* ones from the domain of the *ISP*. They dynamically modify $\mathcal{D}$, with all algorithmic consequences, that is, possible complications if a combinatorial algorithm is used to solve the *ISP*.

In Example 3.2 on the *TCSPP*, assume that we impose $7 \le \mathbf{c}^\mathsf{T}\mathbf{x} \le 15$ within the pricing problem after iteration 3, see Table 3.1 reproduced below as Table 7.1.

	RMP				ISP				
t	primal solution	z_{RMP}	π_0	π_7	$\bar{c}(\pi_7, \pi_0)$	p	c_p	t_p	lb
1	$y_0 = 1$	100.0	100.00	0.00	−97.0	1246	3	18	3.00
2	$y_0 = 0.22, \lambda_{1246} = 0.78$	24.6	100.00	−5.39	−32.9	1356	24	8	−8.33
3	$\lambda_{1246} = 0.6, \lambda_{1356} = 0.4$	11.4	40.80	−2.10	−4.8	13256	15	10	6.60
4	$\lambda_{1246} = \lambda_{13256} = 0.5$	9.0	30.00	−1.50	−2.5	1256	5	15	6.50
5	$\lambda_{13256} = 0.2, \lambda_{1256} = 0.8$	7.0	35.00	−2.00	0.0	-	-	-	7.00
	$x_{12} = 0.8$, $x_{13} = x_{32} = 0.2$, and $x_{25} = x_{56} = 1$								

Table 7.1: Iterations of the column generation algorithm while solving the *MP*.

The imposed lower and upper bounds come from the following observations:

- because all cost data is integral, the current lower bound can be rounded up to $LB = \lceil 6.6 \rceil = 7$;
- the generated path 13256 at iteration 3 is integer feasible for the *ILP* with $(cost, time)$ values $(15, 10)$, hence the upper bound is set to $UB = 15$.

The path 1256 previously generated at the fourth iteration cannot be generated any-more since its cost of 5 is too small. More importantly, the first two generated paths are removed from the *RMP*, the first because its cost of 3 is smaller than *LB*, the second because its cost of 24 is larger than *UB*. Observe that out of the nine possible paths (some already generated, some not), only three satisfy the added *z*-cuts.

$$\begin{array}{ll}\min & 3\lambda_{1246} + \mathbf{14}\lambda_{12456} + 5\lambda_{1256} + \mathbf{13}\lambda_{13246} + 24\lambda_{132456} + \mathbf{15}\lambda_{13256} + 16\lambda_{1346} + 27\lambda_{13456} + 24\lambda_{1356} \\ \text{s.t.} & 18\lambda_{1246} + 14\lambda_{12456} + 15\lambda_{1256} + 13\lambda_{13246} + 9\lambda_{132456} + 10\lambda_{13256} + 17\lambda_{1346} + 13\lambda_{13456} + 8\lambda_{1356} \le 14 \\ & \lambda_{1246} + \lambda_{12456} + \lambda_{1256} + \lambda_{13246} + \lambda_{132456} + \lambda_{13256} + \lambda_{1346} + \lambda_{13456} + \lambda_{1356} = 1 \\ & \lambda_{1246}, \lambda_{12456}, \lambda_{1256}, \lambda_{13246}, \lambda_{132456}, \lambda_{13256}, \lambda_{1346}, \lambda_{13456}, \lambda_{1356} \ge 0.\end{array}$$

Table 7.2 summarizes the modified iterations. The *RMP* is re-solved in iteration $4'$ with a single remaining λ-variable, that is, the path variable $\lambda_{13256} = 1$, hence $\pi_0 = 15.0$ and $\pi_7 = 0$. Solving the *ISP*, we generate the path 13246 with cost 13 and duration 13. This produces an integer solution $x_{13} = x_{32} = x_{25} = x_{56} = 1$ to the *RMP* that sets $UB = 13$. At the same time, the lower bound becomes $LB = \max\{z^\star_{RMP} + \bar{c}(\pi_7, \pi_0), 7\} = \max\{15 - 2, 7\} = 13$ and the column generation algorithm stops with an optimality gap of zero.

Using the *z*-cuts within the *ISP*, an optimal integer solution to the constrained shortest path example has been found without branching, see Figure 7.3. An application is presented in Example 7.9 (Preferential bidding system).

		RMP			*ISP*					
t	primal solution	z_{RMP}	π_0	π_7	$\bar{c}(\pi_7, \pi_0)$	p	c_p	t_p	*LB*	*UB*
1	$y_0 = 1$	100.0	100.00	0.00	−97.0	1246	3	18	3	
2	$y_0 = 0.22, \lambda_{1246} = 0.78$	24.6	100.00	−5.39	−32.9	1356	24	8	$-8.33 \mapsto 3$	
3	$\lambda_{1246} = 0.6, \lambda_{1356} = 0.4$	11.4	40.80	−2.10	−4.8	**13256**	**15**	10	$6.6 \mapsto$ **7**	**15**
4′	$\lambda_{13256} = 1$	15.0	15.00	0.00	−2.0	**13246**	**13**	13	**13**	**13**
	$x^\star_{12} = x^\star_{24} = x^\star_{45} = x^\star_{56} = 1$									

Table 7.2: Solution of the *MP* (and *IMP*) with the dynamically added z-cuts.

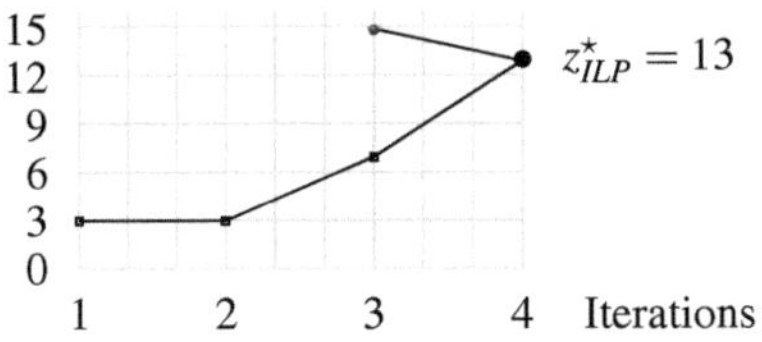

Fig. 7.3: Lower and upper bounds on $z^\star_{ILP}$ with the dynamically added z-cuts.

Discussion

Whenever there is choice, we are in need of advice about what to do. Theory advocates for integrating cuts into the subproblem since this is potentially stronger:

$$\begin{aligned} &\{\mathbf{x} \in \mathbb{R}^n_+ \mid \mathbf{Ax} \geq \mathbf{b}\} \cap \operatorname{conv}\{\mathbf{x} \in \mathbb{Z}^n_+ \mid \mathbf{Dx} \geq \mathbf{d},\ \mathbf{Fx} \geq \mathbf{f}\} \\ \subseteq\ &\{\mathbf{x} \in \mathbb{R}^n_+ \mid \mathbf{Ax} \geq \mathbf{b},\ \mathbf{Fx} \geq \mathbf{f}\} \cap \operatorname{conv}\{\mathbf{x} \in \mathbb{Z}^n_+ \mid \mathbf{Dx} \geq \mathbf{d}\}. \end{aligned} \tag{7.29}$$

However, this may not always be possible. When a cutting plane is linking several subproblems, an integration into the master problem is the way to go. Since they apparently "carry global information," we may be looking for such cuts anyway. When a specialized algorithm for the *ISP* is not compatible with the cutting planes, these must go into the master problem as well.

At this point, we are in the good situation that when aiming for stronger formulations, we do not have to decide for either Dantzig-Wolfe reformulation or the rich body of literature on cutting planes. Even better, since we always work with two models, we have more options to formulate cutting planes. This is discussed next.

Cutting planes on the master variables

We have already seen the cutting planes $\mathbf{Fx} \geq \mathbf{f}$ (7.7d) expressed in the λ-variables as $\sum_{p \in P} \mathbf{f}_p \lambda_p + \sum_{r \in R} \mathbf{f}_r \lambda_r \geq \mathbf{f}$ in (7.10c). However, these came as a byproduct of cuts which are designed to strengthen a model in the original x-variables. Additionally, we may strengthen the model by explicitly deriving information from the integrality of the λ-variables. Example situations are *IMP* and *ISP* ("pattern based") models that are stated without going through a Dantzig-Wolfe reformulation; or when we convexify a binary program; or when we apply discretization and arrive at reformulation *IM*$\ddot{P}$ (7.4).

Now we assume that we know a set of cutting planes $\sum_{p\in\ddot{P}} \mathbf{g}_p\lambda_p + \sum_{r\in\ddot{R}} \mathbf{g}_r\lambda_r \geq \mathbf{h}$ to be added to the linear relaxation of (7.4), the cut coefficients $\mathbf{g}_p, \mathbf{g}_r$ and $\mathbf{h}$ given as integer vectors (scaled if necessary). The resulting *M*$\ddot{P}$ reads as

$$
\begin{aligned}
z^\star_{M\ddot{P}} = \min \quad & \sum_{p\in\ddot{P}} c_p\lambda_p + \sum_{r\in\ddot{R}} c_r\lambda_r && && \text{(7.30a)}\\
\text{s.t.} \quad & \sum_{p\in\ddot{P}} \mathbf{a}_p\lambda_p + \sum_{r\in\ddot{R}} \mathbf{a}_r\lambda_r \geq \mathbf{b} && [\boldsymbol{\pi}_\mathbf{b}] && \text{(7.30b)}\\
& \sum_{p\in\ddot{P}} \mathbf{g}_p\lambda_p + \sum_{r\in\ddot{R}} \mathbf{g}_r\lambda_r \geq \mathbf{h} && [\boldsymbol{\gamma}] && \text{(7.30c)}\\
& \sum_{p\in\ddot{P}} \lambda_p = 1 && [\pi_0] && \text{(7.30d)}\\
& \lambda_p \geq 0 && \forall p\in\ddot{P} && \text{(7.30e)}\\
& \lambda_r \geq 0 && \forall r\in\ddot{R} && \text{(7.30f)}\\
& \sum_{p\in\ddot{P}} \mathbf{x}_p\lambda_p + \sum_{r\in\ddot{R}} \mathbf{x}_r\lambda_r = \mathbf{x} \in \mathbb{R}^n_+. && && \text{(7.30g)}
\end{aligned}
$$

where $\boldsymbol{\pi}_\mathbf{b}, \boldsymbol{\gamma} \geq \mathbf{0}$, and $\pi_0 \in \mathbb{R}$. From the restricted master problem point of view, there is no difference between the strengthened formulations (7.10) and (7.30). The differentiation comes to light in the *ISP*

$$
\begin{aligned}
\bar{c}(\boldsymbol{\pi}_\mathbf{b}, \boldsymbol{\gamma}, \pi_0) = -\pi_0 + \min_{\mathbf{x}\in\mathcal{D}} \quad & c_\mathbf{x} - \boldsymbol{\pi}_\mathbf{b}^\mathsf{T}\mathbf{a}_\mathbf{x} - \boldsymbol{\gamma}^\mathsf{T}\mathbf{g}_\mathbf{x}\\
\text{s.t.} \quad & c_\mathbf{x} = \mathbf{c}^\mathsf{T}\mathbf{x},\ \mathbf{a}_\mathbf{x} = \mathbf{A}\mathbf{x},\ \mathbf{g}_\mathbf{x} = \mathbf{g}(\mathbf{x}),
\end{aligned}
\tag{7.31}
$$

equivalently rewritten as

$$
\begin{aligned}
\bar{c}(\boldsymbol{\pi}_\mathbf{b}, \boldsymbol{\gamma}, \pi_0) = -\pi_0 + \min_{\mathbf{x}\in\mathcal{D}} \quad & \mathbf{c}^\mathsf{T}\mathbf{x} - \boldsymbol{\pi}_\mathbf{b}^\mathsf{T}\mathbf{A}\mathbf{x} - \boldsymbol{\gamma}^\mathsf{T}\mathbf{g}_\mathbf{x}\\
\text{s.t.} \quad & \mathbf{g}_\mathbf{x} = \mathbf{g}(\mathbf{x}),
\end{aligned}
\tag{7.32}
$$

where $\mathbf{g}_\mathbf{x}$ *adequately* computes the column coefficients of the added cuts, as does $\mathbf{c}^\mathsf{T}\mathbf{x}$ for the cost value $c_\mathbf{x}$ and $\mathbf{A}\mathbf{x}$ for the column coefficients $\mathbf{a}_\mathbf{x}$ (see Figure 7.4). However, observe that for some cuts on the master variables, it may not be possible to perform the (linear) substitution for $\mathbf{g}(\mathbf{x})$ in the objective function.

It remains the question of how the $\mathbf{g}(\mathbf{x})$ should "know about" the cut coefficients. In our illustrations below, we restrict ourselves to *rank-1 inequalities*, i.e., those where $\mathbf{g}(\mathbf{x})$ is derived from the column coefficients $\mathbf{a}_\mathbf{x}$ only, i.e., $\mathbf{g}_\mathbf{x} = \mathbf{g}(\mathbf{a}_\mathbf{x})$. It turns out that we can, in concrete cases, encode in a relatively easy way the *semantics* of the cut coefficients using the *extra variables* $\mathbf{g}_\mathbf{x}$ and a system of linear constraints encoding $\mathbf{g}_\mathbf{x} = \mathbf{g}(\mathbf{a}_\mathbf{x})$ in the *ISP*.

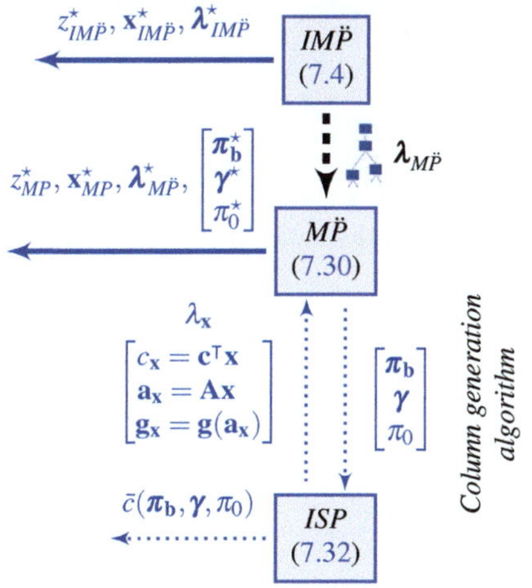

Fig. 7.4: Information flow of the column generation algorithm solving the *MP̈* (7.30) with cuts on the master variables, $\mathbf{g}_{\mathbf{x}}$ given as a function of $\mathbf{a}_{\mathbf{x}}$ in the *ISP* (7.32).

Illustration 7.4 **Chvátal-Gomory cuts**

The *Chvátal-Gomory cuts* are based on a simple observation:

$$\text{If } a \text{ is an integer number and } a \geq b, \text{ then } a \geq \lceil b \rceil. \tag{7.33}$$

Let us write the *IMP̈* using the finite set of variables indexed in $\mathcal{X}$ as

$$\begin{aligned} z^\star_{IM\ddot{P}} = \min \quad & \sum_{\mathbf{x} \in \mathcal{X}} c_{\mathbf{x}} \lambda_{\mathbf{x}} \\ \text{s.t.} \quad & \sum_{\mathbf{x} \in \mathcal{X}} \mathbf{a}_{\mathbf{x}} \lambda_{\mathbf{x}} \geq \mathbf{b} \qquad [\boldsymbol{\pi}] \\ & \lambda_{\mathbf{x}} \in \mathbb{Z}_+ \qquad\qquad \forall \mathbf{x} \in \mathcal{X} \subset \mathbb{Z}^n_+, \end{aligned} \tag{7.34}$$

assuming a single pricing problem. The procedure for defining a Chvátal-Gomory cut of rank-1 on $\sum_{\mathbf{x} \in \mathcal{X}} \mathbf{a}_{\mathbf{x}} \lambda_{\mathbf{x}} \geq \mathbf{b}$ in the *MP̈* is in three steps.

1. Choose a non-negative vector of weights $\mathbf{w} \in \mathbb{R}^m_+$ and left-multiply the system of constraints. This gives $\sum_{\mathbf{x} \in \mathcal{X}} \mathbf{w}^\intercal \mathbf{a}_{\mathbf{x}} \lambda_{\mathbf{x}} \geq \mathbf{w}^\intercal \mathbf{b}$.
2. The following inequality is thus valid: $\sum_{\mathbf{x} \in \mathcal{X}} \lceil \mathbf{w}^\intercal \mathbf{a}_{\mathbf{x}} \rceil \lambda_{\mathbf{x}} \geq \mathbf{w}^\intercal \mathbf{b}$.
3. Because $a = \sum_{\mathbf{x} \in \mathcal{X}} \lceil \mathbf{w}^\intercal \mathbf{a}_{\mathbf{x}} \rceil \lambda_{\mathbf{x}}$ is an integer number when $\lambda_{\mathbf{x}} \in \mathbb{Z}_+$, $\forall \mathbf{x} \in \mathcal{X}$, we use observation (7.33) with $b = \mathbf{w}^\intercal \mathbf{b}$ to obtain the cut

$$\sum_{\mathbf{x} \in \mathcal{X}} \lceil \mathbf{w}^\intercal \mathbf{a}_{\mathbf{x}} \rceil \lambda_{\mathbf{x}} \geq \lceil \mathbf{w}^\intercal \mathbf{b} \rceil. \tag{7.35}$$

In the case of an inequality system $\sum_{\mathbf{x} \in \mathcal{X}} \mathbf{a}_{\mathbf{x}} \lambda_{\mathbf{x}} \leq \mathbf{b}$, the observation $a \leq b$, where a is an integer number, leads to $a \leq \lfloor b \rfloor$, and the cut is given by

$$\sum_{\mathbf{x}\in\mathcal{X}} \lfloor \mathbf{w}^{\mathsf{T}}\mathbf{a}_{\mathbf{x}} \rfloor \lambda_{\mathbf{x}} \leq \lfloor \mathbf{w}^{\mathsf{T}}\mathbf{b} \rfloor. \tag{7.36}$$

If we combine equality constraints, both versions are valid. Assume we are given L cuts in (7.30c) expressed by (7.35) with associated $\boldsymbol{\gamma} \in \mathbb{R}_{+}^{|L|}$. For a row-index $\ell \in L$, coefficient $g_{\ell\mathbf{x}}$ (similarly for right-side h_ℓ) is calculated using the ceiling function taking vector $\mathbf{a}_{\mathbf{x}}$ as argument together with a weight vector $\mathbf{w}_\ell$, that is,

$$g_{\ell\mathbf{x}} = \lceil \mathbf{w}_\ell^{\mathsf{T}}\mathbf{a}_{\mathbf{x}} \rceil \qquad \forall \ell \in L, \mathbf{x} \in \mathcal{X} \tag{7.37a}$$

$$h_\ell = \lceil \mathbf{w}_\ell^{\mathsf{T}}\mathbf{b} \rceil \qquad \forall \ell \in L. \tag{7.37b}$$

The *ISP* (7.32) becomes

$$\begin{aligned} \bar{c}(\boldsymbol{\pi}_{\mathbf{b}}, \boldsymbol{\gamma}, \pi_0) = -\pi_0 \ + \ \min_{\mathbf{x}\in\mathcal{D}} \quad & \mathbf{c}^{\mathsf{T}}\mathbf{x} - \boldsymbol{\pi}_{\mathbf{b}}^{\mathsf{T}}\mathbf{A}\mathbf{x} - \boldsymbol{\gamma}^{\mathsf{T}}\mathbf{g}_{\mathbf{x}} \\ \text{s.t.} \quad & \mathbf{g}_{\mathbf{x}} = \left[\lceil \mathbf{w}_\ell^{\mathsf{T}}\mathbf{a}_{\mathbf{x}} \rceil\right]_{\ell\in L}, \end{aligned} \tag{7.38}$$

where $\left[\lceil \mathbf{w}_\ell^{\mathsf{T}}\mathbf{a}_{\mathbf{x}} \rceil\right]_{\ell\in L}$ denotes the Chvátal-Gomory function of rank 1. Observe that we are lucky regarding the *ceiling* and *floor* functions. Indeed, $y = \lceil x \rceil$ can be computed using the integer linear inequality $y \leq x + 1 - \varepsilon$, $y \in \mathbb{Z}$, for a small value $\varepsilon > 0$ and $g_{\ell\mathbf{x}} = \lceil \mathbf{w}_\ell^{\mathsf{T}}\mathbf{a}_{\mathbf{x}} \rceil$ in (7.37a) and (7.38) is replaced by

$$\begin{aligned} & g_{\ell\mathbf{x}} \leq \mathbf{w}_\ell^{\mathsf{T}}\mathbf{a}_{\mathbf{x}} + 1 - \varepsilon_\ell && \forall \ell \in L, \mathbf{x} \in \mathcal{X} \\ & g_{\ell\mathbf{x}} \in \mathbb{Z} && \forall \ell \in L, \mathbf{x} \in \mathcal{X}. \end{aligned} \tag{7.39}$$

Note that the non-negative sign of the dual values $\boldsymbol{\gamma}$ makes it attractive to have $g_{\ell\mathbf{x}}$ as large as possible. Similarly, $y = \lfloor x \rfloor$ can be replaced by $y \geq x - 1 + \varepsilon$, $y \in \mathbb{Z}$.

Illustration 7.5 Clique cuts for the set partitioning problem

In general mixed-integer programs, clique cuts can be derived from the so-called *conflict graph*. This graph carries information about pairs of binary variables which must not attain certain values simultaneously. We consider a basic version here and restrict the discussion to the linear relaxation of the set partitioning model for which the *MP* is given as

$$\begin{aligned} z_{MP}^{\star} = \min \quad & \sum_{\mathbf{x}\in\mathcal{X}} c_{\mathbf{x}}\lambda_{\mathbf{x}} \\ \text{s.t.} \quad & \sum_{\mathbf{x}\in\mathcal{X}} a_{i\mathbf{x}}\lambda_{\mathbf{x}} = 1 \quad [\pi_i \in \mathbb{R}] \quad i \in \{1, \ldots, m\} \\ & \lambda_{\mathbf{x}} \geq 0 \qquad\qquad \forall \mathbf{x} \in \mathcal{X}, \end{aligned} \tag{7.40}$$

where $\mathcal{X} = \{\mathbf{x}_p\}_{p\in P}$ is the set of extreme points of $\operatorname{conv}(\mathcal{D})$, $\mathcal{D} = \{\mathbf{x} \in \mathbb{Z}^n \mid \mathbf{D}\mathbf{x} \geq \mathbf{d}\}$, and the binary variable $\lambda_{\mathbf{x}}$ takes value 1 if and only if the binary column $\mathbf{a}_{\mathbf{x}} = [a_{i\mathbf{x}}]_{i=1,\ldots,m}$ is selected in the partition of the m rows. We assume that every column is computed by a linear function, say $\mathbf{a}_{\mathbf{x}} = \mathbf{A}\mathbf{x}$, as well as the cost $c_{\mathbf{x}}$ given by $\mathbf{c}^{\mathsf{T}}\mathbf{x}$.

The $\mathbf{x}$-vectors are solutions of the *ISP* given by

$$\begin{aligned} \bar{c}(\boldsymbol{\pi}) = \min_{\mathbf{x} \in \mathcal{D}} \quad & c_{\mathbf{x}} - \boldsymbol{\pi}^\mathsf{T} \mathbf{a}_{\mathbf{x}} \\ \text{s.t.} \quad & c_{\mathbf{x}} = \mathbf{c}^\mathsf{T}\mathbf{x}, \ \mathbf{a}_{\mathbf{x}} = \mathbf{A}\mathbf{x}. \end{aligned} \tag{7.41}$$

We define the *conflict graph* $G = (V, E)$ of (7.40) as follows. There is a node for every generated variable $\lambda_{\mathbf{x}}$, $\mathbf{x} \in \mathcal{X}' = \{\mathbf{x}_p\}_{p \in P'}$; we therefore identify $V = \mathcal{X}'$. Whenever it is infeasible to set $\lambda_{\mathbf{x}_1} = \lambda_{\mathbf{x}_2} = 1$ for some $\mathbf{x}_1 \neq \mathbf{x}_2$—this is called a *conflict*—we have an edge $\{\mathbf{x}_1, \mathbf{x}_2\} \in E$. Note that this is equivalent to $\mathbf{x}_1^\mathsf{T}\mathbf{x}_2 \geq 1$. Every clique $Q \subseteq V$ in G represents a subset of master variables, at most one of which can be set to value 1 in any integer solution to (7.40). This immediately gives the associated clique cut

$$\sum_{\mathbf{x} \in Q} \lambda_{\mathbf{x}} \leq 1 \qquad [\gamma_Q \leq 0]. \tag{7.42}$$

A negative reduced cost variable $\lambda_{\mathbf{x}}$ is either part of the clique cut (7.42) or not. Thus, in the *ISP*, we introduce a variable $g_Q \in \{0, 1\}$ for the binary cut coefficient and it appears in the additional non-negative term $-\gamma_Q g_Q$ in the objective function. We must ensure that $g_Q = 1$ if and only if $\mathbf{x}$ enlarges the clique Q, in which case $\mathbf{x}$ is in conflict with *all* $\mathbf{y} \in Q$. Let us introduce variable $z_{Q\mathbf{y}} \in \{0, 1\}$ to represent whether $\mathbf{x}$ conflicts with $\mathbf{y} \in Q$. For a generated $\mathbf{x} \in \mathcal{D}$, the correct binary values of g_Q and $z_{Q\mathbf{y}}$, $\forall \mathbf{y} \in Q$, can be set, e.g., using the constraints

$$\sum_{i=1}^{m} y_i x_i \leq \Big(\sum_{i=1}^{m} y_i\Big) z_{Q\mathbf{y}} \qquad \forall \mathbf{y} \in Q \tag{7.43a}$$

$$\sum_{\mathbf{y} \in Q} z_{Q\mathbf{y}} - |Q| + 1 \leq g_Q \leq \frac{1}{|Q|} \sum_{\mathbf{y} \in Q} z_{Q\mathbf{y}}. \tag{7.43b}$$

The constraints (7.43a) examine one by one the possible conflicts with every $\mathbf{y} \in Q$, fixing $z_{Q\mathbf{y}} = 1$ if $\sum_{i=1}^{m} y_i x_i \geq 1$, otherwise 0 if there is no conflict with $\mathbf{y}$ as it also sets $g_Q = 0$ in (7.43b) when $\sum_{\mathbf{y} \in Q} z_{Q\mathbf{y}} < |Q|$. These $|Q| + 2$ constraints are added to the *ISP* (7.41) for each Q for which we add a clique cut (7.42) to the *MP* (7.40). This extra burden can be handled, also heuristically, by tailoring to the specific application. Spoorendonk and Desaulniers (2010) present ideas for the *VRPTW*. Note that all the above also works for set packing master problems.

For the separation of a most violated clique cut, one puts a weight of $\lambda_{\mathbf{x}}$ to node $\mathbf{x} \in V$ and solves a maximum weighted clique problem in G. This is itself a hard problem, for which, however, good heuristics exist.

Note 7.6 (How much is too much?) As always with cutting planes, we may find (too) many of them. Then we have to keep an eye on the tradeoff between strengthening the relaxation and the computational burden this entails for re-optimizing a larger restricted master problem. One may first collect violated cuts in a pool, and then select from the pool according to some standard cut selection criteria.

Discussion

We have derived cutting planes in both ways, based on the integrality of the original or the master variables. However, we see that it is not the question whether cutting planes are formulated in the master variables or not (this is possible in both cases), but whether they can be expressed by reformulation of the original variables *alone* or not. More precisely, the former is exactly the special case of the latter when master cut coefficients can be expressed as in (7.11), that is, the values of $\mathbf{f}_\mathbf{x}$ are given by $\mathbf{Fx}$, linearly in $\mathbf{x}$.

Conversely, the cut coefficients $\mathbf{g}_\mathbf{x} = \mathbf{g}(\mathbf{a}_\mathbf{x})$ in Illustration 7.4 (Chvátal-Gomory cuts) are computed using linear inequalities in $\mathbf{x} \in \mathbb{Z}^n_+$ and $\mathbf{g}_\mathbf{x} \in \mathbb{Z}^{|L|}_+$ (similarly for Illustration 7.5, Clique cuts for the set partitioning problem). When we are not able to express $\mathbf{g}(\mathbf{x})$ linearly in $\mathbf{x}$, we need *extra variables* $\mathbf{g}$ and a system of constraints for $\mathbf{g}_\mathbf{x} = \mathbf{g}(\mathbf{x})$, say $\mathbf{Fx} + \mathbf{Gg} \geq \mathbf{f}$ for $\mathbf{x}, \mathbf{g} \in \mathbb{Z}^{n+|L|}_+$, i.e., we construct an *extended formulation* of the *ISP* (see Definitions 1.23–1.24, p. 25).

In the spirit of reverse engineering a compact formulation, we find the *extended compact formulation ILP* (7.44) from which a strengthened *MP* results by Dantzig-Wolfe reformulation:

$$
\begin{aligned}
z^\star_{ILP} = \min \quad & \mathbf{c}^\mathsf{T}\mathbf{x} \\
\text{s.t.} \quad & \mathbf{Ax} \geq \mathbf{b} \\
& \mathbf{Dx} \geq \mathbf{d} \\
& \mathbf{Fx} + \mathbf{Gg} \geq \mathbf{f} \\
& \mathbf{x}, \mathbf{g} \in \mathbb{Z}^{n+|L|}_+,
\end{aligned}
\tag{7.44}
$$

with $|L|$ extra variables in vector $\mathbf{g}$. As a side effect, enabled by discussions about cutting planes, we may find inspiration for different compact formulations for an optimization problem. Conversely, considering extended formulations from the literature may motivate finding cutting planes in master variables.

A main takeaway of this section is that cutting planes formulated in the original variables only cannot be stronger than those formulated in original and extra variables. We keep our observations in mind when we move on to branching.

7.3 Branching

At each node of a branch-and-price tree, the master problem is solved by column generation. If $\mathbf{x}^\star_{MP}$ in (7.5) is not integer (see Integrality Test) and $z^\star_{MP}$ is less than the incumbent's solution value *UB*, we need to make decisions. Even though, in principle, we can arrive at integer solutions by only adding cutting planes, this is not considered efficient. Instead, branching takes place: we split the set of integer solutions into the feasible domains of created nodes, thereby eliminating fractional solutions.

Note 7.7 (Know your branch-and-bound.) For standard branch-and-bound, there is a huge body of literature on branching rules. It is there for a reason: when we make, e.g., a 2-way branching decision "without any effect," we essentially only double the work. It is debatable what an "effect" is, but intuitively, branching should lead to some improvement, primal or dual, like arriving at integer solutions or obtaining (significantly) better local dual bounds in *all* created nodes. The latter is also known as aiming for a *balanced* search tree. This may help in improving the global dual bound. In branch-and-price, solving a node is typically (much) more costly than in branch-and-bound. Thus, branching decisions can become even more critical. Since branching interacts with pricing, we must also keep an eye on implications to the *ISP*. In what follows, we make several suggestions for finding *branching candidates*, that is, values that are fractional but should not be, so that we can branch on them. Check Ranking the candidates later for how to select a candidate when there are many.

Just as we had two options for finding cutting planes, we have the same choices for branching: decisions on the original x-variables *and/or* the master λ-variables. Both have been used with success in the literature, and we present them in turn.

Branching on the original variables

Branching on the "original" x-variables means decisions based on the fractional components of $\mathbf{x}^\star_{MP}$. Whenever $x^\star_j \notin \mathbb{Z}$ for some variable x_j, we can branch on it

$$x_j \leq \lfloor x^\star_j \rfloor \quad \text{or} \quad x_j \geq \lceil x^\star_j \rceil. \tag{7.45}$$

This is called a *variable disjunction* and it is a branch-and-bound classic. When the master problem is not aggregated, this suffices to arrive at an integer solution to the *ILP*. More generally, integer combinations of x-variables can be used for branching. When $\mathbf{f}^\intercal \mathbf{x}^\star_{MP} = \alpha \notin \mathbb{Z}$ for some $\mathbf{f} \in \mathbb{Z}^n$, we can impose the decisions

$$\mathbf{f}^\intercal \mathbf{x} \leq \lfloor \alpha \rfloor \quad \text{or} \quad \mathbf{f}^\intercal \mathbf{x} \geq \lceil \alpha \rceil. \tag{7.46}$$

Both versions (7.45) and (7.46) are special cases of cutting planes, formulated in the original variables of the *ILP* (7.1). We thus already know how to handle this. We have the two options that branching explicitly splits the *ISP* domain or not. Assume a grouping of the constraints of the *ILP* in sets $\mathcal{A}$ and $\mathcal{D}$.

- Either we add the constraints (7.46) to the set $\mathcal{A}$ like in (7.8). The domain $\mathcal{D}$ of the *ISP* is not modified. With $\mathcal{X}$ appropriately defined, in the down-branch, $\mathbf{f}^\intercal \mathbf{x} \leq \lfloor \alpha \rfloor$ appears in the *MP* as

$$\sum_{\mathbf{x} \in \mathcal{X}} f_{\mathbf{x}} \lambda_{\mathbf{x}} \leq \lfloor \alpha \rfloor \qquad [\gamma_1] \tag{7.47}$$

with the associated dual value $\gamma_1 \leq 0$, and $f_{\mathbf{x}} = \mathbf{f}^\mathsf{T}\mathbf{x}$, $\forall \mathbf{x} \in \mathcal{X}$. Then the objective function of the *ISP* is modified with the term $-\gamma_1 f_{\mathbf{x}}$, i.e., a positive slope on $f_{\mathbf{x}}$ favoring lower values as requested in (7.47). In the up-branch, we have a similar behavior:

$$\sum_{\mathbf{x}\in\mathcal{X}} f_{\mathbf{x}}\lambda_{\mathbf{x}} \geq \lceil\alpha\rceil \qquad [\gamma_2] \tag{7.48}$$

in the *MP*, where $\gamma_2 \geq 0$ impacts the objective function of the *ISP* as $-\gamma_2 f_{\mathbf{x}}$, i.e., a negative slope on $f_{\mathbf{x}}$ favoring the larger values for (7.48).

Although the domain of the *ISP* is not modified, new variables in the objective function may change the character of the pricing problem, see Note 7.3 (*A misinterpretation*).

- Or we explicitly split the set $\mathcal{D}$, and impose the branching constraints (7.46) like in (7.14); the domain of the *ISP* changes to:

$$\begin{aligned} &\{\mathbf{x} \in \mathcal{D} \mid \mathbf{f}^\mathsf{T}\mathbf{x} \leq \lfloor\alpha\rfloor\} \quad \text{in the down-branch node,} \\ &\{\mathbf{x} \in \mathcal{D} \mid \mathbf{f}^\mathsf{T}\mathbf{x} \geq \lceil\alpha\rceil\} \quad \text{in the up-branch node.} \end{aligned} \tag{7.49}$$

  Note again that for the branching decision to take effect, that is, to cut off the current fractional *MP* solution, all master variables that are incompatible with the branching decision must be eliminated from the *RMP*.

If one puts the relation $\mathbf{x} = \sum_{p\in P'} \mathbf{x}_p\lambda_p + \sum_{r\in R'} \mathbf{x}_r\lambda_r$ with integrality requirements on $\mathbf{x}$ in the *MP* (also known as the *explicit master*), a solver automatically branches on original variables, however, without "telling" the pricing problem. The latter must be done manually by transferring the local bounds on the $\mathbf{x}$-variables in the explicit master problem to the $\mathbf{x}$-variables in the *ISP*.

Since branching constraints are "convexified" in (7.49), the second option is usually preferred in terms of strength of the dual bound from the reformulation. Less interference with the *ISP* in an implementation may speak for the first option.

In Example 7.1, we look at x-branching decisions to solve our long lasting Time constrained shortest path problem. In some situations, it can be useful to split into more than two branches, as shown in the following Illustration 7.6.

Illustration 7.6 **x-branching for the** *VRPTW*

Recall the two-index formulation (7.21) for the *VRPTW*. Here are some possible branching ways.

- We have already seen a global one on the *ILP*: branching (or cutting) on the number of vehicles if the current value $x^\star_{do}$ is fractional. If the cost c_{do} is large enough, the number of vehicles is already minimized and only a rounding cut is added: $\lceil x^\star_{do}\rceil \leq x_{do} \leq \kappa$. Otherwise, two branches are created:

$$x_{do} \leq \lfloor x^\star_{do}\rfloor \text{ and } \lceil x^\star_{do}\rceil \leq x_{do} \leq \kappa. \tag{7.50}$$

- Another type of decisions concerns a single binary arc-flow variable of the *ISP*. Any fractional one, say x_{ij}, $(i,j) \in A$, can be set to 0 or 1 on the domain $\mathcal{D}$. This also generally implies other restrictions on the network structure, for example the elimination of a number of useless arcs or time interval reductions, see Example 7.1 (Time constrained shortest path problem) for some details.
- If a customer i is visited by more than one route, then the solution in the x_{ij}-variables to formulation (7.21) is necessarily fractional. This may also open opportunities for imposing decisions on the other type of variables present in the *ISP*, that is, the start of service time t_i for $i \in C$. If we assume integer data, two branches are created:

$$t_i \leq \lfloor t_i^\star \rfloor \quad \text{and} \quad \lfloor t_i^\star \rfloor + 1 \leq t_i. \tag{7.51}$$

A way to determine if customer $i \in C$ is a candidate for branching, and hence compute $t_i^\star$, is presented in Gélinas et al. (1995, Proposition 1). Recursively compute the *maximum earliest* and *minimum latest* service times at customer i using the routes visiting it, and denote them as L_i and U_i, respectively. Then i is a candidate if and only if $L_i > U_i$. By choosing $t_i^\star$ such that $U_i \leq t_i^\star < L_i$, one of the routes of the current solution is not feasible in each branch. These decisions create a new time window at customer i on each branch and are, therefore, easily treated within the *ISP*.
- Our final suggestion is derived from a fractional λ-variable in the solution of the *MP*. For example, assume that the value $\lambda^\star_{12345}$ is fractional for the *od*-path

$$o \rightsquigarrow 1 \rightsquigarrow 2 \rightsquigarrow 3 \rightsquigarrow 4 \rightsquigarrow 5 \rightsquigarrow d.$$

Figure 7.5 depicts the natural way of branching on this variable with two branches. As noted before, fixing λ_{12345} at 1 is easy but this is not the case when fixing it at 0 because the pricing may re-generate that positive basic variable.

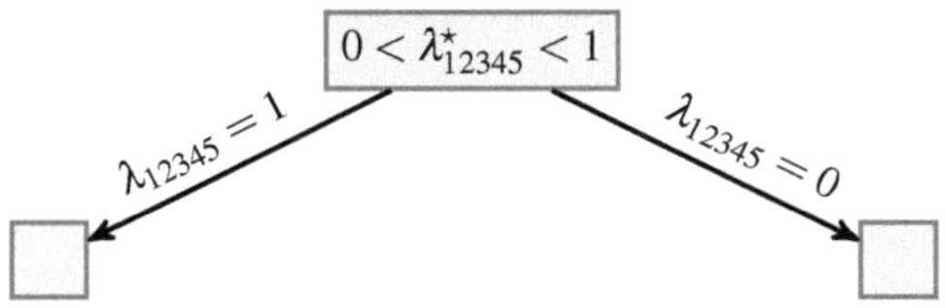

Fig. 7.5: The natural way to branch on a fractional binary λ-variable.

An alternative is to generalize the branching on a single arc-flow variable to a larger set of such variables: this creates multiple x_{ij}-branches as in Desrosiers et al. (1984). Selecting the four binary variables x_{12}, x_{23}, x_{34}, and x_{45} amongst those involved in the above *od*-path, Figure 7.6 depicts a possible partition into only five branches of the $2^4 = 16$ configurations.

1. The branch on the left fixes all four variables at 1. Observe that this does not correspond to $\lambda_{12345} = 1$ unless, for example, customer 1 has no predecessors and 5 has no successors, or if the capacity is exceeded with the addition of any customer. The other four branches are described next.
2. A branch that fixes $x_{12} = x_{23} = x_{34} = 1$ while the last variable x_{45} in the sequence takes value 0. This corresponds to one configuration.
3. A branch that fixes x_{12} and x_{23} to 1 while $x_{34} = 0$ and $x_{45} \in \{0,1\}$. This branch comprises 2 configurations.
4. The next to the last branch fixes $x_{12} = 1$ and $x_{23} = 0$. As $x_{34}, x_{45} \in \{0,1\}$, it comprises 4 configurations.
5. The last branch comprises 8 configurations by only fixing $x_{12} = 0$, leaving free the three other variables.

Although presented here in the context of a fractional path-variable, the more-than-two-branches strategy can be done on any subset of at least two fractional arc-variables.

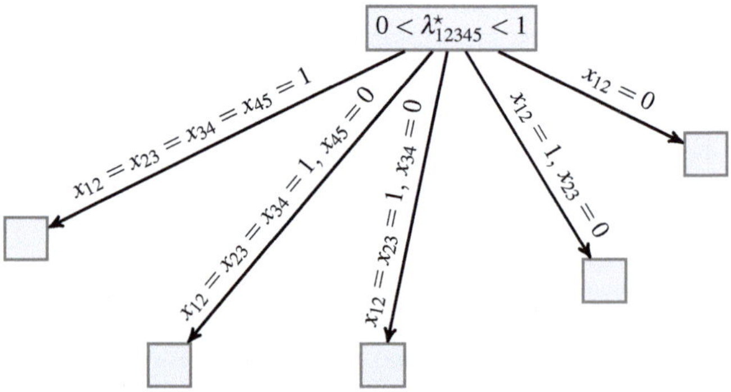

Fig. 7.6: A branching partition for the sequence $1 \rightsquigarrow 2 \rightsquigarrow 3 \rightsquigarrow 4 \rightsquigarrow 5$.

Note 7.8 (It's all about timing) Even though we wrote about *eliminating* variables from the master problem, this is not actually done in an implementation. Remember that there is "only one" linear program maintained in the entire branch-and-price tree, which is locally modified in every node. If variables would be deleted in one node, it would in fact be deleted from all nodes. This is why one works, e.g. with local bounds on the variables that fix them to zero. This brings us to the question when these bounds should be applied. And the answer lies in the above: not when a branching decision is made, but when a node or one of its descendants is visited to be processed.

Identical subproblems and disaggregation

In the case of aggregated master variables,

$$\lambda_{\mathbf{x}} = \sum_{k\in K} \lambda^k_{\mathbf{x}}, \qquad \forall \mathbf{x} \in \mathcal{X}, \tag{7.52}$$

like in (4.53), we have direct access only to $\mathbf{x}_{MP} = \sum_{k\in K} \mathbf{x}^k_{MP}$, the *aggregated original* variables which we compute as $\mathbf{x}_{MP} = \sum_{\mathbf{x}\in\mathcal{X}} \mathbf{x}\lambda_{\mathbf{x}}$. These are sometimes useful, like for the *VRPTW* where they can take fractional values in the formulation (5.7), see also Desaulniers et al. (1998a). Sometimes they are not useful at all. Let us illustrate this with the help of the bin packing problem, in which we have to pack n items of size w_i, $i \in \{1,\ldots,n\}$, into $|K|$ identical bins of capacity W each. The number of used bins should be minimized and the textbook *ILP* reads as follows:

$$z^\star_{ILP} = \min \quad \sum_{k\in K} x^k_0 \tag{7.53a}$$

$$\text{s.t.} \quad \sum_{k\in K} x^k_i = 1 \qquad \forall i \in \{1,\ldots,n\} \tag{7.53b}$$

$$\sum_{i=1}^{n} w_i x^k_i \le W x^k_0 \qquad \forall k \in K \tag{7.53c}$$

$$x^k_i \in \{0,1\} \qquad \forall i \in \{0,\ldots,n\}, k \in K. \tag{7.53d}$$

A Dantzig-Wolfe reformulation by convexification of (7.53c)–(7.53d) and aggregation over K leads to the classic "pattern based" model, with set P of packing patterns. The binary parameter a_{ip} indicates whether pattern p contains item i or not:

$$\begin{aligned} z^\star_{IMP} = \min \quad & \sum_{p\in P} \lambda_p \\ \text{s.t.} \quad & \sum_{p\in P} a_{ip}\lambda_p = 1 \qquad \forall i \in \{1,\ldots,n\} \\ & \lambda_p \in \{0,1\} \quad \forall p \in P. \end{aligned} \tag{7.54}$$

We have not only deliberately lost the index k in the aggregated master variables $\lambda_p = \sum_{k\in K} \lambda^k_p$, but also does the *ISP* use binary variables x_i for every item $i \in \{1,\ldots,n\}$, without any reference to a specific k. These variables, as representatives for *some* bin, can be seen as aggregated original variables

$$x_i = \sum_{k\in K} x^k_i = \sum_{k\in K}\sum_{p\in P} a_{ip}\lambda^k_p \quad \left(= \sum_{p\in P} a_{ip}\Big(\sum_{k\in K} \lambda^k_p\Big) = \sum_{p\in P} a_{ip}\lambda_p = 1 \right), \tag{7.55}$$

that reflect how often we pack item i in all selected patterns. For this model, the answer is easy and comes directly from the set partitioning constraints in (7.54): every item is packed exactly once. The aggregated original variables are integer, regardless of the, potentially fractional, values of the aggregated master λ_p-variables. And this means that we cannot branch on the aggregated x_i-values.

For $\mathbf{x} \in \mathcal{X}$, disaggregation of $\lambda_{\mathbf{x}}$ via the set $\{\lambda_{\mathbf{x}}^k\}_{k \in K}$, into $\{\mathbf{x}_{MP}^k\}_{k \in K}$, is fail-safe because branching on the $\mathbf{x}_{MP}^k$, $k \in K$, finally produces an integer solution to the *ILP*. However, note again from the options we discussed in Chapter 4 (p. 206) that a mapping from the aggregated master variables $\lambda_{\mathbf{x}}$ to the disaggregated $\{\lambda_{\mathbf{x}}^k\}_{k \in K}$, is not unique. We do, for example, prefer the solution from the zero-cost balanced transportation problem (4.62) over the symmetric "even split" disaggregation in (4.57): The latter may produce a completely fractional solution while the former is likely "more λ-integral." In fact, when $\boldsymbol{\lambda}_{RMP}^{\star}$ is integral, the solution $\{\boldsymbol{\lambda}_{RMP}^{k\star}\}_{k \in K}$ to the transportation problem (4.62) is integer, and so is $\{\mathbf{x}_{MP}^{k\star}\}_{k \in K}$. This is also important insofar that a useful disaggregation is needed at the very latest when we produce a solution to the *ILP*, where integer $\{\mathbf{x}_{MP}^{k\star}\}_{k \in K}$ values have to be put on the table.

On the downside of disaggregation during branching, it (at least partially) re-introduces the symmetry in index k that we wanted to eliminate by performing the aggregation in the first place: When we have branched on variables $\mathbf{x}^k$, $k \in K' \subseteq K$, already, we need to differentiate between $|K'|$ pricing problems ISP^k, $k \in K'$, and one more *ISP* "without" index k. An alternative is to formulate a different *ILP* without the index k, as discussed in Extended compact and subproblem formulations or by using the results of Propositions 4.16 or 4.17, respectively, on *ISP* formulations defined on a polyhedral cone or having the integrality property. For the bin packing problem, we can have a network-based formulation, see Example 4.1 on the knapsack problem and Example 4.2 on the cutting stock problem.

And yet another alternative (drum roll!) is to read on: we branch on the λ-variables.

Branching on the master variables

Similarly to what we did for cutting planes, we may derive branching decisions explicitly from the integrality of master variables. In a series of papers, François Vanderbeck (2000, 2005, 2011) develops many general ideas for this, and the following content is largely inspired by his œuvre.

Fig. 7.7: François Vanderbeck, a great contributor to generic branch-and-price algorithmics and implementation, with an admirer (Aussois, France, 2019-01-07).

Let us write again the $IM\ddot{P}$ using the finite set of variables indexed in $\mathcal{X}$ as

$$
\begin{aligned}
z^{\star}_{IM\ddot{P}} = \min \quad & \sum_{\mathbf{x}\in\mathcal{X}} c_{\mathbf{x}}\lambda_{\mathbf{x}} \\
\text{s.t.} \quad & \sum_{\mathbf{x}\in\mathcal{X}} \mathbf{a}_{\mathbf{x}}\lambda_{\mathbf{x}} \geq \mathbf{b} \qquad [\boldsymbol{\pi}] \\
& \lambda_{\mathbf{x}} \in \mathbb{Z}_{+} \qquad\qquad \forall \mathbf{x} \in \mathcal{X} \subset \mathbb{Z}^{n}_{+},
\end{aligned}
\tag{7.56}
$$

assuming a single pricing problem. The notation and discussion can be adapted to other cases like block-diagonal structures with identical or non-identical blocks.

When the master problem solution $\boldsymbol{\lambda}^{\star}_{RMP}$ is fractional, it seems natural to branch on a single $\lambda_{\mathbf{x}}$-variable, but this is not a good idea in general. For an intuition, think of the binary case. When $\lambda^{\star}_{\mathbf{x}} \notin \{0,1\}$ for some $\mathbf{x} \in \mathcal{X}$, the up-branch ($\lambda^{\star}_{\mathbf{x}} = 1$) fixes $\mathbf{x}$, $c_{\mathbf{x}}$ and $\mathbf{a}_{\mathbf{x}}$, as part of the *ILP* and $IM\ddot{P}$ solutions, with a likely impact on the dual bound. On the down-branch ($\lambda^{\star}_{\mathbf{x}} = 0$), however, only one particular $\mathbf{x}$ out of so many is forbidden, and a "similar" variable can take the role of $\lambda_{\mathbf{x}}$ in the *RMP* solution. This down-branch has essentially no effect, in particular not on the dual bound, yielding an unbalanced tree. Worse, in the pricing problem, one has to forbid the re-generation of $\mathbf{x}$, which is difficult, if not impossible to accomplish.

Branching on partitioning $\mathcal{X}$ with a single hyperplane

We assume in the rest of this section that the reformulated domain $\mathcal{D}$ is defined either as a polytope or a polyhedral cone (see subsection Polytope and polyhedral cone in Chapter 4). For both domain types, it is therefore not necessary for the index-set $\mathcal{X}$ to use the binary parameter to differentiate between points and rays.

Let us develop ideas behind branching on constraints in the $\lambda_{\mathbf{x}}$-variables. Classical (original x-variables) branching recursively splits the fractional solution space in two (or more) parts, and as an intended effect, partitions the integer points. The integrality of a solution in the end of the branching process is guaranteed by explicitly eliminating the "space between integer $\mathbf{x}$ points," the fractionality in $\mathbf{x}$. Now, Theorem 4.2 (Hilbert-Giles-Pulleyblank) says that we combine an integer $\mathbf{x}_{IM\ddot{P}}$ solution from an integer number of integer points and rays in $\mathcal{X}$. Remember that the coefficients of this combination are precisely our $\lambda_{\mathbf{x}}$-variables. We conclude that for every subset of integer points and rays in $\mathcal{X}$, the sum of the corresponding $\lambda_{\mathbf{x}}$-variables is integer in any integer solution $\mathbf{x}_{IM\ddot{P}}$. This is true regardless of whether we have one or several pricing problems, whether an aggregation of identical pricing problems was performed or not, and whether the pricing problems are bounded or not.

This motivates the following. We identify a condition on $\mathbf{x}$ that splits in two the vectors in $\mathcal{X}$ such that the *sum* of the corresponding $\lambda_{\mathbf{x}}$-variables on one part is fractional, and then we branch on that sum. This eliminates the fractionality in the space of the $\lambda_{\mathbf{x}}$-variables directly, which implicitly removes the fractionality in $\mathbf{x}$. Formulating the split condition in $\mathbf{x}$ allows us to "check" it in the pricing problem.

We first see that it is always possible to find a single *linear* constraint in $\mathbf{x}$ that separates a fractional master solution.

Proposition 7.1. (Vanderbeck, 2000, Proposition 2, p. 118) *Given a fractional solution $\boldsymbol{\lambda}^\star_{RMP}$, there exists a hyperplane $\mathbf{f}^\intercal\mathbf{x} = f$, $(\mathbf{f}, f) \in \mathbb{Z}^{n+1}$, such that*

$$\beta = \sum_{\mathbf{x}\in\mathcal{X}:\mathbf{f}^\intercal\mathbf{x}\geq f} \lambda^\star_{\mathbf{x}} \quad \textit{is fractional.} \tag{7.57}$$

Proof. Let $\mathcal{F} = \{\mathbf{x} \in \mathcal{X} \mid \lambda^\star_{\mathbf{x}} - \lfloor\lambda^\star_{\mathbf{x}}\rfloor > 0\} \subset \mathbb{Z}^n_+$ denote the index-set of the fractional variables of $\boldsymbol{\lambda}^\star_{RMP}$. Observe that $\mathcal{F}$ also corresponds to the set of extreme points of $\text{conv}(\mathcal{F})$. For an arbitrary but fixed fractional variable $\lambda^\star_q$, $\mathbf{x}_q \in \mathcal{F}$, define the subset $\mathcal{F}' = \mathcal{F} \setminus \{\mathbf{x}_q\}$. There exists a hyperplane separating $\mathbf{x}_q$ from the rational polytope $\text{conv}(\mathcal{F}')$. That is, for some hyperplane expressed by $\mathbf{f}^\intercal\mathbf{x} = f$, $(\mathbf{f}, f) \in \mathbb{Z}^{n+1}$, we have $\mathbf{x}_q$ on one side, and the other points on the other side:

$$\begin{aligned} \mathbf{f}^\intercal\mathbf{x}_q &\geq f \\ \mathbf{f}^\intercal\mathbf{x} &\leq f - 1, \qquad \forall \mathbf{x} \in \text{conv}(\mathcal{F}') \cap \mathbb{Z}^n_+. \end{aligned} \tag{7.58}$$

The set of variables in $\sum_{\mathbf{x}\in\mathcal{X}:\mathbf{f}^\intercal\mathbf{x}\geq f} \lambda^\star_{\mathbf{x}}$ whose index-vector $\mathbf{x}$ satisfies the prescribed condition $\mathbf{f}^\intercal\mathbf{x} \geq f$ includes $\lambda^\star_q$ by construction as the only fractional one, but potentially also a number of integer variables, either positive or zero valued. Because

$$\sum_{\mathbf{x}\in\mathcal{X}:\mathbf{f}^\intercal\mathbf{x}\geq f} \lambda^\star_{\mathbf{x}} - \left\lfloor \sum_{\mathbf{x}\in\mathcal{X}:\mathbf{f}^\intercal\mathbf{x}\geq f} \lambda^\star_{\mathbf{x}} \right\rfloor = \lambda^\star_q - \lfloor\lambda^\star_q\rfloor > 0, \text{ then}$$

$$\beta = \sum_{\mathbf{x}\in\mathcal{X}:\mathbf{f}^\intercal\mathbf{x}>f} \lambda^\star_{\mathbf{x}} = \lambda^\star_q - \lfloor\lambda^\star_q\rfloor + \left\lfloor \sum_{\mathbf{x}\in\mathcal{X}:\mathbf{f}^\intercal\mathbf{x}\geq f} \lambda^\star_{\mathbf{x}} \right\rfloor \quad \text{is fractional.} \qquad \square$$

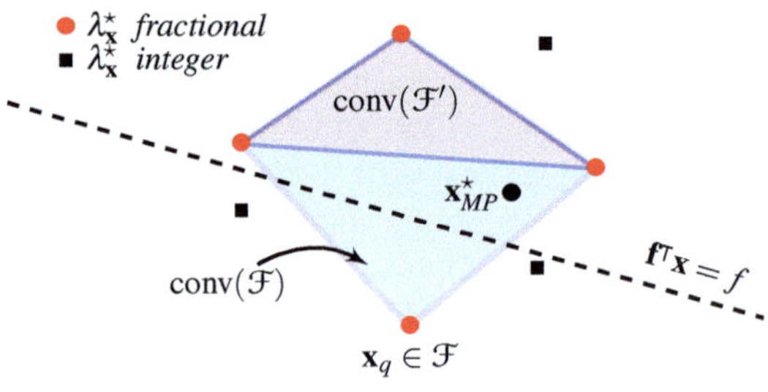

Fig. 7.8: Solution $\mathbf{x}^\star_{MP}$ as a positive combination of seven positive $\lambda^\star_{\mathbf{x}}$-variables.

Let us illustrate some elements of the proof using Figure 7.8. The optimal solution $\mathbf{x}^\star_{MP} \in \mathbb{R}^n_+$ is here a positive combination of seven positive $\lambda^\star_{\mathbf{x}}$-variables, either fractional (•) or integer ones (▪). There are four fractional variables and $\mathbf{x}_q$ is selected as the lowest one in $\mathcal{F}$. It is visually easy to *draw* a hyperplane $\mathbf{f}^\intercal\mathbf{x} = f$ separating $\mathbf{x}_q$ from the polytope $\text{conv}(\mathcal{F}')$ derived from the three other fractional variables.

Let $\mathbf{f}^\mathsf{T}\mathbf{x} \geq f$ be the closed half-space below the hyperplane such that $\mathbf{f}^\mathsf{T}\mathbf{x}_q \geq f$. In $\sum_{\mathbf{x}\in\mathcal{X}:\mathbf{f}^\mathsf{T}\mathbf{x}\geq f} \lambda^\star_\mathbf{x}$, only three positive variables appear: the fractional variable $\lambda^\star_q$ and two integer-valued. This sum is obviously fractional, i.e., $\sum_{\mathbf{x}\in\mathcal{X}:\mathbf{f}^\mathsf{T}\mathbf{x}\geq f} \lambda^\star_\mathbf{x} = \beta \notin \mathbb{Z}_+$.

Master problem modifications. Once we have identified a condition $\mathbf{f}^\mathsf{T}\mathbf{x} \geq f$ such that the sum in (7.57) is fractional, we branch on that sum, i.e., on the number of all $\lambda_\mathbf{x}$-variables that correspond to $\mathbf{x}$ which satisfy the condition:

$$\sum_{\mathbf{x}\in\mathcal{X}:\mathbf{f}^\mathsf{T}\mathbf{x}\geq f} \lambda_\mathbf{x} \leq \lfloor \beta \rfloor \quad \text{or} \quad \sum_{\mathbf{x}\in\mathcal{X}:\mathbf{f}^\mathsf{T}\mathbf{x}\geq f} \lambda_\mathbf{x} \geq \lceil \beta \rceil. \tag{7.59}$$

We write the respective master problems as follows

$$\begin{array}{lll|lll}
\min & \sum_{\mathbf{x}\in\mathcal{X}} c_\mathbf{x}\lambda_\mathbf{x} & & \min & \sum_{\mathbf{x}\in\mathcal{X}} c_\mathbf{x}\lambda_\mathbf{x} & \\
\text{s.t.} & \sum_{\mathbf{x}\in\mathcal{X}} \mathbf{a}_\mathbf{x}\lambda_\mathbf{x} \geq \mathbf{b} & [\boldsymbol{\pi} \geq \mathbf{0}] & \text{s.t.} & \sum_{\mathbf{x}\in\mathcal{X}} \mathbf{a}_\mathbf{x}\lambda_\mathbf{x} \geq \mathbf{b} & [\boldsymbol{\pi} \geq \mathbf{0}] \\
& \sum_{\mathbf{x}\in\mathcal{X}} g_\mathbf{x}\lambda_\mathbf{x} \leq \lfloor \beta \rfloor & [\gamma_1 \leq 0] & & \sum_{\mathbf{x}\in\mathcal{X}} g_\mathbf{x}\lambda_\mathbf{x} \geq \lceil \beta \rceil & [\gamma_2 \geq 0] \\
& \lambda_\mathbf{x} \geq 0 & \forall \mathbf{x} \in \mathcal{X} & & \lambda_\mathbf{x} \geq 0 & \forall \mathbf{x} \in \mathcal{X}
\end{array} \tag{7.60}$$

with a binary coefficient $g_\mathbf{x}$ for all $\mathbf{x} \in \mathcal{X}$ that takes value 1 if and only if the condition $\mathbf{f}^\mathsf{T}\mathbf{x} \geq f$ is satisfied.

Carefully note in (7.47)–(7.48) that the coefficients $f_\mathbf{x}$, $\forall \mathbf{x} \in \mathcal{X}$, of the $\lambda_\mathbf{x}$-variables in x-branching are not restricted to binary values.

Pricing problem modifications. As we already know from our discussion on cutting planes, the pricing problem is responsible for adequately producing the $g_\mathbf{x}$ coefficients. Consider the down-branch with a dual value $\gamma_1 \leq 0$. Similarly to (7.31), the *ISP* is modified to

$$\begin{aligned}
\bar{c}(\boldsymbol{\pi}, \gamma_1) = \quad & \min_{\mathbf{x}\in\mathcal{X}} \quad c_\mathbf{x} - \boldsymbol{\pi}^\mathsf{T}\mathbf{a}_\mathbf{x} - \gamma_1 g_\mathbf{x} \\
& \text{s.t.} \quad c_\mathbf{x} = \mathbf{c}^\mathsf{T}\mathbf{x},\ \mathbf{a}_\mathbf{x} = \mathbf{A}\mathbf{x},\ g_\mathbf{x} = g(\mathbf{x}).
\end{aligned} \tag{7.61}$$

We assume that we know lower and upper bounds on $\mathbf{f}^\mathsf{T}\mathbf{x}$, say $\ell \leq \mathbf{f}^\mathsf{T}\mathbf{x} \leq u$. Then the binary coefficient computes as

$$g(\mathbf{x}) = \begin{cases} 1 & \text{if} \quad f \leq \mathbf{f}^\mathsf{T}\mathbf{x} \leq u \\ 0 & \text{if} \quad \ell \leq \mathbf{f}^\mathsf{T}\mathbf{x} \leq f-1 \end{cases}. \tag{7.62}$$

One way to express $g(\mathbf{x})$ is given by the two inequalities in (7.63): the interval $[f, u]$ is activated with $g_\mathbf{x} = 1$, and otherwise interval $[\ell, f-1]$ with $g_\mathbf{x} = 0$:

$$\ell(1-g_\mathbf{x}) + f g_\mathbf{x} \leq \mathbf{f}^\mathsf{T}\mathbf{x} \leq u g_\mathbf{x} + (1-g_\mathbf{x})(f-1), \qquad g_\mathbf{x} \in \{0,1\}. \tag{7.63}$$

The same set of inequalities (7.63) is used for the pricing in the up-branch, which only slightly differs from (7.61) in the objective function, with dual value $\gamma_2 \geq 0$:

$$\begin{aligned}\bar{c}(\boldsymbol{\pi},\gamma_2) = \quad & \min_{\mathbf{x}\in\mathcal{X}} \quad c_{\mathbf{x}} - \boldsymbol{\pi}^\intercal \mathbf{a}_{\mathbf{x}} - \gamma_2 g_{\mathbf{x}} \\ & \text{s.t.} \quad c_{\mathbf{x}} = \mathbf{c}^\intercal \mathbf{x},\ \mathbf{a}_{\mathbf{x}} = \mathbf{A}\mathbf{x},\ g_{\mathbf{x}} = g(\mathbf{x}).\end{aligned} \tag{7.64}$$

In Exercise 7.5, we develop an alternative way of writing (7.63) as

$$g(\mathbf{x}) = \left\lceil \frac{\mathbf{f}^\intercal \mathbf{x} - (f-1)}{u-\ell} \right\rceil, \tag{7.65}$$

the linearization of which is given by (7.39) with an additional integer variable. When we exploit the sign of the dual values γ_1 and γ_2 in their respective pricing problem (i.e., in their up- or down-branch node), we develop a third way where not both inequalities in (7.63) are needed, see Vanderbeck (2000, eqs. (17)–(18)) and Exercise 7.4:

$$\begin{aligned} & g_{\mathbf{x}} \geq \frac{\mathbf{f}^\intercal \mathbf{x} - f + 1}{u - f + 1}, \quad g_{\mathbf{x}} \in \{0,1\} \quad \text{in the } \textit{ISP} \text{ (7.61) for the down-branch;} \\ & g_{\mathbf{x}} \leq \frac{\mathbf{f}^\intercal \mathbf{x} - \ell}{f - \ell}, \qquad g_{\mathbf{x}} \in \{0,1\} \quad \text{in the } \textit{ISP} \text{ (7.64) for the up-branch.} \end{aligned} \tag{7.66}$$

In all these variants for $g(\mathbf{x})$, the *ISP* modifications can be expressed as linear constraints and potentially extra variables. We wish to emphasize that this λ-branching does *not need* to explicitly partition, at every node of the search tree, the domain $\mathcal{X}$ of the *ISP*. But it could. Here is a fourth option for generating the two types of columns at a node, with an exponentially growing number of types when additional λ-branching decisions are imposed.

At a given branching node, say the down-branch, the *RMP* derived from (7.60) contains both types of columns, that is, those of the first type ($g_{\mathbf{x}} = 1$) satisfying $f \leq \mathbf{f}^\intercal \mathbf{x} \leq u$ and those of the second type ($g_{\mathbf{x}} = 0$) where $\ell \leq \mathbf{f}^\intercal \mathbf{x} \leq f - 1$. Given these non-identical types, the pricing step can *also* be solved using *two* pricing problems, say ISP_1 and ISP_2, where the domains of $\mathbf{x}$ are

$$\begin{aligned} \mathcal{X}_1 &= \{\mathbf{x} \in \mathcal{X} \mid \mathbf{f}^\intercal \mathbf{x} \in [f, u]\} \qquad \text{in } \mathit{ISP}_1, \\ \mathcal{X}_2 &= \{\mathbf{x} \in \mathcal{X} \mid \mathbf{f}^\intercal \mathbf{x} \in [\ell, f-1]\} \quad \text{in } \mathit{ISP}_2. \end{aligned} \tag{7.67}$$

The two pricing problems in this down-branch node become

$$\begin{aligned} \bar{c}(\boldsymbol{\pi},\gamma_1) &= \min_{\mathbf{x}\in\mathcal{X}_1} \quad \mathbf{c}^\intercal \mathbf{x} - \boldsymbol{\pi}^\intercal \mathbf{A}\mathbf{x} - \gamma_1 \quad \text{in } \mathit{ISP}_1, \\ \bar{c}(\boldsymbol{\pi},\gamma_1) &= \min_{\mathbf{x}\in\mathcal{X}_2} \quad \mathbf{c}^\intercal \mathbf{x} - \boldsymbol{\pi}^\intercal \mathbf{A}\mathbf{x} \qquad\quad \text{in } \mathit{ISP}_2. \end{aligned} \tag{7.68}$$

With q λ-branching decisions already made, this could create up to 2^q types for the columns, i.e, 2^q specialized subsets $\mathcal{X}_1, \mathcal{X}_2, \ldots, \mathcal{X}_{2^q} \subset \mathcal{X}$ instead of a single *ISP* imposing on $\mathcal{X}$ up to q sets of linear constraints, in up to q extra binary variables. The shear number of types would quickly get out of hand, but it does not have to if we take some care. When we formulate as second branching condition one that either splits $\mathcal{X}_1$ or $\mathcal{X}_2$ in (7.67) (and keeps the respective other one *intact*), we only have a partition of $\mathcal{X}$ into three parts. In general, we produce only a *linear* number of subsets $\mathcal{X}_1, \mathcal{X}_2, \ldots, \mathcal{X}_{q+1} \subset \mathcal{X}$ after having branched q times. Vanderbeck (2011), in his *generic* branching scheme, calls this a *nested partition* of $\mathcal{X}$, and describes conditions for branching that accomplish it. We note that in practice, we know how to manage a large number of *ISP*s by solving, exactly or not, a small number of them at every column generation iteration, then updating the dual values before solving for another small number of types (as in a partial pricing strategy); see for example Gamache et al. (1999) who dealt with 1131 subproblems decades ago.

Finally note that if we are given two column types and $\sum_{\mathbf{x}\in\mathcal{X}:\mathbf{f}^\intercal\mathbf{x}\geq f} \lambda_{\mathbf{x}} \leq 2$ for the first type, if the total sum of variables $\sum_{\mathbf{x}\in\mathcal{X}} \lambda_{\mathbf{x}}$ is quite large, you should probably call the ISP_2 more often to generate from this specialized generator the numerous requested columns of the second type.

The proof of Proposition 7.1 does not show how to construct or choose a separating hyperplane $\mathbf{f}^\intercal\mathbf{x} \geq f$, let alone one that is *good for the branching tree*. In practice, we could make use of problem knowledge, e.g., enforce a cardinality constraint on each branch for the number of binary $\lambda_{\mathbf{x}}$-variables satisfying, for example, a threshold on the cost or on the number of visited customers in *VRPTW* applications.

Illustration 7.7 **λ-branching for the** *VRPTW*

Let us consider again the Illustration 7.2 with the fractional master problem solution $\lambda^\star_{132} = \lambda^\star_{134} = \lambda^\star_{24} = 0.5$ of total cost $z^\star_{MP} = 65.5$, with respective route costs 37, 48, and 46. Let the route cost be computed as $c_{\mathbf{x}} = \sum_{(i,j)\in A} c_{ij}x_{ij}$ and the number of visited customers as $n_{\mathbf{x}} = \sum_{i\in C}\sum_{(i,j)\in A} x_{ij}$. Both are linear functions in the x_{ij}-variables.

$\mathbf{x}$	132	134	24
$\lambda^\star_{\mathbf{x}}$	0.5	0.5	0.5
$c_{\mathbf{x}}$	37	48	46
$n_{\mathbf{x}}$	3	3	2

By inspection, we find some fractional-valued sums for branching, each isolating one fractional-valued variable from the others:

$$\sum_{\mathbf{x}\in\mathcal{F}:\,c_{\mathbf{x}}\leq 37} \lambda^\star_{\mathbf{x}} = 0.5\ (= \lambda^\star_{132}); \quad \sum_{\mathbf{x}\in\mathcal{F}:\,c_{\mathbf{x}}\geq 48} \lambda^\star_{\mathbf{x}} = 0.5\ (= \lambda^\star_{134}); \quad \sum_{\mathbf{x}\in\mathcal{F}:\,n_{\mathbf{x}}\leq 2} \lambda^\star_{\mathbf{x}} = 0.5\ (= \lambda^\star_{24}).$$

As a fourth suggestion, we observe that the sum of the generated variables, indeed the three fractional ones, happens to be fractional as well: $\sum_{\mathbf{x}\in R'} \lambda^\star_{\mathbf{x}} = 1.5$.

Because a $\boldsymbol{\lambda}^\star_{IM\dot{P}}$-solution is given by $\lambda^\star_{32} = \lambda^\star_{14} = 1$, with respective costs 30 and 44, this optimal solution appears in the up-branch ($\sum_{\mathbf{x}\in\mathcal{X}:\,c_{\mathbf{x}}\leq 37} \lambda_{\mathbf{x}} \geq 1$) for the first suggestion, and also the up-branch ($\sum_{\mathbf{x}\in\mathcal{X}} \lambda_{\mathbf{x}} \geq 2$) for the fourth one. Interestingly, such branching decisions are appropriate in case of vehicles of homogeneous or different capacities, i.e., identical blocks or not.

Branching on lower and upper bounds on the x-variables

Splitting off only one or few points from $\mathcal{X}$ as done in the proof of Proposition 7.1 is likely not a good idea for a balanced branching decision. Probably more "impactful," and yet the simplest hyperplanes one can think of are inspired by branching on original variables, namely bounds on a single x_j variable as in (7.45).

Assume that $x_j^\star$ is fractional in $\mathbf{x}_{MP}^\star$. In the format of Proposition 7.1, we formulate the condition $x_j \geq \lceil x_j^\star \rceil$ on $\mathbf{x} \in \mathcal{X}$ as $\mathbf{e}_j^\intercal \mathbf{x} \geq \lceil x_j^\star \rceil$, where $\mathbf{e}_j$ denotes the j-th unit vector of dimension n. This condition splits the set of master variables in two, that is, the set $\mathcal{X}$ is partitioned into $\{\mathbf{x} \in \mathcal{X} \mid \mathbf{e}_j^\intercal \mathbf{x} \geq \lceil x_j^\star \rceil\}$ and $\{\mathbf{x} \in \mathcal{X} \mid \mathbf{e}_j^\intercal \mathbf{x} \leq \lfloor x_j^\star \rfloor\}$. If $\sum_{\mathbf{x} \in \mathcal{X}: \mathbf{e}_j^\intercal \mathbf{x} \geq \lceil x_j^\star \rceil} \lambda_\mathbf{x}^\star = \beta$ is fractional, then the down-branch master problem becomes

$$
\begin{aligned}
z_{MP}^\star = \min \quad & \sum_{\mathbf{x} \in \mathcal{X}: \mathbf{e}_j^\intercal \mathbf{x} \geq \lceil x_j^\star \rceil} c_\mathbf{x} \lambda_\mathbf{x} + \sum_{\mathbf{x} \in \mathcal{X}: \mathbf{e}_j^\intercal \mathbf{x} \leq \lfloor x_j^\star \rfloor} c_\mathbf{x} \lambda_\mathbf{x} \\
\text{s.t.} \quad & \sum_{\mathbf{x} \in \mathcal{X}: \mathbf{e}_j^\intercal \mathbf{x} \geq \lceil x_j^\star \rceil} \mathbf{a}_\mathbf{x} \lambda_\mathbf{x} + \sum_{\mathbf{x} \in \mathcal{X}: \mathbf{e}_j^\intercal \mathbf{x} \leq \lfloor x_j^\star \rfloor} \mathbf{a}_\mathbf{x} \lambda_\mathbf{x} \geq \mathbf{b} && [\boldsymbol{\pi} \geq \mathbf{0}] \\
& \sum_{\mathbf{x} \in \mathcal{X}: \mathbf{e}_j^\intercal \mathbf{x} \geq \lceil x_j^\star \rceil} \lambda_\mathbf{x} \leq \lfloor \beta \rfloor && [\gamma_1 \leq 0] \\
& \lambda_\mathbf{x} \geq 0 && \forall \mathbf{x} \in \mathcal{X}.
\end{aligned}
\tag{7.69}
$$

Very similarly, the up-branch master problem reads as

$$
\begin{aligned}
z_{MP}^\star = \min \quad & \sum_{\mathbf{x} \in \mathcal{X}: \mathbf{e}_j^\intercal \mathbf{x} \geq \lceil x_j^\star \rceil} c_\mathbf{x} \lambda_\mathbf{x} + \sum_{\mathbf{x} \in \mathcal{X}: \mathbf{e}_j^\intercal \mathbf{x} \leq \lfloor x_j^\star \rfloor} c_\mathbf{x} \lambda_\mathbf{x} \\
\text{s.t.} \quad & \sum_{\mathbf{x} \in \mathcal{X}: \mathbf{e}_j^\intercal \mathbf{x} \geq \lceil x_j^\star \rceil} \mathbf{a}_\mathbf{x} \lambda_\mathbf{x} + \sum_{\mathbf{x} \in \mathcal{X}: \mathbf{e}_j^\intercal \mathbf{x} \leq \lfloor x_j^\star \rfloor} \mathbf{a}_\mathbf{x} \lambda_\mathbf{x} \geq \mathbf{b} && [\boldsymbol{\pi} \geq \mathbf{0}] \\
& \sum_{\mathbf{x} \in \mathcal{X}: \mathbf{e}_j^\intercal \mathbf{x} \geq \lceil x_j^\star \rceil} \lambda_\mathbf{x} \geq \lceil \beta \rceil && [\gamma_2 \geq 0] \\
& \lambda_\mathbf{x} \geq 0 && \forall \mathbf{x} \in \mathcal{X},
\end{aligned}
\tag{7.70}
$$

where the binary coefficients $g_\mathbf{x}$ in (7.60) reflect the condition

$$
g_\mathbf{x} = 1 \Leftrightarrow \mathbf{e}_j^\intercal \mathbf{x} \geq \lceil x_j^\star \rceil, \qquad \forall \mathbf{x} \in \mathcal{X}. \tag{7.71}
$$

When we assume that $x_j \in [\ell_j, u_j]$, relation (7.71) can be implemented in the *ISP*, e.g., along the lines of Exercise 7.4:

$$
g_\mathbf{x} \geq \frac{x_j - \lfloor x_j^\star \rfloor}{u_j - \lfloor x_j^\star \rfloor} \left(= \frac{x_j - \lceil x_j^\star \rceil + 1}{u_j - \lceil x_j^\star \rceil + 1} \right), \qquad g_\mathbf{x} \in \{0,1\} \text{ in (7.61)}; \tag{7.72}
$$

$$
g_\mathbf{x} \leq \frac{x_j - \ell_j}{\lceil x_j^\star \rceil - \ell_j}, \qquad g_\mathbf{x} \in \{0,1\} \text{ in (7.64)}. \tag{7.73}
$$

In the binary case $x_j \in \{0,1\}$, this significantly simplifies to

$$g_{\mathbf{x}} \geq x_j, \qquad g_{\mathbf{x}} \in \{0,1\}, \text{ in (7.61) with slope } -\gamma_1 \geq 0 \text{ on } g_{\mathbf{x}}, \tag{7.74}$$

$$g_{\mathbf{x}} \leq x_j, \qquad g_{\mathbf{x}} \in \{0,1\}, \text{ in (7.64) with slope } -\gamma_2 \leq 0 \text{ on } g_{\mathbf{x}}. \tag{7.75}$$

That is, we can use $g_{\mathbf{x}} = x_j$ and there is in fact no need for an extra binary variable.

So far so good, but the crucial condition is

$$\text{“If } \sum_{\mathbf{x}\in\mathcal{X}:\mathbf{e}_j^\mathsf{T}\mathbf{x}\geq\lceil x_j^\star\rceil} \lambda_{\mathbf{x}}^\star = \beta \text{ is fractional.”}$$

When we restrict ourselves to a subclass of hyperplanes, e.g., the simple bounds above, we may be unable to guarantee the existence of a hyperplane (from that subclass) that cuts off a fractional $\boldsymbol{\lambda}_{RMP}^\star$-solution, as observed by Vanderbeck (2000, p. 119). He shows, however, that we can always accomplish the separation using a very small *set* of lower and upper bounds on the x-variables.

Finding fractional β. Given an optimal fractional master problem solution, the general idea is to iteratively impose lower or upper bounds on x_j-variables, i.e., on a number of components of $\mathbf{x} \in \mathcal{X}$, until we are able to branch on the resulting sum of master variables. We introduce some notation first.

Definition 7.2. Let $\bar{J} \subseteq \{1,\ldots,n\}$ denote the index-set of the x_j-variables on which we impose an upper bound, i.e., $\mathbf{e}_j^\mathsf{T}\mathbf{x} \leq \lfloor v_j \rfloor$ for some $v_j \in [\ell_j, u_j]$. Similarly, we denote by $\underline{J} \subseteq \{1,\ldots,n\}$ the index-set of the x_j-variables on which we impose a lower bound $\mathbf{e}_j^\mathsf{T}\mathbf{x} \geq \lfloor v_j \rfloor + 1$. The number of bound constraints is $|\bar{J}| + |\underline{J}|$. The set of points from $\mathcal{X}$ that satisfy the entire collection of bounds is

$$\mathcal{B} = \{\mathbf{x} \in \mathcal{X} \mid \mathbf{e}_j^\mathsf{T}\mathbf{x} \leq \lfloor v_j \rfloor, \forall j \in \bar{J};\ \mathbf{e}_j^\mathsf{T}\mathbf{x} \geq \lfloor v_j \rfloor + 1, \forall j \in \underline{J}\} \subseteq \mathbb{Z}_+^n. \tag{7.76}$$

We denote (again) by $\mathcal{F} \subseteq \mathcal{X}$ the index-set of fractional variables of $\boldsymbol{\lambda}_{RMP}^\star$. It can happen that the sum of master variables satisfying a condition is integer, but some of the variables are still fractional. We therefore look explicitly at the latter.

Definition 7.3. The *fractionality of* $\boldsymbol{\lambda}_{RMP}^\star$ is given by

$$F = \sum_{\mathbf{x}\in\mathcal{F}} (\lambda_{\mathbf{x}}^\star - \lfloor \lambda_{\mathbf{x}}^\star \rfloor) \geq 0. \tag{7.77}$$

The *fractionality of* $\boldsymbol{\lambda}_{RMP}^\star$ *with respect to* $\mathbf{x} \in \mathcal{B}$ is computed as

$$F_{\mathcal{B}} = \sum_{\mathbf{x}\in\mathcal{F}\cap\mathcal{B}} (\lambda_{\mathbf{x}}^\star - \lfloor \lambda_{\mathbf{x}}^\star \rfloor) \geq 0. \tag{7.78}$$

When $F = 0$, we are done, so the following is relevant for $F > 0$.

When F is integer and we impose the bound constraints $\mathbf{x} \in \mathcal{B}$, two cases can occur: $F_{\mathcal{B}}$ is either *fractional* or *integer*.

1. $a < F_{\mathcal{B}} < a+1$, $a \in \mathbb{Z}_+$. Expression (7.78) re-writes as

$$\sum_{\mathbf{x}\in\mathcal{F}\cap\mathcal{B}} \lfloor \lambda^\star_{\mathbf{x}} \rfloor + a < \sum_{\mathbf{x}\in\mathcal{F}\cap\mathcal{B}} \lambda^\star_{\mathbf{x}} < \sum_{\mathbf{x}\in\mathcal{F}\cap\mathcal{B}} \lfloor \lambda^\star_{\mathbf{x}} \rfloor + a + 1, \tag{7.79}$$

   that is, the sum $\sum_{\mathbf{x}\in\mathcal{F}\cap\mathcal{B}} \lambda^\star_{\mathbf{x}} = \beta$ is guaranteed to be fractional. This is the desired case in which we can branch as before:

$$\sum_{\mathbf{x}\in\mathcal{X}\cap\mathcal{B}} \lambda_{\mathbf{x}} \leq \lfloor \beta \rfloor \quad \text{or} \quad \sum_{\mathbf{x}\in\mathcal{X}\cap\mathcal{B}} \lambda_{\mathbf{x}} \geq \lceil \beta \rceil. \tag{7.80}$$

   Observe that this is also valid for $\mathcal{B} = \mathcal{X}$ (no bound constraints), i.e., when F is fractional.

2. $F_{\mathcal{B}} \in \mathbb{Z}_+$. We are unfortunate in this case as the fractional values of the relevant $\lambda_{\mathbf{x}}$-variables add up to a positive integer. Then we cannot branch as above. However, there exist fractional $\lambda^\star_{\mathbf{x}_1}, \lambda^\star_{\mathbf{x}_2} > 0$ indexed by $\mathbf{x}_1, \mathbf{x}_2 \in \mathcal{F}$, such that $x_{j1} \neq x_{j2}$, say $x_{j1} < x_{j2}$, for some $j \in \{1, \ldots, n\}$. Define $v_j = \frac{x_{j1}+x_{j2}}{2}$ and observe that $x_{j1} \leq \lfloor v_j \rfloor < \lfloor v_j \rfloor + 1 \leq x_{j2}$. With this, we re-write $F_{\mathcal{B}}$ as the result of two positive sums,

$$F_{\mathcal{B}} = \underbrace{\sum_{\mathbf{x}\in\mathcal{F}\cap\mathcal{B}\cap\{x_j \leq \lfloor v_j \rfloor\}} (\lambda^\star_{\mathbf{x}} - \lfloor \lambda^\star_{\mathbf{x}} \rfloor)}_{\geq (\lambda^\star_{\mathbf{x}_1} - \lfloor \lambda^\star_{\mathbf{x}_1} \rfloor) > 0} + \underbrace{\sum_{\mathbf{x}\in\mathcal{F}\cap\mathcal{B}\cap\{x_j \geq \lfloor v_j \rfloor + 1\}} (\lambda^\star_{\mathbf{x}} - \lfloor \lambda^\star_{\mathbf{x}} \rfloor)}_{\geq (\lambda^\star_{\mathbf{x}_2} - \lfloor \lambda^\star_{\mathbf{x}_2} \rfloor) > 0} \tag{7.81}$$

   where not both sums can be larger than $F_{\mathcal{B}}/2$. In other words, imposing one more bound, i.e., considering $\mathcal{B} \cap \{x_j \leq \lfloor v_j \rfloor\}$ or $\mathcal{B} \cap \{x_j \geq \lfloor v_j \rfloor + 1\}$, reduces the fractionality of $\boldsymbol{\lambda}^\star_{RMP}$ with respect to $\mathcal{B}$ by at least a factor of two.

This naturally suggests the following separation procedure. Start with "no bound constraints," i.e., $\mathcal{B} = \mathcal{X}$. As long as $F_{\mathcal{B}}$ is a positive integer, identify a variable x_j and an appropriate value $v_j \in [\ell_j, u_j]$, and impose an additional bound, i.e.,

$$\mathcal{B}_1 = \mathcal{B} \cap \{x_j \leq \lfloor v_j \rfloor\} \quad \text{or} \quad \mathcal{B}_2 = \mathcal{B} \cap \{x_j \geq \lfloor v_j \rfloor + 1\}.$$

Then, either $F_{\mathcal{B}_1}$ and $F_{\mathcal{B}_2}$ are both fractional in which case we have two branching opportunities, or both are integer and we replace $\mathcal{B}$ by $\mathcal{B}_1$ or $\mathcal{B}_2$, whichever results in the smallest fractionality. Since the fractionality is at least halved in each iteration, $F_{\mathcal{B}}$ eventually goes between 0 and 1 (this is the worst case), and this inductively proves the following result:

Proposition 7.2. (Vanderbeck, 2000, Proposition 3, p. 119) *Given $\boldsymbol{\lambda}^\star_{RMP}$ with fractionality $0 < F < 2^t$ for an integer $t \geq 1$, there exist at most t bound constraints on the original x-variables with $\mathcal{B}$ as defined in* (7.76) *such that*

$$\beta = \sum_{\mathbf{x}\in\mathcal{F}\cap\mathcal{B}} \lambda^\star_{\mathbf{x}} \text{ is fractional.} \tag{7.82}$$

It is notable that Vanderbeck (2000, p. 120) states that *"in our experiments, we seldom need to consider sets [of bound constraints] of cardinality greater than one."* This suggests that the case of an "integer fractionality" rarely happens. Moreover, when we consider fractional *basic* master problem solutions, at most m $\lambda^\star_{\mathbf{x}}$-variables can be positive, where m is the number of constraints in the *RMP*. Therefore, no more than m variables can be fractional, limiting $2^t < m$. This gives a small cardinality $|\bar{J}| + |\underline{J}|$ of less than $\log m$ as a corollary.

Pricing problem modifications. After branching, the new master problems exactly follow (7.60) where binary $g_{\mathbf{x}} = 1 \Leftrightarrow \mathbf{x} \in \mathcal{B}$, $\forall \mathbf{x} \in \mathcal{X}$. This needs to be reflected in the pricing. In both *ISP*s, we use a binary variable indicating whether or not a bound constraint is satisfied: $\bar{y}_j = 1 \Leftrightarrow x_j \leq \lfloor v_j \rfloor, \forall j \in \bar{J}$, and $\underline{y}_j = 1 \Leftrightarrow x_j \geq \lfloor v_j \rfloor + 1, \forall j \in \underline{J}$. In (7.83), we have an overview of the conditions on the two pricing problems that, for now, only differ in the slope of $g_{\mathbf{x}}$ in the objective function:

$$
\begin{array}{l|l}
\bar{c}(\boldsymbol{\pi}, \gamma_1) = \min\limits_{\mathbf{x} \in \mathcal{X}} \ \mathbf{c}^\mathsf{T}\mathbf{x} - \boldsymbol{\pi}^\mathsf{T}\mathbf{A}\mathbf{x} - \gamma_1 g_{\mathbf{x}} & \bar{c}(\boldsymbol{\pi}, \gamma_2) = \min\limits_{\mathbf{x} \in \mathcal{X}} \ \mathbf{c}^\mathsf{T}\mathbf{x} - \boldsymbol{\pi}^\mathsf{T}\mathbf{A}\mathbf{x} - \gamma_2 g_{\mathbf{x}}
\end{array}
$$

$$
\begin{aligned}
\text{s.t.} \quad & g_{\mathbf{x}} = 1 \ \Leftrightarrow \ \sum_{j \in \bar{J}} \bar{y}_j + \sum_{j \in \underline{J}} \underline{y}_j = |\bar{J}| + |\underline{J}| \\
& g_{\mathbf{x}} \in \{0,1\} \\
& \bar{y}_j \in \{0,1\} \qquad \forall j \in \bar{J} \\
& \bar{y}_j \in \{0,1\} \qquad \forall j \in \underline{J}.
\end{aligned} \tag{7.83}
$$

- In the down-branch where $-\gamma_1 \geq 0$, the binary $g_{\mathbf{x}}$ naturally takes value 0, so do the $\bar{y}_j$- and $\underline{y}_j$-variables: We need to determine if all these take value 1.
- In the up-branch, $-\gamma_2 \leq 0$ and $g_{\mathbf{x}} = 1$ is preferred (as well as the $\bar{y}_j$- and $\underline{y}_j$-variables at 1): Is there a y-variable taking value 0?

It remains to express an appropriate set of linear constraints for defining binary $g_{\mathbf{x}} = 1 \Leftrightarrow \mathbf{x} \in \mathcal{X} \cap \mathcal{B}$ in each *ISP*, together with binary y-variables. This is presented next in (7.84) while Exercise 7.7 asks the reader to mathematically justify these sets. The special case of binary x_j-variables is considered in Exercise 7.8.

$$
\begin{array}{lr|lr}
g_{\mathbf{x}} \geq 1 + (\sum_{j \in \bar{J}} \bar{y}_j + \sum_{j \in \underline{J}} \underline{y}_j) - (|\bar{J}| + |\underline{J}|) & & g_{\mathbf{x}} \leq \bar{y}_j & \forall j \in \bar{J} \\
& & g_{\mathbf{x}} \leq \underline{y}_j & \forall j \in \underline{J} \\
\bar{y}_j \geq \dfrac{\lfloor v_j \rfloor + 1 - x_j}{\lfloor v_j \rfloor + 1 - \ell_j} & \forall j \in \bar{J} & \bar{y}_j \leq \dfrac{u_j - x_j}{u_j - \lfloor v_j \rfloor} & \forall j \in \bar{J} \\
\underline{y}_j \geq \dfrac{x_j - \lfloor v_j \rfloor}{u_j - \lfloor v_j \rfloor} & \forall j \in \underline{J} & \underline{y}_j \leq \dfrac{x_j - \ell_j}{\lfloor v_j \rfloor + 1 - \ell_j} & \forall j \in \underline{J}.
\end{array} \tag{7.84}
$$

Note that, in particular on the down-branch, it is not clear that there should be any positive $\lambda_{\mathbf{x}}$-variable which satisfies the bound constraints of the parent node. We therefore, in general, cannot just refine set $\mathcal{B}$. One remedy is to start the search for bound constraints from scratch in every node. Note again that no points from $\mathcal{X}$ need to be explicitly excluded from the *ISP* domain through branching.

There may be many different sets $\mathcal{B}$ that lead to a fractional sum of master variables for branching. A smaller number $|\bar{J}| + |\underline{J}|$ of variable bounds reduces the overhead of pricing problem modifications. Furthermore, it may lead to larger sets $\mathcal{B}$ which, intuitively, may have a larger impact, e.g., on the dual bound improvement. Vanderbeck (2000, 2011) proposes several enumeration schemes of varying complexity for finding a set of bounds. Since we are free to choose the order in which we consider the x_j-variables on which to impose bounds, we may start with those that have larger branching priority.

Ryan-Foster rule for set partitioning master problems

A discussion about branching rules on master variables would not be complete without presenting the oldest known general rule, the one by Ryan and Foster (1981). It only works on the important special case of set partitioning *IMP*. Let the *MP* be given as its linear relaxation, where for $i \in \{1, \ldots, m\}$ and $\mathbf{x} \in \mathcal{X}$, the binary parameter $a_{i\mathbf{x}}$ takes value 1 if row i belongs to the column indexed by $\mathbf{x}$ of cost $c_{\mathbf{x}}$:

$$
\begin{aligned}
z^{\star}_{MP} = \min \quad & \sum_{\mathbf{x} \in \mathcal{X}} c_{\mathbf{x}} \lambda_{\mathbf{x}} \\
\text{s.t.} \quad & \sum_{\mathbf{x} \in \mathcal{X}} a_{i\mathbf{x}} \lambda_{\mathbf{x}} = 1 \quad \forall i \in \{1, \ldots, m\} \\
& \lambda_{\mathbf{x}} \geq 0 \quad \forall \mathbf{x} \in \mathcal{X}.
\end{aligned}
\tag{7.85}
$$

For example, $a_{i\mathbf{x}} = 1$ represents a customer i visited in route $\mathbf{x}$ in the *VRPTW*, or a flight i operated in schedule $\mathbf{x}$ in a crew pairing problem, these *tasks* being performed exactly once. Our presentation of the rule more or less follows that of Barnhart et al. (1998). It is based on this result: *For a fractional basic solution to the MP* (7.85), *there exist two rows r and s such that*

$$
0 < \sum_{\mathbf{x} \in \mathcal{X} :\, a_{r\mathbf{x}} = a_{s\mathbf{x}} = 1} \lambda^{\star}_{\mathbf{x}} < 1. \tag{7.86}
$$

Interpreted in the *VRPTW*, if (7.86) holds, then customers r and s are each visited exactly once as requested by the partitioning constraints, but partly when they are together on a route, partly when they are not. This entails a fractional basic $\boldsymbol{\lambda}^{\star}_{RMP}$-solution when the *MP* is solved by column generation. The branching rule imposes that in an integer solution, the customers are either in the same route or in two different ones, suggesting that this can be handled by modifying the *ISP*. The Ryan-Foster condition on $\mathbf{x}$ in (7.86) can be formulated as a separating hyperplane in the language of Proposition 7.1, see for the pleasure Exercise 7.9. It is also a special case of Proposition 7.2 with lower bounds on variables x_r and x_s.

To show that a fractional basic solution implies (7.86), let $\lambda_k^\star$ be fractional (for $\mathbf{x} = \mathbf{x}_k$) in $\boldsymbol{\lambda}_{RMP}^\star$ and select any row r such that $a_{rk} = 1$. Recall that $1 = \sum_{\mathbf{x}\in\mathcal{X}} a_{r\mathbf{x}}\lambda_\mathbf{x}^\star$: hence there exists another fractional basic variable $\lambda_\ell^\star$ such that $a_{r\ell} = 1$. Because there are no identical columns in the basis of the *RMP*, there must exist a row s such that $a_{sk} \neq a_{s\ell}$, i.e., either $a_{sk} = 1$ or $a_{s\ell} = 1$, but not both. Hence

$$1 = \sum_{\mathbf{x}\in\mathcal{X}} a_{r\mathbf{x}}\lambda_\mathbf{x}^\star = \sum_{\mathbf{x}\in\mathcal{X}:\, a_{r\mathbf{x}}=1} \lambda_\mathbf{x}^\star > \sum_{\mathbf{x}\in\mathcal{X}:\, a_{r\mathbf{x}}=a_{s\mathbf{x}}=1} \lambda_\mathbf{x}^\star > 0,$$

where the first inequality comes from the sum on the right that includes either fractional variable λ_k or λ_ℓ, but not both. Such a pair (r,s) leads to branches *married* or *divorced*,

$$\sum_{\mathbf{x}\in\mathcal{X}:\, a_{r\mathbf{x}}=a_{s\mathbf{x}}=1} \lambda_\mathbf{x} = 1 \quad \text{or} \quad \sum_{\mathbf{x}\in\mathcal{X}:\, a_{r\mathbf{x}}=a_{s\mathbf{x}}=1} \lambda_\mathbf{x} = 0, \tag{7.87}$$

that is, rows r and s are covered by the same column on the left or by two columns on the right. The literature calls these the *same* and *differ* branches. An optimal solution on the left branch uses columns having $a_{r\mathbf{x}} = a_{s\mathbf{x}} = 0$ or $a_{r\mathbf{x}} = a_{s\mathbf{x}} = 1$, while on the right, feasible columns have $a_{r\mathbf{x}} = a_{s\mathbf{x}} = 0$, or $a_{r\mathbf{x}} = 0$, $a_{s\mathbf{x}} = 1$, or $a_{r\mathbf{x}} = 1$, $a_{s\mathbf{x}} = 0$. Note that covering neither r nor s is allowed in both branches. For this rule, the *ISP* domain is explicitly partitioned by branching. Note also that, except possibly as a post-processing step, no disaggregation takes place, even though set partitioning master problems often result from aggregation.

The implementation of the Ryan-Foster branching rule depends on the application. For additional details, see *Rule* 3 (p. 484) below and the Examples 7.3 and 7.4 that discuss how to implement the branching decisions in the *ISP*.

Note 7.9 (To **B** *or not to* **B**.*)* Although the validity of the Ryan-Foster rule is presented for a *basic* solution $\boldsymbol{\lambda}_{RMP}^\star$, it remains valid for an interior point solution. Indeed, there are no identical columns in any *RMP*.

Branching on inter-tasks

Because set partitioning *IMP*s are important models, the Ryan and Foster's idea has been used many times and even adapted to ease the solution of the *ISP* for specific cases. A popular adaptation, called *branching on inter-tasks or on follow-ons* (see, e.g., Desrochers and Soumis, 1989; Irnich and Desaulniers, 2005), targets applications such as the *VRPTW* or the crew pairing problem (*CPP*, see Example 5.3 but without the base constraints), where the *ISP* is an *ESPPRC* or a *SPPRC* solved by a labeling algorithm (see Labeling algorithm). It exploits the fact that a path covers the rows in a given order, which is irrelevant for the Ryan-Foster rule presented above.

To describe the inter-task branching, we assume that each set partitioning constraint in the *MP* (7.85) enforces the covering of a task i, e.g., a customer in the *VRPTW* or a flight in the *CPP*. Furthermore, in the *ISP*, we define a binary variable x_i for each task $i \in \{1,\ldots,m\}$ that indicates whether or not task i is covered by a path and an extra binary variable $y_{r\to s}$ for each pair of tasks r and s on which an inter-task decision is imposed. This variable takes value 1 if tasks r and s are covered consecutively in a path (r before s) and 0 otherwise.

Using a reasoning similar to that presented above for the Ryan-Foster rule, we can show the following result: *For a fractional basic solution to the MP* (7.85)*, there exist two rows r and s such that*

$$0 < \sum_{\mathbf{x}\in\mathcal{X}:x_r=x_s=y_{r\to s}=1} \lambda_{\mathbf{x}}^\star < 1. \tag{7.88}$$

Hence, it is possible to branch on this sum of λ-variables:

$$\sum_{\mathbf{x}\in\mathcal{X}:x_r=x_s=y_{r\to s}=1} \lambda_{\mathbf{x}} = 1 \quad \text{or} \quad \sum_{\mathbf{x}\in\mathcal{X}:x_r=x_s=y_{r\to s}=1} \lambda_{\mathbf{x}} = 0, \tag{7.89}$$

where, on the up-branch (left), rows r and s must be covered consecutively and in this order by the same path and on the down-branch (right), they must be covered either by two paths or by the same path but not consecutively in this order. These branching decisions are also imposed directly in the *ISP*, i.e., no constraints are added to the master problem.

Note 7.10 (Inter-task branching is not Ryan-Foster branching.) In the literature, a few papers have mistakenly mentioned that inter-task (or follow-on) branching is a special case of Ryan-Foster branching. This is not true! Indeed, we can observe that the set of columns involved in (7.88) is, in general, only a subset of the columns in (7.86): columns (paths) covering r and s but not consecutively and in the right order (i.e., columns for which $x_r = x_s = 1, y_{r\to s} = 0$) are not considered in (7.88). When the *ISP* is an *ESPPRC* or a *SPPRC* solved by a labeling algorithm, handling Ryan-Foster branching decisions in the *ISP* makes it much harder to solve. This is not the case with inter-task decisions as explained next.

To implement inter-task branching, we consider the following two cases.

Case 1: *Direct connection network.*

In a direct connection network, all nodes, except the source and the sink, are associated with a single task like in the *VRPTW*. Consequently, two tasks r and s can be covered consecutively along a path if and only if the arc linking the task nodes r and s is traversed. In a fractional solution, the pair of customers r and s can be selected if the corresponding arc-flow variable $x_{rs}^\star$ is fractional, that is,

$$0 < x_{rs}^\star = \sum_{\mathbf{x}\in\mathcal{X}} g_{\mathbf{x}}\lambda_{\mathbf{x}}^\star < 1,$$

where $g_{\mathbf{x}} = x_{rs} = 1$ in the paths containing arc (r,s), and 0 otherwise. The up-branch $\sum_{\mathbf{x}\in\mathcal{X}} x_{rs}\lambda_{\mathbf{x}}^\star = 1$ keeps the arc (r,s) but removes all other arcs out of node r and into node s. The down-branch $\sum_{\mathbf{x}\in\mathcal{X}} x_{rs}\lambda_{\mathbf{x}}^\star = 0$ simply removes the arc (r,s).

Case 2: *Arbitrary network.*

In many applications, the network underlying the *ISP* is not a direct connection network. For example, in the electric *VRPTW* (see Desaulniers et al., 2016a), the network includes nodes representing stations where the vehicles can recharge their battery in the middle of their route while, in the *CPP*, the tasks are modeled as flight arcs but the network also contains arcs representing, e.g., connections between two flights or rests between two working days (see Figure 5.7). In such networks, two tasks can be covered consecutively in a path but using a sequence of arcs to link them, i.e., a single arc-flow variable might not be sufficient to determine the consecutiveness. Hence, to branch on a fractional sum of λ-variables

$$0 < \sum_{\mathbf{x} \in \mathcal{X}} g_{\mathbf{x}} \lambda^\star_{\mathbf{x}} < 1,$$

where $g_{\mathbf{x}} = 1 \Leftrightarrow x_r = x_s = y_{r \to s} = 1$ for a given pair of tasks r and s, we can add a new component to the labels to compute $g_{\mathbf{x}}$ in the labeling algorithm used to solve the *ISP*. This implies multiple additional components when there are several inter-task decisions applicable at a branch-and-bound node, substantially increasing the complexity of the labeling algorithm. Nevertheless, by exploiting the consecutiveness condition, this increase in complexity can be controlled by considering a single new label component that only keeps track of the last task covered; see Exercise 7.10.

Note 7.11 (Sufficiency of inter-task branching.) For pure set partitioning *IMP*s with an *ESPPRC* or *SPPRC* pricing problem, branching on inter-tasks is sufficient to fully explore the search tree, even if a sequence of tasks can be represented by several paths in the *ISP* (in this case, a single one of them can appear in the *RMP* basis, namely, one with the least cost). On the other hand, for set partitioning problems with side constraints, the inter-task branching remains applicable but it might not be sufficient to guarantee a complete tree exploration. In this case, it can be complemented with, e.g., branching decisions on arc-flow variables or other λ-branchings. A similar remark applies to the Ryan-Foster branching.

Some practical rules on the sum of binary λ-variables

Simple rules may be easier to tailor to a specific situation. It follows a small collection of such simple rules on the sum of certain sets of $\lambda_{\mathbf{x}}$-variables derived from constraints in the $\mathbf{x}$-variables. These are all inspired by the discussions above.

Rule 1 $\displaystyle\sum_{\mathbf{x} \in \mathcal{X}: c_1 \le \mathbf{c}^\intercal \mathbf{x} \le c_2} \lambda_{\mathbf{x}} \in \mathbb{Z}_+.$

This is a special case of Proposition 7.1 which we have already sketched in Illustration 7.7. As we compute the cost $c_{\mathbf{x}}$ for every variable $\lambda_{\mathbf{x}}$ anyway, a natural hyperplane turns out to be in terms of $\mathbf{c}^\intercal \mathbf{x}$. We assume $\mathbf{c}$ integer, $\mathbf{c}^\intercal \mathbf{x} \in [\ell, u]$, and that a sensible range of cost coefficients occurs. In Exercise 7.13, we ask for a way to *possibly* find a set $\mathcal{B} = \{\mathbf{x} \in \mathcal{X} \mid c_1 \le \mathbf{c}^\intercal \mathbf{x} \le c_2\}$ such that $\sum_{\mathbf{x} \in \mathcal{X} \cap \mathcal{B}} \lambda^\star_{\mathbf{x}} = \beta$ is fractional. If it occurs, the master in the down- and up-branches write similarly as (7.69)–(7.70) and $\mathbf{x} \in \mathcal{B}$ can be implemented in the pricing as in Pricing problem modifications (page 473). Meaningful functions other than the cost, e.g., resource consumption, number of items, etc., can be used in the same way. Of course, this also works for more than one pricing problem.

Rule 2 $\sum_{\mathbf{x}\in\mathcal{X}} \lambda_{\mathbf{x}} \in \mathbb{Z}_+$, adapted for a $|K|$-block structure as $\sum_{\mathbf{x}^k\in\mathcal{X}^k} \lambda_{\mathbf{x}^k} \in \mathbb{Z}_+, \forall k \in K$.

A request that directly follows from the way we apply a decomposition is to have an integer number of columns taken from each pricing problem. Our fourth suggestion in Illustration 7.7 (λ-branching for the *VRPTW*, p. 475) is an example. A related application is the multiple depot vehicle scheduling problem (*MDVSP*), where each depot uses an integer number of vehicles. If such a branching decision is applied to the usual convexity constraint $\sum_{\mathbf{x}^k\in\mathcal{X}^k} \lambda_{\mathbf{x}^k} = 1$ associated with block k, this means that the corresponding ISP^k is used or not. For this rule, there are no modifications to the domain of the ISP^ks.

Rule 3 $\sum_{\mathbf{x}\in\mathcal{X}:x_r=x_s=1} \lambda_{\mathbf{x}} \in \mathbb{Z}_+$, where (x_r, x_s) is a pair of original binary variables.

This generalizes the Ryan-Foster rule. Assuming $0 \leq \sum_{\mathbf{x}\in\mathcal{X}} \lambda_{\mathbf{x}} \leq U$, we describe four special cases. where the introduction in the *ISP* of a binary variable, say y_{rs}, linking the pair (x_r, x_s) may be required. In that case, the condition is interpreted as $x_r = x_s = 1 \Leftrightarrow y_{rs} = 1$.

Rule 3.1 $\sum_{\mathbf{x}\in\mathcal{X}:x_r=x_s=1} \lambda_{\mathbf{x}} \leq \lfloor\beta\rfloor = 0$.

The generated columns satisfying the condition $x_r = x_s = 1$ are removed from the *RMP*, the constraint $x_r + x_s \leq 1$ is added to the *ISP*, and no columns satisfying the condition can be newly generated.

Rule 3.2 $\sum_{\mathbf{x}\in\mathcal{X}:x_r=x_s=1} \lambda_{\mathbf{x}} \leq \lfloor\beta\rfloor$, where $\lfloor\beta\rfloor \geq 1$.

The configuration $x_r = x_s = 1, y_{rs} = 0$ must be eliminated, i.e., $y_{rs} \geq x_r + x_s - 1$ in the *ISP*. Observe that if $x_r + x_s \leq 1$, then $y_{rs} = 0$ because of the non-negative slope of $-\gamma_1 g_{\mathbf{x}}$, $\gamma_1 \leq 0$, in the objective function of $\bar{c}(\boldsymbol{\pi}, \gamma_1)$. Finally, the binary coefficient $g_{\mathbf{x}}$ is computed as $g_{\mathbf{x}} = g(\mathbf{x}) = y_{rs}$ in the *ISP*: this means that the extra variable y_{rs} can be discarded from the formulation and replaced by $g_{\mathbf{x}}$.

Rule 3.3 $\sum_{\mathbf{x}\in\mathcal{X}:x_r=x_s=1} \lambda_{\mathbf{x}} \geq \lceil\beta\rceil = U$.

The generated columns satisfying $x_r + x_s \leq 1$ are removed from the *RMP* because all the U columns must satisfy $x_r = x_s = 1$. The constraint $x_r + x_s \geq 2$ is added to the *ISP*, or the condition $x_r = x_s = 1$ is implemented in an appropriate way for specific applications. Finally, the coefficient $g_{\mathbf{x}} = 1, \forall \mathbf{x} \in \mathcal{X}$, i.e., $\sum_{\mathbf{x}\in\mathcal{X}} \lambda_{\mathbf{x}} = U$.

Rule 3.4 $\sum_{\mathbf{x}\in\mathcal{X}:x_r=x_s=1} \lambda_{\mathbf{x}} \geq \lceil\beta\rceil$, where $\lceil\beta\rceil \leq U - 1$.

The constraint $2y_{rs} \leq x_r + x_s$ is added to the *ISP* and $g_{\mathbf{x}} = y_{rs}$. Alternatively, we can rather impose the two constraints $y_{rs} \leq x_r$ and $y_{rs} \leq x_s$. Observe that if $x_r + x_s \leq 1$, then the binary variable $y_{rs} = 0$ because $y_{rs} \leq \frac{x_r + x_s}{2} \leq 1/2$.

Given $x_r = x_s = 1$ written as $x_r + x_s \geq 2$, constraints $y_{rs} \geq x_r + x_s - 1$ in *Rule* 3.2 and $2y_{rs} \leq x_r + x_s$ in *Rule* 3.4 are derived from (7.66), see Exercise 7.14.

Illustration 7.8 **λ-branching for the** ***VRPTW*** **(cont.)**

Recall the *VRPTW* example in Illustration 7.7 (initially in Illustration 5.1) with four customers. Applying column generation to solve the *MP* (with an *ESPPTWC* pricing problem), we obtain the optimal $\lambda^\star_{RMP}$-solution

$$\lambda^\star_{132} = \lambda^\star_{134} = \lambda^\star_{24} = 0.5 \text{ and } \lambda^\star_r = 0 \text{ for all other routes } r \in R'.$$

The fractionality of this solution is $F = \sum_{r\in\mathcal{F}}(\lambda^\star_r - \lfloor\lambda^\star_r\rfloor) = 0.5 \times 3 = 1.5$. Using the first two routes, where $x_{32,1} = 1$ in the first and $x_{32,2} = 0$ in the second, we obtain an average value $v_{32} = \frac{x_{32,1} + x_{32,2}}{2} = 0.5$. Because $\lfloor v_{32}\rfloor = 0$, $\lfloor v_{32}\rfloor + 1 = 1$, and x_{32} a binary variable, F splits in two sums, as in (7.81):

$$\begin{aligned} F = & \sum_{r\in\mathcal{F}:x_{32}=0} (\lambda^\star_r - \lfloor\lambda^\star_r\rfloor) & + & \sum_{r\in\mathcal{F}:x_{32}=1} (\lambda^\star_r - \lfloor\lambda^\star_r\rfloor) \\ = & \underbrace{\lambda^\star_{134} + \lambda^\star_{24}}_{=1} & + & \underbrace{\lambda^\star_{132}}_{=0.5} \end{aligned}$$

where the fractionality of the second $\sum_{r\in\mathcal{F}:x_{32}=1} \lambda^\star_r = 0.5$, see (7.79)–(7.80), allows for the branching decisions

$$\sum_{r\in R:x_{32}=1} \lambda_r = 0 \quad or \quad \sum_{r\in R:x_{32}=1} \lambda_r \geq 1.$$

Because the *MP* is the linear relaxation of a set partitioning problem, the up-branch is replaced by an equality to 1. The optimal solution $\lambda^\star_{32} = \lambda^\star_{14} = 1$ appears in this branch. We see that this branching on the master variables is in fact a branching on an original variable.

Note 7.12 (The sky is the limit.) Only very few are in need of generally applicable branching rules, but they typically look for strategies that work well in their particular application. In that regard, the above serves as a starting point, and tailoring may be beneficial or even necessary for performance. Consider, for instance, the above *Rule* 1 which formulates a condition that works for every (!) *ISP* under the mild condition that costs of master variables are not all identical. For a specific application, research and experimentation may lead to a customization of such a rule. One question could be how we decide about the "split points," here c_1 and c_2, what gives an "interesting" condition? A suggestion is to order the positive fractional $\lambda^\star_{\mathbf{x}}$-values by some criterion (say cost), and start summing until we are "happy" with the sum. (Another strategy is presented in Exercise 7.13.) In practice, we may not be able to prove that we actually find a fractional sum, then we may not care about theory, and try alternative branching strategies in x and λ, since we have many.

Ranking the candidates

As we announced in Note 7.7, we discussed mainly the first part of a branching rule, the *identification* of branching candidates. Often, however, there are many candidates, many values that are fractional but should be integer. Then, one unavoidably faces the second part, the *ranking* of the candidates. There are general ideas one should have at least heard of, most importantly, branching priorities, pseudo-costs, and strong branching.

It is usually clear from the application context what decisions are most important because of their role in the solution structure or the cost. Global design decisions (use a vehicle or not, open a facility, build infrastructure, and other fixed cost components of the objective function) can have a larger impact than local decisions (like routing, servicing, or other variable cost components). Branch on most important information first, and when there is a difference, in the direction that has greater consequences for the solution and/or its value (usually, the down-branch is weaker than the up-branch). This is immediately realizable for original variable branching. Also master variable branching can be guided by the same thoughts, by accordingly defining the conditions on how to find a fractional sum. This can lead to a "hierarchy" of branching rules, where lower priority rules are called when higher priority ones are failing.

A problem-independent ranking of branching candidates can be done through *pseudo-costs*. For every candidate and every branching direction, they estimate the resulting impact on the dual bound, based on the previous branching on that candidate. One difficulty with pseudo-costs is that we initially have no branching history. A standard approach then is to tentatively branch on (some) candidates, (partly) solve the resulting branching node, assess the impact on the dual bound (maybe combined in both branches), and select a most promising candidate. This *strong branching* strategy can be very expensive, but may be helpful, see Acceleration techniques. As soon as enough historical branching information is collected, or from a certain level of the tree on, one can fall back to pseudo-cost branching (these variants are called reliability and hybrid branching).

We do not recommend a rule which is likely the most popular in research papers: branching on a most fractional variable, that is, one whose fractional part is closest to 0.5. This *most infeasible branching* is maybe the simplest that comes to mind, and easy to implement. There may be cases where this works well, but experimentation tells us that in general "most fractional" is the worst one can do: this rule is no better than random branching and should be avoided (Achterberg et al., 2005).

The other extreme, least fractional branching, selects a candidate closest to an integer. The two branches can have a very unbalanced impact on the dual bound, but one branch potentially leads to integer solutions quickly. It is therefore mainly used in diving, see Primal heuristics below.

7.4 Convexification vs. discretization revisited

It is clear that in general, in one way or another, we need access to the interior points $\ddot{I}$ of $\ddot{P}$ in the original *ISP* domain. Otherwise, we may miss optimal or even only feasible integer solutions to the *ILP*, see Figure 7.9.

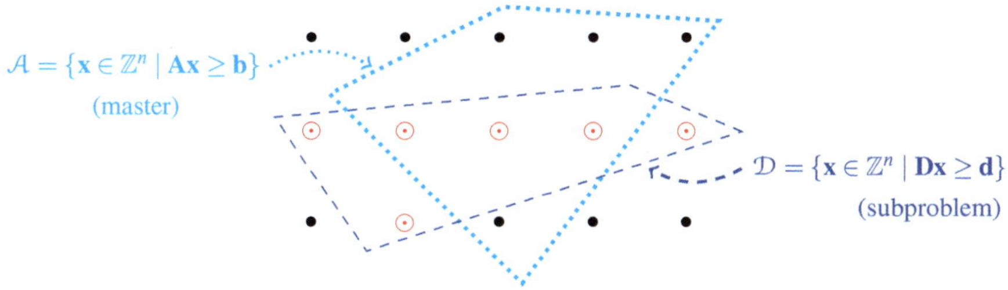

Fig. 7.9: Constraints of the *ILP*, where the integer points for the *ISP* are marked.

"Access" here means that we need a way of representation, and we actually got to know two ways: convexification and discretization. But does it really matter? When you ask a researcher whether they have reformulated their *ILP* by convexification or by discretization, they often cannot tell you. Part of the answer is that for binary problems, and many problems are binary, the decomposition approaches are literally identical. Also, in general, we have observed that the *RMP*s are the same in both approaches, see Proposition 4.2 and Note 4.4. However, this is only true for the root node and branching makes the differences between the two concepts visible.

Convexification requests branching on original variables for the simple reason that the master variables are not required to be integer. Conversely, branching on original variables can be seen as a concept that induces a convexification. How is that? Let us consider the *ISP* domain $\mathcal{D}$ and how it is impacted by branching. Convexification relies on extreme points (and rays) of conv($\mathcal{D}$). When we branch on original variables and impose the decision in the *ISP*, we *explicitly* restrict $\mathcal{D}$ in each child node, giving rise to *new* extreme points that may have been interior to (some face of) conv($\mathcal{D}$) in the parent (in particular, the root) node, see Figure 7.10a. In fact, the *ISP* domain is different in *every* node of the branch-and-price tree and one could speak of a *dynamic convexification*.

Also, when we impose the x-branching decision on the master problem domain $\mathcal{A}$, this creates an explicit boundary that we can hit when optimizing a linear objective function, to reach points that are interior to $\mathcal{D}$, see Figure 7.10b. Note that in this case, it is imperative that we do not only generate but also keep variables in the master problem that correspond to integer points of $\mathcal{D}$ which are infeasible for $\mathcal{A}$. The points of $\mathcal{A} \cap \mathcal{D}$ interior to $\mathcal{D}$ need to be represented by a (strict) convex combination of extreme points (and rays) of conv($\mathcal{D}$), thus again, convexification comes to the surface (see also p. 109 in Chapter 3).

(a) x-decision in subproblem (b) x-decision in master

Fig. 7.10: Domains from Figure 7.9, after an x-branching decision.

Branching on λ-variables in general requests the discretization approach because of the integrality requirement that is needed on the master variables. Again, conversely, discretization suggests to branch on master variables and to handle the branching decisions in the *ISP* without explicitly restricting its domain. In fact, also here the domain of the *ISP* is different in every node of the tree, but using, e.g., (7.63) or (7.84), we *relax* it with every branching decision. The *restriction happens only implicitly*: by introducing extra variables (and constraints), we lift the *ISP* domain to an extended formulation, and by projecting to the original variable space, we arrive at interior points of $\mathcal{D}$, see the following Illustration 7.9.

Illustration 7.9 **What you eat is what you are**

The difference between convexification and discretization becomes visible through branching, and at the same time it vanishes in branching: both approaches give access to interior points of the *ISP* domain. However, it is fair to say that *how you branch determines how you decompose(d).*

Too see this, consider the *ILP*

$$
\begin{aligned}
z^{\star}_{ILP} = \min \quad & -x_1 + x_2 && \text{(7.90a)}\\
\text{s.t.} \quad & 2x_1 + x_2 \leq 3 && \text{(7.90b)}\\
& x_1 \leq 2 && \text{(7.90c)}\\
& x_1 \in \mathbb{Z}_+ && \text{(7.90d)}\\
& x_2 \in \mathbb{Z}_+, && \text{(7.90e)}
\end{aligned}
$$

the optimal solution of which is $x^{\star}_1 = 1$, $x^{\star}_2 = 0$, and $z^{\star}_{ILP} = -1$. We reformulate the *ILP* (7.90) with the grouping of constraints as

$$
\begin{aligned}
\mathcal{A} &= \{x_1, x_2 \in \mathbb{Z}_+ \mid 2x_1 + x_2 \leq 3\} && \text{(7.91a)}\\
\mathcal{D} &= \{x_1 \in \mathbb{Z}_+ \mid x_1 \leq 2\}, && \text{(7.91b)}
\end{aligned}
$$

and do not tell you how, by convexification or by discretization. The domain $\mathcal{D}$ of the *ISP* contains three points (in fact, scalars): $x_1 = 0, 1$, and 2, i.e., $\mathcal{D} = \{0, 1, 2\}$.

- Convexification of $\mathcal{D}$ would give us the extreme points $\{x_p\}_{p\in P} = \{0,2\}$ with master variables $\lambda_0, \lambda_2 \geq 0$, $\lambda_0 + \lambda_2 = 1$, and $x_1 = 0\lambda_0 + 2\lambda_2$.
- Discretization of $\mathcal{D}$ would give us the points $\{x_p\}_{p\in \ddot{P}} = \{0,1,2\}$ with master variables $\lambda_0, \lambda_1, \lambda_2 \in \{0,1\}$, $\lambda_0 + \lambda_1 + \lambda_2 = 1$, and $x_1 = 0\lambda_0 + 1\lambda_1 + 2\lambda_2$.

Note that x_2 is a static variable here, it remains in the master problem as is. As λ_1 is defined for an interior point, we cannot generate it in the *ISP*, and the final (root node) *RMP*, either way, could look like this (remember Proposition 4.2):

$$\begin{aligned} z^\star_{RMP} = \min \quad & -0\lambda_0 - 2\lambda_2 + x_2 \\ \text{s.t.} \quad & 0\lambda_0 + 4\lambda_2 + x_2 \leq 3 \\ & \lambda_0 + \lambda_2 = 1 \\ & \lambda_0, \lambda_2, x_2 \geq 0, \end{aligned} \tag{7.92}$$

with an optimal solution $\lambda_0^\star = 0.25$, $\lambda_2^\star = 0.75$, and $x_2^\star = 0$, resulting in $z^\star_{RMP} = -1.5$ and $x_1^\star = 1.5$.

Option 1: Branch on original variables.
We can branch on $x_1 = 1.5$, imposing either $x_1 \leq 1$ or $x_1 \geq 2$ in the *ISP*.

- In the down-branch, the subproblem domain is $\mathcal{D}_1 = \{x_1 \in \mathbb{Z}_+ \mid x_1 \leq 1\}$. This gives ISP_1 with extreme points $\{x_p\}_{p\in P_1} = \{0,1\}$, $x_1 = 0\lambda_0 + 1\lambda_1$, $\lambda_0 + \lambda_1 = 1$, and $\lambda_0, \lambda_1 \geq 0$. We would remove λ_2 from the *RMP*, generate the extreme point $x_1 = 1$, and corresponding variable λ_1,

$$\begin{aligned} z_{RMP} = \min \quad & -0\lambda_0 - 1\lambda_1 + x_2 \\ \text{s.t.} \quad & 0\lambda_0 + 2\lambda_1 + x_2 \leq 3 \\ & \lambda_0 + \lambda_1 = 1 \\ & \lambda_0, \lambda_1, x_2 \geq 0, \end{aligned} \tag{7.93}$$

  and obtain the optimal integer solution $\lambda_0 = 0$, $\lambda_1 = 1$, $x_2 = 0$, resulting in $z_{RMP} = -1$ and $x_1 = 1$. Note that the new extreme point $x_1 = 1$ of the subproblem domain $\mathcal{D}_1$ is in fact an *interior* point of $\mathcal{D}$. In order to "reach it" we have applied a (re-)convexification, we have adjusted the previous convexification.
- In the up-branch, the subproblem domain becomes $\mathcal{D}_2 = \{x_1 \in \mathbb{Z}_+ \mid x_1 = 2\}$ with extreme point $\{x_p\}_{p\in P_2} = \{2\}$. We would remove λ_0 from the *RMP* (7.92) and observe that the unique extreme point has already been generated. We then find that the resulting *RMP* is infeasible as $\lambda_2 \leq 3/4$ in the first constraint and $\lambda_2 = 1$ from the convexity constraint.

Option 2: Branch on master variables.
We can alternatively branch, e.g., on the fractional sum $\sum_{p\in\ddot{P}:x_1\geq 1} \lambda_p = 0.75$. The branching constraints in the *RMP*s are

$$\sum_{p\in\ddot{P}} g_p\lambda_p \leq 0 \quad [\gamma_1 \leq 0] \qquad \text{or} \qquad \sum_{p\in\ddot{P}} g_p\lambda_p \geq 1 \quad [\gamma_2 \geq 0], \tag{7.94}$$

with $g_p = 1 \Leftrightarrow x_{1p} \geq 1, \forall p \in \ddot{P}$. This condition is linearized as in (7.63) and imposed in the *ISP*s with domain $\mathcal{D}_d = \left\{ \begin{bmatrix} x_1 \\ g \end{bmatrix} \in \mathbb{Z}_+ \times \{0,1\} \,\middle|\, x_1 \leq 2,\ g \leq x_1 \leq 2g \right\}$ which is the same in both branches.

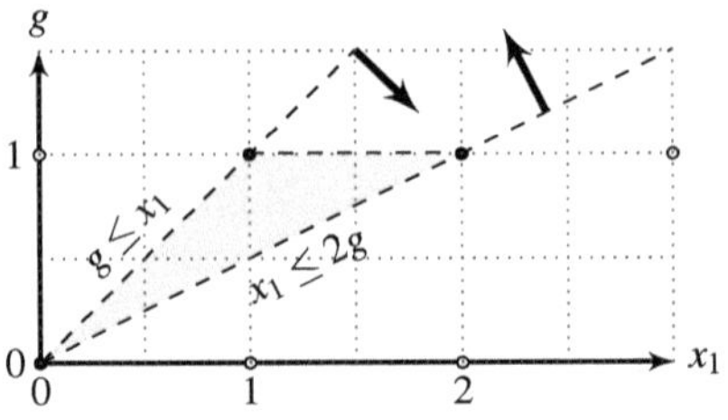

Fig. 7.11: The same domain for the *ISP*s in both the up- and down-branches.

Figure 7.11 shows the integer domain $\mathcal{D}_d = \left\{ \begin{bmatrix} 0 \\ 0 \end{bmatrix}, \begin{bmatrix} 1 \\ 1 \end{bmatrix}, \begin{bmatrix} 2 \\ 1 \end{bmatrix} \right\}$ of the resulting pricing problem: all of these are extreme points of an extended formulation of the original set $\mathcal{D}$. We see that, when projected back into the original variable space (which is only variable x_1 here), we can obtain all of $\mathcal{D} = \{0,1,2\}$ in both branches, in particular also the interior point $x_1 = 1$ and the corresponding variable λ_1. Because $x_1 = 1$ really is an interior point of the original $\mathcal{D}$ here, the decomposition concept apparently must be discretization.

Taking benefit from the slightly different objective functions in the *ISP*s, we may alternatively linearize the logical condition using (7.66). In the *ISP*s, this saves us one linear constraint per branch, and now results in two different extended formulations of the original *ISP* domain $\mathcal{D}$, as illustrated in Figure 7.12. Note again that (here only in the up-branch) an interior point of $\mathcal{D}$ can be generated.

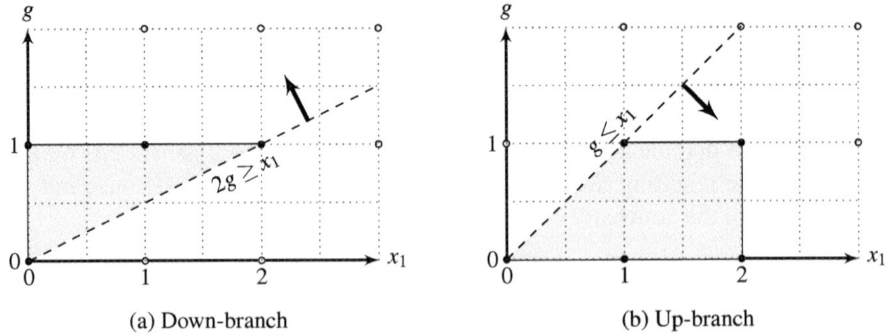

Fig. 7.12: Domains of the *ISP*s derived from (7.66).

En résumé: we *must* arrive and we *do* arrive (by branching) at interior points of the domain $\mathcal{D}$ of the (root node) *ISP*. The way *how* we arrive there determines whether we are in the convexification or in the discretization framework: by creating new extreme points in the variables of the original *ISP* domain formulation, or by creating new extreme points in the variables of an extended *ISP* domain formulation. In the end, the difference between the two concepts is mainly one of presentation: interior points are "there" right from the beginning or they "appear" through branching. In both cases, branching is necessary to actually access them.

It is true that also in λ-variable branching, we can explicitly restrict the *ISP* domain, that is, we create more than one subproblem like in (7.67). But then again, we do not lose any points from the original $\mathcal{D}$ (after projection) and also the *ISP*s of different flavors create integer points interior to $\mathcal{D}$, as just seen in Illustration 7.9.

Discretization is more general than convexification in that every branching idea in the latter works in the former, while the reverse is not true. Yet, the practitioner may still not care about the distinction. If one, maybe not even knowingly, starts with a convexification and then "accidentally" branches on fractional sums of master variables, the approach may automatically turn into one of discretization. Interior points that may be missing "in the concept" are generated "in practice." Conversely, branching on original variables, and thereby explicitly restricting the *ISP* domain, is possible also in discretization. In that sense, both decomposition schemes are very robust against incorrectly using them. The theoretician, of course, wishes to ensure that their presentation is unified and correct, and they have to decide for a framework in which to present their work.

So, *what kind of decomposition person you are* may depend on the branching rules you prefer; branching on original variables may signal your secret love for convexification; directly enforcing integrality of sums of master variables may sort you into the discretization team. When you branch like the Red Hot Chili Peppers, it is Californication. To answer our initial question, in the end it doesn't matter. You can freely mix, you can and you will generate points interior to $\mathcal{D}$ when this is needed, and this is because of branching.

Final remarks

The section on branching could have been very short if we had only non-identical pricing problems. In that case, branching on original variables directly works and always leads to integer solutions. Pricing problem modifications are usually minor, and the impact on the solution structure, and thus on the dual bound, is likely good.

When we have an aggregation of several identical pricing problems, however, things may get more involved. The aggregated original variable values may not provide a "rich enough" information to branch on (they need not be fractional even when we do not have an integer solution to the *ILP* yet). In our discussion on cutting planes, we observed that using an extended formulation, in particular extra variables, in the *ISP* allows us to express more and stronger cuts than when formulating them in original variables only. The situation is similar in branching. We have seen examples, like the Ryan-Foster rule, the more general *Rule* 3, or the Branching on inter-tasks, which can be interpreted as branching on auxiliary variables. They require changes, if only mild, to the *ISP*, but sometimes we are lucky and the "extra variables" are already there. Vanderbeck and Wolsey (2010) call this an extended formulation that is implicit in the solution of the *ISP*. An example is the network flow based *ISP* in routing problems (where a "solution" is an incidence vector of visited customers) in which we can branch, e.g., on arc-flow variables, summed over all vehicles, see Illustration 7.6 (x-branching for the *VRPTW*). Similarly, the knapsack *ISP* of cutting stock or bin packing, in its network-based formulation (see Example 4.1) allows for branching on arc flows as well. In the words of Vanderbeck and Wolsey (2010), these extra variables "offer a larger spectrum of branching objects" than the (aggregated) original variables alone.

A drawback of original variable branching is that a potentially necessary disaggregation brings us back issues with symmetry. Since branching on original variables must explicitly *name* the variable(s) to branch on, we cannot avoid the re-introduction of index k, at least for some variables: the pricing problems re-differentiate into pricing problems belonging to a certain index k. In this situation, λ-branching on fractional sums of master variables has the potential for allowing us to avoid the re-introduction of symmetry: the disaggregation does not have to establish a one-to-one correspondence between master and original variables, like suggested in the Integrality Test.

We have not explicitly discussed the mixed-integer case but the generalization is natural as one (usually) only branches on values that should be integer. In this case, conceptually, convexification applied to continuous variables and discretization used for discrete variables may even mix (Vanderbeck and Savelsbergh, 2006).

Note that tree search strategies, that is, node selection rules like those we discussed on p. 29 for standard branch-and-bound can be relevant in branch-and-price as well, in particular in conjunction with branching rules, see Example 7.2 (Semi-assignment branching), or Primal heuristics.

Fig. 7.13: Eduardo Uchoa and Marcus Poggi de Aragão (Montréal, 2023-05-17).

7.5 Good to Know

This section covers two topics. The first is a method for eliminating flow variables when the *ISP* is a shortest path problem with resource constraints. This is done without compromising optimality of the branch-and-price approach. The second presents a wide range of acceleration techniques.

Arc-flow variable fixing by reduced cost

In many branch-and-price applications, the *ISP* solutions consist of directed paths in a network. The arc elimination, or *variable fixing to zero*, can lead to a substantial reduction of the network size, hence a speedup of the pricing and overall solution process. Irnich et al. (2010) propose a method to do so in this context, with a special attention to variants of the shortest path problem with resource constraints. Our presentation in six points below is intuitive. We refer the reader to the original paper for details.

1. Solving the *LP* (3.1) provides an optimal dual solution $\begin{bmatrix}\sigma^\star_{\mathbf{b}}\\ \sigma^\star_{\mathbf{d}}\end{bmatrix}$ and the Variable fixing by reduced cost (p. 30) can be applied. Proposition 3.1 (Walker, 1969) tells us that this is the same with $\begin{bmatrix}\boldsymbol{\pi}^\star_{\mathbf{b}}\\ \boldsymbol{\pi}^\star_{\mathbf{d}}\end{bmatrix}$ obtained from a Dantzig-Wolfe reformulation, where $\boldsymbol{\pi}^\star_{\mathbf{b}}$ comes from the *MP* while $\boldsymbol{\pi}^\star_{\mathbf{d}}$ is retrieved from the *SP*, a linear program.
2. Next, assume that we want to solve the integer program *ILP* (4.1) using a Dantzig-Wolfe reformulation, where the *ISP* is first defined as a (standard) shortest path problem on an acyclic network $G = (N, A)$. This *ISP* possesses the integrality property, thus $LB = z^\star_{MP} = z^\star_{LP}$. As such, the variable fixing technique stated in the previous paragraph applies. Indeed, the negative of the shortest path labels obtained from a forward dynamic programming (DP) algorithm are optimal dual values for $\boldsymbol{\pi}^\star_{\mathbf{d}} = [\pi^\star_i]_{i\in N}$ (e.g,. see Ahuja et al., 1993, p. 136).
3. The *critical path method* for project scheduling (Elmaghraby, 1977, Chapter 1) looks for a longest path on an acyclic network and is very similar to the above *ISP*. It is common practice to compute, for every activity to be scheduled, not only a single optimal starting time but rather an optimal interval given by the earliest and latest starting times. The earliest times are obtained using a forward DP algorithm while the largest come from using a backward DP algorithm.
4. Interestingly, the same can be done for the dual vector $\boldsymbol{\pi}^\star_{\mathbf{d}} = [\pi^\star_i]_{i\in N}$ retrieved from the *ISP*, that is, computing an optimal dual interval $[E^\star_i, L^\star_i]$ for $\pi^\star_i$, $\forall i \in N$. At optimality of the *MP* and given the adjusted costs $\tilde{c}_{ij}$, $(i, j) \in A$, that depend on $\boldsymbol{\pi}^\star_{\mathbf{b}}$, the reduced cost of x_{ij} is given by $\bar{c}_{ij} = \tilde{c}_{ij} - \pi_i + \pi_j \geq 0$. When there are multiple optimal solutions $\boldsymbol{\pi}^\star_{\mathbf{d}}$, this reduced cost can vary and its maximum is obtained by subtracting the smallest $\pi^\star_i$ and adding the largest $\pi^\star_j$, i.e.,

$$\bar{c}^{\star}_{ij} = \max_{\boldsymbol{\pi}^{\star}_{\mathbf{d}}} \bar{c}_{ij} \quad = \tilde{c}_{ij} - E^{\star}_{i} + L^{\star}_{j}. \tag{7.95}$$

If $\bar{c}^{\star}_{ij} > UB - LB$, where UB is an upper bound on $z^{\star}_{ILP}$, then $x^{\star}_{ij} = 0$ in every optimal solution $\mathbf{x}^{\star}_{ILP}$, i.e., x_{ij} can be fixed to 0.

5. Irnich et al. (2010) suggest computing these intervals *simultaneously*, by using a bidirectional search that first solves the *ISP* using a forward-labeling algorithm (to produce all $E^{\star}_{i}$ values) and then solves it again using a backward-labeling algorithm (to derive all $L^{\star}_{i}$ values). With this technique, the $\bar{c}^{\star}_{ij}$ are computed from different optimal dual solutions, for all $(i, j) \in A$.
6. Technical adjustments are required to adapt this technique to a more complex *ISP* defined as a shortest path problem with resource constraints which can also be represented on an acyclic state-space network. Applications are found in Irnich et al. (2010) for the *VRPTW* and Pecin et al. (2017b) for the capacitated *VRP*. An extension for two-arc sequences appears in Desaulniers et al. (2020a), applied to the *VRPTW* and four variants of the electric *VRPTW*.

Note that fixing an arc-flow variable x_{ij} to 0 corresponds to fixing to 0 all path-flow variables λ_p, $p \in P_{ij}$, where P_{ij} is the set of feasible paths traversing arc (i, j). It can be shown that the maximum reduced cost $\bar{c}^{\star}_{ij}$ is equal to the minimum reduced cost $\bar{c}_p$ (as a function of $\boldsymbol{\pi}^{\star}_{\mathbf{b}}$ and $\pi^{\star}_{0}$) of a path $p \in P_{ij}$, i.e.,

$$\bar{c}^{\star}_{ij} = \min_{p \in P_{ij}} \bar{c}_p. \tag{7.96}$$

Because the lower bound LB on the optimal value $z^{\star}_{ILP}$ does not have to be equal to $z^{\star}_{MP}$, the Lagrangian bound can also be used and, moreover, at any iteration of the column generation algorithm. Pessoa et al. (2010) propose to use such a bound in a variable fixing procedure similar to the one above. Their computational experiments are carried out on an *arc-time-indexed* formulation for parallel machine scheduling problems, where the columns are also generated by solving a shortest path problem. The lower bound is additionally computed using a dual stabilization approach and the fixing procedure is invoked only when there is a new stability center.

Acceleration techniques

Branch-and-price algorithms are quite sophisticated because they involve many components and concepts. To make them as efficient as possible, they are often complemented with various acceleration techniques. However, instead of implementing all existing techniques for solving a particular application, we rather recommend to start by implementing a basic branch-and-price algorithm (using a framework, see Beginner's guide to an implementation) and run computational experiments to identify the algorithm weaknesses on a representative set of relatively large test instances. To analyze the algorithm's performance, we consider three main indicators:

number of column generation iterations, *RMP* time versus *ISP* time over all iterations, and number of branch-and-bound nodes. In the following, we discuss different acceleration techniques that preserve the exactness of the branch-and-price algorithm and that can be applied when these indicators highlight an undesirable behavior. Several of them are surveyed in Desaulniers et al. (2002), together with additional ones.

Large number of column generation iterations

In a branch-and-price algorithm, the linear relaxation at the root node typically requires the largest number of column generation iterations among all linear relaxations solved in the search tree. Although it is difficult to determine what should be the ideal number of column generation iterations for a particular problem instance, we can sometimes observe an excessive number of iterations. To reduce this number (at the root node or any other node), we can use one or several of the following techniques.

- *Generating multiple columns per iteration*
  An obvious way to reduce the total number of column generation iterations is to try to generate more than one column per iteration. This may be possible when there are several subproblems or when the pricing algorithm can return multiple negative reduced cost solutions, for exemple, when the *ISP* is solved by dynamic programming, branch-and-cut, or a local search heuristic. In general, adding a few dozen to a few hundred columns to the *RMP* greatly speeds up the column generation process. However, adding too many columns at once may put too much burden on the *RMP* and have a negative impact on the overall solution process.
- *Complementary columns*
  As suggested in Note 2.25, it might be beneficial to generate columns that impact (or cover) different constraints. To achieve this, it might even be profitable to solve a pricing subproblem more than once in the same iteration, slightly altering it to ensure that the columns produced are complementary, e.g., by forbidding to cover the rows covered in previously generated columns. Keep in mind that the columns generated this way are typically chosen to help feasibility or optimality of the *ILP*, which is not the same as helping the feasibility or optimality of the *MP*, as optimality criteria are different.
- *Dual variable strategies*
  As discussed in Chapter 6, controlling the dual variables can help to generate better columns and improve column generation convergence. Therefore, all proposed tools (dual inequalities, dual boxes, dual variable stabilization, dual variable smoothing, etc.) can be applied to reduce the number of iterations. If you know your problem well, heuristics for finding good dual solutions may come to mind.
- *Early branching*
  When a good Lagrangian bound is available, column generation can always be

stopped early to avoid a long tailing-off at any node of the search tree (see Note 2.18). In this case, column generation is stopped even if there are still columns with a negative reduced cost. Then, the solution process continues with pruning, cutting, or branching at this node. Observe that early termination is commonly used in conjunction with Lagrangian relaxation methods, which we know, most of the time, approximately solves the *AMP* (see Alternative master problem and the proof of Proposition 6.2). One may even branch without a "good" bound available since one always inherits a bound from the parent node; actually, such branching happens regularly in branch-and-bound when solving the LP fails, e.g., because of numerical difficulties. Alternatively, one may want to stop column generation early because it seems obvious that it will lead to a fractional solution, e.g., with a fractional number of vehicles. Branching on the corresponding entity can, thus, be performed right away.

Most of the time spent solving the *RMP*

In a branch-price-and-cut algorithm, most of the computation time is often devoted to the column generation process, i.e., solving the *RMP* and the *ISP* in each column generation iteration. To reduce the total time as much as possible, the effort should concentrate on the most time-consuming part. When the *RMP* requires more time over all iterations than the *ISP*, the following techniques can be considered. Note that they can still be applied in the opposite case but with less overall impact.

- *Degeneracy*
  When the *RMP* is subject to high degeneracy, then several options can be envisaged. First, we can try *LP* solvers that are less prone to degeneracy such as the dual simplex algorithm or the barrier algorithm. Second, all dual variable strategies proposed in Chapter 6 can also help to solve each individual *RMP* more efficiently. Third, at some iterations, a subgradient algorithm can be used to adjust the dual values like in a Lagrangian relaxation method. Finally, if the *MP* contains a large number of set partitioning constraints, a dynamic constraint aggregation method (DCA, El Hallaoui et al., 2005, 2008, 2010) can be applied to take advantage of degeneracy.

- *High-density matrix*
  When the coefficient matrix of the *MP* constraints has a high density, solving the *RMP* might be highly time-consuming due to the numerous matrix operations that are required. In some cases, it might be possible to reduce the matrix density by changing the *MP* formulation. For example, for an integrated bus and driver problem, Haase et al. (2001) replace subsets of dense task covering constraints by corresponding subsets of sparse flow conservation constraints. Because the tasks of each former subset are most likely performed by the same bus driver in a predefined order, most generated columns covering one task in this subset also cover all the others. The corresponding flow conservation constraints simply

ensure that, if a driver leaves the sequence of tasks in the middle of it, then another driver enters it to continue the sequence.

- *Constraint relaxation/generation*
  In some applications, there might be a relatively large set of constraints in the *MP* that are necessary but have little chance of being violated if they were relaxed. In that case, it is often better to remove them a priori and add them to the *MP* dynamically only if they are violated by the solution of a linear relaxation in the search tree. These so-called "lazy constraints" should also be checked to determine the feasibility of any computed integer solution.

- *Column management*
  After a certain number of iterations, the number of columns present in the *RMP* might become large, hindering the *RMP* re-optimization process. In this case, a column management strategy can be implemented to limit the *RMP* size as mentioned in Pivot rules and column management. For instance, when the number of columns in the *RMP* reaches a predefined maximum number, say $3m$ where m is the number of rows, then a certain number of columns, say m, is removed from the *RMP*. These columns can be completely discarded or put in a pool for future direct pricing (Savelsbergh and Sol, 1998). In both cases, these columns can be generated anew by the pricing algorithm if their reduced cost becomes negative in subsequent iterations. The columns to be removed from the *RMP* can be selected based on their reduced cost or on the number of iterations since they have been in the *RMP* basis. Note that there should be a minimum number of column generation iterations (say, 5) between two consecutive *RMP* column removal operations. Otherwise, the column generation process might run into some convergence issues. Consequently, the number of columns generated per iteration should not be too large in this case.

- *Column selection*
  When a very large number of columns is generated in each iteration, then the *RMP* might rapidly become cluttered with too many variables as discussed in the previous point. In this case, it might be advisable to select a subset of the generated columns at each iteration and add only this subset to the *RMP*. For example, when the *MP* contains set partitioning constraints, columns can be selected by groups of orthogonal (complementary) columns with respect to these constraints (Savelsbergh and Sol, 1998; Desaulniers et al., 2002). A machine learning model can also be used to select the columns to add to the *RMP* (Morabit et al., 2021).

Most of the time spent solving the *ISP*

When the *ISP* requires more time over all column generation iterations than the *RMP*, the following strategies can be considered.

- *Partial pricing*
  Partial pricing consists in restricting the pricing to a subset of the feasible columns

(i.e., restricting the domain of the *ISP*), reducing the pricing time in a column generation iteration. To ensure the exactness of a column generation algorithm, full pricing must, however, be performed at least in its last iteration. Partial pricing is essential to any primal simplex algorithm implementation and so for any column generation algorithm implementation. On the other hand, there are many ways to temporarily restrict the domain of the *ISP*. In particular, when there are multiple *ISP*s, not all *ISP*s have to be solved in each iteration (see Heuristic pricing). For example, we can stop solving the *ISP*s if a small number of *ISP*s (say, 3) have been successful at producing negative reduced cost columns; or if at least a small number of *ISP*s (say, 2) have failed to yield such columns and at least one has been successful. Another partial pricing strategy consists in fixing the values of certain variables. For instance, when the *ISP* is a shortest path problem, arcs can be removed from the network (i.e., the associated flow variables are fixed to 0) to speed up the pricing algorithm (Dumas et al., 1991; Fukasawa et al., 2006; Desaulniers et al., 2008). The fixed variables may be selected according to various criteria (e.g., their cost, their adjusted cost, etc., or using a machine learning model as in Quesnel et al., 2022; Morabit et al., 2023) and this selection may vary from one iteration to another. Note that this partial pricing technique also falls into the heuristic pricing category discussed next.

- *Heuristic pricing*
  As mentioned in Heuristic pricing, there is no need to solve the *ISP* to optimality in each column generation iteration, as long as negative reduced cost columns are found. Hence, any heuristic and even several of them can be invoked at each iteration to solve the *ISP* and potentially deliver many columns in a fraction of the time required by an exact pricing algorithm. The heuristic used is often derived from the exact pricing algorithm. For example, for an *ESPPRC* subproblem (see Elementary Shortest Path Problem with Resource Constraints) that is solved by a labeling algorithm (see Labeling algorithm for the *ESPPTWC*), it can be made heuristic by considering only a subset of the arcs, using a heuristic dominance rule, or limiting the number of labels associated with each node (Dumas et al., 1991; Fukasawa et al., 2006; Desaulniers et al., 2008). Also, when the *ISP* is an *ILP* or a *MILP*, it can be solved using a truncated (heuristic) branch-and-cut algorithm by imposing a time or node limit, an optimality tolerance, or a maximum number of integer solutions with a negative reduced cost found. Alternatively, ad hoc heuristics, exploiting the *ISP* structure, can also be applied, such as local search or tabu search (Savelsbergh and Sol, 1998; Desaulniers et al., 2008). Note that some dual bound computations depend on an optimal solution of the pricing problems.

- *Variable fixing*
  The variable fixing technique described in Variable fixing by reduced cost and Arc-flow variable fixing by reduced cost can help to speed up the solution of the *ISP* as it can fix the value of a large subset of its variables (see Irnich et al., 2010; Pecin et al., 2017b; Desaulniers et al., 2020a). This technique is only applicable when a good upper bound *UB* on the optimal value $z^\star_{IMP}$ is known.

- *Relaxed ISP*
  As discussed in Relaxed pricing, relaxed master, when the computational complexity of the *ISP* is high, it might be advantageous to generate columns using a relaxation of the *ISP*, yielding a relaxed *MP* and possibly weaker lower bounds. This technique is often used in branch-and-price algorithms for solving vehicle routing problems, where the pricing subproblem is a strongly $\mathcal{NP}$-hard elementary shortest path problem with resource constraints (see Chapter 5). In this case, the *ISP* is replaced by a relaxed shortest path problem such as the *ng*-shortest path problem with resource constraints (see *ng-SPPTWC* relaxation). Thus, columns associated with inadmissible objects can be generated and must be excluded from the *MP* optimal solutions in the branch-and-bound tree. Note that this technique may allow to substantially reduce the time devoted to the *ISP*, but might result in a much larger search tree to be explored. Consequently, a compromise must be reached between the speed up obtained by relaxing the *ISP* and the quality of the resulting lower bounds.

- *Rolling back on cuts*
  As discussed in Cutting planes on the master variables, adding cuts defined on the master variables may impact the *ISP* formulation, requiring extra variables and constraints. The *ISP* may then become much more difficult to solve as cuts are generated. A typical example for this occurs when subset-row inequalities are applied in a branch-price-and-cut algorithm for the *VRPTW* (see Example 7.8 *VRPTW* and Chvátal-Gomory rank-1 cuts) and the *ISP* is solved using a labeling algorithm similar to that described in Labeling algorithm for the *ESPPTWC*. For this case, the dominance rule becomes less effective at discarding labels due to the handling of the cut dual variables and the *ISP* may consume more than 95% of the total computational time. One option to avoid this burden is to impose a preset limit on the number of cuts that can be generated. This approach may, however, be too conservative as this limit may be reached without inducing an *ISP* that is too difficult to solve. Pecin et al. (2017b) proposed a dynamic alternative that does not rely on a maximum number of cuts but on a difficulty threshold for solving the *ISP* (e.g., a maximum number of labels generated). After adding a round of cuts to the *MP*, a rollback procedure is triggered if this threshold is reached when solving the next *ISP*. This procedure stops the labeling algorithm, removes the last round of cuts and imposes branching decisions instead of adding cuts.

Large number of branch-and-bound nodes

When the number of nodes in the search tree is large, the following techniques can be applied to reduce this number. The impact on the total computation time of each of these techniques depends on the time it consumes versus the time saved by exploring a reduced number of nodes.

- *Lower bound strengthening*
  It is well known that strengthening the lower bounds typically yields smaller

branch-and-bound trees. This can be achieved in different ways. A popular one is to generate cuts as discussed in Section 7.2. Variable fixing, as presented in Variable fixing by reduced cost, can also help to reduce the size of the search tree by yielding stronger lower bounds and more efficient cuts (see Irnich et al., 2010; Desaulniers et al., 2020a). Finally, the definition of the pricing subproblem influences the quality of the lower bounds obtained. Using an *ISP* that does not possess the integrality property yields stronger bounds than solving the *ILP* directly (see Section 4.3). Moreover, integrating more decisions in the *ISP* can produce tighter lower bounds as demonstrated for two different vehicle routing problems in Desaulniers (2010) and Desaulniers et al. (2016b). In the former, the split delivery vehicle routing problem with time windows is addressed and the quantity delivered to each customer visited along a route is a decision made in the *ISP*. In the latter that tackles the inventory routing problem, the *ISP* also determines the quantity delivered to each visited customer in a route, but further specifies how this quantity is used to cover the demands of the customer over the next periods.

- *Strong branching*
Strong branching (see Achterberg et al., 2005) is a technique used in branch-and-bound algorithms that can also be applied to branch-and-price algorithms. Given a subset C of candidate branching decision pairs that can be used to separate a node in the search tree, it evaluates them all to determine which pair of decisions to select. An evaluation consists of computing for each decision (i.e., for each potential child node) an estimate of the objective value deterioration (or, positively speaking, dual bound improvement) if it was imposed. The retained candidate is, then, selected using a rule based on these estimates (for instance, a max-min rule). In a branch-and-price algorithm, the evaluation can be quite time-consuming. Therefore, Pecin et al. (2017b) propose to perform it in two phases. In the first phase, all candidates in C are briefly evaluated by only reoptimizing the *RMP* after imposing each decision. Based on these coarse evaluations, a subset C' of C is selected. In the second phase, the evaluation of each candidate in C' is refined by applying column and cut generation for each potential decision but using only heuristic pricing.

- *Column enumeration*
Introduced by Baldacci et al. (2008) for the capacitated vehicle routing problem, column enumeration is based on the reduced cost criterion applied for variable fixing. Given an upper bound $\bar{z}$ on the optimal value $z^\star_{IMP}$ and a lower bound $\underline{z}$ derived from a dual solution $\boldsymbol{\pi}$ to the *MP* possibly tightened with cuts, the procedure tries to enumerate *all* columns whose reduced cost with respect to $\boldsymbol{\pi}$ is less than or equal to $\bar{z} - \underline{z}$ (the other columns can all be fixed to their zero lower bound). If the enumeration is successful, then an *IMP* restricted to the subset of enumerated variables is solved using a standard branch-and-cut solver and its optimal solution in binary λ-variables, if any, is also optimal for the whole problem. Otherwise (i.e., when memory lacks during enumeration), the procedure simply fails. In their paper, Baldacci et al. (2008) compute $\boldsymbol{\pi}$ and $\underline{z}$ using a dual ascent method based on column and cut generation, enumerate the subset of routes us-

ing dynamic programming, and apply this column (route) enumeration procedure only at the root node, possibly leaving some instances unsolved when enumeration fails.

To fix this issue, Pessoa et al. (2009) develop a hybrid algorithm combining column enumeration and branch-and-price. After solving the *MP* at each node of the search tree, column enumeration is attempted and, if it seems too demanding according to some predefined criteria, it is aborted and branching is rather performed at this node. This algorithm can be further enhanced by using a column pool when a large number of columns can be enumerated but the resulting restricted *IMP* cannot be solved in a reasonable amount of time (see Contardo and Martinelli, 2014).

- *Preprocessing*
  A key element of being able to solve large instances is making them smaller. Branch-and-bound algorithms have a lot of so-called preprocessing procedures that eliminate variables (like in Arc-flow variable fixing by reduced cost) or redundant constraints, tighten bounds, strengthen coefficients, etc. This is no different in branch-and-price. Use as much problem knowledge as possible to reduce the size of the instance to solve. For instance, in *VRPTW* applications, try to reduce wide time windows in order to eliminate impossible visiting times (see Desrosiers et al., 1995, Section 3.2. *Time window reduction and arc elimination*).

7.6 More to Know

We start this section with ideas for primal heuristics and then extend our discussion on Chvátal-Gomory cuts to rank 2. The first topic should indeed be relevant for *every* branch-and-price person, the second is rather for connoisseurs. We close with some thoughts about an implementation. Even though we place this item last, it is indispensable in practice, of course.

Primal heuristics

When students first learn about the standard branch-and-bound algorithm, they often think that integer solutions are obtained by branching. While this is true, of course, this is usually not the *main source* of them. Instead, branch-and-bound solvers have problem-specific and/or general primal heuristics implemented to obtain integer feasible solutions, hopefully early in the process. This is also true for branch-and-price. Finding a good primal bound rapidly can certainly help prune nodes in the search tree, but it can also trigger substantial speedups when applying complementary techniques like variable fixing by reduced cost (see Arc-flow variable fixing by reduced cost) and column enumeration (see Large number of branch-and-bound nodes).

Inventing heuristics for the particular problem at hand is a natural idea. One can find inspiration for this in the many general ideas that exist for primal heuristics. Several are inherited from classical branch-and-bound implementations, and some can even exploit the particular branch-and-price context of having a decomposition available. We sketch a handful of the most popular themes.

- *Restricted master heuristic ("price-and-branch")*
  One of the simplest ideas often happens almost "automatically" when you have not implemented a branching rule (yet). In *price-and-branch*, column generation is only applied at the root node to solve the initial master problem. The final *RMP* is then transformed into a *MILP* and solved using a standard *MILP* solver. This transformation is straightforward if the integrality requirements can be imposed directly on the master variables $\boldsymbol{\lambda}$ (just change the variables' type from continuous to integer). Otherwise, the variables $\mathbf{x}$ of the compact formulation together with the constraints linking $\mathbf{x}$ with $\boldsymbol{\lambda}$ must be added to the *RMP* before imposing the integrality requirements on $\mathbf{x}$. The simplicity of this heuristic comes at a price: the quality of the feasible solution obtained depends on the subset of columns generated. Given that these columns were generated for solving the *MP* but not directly the *IMP*, they are often insufficient to compute a high-quality integer solution. Hence, it is important to generate a sufficiently large subset of columns. Finally, note that *milpping* the *RMP* can be done at any column generation iteration and not necessarily only at the last one at the root node.
- *Complementary pricing*
  We emphasized previously that the columns we generate follow an optimality criterion of *linear* programs, not one of *integer* programs. However, we can always, at least additionally, generate columns with the explicit hope that they fit well in a good *IMP* solution. One such idea is *complementary pricing*, see Acceleration techniques above. It works well in set partitioning master problems: when a column is generated, we temporarily fix the new variable to 1, thus covering certain rows already. Then we call the pricing problem again on the reduced size problem to generate columns that cover the remaining rows, and repeat. The aim is to produce column vectors that "complement well" to an integer feasible solution. The idea can be generalized to other kinds of master problems, plus it is usually not hard to implement. While this is in fact a pricing strategy, see Pivot rules and column management, it is relevant also in the primal heuristics context, of course. This reminds us that, often, everything is connected with everything else.
- *Diving heuristics*
  In an exact branch-and-price algorithm, columns are generated in every node of the search tree. The newly generated columns were not useful to solve the *MP* at the parent node but some of them are complementary to the solution obtained after imposing a branching decision. The main weakness of the restricted master heuristic described above is that it does not possess this recourse mechanism. A *diving heuristic* exploits this recourse option, but restricts the exploration of the search tree to a single branch in general. It just dives in the tree from the root to

a single leaf node, generating columns in each node. In general, it finds just one feasible integer solution in this leaf node.

Given that a single child node is created when the current master problem solution is fractional, certain branching decisions that cannot be applied in an exact algorithm might become applicable. For instance, branching on an integer master variable $\lambda_{\mathbf{x}}$ that takes a fractional value $\lambda^{\star}_{\mathbf{x}}$ in the current *MP* solution is not practical in an exact algorithm because the down-branch decision $\lambda_{\mathbf{x}} \leq \lfloor \lambda^{\star}_{\mathbf{x}} \rfloor$ is usually weak and difficult to implement (see Branching on the master variables). However, in a diving heuristic, we can select the up-branch decision $\lambda_{\mathbf{x}} \geq \lceil \lambda^{\star}_{\mathbf{x}} \rceil$ to create a single node. This strong decision is easy to implement and can help to rapidly reduce the solution space. On the other hand, imposing $\lambda_{\mathbf{x}} \geq 1$ when $0 < \lambda^{\star}_{\mathbf{x}} < 1$ is not very close to 1 (e.g., $\lambda^{\star}_{\mathbf{x}} \leq 0.6$) may be a bad decision that is not reversible because backtracking in the search tree is not allowed. As this situation often arises, we should consider alternative decision types that are defined on the original x-variables or equivalently on a sum of the master variables, and which are less aggressive. For example, for a vehicle routing problem like the *VRPTW* with binary arc-flow variables x_{ij} in the compact formulation and binary route variables λ_p in the extended formulation, we can choose to fix $\lambda_p = 1$ for a route p or $x_{ij} = 1$ for an arc (i, j). The former decision type should be favored only if λ_p takes a sufficiently large fractional value. The latter type is less constraining and has, therefore, less impact on the dual bound. It can be imposed even if the value of x_{ij} is not too close to 1.

To further accelerate the search for an integer solution, multiple up-branch decisions can be imposed simultaneously, which means that columns are not generated between these decisions. This is often acceptable when the set of decisions is defined on variables taking values very close to the imposed lower bounds (e.g., their fractional parts are all greater than or equal to 0.9).

For certain problems, a single dive in the tree might be risky as it may lead to an integer solution of a doubtful quality, or worse to no feasible solution at all. To prevent this, we can explore more than one branch of the tree. This is accomplished by creating more than one child node at some nodes of the tree, typically, when the best candidate decision is rather uncertain (e.g., imposing a lower bound on a master variable that has a fractional part less than 0.7). In this case, multiple nodes are not created by branching, but rather by defining up-branches for different decisions. For example, let us assume that the fractional parts of the three best candidate master variables $\lambda_{\mathbf{x}_1}$, $\lambda_{\mathbf{x}_2}$, and $\lambda_{\mathbf{x}_3}$ are given by 0.7, 0.67, and 0.65, respectively. In this case, three child nodes can be created, one for each decision $\lambda_{\mathbf{x}_1} \geq \lceil \lambda^{\star}_{\mathbf{x}_1} \rceil$, $\lambda_{\mathbf{x}_2} \geq \lceil \lambda^{\star}_{\mathbf{x}_2} \rceil$, and $\lambda_{\mathbf{x}_3} \geq \lceil \lambda^{\star}_{\mathbf{x}_3} \rceil$, where $\lambda^{\star}_{\mathbf{x}_i}$, $i = 1,2,3$, is the variable value in the current *MP* solution. Compared to a pure diving heuristic with a single branch, this diving heuristic with backtrackings can only lead to a better final solution. The creation of multiple nodes must be used with parsimony to limit the computational time. Other diving heuristic variants are described in Sadykov et al. (2019), Lübbecke and Puchert (2013), and Quesnel et al. (2017).

- *Rounding and sub-MILP heuristics*
  Simply rounding a fractional *RMP* solution can be totally useless, but also surprisingly successful. In any case, this is very cheap and always worth a try. However, one can do the rounding more intelligently, but thus also more expensive: Fix all the λ-variables to their current values when these are already integer. Then solve the resulting sub-MILP using branch-and-bound. This can be seen as an optimal rounding. One can also try to fix a smaller, potentially more meaningful subset of λ-variables: for instance, those variables, in which the *RMP* solution and an incumbent (integer feasible) solution agree. Similarly, one can fix the values of x-variables when "many" columns of positive λ-variables agree in the respective components. We refer to Lübbecke and Puchert (2013) for inspiration.
- *Large neighborhood search*
  A sub-MILP searches the neighborhood of a fractional or incumbent solution. This can be generalized to a true *large neighborhood search* (LNS, Shaw, 1998). The LNS procedure starts with an initial solution that can be feasible or not. Then, at each iteration, it destroys part of the current solution before repairing it using branch-and-price as a re-optimization tool. Various destruction operators can be employed throughout the iterations (Røpke and Pisinger, 2006). These operators can work directly on the master variables (e.g., on schedule variables as in Pepin et al., 2009) or on objects appearing in the compact model or *ISP* (e.g., on subsets of customer nodes as in Prescott-Gagnon et al., 2009). An appreciated feature of this heuristic is that it provides a sequence of improving solutions throughout the solution process and can, thus, be stopped at almost any time with a feasible integer solution in hand. The (expensive) re-optimization step by branch-and-price can, and should, be done by a branch-and-price heuristic. Any of those mentioned above could apply. In that sense, LNS can be considered as a hybrid heuristic, see the next item.

Fig. 7.14: Elina Rönnberg (2023-05-17) and Stefan Irnich (2024-07-30) in Montréal.

Maher and Rönnberg (2023) propose a destroy-and-repair scheme for general master problems which include set covering, packing, or partitioning constraints. They adapt the pricing problem so that it does not (only) generate columns that are favorable for optimality of the *RMP MP*, but for the *IMP* itself.

- *Hybrid heuristics*
  Hybrid heuristics combine different heuristic or exact algorithms. One can imagine that the options here are countless. We give two examples. The first one is called *Diving with sub-MILP* by Sadykov et al. (2019) and combines a pure diving heuristic with the restricted master heuristic. It starts by applying a diving heuristic with a single branch to obtain, if any, a feasible integer solution. Then, all columns generated throughout the process are gathered in an augmented *RMP*, which is *milped* and solved using a standard *MILP* solver. Given the augmented set of columns considered, this heuristic can hopefully find a better solution than the restricted master heuristic.

  The second, which can be called *Dive-first, then-branch*, mixes a diving heuristic with an exact branch-and-price algorithm. It begins by a dive in one or multiple branches as discussed above. The dive in a branch is stopped when only highly uncertain candidate decisions are available and the size of the residual problem (i.e., the initial problem restricted by all applicable decisions) is relatively small. One may also use the number of fractional-valued variables in the current *MP* solution to evaluate this size. The search in this branch is then replaced by the exploration of the full sub-tree using the exact branch-and-price algorithm which should be relatively fast for solving the residual problems.

We are in the comfortable situation that we have two formulations in our hands, the *ILP* and the *IMP*. We can therefore run heuristics on (the variables of) both of them (see, e.g, Lübbecke and Puchert, 2012, 2013). One may, for instance, run the primal heuristics of a standard solver on the *ILP* and use the resulting heuristic solution, if one was found, for the initialization of the *RMP*. It shows again, that a good general knowledge about modern branch-and-bound implementations is helpful, see Note 7.7. In the context of primal heuristics, having some background in meta-heuristic ideas like local search, evolutionary algorithms, etc., and how they can be hybridized with exact methods into *matheuristics* certainly helps, too. Let us mention that ideas for primal heuristics for the entire problem can be useful for heuristic pricing as well, as the pricing solves only a part of the same problem.

It is not only that primal heuristics help the exact branch-and-price algorithm. Often, practical-sized instances are too large to be solved to optimality anyway. For example, Barnhart et al. (1998, p. 318) report that Vance (1993) found more than five million crew pairings in a daily problem with 253 flights. You can imagine the size of an instance with 1000 flights per day over a weekly horizon. Then, imposing a time, node, or gap limit turns the exact approach into a heuristic. In branch-and-price, we can then be more sloppy in many places. For instance, there is no requirement to solve the master problem to optimality in each node. In particular, the *ISP* does not have to be solved exactly at any column generation iteration. Hence, we can apply various pricing heuristics that can be simply defined by removing a priori some variables from the *ISP* formulation or by heuristically speeding up the pricing algorithm (for instance, by using an aggressive dominance rule in a labeling algorithm for solving an *ESPPRC* or a *SPPRC*– see Chapter Vehicle Routing and Crew Scheduling Problems). Alternatively, a pure heuristic such as a greedy heuristic or tabu search can also be used. Finally, to avoid the column generation tailing-off at each node, column generation can be halted prematurely when the *RMP* optimal value does not decrease sufficently in a predetermined number of iterations (see also *Early branching*, p. 495). All this can result in stand-alone branch-and-price heuristics which are used in the industry to compute high-quality feasible solutions for huge instances.

Note 7.13 (You do not solve an ILP by column generation.) In many articles, one reads about "column generation approaches to this or that integer program." After having read 506 pages of this book you know that this cannot be true: The column generation algorithm solves *linear* programs only. Often, these authors neither discuss nor implement a branching rule, and thus do not present a true branch-and-price approach. Instead, typically, heuristic integer solutions are obtained using what we called above the *restricted master* or *price-and-branch* method.

Chvátal-Gomory cuts of higher ranks

Recall Illustration 7.4 on the Chvátal-Gomory cuts of rank-1. We here consider the case of rank-2 cuts whereupon it becomes clear how to extend to cuts of any higher rank. Our goal is to add to the *MP* two sets of cuts:

$$z^\star_{MP} = \min \quad \sum_{\mathbf{x}\in\mathcal{X}} c_{\mathbf{x}}\lambda_{\mathbf{x}} \tag{7.97a}$$

$$\text{s.t.} \quad \sum_{\mathbf{x}\in\mathcal{X}} \mathbf{a}_{\mathbf{x}}\lambda_{\mathbf{x}} \geq \mathbf{b} \qquad [\boldsymbol{\pi}_{\mathbf{b}} \in \mathbb{R}^m_+] \tag{7.97b}$$

$$\sum_{\mathbf{x}\in\mathcal{X}} \mathbf{g}^1_{\mathbf{x}}\lambda_{\mathbf{x}} \geq \mathbf{h}^1 \qquad [\boldsymbol{\gamma}^1 \in \mathbb{R}^{|L^1|}_+] \tag{7.97c}$$

$$\sum_{\mathbf{x}\in\mathcal{X}} \mathbf{g}^2_{\mathbf{x}}\lambda_{\mathbf{x}} \geq \mathbf{h}^2 \qquad [\boldsymbol{\gamma}^2 \in \mathbb{R}^{|L^2|}_+] \tag{7.97d}$$

$$\lambda_{\mathbf{x}} \geq 0 \qquad \forall \mathbf{x}\in\mathcal{X}, \tag{7.97e}$$

where the $|L^1|+|L^2|$ additional constraints are rank-1 and rank-2 cuts with respective dual vectors $\boldsymbol{\gamma}^1$ and $\boldsymbol{\gamma}^2$. A column for $\mathbf{x}\in\mathcal{X}$ is composed of $c_{\mathbf{x}} = \mathbf{c}^\intercal\mathbf{x}$, $\mathbf{a}_{\mathbf{x}} = \mathbf{A}\mathbf{x}$, $\mathbf{g}^1_{\mathbf{x}} = [g^1_{\ell\mathbf{x}}]_{\ell\in L^1}$ and $\mathbf{g}^2_{\mathbf{x}} = [g^2_{\ell\mathbf{x}}]_{\ell\in L^2}$. Let $\mathbf{h}^1 = [h^1_\ell]_{\ell\in L^1}$ and $\mathbf{h}^2 = [h^1_\ell]_{\ell\in L^2}$ on the right-side. Given weight vectors $\mathbf{w}^1_\ell \in \mathbb{R}^m_+$ for $\ell\in L^1$ and $\mathbf{w}^2_\ell \in \mathbb{R}^{m+|L^1|}$ for $\ell \in L^2$, the cut coefficients are computed as

$$g^1_{\ell\mathbf{x}} = \lceil \mathbf{w}^{1\intercal}_\ell \mathbf{a}_{\mathbf{x}} \rceil \qquad \forall \ell\in L^1, \mathbf{x}\in\mathcal{X} \tag{7.98a}$$

$$h^1_\ell = \lceil \mathbf{w}^{1\top}_\ell \mathbf{b} \rceil \qquad \forall \ell \in L^1 \tag{7.98b}$$

$$g^2_{\ell\mathbf{x}} = \left\lceil \mathbf{w}^{2\top}_\ell \begin{bmatrix} \mathbf{a}_\mathbf{x} \\ \mathbf{g}^1_\mathbf{x} \end{bmatrix} \right\rceil \qquad \forall \ell \in L^2, \mathbf{x} \in \mathcal{X} \tag{7.98c}$$

$$h^2_\ell = \left\lceil \mathbf{w}^{2\top}_\ell \begin{bmatrix} \mathbf{b} \\ \mathbf{h}^1 \end{bmatrix} \right\rceil \qquad \forall \ell \in L^2. \tag{7.98d}$$

That is, the functions for calculating the rank-2 cut coefficients depend on both $\mathbf{a}_\mathbf{x}$ and the rank-1 coefficients $\mathbf{g}^1_\mathbf{x}$ on the left-side, similarly on $\mathbf{b}$ and $\mathbf{h}^1$ on the right-side. Because $\boldsymbol{\gamma}^1, \boldsymbol{\gamma}^2 \leq \mathbf{0}$, the ceiling functions can be computed as in (7.39), that is, using integer linear inequalities. Hence, the *ISP* becomes

$$\begin{aligned} \bar{c}(\boldsymbol{\pi}_\mathbf{b}, \boldsymbol{\gamma}^1, \boldsymbol{\gamma}^2) = \min_{\mathbf{x} \in \mathcal{X}} \quad & \mathbf{c}^\top \mathbf{x} - \boldsymbol{\pi}^\top_\mathbf{b} \mathbf{A}\mathbf{x} - \boldsymbol{\gamma}^{1\top} \mathbf{g}^1_\mathbf{x} - \boldsymbol{\gamma}^{2\top} \mathbf{g}^2_\mathbf{x} & \\ \text{s.t.} \quad & g^1_{\ell\mathbf{x}} \leq \mathbf{w}^{1\top}_\ell \mathbf{A}\mathbf{x} + 1 - \varepsilon^1_\ell & \forall \ell \in L^1 \\ & g^2_{\ell\mathbf{x}} \leq \mathbf{w}^{2\top}_\ell \begin{bmatrix} \mathbf{A}\mathbf{x} \\ \mathbf{g}^1_\mathbf{x} \end{bmatrix} + 1 - \varepsilon^2_\ell & \forall \ell \in L^2 \\ & \mathbf{g}^1_\mathbf{x} \in \mathbb{Z}^{L^1}, \ \mathbf{g}^2_\mathbf{x} \in \mathbb{Z}^{L^2}. & \end{aligned} \tag{7.99}$$

Repeating the procedure, we can observe that higher rank cuts depend on lower rank cuts to compute adequately the cut coefficients.

Beginner's guide to an implementation

> There is nothing so practical as a good theory.
>
> Kurt Lewin (1951)

When you have read this book, it is likely that you wish to actually *do* some branch-price-and-cut, and use it to solve some real optimization problem. And this unavoidably brings us to an implementation. We have given some specific details at several places, however, we spared a very practical question: where to even start? An answer depends, of course, on your background and goals. We assume that you neither develop a generic branch-price-and-cut solver that can be applied to any problem (because only very few people on earth do this, and the probability that you are one of them is small), nor that you have an existing code, maybe "from your last branch-and-price project" (because then you already know where to start). That is, you are a beginner. Nice! We recommend two phases, the first of which is about creating your setup and doing column generation only, and consists of a few steps.

Step one is solely on paper. Don't skip this. Write down a compact (the "original") MILP model for your problem. We think of it as a formal problem description.

Be clear about all the parameters, and how precisely a solution looks like. Rigor and a good organization here helps in every step downstream.

Step two, choose a MILP solver, proprietary or open source, whatever is available to you, and a programming language, e.g., `Python`. Every solver offers an API to current languages, and usually they come with modeling examples. Now implement your model. The modeling part itself literally translates your model from paper into one written in your programming language. You create some kind of `model` object (which is initially empty), define `variables` and `constraints` (there are methods to `add` them to your `model`), and finally you call the `optimize` method. Learn how to access a solution and print it in an informative way. Since variables and constraints are defined over index sets, the modeling part benefits from precision in your earlier work on paper. The data part is trickier, you have to fill the abstract coefficients in the model with life, and need instances of your problem for this. It may sound obvious, but without data, we cannot compute anything. Acquiring data can be the most time consuming, most boring or most exciting, maybe the most critical phase of the entire project.

The importance of not doing any column generation at first is in the setup that you accomplish: install a solver, learn the basics of a programming language, produce a *pipeline* from the data sources into your model, and from the solution to your model into something interpretable, maybe a visualization or a table. This step also serves as a benchmark: how large (or small) instances can be solved in acceptable time? We could also use an algebraic modeling language for this experiment. Maybe you have done all of this already, and state-of-the-art branch-and-cut (that is built into MILP solvers) left you unsatisfied with the results. Maybe this brought you to this book. This is good, you are not a beginner any more, we can proceed.

Step three is column generation. Again, preparing everything meticulously on paper is our friend. Reformulate your compact model according to Dantzig and Wolfe, let Chapter 4 inspire you. Maybe after aggregation, this results into an *MP* and one or several *ISP*s. Some skip the Dantzig-Wolfe reformulation step and directly formulate a master and pricing problems, but this may conceal connections which are "there" anyway, and this knowledge is likely needed in branching and cutting later. In any case, we work with two communicating `models` now, see Fig. 2.2. The `variable` type in the master `model` should be continuous, and it is likely that this type is integer (or binary) in your *ISP*s. We need methods for obtaining the values of the dual variables (remember, there is one for every constraint in the master, so the dual value may be an attribute of the `constraint` object). Also, the columns that are returned from the *ISP*s need to be added to the *RMP*: Create a new `variable` in the master `model` and `change` its coefficients in the respective master constraints according to the values of the solution obtained from the pricing `model`. You have to write a loop for the column generation algorithm according to Algorithm 2.1 and you may have to think about initialization. We promise that through the implementation (and the debugging!), you understand the theory much better. Maybe you replace the *ISP* MILP `model` with a specialized solver for the pricing later. You may wish to add cutting planes which are derived from the original model, but maybe keep it simple first. Integer solutions are heuristically obtained by switching the `variable` type to integer (or binary) and `optimize` the *MP* with branch-and-bound. This completes phase one.

Doing true branch-and-price, which is phase two of an implementation, can be more involved. It is likely that one needs more programming skills than in the previous phase. A main reason for this is that standard, in particular commercial solvers, do not give us control over how the relaxation is solved in every node of the search tree (in contrast to "callback" functions that allow a user to implement, e.g., an own separation of cutting planes, there is no such form of access to the pricing). If you use an open source solver framework, however, your column generation code from above can be "hooked" into the solving loop to realize precisely this control. Depending on the framework, you may only have to put the code for your *ISP*s in an appropriate place (and the *RMP* is then re-optimized automatically as part of the usual solving process). Now the exchange of information between master and pricing problems likely manifests itself in the implementation of a predefined interface. We can give much less general advise here, except: use a framework that takes care of managing the branch-and-bound process for you. It is not a secret that, e.g., `SCIP` is well-suited for this.

We should not hide that even a large investment in an implementation may turn out to be fruitless in the end. Your optimization problem, at this moment, may be too hard to be solved to proven optimality in acceptable time. After all, this is what drives research. Therefore, before you embark on an implementation, you may wish to have some prediction about a success probability. This is one motivation of the `GCG`[1] project. What is needed is only the original model (the first steps of phase one), the remainder is done automatically by the solver: reformulation, pricing, cutting, branching. Improvement and tailoring like using a specialized pricing solver is still possible, it may even be necessary. But when you are lucky, you may have done full branch-price-and-cut without having to go through an implementation.

7.7 Examples

In Example 7.1 (Time constrained shortest path problem), we use our creative freedom in proposing various branching decisions on the x-variables, integrated either to the *ISP* or the *IMP*. In the second example (Semi-assignment branching), we present a strategy that can be useful for problems such as the generalized assignment, capacitated p-median, unrelated parallel machines scheduling, and similar ones. In Example 7.3 (Ryan-Foster branching for vertex coloring), we use a specialized λ-branching, and decisions are imposed in the set partitioning master problem. In the next one, the Ryan-Foster branching for bin packing makes the *ISP* harder. Continuing with the Edge coloring and odd-circuit cuts helps understanding the use of indicator variables in the *ISP* (which are useful in λ-branching as well).

[1] `https://gcg.or.rwth-aachen.de/`

Example 7.6 (k–path cuts for the *VRPTW*) describes one of the first efficient x-cuts for the *VRPTW* whereas *VRPTW* and non-robust rounded capacity cuts improves on the capacity cuts of Illustration 7.2. For the *VRPTW* and Chvátal-Gomory rank-1 cuts, we present a rather easy implementation of the pricing problem. The set of examples is completed by two large-scale applications. The Preferential bidding system for a monthly airline crew assignment problem makes use of the z-cuts while the Branch-first, Cut-second strategy for locomotive assignment in freight transportation combines branching and cutting on the x-variables. For these two, we have to say: *If we can find good-quality integer solutions, we are very happy.*

Example 7.1 Time constrained shortest path problem

✎ We use various branching rules derived from a compact formulation. The freedom of choosing the branching decisions is a powerful tool when properly applied.

Given is the *ILP* (4.33) for the *time constrained shortest path problem* (*TCSPP*), reformulated as the *IMP* (4.35), for which we want to find an optimal integer solution $\mathbf{x}^{\star}_{ILP}$. The column generation algorithm rather gives us a solution for the *MP*, where $\mathbf{x}^{\star}_{MP}$ is indeed fractional. For reference, we have accomplished all this in Illustration 4.5 and Example 3.2.

From Table 3.1, we get $\lambda_{13256} = 0.2$, $\lambda_{1256} = 0.8$, and $z^{\star}_{MP} = 7$. This is obtained by solving the following *RMP*, where y_0 is an artificial variable in the convexity constraint with a big-M cost equal to 100:

$$\begin{array}{lllllll}
\min 100y_0 + & 3\lambda_{1246} & + 24\lambda_{1356} & + 15\lambda_{13256} & + \ 5\lambda_{1256} & & \\
\text{s.t.} & 18\lambda_{1246} & + \ 8\lambda_{1356} & + 10\lambda_{13256} & + 15\lambda_{1256} & \leq 14 & [\pi_7] \\
\quad y_0 + & \lambda_{1246} & + \ \lambda_{1356} & + \ \lambda_{13256} & + \ \lambda_{1256} & = \ 1 & [\pi_0] \\
\quad y_0, & \lambda_{1246}, & \lambda_{1356}, & \lambda_{13256}, & \lambda_{1256} & \geq \ 0. &
\end{array}$$

In the compact formulation, this λ-solution corresponds to the arc-flows $x_{12} = 0.8$, $x_{13} = x_{32} = 0.2$, and $x_{25} = x_{56} = 1$, see Figure 7.15. Let us instantiate a branch-and-bound tree whose root node, denoted **BB0**, captures all this information.

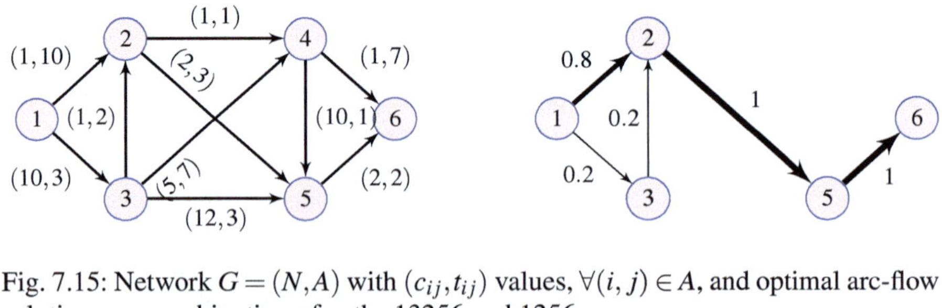

Fig. 7.15: Network $G = (N, A)$ with (c_{ij}, t_{ij}) values, $\forall (i, j) \in A$, and optimal arc-flow solution as a combination of paths 13256 and 1256.

Since the x-variables are required integer, we know that any combination of their values must also be. We can therefore consider branching on the sum of any subset of arc-flow variable values. Immediate examples are the single arc-flow $x_{12} = 0.8$ and the total sum $\sum_{(i,j)\in A} x_{ij} = 3.2$.

Let us instead focus our first decisions on $x_{13} + x_{32} = 0.4$ (just because we can): the down-branch imposes $x_{13} + x_{32} = 0$ whereas we have $x_{13} + x_{32} \geq 1$ in the up-branch. Table 7.3 summarizes the procedure, where $\bar{c}(\cdot)$ denotes the minimum reduced cost obtained by the *ISP* for the appropriate dual vector that is modified in dimension and values during the search.

Node	*RMP* solution	$z^\star_{MP}$	π_0	π_7	π_8	$\bar{c}(\cdot)$	p	c_p	t_p
BB1: BB0 and $x_{13} + x_{32} = 0$									
BB1.1	$y_0 = 1/15, \lambda_{1256} = 14/15$	11.3	100	-6.33	-	0			
BB1.2	M increased to 1000								
	$y_0 = 1/15, \lambda_{1256} = 14/15$	71.3	1000	-66.33	-	-57.3	12456	14	14
BB1.3	$\lambda_{12456} = 1$	**14**	1000	-70.43	-	0			
	Integer arc-flows: $x_{12} = x_{24} = x_{45} = x_{56} = 1$								
BB2: BB0 and $x_{13} + x_{32} \geq 1$									
BB2.1	$\lambda_{1246} = \lambda_{13256} = 0.5$	9	15	-0.67	3.33	0			
	Fractional arc-flows: $x_{12} = x_{13} = x_{24} = x_{25} = x_{32} = x_{46} = x_{56} = 0.5$								
BB3: BB2 and $x_{12} = 0$									
BB3.1	$\lambda_{13256} = 1$	**15**	15	0	0	-2	13246	13	13
BB3.2	$\lambda_{13246} = 1$	**13**	13	0	0	0			
	Integer arc-flows: $x_{13} = x_{32} = x_{24} = x_{46} = 1$								
BB4: BB2 and $x_{12} = 1$									
BB4.1	Infeasible								

Table 7.3: Details of the branch-and-bound tree.

BB1: **BB0** and $x_{13} + x_{32} = 0$

Housekeeping: We 1) discard λ_{1356} and λ_{13256} from the *RMP* and 2) remove the arcs $(1,3)$ and $(3,2)$ from G. As a consequence, node 3 has no more predecessors so it is discarded as well along with the arcs $(3,4)$ and $(3,5)$. In iteration BB1.1, we therefore start by solving the *RMP*

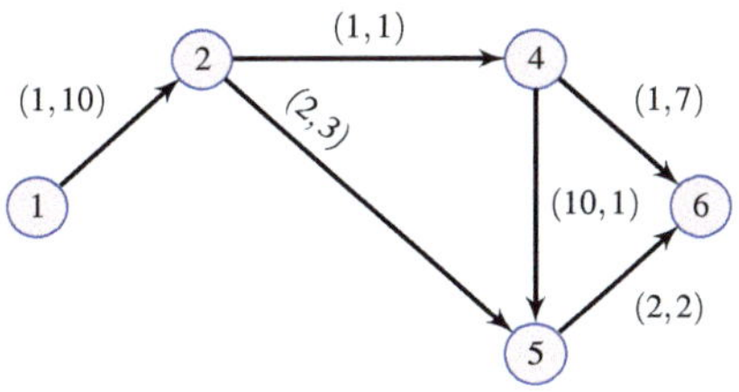

$$
\begin{array}{lllll}
\min 100y_0 & + & 3\lambda_{1246} & + & 5\lambda_{1256} \\
\text{s.t.} & & 18\lambda_{1246} & + & 15\lambda_{1256} \leq 14 \quad [\pi_7] \\
y_0 & + & \lambda_{1246} & + & \lambda_{1256} = \ \ 1 \quad [\pi_0] \\
y_0, & & \lambda_{1246}, & & \lambda_{1256} \geq \ \ 0\,.
\end{array}
$$

The optimal solution with the artificial variable $y_0 = 1/15$ and $\lambda_{1256} = 14/15$ is infeasible for the *MP*, with $z^\star_{MP} = 11.33$. The dual values are $\pi_0 = 100$ and $\pi_7 = -6.33$, and given these, no columns with negative reduced cost can be generated. *Are we ready to prune node **BB1** based on infeasibility?*

Big-M is not big enough. Here we face a drawback of the big-M approach to initialize the column generation algorithm. Indeed,

$$z^\star_{MP} = M(1/15) + 5\,(14/15), \quad \pi_0 = M, \text{ and } \pi_7 = 5 - z^\star_{MP}.$$

Path 12456 is feasible, its cost is 14 as well as its duration, and its reduced cost becomes negative when M is large enough:

$$\bar{c}_{12456} = 14 - 14\pi_7 - \pi_0 < 0 \;\Leftrightarrow\; M > 140.$$

Even though $z^\star_{LP} = 7$ and $z^\star_{ILP} = 13$, the initially chosen constant $M = 100$ is unfortunately too small so we increase it, say to $M = 1000$:

$$\begin{array}{llllll} \min & 1000y_0 & + & 3\lambda_{1246} & + & 5\lambda_{1256} \\ \text{s.t.} & & & 18\lambda_{1246} & + & 15\lambda_{1256} \le 14 \quad [\pi_7] \\ & y_0 & + & \lambda_{1246} & + & \lambda_{1256} = 1 \quad [\pi_0] \\ & y_0, & & \lambda_{1246}, & & \lambda_{1256} \ge 0\,. \end{array}$$

The *RMP* returns $y_0 = 1/15$, $\lambda_{1256} = 14/15$, and $z^\star_{MP} = 71.3$, together with $\pi_0 = 1000$ and $\pi_7 = -66.33$. The *ISP* returns path 12456 with reduced cost -57.33, cost 14, and duration 14.

In BB1.3, the *RMP* finds $\lambda_{12456} = 1$, an integer solution $x_{12} = x_{24} = x_{45} = x_{56} = 1$ with $z^\star_{MP} = 14$ (an upper bound *UB* on $z^\star_{ILP}$). The dual values are $\pi_0 = 1000$ and $\pi_7 = -70.43$, and no new variable is generated. Node **BB1** can be pruned by integrality.

BB2: BB0 and $x_{13} + x_{32} \ge 1$

Housekeeping: The new constraint is handled in the *MP*. It does not forbid any existing columns but does add row coefficients. The domain of the *ISP* is not modified but the objective function must be adapted with the new dual variable.

In more details, this constraint is added to the *ILP* (4.33) together with the associated dual variable π_8. There is no change in the original network so we exploit again the path structure of $\mathcal{D}$ which means that $x_{13} + x_{32} \ge 1$ is a robust cut. We then go through the reformulation process in terms of its extreme points, and obtain a new *MP* to work with. The resulting new constraint in the successive *RMP*s is

$$\sum_{p\in P} (x_{13p} + x_{32p})\lambda_p \ge 1 \qquad [\pi_8]. \tag{7.100}$$

The value $(x_{13p} + x_{32p})$ reflects how often the arcs $(1,3)$ and $(3,2)$ are used to compute the coefficient of λ_p in (7.100): zero for λ_{1246} and λ_{1256}, one for λ_{1356}, and two for λ_{13256}.

The *RMP* at the start of iteration BB2.1 is

$$\begin{array}{llllllll}
\min 1000y_0 & +1000y_8 & +\ 3\lambda_{1246} & +24\lambda_{1356} & +15\lambda_{13256} & +\ 5\lambda_{1256} & & \\
\text{s.t.} & & 18\lambda_{1246} & +\ 8\lambda_{1356} & +10\lambda_{13256} & +15\lambda_{1256} & \leq 14 & [\pi_7] \\
 & y_8 & & +\ \lambda_{1356} & +\ 2\lambda_{13256} & & \geq 1 & [\pi_8] \\
y_0 & & +\ \lambda_{1246} & +\ \lambda_{1356} & +\ \lambda_{13256} & +\ \lambda_{1256} & = 1 & [\pi_0] \\
y_0, & y_8, & \lambda_{1246}, & \lambda_{1356}, & \lambda_{13256}, & \lambda_{1256} & \geq 0, &
\end{array}$$

where the artificial variable y_8 ensures feasibility of the new constraint irrespective of $\mathcal{X}'$. Solving it, we obtain $z^{\star}_{MP} = 9$, $\lambda_{1246} = \lambda_{13256} = 0.5$, $\pi_0 = 15$, $\pi_7 = -0.67$, and $\pi_8 = 3.33$. The modified *ISP* is given by

$$\bar{c}(\pi_7, \pi_8, \pi_0) = \min_{\mathbf{x}\in\mathcal{D}} \sum_{(i,j)\in A} c_{ij}x_{ij} - \pi_7\Big(\sum_{(i,j)\in A} t_{ij}x_{ij}\Big) - \pi_8(x_{13} + x_{32}) - \pi_0. \quad (7.101)$$

For the above dual values, no path of negative reduced cost exists. The fractional arc-flow solution derived from $\lambda_{1246} = \lambda_{13256} = 0.5$ is $x_{12} = x_{13} = x_{24} = x_{25} = x_{32} = x_{46} = x_{56} = 0.5$. We arbitrarily choose the first fractional-valued variable $x_{12} = 0.5$ to branch on.

BB3: **BB2** and $x_{12} = 0$

Housekeeping: We discard λ_{1246} and λ_{1256} from the *RMP*. We then drop arc $(1,2)$ from G. In iteration BB3.1, we therefore start by solving the following *RMP*:

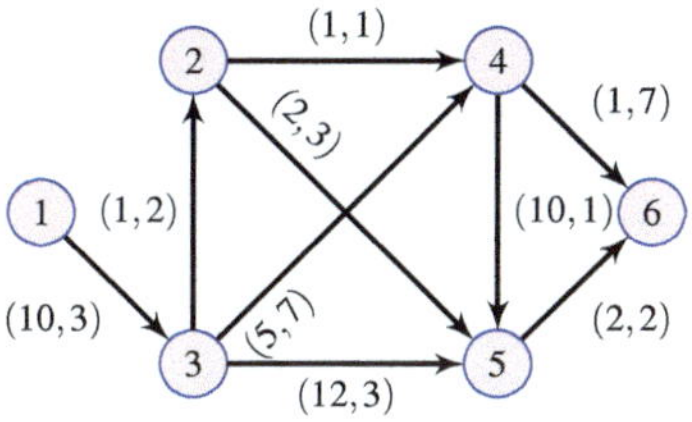

$$\begin{array}{llllll}
\min 1000y_0 & +1000y_8 & +24\lambda_{1356} & +15\lambda_{13256} & & \\
\text{s.t.} & & 8\lambda_{1356} & +10\lambda_{13256} & \leq 14 & [\pi_7] \\
 & y_8\ + & \lambda_{1356} & +\ 2\lambda_{13256} & \geq 1 & [\pi_8] \\
y_0 & & +\ \lambda_{1356} & +\ \lambda_{13256} & = 1 & [\pi_0] \\
y_0, & y_8, & \lambda_{1356}, & \lambda_{13256} & \geq 0. &
\end{array}$$

Two column generation iterations are needed to solve the *MP* at this child node. The dual values we get at BB3.1 are $\pi_0 = 15$ and $\pi_7 = \pi_8 = 0$. The corresponding primal solution $\lambda_{13256} = 1$ of cost 15 is anecdotally integer. This may or may not be relevant to keep track of depending on how difficult it is to find an integer solution at all in the specific application. The path 13246 of reduced cost -2, cost 13, and duration 13 is generated and used in the next iteration BB3.2.

The solution of the *RMP* is integer feasible with variable $\lambda_{13246} = 1$ and a new best integer solution with cost 13 is identified. The dual values are $\pi_0 = 13$ and $\pi_7 = \pi_8 = 0$ for which no negative reduced cost path exists, that is, $\bar{c}(\pi_0, \pi_7, \pi_8) = 0$. Node **BB3** can be pruned by integrality.

BB4: **BB2** and $x_{12} = 1$

Housekeeping: Imposing $x_{12} = 1$ implies setting x_{13} to zero, that is, removing arc $(1,3)$. This implies that node 3 can be discarded as well as its three exiting arcs. We end up with the same network structure as the one for $x_{13} + x_{32} = 0$. Therefore, combining $x_{12} = 1$ with the constraint $x_{13} + x_{32} \geq 1$ introduced in node **BB2** means that there is no solutions for the *MP* so node **BB4** can be pruned by infeasibility. Exercise 7.15 asks the reader to describe how this reasoning plays out from a digital perspective.

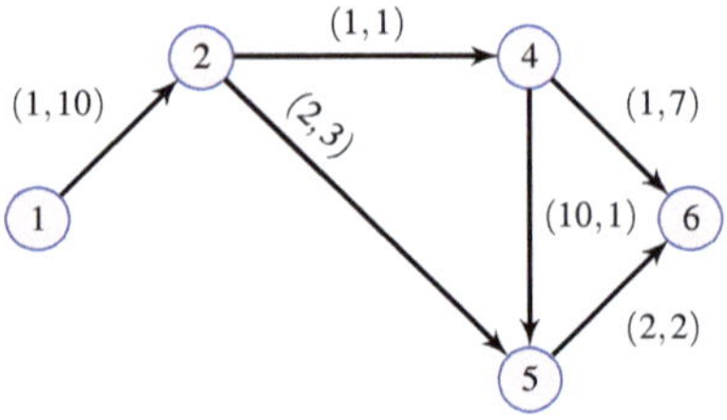

To summarize, a total of five nodes are explored in this branch-and-bound tree (Figure 7.16). It involves decisions incorporated as constraints into the *ISP* as well as in the *ILP*. It produces integer solutions in terms of the original arc-flow variables in two nodes, fractional solutions in two others, and results in infeasibility in one node. An optimal integer solution is found in node **BB3** at a cost $z^{\star}_{ILP} = 13$.

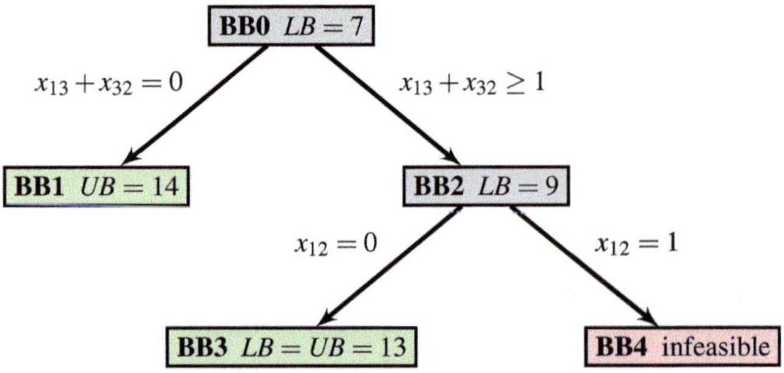

Fig. 7.16: A branch-and-bound tree for the *TCSPP*.

Example 7.2 ***Semi-assignment branching***

✎ Branching on original variables, ranking the candidates, and a rule that works for many similar problems.

We revisit the Capacitated p-median problem that we considered already in Example 6.2 on page 411. Each customer in a set N has to be assigned to exactly one *cluster center* ("median") from a set $K \subseteq N$, such that exactly p clusters result. We repeat for convenience the compact *ILP* (6.89) which contains the remaining notation for costs and capacities:

$$
\begin{aligned}
z^\star_{ILP} = \min \quad & \sum_{k\in K}\sum_{i\in N} c^k_i x^k_i && && \text{(7.102a)}\\
\text{s.t.} \quad & \sum_{k\in K} x^k_i = 1 && \forall i\in N && \text{(7.102b)}\\
& \sum_{k\in K} y^k = p && && \text{(7.102c)}\\
& \sum_{i\in N} d_i x^k_i - D^k y^k \le 0 && \forall k\in K && \text{(7.102d)}\\
& x^k_i \in \{0,1\} && \forall k\in K, i\in N && \text{(7.102e)}\\
& y^k \in \{0,1\} && \forall k\in K. && \text{(7.102f)}
\end{aligned}
$$

Ceselli and Righini (2005) convexify the capacity constraints (7.102d), that is,

$$
\mathcal{D}^k = \left\{ \begin{bmatrix} \mathbf{x}^k \\ y^k \end{bmatrix} \in \{0,1\}^{n+1} \;\middle|\; \sum_{i\in N} d_i x^k_i \le D^k y^k \right\}, \quad \forall k\in K. \tag{7.103}
$$

This results in the *IMP* (6.92) and an ISP^k (6.93) for every $k\in K$ (reproduced here)

$$
\bar{c}^k(\boldsymbol{\pi},\gamma,\pi^k_0) = -\pi^k_0 + \min_{\left[\begin{smallmatrix}\mathbf{x}^k\\ y^k\end{smallmatrix}\right]\in\mathcal{D}^k} \sum_{i\in N}(c^k_i - \pi_i)x^k_i - \gamma y^k, \tag{7.104}
$$

which can be transformed into a knapsack problem. We consider two strategies to branch on original variable values $\mathbf{x}^\star_{RMP} = [x^{k\star}_i]_{k\in K, i\in N}$ and $\mathbf{y}^\star_{RMP} = [y^{k\star}]_{k\in K}$.

Strategy 1. We first put a higher priority on opening medians than on assigning customers, thus prefer branching on y- over branching on x-variables. Branching candidates are all y^k for which $y^{k\star}$ is fractional. They are ranked by non-increasing value of

$$
\tau^k = \min_{\left[\begin{smallmatrix}\mathbf{x}^k\\ y^k\end{smallmatrix}\right]\in\mathcal{D}^k} \sum_{i\in N}(c^k_i - \pi_i)x^k_i = \bar{c}^k(\boldsymbol{\pi},\gamma,\pi^k_0) + \pi^k_0 + \gamma y^k \tag{7.105}
$$

and we pick a $\bar{k}$ for which this value is largest (most *unpromising* median). Branching on whether to open median $\bar{k}$ or not, that is, $y^{\bar{k}} = 1$ or $y^{\bar{k}} = 0$, is performed as follows: in the up-branch, add to the *MP* the constraint $\sum_{p\in P^{\bar{k}}} y^{\bar{k}}_p \lambda^{\bar{k}}_p = 1$; in the down-branch, remove the corresponding *ISP*. Ceselli and Righini (2005) explore the down-branch first and select nodes in a depth-first manner.

Once p medians are fixed to be used, or $|K| - p$ medians are fixed not to be used, branching on binary x^k_i-assignments takes place. Customers are sorted in non-increasing order of $|K_i|$, where $K_i \subseteq K$ is the set of medians to which i is fractionally assigned (most *undecided* customers first). Select a highest ranked customer $\bar{\imath}$ and create p child nodes, one for each used median. In such a node, customer $\bar{\imath}$ is assigned to the respective median k', thus fixing $x^{k'}_{\bar{\imath}} = 1$ in $ISP^{k'}$ and $x^k_{\bar{\imath}} = 0$ in all ISP^k with $k \neq k'$. The pricing problems remain knapsack problems, of smaller size.

Strategy 2. This second strategy refines a proposal by Savelsbergh (1997) for the generalized assignment problem, and is even more interesting (and according to Ceselli and Righini (2005), much more effective). For each customer $i \in N$, consider again the set $K_i \subseteq K$ of all the medians to which i is fractionally assigned. For each i, we create two disjoint subsets K_i^+ and K_i^- by alternately assigning to them the medians in K_i in the order of non-increasing (fractional) value $x_i^{k\star}$ of the assignment variable. The aim is to create a partition of K_i into two subsets which are balanced in fractional assignment to i. Also, the medians in $\bar{K}_i$ to which i is *not* assigned, are partitioned into two subsets $\hat{K}_i^+$ and $\hat{K}_i^-$ of (almost) equal cardinality.

Again, a customer $\bar{\imath}$ is selected for which $|K_{\bar{\imath}}|$ is largest, that is, for which the number of medians to which $\bar{\imath}$ is fractionally assigned is largest. Then branching on the selected semi-assignment constraint $\sum_{k \in K} x_{\bar{\imath}}^k = 1$ in (7.102b) imposes that $\bar{\imath}$ is not assigned to the medians in $K_{\bar{\imath}}^+ \cup \hat{K}_{\bar{\imath}}^+$ in one branch or not to the medians in $K_{\bar{\imath}}^- \cup \hat{K}_{\bar{\imath}}^-$ in the other branch. In the *ILP*, this is done by forcing

$$\text{either} \quad \sum_{k \in K_{\bar{\imath}}^- \cup \hat{K}_{\bar{\imath}}^-} x_{\bar{\imath}}^k = 0 \quad \text{or} \quad \sum_{k \in K_{\bar{\imath}}^+ \cup \hat{K}_{\bar{\imath}}^+} x_{\bar{\imath}}^k = 0, \tag{7.106}$$

or, in terms of the master variables in the *MP*, by

$$\text{either} \quad \sum_{k \in K_{\bar{\imath}}^- \cup \hat{K}_{\bar{\imath}}^-} \sum_{p \in P^k} x_{\bar{\imath}p}^k \lambda_p^k = 0 \quad \text{or} \quad \sum_{k \in K_{\bar{\imath}}^+ \cup \hat{K}_{\bar{\imath}}^+} \sum_{p \in P^k} x_{\bar{\imath}p}^k \lambda_p^k = 0. \tag{7.107}$$

Note that both branches can be implemented by "eliminating" the corresponding variables from the *RMP* and *ISP*s by locally setting them to zero (this reduces the size of the resulting knapsack pricing problems.)

We have semi-assignment constraints in many classical combinatorial optimization models. Sometimes, the subproblems are identical, like in bin packing, vertex coloring, etc. Then, the Ryan and Foster (1981) branching rule (p. 480) is most suited for the resulting set partitioning *IMP*. When the subproblems are different, however, like also for generalized assignment, unrelated parallel machines scheduling, etc., semi-assignment branching is an option worth considering.

Example 7.3 **Ryan-Foster branching for vertex coloring**

✎ Also for master variable branching, we may be able to leave the pricing problem structure intact after branching.

In this example we wish to see the Ryan and Foster (1981) branching rule (p. 480) in action. The rule needs a set partitioning type *IMP* to work. The latter is a very common model and, among others, arises from Dantzig-Wolfe reformulation of *ILP*s that contain *assignment* requirements: assign tasks like nodes, items, customers, flights, etc. to exactly one resource like color, bin, vehicle, crew member, etc.

The vertex coloring problem from Illustration 7.1 fits into this scheme. We formulated an *ILP* in (7.16) that assigns every node $i \in N$ of a graph $G = (N, E)$ exactly one color from a set K. A Dantzig-Wolfe reformulation, e.g., *MP* (7.18), and aggregation over the colors K bring us to a set partitioning program:

$$\begin{aligned} z^\star_{IMP} = \min \quad & \sum_{p\in P} \lambda_p \\ \text{s.t.} \quad & \sum_{p\in P} a_{ip}\lambda_p = 1 \qquad [\pi_i] \quad \forall i \in N \\ & \lambda_p \in \{0,1\} \qquad \forall p \in P. \end{aligned} \tag{7.108}$$

The columns $\mathbf{a}_p = [a_{ip}]_{i\in N}$, $p \in P$, encode (in binary) subsets of nodes that all may receive the same color. That is, for every $p \in P$, whenever $a_{ip} = a_{jp} = 1$, there is no edge (i, j) in set E. In other words, the $\mathbf{a}_p$ are incidence vectors of independent (or stable) sets. An interpretation of (7.108) is thus: we seek a partition of the node set N into a minimum number of independet sets. Some authors directly start here.

For solving the *MP* by column generation, we need to price independent sets where nodes are weighted by the dual values $\boldsymbol{\pi} = [\pi_i]_{i\in N}$:

$$\begin{aligned} \bar{c}(\boldsymbol{\pi}) = \min \quad & x_0 - \sum_{i\in N} \pi_i x_i \\ \text{s.t.} \quad & x_i + x_j \le x_0 \qquad \forall (i,j) \in E \\ & x_0 \in \{0,1\} \\ & x_i \in \{0,1\} \quad \forall i \in N. \end{aligned} \tag{7.109}$$

That is, the *ISP* is a maximum weighted independent set problem. Again, we have the variable x_0 only for correctly reflecting a zero cost of the empty independent set.

Now, assume that we have a fractional solution $\boldsymbol{\lambda}^\star_{MP}$ to the *MP*. The Ryan-Foster rule says that there exist two rows r and s such that we can branch as

$$\sum_{p\in P:\, a_{rp}=a_{sp}=1} \lambda_p = 1 \quad (\textit{same}) \qquad \text{or} \qquad \sum_{p\in P:\, a_{rp}=a_{sp}=1} \lambda_p = 0 \quad (\textit{differ}). \tag{7.110}$$

Barring the removal of columns that violate the branching decision, we can completely implement these two branches in the *ISP*: in the *same* branch, both nodes, r and s (or none of them) must appear in the independent set ($x_r = x_s$); in the *differ* branch, at most one of the two nodes can be part of the independent set ($x_r + x_s \le 1$). Even better, also after these modifications, the *ISP* remains a weighted independent set problem on a slightly altered graph, see Figure 7.17: For the *same* branch, nodes r and s are contracted into one. That is, a new node rs of weight $\pi_r + \pi_s$ is created, and all edges previously incident to r or s become incident to rs. The *differ* branch is even easier, we simply insert a new edge (r,s). A combinatorial algorithm that directly works on the weighted graph remains applicable.

One may ask how we select the pair when there are many (r,s) for which the sum

$$\sum_{p\in P:\, a_{rp}=a_{sp}=1} \lambda_p = \beta \ \text{ is fractional.} \tag{7.111}$$

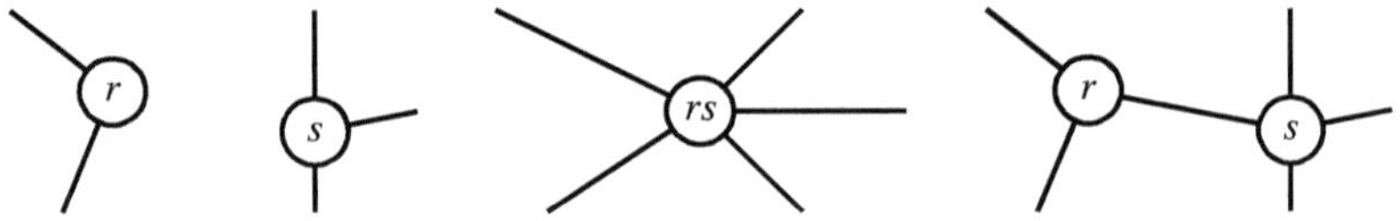

(a) Detail of graph G (b) *Same* branch (c) *Differ* branch

Fig. 7.17: Respecting *same* and *differ* branches directly in the graph.

Mehrotra and Trick (1996) start from a variable λ_p closest to 0.5 and take r as the first row with $a_{rp} = 1$. Then find another variable λ_q with $a_{rq} = 1$ and select s as one row with $a_{sp} \neq a_{sq}$. This rule appears to be a popular one, in particular since it is not problem specific. Other authors tailored the selection to the problem at hand, always with the aim that both branches have a noticeable impact on the solution structure and thus the dual bound. Finally note that for the very similar edge coloring problem in Example 2.3, we do not know how to enforce directly in the graph that two edges are taken together in a matching, neither that at most one edge is taken.

Example 7.4 Ryan-Foster branching for bin packing

✎ Branching can make your pricing problem harder.

This example is related to Example 7.3, which should be read first. We consider the bin packing problem that we introduced in *ILP* (7.53) and for which we also have a set partitioning *IMP* (7.54). The columns $\mathbf{a}_p = [a_{ip}]_{i \in \{1,\dots,n\}}$, $p \in P$, are incidence vectors of subsets of items that together fit into one bin. The *ISP* therefore is a binary knapsack problem. We again associate the dual values $\boldsymbol{\pi} = [\pi_i]_{i \in \{1,\dots,n\}}$ with the set partitioning constraints, and thus, with the items. Similarly to (2.32), we price the packing patterns as:

$$\begin{aligned} \bar{c}(\boldsymbol{\pi}) = \min \quad & x_0 - \sum_{i=1}^{n} \pi_i x_i \\ \text{s.t.} \quad & \sum_{i=1}^{n} w_i x_i \leq W x_0 \\ & x_i \in \{0,1\} \quad \forall i \in \{0,\dots,n\}. \end{aligned} \tag{7.112}$$

According to the Ryan-Foster rule, for a fractional solution to the *MP*, we find two items r and s so that we can branch exactly as in (7.87). What does this mean in the binary knapsack problem? The *same* branch glues items r and s together into a "super item" rs of profit $\pi_r + \pi_s$ that we can either pack or not. This makes the knapsack problem even a tiny bit easier because there is one item less. On the *differ* branch, however, we introduce a *conflict* between items r and s that cannot be packed together. While one branch remains the weakly $\mathcal{NP}$-hard binary knapsack problem, the other branch becomes the strongly $\mathcal{NP}$-hard binary knapsack problem with conflicts. The latter can be solved as an integer program, of course.

Example 7.5 Edge coloring and odd-circuit cuts

✎ Extra variables and constraints in the *ISP* help us capture the semantics of master cut coefficients.

Let us formulate the *ECP* defined on an undirected graph $G = (N, E)$ as an integer program of covering (actually partitioning) edges by matchings (see Example 2.3):

$$\begin{aligned} z^{\star}_{IMP} = \min \quad & \sum_{\mathbf{x}\in\mathcal{X}} \lambda_{\mathbf{x}} \\ \text{s.t.} \quad & \sum_{\mathbf{x}\in\mathcal{X}} a_{e\mathbf{x}}\lambda_{\mathbf{x}} \geq 1 \qquad [\pi_e] \quad \forall e \in E \\ & \lambda_{\mathbf{x}} \in \{0,1\} \qquad\qquad \forall \mathbf{x} \in \mathcal{X}, \end{aligned} \tag{7.113}$$

where $\mathcal{X}$ denotes the set of incidence vectors $\mathbf{x} = [x_e]_{e\in E}$ of matchings. Like in (2.36), we define $\mathcal{X} = \left\{\mathbf{x} \in \{0,1\}^{|E|} \mid \sum_{e\in\delta(\{i\})} x_e \leq 1,\ \forall i \in N\right\}$, where the constraints ensure that, for each node i, at most one incident edge is selected. Let $a_{e\mathbf{x}} = x_e$, $\forall e \in E$; then the *ISP* reads as

$$\min_{\mathbf{x}\in\mathcal{X}} \, (1 - \sum_{e\in E} \pi_e x_e) \;\equiv\; \max_{\mathbf{x}\in\mathcal{X}} \sum_{e\in E} \pi_e x_e. \tag{7.114}$$

Nemhauser and Park (1991) use a combinatorial observation on *odd circuits*, which are *cycles* with an odd number of edges: we need at least three colors (see Figure 7.18).

Fig. 7.18: Taken alone, any odd circuit requires exactly 3 colors (or matchings).

In other words, we need at least three matchings to partition the edges of an odd cycle in G, and this can be formulated as an inequality in the λ-variables. Let $C \subseteq E$ denote the set of edges of an odd circuit. The corresponding *odd-circuit cut* reads as

$$\sum_{\mathbf{x}\in\mathcal{X}:\sum_{e\in C} x_e \geq 1} \lambda_{\mathbf{x}} \geq 3. \tag{7.115}$$

The summation is over all matchings which have non-empty intersection with C. This is the *semantics* of the binary cut coefficient $g_{C\mathbf{x}}$ which equals one if and only if $\sum_{e\in C} x_e \geq 1$. The cut (7.115) can thus be stated as

$$\sum_{\mathbf{x}\in\mathcal{X}} g_{C\mathbf{x}}\lambda_{\mathbf{x}} \geq 3 \quad [\beta_C], \tag{7.116}$$

with the corresponding dual variable β_C. Of course, one does not statically add all odd-circuit cuts $\mathcal{C}$, but dynamically separates violated ones as needed. Let us denote by $\mathcal{C}'$ the set of odd circuits for which we have added an inequality (7.116) to the *MP*. In the *ISP*, we have to ensure that the cut coefficients $g_{C\mathbf{x}}$, $\forall C \in \mathcal{C}'$, are appropriately computed. This can be done using linear inequalities in the original x_e-variables and the extra ones in $\mathbf{g} = [g_C]_{C\in\mathcal{C}'}$:

$$\max_{\mathbf{x}\in\mathcal{X},\, \mathbf{g}\in\{0,1\}^{|\mathcal{C}'|}} \left\{ \sum_{e\in E} \pi_e x_e + \sum_{C\in\mathcal{C}'} \beta_C g_C \;\middle|\; g_C \leq \sum_{e\in C} x_e, \;\; \forall C \in \mathcal{C}' \right\}. \tag{7.117}$$

Since a β_C dual variable is non-negative, the g_C indicator variable is encouraged to be positive. However, the inequality allows g_C to be non-zero only if matching $\mathbf{x}$ has a non-empty intersection with C, as required. The *ISP* (7.117) can be solved to optimality by a branch-and-cut algorithm. There remains to devise a separation routine for recognizing violated odd circuit constraints from a fractional solution $\boldsymbol{\lambda}$, i.e.,

$$\exists C \in \mathcal{C} \setminus \mathcal{C}' \;\text{ such that } \sum_{\mathbf{x}\in\mathcal{X}' : \sum_{e\in C} x_e \geq 1} \lambda_{\mathbf{x}} < 3. \tag{7.118}$$

We point to Nemhauser and Park (1991, §5) for one such method on 3-regular graphs. In any case, these odd-circuit cuts are not sufficient to guarantee we reach integrality so be prepared to devise alternative cutting and branching decisions.

*Example 7.6 **k–path cuts for the** VRPTW*

✎ These valid inequalities are amongst the first producing better lower bounds for the vehicle routing problem with time windows.

Various families of valid inequalities can be considered for the *VRPTW* (see Illustration 7.2 for the problem statement, a two-index arc-flow formulation and a Dantzig-Wolfe reformulation, and such inequalities). In the literature, most of them, such as the k-path cuts proposed by Kohl et al. (1999), are defined as linear combinations of the x_{ij}-variables and easily rewritten in terms of the λ-variables. In this case, their treatment does not pose any difficulties with regard to the complexity of the pricing problem as their dual values affect only the arc costs.

We first formulate the general k-path inequalities and later focus on the specific 2-path type for the *VRPTW*. Let S be a non-empty subset of customers. A k-path inequality writes as

$$\sum_{i\in N\setminus S} \sum_{j\in S} x_{ij} \geq k(S), \qquad S \subseteq C, |S| \geq 1, \tag{7.119}$$

where $k(S)$ is a *lower bound* on the number of vehicles required to service the customers in S. Some such lower bounds are given by

- lb_1 = optimal objective value of a bin packing problem with items of length q_i, $\forall i \in S$, and bins of length Q ($\mathcal{NP}$-hard).
- $lb_2 = \lceil \sum_{i\in S} q_i / Q \rceil$ (linear time but weaker).

- lb_3 = optimal objective value of a vehicle scheduling problem that minimizes the number of vehicles required to service the customers in S considering only the time windows ($\mathcal{NP}$-hard, see Desrosiers et al., 1988b).
- $lb_4 = 2$ if the traveling salesperson problem with time windows (*TSPTW*) for S is infeasible and 1 otherwise (still $\mathcal{NP}$-hard but easier to compute).

In practice, $k(S)$ is set to $\max\{lb_2, lb_4\}$. In general, if time windows are more constraining than vehicle capacity and S is not very large, then $lb_2 \leq lb_4$ and $k(S) = lb_4$; in that case, the 2-path cuts are more relevant.

Let $(\mathbf{x}^\star_{MP}, \mathbf{t}^\star_{MP})$ be fractional in the x-variables. There are no efficient algorithms for finding violated k-path inequalities, yet. We rather proceed with the enumeration of several relatively small subsets of customers. For such a subset S, here is a possible strategy using lb_2 and lb_4:

1. If $\sum_{i\in N\setminus S}\sum_{j\in S} x^\star_{ij} < lb_2$,
   then a cut is found with $k(S) = lb_2$.
2. If $1 < \sum_{i\in N\setminus S}\sum_{j\in S} x^\star_{ij} < 2$ and the *TSPTW* is infeasible,
   then a cut is found with $k(S) = lb_4 = 2$.

Although the *TSPTW* is $\mathcal{NP}$-hard, $|S|$ is selected rather small in practice such that this optimization problem is easily solved by dynamic programming. For example, Dumas et al. (1995) point out that the *TSPTW* can be seen as an elementary shortest path problem with time windows further constrained by the fact that *all* nodes must be visited. This provides an efficient elimination rule: given a subset of visited customers, if the earliest unvisited one is unreachable, the corresponding state is discarded. Therefore, we have a relatively fast, though not polynomial, algorithm to determine whether $k(S) \geq 2$ and possibly identify a set of customers S which requires at least two vehicles to be serviced, but is currently serviced by strictly less than two. The number of subsets S that can be enumerated is, however, exponential and some rules (e.g., on the cardinality of S) are necessary to restrict the enumeration process (see Desaulniers et al., 2008, for further details).

A k-path inequality for subset S is added to the *MP* (7.23) as

$$\sum_{r\in R}\sum_{(i,j)\in A_{do}} \alpha_{ij} x_{ij,r} \lambda_r \geq k(S), \qquad [\pi_S \geq 0] \tag{7.120}$$

where $\alpha_{ij} = 1$ if $i \in N \setminus S, j \in S$, and 0 otherwise. In the *ISP*, the adjusted cost $\tilde{c}_{ij}$ of arc $(i,j) \in A_{do}$ becomes

$$\tilde{c}_{ij} = \begin{cases} c_{ij} - \pi_K & \text{for } (i,j) = (d,o) \\ c_{ij} - \alpha_{ij}\pi_S & \forall (i,j) \in A \text{ such that } i = o \\ c_{ij} - \pi_i - \alpha_{ij}\pi_S & \forall (i,j) \in A \text{ such that } i \neq o \end{cases} . \tag{7.121}$$

Figure 7.19 taken from Kohl et al. (1999) illustrates the efficiency of the 2-path cuts for Solomon's instance RC101.25 in which the 25 customers are located in three

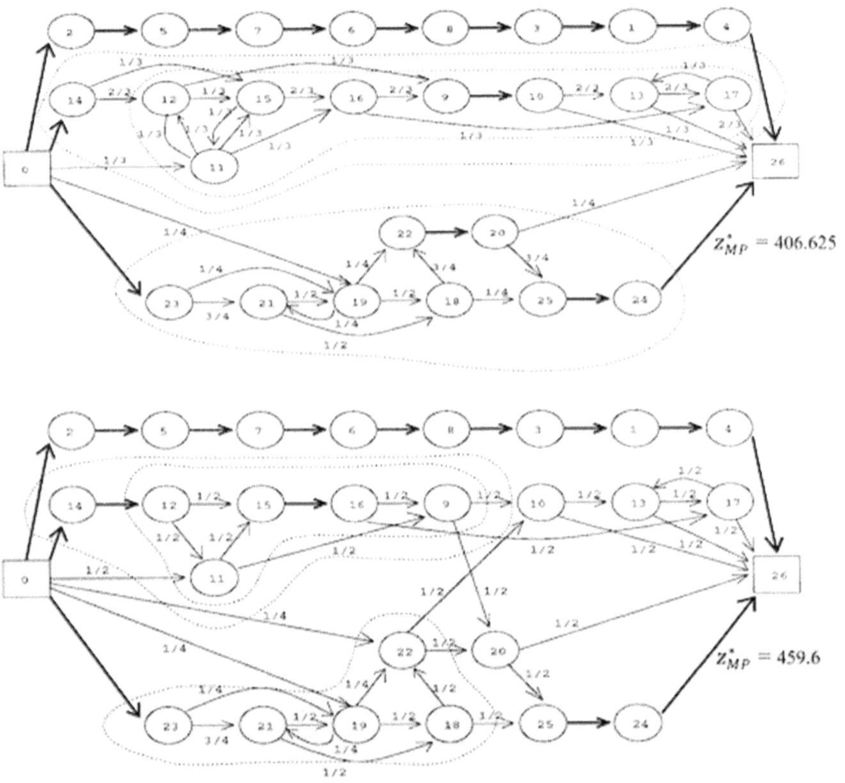

Fig. 7.19: Illustration of the 2-path cuts.

clusters. The fractional solution in the top part of the figure uses $43/12 \approx 3.583$ vehicles and $z^{\star}_{MP} = 406.625$. The three clusters are serviced by respectively 1, $4/3$, and $5/4$ vehicles. Three subsets S (surrounded by dotted lines) are identified for which $\sum_{i\in N\setminus S}\sum_{j\in S} x_{ij} < 2$ but requires at least 2 vehicles. After the insertion of three cuts, the lower bound increases to 459.03125 while the number of vehicles goes up to $31/8 = 3.875$ (this solution is not displayed). Three new cuts are inserted and the solution appears in the bottom part: it requires 4 vehicles and costs 459.6. After insertion of three more cuts, an optimal integer solution with an objective value of 461.1 is obtained without any branching.

Example 7.7 VRPTW ***and non-robust rounded capacity cuts***

✎ A non-robust version of these cuts is stronger but requires modifications to the pricing algorithm.

In Illustration 7.2, we have presented a robust version of the rounded capacity cuts that writes as follows:

$$\sum_{r \in R} n_{Sr} \lambda_r \geq LB(S) \qquad [\pi_S] \qquad \forall S \subseteq C \text{ such that } |S| \geq 2, \tag{7.122}$$

where $LB(S)$ is a lower bound on the number of vehicles needed to serve all customers in subset S and the coefficient n_{Sr} is equal to the number of times that route $r \in R$ enters subset S, i.e., traverses an arc $(i, j) \in \delta^-(S)$. This version of the cuts can easily be handled in the column generation algorithm by simply transferring the dual values π_S of these cuts to the adjusted cost $\tilde{c}_{ij}$ of the arc-flow variables in the *ISP*:

$$\tilde{c}_{ij} = \begin{cases} c_{ij} - \pi_K & \text{for } (i,j) = (d,o) \\ c_{ij} - \sum_{S \in \mathcal{S}_{ij}} \pi_S & \forall (i,j) \in A \text{ such that } i = o \\ c_{ij} - \pi_i - \sum_{S \in \mathcal{S}_{ij}} \pi_S & \forall (i,j) \in A \text{ such that } i \neq o, \end{cases} \tag{7.123}$$

where $\mathcal{S}_{ij}$ is the set of subsets S such that $(i, j) \in \delta^-(S)$.

Because the coefficient n_{Sr} can be greater than 1, the left-hand side in (7.122) may overestimate the number of vehicles used to serve the customers in S in the current solution, yielding *weak* rounded capacity cuts. A stronger version of these cuts, introduced by Baldacci et al. (2008) for the capacitated *VRP*, expresses as follows:

$$\sum_{r \in R} g_{Sr} \lambda_r \geq LB(S) \qquad [\pi_S] \qquad \forall S \subseteq C \text{ such that } |S| \geq 2, \tag{7.124}$$

where g_{Sr} is a binary parameter taking value 1 if route r visits at least one customer in S and 0 otherwise. In this case, the left-hand side really counts the number of vehicles used to serve the customers in S. These strong rounded capacity cuts are defined on the master variables because there are no equivalent cuts defined on the x-variables of the compact formulation (7.21). In this case, the *ISP* cannot be obtained by transferring the dual values of (7.124) in the adjusted arc costs as in (7.123) because the dual π_S cannot be subtracted more than once along a route even if this route enters S multiple times.

Let $\mathcal{S}$ be the set of customer subsets S for which a strong rounded capacity inequality (7.124) has been generated at a given column generation iteration and $\pi_S > 0$. Using the initial grouping of the constraints (7.22) where these cuts are integrated into set $\mathcal{A}$, the modified *ISP* at that iteration is given by

$$\bar{c}(\boldsymbol{\pi}) = -\pi_K + \min_{\mathbf{x} \in \mathcal{X}} \quad c_{\mathbf{x}} - \sum_{i \in C} \pi_i a_{i\mathbf{x}} - \sum_{S \in \mathcal{S}} \pi_S g_{S\mathbf{x}} \tag{7.125a}$$

$$\text{s.t.} \quad c_{\mathbf{x}} = \sum_{(i,j) \in A_{do}} c_{ij} x_{ij} \tag{7.125b}$$

$$a_{i\mathbf{x}} = \sum_{j:(i,j) \in A} x_{ij} \qquad \forall i \in C \tag{7.125c}$$

$$g_{S\mathbf{x}} = \left\lceil \sum_{(i,j) \in \delta^-(S)} x_{ij} / \left|\delta^-(S)\right| \right\rceil \qquad \forall S \in \mathcal{S}. \tag{7.125d}$$

Non-linear function $\lceil \sum_{(i,j)\in\delta^-(S)} x_{ij} / |\delta^-(S)| \rceil$ takes value 1 if at least one arc $(i,j) \in \delta^-(S)$ is used in the *ISP* solution and 0 otherwise. Given that $\pi_S > 0$ and the objective function of (7.125) favors larger values of $g_{S\mathbf{x}}$, it can be linearized as follows

$$\begin{aligned} g_{S\mathbf{x}} \leq \sum_{(i,j)\in\delta^-(S)} x_{ij} \quad \forall S \in \mathcal{S} \\ 0 \leq g_{S\mathbf{x}} \leq 1 \quad \forall S \in \mathcal{S}. \end{aligned} \tag{7.126}$$

Setting $x_{do} = 1$ in $\mathcal{D}$, the *ISP* remains an *ESPPRC* that can be solved by a labeling algorithm similar to that described in Labeling algorithm for the *ESPPTWC*. A label

$$E_p = (T_p^{rCost}, T_p^{time}, T_p^{load}, [T_p^{cust_i}]_{i\in C}, [T_p^{rccs}]_{S\in\mathcal{S}^+}) \tag{7.127}$$

representing a partial path $p = (i_0 = o, i_1, \ldots, i_m = j)$ contains a binary resource component T_p^{rccs} for each generated rounded capacity cut (7.124) associated with a subset $S \in \mathcal{S}^+$, where $\mathcal{S}^+ = \{S \in \mathcal{S} \mid \pi_S > 0\}$. When extending E_p along an arc $(j,h) \in A$ to yield a new path p', the values $T_{p'}^{rccs}$ of these components in label $E_{p'}$ are computed using the REF

$$T_{p'}^{rccs} = f_{jh}^{rccs}(T_p^{rccs}) = \begin{cases} 1 & \text{if } (j,h) \in \delta^-(S) \\ T_p^{rccs} & \text{otherwise} \end{cases} \quad \forall S \in \mathcal{S}^+. \tag{7.128}$$

Moreover, the REF (5.18a) for the *rCost* component is replaced by

$$T_{p'}^{rCost} = f_{jh}^{rCost}(T_p^{rCost}, [T_p^{rccs}]_{S\in\mathcal{S}^+}) = T_p^{rCost} + \tilde{c}_{jh} - \sum_{S\in\mathcal{S}^+} \pi_S (T_{p'}^{rccs} - T_p^{rccs}), \tag{7.129}$$

that is, a dual value π_S is subtracted from the path reduced cost only if (j,h) is the first arc of the path entering S (or, equivalently, if $T_{p'}^{rccs} = 1$ and $T_p^{rccs} = 0$). The other REFs do not change. On the other hand, given that the REFs (7.128)–(7.129) are non-decreasing, the dominance rule (5.19) augmented with the conditions

$$T_p^{rccs} \leq T_{p'}^{rccs}, \quad \forall S \in \mathcal{S}^+ \tag{7.130}$$

can be applied to eliminate labels. Clearly, handling the duals of these non-robust cuts in the *ISP* reduces the number of labels that can be dominated and, thus, increases the difficulty of solving the pricing subproblem.

To improve this dominance rule, we can drop the conditions (7.128) and replace the condition (5.19a) on the reduced cost by

$$T_p^{rCost} \leq T_{p'}^{rCost} - \sum_{S\in\mathcal{S}_{10}^+} \pi_S, \tag{7.131}$$

where $\mathcal{S}_{10}^+ = \{S \in \mathcal{S}^+ \mid T_p^{rccs} = 1, T_{p'}^{rccs} = 0\}$. To summarize, a label E_p dominates a label $E_{p'}$ ending at the same node if the following conditions are met:

$$T_p^{rCost} \leq T_{p'}^{rCost} - \sum_{S \in \mathcal{S}_{10}^+} \pi_S \tag{7.132a}$$

$$T_p^{time} \leq T_{p'}^{time} \tag{7.132b}$$

$$T_p^{load} \leq T_{p'}^{load} \tag{7.132c}$$

$$T_p^{cust_i} \leq T_{p'}^{cust_i}, \qquad \forall i \in C. \tag{7.132d}$$

Note 7.14 (Weak or strong?) Several efficient separation heuristics have been developed to separate the weak version of the rounded capacity inequalities (see, e.g., Lysgaard et al., 2004). On the other hand, no specific separation algorithms have been designed for the strong version. As any violated weak inequality implies a violated strong inequality, the separation approach consists in searching for violated weak inequalities only and convert them into strong ones whenever useful. More precisely, weak cuts are first added to the *MP* to minimize the impact on the *ISP* solution process. Then, after re-optimizing the *MP*, a weak rounded capacity cut that was previously added can be lifted to its strong version if this version remains violated by the current *MP* solution.

Example 7.8 VRPTW ***and Chvátal-Gomory rank-1 cuts***

✎ We present an application that turns out to be rather easy to implement.

Consider the set partitioning formulation (7.133) of the *VRPTW* (see Illustration 7.2 for the parameter definitions) for which the number of available vehicles is sufficiently large, hence not binding, and the corresponding constraint is discarded:

$$\begin{aligned} z_{IMP}^{\star} = \min \quad & \sum_{\mathbf{x} \in \mathcal{X}} c_{\mathbf{x}} \lambda_{\mathbf{x}} \\ \text{s.t.} \quad & \sum_{\mathbf{x} \in \mathcal{X}} a_{i\mathbf{x}} \lambda_{\mathbf{x}} = 1 \qquad [\pi_i] \quad \forall i \in C \\ & \lambda_{\mathbf{x}} \in \{0,1\} \qquad \forall \mathbf{x} \in \mathcal{X}. \end{aligned} \tag{7.133}$$

As already defined in (7.22b), a route $\mathbf{x} \in \mathcal{X}$ is an *od*-path scaled with $x_{do} = 1$, of cost $c_{\mathbf{x}} = \sum_{(i,j) \in A_{do}} c_{ij} x_{ij}$, and binary coefficients $a_{i\mathbf{x}} = \sum_{j:(i,j) \in A} x_{ij}$, $\forall i \in C$. Assume known a Chvátal-Gomory rank-1 cut, i.e., we know a vector of non-negative rational weights $[w_i]_{i \in C}$ on the columns $[a_{i\mathbf{x}}]_{i \in C}$, $\mathbf{x} \in \mathcal{X}$, and the linear relaxation of (7.133) is augmented with constraint

$$\sum_{\mathbf{x} \in \mathcal{X}} g_{\mathbf{x}} \lambda_{\mathbf{x}} \leq \left\lfloor \sum_{i \in C} w_i \right\rfloor, \qquad [\gamma \leq 0], \tag{7.134}$$

where $g_{\mathbf{x}}$ is computed in the modified *ISP*:

$$\bar{c}(\boldsymbol{\pi}, \gamma) = \min_{\mathbf{x} \in \mathcal{X}} \qquad c_{\mathbf{x}} - \sum_{i \in C} \pi_i a_{i\mathbf{x}} - \gamma g_{\mathbf{x}} \tag{7.135a}$$

$$\text{s.t.} \qquad c_{\mathbf{x}} = \sum_{(i,j)\in A_{do}} c_{ij}x_{ij} \tag{7.135b}$$

$$a_{i\mathbf{x}} = \sum_{j:(i,j)\in A} x_{ij} \qquad \forall i \in C \tag{7.135c}$$

$$g_{\mathbf{x}} = \left\lfloor \sum_{i\in C} w_i a_{i\mathbf{x}} \right\rfloor. \tag{7.135d}$$

Function $\lfloor \sum_{i\in C} w_i a_{i\mathbf{x}} \rfloor$ is non-linear but $\sum_{i\in C} \pi_i a_{i\mathbf{x}}$ and $\sum_{i\in C} w_i a_{i\mathbf{x}}$ are linear expressions computed in the same way. Moreover, it is only on the last arc reaching node d that we use the floor function. The *ISP* (7.135) can still be regarded as an *elementary shortest path problem with resource constraints* with an additional resource for every Chvátal-Gomory cut. In this *ISP*, the adjusted cost $\tilde{c}_{ij}$ of arc $(i,j) \in A$ is still computed as $\tilde{c}_{ij} = c_{ij} - \pi_i$ and does not depend on γ. This dual value is treated as discussed below. Let a partial path p ending at node $j \in C$ be represented by the label

$$E_p = (T_p^{rCost}, T_p^{time}, T_p^{load}, [T_p^{cust_i}]_{i\in C}, T_p^{cgc}) \tag{7.136}$$

consisting of the current resource values from o to j, i.e., the reduced cost T_p^{rCost}, the earliest start of service time T_p^{time}, the vehicle load T_p^{load}, the vector of visited customers $[T_p^{cust_i}]_{i\in C}$, and the cut resource T_p^{cgc}. At the source o, all components of E_o are initialized to zero. Extending label E_p on arc $(j,h) \in A, h \neq d$ to create a partial path p' represented by label $E_{p'}$ (if feasible considering the resource limitations) is performed by the following REFs:

$$T_{p'}^{rCost} = f_{jh}^{rCost}(T_p^{rCost}) = T_p^{rCost} + \tilde{c}_{jh} \tag{7.137a}$$

$$T_{p'}^{time} = f_{jh}^{time}(T_p^{time}) = \max\{a_h, T_p^{time} + t_{jh}\} \tag{7.137b}$$

$$T_{p'}^{load} = f_{jh}^{load}(T_p^{load}) = T_p^{load} + q_h \tag{7.137c}$$

$$T_{p'}^{cust_i} = f_{jh}^{cust_i}(T_p^{cust_i}) = \begin{cases} T_p^{cust_i} + 1 & \text{if } h = i \\ T_p^{cust_i} & \text{otherwise} \end{cases} \qquad \forall i \in C \tag{7.137d}$$

$$T_{p'}^{cgc} = f_{jh}^{time}(T_p^{cgc}) = T_p^{cgc} + w_h. \tag{7.137e}$$

On an arc (j,d), the extension functions are slightly different:

$$T_{p'}^{rCost} = f_{jd}^{rCost}(T_p^{rCost}, T_p^{cgc}) = T_p^{rCost} + \tilde{c}_{jh} - \gamma \lfloor T_p^{cgc} \rfloor \tag{7.138a}$$

$$T_{p'}^{time} = f_{jd}^{time}(T_p^{time}) = \max\{a_d, T_p^{time} + t_{jd}\} \tag{7.138b}$$

$$T_{p'}^{load} = f_{jd}^{load}(T_p^{load}) = T_p^{load} \tag{7.138c}$$

$$T_{p'}^{cust_i} = f_{jd}^{cust_i}(T_p^{cust_i}) = T_p^{cust_i} \qquad \forall i \in C \tag{7.138d}$$

$$T_{p'}^{cgc} = f_{jd}^{time}(T_p^{cgc}) = \lfloor T_p^{cgc} \rfloor. \tag{7.138e}$$

From j to d, the reduced cost $T^{rCost}_{p'}$ not only depends on T^{rCost}_{p} but also on T^{cgc}_{p}, the value of the resource computing the $g_{\mathbf{x}}$-coefficient, see (7.138a). The two-dimensional REF $f^{rCost}_{jd}(T^{rCost}_{p}, T^{cgc}_{p})$ is non-decreasing with respect to each variable with a unitary slope for T^{rCost}_{p} and a non-negative slope $-\gamma$ for $\lfloor T^{cgc}_{p} \rfloor$, the floor function being also non-decreasing. All extension functions in (7.137)–(7.138) are non-decreasing with respect to each of their variables and, therefore, the classical dominance rule between labels can be used to discard dominated ones.

Jepsen et al. (2008) introduced the so-called subset-row inequalities, which are Chvátal-Gomory rank-1 cuts, but implemented only a special case of them that is expressed with exactly three non-zero weights equal to $1/2$. Their computational results show that this strategy significantly improves the lower bound and, in many cases, makes it possible to prove optimality in the root node. They also used an improved dominance rule by exploiting the step-like structure of the objective function of the *ISP*. Subset-row inequalities involving more than three non-zero weights have been studied in Petersen et al. (2008) and Pecin et al. (2017c,b). To limit the impact of handling these cut dual values in the *ISP*, Pecin et al. (2017a,b) have proposed weaker versions that are called *limited-memory subset-row inequalities*.

Subset-row inequalities have been applied in branch-price-and-cut algorithms for other problem types that involve set-packing or set-partitioning constraints, namely, for computer vision applications (Yarkony et al., 2020), bin packing (Wei et al., 2020), and vector packing (Wang et al., 2022), among others.

Example 7.9 ***Preferential bidding system***

✎ We present a real-life application for the z-cuts. We exploit lexicographic optimization properties to discard significant portions of the solution space that are irrelevant for integrality optimality. These cuts allow us to rapidly improve the dual bound on integer optimality before any branching even occurs.

Opportunities for optimization in the airline industry are legion. Crew scheduling is a prominent example of operational and tactical planning. The importance of this problem can be simply explained by the fact that crew costs are the second largest expense after fuel costs. Because it is so complex, it is typically tackled via two sequential problems: crew *pairing* and crew *assignment*. In the crew pairing problem, we aim to pair all flights to anonymous crews at minimal cost. To accomplish this, each flight is decomposed into a sequence of possibly shorter flights that correspond to each nonstop departure/arrival portion. This breakdown increases flexibility as crew pairing can be done at this finer granularity. In the simplest of terms, a pairing consists of a series of flights intertwined with enough but not too much in-between time to be feasible. In reality, all legal regulations have to be met as well as conditions specified by the collective agreements. (For more definitions, see Crew pairing problem with base constraints.) In any case, we especially point to Figure 5.7 to appreciate how time is embedded into the network structure.

In the crew assignment problem, we get rid of the anonymousness by assigning explicit crew members to every pairing we found in the first step. Pairings are themselves sequenced in time to produce a schedule for a crew member, say for the next month. A set of schedules, called a roster, is feasible if it covers all pairings and regulations/collective agreements are still taken into account. There are three main ways to create a roster: bidline, rostering, and preferential bidding.

Let us focus on a simpler crew variant in which we only take care of the pilots with same qualifications such that any of them can cover any schedule. In a bidline process, schedules are first generated anonymously before each pilot, from the most senior to the most junior, selects his or her preferred schedule amongst the remaining ones. For rostering, we generate personalized schedules that must satisfy a list of predefined activities for each pilot with the objective of balancing as much as possible some schedule characteristics (e.g., number of worked hours or days off) between the pilots. Preferential bidding is similar to rostering but aims at maximizing how much a schedule fulfills the desiderata of each pilot, favoring the pilots in order of seniority. We learn from Gamache et al. (1998) that such a preferential bidding system is indeed in place at Air Canada.

Interestingly, we can greatly exploit the switch to a preference-based objective function if the policy favors seniority above all else. More specifically, each pilot k in a set K assigns *integer* weights to various preference indicators that we use to measure the score s^k of a schedule $\mathbf{x}^k \in \mathcal{D}^k$, say $s^k = f^k(\mathbf{x}^k)$. Respecting the seniority policy, we get a lexicographic optimization problem (Ehrgott, 2005) for which the contribution of every function $f^k(\mathbf{x}^k)$ is more important than those that follow. The rest of the model has a block-diagonal structure over the $|K|$ pilots.

In principle, a lexicographic problem can be solved sequentially by optimizing one objective at a time and then imposing higher-ranked optimal objective values as constraints in the following lower-ranked objectives. Let us not forget that we face column generation and integrality at every step of the way which means that we do not want to get lost in branching decisions before having a chance to find a decent solution.

- The *ILP* has a block-diagonal structure with partitioning constraints on the pairings as the linking constraints and bounded domains $\mathcal{D}^k$, $\forall k \in K$, each one appearing in a dedicated pricing problem. Indeed, the ISP^k for finding the maximum score for pilot k is a *longest* path problem with resource constraints defined on an acyclic network. The *IMP* comprises partitioning constraints for both the pairings and pilots (indeed, expressed by the convexity constraints for the latter).
- The first integer master pilot-problem (denoted IMP^1) is solved while only considering the bids of the most senior pilot in the objective function. In other words, we solve the IMP^1 to integer optimality by using $|K|$ pricing problems such that all pairings are assigned and all pilots have a feasible schedule.

  Let $\mathcal{P}$ denote the set of pairings and define the binary parameter a^k_{ip} taking value 1 if and only if pairing $i \in \mathcal{P}$ belongs to schedule $\mathbf{x}^k_p \in \mathcal{D}^k$, for $k \in K$, $p \in P^k$. Morever, let $s^k_p = f^k(\mathbf{x}^k_p)$ denote the score of schedule $\mathbf{x}^k_p$. Then

$$
\begin{aligned}
z^\star_{IMP^1} = \max \quad & \sum_{p \in P^1} s^1_p \lambda^1_p \\
\text{s.t.} \quad & \sum_{k \in K} \sum_{p \in P^k} a^k_{ip} \lambda^k_p = 1 \qquad \forall i \in \mathcal{P} \\
& \sum_{p \in P^k} \lambda^k_p = 1 \qquad \forall k \in K \\
& \lambda^k_p \in \{0,1\} \quad \forall k \in K, p \in P^k.
\end{aligned} \tag{7.139}
$$

- Then, the second integer master pilot-problem (IMP^2) is solved for maximizing the score of the second most senior pilot but without deteriorating the score of the first, i.e., maximizing $\sum_{p \in P^2} s^2_p \lambda^2_p$ but additionally imposing $\sum_{p \in P^1} c^1_p \lambda^1_p \geq z^\star_{IMP^1}$:

$$
\begin{aligned}
z^\star_{IMP^2} = \max \quad & \sum_{p \in P^2} s^2_p \lambda^2_p \\
\text{s.t.} \quad & \sum_{p \in P^1} s^1_p \lambda^1_p \geq z^\star_{IMP^1} \\
& \sum_{k \in K} \sum_{p \in P^k} a^k_{ip} \lambda^k_p = 1 \qquad \forall i \in \mathcal{P} \\
& \sum_{p \in P^k} \lambda^k_p = 1 \qquad \forall k \in K \\
& \lambda^k_p \in \{0,1\} \quad \forall k \in K, p \in P^k.
\end{aligned} \tag{7.140}
$$

  This is what the authors refer to as a residual pilot-problem. We can even set the schedule $\mathbf{x}^{1\star}$ for pilot 1 in stone (potentially turning the overall algorithm into a heuristic) in which case there remains only $|K| - 1$ pricing problems and the objective value constraint becomes redundant.
- And so on for the third integer master pilot-problem (IMP^3), imposing both

$$
\sum_{p \in P^1} s^1_p \lambda^1_p \geq z^\star_{IMP^1} \quad \text{and} \quad \sum_{p \in P^2} s^2_p \lambda^2_p \geq z^\star_{IMP^2},
$$

  and the remaining (residual) pilot-problems. A simple sum tells us that a total of $\frac{|K|(|K|+1)}{2}$ pricing problems have to be considered in the (possibly heuristic) sequential process.
- Of course, since there can exist alternative same-score schedules, allowing retroactive decisions potentially yields better overall preferences at the cost of having to deal with $|K|^2$ pricing problems. The computation results suggest that cutting this corner is a reasonable decision.

In the linear relaxation of the IMP^k, the maximum score for pilot k is a convex combination of scores which is calculated as

$$z^{\star}_{MP^k} = \max \sum_{p \in P^k} s^k_p \lambda^k_p, \tag{7.141}$$

where $\sum_{p\in P^k} \lambda^k_p = 1$ is required by the convexity constraint. If these scores are not all equal, then some are strictly greater than the maximized objective value $z^{\star}_{MP^k}$. Hence, these large score-schedules cannot appear in any optimal integer solution for pilot k. Figure 7.20 illustrates a fractional λ-solution in the MP^k.

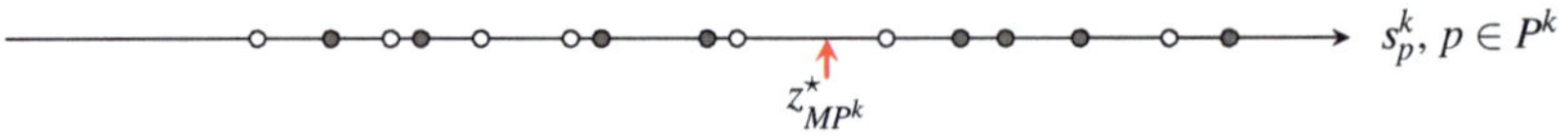

Fig. 7.20: Possible score-schedules (o and •) and selected ones (•) for an optimal convex combination of schedules in the MP^k.

Because an integer solution for the IMP^k assigns only one schedule to pilot k, the z-cut in $\mathbf{x}^k$-variables

$$f^k(\mathbf{x}^k) \leq UB^k = \lfloor z^{\star}_{MP^k} \rfloor \tag{7.142}$$

removes all schedules with a score larger than UB^k. The floor function can be used as we know the optimum must be an integer score given the integral weight inputs. Those already generated are discarded from the current RMP^k. Such a cut is imposed on the subproblem domain $\mathcal{D}^k$ and implemented as an additional resource in the constrained longest path problem. This maximum-score resource is treated as a capacity constraint.

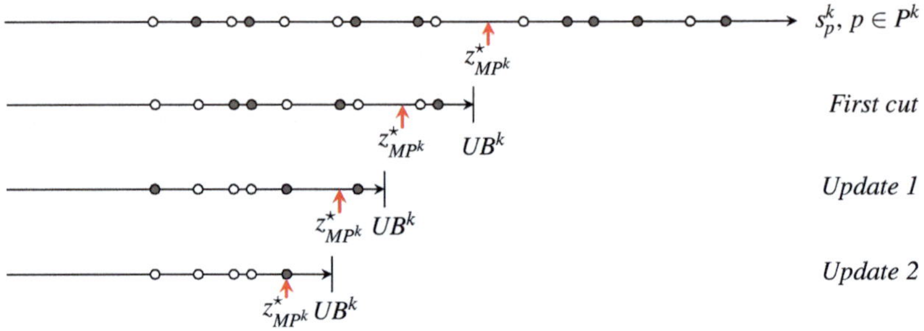

Fig. 7.21: Successive impact on $\mathcal{D}^k$ of the dynamic z-cut $f^k(\mathbf{x}^k) \leq UB^k$.

Although a cut on the objective function imposed in the ILP is usually not all that efficient in practice, only a few are needed in this application to obtain a convex combination of equal maximum-score schedules for pilot k by dynamically updating UB^k. Figure 7.21 illustrates the successive elimination of integer points from the domain of the ISP^k with the use of three cuts (i.e., two upper bound updates). At that point, classical decisions on the binary flow variables are used for the other pilots. There is no need to branch on pilot k because its solution automatically becomes integer when all other pilots have assigned schedules.

Gamache et al. (1998) report solving 24 monthly problems giving rise to 710 pilot-problems. The largest *ILP* involves 568 pairings and 108 pilots; this means that a total of $108(109)/2 = 5886$ pricing problems are solved (72 % of the computation time) during the solution process that lasts 5.7 hours on a HP9000/735 with 128 MB of RAM, more than two decades ago.

Amongst these 24 *ILP*s, 15 need such z-cuts on 51 IMP^k. The average number of cuts on these 51 pilot-problems is only 2.3, with a maximum of 7. The average integrality gap is 15 %, with a maximum of 99 %: branching decisions on the flow variables alone cannot be expected to handle such large gaps. We emphasize that if any of the pilot-problems cannot be solved, then neither can the original *ILP*.

Example 7.10 ***Branch-first, Cut-second strategy***

✎ Branching decisions are taken such that, on one branch, the facets of the polyhedron describing a subset of constraints are easily derived.

The problem considered in this example is to assign locomotives of different types to trains to operate a pre-planned train schedule. Each train may be operated by several locomotives of different types (*heterogeneous consists*). The following text is a simplified version inspired by Ziarati et al. (1997, 1999b).

The proposed *ILP* is an integer multi-commodity flow problem with operational constraints, where each locomotive type defines a commodity. In practice, we observe integrality gaps well above 5 % on this formulation. An important reason is that the train covering constraints are often expressed in terms of horsepower or tonnage requirements rather than a number of locomotives. Therefore the solution of the *LP* is strongly fractional.

A time-space network similar to an airline time-line network is constructed for each commodity. Arcs represent activities such as trains, waiting, and so on (Ziarati et al., 1997, Fig. 1). For each commodity $k \in K$, let $G^k = (V^k, A^k_{do})$ be the associated network, where $V^k = N^k \cup \{o^k, d^k\}$ is the node set and $A^k_{do} = A^k \cup \{(d^k, o^k)\}$ the arc set. There are n^k available locomotives of type k at the origin node o^k.

Let W be the set of trains and set binary coefficient $a^k_{w,ij}$ to 1 if arc $(i,j) \in A^k$ covers train $w \in W$, 0 otherwise. For every train covered by active locomotives, pull power requirements are expressed in terms of the minimum number of locomotives, the requested horsepower or tonnage, or some combination. Let W^n, W^p, W^q be, respectively, the trains for which the corresponding measure is used ($W = W^n \cup W^p \cup W^q$). We utilize the following parameters:

n_w: minimum number of locomotives necessary to cover train $w \in W^n$,

p^k: operating power of locomotive type k,

p_w: requested power for train $w \in W^p$,

q^k_{ij}: total weight that may be pulled by a locomotive of type k on arc (i,j),

q_w: tonnage of train $w \in W^q$.

Integer flow variables in block k are x_{ij}^k and y_{ij}^k, respectively indicating the number of active and deadheading locomotives of type k covering arc $(i,j) \in A^k$. The cost of an active locomotive is denoted c_{ij}^k, that of a deadhead d_{ij}^k. The *ILP* with a block-diagonal structure writes as

$$z_{ILP}^\star = \min \sum_{k\in K} \sum_{(i,j)\in A^k} (c_{ij}^k x_{ij}^k + d_{ij}^k y_{ij}^k) \tag{7.143a}$$

$$\text{s.t.} \quad \sum_{k\in K} \sum_{j:(i,j)\in A^k} a_{w,ij}^k x_{ij}^k \geq n_w \qquad \forall w \in W^n \tag{7.143b}$$

$$\sum_{k\in K} \sum_{j:(i,j)\in A^k} p^k a_{w,ij}^k x_{ij}^k \geq p_w \qquad \forall w \in W^p \tag{7.143c}$$

$$\sum_{k\in K} \sum_{j:(i,j)\in A^k} q_{ij}^k a_{w,ij}^k x_{ij}^k \geq q_w \qquad \forall w \in W^q \tag{7.143d}$$

$$x_{do}^k \leq n^k \qquad \forall k \in K \tag{7.143e}$$

$$\sum_{j:(i,j)\in A^k} (x_{ij}^k + y_{ij}^k) - \sum_{j:(j,i)\in A^k} (x_{ji}^k + y_{ji}^k) = \begin{cases} x_{do}^k & \forall k \in K, i = o^k \\ 0 & \forall k \in K, i \in N^k \\ -x_{do}^k & \forall k \in K, i = d^k \end{cases} \tag{7.143f}$$

$$x_{ij}^k \in \mathbb{Z}_+ \qquad \forall k \in K, (i,j) \in A_{do}^k. \tag{7.143g}$$

The objective function (7.143a) minimizes the total cost to cover all trains. Constraints (7.143b)–(7.143d) represent train coverage requirements. The set (7.143e) makes at most n^k locomotives of type k available at o^k whereas (7.143f) provides the flow conservation equations at every node $i \in V^k$, for all $k \in K$. We group the covering constraints and the upper bound on x_{do}^k in $\mathcal{A}$ whereas $\mathcal{D}^k$ contains the flow conservation equations for type k:

$$\mathcal{A} = \left\{ \begin{bmatrix} \mathbf{x}^k \\ \mathbf{y}^k \\ x_{do}^k \end{bmatrix} \in \mathbb{Z}_+^{|A^k|} \times \mathbb{Z}_+^{|A^k|} \times \mathbb{Z}_+ \;\middle|\; \text{(7.143b)–(7.143e)} \right\} \tag{7.144a}$$

$$\mathcal{D}^k = \left\{ \begin{bmatrix} \mathbf{x}^k \\ \mathbf{y}^k \\ x_{do}^k \end{bmatrix} \in \mathbb{Z}_+^{|A^k|} \times \mathbb{Z}_+^{|A^k|} \times \mathbb{Z}_+ \;\middle|\; \text{(7.143f) for type } k \right\}, \quad \forall k \in K. \tag{7.144b}$$

The set $\mathcal{D}^k$ is a polyhedral cone with the unique zero extreme point and the set of extreme rays indexed by R^k. Such a ray is a cycle formed by a path from o^k to d^k and returning arc (d^k, o^k). We represent it with a unit-flow on the selected arcs, i.e., we set $x_{do}^k = 1$. The Dantzig-Wolfe reformulation becomes

$$z_{IMP}^\star = \min \sum_{k\in K} \sum_{r\in R^k} c_r^k \lambda_r^k \tag{7.145a}$$

$$\text{s.t.} \quad \sum_{k\in K} \sum_{r\in R^k} n_{w,r}^k \lambda_r \geq p_w \qquad \forall w \in W^n \tag{7.145b}$$

$$\sum_{k\in K}\sum_{r\in R^k} p^k_{w,r}\lambda_r \geq p_w \qquad \forall w \in W^p \qquad (7.145c)$$

$$\sum_{k\in K}\sum_{r\in R^k} q^k_{w,r}\lambda_r \geq p_w \qquad \forall w \in W^q \qquad (7.145d)$$

$$\sum_{r\in R^k} \lambda^k_r \leq n^k \qquad \forall k \in K \qquad (7.145e)$$

$$\lambda^k_r \geq 0 \qquad \forall k \in K, r \in R^k \qquad (7.145f)$$

$$\sum_{r\in R^k} \begin{bmatrix} \mathbf{x}^k_r \\ \mathbf{y}^k_r \\ 1 \end{bmatrix} \lambda^k_r = \begin{bmatrix} \mathbf{x}^k \\ \mathbf{y}^k \\ x^k_{do} \end{bmatrix} \geq \begin{bmatrix} \mathbf{0} \\ \mathbf{0} \\ 0 \end{bmatrix}, \textit{ integer} \quad \forall k \in K, \qquad (7.145g)$$

where, for an extreme ray $[\mathbf{x}^k_r, \mathbf{y}^k_r, 1]^\intercal$, $r \in R^k$,

$$c^k_r = \sum_{(i,j)\in A^k} (c^k_{ij,r} x^k_{ij} + d^k_{ij} y^k_{ij,r}), \qquad n^k_{w,r} = \sum_{j:(i,j)\in A} a^k_{w,ij} x^k_{ij,r},$$

$$p^k_{w,r} = \sum_{j:(i,j)\in A} p^k a^k_{w,ij} x^k_{ij,r}, \qquad q^k_{w,r} = \sum_{j:(i,j)\in A} q^k_{ij} a^k_{w,ij} x^k_{ij,r}.$$

In the applications described in Ziarati et al. (1997, 1999b), there are 26 locomotive types at Canadian National North America railway company (CN), yet on average only 2.3 are used per train. This provides the rationale for developing the *Branch-first, Cut-second* strategy exploiting the facets defined for at most two types of locomotives.

Facets. In general, it is unlikely to find a mix of locomotives that exactly matches the requested power p_w or tonnage q_w for train w, let alone both at the same time. We take advantage of this by adjusting the power or tonnage requests ad hoc with respect to fractional solutions. For example, consider two locomotive types represented by the integer variables x_1 and x_2 in Figure 7.22-*Left*: 3000 and 4000 hp, respectively, and a requested power of $3000x_1 + 4000x_2 \geq 8500$. The smallest hp-amounts are 12 000 at $(0,3)$, 11 000 at $(1,2)$, 10 000 at $(2,1)$, and 12 000 at $(4,0)$. Then 10 000 hp becomes the requested lifted power: $3000x_1 + 4000x_2 \geq 10000$. The solution $(2,1)$ is obviously integer but the extreme points $(0, {}^{10}/_{4})$ and $({}^{10}/_{3}, 0)$ are not. Indeed, the set of feasible solutions is bounded from below by the two line segments passing through the points (4,0)–(2,1) and (2,1)–(0,3). We can then replace the original constraint $3000x_1 + 4000x_2 \geq 8500$ specifying the requested pull power by the two facets $x_1 + x_2 \geq 3$ and $x_1 + 2x_2 \geq 4$.

Taking into account all covering constraints (7.143b)–(7.143d), for a given train, Figure 7.22-*Right* depicts all possible integer solutions of at least one and at most four locomotives of two different types, i.e., $1 \leq x_1, x_2 \leq 4$. Given that zero-facets are defined as $x_1, x_2 \geq 0$, it is easy to see that at most three non-zero facets are sufficient to describe the lower envelope of the convex hull of the integer solutions. This is the case for up to 6 locomotives (Ziarati et al., 1999b, Propositions 2 and 3), which is in practice the maximum number of active locomotives pulling a train.

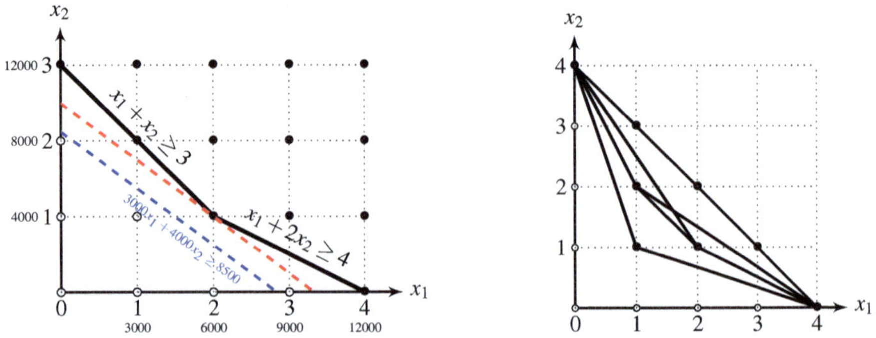

Fig. 7.22: *Left*: power request 8500 lifted to 10000 or $3000x_1 + 4000x_2 \geq 8500$ replaced by two facets. *Right*: non-zero facets with at most 4 locomotives of 2 types.

Branch-first. Assume that for a given train w, the *MP* solution at a certain branching node uses exactly two types of locomotives (let this set be denoted K_w), with fractional values for $a^k_{w,ij}x^k_{ij}, k \in K_w$. A possible set of decisions is to restrict the covering of this train by the actual two locomotives types or accept at least one from another type:

$$\sum_{k \in K \setminus K_w} \sum_{j:(i,j) \in A^k} a^k_{w,ij} x^k_{ij} = 0 \quad or \quad \sum_{k \in K \setminus K_w} \sum_{j:(i,j) \in A^k} a^k_{w,ij} x^k_{ij} \geq 1. \tag{7.146}$$

The proposed branching decisions create an unbalanced search tree. On the down-branch, the actual solution is still feasible since only the two locomotive types in K_w are used in the *MP* solution. For this train w, this is equivalent to replacing constraint (7.143b) by

$$\sum_{k \in K_w} \sum_{j:(i,j) \in A^k} a^k_{w,ij} x^k_{ij} \geq n_w. \tag{7.147}$$

Cut-second. On the down-branch, we can impose the facets describing the lower envelope of the polyhedron of the covering constraints (7.143b)–(7.143d) for the locomotives of types in K_w, that is, replacing one to three constraints for train w by at most three facets. Let K_w comprises types 1 and 2: any 2-dimensional facet of the form $x_2 \geq ax_1 + b$ writes as

$$-a \sum_{j:(i,j) \in A^1} a^1_{w,ij} x^1_{ij} + \sum_{j:(i,j) \in A^2} a^2_{w,ij} x^2_{ij} \geq b. \tag{7.148}$$

On the up-branch, the locomotive types in $K \setminus K_w$ are not currently being used for train w. The new *RMP* solution can be quite costly if the alternative locomotive types are marginally expensive to cover train w.

Selection of w. At a given branching node, let $W_2 \subseteq W$ be the subset of trains using at most two types of locomotives with fractional x-values in the *MP* solution. If a single type is used for a train w, each of the other types is evaluated by computing the marginal cost of forcing their use. The least-cost type is then coupled with the currently used type to form K_w. If $W_2 = \emptyset$, the heuristic branching rules of Ziarati et al. (1997) are applied, that is, selecting the λ-variables having fractional parts greater than or equal to 0.8 and fixing to the next integer those producing the minimal horsepower excess. Otherwise $W_2 \neq \emptyset$ and we compute for every train w in this set the marginal cost, say $\hat{C}^k_w$, of assigning a third locomotive type $k \in K \setminus K_w$ to w and we save the minimum value. The marginal cost $\hat{C}^k_w$ is computed by solving a shortest path problem passing through arc $(i,j) \in A^k$ for which $a^k_{w,ij} = 1$. Since we only need an underestimate, it is derived from the current solution of subproblem k in the column generation process. Then

$$\hat{C}^{min}_w = \min_{w \in K \setminus K_w} \hat{C}^k_w, \qquad \forall w \in W_2, \tag{7.149}$$

and the selected train corresponds to the largest value:

$$w \in \arg\max_{w \in W_2} \hat{C}^{min}_w. \tag{7.150}$$

In other words, for the selected train w, we impose the use of at most two locomotive types on the down-branch while on the up-branch, the likelihood of adding a third type of locomotive is marginally costly.

Ziarati et al. (1999b) report computational results on a weekly scheduling problem using data from the CN. In addition to the 26 locomotive types covering 1988 trains, it involves 171 critical locomotives that need maintenance (each subproblem is solved using a shortest path problem with time windows), 18 shop maintenance restrictions, and 26 power change point demands. The *ILP* with $26 + 171 = 197$ different blocks also comprises a total of 4000 train coverage constraints. Some results are presented in Table 7.4.

Solution approach	Number of locomotives	Horsepower (hp)	time (hours)
CN (manual)	1090	3 116 000	–
B&B without facets	1024	2 939 000	2.45
B&B with facets	1013	2 902 500	3.27

Table 7.4: A CN weekly scheduling problem involving 26 locomotive types and 171 critical locomotives for the covering of 1988 trains.

The reduction of 11 locomotives from 1024 to 1013 translates into savings of 1.1 % (more than \$30 M). The decrease in the power consumed is also of 1.2 % and the total cpu time increases by only 33 % by introducing the facets, from 2.45 to 3.27 hours (on a workstation HP9000/735). Similar results were obtained on a 1794-train problem: a decrease of 14 locomotives from 1044 to 1030 for a cpu time increase of 24 % (from 42 to 52 minutes on a SUN Ultra-2/2300 workstation).

7.8 Reference notes

Introduction In the first two decades of the GENCOL team on experiments with large scale applications, e.g., school bus routing, bus drivers scheduling, transportation of persons with reduced mobility, crew pairing, monthly rostering and preferential biding system, locomotive and car assignment, etc., the main difficulties always came from the design of branching and cutting rules: they were *different* in all applications. The rules working previously were not anymore efficient in the new application. This is indeed typical of integer programming applications, where *specialized decision rules* are often more efficient than those used by general solvers.

Section 7.1 The simple necessity to check an original or master problem solution for integrality, formally promoted by Vanderbeck (2011); Vanderbeck and Wolsey (2010), reminds us of many topics seen in Chapter 4, in particular aggregation and disaggregation of subproblems.

Section 7.2 The theoretical material of the subsections Cutting planes on the original variables and Cutting planes on the master variables is largely inspired by Desaulniers et al. (2011). The design of z-cuts within the *ISP* in Illustration 7.3 come from a one-hour discussion between François Soumis and Jacques so as to efficiently find optimal *integer* solutions to the many ISP^k, $k \in K$, in the Preferential bidding system, where $|K|$ is the number of pilots considered in the monthly schedules (Example 7.9). Lübbecke et al. (2021) eliminate solutions from the *ISP*s which cannot be part of an optimal *IMP* solution using cuts from Benders-like arguments.

For the Clique cuts for the set partitioning problem in Illustration 7.5, we are the first to write this down in this generality. The specialized idea for the *VRPTW* is from Spoorendonk and Desaulniers (2010).

Section 7.3 The general strategy of Branching on the original variables (x-branching) starts with Desrosiers et al. (1984) for a network-based *ILP* formulation of a vehicle routing problem with time window constraints. For aggregated master problems, Villeneuve et al. (2005) suggest to disaggregate and branch on original variables, thereby partially re-introducing the symmetry in pricing problems. As pointed out in Chapter 4, note that there are alternative compact models without a block-diagonal structure if the *ISP* formulation is defined on a cone (Proposition 4.16) or if it possesses the integrality property (Proposition 4.17).

Illustration 7.6 (x-branching for the *VRPTW*) presents various strategies for vehicle routing applications. Branching on the resource variables as in (7.51) is first found in Gélinas et al. (1995), where decisions are taken on the time or capacity values rather than on the network flow variables of a routing problem with backhauling. This branching type has been re-used with success by Pessoa et al. (2021) for solving bin packing models. The strategy involving multiple x_{ij}-branches in Figure 7.6 originally appears in Bellmore and Malone (1971), later adapted by Carpaneto and Toth (1980) for the asymmetric *TSP*.

The strategy of Branching on the master variables (λ-branching) first appears in Ryan and Foster (1981) for a set partitioning *IMP* formulation, that is, a few years before that on Branching on the original variables. In full generality, it was later developed by François Vanderbeck (2000, 2005) and we hope having well summarized his work. Extensions to the mixed-integer case appear in Vanderbeck and Savelsbergh (2006). The latest proposal of a generic λ-branching (Vanderbeck, 2011) goes beyond our presentation. The alternative way of writing (7.63) using a floor function in (7.65) is new. We recall that the x- and λ-branchings are complementary and the best of each should be used in appropriate situations. Advantages and disadvantages of certain strategies are discussed by Vanderbeck (2000) and Vanderbeck and Wolsey (2010).

Successful applications of strong branching were reported e.g., by Stefan Røpke at the *Column Generation 2012* workshop (see Figure 5.9). The question whether to generate columns (and how many) in the linear programming subproblems led to the notion of hierarchical or multi-phase branching (e.g., Pecin et al., 2017b).

Referring to the Brazilian friends in Figure 7.13, Eduardo is a former doctoral student (PUC-Rio, 2001) of Marcus who previously obtained his PhD under the supervision of Professor Brigitte Jaumard (Polytechnique Montréal, 1993).

Good to Know As already mentioned in Arc-flow variable fixing by reduced cost, applications of this strategy are found in Irnich et al. (2010) for the *VRPTW* and Pecin et al. (2017b) for the capacitated *VRP*. An extension for two-arc sequences appears in Desaulniers et al. (2020a), applied to the *VRPTW* and four variants of the electric *VRPTW*.

A lot has been written on Acceleration techniques, indeed ideas spread over hundreds of various application papers. Let us here concentrate on a single item, the *degeneracy of the RMP*. Dynamic constraint aggregation (DCA) and dual variable stabilization are two methods that reduce the negative impact of degeneracy. The first uses a projection to reduce the primal space (the number of constraints of the *RMP*, hence the size of the *working basis*, decreases to the number of positive basic variables) whereas the second acts in the dual space by providing relevant restrictions. Benchimol et al. (2012) develop a *stabilized dynamic constraint aggregation* that simultaneously combines both for solving set partitioning problems. On highly degenerate *MDVSP* instances, it reduces the average CPU time of the *RMP* by a factor of up to 7 with respect to the best of the two combined methods. By going back to the linear algebra framework from which emanates the so-called concept of subspace basis, Gauthier et al. (2016) establish tight bounds between three tools for dealing with primal degeneracy, that is, the *dynamic constraint aggregation*, the *positive edge* rule, and the *improved primal simplex* (see Positive edge rule and Improved Primal Simplex algorithm in Chapter 3 for a short description of the last two). Combining the first and the last tools in the context of column generation, Desrosiers et al. (2014) propose a *row-reduced column generation algorithm.*

More to Know When it comes to an implementation, at least three viewpoints are interesting: the theoretical background, the technical setup, and the "tricks" that bring efficiency. Sometimes the three are hard to distinguish from one another. Most of the practical advise given by Vanderbeck (2005) is in fact theory. Descriptions on general branch-and-price frameworks are very helpful, too (e.g., Pessoa et al., 2020, 2021; Sadykov and Vanderbeck, 2021), also because they come with examples.

The major contributors to the GCG project (Gamrath and Lübbecke, 2010) over the years have been Gerald Gamrath (who laid out the original "two communicating trees" design which is still intact today), Martin Bergner, Michael Bastubbe, Christian Puchert, Jonas Witt, Steve Maher, Erik Mühmer, and many more. Marco is the one with the vision who keeps it all together.

Examples
Example 7.5 Edge coloring and odd-circuit cuts is inspired by Desaulniers et al. (2011). For the other examples, the text already includes the main references.

Exercises

7.1 François Vanderbeck
Who is François Vanderbeck, chair of the organization committee in charge of the logistics for the 23rd ISMP (International Symposium on Mathematical Programming) held in Bordeaux (July 1–6, 2018), cited or mentioned close to 50 times in this book?

7.2 B&P: Wikipedia
At the time of publication, the entry about *branch-and-price* on Wikipedia supports the explanation of this method using Figure 7.23. What is wrong with it?

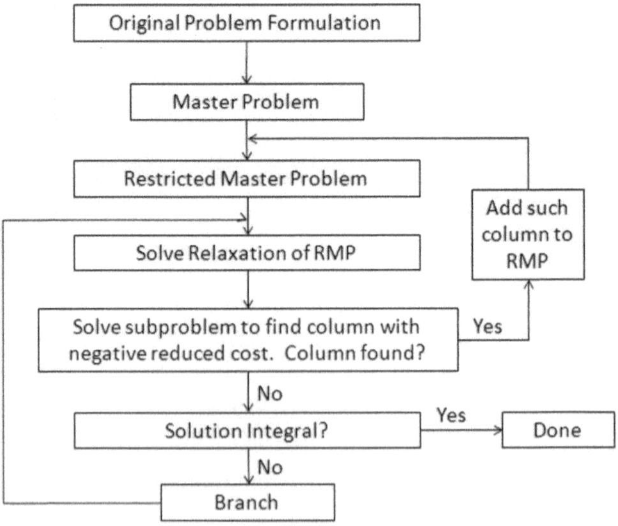

Fig. 7.23: Wikipedia's misleading flowchart for branch-and-price, 2024.04.01.

7.3 B&P: early branching
Consider a minimization problem and a branch-and-bound tree in which some nodes are not solved to optimality, that is, more specifically, an early branching strategy is used such that the column generation process is stopped even if there are still columns with a negative reduced cost (see *Early branching*, p. 495). Consider the cases where the Lagrangian lower bound is computed at a node and that where it is not, for example, because the subproblems are not all solved. Adapt the Branch-and-Bound method of Section 1.5, in particular, the definition of *node value* (p. 29).

7.4 *ISP* modifications for a λ-branching
By exploiting the sign of $\gamma_1 \leq 0$ and $\gamma_2 \geq 0$ in their respective pricing problem, show that (7.63) can be replaced by

(a) $g_{\mathbf{x}} \geq \frac{\mathbf{f}^{\mathsf{T}}\mathbf{x}-f+1}{u-f+1}, \; g_{\mathbf{x}} \in \{0,1\}$ in the *ISP* (7.61).

(b) $g_{\mathbf{x}} \leq \frac{\mathbf{f}^{\mathsf{T}}\mathbf{x}-\ell}{f-\ell}, \quad g_{\mathbf{x}} \in \{0,1\}$ in the *ISP* (7.64).

7.5 *ISP* modifications for a λ-branching: alternative function $g(\mathbf{x})$
Show that $g(\mathbf{x})$ in the *ISP*s (7.61) and (7.64) can take the form $g(\mathbf{x}) = \left\lceil \frac{\mathbf{f}^{\mathsf{T}}\mathbf{x}-(f-1)}{u-\ell} \right\rceil$.

7.6 A binary x-branching decision imposed in $\mathcal{A}$ rather than $\mathcal{D}$
Let a Dantzig-Wolfe reformulation of the *ILP* (7.1) be based on the grouping $\mathcal{A}$ and $\mathcal{D}$ in (7.2), where conv($\mathcal{D}$) is a polytope. Assume that the branching decision on the binary variable x_j is imposed in $\mathcal{A}$ rather than $\mathcal{D}$. Describe the modifications on the *RMP* and *ISP* in the down- and up-branches.

7.7 Bound constraints on several x-variables: underlying mathematics
Mathematically justify the added constraints to the *ISP*s in (7.84) for
(a) the down-branch; (b) the up-branch.

7.8 Bound constraints on several x-variables: binary case
Refresh the set of constraints in (7.84) if $\mathbf{x} \in \{0,1\}^n$. Show that it simplifies to

$$g_{\mathbf{x}} \geq 1 - \sum_{j \in \bar{J}} x_j + \sum_{j \in \underline{J}} x_j - (|\bar{J}| + |\underline{J}|) \quad \Bigg| \quad \begin{array}{ll} g_{\mathbf{x}} \leq 1 - x_j & \forall j \in \bar{J} \\ g_{\mathbf{x}} \leq x_j & \forall j \in \underline{J}. \end{array} \tag{7.151}$$

7.9 Ryan-Foster separating hyperplane for the *VRPTW*
(a) Given the Ryan-Foster rule in (7.86), how does a separating hyperplane for the *VRPTW* in the language of Proposition 7.1 look like?

(b) Generalize it for an *ILP* (7.1) whose *IMP* reformulation is a set partitioning model.

7.10 Handling inter-task branching decisions in labeling algorithm
We consider applying Branching on inter-tasks when the *ISP* is a *SPPRC* defined on an acyclic network $G = (N, A)$ (with source and sink nodes o and d) such as that in Figure 5.7. Let $W = \{1, \ldots, m\}$ be the set of tasks to cover. In network G, each task $w \in W$ is associated with a single arc and each arc represents at most one task. Let

w_{ij} be the task associated with arc $(i, j) \in A$, including the fictitious task *NIL* if the arc does not represent a real task in W.

Assume that, at the current branch-and-bound node, multiple inter-task decisions are applicable and compatible (i.e., they do not contradict): some are from up-branches and stored as follow-on task pairs (r,s) in a set F, others are from down-branches and stored as do-not-follow-on task pairs (r,s) in a set D.

(a) Describe a labeling algorithm that takes into account these inter-task branching decisions using a single additional label component which keeps track of the last task covered in a path. Focus on this additional component by presenting its extension functions and its roles in label feasibility and dominance.

For convenience, use the following notation:

- For $w \in W$, let $f_w = s$ be the task that must be covered immediately after w if $\exists s \in W$ such that $(w,s) \in F$; otherwise, set $f_w = NIL$.
- In the labeling algorithm, a label E_p representing a partial path p contains an additional component T_p^{last} that indicates the last task covered in p.

(b) In an acyclic network, we can easily compute a priori a subset of unreachable tasks $U_i \subseteq W$ from a node $i \in N$, i.e., tasks that cannot appear in any feasible extension of a path ending in node i. How can you improve the algorithm proposed in (a) by considering unreachable tasks?

7.11 Finite number of iterations in Proposition 7.2

Can the bound constraints $x_j \leq \lfloor v_j \rfloor$ or $x_j \geq \lfloor v_j \rfloor + 1$ already exist in set $\mathcal{B}$ in (7.81)?

7.12 λ-branching: single bound constraint on x_j

Let $x_j \in \mathbb{Z}_+$ take values between 0 and $u_j > 2$ in the *ILP*, with $x_j^\star \notin \mathbb{Z}_+$ in the final *RMP*. Let us refer to (7.59) and assume that $\sum_{\mathbf{x} \in \mathcal{X}: x_j \geq 2} \lambda_{\mathbf{x}}^\star = \beta$ is fractional. Derive a λ-branching, providing the modifications to the *MP*s (7.60) and *ISP*s (7.66).

7.13 λ-branching: interval constraint on the cost function

Assume that $\mathbf{c}$ in *Rule* 1 (p. 483) is integer and that we have cost coefficients $\{c_{\mathbf{x}}\}_{\mathbf{x} \in \mathcal{X}}$, not all identical, ranging between lower and upper bounds ℓ and u, respectively. Given $\lambda_{RMP}^\star$ fractional, show how to *possibly* find a set $\mathcal{B} = \{\mathbf{x} \in \mathcal{X} \mid c_1 \leq \mathbf{c}^\intercal \mathbf{x} \leq c_2\}$ such that $\sum_{\mathbf{x} \in \mathcal{X} \cap \mathcal{B}} \lambda_{\mathbf{x}}^\star = \beta$ is fractional.

7.14 λ-branching: ISP modifications for *Rules* 3.2 and 3.4

Given the condition $x_r = x_s = 1$ re-written as $x_r + x_s \geq 2$ in page 484, show that the constraints $y_{rs} \geq x_r + x_s - 1$ in *Rule* 3.2 (down-branch) and $2y_{rs} \leq x_r + x_s$ in *Rule* 3.4 (up-branch) can be derived from (7.66).

7.15 Time constrained shortest path problem: thinking like a computer

In Example 7.1, we describe a branch-and-bound for the *TCSPP* but we skip or bend a lot of technical explanations regarding implementation. Let us take a snapshot of the internal computer representation of the *RMP* at node **BB4**. In particular, one has to think that variables and constraints are passively suppressed. Describe exactly the optimization program that the solver solves for the *RMP* after housekeeping and explains how infeasibility is detected.

7.16 Minimum number of vehicles for the VRPTW
Give a compact *ILP* formulation and propose branching and cutting strategies to only determine the minimum number of vehicles for the *VRPTW*.

7.17 Validity of a dominance rule for the VRPTW with strong capacity cuts
Show that the dominance rule (7.132) is valid.

7.18 Comparison of dominance rules for the VRPTW with strong capacity cuts
Show that the dominance rule (7.132) is stronger (i.e., cannot yield less dominated labels) than the dominance rule (5.19) and (7.130).

7.19 Arc-flow variable fixing for the ***VRPTW***
Consider the *VRPTW* example in Illustration 5.1 with $C = \{1,2,3,4\}$, $q_1 = q_3 = 1$, $q_2 = q_4 = 2$, $Q = 5$, and $z^\star_{IMP} = 74$. Its *MP* is the linear relaxation of (5.8) without the constraint (5.8c) (the number of available vehicles is unconstraining). Let us assume that a heuristic solution of cost 75 is known and provides an upper bound $UB = 75$ on $z^\star_{IMP}$.

(a) After solving the *MP* by column generation (with an *ESPPRC* pricing problem), we obtain the optimal primal-dual solution pair:

$$\lambda^\star_{132} = \lambda^\star_{134} = \lambda^\star_{24} = 0.5, \;\; \lambda^\star_r = 0 \;\text{ for all other routes } r \in R,$$

and

$$\pi^\star_1 = 13.5,\; \pi^\star_2 = 17.5,\; \pi^\star_3 = 6,\; \pi^\star_4 = 28.5,$$

with a cost $z^\star_{MP} = LB = 65.5$. The following table lists the non-dominated labels obtained at nodes 1 and 2 when applying a forward-labeling algorithm to solve the *ISP* defined for this dual solution.

Node	Forward labels $(T^{rCost}, T^{time}, T^{load}, [T^{uCust_h}]_{h\in C})$
1	$(3,2,1,[1,0,0,0])$
2	$(15,8,2,[1,1,0,0]), (8.5,12,3,[1,1,0,1]), (2.5,15,4,[1,1,1,1])$

Next, executing a backward-labeling algorithm yields at nodes 2 and 3 the non-dominated labels listed in the following table.

Node	Backward labels $(B^{rCost}, B^{lTime}, B^{mLoad}, [B^{uCust_h}]_{h\in C})$
2	$(-2.5,18,3,[0,1,0,1]), (-15,9,1,[1,1,1,1])$
3	$(12,25,4,[0,0,1,0]), (-5.5,12,2,[0,0,1,1])$

In these backward labels, resources *rCost* and $uCust_j$, $j \in C$, act like the corresponding resources in the forward labels. Resource *lTime* provides the latest time at which the service at the corresponding node can start, whereas resource *mLoad* indicates the maximum load that can be collected/delivered before this node, i.e., the vehicle capacity minus the total load of the customers visited

along the path. For instance, label $(-5.5, 12, 2, [0, 0, 1, 1])$ at node 3 represents the path 34d traversed backwardly. To be feasibly concatenated with it, any forward path must arrive at node 3 no later than time 12 and with a load not exceeding 2.
For arcs $(i, j) \in \{(1,2),(2,3)\}$ (for which $t_{12} = 10$, $t_{23} = 5$, $\tilde{c}_{12} = 5.5$, and $\tilde{c}_{23} = -13.5$), use the above forward and backward labels to compute $\bar{c}^{\star}_{ij}$, the minimum reduced cost of a path traversing (i, j), and determine if x_{ij} can be fixed to 0.

(b) Same question as the previous one except that we have added to the *MP* the capacity cut (7.26) for $S = C$. In this case, we obtain the following optimal primal-dual solution pair:

$$\lambda^{\star}_{1} = \lambda^{\star}_{132} = \lambda^{\star}_{24} = \lambda^{\star}_{34} = 0.5, \;\; \lambda^{\star}_{r} = 0 \;\text{ for all other routes } r \in R,$$

and

$$\pi^{\star}_{1} = 7,\ \pi^{\star}_{2} = 9.5,\ \pi^{\star}_{3} = 4.5,\ \pi^{\star}_{4} = 20.5,\ \psi^{\star} = 16$$

with $z^{\star}_{MP} = LB = 73.5$, where ψ is the dual variable associated with the cut. The following two tables provide the forward and backward labels at the appropriate nodes.

Node	Forward labels $(T^{rCost}, T^{time}, T^{load}, [T^{uCust_h}]_{h\in C})$
1	$(-13, 2, 1, [1, 0, 0, 0])$
2	$(-1, 8, 2, [1, 1, 0, 0]), (-5.5, 14, 3, [1, 1, 1, 1])$

Node	Backward labels $(B^{rCost}, B^{lTime}, B^{mLoad}, [B^{uCust_h}]_{h\in C})$
2	$(5.5, 18, 3, [0, 1, 0, 1]), (1, 9, 1, [1, 1, 1, 1])$
3	$(13.5, 25, 4, [0, 0, 1, 0]), (4, 14, 2, [0, 1, 1, 0])$

The adjusted costs of the arcs $(1,2)$ and $(2,3)$ are $\tilde{c}_{12} = 12$, and $\tilde{c}_{23} = -5.5$.

7.20 Branching on a resource interval
Let the value of a certain resource in a constrained shortest *od*-path problem be restricted to an interval, for example, the number of flight credits for a pilot must be between 70 and 78 in a monthly crew assignment problem. How is this resource, say F, implemented in the corresponding pricing problem if we only want to use non-decreasing resource extension functions?

7.21 B&P&C strategies for various applications
We provide the solution to all exercises in the book *except this one.* The applications listed below can be found in the literature, each one with some branching or cutting decisions. Since their publication, theory and practice have evolved. Therefore, propose and discuss various strategies that might be efficient in finding optimal (or near-optimal) integer solutions.

(a) One-dimensional cutting stock problem (Chapter 2)

(b) Aircraft routing with schedule synchronization (Chapter 2)
(c) Scene selection problem (Chapter 4)
(d) Design of balanced student teams (Chapter 4)
(e) Multiple depot vehicle scheduling problem (Chapter 4)
(f) Pickup and delivery problem with time windows (Chapter 5)
(g) Crew pairing problem with base constraints (Chapter 5)
(h) Capacitated p-median problem (Chapter 6)
(i) Generalized assignment problem (Chapter 6)

8 Conclusion

> Rien de beau ne peut se résumer.
>
> *Rhumbs*
> Paul Valéry

> Mmm num ba de
> Dum bum ba be
> Doo buh dum ba beh beh
>
> *Under Pressure*
> Queen and David Bowie

Abstract We have reached the final chapter to conclude this great and hazardous adventure of writing a book on the column generation algorithm, Dantzig-Wolfe reformulations, and resolution of integer linear programs, indeed, the branch-and-price approach. We begin with a brief history of the research activities carried out over five decades by the Montréal group `GENCOL`, then look to the future.

Contents

Outroduction . . . 546
8.1 GENCOL . . . 546
First decade (1980s) . . . 547
Second decade (1990s) . . . 548
Third decade (2000s) . . . 549
Fourth decade (2010s) . . . 550
Fifth decade (2020s) . . . 551
8.2 Branch-and-Price: *More to Know* . . . 552
Exercises . . . 553

Acronyms

`FORTRAN` FORmula TRANslation . . . 547
SPPTW shortest path problem with time windows . . . 547

J. Desrosiers et al., *Branch-and-Price*,
https://doi.org/10.1007/978-3-031-96917-1_8

SPPRC	shortest path problem with resource constraints	548
VRPTW	vehicle routing problem with time windows	549
ISUD	integral simplex using decomposition	550
ESPPRC	elementary shortest path problem with resource constraints	551
DOI	dual-optimal inequality	552
SPP	set partitioning problem	552

My next guest does not need an introduction.

He needs an ...
Marco, Guy, and Jean Bertrand

... Outroduction

The conclusion at last. Whether column generation is part of your daily work routine or you rather appreciate its value as an enthusiastic observer, we can only encourage giving this book another read. Before we turn the final page, Section 8.1 connects the dot between Jacques' career debut and where we stand today. With the utmost modesty and respect towards all those who accompanied and supported him professionally and emotionally, let this book be his legacy. It is the end of a journey but certainly not that of column generation. We thus leave space for the story to unfold with general perspectives for the future. These are summarized, along with a few others we have raised in the course of this book, in Section 8.2 as suggestions for those with an interest in writing.

8.1 GENCOL

The `GENCOL` team—that is the name of the research team supervised since 1981 by François Soumis, Jacques Desrosiers, and Guy Desaulniers at the GERAD[1] research center in Montréal—has revived the column generation method which is now considered a major tool in the solution of large-scale integer linear programs. Indeed, from a negligible number of column generation papers published fifty years ago, we see many more of them every year as illustrated in Figure 8.1 obtained by the topic search *"column generation" or "branch-and-price" or "Dantzig-Wolfe"* on the Web of Science[2].

Since 2005, members of the team edited a book on column generation (Desaulniers et al., 2005) and organized either an international school (Montréal 2006,

[1] `https://gerad.ca`

[2] `https://webofscience.com`

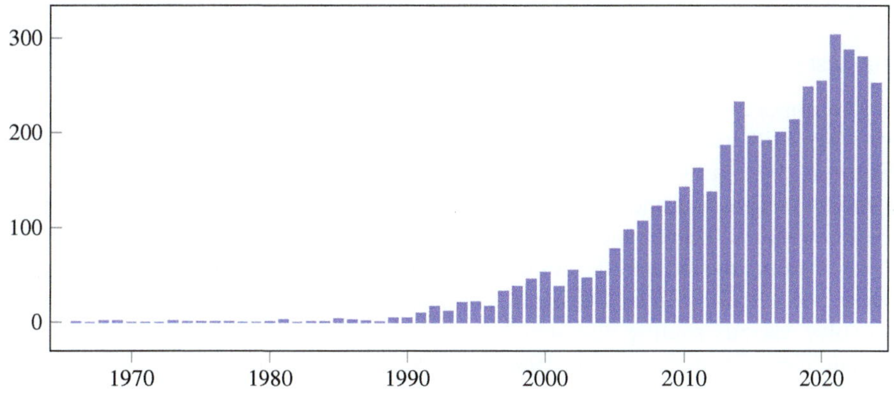

Fig. 8.1: Publications on column generation every year from 1966 to 2024.

Darmstadt 2010, Paris 2014, 2018, and 2025 with the help of Roberto Wolfler Calvo) or a workshop (Aussois 2008, Bromont 2012, Búzios 2016, Montréal 2023). Let us highlight some marking events that occurred for the GENCOL team in the last four decades and what the future looks like for the fifth one. Note that an earlier text already appeared in Desrosiers (2010).

First decade (1980s)

At the Sixth European Congress on Operational Research (Euro VI) held in Vienna (1983), Jacques Desrosiers, François Soumis, and Martin Desrochers are awarded with the *best scientific contribution paper* for "Routing with Time Windows by Column Generation" (Desrosiers et al., 1984). Several school bus transportation problems with up to 151 trips are solved. This paper is the result of three years of research, the first month devoted to design and code the pricing problem in FORTRAN (the first high-level programming language), the rest of the time to numerically stabilize that particular column generation framework using the open source code LANDP by Ailsa Land and Suzan Powell, see Land and Powell (1973). It is one of the first efficient applications of column generation integrating branch-and-bound decisions and cutting planes to reach integer solutions. The algorithm for the shortest path problem with time windows (*SPPTW*) is designed in November 1981, published in Desrosiers et al. (1983), and the improved and re-optimized versions appeared some years later in Desrochers and Soumis (1988b,c).

A key point of this stream of research is to recognize that the sets of incoming/outgoing constraints for visiting a node k exactly once in the directed network $G = (N, A)$ can be separated in the master and pricing problems. That is, we can keep one of $\sum_{j:(k,j)\in A} x_{kj} = 1$ or $\sum_{i:(i,k)\in A} x_{ik} = 1$ in the master problem and use a transformation of their combination in the pricing problem, e.g., their difference as

$\sum_{i:(i,k)\in A} x_{ik} - \sum_{j:(k,j)\in A} x_{kj} = 0$. These are flow conservation constraints found in the shortest path formulations of the *ISP*.

An important breakthrough is the generalization of *SPPTW* to the shortest path problem with resource constraints (*SPPRC*) (Desrochers, 1986; Desrochers and Soumis, 1991; Irnich and Desaulniers, 2005). These constraints are treated by dynamic programming and enable the handling of non-decreasing extension functions between states. These functions, which can be discontinuous and non-convex, are used to model the industrial rules that are defined by administrators and lawyers (see Desrochers et al., 1992b). During this first decade, the team uses column generation to solve various routing and scheduling integer programming problems: pick-up and delivery (Desrosiers et al., 1986; Dumas et al., 1991), transportation of persons with reduced mobility (Desrosiers et al., 1988a), and bus driver scheduling (Soumis et al., 1984; Desrochers and Soumis, 1988a), amongst others. Two surveys are published in this early period, see Desrochers et al. (1988) and Solomon and Desrosiers (1988).

International commercialization starts with GIRO Inc.[3] at the end of the '80s: In over 300 cities worldwide, their `HASTUS` software system is now used by transit authorities ranging in size from 20 to over 6500 vehicles/rail cars, including installations in Montréal, Los Angeles, Chicago, New York, Barcelona, Brussels, Geneva, Stockholm, Vienna, Hamburg, Singapore, Tokyo, and Sydney.

Second decade (1990s)

While Irvin Lustig, Roy Marsten, and David Shanno present an interior-point method capable of solving a linear program involving 4 million variables, the `GENCOL` team shows in the same ORSA/TIMS meeting (1992) that an airline crew pairing problem with over 190 millions of millions of variables is solved to optimality using column generation. The solution improves by 6 % those computed by the best available software systems.

AD OPT Technologies[4] starts the commercialization of the `Altitude` software suite for the airline industry (Jacques Desrosiers and Yvan Dumas are partners in this startup co-founded in 1987 by François Soumis and four OR researchers). In 1993, it acquires Volvo's Carmen division, which provides pilot schedules for SAS and Lufthansa. The company goes public on the Toronto Stock Exchange market in June 1999, carving out a place for itself in a market already occupied by major companies such as IBM, Unisys, and AT&T. The core engine of this suite is `GENCOL`, the column generation solver developed at GERAD and already utilized by GIRO. The acronym comes from the French expression *GÉNération de COLonnes*. This is a success story that is in particular recognized by the prize *Excellence in Innovative Partnership, University-Industries Synergy*, awarded by the Natural Sciences

[3] `https://giro.ca`

[4] `https://ad-opt.com`

and Engineering Research Council of Canada and the Conference Board of Canada (Vancouver, 1997).

Some notable papers for this time period are for multi-depot vehicle routing (Ribeiro and Soumis, 1994; Desaulniers et al., 1998b), crew pairing (Desaulniers et al., 1997b), aircraft routing and scheduling (Desaulniers et al., 1997c), monthly schedules at Air Canada (Gamache et al., 1998) and Air France (Gamache et al., 1999), simultaneous aircraft routing and crew scheduling (Cordeau et al., 2001b), and crew recovery (Stojković et al., 1998, 2002).

Applications in the rail industry include Ziarati et al. (1997, 1999b,a) for freight trains and Cordeau et al. (2000, 2001a); Lingaya et al. (2002) for passenger trains. This decade also sees the first attempts to solve the vehicle routing problem with time windows (*VRPTW*) using a branch-and-price approach (Desrochers et al., 1992a). In a certain way, it is a failure on these small 25- to 100-customer instances compared to those in the airline industry, but this paper opens the way to fundamental research, e.g., on cutting plane generation (Kohl et al., 1999).

Three survey papers on column generation and branch-and-price applied to vehicle routing and crew scheduling problems appear during that period: Desrosiers et al. (1995), Soumis (1997), and Desaulniers et al. (1998a).

Third decade (2000s)

Together with Bombardier Flexjet[5], the team is a finalist in Boston for the INFORMS 2004 *Franz Edelman Award* for Achievement in Operations Research and the Management Sciences. The three-year project with AD OPT Technologies helped Flexjet, a division of Bombardier operating a fleet of more than 100 business jets, to achieve operational profitability and improve its scheduling process (Hicks et al., 2005). In the first two years of operation, the new optimization system generated savings of US$54 million. FlexJet became profitable and was subsequently sold for US$500 million with a firm aircraft order valued at US$1.5 billion when Bombardier needed funds to develop the C Series.

With about 400 large-scale worldwide industrial implementations using GENCOL and hundreds of academic research projects published in over 150 papers by the team, the column generation method is mathematically better understood. This is true for its virtues, but also for its shortcomings. At that point in time, branch-and-price (Barnhart et al., 1998), that is, column generation combined with a branch-and-bound search tree, becomes a powerful tool in solving large-scale integer programs. Under mild assumptions, it is shown that a compact formulation *exists* for any column generation scheme, hence branching and cutting decisions can always be defined on the variables of this formulation. It has a block diagonal structure with identical subproblems, each of which contributes only one column in an in-

---

[5] `https://flexjet.com`

teger solution (Villeneuve et al., 2005). Single block special cases are presented in Propositions 4.16 for and 4.17.

Dantzig-Wolfe reformulation and column generation, both devised for linear programs, are the main characters of a success story in large-scale integer programming. Lübbecke and Desrosiers (2005) outline and relate these approaches. The authors emphasize the growing understanding of the dual point of view, which brings considerable progress to the column generation theory and practice.

In fact, the `GENCOL` team develops during that time period *dual variable stabilization* techniques to reduce the impact of the well-known tailing-off effect seen in column generation (du Merle et al., 1999; Oukil et al., 2007; Ben Amor et al., 2009) and introduces the *dynamic constraint aggregation* procedure to dynamically reduce the row-size of set partitioning master problems (El Hallaoui et al., 2005, 2008, 2010). The latter procedure is elegantly generalized to the *improved primal simplex* algorithm (El Hallaoui et al., 2011), allowing for a better treatment of degeneracy in linear programming. Degeneracy, which has been seen as the major drawback of the primal simplex and column generation algorithms, can be exploited advantageously. As a by-product, the primal and dual versions of the *positive edge rule* (Towhidi et al., 2014) is implemented within COIN-OR's `Clp` by John Forrest (see Figure 3.22). Moreover, this research on the dual side leads to an efficient crossover method, from an optimal interior to extreme point solution (Ben Amor et al., 2006a).

Fourth decade (2010s)

The fourth decade sees the use of new pricing problems based on alternative optimality conditions (Desrosiers et al., 2014; Gauthier et al., 2016, 2018). From a dual point of view, we essentially optimize all or a subset of the dual variables to maximize the smallest reduced cost. From a primal point of view, we look for an improving combination of variables, also called an augmenting cycle of negative cost (as in solving network flow problems). Furthermore, the nature of the pricing problem is such that it identifies combinations containing few variables. This approach is an extension of the *improved primal simplex* algorithm (El Hallaoui et al., 2011). The *integral simplex using decomposition (*ISUD*) algorithm* extends upon this pricing problem in an effort to maintain integer solutions while solving the linear relaxation (Zaghrouti et al., 2014; Rosat et al., 2017a). The former fosters integrality by imposing that negative reduced cost columns must also lead to neighboring integer solutions. It effectively alters the solution paradigm since the integrality requirement is now solely the responsibility of the pricing problem. It has been developed in the context of set partitioning models when columns are all known a priori. *ISUD* is then generalized to the case with a very large number of columns that are generated dynamically to yield the *integral column generation algorithm* (Tahir et al., 2019) that can also handle side constraints (Tahir et al., 2022).

In parallel to these theoretical developments, the GENCOL team develops branch-price-and-cut algorithms for solving to optimality various *rich* vehicle routing problems: *split-delivery* (Desaulniers, 2010; Archetti et al., 2011), *stochastic demands* (Gauvin et al., 2014), *pickup and delivery with time windows and loading constraints* (Cherkesly et al., 2015, 2016), *inventory-routing* (Desaulniers et al., 2016b), *stochastic service times* (Errico et al., 2016, 2018), and *electric* (Desaulniers et al., 2016a), among others. For these applications, the pricing problems can still be seen as *elementary shortest path problems with resource constraints* (*ESPPRC*s) but with more complex label definitions, resource extension functions, and dominance rules.

In 2014, François Soumis receives the *Lionel-Boulet Prize*, the highest distinction awarded by the Québec government to a researcher, for his entire career, who distinguished himself by his inventions, innovations, leadership in scientific development, and his contribution to the economic growth of Québec. One year later, Jacques Desrosiers wins the *Pierre Laurin Award* at HEC Montréal recognizing research conducted over his entire career whereas AD OPT receives from the Canadian Operational Research Society the *Omond Solandt Award* for Excellence in Operations Research.

Fifth decade (2020s)

While Jacques is happily heading towards retirement, Guy is still doubly active, as usual, as is François who received from Polytechnique Montréal the *Prix d'excellence en recherche et innovation 2023*. New collaborators joined the team in the last 20 years such as Marco Lübbecke (RWTH Aachen University), Issmail El Hallaoui (Polytechnique Montréal), Claudio Contardo (Concordia University), and Jean Bertrand Gauthier who went private. More recently, the team has grown with the arrival of Fausto Errico (École de technologie supérieure) together with Marilène Cherkesly and Frédéric Quesnel (Université du Québec à Montréal).

Fundamental research still has its place, as we can see with improved dynamic programming algorithms for solving shortest path problems with resource constraints (Himmich et al., 2020, 2023) and also Altman et al. (2023) for the fragility-constrained *VRPTW*, and a selective pricing mechanism for a variant of the *ng*-route relaxation in which the neighborhoods are associated with arcs instead of nodes (Costa et al., 2021). Another stream of research is about alternative stabilization techniques such as those described in Haghani et al. (2022) and Costa et al. (2022).

For sure, the future looks like *larger instances*, again. This has always been the case for over 40 years, one such application being solving an industrial monthly crew pairing problem instance with 46 588 flights (Desaulniers et al., 2020b). The solution approach combines, among others, column generation with dynamic constraint aggregation, hence efficiently managing the degeneracy issues of the set partitioning master problem. When embedded in a rolling horizon procedure, the aggregation of the constraints allows to consider wider time periods and therefore yields better solutions.

The future also looks like *machine learning*, a strong tendency, see Kruber et al. (2017) and Bengio et al. (2021). For example, Yaakoubi et al. (2020) use machine learning and mathematical programming to solve larger crew pairing problems, Morabit et al. (2021, 2023) accelerate the column generation algorithm with a better column or arc selection, Tahir et al. (2021) combine an integral column-generation algorithm and high probability flight connections for solving the crew pairing problem, and Quesnel et al. (2022) solve a personalized crew rostering problem, where the pairings to include in each pricing problem are chosen by a deep neural network trained on historical data.

Another trend getting more and more traction is dealing with *stochastic* models. This is seen in Errico et al. (2018) for solving the *VRPTW* with stochastic service times by a branch-price-and-cut algorithm. The authors adapt the dynamic programming algorithm of the pricing problem to account for the probabilistic resource consumption by extending the label dimension and providing new dominance rules. Another one is for solving a version of the capacitated vehicle routing problem where the travel times are assumed to be uncertain and statistically correlated (Rostami et al., 2021).

8.2 Branch-and-Price: *More to Know*

The following is a short list of advanced subjects, more or less tied to column generation, that might be covered in a second book (and a third one as *Much more to Know*), by various authors hopefully following the notation introduced in this one:

1. Volume algorithm
2. Bundle methods
3. Dual heuristics
4. Lagrangean decomposition
5. Smooth and flexible *DOI*s

6. Improved primal simplex
7. Positive edge rule
8. Constraints aggregation for the *SPP*

9. Row-reduced column generation
10. Stabilized dynamic constraints aggregation for the *SPP*

11. Nested column generation
12. Benders decomposition
13. Combination of Dantzig-Wolfe and Benders decompositions

14. Integral simplex
15. Integral column generation
16. Constraint aggregation and integral simplex for the *SPP*

17. Cycle-canceling algorithms
18. Vector space decomposition
19. Interior point algorithms
20. Non-linear decomposition

21. Machine learning
22. Integration of stochastic aspects

23. Implementation
24. History of `SCIP`
25. History of `BaPCoD`

Exercises

8.1 François Soumis

Who is François Soumis, a former director of the GERAD research center from 1992 to 1996 and co-author in Desrochers and Soumis (1991), a paper for which we provide below the main contribution?

> *In all dynamic programming problems, the aim is to solve the recurrence equations for all states. The optimal solution is a set of paths going from the initial states to the final ones. This paper studies the implementation of the shortest path problem with resource constraints. The authors compare two classes of algorithms:* reaching *and* pulling. *The main conclusion is that the computation times of an implementation of the pulling algorithm does not grow with the number of states but is rather proportional to the number of feasible paths in the solution.*

8.2 Undivided attention

Find all mistakes in this book.

Und wenn die Frage jemals fällt:
"Was ist es, das am Ende zählt?"
Dann wird die Antwort immer sein
Dass man nicht solche Fragen stellt
Doch wenn ich ehrlich zu mir bin
Dann macht nur die Erkenntnis Sinn:
Das Ende ist nur'n Meilenstein
Und wichtig ist der Weg dahin

Die Parade
von Brücken (2015)

Solutions

Exercises of Chapter 1

1.1 George Dantzig

> In 1939 Dantzig enrolled in the Ph.D. program of the Berkeley Mathematics Department where Neyman's professorship was located. Dantzig took only two courses from Neyman, but in one of them he had a remarkable experience that was to become a famous legend. Arriving late to one of Neyman's classes, Dantzig saw two problems written on the blackboard and mistook them for a homework assignment. He found them more challenging than usual, but managed to solve them and submitted them directly to Neyman. As it turned out, these problems were actually two open questions in the theory of mathematical statistics. Dantzig's 57-page Ph.D. thesis [*I Complete Form of the Neyman-Pearson Lemma; II On the Non-Existence of Tests of "Student's" Hypothesis Having Power Functions Independent of Sigma*, 1943] was composed of his solutions to these two problems. – Cottle et al. (2007)

1.2 DDL1x

DDL1x was an abbreviated way to reference to the forthcoming book. In the alphabetic order of the author's name: Desaulniers, Desrosiers, and Lübbecke. This is followed by the *expected* year of publication: 2015, or 2016, ... and so on.

Because the project was a bit stalling, Jean Bertrand joined the team a few months after completing his doctoral dissertation. Then, the *Branch-and-Price* team structure took place: a master coordinator (JD) and three text generators (MEL, GD, and JBG).

1.3 Farkas' lemma

Following the proposed *hint*, let $\mathbf{y}$ be the vector of artificial variables in the *Phase I* of the primal simplex algorithm. The primal-dual pair reads as follows:

$$\begin{aligned} z^\star_{LP} = \min \quad & \mathbf{1}^\mathsf{T}\mathbf{y} \\ \text{s.t.} \quad & \mathbf{A}\mathbf{x} - \mathbf{s} + \mathbf{y} = \mathbf{b} \quad [\boldsymbol{\pi}] \\ & \mathbf{x},\ \mathbf{s},\ \mathbf{y} \geq \mathbf{0} \end{aligned} \qquad \Bigg| \qquad \begin{aligned} z^\star_{LD} = \max \quad & \mathbf{b}^\mathsf{T}\boldsymbol{\pi} \\ \text{s.t.} \quad & \mathbf{A}^\mathsf{T}\boldsymbol{\pi} \leq \mathbf{0} \quad [\mathbf{x}] \\ & -\boldsymbol{\pi} \leq \mathbf{0} \quad [\mathbf{s}] \\ & \boldsymbol{\pi} \leq \mathbf{1} \quad [\mathbf{y}] \end{aligned}$$

The *LP* formulation is feasible ($\mathbf{x} = \mathbf{0}, \mathbf{s} = \mathbf{0}, \mathbf{y} = \mathbf{b}$) and the optimal objective value is finite, either $z^\star_{LP} = 0$ or $z^\star_{LP} > 0$. At the termination of the primal simplex algorithm, all the dual constraints are satisfied, that is, $\mathbf{A}^\mathsf{T}\boldsymbol{\pi} \leq \mathbf{0}$, $\mathbf{0} \leq \boldsymbol{\pi} \leq \mathbf{1}$. By *strong duality*, $z^\star_{LP} = z^\star_{LD}$ and we therefore have:

$$\begin{aligned} \mathbf{1}^\mathsf{T}\mathbf{y} = \boldsymbol{\pi}^\mathsf{T}\mathbf{b} = 0 \quad &\Leftrightarrow \quad LP \text{ is feasible} \quad &&\Leftrightarrow \quad \mathbf{A}\mathbf{x} \geq \mathbf{b},\ \mathbf{x} \geq \mathbf{0} \\ \mathbf{1}^\mathsf{T}\mathbf{y} = \boldsymbol{\pi}^\mathsf{T}\mathbf{b} > 0 \quad &\Leftrightarrow \quad LP \text{ is infeasible} \quad &&\Leftrightarrow \quad \mathbf{A}^\mathsf{T}\boldsymbol{\pi} \leq \mathbf{0},\ \boldsymbol{\pi} \geq \mathbf{0},\ \boldsymbol{\pi}^\mathsf{T}\mathbf{b} > 0. \end{aligned}$$

J. Desrosiers et al., *Branch-and-Price*,
https://doi.org/10.1007/978-3-031-96917-1

1.4 Infeasible primal-dual pair of linear programs

(a) Let $c \in \mathbb{R}$ and $a, b > 0$ such that the following *LP* is clearly infeasible:

$$\begin{array}{rl|rl} z^\star_{LP} = \min cx & & z^\star_{LD} = \max b\pi & \\ \text{s.t.} -ax \geq b & [\pi] & \text{s.t.} -a\pi \leq c & [x] \\ x \geq 0 & & \pi \geq 0. & \end{array}$$

In the dual formulation, $-a\pi \leq 0$ by construction such that the *LD* is feasible for any value of c. Indeed, if $c \geq 0$, then the inequality is trivially fulfilled. Otherwise, there exists some $\pi > 0$ such that $-a\pi = c$. One may then incorporate an arbitrary number of constraints in the primal for which all dual variables at value 0 would produce a feasible dual with respect to π.

(b) The key to conceiving a simple example is to realize that a symmetric matrix $\mathbf{A}$ produces an equivalent left-hand side system in both formulations. The task is then accomplished by using appropriate $\mathbf{c}$ and $\mathbf{b}$ signed values to have conflicting constraints in both formulations.

$$\begin{array}{rrrll|rrrll} z^\star_{LP} = \min & -x_1 - & x_2 & & & z^\star_{LD} = \max & \pi_1 + & \pi_2 & & \\ \text{s.t.} & -x_1 & & \geq 1 & [\pi_1] & \text{s.t.} & -\pi_1 & & \leq -1 & [x_1] \\ & & x_2 & \geq 1 & [\pi_2] & & & \pi_2 & \leq -1 & [x_2] \\ & x_1, & x_2 & \geq 0 & & & \pi_1, & \pi_2 & \geq 0. & \end{array}$$

1.5 Direction in the primal simplex algorithm

Without loss of generality, let B be an index-set from 1 to m and indices of N' be numbered from $m+2$ up to n.

$$\mathbf{A}\begin{bmatrix} -\bar{\mathbf{a}}_\ell \\ 1 \\ \mathbf{0} \end{bmatrix} = \begin{bmatrix} \mathbf{A}_B & \mathbf{a}_\ell & \mathbf{A}_{N'} \end{bmatrix} \begin{bmatrix} -\mathbf{A}_B^{-1}\mathbf{a}_\ell \\ 1 \\ \mathbf{0} \end{bmatrix} = \left(\mathbf{A}_B(-\mathbf{A}_B^{-1}\mathbf{a}_\ell) + \mathbf{a}_\ell + \mathbf{A}_{N'}\mathbf{0}\right) = \mathbf{0}.$$

1.6 Sensitivity range for an increase of the lower bound

The non-basic variable x_ℓ is simply considered as an entering variable and its maximum value $\Delta\ell_j$ is computed using the minimum ratio rule:

$$\Delta x_\ell = \min_{i \in \{1,\dots,m\} | \bar{a}_{i\ell} > 0} \frac{\bar{b}_i}{\bar{a}_{i\ell}}.$$

1.7 Lost in translation?

Subtracting the vector $\mathbf{s}$ of surplus variables to the constraints of the *LP* (1.2) and taking its dual recovers the signed dual variables $\boldsymbol{\pi} \geq \mathbf{0}$:

$$\begin{array}{rrrll|rll} z^\star_{LP} = \min & \mathbf{c}^\mathsf{T}\mathbf{x} & + \mathbf{0}^\mathsf{T}\mathbf{s} & & & z^\star_{LD} = \max & \mathbf{b}^\mathsf{T}\boldsymbol{\pi} & \\ \text{s.t.} & \mathbf{A}\mathbf{x} & - \quad \mathbf{s} & = \mathbf{b} & [\boldsymbol{\pi}] & \text{s.t.} & \mathbf{A}^\mathsf{T}\boldsymbol{\pi} \leq \mathbf{c} & [\mathbf{x}] \\ & \mathbf{x}, & \mathbf{s} & \geq \mathbf{0} & & & -\boldsymbol{\pi} \leq \mathbf{0} & [\mathbf{s}]. \end{array}$$

1.8 Primal simplex and strong duality

Introducing the vector $\mathbf{s}$ of surplus variables in the constraints of the *LP* (1.2), the primal-dual pair writes as

$$
\begin{array}{llll|llll}
z^\star_{LP} = \min & \mathbf{c}^\mathsf{T}\mathbf{x} & + \ \mathbf{0}^\mathsf{T}\mathbf{s} & & z^\star_{LD} = \max & \mathbf{b}^\mathsf{T}\boldsymbol{\pi} & \\
\text{s.t.} & \mathbf{A}\mathbf{x} & - \ \mathbf{I}_m\mathbf{s} = \mathbf{b} & [\boldsymbol{\pi}] & \text{s.t.} & \mathbf{A}^\mathsf{T}\boldsymbol{\pi} \le \mathbf{c} & [\mathbf{x}] \\
& \mathbf{x}, & \mathbf{s} \ge \mathbf{0} & & & -\mathbf{I}_m^\mathsf{T}\boldsymbol{\pi} \le \mathbf{0} & [\mathbf{s}].
\end{array}
$$

Assume the primal formulation *LP* is feasible and bounded. Then the simplex algorithm gives optimal primal solutions as $\mathbf{x}^\star$ and $\mathbf{s}^\star$ toghether with a dual one $\boldsymbol{\pi}^\star$. Let $\bar{\mathbf{c}}_\mathbf{x}^\mathsf{T} = \mathbf{c}^\mathsf{T} - \boldsymbol{\pi}^{\star\mathsf{T}}\mathbf{A}$ and $\bar{\mathbf{c}}_\mathbf{s}^\mathsf{T} = \mathbf{0}^\mathsf{T} + \boldsymbol{\pi}^{\star\mathsf{T}}\mathbf{I}_m$ denote the *non-negative* reduced cost vectors at the stopping rule of the algorithm; this ensure the feasibility of $\boldsymbol{\pi}^\star$. Then

$$z^\star_{LP} = \mathbf{c}^\mathsf{T}\mathbf{x}^\star + \mathbf{0}^\mathsf{T}\mathbf{s}^\star = \boldsymbol{\pi}^{\star\mathsf{T}}(\mathbf{A}\mathbf{x}^\star - \mathbf{I}_m\mathbf{s}^\star) + \bar{\mathbf{c}}_\mathbf{x}^\mathsf{T}\mathbf{x}^\star + \bar{\mathbf{c}}_\mathbf{s}^\mathsf{T}\mathbf{s}^\star = \boldsymbol{\pi}^{\star\mathsf{T}}\mathbf{b} + \bar{\mathbf{c}}_\mathbf{x}^\mathsf{T}\mathbf{x}^\star + \bar{\mathbf{c}}_\mathbf{s}^\mathsf{T}\mathbf{s}^\star.$$

It remains to show that $\bar{\mathbf{c}}_\mathbf{x}^\mathsf{T}\mathbf{x}^\star + \bar{\mathbf{c}}_\mathbf{s}^\mathsf{T}\mathbf{s}^\star = 0$. This is true because each term only comprises a set of basic variables with 0-valued reduced cost and a complementary set of non-basic variables taking value 0.
The argument in terms of the dual problem being feasible in Proposition 1.6 (*Strong duality*) is analogous.

1.9 Lower bound on the optimum

Let $\mathbf{x}^\star$ denote an optimal solution to (1.2) and the reduced cost vector be defined as usual as $\bar{\mathbf{c}}^\mathsf{T} = \mathbf{c}^\mathsf{T} - \boldsymbol{\pi}^\mathsf{T}\mathbf{A}$, hence $\bar{\mathbf{c}}^\mathsf{T} \ge \bar{c}(\boldsymbol{\pi})\mathbf{1}^\mathsf{T}$. Then we have

$$z^\star_{LP} = \mathbf{c}^\mathsf{T}\mathbf{x}^\star = (\boldsymbol{\pi}^\mathsf{T}\mathbf{A} + \bar{\mathbf{c}}^\mathsf{T})\mathbf{x}^\star = \boldsymbol{\pi}^\mathsf{T}\mathbf{A}\mathbf{x}^\star + \bar{\mathbf{c}}^\mathsf{T}\mathbf{x}^\star \ \ge \ \boldsymbol{\pi}^\mathsf{T}\mathbf{b} \ + \ \kappa\bar{c}(\boldsymbol{\pi}), \tag{8.1}$$

because

- $\boldsymbol{\pi}^\mathsf{T}\mathbf{A}\mathbf{x}^\star \ge \boldsymbol{\pi}^\mathsf{T}\mathbf{b}$ by feasibility of $\mathbf{x}^\star$ in (1.2),
- $\bar{\mathbf{c}}^\mathsf{T}\mathbf{x}^\star \ge \bar{c}(\boldsymbol{\pi})\mathbf{1}^\mathsf{T}\mathbf{x}^\star \ge \kappa\bar{c}(\boldsymbol{\pi})$ as $\bar{c}(\boldsymbol{\pi}) \le 0$ and $\mathbf{1}^\mathsf{T}\mathbf{x}^\star \le \kappa$.

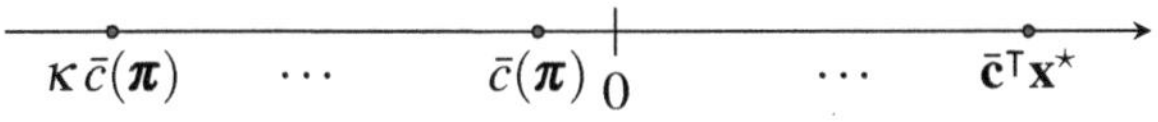

1.10 Optimizing π for maximizing the smallest reduced cost

(a) Let $\mu = \min_{j=1,\dots,n} c_j - \boldsymbol{\pi}^\mathsf{T}\mathbf{a}_j$ denote the smallest reduced cost, μ being an unrestricted variable whereas $\boldsymbol{\pi} \in \mathbb{R}^m$ is also unknown. Then

$$
\begin{aligned}
\max_{\boldsymbol{\pi}} \min_{j=1,\dots,n} \ c_j - \boldsymbol{\pi}^\mathsf{T}\mathbf{a}_j = \max \quad & \mu \\
\text{s.t.} \quad & \mu \le c_j - \boldsymbol{\pi}^\mathsf{T}\mathbf{a}_j \quad [y_j] \quad \forall j \in \{1,\dots,n\}.
\end{aligned}
$$

Transferring the π-variables on the left-hand side, we obtain

$$
\begin{aligned}
\max_{\boldsymbol{\pi}} \min_{j=1,\dots,n} \ c_j - \boldsymbol{\pi}^\mathsf{T}\mathbf{a}_j = \max \quad & \mu \\
\text{s.t.} \quad & \mu + \boldsymbol{\pi}^\mathsf{T}\mathbf{a}_j \le c_j \quad [y_j] \quad \forall j \in \{1,\dots,n\}.
\end{aligned}
$$

(b) In the above formulation (a), $y_j \geq 0$, $j \in \{1, \dots, n\}$, acts as a dual variable so the linear programming dual reads as

$$\max_{\boldsymbol{\pi}} \min_{j=1,\dots,n} c_j - \boldsymbol{\pi}^\mathsf{T}\mathbf{a}_j = \min \sum_{j\in\{1,\dots,n\}} c_j y_j$$
$$\text{s.t.} \sum_{j\in\{1,\dots,n\}} y_j = 1 \quad [\mu]$$
$$\sum_{j\in\{1,\dots,n\}} \mathbf{a}_j y_j = 0 \quad [\boldsymbol{\pi}]$$
$$y_j \geq 0 \qquad \forall j \in \{1, \dots, n\}.$$

1.11 Integrality property and algorithm selection bias

The answer is yes. From the equivalence between Proposition 1.9 and Definition 1.25, we know that every extreme point of the feasible region is integer. Since the primal simplex algorithm restricts the search to such solutions, it terminates with an optimal integer solution.

1.12 Unbalanced capacitated minimum cost flow problem

Recall the formulation (1.45) rewritten as

$$\min \sum_{(i,j)\in A} c_{ij} x_{ij}$$
$$\text{s.t.} \sum_{j:(i,j)\in A} x_{ij} - \sum_{j:(j,i)\in A} x_{ji} = b_i \quad \forall i \in S$$
$$\sum_{j:(i,j)\in A} x_{ij} - \sum_{j:(j,i)\in A} x_{ji} = b_i \quad \forall i \in D$$
$$\sum_{j:(i,j)\in A} x_{ij} - \sum_{j:(j,i)\in A} x_{ji} = 0 \quad \forall i \in N \setminus \{S \cup D\}$$
$$0 \leq x_{ij} \leq u_{ij} \quad \forall (i,j) \in A.$$

Take a look at the accompanying figure below, where the node 0 added to G is connected to the set $S = \{s_1, \dots, s_5\}$ of supply nodes by incoming arcs and to the set $D = \{d_1, \dots, d_4\}$ of demand nodes by outgoing arcs. The net flow at node 0 mimics flow conservation as $-\sum_{i\in D} b_i - \sum_{i\in S} b_i = b_0 \in \mathbb{Z}$.

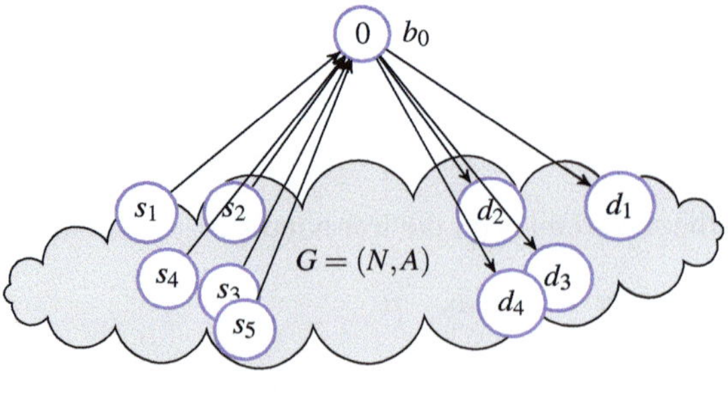

Such a network results from

- adding an artificial variable $x_{i0} \geq 0$ for each supply constraint at $i \in S$,
- subtracting an artificial variable $x_{0i} \geq 0$ for each demand constraint at $i \in D$:

$$\begin{array}{llll} x_{i0} + \sum_{j:(i,j)\in A} x_{ij} - \sum_{j:(j,i)\in A} x_{ji} & & = b_i & \forall i \in S \\ \sum_{j:(i,j)\in A} x_{ij} - \sum_{j:(j,i)\in A} x_{ji} & - x_{0i} & = b_i & \forall i \in D \\ \sum_{j:(i,j)\in A} x_{ij} - \sum_{j:(j,i)\in A} x_{ji} & & = 0 & \forall i \in N \setminus \{S \cup D\} \end{array}$$

Setting $x_{i0} = b_i$, $\forall i \in S$, and $x_{0i} = -b_i > 0$, $\forall i \in D$, provides a *feasible* solution. The flow conservation constraint at node 0 is redundant but can be retrieved by adding the constraints for all $i \in N$ and taking the negative of this sum. Because each x_{ij}-variable appears twice for $(i,j) \in A$, once with a $+1$ coefficient, once with a -1 coefficient, it remains only $-(\sum_{i\in S} x_{i0} - \sum_{i\in D} x_{0i}) = -(\sum_{i\in D} b_i + \sum_{i\in S} b_i) = b_0$. Therefore, $\sum_{i\in D} b_i + \sum_{i\in S} b_i + b_0 = 0$ and the new network is also *balanced.*

A set of possible [cost; upper bound] values appears next:

$$\begin{array}{lll} [M;+b_i], & \text{for all arcs } (i,0), & i \in S; \\ [M;-b_i], & \text{for all arcs } (0,j), & j \in D. \end{array}$$

The cost of the artificial variables is set to big-M. Although there are usually no upper bounds on such variables, the proposed ones are sufficient for providing an initial solution: for each $i \in S$, $x_{i0} = b_i$, for each $i \in D$, $x_{0j} = -b_i > 0$, whereas $x_{ij} = 0$, $\forall (i,j) \in A$. In any optimal solution, the units available from $i \in S$ that do not flow through G go to the extra node on arc $(i,0)$ and there are at most b_i such units. All units reaching node 0 are redirected to the demand nodes, with at most $-b_j$ requested units to node $j \in D$ on arc $(0,j)$.

Special case. If there are no upper bounds on the arcs of A in G, as in the classical formulation of the transportation problem, it is known up-front whether the total demand *or* total supply is larger, so we need only one of the two sets of arcs. Moreover, a cost value that better reflects reality can be assigned to an arc, for example, the penalty cost c_{0j} incurred for a back-order delivery at a demand node $j \in D$, or the storage cost c_{i0} at a supply node $i \in S$. If zero-costs are assigned to the additional variables, these are in fact simply considered as slack or suplus variables.

1.13 Totally unimodular matrices with consecutive ones

(a) If $\mathbf{A}$ contains a column of 0's, then the determinant of $\mathbf{A}$ is 0. Otherwise, let $\mathbf{C}$ be the matrix obtained by subtracting from each row of $\mathbf{A}$ except its first, its previous row. All columns of $\mathbf{C}$ contains a single 1 if $a_{mj} = 1$, or a single 1 and a single -1 if $a_{mj} = 0$. We can easily check that conditions (i)–(iii) are satisfied for $\mathbf{C}$ by setting $M_1 = \{1, \ldots, m\}$ and $M_2 = \emptyset$. Consequently, $\mathbf{C}$ is totally unimodular and its determinant is, thus, equal to 0, $+1$, or -1. Given that $\mathbf{C}$ is obtained from $\mathbf{A}$ using determinant-invariant line operations, the determinant of $\mathbf{A}$ is also 0, $+1$, or -1.

(b) Let $\bar{\mathbf{A}}$ be any square submatrix of $\mathbf{A}$. We observe that $\bar{\mathbf{A}}$ has also the consecutive 1's property. Hence, using the result in (a), the determinant of $\bar{\mathbf{A}}$ is 0, +1, or −1. Consequently, $\mathbf{A}$ is totally unimodular.

1.14 Variable fixing at upper bound

If x_j is at its upper bound in $\mathbf{x}^\star_{LP}$, then its reduced cost $\bar{c}_j \leq 0$. By the sensitivity analysis for x_j at its upper bound u_j, we find in (1.42) that $z^\star_{LP}(\Delta u_j) = z^\star_{LP} + \bar{c}_j\, \Delta u_j$. Reducing the value of x_j by 1 would, thus, induce a cost increase of at least $-\bar{c}_j$. Denoting $LB = z^\star_{LP}$, we deduce the following sufficient condition:

if $-\bar{c}_j > UB - LB$ (or equivalently, $\bar{c}_j < LB - UB$), x_j can be fixed at u_j.

1.15 Formulations for the minimum-weight perfect matching problem

(a) By observation, we can easily find that
i) $x_{12} = x_{35} = x_{46} = 1$ and $x_e = 0$ for all other edges $e \in E$ is the unique optimal integer solution of cost $z^\star_{ILP} = 102$;
ii) $x_{12} = x_{13} = x_{23} = x_{45} = x_{46} = x_{56} = 0.5$ and $x_e = 0$ for all other edges $e \in E$ is the unique optimal linear relaxation solution of cost $z^\star_{LP} = 3$;
iii) the relative integrality gap is $(z^\star_{ILP} - z^\star_{LP})/|z^\star_{ILP}| = (102-3)/102 = 97\%$.

(b) Proof by contradiction. Consider a feasible integer solution $\hat{\mathbf{x}}$ to (1.57) and a subset of nodes S of odd cardinality. Assume that the corresponding blossom inequality (1.58) is violated by $\hat{\mathbf{x}}$, i.e.,

$$\sum_{e \in E(S)} \hat{x}_e > \frac{1}{2}(|S| - 1).$$

Given that $\hat{\mathbf{x}}$ is integer and $|S|$ is odd, it ensues that

$$\sum_{e \in E(S)} \hat{x}_e \geq \frac{1}{2}(|S| + 1).$$

Because each edge $e \in E(S)$ is incident to two nodes in S, the selected edges in $E(S)$ are, thus, incident to a total of at least $|S| + 1$ nodes. Consequently, S contains a node that is incident to at least two selected edges, which contradicts the feasibility of $\hat{\mathbf{x}}$.

(c) Let $S_1 = \{1, 2, 3\}$. For the optimal linear relaxation solution of (a), we have

$$\sum_{e \in E(S_1)} x_e = x_{12} + x_{13} + x_{23} = 1.5 \not\leq \frac{1}{2}(|S_1| - 1) = 1.$$

Hence, the corresponding blossom inequality is violated by this solution. Similarly, the blossom inequality for $S_2 = \{4, 5, 6\}$ is also violated by it.

(d) To show that $\mathcal{P}_2$ is stronger than $\mathcal{P}_1$, we need to prove that $\mathcal{P}_2 \subset \mathcal{P}_1$. Given that $\mathcal{P}_2$ is obtained from $\mathcal{P}_1$ by adding the blossom inequalities (1.58), we get that $\mathcal{P}_2 \subseteq \mathcal{P}_1$. The previous example has shown that, for certain problem instances, there are solutions in $\mathcal{P}_1$ that do not belong to $\mathcal{P}_2$. Thus, $\mathcal{P}_2 \subset \mathcal{P}_1$.

(e) Let y_i, $i \in N$, be a binary variable equal to 1 if $i \in S$ and 0 otherwise. Let w_e, $e \in E$, be a binary variable equal to 1 if $e \in E(S)$ and 0 otherwise. Finally, let $k \in \mathbb{Z}_+$ be an integer variable such that $|S| = 2k+1$ is odd. The separation *ILP* is given by

$$
\begin{aligned}
z^\star_{sep} = \max \quad & \sum_{e \in E} x^\star_e w_e - k \\
\text{s.t.} \quad & w_e \le y_{i_e} && \forall e \in E \\
& w_e \le y_{j_e} && \forall e \in E \\
& \sum_{i \in N} y_i = 2k+1 \\
& y_i \in \{0,1\} && \forall i \in N \\
& w_e \in \{0,1\} && \forall e \in E \\
& k \in \mathbb{Z}_+
\end{aligned}
$$

where i_e and j_e denote the end nodes of edge e. The first two constraint sets ensure that an edge e belong to $E(S)$ only if its end nodes i_e and j_e are part of S. The third constraint imposes the odd cardinality condition.

If $z^\star_{sep} > 0$, a violated inequality is obtained for the subset S composed of the nodes $i \in N$ such that $y_i = 1$ in the computed solution. Note that we can restrict this *ILP* to the edges for which $0 < x^\star_e < 1$ and the nodes with at least two incident edges in the linear relaxation solution $\mathbf{x}^\star_{LP}$.

(f) Let S be a node subset of odd cardinality. Summing the main constraints in model (1.57) associated with the nodes $i \in S$, we obtain

$$
\sum_{i \in S} \sum_{e \in \delta(i)} x_e = |S| \qquad \text{or, equivalently,} \qquad \sum_{e \in \delta(S)} x_e + 2 \sum_{e \in E(S)} x_e = |S|
$$

because each edge in $\delta(S)$ is incident to one node in S and each edge in $E(S)$ is incident to two nodes in S. Using the latter equation, we start from inequalities (1.59) to obtain inequalities (1.58):

$$
\begin{aligned}
& \sum_{e \in \delta(S)} x_e \ge 1 \\
\Leftrightarrow \quad & |S| - 2 \sum_{e \in E(S)} x_e \ge 1 \\
\Leftrightarrow \quad & \sum_{e \in E(S)} x_e \le \frac{1}{2}(|S| - 1).
\end{aligned}
$$

Separating blossom inequalities of the form (1.59) corresponds to finding a cut set $\delta(S)$ in G of minimum weight, where $|S|$ is odd and the weight of an edge e is equal to $x^\star_e$.

Exercises of Chapter 2

2.1 Ralph Gomory

Gomory is one of those people with a fascinating biography. Here is a potpourri of excerpts from various sources which probably does not do justice to his legacy.

> Gomory originally studied physics at Williams College but was drawn to mathematics there, and wrote a Princeton Ph.D. thesis on nonlinear differential equations. After completing his Ph.D. he joined the U.S. Navy as an officer and was assigned to the Physics Branch of Office of Naval Research in DC. There he discovered the Operations Research Group (ORG) and learned linear programming from Alan Goldman. – INFORMS
>
> He is a mathematical engineer in the true sense of the word: mathematician and engineer. Accordingly he is member of both the National Academy of Sciences and the National Academy of Engineering. His most influential result is the cutting plane method for integer linear programming; it bares his name. – Technische Universiteit Eindhoven
>
> After his career in the corporate world, Gomory became the president of the Alfred P. Sloan Foundation, where he oversaw programs dedicated to broadening public understanding in three key areas: the economic importance of science and research; the effects of globalization on the United States; and the role of technology in education.
>
> Gomory has written extensively on the nature of technology development, industrial competitiveness, models of international trade, social issues under current economics and law, and the function of the corporation in a globalizing world. – Wikipedia

2.2 Finiteness of $\mathcal{X}$ and $z^\star_{MP}$

(a) If an infinite number of constraints is required to express the domain of the dual of (2.1), the said domain is not a polyhedron, hence the dual formulation is not a linear program. Since we assume that the primal *is* a linear program, it must be expressed with $|\mathcal{X}| < \infty$ variables.

(b) $\sum_{\mathbf{x}\in\mathcal{X}} \lambda_{\mathbf{x}} \leq \kappa$, where κ is a positive and finite number. Alternatively, with an upper bound $u_{\mathbf{x}}$ on every variable: $\lambda_{\mathbf{x}} \leq u_{\mathbf{x}}, \forall x \in \mathcal{X}$.

2.3 Cutting stock problem with rolls of different width

(a) Assume an optimal solution to the *MP* comprises a pattern $\mathbf{x}^{k_2}$ such that

$$\sum_{i=1}^{m} w_i x_i^{k_2} \leq W^{k_1} < W^{k_2}.$$

The cost or loss $(W^{k_2} - \sum_{i=1}^{m} w_i x_i^{k_2})$ of such a pattern is computed relatively to W^{k_2} while it would be smaller if computed relatively to W^{k_1}. Hence a contradiction on the optimality of the solution and constraint $\sum_{i=1}^{m} w_i x_i^{k_2} \geq W^{k_1} + 1$ can be added to the ISP^{k_2}.

(b) The *IMP* becomes

$$\begin{aligned} z^\star_{IMP} = \min \quad & \sum_{k\in K} \sum_{\mathbf{x}^k\in\mathcal{X}^k} c_{\mathbf{x}^k} \lambda_{\mathbf{x}^k} \\ \text{s.t.} \quad & \sum_{k\in K} \sum_{\mathbf{x}^k\in\mathcal{X}^k} a_{i\mathbf{x}^k} \lambda_{\mathbf{x}^k} = b_i \qquad [\pi_i] \qquad \forall i \in \{1,\ldots,m\} \end{aligned}$$

$$\sum_{\mathbf{x}^k \in \mathcal{X}^k} d_{\mathbf{x}^k} \lambda_{\mathbf{x}^k} \le n^k \qquad [\pi^k] \qquad \forall k \in K$$
$$\lambda_{\mathbf{x}^k} \in \mathbb{Z}_+ \qquad \forall k \in K, \mathbf{x}^k \in \mathcal{X}^k,$$

where $\mathcal{X}^k$ denotes the set of cutting patterns for the roll width W^k and $d_{\mathbf{x}^k}$ is a parameter that takes value 1 whenever pattern $\mathbf{x}^k$ uses a roll of width W^k, 0 otherwise. We need a new encoding function $d_{\mathbf{x}^k} = x_0^k$ and modify the objective function of the *ISP*^{k} (2.34) as

$$\begin{aligned}
\bar{c}^k(\boldsymbol{\pi}, \pi^k) = \min \quad & c_{\mathbf{x}^k} - \sum_{i=1}^m \pi_i a_{i\mathbf{x}^k} - \pi^k d_{\mathbf{x}^k} \\
\text{s.t.} \quad & \sum_{i=1}^m w_i x_i^k \le W^k x_0^k \\
& x_i^k \le b_i && \forall i \in \{1, \ldots, m\} \\
& x_0^k \in \{0, 1\} \\
& x_i^k \in \mathbb{Z}_+ && \forall i \in \{1, \ldots, m\} \\
& c_{\mathbf{x}^k} = W^k x_0^k - \sum_{i=1}^m w_i x_i^k \\
& a_{i\mathbf{x}^k} = x_i^k && \forall i \in \{1, \ldots, m\} \\
& d_{\mathbf{x}^k} = x_0^k.
\end{aligned}$$

2.4 Cutting stock problem with rolls of different width: a single subproblem

(a) The *ISP* reads as

$$\begin{aligned}
\bar{c}(\boldsymbol{\pi}) = \min \quad & c_{\mathbf{x}} - \sum_{i=1}^m \pi_i a_{i\mathbf{x}} \\
\text{s.t.} \quad & \sum_{i=1}^m w_i x_i^k \le W^k && \forall k \in K \\
& x_i^k \in \mathbb{Z}_+ && \forall k \in K,\ i \in \{1, \ldots, m\} \\
& c_{\mathbf{x}} = \sum_{k \in K} W^k - \sum_{i=1}^m w_i x_i^k \\
& \mathbf{a}_{i\mathbf{x}} = \sum_{k \in K} x_i^k, && \forall i \in \{1, \ldots, m\}.
\end{aligned}$$

This is equivalent to

$$\begin{aligned}
\sum_{k \in K} W^k + \max \quad & \sum_{k \in K} \sum_{i=1}^m (w_i + \pi_i) x_i^k \\
\text{s.t.} \quad & \sum_{i=1}^m w_i x_i^k \le W^k \quad \forall k \in K \\
& x_i^k \in \mathbb{Z}_+ \quad \forall k \in K,\ i \in \{1, \ldots, m\}.
\end{aligned}$$

(b) We see the competitive pricing problems between the usage of different roll widths in a more direct angle. In this "multiple knapsack problem," units of item i are favored/deterred by the same value $w_i + \pi_i$ regardless of which knapsack we pack them in. In fact, putting conflicting widths W^k aside, we see that we are likely to reproduce packing choices in every knapsack. We thus end up with a column $\lambda_{\mathbf{x}}$ using more or less the same pattern $|K|$ times. This is exactly the warning we give in Note 2.13.
We have a much more expensive pricing problem which produces basically the same information as the simple knapsack problem.
Amongst other inconveniences of this formulation, the interpretation of column $\mathbf{a}_{\mathbf{x}}$ is not very intuitive. It corresponds to a merged list of cutting patterns, one for each roll width. What is the interpretation of a λ-variable with a fractional value then?

2.5 Farley's lower bound

(a) We show that for any $\boldsymbol{\pi} \geq \mathbf{0}$, the vector $\bar{\boldsymbol{\pi}} = \frac{\boldsymbol{\pi}}{1-\bar{c}(\boldsymbol{\pi})}$ is dual feasible in Corollary 2.2, that is, $\bar{\boldsymbol{\pi}} \geq \mathbf{0}$ and $1 - \bar{\boldsymbol{\pi}}^\intercal \mathbf{a}_{\mathbf{x}} \geq 0, \forall \mathbf{x} \in \mathcal{X}$:

$$1 - \left(\frac{\boldsymbol{\pi}}{1-\bar{c}(\boldsymbol{\pi})}\right)^\intercal \mathbf{a}_{\mathbf{x}} \geq 0 \Leftrightarrow 1 - \bar{c}(\boldsymbol{\pi}) - \boldsymbol{\pi}^\intercal \mathbf{a}_{\mathbf{x}} \geq 0 \Leftrightarrow \bar{c}(\boldsymbol{\pi}) \leq 1 - \boldsymbol{\pi}^\intercal \mathbf{a}_{\mathbf{x}},$$

where the first equivalence is true because $\bar{c}(\boldsymbol{\pi}) \leq 0$ so the denominator is always positive (non-negativity of $\bar{\boldsymbol{\pi}}$ holds as well) and the second equivalence yields an inequality that holds by definition of the minimum reduced cost $\bar{c}(\boldsymbol{\pi})$.

(b) We compare $\frac{\boldsymbol{\pi}^\intercal \mathbf{b}}{1-\bar{c}(\boldsymbol{\pi})} \leq z^\star_{MP}$ to $l(\boldsymbol{\pi}) \cdot \boldsymbol{\pi}^\intercal \mathbf{b} \leq z^\star_{MP}$ and show that $\frac{1}{1-\bar{c}(\boldsymbol{\pi})} = l(\boldsymbol{\pi})$ given $c_{\mathbf{x}} = 1, \forall \mathbf{x} \in \mathcal{X}$. Starting from the right-hand side, we have

$$l(\boldsymbol{\pi}) = \min_{\mathbf{x} \in \mathcal{X} | \boldsymbol{\pi}^\intercal \mathbf{a}_{\mathbf{x}} > 0} \frac{1}{\boldsymbol{\pi}^\intercal \mathbf{a}_{\mathbf{x}}} = \frac{1}{\max_{\mathbf{x} \in \mathcal{X}} \boldsymbol{\pi}^\intercal \mathbf{a}_{\mathbf{x}}}$$

which is equal to the left-hand side since the denominators are equal, i.e.,

$$\max_{\mathbf{x} \in \mathcal{X}} \boldsymbol{\pi}^\intercal \mathbf{a}_{\mathbf{x}} = -\min_{\mathbf{x} \in \mathcal{X}} (-\boldsymbol{\pi}^\intercal \mathbf{a}_{\mathbf{x}}) = 1 - \min_{\mathbf{x} \in \mathcal{X}} (1 - \boldsymbol{\pi}^\intercal \mathbf{a}_{\mathbf{x}}) = 1 - \bar{c}(\boldsymbol{\pi}).$$

(c) We show that their difference is positive under the given assumptions $\bar{c}(\boldsymbol{\pi}) < 0$ and $z_{RMP} = \boldsymbol{\pi}^\intercal \mathbf{b}$. The trick is to realize that the objective function implies that $\kappa = z^\star_{MP} > 0$ by definition (2.10). At any point during the column generation process, the tightest *known value* for κ is $z_{RMP} \geq z^\star_{MP}$ such that a lower bound can be computed as

$$z_{RMP} + z_{RMP} \cdot \bar{c}(\boldsymbol{\pi}) = z_{RMP}(1 + \bar{c}(\boldsymbol{\pi})).$$

The desired result follows from

$$\frac{z_{RMP}}{1-\bar{c}(\boldsymbol{\pi})} - z_{RMP}(1 + \bar{c}(\boldsymbol{\pi})) = z_{RMP} \left(\frac{\bar{c}(\boldsymbol{\pi})^2}{1-\bar{c}(\boldsymbol{\pi})}\right) > 0.$$

2.6 Sufficient optimality condition

Figure 2.4 provides such evidence. We found the optimal objective value at iteration 736 and produce a sequence of $952 - 736 = 216$ negative reduced cost columns with no impact thereafter (degenerate pivots). Each new column does however modify the dual values until $\bar{c}(\boldsymbol{\pi}) \geq 0$ is fulfilled.

2.7 Single depot vehicle scheduling problem: adjusted arc costs

The adjusted arc costs are defined on network $G = (V, A)$, with the node set $V = N \cup \{o, d\}$ and arc set $A = I \cup (\{o\} \times N) \cup (N \times \{d\}) \cup \{(o, d)\}$, see the figure below.

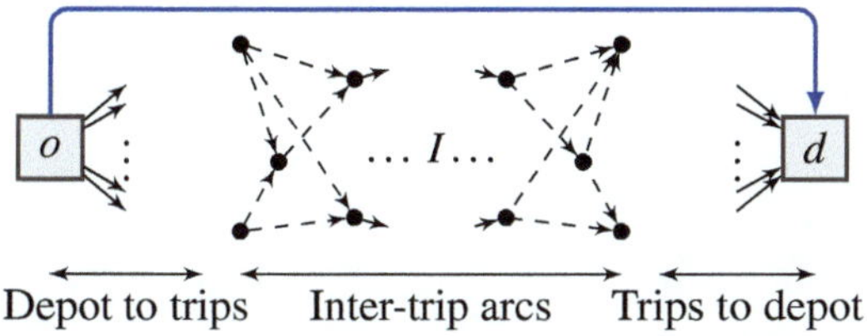

$$\begin{aligned}\sum_{(i,j)\in A} \tilde{c}_{ij}x_{ij} &= \sum_{(i,j)\in A} c_{ij}x_{ij} - \sum_{i\in N}\pi_i\Big(\sum_{j:(i,j)\in A} x_{ij}\Big) - \pi_o\Big(\sum_{j:(o,j)\in A} x_{oj}\Big)\\ &= \sum_{i\in N}\sum_{j:(i,j)\in A} c_{ij}x_{ij} - \sum_{i\in N}\sum_{j:(i,j)\in A}\pi_i x_{ij} + \sum_{j:(o,j)\in A} c_{oj}x_{oj} - \sum_{j:(o,j)\in A}\pi_o x_{oj}\\ &= \sum_{i\in N}\sum_{j:(i,j)\in A}(c_{ij}-\pi_i)x_{ij} + \sum_{j:(o,j)\in A}(c_{oj}-\pi_o)x_{oj}\end{aligned}$$

Hence $\tilde{c}_{ij} = c_{ij} - \pi_i, \forall i \in N, (i,j) \in A$ and $\tilde{c}_{oj} = c_{oj} - \pi_o, \forall (o,j) \in A$.

2.8 Single depot vehicle scheduling problem: arc-flow formulation

(a) Every trip $i \in N$ is represented by two nodes for the start (in gray) and end (in black) of service; these are linked by an arc with $x_i = 1$ (indeed, lower and upper bounds equal to 1). The cost of such an arc can be set to zero or to the fixed cost c_i incurred to cover that trip; since each trip has to be covered exactly once, the sum of the costs of all the trips is a constant. The representation with two nodes is natural in the airline area, for example, the flight Montréal – Frankfurt is represented by the two city-nodes.

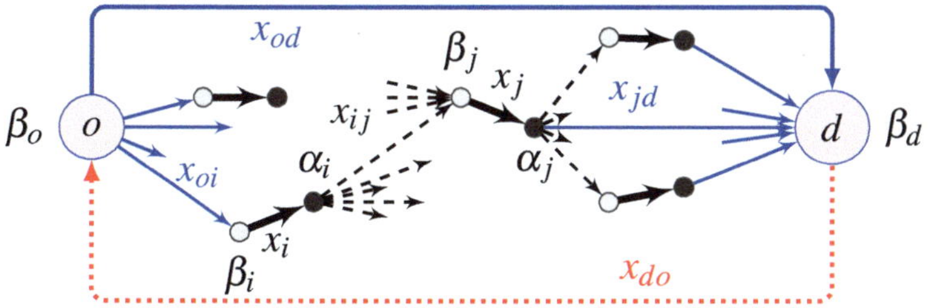

(b) $\bar{c}_{od} = c_{od} - \beta_o + \beta_d$; $\quad \bar{c}_{do} = -\beta_d + \beta_o$; $\quad \bar{c}_i = c_i - \beta_i + \alpha_i, \forall i \in N$; $\bar{c}_{ij} = c_{ij} - \alpha_i + \beta_j, \forall (i,j) \in A$.

2.9 Multiple depot vehicle scheduling problem

(a) The *IMP* formulation is

$$
\begin{aligned}
z^\star_{IMP} = \min \quad & \sum_{k\in K}\sum_{\mathbf{x}^k\in\mathcal{X}^k} c_{\mathbf{x}^k}\lambda_{\mathbf{x}^k} && \\
\text{s.t.} \quad & \sum_{k\in K}\sum_{\mathbf{x}^k\in\mathcal{X}^k} a_{i\mathbf{x}^k}\lambda_{\mathbf{x}^k} = 1 && [\pi_i] \quad \forall i\in N\\
& \sum_{\mathbf{x}^k\in\mathcal{X}^k} \lambda_{\mathbf{x}^k} \le v^k && [\pi_0^k] \quad \forall k\in K\\
& \lambda_{\mathbf{x}^k} \text{ binary} && \forall k\in K,\ \mathbf{x}^k\in\mathcal{X}^k.
\end{aligned}
$$

(b) A multi-commodity formulation, with commodity k being associated to depot k, is

$$
\begin{aligned}
z^\star_{ILP} = \min \quad & \sum_{k\in K}\sum_{(i,j)\in A^k_{do}} c^k_{ij}x^k_{ij} && \\
\text{s.t.} \quad & \sum_{k\in K}\sum_{j:(i,j)\in A^k} x^k_{ij} = 1 && \forall i\in N\\
& \sum_{j:(i,j)\in A^k_{do}} x^k_{ij} - \sum_{j:(j,i)\in A^k_{do}} x^k_{ji} = 0 && \forall k\in K, i\in N^k\\
& 0 \le x^k_{d^ko^k} \le v^k && \forall k\in K\\
& x^k_{ij}\in\{0,1\} && \forall k\in K, (i,j)\in A^k\setminus\{(d^k,o^k)\}.
\end{aligned}
$$

Note that $x^k_{d^ko^k}\in\mathbb{Z}_+$, $\forall k\in K$, as a consequence of the flow conservation equations and all other flow variables being binary.

(c) A possible network for 2 depots is illustrated below. The minimum number of buses is obtained as

$$v^\star = \sum_{k\in K} v^k - \max\sum_{k\in K} x^k_{o^kd^k},$$

subject to the network constraints. In this new formulation, there is no travel costs and the *depot restriction constraints* can be satisfied a posteriori by reassigning the routes to appropriate depots.

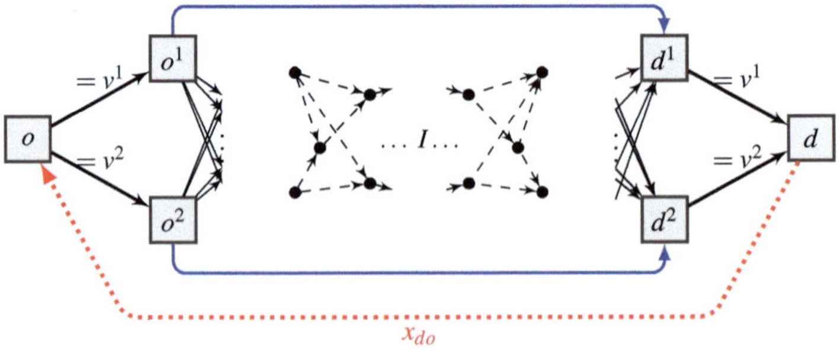

2.10 Maximum flow problem: arc-flow and path-flow formulations

(a) The maximum flow problem between o and d is a specialized form of the general flow formulation with zero-supply and zero-demand vectors. Let x_{ij} be the number of units traveling on arc $(i,j) \in A_{do}$. An arc-flow formulation is

$$\begin{aligned} z^\star_{LP} = \max \quad & x_{do} \\ \text{s.t.} \quad & \sum_{j:(i,j)\in A_{do}} x_{ij} - \sum_{j:(j,i)\in A_{do}} x_{ji} = 0 \quad && \forall i \in V \\ & 0 \le x_{ij} \le u_{ij} && \forall (i,j) \in A. \end{aligned}$$

There is a flow conservation equation per node in N and a pair of bounds for each arc in A. Variable x_{do} is maximized in the objective function.

(b) Let $\mathcal{X}$ be the finite set of feasible od-paths in $G_{do} = (V, A_{do})$, where path $\mathbf{x} \in \mathcal{X}$ is encoded in the *MP* as the binary vector $\mathbf{a_x} = [a_{ij\mathbf{x}}]_{(i,j)\in A}$ with $a_{ij\mathbf{x}} = 1$ if and only if path $\mathbf{x}$ uses arc (i,j). The *MP* formulation reads as

$$\begin{aligned} z^\star_{MP} = \max \quad & \sum_{\mathbf{x}\in\mathcal{X}} \lambda_\mathbf{x} \\ \text{s.t.} \quad & \sum_{\mathbf{x}\in\mathcal{X}} \mathbf{a_x}\lambda_\mathbf{x} \le \mathbf{u} \quad [\boldsymbol{\pi}] \\ & \lambda_\mathbf{x} \ge \mathbf{0}, \end{aligned}$$

where $\mathbf{u} = [u_{ij}]_{(i,j)\in A}$ is the vector of upper bounds and $\boldsymbol{\pi} = [\pi_{ij}]_{(i,j)\in A}$ is the non-negative dual vector. Regarding the *SP*, it can be seen as the longest path problem from o to d with objective function $\max_{\mathbf{x}\in\mathcal{X}} 1 - \boldsymbol{\pi}^\intercal \mathbf{a_x}$, where $\mathbf{a_x} = \mathbf{x}$. Alternatively, it can be formulated on G_{do} as

$$\begin{aligned} \bar{c}(\boldsymbol{\pi}) = \max \quad & c_\mathbf{x} - \boldsymbol{\pi}^\intercal \mathbf{a_x} \\ \text{s.t.} \quad & \sum_{j:(o,j)\in A} x_{oj} = x_{do} \\ & \sum_{j:(i,j)\in A} x_{ij} - \sum_{j:(j,i)\in A} x_{ji} = 0 && \forall i \in N \\ & \sum_{i:(i,d)\in A} x_{id} = x_{do} \\ & x_{ij} \in \{0,1\} && \forall (i,j) \in A_{do} \\ & c_\mathbf{x} = x_{do} \\ & \mathbf{a_x} = \mathbf{x} = [x_{ij}]_{(i,j)\in A}. \end{aligned}$$

(c) In any *RMP* solution, either an arc $(i,j) \in A$ is saturated or not.

- In the latter case, $x_{ij} < u_{ij}$ and $\pi_{ij} = 0$.
- Otherwise the arc is saturated and π_{ij} gives the impact on the maximum flow (that is, $z^\star_{MP}$ or equivalently $z^\star_{LP}$) if u_{ij} is increased by 1. Then, either

the minimal cut remains unchanged and $\pi_{ij} = 0$ or it allows for one additional unit of flow and $\pi_{ij} = 1$. Hence $\pi_{ij} \in \{0, 1\}$.

(d) Given $\boldsymbol{\pi} \geq \mathbf{0}$ and $\bar{c}(\boldsymbol{\pi}) > 0$, we show that $\bar{c}(\boldsymbol{\pi}) = 1$.

- If $x_{do} = 1$, then $1 > \sum_{(i,j)\in A} \pi_{ij} x_{ij}$. An optimized *od*-path $\mathbf{x}$ must go through a subset $A_{\mathbf{x}} \subset A$ of arcs for which the dual values are $\pi_{ij} = 0$, $\forall (i,j) \in A_{\mathbf{x}}$, hence $\boldsymbol{\pi}^{\mathsf{T}} \mathbf{a}_{\mathbf{x}} = 0$ and $\bar{c}(\boldsymbol{\pi}) = x_{do} = 1$.
- The case $x_{do} = 0$ is in contradiction with $\bar{c}(\boldsymbol{\pi}) > 0$. If we assume it to be true, then $x_{ij} = 0$, $\forall (i,j) \in A$, $\mathbf{a}_{\mathbf{x}} = \mathbf{x} = \mathbf{0}$, and $\bar{c}(\boldsymbol{\pi}) = x_{do} = 0$.

(e) At optimality of the *MP*, there is no $\lambda_{\mathbf{x}}$-variables with positive reduced cost, hence $\bar{c}(\boldsymbol{\pi}^\star) = 0$.

- This can be achieved in $\bar{c}(\boldsymbol{\pi}^\star) = x_{do} - \boldsymbol{\pi}^{\star\mathsf{T}} \mathbf{x}$ with $\begin{bmatrix} x_{do} \\ \mathbf{x} \end{bmatrix} = \begin{bmatrix} 0 \\ \mathbf{0} \end{bmatrix}$.
- Alternatively, any *od*-path $\mathbf{x} \in \mathcal{X}$ whose variable $\lambda_{\mathbf{x}}$ is positive in the *RMP* solution has a zero reduced cost by complementary slackness.

2.11 Maximizing the smallest reduced cost within column generation

(a) Letting $\mu = \min_{\mathbf{x}\in\mathcal{X}} c_{\mathbf{x}} - \boldsymbol{\pi}^{\mathsf{T}} \mathbf{a}_{\mathbf{x}}$, the *SP* becomes

$$
\begin{aligned}
\bar{c}^\star = \max_{\boldsymbol{\pi}\in\mathbb{R}^m} \min_{\mathbf{x}\in\mathcal{X}} \bar{c}_{\mathbf{x}} \quad &= \quad \max_{\boldsymbol{\pi}\in\mathbb{R}^m} \quad \mu \\
&\qquad\ \text{s.t.} \quad \mu \leq c_{\mathbf{x}} - \boldsymbol{\pi}^{\mathsf{T}} \mathbf{a}_{\mathbf{x}} \quad [\theta_{\mathbf{x}}] \quad \forall \mathbf{x} \in \mathcal{X} \\
&= \quad \max_{\boldsymbol{\pi}\in\mathbb{R}^m} \quad \mu \\
&\qquad\ \text{s.t.} \quad \mu + \boldsymbol{\pi}^{\mathsf{T}} \mathbf{a}_{\mathbf{x}} \leq c_{\mathbf{x}} \quad [\theta_{\mathbf{x}}] \quad \forall \mathbf{x} \in \mathcal{X}.
\end{aligned}
$$

(b) Given $\theta_{\mathbf{x}} \geq 0$, $\forall \mathbf{x} \in \mathcal{X}$, the dual formulation of the above is

$$
\begin{aligned}
\bar{c}^\star = \min \quad & \sum_{\mathbf{x}\in\mathcal{X}} c_{\mathbf{x}} \theta_{\mathbf{x}} \\
\text{s.t.} \quad & \sum_{\mathbf{x}\in\mathcal{X}} \mathbf{a}_{\mathbf{x}} \theta_{\mathbf{x}} = \mathbf{0} \quad [\boldsymbol{\pi}] \\
& \sum_{\mathbf{x}\in\mathcal{X}} \theta_{\mathbf{x}} = 1 \quad [\mu] \\
& \theta_{\mathbf{x}} \geq 0 \qquad \forall \mathbf{x} \in \mathcal{X}.
\end{aligned}
$$

(c) This *SP* looks for a minimum cost normalized combination of columns summing up to $\mathbf{0}$. For a network flow problem, this corresponds to finding a minimum mean-cost cycle because the variables take value $1/|W|$, for a cycle comprising $|W|$ arcs. For a network given by $G = (N, A)$ with arc-flow vector $\mathbf{x} = [x_{ij}]_{(i,j)\in A}$, the above formulation becomes

$$
\begin{aligned}
\bar{c}^\star = \min \quad & \sum_{(i,j)\in A} c_{ij} x_{ij} \\
\text{s.t.} \quad & \sum_{j:(i,j)\in A} x_{ij} - \sum_{j:(j,i)\in A} x_{ji} = 0 \quad [\pi_i] \quad \forall i \in N \\
& \sum_{(i,j)\in A} x_{ij} = 1 \quad [\mu] \\
& x_{ij} \geq 0 \qquad \forall (i,j) \in A.
\end{aligned}
$$

2.12 Tailing-off effect
Stop solving the *MP* before optimality is reached and start the branching process. This early stopping can be based on the computation of a lower bound on $z^\star_{MP}$, or if the value of z_{RMP} does not change sufficiently in a certain number of iterations. Indeed, branch-and-bound can be started at any time, even while heuristically solving the *MP*; we only need a branching rule. See the discussion on early termination in Note 2.18.

2.13 Transforming a set partitioning problem into a set covering problem
In column generation notation, x_j, $j \in \{1,\dots,n\}$, becomes $\lambda_{\mathbf{x}}$, $\mathbf{x} \in \mathcal{X}$. Any upper bound $UB \geq \sum_{\mathbf{x}\in\mathcal{X}} c_{\mathbf{x}}\lambda^\star_{\mathbf{x}}$ on the optimal objective value can be selected. Then (2.60) reads as $\sum_{\mathbf{x}\in\mathcal{X}} C_{\mathbf{x}}\lambda^\star_{\mathbf{x}} \leq UB(m+1)$ while (2.61) becomes $\sum_{\mathbf{x}\in\mathcal{X}} C_{\mathbf{x}}\lambda^\bullet_{\mathbf{x}} > UB(m+1)$.

2.14 Set covering vs. set partitioning

(a) Let $\boldsymbol{\lambda}^\star_{SCP} = [\lambda^\star_{\mathbf{x}}]_{\mathbf{x}\in\mathcal{X}}$ be an optimal solution to the *SCP* under the given conditions and assume it over-covers the right-hand side vector **1** (otherwise we trivially have $\boldsymbol{\lambda}^\star_{SPP} = \boldsymbol{\lambda}^\star_{SCP}$). Select a variable $\lambda^\star_{\mathbf{x}} = 1$ whose column $\mathbf{a}_{\mathbf{x}}$ induces an over-covering on some row i. Let $R'_{\mathbf{x}} = R_{\mathbf{x}} \setminus \{i\}$.

- If $c(R'_{\mathbf{x}}) < c(R_{\mathbf{x}})$, then $\boldsymbol{\lambda}^\star_{SCP}$ cannot be optimal. Hence $c(R'_{\mathbf{x}}) = c(R_{\mathbf{x}})$ and the use of $R'_{\mathbf{x}}$ rather than $R_{\mathbf{x}}$ reduces the over-covering on row i.
- Repeat this process until there is no over-covering, hence producing an optimal solution $\boldsymbol{\lambda}^\star_{SPP}$ to the set partitioning problem.

(b) From the primal point of view, finding feasible solutions is much easier with the *SCP* formulation because over-covering is permitted. Moreover, there exists fast heuristics to solve the *SCP*. This is particularly important for the *Phase I* to rapidly find relatively good dual variables.
From the dual point of view, recall that the optimality conditions are based on the reduced costs, hence on the dual vector. *Restricting* to $\boldsymbol{\pi} \geq \mathbf{0}$ in the *SCP* rather than $\boldsymbol{\pi} \in \mathbb{R}^m$ in the *SPP relaxes* the primal formulation and this strategy has proven to be much faster in practice to solve the linear relaxation. This is also the motivation for stabilization techniques used in numerous implementations of the column generation algorithm, that is, looking for a better control over likely dual variable values.

2.15 Cutting stock problem: integer optimum bound
We can round up each of the positive $\lambda^\star_{\mathbf{x}}$-values to the nearest integer. This solution is feasible and has value $UB = \sum_{\mathbf{x}\in\mathcal{X}'} \lceil \lambda^\star_{\mathbf{x}} \rceil$. (If we also assume an optimal *basic* solution to the *MP*, there are at most m such positive variables and $UB < z^\star_{MP} + m$.)

2.16 Cutting stock problem with equality constraints

(a) The *ISP* (2.30) is here given on a one-dimensional domain as

$$\begin{aligned} \bar{c}(\pi_1) = \min \quad & 1 - \pi_1 x_1 \\ \text{s.t.} \quad & 2x_1 \le 11 \\ & x_1 \in \mathbb{Z}_+. \end{aligned}$$

There are six feasible solutions given by $x_1 \in \{0,\ldots,5\}$. Each has a cost encoding of 1 whereas the respective column encoding is given by the value of x_1.

(b) Including the *empty* pattern ($x_1 = 0$) that does not select the requested item, the *IMP* writes as

$$\begin{aligned} z^\star_{IMP} = \min \quad & \lambda_0 + \lambda_1 + \lambda_2 + \lambda_3 + \lambda_4 + \lambda_5 \\ \text{s.t.} \quad & \lambda_1 + 2\lambda_2 + 3\lambda_3 + 4\lambda_4 + 5\lambda_5 = 4 \quad [\pi_1 \in \mathbb{R}] \\ & \lambda_0, \lambda_1, \lambda_2, \lambda_3, \lambda_4, \lambda_5 \in \mathbb{Z}_+ . \end{aligned}$$

(c) The unique optimal integer solution is $z^\star_{IMP} = 1$ with $\lambda^\star_4 = 1$.

(d) The domain $\text{conv}(\mathcal{X})$ contains two extreme points: $x_1 = 0$ and $x_1 = 5$.

(e) No, there is an obvious difference between the six columns in the *IMP* and the two extreme points which cannot produce λ_4.

(f) Solving by column generation goes as follows:

- *Phase 1* starts with an artificial variable of cost big-$M > 0$. Then $\pi_1 = M$, i.e., if the demand increases by 1 unit, the value of z_{RMP} goes up by M.
- The *ISP* generates $\begin{bmatrix}1\\x_1\end{bmatrix} = \begin{bmatrix}1\\5\end{bmatrix}$ with $\bar{c}(M) = 1 - M(5) < 0$. Hence $z_{RMP} = 0.8$ with solution $\lambda_5 = 0.8$ whereas $\pi_1 = 0.2$.
- The next iteration gives $\bar{c}(0.2) = 1 - 0.2(5) = 0$, again with $x_1 = 5$. We terminate with an optimal solution $z^\star_{MP} = 0.8$ using $\lambda^\star_5 = 0.8$.

(g) There are now only five feasible solutions to $\mathcal{X} = \{x_1 \in \mathbb{Z}_+ \mid 2x_1 \le 11, x_1 \le 4\}$ given by $x_1 \in \{0,\ldots,4\}$ and the variable λ_5 is discarded from the *IMP*. The optimal solution to the *MP* is already integer: $z^\star_{MP} = 1$ using $\lambda^\star_4 = 1$. Observe that $z^\star_{MP} \le z^\star_{MP(R)} \le z^\star_{IMP} = z^\star_{IMP(R)} = z^\star_{ILP}$, where (R) indicates that over-packing solutions are removed from the *ISP*.

2.17 Edge coloring problem with equality constraints

(a) The *ISP* is given by

$$\begin{aligned} \bar{c}(\boldsymbol{\pi}) = \min \quad & x_0 - \sum_{e \in E} \pi_e x_e \\ \text{s.t.} \quad & \sum_{e \in \delta(\{i\})} x_e \le x_0 \qquad \forall i \in N \\ & x_0 \in \{0,1\} \\ & x_e \in \{0,1\} \quad \forall e \in E. \end{aligned}$$

There are ten matchings: a 0-edge matching with $\mathbf{x} = 0$ and $x_0 = 0$ as well as five 1-edge matchings and, illustrated below, four 2-edge matchings.

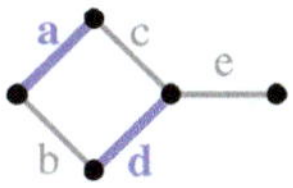

(b) Dropping the 0-edge or empty matching, the *IMP* formulation writes as

$$
\begin{array}{rllllllllll}
z^\star_{IMP} = \min & \lambda_a + \lambda_b + \lambda_c + \lambda_d + \lambda_e & + & \lambda_{ad} + \lambda_{ae} + \lambda_{bc} + \lambda_{be} & & \\
\text{s.t.} & \lambda_a & + & \lambda_{ad} + \lambda_{ae} & = 1 & [\pi_a] \\
& \lambda_b & & + \lambda_{bc} + \lambda_{be} & = 1 & [\pi_b] \\
& \lambda_c & & + \lambda_{bc} & = 1 & [\pi_c] \\
& \lambda_d & + & \lambda_{ad} & = 1 & [\pi_d] \\
& \lambda_e & & + \lambda_{ae} \quad + \lambda_{be} & = 1 & [\pi_e] \\
& \lambda_a, \ldots & & \ldots \lambda_{be} \in \{0,1\}. & &
\end{array}
$$

(c) Three colors are sufficient. The possible optimal solutions are respectively given by $\lambda_{ad} = \lambda_{bc} = \lambda_e = 1$, $\lambda_{ae} = \lambda_{bc} = \lambda_d = 1$, and $\lambda_{ad} = \lambda_{be} = \lambda_c = 1$ plus all 3-color permutations.

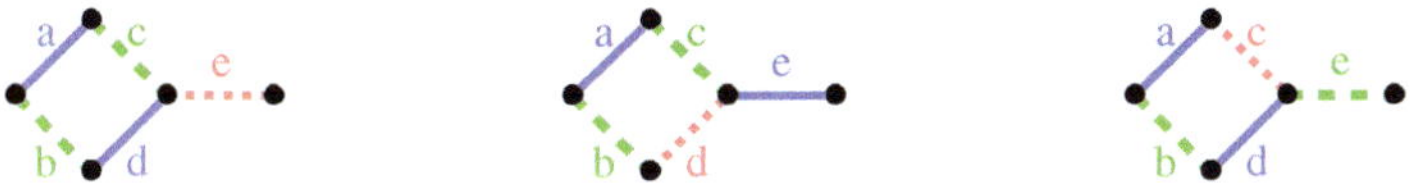

(d) All matchings listed in a) are extreme points of $\text{conv}(\mathcal{X})$. Indeed, we see by the objective function that a 1-edge matching has a different reduced cost than a 2-edge matching covering the same edge.

(e) Yes, there is no difference between the columns of the *IMP* and those using the extreme points of $\text{conv}(\mathcal{X})$.

(f) The initial artificial solution implies that $\pi_a = \pi_b = \ldots = \pi_e = M > 1$, hence the four 2-edge matchings have reduced costs equal to $1 - 2M$ while it is only $1 - M$ for the 1-edge matchings. Assuming we enter in the basis all the four maximal matchings, then any of the *artificial* variables covering rows $\{c,d,e\}$ can complete the basis. Let us take the artificial variable with an entry 1 in row c as the last basic column while similar results are obtained with the two other selections. The basis $\mathbf{A}_B$ and its inverse $\mathbf{A}_B^{-1}$ are

$$\mathbf{A}_B = \begin{bmatrix} 1 & 1 & 0 & 0 & 0 \\ 0 & 0 & 1 & 1 & 0 \\ 0 & 0 & 1 & 0 & 1 \\ 1 & 0 & 0 & 0 & 0 \\ 0 & 1 & 0 & 1 & 0 \end{bmatrix} \qquad \mathbf{A}_B^{-1} = \begin{bmatrix} 0 & 0 & 0 & 1 & 0 \\ 1 & 0 & 0 & -1 & 0 \\ 1 & 1 & 0 & -1 & -1 \\ -1 & 0 & 0 & 1 & 1 \\ -1 & -1 & 1 & 1 & 1 \end{bmatrix}$$

and $\boldsymbol{\pi}^\mathsf{T} = \mathbf{c}_B^\mathsf{T}\mathbf{A}_B^{-1} = [1,\ 1,\ 1,\ 1,\ M]\mathbf{A}_B^{-1} = [1-M,\ \ 1-M,\ \ M,\ \ M,\ \ M]$. Pricing the non-basic variables, that is, the five 1-edge matchings, we obtain

$$\bar{c}_a = \bar{c}_b = M > 0 \ \text{ and } \ \bar{c}_c = \bar{c}_d = \bar{c}_e = 1 - M < 0.$$

Selecting λ_c to enter the basis (replacing the artificial variable), the current basic solution is $z_{RMP} = 3$ with

$$\lambda_{ad} = \lambda_{be} = \lambda_c = 1,\ \lambda_{ae} = \lambda_{bc} = 0,\ \text{ and } \ \boldsymbol{\pi}^\mathsf{T} = [0,\ 0,\ 1,\ 1,\ 1].$$

Pricing again the non-basic variables, we prove optimality of the current solution because

$$\bar{c}_a = \bar{c}_b = 1 \ \text{ and } \ \bar{c}_d = \bar{c}_e = 0.$$

Note that the *MP* finds an integer λ-solution ... by chance.

2.18 Tolerance on the schedule synchronization constraints

We can use one bounded surplus and one bounded slack variable for every flight, or simply a single variable $v_i \in [-2,2]$. Using the revised model (p. 90) with constraints (2.53),

$$\begin{aligned}
z^\star_{IMP} = \min \quad & \sum_{k\in K}\sum_{p\in P^k} c^k_p \lambda^k_p \\
\text{s.t.} \quad & \sum_{k\in K}\sum_{p\in P^k} a^k_{ip}\lambda^k_p && = 1 && \forall i \in N \\
& \sum_{k\in K}\sum_{p\in P^k} b^k_{ip}\lambda^k_p \quad -t_{m_i} + v_i && = 0 && \forall i \in N \\
& -2 \le v_i \le 2 && && \forall i \in N \\
& \sum_{p\in P^k} \lambda^k_p && = 1 && \forall k \in K \\
& \lambda^k_p \ge 0 && && \forall k \in K, p \in P^k \\
& \sum_{p\in P^k} x^k_{ijp}\lambda^k_p && = x^k_{ij} \in \{0,1\} && \forall k \in K, (i,j) \in A^k.
\end{aligned}$$

2.19 Degenerate or small step-length pivots

(a) Using $X \sim H(n, 20, 200)$, $\Pr(X \geq 1)$ for a degenerate pivot:

n	5	10	15	20	25	30	35	40	45	50
%	41.3	66.0	80.6	**89.1**	94.0	96.8	98.3	99.1	99.5	99.8

(b) Using $X \sim H(n, 20, 500)$, $\Pr(X \geq 1)$ for a degenerate or small step-length pivot:

n	5	10	15	20	25	30	35	40	45	50
%	18.5	33.8	46.3	56.5	64.9	71.7	77.3	81.8	85.4	88.4

2.20 Pseudo-code practice

Whenever the *RMP* is feasible, we have $z_{RMP} < \infty$ and we perform column generation as usual. Otherwise, Farkas pricing intervenes by destroying the certificate of infeasibility found in the *RMP* or proving that the *MP* is indeed infeasible. This pseudo-code also handles infeasibility of the *ISP* which obviously implies that of the *MP* because there do not exist any columns at all. In this case, we go through the 'else' condition and correctly break by infeasibility since $F(\boldsymbol{\pi}) = -\infty$.

Algorithm: The column generation algorithm using Farkas pricing.

input : *RMP*, *SP*
output : Certificate of optimization
initialization : $\mathcal{X}' \leftarrow \emptyset$

loop
  $z_{RMP}, \boldsymbol{\lambda}_{RMP}, \boldsymbol{\pi} \leftarrow RMP$
  if $z_{RMP} < \infty$
    $\bar{c}(\boldsymbol{\pi}), \mathbf{x}, c_{\mathbf{x}}, \mathbf{a}_{\mathbf{x}} \leftarrow SP$
    if $\bar{c}(\boldsymbol{\pi}) \geq 0$
      break *by optimality of the MP*
  else
    $F(\boldsymbol{\pi}), \mathbf{x}, c_{\mathbf{x}}, \mathbf{a}_{\mathbf{x}} \leftarrow$ Farkas pricing
    if $F(\boldsymbol{\pi}) \leq 0$
      break *by infeasibility of the MP*
  $\mathcal{X}' \leftarrow \mathcal{X}' \cup \{\mathbf{x}\}$
return $\boldsymbol{\lambda}_{RMP}$, $\boldsymbol{\pi}$, and z_{RMP}

Exercises of Chapter 3

3.1 Philip Wolfe

Before joining Princeton, Wolfe had been offered a higher-paid position at RAND, which he turned down. But in 1957 RAND doubled their earlier offer and Wolfe went back to California, where he was assigned to RAND's computing group. There he was associated with George Dantzig, Ray Fulkerson, and Lloyd Shapley, and worked on ways to improve the simplex algorithm. In particular, with George Dantzig he developed a decomposition method to solve linear programming problems that had only a few constraints linking variables of several smaller LP problems and, for other problems, to generate LP matrix columns as they were needed in the computation. – INFORMS

3.2 Column generation vs. Dantzig-Wolfe decomposition

Column generation is a solution method, thus the term *algorithm*. The Dantzig-Wolfe decomposition is a principle that leads to an equivalent mathematical model: Dantzig-Wolfe *reformulation*. The confusion arises from the fact that the *MP* obtained from a Dantzig-Wolfe reformulation is often solved by column generation.

At the time of writing, the entry on Wikipedia about Dantzig-Wolfe decomposition perpetuates this confusion with its leading statement "Dantzig-Wolfe decomposition is an algorithm [...]." Feel free to also browse through the implementation paragraphs and compare with Note 2.14.

3.3 Bounded variables in the compact formulation

The domain of $\mathcal{D}$ is obviously bounded so we only have a set of $|P|$ extreme points. Since every extreme point $\mathbf{x}_p$, $p \in P$, of $\mathcal{D}$ fulfills $\boldsymbol{\ell} \leq \mathbf{x}_p \leq \mathbf{u}$, any convex combination of those extreme points necessarily fulfills the bounds which means they are indeed redundant in $\mathcal{A}$.

The *MP* thus reads no differently than what we are used to, i.e., reformulation (3.11)_left which we repeat for convenience:

$$
\begin{aligned}
z^\star_{MP} = \min \quad & \sum_{p\in P} c_p\lambda_p \\
\text{s.t.} \quad & \sum_{p\in P} \mathbf{a}_p\lambda_p \geq \mathbf{b} && [\boldsymbol{\pi}_\mathbf{b}] \\
& \sum_{p\in P} \lambda_p = 1 && [\pi_0] \\
& \lambda_p \geq 0 && \forall p \in P \\
& \sum_{p\in P} \mathbf{x}_p\lambda_p = \mathbf{x}.
\end{aligned}
$$

In particular, we still simply have non-negativity conditions on the λ-variables. We reach the same conclusion if the bounds are kept only in the set $\mathcal{A}$, see Decomposition theorem of a network flow solution.

3.4 Duplicated cost and column variables

(a) Similarly to (3.3),

$$\begin{aligned}
\sum_{p\in P} \mathbf{x}_p\lambda_p &+ \sum_{r\in R} \mathbf{x}_r\lambda_r = \mathbf{x}\\
\sum_{p\in P} c_p\lambda_p &+ \sum_{r\in R} c_r\lambda_r = c_{\mathbf{x}}\\
\sum_{p\in P} \mathbf{a}_p\lambda_p &+ \sum_{r\in R} \mathbf{a}_r\lambda_r = \mathbf{a}_{\mathbf{x}}\\
\sum_{p\in P} \lambda_p &= 1\\
\lambda_p \geq 0, &\qquad \lambda_r \geq 0 \quad \forall p\in P,\ r\in R,
\end{aligned}$$

and the reformulation is the same as (3.9).

(b) As in (3.15):

$$\begin{aligned}
\bar{c}(\boldsymbol{\pi}_{\mathbf{b}}, \pi_0) = \quad -\pi_0 + \min \quad & c_{\mathbf{x}} - \boldsymbol{\pi}_{\mathbf{b}}^{\mathsf{T}}\mathbf{a}_{\mathbf{x}}\\
\text{s.t.} \quad & \mathbf{Dx} \geq \mathbf{d}\\
& c_{\mathbf{x}} = \mathbf{c}^{\mathsf{T}}\mathbf{x},\ \ \mathbf{a}_{\mathbf{x}} = \mathbf{Ax},\ \ \mathbf{x}\in\mathbb{R}^n_+.
\end{aligned}$$

3.5 Dual formulations for the *MP*

The respective dual formulations are

(a)

$$\begin{aligned}
z^\star_{DMP} = \max \ \ & \mathbf{b}^{\mathsf{T}}\boldsymbol{\pi}_{\mathbf{b}} + \pi_0\\
\text{s.t.} \ \ & \mathbf{a}_p^{\mathsf{T}}\boldsymbol{\pi}_{\mathbf{b}} + \pi_0 \leq c_p \quad [\lambda_p] \quad \forall p\in P\\
& \mathbf{a}_r^{\mathsf{T}}\boldsymbol{\pi}_{\mathbf{b}} \leq c_r \quad [\lambda_r] \quad \forall r\in R\\
& \boldsymbol{\pi}_{\mathbf{b}} \geq \mathbf{0}, \quad \pi_0 \in \mathbb{R}.
\end{aligned}$$

(b)

$$\begin{aligned}
z^\star_{DMP} = \max \ \ & \mathbf{b}^{\mathsf{T}}\boldsymbol{\pi}_{\mathbf{b}} + \sum_{k\in K} \pi_0^k\\
\text{s.t.} \ \ & \mathbf{a}_p^{k\mathsf{T}}\boldsymbol{\pi}_{\mathbf{b}} + \pi_0^k \leq c_p^k \quad [\lambda_p^k] \quad \forall k\in K,\ p\in P^k\\
& \mathbf{a}_r^{k\mathsf{T}}\boldsymbol{\pi}_{\mathbf{b}} \leq c_r^k \quad [\lambda_r^k] \quad \forall k\in K,\ r\in R^k\\
& \boldsymbol{\pi}_{\mathbf{b}} \geq \mathbf{0}\\
& \pi_0^k \in \mathbb{R} \qquad \forall k\in K.
\end{aligned}$$

3.6 2D illustration: role inversion

$$\begin{aligned}
\mathcal{A} &= \{x_1, x_2 \geq 0 \mid -x_1 + 4x_2 \leq 8,\ x_2 \geq 2\}\\
\mathcal{D} &= \{x_1, x_2 \geq 0 \mid 3x_1 + 4x_2 \geq 24,\ x_1 \leq 10,\ x_2 \leq 6\}.
\end{aligned}$$

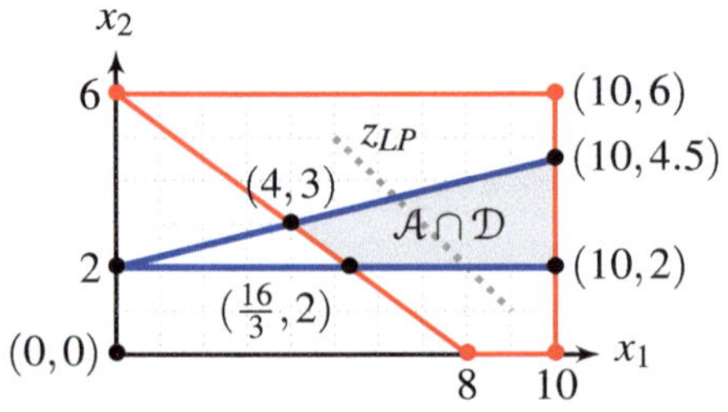

(a) $\mathcal{X}$ only comprises the extreme points $(0,6)$, $(10,6)$, $(10,0)$ and $(8,0)$.

(b) $c_{\mathbf{x}} = x_1 + x_2$ and $\mathbf{a}_{\mathbf{x}} = \begin{bmatrix} -x_1 + 4x_2 \\ x_2 \end{bmatrix}$.

(c) Let index $p \in \{1, \ldots, 4\}$ represent the extreme points of $\mathcal{D}$. The domain of the *MP* comprises three constraints, including the convexity one:

$$\begin{array}{rrrrrrrrrl} & \mathbf{x}_p\!: & (0,6) & & (10,6) & & (10,0) & & (8,0) & \\ z^\star_{MP} = & \min & 6\lambda_1 & + & 16\lambda_2 & + & 10\lambda_3 & + & 8\lambda_4 & \\ & \text{s.t.} & 24\lambda_1 & + & 14\lambda_2 & - & 10\lambda_3 & - & 8\lambda_4 \le 8 & [\pi_1] \\ & & 6\lambda_1 & + & 6\lambda_2 & & & & \ge 2 & [\pi_4] \\ & & \lambda_1 & + & \lambda_2 & + & \lambda_3 & + & \lambda_4 = 1 & [\pi_0] \\ & & \lambda_1, & & \lambda_2, & & \lambda_3, & & \lambda_4 \ge 0. & \end{array}$$

The *SP* writes as $\bar{c}(\boldsymbol{\pi}_{\mathbf{b}}, \pi_0) = -\pi_0 + \min_{\mathbf{x}\in\mathcal{D}} c_{\mathbf{x}} - [\pi_1, \pi_4]\mathbf{a}_{\mathbf{x}}$, where

$$c_{\mathbf{x}} = x_1 + x_2 \qquad \text{and} \qquad \mathbf{a}_{\mathbf{x}} = \begin{bmatrix} -x_1 + 4x_2 \\ x_2 \end{bmatrix}.$$

$$\begin{array}{rrrrl} \bar{c}(\pi_1, \pi_4, \pi_0) = -\pi_0 & + \min & (1+\pi_1)x_1 + & (1-4\pi_1-\pi_4)x_2 & \\ & \text{s.t.} & 3x_1 + & 4x_2 \ge 24 & [\pi_2] \\ & & x_1 & \le 10 & [\pi_3] \\ & & & x_2 \le \ 6 & [\pi_5] \\ & & x_1, & x_2 \ge 0. & \end{array}$$

(d) Solving the *MP*

- directly gives $z^\star_{MP} = 7$ with primal values $\boldsymbol{\lambda}^\star_{MP} = (0.5,\ 0,\ 0,\ 0.5)$ and dual vector $[\boldsymbol{\pi}^\star_{\mathbf{b}}, \pi^\star_0] = [\pi^\star_1, \pi^\star_4, \pi^\star_0] = [-1/16,\ 0,\ 7.5]$. For this optimal dual vector, the *SP* becomes

$$\begin{array}{rrrrl} \bar{c}(-1/16,\ 0,\ 7.5) = -7.5 & + \min & 15/16 x_1 & + \ 1.25x_2 & \\ & \text{s.t.} & 3x_1 & + \ 4x_2 \ge 24 & [\pi_2] \\ & & x_1 & \le 10 & [\pi_3] \\ & & & x_2 \le \ 6 & [\pi_5] \\ & & x_1, & x_2 \ge 0, & \end{array}$$

where $\bar{c}(\boldsymbol{\pi}^\star_{\mathbf{b}}, \pi^\star_0) = 0$ at both extreme points $(0,6)$ and $(8,0)$ with dual vector $[\pi^\star_2, \pi^\star_3, \pi^\star_5] = [5/16,\ 0,\ 0]$.

- by column generation using Farkas pricing to initialize with objective function here given by $\bar{c}(\boldsymbol{\pi}_{\mathbf{b}}, \pi_0) - c_{\mathbf{x}} = -\pi_0 + \min \pi_1 x_1 - (4\pi_1 + \pi_4)x_2$. We need two iterations to reach feasibility, one to improve the initial solution and one more to prove optimality.

		MP		*SP*		
t	*RMP* solution	z_{RMP}	$[\pi_1\ \pi_4\ \pi_0]$	$\bar{c}(\boldsymbol{\pi}_{\mathbf{b}}, \pi_0)$	$\mathbf{x}_p$	*lb*
1	-	-	$[0\ 1\ 0]$	−24.00	$[0\ 6]$	-
2	-	-	$[-1\ 0\ 24]$	−2.00	$[10\ 0]$	-
3	$\lambda_1 = 0.529, \lambda_3 = 0.471$	7.88	$[-0.1176\ 0\ 8.8235]$	−1.76	$[8\ 0]$	6.12
4	$\lambda_1 = 0.5, \lambda_4 = 0.5$	7.00	$[-0.0625\ 0\ 7.5000]$	0.00	$[0\ 6]$	7.00

(e) Verification of primal-dual solutions:

- $(x^\star_1, x^\star_2) = \sum_{p=1}^{4} \mathbf{x}_p \lambda^\star_p = (0,6)(0.5) + (8,0)(0.5) = (3,4)$.
- $[\pi^\star_1, \pi^\star_2, \pi^\star_3, \pi^\star_4, \pi^\star_5] = [-1/16,\ 5/16,\ 0,\ 0,\ 0]$.

3.7 2D illustration: maximization

(a) $z^\star_{LP} = 14.5$, $\mathbf{x}^\star_{LP} = (10, 4.5)$, and $\boldsymbol{\sigma}^{\star\mathsf{T}} = [0.25,\ 0,\ 1.25,\ 0,\ 0]$.

(b) Let index $p = 1$ represent the extreme point $(0,2)$ in $\mathcal{D}$ whereas $r = 2$ and $r = 3$ are used for the extreme rays $(4,1)$ and $(1,0)$, respectively. After substitution in the constraints of $\mathcal{A}$ and the objective function, the *MP* comprises four constraints, where the row indices correspond to those of the dual variables $\boldsymbol{\pi}^{\mathsf{T}}_{\mathbf{b}} = [\pi_2, \pi_3, \pi_5]$, π_0 being reserved for the convexity constraint (here involving only variable λ_1):

$$\begin{array}{rrrrrrl}
 & \mathbf{x}: & (0,2) & & (4,1) & & (1,0) \\
z^\star_{MP} = & \max & 2\lambda_1 & + & 5\lambda_2 & + & \lambda_3 \\
 & \text{s.t.} & 8\lambda_1 & + & 16\lambda_2 & + & 3\lambda_3 \geq 24 \quad [\pi_2 \leq 0] \\
 & & & & 4\lambda_2 & + & \lambda_3 \leq 10 \quad [\pi_3 \geq 0] \\
 & & 2\lambda_1 & + & \lambda_2 & & \leq \ 6 \quad [\pi_5 \geq 0] \\
 & & \lambda_1 & & & & = \ 1 \quad [\pi_0 \in \mathbb{R}] \\
 & & \lambda_1, & & \lambda_2, & & \lambda_3 \geq 0.
\end{array}$$

- $\boldsymbol{\lambda}^\star_{MP} = (1,\ 2.5,\ 0)$ with objective value $z^\star_{MP} = 14.5 = z^\star_{LP}$;
- $\mathbf{x}^\star_{MP} = (0,2)\lambda^\star_1 + (4,1)\lambda^\star_2 + (1,0)\lambda^\star_3 = (10, 4.5) = \mathbf{x}^\star_{LP}$;
- $\boldsymbol{\pi}^{\star\mathsf{T}}_{\mathbf{b}} = [\pi^\star_2, \pi^\star_3, \pi^\star_5] = [0,\ 1.25,\ 0]$ together with $\pi^\star_0 = 2$.

- The *SP*, with the appropriate dual vector $\boldsymbol{\pi}_{\mathbf{b}}$, writes as

$$\bar{c}(\boldsymbol{\pi}_{\mathbf{b}}, \pi_0) = -\pi_0 + \max_{\mathbf{x} \in \mathcal{D}} c_{\mathbf{x}} - \begin{bmatrix} \pi_2 \\ \pi_3 \\ \pi_5 \end{bmatrix}^{\mathsf{T}} \mathbf{a}_{\mathbf{x}}, \text{ where } c_{\mathbf{x}} = x_1 + x_2 \text{ and } \mathbf{a}_{\mathbf{x}} = \begin{bmatrix} 3x_1 + 4x_2 \\ x_1 \\ x_2 \end{bmatrix},$$

that is,

$$
\begin{array}{llllr}
-\pi_0 + \max & (1-3\pi_2-\pi_3)x_1 & + (1-4\pi_2-\pi_5)x_2 & & \\
\text{s.t.} & -x_1 & + 4x_2 \le 8 & [\pi_1 \ge 0] \\
 & & x_2 \ge 2 & [\pi_4 \le 0] \\
 & x_1 \ge 0. & & &
\end{array}
$$

- Given $\boldsymbol{\pi}_{\mathbf{b}}^\star$ and $\pi_0^\star$, the *SP* becomes

$$
\begin{array}{lllll}
\bar{c}(0,\ 1.25,\ 0,\ 2) = -2 + \max & -{}^1\!/\!{}_4 x_1 & + \ x_2 & & \\
\text{s.t.} & -x_1 & + 4x_2 \le 8 & [\pi_1 \ge 0] \\
 & & x_2 \ge 2 & [\pi_4 \le 0] \\
 & x_1 \ge 0, & & &
\end{array}
$$

where $\bar{c}(\boldsymbol{\pi}_{\mathbf{b}}^\star, \pi_0^\star) = 0$ at the extreme point $(0,2)$ with $[\pi_1^\star, \pi_4^\star] = [0.25,\ 0]$.

3.8 Minimum reduced cost of zero at optimality of the *MP*
Because the *MP* needs not be solved by a simplex-type algorithm, the zero reduced cost of the basic variables is not a valid argument. The result follows from the positiveness of at least one λ_p-variable in the convexity constraint $\sum_{p\in P}\lambda_p = 1$ and the optimality conditions in Proposition 1.7: $\lambda_p > 0 \Rightarrow \bar{c}_p = 0$.

3.9 The Dantzig-Wolfe lower bound does not depend on π_0

$$
\begin{aligned}
z_{RMP} + \bar{c}(\boldsymbol{\pi}_{\mathbf{b}}, \pi_0) &= (\boldsymbol{\pi}_{\mathbf{b}}^\mathsf{T}\mathbf{b} + \pi_0) + (-\pi_0 + \min_{\mathbf{x}\in\mathcal{D}} (\mathbf{c}^\mathsf{T} - \boldsymbol{\pi}_{\mathbf{b}}^\mathsf{T}\mathbf{A})\mathbf{x}) \\
&= \boldsymbol{\pi}_{\mathbf{b}}^\mathsf{T}\mathbf{b} + \min_{\mathbf{x}\in\mathcal{D}} (\mathbf{c}^\mathsf{T} - \boldsymbol{\pi}_{\mathbf{b}}^\mathsf{T}\mathbf{A})\mathbf{x}.
\end{aligned}
$$

3.10 All constraints in the pricing problem
(a) The *MP* is adapted from (3.9):

$$
\begin{array}{llll}
z_{MP}^\star = \min & \sum_{p\in P} c_p\lambda_p + \sum_{r\in R} c_r\lambda_r & & \\
\text{s.t.} & \sum_{p\in P} \lambda_p = 1 & [\pi_0] & \\
 & \lambda_p \ge 0, \quad \lambda_r \ge 0 & & \forall p \in P,\ r \in R \\
 & \sum_{p\in P} \mathbf{x}_p\lambda_p + \sum_{r\in R} \mathbf{x}_r\lambda_r = \mathbf{x}. & &
\end{array}
$$

(b) The *SP* is adapted from (3.15):

$$
\begin{array}{lll}
\bar{c}(\pi_0) = -\pi_0 + \min & \mathbf{c}^\mathsf{T}\mathbf{x} & \\
\text{s.t.} & \mathbf{A}\mathbf{x} \ge \mathbf{b} & [\boldsymbol{\pi}_{\mathbf{b}}] \\
 & \mathbf{D}\mathbf{x} \ge \mathbf{d} & [\boldsymbol{\pi}_{\mathbf{d}}] \\
 & \mathbf{x} \ge \mathbf{0}. &
\end{array}
$$

(c) The *MP* only contains the convexity constraint with associated dual variable π_0. A single artificial variable, say y_0, is needed to start, with a large cost $M > 0$ in the objective function. The first *RMP* reads as

$$\begin{aligned} z_{RMP} = \min \quad & M y_0 \\ \text{s.t.} \quad & y_0 = 1 \quad [\pi_0] \\ & y_0 \geq 0. \end{aligned}$$

Obviously, $y_0 = 1$, $z_{RMP} = M$, and $\pi_0 = M$ in the optimal solution of this first *RMP*. Assuming $M > z^\star_{LP}$, the first call to the *SP* finds

- $\bar{c}(\pi_0) = -M + \mathbf{c}^\mathsf{T}\mathbf{x}^\star = -M + z^\star_{LP} < 0$
- at an extreme point $\mathbf{x}^\star_{LP}$ (because $z^\star_{LP}$ is finite),
- as well as $\boldsymbol{\pi}^\star_\mathbf{b} = \boldsymbol{\sigma}^\star_\mathbf{b}$ and $\boldsymbol{\pi}^\star_\mathbf{d} = \boldsymbol{\sigma}^\star_\mathbf{d}$.

The lower bound (3.32) on $z^\star_{LP}$ reads as $z_{RMP} + \bar{c}(\pi_0) = M - M + z^\star_{LP} = z^\star_{LP}$. We thus terminate with an optimal primal-dual solution for the *LP* found in the *SP*.
• If the lower bound criterion is not used, it takes a second iteration of the *RMP* and *SP*. Let $p = 1$, $\mathbf{x}_1 = \mathbf{x}^\star_{LP}$, $c_1 = z^\star_{LP}$. The second *RMP* reads as

$$\begin{aligned} z_{RMP} = \min \quad & c_1\lambda_1 \\ \text{s.t.} \quad & \lambda_1 = 1 \quad [\pi_0] \\ & \lambda_1 \geq 0 \\ & \mathbf{x}_1\lambda_1 = \mathbf{x}, \end{aligned}$$

for which the optimal solution is $\lambda_1 = 1$, $z_{RMP} = c_1 = z^\star_{LP}$, and $\pi_0 = z^\star_{LP}$. The *SP* again finds the same extreme point $\mathbf{x}_1$ and $\bar{c}(\pi_0) = -z^\star_{LP} + c_1 = 0$.

3.11 Generating variable $\lambda_\mathbf{0}$

(a) Initialize the *RMP* with artificial variables of cost big-*M* as needed, one being y_0 in the convexity constraint. Then $\pi_0 = M$ and $\bar{c}_\mathbf{x} = -\pi_0 + (\mathbf{c}^\mathsf{T} - \boldsymbol{\pi}^\mathsf{T}_\mathbf{b}\mathbf{A})\mathbf{x}$, $\forall \mathbf{x} \in \mathcal{X}$. Therefore, the reduced cost of $\lambda_\mathbf{0}$ is $\bar{c}_\mathbf{0} = -\pi_0 < 0$. Although there may be other $\lambda_\mathbf{x}$-variables with negative reduced costs, it is possible that $\lambda_\mathbf{0}$ dominates all of them such that the *ISP* indeed generates it. This variable then enters the basis in the *MP* with a contribution of zero but it does also change the dual values for the next iteration.

(b) If there are several pricing problems, a variable $\lambda^k_\mathbf{0}$ may be generated after some other extreme points $\mathbf{x}^k$ have been found. We can see this from the complicating constraints that can be covered by extreme points of any block. It may happen that all previously generated columns of a block k are no longer cost effective and it is better to do nothing with $\lambda^k_\mathbf{0} = 1$.

(c) In the case of a polyhedral cone, there is only one extreme point per domain $\mathcal{D}^k$, $k \in K$. Each variable $\lambda^k_\mathbf{0}$ therefore has to be generated otherwise we cannot fulfill the k-th convexity constraint. Once again, we can alternatively remove the convexity constraints from the *MP* and only generate extreme rays.

3.12 Time constrained shortest path problem: duality

The figure presents side-by-side the dual domains for the *MP*s with a level curve for their respective objective functions in dotted lines (Not all line-equations are identified).

$$z^\star_{MP} = 7 = 14\pi^\star_7 + \pi^\star_0|_{(-2,\,35)} = \mu^\star|_{(-2,\,7)}.$$

Remark: *Dual formulations* (3.107) *and* $(3.108)_{right}$ *of the master problem are equivalent. The second dual formulation, the one based on the alternative MP formulation, is used in Lagrangian relaxation, not the first.*

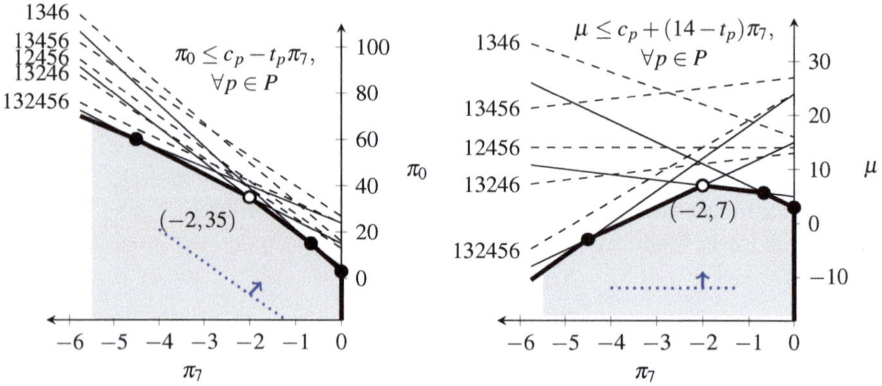

(a) Domain of the dual of the *MP*. (b) Domain of the dual of the alternative *MP*.

3.13 Time constrained shortest path problem: circulation pricing problem

(a) The set $\{\mathbf{x}_p\}_{p\in P}$ contains the single extreme point $\mathbf{0}$. Any extreme ray of set $\{\mathbf{x}_r\}_{r\in R}$ can be represented as a unit-cycle composed of a path from 1 to 6 completed by the returning arc $(6,1)$ at value 1; other arc-components are zero. These cycles are in a one-to-one correspondence with the paths from 1 to 6.

(b) Because the extreme rays are scaled with $x_{61} = 1$, we show that the *MP* remains the same as (3.81) in Example 3.2, except that index set P is replaced by R.

- First, the variable $\lambda_{\mathbf{0}}$ is discarded from the *MP* as well as the convexity constraint $\lambda_{\mathbf{0}} = 1$ on the unique extreme point.
- Second, we have to reformulate both constraints of $\mathcal{A}$, in particular $x_{61} = 1$ in (3.109d), when substituting the extreme rays indexed in R.
  This gives back an equivalent "convexity" constraint on the λ_r-variables.

$$
\begin{aligned}
z^\star_{MP} = \min \quad & \sum_{r\in R} c_r\lambda_r \\
\text{s.t.} \quad & \sum_{r\in R} t_r\lambda_r \le 14 && [\pi_7 \le 0] \\
& \sum_{r\in R} \lambda_r = 1 && [\pi_{61} \in \mathbb{R}] \\
& \lambda_r \ge 0 && \forall r \in R \\
& \sum_{r\in R} x_{ijr}\lambda_r = x_{ij} && \forall (i,j) \in A_{do},
\end{aligned}
$$

The *SP* is a minimum cost circulation problem:

$$
\begin{aligned}
\bar{c}(\pi_7, \pi_{61}) = \quad & \min \; c_{\mathbf{x}} - [\pi_7, \pi_{61}]\mathbf{a}_{\mathbf{x}} \\
\text{s.t.} \quad & \sum_{j:(i,j)\in A_{do}} x_{ij} - \sum_{j:(j,i)\in A_{do}} x_{ji} = 0 \;\; [\pi_i] \;\; \forall i \in \{1,\ldots,6\} \\
& x_{ij} \geq 0, \; \forall (i,j) \in A_{do} \\
& c_{\mathbf{x}} = \sum_{(i,j)\in A_{do}} c_{ij} x_{ij} \\
& \mathbf{a}_{\mathbf{x}} = \begin{bmatrix} \sum_{(i,j)\in A_{do}} t_{ij} x_{ij} \\ x_{61} \end{bmatrix}.
\end{aligned}
$$

3.14 Aircraft routing

(a) • Objective function: min $\sum_{(i,j)\in Night} x_{ij}$,
i.e., the sum of the three night arcs, one per city. Other possibilities exist, such as the sum of the ground and flight arcs at any specific moment.
• Bounds: $x_{ij} = 1, \; \forall (i,j) \in Flight$; $\quad x_{ij} \geq 0, \; \forall (i,j) \in Ground$.

(b) 6 aircraft on 5 cycles: 3 aircraft at **A**, 2 at **B** and 1 at **C**. The decomposition into cycles is not unique, see Proposition 3.4.

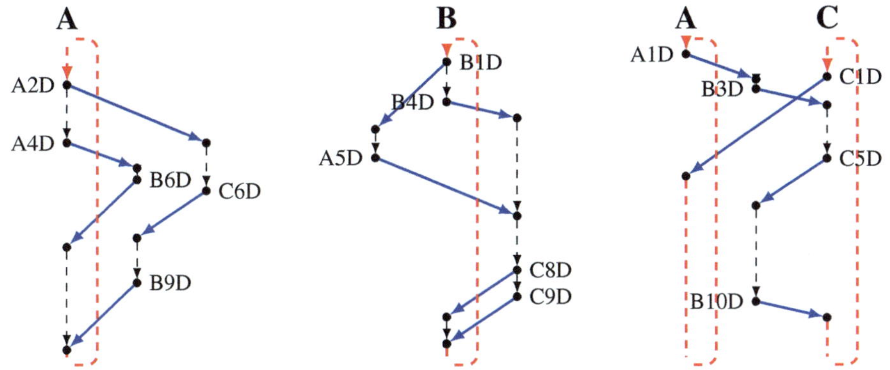

(a) Two aircraft cycles from city **A**. (b) Two aircraft cycles from city **B**. (c) A cycle with two aircraft from cities **A** and **C**.

(c) The aircraft at the beginning of the day in city **C** operates 5 legs over a period of two days, 1 on the first day, 4 on the second. Several solutions are possible, and they look like

- flight C1 to **A**, night arc in **A**;
- A1 to **B**, B3 or B4 to **C**, C5 or C6 to **B**, B10 back to **C**, night arc in **C**.

The schedule repeats itself after two days.

(d) The time of departure in the city of origin remains the same but the arrival time is modified according to the difference between the time zones.

(e) Time-space network as a *SDVSP*:

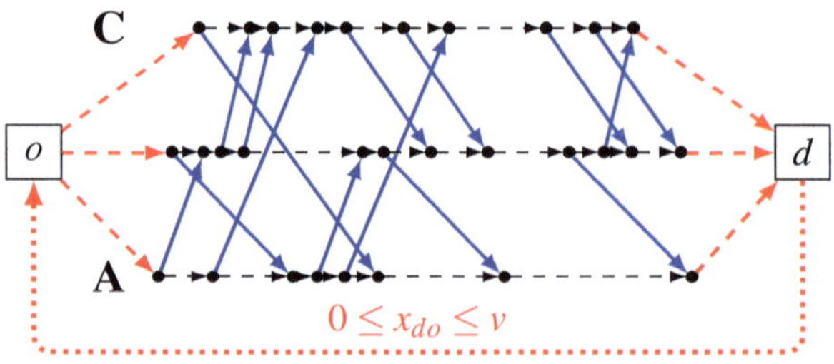

Because the number of departures and arrivals are equal in every city, we do not have to impose the equality constraints $x_{oA} = x_{Ad}$, $x_{oB} = x_{Bd}$, and $x_{oC} = x_{Cd}$.

(f) We can group a few successive nodes at an airport. Nodes are examined in chronological order and a new group is started each time an arrival node is encountered, provided that in the current group there already exists a departure node. This aggregation rule ensures that a flight leaving an airport utilizes an aircraft already stationed (Soumis et al., 1980). The reduction is illustrated for city **A** followed by that in **B**.

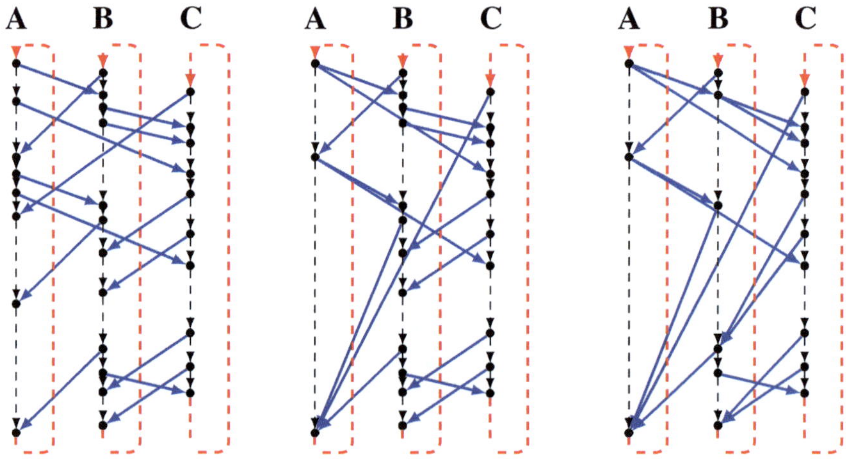

(a) Original network. (b) Reduction performed in **A**. (c) Reduction in **A** and **B**.

3.15 Aircraft routing: reformulation and column generation

The set $\mathcal{X} = \{\mathbf{0}\} \cup \{\mathbf{x}_r\}_{r \in R}$, where $\mathbf{x}_r \in \mathcal{D}$ is an extreme ray scaled to one unit of flow, a directed cycle on the network of Figure 3.18. Discarding index $\mathbf{0}$, the *MP* reads as

$$\begin{aligned} z^{\star}_{MP} = \min \quad & \sum_{r \in R} c_r \lambda_r \\ \text{s.t.} \quad & \sum_{r \in R} a_{ij,r} \lambda_r = 1 \quad [\pi_{ij}] \quad \forall (i,j) \in \mathit{Flight} \\ & \lambda_r \geq 0 \qquad \forall r \in R, \end{aligned}$$

where c_r counts the number of night arcs in cycle $\mathbf{x}_r$ and $a_{ij,r} = 1$ if flight (i,j) belongs to the cycle, 0 otherwise.

Let $\boldsymbol{\pi} = [\pi_{ij}]_{(i,j)\in Flight}$. The *SP* with domain $\mathcal{D}$ (3.113b) writes as

$$\begin{aligned}
\bar{c}(\boldsymbol{\pi}) = \min \quad & c_{\mathbf{x}} - \sum_{(i,j)\in Flight} \pi_{ij} a_{ij,\mathbf{x}} & \\
\text{s.t.} \quad & \sum_{j:(i,j)\in A} x_{ij} - \sum_{j:(j,i)\in A} x_{ji} = 0 & \forall i \in N \\
& x_{ij} \geq 0 & \forall (i,j) \in A \\
& c_{\mathbf{x}} = \sum_{(i,j)\in Night} x_{ij} & \\
& a_{ij,\mathbf{x}} = x_{ij} & \forall (i,j) \in Flight.
\end{aligned}$$

3.16 Linear relaxation of the cutting stock problem

(a) The first inequality system can be consolidated into a single equation, that is,

$$\sum_{k\in K} x_i^k \geq b_i,\ \forall i \in \{1,\dots,m\} \ \Rightarrow \sum_{i=1}^{m}\sum_{k\in K} w_i x_i^k \geq \sum_{i=1}^{m} w_i b_i.$$

The same consolidation idea can be applied on the second inequality system as

$$\sum_{i=1}^{m} w_i x_i^k \leq W x_0^k,\ \forall k \in K \ \Rightarrow \sum_{k\in K}\sum_{i=1}^{m} w_i x_i^k \leq \sum_{k\in K} W x_0^k.$$

Merging these two results yields $\sum_{i=1}^m w_i b_i \leq \sum_{k\in K}\sum_{i=1}^m w_i x_i^k \leq \sum_{k\in K} W x_0^k$ which corresponds to the requested lower bound on $z_{LP}^\star$ when dividing the left and right sides by W: $\sum_{i=1}^m w_i b_i / W \leq \sum_{k\in K} x_0^k$.

Now assign the following values for the variables:

$$\begin{aligned}
x_0^k &= \frac{\sum_{i=1}^m w_i b_i}{W|K|}, & \forall k \in K; \\
x_i^k &= \frac{b_i}{|K|}, & \forall k \in K,\ i \in \{1,\dots,m\}.
\end{aligned}$$

Then, the values $x_i^k = b_i/|K|$ fulfill the first inequality system $\sum_{k\in K} x_i^k \geq b_i$, $\forall i \in \{1,\dots,m\}$, and, for all $k \in K$, the inequalities $\sum_{i=1}^m w_i x_i^k \leq W x_0^k$ are as well satisfied. Since all bound restrictions are also satisfied, we have a primal solution that reaches the lower bound, hence $z_{LP}^\star = \sum_{k\in K} x_0^k = \sum_{i=1}^m w_i b_i / W$.

(b) First note that for all $k \in K$, the dual variable σ_0^k associated with the second system of constraints $\sum_{i=1}^m w_i x_i^k - W x_0^k \leq 0$ does not contribute at all to the objective value in the dual formulation. Second, $\sigma_i^\star = w_i/W$, $\forall i \in \{1,\dots,m\}$, provides the same solution cost as that of the primal:

$$\sum_{i=1}^{m} b_i \pi_i^\star = \sum_{i=1}^{m} b_i \frac{w_i}{W} = z_{LP}^\star.$$

There remains to complete the solution and tackle the feasibility aspects. An optimal dual solution implies that the reduced cost of all the variables is greater than or equal to zero:

$$\begin{aligned}\bar{c}_0^k &= \quad 1 + W\sigma_0^k \geq 0, \quad \forall k \in K;\\ \bar{c}_i^k &= \quad -\sigma_i - w_i\sigma_0^k \geq 0, \quad \forall k \in K,\ i \in \{1,\ldots,m\}.\end{aligned}$$

The first set implies $\sigma_0^k \geq -1/W$, $\forall k \in K$, whereas the second, combined with $\sigma_i^\star = w_i/W$, $\forall i \in \{1,\ldots,m\}$, implies $\sigma_0^k \leq -1/W$, $\forall k \in K$. Hence the required values are $\sigma_0^{k*} = -1/W$, $\forall k \in K$, to complete an optimal dual solution.

(c) **1-** Round up: $\lceil z_{LP}^\star \rceil$. **2-** Integer requirements in the subproblems:

$$\mathcal{D}^k = \left\{ \mathbf{x}^k \in \mathbb{Z}_+^{m+1} \;\middle|\; \sum_{i=1}^m w_i x_i^k \leq W x_0^k,\ x_0^k \leq 1 \right\}, \ \forall k \in K.$$

3- A compact *ILP* formulation with a better linear relaxation bound $z_{LP}^\star$. Forthcoming in Example 4.2, a Network-based compact formulation for the *CSP*.

3.17 PS, IPS, and MMCC: a dual point of view

Let us dualize the proposed pricing problems (3.119) and witness how a normalization constraint naturally appears on a subset of the variables. We are thus looking for a convex combination of these columns whose impact on the solution is later determined on the other variables that are not part of this subset. Pay attention to the presence of an equality system in the proposed pricing problems.

- In **PS**, we have a system of m equalities and m basic variables which immediately leads to the well-known $\boldsymbol{\pi}^\mathsf{T} = \mathbf{c}_B^\mathsf{T}\mathbf{A}_B^{-1}$ with zero degrees of freedom.
- In **IPS**, we rather have $m - |F| \geq 0$ degrees of freedom which means that $\boldsymbol{\pi}$ breaks down in $|F|$ fixed values and $m - |F|$ optimized ones.
- In **MMCC**, there are no equalities and we have m degrees of freedom such that all dual values are optimized.

(a) For **PS**, the dual writes as

$$\begin{aligned}\bar{c}(\boldsymbol{\pi}) = \min \quad & \sum_{j\in B} c_j y_j + \sum_{j\in N} c_j y_j \\ & \sum_{j\in B} \mathbf{a}_j y_j + \sum_{j\in N} \mathbf{a}_j y_j = \mathbf{0} \quad [\boldsymbol{\pi} \in \mathbb{R}^m] \\ & \sum_{j\in N} y_j = 1 \quad [\mu \in \mathbb{R}] \\ & y_j \in \mathbb{R} \qquad \forall j \in B \\ & y_j \geq 0 \qquad \forall j \in N.\end{aligned}$$

The optimized y-direction allows all basic variables to increase or decrease as $y_j \in \mathbb{R}$, $\forall j \in B$. If the impact on any basic variable already at zero is a decrease, a degenerate pivot unfortunately occurs. The top rows are fully determined so an extreme point can only have one variable in N. Perhaps it is easier to see it when we explicitly impose $\boldsymbol{\pi}^\intercal = \mathbf{c}_B^\intercal \mathbf{A}_B^{-1}$ from the first condition $\bar{c}_j = 0, \forall j \in B$, in (3.119) for **PS**. This leads to the simplified form

$$\begin{aligned} \bar{c}(\boldsymbol{\pi}) = \max \quad & \mu \\ \text{s.t.} \quad & \mu \leq \bar{c}_j \quad [y_j \geq 0] \quad \forall j \in N, \end{aligned}$$

where μ is unrestricted. The dual program reads as

$$\begin{aligned} \bar{c}(\boldsymbol{\pi}) = \min \quad & \sum_{j \in N} \bar{c}_j y_j \\ \text{s.t.} \quad & \sum_{j \in N} y_j = 1 \quad [\mu] \\ & y_j \geq 0 \qquad \forall j \in N. \end{aligned}$$

This program looks for a convex combination of variables, where an extreme point solution selects a single non-basic variable with the smallest reduced cost.

(b) For **IPS**, the pricing problem in y-variables writes as

$$\begin{aligned} \bar{c}(\boldsymbol{\pi}) = \min \quad & \sum_{j \in F} c_j y_j + \sum_{j \in L} c_j y_j \\ & \sum_{j \in F} \mathbf{a}_j y_j + \sum_{j \in L} \mathbf{a}_j y_j = \mathbf{0} \quad [\boldsymbol{\pi} \in \mathbb{R}^m] \\ & \sum_{j \in L} y_j = 1 \quad [\mu \in \mathbb{R}] \\ & y_j \in \mathbb{R} \qquad \forall j \in F \\ & y_j \geq 0 \qquad \forall j \in L. \end{aligned}$$

Recall that the system in $\boldsymbol{\pi}$ is partially determined so we are indeed looking for a convex combination of variables. The y-direction allows the free variables ($j \in F$) to either increase or decrease which cannot lead to a degenerate pivot. Such a direction always occurs on an edge (Raymond et al., 2010b).

(c) For **MMCC**, the pricing problem has another interpretation with an absolute value, i.e., $\mu \leq -|c_j - \boldsymbol{\pi}^\intercal \mathbf{a}_j|, \forall j \in F$. In other words, we aim to reduce the disparity in signed reduced cost values of the positive variables until they reach 0 by the complementary slackness optimality conditions. The extra variable y_{j+n}, $j \in \{i, \ldots, n\}$, is the result of linearizing this constraint and echoes the forward and backward directions of the free variables. The dual writes as

$$
\begin{aligned}
\bar{c}(\boldsymbol{\pi}) = \min \quad & \sum_{j\in F} -c_j y_{j+n} + \sum_{j\in F\cup L} c_j y_j \\
\text{s.t.} \quad & \sum_{j\in F} -\mathbf{a}_j y_{j+n} + \sum_{j\in F\cup L} \mathbf{a}_j y_j = \mathbf{0} \quad [\boldsymbol{\pi} \in \mathbb{R}^m] \\
& \sum_{j\in F} y_{j+n} + \sum_{j\in F\cup L} y_j = 1 \quad [\mu \in \mathbb{R}] \\
& y_{j+n} \geq 0 \qquad \forall j \in F \\
& y_j \geq 0 \qquad \forall j \in F \cup L.
\end{aligned}
$$

Recall that the system in $\boldsymbol{\pi}$ is completely free so we are indeed looking for a convex combination of variables. The y-direction cannot be degenerate since there are no other variables to account for. Gauthier et al. (2018) and Gauthier and Desrosiers (2022) show that it can be on an edge, a face, or interior.

Exercises of Chapter 4

4.1 Hermann Minkowski

Minkowski's original mathematical interests were in pure mathematics and he spent much of his time investigating quadratic forms and continued fractions. His most original achievement, however, was his 'geometry of numbers' which he initiated in 1890. *Geometrie der Zahlen* was first published in 1910 but the first 240 pages (of the 256) appeared as the first section in 1896.
– MacTutor

Minkowski is perhaps best known for his work in relativity, in which he showed in 1907 that his former student Albert Einstein's special theory of relativity (1905) could be understood geometrically as a theory of four-dimensional space–time, since known as the "Minkowski spacetime."
– Wikipedia

4.2 Alternative decomposition of the 2D illustration

Given are $\mathcal{A}$ and $\mathcal{D}$ in Figure 4.1.

(a) $\text{conv}(\mathcal{A})$.

(b) Domain of the *MP* $\{\mathbf{x} \in \mathbb{R}^2_+ \mid \mathbf{Dx} \geq \mathbf{d}\} \cap \text{conv}(\mathcal{A})$.

(c) Domain of the *MP* compared to $\text{conv}(\mathcal{A} \cap \mathcal{D})$, that of the *ILP* (and *IMP*).

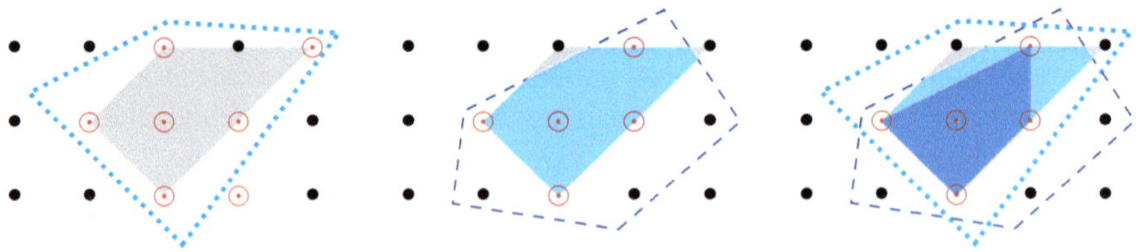

(a) (b) (c)

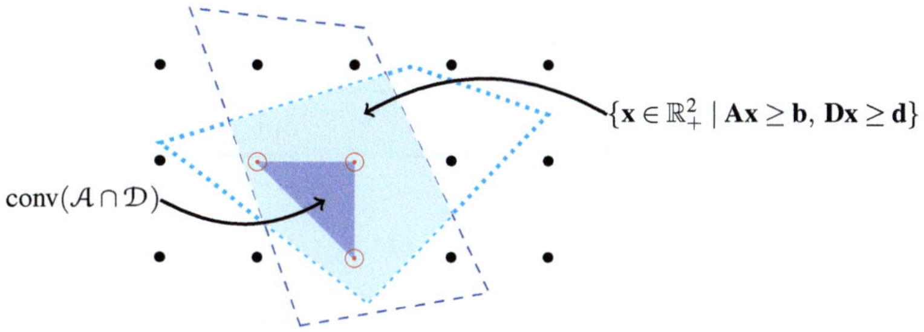

4.3 2D convexification practice

(a) Domain $\{\mathbf{x} \in \mathbb{R}^2_+ \mid \mathbf{Ax} \geq \mathbf{b},\, \mathbf{Dx} \geq \mathbf{d}\}$ of the *LP* in light color. Integer hull $\text{conv}(\mathcal{A} \cap \mathcal{D})$ of the *ILP* in dark color.

(b) Reformulation based on the integer set $\mathcal{D} = \{\mathbf{x} \in \mathbb{Z}^2_+ \mid \mathbf{Dx} \geq \mathbf{d}\}$.

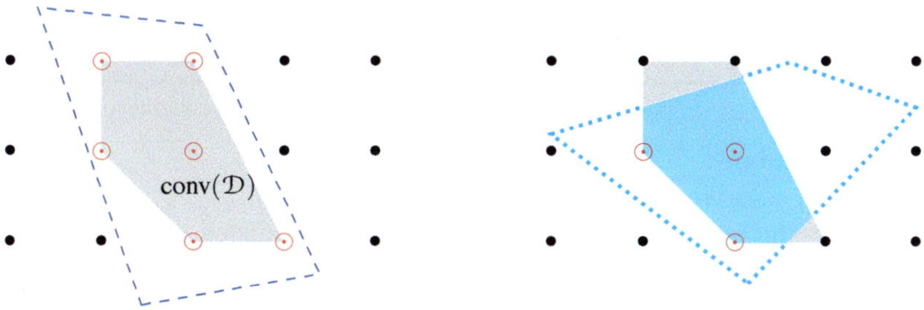

(a) Integer hull $\text{conv}(\mathcal{D})$. (b) Domain $\{\mathbf{x} \in \mathbb{R}^2_+ \mid \mathbf{Ax} \geq \mathbf{b}\} \cap \text{conv}(\mathcal{D})$.

(c) Reformulation based on the integer set $\mathcal{A} = \{\mathbf{x} \in \mathbb{Z}^2_+ \mid \mathbf{Ax} \geq \mathbf{b}\}$.

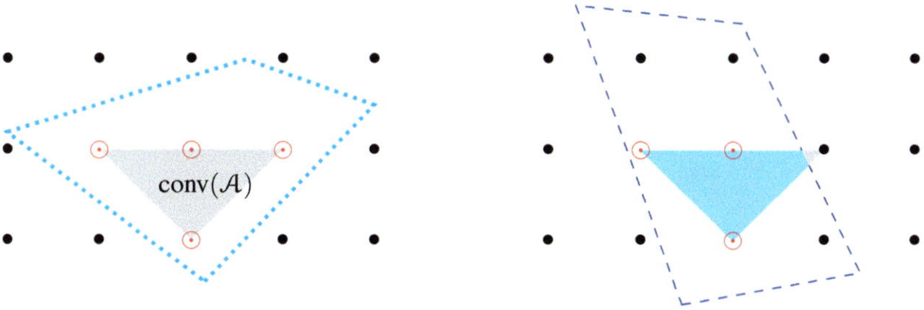

(a) Integer hull $\text{conv}(\mathcal{A})$. (b) Domain $\{\mathbf{x} \in \mathbb{R}^2_+ \mid \mathbf{Dx} \geq \mathbf{d}\} \cap \text{conv}(\mathcal{A})$.

4.4 Reformulation by discretization

(a)

$$\begin{aligned}(17,5) &= (0,2)+5(1,0)+3(4,1)\\ &= \underbrace{(0,2)+0(1,0)+0(4,1)}_{(0,2)\in\mathcal{T}} + \underbrace{5(1,0)+3(4,1)}_{\text{integer combination of } \mathbf{x}_r,\ r\in\ddot{R}}\end{aligned}$$

$$\begin{aligned}(9,4) &= (0,2)+1(1,0)+2(4,1)\\ &= \underbrace{(0,2)+0(1,0)+0(4,1)}_{(0,2)\in\mathcal{T}} + \underbrace{1(1,0)+2(4,1)}_{\text{integer combination of } \mathbf{x}_r,\ r\in\ddot{R}}\end{aligned}$$

(b)

$$\begin{aligned}(6,3) &= (0,2)+1(2,0)+\frac{1}{2}(8,2)\\ &= \underbrace{(0,2)+0(2,0)+\frac{1}{2}(8,2)}_{(4,3)\in\mathcal{T}} + \underbrace{1(2,0)+0(8,2)}_{\text{integer combination of } \mathbf{x}_r,\ r\in\ddot{R}}\end{aligned}$$

$$\begin{aligned}(7,3) &= (0,2)+\frac{3}{2}(2,0)+\frac{1}{2}(8,2)\\ &= \underbrace{(0,2)+\frac{1}{2}(2,0)+\frac{1}{2}(8,2)}_{(5,3)\in\mathcal{T}} + \underbrace{1(2,0)+0(8,2)}_{\text{integer combination of } \mathbf{x}_r,\ r\in\ddot{R}}\end{aligned}$$

4.5 Trick question

The answer is negative, but why?

Intentionally, the question does not mention the reformulated programs, the *IMP* and $\ddot{IMP}$, but only the integer set $\mathcal{D}$ that is reformulated in the theorems' proofs. Those two integer masters are however where convexification appears through the convexity constraint. Or, more precisely, in their linear relaxations we solve until integer optimality is reached. Moreover, full convexification of $\mathcal{D}$ is only possible if *all* extreme points and extreme rays of $\text{conv}(\mathcal{D})$ are generated, which is never the case in practice. It is in fact the beauty of a Dantzig-Wolfe reformulation that the *MP* finds itself implicitly working on this integer hull without us having to know what it is.

4.6 Alternative expression for the Dantzig-Wolfe lower bound

We have

$$\bar{c}^k(\boldsymbol{\pi}_\mathbf{b}, \pi_0^k) = -\pi_0^k + \min_{\mathbf{x}^k\in\mathcal{D}^k} (\mathbf{c}^{k\intercal} - \boldsymbol{\pi}_\mathbf{b}^\intercal \mathbf{A}^k)\mathbf{x}^k$$

by definition (4.45) and

$$z_{RMP} = \boldsymbol{\pi}_\mathbf{b}^\intercal \mathbf{b} + \sum_{k\in K} \pi_0^k$$

by strong duality. The terms π_0^k, $k \in K$, therefore cancel out to

$$\boldsymbol{\pi}_\mathbf{b}^\intercal \mathbf{b} + \sum_{k\in K} \min_{\mathbf{x}^k\in\mathcal{D}^k} (\mathbf{c}^{k\intercal} - \boldsymbol{\pi}_\mathbf{b}^\intercal \mathbf{A}^k)\mathbf{x}^k \le z_{MP}^\star.$$

4.7 Optimality test for the compact formulation
Because $\mathbf{x}_p$ is feasible for the *ILP*, it provides an upper bound: $z^\star_{ILP} \le \mathbf{c}^\intercal \mathbf{x}_p = c_p$. For $|K| = 1$ in the lower bound expression on $z^\star_{MP}$ in (4.68), we also have

$$\boldsymbol{\pi}_\mathbf{b}^\intercal \mathbf{b} + \pi_0 + \bar{c}(\boldsymbol{\pi}_\mathbf{b}, \pi_0) \le z^\star_{MP} \le z^\star_{IMP} = z^\star_{ILP},$$

where, for the optimal solution $\mathbf{x}_p$ to the *ISP*, $\bar{c}(\boldsymbol{\pi}_\mathbf{b}, \pi_0) = c_p - \boldsymbol{\pi}_\mathbf{b}^\intercal \mathbf{a}_p - \pi_0$ by (4.12). Hence,

$$\boldsymbol{\pi}_\mathbf{b}^\intercal \mathbf{b} + c_p - \boldsymbol{\pi}_\mathbf{b}^\intercal \mathbf{a}_p = c_p + \boldsymbol{\pi}_\mathbf{b}^\intercal (\mathbf{b} - \mathbf{a}_p) \le z^\star_{ILP} \le c_p,$$

where $\boldsymbol{\pi}_\mathbf{b}^\intercal (\mathbf{b} - \mathbf{a}_p) = 0$ by assumption. Therefore, $z^\star_{ILP} = c_p$ and $\mathbf{x}^\star_{ILP} = \mathbf{x}_p$.

• *Observe that the value of π_0 is useless in this proof.*

4.8 Identical subproblems: solving an aggregated compact formulation
Compared to the extreme points and extreme rays of $\mathcal{D}$ in (4.49), the index-set $\mathcal{Y}$ associated with $\mathcal{D}_\mathbf{y}$ is

$$\mathcal{Y} = \{\mathbf{y}_p\}_{p \in P} \cup \{\mathbf{y}_r\}_{r \in R} = \{|K|\mathbf{x}_p\}_{p \in P} \cup \{|K|\mathbf{x}_r\}_{r \in R}.$$

Let the Minkowski-Weyl substitution be given by

$$\sum_{p \in P} \mathbf{y}_p \theta_p + \sum_{r \in R} \mathbf{y}_r \theta_r = \mathbf{y}, \quad \sum_{p \in P} \theta_p = 1, \quad \theta_p, \theta_r \ge 0, \ \forall p \in P, r \in R,$$

$$\text{i.e.,} \quad \sum_{p \in P} |K|\mathbf{x}_p \theta_p + \sum_{r \in R} |K|\mathbf{x}_r \theta_r = \mathbf{y}, \quad \sum_{p \in P} \theta_p = 1, \quad \theta_p, \theta_r \ge 0, \ \forall p \in P, r \in R,$$

and define $\lambda_p = |K|\theta_p$, $\lambda_r = |K|\theta_r$, $\forall p \in P, r \in R$. The substitution becomes

$$\sum_{p \in P} \mathbf{x}_p \lambda_p + \sum_{r \in R} \mathbf{x}_r \lambda_r = \mathbf{y}, \quad \sum_{p \in P} \lambda_p = |K|, \quad \lambda_p, \lambda_r \ge 0, \ \forall p \in P, r \in R,$$

and the resulting reformulation of the *aggregated compact formulation* (4.183) is

$$\begin{aligned}
z^\star_{IMP} = \min \ & \sum_{p \in P} c_p \lambda_p + \sum_{r \in R} c_r \lambda_r \\
\text{s.t.} \ & \sum_{p \in P} \mathbf{a}_p \lambda_p + \sum_{r \in R} \mathbf{a}_r \lambda_r \ge \mathbf{b} \\
& \sum_{p \in P} \lambda_p = |K| \\
& \lambda_p \ge 0, \qquad \lambda_r \ge 0 \qquad \forall p \in P, \ r \in R \\
& \sum_{p \in P} \mathbf{x}_p \lambda_p + \sum_{r \in R} \mathbf{x}_r \lambda_r = \mathbf{y} \in \mathbb{Z}^n_+.
\end{aligned}$$

The linear relaxation of the above is obviously the *MP* (4.56). Note that if $\mathcal{D}_\mathbf{y}$ is a polyhedral cone, i.e., $\mathbf{d} = \mathbf{0}$ in (4.183), the *ILP–IMP* special case derived in Proposition 4.11 is validated, that is, index-set K is useless in the compact formulation.

4.9 Identical subproblems: lexicographic ordering of the extreme points
The idea behind the disaggregation rule (4.184) is that the value of λ_p^k is equal to the minimum between the residual supply at node p and the residual demand at node k. As such, recursively compute for $p = 1, \ldots, m$,

$$\lambda_p^k = \min\left\{\lambda_p - \sum_{j=1}^{k-1} \lambda_p^j,\, 1 - \sum_{i=1}^{p-1} \lambda_i^k\right\}, \qquad \text{for } k = 1, \ldots, |K|.$$

It remains to show that, while distributing λ_p, the residual demand at k is also given by $\min\{1, (k - \sum_{i=1}^{p-1} \lambda_i)^+\}$. Indeed, it depends on the total supply available from λ_1 up to λ_{p-1}.

- If $\sum_{i=1}^{p-1} \lambda_i \geq k$, this is sufficient to satisfy all the unit-demands of destination nodes $j = 1, \ldots, k$ and hence the demand at k is $0 = (k - \sum_{i=1}^{p-1} \lambda_i)^+$.
- Otherwise, $(k - \sum_{i=1}^{p-1} \lambda_i) > 0$, either greater than 1, or not, and the demand at k is equal to $\min\{1, (k - \sum_{i=1}^{p-1} \lambda_i)^+\}$.

4.10 Not all blocks are used
The ISP^k is solved over the domain $\mathcal{D}_{\mathbf{0}}^k$ in (4.186b), where we recall that $x_0^k \geq 0$.

- If $x_0^k = 0$, we have $\mathbf{x}^k = \mathbf{0}$ and we obtain the zero-vector $\left[\begin{smallmatrix}0\\ \mathbf{0}\end{smallmatrix}\right]$ for that block.
- Otherwise, a negative reduced cost extreme ray is obtained, encoded with x_0^k=1, equivalent to an extreme point of conv($\mathcal{D}^k$), i.e., the set $\{\mathbf{x}_p\}_{p\in P}$ is in a one-to-one correspondence with $\left\{\left[\begin{smallmatrix}1\\ \mathbf{x}_r\end{smallmatrix}\right]\right\}_{r\in R}$.

With the grouping (4.186) in a Dantzig-Wolfe reformulation, the *IMP* writes as

$$\begin{aligned}
\min \quad & \sum_{k\in K} [c_0^k\ \mathbf{c}^{k\intercal}] \begin{bmatrix}0\\ \mathbf{0}\end{bmatrix} \lambda_0^k + \sum_{k\in K}\sum_{r\in R^k} [c_0^k\ \mathbf{c}^{k\intercal}] \begin{bmatrix}1\\ \mathbf{x}_r^k\end{bmatrix} \lambda_r^k \\
\text{s.t.} \quad & \sum_{k\in K} [0\ \mathbf{A}^k] \begin{bmatrix}0\\ \mathbf{0}\end{bmatrix} \lambda_0^k + \sum_{k\in K}\sum_{r\in R^k} [0\ \mathbf{A}^k] \begin{bmatrix}1\\ \mathbf{x}_r^k\end{bmatrix} \lambda_r^k \geq \mathbf{b} \\
& \lambda_0^k = 1 \qquad \forall k \in K \\
& \lambda_r^k \geq 0 \quad \forall k \in K, r \in R^k \\
& \begin{bmatrix}0\\ \mathbf{0}\end{bmatrix} \lambda_0^k + \sum_{r\in R^k} \begin{bmatrix}1\\ \mathbf{x}_r^k\end{bmatrix} \lambda_r^k = \begin{bmatrix}x_0^k\\ \mathbf{x}^k\end{bmatrix} \in \{0,1\} \times \mathbb{Z}_+^{n^k} \ \forall k \in K.
\end{aligned}$$

For every k, λ_0^k is discarded as well as the convexity constraint $\lambda_0^k = 1$. The *IMP*, where we compute the vector product for the coefficients of the λ_r^k-variables, is

$$\begin{aligned} z^\star_{IMP} = \min \quad & \sum_{k \in K} \sum_{r \in R^k} (c_0^k + c_r^k \lambda_r^k) \\ \text{s.t.} \quad & \sum_{k \in K} \sum_{r \in R^k} \mathbf{a}_r^k \lambda_r^k \geq \mathbf{b} \\ & \lambda_r^k \geq 0 \qquad \forall k \in K, r \in R^k \\ & \sum_{r \in R^k} \lambda_r^k = x_0^k \in \{0,1\} \quad \forall k \in K \\ & \sum_{r \in R^k} \mathbf{x}_r^k \lambda_r^k = \mathbf{x}^k \in \mathbb{Z}_+^{n^k} \qquad \forall k \in K, \end{aligned}$$

for which, also removing all the constraints on the x-variables, the *MP* reads as the linear relaxation of (4.84):

$$\begin{aligned} z^\star_{MP} = \min \quad & \sum_{k \in K} \sum_{r \in R^k} (c_0^k + c_r^k \lambda_r^k) \\ \text{s.t.} \quad & \sum_{k \in K} \sum_{r \in R^k} \mathbf{a}_r^k \lambda_r^k \geq \mathbf{b} \\ & \sum_{r \in R^k} \lambda_r^k \leq 1 \quad \forall k \in K \\ & \lambda_r^k \geq 0 \quad \forall k \in K, r \in R^k. \end{aligned}$$

Notice the re-appearance of the *less-than-or-equal-to-one* inequality constraints.

4.11 Binary knapsack problem

The table below contains the original data with an additional row providing the ratio u_i/w_i, that is, the utility per unit of size. The first item is the most profitable followed by the second, and so on. The optimal solution to the *ILP* formulation (4.103)

$$z^\star_{ILP} = \max \sum_{i=1}^{4} u_i x_i \quad \text{s.t.} \sum_{i=1}^{4} w_i x_i \leq W, \; x_i \in \{0,\, 1\}, \; \forall i \in \{1,\ldots,4\},$$

is $z^\star_{ILP} = 60$, where $x_1^\star = x_2^\star = x_4^\star = 1$. For the linear relaxation, $z^\star_{LP} = 62$, where $x_1^\star = x_2^\star = 1$ for a load of 6 units completed with $x_3^\star = 1/3$. Hence $z^\star_{LP} \neq z^\star_{ILP}$ for this data set and the above knapsack formulation does not possess the integrality property.

i	1	2	3	4
u_i	20	36	18	4
w_i	2	4	3	1
u_i/w_i	*10*	*9*	*6*	*4*
$W = 7$				

4.12 Cutting stock problem: λ-integrality

The domain $\mathcal{D}_K$ in (4.113) is bounded and the same for all pricing problems:

$$\begin{bmatrix} x_0^k \\ \mathbf{x}^k \end{bmatrix} \in \mathcal{D}_K = \left\{ \begin{bmatrix} x_0 \\ \mathbf{x} \end{bmatrix} \in \{0,1\} \times \mathbb{Z}_+^m \;\middle|\; \sum_{i=1}^{m} w_i x_i \leq W x_0 \right\}, \qquad \forall k \in K.$$

Therefore we can use the discretization approach and the *IMP̈* (4.60), with the variables indexed in $\ddot{P}$ only, reads as

$$
\begin{aligned}
z^\star_{IM\ddot{P}} = \min \quad & \sum_{p \in \ddot{P}} c_p \lambda_p \\
\text{s.t.} \quad & \sum_{p \in \ddot{P}} \mathbf{a}_p \lambda_p \geq \mathbf{b} \\
& \sum_{p \in \ddot{P}} \lambda_p = |K| \\
& \lambda_p \in \mathbb{Z}_+, && \forall p \in \ddot{P} \\
& \lambda_p = \sum_{k \in K} \lambda_r^k && \forall p \in \ddot{P} \\
& \sum_{p \in \ddot{P}} \lambda_p^k = 1 && \forall k \in K \\
& \lambda_p^k \in \{0,1\} && \forall k \in K, p \in \ddot{P} \\
& \sum_{p \in \ddot{P}} \mathbf{x}_p \lambda_p^k = \mathbf{x}^k \in \mathbb{Z}_+^n && \forall k \in K,
\end{aligned}
$$

where the aggregated λ_p-variables become non-negative integers because they are computed as the sum over $k \in K$ of the binary λ_p^k-variables.

4.13 Time constrained shortest path problem: nine reformulations

(a) Let the *six path constraints* refer to (4.131b)–(4.131d), the *four flow conservation constraints* to (4.131c), the *duration constraint* to (4.131e), and *all the seven constraints* to (4.131b)–(4.131e) of the *ILP* (4.131).

$$
\begin{aligned}
&\mathcal{D}_1 = \{\mathbf{x} \in \{0,1\}^{|A|} \mid \text{(4.131b)–(4.131d)}\} && \mathcal{A}_1 = \{\mathbf{x} \in \{0,1\}^{|A|} \mid \text{(4.131e)}\} \\
&\mathcal{D}_2 = \mathcal{D}_1 \cap \{3 \leq \sum_{(i,j)\in A} x_{ij} \leq 5\} && \mathcal{A}_2 = \{\mathbf{x} \in \{0,1\}^{|A|} \mid \text{(4.131e)}\} \\
&\mathcal{D}_3 = \{\mathbf{x} \in \{0,1\}^{|A|} \mid \text{(4.131e)}\} && \mathcal{A}_3 = \{\mathbf{x} \in \{0,1\}^{|A|} \mid \text{(4.131b)–(4.131d)}\} \\
&\mathcal{D}_4 = \mathcal{D}_3 \cap \{3 \leq \sum_{(i,j)\in A} x_{ij} \leq 5\} && \mathcal{A}_4 = \{\mathbf{x} \in \{0,1\}^{|A|} \mid \text{(4.131b)–(4.131d)}\} \\
&\mathcal{D}_5 = \mathcal{D}_3 \cap \{\sum_{j=2}^{3} x_{1j} = \sum_{i=4}^{5} x_{i6} = 1\} && \mathcal{A}_5 = \{\mathbf{x} \in \{0,1\}^{|A|} \mid \text{(4.131c)}\} \\
&\mathcal{D}_6 = \mathcal{D}_4 \cap \mathcal{D}_5 && \mathcal{A}_6 = \{\mathbf{x} \in \{0,1\}^{|A|} \mid \text{(4.131c)}\} \\
&\mathcal{D}_7 = \mathcal{D}_6 \cap \{\sum_{i:(i,j)\in A} x_{ij} \leq 1, \forall j \in \{2,\ldots,5\}\} && \mathcal{A}_7 = \{\mathbf{x} \in \{0,1\}^{|A|} \mid \text{(4.131c)}\} \\
&\mathcal{D}_8 = \mathcal{D}_6 \cap \{\sum_{j:(i,j)\in A} x_{ij} \leq 1, \forall i \in \{2,\ldots,5\}\} && \mathcal{A}_8 = \{\mathbf{x} \in \{0,1\}^{|A|} \mid \text{(4.131c)}\} \\
&\mathcal{D}_9 = \{\mathbf{x} \in \{0,1\}^{|A|} \mid \textit{all constraints}\} && \mathcal{A}_9 = \{\mathbf{x} \in \{0,1\}^{|A|}\}.
\end{aligned}
$$

(b) Let $\{\mathbf{x}_p\}_{p\in P}$ be the set of extreme points of $\operatorname{conv}(\mathcal{D}_9)$ and $c_p = \mathbf{c}^\intercal \mathbf{x}_p$, $\forall p \in P$. The *IMP* comprises the convexity constraint and the bounds on the λ- and x-variables:

$$\begin{aligned} z^\star_{IMP} = \min \quad & \sum_{p\in P} c_p \lambda_p \\ \text{s.t.} \quad & \sum_{p\in P} \lambda_p = 1 && [\pi_0] \\ & \lambda_p \geq 0 && \forall p \in P \\ & \sum_{p\in P} \mathbf{x}_p \lambda_p = \mathbf{x} \in \{0,1\}^{|A|}. \end{aligned}$$

4.14 Reformulation of the scene selection problem

(a) Consider the solution

$$\begin{aligned} x_i^k &= \frac{1}{m} && \forall i \in N,\ k \in K \\ y_j^k &= \frac{1}{m} && \forall j \in A,\ k \in K \end{aligned}$$

with objective value $\sum_{k=1}^m \sum_{j\in A} c_j y_j^k = \sum_{j\in A} c_j (\sum_{k=1}^m y_j^k) = \sum_{j\in A} c_j$.

• First, we show that the proposed solution is feasible. As the set K contains m days, it holds in the *assignment* constraints (4.159b) that

$$\sum_{k=1}^m x_i^k = m(\frac{1}{m}) = 1, \qquad \forall i \in N.$$

To have a feasible program, we need at least $\lceil |N|/W \rceil$ days, hence $m \geq |N|/W$, or equivalently $1/m \leq W/|N|$. Therefore

$$\sum_{i\in N} x_i^k = |N|(\frac{1}{m}) \leq |N|(\frac{W}{|N|}) = W, \qquad \forall k \in K,$$

and the *capacity* constraints (4.159c) are satisfied. The last set of *actor* constraints in (4.159d) is obviously satisfied as

$$a_{ij}(\frac{1}{m}) \leq (\frac{1}{m}), \qquad \forall k \in K, i \in N, j \in A.$$

• Second, we show that $\sum_{j\in A} c_j$ is a lower bound for

$$\begin{aligned} \sum_{k=1}^m \sum_{j\in A} c_j y_j^k = \sum_{j\in A} c_j (\sum_{k\in K} y_j^k) \\ 0 \leq y_j^k \leq 1 && \forall k \in K,\ j \in A \\ 0 \leq x_i^k \leq 1 && \forall k \in K,\ i \in N. \end{aligned}$$

For each actor j, there is at least one scene i for which $a_{ij} = 1$, say $\hat{\imath} \in N$. Hence in (4.159d):

$$\sum_{k\in K} y_j^k \geq \sum_{k\in K} a_{\hat{\imath}j} x_{\hat{\imath}}^k = a_{\hat{\imath}j} \sum_{k\in K} x_{\hat{\imath}}^k = 1,$$

the right-hand side being equal to 1 due to the first constraint set $\sum_{k\in K} x_{\hat{\imath}}^k = 1$ in (4.159b) and $a_{\hat{\imath}j} = 1$, hence

$$\sum_{j\in A} c_j (\sum_{k\in K} y_j^k) \geq \sum_{j\in A} c_j.$$

(b) Days in K are identical so that $\mathcal{D}^k = \mathcal{D}, \forall k \in K$, where $\mathcal{D}$ is given by the x_i and y_j binary variables satisfying

$$\sum_{i\in N} x_i \leq W$$
$$a_{ij} x_i \leq y_j \quad \forall i \in N,\ j \in A.$$

The set of extreme points is $\ddot{\mathcal{D}} = \left\{ \begin{bmatrix} \mathbf{x}_p \\ \mathbf{y}_p \end{bmatrix} \in \{0,1\}^{|N|+|A|} \right\}_{p\in\ddot{P}}$.

For a *day*-pattern indexed by $p \in \ddot{P}$, component $x_{ip} = 1$ indicates that scene $i \in N$ is shot whereas component $y_{jp} = 1$ indicates that actor $j \in A$ has to be present (at a cost c_j). Observe that the zero-pattern exists, that is, $\begin{bmatrix} \mathbf{0} \\ \mathbf{0} \end{bmatrix} \in \mathcal{D}$, a free day without any actor being present and scene shot.

(c) We here face identical subproblems, where the common domain is defined for binary variables. Following Proposition 4.10 restricted to extreme points, the $IM\ddot{P}$ becomes

$$\begin{aligned} z^\star_{IM\ddot{P}} = \min \quad & \sum_{p\in\ddot{P}} (\sum_{j\in A} c_j y_{jp}) \lambda_p \\ \text{s.t.} \quad & \sum_{p\in\ddot{P}} x_{ip} \lambda_p = 1 && [\pi_i \in \mathbb{R}] \quad \forall i \in N \\ & \sum_{p\in\ddot{P}} \lambda_p = |K| && [\pi_0 \in \mathbb{R}] \\ & \lambda_p \in \mathbb{Z}_+ && \forall p \in \ddot{P}. \end{aligned}$$

(d) Let $\boldsymbol{\pi} = [\pi_i]_{i\in N}$. The *ISP* writes as

$$\begin{aligned} \bar{c}(\boldsymbol{\pi}, \pi_0) = -\pi_0 + \min \quad & \sum_{j\in A} c_j y_j - \sum_{i\in N} \pi_i x_i \\ \text{s.t.} \quad & \sum_{i\in N} x_i \leq W \\ & a_{ij} x_i \leq y_j && \forall i \in N,\ j \in A \\ & x_i, y_j \in \{0,1\} && \forall i \in N, j \in A. \end{aligned}$$

(e) If we first select an actor $j \in A$, we still have the choice regarding the scenes $i \in N$ such that $a_{ij} = 1$ and $\sum_{i \in N} x_i \leq 5$.
If we rather select a scene $i \in N$, then $x_i = 1$ in $\mathfrak{D}$ and we need all the actors $j \in A$ for which $a_{ij} = 1$. The number of *scene selection* patterns is

$$\sum_{s=0}^{5} \binom{19}{s} = 1 + 19 + 171 + 969 + 3\,876 + 11\,628 = 16\,664,$$

including that with $s = 0$.

(f) With the data of Example 4.7, Table 4.10 shows

$$z^\star_{MP} = 330\,405.41 \gg z^\star_{LP} = 137\,739.00.$$

The formulation of the *ISP* does not possess the integrality property and, most of the time, $z^\star_{MP} \gg z^\star_{LP}$.

4.15 Design of balanced student teams

(a) As stated, an optimal solution to the *IMP* (4.168) is composed of 6 teams for which the corresponding column-vectors are $\mathbf{b}_1, \ldots, \mathbf{b}_6$. In a sequential approach for the second semester, remove from the *IMP* all teams with two or more students grouped together in the first semester. That is, given the teams of 4 and 5 students, discard all teams indexed by $p \in P^4$ and $p' \in P^5$ for which

$$\exists t \in \{1,\ldots,6\} : \mathbf{b}_t^\mathsf{T} \mathbf{a}_p \geq 2; \qquad \exists t \in \{1,\ldots,6\} : \mathbf{b}_t^\mathsf{T} \mathbf{a}_{p'} \geq 2.$$

Equivalently, we keep all teams indexed by p and p' such that

$$\mathbf{b}_t^\mathsf{T} \mathbf{a}_p \leq 1, \quad \forall t \in \{1,\ldots,6\}; \qquad \mathbf{b}_t^\mathsf{T} \mathbf{a}_{p'} \leq 1, \quad \forall t \in \{1,\ldots,6\}.$$

(b) One modeling option is to encode the double semester directly in the columns. For every team of size ℓ, we create $O(|P^\ell|)$ columns which represent every possibility that this team could become in the second semester. This means that we now have $O(\sum_{\ell \in L} |P^\ell|^2)$ columns, i.e., $2\,228\,162\,256 = 11\,016^2 + 45\,900^2$, from which we can filter out all infeasible combinations similarly to what we have seen in (a), i.e., the new set of columns is composed of

$$\forall \ell \in \{4,5\} \qquad \bigcup_{p \in \ddot{P}^\ell} \left\{ \begin{bmatrix} \mathbf{a}_p^\ell \\ \mathbf{a}_{p'}^\ell \end{bmatrix} \right\}_{p' \in \ddot{P}^\ell | \mathbf{a}_{p'}^\mathsf{T} \mathbf{a}_p^\ell \leq 1}.$$

This filter works column by column and is therefore much less aggressive than when an optimal 1-semester solution is known. An obvious consequence of a more difficult problem to solve. The actual number of combinations is $1\,710\,408\,420 = 108\,984\,960 + 1\,601\,423\,460$. To be fair, one needs to come up with more filtering because this is still too many columns even by a modern solver standard. In our experiment, we can fairly easily solve instances with up to 80 people and team size of 7.

4.16 Secret ballot

The answers are based on the following table, where the vote patterns in $\mathcal{D}^a$ are presented according to the various rules identified in the last three rows. The first 10 patterns are those for the 1-decimal digit data and results. The two additional patterns on the right are valid if the percentage shares of every shareholder is given with two decimal digits while the results are rounded to only one.

Voters	Shares (%) 2-digits	Shares (%) 1-digit	Patterns (#) 1	2	3	4	5	6	7	8	9	10	11	12
9	4.51	4.5	1											
10	4.24	4.2		1										
11	3.61	3.6												
12	3.07	3.1			1	1								
13	2.66	2.7					1	1	1				1	
14	2.37	2.4								1				1
15	1.49	1.5					1				1		1	1
[a]16	1.41	1.4			1			1		1	1	1		
17	1.32	1.3									1	1		
18	1.13	1.1				1			1			1		
19	0.36	0.4						1	1	1		1	1	1
20	0.26	0.3		1		1	1		1	1	1	1		1
		1-digit	**4.5**	**4.5**	**4.5**	**4.5**	**4.5**	**4.5**	**4.5**	**4.5**	**4.5**	**4.5**	4.6	4.6
		2-digits	4.51	4.50	**4.48**	4.46	4.41	4.43	4.41	4.40	**4.48**	**4.48**	4.51	**4.48**
		Rounded	**4.5**	**4.5**	**4.5**	**4.5**	4.4	4.4	4.4	4.4	**4.5**	**4.5**	**4.5**	**4.5**
		No. of voters	1	**2**	**2**	3	3	3	4	4	4	5	3	4

(a) For the two decimal digits results, we simply use the more precise data in the knapsack equality constraints, i.e.,

$$\sum_{i=1}^{20} p_i x_i^k = b_2^k \quad \forall k \in K.$$

The number of possible voting patterns with $b_2^a = 4.48\,\%$ in $\mathcal{D}^a$ is only 4: #3, #9, #10, and #12.

For the rounded results, each knapsack equality constraint becomes an interval system, i.e.,

$$b^k - 0.05 \le \sum_{i=1}^{20} p_i x_i^k \le b^k + 0.04, \qquad \forall k \in K.$$

There are 8 possible voting patterns with $\sum_{i=1}^{20} p_i x_i^a$ around $4.5\,\%$ in $\mathcal{D}^a$: #1 (4.51), #2 (4.50), #3 (4.48), #4 (4.46), #9 (4.48), #10 (4.48), #11(4.51), and #12 (4.48). *We observe that more precision means less freedom in the voting patterns.*

(b) Candidate a, i.e., shareholder 16 voting for him or herself, appears in 5 patterns: #3, #6, #8, #9, and #10.

(c) There are only two possible voting patterns for candidate a: #2 and #3.

4.17 Cutting stock with rolls of different widths: compact formulation

(a) Starting with the structure of the ISP^k (2.34), a compact formulation is

$$
\begin{aligned}
z^\star_{ILP} = \min \quad & \sum_{k\in K}(W^k x_0^k - \sum_{i=1}^m w_i x_i^k) \\
\text{s.t.} \quad & \sum_{k\in K} x_i^k = b_i && \forall i \in \{1,\dots,m\} \\
& \sum_{i=1}^m w_i x_i^k \le W^k x_0^k && \forall k \in K \\
& x_i^k \le b_i x_0^k && \forall k \in K,\ i \in \{1,\dots,m\} \\
& x_0^k \in \{0,1\} && \forall k \in K \\
& x_i^k \in \mathbb{Z}_+ && \forall k \in K,\ i \in \{1,\dots,m\}.
\end{aligned}
$$

Compared to (2.34), note the presence of the variable x_0^k in $x_i^k \le b_i x_0^k$.

(b) Let $\mathbf{x}^k = [x_i^k]_{i\in\{1,\dots,m\}}$, $\forall k \in K$, and group the constraints as

$$
\mathcal{A} = \left\{ \left\{ x_0^k \in \{0,1\},\ \mathbf{x}^k \in \mathbb{Z}_+^m \right\}_{k\in K} \;\middle|\; \sum_{k\in K} x_i^k = b_i,\ \forall i \in \{1,\dots,m\} \right\}
$$

$$
\mathcal{D}^k = \left\{ x_0^k \ge 0,\ \mathbf{x}^k \in \mathbb{Z}_+^m \;\middle|\; \begin{array}{l} \sum_{i=1}^m w_i x_i^k \le W^k x_0^k \\ x_i^k \le \ b_i x_0^k \quad \forall i \in \{1,\dots,m\} \end{array} \right\}, \quad \forall k \in K.
$$

Observe that in this grouping, $x_0^k \ge 0$ is not requested to binary values in $\mathcal{D}^k$. Consequently, $\text{conv}(\mathcal{D}^k)$ is a polyhedral cone for which the set of extreme rays can be represented with the integer scaled value $x_0^k = 1$. Using the discretization approach on polyhedral cones (Section Polytope and polyhedral cone, p. 187), the reformulation writes as

$$
\begin{aligned}
z^\star_{IM\ddot{P}} = \min \quad & \sum_{k\in K} \sum_{r\in \ddot{R}_0^k} c_r^k \lambda_r^k \\
\text{s.t.} \quad & \sum_{k\in K} \sum_{r\in \ddot{R}_0^k} a_{ir}^k \lambda_r^k \ge b_i \quad [\pi_i] && \forall i \in \{1,\dots,m\} \\
& \lambda_r^k \in \mathbb{Z}_+ && \forall k \in K,\ r \in \ddot{R}_0^k,
\end{aligned}
$$

where $c_r^k = W^k - \sum_{i=1}^m w_i x_{ir}^k$, $\forall k \in K$, and $a_{ir}^k = x_{ir}^k$, $\forall i \in \{1,\dots,m\}$.

An optimal solution in x-variables is computed *a posteriori* as

$$
\sum_{r\in \ddot{R}^k} \mathbf{x}_r^k \lambda_r^{k\star} = \mathbf{x}^{k\star} \in \mathbb{Z}_+^n, \qquad \forall k \in K.
$$

4.18 Aircraft routing with schedule synchronization: compact formulation
A compact formulation is

$$
\begin{aligned}
z^\star_{ILP} = \min \; & \sum_{k\in K} \sum_{(i,j)\in A^k} c^k_{ij} x^k_{ij} \\
\text{s.t.} \; & \sum_{k\in K} \sum_{j:(i,j)\in A^k} x^k_{ij} = 1 && \forall i \in N \\
& \sum_{k\in K} (t^k_i - a_i x^k_i) - t_{m_i} = 0 && \forall i \in N \\
& \begin{bmatrix} \mathbf{x}^k \\ \mathbf{t}^k \\ [x^k_i]_{i\in N} \end{bmatrix} \in \mathcal{D}^k && \forall k \in K,
\end{aligned}
$$

where $\mathcal{D}^k = \left\{ \begin{bmatrix} \mathbf{x}^k \\ \mathbf{t}^k \\ [x^k_i]_{i\in N} \end{bmatrix} \in \{0,1\}^{|A^k|} \times \mathbb{R}^{|N|}_+ \times \{0,1\}^{|N|} \;\middle|\; (2.45) \right\}$.

4.19 Single depot vehicle scheduling problem: compact formulations

(a) Let $K = \{1, \ldots, v\}$ and for a given $k \in K$, consider the following network $G^k = (V^k, A^k)$ with node set $V^k = N \cup \{o^k, d^k\}$ and arc set $A^k = I \cup (\{o^k\} \times N) \cup (N \times \{d^k\}) \cup \{(o^k, d^k)\}$, where I is the common set of *inter-trip arcs*.

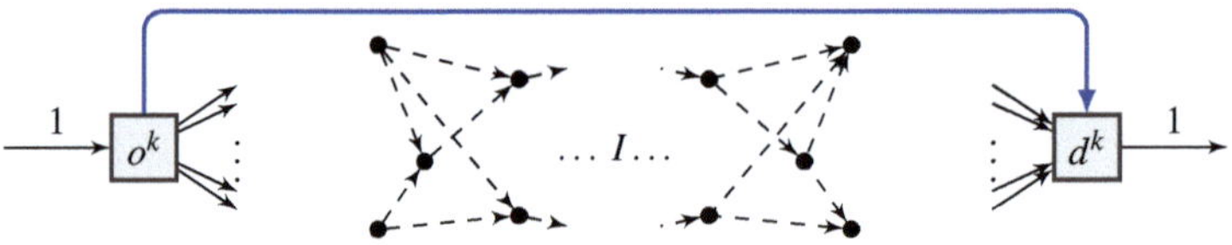

Let the binary variable x^k_{ij} be the flow through arc $(i,j) \in A^k$. Introducing the binary variables x^k_0, $\forall k \in K$, a multi-commodity network flow formulation derived from Proposition 4.15 is given as

$$
\begin{aligned}
z^\star_{ILP} = \min \; & \sum_{k\in K} \sum_{(i,j)\in A^k} c_{ij} x^k_{ij} \\
\text{s.t.} \; & \sum_{k\in K} \sum_{j:(i,j)\in A^k} x^k_{ij} = 1 && \forall i \in N \\
& \sum_{j:(o^k,j)\in A^k} x_{o^k j} = x^k_0 && \forall k \in K \\
& \sum_{j:(i,j)\in A^k} x^k_{ij} - \sum_{j:(j,i)\in A^k} x^k_{ji} = 0 && \forall k \in K, i \in N \\
& - \sum_{j:(j,d^k)\in A^k} x_{jd^k} = -x^k_0 && \forall k \in K \\
& x^k_0, x^k_{ij} \in \{0,1\} && \forall k \in K, (i,j) \in A^k.
\end{aligned}
$$

Obviously, the binary variable x^k_0 can be renamed as x^k_{do}, $\forall k \in K$.

(b) Because the *ISP* (4.188) possesses the integrality property (Condition **1**) and the **0**-vector is not a valid x-solution to the partitioning constraints (4.187b) (Condition **2**), Proposition 4.17 is used to derive the following *ILP*:

$$\begin{aligned} z^\star_{ILP} = \min \quad & \sum_{(i,j)\in A} c_{ij}x_{ij} \\ \text{s.t.} \quad & \sum_{j:(i,j)\in A} x_{ij} = 1 \qquad \forall i \in N \\ & \sum_{j:(o,j)\in A} x_{oj} \leq v \\ & \sum_{j:(o,j)\in A} x_{oj} = x_0 \\ \sum_{j:(i,j)\in A} x_{ij} - & \sum_{j:(j,i)\in A} x_{ji} = 0 \qquad \forall i \in N \\ - & \sum_{i:(i,d)\in A} x_{id} = -x_0 \\ & x_0, x_{ij} \geq 0 \qquad \forall (i,j) \in A, \end{aligned}$$

where the bounding constraints $x_{ij} \leq x_0$, $\forall (i,j) \in A$, are redundant. Again, x_0 can be renamed as x_{do}, as in the figure below.

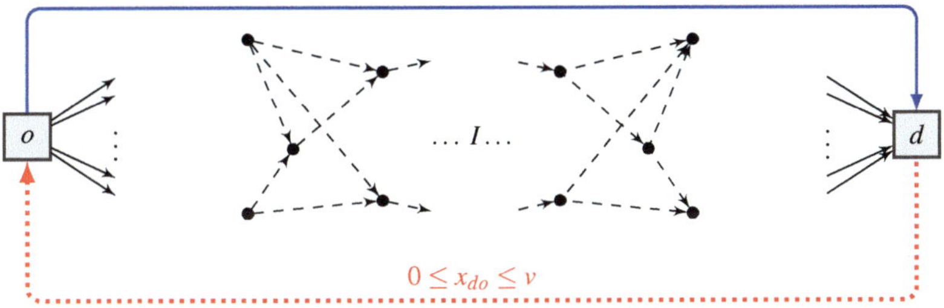

The definition by Geoffrion (1974, p. 89) states that *the integrality gap is zero* for *any cost* coefficients. A Dantzig-Wolfe reformulation *IMP* of the above *ILP*, expressed as a mixed-integer linear program formulated in terms of continuous λ- and integer x-variables, also possesses the integrality property by Proposition 4.6 . This proposition shows that $z^\star_{LP} \leq z^\star_{MP} \leq z^\star_{IMP}$, where $z^\star_{IMP} = z^\star_{ILP} = z^\star_{LP}$. Hence $z^\star_{IMP} = z^\star_{MP}$ yields a zero-integrality gap. Indeed, no matter the way we group the constraints of the *ILP*, the x-domain of the arc-flow variables remains the same in any *IMP*. See also Note 4.9.

4.20 Multiple depot vehicle scheduling problem

(a) $|N|$ in (4.189b); $|K|$ in (4.189c); $|K| \times (|N|+2)$ in (4.189d).

(b) For all $k \in K$, let $\mathbf{x}^k = [x^k_{ij}]_{(i,j)\in A^k_{do}}$.

$$\mathcal{A} = \left\{ \left\{ \mathbf{x}^k \in \mathbb{Z}_+^{|A^k_{do}|} \right\}_{k \in K} \;\middle|\; (4.189\text{b})\text{–}(4.189\text{c}) \right\}$$

$$\mathcal{D}^k = \left\{ \mathbf{x}^k \in \mathbb{Z}_+^{|A^k_{do}|} \;\middle|\; (4.189\text{d}) \right\}, \qquad \forall k \in K.$$

(c) $\text{conv}(\mathcal{D}^k)$ is a polyhedral cone with $\mathbf{0}$ as the single extreme point (flow conservation equations for all nodes in N^k), hence $\mathcal{X}^k = \{\mathbf{0}\} \cup \{\mathbf{x}^k_r\}_{r \in R^k}$ and $\mathcal{X} = \cup_{k \in K} \mathcal{X}^k$.

(d) For all $k \in K$, we discard the $\mathbf{0}$-vector together with $\lambda^k_{\mathbf{0}}$ and its convexity constraint $\lambda^k_{\mathbf{0}} = 1$. The *IMP* writes as

$$\begin{aligned}
z^\star_{IMP} = \min \quad & \sum_{k \in K} \sum_{r \in R^k} c^k_r \lambda^k_r \\
\text{s.t.} \quad & \sum_{k \in K} \sum_{r \in R^k} a^k_{ir} \lambda^k_r = 1 && [\pi_i] \quad \forall i \in N \\
& \sum_{r \in R^k} a^k_r \lambda^k_r \le v^k && [\pi^k_0] \quad \forall k \in K \\
& \lambda^k_r \ge 0 && \forall k \in K, r \in R^k \\
& \sum_{r \in R^k} \mathbf{x}^k_r \lambda^k_r = \mathbf{x}^k \in \mathbb{Z}_+^{|A^k_{do}|},
\end{aligned}$$

$$\text{where } c^k_r = \sum_{(i,j) \in A^k_{do}} c^k_{ij} x^k_{ij,r}, \; a^k_{ir} = \sum_{j:(i,j) \in A^k} x^k_{ij,r},$$
$$\text{and } a^k_r = x^k_{do,r} = 1, \quad \forall k \in K, r \in R^k.$$

Row-size of the IMP: $|N| + |K|$, the number of trips and depots.

(e) Let $\boldsymbol{\pi} = [\pi_i]_{i \in N}$. The ISP^k writes as

$$\begin{aligned}
\bar{c}^k(\boldsymbol{\pi}, \pi^k_0) = \min \quad & c^k - \sum_{i \in N} \pi_i a^k_i - a^k \pi^k_0 \\
\text{s.t.} \quad & \sum_{j:(i,j) \in A^k_{do}} x^k_{ij} - \sum_{j:(j,i) \in A^k_{do}} x^k_{ji} = 0 && \forall i \in N^k \\
& x^k_{ij} \in \mathbb{Z}_+ && \forall (i,j) \in A^k_{do} \\
& c^k = \sum_{(i,j) \in A^k_{do}} c^k_{ij} x^k_{ij} \\
& a^k_i = \sum_{j:(i,j) \in A^k} x^k_{ij} \\
& a^k = x^k_{do} \quad \text{(scaled to 1).}
\end{aligned}$$

(f) Solving the ISP^k given $a^k = x^k_{do} = 1$ results in solving a shortest path problem from o^k to d^k in an acyclic network; dynamic programming complexity of

$O(|A^k_{do}|)$, see Ahuja et al. (1993, § 4.4). From the *tree*-solution of the ISP^k, bring to the *RMP* a subset of columns, ideally partitioning the covered trips.

(g) The formulation of every ISP^k possesses the integrality property: $z^\star_{LP} = z^\star_{MP}$ by Proposition 4.13.

(h) Let us start with the network G_{do} on which we define the *SDVSP*, a network flow problem, where for each trip $i \in N$, the costs *depot-to-trip* and *trip-to-depot* are computed as $c_{oi} = \min_{k\in K} c_{o^k i}$ and $c_{id} = \min_{k\in K} c_{id^k}$. Moreover, the cost c_v of a vehicle is assigned to arc (d,o) whereas arc (o,d) is removed.

$$
\begin{aligned}
z^\star_{SDVSP} = \min \quad & \sum_{(i,j)\in A_{do}} c_{ij}x_{ij} \\
\text{s.t.} \quad & \sum_{j:(i,j)\in A} x_{ij} = 1 && [\alpha_i] \quad \forall i \in N \\
\sum_{j:(i,j)\in A_{do}} x_{ij} - & \sum_{j:(j,i)\in A_{do}} x_{ji} = 0 && [\beta_i] \quad \forall i \in N \cup \{o,d\} \\
& x_{ij} \geq 0 && \forall (i,j) \in A_{do}.
\end{aligned}
$$

Given an (integer) optimal solution for the *SDVSP*:

1. Retrieve the $v^\star$ *od*-paths. The dual value α_i is a good estimation of π_i and can be used in a Stabilized column generation strategy, see p. 398.
2. $\forall p \in \{1,\ldots,v^\star\}$, $k \in K$, compute the cost c^k_p of path p assigned to depot k.
3. *Optional.* Solve a network problem for finding a least-cost *paths-to-depots assignment* (illustrated below). This provides an upper bound on $z^\star_{IMP}$.

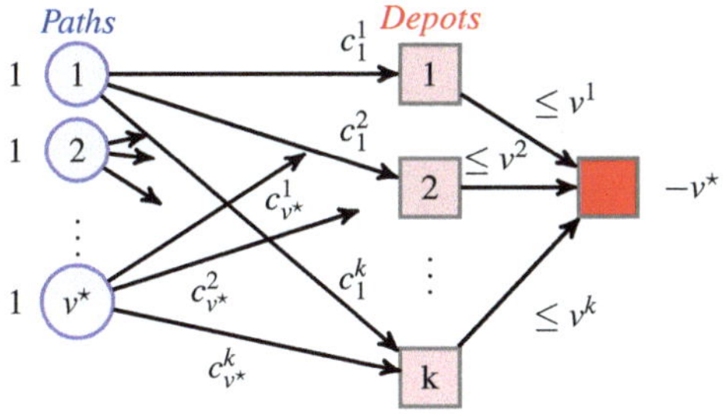

4. Initialize the *RMP* with λ^k_p, $\forall p \in \{1,\ldots,v^\star\}$, $k \in K$, each with its cost and column encodings. Add the constraint $\sum_{k\in K}\sum_{r\in R^k} \lambda^k_r = v^\star$ (or $\geq v^\star$) with dual variable π_v. Modify the ISP^k with this additional dual value:

$$\bar{c}^k(\boldsymbol{\pi}, \pi^k_0, \pi_v) = \min_{\mathcal{D}^k} \sum_{(i,j)\in A^k_{do}} c^k_{ij}x^k_{ij} - \sum_{i\in N} \pi_i \Big(\sum_{j:(i,j)\in A^k} x^k_{ij}\Big) - \pi^k_0 - \pi_v.$$

4.21 Useless Dantzig-Wolfe reformulations
We see two additional situations: $\mathcal{D} = \{\mathbf{x} \in \mathbb{R}^n_+\}$ and $\mathcal{D} = \{\mathbf{x} \in [0,1]^n \mid \sum_{j=1}^n x_j = 1\}$. The proofs are similar to that of Proposition 4.5.

• In the first case, $\mathcal{D}$ is reformulated with extreme rays represented by the orthogonal unit-vectors $\mathbf{x}_r = \mathbf{e}_r, \forall r \in \{1,\ldots,n\}$, together with the $\mathbf{0}$ removable extreme point. We express $\mathbf{x}$ as

$$\sum_{j=1}^{n} \mathbf{e}_j \lambda_j = \mathbf{x} \geq \mathbf{0}.$$

Writing this equation component-wise, we get $\lambda_1 = x_1, \ldots, \lambda_n = x_n$ which simply renames the x-variables. This Dantzig-Wolfe reformulation gives back the original *LP* formulation.

• In the second case, we reformulate $\mathcal{D}$ which is the convex hull of the n unit vectors in $\mathbb{R}^n_+$. The set of extreme points of $\mathcal{D}$ is $\{\mathbf{e}_1, \ldots, \mathbf{e}_n\}$. We express $\mathbf{x}$ as

$$\sum_{j=1}^{n} \mathbf{e}_j \lambda_j = \mathbf{x} \in [0,1]^n, \qquad \sum_{j=1}^{n} \lambda_j = 1, \qquad \lambda_j \geq 0, \ j = 1, \ldots, n.$$

Writing the first equation component-wise, we also get $\lambda_1 = x_1, \ldots, \lambda_n = x_n$. Finally, the introduced convexity constraint is exactly the original constraint in $\mathcal{D}$.

Exercises of Chapter 5

5.1 Martin Desrochers
Martin Desrochers is a former PhD student (1983–1986) of François Soumis. The title of his dissertation is: *La fabrication d'horaires de travail pour les conducteurs d'autobus par une méthode de génération de colonnes*, Université de Montréal. Amongst his contributions, we can cite the design of algorithms for the *shortest path problem with time windows* (Desrochers and Soumis, 1988c,b). He is best known for the algorithmic design of the *shortest path problem with resource constraints* (Desrochers, 1988).

With Paul Pelletier, Yvan Dumas and Michel Sauvé, Martin Desrochers is behind the development and implementation of the first version of the `GENCOL` solver (1981–1987, ***gén**ération de **col**onnes*, in French) at the heart of two commercial optimization systems. Indeed, the findings of his dissertation are part of `HASTUS`, a modular software distributed by GIRO (Montréal) for bus, metro, tram and passenger rail operations, used all around the world. The same is true for the `ALTITUDE` suite of crew planning solutions for airline companies distributed by AD OPT, the Montréal division of IBS Software.

5.2 Zero-objects for the *VRPTW*
(a) Following formulation (4.81) of Section Not all blocks are used, add the binary variables $x^k_{do}, \forall k \in K$, and write the new compact formulation as

Martin Desrochers and Jacques (Vancouver, Canada, May 1989).

$$
\begin{aligned}
z^{\star}_{ILP} = \min \quad & \sum_{k\in K}\sum_{(i,j)\in A} c_{ij}x^k_{ij} & \\
\text{s.t.} \quad & \sum_{k\in K}\sum_{j:(i,j)\in A} x^k_{ij} = 1 & \forall i \in C \\
& \sum_{j:(i,j)\in A} x^k_{ij} - \sum_{j:(j,i)\in A} x^k_{ji} = \begin{cases} x^k_{do} & \text{for } i = o \\ 0 & \forall i \in C \\ -x^k_{do} & \text{for } i = d \end{cases} & \forall k \in K \\
& \sum_{i\in C}\sum_{j:(i,j)\in A} q_i\, x^k_{ij} \le Q x^k_{do} & \forall k \in K \\
& a_i x^k_{do} \le t^k_i \le b_i x^k_{do} & \forall k \in K, i \in \{o,d\} \\
& a_i(\sum_{j:(i,j)\in A} x^k_{ij}) \le t^k_i \le b_i(\sum_{j:(i,j)\in A} x^k_{ij}) & \forall k \in K, i \in C \\
& x^k_{ij}(t^k_i + t_{ij} - t^k_j) \le 0 & \forall k \in K, (i,j) \in A \\
& x^k_{ij}, x^k_{do} \in \{0,1\} & \forall k \in K, (i,j) \in A.
\end{aligned}
$$

When $x^k_{do} = 0$, we have $x^k_{ij} = 0$, $\forall (i,j) \in A$, such that the vehicle is empty and $t^k_i = 0$, $\forall i \in C \cup \{o,d\}$. Otherwise $x^k_{do} = 1$ and we compute an *od*-path constrained by the load and time window constraints.

(b) For $k \in K$, let the new set of arcs be $A_{do} = A \cup \{(d,o)\}$ and $\mathbf{x}^k = [x^k_{ij}]_{(i,j)\in A_{do}}$. A possible grouping is

$$\mathcal{A} = \left\{ \left\{ \mathbf{x}^k \in \{0,1\}^{|A_{do}|} \right\}_{k \in K} \mid \sum_{k \in K} \sum_{j:(i,j) \in A} x_{ij}^k = 1, \ \forall i \in C \right\}$$
$$\mathcal{D}^k = \left\{ \mathbf{x}^k \in \{0,1\}^{|A_{do}|} \mid \mathbf{D}^k \mathbf{x}^k \geq \mathbf{d}^k \right\}, \qquad \forall k \in K,$$

where the set $\mathbf{D}^k \mathbf{x}^k \geq \mathbf{d}^k$ denotes again the constraints associated with vehicle k. Because all domains $\mathcal{D}^k$ are identical, we can aggregate them into a single one by omitting index k:

$$\sum_{j:(i,j) \in A} x_{ij} - \sum_{j:(j,i) \in A} x_{ji} = \begin{cases} x_{do} & \text{for } i = o \\ 0 & \forall i \in C \\ -x_{do} & \text{for } i = d \end{cases}$$
$$\sum_{i \in C} \sum_{j:(i,j) \in A} q_i \, x_{ij} \leq Q x_{do}$$
$$a_i x_{do} \leq \ t_i \leq b_i x_{do} \qquad \forall i \in \{o,d\}$$
$$a_i \Big(\sum_{j:(i,j) \in A} x_{ij} \Big) \leq \ t_i \leq b_i \Big(\sum_{j:(i,j) \in A} x_{ij} \Big) \qquad \forall i \in C$$
$$x_{ij}(t_i + t_{ij} - t_j) \leq 0 \qquad \forall (i,j) \in A$$
$$x_{ij} \in \{0,1\} \qquad \forall (i,j) \in A_{do}.$$

(c) Adapting the *IMP* (5.7), the *MP* becomes

$$z^\star_{MP} = \min \sum_{p \in P} c_p \lambda_p$$
$$\text{s.t.} \ \sum_{p \in P} a_{ip} \lambda_p = 1 \qquad \forall i \in C$$
$$\sum_{p \in P} \lambda_p \leq |K|$$
$$\lambda_p \geq 0 \qquad \forall p \in P.$$

5.3 Two-index arc-flow formulation for the *VRPTW*
Let $\mathcal{R} = \{time, load\}$.
For $r = time$, set $a_i^{time} = a_i$, $b_i^{time} = b_i$, $\forall i \in N$, and $t_{ij}^{time} = t_{ij}$, $\forall (i,j) \in A$.
For $r = load$, set $a_i^{load} = 0$, $b_i^{load} = Q$, $\forall i \in N$, and $t_{ij}^{load} = q_j$, $\forall (i,j) \in A$.

$$
\begin{aligned}
z^\star_{ILP} = \min \quad & \sum_{(i,j)\in A} c_{ij}x_{ij} & \\
\text{s.t.} \quad & \sum_{j:(i,j)\in A} x_{ij} = 1 & \forall i \in C \\
& \sum_{j:(i,j)\in A} x_{ij} - \sum_{j:(j,i)\in A} x_{ji} = \begin{cases} |K| & \text{for } i = o \\ 0 & \forall i \in C \\ -|K| & \text{for } i = d \end{cases} & \\
& a^r_i\Big(\sum_{j:(i,j)\in A} x_{ij}\Big) \le t^r_i \le b^r_i\Big(\sum_{j:(i,j)\in A} x_{ij}\Big) & \forall r \in \mathcal{R}, i \in N \\
& x_{ij}(t^r_i + t^r_{ij} - t^r_j) \le 0 & \forall r \in \mathcal{R}, (i,j) \in A \\
& x_{od} \ge 0, \text{ integer} & \\
& x_{ij} \in \{0,1\} & \forall (i,j) \in A \setminus \{(o,d)\}.
\end{aligned}
$$

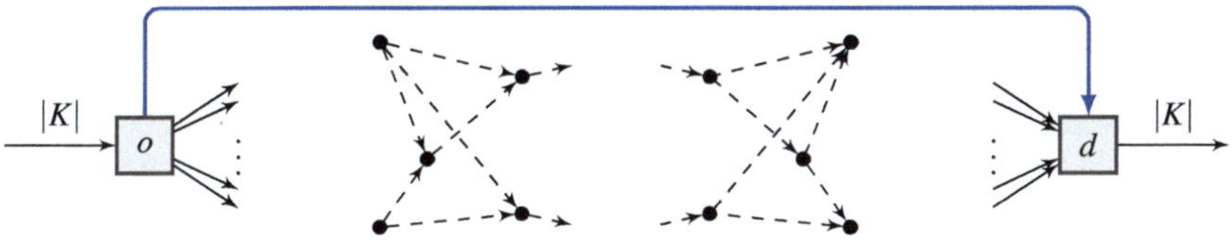

5.4 Tightness of the linear relaxations for the *VRPTW*
Given that $q_i = 10$ for all $i \in C$ and $Q = 15$, any feasible route in P visits at most one customer. Thus, P contains only four routes, denoted p_j, $j \in \{0,1,2,3\}$, where p_0 visits no customer and p_j, $j = 1,2,3$, visits only customer j. An optimal solution to the linear relaxation of (5.8) is given by: $\lambda_{p_j} = 1$, $j = 1,2,3$, and $\lambda_{p_0} = 0$. Its cost is $z^\star_{MP} = 6$.

For model (5.1), let $K = \{k_1, k_2, k_3\}$. An optimal solution to its linear relaxation is given by: $x^{k_j}_{o1} = x^{k_j}_{12} = x^{k_j}_{23} = x^{k_j}_{3d} = x^{k_j}_{od} = 0.5$ for $j = 1,2$ and all other variables equal to 0. Its cost is $z^\star_{LP} = 4$. Thus, $z^\star_{LP} < z^\star_{MP}$, showing that model (5.8) has a tighter linear relaxation than that of model (5.1).

5.5 Dominance rule for the *ESPPTWC*
Like in Definition 5.3, consider two feasible partial paths p and p', both ending at node $j \in N$. Consider also a feasible extension χ' of p' that ends at node $\ell \in N$ and would yield labels

$$E_{p\oplus\chi'} = \left(T^{rCost}_{p\oplus\chi'}, T^{time}_{p\oplus\chi'}, T^{load}_{p\oplus\chi'}, [T^{cust_i}_{p\oplus\chi'}]_{i\in C}\right)$$

and

$$E_{p'\oplus\chi'} = \left(T^{rCost}_{p'\oplus\chi'}, T^{time}_{p'\oplus\chi'}, T^{load}_{p'\oplus\chi'}, [T^{cust_i}_{p'\oplus\chi'}]_{i\in C}\right)$$

if used to extend paths p and p', respectively.

Because $p' \oplus \chi'$ is feasible, we deduce that

$$\begin{aligned} T^{time}_{p' \oplus \chi'} &\leq b^{time}_{\ell} \\ T^{load}_{p' \oplus \chi'} &\leq b^{load}_{\ell} \\ T^{cust_i}_{p' \oplus \chi'} &\leq b^{cust_i}_{\ell} \quad \forall i \in C. \end{aligned}$$

Because all REFs are non-decreasing with respect to each variable and the composition of non-decreasing functions is non-decreasing, we get that

$$\begin{aligned} T^{time}_{p \oplus \chi'} &\leq T^{time}_{p' \oplus \chi'} \leq b^{time}_{\ell} \\ T^{load}_{p \oplus \chi'} &\leq T^{load}_{p' \oplus \chi'} \leq b^{load}_{\ell} \\ T^{cust_i}_{p \oplus \chi'} &\leq T^{cust_i}_{p' \oplus \chi'} \leq b^{cust_i}_{\ell} \quad \forall i \in C, \end{aligned}$$

which means that $p \oplus \chi'$ is also feasible. Furthermore, due again to the non-decreasing property of the REFs, we find that $T^{rCost}_{p \oplus \chi'} \leq T^{rCost}_{p' \oplus \chi'}$, which is equivalent to $c_{p \oplus \chi'} \leq c_{p' \oplus \chi'}$. Therefore, p dominates p' according to Definition 5.3.

5.6 Dominance of labels by a dominated label
Observe that the set $\mathcal{U}_h \cup \mathcal{P}_h$ always contains labels that do not dominate each other. Let E_{p^*} be a label in $\mathcal{U}_h \cup \mathcal{P}_h$ that dominates $E_{p'}$ in Step 6. Therefore,

$$\begin{aligned} T^{rCost}_{p^*} &\leq T^{rCost}_{p'} \\ T^{time}_{p^*} &\leq T^{time}_{p'} \\ T^{load}_{p^*} &\leq T^{load}_{p'} \\ T^{uCust_i}_{p^*} &\leq T^{uCust_i}_{p'} \quad \forall i \in C. \end{aligned}$$

Because E_{p^*} does not dominate any other label in $\mathcal{U}_h \cup \mathcal{P}_h$, these inequalities imply that $E_{p'}$ cannot either. It may only dominate E_{p^*} if both labels are equal (one of them must, however, be kept).

5.7 Dominance of processed labels by a non-dominated label
Observe that $h \in C$ and $t_{jh} > 0$ under the assumption. Because all labels $E_{p''}$ in $\mathcal{P}_h$ have been processed before E_p, then $T^{time}_{p''} \leq T^{time}_{p} < T^{time}_{p} + t_{jh} \leq T^{time}_{p'}$. Therefore, according to condition (5.19b) of the dominance rule, label $E_{p'}$ cannot dominate $E_{p''}$.

5.8 Omitting the *load* resource in the *SPPTWC*
Label E_4 would be dominated by label E_8. Consequently, E_4 would not be extended and, thus, labels E_{28}, E_{29}, E_{37}, E_{38}, and E_{39} would not be generated.

5.9 Revisiting the time-constrained shortest path problem

(a) The resource set includes only two resources: $\mathcal{R} = \{cost, time\}$. Note that, because network $G = (N,A)$ is acyclic, there is no need to define additional resources to ensure path elementarity. There are only resource windows for resource $r = time$: $[a_i^{time}, b_i^{time}] = [0, 14]$ for all $i \in N$.
Let $E_p = (T_p^{cost}, T_p^{time})$ be a label representing a path p ending at a node h. Extending E_p along an arc $(h, j) \in A$ produces a new label $E_{p'} = (T_{p'}^{cost}, T_{p'}^{time})$ whose components are computed using the following REFs:

$$T_{p'}^{cost} = f_{hj}^{cost}(T_p^{cost}) = T_p^{cost} + c_{hj}$$
$$T_{p'}^{time} = f_{hj}^{time}(T_p^{time}) = T_p^{time} + t_{hj},$$

where the time resource REFs exploit the facts that $a_i^{time} = 0$, for all $i \in N$, and $t_{hj} > 0$ for all $(h, j) \in A$.

(b) Given that the network $G = (N,A)$ is acyclic, the nodes in N can be first sorted on topological order and the labels processed according to this order (i.e., all labels associated with a node are extended consecutively before moving on to the labels of the next node).
For Example 3.2, a topological sorting of the nodes in N gives the following order: $(1,3,2,4,5,6)$. Starting with label $E_0 = (0,0)$ at node 1, the labeling algorithm generates the labels listed on page 607. Each row in this table indicates the extensions of a label (column xLbl) to create new labels (columns nLbl). For each new label, column F/I specifies if it is feasible (F) or infeasible (I), while column dLbl identifies a label that dominates it whenever it is the case. From the labels highlighted in bold, we deduce that $(1,3,2,4,6)$ is an optimal path of cost 13.

	Node 3			Node 2			Node 4			Node 5			Node 6		
xLbl	nLbl	F/I	dLbl	nLbl	F/I	dLbl	nLbl	F/I	dLbl	nLbl	F/I	dLbl	nLbl	F/I	dLbl
$\mathbf{E_0 = (0,0)}$	$\mathbf{E_1 = (10,3)}$	F	-	$E_2 = (1,10)$	F	-									
$E_1 = (10,3)$				$\mathbf{E_3 = (11,5)}$	F	-	$E_4 = (15,10)$	F	E_8	$E_5 = (22,6)$	F	-			
$E_2 = (1,10)$							$E_6 = (2,11)$	F	-	$E_7 = (3,13)$	F	-			
$E_3 = (11,5)$							$\mathbf{E_8 = (12,6)}$	F	-	$E_9 = (13,8)$	F	-			
$E_6 = (2,11)$										$E_{10} = (12,12)$	F	-	$E_{11} = (3,18)$	I	-
$E_8 = (12,6)$										$E_{12} = (22,7)$	F	E_5	$\mathbf{E_{13} = (13,13)}$	F	-
$E_5 = (22,6)$													$E_{14} = (24,8)$	F	-
$E_7 = (3,13)$													$E_{15} = (5,15)$	I	-
$E_9 = (13,8)$													$E_{16} = (15,10)$	F	-
$E_{10} = (12,12)$													$E_{17} = (14,14)$	F	E_{13}

Exercise 5.9: Generated labels.

5.10 Two-cycle elimination

We present two solutions.

First solution: To account for the elimination of 2-cycles, a new component is considered in the labels. This component is not quantitative. It stores the next-to-last node of the path represented by the label if this node exists and is set to NIL otherwise. For a path $p = (i_0 = o, i_1, \ldots, i_m)$, this component, denoted T_p^{ntln}, is given by

$$T_p^{ntln} = \begin{cases} i_{m-1} & \text{if } m \geq 1 \\ NIL & \text{otherwise.} \end{cases}$$

In label E_0, we set $T_0^{ntln} = NIL$. When extending a path $p = (i_0 = o, i_1, \ldots, i_m = j)$ along an arc $(j,h) \in A$ to create a new path $p' = (i_0 = o, i_1, \ldots, i_m = j, h)$ represented by $E_{p'} = (T_{p'}^{rCost}, T_{p'}^{time}, T_{p'}^{load}, T_{p'}^{ntln})$, the REF $f_{jh}^{ntln}(\cdot)$ for this new resource is given by

$$T_{p'}^{ntln} = f_{jh}^{ntln}() = j.$$

However, to avoid a 2-cycle, path p, represented by $E_p = (T_p^{rCost}, T_p^{time}, T_p^{load}, T_p^{ntln})$, cannot be extended along (j,h) if $T_p^{ntln} = h$. According to Definition 5.3, label E_p cannot dominate a label that can be directly extended to node T_p^{ntln} because E_p cannot. Therefore, a label E_p can only dominate a label $E_{p'}$ associated with the same node if the condition $T_p^{ntln} = T_{p'}^{ntln}$ also holds.

Second solution: The latter condition is very restrictive. Indeed, one can observe that, if all dominance conditions are met beside this one, then the label resulting from an extension of label E_p along an arc $(j,h') \in A$ with $h' \neq h$ dominates that resulting from the corresponding extension of label $E_{p'}$. Therefore, it is not worth extending $E_{p'}$ along any arc (j,h') with $h' \neq h$. When all dominance conditions hold including (resp. except) $T_p^{ntln} = T_{p'}^{ntln}$, we say that E_p *strongly* (resp. *weakly*) dominates $E_{p'}$. When two labels E_{p_1} and E_{p_2} with $T_{p_1}^{ntln} \neq T_{p_2}^{ntln}$ weakly dominate a label $E_{p'}$, label $E_{p'}$ can be discarded.

To apply strong and weak dominance, the labeling algorithm needs to be modified as follows. First, for each label E_p representing a path $p = (i_0 = o, i_1, \ldots, i_m)$, the T_p^{ntln} component needs to be considered as described in the first solution above, except for the dominance condition. Second, the labels must include another new non-quantitative component, denoted T_p^{extN}, which indicates the single node to which the label can be extended if it has already been weakly dominated and is equal to NIL otherwise. In the initial label E_0, $T_0^{extN} = NIL$. Furthermore, when creating a new label $E_{p'}$ by extending a label E_p, $T_{p'}^{extN}$ is initially set to NIL. Dominance works as follows. Given two labels $E_p = (T_p^{rCost}, T_p^{time}, T_p^{load}, T_p^{ntln}, T_p^{extN})$ and $E_{p'} = (T_{p'}^{rCost}, T_{p'}^{time}, T_{p'}^{load}, T_{p'}^{ntln}, T_{p'}^{extN})$ associated with the same node and such that

$$\begin{aligned} T_p^{rCost} &\leq T_{p'}^{rCost} \\ T_p^{time} &\leq T_{p'}^{time} \\ T_p^{load} &\leq T_{p'}^{load}, \end{aligned}$$

then $E_{p'}$ is dominated if $T_p^{ntln} = T_{p'}^{ntln}$ or if $T_{p'}^{extN} \neq NIL$ and $T_p^{ntln} \neq T_{p'}^{extN}$. Otherwise, $T_{p'}^{extN} = T_p^{ntln}$. Finally, a label $E_p = (T_p^{rCost}, T_p^{time}, T_p^{load}, T_p^{ntln}, T_p^{extN}) \neq E_0$ cannot be extended to node T_p^{ntln} and can only be extended to node T_p^{extN} if $T_p^{extN} \neq NIL$.

5.11 Backward labeling for the *ESPPTWC*

Each partial path $p = (i_m = j, \ldots, i_1, i_0 = d)$ from a node $j \in N$ to node d is represented by a label

$$E_p^{bw} = (T_p^{bw\_rCost}, T_p^{bw\_time}, T_p^{bw\_rCap}, [T_p^{bw\_cust_i}]_{i \in C})$$

associated with node j and whose components are defined as follows:

$T_p^{bw\_rCost}$: Reduced cost of path p,
i.e., $T_p^{bw\_rCost} = \sum_{\ell=1}^{m} \tilde{c}_{i_\ell, i_{\ell-1}}$;

$T_p^{bw,time}$: Latest feasible time at node j using path p.
Its value can be computed recursively as

$$\begin{aligned} T_{p0}^{bw\_time} &= b_{i_0} \\ T_{p\ell}^{bw\_time} &= \min\{b_{i_\ell}, T_{p,\ell-1}^{bw\_time} - t_{i_\ell, i_{\ell-1}}\}, \qquad \ell = 1, \ldots, m \\ T_p^{bw\_time} &= T_{pm}^{bw\_time}; \end{aligned}$$

$T_p^{bw\_rCap}$: Residual capacity required at node j to complete path p,
i.e., $T_p^{bw\_rCap} = Q - \sum_{\ell=1}^{m} q_{i_\ell}$;

$T_p^{bw\_cust_i}$, $i \in C$: Indicator equal to 1 if customer i is visited along path p,
i.e., if there exists $\ell \in \{1, \ldots, m\}$ such that $i_\ell = i$, and 0 otherwise.

Thus, the backward resource set $\mathcal{R}^{bw}$ is defined by

$$\mathcal{R}^{bw} = \{bw\_rCost, bw\_time, bw\_rCap, [bw\_cust_i]_{i \in C}\}.$$

At a node $j \in N$, the resource windows for the resources in $\mathcal{R}^{bw} \setminus \{bw\_rCost\}$ are given by

$$\begin{aligned} T_p^{bw\_time} &\in [a_j, b_j] \\ T_p^{bw\_rCap} &\in [0, Q] \\ T_p^{bw\_cust_i} &\in [0, 1], \quad \forall i \in C. \end{aligned}$$

The initial label $E_0^{bw} = (0, b_d, Q, [0]_{i \in C})$ is associated with node d. To create a new partial path $p' = (h, i_m = j, \ldots, i_1, i_0 = d)$ from path $p = (i_m = j, \ldots, i_1, i_0 = d)$, label E_p^{bw} is extended backwardly along the arc $(h, j) \in A$ to yield a new label

$$E_{p'} = (T_{p'}^{bw\_rCost}, T_{p'}^{bw\_time}, T_{p'}^{bw\_rCap}, (T_{p'}^{bw\_cust_i})_{i \in C})$$

using the following backward REFs:

$$\begin{aligned}
T_{p'}^{bw_rCost} &= f_{hj}^{bw_rCost}(T_p^{bw_rCost}) &&= T_p^{bw_rCost} + \tilde{c}_{hj} \\
T_{p'}^{bw_time} &= f_{hj}^{bw_time}(T_p^{bw_time}) &&= \min\{b_h, T_p^{bw_time} - t_{hj}\} \\
T_{p'}^{bw_rCap} &= f_{hj}^{bw_rCap}(T_p^{bw_rCap}) &&= T_p^{bw_rCap} - q_h \\
T_{p'}^{bw_cust_i} &= f_{hj}^{bw_cust_i}(T_p^{bw_cust_i}) &&= \begin{cases} T_p^{bw_cust_i} + 1 \text{ if } h = i \\ T_p^{bw_cust_i} \quad \text{otherwise} \end{cases} \qquad \forall i \in C.
\end{aligned}$$

A label E_p^{bw} dominates a label $E_{p'}^{bw}$ associated with the same node if the following conditions hold:

$$\begin{aligned}
T_p^{bw_rCost} &\le T_{p'}^{bw_rCost} \\
T_p^{bw_time} &\ge T_{p'}^{bw_time} \\
T_p^{bw_rCap} &\ge T_{p'}^{bw_rCap} \\
T_p^{bw_cust_i} &\le T_{p'}^{bw_cust_i}, \qquad \forall i \in C.
\end{aligned}$$

Observe the $\ge$ inequalities for the *bw_time* and *bw_rCap* resources which ensue from the fact that larger values for the *bw_time* and *bw_rCap* resources are preferred.

5.12 Bidirectional labeling for the *ESPPTWC*
The path $p \oplus p'$ is feasible if and only if

$$\begin{aligned}
T_p^{time} &\le T_{p'}^{bw_time} \\
T_p^{load} &\le T_{p'}^{bw_rCap} \\
T_p^{cust_i} + T_{p'}^{bw_cust_i} &\le 1, \qquad \forall i \in C \setminus \{j\}.
\end{aligned}$$

5.13 Unreachable customers in the *ng-SPPTWC*
When a customer is unreachable for a path p, it will also be unreachable for all extensions of p. This is not necessarily the case for a customer $i \in C$ that cannot be reached from a direct extension of a label E_p because $T_p^{ngCust_i} = 1$. Indeed, depending on the customer neighborhoods NG_i, $i \in C$, it might be possible to reach customer i using two or more arcs. Consequently, both types of information must be stored and treated separately in the labels. Furthermore, the definition of unreachability must be revised as follows. For a path ending at node j and represented by label $E_p = (T_p^{rCost}, T_p^{time}, T_p^{load}, [T_p^{ngCust_i}]_{i \in C})$, the customer $i \in C$ is said to be unreachable if and only if one of the following three conditions is met:

$$(j,i) \notin A; \qquad T_p^{time} + t_{ji} > b_i; \qquad T^{load} + q_i > Q.$$

Both information can, however, be used simultaneously to improve dominance by replacing conditions (5.29) with

$$T_p^{ngCust_i} \le \max\{T_{p'}^{ngCust_i}, T_{p'}^{uCust_i}\}, \qquad \forall i \in C \cap NG_h.$$

5.14 Applying the *ng-SPPTWC* labeling algorithm
The labels generated by the algorithm are displayed on page 612. A label E_p is represented by a vector of five components, where, to be concise, the last two components indicate the subset of customers i for which $T_p^{ngCust_i} = 1$ (this subset is called the memory of E_p) and the subset of customers i for which $T_p^{uCust_i} = 1$, i.e., which are unreachable. The table only lists the feasible labels and specifies in columns dLbl the labels which have dominated the labels in columns nLbl. Note that the extensions of a label E_p yielding infeasible paths can be identified a priori by checking the memory of E_p and its unreachable customers and, thus, do not have to be performed.

From the labels highlighted in bold, we deduce that an optimal solution is given by path $(o, 1, 3, 2, d)$ which has a reduced cost of -37. Given that this path is elementary, it would also be an optimal solution to the corresponding *ESPPTWC*.

5.15 Non-decreasing REFs assumption for the *ESPPRC*
When some REFs are non-decreasing, we cannot guarantee that the conditions (5.35) are sufficient to ensure dominance according to Definition 5.3. Indeed, consider a resource $r \in \mathcal{R}$ for which some REFs f_{ij}^r, $(i, j) \in A$ are not non-decreasing. Let p and p' be two partial paths ending at the same node and represented by the labels $E_p = \left(T_r^p\right)_{r\in\mathcal{R}}$ and $E_{p'} = \left(T_r^{p'}\right)_{r\in\mathcal{R}}$, respectively. Furthermore, let χ' be a feasible extension of p'. Even if $T_p^r \le T_{p'}^r$, there is no guarantee that $T_{p\oplus\chi'}^r \le T_{p'\oplus\chi'}^r$ because some REFs f_{ij}^r, $(i, j) \in A$, are not non-decreasing. Consequently, χ' might be an infeasible extension of p or $c_{p\oplus\chi'}$ might be greater than $c_{p'\oplus\chi'}$.

5.16 Composition and addition of non-decreasing functions
Let $\mathbf{x}_1, \mathbf{x}_2 \in \mathbb{R}^n$ such that $\mathbf{x}_1 \le \mathbf{x}_2$. Then $f(\mathbf{x}_1) \le f(\mathbf{x}_2)$ and $\mathbf{y}_1 = g(\mathbf{x}_1) \le g(\mathbf{x}_2) = \mathbf{y}_2$.

(a) Therefore, $f(\mathbf{y}_1) \le f(\mathbf{y}_2)$, that is, $f \circ g(\mathbf{x}_1) \le f \circ g(\mathbf{x}_2)$.
(b) Adding up the two inequalities gives $f(\mathbf{x}_1) + g(\mathbf{x}_1) \le f(\mathbf{x}_2) + g(\mathbf{x}_2)$, i.e., $(f+g)(\mathbf{x}_1) \le (f+g)(\mathbf{x}_2)$.

5.17 Customers with both a pickup and a delivery
It would not be valid to use a single node because the quantity delivered at a customer that would not be accounted for (q_i^D if $q_i^P \ge q_i^D$ and q_i^P otherwise) comes from the depot and, thus, limits the total quantity that can be onboard the vehicle before this customer. Similarly, the quantity picked up at a customer that would not be accounted for (q_i^D if $q_i^P \ge q_i^D$ and q_i^P otherwise) must be transported to the depot and, thus, limits the total quantity that can be onboard the vehicle after this customer.

5.18 Unreachable customers for the *SPDP*
Let $p = (i_0 = o, i_1, \ldots, i_m)$ be a partial path in G that is represented by a label E_p with components T_p^{time} and T_p^{maxL} amongst others. A customer $i \in C$ is unreachable from p if at least one of the following conditions hold:

1. $\exists h \in \{1, \ldots, m\}$ such that $i_h = i$;
2. $i \in C^P$ and $T_p^{loadP} + q_i > Q$;
3. $i \in C^D$ and $T_p^{maxL} + q_i > Q$.

	Node 1			Node 2		
xLbl	nLbl	F/I	dLbl	nLbl	F/I	dLbl
$\mathbf{E_0 = (0,0,0,\emptyset,\emptyset)}$	$\mathbf{E_1 = (3,2,1,\{1\},\{1\})}$	F	-	$E_2 = (15,8,2,\{2\},\{1,2\})$	F	-
$E_1 = (3,2,1,\{1\},\{1\})$				$E_6 = (0,12,3,\{2\},\{1,2,4\})$	F	-
$E_2 = (15,8,2,\{2\},\{1,2\})$						
$E_3 = (12,10,1,\{3\},\{1,3\})$				$E_{13} = (-17,14,3,\{2,3\},\{1,2,4\})$	F	-
$E_7 = (-3,11,2,\{3\},\{1,3\})$				$\mathbf{E_{16} = (-32,15,4,\{2,3\},\{1,2,4\})}$	F	-
$E_6 = (0,12,3,\{2\},\{1,2,4\})$						
$E_{13} = (-17,14,3,\{2,3\},\{1,2,4\})$						
$E_{16} = (-32,15,4,\{2,3\},\{1,2,4\})$						
$E_4 = (20,16,2,\{4\},\{1,2,4\})$						
$E_{14} = (-11,16,3,\{3,4\},\{1,2,4\})$						
$E_{17} = (-26,17,4,\{3,4\},\{1,2,4\})$						
$E_{19} = (-16,17,4,\{3\},C)$						
$E_{23} = (3,18,3,\{3,4\},C)$						

Node 3			Node 4			Node d		
nLbl	F/I	dLbl	nLbl	F/I	dLbl	nLbl	F/I	dLbl
$E_3 = (12,10,1,\{3\},\{1,3\})$	F	-	$E_4 = (20,16,2,\{4\},\{1,2,4\})$	F	-	$E_5 = (0,0,0,\emptyset,C)$	F	-
$\mathbf{E_7 = (-3,11,2,\{3\},\{1,3\})}$	F	-	$E_8 = (2,16,3,\{4\},\{1,4\})$	F	-	$E_9 = (1,12,1,\emptyset,C)$	F	E_5
$E_{10} = (-1,13,3,\{3\},\{1,3,4\})$	F	E_7	$E_{11} = (6,17,4,\{4\},\{1,2,4\})$	F	-	$E_{12} = (10,20,2,\emptyset,C)$	F	E_5
			$E_{14} = (-11,16,3,\{3,4\},\{1,2,4\})$	F	-	$E_{15} = (-2,24,1,\emptyset,C)$	F	-
			$E_{17} = (-26,17,4,\{3,4\},\{1,2,4\})$	F	-	$E_{18} = (-17,25,2,\emptyset,C)$	F	-
$E_{19} = (-16,17,4,\{3\},C)$	F	-				$E_{20} = (-5,24,3,\emptyset,C)$	F	-
						$E_{21} = (-22,26,3,\emptyset,C)$	F	-
						$\mathbf{E_{22} = (-37,27,4,\emptyset,C)}$	F	-
$E_{23} = (3,18,3,\{3,4\},C)$	F	-				$E_{24} = (15,34,2,\emptyset,C)$	F	E_5
						$E_{25} = (-16,34,3,\emptyset,C)$	F	E_{18}
						$E_{26} = (-31,35,4,\emptyset,C)$	F	E_{22}
						$E_{27} = (-30,31,4,\emptyset,C)$	F	E_{22}
						$E_{39} = (-11,32,3,\emptyset,C)$	F	E_{18}

Exercise 5.14: Labels generated for the *ng-SPPTWC*.

5.19 *SPDP* is a special case of the *PDP*

To model the *SPDP* as a *PDP*, we define a transportation request for each customer $i \in C^P \cup C^D$ as follows. The request u_i for a customer $i \in C^P$ is represented by a pickup node $u_i^+ = i$ and a delivery node u_i^- that is adjacent to the destination depot node d. On the contrary, the request u_i for a customer $i \in C^D$ is represented by a pickup node u_i^+ adjacent to the origin depot node o and a delivery node $u_i^- = i$.

Assuming that the customers $i \in C^P$ are numbered from 1 to $|C^P|$, the only arcs exiting node u_i^- for $i \in C^P$ are

$$(u_i^-, u_{i+1}^-), (u_i^-, u_{i+2}^-), \ldots, (u_i^-, u_{|C^P|}^-), \text{ and } (u_i^-, d).$$

They all have a zero-cost. Similarly, assuming that the customers in $i \in C^D$ are numbered from 1 to $|C^D|$, the only arcs entering node u_i^+ for $i \in C^D$ are

$$(u_1^+, u_i^+), (u_2^+, u_i^+), \ldots, (u_{i-1}^+, u_i^+), \text{ and } (o, u_i^+).$$

They also have a zero-cost. With this modeling, any route starts with a (possibly empty) sequence of pickup nodes u_i^+, $i \in C^D$, that represent the merchandise originating from the depot and delivered to customers in C^D. It is followed by a mixed sequence of pickup nodes u_i^+, $i \in C^P$ and delivery nodes u_i^-, $i \in C^D$, in any ordered which may not contain both node types. Finally, it ends with a (possibly empty) sequence of delivery nodes u_i^-, $i \in C^P$.

5.20 The vehicle routing problem with backhauls

As for the *VRPTW*, the network $G = (N, A)$ contains a source node o, a sink node d, and one node for each customer $i \in C$, i.e., $N = C \cup \{o, d\}$. The arc set A contains:

- All arcs (o, j) such that $j \in N \setminus \{o\}$;
- All arcs (i, d) such that $i \in N \setminus \{d\}$;
- All arcs $(i, j) \in C^D \times C^D$ such that $q_i + q_j \leq Q$;
- All arcs $(i, j) \in C^P \times C^P$ such that $q_i + q_j \leq Q$;
- All arcs $(i, j) \in C^D \times C^P$.

Because A contains no arcs from a pickup node to a delivery node, a delivery cannot be performed after a pickup in any $o-d$ path. The adjusted cost of an arc $(i, j) \in A$ is given by $\tilde{c}_{ij} = c_{ij} - \pi_i$.

The set of resources is given by $\mathcal{R} = \{rCost, load, [uCust_i]_{i \in C}\}$. Resource *rCost* stores the reduced cost and is managed similarly as for the *VRPTW*. Resource *load* is used to impose vehicle capacity as follows. We set the resource windows $[a_i^{load}, b_i^{load}] = [0, Q]$ for all $i \in N$ and use the following REFs when extending a label $E_p = (T_p^{rCost}, T_p^{load}, [T_p^{uCust_i}]_{i \in C})$ along an arc $(j, h) \in A$:

$$f_{jh}^{load}(T_p^{load}) = \begin{cases} T_p^{load} + q_h, & \text{if } (j,h) \in A \setminus (C^D \times C^P) \\ q_h, & \text{if } (j,h) \in C^D \times C^P, \end{cases}$$

assuming that $q_d = 0$. We highlight that, for increased efficiency, a single resource is used to ensure that vehicle capacity is not exceeded for both the deliveries and the pickups.

The unreachable customer resources $[uCust_i]_{i \in C}$ are used to ensure path elementarity. For each customer $i \in C$, a resource window $[0, 1]$ is associated with each node $i \in N$. For all customers $i \in C$, the following REFs are applied to compute these resource values when extending a label $E_p = (T_p^{rCost}, T_p^{load}, [T_p^{uCust_i}]_{i \in C})$ along an arc $(j, h) \in A$:

$$f_{jh}^{uCust_i}(T_p^{load}, T_p^{uCust_i}) = \begin{cases} T_p^{uCust_i} + 1, & \text{if } h = i \\ 0, & \text{if } h \neq i, (j,h) \in C^D \times C^P \\ \max\{T_p^{uCust_i}, U_{jh}^i(T_p^{load})\}, & \text{otherwise,} \end{cases}$$

where

$$U^i_{jh}(T^{load}_p) = \begin{cases} 1 \text{ if } (h,i \in C^D \vee h,i \in C^P) \wedge (f^{load}_{jh}(T^{load}_p) + q_i > Q) \\ 0 \text{ otherwise.} \end{cases}$$

We observe that, to increase label elimination, all resources $uCust_i$, $i \in C^D$, take value 0 at pickup nodes. Indeed, all these resource values are reset to 0 when traversing an arc $(j,h) \in C^D \times C^P$.

With this resource set and these non-decreasing REFs (with respect to each variable), the *standard* dominance rule can be applied: $T^r_p \le T^r_{p'}$, $\forall r \in \mathcal{R}$.

5.21 Resource validity for the *PDPTW*

The resource req_u is bounded by $[0,1]$ at all nodes. As it increases by 1 each time that node u^+ is visited and never changes otherwise, request u cannot be picked up more than once.

When request u is picked up (i.e., node u^+ is visited), resource on_u becomes equal to 1. Because the corresponding resource window is $[0,0]$ at the sink node, the value of on_u must decrease afterwards. The only way to do so is to visit the delivery node u^-, i.e., to deliver request u.

Finally, observe that the resources on_u are not sufficient to ensure that deliveries are performed at most once because the REFs $f^{on_u}_{ij}(\mathbf{t}_i)$ only provide lower bounds on the resource values $t^{on_u}_j$ in the *ESPPRC* (see Note 5.10). Without the resources $notOn_u$, it might be possible to perform a delivery without having picked up the request or to perform several times the same delivery. For a request u, the resource $notOn_u$ plays a role symmetric to that of resource on_u. Its value is equal to 1 at the source node and becomes equal to 0 only at pickup node u^+ (which can be visited at most once). Because visiting node u^- increases the value of $notOn_u$ by 1 and the resource window for this node and resource is $[1,1]$, it is impossible to perform the same delivery multiple times.

5.22 Unreachable requests for the *PDPTW*

Let $p = (i_0 = o, i_1, \ldots, i_m)$ be a partial path in G that is represented by a label E_p with components T^{time}_p and $[T^{on_u}_p]_{u \in U}$ amongst others. A request $v \in U$ is said to be unreachable from p if at least one of the following conditions hold:

1. $\exists h \in \{1,\ldots,m\}$ such that $i_h = v^-$;
2. $T^{on_v} = 1$ and $T^{time}_p + t_{i_m,v^-} > b_{v^-}$;
3. $T^{on_v} = 0$ and $T^{time}_p + t_{i_m,v^+} + t_{v^+,v^-} > b_{v^-}$.

Note that more complex conditions (taking into account, e.g., the time windows of the other onboard requests) might also be considered in conditions 2 and 3. However, checking them might be too time-consuming to be worthwhile.

5.23 Dominance only for same onboard requests in the *PDPTW*

Let p and p' be two partial paths ending at the same node and denote by $[T^{on_u}_j]_{u \in U}$ and $[T^{notOn_u}_j]_{u \in U}$ the *on* and *notOn* components of their associated labels E_j, $j = p, p'$. From the REFs (5.46) and (5.47), we can easily deduce that $T^{on_u}_j + T^{notOn_u}_j = 1$

for all $u \in U$ and $j \in \{p, p'\}$, i.e., if $T_j^{on_u} = 0$, then $T_j^{notOn_u} = 1$ and vice versa. Consequently, in rule (5.35), the dominance conditions

$$T_p^{on_u} \leq T_{p'}^{on_u}, \qquad T_p^{notOn_u} \leq T_{p'}^{notOn_u},$$

for request u can be both satisfied if and only if $T_p^{on_u} = T_{p'}^{on_u}$ and $T_p^{notOn_u} = T_{p'}^{notOn_u}$. This is equivalent to saying that these conditions are met for all requests $u \in U$ if and only if $O_p = O_{p'}$.

5.24 Pairing feasibility constraints

The structure of a network G^{wb} ensures that

- A pairing starts and ends at the same base because the *sop* and *eop* arcs are linked to departure and arrival nodes at base b only.
- A pairing spans at most $\bar{n}^D$ days because the network contains *flight* arcs representing flights in W_w only.
- A connection between two consecutive flights in a duty lasts at least $\underline{t}^{CNX}$ because the time associated with an *arrival* node is equal to the flight arrival time plus $\underline{t}^{CNX}$.
- A rest period between two consecutive duties lasts at least $\underline{t}^{RST}$ due to the construction of the *rest* arcs.
- A rest period cannot occur at base b because there are no rest arcs associated with this base.

5.25 Minimum and maximum resource constraints for the *CPPBC*

(a) The dominance rule would not be valid. Indeed, the condition $T_p^{dtyMax} \leq T_{p'}^{dtyMax}$ would favor paths with smaller durations, i.e., paths that could potentitally not be extended to reach the minimum duty duration $\underline{t}^{DTY}$.

(b) According to the REFs (5.57d)–(5.57e) and the resource windows in Table 5.4, we can deduce that

$$T_q^{dtyMin} = \begin{cases} \underline{t}^{DTY} - T_q^{dtyMax} & \text{if } T_q^{dtyMax} < \underline{t}^{DTY} \\ 0 & \text{otherwise} \end{cases} \qquad q = p, p'.$$

Therefore, when the dominance condition $T_p^{dtyMax} \leq T_{p'}^{dtyMax}$ holds, we get that $T_p^{dtyMin} > T_{p'}^{dtyMin}$ unless $T_p^{dtyMax} = T_{p'}^{dtyMax} < \underline{t}^{DTY}$ (in which case $T_p^{dtyMin} = T_{p'}^{dtyMin} > 0$) or $T_p^{dtyMax} \geq \underline{t}^{DTY}$ (in which case $T_p^{dtyMin} = T_{p'}^{dtyMin} = 0$).

5.26 New pairing constraints for the *CPPBC*

(a) In each subproblem ISP^{wb}, a new resource *flyTime* is added with a resource window $[0,0]$ at the source node and $[0, \bar{t}^{FLY}]$ at all other nodes. When extending a label along an arc $(i, j) \in A^{wb}$, the REF for this resource is given by:

$$f_{ij}^{flyTime}(\mathbf{t}_i) = \begin{cases} t_i^{flyTime} + F_{ij} & \text{if } (i,j) \in A^{FLY} \\ 0 & \text{if } (i,j) \in A^{RST} \\ t_i^{flyTime} & \text{otherwise.} \end{cases}$$

(b) We introduce a new type of arcs and a new resource.

► **The new arc type**, called *srf (short-rest-and-flight)*, represents a rest period of a duration less than $\underline{t}^{RST}$ followed by a start of a duty and a flight. In a network G^{wb}, $w \in W$, $b \in B$, an *srf* arc links the arrival nodes of two flights f_1 and f_2 if

- the arrival airport of f_1 is the same as the departure airport of f_2;
- this airport is not base b;
- the difference between the departure time of f_2 and the arrival time of f_1 falls in the semi-open interval $[\underline{\underline{t}}^{RST}, \underline{t}^{RST})$.

Such an arc (i, j) is associated with an elapsed time E_{ij} equal to the difference between the arrival times of f_1 and f_2, a flying time F_{ij} equal to the flying time of flight f_2, and an associated flight $f(i, j) = f_2$. The subset of these *srf* arcs is denoted A^{SRF}. When extending a label along an arc $(i, j) \in A^{SRF}$, the REFs for the resources in $\mathcal{R}$ are defined as follows:

$$f_{ij}^{rCost}(\mathbf{t}_i) = \max\{f_{ij}^{costF}(\mathbf{t}_i), f_{ij}^{costS}(\mathbf{t}_i)\}$$

$$f_{ij}^{costF}(\mathbf{t}_i) = t_i^{costF} + F_{ij}(1 - \pi_b) - \pi_{f(i,j)}$$

$$f_{ij}^{costS}(\mathbf{t}_i) = t_i^{costS} + \alpha E_{ij} - \pi_{f(i,j)} - F_{ij}\pi_b$$

$$f_{ij}^{dtyMin}(\mathbf{t}_i) = \max\{\underline{t}^{DTY} - F_{ij} - \underline{t}^{CNX}, 0\}$$

$$f_{ij}^{dtyMax}(\mathbf{t}_i) = F_{ij} + \underline{t}^{CNX}$$

$$f_{ij}^{cnxMax}(\mathbf{t}_i) = \underline{t}^{CNX}$$

$$f_{ij}^{rstMax}(\mathbf{t}_i) = 0.$$

► **The new resource** *shRst* counts the number of short rest periods taken. When it takes value 1, no more rests are allowed before completing the pairing. Its resource windows are: $[0,0]$ at the source node and all to-flight nodes; and $[0,1]$ at all other nodes. When extending a label along an arc $(i, j) \in A^{wb}$, the REF for this resource is given by:

$$f_{ij}^{shRst}(\mathbf{t}_i) = \begin{cases} t_i^{shRst} + 1 & \text{if } (i,j) \in A^{SRF} \\ t_i^{shRst} & \text{otherwise.} \end{cases}$$

Exercises of Chapter 6

6.1 Joseph-Louis Lagrange

Joseph-Louis Lagrange is usually considered to be a French mathematician, but the Italian Encyclopædia refers to him as an Italian mathematician. They certainly have some justification in this claim since Lagrange was born in Turin and baptised in the name of Giuseppe Lodovico Lagrangia.

– MacTutor

6.2 Nature of the *LDP*

The *LDP* can be written as a linear program, see formulation (6.29). Recall that it is equivalent to another linear program, the *MP* of a Dantzig-Wolfe reformulation of the *ILP*.

6.3 About the *AMP*

(a) Given the restricted masters *RMP* and *ARMP* solved on the same set of λ_p-variables, $p \in P'$, we have $z_{RMP} = z_{ARMP}$. Moreover, $\boldsymbol{\pi}_{\mathbf{b}}^{\mathsf{T}}\mathbf{b} + \pi_0 = \mu$ by duality. Then, the reduced cost of a λ_p-variable is computed the same way:

$$\bar{c}_p = c_p - \boldsymbol{\pi}_{\mathbf{b}}^{\mathsf{T}}\mathbf{a}_p - \pi_0 \qquad \text{with the } MP$$
$$\bar{c}_p = c_p - \boldsymbol{\pi}_{\mathbf{b}}^{\mathsf{T}}(\mathbf{a}_p - \mathbf{b}) - \mu = c_p - \boldsymbol{\pi}_{\mathbf{b}}^{\mathsf{T}}\mathbf{a}_p - (\mu - \boldsymbol{\pi}_{\mathbf{b}}^{\mathsf{T}}\mathbf{b}) \qquad \text{with the } AMP.$$

(b) By (6.11) and Proposition 6.1, the Lagrangian bound of the *AMP* is

$$\boldsymbol{\pi}_{\mathbf{b}}^{\mathsf{T}}\mathbf{0} + \min_{\mathbf{x}\in\mathcal{D}} \mathbf{c}^{\mathsf{T}}\mathbf{x} + \boldsymbol{\pi}_{\mathbf{b}}^{\mathsf{T}}(\mathbf{0} - (\mathbf{A}\mathbf{x} - \mathbf{b})) \le z^{\star}_{AMP} = z^{\star}_{MP},$$

which is obviously the same expression as $\boldsymbol{\pi}_{\mathbf{b}}^{\mathsf{T}}\mathbf{b} + \min_{\mathbf{x}\in\mathcal{D}} \mathbf{c}^{\mathsf{T}}\mathbf{x} - \boldsymbol{\pi}_{\mathbf{b}}^{\mathsf{T}}\mathbf{A}\mathbf{x} \le z^{\star}_{MP}$. It is indeed independent of π_0 or μ.

6.4 *AMP* formulations for a block-diagonal structure

Given that $\sum_{k\in K}\sum_{p\in P^k}\lambda_p^k = |K|$, we have $\mathbf{b} = \sum_{k\in K}\sum_{p\in P^k}\lambda_p^k \frac{\mathbf{b}}{|K|}$. Let us write the primal formulation of the *AMP* followed by its dual counterpart.

$$\begin{aligned}
z^{\star}_{AMP} = \min \quad & \sum_{k\in K}\sum_{p\in P^k} c_p^k\lambda_p^k + \sum_{k\in K}\sum_{r\in R^k} c_r^k\lambda_r^k \\
\text{s.t.} \quad & \sum_{k\in K}\sum_{p\in P^k}\left(\mathbf{a}_p^k - \frac{\mathbf{b}}{|K|}\right)\lambda_p^k + \sum_{k\in K}\sum_{r\in R^k}\mathbf{a}_r^k\lambda_r^k \ge \mathbf{0} \quad [\boldsymbol{\pi}_{\mathbf{b}}] \\
& \sum_{p\in P^k}\lambda_p^k = 1 \quad [\mu^k] \quad \forall k\in K \\
& \lambda_p^k \ge 0 \quad \forall k\in K, p\in P^k \\
& \lambda_r^k \ge 0 \quad \forall k\in K, r\in R^k,
\end{aligned}$$

$$\begin{aligned}
z^{\star}_{AMP} = \max \quad & \sum_{k\in K}\mu^k \\
\text{s.t.} \quad & \left(\mathbf{a}_p^k - \frac{\mathbf{b}}{|K|}\right)^{\mathsf{T}}\boldsymbol{\pi}_{\mathbf{b}} + \mu^k \le c_p^k \quad [\lambda_p^k] \quad \forall k\in K, p\in P^k \\
& \mathbf{a}_r^{k\mathsf{T}}\boldsymbol{\pi}_{\mathbf{b}} \le c_r^k \quad [\lambda_r^k] \quad \forall k\in K, r\in R^k \\
& \boldsymbol{\pi}_{\mathbf{b}} \ge \mathbf{0} \\
& \mu^k \in \mathbb{R} \quad \forall k\in K.
\end{aligned}$$

6.5 Dantzig-Wolfe vs. Lagrange for a block-diagonal structure

(a) Given $\boldsymbol{\pi}_{\mathbf{b}} \geq \mathbf{0}$, we start with the *ISP* (6.17) given by

$$LR(\boldsymbol{\pi}_{\mathbf{b}}) = \boldsymbol{\pi}_{\mathbf{b}}^{\mathsf{T}}\mathbf{b} + \sum_{k\in K} \min_{\mathbf{x}^k \in \mathcal{D}^k} (\mathbf{c}^{k\mathsf{T}} - \boldsymbol{\pi}_{\mathbf{b}}^{\mathsf{T}}\mathbf{A}^k)\mathbf{x}^k.$$

Let the *ISP* be solved with an algorithm that returns $\mathbf{x}^k$ as either an extreme point or extreme ray of $\operatorname{conv}(\mathcal{D}^k)$, respectively indexed in P^k and R^k, $\forall k \in K$. The Lagrangian bound is either $-\infty$ if there exists an optimal extreme ray such that $c_r^k - \mathbf{a}_r^{k\mathsf{T}}\boldsymbol{\pi}_{\mathbf{b}} < 0$, $r \in R^k$, $k \in K$, otherwise finite and given by

$$LR(\boldsymbol{\pi}_{\mathbf{b}}) = \mathbf{b}^{\mathsf{T}}\boldsymbol{\pi}_{\mathbf{b}} + \sum_{k\in K} \min_{p\in P^k} (c_p^k - \mathbf{a}_p^k)^{\mathsf{T}}\boldsymbol{\pi}_{\mathbf{b}}.$$

Looking for a finite bound, we restrain $\boldsymbol{\pi}_{\mathbf{b}}$ with $c_r^k - \mathbf{a}_r^{k\mathsf{T}}\boldsymbol{\pi}_{\mathbf{b}} \geq 0$, $\forall k \in K$, $r \in R^k$, which we know must hold for any optimal $\boldsymbol{\pi}_{\mathbf{b}}^\star$, see Note 6.3. The *LDP* is thus restated as

$$\begin{aligned} z_{LDP}^\star = \max_{\boldsymbol{\pi}_{\mathbf{b}}} \quad & \mathbf{b}^{\mathsf{T}}\boldsymbol{\pi}_{\mathbf{b}} + \sum_{k\in K} \min_{p\in P^k} (c_p^k - \mathbf{a}_p^k)^{\mathsf{T}}\boldsymbol{\pi}_{\mathbf{b}} \\ \text{s.t.} \quad & c_r^k - \mathbf{a}_r^{k\mathsf{T}}\boldsymbol{\pi}_{\mathbf{b}} \geq 0 \quad \forall k \in K, r \in R^k \\ & \boldsymbol{\pi}_{\mathbf{b}} \geq \mathbf{0}. \end{aligned}$$

We reformulate the inner optimization $\min_{p\in P^k} (c_p^k - \mathbf{a}_p^k)^{\mathsf{T}}\boldsymbol{\pi}_{\mathbf{b}}$ as

$$\begin{aligned} \max \quad & \pi_0^k \\ \text{s.t.} \quad & \pi_0^k \leq (c_p^k - \mathbf{a}_p^k)^{\mathsf{T}}\boldsymbol{\pi}_{\mathbf{b}} \quad \forall p \in P^k. \end{aligned}$$

Combining the last two formulations, we arrive at a reformulation of the *LDP* as a linear program in $\pi_0^k \in \mathbb{R}$, $\forall k \in K$, and $\boldsymbol{\pi}_{\mathbf{b}} \geq \mathbf{0}$, where the corresponding dual λ^k-variables associated with the constraints appear in brackets:

$$\begin{aligned} z_{LDP}^\star = \max \quad & \mathbf{b}^{\mathsf{T}}\boldsymbol{\pi}_{\mathbf{b}} + \sum_{k\in K} \pi_0^k \\ \text{s.t.} \quad & \mathbf{a}_p^{k\mathsf{T}}\boldsymbol{\pi}_{\mathbf{b}} + \pi_0^k \leq c_p^k \quad [\lambda_p^k] \quad \forall k \in K, p \in P^k \\ & \mathbf{a}_r^{k\mathsf{T}}\boldsymbol{\pi}_{\mathbf{b}} \leq c_r^k \quad [\lambda_r^k] \quad \forall k \in K, r \in R^k \\ & \boldsymbol{\pi}_{\mathbf{b}} \geq \mathbf{0} \\ & \pi_0^k \in \mathbb{R} \quad \forall k \in K. \end{aligned}$$

Dualizing, we obtain the formulation of the *MP*, hence, $z_{LDP}^\star = z_{MP}^\star$:

$$\begin{array}{llll} z^\star_{LDP} = \min & \sum\limits_{k\in K}\sum\limits_{p\in P^k} c^k_p\lambda^k_p & +\sum\limits_{k\in K}\sum\limits_{r\in R^k} c^k_r\lambda^k_r & \\ \text{s.t.} & \sum\limits_{k\in K}\sum\limits_{p\in P^k} \mathbf{a}^k_p\lambda^k_p & +\sum\limits_{k\in K}\sum\limits_{r\in R^k} \mathbf{a}^k_r\lambda^k_r \geq \mathbf{b} \;\; [\boldsymbol{\pi}_\mathbf{b}] & \\ & \sum\limits_{p\in P^k} \lambda^k_p & = 1 \;\; [\pi^k_0] & \forall k \in K \\ & \lambda^k_p \geq 0 & & \forall k \in K, p \in P^k \\ & & \lambda^k_r \geq 0 & \forall k \in K, r \in R^k. \end{array}$$

(b) Starting with the *ILP* (6.16), we rewrite the *ISP* (6.17) for $\boldsymbol{\pi}_\mathbf{b} \geq \mathbf{0}$ as

$$\begin{array}{ll} LR(\boldsymbol{\pi}_\mathbf{b}) = \min & \sum\limits_{k\in K} \mathbf{c}^{k\mathsf{T}}\mathbf{x}^k + \boldsymbol{\pi}_\mathbf{b}^\mathsf{T}\left(\frac{\mathbf{b}}{|K|} - \mathbf{A}^k\mathbf{x}^k\right) \\ \text{s.t.} & \mathbf{x}^k \in \mathcal{D}^k, \quad \forall k \in K, \end{array}$$

which makes the objective function separable in index k, that is,

$$LR(\boldsymbol{\pi}_\mathbf{b}) = \sum_{k\in K} \min_{\mathbf{x}^k\in\mathcal{D}^k} \mathbf{c}^{k\mathsf{T}}\mathbf{x}^k + \boldsymbol{\pi}_\mathbf{b}^\mathsf{T}\left(\frac{\mathbf{b}}{|K|} - \mathbf{A}^k\mathbf{x}^k\right).$$

Let again the *ISP* be solved with an algorithm that returns $\mathbf{x}^k$, $\forall k \in K$, as either an extreme point or an extreme ray of $\text{conv}(\mathcal{D}^k)$. The Lagrangian bound is either $-\infty$ if there exists an extreme ray solution such that $c^k_r - \mathbf{a}^{k\mathsf{T}}_r\boldsymbol{\pi}_\mathbf{b} < 0$, $r \in R^k$, $k \in K$, otherwise finite and given by

$$LR(\boldsymbol{\pi}_\mathbf{b}) = \sum_{k\in K} \min_{p\in P^k} c^k_p + \left(\frac{\mathbf{b}}{|K|} - \mathbf{a}^k_p\right)^\mathsf{T}\boldsymbol{\pi}_\mathbf{b}.$$

Looking for a finite bound, we restrain $\boldsymbol{\pi}_\mathbf{b}$ with $c^k_r - \mathbf{a}^{k\mathsf{T}}_r\boldsymbol{\pi}_\mathbf{b} \geq 0$, $\forall k \in K, r \in R^k$, which we know must hold for any optimal $\boldsymbol{\pi}^\star_\mathbf{b}$, see Note 6.3. The *LDP* is thus restated as

$$\begin{array}{lll} z^\star_{LDP} = \max\limits_{\boldsymbol{\pi}_\mathbf{b}} & \sum\limits_{k\in K} \min\limits_{p\in P^k} c^k_p + \left(\frac{\mathbf{b}}{|K|} - \mathbf{a}^k_p\right)^\mathsf{T}\boldsymbol{\pi}_\mathbf{b} & \\ \text{s.t.} & c^k_r - \mathbf{a}^{k\mathsf{T}}_r\boldsymbol{\pi}_\mathbf{b} \geq 0 & \forall k \in K, r \in R^k \\ & \boldsymbol{\pi}_\mathbf{b} \geq \mathbf{0}. & \end{array}$$

We reformulate the inner optimization $\min_{p\in P^k} c^k_p + (\frac{\mathbf{b}}{|K|} - \mathbf{a}^k_p)^\mathsf{T}\boldsymbol{\pi}_\mathbf{b}$ as

$$\begin{array}{ll} \max & \mu^k \\ \text{s.t.} & \mu^k \leq c^k_p + \left(\frac{\mathbf{b}}{|K|} - \mathbf{a}^k_p\right)^\mathsf{T}\boldsymbol{\pi}_\mathbf{b} \quad \forall p \in P^k. \end{array}$$

Combining the last two formulations, we arrive at a reformulation of the *LDP* as a linear program in $\mu^k \in \mathbb{R}$, $\forall k \in K$, and $\boldsymbol{\pi}_\mathbf{b} \geq \mathbf{0}$, where the corresponding dual λ^k-variables associated with the constraints appear in brackets:

$$
\begin{aligned}
z^\star_{LDP} = \max \quad & \sum_{k\in K} \mu^k \\
\text{s.t.} \quad & (\mathbf{a}_p^k - \frac{\mathbf{b}}{|K|})^\mathsf{T} \boldsymbol{\pi}_\mathbf{b} + \mu^k \leq c_p^k \quad [\lambda_p^k] \quad \forall k \in K, p \in P^k \\
& \mathbf{a}_r^{k\mathsf{T}} \boldsymbol{\pi}_\mathbf{b} \leq c_r^k \quad [\lambda_r^k] \quad \forall k \in K, r \in R^k \\
& \boldsymbol{\pi}_\mathbf{b} \geq \mathbf{0} \\
& \mu^k \in \mathbb{R} \quad \forall k \in K.
\end{aligned}
$$

Dualizing, we obtain the alternative Dantzig-Wolfe master problem:

$$
\begin{aligned}
z^\star_{LDP} = \min \quad & \sum_{k\in K}\sum_{p\in P^k} c_p^k \lambda_p^k + \sum_{k\in K}\sum_{r\in R^k} c_r^k \lambda_r^k \\
\text{s.t.} \quad & \sum_{k\in K}\sum_{p\in P^k} (\mathbf{a}_p^k - \frac{\mathbf{b}}{|K|}) \lambda_p^k + \sum_{k\in K}\sum_{r\in R^k} \mathbf{a}_r^k \lambda_r^k \geq \mathbf{0} \quad [\boldsymbol{\pi}_\mathbf{b}] \\
& \sum_{p\in P^k} \lambda_p^k = 1 \quad [\mu^k] \quad \forall k \in K \\
& \lambda_p^k \geq 0 \quad \forall k \in K, p \in P^k \\
& \lambda_r^k \geq 0 \quad \forall k \in K, r \in R^k,
\end{aligned}
$$

hence, $z^\star_{LDP} = z^\star_{AMP} = z^\star_{MP}$.

6.6 *VRPTW*: Dantzig-Wolfe vs. Lagrange
The sequences of dual multipliers differ. The *MP* solution process by column generation starts with costly artificial variables, hence favoring paths with a large number of visited customers in the *ISP* and also possible cycles. On the contrary, the subgradient algorithm for the *LDP* rather starts with zero or small dual values: such a strategy reduces both the number of visited nodes and the cycles and is on average faster per iteration.

6.7 Yes or No? And why?

(a) No in general, except by chance if the selected extreme point solution is integer, or if the *ILP* formulation possesses the integrality property. We can however indirectly solve the *ILP* with the primal simplex algorithm embedded in a branch-and-bound tree.

(b) The column generation algorithm is the primal simplex method for solving a linear program with a huge number of variables, hence the same answer as (a).

(c) The Dantzig-Wolfe decomposition is a reformulation of the *LP* or *ILP*; it cannot solve anything, see Figures 6.6 and 6.14.

(d) The *LDP* can be seen as the dual of the *MP*, hence it is a relaxation of the original problem. Furthermore, it is not an algorithm so it does not solve anything, see Figures 6.6 and 6.14.

6.8 Ready to answer?

(a) A dual optimal solution $\boldsymbol{\pi}_{\mathbf{b}}^{\star}$ alone is of no use as we need actionable information in $\mathbf{x}$ or $\boldsymbol{\lambda}$ to formulate a strategy.

(b) We see in Illustration 6.10 that a dual box on π_0 can improve the initial objective value but most of the help comes from using $\boldsymbol{\pi}_{\mathbf{b}}^{\star}$. In fact, in Illustration 6.11 we do not bother at all to test dual boxes on the π_0^k, $k \in K$, and rather concentrate again on $\boldsymbol{\pi}_{\mathbf{b}}^{\star}$.

(c) The black on white answer is yes but only because a Dantzig-Wolfe reformulation *IMP* is equivalent to the original problem *ILP* whereas a Lagrangian relaxation is only equivalent to the *MP* at the root node of a Dantzig-Wolfe reformulation. With that said, even focusing on said root node, we have seen some arguments as to why a Lagrangian relaxation may be solved faster. For example, the dual values used at initialization may be easier to deal with in the pricing problem.

(d) We can collect the subgradients we find and post-process them in a master problem. Since the latter provides analogous solutions to those one would find in the *MP* of a Dantzig-Wolfe reformulation, this means that it is neither easier, nor harder to find a solution to the compact formulation than what we would observe in the branch-and-price tree.

Sometimes, as in Example 6.5 Balancing printed circuit board assembly line systems, a solution to the *ISP* can be transformed into a feasible one for the *ILP*, hence providing an upper bound on top of a lower bound. In fact, one could argue that a Lagrangian relaxation can be preferable to a Dantzig-Wolfe reformulation whenever integer solutions are indeed *easy* to recover from collected subgradients and their quality is good enough such that we do not have to deal with a full-blown master problem implementation.

(e) We can incorporate dual information in the master problem formulation to recover efficiently a primal solution in λ, x-variables using dual boxes. Furthermore, we can use stabilization and work our way to optimality even if we start with incorrect dual information.

6.9 Using optimal Lagrangian multipliers with dual boxes

If $|K| = 1$, we can deduce $\pi_0^{\star} = z_{LDP}^{\star} - \boldsymbol{\pi}_{\mathbf{b}}^{\star\mathsf{T}}\mathbf{b}$. Since this is dual optimal, it leads to $\bar{c}(\boldsymbol{\pi}_{\mathbf{b}}^{\star}, \pi_0^{\star}) = 0$ such that we cannot generate any columns.
We could also use M as a placeholder for π_0 in $\bar{c}(\boldsymbol{\pi}_{\mathbf{b}}^{\star}, M) < 0$ which generates a negative reduced cost column and then modifies all dual values upon solving the *RMP*, thus losing the dual optimal information $\boldsymbol{\pi}_{\mathbf{b}}^{\star}$. The big-$M$ strategy obviously holds no better for $|K| > 1$.

Proposition 6.10 and Note 6.18 provide a more meaningful way to exploit $\boldsymbol{\pi}_{\mathbf{b}}^{\star}$. Indeed, it is hard to imagine a more to the point approach because we basically only generate the columns we need to match this dual optimal solution. In fact, it makes the most sense in stabilization since in general we have no guarantee that Lagrangian multipliers $\boldsymbol{\pi}_{\mathbf{b}}$ are optimal in which case we do likely use "roomier" dual boxes at initialization.

6.10 Stabilization parameters: initialization and update

(a) In Corollary 6.3, we establish that the MP_{stab} solves the MP for any open interval and $\boldsymbol{\varepsilon}_1, \boldsymbol{\varepsilon}_2 > \mathbf{0}$. When we initialize the RMP_{stab} with $\mathcal{X}' = \emptyset$, the values of $\boldsymbol{\varepsilon}_1, \boldsymbol{\varepsilon}_2$ must be *large enough* to ensure primal feasibility in the y-variables. The primal and dual stabilized problems respectively read as (without the 0-index)

$$\begin{array}{llllr}
\min & -\boldsymbol{\delta}_1^\mathsf{T}\mathbf{y}_1 & +\boldsymbol{\delta}_2^\mathsf{T}\mathbf{y}_2 & & \\
\text{s.t.} & -\mathbf{y}_1 & +\mathbf{y}_2 = \mathbf{b} & & [\boldsymbol{\pi}] \\
& \mathbf{y}_1 \leq \boldsymbol{\varepsilon}_1 & & & [-\mathbf{w}_1 \leq \mathbf{0}] \\
& & \mathbf{y}_2 \leq \boldsymbol{\varepsilon}_2 & & [-\mathbf{w}_2 \leq \mathbf{0}] \\
& \mathbf{y}_1 \geq \mathbf{0}, & \mathbf{y}_2 \geq \mathbf{0} & &
\end{array}$$

and

$$\begin{array}{lllllr}
\max & \mathbf{b}^\mathsf{T}\boldsymbol{\pi} & -\boldsymbol{\varepsilon}_1^\mathsf{T}\mathbf{w}_1 & -\boldsymbol{\varepsilon}_2^\mathsf{T}\mathbf{w}_2 & & \\
\text{s.t.} & -\boldsymbol{\pi} & -\mathbf{w}_1 & & \leq -\boldsymbol{\delta}_1 & [\mathbf{y}_1 \geq \mathbf{0}] \\
& \boldsymbol{\pi} & & -\mathbf{w}_2 & \leq \boldsymbol{\delta}_2 & [\mathbf{y}_2 \geq \mathbf{0}] \\
& & \mathbf{w}_1 \geq \mathbf{0}, & \mathbf{w}_2 \geq \mathbf{0}. & &
\end{array}$$

Assuming non-binding $\boldsymbol{\varepsilon}_1, \boldsymbol{\varepsilon}_2$, then $\mathbf{w}_1 = \mathbf{w}_2 = \mathbf{0}$ and the first iteration yields dual values at either end of the interval, that is,

$$\pi_i = \delta_{i2} \text{ if } b_i \geq 0 \quad \text{and} \quad \pi_i = \delta_{i1} \text{ if } b_i < 0, \quad \forall i \in \{1, \ldots, m\}.$$

Primal feasibility, at the same cost, is achieved by setting

$$y_{i2} = b_i \text{ if } b_i \geq 0 \quad \text{and} \quad y_{i1} = -b_i \text{ if } b_i < 0, \quad \forall i \in \{1, \ldots, m\}.$$

This follows the statement of Proposition 6.11 and intuitively corresponds to the behavior we describe for dual optimal inequalities in Note 6.18, where we then do not solve the pricing problem with $\boldsymbol{\pi}^\star$ as long as the $RMP_{\boldsymbol{\delta}}$ does not contain an optimal primal solution. Numeric precision issues in the pricing suggest that the $\boldsymbol{\delta}$ values should also be roomy enough not to behave as $\boldsymbol{\pi}^\star$.

Perhaps we can add a little more intuition in applications with non-negative coefficients $\mathbf{a_x}$ and $\mathbf{b}$. The RMP_{stab} with $\mathcal{X}' = \emptyset$ can be initialized with $\boldsymbol{\varepsilon}_2 = \mathbf{b}$ whereas the pertinence of $\boldsymbol{\varepsilon}_1 = \mathbf{b}$ becomes visible as the content of $\mathcal{X}'$ increases along with our ability to over-cover the right-hand side. The delicate tango we see between $\boldsymbol{\lambda}_\mathbf{x}$- and $\mathbf{y}$-variables is the essence of stabilization. Its rhythm is dictated by the current parameter values $\boldsymbol{\varepsilon}$ and $\boldsymbol{\delta}$. Indeed, the stabilization variables $\mathbf{y}_1$ and $\mathbf{y}_2$ are there to absorb primal infeasibility but by restricting their influence in size ($\boldsymbol{\varepsilon}$) and weight ($\boldsymbol{\delta}$), we more or less favor the use of columns already present in $\mathcal{X}'$. The more accurate $\hat{\boldsymbol{\pi}}$ is, the more we expect to have relevant columns in $\mathcal{X}'$ because those are precisely the type of columns we generate.

(b) In Algorithm 6.2, we are actually free to update the $\boldsymbol{\delta}$ and $\boldsymbol{\varepsilon}$ parameters whenever we want. This may especially be worth it right after initialization to rapidly reassess more appropriate values than $\boldsymbol{\varepsilon} = \mathbf{b}$. After all, these are conceptually perturbation parameters.

A general guideline for $\boldsymbol{\delta}$ can be to reduce the dual box if $\boldsymbol{\delta}_1 < \boldsymbol{\pi} < \boldsymbol{\delta}_2$, e.g., $\boldsymbol{\delta}_1^2 = 0.5\boldsymbol{\delta}_1^1$, and to increase it otherwise, e.g., $\boldsymbol{\delta}_1^2 = 1.5\boldsymbol{\delta}_1^1$.

Similarly for $\boldsymbol{\varepsilon}$, we augment the penalty when we are in the dual box, e.g., The idea here is that if the dual box is good, the value of $\boldsymbol{\varepsilon}$ is mostly irrelevant by Corollary 6.3 whereas we really do want the perturbation effect when the dual information is not as potent.

Note 6.21 warns us that these parameters must not reduce the dual intervals to single points so it does not hurt to post-process some hard limits like 10^{-3} to avoid numerical instability traps as observed in Illustration 6.11 (see the discussion regarding the last row of Table 6.6). In any case, one must always have a fail-safe test to ensure that the infeasibility of the RMP_{stab} does not come from $\boldsymbol{\varepsilon}$. Furthermore, we can obviously make granular operations per component, and non-symmetrical updates.

6.11 A crossover method for finding a basic $\mathbf{x}^\star_{LP}$
For $i \in \{1,\ldots,m\}$, we have $\pi_i^\star - \delta_i \le \pi_i \le \pi_i^\star + \delta_i$. At optimality, we can fix the variable x_j to zero if the *smallest* possible value of its reduced cost is positive, that is, if

$$\min_{\boldsymbol{\pi}\in[\boldsymbol{\pi}^\star-\boldsymbol{\delta},\boldsymbol{\pi}^\star+\boldsymbol{\delta}]} c_j - \boldsymbol{\pi}^\intercal \mathbf{a}_j > 0.$$

This is equivalent to

$$\begin{aligned} & c_j - \sum_{i:a_{ij}\ge 0} (\pi_i^\star + \delta_i)a_{ij} - \sum_{i:a_{ij}<0} (\pi_i^\star - \delta_i)a_{ij} > 0 \\ \Leftrightarrow\quad & c_j - \sum_{i=1}^{m} \pi_i^\star a_{ij} - \sum_{i:a_{ij}\ge 0} \delta_i a_{ij} - \sum_{i:a_{ij}<0} \delta_i(-a_{ij}) > 0 \\ \Leftrightarrow\quad & c_j - \sum_{i=1}^{m} \pi_i^\star a_{ij} > \sum_{i=1}^{m} \delta_i |a_{ij}|. \end{aligned}$$

6.12 Symmetric *TSP*: a compact formulation without block-index k
Because $\sum_{p\in P} \lambda_p = 1$, Proposition 4.15 applies with $|K| = 1$, hence there is no need for a block-index. Moreover, as the single block is obviously used, the corresponding variable x_0 is set to 1 and disappears.

6.13 Asymmetric and symmetric *TSP*: flow conservation constraints
• Subtracting the assignment constraints, that is, (6.110b) minus (6.110c) in the *ATSP* formulation (6.110) gives

$$\sum_{j:(k,j)\in A} x_{kj} - \sum_{i:(i,k)\in A} x_{ik} = 0, \qquad \forall k \in N.$$

• There is always at least one contradiction in particular at node n. Assume that an optimal solution for the *STSP* contains the sequence $1-3-2$: it has degree 2 at node 3 in both formulations (6.111) and (6.119), using arcs $(1,3)$ and $(2,3)$, where $i < j$, $(i,j) \in E$. In this case, the flow conservation constraint at node 3 is not satisfied:

$$\sum_{j:(3,j)\in E} x_{3j} - \sum_{i:(i,3)\in E} x_{i3} = 0 - 2 \neq 0.$$

6.14 Printed circuit boards: Lagrangian multipliers

(a) For $i \in N$, $k \in K$, let $-a^k_{i,t} \geq 0$ be the entry of the subgradient direction at iteration $t \geq 1$ and $\beta^k_{i,t+1}$ denote the value of the multiplier *before the normalization.*

$$\begin{aligned} \beta^k_{i,t+1} &= \pi^k_{i,t} + \theta_t(-a^k_{i,t}) && \forall i \in N, k \in K \\ \pi^k_{i,t+1} &= \left(\frac{\beta^k_{i,t+1}}{\sum_{k\in K} \beta^k_{i,t+1}}\right) p_i && \forall i \in N, k \in K. \end{aligned}$$

(b) A multiplier decreases from one iteration to the next, $\pi^k_{i,t+1} < \pi^k_{i,t}$, if and only if $\left(\frac{\beta^k_{i,t+1}}{\sum_{k\in K} \beta^k_{i,t+1}}\right) < \left(\frac{\pi^k_{i,t}}{p_i}\right)$, that is, if the contribution of $\beta^k_{i,t+1}$ relatively to the sum $\sum_{k\in K} \beta^k_{i,t+1}$ over all the machines is less than the contribution of $\pi^k_{i,t}$ compared to $\sum_{k\in K} \pi^k_{i,t} = p_i$.

(c) Because the multipliers are initialized at values greater than 0, $\beta^k_{i,t+1} > 0$, hence also $\pi^k_{i,t+1} > 0$, $\forall t \geq 1$, even if $\sum_{j\in J_i} \sum_{s\in S^k} t_{js}(\sum_{p\in P} x_{js,p}\lambda_p) < t_i$.

6.15 Warehouse location problem

(a) The Lagrangian subproblem obtained by relaxing $\sum_{k\in K} x^k_i = 1$, $\forall i \in N$, is

$$\begin{aligned} LR(\boldsymbol{\pi}) = \min \quad & \sum_{i\in N}\sum_{k\in K} c^k_i x^k_i + \sum_{k\in K} F^k y^k + \sum_{i\in N} \pi_i (1 - \sum_{k\in K} x^k_i) \\ \text{s.t.} \quad & \sum_{i\in N} d_i x^k_i \leq D^k y^k \quad \forall i \in N \\ & 0 \leq x^k_i \leq 1 \qquad \forall i \in N, k \in K \\ & y^k \in \{0,1\} \qquad \forall k \in K, \end{aligned}$$

where the objective function can be rewritten as

$$\min \sum_{i\in N} \pi_i + \min \sum_{i\in N}\sum_{k\in K} (c^k_i - \pi_i) x^k_i + \sum_{k\in K} F^k y^k.$$

The above formulation does not possess the integrality property and it must be solved as a mixed-integer linear program, or by a specialized algorithm. As such, the best lower bound $z^\star_{LDP} \geq z^\star_{LP}$.

(b) The Lagrangian subproblem obtained by relaxing $\sum_{i\in N} d_i x_i^k \leq D^k y^k$, $\forall k \in K$, is

$$\begin{aligned} LR(\boldsymbol{\omega}) = \min \quad & \sum_{i\in N}\sum_{k\in K} c_i^k x_i^k + \sum_{k\in K} F^k y^k + \sum_{k\in K} \omega^k (D^k y^k - \sum_{i\in N} d_i x_i^k) \\ \text{s.t.} \quad & \sum_{k\in K} x_i^k = 1 && \forall i \in N \\ & 0 \leq x_i^k \leq 1 && \forall i \in N, k \in K \\ & y_j \in \{0,1\} && \forall k \in K, \end{aligned}$$

where the objective function can be rewritten as

$$\min \sum_{i\in N}\sum_{k\in K} (c_i^k - \omega^k d_i) x_i^k + \sum_{k\in K} (F^k + \omega^k D^k) y^k.$$

It is separable in x- and y-variables.

- For the continuous x-variables, a solution is obtained by inspection for each i, fixing $x_i^\ell = 1$ for any warehouse $\ell \in K$ with the smallest adjusted coefficient, i.e., $\ell \in \arg\min_{k\in K} (c_i^k - \omega^k d_i)$.
- For the binary y-variables, a solution is also obtained by inspection for each k, fixing $y^k = 0$ if $F^k + \omega^k D^k > 0$ and equal to 1 otherwise.
- This formulation hence possesses the integrality property and the best lower bound is $z^\star_{LDP} = \max_{\boldsymbol{\omega}\in\mathbb{R}_-^{|K|}} LR(\boldsymbol{\omega}) = z^\star_{LP}$.

(c) Once again, the Lagrangian subproblem with objective value $LR(\boldsymbol{\pi})$ does not possess the integrality property while that with $LR(\boldsymbol{\omega})$ still has so

$$z^\star_{LP} = \max_{\boldsymbol{\omega}\in\mathbb{R}_-^{|K|}} LR(\boldsymbol{\omega}) \leq \max_{\boldsymbol{\pi}\in\mathbb{R}^{|N|}} LR(\boldsymbol{\pi}) \leq z^\star_{ILP}.$$

As seen in the Generalized assignment problem with the grouping of constraints in (4.150), another answer that relies on Proposition 4.5 directly tells us that a Dantzig-Wolfe reformulation with

$$\mathcal{D}_i = \{\mathbf{x}_i \in \{0,1\}^{|K|} \mid \sum_{k\in K} x_i^k = 1\}, \qquad \forall i \in N$$

in the pricing problems gives us back exactly the compact formulation, that is, the *IMP* is the *ILP*. Therefore,

$$\max_{\boldsymbol{\omega}\in\mathbb{R}_-^{|K|}} LR(\boldsymbol{\omega}) = z^\star_{MP} = z^\star_{LP}.$$

6.16 Dual inequalities for a two-echelon vehicle routing problem

(a) Recall that a dual variable indicates the rate of change of the optimal objective value with respect to a change in the right-hand side of the corresponding constraint, see the Sensitivity analysis (p. 11). Because $R^{k_1} \subseteq R^{k_2}$, each variable $\lambda_r^{k_1}$, $r \in R^{k_1}$, in the k_1-constraint (6.145c) has an equivalent variable $\lambda_r^{k_2}$ in the k_2-constraint (6.145c) and the latter constraint possibly involves additional variables, the k_2-constraint offers more leeway to react to a change in its right-hand side. Therefore, its dual variable should take a larger absolute value at optimality, that is, $\pi^{k_2} \leq \pi^{k_1}$ because $\pi^k \leq 0$ for all $k \in K$.

(b) We replace the constraints (6.145c) for k_1 and k_2 by

$$\begin{aligned} \sum_{r \in R^{k_1}} d_r \lambda_r^{k_1} - u^{k_1 k_2} &\leq Q^1 y^{k_1} & [\pi^{k_1} \leq 0] \\ \sum_{r \in R^{k_2}} d_r \lambda_r^{k_2} + u^{k_1 k_2} &\leq Q^1 y^{k_2} & [\pi^{k_2} \leq 0] \\ u^{k_1 k_2} &\geq 0. \end{aligned}$$

(c) Variable $u^{k_1 k_2}$ allows to transfer some load from first-echelon route k_1 to route k_2, i.e., the demand of one or several second-echelon routes that is supposed to be supplied by k_1 can rather be (partially) supplied by k_2. Its value indicates the quantity transferred. It can speed up column generation because, if variable $\lambda_r^{k_1}$ is already generated, there is no need to generate $\lambda_r^{k_2}$. Indeed, variable $u^{k_1 k_2}$ can transfer the demand of second-echelon route r from k_1 to k_2.

(d) Constraints (6.145c) are replaced by

$$\begin{aligned} \sum_{r \in R^k} d_r \lambda_r^k + \sum_{(k_1,k) \in F} u^{k_1 k} - \sum_{(k,k_2) \in F} u^{k k_2} &\leq Q^1 y^k \quad [\pi^k \leq 0] \quad \forall k \in K \\ u^{k_1 k_2} &\geq 0 \quad \forall (k_1, k_2) \in F. \end{aligned}$$

(e)
- No positive u-variables:
  There is no load transfer and, thus, solution $(\tilde{\boldsymbol{\lambda}}, \tilde{\mathbf{y}})$ is integer and feasible.
- Only the variable $u^{k_1 k_2} = \tilde{u}^{k_1 k_2}$ is positive:
  In that case (and the next one), it might be impossible to recover primal feasibility, i.e., the *DDOI*s are incompatible. Indeed, there exists an equivalent feasible solution if and only if there is a second-echelon route $r \in R^{k_1}$ such that $\tilde{\lambda}_r^{k_1} = 1$ and $d_r = \tilde{u}^{k_1 k_2}$.
- At least two u-variables take positive values:
  Conditions can be defined for some specific cases. However, in general, determining whether there exists an equivalent solution is a difficult combinatorial problem, especially when the same first-echelon route k acts as a receiver and a giver in different load transfers, i.e., when there are two positive-valued variables $u^{k_1 k}$ and $u^{k k_2}$.

Exercises of Chapter 7

7.1 François Vanderbeck

This François holds a master's degree in operations research from MIT ORC. He continued his studies in the same field and obtained a PhD (1994) at the CORE (Université Catholique de Louvain, Belgique) under the supervision of professor Laurence Wolsey. At the time of writing, he is CEO (and co-founder) of Atoptima.

He is known for reformulation and decomposition methods of linear integer programs "*so as to tackle large scale applications arising in planning, scheduling, logistics and routing problems, as well as cutting and packing problems, or production and inventory control*" (see Institut de Mathématiques de Bordeaux).

As seen in this chapter, François-λ is well renowned for his contributions to branch-and-price via numerous branching strategies derived from the master variables in the discretization approach. While he was Professor in Operations Research at the Université de Bordeaux, he developed `BaPCod` (swmath.org/software/9871), a generic branch-and-price code (Sadykov and Vanderbeck, 2021). It solves mixed-integer linear programs by the application of a Dantzig-Wolfe reformulation on a compact model. It is actually available under the name Coluna, as an open-source platform.

He is a good friend of Jacques, Marco, Guy, Eduardo, and many others. Moreover, he likes red wine from France and maple syrup from Québec.

7.2 B&P: Wikipedia

Contrary to what is suggested by the 'done' box, finding an integer solution is not enough to conclude branch-and-price. Indeed, at this point, we have to update the incumbent integer solution. We can only conclude branch-and-price when we can prove that this upper bound cannot be improved. (Note also that according to our terminology, the *MP* and *RMP* are linear programs, not integer ones)

7.3 B&P: early branching

In principle, to branch, we do not need to solve a node at all, which amounts to making a random branching decision. Hopefully, solving a node approximately can be enough to put forth a meaningful decision, either branching or cutting (and even heuristically fixing). Let us redefine the *node value* as follows:

1. *If the node is solved to optimality*: the objective value of the linear relaxation, as in § 1.5; this is a lower bound on an optimal integer solution for that node.
2. *Else and if available*: any other lower bound such as the Lagrangian bound.
3. *Otherwise*: the current objective value of the *RMP* (which is not a lower bound on an optimal integer solution for that node).

- As in the Branch-and-Bound method of § 1.5, before being solved, every child node inherits the node value of its parent. The node values keep their monotonic behavior in the above first two cases, i.e., a lower bound is used.
  Otherwise, if a child node is assigned a node value that is less than that of its parent node, the latter is refreshed to the child node value (this retroactive monotonicity impacts the node selection strategy).

- **B&P** stops when $LB = UB$, which means that optimality is guaranteed when every leaf node is assigned a pruning state. In particular, pruning by bound can occur whenever the relative optimality gap (see Optimality gap in Chapter 1) guarantees that the lower bound at a node is greater-than-or-equal to UB.
- When applying early branching, column generation may stop with an integer *RMP* solution whose cost differs from the computed lower bound, if any. The corresponding leaf node cannot be pruned by integrality. If it cannot be pruned by bound either, a specialized branching decision or cutting plane must be imposed to get rid of this integer solution. Alternatively, column generation may be restarted to sufficiently improve the lower bound and possibly solve the node to optimality.

7.4 ***ISP*** **modifications for a λ-branching**

(a) Given $\gamma_1 \leq 0$ in the *ISP* (7.61), the slope $-\gamma_1$ is non-negative and $g_{\mathbf{x}}$ can be set to 0 unless forced to be 1 if $\mathbf{f}^\mathsf{T}\mathbf{x} > f-1$, as imposed by

$$g_{\mathbf{x}} = \begin{cases} 1 & \text{if} \quad f \leq \mathbf{f}^\mathsf{T}\mathbf{x} \leq u \\ 0 & \text{if} \quad \ell \leq \mathbf{f}^\mathsf{T}\mathbf{x} \leq f-1 \end{cases}.$$

Enforcing the upper bound inequality in (7.63), $(\mathbf{f}^\mathsf{T}\mathbf{x} - f + 1) \leq g_{\mathbf{x}}(u - f + 1)$, this results in

$$g_{\mathbf{x}} = 1 \;\Leftrightarrow\; \frac{\mathbf{f}^\mathsf{T}\mathbf{x} - f + 1}{u - f + 1} > 0 \;\Leftrightarrow\; \mathbf{f}^\mathsf{T}\mathbf{x} > f - 1 \;\Leftrightarrow\; \mathbf{f}^\mathsf{T}\mathbf{x} \geq f.$$

(b) Given $\gamma_2 \geq 0$ in the *ISP* (7.64), the slope $-\gamma_2$ is non-positive and $g_{\mathbf{x}}$ can be set to 1 unless forced to be 0 if $\mathbf{f}^\mathsf{T}\mathbf{x} < f$. Enforcing the lower bound inequality in (7.63), $(f - \ell)g_{\mathbf{x}} \leq (\mathbf{f}^\mathsf{T}\mathbf{x} - \ell)$, this imposes

$$g_{\mathbf{x}} = 0 \;\Leftrightarrow\; \frac{\mathbf{f}^\mathsf{T}\mathbf{x} - \ell}{f - \ell} < 1 \;\Leftrightarrow\; \mathbf{f}^\mathsf{T}\mathbf{x} < f \;\Leftrightarrow\; \mathbf{f}^\mathsf{T}\mathbf{x} \leq f - 1.$$

7.5 ***ISP*** **modifications for a λ-branching: alternative function $g(\mathbf{x})$**

Let us start with the set of inequalities in (7.63):

$$\ell(1 - g_{\mathbf{x}}) + f g_{\mathbf{x}} \leq \mathbf{f}^\mathsf{T}\mathbf{x} \leq u g_{\mathbf{x}} + (1 - g_{\mathbf{x}})(f - 1), \qquad g_{\mathbf{x}} \in \{0, 1\}.$$

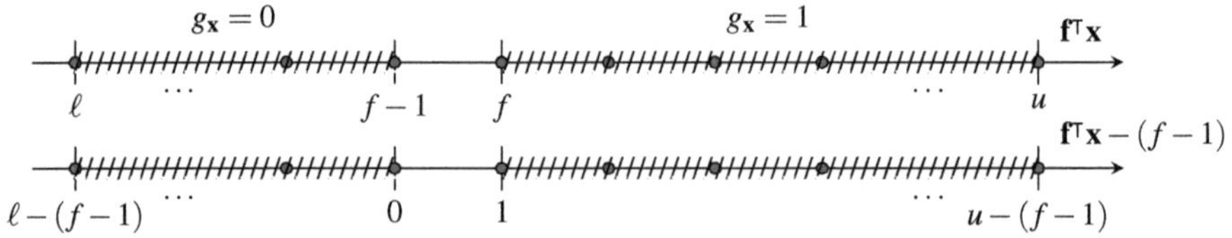

Subtracting $(f-1)$ from $\mathbf{f}^\mathsf{T}\mathbf{x}$ gives $g_{\mathbf{x}} = 1$ if $1 \leq \mathbf{f}^\mathsf{T}\mathbf{x} - (f-1) \leq u - (f-1)$ and $g_{\mathbf{x}} = 0$ if $\ell - (f-1) \leq \mathbf{f}^\mathsf{T}\mathbf{x} - (f-1) \leq 0$. Dividing the left-shifted function by $u - \ell$, assuming $\ell \leq f-1$ and $f \leq u$, we obtain the proposed ceiling function

$$g(\mathbf{x}) = \left\lceil \frac{\mathbf{f}^\intercal\mathbf{x}-(f-1)}{u-\ell} \right\rceil = \begin{cases} 1 & \text{if} \quad 0 < \frac{\mathbf{f}^\intercal\mathbf{x}-(f-1)}{u-\ell} \leq 1 \\ 0 & \text{if} \quad -1 < \frac{\mathbf{f}^\intercal\mathbf{x}-(f-1)}{u-\ell} \leq 0 \end{cases}.$$

7.6 A binary x-branching decision imposed in $\mathcal{A}$ rather than $\mathcal{D}$

Down-branch: $x_j \leq 0$ appears in the *RMP* as $\sum_{\mathbf{x}\in\mathcal{X}} x_j\lambda_\mathbf{x} \leq 0$ with the associated dual variable $\gamma_1 \leq 0$. The objective function of the *ISP* is modified with the term $-\gamma_1 x_j$, i.e., a positive slope on x_j, hence favoring $x_j = 0$ as requested.

Up-branch: we have a similar behavior, that is, $x_j \geq 1$ becomes $\sum_{\mathbf{x}\in\mathcal{X}} x_j\lambda_\mathbf{x} \geq 1$ with $\gamma_2 \geq 0$ in the *RMP*. This impacts the objective function of the *ISP* with the term $-\gamma_2 x_j$, i.e., a negative slope on x_j which favors $x_j = 1$ in that branch.

This is similar to the λ-branching in (7.69)–(7.70) with $\sum_{\mathbf{x}\in\mathcal{X}:\mathbf{e}_j^\intercal\mathbf{x}\geq\lceil x_j^\star\rceil} \lambda_\mathbf{x}^\star = \beta$ fractional.

7.7 Bound constraints on several x-variables: underlying mathematics

(a) *Down-branch*: Given the slope $-\gamma_1 \geq 0$ on $g_\mathbf{x}$, we must detect in (7.84) when $g_\mathbf{x} = 1$, i.e., when $\bar{y}_j = 1, \forall j \in \bar{J}$, and $\underline{y}_j = 1, \forall j \in \underline{J}$. The value of $g_\mathbf{x}$ follows with the constraint $g_\mathbf{x} \geq 1 + (\sum_{j\in\bar{J}}\bar{y}_j + \sum_{j\in\underline{J}}\underline{y}_j) - (|\bar{J}| + |\underline{J}|)$, where the total inner sum equals $|\bar{J}| + |\underline{J}|$ if and only if all the y_j-variables take value 1.

• For the set $\bar{J}$, we derive $\bar{f}(x_j, v_j) \in (0,1]$ such that $\bar{y}_j \geq \bar{f}(x_j, v_j) \Rightarrow \bar{y}_j = 1$.

$$\begin{aligned} j \in \bar{J} : \bar{y}_j = 1 &\Leftrightarrow \ell_j \leq x_j \leq \lfloor v_j \rfloor \\ &\Leftrightarrow -\lfloor v_j \rfloor \leq -x_j \leq -\ell_j \\ &\Leftrightarrow \lfloor v_j \rfloor + 1 - \lfloor v_j \rfloor \leq \lfloor v_j \rfloor + 1 - x_j \leq \lfloor v_j \rfloor + 1 - \ell_j \\ &\Leftrightarrow 0 < \frac{1}{\lfloor v_j \rfloor + 1 - \ell_j} \leq \frac{\lfloor v_j \rfloor + 1 - x_j}{\lfloor v_j \rfloor + 1 - \ell_j} \leq 1; \\ &\blacktriangleright \bar{f}(x_j, v_j) = \frac{\lfloor v_j \rfloor + 1 - x_j}{\lfloor v_j \rfloor + 1 - \ell_j}. \end{aligned}$$

• For $\underline{J}$, we derive $\underline{f}(x_j, v_j) \in (0,1]$ such that $\underline{y}_j \geq \underline{f}(x_j, v_j) \Rightarrow \underline{y}_j = 1$.

$$\begin{aligned} j \in \underline{J} : \underline{y}_j = 1 &\Leftrightarrow \lfloor v_j \rfloor + 1 \leq x_j \leq u_j \\ &\Leftrightarrow 1 \leq x_j - \lfloor v_j \rfloor \leq u_j - \lfloor v_j \rfloor \\ &\Leftrightarrow 0 < \frac{1}{u_j - \lfloor v_j \rfloor} \leq \frac{x_j - \lfloor v_j \rfloor}{u_j - \lfloor v_j \rfloor} \leq 1; \\ &\blacktriangleright \underline{f}(x_j, v_j) = \frac{x_j - \lfloor v_j \rfloor}{u_j - \lfloor v_j \rfloor}. \end{aligned}$$

(b) *Up-branch*: Given the slope $-\gamma_2 \leq 0$, $g_\mathbf{x} = 1$ is the preferred choice in the *ISP*. Hence, we must detect in (7.84) when $g_\mathbf{x} = 0$. This follows from the first two added constraints $g_\mathbf{x} \leq \bar{y}_j, \forall j \in \bar{J}$, and $g_\mathbf{x} \leq \underline{y}_j, \forall j \in \underline{J}$, if at least one of the y_j-variable equals 0.

- For the set $\bar{J}$, we derive $\underline{f}(x_j, v_j) \in [0,1)$ such that $\bar{y}_j \leq \bar{f}(x_j, v_j) \Rightarrow \bar{y}_j = 0$.

$$\begin{aligned} j \in \bar{J} : \bar{y}_j = 0 &\Leftrightarrow \lfloor v_j \rfloor + 1 \leq x_j \leq u_j \\ &\Leftrightarrow -u_j \leq -x_j \leq -\lfloor v_j \rfloor - 1 \\ &\Leftrightarrow 0 \leq u_j - x_j \leq u_j - \lfloor v_j \rfloor - 1 \\ &\Leftrightarrow 0 \leq \frac{u_j - x_j}{u_j - \lfloor v_j \rfloor} \leq 1 - \frac{1}{u_j - \lfloor v_j \rfloor} < 1; \\ &\blacktriangleright\ \bar{f}(x_j, v_j) = \frac{u_j - x_j}{u_j - \lfloor v_j \rfloor}. \end{aligned}$$

- For $\underline{J}$, we derive $\underline{f}(x_j, v_j) \in [0,1)$ such that $\underline{y}_j \leq \underline{f}(x_j, v_j) \Rightarrow \underline{y}_j = 0$.

$$\begin{aligned} j \in \underline{J} : \underline{y}_j = 0 &\Leftrightarrow \ell_j \leq x_j \leq \lfloor v_j \rfloor \\ &\Leftrightarrow 0 \leq x_j - \ell_j \leq \lfloor v_j \rfloor - \ell_j \\ &\Leftrightarrow 0 \leq \frac{x_j - \ell_j}{\lfloor v_j \rfloor + 1 - \ell_j} \leq \frac{\lfloor v_j \rfloor - \ell_j}{\lfloor v_j \rfloor + 1 - \ell_j} < 1; \\ &\blacktriangleright\ \underline{f}(x_j, v_j) = \frac{x_j - \ell_j}{\lfloor v_j \rfloor + 1 - \ell_j}. \end{aligned}$$

7.8 Bound constraints on several x-variables: binary case
For a binary x_j, $\ell_j = \lfloor v_j \rfloor = 0$, $u_j = 1$; the set of constraints in (7.84) becomes

$$\begin{array}{ll|ll} g_{\mathbf{x}} \geq 1 + (\sum_{j \in \bar{J}} \bar{y}_j + \sum_{j \in \underline{J}} \underline{y}_j) - (|\bar{J}| + |\underline{J}|) & & g_{\mathbf{x}} \leq \bar{y}_j & \forall j \in \bar{J} \\ & & g_{\mathbf{x}} \leq \underline{y}_j & \forall j \in \underline{J} \\ \bar{y}_j \geq 1 - x_j & \forall j \in \bar{J} & \bar{y}_j \leq 1 - x_j & \forall j \in \bar{J} \\ \underline{y}_j \geq x_j & \forall j \in \underline{J} & \underline{y}_j \leq x_j & \forall j \in \underline{J}. \end{array}$$

Following the solution of Exercise 7.7, the y-variables are eliminated:

$$\begin{array}{rll} \textit{Down-branch} & j \in \bar{J} : \bar{y}_j = 1 \Leftrightarrow x_j = 0; & \bar{y}_j = 1 - x_j; \\ & j \in \underline{J} : \underline{y}_j = 1 \Leftrightarrow x_j = 1; & \underline{y}_j = x_j. \\ \textit{Up-branch} & j \in \bar{J} : \bar{y}_j = 0 \Leftrightarrow x_j = 1 & \bar{y}_j = 1 - x_j; \\ & j \in \underline{J} : \underline{y}_j = 0 \Leftrightarrow x_j = 0; & \underline{y}_j = x_j. \end{array}$$

Therefore, the set of constraints in (7.84) simplifies to

$$\begin{array}{l|ll} g_{\mathbf{x}} \geq 1 - \sum_{j \in \bar{J}} x_j + \sum_{j \in \underline{J}} x_j - (|\bar{J}| + |\underline{J}|) & g_{\mathbf{x}} \leq 1 - x_j & \forall j \in \bar{J} \\ & g_{\mathbf{x}} \leq x_j & \forall j \in \underline{J}. \end{array}$$

7.9 Ryan-Foster separating hyperplane for the *VRPTW*
(a) The Ryan-Foster rule in (7.86) reads as $0 < \sum_{\mathbf{x} \in \mathcal{X} : a_{r\mathbf{x}} = a_{s\mathbf{x}} = 1} \lambda_{\mathbf{x}} < 1$. Following the expression of $a_{r\mathbf{x}}, a_{s\mathbf{x}}$ in (5.6), $\mathbf{f}^{\intercal}\mathbf{x} \geq f$ is given by

$$\sum_{j:(r,j)\in A} x_{rj} + \sum_{j:(s,j)\in A} x_{sj} \geq 2.$$

(b) For an *ILP* (7.1) reformulated as a set partitioning model, it becomes

$$\mathbf{A}_{r*}\mathbf{x} + \mathbf{A}_{s*}\mathbf{x} \geq 2,$$

where $\mathbf{A}_{r*}$ refers to the r-th row of matrix $\mathbf{A}$ (see **Notation** in Chapter 1, page 3).

7.10 Handling inter-task branching decisions in labeling algorithm

(a) For $w \in W$, let D_w be the set of tasks of W that cannot be covered immediately after w, i.e.,

$$D_w = \begin{cases} W \setminus \{f_w\} & \text{if } f_w \in W \\ \{s \in W \mid (w,s) \in D\} & \text{otherwise.} \end{cases}$$

The first case specifies that, if w is the first task of a follow-on decision, then w cannot be followed immediately by any task in W except f_w. The second case ensues from the applicable do-not-follow-on decisions for task w. We also define $D_{NIL} = \emptyset$.
The labeling algorithm starts with an initial label E_0, $T_0^{last} = NIL$. A path p ending at node $h \in N$ and represented by label E_p can be extended along an arc (h, j) only if $w_{hj} \notin D_{T_p^{last}}$. If the path can be extended, it generates a new path p' with a label $E_{p'}$ associated with node j, where $T_{p'}^{last}$ is computed as follows:

$$T_{p'}^{last} = \begin{cases} T_p^{last} & \text{if } w_{hj} = NIL \\ w_{jh} & \text{otherwise.} \end{cases}$$

If $j = d$ and $f_{T_{p'}^{last}} \neq NIL,$ then p' is discarded.
Finally, the following condition is added to the dominance rule that applies to two labels representing paths p and p' ending at the same node:

$$D_{T_p^{last}} \subseteq D_{T_{p'}^{last}}.$$

(b) Assuming that paths p and p' end at node j, the previous dominance condition can be replaced by

$$D_{T_p^{last}} \subseteq D_{T_{p'}^{last}} \cup U_j.$$

7.11 Finite number of iterations in Proposition 7.2

The answer is **No**, although *values* $\lfloor v_j \rfloor$ or $\lfloor v_j \rfloor + 1$ may already exist.

For example, assume that the current integer lower and upper bounds on x_j, including those from $\mathcal{B}$, are given by $x_j \geq \ell_j$ and $x_j \leq u_j$, and such that $u_j - \ell_j = 1$, i.e., separated by one unit.
Then $\ell_j = x_{j1} < x_{j2} = u_j$ is a possible choice for which the average is

$$v_j = \ell_j + 1/2 = u_j - 1/2.$$

Then $x_j \leq \ell_j = \lfloor v_j \rfloor$ is imposed in the left-sum of (7.81) whereas $x_j \geq u_j = \lfloor v_j \rfloor + 1$ appears in the right-sum. More generally, given $\ell_j < u_j$:

- If $v_j = \ell_j + 1/2$, then $x_j \leq \ell_j$ appears in the left-sum (in addition to $x_j \geq \ell_j$).
- If $v_j = u_j - 1/2$, then $x_j \geq u_j$ appears in the right-sum (in addition to $x_j \leq u_j$).
- Hence, a bound constraint cannot be imposed twice. In fact, a bound *value* can only be used in opposite inequalities, which also fixes the variable x_j.

7.12 λ-branching: single bound constraint on x_j

As in (7.60), two branches are created with $\sum_{\mathbf{x}\in\mathcal{X}} g_{\mathbf{x}}\lambda_{\mathbf{x}} \leq \lfloor \beta \rfloor$ and $\sum_{\mathbf{x}\in\mathcal{X}} g_{\mathbf{x}}\lambda_{\mathbf{x}} \geq \lceil \beta \rceil$ incorporated to their respective *MP* formulation, where the binary coefficient $g_{\mathbf{x}}$ takes value 1 if and only if $x_j \geq 2$.

Following (7.66) with $\mathbf{f}^\mathsf{T}\mathbf{x} = \mathbf{e}_j^\mathsf{T}\mathbf{x} = x_j$ (where $\mathbf{e}_j$ is the j-th unit vector of dimension n), $f = 2$, $\ell = 0$, and $u = u_j$, the *ISP* is modified by adding

- $g_{\mathbf{x}} \geq \frac{\mathbf{f}^\mathsf{T}\mathbf{x}-f+1}{u_j-f+1} = \frac{x_j-2+1}{u_j-2+1} = \frac{x_j-1}{u_j-1}$, $\quad g_{\mathbf{x}} \in \{0,1\}$ in the down-branch;
- $g_{\mathbf{x}} \leq \frac{\mathbf{f}^\mathsf{T}\mathbf{x}-\ell_j}{f-\ell_j} = \frac{x_j-0}{2-0} = \frac{x_j}{2}$, $\quad g_{\mathbf{x}} \in \{0,1\}$ in the up-branch.

7.13 λ-branching: interval constraint on the cost function

The fractionality of $\lambda^\star_{RMP}$ is given by $F = \sum_{\mathbf{x}\in\mathcal{F}}(\lambda^\star_{\mathbf{x}} - \lfloor \lambda^\star_{\mathbf{x}} \rfloor)$. Rank the costs in non-decreasing order for this index-set $\mathcal{F}$ of variables, say $\ell \leq c_{\mathbf{x}_1} \leq \cdots \leq c_{\mathbf{x}_{|\mathcal{F}|}} \leq u$. When we are lucky, and $c_{\mathbf{x}_1} < c_{\mathbf{x}_{|\mathcal{F}|}}$, the following works. Compute v such that $c_{\mathbf{x}_1} \leq \lfloor v \rfloor < \lfloor v \rfloor + 1 \leq c_{\mathbf{x}_{|\mathcal{F}|}}$, either by taking the average for $\mathbf{x} \in \mathcal{F}$, or the median, or simply $v = \frac{c_{\mathbf{x}_1}+c_{\mathbf{x}_{|\mathcal{F}|}}}{2}$. Then F, as in (7.81), is the result of two sums:

$$F = \underbrace{\sum_{\mathbf{x}\in\mathcal{F}:c_{\mathbf{x}_1}\leq c_{\mathbf{x}}\leq\lfloor v\rfloor}(\lambda^\star_{\mathbf{x}} - \lfloor \lambda^\star_{\mathbf{x}} \rfloor)}_{\geq (\lambda^\star_{\mathbf{x}_1} - \lfloor \lambda^\star_{\mathbf{x}_1} \rfloor) > 0} + \underbrace{\sum_{\mathbf{x}\in\mathcal{F}:\lfloor v\rfloor+1\leq c_{\mathbf{x}}\leq c_{\mathbf{x}_{|\mathcal{F}|}}}(\lambda^\star_{\mathbf{x}} - \lfloor \lambda^\star_{\mathbf{x}} \rfloor)}_{\geq (\lambda^\star_{\mathbf{x}_{|\mathcal{F}|}} - \lfloor \lambda^\star_{\mathbf{x}_{|\mathcal{F}|}} \rfloor) > 0}$$

and the fractionality of one of these sums cannot exceed $F/2$.

If F is fractional at the start, at least one of these sums is also fractional and this gives us a branching interval on $\mathbf{c}^\mathsf{T}\mathbf{x}$, e.g. for the first sum:

$$\sum_{\mathbf{x}\in\mathcal{X}:\ell\leq c_{\mathbf{x}}\leq\lfloor v\rfloor} \lambda_{\mathbf{x}} = \beta \text{ fractional.}$$

Otherwise, F takes an integer value and, either both sums are fractional (with two branching opportunities) or both are integer. In the later case, we split again using the $c_{\mathbf{x}}$ values in the lower or the upper interval until we find a fractional sum. Although there are many ways to split, interesting intervals in the end are one of these two forms: $\mathcal{B} = \{\mathbf{x} \in \mathcal{X} \mid \ell \leq \mathbf{c}^\mathsf{T}\mathbf{x} \leq c\}$ or $\mathcal{B} = \{\mathbf{x} \in \mathcal{X} \mid c \leq \mathbf{c}^\mathsf{T}\mathbf{x} \leq u\}$.

7.14 λ-branching: ISP modifications for *Rules* 3.2 and 3.4
Expression $\mathbf{f}^\intercal\mathbf{x} \geq f$ is given by $x_r + x_s \geq 2$, where $x_r + x_s \in [0,2]$. Moreover, in both *Rules* 3.2 and 3.4, $g_\mathbf{x} = y_{rs}$. Therefore, (7.66) becomes

$$\begin{aligned}
y_{rs} &\geq \frac{x_r + x_s - 2 + 1}{2 - 2 + 1} = x_r + x_s - 1, \quad y_{rs} \in \{0,1\} \quad \text{in the down-branch;} \\
y_{rs} &\leq \frac{x_r + x_s - 0}{2 - 0} = \frac{x_r + x_s}{2}, \quad y_{rs} \in \{0,1\} \quad \text{in the up-branch.}
\end{aligned}$$

7.15 Time constrained shortest path problem: thinking like a computer
We start by making an exhaustive list of the *variables and constraints created* at any point before node **BB4** is processed, see the *RMP* solved at **BB0** (p. 510) and branch-and-bound process.

- Node **BB0**: artificial variable y_0 and path-variables λ_{1246}, λ_{1356}, λ_{13256}, λ_{1256}; constraint $x_{13} + x_{32} = 0$ and artificial variable y_8 together with $x_{13} + x_{32} \geq 1$.
- Node **BB1**: path-variable λ_{12456}.
- Node **BB2**: constraints $x_{12} = 0$ and $x_{12} = 1$.
- Node **BB3**: path-variable λ_{13246}.

During housekeeping for node **BB4**, we activate constraints $x_{13} + x_{32} \geq 1$ and $x_{12} = 1$. We can then set the variable bounds to $[0, \infty)$ for y_0, λ_{1246}, λ_{1256}, y_8, and λ_{12456} whereas the bounds must be set to $[0,0]$ for λ_{1356}, λ_{13256}, and λ_{13246}. The *RMP* we deal with after housekeeping is

$$\begin{array}{lllllll}
\min 1000y_0 & +1000y_8 & +\ 3\lambda_{1246} & +\ 5\lambda_{1256} & +\ 14\lambda_{12456} & & \\
\text{s.t.} & & 18\lambda_{1246} & +\ 15\lambda_{1256} & +\ 14\lambda_{12456} & \leq 14 & [\pi_7] \\
 & y_8 & & & & \geq 1 & [\pi_8] \\
y_0 & & +\ \lambda_{1246} & +\ \lambda_{1256} & +\ \lambda_{12456} & = 1 & [\pi_0] \\
y_0, & y_8, & \lambda_{1246}, & \lambda_{1256}, & \lambda_{12456} & \geq 0. &
\end{array}$$

An optimal solution is $y_8 = \lambda_{12456} = 1$ with cost 1014 together with dual values $\pi_7 = -70.42857$ and $\pi_8 = \pi_0 = 1000$. Solving the *ISP* produces path 12456 at reduced cost $\bar{c}_{12456} = 14 - \pi_7(14) - \pi_8(0) - \pi_0 = 0$. We confirm that big-$M$ is large enough thus proving optimality of the *MP*. Since there remains at least one positive artificial y-variable, we conclude that node **BB4** can be pruned by infeasibility.

A couple more words of wisdom are in order. We conclude by noting that not only are the way the cuts identified application-specific, but also the way we handle them during housekeeping depends on what we know about the problem. In particular, we have handled the branch $x_{13} + x_{32} = 0$ in the *ISP* but could also have imposed it in the *IMP* in a generic fashion.

7.16 Minimum number of vehicles for the VRPTW

- The two-index arc-flow formulation (7.21) is a good choice for the *ILP*, defined on the network $G_{do} = (N, A_{do})$ (or the similar one from Exercise 5.3). The important aspect to consider is the objective function, say $z = \sum_{(i,j)\in A_{do}} c_{ij}x_{ij}$.

The number of vehicles utilized is given by the value of x_{do} in an optimal solution. Defining $z = x_{do}$ alone is the worst objective function because the optimization process does not look for "optimal routes." Therefore, the traveling costs c_{ij}, $(i,j) \in A$, should be kept while c_{do} is set to a large value, say big-M:

$$z = M x_{do} + \sum_{(i,j)\in A} c_{ij}x_{ij}.$$

- The various branching and cutting strategies described in this chapter can be applied to the reader's choice. The stopping rule is however based on an early termination: from the usual dual bound, derive an integer one on the number of vehicles only, compare it to the primal integer bound on the number of vehicles only, and stop when the equality is reached.
- For an early implementation on school bus scheduling problems, see Desrosiers et al. (1988b) presenting an approach using the augmented Lagrangian method.

7.17 Validity of a dominance rule for the VRPTW with strong capacity cuts
According to Definition 5.3, we need to show that, for every feasible extension χ' of p', $p \oplus \chi'$ is feasible and $\bar{c}_{p\oplus\chi'} \leq \bar{c}_{p'\oplus\chi'}$. Let χ' be a feasible extension of p'.
• Given (7.132b)–(7.132d) and that the REFs (5.18b)–(5.18d) are non-decreasing, χ' is also a feasible extension of p, i.e., $p \oplus \chi'$ is feasible.
• Now, let

$$\mathcal{S}^+(\chi') = \{S \in \mathcal{S}^+ \mid \exists (i,j) \in \chi' \text{ such that } S \in \mathcal{S}_{ij}\},$$
$$\mathcal{S}^+_{u_1u_2}(\chi') = \{S \in \mathcal{S}^+(\chi') \mid T^{rccs}_p = u_1, T^{rccs}_{p'} = u_2\} \text{ for } u_1, u_2 \in \{0,1\}, \text{ and}$$
$$\bar{c}_{\chi'} = \textstyle\sum_{(i,j)\in\chi'} \bar{c}_{ij}.$$

We get

$$\begin{aligned}
\bar{c}_{p\oplus\chi'} &= T^{rCost}_p + \bar{c}_{\chi'} - \sum_{S\in\mathcal{S}^+_{00}(\chi')} \pi_S - \sum_{S\in\mathcal{S}^+_{01}(\chi')} \pi_S && \text{(a)}\\
&\leq T^{rCost}_{p'} - \sum_{S\in\mathcal{S}^+_{10}} \pi_S + \bar{c}_{\chi'} - \sum_{S\in\mathcal{S}^+_{00}(\chi')} \pi_S - \sum_{S\in\mathcal{S}^+_{01}(\chi')} \pi_S && \text{(b)}\\
&\leq T^{rCost}_{p'} - \sum_{S\in\mathcal{S}^+_{10}} \pi_S + \bar{c}_{\chi'} - \sum_{S\in\mathcal{S}^+_{00}(\chi')} \pi_S && \text{(c)}\\
&\leq T^{rCost}_{p'} + \bar{c}_{\chi'} - \sum_{S\in\mathcal{S}^+_{00}(\chi')} \pi_S - \sum_{S\in\mathcal{S}^+_{10}(\chi')} \pi_S && \text{(d)}\\
&= \bar{c}_{p'\oplus\chi'}, && \text{(e)}
\end{aligned}$$

where (b) is derived from (7.132a), (c) from the fact that $\pi_S > 0$ for all $S \in \mathcal{S}^+$, and (d) from $\mathcal{S}^+_{10}(\chi') \subseteq \mathcal{S}^+_{10}$. Hence, $\bar{c}_{p\oplus\chi'} \leq \bar{c}_{p'\oplus\chi'}$ which completes the proof.

7.18 Comparison of dominance rules for the VRPTW with strong capacity cuts
We need to prove that any label $E_{p'}$ declared to be dominated by a label E_p according to rule (5.19) and (7.130) is also identified as such by rule (7.132).
According to the first rule, we have that

$$\begin{aligned}
T^{rCost}_p &\leq T^{rCost}_{p'} && \text{(a)}\\
T^{rccs}_p &\leq T^{rccs}_{p'}, \quad \forall S \in \mathcal{S}. && \text{(b)}
\end{aligned}$$

From (b), we deduce that $\mathcal{S}^+_{10} = \emptyset$. Combined with (a), it thus implies (7.132a). Given that (7.132b)–(7.132d) correspond to (5.19b)–(5.19d), rule (7.132) is also satisfied.

7.19 Arc-flow variable fixing for the *VRPTW*

(a) To compute $\bar{c}^{\star}_{ij}$ for an arc (i,j), we must consider all concatenations of a forward label $(T_i^{rCost}, T_i^{time}, T_i^{load}, [T_i^{uCust_h}]_{h\in C})$ at node i, arc (i,j), and a backward label $(B_i^{rCost}, B_i^{lTime}, B_j^{mLoad}, [B_j^{uCust_h}]_{h\in C})$ at node j that yield a feasible path and select the least cost one. Such a concatenation is feasible if

$$\begin{aligned} T_i^{time} + t_{ij} &\leq B_j^{lTime} \\ T_i^{load} &\leq B_j^{mLoad} \\ T_i^{uCust_h} + B_j^{uCust_h} &\leq 1 \qquad \forall h \in C. \end{aligned}$$

Its reduced cost is equal to $T_i^{rCost} + \tilde{c}_{ij} + B_j^{rCost}$.

• For arc $(1,2)$, we can concatenate forward label $(3,2,1,[1,0,0,0]$, arc $(1,2)$, and backward label $(-2.5,18,3,[0,1,0,1])$ to yield a feasible path of reduced cost 6. The other possible concatenation of forward label $(3,2,1,[1,0,0,0]$, arc $(1,2)$, and backward label $(-15,9,1,[1,1,1,1])$ is infeasible because of the time condition $(2+10 \not\leq 9)$ and the condition on customer 1. Therefore, $\bar{c}^{\star}_{12} = 6$. Given that $\bar{c}^{\star}_{12} = 6 \not> UB - LB = 9.5$, variable x_{12} cannot be fixed to 0.

• For arc $(2,3)$, there are 6 possible concatenations to check. Only two of them are feasible, namely, forward label $(15,8,2,[1,1,0,0])$, arc $(2,3)$, and backward label $(12,25,4,[0,0,1,0])$, and forward label $(8.5,12,3,[1,1,0,1])$, arc $(2,3)$, and backward label $(12,25,4,[0,0,1,0])$. The first option yields a reduced cost of 13.5 and the second of 7. Hence, $\bar{c}^{\star}_{23} = \min\{13.5, 7\} = 7$ and x_{23} cannot be fixed to 0 because $\bar{c}^{\star}_{23} = 7 \not> UB - LB = 9.5$.

(b) For arc $(1,2)$, there are 2 possible concatenations but only the concatenation of the forward label $(-13,2,1,[1,0,0,0])$, arc $(1,2)$, and the backward label $(5.5,18,3,[0,1,0,1])$ is feasible, with a reduced cost of 4.5. Thus, $\bar{c}^{\star}_{12} = 4.5 > UB - LB = 1.5$ and x_{12} can be fixed to 0.

For arc $(2,3)$, there are also 2 possible concatenations and only the concatenation of the forward label $(-1,8,2,[1,1,0,0])$, arc $(2,3)$, and backward label $(13.5,25,4,[0,0,1,0])$ is feasible, with a reduced cost of 7. Because $\bar{c}^{\star}_{23} = 7 > UB - LB = 1.5$, x_{23} can be fixed to 0.

This shows that adding a cut to substantially raise the lower bound can help to eliminate arc-flow variables and reduce the size of the *ISP*.

7.20 Branching on a resource interval

The following solution is adapted from Gamache et al. (1998, pp. 250–251).

- The upper bound 78 is like the vehicle capacity in the *VRPTW*, and as such, F is used to cumulate the number of flight credits. Let T_i^F denote the resource value at node i and f_{ij} be the number of flight credits on arc (i,j). If this arc is used in an *od*-path, the F-resource consumption is updated from i to j by $T_j^F = T_i^F + f_{ij}$, provided T_j^F is less than or equal to 78.

- The lower bound 70 requires an additional resource, say $negF$, which is the negative of F. Then the restriction becomes $-F = negF \leq -70$ and the lower bound is also treated as a capacity restriction. Resource $negF$ is updated from i to j by $T_j^{negF} = T_i^{negF} - f_{ij}$, where the negative upper bound -70 is only enforced at the sink node d. If $T_d^{negF} > -70$, then the number of flight credits is insufficient and the corresponding path is discarded.

  An unfortunate outcome of having these two opposite resources is that, for two partial paths reaching node i with different flight credit values, say $T_{i,1}^F < T_{i,2}^F$, the dominance rule never applies. Indeed, $T_{i,1}^F < T_{i,2}^F \Leftrightarrow T_{i,1}^{negF} > T_{i,2}^{negF}$. To increase the possibility of dominance once the lower bound is reached, we can replace the above extension function for resource $negF$ by

$$T_j^{negF} = \max\{-70, T_i^{negF} - f_{ij}\}.$$

7.21 B&P&C strategies for various applications
Compare your strategies with those in the suggested references.

(a) One-dimensional cutting stock problem
Desaulniers et al. (2011)

(b) Aircraft routing with schedule synchronization
Ioachim et al. (1999)

(c) Scene selection problem
Jans and Desrosiers (2010)

(d) Design of balanced student teams
Desrosiers et al. (2005)

(e) Multiple depot vehicle scheduling problem
Ribeiro and Soumis (1994)

(f) Pickup and delivery problem with time windows
Dumas et al. (1991); Røpke and Cordeau (2009)

(g) Crew pairing problem with base constraints
Quesnel et al. (2017)

(h) Capacitated p-median problem
Ceselli and Righini (2005)

(i) Generalized assignment problem
Savelsbergh (1997); Barnhart et al. (1998)

Exercises of Chapter 8

8.1 François Soumis

This François holds a master's degree in mathematics (topology) from the Département de mathématiques et de statistique (Université de Montréal). He continued his studies at the Département d'informatique et de recherche opérationnelle and obtained a PhD (1979) in operations research under the supervision of Professors Jacques A. Ferland and Jean-Marc Rousseau (co-founder of GIRO).

Since the start of the GENCOL team with Jacques in 1981, François-x is known to largely prefer the convexification approach over the discretization one (although he is always open to new ideas). At the time of writing, he has supervised or co-supervised 107 master's and 68 doctoral students, his contributions to decomposition methods (such as Dantzig-Wolfe and Benders) are numerous, both on theory and efficient resolution of large-scale applications. To name a few, consider the following list where his inspiration is decisive, with the theoretical innovations on the left and applications on the right:

Constrained shortest path algorithms	Urban transit systems
k-path cuts and z-cuts	Telecommunication networks
A crossover method	Automated vehicles
Branch-first, Cut-second strategy	Airline crew scheduling
Existence of a compact formulation	Aircraft routing
Constraints aggregation for the *SPP*	Flight planing
Improved primal simplex	Production scheduling
Positive edge rule	Real-time control of mining trucks
Integral simplex for the *SPP*	Locomotive assignment
Lagrangian decomposition	Shift scheduling
Primal Benders decomposition	Fractional aircraft ownership operations

In the late 1980s, Montreal-based GIRO, which today supplies bus and metro scheduling software to over 300 cities worldwide, funded the university team to develop a commercial version of `GENCOL`. This led to the creation of the `Crew-Opt` optimizer, subsequently integrated into the `HASTUS` software, a world reference in the scheduling of public transit services.

In 1987, François co-founded AD OPT Technologies to put the fruits of his research into practice. The company creates and markets a real-time planning and optimization system for trucks in open-pit mines. The system is implemented in six mines in Québec, Brazil, India, and the United States.

Marketing of air transport products began in the early 1990s, with UPS as its first customer, followed by Air Transat, Air Canada, Quantas, FedEx, etc. Over the years, AD OPT has done business with some thirty airlines, while developing the `Altitude` software suite:

- `Pairing`
  Builds cost-effective pairings meeting all rules, regulations, and fatigue targets.
- `Rostering`
  Creates monthly schedules for pilots and flight attendants.
- `PBS` (Preferential Bidding System)
  Takes into account crew member preferences, collective agreements, and preassigned tasks.
- `BLISS` (BidLine Integrated Scheduling System)
  Builds quality lines by offering user-driven, flexible line generation.
- `Intuitive Crew Interface`
  Allows crew members to enter preferences and requests through web and mobile devices for upcoming planning periods.
- `Insight`
  Manages short-term and long-term planning requirements for every applicable category of crew member.

The company also designs schedules for shift workers in a wide range of industries such as retail, factories, hospitals, and service centers (`ShiftLogic`).

From 1992 to 1996, François is the director of the GERAD research center whose members come from the seven universities in Montréal: specialists in data and decision sciences, computer scientists, applied mathematicians, and mathematical engineers. Moreover, and for several years, the `GENCOL` team comprises up to 35 students, programmers, professors, and researchers.

In 1999, AD OPT enters the Toronto Stock Exchange and, with its 250 employees, is subsequently acquired by Kronos in 2004, a world leader in workforce management solutions. In 2019, the AD OPT division is sold to the aviation and transportation IT services provider IBS Software.

From 2014 to 2017, François is the founding director of IVADO (Institut de valorisation des données), which is now a huge research organization in artificial intelligence (AI) and optimization.

> IVADO is an interdisciplinary, cross-sectoral research, training and knowledge mobilization consortium whose mission is to develop and promote a robust, reasoning and responsible AI. Led by Université de Montréal with four university partners (Polytechnique Montréal, HEC Montréal, Université Laval and McGill University), IVADO brings together research centres, government bodies and industry members to co-build ambitious cross-sectoral initiatives with the goal of fostering a paradigm shift for AI and its adoption. – IVADO

We could go on with his achievements but summarize with a potpourri that highlights the extent of his career. The *prize Lionel-Boulet* is "awarded by the Québec government since 1999 to a person who has led an outstanding career in industrial research." Those who had the opportunity to discuss with François know that this is a well-deserved recognition, not only because of his exceptional research skills and strong business acumen, but also his lovable nature.

- programmer at the computing center (Université de Montréal)
- IT analyst at Sainte-Justine Hospital in Montréal

- researcher at HEC Montréal
- professor in Rwanda for the Canadian International Development Agency
- gentleman farmer selling tomatoes at 4 in the morning at the market place
- professor at Polytechnique Montréal (polymtl.ca/expertises/en/soumis-francois)
- artist-sculptor
- laureate of the Prix du Québec Lionel-Boulet (2014)

By the way, he is a formidable squash/chess/backgammon player who we have learned by now not to bet against. He also loves French wines and once made with Jacques two gallons of maple syrup of dubious dark color over an open wood fire (Spring 1981). Did we mention that he and Jacques are good friends? Well, the same is true for Guy who happens to have been supervised for his PhD by François. In the midst of this reminiscence, we cannot help but mention a few others: Mirela, Goran, and the late Egon (1922–2019).

At the age of 78, François sadly passed away on February 14, 2025, just a few days before handing over the final version of the manuscript to the publisher. We have kept the present tense in the above to emphasize his immense contribution to operations research, today and in the future. We finalize this short addendum by quoting José Manuel Valério de Carvalho:

"François was brilliant, a genius,
who contributed many complex and rich ideas.
[...] *We will all miss his insights."*

8.2 Undivided attention

Look out for an (online) errata appendix and do not hesitate to drop us a line should it prove incomplete.

References

Tobias Achterberg. *Constraint Integer Programming.* PhD thesis, Technische Universität Berlin, Fakultät II - Mathematik und Naturwissenschaften, Berlin, July 2007. doi:10.14279/depositonce-1634.

Tobias Achterberg. SCIP: Solving constraint integer programs. *Mathematical Programming Computation*, 1(1):1–41, January 2009. doi:10.1007/s12532-008-0001-1.

Tobias Achterberg, Thorsten Koch, and Alexander Martin. Branching rules revisited. *Operations Research Letters*, 33(1):42–54, January 2005. doi:10.1016/j.orl.2004.04.002.

Ravindra Kumar Ahuja, Thomas Lee Magnanti, and James Berger Orlin. *Network Flows: Theory, Algorithms, and Applications*. Prentice Hall, Upper Saddle River, February 1993.

Clément Altman, Guy Desaulniers, and Fausto Errico. The fragility-constrained vehicle routing problem with time windows. *Transportation Science*, 57(2):552–572, March–April 2023. doi:10.1287/trsc.2022.1168.

Leif H. Appelgren. A column generation algorithm for a ship scheduling problem. *Transportation Science*, 3(1):53–68, February 1969. doi:10.1287/trsc.3.1.53.

Leif H. Appelgren. Integer programming methods for a vessel scheduling problem. *Transportation Science*, 5(1):64–78, February 1971. doi:10.1287/trsc.5.1.64.

David Applegate, Robert Bixby, Vašek Chvátal, and William Cook. Implementing the Dantzig-Fulkerson-Johnson algorithm for large traveling salesman problems. *Mathematical Programming*, 97(1):91–153, July 2003. doi:10.1007/s10107-003-0440-4.

Claudia Archetti, Mathieu Bouchard, and Guy Desaulniers. Enhanced branch and price and cut for vehicle routing with split deliveries and time windows. *Transportation Science*, 45(3):285–298, 2011. doi:10.1287/trsc.1100.0363.

Larry Armijo. Minimization of functions having Lipschitz continuous first partial derivatives. *Pacific Journal of Mathematics*, 16(1):1–3, January 1966. doi:10.2140/pjm.1966.16.1.

Arjang A. Assad and Saul Irving Gass. *Profiles in Operations Research: Pioneers and Innovators*, volume 147 of *International Series in Operations Research & Management Science*. Springer, New York, 2011. doi:10.1007/978-1-4419-6281-2.

David Avis, Bohdan Kaluzny, and David Titley-Péloquin. Visualizing and constructing cycles in the simplex method. *Operations Research*, 56(2):512–518, April 2008. doi:10.1287/opre.1070.0474.

Cevdet Aykanat, Ali Pinar, and Ümit Veysel Çatalyürek. Permuting sparse rectangular matrices into block-diagonal form. *SIAM Journal on Scientific Computing*, 25(6):1860–1879, 2004. doi:10.1137/S1064827502401953.

David A. Bader and Guojing Cong. Fast shared-memory algorithms for computing the minimum spanning forest of sparse graphs. *Journal of Parallel and Distributed Computing*, 66(11):1366–1378, November 2006. doi:10.1016/j.jpdc.2006.06.001.

Roberto Baldacci, Nicos Christofides, and Aristide Mingozzi. An exact algorithm for the vehicle routing problem based on the set partitioning formulation with additional cuts. *Mathematical Programming*, 115(2):351–385, August 2008. doi:10.1007/s10107-007-0178-5.

Roberto Baldacci, Aristide Mingozzi, and Roberto Roberti. New route relaxation and pricing strategies for the vehicle routing problem. *Operations Research*, 59(5):1269–1283, September–October 2011. doi:10.1287/opre.1110.0975.

Roberto Baldacci, Aristide Mingozzi, and Roberto Roberti. Recent exact algorithms for solving the vehicle routing problem under capacity and time window constraints. *European Journal of Operational Research*, 218(1):1–6, April 2012. doi:10.1016/j.ejor.2011.07.037.

Francisco Barahona and Ranga Anbil. The volume algorithm: producing primal solutions with a subgradient method. *Mathematical Programming*, 87(3):385–399, May 2000. doi:10.1007/s101070050002.

J. Desrosiers et al., *Branch-and-Price*,
https://doi.org/10.1007/978-3-031-96917-1

Cynthia Barnhart, Levent Hatay, and Ellis Lane Johnson. Deadhead selection for the long-haul crew pairing problem. *Operations Research*, 43(3):491–499, May–June 1995. doi:10.1287/opre.43.3.491.

Cynthia Barnhart, Ellis Lane Johnson, George Lann Nemhauser, Martin W.P. Savelsbergh, and Pamela Hatch Vance. Branch-and-Price: Column generation for solving huge integer programs. *Operations Research*, 46(3):316–329, May–June 1998. doi:10.1287/opre.46.3.316.

Cynthia Barnhart, Amy Mainville Cohn, Ellis Lane Johnson, Diego Klabjan, George Lann Nemhauser, and Pamela Hatch Vance. Airline crew scheduling. In Randolph William Hall, editor, *Handbook of Transportation Science*, chapter 14, pages 517–560. Springer, Boston, 2003. doi:10.1007/0-306-48058-1_14.

Richard Harold Bartels and Gene Howard Golub. The simplex method of linear programming using LU decomposition. *Communications of the ACM*, 12(5):266–268, May 1969. doi:10.1145/362946.362974.

Saverio Basso and Alberto Ceselli. Distributed asynchronous column generation. *Computers & Operations Research*, 146:105894, October 2022. doi:10.1016/j.cor.2022.105894.

Saverio Basso and Alberto Ceselli. A data driven Dantzig-Wolfe decomposition framework. *Mathematical Programming Computation*, 15(1):153–194, March 2023. doi:10.1007/s12532-022-00230-4.

Michael Bastubbe, Marco E. Lübbecke, and Jonas Timon Witt. A computational investigation on the strength of Dantzig-Wolfe reformulations. In Gianlorenzo D'Angelo, editor, *Experimental Algorithms*, volume 103 of *Leibniz International Proceedings in Informatics (LIPIcs)*, pages 11:1–11:12, Germany, May 2018. Schloss Dagstuhl – Leibniz-Zentrum für Informatik. doi:10.4230/LIPIcs.SEA.2018.11.

Steven Baum. *Integral Near-Optimal Solutions to Certain Classes of Linear Programming Problems*. PhD thesis, Cornell University, Ithaca, NY, USA, September 1977. No. 360.

Steven Baum and Leslie Earl Trotter, Jr. Finite checkability for integer rounding properties in combinatorial programming problems. *Mathematical Programming*, 22(1):141–147, December 1982. doi:10.1007/BF01581034.

Evelyn Martin Lansdowne Beale. An alternative method for linear programming. *Mathematical Proceedings of the Cambridge Philosophical Society*, 50(4):513–523, October 1954. doi:10.1017/S0305004100029650.

Richard Ernest Bellman. *Dynamic Programming*. Princeton University Press, Princeton, 1957.

Mandell Bellmore and John C. Malone. Pathology of traveling-salesman subtour-elimination algorithms. *Operations Research*, 19(2):278–307, April 1971. doi:10.1287/opre.19.2.278.

Gleb Belov and Guntram Scheithauer. A cutting plane algorithm for the one-dimensional cutting stock problem with multiple lengths. *European Journal of Operational Research*, 141(2):274–294, September 2002. doi:10.1016/S0377-2217(02)00125-X.

Gleb Belov and Guntram Scheithauer. A branch-and-cut-and-price algorithm for one-dimensional stock cutting and two-dimensional two-stage cutting. *European Journal of Operational Research*, 171(1):85–106, May 2006. doi:10.1016/j.ejor.2004.08.036.

Hatem M. T. Ben Amor and Jacques Desrosiers. A proximal trust-region algorithm for column generation stabilization. *Computers & Operations Research*, 33(4):910–927, April 2006. doi:10.1016/j.cor.2004.08.003.

Hatem M. T. Ben Amor and José Manuel Valério de Carvalho. Cutting stock problems. In Guy Desaulniers, Jacques Desrosiers, and Marius Mihai Solomon, editors, *Column Generation*, chapter 5, pages 131–161. Springer, New York, April 2005. doi:10.1007/0-387-25486-2_5.

Hatem M. T. Ben Amor, Jacques Desrosiers, and François Soumis. Recovering an optimal LP basis from an optimal dual solution. *Operations Research Letters*, 34(5):569–576, September 2006a. doi:10.1016/j.orl.2005.10.001.

Hatem M. T. Ben Amor, Jacques Desrosiers, and José Manuel Valério de Carvalho. Dual-optimal inequalities for stabilized column generation. *Operations Research*, 54(3):454–463, June 2006b. doi:10.1287/opre.1060.0278.

Hatem M. T. Ben Amor, Jacques Desrosiers, and Antonio Frangioni. On the choice of explicit stabilizing terms in column generation. *Discrete Applied Mathematics*, 157(6):1167–1184, March 2009. doi:10.1016/j.dam.2008.06.021.

Pascal Benchimol, Guy Desaulniers, and Jacques Desrosiers. Stabilized dynamic constraint aggregation for solving set partitioning problems. *European Journal of Operational Research*, 223 (2):360–371, December 2012. doi:10.1016/j.ejor.2012.07.004.

Jacobus Franciscus Benders. Partitioning procedures for solving mixed-variables programming problems. *Numerische Mathematik*, 4(1):238–252, December 1962. doi:10.1007/BF01386316.

Yoshua Bengio, Andrea Lodi, and Antoine Prouvost. Machine learning for combinatorial optimization: A methodological tour d'horizon. *European Journal of Operational Research*, 290 (2):405–421, April 2021. doi:10.1016/j.ejor.2020.07.063.

Martin Bergner, Alberto Caprara, Alberto Ceselli, Fabio Furini, Marco E. Lübbecke, Enrico Malaguti, and Emiliano Traversi. Automatic Dantzig-Wolfe reformulation of mixed integer programs. *Mathematical Programming*, 149(1):391–424, February 2015. doi:10.1007/s10107-014-0761-5.

Alan Albert Bertossi, Paolo Carraresi, and Giorgio Gallo. On some matching problems arising in vehicle scheduling models. *Networks*, 17(3):271–281, 1987. doi:10.1002/net.3230170303.

Dimitri Panteli Bertsekas. *Nonlinear Programming*. Athena Scientific, Belmont, 1995.

Dimitris John Bertsimas and John Nikolaos Tsitsiklis. *Introduction to Linear Optimization*. Athena Scientific, Belmont, 1997.

Dimitris John Bertsimas and Robert Weismantel. *Optimization over integers*. Athena Scientific, Belmont, 2005.

Ksenia Bestuzheva, Mathieu Besançon, Wei-Kun Chen, Antonia Chmiela, Tim Donkiewicz, Jasper van Doornmalen, Leon Eifler, Oliver Gaul, Gerald Gamrath, Ambros Gleixner, Leona Gottwald, Christoph Graczyk, Katrin Halbig, Alexander Hoen, Christopher Hojny, Rolf van der Hulst, Thorsten Koch, Marco Lübbecke, Stephen John Maher, Frederic Matter, Erik Mühmer, Benjamin Müller, Marc Emanuel Pfetsch, Daniel Rehfeldt, Steffan Schlein, Franziska Schlösser, Felipe Serrano, Yuji Shinano, Boro Sofranac, Mark Turner, Stefan Vigerske, Fabian Wegscheider, Philipp Wellner, Dieter Weninger, and Jakob Witzig. The SCIP Optimization Suite 8.0. arXiv, March 2021. doi:10.48550/ARXIV.2112.08872.

Robert Bixby and Edward Rothberg. Progress in computational mixed integer programming–A look back from the other side of the tipping point. *Annals of Operations Research*, 149(1): 37–41, January 2007. doi:10.1007/s10479-006-0091-y.

Robert Eugene Bixby. Solving real-world linear programs: A decade and more of progress. *Operations Research*, 50(1):3–15, February 2002. doi:10.1287/opre.50.1.3.17780.

Robert Eugene Bixby. A brief history of linear and mixed-integer programming computation. In Martin Grötschel, editor, *Optimization Stories*, Extra Volume ISMP, pages 107–121, Bielefeld, 2012. Documenta Mathematica, Journal der Deutschen Mathematiker-Vereinigung. ISBN 978-3-936609-58-5.

Robert Eugene Bixby. *Computational Progress in Linear and Mixed Integer Programming*. 12th Congress of ECCO, Barcelona, Spain, February 15–18, 2017. URL `https://ecco2017.euro-online.org/en/resources/files/invited-speakers/robertbixby.pdf`.

Robert Eugene Bixby, John W. Gregory, Irvin Jay Lustig, Roy Earl Marsten, and David Francis Shanno. Very large-scale linear programming: A case study in combining interior point and simplex methods. *Operations Research*, 40(5):885–897, September–October 1992. doi:10.1287/opre.40.5.885.

Natashia Boland, John Dethridge, and Irina Dumitrescu. Accelerated label setting algorithms for the elementary resource constrained shortest path problem. *Operations Research Letters*, 34(1): 58–68, January 2006. doi:10.1016/j.orl.2004.11.011.

Otakar Borůvka. O jistém problému minimálním. *Práce Moravské Přírodovědecké Společnosti*, 3:37–58, 1926. (in Czech and German).

Olivier Briant, Claude Lemaréchal, Philippe Meurdesoif, Sophie Michel, Nancy Perrot, and François Vanderbeck. Comparison of bundle and classical column generation. *Mathematical Programming*, 113(2):299–344, December 2008. doi:10.1007/s10107-006-0079-z.

E. Rod Butchers, Paul R. Day, Andrew P. Goldie, Stephen Miller, Jeff A. Meyer, David Murray Ryan, Amanda C. Scott, and Chris A. Wallace. Optimized crew scheduling at Air New Zealand. *Interfaces*, 31(1):30–56, January–February 2001. doi:10.1287/inte.31.1.30.9688.

Giorgio Carpaneto and Paolo Toth. Some new branching and bounding criteria for the asymmetric travelling salesman problem. *Management Science*, 26(7):736–743, July 1980. doi:10.1287/mnsc.26.7.736.

Alberto Ceselli and Giovanni Righini. A branch-and-price algorithm for the capacitated p-median problem. *Networks*, 45(3):125–142, May 2005. doi:10.1002/net.20059.

Abraham Charnes and William Wager Cooper. The strong Minkowski-Farkas-Weyl theorem for vector spaces over ordered fields. *Proceedings of the National Academy of Sciences*, 44(9): 914–916, September 1958. doi:10.1073/pnas.44.9.914.

Abraham Charnes and William Wager Cooper. On some works of Kantorovich, Koopmans and others. *Management Science*, 8(3):246–263, 1962. doi:10.1287/mnsc.8.3.246.

Elliott Ward Cheney, Jr. and Allen Abbey Goldstein. Newton's method for convex programming and Tchebycheff approximation. *Numerische Mathematik*, 1(1):253–268, December 1959. doi:10.1007/BF01386389.

Marilène Cherkesly, Guy Desaulniers, and Gilbert Laporte. Branch-price-and-cut algorithms for the pickup and delivery problem with time windows and last-in-first-out loading. *Transportation Science*, 49(4):752–766, 2015. doi:10.1287/trsc.2014.0535.

Marilène Cherkesly, Guy Desaulniers, Stefan Irnich, and Gilbert Laporte. Branch-price-and-cut algorithms for the pickup and delivery problem with time windows and multiple stacks. *European Journal of Operational Research*, 250(3):782–793, 2016. doi:10.1016/j.ejor.2015.10.046.

Vašek Chvátal. *Linear Programming*. Series of Books in the Mathematical Sciences. W. H. Freeman, New York, 1983.

Jens Vinther Clausen, Richard Lusby, and Stefan Røpke. Consistency cuts for Dantzig-Wolfe reformulations. *Operations Research*, 70(5):2883–2905, September–October 2022. doi:10.1287/opre.2021.2160.

Amy Mainville Cohn and Cynthia Barnhart. Improving crew scheduling by incorporating key maintenance routing decisions. *Operations Research*, 51(3):387–396, May–June 2003. doi:10.1287/opre.51.3.387.14959.

Michele Conforti, Gérard Cornuéjols, and Giacomo Zambelli. *Integer programming*, volume 271 of *Graduate Texts in Mathematics*. Springer, Cham, 2014. doi:10.1007/978-3-319-11008-0.

Claudio Contardo and Rafael Martinelli. A new exact algorithm for the multi-depot vehicle routing problem under capacity and route length constraints. *Discrete Optimization*, 12:129–146, May 2014. doi:10.1016/j.disopt.2014.03.001.

Claudio Contardo, Guy Desaulniers, and François Lessard. Reaching the elementary lower bound in the vehicle routing problem with time windows. *Networks*, 65(1):88–99, January 2015. doi:10.1002/net.21594.

Jean-François Cordeau, François Soumis, and Jacques Desrosiers. A Benders decomposition approach for the locomotive and car assignment problem. *Transportation Science*, 34(2):133–149, May 2000. doi:10.1287/trsc.34.2.133.12308.

Jean-François Cordeau, Guy Desaulniers, Norbert Lingaya, François Soumis, and Jacques Desrosiers. Simultaneous locomotive and car assignment at VIA Rail Canada. *Transportation Research Part B: Methodological*, 35(8):767–787, September 2001a. doi:10.1016/S0191-2615(00)00022-9.

Jean-François Cordeau, Goran Stojković, François Soumis, and Jacques Desrosiers. Benders decomposition for simultaneous aircraft routing and crew scheduling. *Transportation Science*, 35 (4):375–388, November 2001b. doi:10.1287/trsc.35.4.375.10432.

Luciano Costa, Claudio Contardo, and Guy Desaulniers. Exact branch-price-and-cut algorithms for vehicle routing. *Transportation Science*, 53(4):946–985, July–August 2019. doi:10.1287/trsc.2018.0878.

Luciano Costa, Claudio Contardo, Guy Desaulniers, and Diego Galindo Pecin. Selective arc-*ng* pricing for vehicle routing. *International Transactions in Operational Research*, 28(5):2633–2690, September 2021. doi:10.1111/itor.12911.

Luciano Costa, Claudio Contardo, Guy Desaulniers, and Julian Yarkony. Stabilized column generation via the dynamic separation of aggregated rows. *INFORMS Journal on Computing*, 34(2): 1141–1156, March–April 2022. doi:10.1287/ijoc.2021.1094.

Richard Warren Cottle, Ellis Lane Johnson, and Roger Jean-Baptiste Robert Wets. George B. Dantzig (1914–2005). *Notices of the American Mathematical Society*, 54(3):344–362, 2007.

George Bernard Dantzig. *Linear programming and extensions*. Princeton University Press, Princeton, 1963. doi:10.7249/R366.

George Bernard Dantzig. *Applications et prolongements de la programmation linéaire*. Dunod, Paris, 1966. French translation and adaptation of Dantzig (1963) by Elio Ventura.

George Bernard Dantzig. Reminiscences about the origins of linear programming. *Operations Research Letters*, 1(2):43–48, 1982. doi:10.1016/0167-6377(82)90043-8.

George Bernard Dantzig. Origins of the simplex method. In *A History of Scientific Computing*, pages 141–151. Association for Computing Machinery, New York, 1990. doi:10.1145/87252.88081.

George Bernard Dantzig and Mukund Narain Thapa. *Linear Programming 1: Introduction*. Springer, Berlin Heidelberg, 1997. doi:10.1007/b97672.

George Bernard Dantzig and Mukund Narain Thapa. *Linear Programming 2: Theory and Extensions*. Springer, New York, 2003. doi:10.1007/b97283.

George Bernard Dantzig and Philip Wolfe. The decomposition algorithm for linear programs. *Econometrica*, 29(4):767–778, October 1961. doi:10.2307/1911818.

George Bernard Dantzig and Philip Starr Wolfe. Decomposition principle for linear programs. *Operations Research*, 8(1):101–111, February 1960. doi:10.1287/opre.8.1.101.

George Bernard Dantzig, Delbert Ray Fulkerson, and Selmer Martin Johnson. Solution of a large-scale traveling-salesman problem. *Journal of the Operations Research Society of America*, 2 (4):393–410, November 1954. doi:10.1287/opre.2.4.393.

George Bernard Dantzig, Lester Randolph Ford, and Delbert Ray Fulkerson. A primal-dual algorithm for linear programs. In Harold William Kuhn and Albert William Tucker, editors, *Linear Inequalities and Related Systems*, Annals of Mathematics Study, No. 38, pages 171–182. Princeton University Press, Princeton, 1956.

Vinícius Loti de Lima, Cláudio Alves, François Clautiaux, Manuel Iori, and José Manuel Valério de Carvalho. Arc flow formulations based on dynamic programming: Theoretical foundations and applications. *European Journal of Operational Research*, 296(1):3–21, January 2022. doi:10.1016/j.ejor.2021.04.024.

Zeger Degraeve and Raf Jans. A new Dantzig-Wolfe reformulation and branch-and-price algorithm for the capacitated lot-sizing problem with setup times. *Operations Research*, 55(5):909–920, September–October 2007. doi:10.1287/opre.1070.0404.

Maxence Delorme and Manuel Iori. Enhanced pseudo-polynomial formulations for bin packing and cutting stock problems. *INFORMS Journal on Computing*, 32(1):101–119, Winter 2020. doi:10.1287/ijoc.2018.0880.

Maxence Delorme, Manuel Iori, and Silvano Martello. Bin packing and cutting stock problems: Mathematical models and exact algorithms. *European Journal of Operational Research*, 255 (1):1–20, November 2016. doi:10.1016/j.ejor.2016.04.030.

Guy Desaulniers. Branch-and-price-and-cut for the split-delivery vehicle routing problem with time windows. *Operations Research*, 58(1):179–192, January–February 2010. doi:10.1287/opre.1090.0713.

Guy Desaulniers, Jacques Desrosiers, Yvan Dumas, Simon Marc, Brigitte Rioux, Marius Mihai Solomon, and François Soumis. Crew pairing at Air France. *European Journal of Operational Research*, 97(2):245–259, March 1997a. doi:10.1016/S0377-2217(96)00195-6.

Guy Desaulniers, Jacques Desrosiers, Yvan Dumas, Simon Marc, Brigitte Rioux, Marius Mihai Solomon, and François Soumis. Crew pairing at Air France. *European Journal of Operational Research*, 97(2):245–259, March 1997b. doi:10.1016/S0377-2217(96)00195-6.

Guy Desaulniers, Jacques Desrosiers, Yvan Dumas, Marius Mihai Solomon, and François Soumis. Daily aircraft routing and scheduling. *Management Science*, 43(6):841–855, June 1997c. doi:10.1287/mnsc.43.6.841.

Guy Desaulniers, Jacques Desrosiers, Yvan Dumas, Marius Mihai Solomon, and François Soumis. Daily aircraft routing and scheduling. *Management Science*, 43(6):841–855, June 1997d. doi:10.5555/3059745.3059753.

Guy Desaulniers, Jacques Desrosiers, Irina Ioachim, Marius Mihai Solomon, François Soumis, and Daniel Villeneuve. A unified framework for deterministic time constrained vehicle routing and crew scheduling problems. In Teodor Gabriel Crainic and Gilbert Laporte, editors, *Fleet Management and Logistics*, chapter 3, pages 57–93. Springer, New York, June 1998a. doi:10.1007/978-1-4615-5755-5_3.

Guy Desaulniers, June Lavigne, and François Soumis. Multi-depot vehicle scheduling problems with time windows and waiting costs. *European Journal of Operational Research*, 111(3): 479–494, February 1998b. doi:10.1016/S0377-2217(97)00363-9.

Guy Desaulniers, Jacques Desrosiers, and Marius Mihai Solomon. *Accelerating Strategies in Column Generation Methods for Vehicle Routing and Crew Scheduling Problems*, pages 309–324. Springer, Boston, 2002. doi:10.1007/978-1-4615-1507-4_14.

Guy Desaulniers, Jacques Desrosiers, and Marius Mihai Solomon, editors. *Column Generation*. Springer, Boston, April 2005. doi:10.1007/b135457.

Guy Desaulniers, François Lessard, and Ahmed Hadjar. Tabu search, partial elementarity, and generalized k-path inequalities for the vehicle routing problem with time windows. *Transportation Science*, 42(3):387–404, March 2008. doi:10.1287/trsc.1070.0223.

Guy Desaulniers, Jacques Desrosiers, and Simon Spoorendonk. Cutting planes for branch-and-price algorithms. *Networks*, 58(4):301–310, December 2011. doi:10.1002/net.20471.

Guy Desaulniers, Oli B. G. Madsen, and Stefan Røpke. The vehicle routing problem with time windows. In Paolo Toth and Daniele Vigo, editors, *Vehicle routing: Problems, Methods, and Applications*, MOS-SIAM Series on Optimization, chapter 5, pages 119–159. SIAM, Philadelphia, 2nd edition, 2014. doi:10.1137/1.9781611973594.ch5.

Guy Desaulniers, Fausto Errico, Stefan Irnich, and Michael Schneider. Exact algorithms for electric vehicle-routing problems with time windows. *Operations Research*, 64(6):1388–1405, 2016a. doi:10.1287/opre.2016.1535.

Guy Desaulniers, Jørgen Glomvik Rakke, and Leandro Callegari Coelho. A branch-price-and-cut algorithm for the inventory-routing problem. *Transportation Science*, 50(3):1060–1076, August 2016b. doi:10.1287/trsc.2015.0635.

Guy Desaulniers, Timo Gschwind, and Stefan Irnich. Variable fixing for two-arc sequences in branch-price-and-cut algorithms on path-based models. *Transportation Science*, 54(5):1170–1188, September–October 2020a. doi:10.1287/trsc.2020.0988.

Guy Desaulniers, François Lessard, Mohammed Saddoune, and François Soumis. Dynamic constraint aggregation for solving very large-scale airline crew pairing problems. *SN Operations Research Forum*, 1(3):1–23, September 2020b. doi:10.1007/s43069-020-00016-1.

Guy Desaulniers, François Lessard, Mohammed Saddoune, and François Soumis. Dynamic constraint aggregation for solving very large-scale airline crew pairing problems. *SN Operations Research Forum*, 1(3):19–41, August 2020c. doi:10.1007/s43069-020-00016-1.

Martin Desrochers. *La fabrication d'horaires de travail pour les conducteurs d'autobus par une méthode de génération de colonnes*. PhD thesis, Université de Montréal, Centre de Recherche sur les Transports, Montreal, QC, Canada, 1986. Publication #470 (in French).

Martin Desrochers. An algorithm for the shortest path problem with resource constraints. Les Cahiers du GERAD G-88-27, HEC Montréal, Montreal, QC, Canada, 1988.

Martin Desrochers and François Soumis. CREW-OPT: Crew scheduling by column generation. In Joachim R. Daduna and Anthony Wren, editors, *Computer-Aided Transit Scheduling*, volume 308 of *Lecture Note in Economics Mathematical Systems*, pages 83–90. Springer, Berlin Heidelberg, 1988a. doi:10.1007/978-3-642-85966-3_8.

Martin Desrochers and François Soumis. A reoptimization algorithm for the shortest path problem with time windows. *European Journal of Operational Research*, 35(2):242–254, February 1988b. doi:10.1016/0377-2217(88)90034-3.

Martin Desrochers and François Soumis. A generalized permanent labelling algorithm for the shortest path problem with time windows. *INFOR: Information Systems and Operational Research*, 26(3):191–212, January 1988c. doi:10.1080/03155986.1988.11732063.

Martin Desrochers and François Soumis. A column generation approach to the urban transit crew scheduling problem. *Transportation Science*, 23(1):1–13, 1989. doi:10.1287/trsc.23.1.1.

Martin Desrochers and François Soumis. Implantation et complexité des techniques de programmation dynamique dans les méthodes de confection de tournées et d'horaires. *RAIRO - Operations Research*, 25:291–310, January 1991. doi:10.1051/ro/1991250302911. (in French).

Martin Desrochers, Jan Karel Lenstra, Martin W. P. Savelsbergh, and François Soumis. Vehicle routing with time windows: Analysis and algorithms. *TIMS Studies in the Management Sciences*, 16:65–85, 1988.

Martin Desrochers, Jacques Desrosiers, and Marius Mihai Solomon. A new optimization algorithm for the vehicle routing problem with time windows. *Operations Research*, 40(2):342–354, March–April 1992a. doi:10.1287/opre.40.2.342.

Martin Desrochers, Johanne Gilbert, Michel Sauvé, and François Soumis. CREW-OPT: Subproblem modeling in a column generation approach to urban crew scheduling. In Martin Desrochers and Jean-Marc Rousseau, editors, *Computer-Aided Transit Scheduling*, volume 386 of *Lecture Notes in Economics and Mathematical Systems*, pages 395–406. Springer, Berlin Heidelberg, 1992b. doi:10.1007/978-3-642-85968-7_25.

Jacques Desrosiers. GENCOL: une équipe et un logiciel d'optimisation. In Ivan Lavallée, Alain Bui, and Ider Tseveendorj, editors, *Combinatorial Optimization in Practice*, volume 8.2, pages 61–96. Studia Informatica Universalis, Hermann, Paris, 2010. URL `https://www.editions-hermann.fr/livre/studia-informatica-universalis-n-8-2-ivan-lavallee`. (in French).

Jacques Desrosiers and Marco E. Lübbecke. A primer in column generation. In Guy Desaulniers, Jacques Desrosiers, and Marius Mihai Solomon, editors, *Column Generation*, chapter 1, pages 1–32. Springer, Boston, April 2005. doi:10.1007/0-387-25486-2_1.

Jacques Desrosiers, Paul Pelletier, and François Soumis. Plus court chemin avec contraintes d'horaires. *RAIRO - Operations Research - Recherche Opérationnelle*, 17(4):357–377, 1983. URL `http://numdam.org/item/RO_1983__17_4_357_0`. (in French).

Jacques Desrosiers, François Soumis, and Martin Desrochers. Routing with time windows by column generation. *Networks*, 14(4):545–565, Winter 1984. doi:10.1002/net.3230140406.

Jacques Desrosiers, Yvan Dumas, and François Soumis. A dynamic programming solution of the large-scale single-vehicle dial-a-ride problem with time windows. *American Journal of Mathematical and Management Sciences*, 6(3–4):301–325, 1986. doi:10.1080/01966324.1986.10737198.

Jacques Desrosiers, Yvan Dumas, and François Soumis. The multiple vehicle DIAL-A-RIDE problem. In Joachim R. Daduna and Anthony Wren, editors, *Computer-Aided Transit Scheduling*, volume 308 of *Lecture Note in Economics Mathematical Systems*, pages 15–27. Springer, Berlin Heidelberg, 1988a. doi:10.1007/978-3-642-85966-3_3.

Jacques Desrosiers, Michel Sauvé, and François Soumis. Lagrangian relaxation methods for solving the minimum fleet size multiple traveling salesman problem with time windows. *Management Science*, 34(8):1005–1022, August 1988b. doi:10.1287/mnsc.34.8.1005.

Jacques Desrosiers, Yvan Dumas, Marius Mihai Solomon, and François Soumis. Time constrained routing and scheduling. In Michael Owen Ball, Thomas Lee Magnanti, Clyde Lawrence Monma, and George Lann Nemhauser, editors, *Handbooks in Operations Research and Management Science*, volume 8: Network Routing, chapter 2, pages 35–139. Elsevier, Maryland Heights, 1995. doi:10.1016/S0927-0507(05)80106-9.

Jacques Desrosiers, Nenad Mladenović, and Daniel Villeneuve. Design of balanced MBA student teams. *Journal of the Operational Research Society*, 56(1):60–66, 2005. doi:10.1057/palgrave.jors.2601775.

Jacques Desrosiers, Jean Bertrand Gauthier, and Marco E. Lübbecke. Row-reduced column generation for degenerate master problems. *European Journal of Operational Research*, 236(2): 453–460, July 2014. doi:10.1016/j.ejor.2013.12.016.

Edsger Wybe Dijkstra. A note on two problems in connexion with graphs. *Numerische Mathematik*, 1(1):269–271, December 1959. doi:10.1007/BF01386390.

Olivier du Merle, Daniel Villeneuve, Jacques Desrosiers, and Pierre Hansen. Stabilized column generation. *Discrete Mathematics*, 194(1–3):229–237, January 1999. doi:10.1016/S0012-365X(98)00213-1.

Yvan Dumas, Jacques Desrosiers, and François Soumis. The pickup and delivery problem with time windows. *European Journal of Operational Research*, 54(1):7–22, September 1991. doi:10.1016/0377-2217(91)90319-Q.

Yvan Dumas, Jacques Desrosiers, Éric Gélinas, and Marius Mihai Solomon. An optimal algorithm for the traveling salesman problem with time windows. *Operations Research*, 43(2):367–371, April 1995. doi:10.1287/opre.43.2.367.

Bernard P. Dzielinski and Ralph Edward Gomory. Optimal programming of lot sizes, inventory and labor allocations. *Management Science*, 11(9):874–890, July 1965. doi:10.1287/mnsc.11.9.874.

Jack Edmonds. Maximum matching and a polyhedron with 0,1-vertices. *Journal of Research of the National Bureau of Standards Section B Mathematics and Mathematical Physics*, 69B(1 and 2): 125–130, December 1965. doi:10.6028/jres.069B.013.

Jack Edmonds. Matroids and the greedy algorithm. *Mathematical Programming*, 1(1):127–136, December 1971. doi:10.1007/BF01584082.

Matthias Ehrgott. *Multicriteria Optimization*. Springer, Berlin Heidelberg, 2nd edition, 2005. doi:10.1007/3-540-27659-9.

Issmail El Hallaoui, Daniel Villeneuve, François Soumis, and Guy Desaulniers. Dynamic aggregation of set-partitioning constraints in column generation. *Operations Research*, 53(4):632–645, August 2005. doi:10.1287/opre.1050.0222.

Issmail El Hallaoui, Guy Desaulniers, Abdelmoutalib Metrane, and François Soumis. Bi-dynamic constraint aggregation and subproblem reduction. *Computers & Operations Research*, 35(5): 1713–1724, May 2008. doi:10.1016/j.cor.2006.10.007.

Issmail El Hallaoui, Abdelmoutalib Metrane, François Soumis, and Guy Desaulniers. Multi-phase dynamic constraint aggregation for set partitioning type problems. *Mathematical Programming*, 123(2):345–370, June 2010. doi:10.1007/s10107-008-0254-5.

Issmail El Hallaoui, Abdelmoutalib Metrane, Guy Desaulniers, and François Soumis. An improved primal simplex algorithm for degenerate linear programs. *INFORMS Journal on Computing*, 23(4):569–577, Fall 2011. doi:10.1287/ijoc.1100.0425.

Salah Eldin Elmaghraby. *Activity networks: project planning and control by network models*. John Wiley & Sons, New York, 1977.

Gary Dean Eppen and Richard Kipp Martin. Solving multi-item capacitated lot-sizing problems using variable redefinition. *Operations Research*, 35(6):832–848, November–December 1987. doi:10.1287/opre.35.6.832.

Donald Erlenkotter. A dual-based procedure for uncapacitated facility location. *Operations Research*, 26(6):992–1009, November–December 1978. doi:10.1287/opre.26.6.992.

Fausto Errico, Guy Desaulniers, Michel Gendreau, Walter Rei, and Louis-Martin Rousseau. A priori optimization with recourse for the vehicle routing problem with hard time windows and stochastic service times. *European Journal of Operational Research*, 249(1):55–66, 2016. doi:10.1016/j.ejor.2015.07.027.

Fausto Errico, Guy Desaulniers, Michel Gendreau, Walter Rei, and Louis-Martin Rousseau. The vehicle routing problem with hard time windows and stochastic service times. *EURO Journal on Transportation and Logistics*, 7(3):223–251, September 2018. doi:10.1007/s13676-016-0101-4.

Shimon Even, Alon Itai, and Adi Shamir. On the complexity of timetable and multicommodity flow problems. *SIAM Journal on Computing*, 5(4):691–703, 1976. doi:10.1137/0205048.

Julie C. Falkner and David Murray Ryan. Aspects of bus crew scheduling using a set partitioning model. In Joachim R. Daduna and Anthony Wren, editors, *Computer-Aided Transit Schedul-*

ing, volume 308 of *Lecture Note in Economics Mathematical Systems*, pages 91–103. Springer, Berlin Heidelberg, 1988. doi:10.1007/978-3-642-85966-3_9.

Alan Arthur Farley. A note on bounding a class of linear programming problems, including cutting stock problems. *Operations Research*, 38(5):922–923, September–October 1990. doi:10.1287/opre.38.5.922.

Dominique Feillet. A tutorial on column generation and branch-and-price for vehicle routing problems. *4OR*, 8(4):407–424, June 2010. doi:10.1007/s10288-010-0130-z.

Dominique Feillet, Pierre Dejax, Michel Gendreau, and Cyrille Gueguen. An exact algorithm for the elementary shortest path problem with resource constraints: Application to some vehicle routing problems. *Networks*, 44(3):216–229, August 2004. doi:10.1002/net.20033.

Michael C. Ferris and Jeffrey D. Horn. Partitioning mathematical programs for parallel solution. *Mathematical Programming*, 80:35–61, January 1998. doi:10.1007/BF01582130.

Marshall Lee Fisher. The Lagrangian relaxation method for solving integer programming problems. *Management Science*, 27(1):1–18, December 1981. doi:10.1287/mnsc.1040.0263.

Robert W Floyd. Algorithm 97: Shortest Path. *Communications of the ACM*, 5(6):345, June 1962. doi:10.1145/367766.368168.

Lester Randolph Ford, Jr. and Delbert Ray Fulkerson. Maximal flow through a network. *Canadian Journal of Mathematics*, 8:399–404, 1956. doi:10.4153/CJM-1956-045-5.

Lester Randolph Ford, Jr. and Delbert Ray Fulkerson. A suggested computation for maximal multicommodity network flows. *Management Science*, 5(1):97–101, October 1958. doi:10.1287/mnsc.5.1.97.

John J. H. Forrest and John A. Tomlin. Updated triangular factors of the basis to maintain sparsity in the product form simplex method. *Mathematical Programming*, 2(1):263–278, February 1972. doi:10.1007/BF01584548.

Omar Foutlane, Issmail El Hallaoui, and Pierre Hansen. Distributed integral column generation for set partitioning problems. *Operations Research Forum*, 3(2):27–48, April 2022. doi:10.1007/s43069-022-00136-w.

Antonio Frangioni. Generalized bundle methods. *SIAM Journal on Optimization*, 13(1):117–156, 2002. doi:10.1137/S1052623498342186.

Antonio Frangioni and Enrico Gorgone. Bundle methods for sum-functions with "easy" components: applications to multicommodity network design. *Mathematical Programming*, 145(1): 133–161, February 2014. doi:10.1007/s10107-013-0642-3.

Michael Lawrence Fredman and Robert Endre Tarjan. Fibonacci heaps and their uses in improved network optimization algorithms. In *25th Annual Symposium on Foundations of Computer Science, 1984.*, pages 338–346. IEEE, October 1984. doi:10.1109/SFCS.1984.715934.

Michael Lawrence Fredman and Robert Endre Tarjan. Fibonacci heaps and their uses in improved network optimization algorithms. *Journal of the ACM*, 34(3):596–615, July 1987. doi:10.1145/28869.28874.

Ricardo Fukasawa, Humberto Longo, Jens Lysgaard, Marcus Poggi de Aragão, Marcelo Reis, Eduardo Uchoa, and Renato Fonseca Werneck. Robust branch-and-cut-and-price for the capacitated vehicle routing problem. *Mathematical Programming*, 106(3):491–511, May 2006. doi:10.1007/s10107-005-0644-x.

Michel Gamache, François Soumis, Daniel Villeneuve, Jacques Desrosiers, and Éric Gélinas. The preferential bidding system at Air Canada. *Transportation Science*, 32(3):246–255, August 1998. doi:10.1287/trsc.32.3.246.

Michel Gamache, François Soumis, Gérald Marquis, and Jacques Desrosiers. A column generation approach for large-scale aircrew rostering problems. *Operations Research*, 47(2):247–263, March–April 1999. doi:10.1287/opre.47.2.247.

Gerald Gamrath and Marco E. Lübbecke. Experiments with a generic Dantzig-Wolfe decomposition for integer programs. In Paola Festa, editor, *Experimental Algorithms*, volume 6049 of *Lecture Notes in Computer Science*, pages 239–252, Berlin Heidelberg, 2010. Springer. doi:10.1007/978-3-642-13193-6_21.

Robert Shaun Garfinkel and George Lann Nemhauser. *Integer Programming*. John Wiley & Sons, New York, 1972.

Jean Bertrand Gauthier and Jacques Desrosiers. The minimum mean cycle-canceling algorithm for linear programs. *European Journal of Operational Research*, 298(1):36–44, April 2022. doi:10.1016/j.ejor.2021.09.022.

Jean Bertrand Gauthier, Jacques Desrosiers, and Marco E. Lübbecke. Decomposition theorems for linear programs. *Operations Research Letters*, 42(8):553–557, December 2014. doi:10.1016/j.orl.2014.10.001.

Jean Bertrand Gauthier, Jacques Desrosiers, and Marco E. Lübbecke. About the minimum mean cycle-canceling algorithm. *Discrete Applied Mathematics*, 196:115–134, December 2015. doi:10.1016/j.dam.2014.07.005. Advances in Combinatorial Optimization.

Jean Bertrand Gauthier, Jacques Desrosiers, and Marco E. Lübbecke. Tools for primal degenerate linear programs: IPS, DCA, and PE. *EURO Journal on Transportation and Logistics*, 5(2): 161–204, June 2016. doi:10.1007/s13676-015-0077-5.

Jean Bertrand Gauthier, Jacques Desrosiers, and Marco E. Lübbecke. Vector space decomposition for solving large-scale linear programs. *Operations Research*, 66(5):1376–1389, September–October 2018. doi:10.1287/opre.2018.1728.

Charles Gauvin, Guy Desaulniers, and Michel Gendreau. A branch-cut-and-price algorithm for the vehicle routing problem with stochastic demands. *Computers & Operations Research*, 50: 141–153, October 2014. doi:10.1016/j.cor.2014.03.028.

Hermann Gehring and Jörg Homberger. A parallel hybrid evolutionary metaheuristic for the vehicle routing problem with time windows. In *Proceedings of EUROGEN99*, pages 57–64, New York, 1999. John Wiley & Sons.

Sylvie Gélinas, Martin Desrochers, Jacques Desrosiers, and Marius Mihai Solomon. A new branching strategy for time constrained routing problems with application to backhauling. *Annals of Operations Research*, 61(1):91–109, December 1995. doi:10.1007/BF02098283.

Arthur Minot Geoffrion. Lagrangean relaxation for integer programming. In Michel Louis Balinski, editor, *Approaches to Integer Programming*, pages 82–114. Springer, Berlin Heidelberg, 1974. doi:10.1007/BFb0120690.

Ahmed Ghoniem and Hanif D. Sherali. Complementary column generation and bounding approaches for set partitioning formulations. *Optimization Letters*, 3(1):123–136, January 2009. doi:10.1007/s11590-008-0097-2.

Frederick Richard Giles and William Robert Pulleyblank. Total dual integrality and integer polyhedra. *Linear Algebra and its Applications*, 25:191–196, June 1979. doi:10.1016/0024-3795(79)90018-1.

Paul Carl Gilmore and Ralph Edward Gomory. A linear programming approach to the cutting-stock problem. *Operations Research*, 9(6):849–859, December 1961. doi:10.1287/opre.9.6.849.

Paul Carl Gilmore and Ralph Edward Gomory. A linear programming approach to the cutting stock problem—part II. *Operations Research*, 11(6):863–888, December 1963. doi:10.1287/opre.11.6.863.

Paul Carl Gilmore and Ralph Edward Gomory. The theory and computation of knapsack functions. *Operations Research*, 14(6):1045–1074, December 1966. doi:10.1287/opre.14.6.1045.

Ambros Gleixner, Gregor Hendel, Gerald Gamrath, Tobias Achterberg, Michael Bastubbe, Timo Berthold, Philipp M. Christophel, Kati Jarck, Thosten Koch, Jeffrey Linderoth, Marco Lübbecke, Hand D. Mittelmann, Derya Ozyurt, Ted K. Ralphs, Domenico Salvagnin, and Yuji Shinano. MIPLIB 2017: Data-driven compilation of the 6th mixed-integer programming library. *Mathematical Programming Computation*, 13:443–490, September 2021. doi:10.1007/s12532-020-00194-3.

Fred Glover. Heuristics for integer programming using surrogate constraints. *Decision Sciences*, 8(1):156–166, January 1977. doi:10.1111/j.1540-5915.1977.tb01074.x.

Jean-Louis Goffin and Jean-Philippe Vial. Cutting planes and column generation techniques with the projective algorithm. *Journal of Optimization Theory and Applications*, 65(3):409–429, June 1990. doi:10.1007/BF00939559.

Jean-Louis Goffin and Jean-Philippe Vial. Multiple cuts in the analytic center cutting plane method. *SIAM Journal on Optimization*, 11(1):266–288, 2000. doi:10.1137/S1052623498340266.

Jean-Louis Goffin and Jean-Philippe Vial. Convex nondifferentiable optimization: A survey focused on the analytic center cutting plane method. *Optimization Methods and Software*, 17(5): 805–867, 2002. doi:10.1080/1055678021000060829a.

Jean-Louis Goffin, Alain Haurie, and Jean-Philippe Vial. Decomposition and nondifferentiable optimization with the projective algorithm. *Management Science*, 38(2):284–302, February 1992. doi:10.1287/mnsc.38.2.284.

Jean-Louis Goffin, Alain Haurie, Jean-Philippe Vial, and Dao Li Zhu. Using central prices in the decomposition of linear programs. *European Journal of Operational Research*, 64(3):393–409, February 1993. doi:10.1016/0377-2217(93)90129-B.

Andrew Vladislav Goldberg and Robert Endre Tarjan. Finding minimum-cost circulations by canceling negative cycles. *Journal of the ACM*, 36(4):873–886, October 1989. doi:10.1145/76359.76368.

Ralph Edward Gomory. An algorithm for integer solutions to linear programs. In Robert Lawrence Graves and Philip Wolfe, editors, *Recent Advances in Mathematical Programming*, pages 269–302. McGraw-Hill, New York, 1963.

Jacek Gondzio, Olivier du Merle, Robert Sarkissian, and Jean-Philippe Vial. ACCPM – A library for convex optimization based on an analytic center cutting plane method. *European Journal of Operational Research*, 94(1):206–211, October 1996. doi:10.1016/0377-2217(96)00169-5.

Jacek Gondzio, Pablo González-Brevis, and Pedro Munari. New developments in the primal-dual column generation technique. *European Journal of Operational Research*, 224(1):41–51, January 2013. doi:10.1016/j.ejor.2012.07.024.

Jacek Gondzio, Pablo González-Brevis, and Pedro Munari. Large-scale optimization with the primal-dual column generation method. *Mathematical Programming Computation*, 8(1):47–82, 2016. doi:10.1007/s12532-015-0090-6.

Igor Griva, Stephen Gregory Nash, and Ariela Sofer. *Linear and Nonlinear Optimization*. SIAM, Philadelphia, 2nd edition, 2008.

Martin Grötschel, editor. *Optimization Stories*, Extra Volume ISMP, Bielefeld, 2012. Documenta Mathematica, Journal der Deutschen Mathematiker-Vereinigung. ISBN 978-3-936609-58-5.

Martin Grötschel and Olaf Holland. Solving matching problems with linear programming. *Mathematical Programming*, 33(3):243–259, December 1985. doi:10.1007/BF01584376.

Timo Gschwind and Stefan Irnich. Dual inequalities for stabilized column generation revisited. *INFORMS Journal on Computing*, 28(1):175–194, February 2016. doi:10.1287/ijoc.2015.0670.

Monique Guignard. Lagrangean relaxation. *Top*, 11(2):151–200, December 2003. doi:10.1007/BF02579036.

Monique Guignard and Siwhan Kim. Lagrangean decomposition: A model yielding stronger Lagrangean bounds. *Mathematical Programming*, 39(2):215–228, November 1987. doi:10.1007/BF02592954.

Knut Haase, Guy Desaulniers, and Jacques Desrosiers. Simultaneous vehicle and crew scheduling in urban mass transit systems. *Transportation Science*, 35(3):286–303, August 2001. doi:10.1287/trsc.35.3.286.10153.

Naveed Haghani, Claudio Contardo, and Julian Yarkony. Smooth and flexible dual optimal inequalities. *INFORMS Journal on Optimization*, 4(1):29–44, Winter 2022. doi:10.1287/ijoo.2021.0057.

Mohamed Hamdouni, Guy Desaulniers, and François Soumis. Parking buses in a depot using block patterns: A Benders decomposition approach for minimizing type mismatches. *Computers & Operations Research*, 34(11):3362–3379, November 2007. doi:10.1016/j.cor.2006.02.002.

Michael Held and Richard Manning Karp. The traveling-salesman problem and minimum spanning trees. *Operations Research*, 18(6):1138–1162, November–December 1970. doi:10.1287/opre.18.6.1138.

Michael Held and Richard Manning Karp. The traveling-salesman problem and minimum spanning tree: Part II. *Mathematical Programming*, 1(1):6–25, December 1971. doi:10.1007/BF01584070.

Michael Held, Philip Wolfe, and Harlan Pinkney Crowder. Validation of subgradient optimization. *Mathematical Programming*, 6(1):62–88, December 1974. doi:10.1007/BF01580223.

Richard Hicks, Richard Madrid, Chris Milligan, Robert Pruneau, Mike Kanaley, Yvan Dumas, Benoit Lacroix, Jacques Desrosiers, and François Soumis. Bombardier Flexjet significantly improves its fractional aircraft ownership operations. *Interfaces*, 35(1):49–60, January–February 2005. doi:10.1287/inte.1040.0113.

David Hilbert. Ueber die Theorie der algebraischen Formen. *Mathematische Annalen*, XXXVI(4): 473–534, December 1890. doi:10.1007/BF01208503. (in German).

Ilyas Himmich, Hatem M. T. Ben Amor, Issmail El Hallaoui, and François Soumis. A primal adjacency-based algorithm for the shortest path problem with resource constraints. *Transportation Science*, 54(5):1153–1169, July 2020. doi:10.1287/trsc.2019.0941.

Ilyas Himmich, Issmail El Hallaoui, and François Soumis. A multiphase dynamic programming algorithm for the shortest path problem with resource constraints. *European Journal of Operational Research*, 315(2):470–483, 2023. doi:10.1016/j.ejor.2023.11.047.

Alan Jerome Hoffman. Cycling in the simplex algorithm. *Report No. 2974,* National Bureau of Standards, Washington, DC, 1953.

John N. Hooker and Greger Ottosson. Logic-based Benders decomposition. *Mathematical Programming*, 96(1):33–60, April 2003. doi:10.1007/s10107-003-0375-9.

Irina Ioachim, Sylvie Gélinas, François Soumis, and Jacques Desrosiers. A dynamic programming algorithm for the shortest path problem with time windows and linear node costs. *Networks*, 31(3):193–204, May 1998. doi:10.1002/(SICI)1097-0037(199805)31:3<193::AID-NET6>3.0.CO;2-A.

Irina Ioachim, Jacques Desrosiers, François Soumis, and Nicolas Bélanger. Fleet assignment and routing with schedule synchronization constraints. *European Journal of Operational Research*, 119(1):75–90, November 1999. doi:10.1016/S0377-2217(98)00343-9.

Stefan Irnich and Guy Desaulniers. Shortest path problems with resource constraints. In Guy Desaulniers, Jacques Desrosiers, and Marius Mihai Solomon, editors, *Column Generation*, chapter 2, pages 33–66. Springer, New York, April 2005. doi:10.1007/0-387-25486-2_2.

Stefan Irnich, Guy Desaulniers, Jacques Desrosiers, and Ahmed Hadjar. Path-reduced costs for eliminating arcs in routing and scheduling. *INFORMS Journal on Computing*, 22(2):297–313, Spring 2010. doi:10.1287/ijoc.1090.0341.

John Rolfe Isbell and William Henry Marlow. On an industrial programming problem of Kantorovich. *Management Science*, 8(1):13–17, 1961. doi:10.1287/mnsc.8.1.13.

Raf Jans. Classification of Dantzig-Wolfe reformulations for binary mixed integer programming problems. *European Journal of Operational Research*, 204(2):251–254, July 2010. doi:10.1016/j.ejor.2009.11.014.

Raf Jans and Jacques Desrosiers. Binary clustering problems: Symmetric, asymmetric and decomposition formulations. Les Cahiers du GERAD G-2010-44, HEC Montréal, Montreal, QC, Canada, August 2010.

Tiago Januario, Sebastián Urrutia, Celso Carneiro Ribeiro, and Dominique de Werra. Edge coloring: A natural model for sports scheduling. *European Journal of Operational Research*, 254(1): 1–8, October 2016. doi:10.1016/j.ejor.2016.03.038.

Vojtěch Jarník. O jistém problému minimálním. *Práce Moravské Přírodovědecké Společnosti*, 6 (4):57–63, 1930. (in Czech).

Brigitte Jaumard and François Soumis. Extraction de relation logique et saut de dualité. Technical Report #458, Centre de recherche sur les transports, Université de Montréal, Montreal, QC, Canada, Avril 1986. (in French).

Mads Jepsen, Bjørn Petersen, Simon Spoorendonk, and David Pisinger. Subset-row inequalities applied to the vehicle-routing problem with time windows. *Operations Research*, 56(2):497–511, March–April 2008. doi:10.1287/opre.1070.0449.

William Sylvester Jewell. Optimal flows through networks. Interim Technical Report No. 8, on Fundamental Investigations in Methods of Operations Research, Operations Research Center, MIT, Cambridge, MA, USA, 1958. [Partial reprint: Jewell (1966)].

William Sylvester Jewell. A primal-dual multicommodity flow algorithm. Technical Report ORC-66-24, Operations Research Center, University of California, Berkeley, CA, USA, 1966.

Ellis Lane Johnson. Modeling and strong linear programs for mixed integer programming. In Stein William Wallace, editor, *Algorithms and Model Formulations in Mathematical Programming*, pages 1–43, Berlin Heidelberg, 1989. Springer. doi:10.1007/978-3-642-83724-1.

Michael Jünger, Thomas M. Liebling, Denis Naddef, George Lann Nemhauser, William Robert Pulleyblank, Gerhard Reinelt, Giovanni Rinaldi, and Laurence Alexander Wolsey, editors. *50 Years of Integer Programming 1958–2008: From the Early Years to the State-of-the-Art*. Springer, Berlin Heidelberg, 2010. doi:10.1007/978-3-540-68279-0.

Brian Kallehauge, Jesper Larsen, and Oli B.G. Madsen. Lagrangian duality applied to the vehicle routing problem with time windows. *Computers & Operations Research*, 33(5):1464–1487, May 2006. doi:10.1016/j.cor.2004.11.002.

Leonid Vitaliyevich Kantorovich. Mathematical methods of organizing and planning production. *Leningrad State University*, 1939. (Translated from Russian in Management Science, 6(4):366–422, 1960. doi:10.1287/mnsc.6.4.366).

Leonid Vitaliyevich Kantorovich and Victor Abramovich Zalgaller. *Calculation of a Rational Apportionment of Industrial Materials*. Lenizdat, Leningrad, 1951. (in Russian).

Oded Kariv and Seifollah Louis Hakimi. An algorithmic approach to network location problems. II: The p-medians. *SIAM Journal on Applied Mathematics*, 37(3):539–560, 1979. doi:10.1137/0137041.

Narendra Karmarkar. A new polynomial-time algorithm for linear programming. *Combinatorica*, 4(4):373–395, December 1984. doi:10.1007/BF02579150.

James E. Kelley, Jr. The cutting-plane method for solving convex programs. *Journal of the Society for Industrial and Applied Mathematics*, 8(4):703–712, December 1960. doi:10.1137/0108053.

Leonid Genrikhovich Khachiyan. A polynomial algorithm in linear programming. *Doklady Akademii Nauk SSSR*, 244(5):1093–1096, 1979. (Translated from Russian in Soviet Mathematics Doklady, 20(1):191–194, 1979).

Taghi Khaniyev, Samir Elhedhli, and Fatih Safa Erenaya. Structure detection in mixed-integer programs. *INFORMS Journal on Computing*, 30(3):570–587, October 2018. doi:10.1287/ijoc.2017.0797.

Gerard A. P. Kindervater and Martin W. P. Savelsbergh. Vehicle routing: handling edge exchanges. In Emile Hubertus Leonardus Aarts and Jan Karel Lenstra, editors, *Local Search in Combinatorial Optimization*, chapter 10, pages 337–360. John Wiley & Sons, New York, 1997.

Krzysztof Czesław Kiwiel. A bundle Bregman proximal method for convex nondifferentiable minimization. *Mathematical Programming*, 85(2):241–258, June 1999. doi:10.1007/s101070050056.

Diego Klabjan, Ellis Lane Johnson, George Lann Nemhauser, Eric Gelman, and Srini Ramaswamy. Solving large airline crew scheduling problems: Random pairing generation and strong branching. *Computational Optimization and Applications*, 20(1):73–91, October 2001. doi:10.1023/A:1011223523191.

Diego Klabjan, Ellis Lane Johnson, George Lann Nemhauser, Eric Gelman, and Srini Ramaswamy. Airline crew scheduling with time windows and plane-count constraints. *Transportation Science*, 36(3):337–348, August 2002. doi:10.1287/trsc.36.3.337.7831.

Victor Klee and George James Minty. How good is the simplex algorithm? In Oved Shisha, editor, *Inequalities*, volume III, pages 159–175. Academic Press, New York, 1972.

Donald Ervin Knuth. Robert W Floyd, In Memoriam. *ACM SIGACT News*, 34(4):3–13, 2003. doi:10.1145/954092.954488.

Achim Koberstein and Uwe H. Suhl. Progress in the dual simplex method for large scale LP problems: Practical dual phase 1 algorithms. *Computational Optimization and Applications*, 37 (1):49–65, March 2007. doi:10.1007/s10589-007-9022-3.

Thorsten Koch, Tobias Achterberg, Erling Andersen, Oliver Bastert, Timo Berthold, Robert Eugene Bixby, Emilie Danna, Gerald Gamrath, Ambros M. Gleixner, Stefan Heinz, Andrea Lodi, Hans Mittelmann, Ted Ralphs, Domenico Salvagnin, Daniel Eli Steffy, and Kati Wolter. MIPLIB 2010. *Mathematical Programming Computation*, 3(2):103–163, June 2011. doi:10.1007/s12532-011-0025-9.

Thorsten Koch, Timo Berthold, Jaap Pedersen, and Charlie Vanaret. Progress in mathematical programming solvers from 2001 to 2020. *EURO Journal on Computational Optimization*, 10: 100031, 2022. doi:10.1016/j.ejco.2022.100031.

Niklas Kohl. *Exact methods for time constrained routing and related scheduling problems*. PhD thesis, Technical University of Denmark (DTU), 1995.

Niklas Kohl and Oli B. G. Madsen. An optimization algorithm for the vehicle routing problem with time windows based on Lagrangian relaxation. *Operations Research*, 45(3):395–406, May–June 1997. doi:10.1287/opre.45.3.395.

Niklas Kohl, Jacques Desrosiers, Oli B. G. Madsen, Marius Mihai Solomon, and François Soumis. 2-path cuts for the vehicle routing problem with time windows. *Transportation Science*, 33(1): 101–116, February 1999. doi:10.1287/trsc.33.1.101.

Tamás Koltai and Viola Tatay. A practical approach to sensitivity analysis in linear programming under degeneracy for management decision making. *International Journal of Production Economics*, 131(1):392–398, 2011. doi:10.1016/j.ijpe.2010.04.037.

Tjalling Charles Koopmans. Exchange ratios between cargoes on various routes (non-refrigerating dry cargoes), 1942. Memorandum for the Combined Shipping Adustment Board. Reprinted in *Scientific Papers of Tjalling C. Koopmans*. Berlin: Springer, 1970.

Tjalling Charles Koopmans. Optimum utilization of the transportation system. Reprinted in Supplement to *Econometrica* 17 (July 1949): 136–146., 1947. doi:10.2307/1907301.

Tjalling Charles Koopmans. On the evaluation of Kantorovich's work of 1939. *Management Science*, 8(3):264–265, 1962. doi:10.1287/mnsc.8.3.264.

Markus Kruber, Marco E. Lübbecke, and Axel Parmentier. Learning when to use a decomposition. In Domenico Salvagnin and Michele Lombardi, editors, *Integration of AI and OR Techniques in Constraint Programming*, volume 10335 of *Lecture Notes in Computer Science*, pages 202–210, Cham, May 2017. Springer. doi:10.1007/978-3-319-59776-8_16.

Joseph Bernard Kruskal. On the shortest spanning subtree of a graph and the traveling salesman problem. *Proceedings of the American Mathematical Society*, 7(1):48–50, 1956. doi:10.1090/S0002-9939-1956-0078686-7.

Ailsa H. Land and Susan Powell. *FORTRAN Codes for Mathematical Programming: Linear, Quadratic and Discrete*. John Wiley & Sons, New York, 1973.

Ailsa Horton Land and Alison Grant Doig. An automatic method of solving discrete programming problems. *Econometrica*, 28(3):497–520, July 1960. doi:10.2307/1910129.

Sophie D. Lapierre, Luc Debargis, and François Soumis. Balancing printed circuit board assembly line systems. *International Journal of Production Research*, 38(16):3899–3911, 2000. doi:10.1080/00207540050176076.

Gilbert Laporte. *A Short History of the Traveling Salesman Problem*, 2006. URL `http://neumann.hec.ca/chairedistributique/common/laporte-short.pdf`.

Gilbert Laporte and François Victor Louveaux. The integer L-shaped method for stochastic integer programs with complete recourse. *Operations Research Letters*, 13(3):133–142, April 1993. doi:10.1016/0167-6377(93)90002-X.

Gilbert Laporte, Yves Nobert, and Martin Desrochers. Optimal routing under capacity and distance restrictions. *Operations Research*, 33(5):1050–1073, September–October 1985. doi:10.1287/opre.33.5.1050.

Leon Stephen Lasdon. *Optimization Theory for Large Systems*. Macmillan series in operations research. Macmillan, New York, 1970. [Reprint: Dover Publications, Mineola, 2002].

Sylvie Lavoie, Michel Minoux, and Édouard Odier. A new approach for crew pairing problems with an application to air transportation. *European Journal of Operational Research*, 35(1): 45–58, April 1988. doi:10.1016/0377-2217(88)90377-3.

Eugene Leighton Lawler, Jan Karel Lenstra, Alexander Hendrik George Rinnooy Kan, and David Bernard Shmoys. *The Travelling Salesman Problem: A Guided Tour of Combinatorial Optimization*. Wiley-Interscience series in discrete mathematics and optimization. John Wiley & Sons, Chichester, 1985.

Claude Lemaréchal. Bundle methods in nonsmooth optimization. In Claude Lemaréchal and Robert Mifflin, editors, *Nonsmooth optimization*, volume 3 of *Proceedings of a IIASA Workshop*, pages 79–102, Oxford, 1978. Pergamon Press.

Claude Lemaréchal, Arkadi Nemirovski, and Yurii Nesterov. New variants of bundle methods. *Mathematical Programming*, 69:111–147, July 1995. doi:10.1007/BF01585555.

Carlton Edward Lemke. The dual method of solving the linear programming problem. *Naval Research Logistics Quarterly*, 1:36–47, March 1954. doi:10.1002/nav.3800010107.

Jan Karel Lenstra, Alexander Hendrik George Rinnooy Kan, and Alexander Schrijver, editors. *History of Mathematical Programming: A Collection of Personal Reminiscences*. North-Holland, Amsterdam, 1991. ISBN 0444888187; 9780444888181.

Kuan-Min Lin, Matthias Ehrgott, and Andrea Raith. Integrating column generation in a method to compute a discrete representation of the non-dominated set of multi-objective linear programmes. *4OR*, 15(4):331–357, December 2017. doi:10.1007/s10288-016-0336-9.

Norbert Lingaya, Jean-François Cordeau, Guy Desaulniers, Jacques Desrosiers, and Françcois Soumis. Operational car assignment at VIA Rail Canada. *Transportation Research Part B: Methodological*, 36(9):755–778, November 2002. doi:10.1016/S0191-2615(01)00027-3.

Andreas Löbel. Vehicle scheduling in public transit and Lagrangean pricing. *Management Science*, 44(12):1637–1649, December 1998. doi:10.1287/mnsc.44.12.1637.

Marco E. Lübbecke. Dual variable based fathoming in dynamic programs for column generation. *European Journal of Operational Research*, 162(1):122–125, April 2005. doi:10.1016/j.ejor.2003.05.006.

Marco E. Lübbecke and Jacques Desrosiers. Selected topics in column generation. *Operations Research*, 53(6):1007–1023, November–December 2005. doi:10.1287/opre.1050.0234.

Marco E. Lübbecke and Christian Puchert. Primal heuristics for branch-and-price algorithms. In Diethard Klatte, Hans-Jakob Lüthi, and Karl Schmedders, editors, *Operations Research Proceedings 2011*, pages 65–70, Berlin Heidelberg, April 2012. Springer. doi:10.1007/978-3-642-29210-1_11.

Marco E. Lübbecke and Christian Puchert. Large neighborhood search and diving heuristics based on Dantzig-Wolfe reformulation. Unpublished manuscript, 2013. Chair of Operations Research, RWTH Aachen University, Germany.

Marco E. Lübbecke, Stephen John Maher, and Jonas Timon Witt. Avoiding redundant columns by adding classical Benders cuts to column generation subproblems. *Discrete Optimization*, 39: Article 100626, 2021. doi:10.1016/j.disopt.2021.100626.

Jens Lysgaard, Adam Nicholas Letchford, and Richard William Eglese. A new branch-and-cut algorithm for the capacitated vehicle routing problem. *Mathematical Programming*, 100(2): 423–445, June 2004. doi:10.1007/s10107-003-0481-8.

Thomas Lee Magnanti, Jeremy Frank Shapiro, and Harvey Maurice Wagner. Generalized linear programming solves the dual. *Management Science*, 22(11):1195–1203, July 1976. doi:10.1287/mnsc.22.11.1195.

Stephen John Maher and Elina Rönnberg. Integer programming column generation: accelerating branch-and-price using a novel pricing scheme for finding high-quality solutions in set covering, packing, and partitioning problems. *Mathematical Programming Computation*, 15(3):509–548, September 2023. doi:10.1007/s12532-023-00240-w.

John William Mamer and Richard Dewayne McBride. A decomposition-based pricing procedure for large-scale linear programs: An application to the linear multicommodity flow problem. *Management Science*, 46(5):693–709, May 2000. doi:10.1287/mnsc.46.5.693.12042.

Alan Sussmann Manne. Programming of economic lot sizes. *Management Science*, 4(2):115–135, January 1958. doi:10.1287/mnsc.4.2.115.

Roy Earl Marsten. The use of the boxstep method in discrete optimization. In Michel Louis Balinski and Philip Wolfe, editors, *Nondifferentiable Optimization*, volume 3 of *Mathematical Programming Studies*, pages 127–144. Springer, Berlin Heidelberg, 1975. doi:10.1007/BFb0120702.

Roy Earl Marsten, William Walter Hogan, and Jerry W. Blankenship. The boxstep method for large-scale optimization. *Operations Research*, 23(3):389–405, May–June 1975. doi:10.1287/opre.23.3.389.

Silvano Martello and Paolo Toth. *Knapsack Problems: Algorithms and Computer Implementations*. John Wiley & Sons, New York, 1990.

Rafael Martinelli, Diego Pecin, and Marcus Poggi de Aragão. Efficient elementary and restricted non-elementary route pricing. *European Journal of Operational Research*, 239(1):102–111, November 2014. doi:10.1016/j.ejor.2014.05.005.

Anuj Mehrotra and Michael Alan Trick. A column generation approach for graph coloring. *INFORMS Journal on Computing*, 8(4):344–354, Fall 1996. doi:10.1287/ijoc.8.4.344.

Abdelmoutalib Metrane, François Soumis, and Issmail El Hallaoui. Column generation decomposition with the degenerate constraints in the subproblem. *European Journal of Operational Research*, 207(1):37–44, November 2010. doi:10.1016/j.ejor.2010.05.002.

Tayeb Mhamedi, Henrik Andersson, Marilène Cherkesly, and Guy Desaulniers. A branch-price-and-cut algorithm for the two-echelon vehicle routing problem with time windows. *Transportation Science*, 56(1):245–264, 2022. doi:10.1287/trsc.2021.1092.

Mouad Morabit, Guy Desaulniers, and Andrea Lodi. Machine-learning-based column selection for column generation. *Transportation Science*, 55(4):815–831, July–August 2021. doi:10.1287/trsc.2021.1045.

Mouad Morabit, Guy Desaulniers, and Andrea Lodi. Machine-learning-based arc selection for constrained shortest path problems in column generation. *INFORMS Journal on Optimization*, 5(2):191–210, 2023. doi:10.1287/ijoo.2022.0082.

Pedro Munari and Jacek Gondzio. Using the primal-dual interior point algorithm within the branch-price-and-cut method. *Computers & Operations Research*, 40(8):2026–2036, August 2013. doi:10.1016/j.cor.2013.02.028.

Philip James Neame. *Nonsmooth Dual Methods in Integer Programming*. PhD thesis, University of Melbourne, Department of Mathematics and Statistics, Australia, March 1999.

George Lann Nemhauser. The age of optimization: Solving large-scale real-world problems. *Operations Research*, 42(1):5–13, January–February 1994. doi:10.1287/opre.42.1.5.

George Lann Nemhauser. Column generation for linear and integer programming. In Martin Grötschel, editor, *Optimization Stories*, Extra Volume ISMP, pages 65–73, Bielefeld, 2012. Documenta Mathematica, Journal der Deutschen Mathematiker-Vereinigung. ISBN 978-3-936609-58-5.

George Lann Nemhauser and Sungsoo Park. A polyhedral approach to edge coloring. *Operations Research Letters*, 10(6):315–322, August 1991. doi:10.1016/0167-6377(91)90003-8.

George Lann Nemhauser and Laurence Alexander Wolsey. *Integer and Combinatorial Optimization*. John Wiley & Sons, Chichester, 1988.

Jérémy Omer and François Soumis. A linear programming decomposition focusing on the span of the nondegenerate columns. *European Journal of Operational Research*, 245(2):371–383, September 2015. doi:10.1016/j.ejor.2015.03.019.

Jérémy Omer, Samuel Rosat, Vincent Raymond, and François Soumis. Improved primal simplex: A more general theoretical framework and an extended experimental analysis. *INFORMS Journal on Computing*, 27(4):773–787, Fall 2015a. doi:10.1287/ijoc.2015.0656.

Jérémy Omer, Mehdi Towhidi, and François Soumis. The positive edge pricing rule for the dual simplex. *Computers & Operations Research*, 61:135–142, September 2015b. doi:10.1016/j.cor.2015.03.009.

Temel Öncan, Ismail Kuban Altínel, and Gilbert Laporte. A comparative analysis of several asymmetric traveling salesman problem formulations. *Computers & Operations Research*, 36(3): 637–654, March 2009. doi:10.1016/j.cor.2007.11.008.

Amar Oukil, Hatem M. T. Ben Amor, Jacques Desrosiers, and Hicham El Gueddari. Stabilized column generation for highly degenerate multiple-depot vehicle scheduling problems. *Computers & Operations Research*, 34(3):817–834, March 2007. doi:10.1016/j.cor.2005.05.011.

Manfred Padberg and Giovanni Rinaldi. A branch-and-cut algorithm for the resolution of large-scale symmetric traveling salesman problems. *SIAM Review*, 33(1):60–100, March 1991. doi:10.1137/1033004.

Diego Galindo Pecin, Claudio Contardo, Guy Desaulniers, and Eduardo Uchoa. New enhancements for the exact solution of the vehicle routing problem with time windows. *INFORMS Journal on Computing*, 29(3):489–502, Summer 2017a. doi:10.1287/ijoc.2016.0744.

Diego Galindo Pecin, Artur Pessoa, Marcus Poggi de Aragão, and Eduardo Uchoa. Improved branch-cut-and-price for capacitated vehicle routing. *Mathematical Programming Computation*, 9(1):61–100, March 2017b. doi:10.1007/s12532-016-0108-8.

Diego Galindo Pecin, Artur Pessoa, Marcus Poggi de Aragão, Eduardo Uchoa, and Haroldo Santos. Limited memory rank-1 cuts for vehicle routing problems. *Operations Research Letters*, 45(3): 206–209, May 2017c. doi:10.1016/j.orl.2017.02.006.

Ann-Sophie Pepin, Guy Desaulniers, Alain Hertz, and Dennis Huisman. A comparison of five heuristics for the multiple depot vehicle scheduling problem. *Journal of Scheduling*, 12(1): 17–30, February 2009. doi:10.1007/s10951-008-0072-x.

Artur Pessoa, Eduardo Uchoa, and Marcus Poggi de Aragão. A robust branch-cut-and-price algorithm for the heterogeneous fleet vehicle routing problem. *Networks*, 54(4):167–177, December 2009. doi:10.1002/net.20330.

Artur Pessoa, Eduardo Uchoa, Marcus Poggi de Aragão, and Rosiane de Freitas Rodrigues. Exact algorithm over an arc-time-indexed formulation for parallel machine scheduling problems. *Mathematical Programming Computation*, 2(3):259–290, December 2010. doi:10.1007/s12532-010-0019-z.

Artur Pessoa, Ruslan Sadykov, Eduardo Uchoa, and François Vanderbeck. In-out separation and column generation stabilization by dual price smoothing. In Vincenzo Bonifaci, Camil Demetrescu, and Alberto Marchetti-Spaccamela, editors, *Experimental Algorithms*, volume 7933 of *Lecture Notes in Computer Science*, pages 354–365. Springer, Berlin Heidelberg, 2013. doi:10.1007/978-3-642-38527-8_31.

Artur Pessoa, Ruslan Sadykov, Eduardo Uchoa, and François Vanderbeck. Automation and combination of linear-programming based stabilization techniques in column generation. *INFORMS Journal on Computing*, 30(2):339–360, Spring 2018. doi:10.1287/ijoc.2017.0784.

Artur Pessoa, Ruslan Sadykov, Eduardo Uchoa, and François Vanderbeck. A generic exact solver for vehicle routing and related problems. *Mathematical Programming*, 183:483–523, 2020. doi:10.1007/s10107-020-01523-z.

Artur Pessoa, Ruslan Sadykov, and Eduardo Uchoa. Solving bin packing problems using VRPSolver models. *Operations Research Forum*, 2:article number 20, April 2021. doi:10.1007/s43069-020-00047-8.

Bjørn Petersen, David Pisinger, and Simon Spoorendonk. Chvátal-Gomory rank-1 cuts used in a Dantzig-Wolfe decomposition of the vehicle routing problem with time windows. In Bruce Golden, Saahitya Raghavan, and Edward Wasil, editors, *The Vehicle Routing Problem: Latest Advances and New Challenges*, pages 397–420. Springer, New York, 2008. doi:10.1007/978-0-387-77778-8_18.

Olivier Péton and Jean-Philippe Vial. A brief tutorial on ACCPM. In *Proceedings of the Operational Research Peripatetic Postgraduate Programme*, pages 1–8, September 26–29 2001.

Seth Pettie. A new approach to all-pairs shortest paths on real-weighted graphs. *Theoretical Computer Science*, 312(1):47–74, January 2004. doi:10.1016/S0304-3975(03)00402-X.

Boris Teodorovich Polyak. A general method of solving extremum problems. *Doklady Akademii Nauk SSSR*, 174(1):33–36, 1967. (Translated from Russian in Soviet Mathematics Doklady, 8(3):593–597, 1967).

Boris Teodorovich Polyak. History of mathematical programming in the USSR: analyzing the phenomenon. *Mathematical Programming*, 91:401–416, April 2002. doi:10.1007/s101070100258.

Eric Prescott-Gagnon, Guy Desaulniers, and Louis-Martin Rousseau. A branch-and-price-based large neighborhood search algorithm for the vehicle routing problem with time windows. *Networks*, 54(4):190–204, August 2009. doi:10.1002/net.20332.

Robert Clay Prim. Shortest connection networks and some generalizations. *Bell System Technical Journal*, 36(6):1389–1401, November 1957. doi:10.1002/j.1538-7305.1957.tb01515.x.

Frédéric Quesnel, Guy Desaulniers, and François Soumis. A new heuristic branching scheme for the crew pairing problem with base constraints. *Computers & Operations Research*, 80:159–172, April 2017. doi:10.1016/j.cor.2016.11.020.

Frédéric Quesnel, Alice Wu, Guy Desaulniers, and François Soumis. Deep-learning-based partial pricing in a branch-and-price algorithm for personalized crew rostering. *Computers & Operations Research*, 138(C):105554, February 2022. doi:10.1016/j.cor.2021.105554.

Tomasz Radzik and Andrew Vladislav Goldberg. Tight bounds on the number of minimum-mean cycle cancellations and related results. *Algorithmica*, 11(3):226–242, March 1994. doi:10.1007/BF01240734.

Ragheb Rahmaniani, Teodor Gabriel Crainic, Michel Gendreau, and Walter Rei. The Benders decomposition algorithm: A literature review. *European Journal of Operational Research*, 259 (3):801–817, June 2017. doi:10.1016/j.ejor.2016.12.005.

Vincent Raymond, François Soumis, Abdelmoutalib Metrane, and Jacques Desrosiers. Positive edge: A pricing criterion for the identification of non-degenerate simplex pivots. Les Cahiers du GERAD G-2010-61, HEC Montréal, Montreal, QC, Canada, 2010a.

Vincent Raymond, François Soumis, and Dominique Orban. A new version of the improved primal simplex for degenerate linear programs. *Computers & Operations Research*, 37(1):91–98, January 2010b. doi:10.1016/j.cor.2009.03.020.

John Ker Reid. A sparsity-exploiting variant of the Bartels–Golub decomposition for linear programming bases. *Mathematical Programming*, 24(1):55–69, December 1982. doi:10.1007/BF01585094.

Celso Carneiro Ribeiro and François Soumis. A column generation approach to the multiple-depot vehicle scheduling problem. *Operations Research*, 42(1):41–53, January–February 1994. doi:10.1287/opre.42.1.41.

Celso Carneiro Ribeiro, Sebastián Urrutia, and Dominique de Werra. *Combinatorial Models for Scheduling Sports Tournaments*. EURO Advanced Tutorials on Operational Research. Springer, Cham, January 2023. doi:10.1007/978-3-031-37283-4.

Jürgen Rietz, Guntram Scheithauer, and Johannes Terno. Tighter bounds for the gap and non-IRUP constructions in the one-dimensional cutting stock problem. *Optimization*, 51(6):927–963, 2002. doi:10.1080/0233193021000066545.

Giovanni Righini and Matteo Salani. Symmetry helps: Bounded bi-directional dynamic programming for the elementary shortest path problem with resource constraints. *Discrete Optimization*, 3(3):255–273, September 2006. doi:10.1016/j.disopt.2006.05.007.

Giovanni Righini and Matteo Salani. New dynamic programming algorithms for the resource constrained elementary shortest path problem. *Networks*, 51(3):155–170, May 2008. doi:10.1002/net.20212.

Elina Rönnberg and Torbjörn Larsson. Column generation in the integral simplex method. *European Journal of Operational Research*, 192(1):333–342, January 2009. doi:10.1016/j.ejor.2007.09.037.

Elina Rönnberg and Torbjörn Larsson. All-integer column generation for set partitioning: Basic principles and extensions. *European Journal of Operational Research*, 233(3):529–538, March 2014. doi:10.1016/j.ejor.2013.08.036.

Stefan Røpke and Jean-François Cordeau. Branch and cut and price for the pickup and delivery problem with time windows. *Transportation Science*, 43(3):267–286, August 2009. doi:10.1287/trsc.1090.0272.

Stefan Røpke and David Pisinger. An adaptive large neighborhood search heuristic for the pickup and delivery problem with time windows. *Transportation Science*, 40(4):455–472, November 2006. doi:10.1287/trsc.1050.0135.

Samuel Rosat, Issmail El Hallaoui, François Soumis, and Driss Chakour. Influence of the normalization constraint on the integral simplex using decomposition. *Discrete Applied Mathematics*, 217(Part 1):53–70, January 2017a. doi:10.1016/j.dam.2015.12.015.

Samuel Rosat, Issmail El Hallaoui, François Soumis, and Andrea Lodi. Integral simplex using decomposition with primal cutting planes. *Mathematical Programming*, 166(1–2):327–367, March 2017b. doi:10.1007/s10107-017-1123-x.

Samuel Rosat, Frédéric Quesnel, Issmail El Hallaoui, and François Soumis. Dynamic penalization of fractional directions in the integral simplex using decomposition: Application to aircrew scheduling. *European Journal of Operational Research*, 263(3):1007–1018, December 2017c. doi:10.1016/j.ejor.2017.05.047.

Borzou Rostami, Guy Desaulniers, Fausto Errico, and Andrea Lodi. Branch-price-and-cut algorithms for the vehicle routing problem with stochastic and correlated travel times. *Operations Research*, 69(2):436–455, March 2021. doi:10.1287/opre.2020.2037.

Louis-Martin Rousseau, Michel Gendreau, and Dominique Feillet. Interior point stabilization for column generation. *Operations Research Letters*, 35(5):660–668, September 2007. doi:10.1016/j.orl.2006.11.004.

Kevin Scott Ruland and Ervin Yechiel-Laszlo Rodin. Survey of facial results for the traveling salesman polytope. *Mathematical and Computer Modelling*, 27(8):11–27, April 1998. doi:10.1016/S0895-7177(98)00041-7.

David Murray Ryan. The solution of massive generalized set partitioning problems in aircrew rostering. *Journal of the Operational Research Society*, 43(5):459–467, 1992. doi:10.1057/jors.1992.72.

David Murray Ryan and Julie C. Falkner. On the integer properties of scheduling set partitioning models. *European Journal of Operational Research*, 35(3):442–456, June 1988. doi:10.1016/0377-2217(88)90233-0.

David Murray Ryan and Brian A. Foster. An integer programming approach to scheduling. In Anthony Wren, editor, *Computer Scheduling of Public Transport: Urban Passenger Vehicle and Crew Scheduling*, pages 269–280, Amsterdam, 1981. North-Holland Publishing Company.

David Murray Ryan and Michael Robert Osborne. On the solution of highly degenerate linear programmes. *Mathematical Programming*, 41(1–3):385–392, May 1988. doi:10.1007/BF01580776.

Ruslan Sadykov and François Vanderbeck. Column generation for extended formulations. *EURO Journal on Computational Optimization*, 1(1–2):81–115, March 2013. doi:10.1007/s13675-013-0009-9.

Ruslan Sadykov and François Vanderbeck. BaPCod - a generic branch-and-price code. Technical Report HAL-03340548, Inria Bordeaux Sud-Ouest, November 2021.

Ruslan Sadykov, François Vanderbeck, Artur Pessoa, Issam Tahiri, and Eduardo Uchoa. Primal heuristics for branch and price: The assets of diving methods. *INFORMS Journal on Computing*, 31(2):251–267, Spring 2019. doi:10.1287/ijoc.2018.0822.

Ruslan Sadykov, Eduardo Uchoa, and Artur Pessoa. A bucket graph-based labeling algorithm with application to vehicle routing. *Transportation Science*, 55(1):4–28, January–February 2021. doi:10.1287/trsc.2020.0985.

Domenico Salvagnin. Detecting semantic groups in MIP models. In Claude-Guy Quimper, editor, *Integration of AI and OR Techniques in Constraint Programming*, volume 9676 of *Lecture Notes in Computer Science*, pages 329–341, Cham, 2016. Springer. doi:10.1007/978-3-319-33954-2_24.

Martin Savelsbergh and Marc Sol. Drive: Dynamic routing of independent vehicles. *Operations Research*, 46(4):474–490, April 1998. doi:10.1287/opre.46.4.474.

Martin W.P. Savelsbergh. A branch-and-price algorithm for the generalized assignment problem. *Operations Research*, 45(6):831–841, November–December 1997. doi:10.1287/opre.45.6.831.

Guntram Scheithauer and Johannes Terno. The modified integer round-up property of the one-dimensional cutting stock problem. *European Journal of Operational Research*, 84(3):562–571, August 1995. doi:10.1016/0377-2217(95)00022-I.

Alexander Schrijver. *Theory of Linear and Integer Programming*. John Wiley & Sons, Chichester, 1986.

Ron Shamir. The efficiency of the simplex method: A survey. *Management Science*, 33(3):301–334, March 1987. doi:10.1287/mnsc.33.3.301.

Jeremy Frank Shapiro. *Mathematical Programming: Structures and Algorithms*. John Wiley & Sons, New York, 1979a.

Jeremy Frank Shapiro. A survey of Lagrangean techniques for discrete optimization. In Peter Ladislaw Hammer, Ellis Lane Johnson, and Bernhard Hermann Korte, editors, *Discrete Optimization II*, volume 5 of *Annals of Discrete Mathematics*, pages 113–138. North-Holland Publishing Company, Amsterdam, 1979b. doi:10.1016/S0167-5060(08)70346-7.

Paul Shaw. Using constraint programming and local search methods to solve vehicle routing problems. In Michael Maher and Jean-François Puget, editors, *Principles and Practice of Constraint Programming — CP98*, volume 1520 of *Lecture Notes in Computer Science*, pages 417–431, Berlin Heidelberg, 1998. Springer. doi:10.1007/3-540-49481-2_30.

Hanif D. Sherali and Barbara M. P. Fraticelli. A modification of Benders' decomposition algorithm for discrete subproblems: An approach for stochastic programs with integer recourse. *Journal of Global Optimization*, 22(1):319–342, January 2002. doi:10.1023/A:1013827731218.

Jeffrey Stewart Simonoff. *Smoothing Methods in Statistics*. Springer Series in Statistics. Springer, New York, 1996. doi:10.1007/978-1-4612-4026-6.

Marius Mihai Solomon. Algorithms for the vehicle routing and scheduling problem with time window constraints. *Operations Research*, 35(2):254–265, March–April 1987. doi:10.1287/opre.35.2.254.

Marius Mihai Solomon and Jacques Desrosiers. Survey paper—Time window constrained routing and scheduling problems. *Transportation Science*, 22(1):1–13, February 1988. doi:10.1287/trsc.22.1.1.

François Soumis. Decomposition and column generation. In Mauro Dell'Amico, Francesco Maffioli, and Silvano Martello, editors, *Annotated bibliographies in combinatorial optimization*, pages 115–126. John Wiley & Sons, New York, 1997.

François Soumis, Jacques A. Ferland, and Jean-Marc Rousseau. A model for large-scale aircraft routing and scheduling problems. *Transportation Research Part B: Methodological*, 14(1–2): 191–201, March–June 1980. doi:10.1016/0191-2615(80)90044-2.

François Soumis, Jacques Desrosiers, and Martin Desrochers. Optimal urban bus routing with scheduling flexibilities. In Palle Thoft-Christensen, editor, *System Modelling and Optimization*, volume 59 of *Lecture Notes in Control and Information Sciences*, pages 155–165, Berlin Heidelberg, 1984. Springer. doi:10.1007/BFb0008887.

Simon Spoorendonk and Guy Desaulniers. Clique inequalities applied to the vehicle routing problem with time windows. *INFOR: Information Systems and Operational Research*, 48(1):53–67, 2010. doi:10.3138/infor.48.1.053.

Goran Stojković, François Soumis, Jacques Desrosiers, and Marius Mihai Solomon. An optimization model for a real-time flight scheduling problem. *Transportation Research Part A: Policy and Practice*, 36(9):779–788, November 2002. doi:10.1016/S0965-8564(01)00039-8.

Mirela Stojković, François Soumis, and Jacques Desrosiers. The operational airline crew scheduling problem. *Transportation Science*, 32(3):232–245, August 1998. doi:10.1287/trsc.32.3.232.

Anand Subramanian. *Subject to*. `https://listennotes.com/podcasts/subject-to-anand-subramanian-su0JT1q7ygf/#podcast`, 2023.

Adil Tahir, Guy Desaulniers, and Issmail El Hallaoui. Integral column generation for the set partitioning problem. *EURO Journal on Transportation and Logistics*, 8(5):713–744, June 2019. doi:10.1007/s13676-019-00145-6.

Adil Tahir, Frédéric Quesnel, Guy Desaulniers, Issmail El Hallaoui, and Yassine Yaakoubi. An improved integral column generation algorithm using machine learning for aircrew pairing. *Transportation Science*, 55(6):1411–1429, November–December 2021. doi:10.1287/trsc.2021.1084.

Adil Tahir, Guy Desaulniers, and Issmail El Hallaoui. Integral column generation for set partitioning problems with side constraints. *INFORMS Journal on Computing*, 34(4):2313–2331, 2022. doi:10.1287/ijoc.2022.1174.

Christian Tilk, Ann-Kathrin Rothenbächer, Timo Gschwind, and Stefan Irnich. Asymmetry matters: Dynamic half-way points in bidirectional labeling for solving shortest path problems with resource constraints faster. *European Journal of Operational Research*, 261(2):530–539, September 2017. doi:10.1016/j.ejor.2017.03.017.

Mehdi Towhidi, Jacques Desrosiers, and François Soumis. The positive edge criterion within COIN-OR's CLP. *Computers & Operations Research*, 49:41–46, September 2014. doi:10.1016/j.cor.2014.03.020.

Eduardo Uchoa and Ruslan Sadykov. Kantorovich and Zalgaller (1951): the 0-th Column Generation Algorithm. Technical Report L-2024-1, Cadernos do LOGIS-UFF, Niterói, Brazil, January 2024.

José Manuel Valério de Carvalho. Exact solution of bin-packing problems using column generation and branch-and-bound. *Annals of Operations Research*, 86(0):629–659, January 1999. doi:10.1023/A:1018952112615.

José Manuel Valério de Carvalho. LP models for bin-packing and cutting stock problems. *European Journal of Operational Research*, 141(2):253–273, September 2002. doi:10.1016/S0377-2217(02)00124-8.

José Manuel Valério de Carvalho. Using extra dual cuts to accelerate convergence in column generation. *INFORMS Journal on Computing*, 17(2):175–182, Spring 2005. doi:10.1287/ijoc.1030.0060.

Pascal Van Hentenryck. Constraint and integer programming in OPL. *INFORMS Journal on Computing*, 14(4):345–372, Fall 2002. doi:10.1287/ijoc.14.4.345.2826.

Richard Maurice Van Slyke and Roger Jean-Baptiste Wets. L-shaped linear programs with applications to optimal control and stochastic programming. *SIAM Journal on Applied Mathematics*, 17(4):638–663, 1969. doi:10.1137/0117061.

Pamela Hatch Vance. *Crew scheduling, cutting stock, and column generation: Solving huge integer programs*. PhD thesis, Georgia Institute of Technology, Atlanta, USA, August 1993.

Pamela Hatch Vance, Cynthia Barnhart, Ellis Lane Johnson, and George Lann Nemhauser. Airline crew scheduling: A new formulation and decomposition algorithm. *Operations Research*, 45 (2):188–200, March–April 1997. doi:10.1287/opre.45.2.188.

François Vanderbeck. *Decomposition and column generation for integer programs*. PhD thesis, Université catholique de Louvain, Belgium, 1994.

François Vanderbeck. Computational study of a column generation algorithm for bin packing and cutting stock problems. *Mathematical Programming*, 86(3):565–594, December 1999. doi:10.1007/s101070050105.

François Vanderbeck. On Dantzig-Wolfe decomposition in integer programming and ways to perform branching in a branch-and-price algorithm. *Operations Research*, 48(1):111–128, January–February 2000. doi:10.1287/opre.48.1.111.12453.

François Vanderbeck. Implementing mixed integer column generation. In Guy Desaulniers, Jacques Desrosiers, and Marius Mihai Solomon, editors, *Column Generation*, pages 331–358. Springer, New York, 2005. doi:10.1007/0-387-25486-2_12.

François Vanderbeck. Branching in branch-and-price: a generic scheme. *Mathematical Programming*, 130(2):249–294, December 2011. doi:10.1007/s10107-009-0334-1.

François Vanderbeck and Martin W.P. Savelsbergh. A generic view of Dantzig-Wolfe decomposition in mixed integer programming. *Operations Research Letters*, 34(3):296–306, May 2006. doi:10.1016/j.orl.2005.05.009.

François Vanderbeck and Laurence Alexander Wolsey. Reformulation and decomposition of integer programs. In Michael Jünger, Thomas M. Liebling, Denis Naddef, George Lann Nemhauser, William Robert Pulleyblank, Gerhard Reinelt, Giovanni Rinaldi, and Laurence Alexander Wolsey, editors, *50 Years of Integer Programming 1958–2008: From the Early Years to the State-of-the-Art*, chapter 13, pages 431–502. Springer, Berlin Heidelberg, 2010. doi:10.1007/978-3-540-68279-0_13.

Robert Joseph Vanderbei. *Linear Programming: Foundations and Extensions*, volume 285 of *International Series in Operations Research & Management Science*. Springer, Cham, 5th edition, 2020. doi:10.1007/978-3-030-39415-8.

Daniel Villeneuve. *Logiciel de Génération de Colonnes*. PhD thesis, Polytechnique Montréal, Montreal, QC, Canada, October 1999. (in French).

Daniel Villeneuve, Jacques Desrosiers, Marco E. Lübbecke, and François Soumis. On compact formulations for integer programs solved by column generation. *Annals of Operations Research*, 139(1):375–388, October 2005. doi:10.1007/s10479-005-3455-9.

Vadim Georgievich Vizing. On an estimate of the chromatic class of a p-graph. *Diskretnyi Analiz*, 3:25–30, 1964. (in Russian).

Harvey Maurice Wagner and Thomson McLintock Whitin. Dynamic version of the economic lot size model. *Management Science*, 5(1):89–96, January 1958. doi:10.1287/mnsc.5.1.89.

Warren Elliott Walker. Letter to the editor–A method for obtaining the optimal dual solution to a linear program using the Dantzig-Wolfe decomposition. *Operations Research*, 17(2):368–370, March–April 1969. doi:10.1287/opre.17.2.368.

I-Lin Wang. Multicommodity network flows: A survey, part I: Applications and formulations. *International Journal of Operations Research*, 15(4):145–153, December 2018a. doi:10.6886/IJOR.201812_15(4).0001.

I-Lin Wang. Multicommodity network flows: A survey, part II: Solution methods. *International Journal of Operations Research*, 15(4):155–173, December 2018b. doi:10.6886/IJOR.201812_15(4).0002.

Jiadong Wang and Ted Ralphs. Computational experience with hypergraph-based methods for automatic decomposition in discrete optimization. In Carla Gomes and Meinolf Sellmann, editors, *Integration of AI and OR Techniques in Constraint Programming for Combinatorial Optimization Problems*, volume 7874 of *Lecture Notes in Computer Science*, pages 394–402. Springer, Berlin Heidelberg, 2013. doi:10.1007/978-3-642-38171-3_31.

Ting Wang, Qian Hu, and Andrew Lim. An exact algorithm for two-dimensional vector packing problem with volumetric weight and general costs. *European Journal of Operational Research*, 300(1):20–34, July 2022. doi:10.1016/j.ejor.2021.10.011.

Lijun Wei, Zhixing Luo, Roberto Baldacci, and Andrew Lim. A new branch-and-price-and-cut algorithm for one-dimensional bin-packing problems. *INFORMS Journal on Computing*, 32 (2):428–443, Spring 2020. doi:10.1287/ijoc.2018.0867.

Oliver Weide, David Murray Ryan, and Matthias Ehrgott. An iterative approach to robust and integrated aircraft routing and crew scheduling. *Computers & Operations Research*, 37(5): 833–844, May 2010. doi:10.1016/j.cor.2009.03.024.

Paul Wentges. Weighted Dantzig-Wolfe decomposition of linear mixed-integer programming. *International Transactions in Operational Research*, 4(2):151–162, March 1997. doi:10.1016/S0969-6016(97)00001-4.

Jonas Timon Witt. *Polyhedral aspects of Dantzig-Wolfe reformulation*. PhD thesis, RWTH Aachen University, November 2019. doi:10.18154/RWTH-2020-00547.

Philip Wolfe. Convergence conditions for ascent methods. *SIAM Review*, 11(2):226–235, 1969. doi:10.1137/1011036.

Laurence Alexander Wolsey. *Integer programming*. John Wiley & Sons, Chichester, 1998.

Yassine Yaakoubi, François Soumis, and Simon Lacoste-Julien. Machine learning in airline crew pairing to construct initial clusters for dynamic constraint aggregation. *EURO Journal on Transportation and Logistics*, 9(4):100020, December 2020. doi:10.1016/j.ejtl.2020.100020.

Julian Yarkony, Yossiri Adulyasak, Maneesh Singh, and Guy Desaulniers. Data association via set packing for computer vision applications. *INFORMS Journal on Optimization*, 2(3):167–191, Summer 2020. doi:10.1287/ijoo.2019.0030.

Abdelouahab Zaghrouti, François Soumis, and Issmail El Hallaoui. Integral simplex using decomposition for the set partitioning problem. *Operations Research*, 62(2):435–449, March–April 2014. doi:10.1287/opre.2013.1247.

Abdelouahab Zaghrouti, Issmail El Hallaoui, and François Soumis. Improved integral simplex using decomposition for the set partitioning problem. *EURO Journal on Computational Optimization*, 6(2):185–206, June 2018. doi:10.1007/s13675-018-0098-6.

Koorush Ziarati, François Soumis, Jacques Desrosiers, Sylvie Gélinas, and André Saintonge. Locomotive assignment with heterogeneous consists at CN North America. *European Journal of Operational Research*, 97(2):281–292, March 1997. doi:10.1016/S0377-2217(96)00198-1.

Koorush Ziarati, François Soumis, and Jacques Desrosiers. Locomotive assignment using train delays. In Nigel H. M. Wilson, editor, *Computer-Aided Transit Scheduling*, volume 471 of *Lecture Note in Economics Mathematical Systems*, pages 285–297. Springer, Berlin Heidelberg, 1999a. doi:10.1007/978-3-642-85970-0_14.

Koorush Ziarati, François Soumis, Jacques Desrosiers, and Marius Mihai Solomon. A Branch-first, Cut-second approach for locomotive assignment. *Management Science*, 45(8):1156–1168, August 1999b. doi:10.1287/mnsc.45.8.1156.

Günter Matthias Ziegler. *Lectures on Polytopes*. Springer, New York, 1995. doi:10.1007/978-1-4613-8431-1.